PLASMA CHEMISTRY

This unique book provides a fundamental introduction to all aspects of modern plasma chemistry. The book describes mechanisms and kinetics of chemical processes in plasma, plasma statistics, thermodynamics, fluid mechanics, and electrodynamics, as well as all major electric discharges applied in plasma chemistry. The book considers most of the major applications of plasma chemistry, from electronics to thermal coatings, from treatment of polymers to fuel conversion and hydrogen production, and from plasma metallurgy to plasma medicine. The book can be helpful to engineers, scientists, and students interested in plasma physics, plasma chemistry, plasma engineering, and combustion, as well as in chemical physics, lasers, energy systems, and environmental control. The book contains an extensive database on plasma kinetics and thermodynamics, as well as many convenient numerical formulas for practical calculations related to specific plasma–chemical processes and applications. The book contains a large number of problems and concept questions that are helpful in university courses related to plasma, lasers, combustion, chemical kinetics, statistics and thermodynamics, and high-temperature and high-energy fluid mechanics.

Alexander Fridman is Nyheim Chair Professor of Drexel University and Director of Drexel Plasma Institute. His research focuses on plasma approaches to material treatment, fuel conversion, hydrogen production, biology, medicine, and environmental control. Professor Fridman has more than 35 years of plasma research experience in national laboratories and universities in Russia, France, and the United States. He has published 6 books and 450 papers, chaired several international plasma conferences, and received numerous awards, including the Stanley Kaplan Distinguished Professorship in Chemical Kinetics and Energy Systems, the George Soros Distinguished Professorship in Physics, and the State Prize of the USSR for discovery of selective stimulation of chemical processes in non-thermal plasma.

Plasma Chemistry

Alexander Fridman

Drexel University

CAMBRIDGE
UNIVERSITY PRESS

CAMBRIDGE UNIVERSITY PRESS
Cambridge, New York, Melbourne, Madrid, Cape Town, Singapore, São Paulo, Delhi

Cambridge University Press
32 Avenue of the Americas, New York, NY 10013-2473, USA

www.cambridge.org
Information on this title: www.cambridge.org/9780521847353

First published 2008

Printed in the United States of America.

A catalog record for this publication is available from the British Library.

ISBN-13   978-0-521-84735-3 hardback

To my wife Irene

Contents

Foreword   page xxxix

Preface   xli

1   Introduction to Theoretical and Applied Plasma Chemistry   1
  1.1. Plasma as the Fourth State of Matter   1
  1.2. Plasma in Nature and in the Laboratory   2
  1.3. Plasma Temperatures: Thermal and Non-Thermal Plasmas   4
  1.4. Plasma Sources for Plasma Chemistry: Gas Discharges   5
  1.5. Fundamentals of Plasma Chemistry: Major Components of
      Chemically Active Plasma and Mechanisms of Plasma-Chemical
      Processes   8
  1.6. Applied Plasma Chemistry   9
  1.7. Plasma as a High-Tech Magic Wand of Modern Technology   10

2   Elementary Plasma-Chemical Reactions   12
  2.1. Ionization Processes   12
    2.1.1. Elementary Charged Particles in Plasma   12
    2.1.2. Elastic and Inelastic Collisions and Their Fundamental
        Parameters   13
    2.1.3. Classification of Ionization Processes   14
    2.1.4. Elastic Scattering and Energy Transfer in Collisions of
        Charged Particles: Coulomb Collisions 15
    2.1.5. Direct Ionization by Electron Impact: Thomson Formula 16
    2.1.6. Specific Features of Ionization of Molecules by Electron
        Impact: Frank-Condon Principle and Dissociative
        Ionization   17
    2.1.7. Stepwise Ionization by Electron Impact   18
    2.1.8. Ionization by High-Energy Electrons and Electron Beams:
        Bethe-Bloch Formula   20
    2.1.9. Photo-Ionization Processes   20
    2.1.10. Ionization in Collisions of Heavy Particles: Adiabatic
        Principle and Massey Parameter   21
    2.1.11. Penning Ionization Effect and Associative Ionization 21
  2.2. Elementary Plasma-Chemical Reactions of Positive Ions   22
    2.2.1. Different Mechanisms of Electron–Ion Recombination in
        Plasma   22

Contents

2.2.2. Dissociative Electron–Ion Recombination and Possible Preliminary Stage of Ion Conversion   23

2.2.3. Three-Body and Radiative Electron–Ion Recombination Mechanisms   25

2.2.4. Ion–Molecular Reactions, Ion–Molecular Polarization Collisions, and the Langevin Rate Coefficient   26

2.2.5. Ion–Atomic Charge Transfer Processes and Resonant Charge Transfer   28

2.2.6. Non-Resonant Charge Transfer Processes and Ion–Molecular Chemical Reactions of Positive and Negative Ions   29

2.3. Elementary Plasma-Chemical Reactions Involving Negative Ions   31

2.3.1. Dissociative Electron Attachment to Molecules as a Major Mechanism of Negative Ion Formation in Electronegative Molecular Gases   31

2.3.2. Three-Body Electron Attachment and Other Mechanisms of Formation of Negative Ions   33

2.3.3. Destruction of Negative Ions: Associative Detachment, Electron Impact Detachment, and Detachment in Collisions with Excited Particles   35

2.3.4. Recombination of Negative and Positive Ions   37

2.3.5. Ion–Ion Recombination in Binary Collisions   38

2.3.6. Three-Body Ion–Ion Recombination: Thomson's Theory and Langevin Model   39

2.4. Electron Emission and Heterogeneous Ionization Processes   42

2.4.1. Thermionic Emission: Sommerfeld Formula and Schottky Effect   42

2.4.2. Field Emission of Electrons in Strong Electric Fields: Fowler-Nordheim Formula and Thermionic Field Emission   43

2.4.3. Secondary Electron Emission   45

2.4.4. Photo-Ionization of Aerosols: Monochromatic Radiation   46

2.4.5. Photo-Ionization of Aerosols: Continuous-Spectrum Radiation   49

2.4.6. Thermal Ionization of Aerosols: Einbinder Formula   51

2.4.7. Space Distribution of Electrons and Electric Field Around a Thermally Ionized Macro-Particle   52

2.4.8. Electric Conductivity of Thermally Ionized Aerosols   53

2.5. Excitation and Dissociation of Neutral Particles in Ionized Gases   54

2.5.1. Vibrational Excitation of Molecules by Electron Impact   54

2.5.2. Rate Coefficients of Vibrational Excitation by Electron Impact: Semi-Empirical Fridman Approximation   56

2.5.3. Rotational Excitation of Molecules by Electron Impact   58

2.5.4. Electronic Excitation of Atoms and Molecules by Electron Impact   59

2.5.5. Dissociation of Molecules by Direct Electron Impact   61

2.5.6. Distribution of Electron Energy in Non-Thermal Discharges Between Different Channels of Excitation and Ionization   63

Contents

2.6. Elementary Relaxation Processes of Energy Transfer Involving
Vibrationally, Rotationally, and Electronically Excited Molecules 67
    2.6.1. Vibrational–Translational (VT) Relaxation: Slow Adiabatic
    Elementary Process 67
    2.6.2. Landau–Teller Formula for VT-Relaxation Rate Coefficients 69
    2.6.3. Fast Non-Adiabatic Mechanisms of VT Relaxation 71
    2.6.4. Vibrational Energy Transfer Between Molecules: Resonant
    VV Relaxation 72
    2.6.5. Non-Resonant VV Exchange: Relaxation of Anharmonic
    Oscillators and Intermolecular VV Relaxation 74
    2.6.6. Rotational Relaxation Processes: Parker Formula 76
    2.6.7. Relaxation of Electronically Excited Atoms and Molecules 76
2.7. Elementary Chemical Reactions of Excited Molecules:
Fridman-Macheret α-Model 79
    2.7.1. Rate Coefficient of Reactions of Excited Molecules 79
    2.7.2. Efficiency α of Vibrational Energy in Overcoming
    Activation Energy of Chemical Reactions: Numerical
    Values and Classification Table 81
    2.7.3. Fridman-Macheret α-Model 81
    2.7.4. Efficiency of Vibrational Energy in Elementary Reactions
    Proceeding Through Intermediate Complexes: Synthesis of
    Lithium Hydride 83
    2.7.5. Dissociation of Molecules in Non-Equilibrium Conditions
    with Essential Contribution of Translational Energy:
    Non-Equilibrium Dissociation Factor Z 86
    2.7.6. Semi-Empirical Models of Non-Equilibrium Dissociation of
    Molecules Determined by Vibrational and Translational
    Temperatures 87
    Problems and Concept Questions 89

3 **Plasma-Chemical Kinetics, Thermodynamics,
and Electrodynamics** 92
3.1. Plasma Statistics and Thermodynamics, Chemical and Ionization
Equilibrium, and the Saha Equation 92
    3.1.1. Statistical Distributions: Boltzmann Distribution Function 92
    3.1.2. Equilibrium Statistical Distribution of Diatomic Molecules
    over Vibrational–Rotational States 93
    3.1.3. Saha Equation for Ionization Equilibrium in Thermal Plasma 94
    3.1.4. Dissociation Equilibrium in Molecular Gases 94
    3.1.5. Complete Thermodynamic Equilibrium (CTE) and Local
    Thermodynamic Equilibrium (LTE) in Plasma 95
    3.1.6. Thermodynamic Functions of Quasi-Equilibrium Thermal
    Plasma Systems 95
    3.1.7. Non-Equilibrium Statistics of Thermal and Non-Thermal
    Plasmas 97
    3.1.8. Non-Equilibrium Statistics of Vibrationally Excited
    Molecules: Treanor Distribution 99
3.2. Electron Energy Distribution Functions (EEDFs) in Non-Thermal
Plasma 100
    3.2.1. Fokker-Planck Kinetic Equation for Determination of EEDF 100

Contents

3.2.2. Druyvesteyn Distribution, Margenau Distributions, and
Other Specific EEDF 101
3.2.3. Effect of Electron–Molecular and Electron–Electron
Collisions on EEDF 103
3.2.4. Relation Between Electron Temperature and the Reduced
Electric Field 104
3.2.5. Isotropic and Anisotropic Parts of the Electron
Distribution Functions: EEDF and Plasma Conductivity 104
3.3. Diffusion, Electric/Thermal Conductivity, and Radiation in Plasma 106
3.3.1. Electron Mobility, Plasma Conductivity, and Joule Heating 106
3.3.2. Plasma Conductivity in Crossed Electric and Magnetic
Fields 107
3.3.3. Ion Energy and Ion Drift in Electric Field 109
3.3.4. Free Diffusion of Electrons and Ions; Continuity Equation;
and Einstein Relation Between Diffusion Coefficient,
Mobility, and Mean Energy 109
3.3.5. Ambipolar Diffusion and Debye Radius 110
3.3.6. Thermal Conductivity in Plasma 111
3.3.7. Non-Equilibrium Thermal Conductivity and Treanor Effect
in Vibrational Energy Transfer 112
3.3.8. Plasma Emission and Absorption of Radiation in
Continuous Spectrum and Unsold-Kramers Formula 112
3.3.9. Radiation Transfer in Plasma: Optically Thin and Optically
Thick Plasmas 113
3.4. Kinetics of Vibrationally and Electronically Excited Molecules in
Plasma: Effect of Hot Atoms 114
3.4.1. Fokker-Plank Kinetic Equation for Non-Equilibrium
Vibrational Distribution Functions 114
3.4.2. VT and VV Fluxes of Excited Molecules in Energy Space 115
3.4.3. Non-Equilibrium Vibrational Distribution Functions:
Regime of Strong Excitation 117
3.4.4. Non-Equilibrium Vibrational Distribution Functions:
Regime of Weak Excitation 119
3.4.5. Kinetics of Population of Electronically Excited States in
Plasma 120
3.4.6. Non-Equilibrium Translational Energy Distribution
Functions of Heavy Neutrals: Effect of "Hot" Atoms in Fast
VT-Relaxation Processes 122
3.4.7. Generation of "Hot" Atoms in Chemical Reactions 123
3.5. Vibrational Kinetics of Gas Mixtures, Chemical Reactions, and
Relaxation Processes 124
3.5.1. Kinetic Equation and Vibrational Distributions in Gas
Mixtures: Treanor Isotopic Effect in Vibrational Kinetics 124
3.5.2. Reverse Isotopic Effect in Plasma-Chemical Kinetics 126
3.5.3. Macrokinetics of Chemical Reactions of Vibrationally
Excited Molecules 129
3.5.4. Vibrational Energy Losses Due to VT Relaxation 131
3.5.5. Vibrational Energy Losses Due to Non-Resonance VV
Exchange 132
3.6. Energy Balance and Energy Efficiency of Plasma-Chemical
Processes 132

Contents

3.6.1. Energy Efficiency of Quasi-Equilibrium and Non-Equilibrium Plasma-Chemical Processes   132

3.6.2. Energy Efficiency of Plasma-Chemical Processes Stimulated by Vibrational Excitation of Molecules   133

3.6.3. Energy Efficiency of Plasma-Chemical Processes Stimulated by Electronic Excitation and Dissociative Attachment   134

3.6.4. Energy Balance and Energy Efficiency of Plasma-Chemical Processes Stimulated by Vibrational Excitation of Molecules   134

3.6.5. Components of Total Energy Efficiency: Excitation, Relaxation, and Chemical Factors   136

3.6.6. Energy Efficiency of Quasi-Equilibrium Plasma-Chemical Systems: Absolute, Ideal, and Super-Ideal Quenching   137

3.6.7. Mass and Energy Transfer Equations in Multi-Component Quasi-Equilibrium Plasma-Chemical Systems   137

3.6.8. Transfer Phenomena Influence on Energy Efficiency of Plasma-Chemical Processes   139

3.7. Elements of Plasma Electrodynamics   140

3.7.1. Ideal and Non-Ideal Plasmas   140

3.7.2. Plasma Polarization: Debye Shielding of Electric Field in Plasma   141

3.7.3. Plasmas and Sheaths: Physics of DC Sheaths   142

3.7.4. High-Voltage Sheaths: Matrix and Child Law Sheath Models   144

3.7.5. Electrostatic Plasma Oscillations: Langmuir or Plasma Frequency   145

3.7.6. Penetration of Slow-Changing Fields into Plasma: Skin Effect in Plasma   146

3.7.7. Magneto-Hydrodynamics: "Diffusion" of Magnetic Field and Magnetic Field Frozen in Plasma   146

3.7.8. Magnetic Pressure: Plasma Equilibrium in Magnetic Field and Pinch Effect   147

3.7.9. Two-Fluid Magneto-Hydrodynamics: Generalized Ohm's Law   149

3.7.10. Plasma Diffusion Across Magnetic Field   149

3.7.11. Magneto-Hydrodynamic Behavior of Plasma: Alfven Velocity and Magnetic Reynolds Number   150

3.7.12. High-Frequency Plasma Conductivity and Dielectric Permittivity   151

3.7.13. Propagation of Electromagnetic Waves in Plasma   153

3.7.14. Plasma Absorption and Reflection of Electromagnetic Waves: Bouguer Law: Critical Electron Density   154

Problems and Concept Questions   155

4 Electric Discharges in Plasma Chemistry   **157**

4.1. Fundamentals of Electric Breakdown, Streamer Processes, and Steady-State Regimes of Non-Equilibrium Electrical Discharges   157

4.1.1. Townsend Mechanism of Electric Breakdown and Paschen Curves   157

4.1.2. Spark Breakdown Mechanism: Streamer Concept   159

4.1.3. Meek Criterion of Streamer Formation: Streamer
Propagation Models 163
4.1.4. Streamers and Microdischarges 164
4.1.5. Interaction of Streamers and Microdischarges 166
4.1.6. Monte Carlo Modeling of Interaction of Streamers and
Microdischarges 167
4.1.7. Self-Organized Pattern of DBD Microdischarges due to
Streamer Interaction 168
4.1.8. Steady-State Regimes of Non-Equilibrium Electric
Discharges and General Regimes Controlled by Volume
and Surface Recombination Processes 170
4.1.9. Discharge Regime Controlled by Electron–Ion
Recombination 171
4.1.10. Discharge Regime Controlled by Electron Attachment 172
4.1.11. Non-Thermal Discharge Regime Controlled by
Charged-Particle Diffusion to the Walls: The
Engel-Steenbeck Relation 172
4.2. Glow Discharges 175
4.2.1. General Structure and Configurations of Glow
Discharges 175
4.2.2. Current-Voltage Characteristics of DC Discharges 177
4.2.3. Dark Discharge and Transition from Townsend Dark to
Glow Discharge 178
4.2.4. Current-Voltage Characteristics of Cathode Layer:
Normal Glow Discharge 179
4.2.5. Abnormal, Subnormal, and Obstructed Regimes of Glow
Discharges 181
4.2.6. Positive Column of Glow Discharge 182
4.2.7. Hollow Cathode Glow Discharge 183
4.2.8. Other Specific Glow Discharge Plasma Sources 184
4.2.9. Energy Efficiency Peculiarities of Glow Discharge
Application for Plasma-Chemical Processes 186
4.3. Arc Discharges 187
4.3.1. Classification and Current-Voltage Characteristics of Arc
Discharges 187
4.3.2. Cathode and Anode Layers of Arc Discharges 189
4.3.3. Cathode Spots in Arc Discharges 191
4.3.4. Positive Column of High-Pressure Arcs: Elenbaas-Heller
Equation 193
4.3.5. Steenbeck-Raizer "Channel" Model of Positive
Column 194
4.3.6. Steenbeck-Raizer Arc "Channel" Modeling of Plasma
Temperature, Specific Power, and Electric Field in Positive
Column 196
4.3.7. Configurations of Arc Discharges Applied in Plasma
Chemistry and Plasma Processing 197
4.3.8. Gliding Arc Discharge 200
4.3.9. Equilibrium Phase of Gliding Arc, Its Critical Parameters,
and Fast Equilibrium-to-Non-Equilibrium Transition 204
4.3.10. Gliding Arc Stability Analysis and Transitional and
Non-Equilibrium Phases of the Discharge 205

Contents

4.3.11. Special Configurations of Gliding Arc Discharges: Gliding
Arc Stabilized in Reverse Vortex (Tornado) Flow   207
4.4. Radiofrequency and Microwave Discharges in Plasma Chemistry   209
4.4.1. Generation of Thermal Plasma in Radiofrequency
Discharges   209
4.4.2. Thermal Plasma Generation in Microwave and Optical
Discharges   211
4.4.3. Non-Thermal Radiofrequency Discharges: Capacitive and
Inductive Coupling of Plasma   215
4.4.4. Non-Thermal RF-CCP Discharges in Moderate Pressure
Regimes   216
4.4.5. Low-Pressure Capacitively Coupled RF Discharges   219
4.4.6. RF Magnetron Discharges   222
4.4.7. Non-Thermal Inductively Coupled RF Discharges in
Cylindrical Coil   224
4.4.8. Planar-Coil and Other Configurations of Non-Thermal
Inductively Coupled RF Discharges   226
4.4.9. Non-Thermal Low-Pressure Microwave and Other
Wave-Heated Discharges   229
4.4.10. Non-Equilibrium Plasma-Chemical Microwave Discharges
of Moderate Pressure   231
4.5. Non-Thermal Atmospheric Pressure Discharges   233
4.5.1. Corona Discharges   233
4.5.2. Pulsed Corona Discharges   234
4.5.3. Dielectric Barrier Discharges   237
4.5.4. Special Modifications of DBD: Surface, Packed-Bed, and
Ferroelectric Discharges   239
4.5.5. Spark Discharges   240
4.5.6. Atmospheric Pressure Glow Mode of DBD   241
4.5.7. APGs: Resistive Barrier Discharge   242
4.5.8. One-Atmosphere Uniform Glow Discharge Plasma as
Another Modification of APG   243
4.5.9. Electronically Stabilized APG Discharges   244
4.5.10. Atmospheric-Pressure Plasma Jets   245
4.6. Microdischarges   247
4.6.1. General Features of Microdischarges   247
4.6.2. Micro-Glow Discharge   248
4.6.3. Micro-Hollow-Cathode Discharge   251
4.6.4. Arrays of Microdischarges: Microdischarge
Self-Organization and Structures   252
4.6.5. Kilohertz-Frequency-Range Microdischarges   254
4.6.6. RF Microdischarges   255
4.6.7. Microwave Microdischarges   257
Problems and Concept Questions   257

5 Inorganic Gas-Phase Plasma Decomposition Processes   259
5.1. CO_2: Dissociation in Plasma, Thermal, and Non-Thermal
Mechanisms   259
5.1.1. Fundamental and Applied Aspects of the CO_2 Plasma
Chemistry   259

Contents

5.1.2. Major Experimental Results on CO_2: Dissociation in Different Plasma Systems and Energy Efficiency of the Process   260

5.1.3. Mechanisms of CO_2 Decomposition in Quasi-Equilibrium Thermal Plasma   262

5.1.4. CO_2 Dissociation in Plasma, Stimulated by Vibrational Excitation of Molecules   263

5.1.5. CO_2 Dissociation in Plasma by Means of Electronic Excitation of Molecules   265

5.1.6. CO_2 Dissociation in Plasma by Means of Dissociative Attachment of Electrons   267

5.2. Physical Kinetics of CO_2 Dissociation, Stimulated by Vibrational Excitation of the Molecules in Non-Equilibrium Plasma   268

5.2.1. Asymmetric and Symmetric CO_2 Vibrational Modes   268

5.2.2. Contribution of Asymmetric and Symmetric CO_2 Vibrational Modes into Plasma-Chemical Dissociation Process   269

5.2.3. Transition of Highly Vibrationally Excited CO_2 Molecules into the Vibrational Quasi Continuum   271

5.2.4. One-Temperature Approximation of CO_2 Dissociation Kinetics in Non-Thermal Plasma   273

5.2.5. Two-Temperature Approximation of CO_2 Dissociation Kinetics in Non-Thermal Plasma   274

5.2.6. Elementary Reaction Rates of CO_2 Decomposition, Stimulated in Plasma by Vibrational Excitation of the Molecules   275

5.3. Vibrational Kinetics and Energy Balance of Plasma-Chemical CO_2 Dissociation   276

5.3.1. Two-Temperature Approach to Vibrational Kinetics and Energy Balance of CO_2 Dissociation in Non-Equilibrium Plasma: Major Energy Balance and Dynamic Equations   276

5.3.2. Two-Temperature Approach to Vibrational Kinetics and Energy Balance of CO_2 Dissociation in Non-Equilibrium Plasma: Additional Vibrational Kinetic Relations   277

5.3.3. Results of CO_2 Dissociation Modeling in the Two-Temperature Approach to Vibrational Kinetics   279

5.3.4. One-Temperature Approach to Vibrational Kinetics and Energy Balance of CO_2 Dissociation in Non-Equilibrium Plasma: Major Equations   280

5.3.5. Threshold Values of Vibrational Temperature, Specific Energy Input, and Ionization Degree for Effective Stimulation of CO_2 Dissociation by Vibrational Excitation of the Molecules   281

5.3.6. Characteristic Time Scales of CO_2 Dissociation in Plasma Stimulated by Vibrational Excitation of the Molecules: VT-Relaxation Time   282

5.3.7. Flow Velocity and Compressibility Effects on Vibrational Relaxation Kinetics During Plasma-Chemical CO_2 Dissociation: Maximum Linear Preheating Temperature   283

5.3.8. CO_2 Dissociation in Active and Passive Discharge Zones: Discharge (τ_{eV}) and After-Glow (τ_p) Residence Time   284

Contents

5.3.9. Ionization Degree Regimes of the CO_2 Dissociation Process in Non-Thermal Plasma   285

5.3.10. Energy Losses Related to Excitation of CO_2 Dissociation Products: Hyperbolic Behavior of Energy Efficiency Dependence on Specific Energy Input   286

5.4. Energy Efficiency of CO_2 Dissociation in Quasi-Equilibrium Plasma, and Non-Equilibrium Effects of Quenching Products of Thermal Dissociation   288

5.4.1. Ideal and Super-Ideal Modes of Quenching Products of CO_2 Dissociation in Thermal Plasma   288

5.4.2. Kinetic Evolution of Thermal CO_2 Dissociation Products During Quenching Phase   288

5.4.3. Energy Efficiency of CO_2 Dissociation in Thermal Plasma Under Conditions of Ideal Quenching of Products   289

5.4.4. Vibrational–Translational Non-Equilibrium Effects of Quenching Products of Thermal CO_2 Dissociation in Plasma: Super-Ideal Quenching Mode   290

5.4.5. Maximum Value of Energy Efficiency of CO_2 Dissociation in Thermal Plasma with Super-Ideal Quenching of the Dissociation Products   291

5.4.6. Kinetic Calculations of Energy Efficiency of CO_2 Dissociation in Thermal Plasma with Super-Ideal Quenching   291

5.4.7. Comparison of Thermal and Non-Thermal Plasma Approaches to CO_2 Dissociation: Comments on Products (CO-O_2) Oxidation and Explosion   292

5.5. Experimental Investigations of CO_2 Dissociation in Different Discharge Systems   293

5.5.1. Experiments with Non-Equilibrium Microwave Discharges of Moderate Pressure, Discharges in Waveguide Perpendicular to Gas Flow Direction, and Microwave Plasma Parameters in CO_2 293

5.5.2. Plasma-Chemical Experiments with Dissociation of CO_2 in Non-Equilibrium Microwave Discharges of Moderate Pressure   295

5.5.3. Experimental Diagnostics of Plasma-Chemical Non-Equilibrium Microwave Discharges in Moderate-Pressure CO_2: Plasma Measurements   296

5.5.4. Experimental Diagnostics of Plasma-Chemical Non-Equilibrium Microwave Discharges in Moderate-Pressure CO_2: Temperature Measurements   297

5.5.5. CO_2 Dissociation in Non-Equilibrium Radiofrequency Discharges: Experiments with Inductively Coupled Plasma   299

5.5.6. CO_2 Dissociation in Non-Equilibrium Radiofrequency Discharges: Experiments with Capacitively Coupled Plasma   300

5.5.7. CO_2 Dissociation in Non-Self-Sustained Atmospheric-Pressure Discharges Supported by High-Energy Electron Beams or UV Radiation   302

5.5.8. CO_2 Dissociation in Different Types of Glow Discharges   302

Contents

5.5.9. CO_2 Dissociation in Other Non-Thermal and Thermal
Discharges: Contribution of Vibrational and Electronic
Excitation Mechanisms   304

5.6. CO_2 Dissociation in Special Experimental Systems, Including
Supersonic Stimulation and Plasma Radiolysis   304

5.6.1. Dissociation of CO_2 in Supersonic Non-Equilibrium
Discharges: Advantages and Gasdynamic Characteristics   304

5.6.2. Kinetics and Energy Balance of Non-Equilibrium
Plasma-Chemical CO_2 Dissociation in Supersonic Flow   306

5.6.3. Limitations of Specific Energy Input and CO_2 Conversion
Degree in Supersonic Plasma Related to Critical Heat
Release and Choking the Flow   308

5.6.4. Experiments with Dissociation of CO_2 in Non-Equilibrium
Supersonic Microwave Discharges   308

5.6.5. Gasdynamic Stimulation of CO_2 Dissociation in Supersonic
Flow: "Plasma Chemistry Without Electricity"   309

5.6.6. Plasma Radiolysis of CO_2 Provided by High-Current
Relativistic Electron Beams   310

5.6.7. Plasma Radiolysis of CO_2 in Tracks of Nuclear Fission
Fragments   311

5.6.8. Ionization Degree in Tracks of Nuclear Fission Fragments,
Energy Efficiency of Plasma Radiolysis of CO_2, and
Plasma-Assisted Chemonuclear Reactors   313

5.7. Complete CO_2 Dissociation in Plasma with Production of
Carbon and Oxygen   314

5.7.1. Complete Plasma-Chemical Dissociation of CO_2: Specifics
of the Process and Elementary Reaction Mechanism   314

5.7.2. Kinetics of CO Disproportioning Stimulated in
Non-Equilibrium Plasma by Vibrational Excitation of
Molecules   314

5.7.3. Experiments with Complete CO_2 Dissociation in
Microwave Discharges Operating in Conditions of Electron
Cyclotron Resonance   316

5.7.4. Experiments with Complete CO_2 Dissociation in
Stationary Plasma-Beam Discharge   317

5.8. Dissociation of Water Vapor and Hydrogen Production in
Plasma-Chemical Systems   318

5.8.1. Fundamental and Applied Aspects of H_2O Plasma
Chemistry   318

5.8.2. Kinetics of Dissociation of Water Vapor Stimulated in
Non-Thermal Plasma by Vibrational Excitation of Water
Molecules   319

5.8.3. Energy Efficiency of Dissociation of Water Vapor
Stimulated in Non-Thermal Plasma by Vibrational
Excitation   320

5.8.4. Contribution of Dissociative Attachment of Electrons into
Decomposition of Water Vapor in Non-Thermal
Plasma   322

5.8.5. Kinetic Analysis of the Chain Reaction of H_2O Dissociation
via Dissociative Attachment/Detachment Mechanism   324

Contents

5.8.6. H_2O Dissociation in Thermal Plasma and Quenching of the Dissociation Products: Absolute and Ideal Quenching Modes   325

5.8.7. Cooling Rate Influence on Kinetics of H_2O Dissociation Products in Thermal Plasma: Super-Ideal Quenching Effect   326

5.8.8. Water Dissociation and H_2 Production in Plasma-Chemical System CO_2–H_2O 328

5.8.9. CO-to-H_2 Shift Reaction: Plasma Chemistry of CO–O_2–H_2O Mixture   330

5.9. Experimental Investigations of H_2O Dissociation in Different Discharge Systems   331

    5.9.1. Microwave Discharge in Water Vapor   331

    5.9.2. Plasma-Chemical Experiments with Microwave Discharge in Water Vapor   332

    5.9.3. Dissociation of Water Vapor in Glow Discharges   332

    5.9.4. Dissociation of H_2O with Production of H_2 and H_2O_2 in Supersonic Microwave Discharges   334

    5.9.5. Plasma Radiolysis of Water Vapor in Tracks of Nuclear Fission Fragments   335

    5.9.6. Effect of Plasma Radiolysis on Radiation Yield of Hydrogen Production in Tracks of Nuclear Fission Fragments   336

5.10. Inorganic Gas-Phase Plasma-Chemical Processes of Decomposition of Triatomic Molecules: NH_3, SO_2, and N_2O 336

    5.10.1. Gas-Phase Plasma Decomposition Reactions in Multi-Phase Technologies   336

    5.10.2. Dissociation of Ammonia in Non-Equilibrium Plasma: Mechanism of the Process in Glow Discharge   337

    5.10.3. Mechanism of Formation of Molecular Nitrogen and Hydrogen in Non-Equilibrium Plasma-Chemical Process of Ammonia Dissociation   338

    5.10.4. Plasma Dissociation of Sulfur Dioxide   338

    5.10.5. Destruction and Conversion of Nitrous Oxide in Non-Equilibrium Plasma   340

5.11. Non-Thermal and Thermal Plasma Dissociation of Diatomic Molecules   341

    5.11.1. Plasma-Chemical Decomposition of Hydrogen Halides: Example of HBr Dissociation with Formation of Hydrogen and Bromine   341

    5.11.2. Dissociation of HF, HCl, and HI in Plasma   343

    5.11.3. Non-Thermal and Thermal Dissociation of Molecular Fluorine   344

    5.11.4. Dissociation of Molecular Hydrogen in Non-Thermal and Thermal Plasma Systems   345

    5.11.5. Dissociation of Molecular Nitrogen in Non-Thermal and Thermal Plasma Systems   347

    5.11.6. Thermal Plasma Dissociation of Other Diatomic Molecules (O_2, Cl_2, Br_2)   347

    Problems and Concept Questions   351

Contents

6 Gas-Phase Inorganic Synthesis in Plasma   **355**

6.1. Plasma-Chemical Synthesis of Nitrogen Oxides from Air and
      Nitrogen–Oxygen Mixtures: Thermal and Non-Thermal
      Mechanisms   355

  6.1.1. Fundamental and Applied Aspects of NO Synthesis in Air
          Plasma   355

  6.1.2. Mechanisms of NO Synthesis Provided in Non-Thermal
          Plasma by Excitation of Neutral Molecules: Zeldovich
          Mechanism   356

  6.1.3. Mechanisms of NO Synthesis Provided in Non-Thermal
          Plasma by Charged Particles   358

  6.1.4. NO Synthesis in Thermal Plasma Systems   358

  6.1.5. Energy Efficiency of Different Mechanisms of NO Synthesis
          in Thermal and Non-Thermal Discharge Systems   359

6.2. Elementary Reaction of NO Synthesis Stimulated by Vibrational
      Excitation of Molecular Nitrogen   361

  6.2.1. Limiting Elementary Reaction of Zeldovich Mechanism:
          Adiabatic and Non-Adiabatic Channels of NO Synthesis   361

  6.2.2. Electronically Adiabatic Channel of NO Synthesis
          $O + N_2 \rightarrow NO + N$ Stimulated by Vibrational Excitation
          of Molecular Nitrogen   361

  6.2.3. Electronically Non-Adiabatic Channel of NO Synthesis
          $(O + N_2 \rightarrow NO + N)$: Stages of the Elementary Process
          and Method of Vibronic Terms   363

  6.2.4. Transition Probability Between Vibronic Terms
          Corresponding to Formation of Intermediate $N_2O^*(^1\Sigma^+)$
          Complex   364

  6.2.5. Probability of Formation of Intermediate $N_2O^*(^1\Sigma^+)$
          Complex in Electronically Non-Adiabatic Channel of NO
          Synthesis   365

  6.2.6. Decay of Intermediate Complex $N_2O^*(^1\Sigma^+)$: Second Stage
          of Electronically Non-Adiabatic Channel of NO Synthesis   366

  6.2.7. Total Probability of Electronically Non-Adiabatic Channel
          of NO Synthesis $(O + N_2 \rightarrow NO + N)$ 367

6.3. Kinetics and Energy Balance of Plasma-Chemical NO Synthesis
      Stimulated in Air and O_2–N_2 Mixtures by Vibrational Excitation   367

  6.3.1. Rate Coefficient of Reaction $O + N_2 \rightarrow NO + N$
          Stimulated in Non-Equilibrium Plasma by Vibrational
          Excitation of Nitrogen Molecules   367

  6.3.2. Energy Balance of Plasma-Chemical NO Synthesis:
          Zeldovich Mechanism Stimulated by Vibrational Excitation   368

  6.3.3. Macro-Kinetics of Plasma-Chemical NO Synthesis: Time
          Evolution of Vibrational Temperature   369

  6.3.4. Energy Efficiency of Plasma-Chemical NO Synthesis:
          Excitation and Relaxation Factors   370

  6.3.5. Energy Efficiency of Plasma-Chemical NO Synthesis:
          Chemical Factor   371

  6.3.6. Stability of Products of Plasma-Chemical Synthesis to
          Reverse Reactions in Active Zone of Non-Thermal Plasma   371

Contents

6.3.7. Effect of "Hot Nitrogen Atoms" on Yield of NO Synthesis in Non-Equilibrium Plasma in Air and Nitrogen–Oxygen Mixtures   372

6.3.8. Stability of Products of Plasma-Chemical NO Synthesis to Reverse Reactions Outside of the Discharge Zone   373

6.4. Experimental Investigations of NO Synthesis from Air and N_2–O_2 Mixtures in Different Discharges   374

6.4.1. Non-Equilibrium Microwave Discharge in Magnetic Field Operating in Conditions of Electron Cyclotron Resonance   374

6.4.2. Evolution of Vibrational Temperature of Nitrogen Molecules in Non-Equilibrium ECR: Microwave Discharge During Plasma-Chemical NO Synthesis   376

6.4.3. NO Synthesis in the Non-Equilibrium ECR Microwave Discharge   377

6.4.4. NO Synthesis in Non-Self-Sustained Discharges Supported by Relativistic Electron Beams   378

6.4.5. Experiments with NO Synthesis from Air in Stationary Non-Equilibrium Plasma-Beam Discharge   379

6.4.6. Experiments with NO Synthesis from N_2 and O_2 in Thermal Plasma of Arc Discharges   380

6.4.7. General Schematic and Parameters of Industrial Plasma-Chemical Technology of NO Synthesis from Air   381

6.5. Plasma-Chemical Ozone Generation: Mechanisms and Kinetics   382

6.5.1. Ozone Production as a Large-Scale Industrial Application of Non-Thermal Atmospheric-Pressure Plasma   382

6.5.2. Energy Cost and Energy Efficiency of Plasma-Chemical Production of Ozone in Some Experimental and Industrial Systems   383

6.5.3. Plasma-Chemical Ozone Formation in Oxygen   383

6.5.4. Optimum DBD Microdischarge Strength and Maximization of Energy Efficiency of Ozone Production in Oxygen Plasma   385

6.5.5. Plasma-Chemical Ozone Generation in Air   386

6.5.6. Discharge Poisoning Effect During Ozone Generation in Air Plasma   387

6.5.7. Temperature Effect on Plasma-Chemical Generation and Stability of Ozone   388

6.5.8. Negative Effect of Water Vapor on Plasma-Chemical Ozone Synthesis   389

6.5.9. Effect of Hydrogen, Hydrocarbons, and Other Admixtures on Plasma-Chemical Ozone Synthesis   390

6.6. Experimental and Industrial Plasma-Chemical Ozone Generators   392

6.6.1. Synthesis of Ozone in Dielectric Barrier Discharges as the Oldest and Still Most Successful Approach to Ozone Generation   392

6.6.2. Tubular DBD Ozone Generators and Large Ozone Production Installations   392

6.6.3. Planar and Surface Discharge Configurations of DBD Ozone Generators   394

6.6.4. Synthesis of Ozone in Pulsed Corona Discharges   395

Contents

6.6.5. Peculiarities of Ozone Synthesis in Pulsed Corona with Respect to DBD ... 396

6.6.6. Possible Specific Contribution of Vibrational Excitation of Molecules to Ozone Synthesis in Pulsed Corona Discharges ... 397

6.7. Synthesis of KrF_2 and Other Aggressive Fluorine Oxidizers ... 399

6.7.1. Plasma-Chemical Gas-Phase Synthesis of KrF_2 and Mechanism of Surface Stabilization of Reaction Products ... 399

6.7.2. Physical Kinetics of KrF_2 Synthesis in Krypton Matrix ... 400

6.7.3. Synthesis of KrF_2 in Glow Discharges, Barrier Discharges, and Photo-Chemical Systems ... 401

6.7.4. Synthesis of KrF_2 in Non-Equilibrium Microwave Discharge in Magnetic Field ... 402

6.7.5. Plasma F_2 Dissociation as the First Step in Synthesis of Aggressive Fluorine Oxidizers ... 402

6.7.6. Plasma-Chemical Synthesis of O_2F_2 and Other Oxygen Fluorides ... 403

6.7.7. Plasma-Chemical Synthesis of NF_3 and Other Nitrogen Fluorides ... 404

6.7.8. Plasma-Chemical Synthesis of Xenon Fluorides and Other Fluorine Oxidizers ... 405

6.8. Plasma-Chemical Synthesis of Hydrazine (N_2H_4), Ammonia (NH_3), Nitrides of Phosphorus, and Some Other Inorganic Compounds ... 406

6.8.1. Direct Plasma-Chemical Hydrazine (N_2H_4) Synthesis from Nitrogen and Hydrogen in Non-Equilibrium Discharges ... 406

6.8.2. Hydrazine (N_2H_4) Synthesis from N_2–H_2 Mixture in Non-Self-Sustained Stationary Discharge Supported by Electron Beam ... 407

6.8.3. Kinetics of Hydrazine (N_2H_4) Synthesis from N_2–H_2 Mixture in Non-Thermal Plasma Conditions ... 407

6.8.4. Synthesis of Ammonia in DBD and Glow Discharges ... 408

6.8.5. Plasma-Chemical Synthesis of Nitrides of Phosphorus ... 409

6.8.6. Sulfur Gasification by Carbon Dioxide in Non-Thermal and Thermal Plasmas ... 409

6.8.7. CN and NO Synthesis in CO–N_2 Plasma ... 412

6.8.8. Gas-Phase Synthesis Related to Plasma-Chemical Oxidation of HCl and SO_2 ... 413

Problems and Concept Questions ... 414

7 **Plasma Synthesis, Treatment, and Processing of Inorganic Materials, and Plasma Metallurgy** ... 417

7.1. Plasma Reduction of Oxides of Metals and Other Elements ... 417

7.1.1. Thermal Plasma Reduction of Iron Ore, Iron Production from Oxides Using Hydrogen and Hydrocarbons, and Plasma-Chemical Steel Manufacturing ... 417

7.1.2. Productivity and Energy Efficiency of Thermal Plasma Reduction of Iron Ore ... 419

7.1.3. Hydrogen Reduction of Refractory Metal Oxides in Thermal Plasma and Plasma Metallurgy of Tungsten and Molybdenum ... 420

Contents

7.1.4. Thermal Plasma Reduction of Oxides of Aluminum and Other Inorganic Elements   423

7.1.5. Reduction of Metal Oxides and Production of Metals Using Non-Thermal Hydrogen Plasma   425

7.1.6. Non-Equilibrium Surface Heating and Evaporation Effect in Heterogeneous Plasma-Chemical Processes in Non-Thermal Discharges   426

7.1.7. Non-Equilibrium Surface Heating and Evaporation in Plasma Treatment of Thin Layers of Flat Surfaces: Effect of Short Pulses   427

7.2. Production of Metals and Other Elements by Carbothermic Reduction and Direct Decomposition of Their Oxides in Thermal Plasma   429

    7.2.1. Carbothermic Reduction of Elements from Their Oxides   429

    7.2.2. Production of Pure Metallic Uranium by Carbothermic Plasma-Chemical Reduction of Uranium Oxides   429

    7.2.3. Production of Niobium by Carbothermic Plasma-Chemical Reduction of Niobium Oxides   430

    7.2.4. Double-Stage Carbothermic Thermal Plasma Reduction of Rare and Refractory Metals from Their Oxides   430

    7.2.5. Carbothermic Reduction of Iron from Iron Titanium Oxide Concentrates in a Thermal Plasma Fluidized Bed   431

    7.2.6. Production of Silicon Monoxide by SiO_2 Decomposition in Thermal Plasma   432

    7.2.7. Experiments with SiO_2 Reduction to Pure Silicon Monoxide in High-Temperature Radiofrequency ICP Discharges   433

    7.2.8. Reduction of Aluminum by Direct Thermal Plasma Decomposition of Alumina   434

    7.2.9. Reduction of Vanadium by Direct Plasma Decomposition of Its Oxides, V_2O_5 and V_2O_3 436

    7.2.10. Reduction of Indium and Germanium by Direct Plasma Decomposition of Their Oxides   439

7.3. Hydrogen Plasma Reduction of Metals and Other Elements from Their Halides   440

    7.3.1. Using Halides for Production of Metals and Other Elements from Their Compounds   440

    7.3.2. Plasma-Chemical Production of Boron: Thermal Plasma Reduction of BCl_3 with Hydrogen   441

    7.3.3. Hydrogen Reduction of Niobium from Its Pentachloride ($NiCl_5$) in Thermal Plasma   442

    7.3.4. Hydrogen Reduction of Uranium from Its Hexafluoride (UF_6) in Thermal Plasma   442

    7.3.5. Hydrogen Reduction of Tantalum (Ta), Molybdenum (Mo), Tungsten (W), Zirconium (Zr), and Hafnium (Hf) from Their Chlorides in Thermal Plasma   443

    7.3.6. Hydrogen Reduction of Titanium (Ti), Germanium (Ge), and Silicon (Si) from Their Tetrachlorides in Thermal Plasma   445

    7.3.7. Thermal Plasma Reduction of Some Other Halides with Hydrogen: Plasma Production of Intermetallic Compounds   446

Contents

7.3.8. Hydrogen Reduction of Halides in Non-Thermal Plasma   448

7.4. Direct Decomposition of Halides in Thermal and Non-Thermal Plasma   448

7.4.1. Direct Decomposition of Halides and Production of Metals in Plasma   448

7.4.2. Direct UF_6 Decomposition in Thermal Plasma: Requirements for Effective Product Quenching   449

7.4.3. Direct Decomposition of Halides of Some Alkali and Alkaline Earth Metals in Thermal Plasma   451

7.4.4. Direct Thermal Plasma Decomposition of Halides of Aluminum, Silicon, Arsenic, and Some Other Elements of Groups 3, 4, and 5   457

7.4.5. Direct Thermal Plasma Decomposition of Halides of Titanium (Ti), Zirconium (Zr), Hafnium (Hf), Vanadium (V), and Niobium (Nb)   461

7.4.6. Direct Decomposition of Halides of Iron (Fe), Cobalt (Co), Nickel (Ni), and Other Transition Metals in Thermal Plasma   465

7.4.7. Direct Decomposition of Halides and Reduction of Metals in Non-Thermal Plasma   469

7.4.8. Kinetics of Dissociation of Metal Halides in Non-Thermal Plasma: Distribution of Halides over Oxidation Degrees   470

7.4.9. Heterogeneous Stabilization of Products During Direct Decomposition of Metal Halides in Non-Thermal Plasma: Application of Plasma Centrifuges for Product Quenching   472

7.5. Plasma-Chemical Synthesis of Nitrides and Carbides of Inorganic Materials   472

7.5.1. Plasma-Chemical Synthesis of Metal Nitrides from Elements: Gas-Phase and Heterogeneous Reaction Mechanisms   472

7.5.2. Synthesis of Nitrides of Titanium and Other Elements by Plasma-Chemical Conversion of Their Chlorides   473

7.5.3. Synthesis of Silicon Nitride (Si_3N_4) and Oxynitrides by Non-Thermal Plasma Conversion of Silane (SiH_4)   474

7.5.4. Production of Metal Carbides by Solid-Phase Synthesis in Thermal Plasma of Inert Gases   475

7.5.5. Synthesis of Metal Carbides by Reaction of Solid Metal Oxides with Gaseous Hydrocarbons in Thermal Plasma   475

7.5.6. Gas-Phase Synthesis of Carbides in Plasma-Chemical Reactions of Halides with Hydrocarbons   475

7.5.7. Conversion of Solid Oxides into Carbides Using Gaseous Hydrocarbons Inside of RF-ICP Thermal Plasma Discharge and Some Other Plasma Technologies for Carbide Synthesis   477

7.6. Plasma-Chemical Production of Inorganic Oxides by Thermal Decomposition of Minerals, Aqueous Solutions, and Conversion Processes   477

7.6.1. Plasma Production of Zirconia (ZrO_2) by Decomposition of Zircon Sand ($ZrSiO_4$)   477

7.6.2. Plasma Production of Manganese Oxide (MnO) by Decomposition of Rhodonite ($MnSiO_3$)   478

Contents

7.6.3. Plasma-Chemical Extraction of Nickel from Serpentine
Minerals   482

7.6.4. Production of Uranium Oxide (U_3O_8) by Thermal Plasma
Decomposition of Uranyl Nitrate ($UO_2(NO_3)_2$) Aqueous
Solutions   482

7.6.5. Production of Magnesium Oxide (MgO) by Thermal
Plasma Decomposition of Aqueous Solution or Melt of
Magnesium Nitrate ($Mg(NO_3)_2$)   483

7.6.6. Plasma-Chemical Production of Oxide Powders for
Synthesis of High-Temperature Superconducting
Composites   483

7.6.7. Production of Uranium Oxide (U_3O_8) by Thermal Plasma
Conversion of Uranium Hexafluoride (UF_6) with Water
Vapor   484

7.6.8. Conversion of Silicon Tetrafluoride (SiF_4) with Water
Vapor into Silica (SiO_2) and HF in Thermal Plasma   484

7.6.9. Production of Pigment Titanium Dioxide (TiO_2) by
Thermal Plasma Conversion of Titanium Tetrachloride
($TiCl_4$) in Oxygen   485

7.6.10. Thermal Plasma Conversion of Halides in Production of
Individual and Mixed Oxides of Chromium, Aluminum,
and Titanium   486

7.6.11. Thermal Plasma Treatment of Phosphates: Tricalcium
Phosphate ($Ca_3(PO_4)_2$) and Fluoroapatite ($Ca_5F(PO_4)_3$)   487

7.6.12. Oxidation of Phosphorus and Production of Phosphorus
Oxides in Air Plasma   488

7.7. Plasma-Chemical Production of Hydrides, Borides, Carbonyls,
and Other Compounds of Inorganic Materials   488

7.7.1. Production of Hydrides in Thermal and Non-Thermal
Plasma   488

7.7.2. Non-Thermal Plasma Mechanisms of Hydride Formation
by Hydrogen Gasification of Elements and by
Hydrogenation of Thin Films   489

7.7.3. Synthesis of Metal Carbonyls in Non-Thermal Plasma:
Effect of Vibrational Excitation of CO Molecules on
Carbonyl Synthesis   490

7.7.4. Plasma-Chemical Synthesis of Borides of Inorganic
Materials   492

7.7.5. Synthesis of Intermetallic Compounds in Thermal Plasma   493

7.8. Plasma Cutting, Welding, Melting, and Other High-Temperature
Inorganic Material Processing Technologies   493

7.8.1. Plasma Cutting Technology   493

7.8.2. Plasma Welding Technology   494

7.8.3. About Plasma Melting and Remelting of Metals   495

7.8.4. Plasma Spheroidization and Densification of Powders   495

Problems and Concept Questions   496

8 **Plasma-Surface Processing of Inorganic Materials: Micro- and
Nano-Technologies**   499

8.1. Thermal Plasma Spraying   499

8.1.1. Plasma Spraying as a Thermal Spray Technology   499

Contents

8.1.2. DC-Arc Plasma Spray: Air Plasma Spray   500

8.1.3. DC-Arc Plasma Spray: VPS, LPPS, CAPS, SPS, UPS, and Other Specific Spray Approaches   501

8.1.4. Radiofrequency Plasma Spray   502

8.1.5. Thermal Plasma Spraying of Monolithic Materials   503

8.1.6. Thermal Plasma Spraying of Composite Materials   505

8.1.7. Thermal Spray Technologies: Reactive Plasma Spray Forming   507

8.1.8. Thermal Plasma Spraying of Functionally Gradient Materials   508

8.1.9. Thermal Plasma Spray Modeling   510

8.2. Plasma-Chemical Etching: Mechanisms and Kinetics   510

8.2.1. Main Principles of Plasma Etching as Part of Integrated Circuit Fabrication Technology   510

8.2.2. Etch Rate, Anisotropy, Selectivity, and Other Plasma Etch Requirements   511

8.2.3. Basic Plasma Etch Processes: Sputtering   514

8.2.4. Basic Plasma Etch Processes: Pure Chemical Etching   515

8.2.5. Basic Plasma Etch Processes: Ion Energy-Driven Etching   516

8.2.6. Basic Plasma Etch Processes: Ion-Enhanced Inhibitor Etching   516

8.2.7. Surface Kinetics of Etching Processes; Kinetics of Ion Energy-Driven Etching   517

8.2.8. Discharges Applied for Plasma Etching: RF-CCP Sources, RF Diodes and Triodes, and MERIEs   519

8.2.9. Discharges Applied for Plasma Etching: High-Density Plasma Sources   520

8.2.10. Discharge Kinetics in Etching Processes: Ion Density and Ion Flux   520

8.2.11. Discharge Kinetics in Etching Processes: Density and Flux of Neutral Etchants   521

8.3. Specific Plasma-Chemical Etching Processes   523

8.3.1. Gas Composition in Plasma Etching Processes: Etchants-to-Unsaturates Flux Ratio   523

8.3.2. Pure Chemical F-Atom Etching of Silicon: Flamm Formulas and Doping Effect   523

8.3.3. Ion Energy-Driven F-Atom Etching Process: Main Etching Mechanisms   524

8.3.4. Plasma Etching of Silicon in CF_4 Discharges: Kinetics of Fluorine Atoms   525

8.3.5. Plasma Etching of Silicon in CF_4 Discharges: Kinetics of CF_x Radicals and Competition Between Etching and Carbon Deposition   526

8.3.6. Plasma Etching of Silicon by Cl Atoms   528

8.3.7. Plasma Etching of SiO_2 by F Atoms and CF_x Radicals   529

8.3.8. Plasma Etching of Silicon Nitride (Si_3N_4)   529

8.3.9. Plasma Etching of Aluminum   529

8.3.10. Plasma Etching of Photoresist   530

8.3.11. Plasma Etching of Refractory Metals and Semiconductors   530

8.4. Plasma Cleaning of CVD and Etching Reactors in Micro-Electronics and Other Plasma Cleaning Processes   531

Contents

8.4.1. In Situ Plasma Cleaning in Micro-Electronics and Related Environmental Issues   531

8.4.2. Remote Plasma Cleaning Technology in Microelectronics: Choice of Cleaning Feedstock Gas   532

8.4.3. Kinetics of F-Atom Generation from NF_3, CF_4, and C_2F_6 in Remote Plasma Sources   533

8.4.4. Surface and Volume Recombination of F Atoms in Transport Tube   535

8.4.5. Effectiveness of F Atom Transportation from Remote Plasma Source   538

8.4.6. Other Plasma Cleaning Processes: Passive Plasma Cleaning   539

8.4.7. Other Plasma Cleaning Processes: Active Plasma Cleaning   540

8.4.8. Wettability Improvement of Metallic Surfaces by Active and Passive Plasma Cleaning   541

8.5. Plasma Deposition Processes: Plasma-Enhanced Chemical Vapor Deposition and Sputtering Deposition   541

    8.5.1. Plasma-Enhanced Chemical Vapor Deposition: General Principles   541

    8.5.2. PECVD of Thin Films of Amorphous Silicon   542

    8.5.3. Kinetics of Amorphous Silicon Film Deposition in Silane (SiH_4) Discharges   543

    8.5.4. Plasma Processes of Silicon Oxide (SiO_2) Film Growth: Direct Silicon Oxidation   545

    8.5.5. Plasma Processes of Silicon Oxide (SiO_2) Film Growth: PECVD from Silane–Oxygen Feedstock Mixtures and Conformal and Non-Conformal Deposition Within Trenches   545

    8.5.6. Plasma Processes of Silicon Oxide (SiO_2) Film Growth: PECVD from TEOS–O_2 Feed-Gas Mixtures   547

    8.5.7. PECVD Process of Silicon Nitride (Si_3N_4)   547

    8.5.8. Sputter Deposition Processes: General Principles   548

    8.5.9. Physical Sputter Deposition   548

    8.5.10. Reactive Sputter Deposition Processes   549

    8.5.11. Kinetics of Reactive Sputter Deposition: Hysteresis Effect   550

8.6. Ion Implantation Processes: Ion-Beam Implantation and Plasma-Immersion Ion Implantation   551

    8.6.1. Ion-Beam Implantation   551

    8.6.2. Plasma-Immersion Ion Implantation: General Principles   552

    8.6.3. Dynamics of Sheath Evolution in Plasma-Immersion Ion Implantation: From Matrix Sheath to Child Law Sheath   553

    8.6.4. Time Evolution of Implantation Current in PIII Systems   555

    8.6.5. PIII Applications for Processing Semiconductor Materials   556

    8.6.6. PIII Applications for Modifying Metallurgical Surfaces: Plasma Source Ion Implantation   557

8.7. Microarc (Electrolytic-Spark) Oxidation Coating and Other Microdischarge Surface Processing Systems   557

    8.7.1. Microarc (Electrolytic-Spark) Oxidation Coating: General Features   557

    8.7.2. Major Characteristics of the Microarc (Electrolytic-Spark) Oxidation Process   558

8.7.3. Mechanism of Microarc (Electrolytic-Spark) Oxidation Coating of Aluminum in Sulfuric Acid — 559

8.7.4. Breakdown of Oxide Film and Starting Microarc Discharge — 560

8.7.5. Microarc Discharge Plasma Chemistry of Oxide Coating Deposition on Aluminum in Concentrated Sulfuric Acid Electrolyte — 562

8.7.6. Direct Micropatterning and Microfabrication in Atmospheric-Pressure Microdischarges — 563

8.7.7. Microetching, Microdeposition, and Microsurface Modification by Atmospheric-Pressure Microplasma Discharges — 564

8.8. Plasma Nanotechnologies: Nanoparticles and Dusty Plasmas — 566

8.8.1. Nanoparticles in Plasma: Kinetics of Dusty Plasma Formation in Low-Pressure Silane Discharges — 566

8.8.2. Formation of Nanoparticles in Silane: Plasma Chemistry of Birth and Catastrophic Evolution — 567

8.8.3. Critical Phenomena in Dusty Plasma Kinetics: Nucleation of Nanoparticles, Winchester Mechanism, and Growth of First Generation of Negative Ion Clusters — 570

8.8.4. Critical Size of Primary Nanoparticles in Silane Plasma — 572

8.8.5. Critical Phenomenon of Neutral-Particle Trapping in Silane Plasma — 573

8.8.6. Critical Phenomenon of Super-Small Nanoparticle Coagulation — 575

8.8.7. Critical Change of Plasma Parameters due to Formation of Nanoparticles: $\alpha-\gamma$ Transition — 577

8.8.8. Other Processes of Plasma Production of Nanoparticles: Synthesis of Aluminum Nanopowder and Luminescent Silicon Quantum Dots — 579

8.8.9. Plasma Synthesis of Nanocomposite Particles — 580

8.9. Plasma Nanotechnologies: Synthesis of Fullerenes and Carbon Nanotubes — 581

8.9.1. Highly Organized Carbon Nanostructures: Fullerenes and Carbon Nanotubes — 581

8.9.2. Plasma Synthesis of Fullerenes — 583

8.9.3. Plasma Synthesis of Endohedral Fullerenes — 583

8.9.4. Plasma Synthesis of Carbon Nanotubes by Dispersion of Thermal Arc Electrodes — 584

8.9.5. Plasma Synthesis of Carbon Nanotubes by Dissociation of Carbon Compounds — 584

8.9.6. Surface Modification of Carbon Nanotubes by RF Plasma — 585

Problems and Concept Questions — 586

9 Organic and Polymer Plasma Chemistry — 589

9.1. Thermal Plasma Pyrolysis of Methane and Other Hydrocarbons: Production of Acetylene and Ethylene — 589

9.1.1. Kinetics of Thermal Plasma Pyrolysis of Methane and Other Hydrocarbons: The Kassel Mechanism — 589

9.1.2. Kinetics of Double-Step Plasma Pyrolysis of Hydrocarbons — 591

9.1.3. Electric Cracking of Natural Gas with Production of Acetylene–Hydrogen or Acetylene–Ethylene–Hydrogen Mixtures   591

9.1.4. Other Processes and Regimes of Hydrocarbon Conversion in Thermal Plasma   592

9.1.5. Some Chemical Engineering Aspects of Plasma Pyrolysis of Hydrocarbons   595

9.1.6. Production of Vinyl Chloride as an Example of Technology Based on Thermal Plasma Pyrolysis of Hydrocarbons   596

9.1.7. Plasma Pyrolysis of Hydrocarbons with Production of Soot and Hydrogen   597

9.1.8. Thermal Plasma Production of Acetylene by Carbon Vapor Reaction with Hydrogen or Methane   598

9.2. Conversion of Methane into Acetylene and Other Processes of Gas-Phase Conversion of Hydrocarbons in Non-Thermal Plasmas   598

9.2.1. Energy Efficiency of CH_4 Conversion into Acetylene in Thermal and Non-Thermal Plasmas   598

9.2.2. High-Efficiency CH_4 Conversion into C_2H_2 in Non-Thermal Moderate-Pressure Microwave Discharges   598

9.2.3. Limits of Quasi-Equilibrium Kassel Kinetics for Plasma Conversion of CH_4 into C_2H_2 600

9.2.4. Contribution of Vibrational Excitation to Methane Conversion into Acetylene in Non-Equilibrium Discharge Conditions   601

9.2.5. Non-Equilibrium Kinetics of Methane Conversion into Acetylene Stimulated by Vibrational Excitation   602

9.2.6. Other Processes of Decomposition, Elimination, and Isomerization of Hydrocarbons in Non-Equilibrium Plasma: Plasma Catalysis   603

9.3. Plasma Synthesis and Conversion of Organic Nitrogen Compounds   604

9.3.1. Synthesis of Dicyanogen (C_2N_2) from Carbon and Nitrogen in Thermal Plasma   604

9.3.2. Co-Production of Hydrogen Cyanide (HCN) and Acetylene (C_2H_2) from Methane and Nitrogen in Thermal Plasma Systems   605

9.3.3. Hydrogen Cyanide (HCN) Production from Methane and Nitrogen in Non-Thermal Plasma   606

9.3.4. Production of HCN and H_2 in CH_4–NH_3 Mixture in Thermal and Non-Thermal Plasmas   608

9.3.5. Thermal and Non-Thermal Plasma Conversion Processes in CO–N_2 Mixture   609

9.3.6. Other Non-Equilibrium Plasma Processes of Organic Nitrogen Compounds Synthesis   610

9.4. Organic Plasma Chemistry of Chlorine and Fluorine Compounds   611

9.4.1. Thermal Plasma Synthesis of Reactive Mixtures for Production of Vinyl Chloride   611

9.4.2. Thermal Plasma Pyrolysis of Dichloroethane, Butyl Chloride, Hexachlorane, and Other Organic Chlorine Compounds for Further Synthesis of Vinyl Chloride   612

9.4.3. Thermal Plasma Pyrolysis of Organic Fluorine Compounds   613

Contents

9.4.4. Pyrolysis of Organic Fluorine Compounds in Thermal Plasma of Nitrogen: Synthesis of Nitrogen-Containing Fluorocarbons 614

9.4.5. Thermal Plasma Pyrolysis of Chlorofluorocarbons 614

9.4.6. Non-Thermal Plasma Conversion of CFCs and Other Plasma Processes with Halogen-Containing Organic Compounds 616

9.5. Plasma Synthesis of Aldehydes, Alcohols, Organic Acids, and Other Oxygen-Containing Organic Compounds 617

9.5.1. Non-Thermal Plasma Direct Synthesis of Methanol from Methane and Carbon Dioxide 617

9.5.2. Non-Thermal Plasma Direct Synthesis of Methanol from Methane and Water Vapor 617

9.5.3. Production of Formaldehyde (CH_2O) by CH_4 Oxidation in Thermal and Non-Thermal Plasmas 618

9.5.4. Non-Thermal Plasma Oxidation of Methane and Other Hydrocarbons with Production of Methanol and Other Organic Compounds 619

9.5.5. Non-Thermal Plasma Synthesis of Aldehydes, Alcohols, and Organic Acids in Mixtures of Carbon Oxides with Hydrogen: Organic Synthesis in $CO_2–H_2O$ Mixture 620

9.5.6. Non-Thermal Plasma Production of Methane and Acetylene from Syngas ($CO–H_2$) 621

9.6. Plasma-Chemical Polymerization of Hydrocarbons: Formation of Thin Polymer Films 622

9.6.1. General Features of Plasma Polymerization 622

9.6.2. General Aspects of Mechanisms and Kinetics of Plasma Polymerization 622

9.6.3. Initiation of Polymerization by Dissociation of Hydrocarbons in Plasma Volume 623

9.6.4. Heterogeneous Mechanisms of Plasma-Chemical Polymerization of C_1/C_2 Hydrocarbons 625

9.6.5. Plasma-Initiated Chain Polymerization: Mechanisms of Plasma Polymerization of Methyl Methacrylate 625

9.6.6. Plasma-Initiated Graft Polymerization 626

9.6.7. Formation of Polymer Macroparticles in Volume of Non-Thermal Plasma in Hydrocarbons 627

9.6.8. Plasma-Chemical Reactors for Deposition of Thin Polymer Films 628

9.6.9. Some Specific Properties of Plasma-Polymerized Films 628

9.6.10. Electric Properties of Plasma-Polymerized Films 630

9.6.11. Some Specific Applications of Plasma-Polymerized Film Deposition 631

9.7. Interaction of Non-Thermal Plasma with Polymer Surfaces: Fundamentals of Plasma Modification of Polymers 632

9.7.1. Plasma Treatment of Polymer Surfaces 632

9.7.2. Major Initial Chemical Products Created on Polymer Surfaces During Their Interaction with Non-Thermal Plasma 633

Contents

9.7.3. Kinetics of Formation of Main Chemical Products in Process of Polyethylene Treatment in Pulsed RF Discharges — 634

9.7.4. Kinetics of Polyethylene Treatment in Continuous RF Discharge — 636

9.7.5. Non-Thermal Plasma Etching of Polymer Materials — 636

9.7.6. Contribution of Electrons and Ultraviolet Radiation in the Chemical Effect of Plasma Treatment of Polymer Materials — 637

9.7.7. Interaction of Atoms, Molecules, and Other Chemically Active Heavy Particles Generated in Non-Thermal Plasma with Polymer Materials: Plasma-Chemical Oxidation of Polymer Surfaces — 638

9.7.8. Plasma-Chemical Nitrogenation of Polymer Surfaces — 639

9.7.9. Plasma-Chemical Fluorination of Polymer Surfaces — 640

9.7.10. Synergetic Effect of Plasma-Generated Active Atomic/Molecular Particles and UV Radiation During Plasma Interaction with Polymers — 640

9.7.11. Aging Effect in Plasma-Treated Polymers — 641

9.8. Applications of Plasma Modification of Polymer Surfaces — 641

9.8.1. Plasma Modification of Wettability of Polymer Surfaces — 641

9.8.2. Plasma Enhancement of Adhesion of Polymer Surfaces: Metallization of Polymer Surfaces — 643

9.8.3. Plasma Modification of Polymer Fibers and Polymer Membranes — 645

9.8.4. Plasma Treatment of Textile Fibers: Treatment of Wool — 645

9.8.5. Plasma Treatment of Textile Fibers: Treatment of Cotton and Synthetic Textiles and the Lotus Effect — 648

9.8.6. Specific Conditions and Results of Non-Thermal Plasma Treatment of Textiles — 649

9.8.7. Plasma-Chemical Processes for Final Fabric Treatment — 649

9.8.8. Plasma-Chemical Treatment of Plastics, Rubber Materials, and Special Polymer Films — 654

9.9. Plasma Modification of Gas-Separating Polymer Membranes — 655

9.9.1. Application of Polymer Membranes for Gas Separation: Enhancement of Polymer Membrane Selectivity by Plasma Polymerization and by Plasma Modification of Polymer Surfaces — 655

9.9.2. Microwave Plasma System for Surface Modification of Gas-Separating Polymer Membranes — 656

9.9.3. Influence of Non-Thermal Discharge Treatment Parameters on Permeability of Plasma-Modified Gas-Separating Polymer Membranes — 657

9.9.4. Plasma Enhancement of Selectivity of Gas-Separating Polymer Membranes — 659

9.9.5. Chemical and Structural Modification of Surface Layers of Gas-Separating Polymer Membranes by Microwave Plasma Treatment — 661

9.9.6. Theoretical Model of Modification of Polymer Membrane Surfaces in After-Glow of Oxygen-Containing Plasma of Non-Polymerizing Gases: Lame Equation — 662

Contents

9.9.7. Elasticity/Electrostatics Similarity Approach to
Permeability of Plasma-Treated Polymer Membranes   663

9.9.8. Effect of Cross-Link's Mobility and Clusterization on
Permeability of Plasma-Treated Polymer Membranes   664

9.9.9. Modeling of Selectivity of Plasma-Treated Gas-Separating
Polymer Membranes   666

9.9.10. Effect of Initial Membrane Porosity on Selectivity
of Plasma-Treated Gas-Separating Polymer Membranes   667

9.10. Plasma-Chemical Synthesis of Diamond Films   668

9.10.1. General Features of Diamond-Film Production and
Deposition in Plasma   668

9.10.2. Different Discharge Systems Applied for Synthesis of
Diamond Films   669

9.10.3. Non-Equilibrium Discharge Conditions and Gas-Phase
Plasma-Chemical Processes in the Systems Applied for
Synthesis of Diamond Films   671

9.10.4. Surface Chemical Processes of Diamond-Film Growth
in Plasma   672

9.10.5. Kinetics of Diamond-Film Growth   673

Problems and Concept Questions   674

10   Plasma-Chemical Fuel Conversion and Hydrogen Production   **676**

10.1. Plasma-Chemical Conversion of Methane, Ethane, Propane, and
Natural Gas into Syngas ($CO-H_2$) and Other Hydrogen-Rich
Mixtures   676

10.1.1. General Features of Plasma-Assisted Production of
Hydrogen from Hydrocarbons: Plasma Catalysis   676

10.1.2. Syngas Production by Partial Oxidation of Methane in
Different Non-Equilibrium Plasma Discharges, Application
of Gliding Arc Stabilized in Reverse Vortex (Tornado)
Flow   678

10.1.3. Plasma Catalysis for Syngas Production by Partial
Oxidation of Methane in Non-Equilibrium Gliding Arc
Stabilized in Reverse Vortex (Tornado) Flow   681

10.1.4. Non-Equilibrium Plasma-Catalytic Syngas Production from
Mixtures of Methane with Water Vapor   683

10.1.5. Non-Equilibrium Plasma-Chemical Syngas Production
from Mixtures of Methane with Carbon Dioxide   685

10.1.6. Plasma-Catalytic Direct Decomposition (Pyrolysis) of
Ethane in Atmospheric-Pressure Microwave
Discharges   687

10.1.7. Plasma Catalysis in the Process of Hydrogen Production
by Direct Decomposition (Pyrolysis) of Methane   688

10.1.8. Mechanism of Plasma Catalysis of Direct CH_4
Decomposition in Non-Equilibrium Discharges   689

10.1.9. Plasma-Chemical Conversion of Propane,
Propane–Butane Mixtures, and Other Gaseous
Hydrocarbons to Syngas and Other Hydrogen-Rich
Mixtures   690

Contents

10.2. Plasma-Chemical Reforming of Liquid Fuels into Syngas
(CO–H₂): On-Board Generation of Hydrogen-Rich Gases for
Internal Combustion Engine Vehicles   692
- 10.2.1. Specific Applications of Plasma-Chemical Reforming of
Liquid Automotive Fuels: On-Board Generation of
Hydrogen-Rich Gases   692
- 10.2.2. Plasma-Catalytic Steam Conversion and Partial Oxidation
of Kerosene for Syngas Production   693
- 10.2.3. Plasma-Catalytic Conversion of Ethanol with Production
of Syngas   694
- 10.2.4. Plasma-Stimulated Reforming of Diesel Fuel and Diesel
Oils into Syngas   697
- 10.2.5. Plasma-Stimulated Reforming of Gasoline into
Syngas   698
- 10.2.6. Plasma-Stimulated Reforming of Aviation Fuels into Syngas   698
- 10.2.7. Plasma-Stimulated Partial Oxidation Reforming of
Renewable Biomass: Biodiesel   699
- 10.2.8. Plasma-Stimulated Partial Oxidation Reforming of
Bio-Oils and Other Renewable Biomass into Syngas   700

10.3. Combined Plasma–Catalytic Production of Hydrogen by Partial
Oxidation of Hydrocarbon Fuels   701
- 10.3.1. Combined Plasma–Catalytic Approach Versus Plasma
Catalysis in Processes of Hydrogen Production by Partial
Oxidation of Hydrocarbons   701
- 10.3.2. Pulsed-Corona-Based Combined Plasma–Catalytic System
for Reforming of Hydrocarbon Fuel and Production of
Hydrogen-Rich Gases   702
- 10.3.3. Catalytic Partial Oxidation Reforming of Isooctane   703
- 10.3.4. Partial Oxidation Reforming of Isooctane Stimulated by
Non-Equilibrium Atmospheric-Pressure Pulsed Corona
Discharge   703
- 10.3.5. Reforming of Isooctane and Hydrogen Production in
Pulsed-Corona-Based Combined Plasma–Catalytic
System   704
- 10.3.6. Comparison of Isooctane Reforming in Plasma
Preprocessing and Plasma Postprocessing Configurations
of the Combined Plasma–Catalytic System   706

10.4. Plasma-Chemical Conversion of Coal: Mechanisms, Kinetics, and
Thermodynamics   707
- 10.4.1. Coal and Its Composition, Structure, and Conversion to
Other Fuels   707
- 10.4.2. Thermal Conversion of Coal   708
- 10.4.3. Transformations of Sulfur-Containing Compounds During
Thermal Conversion of Coal   710
- 10.4.4. Transformations of Nitrogen-Containing Compounds
During Thermal Conversion of Coal   711
- 10.4.5. Thermodynamic Analysis of Coal Conversion in Thermal
Plasma   711
- 10.4.6. Kinetic Phases of Coal Conversion in Thermal Plasma   712

Contents

10.4.7. Kinetic Analysis of Thermal Plasma Conversion of Coal:
Kinetic Features of the Major Phases of Coal Conversion
in Plasma ... 714

10.4.8. Coal Conversion in Non-Thermal Plasma ... 715

10.5. Thermal and Non-Thermal Plasma-Chemical Systems for Coal
Conversion ... 716

10.5.1. General Characteristics of Coal Conversion in Thermal
Plasma Jets ... 716

10.5.2. Thermal Plasma Jet Pyrolysis of Coal in Argon, Hydrogen,
and Their Mixtures: Plasma Jet Production of Acetylene
from Coal ... 716

10.5.3. Heating of Coal Particles and Acetylene Quenching
During Pyrolysis of Coal in Argon and Hydrogen Plasma
Jets ... 719

10.5.4. Pyrolysis of Coal in Thermal Nitrogen Plasma Jet with
Co-Production of Acetylene and Hydrogen Cyanide ... 721

10.5.5. Coal Gasification in a Thermal Plasma Jet of Water Vapor ... 721

10.5.6. Coal Gasification by H_2O and Syngas Production in
Thermal Plasma Jets: Application of Steam Plasma Jets and
Plasma Jets of Other Gases ... 722

10.5.7. Coal Gasification in Steam–Oxygen and Air Plasma Jets ... 724

10.5.8. Conversion of Coal Directly in Electric Arcs ... 724

10.5.9. Direct Pyrolysis of Coal with Production of Acetylene
(C_2H_2) in Arc Plasma of Argon and Hydrogen ... 724

10.5.10. Direct Gasification of Coal with Production of Syngas
(H_2–CO) in Electric Arc Plasma of Water Vapor ... 725

10.5.11. Coal Conversion in Non-Equilibrium Plasma of
Microwave Discharges ... 726

10.5.12. Coal Conversion in Non-Equilibrium Microwave
Discharges Containing Water Vapor or Nitrogen ... 728

10.5.13. Coal Conversion in Low-Pressure Glow and Other
Strongly Non-Equilibrium Non-Thermal Discharges ... 730

10.5.14. Plasma-Chemical Coal Conversion in Corona and
Dielectric Barrier Discharges ... 731

10.6. Energy and Hydrogen Production from Hydrocarbons with
Carbon Bonding in Solid Suboxides and without CO_2 Emission ... 732

10.6.1. Highly Ecological Hydrogen Production by Partial
Oxidation of Hydrocarbons without CO_2 Emission:
Plasma Generation of Carbon Suboxides ... 732

10.6.2. Thermodynamics of the Conversion of Hydrocarbons
into Hydrogen with Production of Carbon Suboxides and
without CO_2 Emission ... 732

10.6.3. Plasma-Chemical Conversion of Methane and Coal into
Carbon Suboxide ... 734

10.6.4. Mechanochemical Mechanism of Partial Oxidation of
Coal with Formation of Suboxides ... 735

10.6.5. Kinetics of Mechanochemical Partial Oxidation of Coal to
Carbon Suboxides ... 736

10.6.6. Biomass Conversion into Hydrogen with the Production
of Carbon Suboxides and Without CO_2 Emission ... 737

Contents

10.7. Hydrogen Sulfide Decomposition in Plasma with Production of
Hydrogen and Sulfur: Technological Aspects of Plasma-Chemical
Hydrogen Production   738
  10.7.1. H_2S Dissociation in Plasma with Production of Hydrogen
          and Elemental Sulfur and Its Industrial Applications   738
  10.7.2. Application of Microwave, Radiofrequency, and Arc
          Discharges for H_2S Dissociation with Production of
          Hydrogen and Elemental Sulfur   740
  10.7.3. Technological Aspects of Plasma-Chemical Dissociation of
          Hydrogen Sulfide with Production of Hydrogen and
          Elemental Sulfur   741
  10.7.4. Kinetics of H_2S Decomposition in Plasma   744
  10.7.5. Non-Equilibrium Clusterization in a Centrifugal Field and
          Its Effect on H_2S Decomposition in Plasma with
          Production of Hydrogen and Condensed-Phase Elemental
          Sulfur   745
  10.7.6. Influence of the Centrifugal Field on Average Cluster
          Sizes: Centrifugal Effect Criterion for Energy Efficiency of
          H_2S Decomposition in Plasma   748
  10.7.7. Effect of Additives (CO_2, O_2, and Hydrocarbons) on
          Plasma-Chemical Decomposition of H_2S 749
  10.7.8. Technological Aspects of H_2 Production from Water in
          Double-Step and Multi-Step Plasma-Chemical Cycles   751
  Problems and Concept Questions   753

11 Plasma Chemistry in Energy Systems and Environmental
   Control   755
  11.1. Plasma Ignition and Stabilization of Flames   755
    11.1.1. General Features of Plasma-Assisted Ignition and
            Combustion   755
    11.1.2. Experiments with Plasma Ignition of Supersonic Flows   757
    11.1.3. Non-Equilibrium Plasma Ignition of Fast and Transonic
            Flows: Low-Temperature Fuel Oxidation Versus Ignition   758
    11.1.4. Plasma Sustaining of Combustion in Low-Speed Gas Flows   760
    11.1.5. Kinetic Features of Plasma-Assisted Ignition and
            Combustion   761
    11.1.6. Combined Non-Thermal/Quasi-Thermal Mechanism of
            Flame Ignition and Stabilization: "Zebra" Ignition and
            Application of Non-Equilibrium Magnetic Gliding Arc
            Discharges   763
    11.1.7. Magnetic Gliding Arc Discharge Ignition of Counterflow
            Flame   765
    11.1.8. Plasma Ignition and Stabilization of Combustion of
            Pulverized Coal: Application for Boiler Furnaces   768
  11.2. Mechanisms and Kinetics of Plasma-Stimulated Combustion   770
    11.2.1. Contribution of Different Plasma-Generated Chemically
            Active Species in Non-Equilibrium Plasma Ignition and
            Stabilization of Flames   770
    11.2.2. Numerical Analysis of Contribution of Plasma-Generated
            Radicals to Stimulate Ignition   770

Contents

11.2.3. Possibility of Plasma-Stimulated Ignition Below the Auto-Ignition Limit: Conventional Kinetic Mechanisms of Explosion of Hydrogen and Hydrocarbons   771

11.2.4. Plasma Ignition in H_2–O_2–He Mixtures   773

11.2.5. Plasma Ignition in Hydrocarbon–Air Mixtures   774

11.2.6. Analysis of Subthreshold Plasma Ignition Initiated Thermally: The "Bootstrap" Effect   775

11.2.7. Subthreshold Ignition Initiated by Plasma-Generated Radicals   776

11.2.8. Subthreshold Ignition Initiated by Plasma-Generated Excited Species   778

11.2.9. Contribution of Plasma-Excited Molecules into Suppressing HO_2 Formation During Subthreshold Plasma Ignition of Hydrogen   779

11.2.10. Subthreshold Plasma Ignition of Hydrogen Stimulated by Excited Molecules Through Dissociation of HO_2 781

11.2.11. Subthreshold Plasma Ignition of Ethylene Stimulated by Excited Molecules Effect of NO   783

11.2.12. Contribution of Ions in the Subthreshold Plasma Ignition   784

11.2.13. Energy Efficiency of Plasma-Assisted Combustion in Ram/Scramjet Engines   785

11.3. Ion and Plasma Thrusters   787

11.3.1. General Features of Electric Propulsion: Ion and Plasma Thrusters   787

11.3.2. Optimal Specific Impulse of an Electric Rocket Engine   788

11.3.3. Electric Rocket Engines Based on Ion Thrusters   789

11.3.4. Classification of Plasma Thrusters: Electrothermal Plasma Thrusters   790

11.3.5. Electrostatic Plasma Thrusters   791

11.3.6. Magneto-Plasma-Dynamic Thrusters   791

11.3.7. Pulsed Plasma Thrusters   792

11.4. Plasma Applications in High-Speed Aerodynamics   792

11.4.1. Plasma Interaction with High-Speed Flows and Shocks   792

11.4.2. Plasma Effects on Shockwave Structure and Velocity   793

11.4.3. Plasma Aerodynamic Effects in Ballistic Range Tests   793

11.4.4. Global Thermal Effects: Diffuse Discharges   795

11.4.5. High-Speed Aerodynamic Effects of Filamentary Discharges   795

11.4.6. Aerodynamic Effects of Surface and Dielectric Barrier Discharges: Aerodynamic Plasma Actuators   797

11.4.7. Plasma Application for Inlet Shock Control: Magneto-Hydrodynamics in Flow Control and Power Extraction   798

11.4.8. Plasma Jet Injection in High-Speed Aerodynamics   799

11.5. Magneto-Hydrodynamic Generators and Other Plasma Systems of Power Electronics   799

11.5.1. Plasma Power Electronics   799

11.5.2. Plasma MHD Generators in Power Electronics: Different Types of MHD Generators   800

11.5.3. Major Electric and Thermodynamic Characteristics of MHD Generators   801

Contents

11.5.4. Electric Conductivity of Working Fluid in Plasma MHD Generators 802

11.5.5. Plasma Thermionic Converters of Thermal Energy into Electricity: Plasma Chemistry of Cesium 803

11.5.6. Gas-Discharge Commutation Devices 804

11.6. Plasma Chemistry in Lasers and Light Sources 804

11.6.1. Classification of Lasers: Inversion Mechanisms in Gas and Plasma Lasers and Lasers on Self-Limited Transitions 804

11.6.2. Pulse-Periodic Self-Limited Lasers on Metal Vapors and on Molecular Transitions 805

11.6.3. Quasi-Stationary Inversion in Collisional Gas-Discharge Lasers: Excitation by Long-Lifetime Particles and Radiative Deactivation 806

11.6.4. Ionic Gas-Discharge Lasers of Low Pressure: Argon and He–Ne Lasers 806

11.6.5. Inversion Mechanisms in Plasma Recombination Regime: Plasma Lasers 807

11.6.6. Plasma Lasers Using Electronic Transitions: He–Cd, He–Zn, He–Sr, and Penning Lasers 807

11.6.7. Plasma Lasers Based on Atomic Transitions of Xe and on Transitions of Multi-Charged Ions 808

11.6.8. Excimer Lasers 809

11.6.9. Gas-Discharge Lasers Using Vibrational–Rotational Transitions: CO_2 Lasers 810

11.6.10. Gas-Discharge Lasers Using Vibrational–Rotational Transitions: CO Lasers 811

11.6.11. Plasma Stimulation of Chemical Lasers 811

11.6.12. Energy Efficiency of Chemical Lasers: Chemical Lasers with Excitation Transfer 812

11.6.13. Plasma Sources of Radiation with High Spectral Brightness 814

11.6.14. Mercury-Containing and Mercury-Free Plasma Lamps 815

11.6.15. Plasma Display Panels and Plasma TV 816

11.7. Non-Thermal Plasma in Environmental Control: Cleaning Exhaust Gas of SO_2 and NO_x 817

11.7.1. Industrial SO_2 Emissions and Plasma Effectiveness of Cleaning Them 817

11.7.2. Plasma-Chemical SO_2 Oxidation to SO_3 in Air and Exhaust Gas Cleaning Using Relativistic Electron Beams 818

11.7.3. SO_2 Oxidation in Air to SO_3 Using Continuous and Pulsed Corona Discharges 819

11.7.4. Plasma-Stimulated Liquid-Phase Chain Oxidation of SO_2 in Droplets 820

11.7.5. Plasma-Catalytic Chain Oxidation of SO_2 in Clusters 822

11.7.6. Simplified Mechanism and Energy Balance of the Plasma-Catalytic Chain Oxidation of SO_2 in Clusters 823

11.7.7. Plasma-Stimulated Combined Oxidation of NO_x and SO_2 in Air: Simultaneous Industrial Exhaust Gas Cleaning of Nitrogen and Sulfur Oxides 824

Contents

11.7.8. Plasma-Assisted After Treatment of Automotive Exhaust: Kinetic Mechanism of Double-Stage Plasma-Catalytic NO_x and Hydrocarbon Remediation   825

11.7.9. Plasma-Assisted Catalytic Reduction of NO_x in Automotive Exhaust Using Pulsed Corona Discharge: Cleaning of Diesel Engine Exhaust   827

11.8. Non-Thermal Plasma Treatment of Volatile Organic Compound Emissions, and Some Other Plasma-Ecological Technologies   830

   11.8.1. General Features of the Non-Thermal Plasma Treatment of Volatile Organic Compound Emissions   830

   11.8.2. Mechanisms and Energy Balance of the Non-Thermal Plasma Treatment of VOC Emissions: Treatment of Exhaust Gases from Paper Mills and Wood Processing Plants   830

   11.8.3. Removal of Acetone and Methanol from Air Using Pulsed Corona Discharge   832

   11.8.4. Removal of Dimethyl Sulfide from Air Using Pulsed Corona Discharge   833

   11.8.5. Removal of α-Pinene from Air Using Pulsed Corona Discharge; Plasma Treatment of Exhaust Gas Mixtures   835

   11.8.6. Treatment of Paper Mill Exhaust Gases Using Wet Pulsed Corona Discharge   836

   11.8.7. Non-Thermal Plasma Control of Diluted Large-Volume Emissions of Chlorine-Containing VOCs   839

   11.8.8. Non-Thermal Plasma Removal of Elemental Mercury from Coal-Fired Power Plants and Other Industrial Offgases   843

   11.8.9. Mechanism of Non-Thermal Plasma Removal of Elemental Mercury from Exhaust Gases   844

   11.8.10. Plasma Decomposition of Freons (Chlorofluorocarbons) and Other Waste Treatment Processes Organized in Thermal and Transitional Discharges   845

Problems and Concept Questions   846

12 Plasma Biology and Plasma Medicine   848

12.1. Non-Thermal Plasma Sterilization of Different Surfaces: Mechanisms of Plasma Sterilization   848

   12.1.1. Application of Low-Pressure Plasma for Biological Sterilization   848

   12.1.2. Inactivation of Micro-Organisms by Non-Equilibrium High-Pressure Plasma   850

   12.1.3. Plasma Species and Factors Active for Sterilization: Direct Effect of Charged Particles   851

   12.1.4. Plasma Species and Factors Active for Sterilization: Effects of Electric Fields, Particularly Related to Charged Plasma Particles   854

   12.1.5. Plasma Species and Factors Active for Sterilization: Effect of Reactive Neutral Species   855

   12.1.6. Plasma Species and Factors Active for Sterilization: Effects of Heat   858

Contents

12.1.7. Plasma Species and Factors Active for Sterilization: Effect of Ultraviolet Radiation   858

12.2. Effects of Atmospheric-Pressure Air Plasma on Bacteria and Cells: Direct Versus Indirect Treatment, Surface Versus In-Depth Treatment, and Apoptosis Versus Necrosis   859

    12.2.1. Direct and Indirect Effects of Non-Thermal Plasma on Bacteria   859

    12.2.2. Two Experiments Proving Higher Effectiveness of Direct Plasma Treatment of Bacteria   862

    12.2.3. Surface Versus In-Depth Plasma Sterilization: Penetration of DBD Treatment into Fluid for Biomedical Applications   863

    12.2.4. Apoptosis Versus Necrosis in Plasma Treatment of Cells: Sublethal Plasma Treatment Effects   865

12.3. Non-Thermal Plasma Sterilization of Air Streams: Kinetics of Plasma Inactivation of Biological Micro-Organisms   866

    12.3.1. General Features of Plasma Inactivation of Airborne Bacteria   866

    12.3.2. Pathogen Detection and Remediation Facility for Plasma Sterilization of Air Streams   867

    12.3.3. Special DBD Configuration – the Dielectric Barrier Grating Discharge – Applied in PDRF for Plasma Sterilization of Air Streams   869

    12.3.4. Rapid and Direct Plasma Deactivation of Airborne Bacteria in the PDRF   870

    12.3.5. Phenomenological Kinetic Model of Non-Thermal Plasma Sterilization of Air Streams   871

    12.3.6. Kinetics and Mechanisms of Rapid Plasma Deactivation of Airborne Bacteria in the PDRF   872

12.4. Plasma Cleaning and Sterilization of Water: Special Discharges in Liquid Water Applied for Its Cleaning and Sterilization   874

    12.4.1. Needs and General Features of Plasma Water Treatment: Water Disinfection Using UV Radiation, Ozone, or Pulsed Electric Fields   874

    12.4.2. Electrical Discharges in Water   875

    12.4.3. Mechanisms and Characteristics of Plasma Discharges in Water   876

    12.4.4. Physical Kinetics of Water Breakdown   878

    12.4.5. Experimental Applications of Pulsed Plasma Discharges for Water Treatment   879

    12.4.6. Energy-Effective Water Treatment Using Pulsed Spark Discharges   880

12.5. Plasma-Assisted Tissue Engineering   882

    12.5.1. Plasma-Assisted Regulation of Biological Properties of Medical Polymer Materials   882

    12.5.2. Plasma-Assisted Attachment and Proliferation of Bone Cells on Polymer Scaffolds   884

    12.5.3. DBD Plasma Effect on Attachment and Proliferation of Osteoblasts Cultured over Poly-ε-Caprolactone Scaffolds   885

    12.5.4. Controlling of Stem Cell Behavior on Non-Thermal Plasma Modified Polymer Surfaces   887

12.5.5. Plasma-Assisted Bio-Active Liquid Microxerography,
Plasma Bioprinter 888
12.6. Animal and Human Living Tissue Sterilization 888
12.6.1. Direct Plasma Medicine: Floating-Electrode Dielectric
Barrier Discharge 888
12.6.2. Direct Plasma-Medical Sterilization of Living Tissue Using
FE-DBD Plasma 889
12.6.3. Non-Damage (Toxicity) Analysis of Direct Plasma
Treatment of Living Tissue 890
12.7. Non-Thermal Plasma-Assisted Blood Coagulation 892
12.7.1. General Features of Plasma-Assisted Blood Coagulation 892
12.7.2. Experiments with Non-Thermal Atmospheric-Pressure
Plasma-Assisted In Vitro Blood Coagulation 892
12.7.3. In Vivo Blood Coagulation Using FE-DBD Plasma 893
12.7.4. Mechanisms of Non-Thermal Plasma-Assisted Blood
Coagulation 894
12.8. Plasma-Assisted Wound Healing and Tissue Regeneration 896
12.8.1. Discharge Systems for Air-Plasma Surgery and Nitrogen
Oxide (NO) Therapy 896
12.8.2. Medical Use of Plasma-Generated Exogenic NO 898
12.8.3. Experimental Investigations of NO Effect on Wound
Healing and Inflammatory Processes 899
12.8.4. Clinical Aspects of Use of Air Plasma and Exogenic NO in
Treatment of Wound Pathologies 900
12.8.5. Air Plasma and Exogenic NO in Treatment of
Inflammatory and Destructive Illnesses 904
12.9. Non-Thermal Plasma Treatment of Skin Diseases 906
12.9.1. Non-Thermal Plasma Treatment of Melanoma Skin
Cancer 906
12.9.2. Non-Thermal Plasma Treatment of Cutaneous
Leishmaniasis 908
12.9.3. Non-Equilibrium Plasma Treatment of Corneal Infections 910
12.9.4. Remarks on the Non-Thermal Plasma-Medical Treatment
of Skin 911
Problems and Concept Questions 912

References 915

Index 963

Foreword

Although the public understanding of plasmas may be limited to plasma TVs, low-temperature plasma processes are beginning to enter into a higher level of consciousness due to the importance of plasma in many aspects of technological developments. The use of plasma for industrial purposes began more than 100 years ago with plasma sources used to produce light. Since then, plasma processes have emerged in transforming wide-ranging technologies, including microelectronics, gas lasers, polymers and novel materials, protective coatings, and water purification, and finally found their ubiquitous place in our homes. Plasma systems or plasma-treated materials are now commonly used and can be found in air-cleaning systems; food containers; fruit, meat, and vegetable treatment; fabrics; and medical devices.

In recent years, new application areas of plasma chemistry and plasma processing have been established, such as plasma nanotechnology with the continuous growth of the "dusty plasmas" domain, plasma production and modification of nanotubes, plasma aerodynamics, and plasma ignition and stabilization of flames. With the recent emphasis on alternative energy and environmental concerns, plasma chemistry has revolutionized hydrogen production, biomass conversion, and fuel-cell technology. In the same manner, the use of non-thermal plasmas in biology and medicine will likely "explode" in the coming years for various applications. Plasma is expected to soon be widely used in surgery, decontamination and sterilization of surfaces and devices, and air and water streams, as well as in tissue engineering and direct treatment of skin diseases.

In many of these applications, comprehension of detailed mechanisms, knowledge of the reaction kinetics, and understanding of the production of radicals or excited species are vital for optimization of plasma reactors and plasma processes. A growing number of universities recognize the importance of plasma technology and are preparing future professionals who are cognizant of the latest achievements in practice. Drexel University (more specifically, Drexel Plasma Institute), where the author of the book serves as the Nyheim Chair Professor, is one of the world's leading centers focused on plasma chemistry and engineering. Drexel University today makes a significant contribution to the successful development of both plasma research and plasma education and works closely with the International Plasma Chemistry Society to coordinate international activities in research, education, and outreach.

As a result of the increased interest in low-pressure plasma science, researchers and engineers are faced with the problem of evaluating the broad and varied literature from a common basis of fundamental plasma chemistry. The book that you hold in your hands meets this challenge! This book represents the first comprehensive contribution that presents the fundamentals of plasma chemistry and the scientific basis of most modern applications of plasma technologies. This book is written by my distinguished colleague and friend,

Foreword

Alexander Fridman, who has made outstanding contributions in the development of modern plasma science and engineering, especially in plasma kinetics of excited and charged particles, in the development of novel non-thermal atmospheric-pressure discharges, in fuel conversion and hydrogen production, in plasma sterilization and disinfection, and, more recently, in breakthrough developments in plasma medicine. *Plasma Chemistry* is of unique value to scientists, engineers, and students in the domains of plasma physics, chemistry, and engineering. It is my great pleasure to recommend this excellent work to practitioners, students, and scientists who are interested in the fundamentals and applications of plasma chemistry.

Jean-Michel Pouvesle,
President of the International Plasma Chemistry Society
Director of GREMI, University of Orleans, France
June, 2007

Preface

Plasma chemistry is an area of research that has consumed and inspired more than 35 years of the author's professional activities. During this period, plasma chemistry has become a rapidly growing area of scientific endeavor that holds great promise for practical applications for industrial and medical fields. Plasma has become a ubiquitous element that pervades many aspects of our lives. For example, the public is well aware of plasma TV, fluorescent lamps, and plasma thrusters, as well as popular-culture concepts such as plasma guns and plasma shields from *Star Trek*. Not many are aware, however, that computers, cell phones, and other modern electronic devices are manufactured using plasma-enabled chemical processing equipment; that most of the synthetic fibers used in clothing, photomaterials, and advanced packaging materials are plasma treated; that a significant amount of potable water in the world is purified using ozone-plasma technology; and that many different tools and special surfaces are plasma coated to protect and provide them with new extraordinary properties. The developments in plasma chemistry are enabling tremendous growth in a variety of applications for manufacturing, environmental remediation, and therapeutic and preventive medicine.

The motivation for this book is to provide engineers and scientists with a foundational understanding of the physical and chemical phenomena associated with both thermal and non-thermal discharge plasmas. Students pursuing degrees in electrical, chemical, mechanical, environmental, and materials engineering will find that the applications in plasma and plasma chemistry will have many important bearings in their own disciplinary areas. Therefore, the objectives and challenges of this book are to present the broad extent of basic and applied knowledge on modern plasma chemistry in a comprehensive manner for students, as well as for senior scientists and engineers.

This book also includes detailed problems and inquiries to enhance the conceptual understanding of the diverse plasma chemistry–related topics ranging from nonequilibrium processes to quantum chemistry in a manner that is readily amenable to interactive learning for students and practitioners of the topic. The problems and concept questions have been developed based on the sequence of plasma courses taught by the author at Drexel University. The book also contains extensive data tables and numerical and empirical formulas to help engineers, scientists, and practitioners in calculations of plasma-chemical systems and plasma-chemical processes. The book consists of 12 chapters; the first 4 chapters focus on the fundamental aspects of plasma chemistry, including elementary processes, physical and chemical kinetics of charged and excited plasma particles, and basic physics of gas discharges. The following 8 chapters deal with specific applications of plasma chemistry on practical implementation in areas such as electronics manufacturing, energy systems, fuel conversion, surface treatment, remediation of contaminated air and water, treatment of diseases, and destruction of pathogens.

Preface

The author gratefully acknowledges the support of his family; and support of plasma research by John and Chris Nyheim and the Stanley Kaplan family; support of the Drexel Plasma Institute (DPI) through leaders of Drexel University: Provost Steve Director, Vice Provost Ken Blank, and Dean Selcuk Guceri. The author is grateful to Professor Mun Choi, Dean of the University of Connecticut, for useful recommendations and help in working on the text. The author greatly appreciates the research support provided to DPI by the National Science Foundation, U.S. Department of Energy, U.S. Department of Defense (specifically DARPA, TARDEC Army Research Lab, and Air Force OSR), NASA, and USDA, as well as the support of our long-term industrial sponsors, Chevron, Kodak, Air Products, Georgia Pacific, Applied Materials, and Ceramatec.

For stimulating discussions on the topic of plasma chemistry and immeasurable assistance in development of the book, the author gratefully acknowledges Dr. James Hervonen from the Army Research Lab, as well as all his colleagues and friends from DPI, especially Professors A. Brooks, N. Cernansky, Y. Cho, B. Farouk, G. Friedman, A. Gutsol, R. Knight, T. Miller, G. Palmese, W. Sun, V. Vasilets, and Ph.D. students M. Cooper, G. Fridman, M. Gallagher, S. Gangoli, and D. Staack. Special thanks are addressed to Kirill Gutsol for assistance with numerous illustrations.

1

Introduction to Theoretical and Applied Plasma Chemistry

1.1. PLASMA AS THE FOURTH STATE OF MATTER

Although the term *chemistry* in the title of the book does not require a special introduction, the term *plasma* probably does. Plasma is an ionized gas, a distinct fourth state of matter. "Ionized" means that at least one electron is not bound to an atom or molecule, converting the atoms or molecules into positively charged ions. As temperature increases, molecules become more energetic and transform matter in the sequence: solid, liquid, gas, and finally plasma, which justifies the title "fourth state of matter."

The free electric charges – electrons and ions – make plasma electrically conductive (sometimes more than gold and copper), internally interactive, and strongly responsive to electromagnetic fields. Ionized gas is usually called plasma when it is electrically neutral (i.e., electron density is balanced by that of positive ions) and contains a significant number of the electrically charged particles, sufficient to affect its electrical properties and behavior. In addition to being important in many aspects of our daily lives, plasmas are estimated to constitute more than 99% of the visible universe.

The term *plasma* was first introduced by Irving Langmuir (1928) because the multi-component, strongly interacting ionized gas reminded him of blood plasma. Langmuir wrote: "Except near the electrodes, where there are sheaths containing very few electrons, the ionized gas contains ions and electrons in about equal numbers so that the resultant space charge is very small. We shall use the name **plasma** to describe this region containing balanced charges of ions and electrons." There is usually not much confusion between the fourth state of matter (plasma) and blood plasma; probably the only exception is the process of plasma-assisted blood coagulation, where the two concepts meet.

Plasmas occur naturally but also can be effectively man-made in laboratory and in industry, which provides opportunities for numerous applications, including thermonuclear synthesis, electronics, lasers, fluorescent lamps, and many others. To be more specific, most computer and cell-phone hardware is made based on plasma technologies, not to forget about plasma TV. In this book, we are going to focus on fundamental and practical aspects of plasma applications to chemistry and related disciplines, which probably represent the major part of complete plasma science and engineering.

Plasma is widely used in practice. Plasma offers three major features that are attractive for applications in chemistry and related disciplines: (1) temperatures of at least some plasma components and energy density can significantly exceed those in conventional chemical technologies, (2) plasmas are able to produce very high concentrations of energetic and chemically active species (e.g., electrons, ions, atoms and radicals, excited states, and different wavelength photons), and (3) plasma systems can essentially be far from thermodynamic equilibrium, providing extremely high concentrations of the chemically active species and

Figure 1–1. Benjamin Franklin's first experiments with the atmospheric plasma phenomenon of lightning.

keeping bulk temperature as low as room temperature. These plasma features permit significant intensification of traditional chemical processes, essential increases of their efficiency, and often successful stimulation of chemical reactions impossible in conventional chemistry.

Plasma chemistry today is a rapidly expanding area of science and engineering, with applications widely spread from micro-fabrication in electronics to making protective coatings for aircrafts, from treatment of polymer fibers and films before painting to medical cauterization for stopping blood and wound treatment, and from production of ozone to plasma TVs. Let us start the long journey to theoretical and applied plasma chemistry by getting acquainted with the major natural and man-made plasmas.

1.2. PLASMA IN NATURE AND IN THE LABORATORY

Plasma comprises the majority of the universe: the solar corona, solar wind, nebula, and Earth's ionosphere are all plasmas. The best known natural plasma phenomenon in Earth's atmosphere is lightning. The breakthrough experiments with this natural form of plasma were performed long ago by Benjamin Franklin (Fig. 1–1), which probably explains the special interest in plasma research in Philadelphia, where the author of the book works at the Drexel Plasma Institute (Drexel University).

At altitudes of approximately 100 km, the atmosphere no longer remains non-conducting due to ionization and formation of plasma by solar radiation. As one progresses further into near-space altitudes, the Earth's magnetic field interacts with charged particles streaming from the sun. These particles are diverted and often become trapped by the Earth's magnetic field. The trapped particles are most dense near the poles and account for the aurora borealis (Fig. 1–2). Lightning and the aurora borealis are the most common natural plasmas observed on Earth.

Natural and man-made plasmas (generated in gas discharges) occur over a wide range of pressures, electron temperatures, and electron densities (Fig. 1–3). The temperatures of

Figure 1–2. Aurora borealis.

man-made plasmas range from slightly above room temperature to temperatures comparable to the interior of stars, and electron densities span over 15 orders of magnitude. Most plasmas of practical significance, however, have electron temperatures of 1–20 eV, with electron densities in the range 10^6–10^{18} cm^{-3}. (High temperatures are conventionally expressed in electron volts; 1 eV approximately equals 11,600 K.)

Not all particles need to be ionized in plasma; a common condition in plasma chemistry is for the gases to be only partially ionized. The ionization degree (i.e., ratio of density of major charged species to that of neutral gas) in the conventional plasma–chemical systems is in the range 10^{-7}–10^{-4}. When the ionization degree is close to unity, such plasma is called **completely ionized plasma**. Completely ionized plasmas are conventional for thermonuclear plasma systems: tokomaks, stellarators, plasma pinches, focuses, and so on. Completely ionized plasma and related issues of nuclear fusion and space plasmas are the subject of several books, in particular those of Bittencourt (2004) and Chen (2006); thermonuclear and space plasmas are not the focus of this book. When the ionization degree is low, the plasma is called **weakly ionized plasma**, which is the main focus of plasma chemistry and this book.

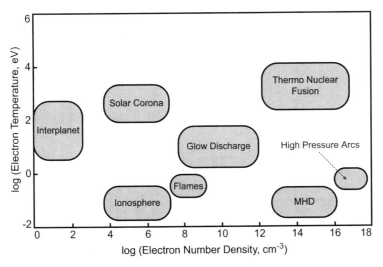

Figure 1–3. Plasma temperatures and densities.

Both natural and man-made laboratory plasmas are quasi-neutral, which means that concentrations of positively charged particles (positive ions) and negatively charged particles (electrons and negative ions) are well balanced. Langmuir was one of the pioneers who studied gas discharges and defined plasma to be a region not influenced by its boundaries. The transition zone between the plasma and its boundaries was termed the plasma **sheath**. The properties of the sheath differ from those of the plasma, and these boundaries influence the motion of the charged particles in this sheath. The particles form an electrical screen for the plasma from influences of the boundary.

1.3. PLASMA TEMPERATURES: THERMAL AND NON-THERMAL PLASMAS

As in any gas, temperature in plasma is determined by the average energies of the plasma particles (neutral and charged) and their relevant degrees of freedom (translational, rotational, vibrational, and those related to electronic excitation). Thus, plasmas, as multi-component systems, are able to exhibit multiple temperatures. In electric discharges common for plasma generation in the laboratory, energy from the electric field is first accumulated by the electrons between collisions and, subsequently, is transferred from the electrons to the heavy particles.

Electrons receive energy from the electric field during their mean free path and, during the following collision with a heavy particle, lose only a small portion of that energy (because electrons are much lighter than the heavy particles). That is why the electron temperature in plasma is initially higher than that of heavy particles. Subsequently, collisions of electrons with heavy particles (Joule heating) can equilibrate their temperatures, unless time or energy are not sufficient for the equilibration (such as in coronas and pulsed discharges) or there is an intensive cooling mechanism preventing heating of the entire gas (such as in wall-cooled low-pressure discharges).

The temperature difference between electrons and heavy neutral particles due to Joule heating in the collisional weakly ionized plasma is conventionally proportional to the square of the ratio of the electric field (E) to the pressure (p). Only in the case of small values of E/p do the temperatures of electrons and heavy particles approach each other. Thus, this is a basic requirement for **local thermodynamic equilibrium** (LTE) in plasma. Additionally, LTE conditions require chemical equilibrium as well as restrictions on the gradients. The LTE plasma follows the major laws of equilibrium thermodynamics and can be characterized by a single temperature at each point of space. Ionization and chemical processes in such plasmas are determined by temperature (and only indirectly by the electric fields through Joule heating). The quasi-equilibrium plasma of this kind is usually called **thermal plasma**. Thermal plasmas in nature can be represented by solar plasma (Fig. 1–4).

Numerous plasmas exist very far from the thermodynamic equilibrium and are characterized by multiple different temperatures related to different plasma particles and different degrees of freedom. It is the electron temperature that often significantly exceeds that of heavy particles ($T_e \gg T_0$). Ionization and chemical processes in such non-equilibrium plasmas are directly determined by electron temperature and, therefore, are not so sensitive to thermal processes and temperature of the gas. The non-equilibrium plasma of this kind is usually called **non-thermal plasma**. An example of non-thermal plasmas in nature is the aurora borealis (Fig. 1–2).

Although the relationship between different plasma temperatures in non-thermal plasmas can be quite sophisticated, it can be conventionally presented in the collisional weakly ionized plasmas as $T_e > T_v > T_r \approx T_i \approx T_0$. Electron temperature ($T_e$) is the highest in the system, followed by the temperature of vibrational excitation of molecules (T_v). The lowest temperature is usually shared in plasma by heavy neutrals (T_0, temperature of translational

Figure 1–4. Solar plasma.

degrees of freedom or simply gas temperature), ions (T_i), as well as rotational degrees of freedom of molecules (T_r). In many non-thermal plasma systems, electron temperature is about 1 eV (about 10,000 K), whereas the gas temperature is close to room temperature.

Non-thermal plasmas are usually generated either at low pressures or at lower power levels, or in different kinds of pulsed discharge systems. The engineering aspects, and application areas are quite different for thermal and non-thermal plasmas. Thermal plasmas are usually more powerful, whereas non-thermal plasmas are more selective. However, these two very different types of ionized gases have many more features in common and both are plasmas.

It is interesting to note that both thermal and non-thermal plasmas usually have the highest temperature (T_e in one case, and T_0 in the other) on the order of magnitude of 1 eV, which is about 10% of the total energy required for ionization (about 10 eV). It reflects the general rule found by Zeldovich and Frank-Kamenetsky for atoms and small molecules in chemical kinetics: the temperature required for a chemical process is typically about 10% of the total required energy, which is the Arrhenius activation energy. A funny fact is that a similar rule (10%) can usually be applied to determine a down payment to buy a house or a new car. Thus, the plasma temperatures can be somewhat identified as the down payment for the ionization process.

1.4. PLASMA SOURCES FOR PLASMA CHEMISTRY: GAS DISCHARGES

Plasma chemistry is clearly the chemistry organized in or with plasma. Thus, a plasma source, which in most laboratory conditions is a gas discharge, represents the physical and engineering basis of the plasma chemistry. For simplicity, an electric discharge can be viewed as two electrodes inserted into a glass tube and connected to a power supply. The tube can be filled with various gases or evacuated. As the voltage applied across the two

Figure 1–5. Glow discharge.

electrodes increases, the current suddenly increases sharply at a certain voltage required for sufficiently intensive electron avalanches. If the pressure is low, on the order of a few torrs, and the external circuit has a large resistance to prohibit a large current, a glow discharge develops. This is a low-current, high-voltage discharge widely used to generate non-thermal plasma. A similar discharge is known by everyone as the plasma source in fluorescent lamps. The glow discharge can be considered a major example of low-pressure, non-thermal plasma sources (see Fig. 1–5).

A non-thermal corona discharge occurs at high pressures (including atmospheric pressure) only in regions of sharply non-uniform electric fields. The field near one or both

Figure 1–6. Corona discharge.

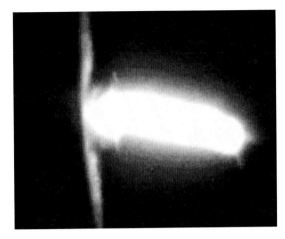

Figure 1–7. Arc discharge.

electrodes must be stronger than in the rest of the gas. This occurs near sharp points, edges, or small-diameter wires, which tend to be low-power plasma sources limited by the onset of electrical breakdown of the gas. However, it is possible to circumvent this restriction through the use of pulsating power supplies. Electron temperature in the corona exceeds 1 eV, whereas the gas remains at room temperature. The corona discharges are, in particular, widely applied in the treatment of polymer materials: most synthetic fabrics applied to make clothing have been treated before dyeing in corona-like discharges to provide sufficient adhesion. The corona discharge can be considered a major example of an atmospheric pressure non-thermal plasma source (see Fig. 1–6).

If the pressure is high, on the order of an atmosphere, and the external circuit resistance is low, a thermal arc discharge can be organized between two electrodes. Thermal arcs usually carry large currents, greater than 1 A at voltages of the order of tens of volts. Furthermore, they release large amounts of thermal energy at very high temperatures often exceeding 10,000 K. The arcs are often coupled with a gas flow to form high-temperature plasma jets. The arc discharges are well known not only to scientists and engineers but also to the general public because of their wide applications in welding devices. The arc discharge can be considered a major example of thermal plasma sources (see Fig. 1–7).

Between other electric discharges widely applied in plasma chemistry, we should emphasize that non-equilibrium, low-pressure radiofrequency discharges play the key roles in sophisticated etching and deposition processes of modern micro-electronics, as well as in treatment of polymer materials. More and more chemical processes have been organized recently in gliding arc discharges (powerful generators of non-equilibrium atmospheric pressure plasma), especially with plasma stabilization in reverse vortex "tornado" flow (see Fig. 1–8). The gliding arc "tornado" discharges provide a unique opportunity of combining the high power typical for arc discharges with the relatively high level of non-equilibrium typical for non-thermal atmospheric pressure discharges.

Between "non-traditional" but very practically interesting discharges, we can point out the non-thermal, high-voltage, atmospheric-pressure, floating-electrode dielectric barrier discharge (FE-DBD), which can use the human body as a second electrode without damaging the living tissue. Such a discharge obviously provides very interesting opportunities for direct plasma applications in biology and medicine (Fig. 1–9). Major discharges applied in plasma chemistry are to be discussed in Chapter 4. More detailed information on the subject can be found in special books focused on plasma physics and engineering, for example those of Roth (1995) and Fridman and Kennedy (2004).

Figure 1–8. Gliding arc discharge stabilized in the reverse vortex ("tornado") gas flow.

1.5. FUNDAMENTALS OF PLASMA CHEMISTRY: MAJOR COMPONENTS OF CHEMICALLY ACTIVE PLASMA AND MECHANISMS OF PLASMA-CHEMICAL PROCESSES

Chemically active plasma is a multi-component system highly reactive due to large concentrations of charged particles (electrons, negative and positive ions), excited atoms and molecules (electronic and vibrational excitation make a major contribution), active atoms and radicals, and UV photons. Each component of the chemically active plasma plays its own specific role in plasma-chemical kinetics. Electrons, for example, are usually first to receive the energy from an electric field and then distribute it between other plasma components and specific degrees of freedom of the system. Changing parameters of the electron gas (density, temperature, electron energy distribution function) often permit control and optimization of plasma-chemical processes.

Ions are charged heavy particles, that are able to make a significant contribution to plasma-chemical kinetics either due to their high energy (as in the case of sputtering and reactive ion etching) or due to their ability to suppress activation barriers of chemical reactions. This second feature of plasma ions results in the so-called ion or plasma catalysis, which is particularly essential in plasma-assisted ignition and flame stabilization, fuel

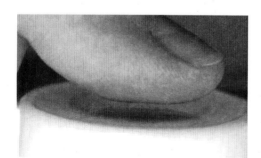

Figure 1–9. Floating-electrode dielectric barrier discharge (FE-DBD) with a finger as a second electrode.

conversion, hydrogen production, exhaust gas cleaning, and even in the direct plasma treatment of living tissue.

The vibrational excitation of molecules often makes a major contribution to plasma-chemical kinetics because the plasma electrons with energies around 1 eV primarily transfer most of the energy in such gases as N_2, CO, CO_2, H_2, and so forth into vibrational excitation. Stimulation of plasma-chemical processes through vibrational excitation permits the highest values of energy efficiency to be reached; see the example with CO_2 dissociation in the next section. Electronic excitation of atoms and molecules can also play a significant role, especially when the lifetime of the excited particles is quite long (as in the case of metastable electronically excited atoms and molecules). As an example, we can mention plasma-generated metastable electronically excited oxygen molecules $O_2(^1\Delta_g)$ (singlet oxygen), which effectively participate in the plasma-stimulated oxidation process in polymer processing, and biological and medical applications.

The contribution of atoms and radicals is obviously significant. As an example, we can point out that O atoms and OH radicals effectively generated in atmospheric air discharges which play a key role in numerous plasma-stimulated oxidation processes. Plasma-generated photons play a key role in a wide range of applications, from plasma light sources to UV sterilization of water.

Plasma is not only a multi-component system, but often a very non-equilibrium one (see Section 1.3). Concentrations of the active species described earlier can exceed those of quasi-equilibrium systems by many orders of magnitude at the same gas temperature. The successful control of plasma permits chemical processes to be directed in a desired direction, selectively, and through an optimal mechanism. Control of a plasma-chemical system requires detailed understanding of elementary processes and the kinetics of the chemically active plasma. The major fundamentals of plasma physics, elementary processes in plasma, and plasma kinetics are to be discussed in Chapters 2 and 3; more details on the subject can be found in Fridman and Kennedy (2004).

1.6. APPLIED PLASMA CHEMISTRY

Applications of plasma technologies today are numerous and involve many industries. High-energy efficiency (energy cost with respect to the minimum determined by thermodynamics), high specific productivity (productivity per unit volume of reactor), and high selectivity may be achieved in plasmas for a wide range of chemical processes. As an example, for CO_2 dissociation in non-equilibrium plasma under supersonic flow conditions, it is possible to selectively introduce up to 90% of the total discharge power in CO production when the vibrational temperature is about 4000 K and the translational temperature is only about 100 K. The specific productivity of such a supersonic reactor achieves 1,000,000 L/h, with power levels up to 1 MW. This plasma process has been examined for fuel production on Mars, where the atmosphere mostly consists of CO_2. On the Earth, it was applied as a plasma stage in a two-step process for hydrogen production from water.

As mentioned in the previous section, the key point for practical use of any chemical process in a particular plasma system is to find the proper regime and optimal plasma parameters among the numerous possibilities intrinsic to systems far from thermodynamic equilibrium. In particular, it is desired to provide high operating power for the plasma chemical reactor together with a high selectivity of energy input while maintaining non-equilibrium plasma conditions. Thermal plasma generators have been designed for many diverse industrial applications covering a wide range of operating power levels from less than 1 kW to over 50 MW. However, in spite of providing sufficient power levels, these generators are not well adapted to the purposes of plasma chemistry, where selective treatment of reactants and high efficiency are required. The main drawback of using thermal plasmas for plasma-chemical applications are that the reaction media become overheated when energy is uniformly consumed by the reagents into all degrees of freedom and, hence, high-energy consumption is required to provide special quenching of the reagents. Because of these drawbacks, the energy efficiency and selectivity of such systems are rather small (only a few of many thermal plasma-chemical processes developed in the first decades of the 20th century remain, e.g., the production of acetylene and other valuable chemicals from light hydrocarbons in Germany).

Recently, energy-efficient and powerful plasma-chemical systems have been developed based on microwave discharges. The skin effect in this case permits simultaneous achievement of a high level of electron density and a high electric field (and hence a high electron temperature as well) in the relatively cold gas. Microwave plasma technology permits dense ($n_e = 10^{13}$ cm^{-3}) non-equilibrium plasmas to be generated ($T_e = $ 1–2 eV, $T_v = $ 3000–5000 K, $T_0 = $ 800–1500 K, and for supersonic flow $T_n \leq 150$ K) at pressures up to 200–300 torr and at power levels reaching 1 MW. Similar plasma parameters can be achieved in recently developed gliding arc discharges (see Fig. 1–8). These energy-effective, non-equilibrium but at the same time powerful plasma systems are optimal for fuel conversion and hydrogen production applications, where minimization of electric energy cost should be achieved together with very high productivities of the process.

Most of the book considers specific plasma applications (Chapters 5–12), widely spread from etching and chemical vapor deposition in micro-electronics to thermal spray coatings; from plasma metallurgy to the production of ozone; from plasma ignition and stabilization of flames to the treatment of synthetic fabrics and other polymer materials; from plasma TVs to sterilization of water and air streams; and from plasma treatment of exhaust gases to direct plasma treatment of burns, ulcers, and other skin diseases. Each plasma process obviously requires understanding of its mechanism, development of the most relevant discharge system, and choice of the most optimal plasma regime. These factors make the journey through applied plasma chemistry not only interesting and exciting but also somewhat long.

1.7. PLASMA AS A HIGH-TECH MAGIC WAND OF MODERN TECHNOLOGY

In many practical applications, plasma technologies compete with other approaches and successfully find their specific niche in modern industry. Such a situation takes place, for example, in thermal plasma deposition of protective coatings, in plasma stabilization of flames, in plasma conversion of fuels, in plasma light sources, in plasma cleaning of exhaust gases, in plasma sterilization of water, and so on. All these plasma technologies are practically interesting, commercially viable, and generally make an important contribution to the development of our society.

The most exciting applications of plasma, however, are related not to the aforementioned technologies but to those which actually have no analogies and no (or almost no) competitors. A good relevant example is plasma applications in micro-electronics, such as etching deep trenches (0.2 μm wide and 4 μm deep) in single crystal silicon, which is so important in the fabrication of integrated circuits. Capabilities of plasma processing in micro-electronics are extraordinary and unique. We probably would not have computers and cell phones like we have now without plasma processing. When all alternatives fail, plasma can still be utilized; plasma chemistry in this case plays the role of the high-tech magic wand of modern technology.

Among other examples, when plasma abilities are extraordinary and unique, we can point out plasma production of ozone where no other technologies are able to challenge plasma for more than 100 years, thermonuclear plasma as a major future source of energy, and low-temperature fuel conversion where hydrogen is produced without CO_2 exhaust, which is

Figure 1–10. FE-DBD plasma device for direct treatment of wounds, skin sterilization, and treatment of skin diseases.

still under research (see Section 10.6). A student of the Drexel Plasma Institute in Fig. 1–10 holds in his hands a pencil-like active 35-kV FE-DBD electrode, which can be safely and directly applied to the human body and opens possibilities to cure diseases that were previously incurable (see Chapter 12). This plasma medical device even looks like a magic wand. Each type of magic, however, requires a well-prepared magician. With these words, we now step from the introduction into the following chapters focused on the fundamentals of plasma chemistry.

2

Elementary Plasma-Chemical Reactions

The yield of a total plasma-chemical process is due to synergistic contributions of numerous different elementary reactions taking place simultaneously in a discharge system. The sequence of transformations of initial chemical substances and electric energy into products and thermal energy is usually referred to as the **mechanism of the plasma-chemical process**. Elementary reaction rates are determined by the micro-kinetic characteristics of individual reactive collisions (like, for example, reaction cross sections or elementary reaction probabilities) as well as by relevant kinetic distribution functions (like the electron energy distribution function [EEDF], or population function of excited molecular states). The elementary reaction rate is actually a result of integration of the reaction cross section or probability over the relevant distribution function and characterizes the energy or excitation state of reactants. We will focus in this chapter mostly on the micro-kinetics of the elementary reactions – on their cross sections and probabilities – assuming, if necessary, conventional Maxwellian or Boltzmann distribution functions. More sophisticated non-Maxwellian and non-Boltzmann kinetic distribution functions typical for strongly non-equilibrium discharge conditions, like the Druyvesteyn EEDF for electrons or Treanor distribution for vibrationally excited molecules, are to be considered in the next chapter.

2.1. IONIZATION PROCESSES

Plasma is an ionized gas. The key process in plasma is ionization, which means conversion of neutral atoms or molecules into **electrons** and **positive ions**. Thus, ionization is the first elementary plasma-chemical processes to be considered.

2.1.1. Elementary Charged Particles in Plasma

Usually the number densities of electrons and positive ions are equal or close in quasi-neutral plasmas, but in "electronegative" gases (like O_2, Cl_2, SF_6, UF_6, $TiCl_4$, etc.) with high electron affinity, negative ions are also effectively formed. Because ionization is the first plasma-chemical reaction to consider, let us first briefly discuss the main features of elementary charged particles in plasma and their elastic and inelastic collisions. Electrons are first in getting energy from electric fields, because of their low mass and high mobility. Electrons then transmit the energy to all other plasma components, providing energy for ionization, excitation, dissociation, and other plasma-chemical processes. The rates of such processes depend on how many electrons have enough energy to do the job. It can be described by means of the **electron energy distribution function (EEDF)** $f(\varepsilon)$, which is the probability density for an electron to have energy ε.

The EEDF strongly depends on electric field and gas composition in plasma (especially in non-thermal discharges) and often can be very far from the equilibrium distribution.

The non-equilibrium EEDF for different discharge systems and different plasma conditions will be discussed in Chapter 3. Sometimes, however (even in non-equilibrium plasmas), the EEDF is determined mostly by the electron temperature T_e and, therefore, can be described by the quasi-equilibrium Maxwell-Boltzmann distribution function:

$$f(\varepsilon) = 2\sqrt{\varepsilon/\pi(kT_e)^3}\exp(-\varepsilon/kT_e), \tag{2-1}$$

where k is a Boltzmann constant (when temperature is given in energy units, then $k = 1$ and can be omitted). The **mean electron energy**, which is the first moment of the distribution function, in this case is proportional to temperature in the conventional way:

$$\langle\varepsilon\rangle = \int\limits_0^\infty \varepsilon f(\varepsilon)d\varepsilon = \frac{3}{2}T_e. \tag{2-2}$$

Numerically, in most plasmas under consideration, the mean electron energy is from 1 to 5 eV.

Atoms or molecules lose their electrons in the ionization process and form positive ions. In hot thermonuclear plasmas, the ions are multi-charged, but in relatively cold plasmas of technological interest their charge is usually equal to $+1e(1.6 \cdot 10^{-19}\,\text{C})$. Ions are heavy particles, so usually they cannot receive high energy directly from electric fields because of intensive collisional energy exchange with other plasma components. The collisional nature of the energy transfer results in the fact that the ion energy distribution function at lower pressures is often not far from the Maxwellian one (2-1), with an ion temperature T_i close to neutral gas temperature T_0.

Electron attachment leads to formation of negative ions with charge $-1e(1.6 \cdot 10^{-19}\,\text{C})$. Attachment of another electron and formation of multi-charged negative ions is actually impossible in the gas phase because of electric repulsion. Negative ions are also heavy particles, so usually their energy balance is due not to electric fields but to collisional processes. The energy distribution functions for negative ions, similar to that for the positive ones, are not far from the Maxwellian distributions (2-1) at pressures that are not too low. Their temperatures in this case are also close to those of a neutral gas.

2.1.2. Elastic and Inelastic Collisions and Their Fundamental Parameters

Elementary processes can be generally subdivided into two classes – elastic and non-elastic. The **elastic collisions** are those in which the internal energies of colliding particles do not change; therefore, total kinetic energy is conserved as well. Actually, the elastic processes result only in geometric scattering and redistribution of kinetic energy. All other collisions, like ionization, for example, are inelastic. Most **inelastic collisions**, like ionization, result in energy transfer from the kinetic energy of colliding partners into internal energy. Sometimes, however, the internal energy of excited atoms or molecules can be transferred back into kinetic energy (in particular, into kinetic energy of plasma electrons). These elementary processes are usually referred to as **superelastic collisions**.

The elementary processes can be described in terms of six major collision parameters: cross section, probability, mean free path, interaction frequency, reaction rate, and finally reaction rate coefficient. The most fundamental characteristic of all elementary processes is the **cross section**, which can be interpreted as an imaginary circle of area σ, moving together with one of the collision partners. If the center of the other collision partner crosses the circle, then the elementary reaction takes place.

If two colliding particles can be considered hard elastic spheres of radii r_1 and r_2, their collisional cross section is equal to $\pi(r_1 + r_2)^2$. Obviously, interaction radius and cross section can exceed corresponding geometric sizes because of the long-distance nature of forces acting between electric charges and dipoles. On the other hand, if only a few out of many collisions result in a chemical reaction, the cross section is considered to be less than geometric. The ratio of the inelastic collision cross section to the corresponding cross section of elastic collision in the same condition sometimes is called the dimensionless **probability of the elementary process**.

The **mean free path** λ of one collision partner A with respect to the elementary process A + B with another collision partner, B, can be calculated as

$$\lambda = 1/n_{\mathrm{B}}\sigma, \tag{2-3}$$

where n_{B} is the number density (concentration) of the particles B. The **interaction frequency** ν of one collision partner A with another collision partner B can be defined as the ratio of their relative velocity v to the mean free path λ:

$$\nu_{\mathrm{A}} = n_{\mathrm{B}}\sigma v. \tag{2-4}$$

Actually, this relation should be averaged, taking into account the velocity distribution function $f(v)$ and the dependence of cross section σ on the collision partners' velocity:

$$\nu_{\mathrm{A}} = n_{\mathrm{B}} \int \sigma(v)vf(v)dv = \langle \sigma v \rangle n_{\mathrm{B}}. \tag{2-5}$$

The number of elementary processes, w, which take place per unit volume per unit time is called the elementary **reaction rate**. For bimolecular processes A + B, the reaction rate can be calculated by multiplication of the interaction frequency of partner A with partner B, ν_{A}, and by the number of particles A in the unit volume (which is their number density, n_{A}):

$$w_{\mathrm{A+B}} = \nu_{\mathrm{A}} n_{\mathrm{A}} = \langle \sigma v \rangle n_{\mathrm{A}} n_{\mathrm{B}}. \tag{2-6}$$

The factor $\langle \sigma v \rangle$ is the so-called **reaction rate coefficient**, which can be calculated for bimolecular reactions as

$$k_{\mathrm{A+B}} = \int \sigma(v)vf(v)dv = \langle \sigma v \rangle. \tag{2-7}$$

In contrast to the reaction cross section σ, which is a function of the partners' energy, the reaction rate coefficient k is an integral factor, which includes information on the energy distribution functions and depends on temperatures or mean energies of the collision partners. Formula (2–7) establishes a relation between micro-kinetics concerned with elementary processes and macro-kinetics, which takes into account real energy distribution functions.

2.1.3. Classification of Ionization Processes

Mechanisms of ionization can be very different in different plasma-chemical systems and may be subdivided generally into the following five groups.

1. **Direct ionization by electron impact** is ionization of neutral and previously unexcited atoms, radicals, or molecules by an electron whose energy is high enough to provide the ionization act in one collision. These processes are the most important in cold or non-thermal discharges, where electric fields and therefore electron energies are quite high, but where the excitation level of neutral species is relatively moderate.
2. **Stepwise ionization by electron impact is** ionization of preliminary excited neutral species. These processes are important mostly in thermal or energy-intense discharges, when the ionization degree n_e/n_0 as well as the concentration of highly excited neutral species is quite high.
3. **Ionization by collision of heavy particles** takes place during ion–molecule or ion–atom collisions, as well as in collisions of electronically or vibrationally excited species, when

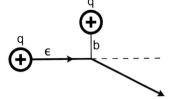

Figure 2–1. Illustration of Coulomb collisions of charged particles.

the total energy of the collision partners exceeds the ionization potential. The chemical energy of colliding neutral species can also contribute to ionization in the so-called associative ionization processes.

4. **Photo-ionization** takes place in collisions of neutrals with photons, which result in the formation of an electron–ion pair. Photo-ionization is mostly important in thermal plasmas and in some mechanisms of propagation of non-thermal discharges.

5. **Surface ionization (electron emission)** is provided by electron, ion, and photon collisions with different surfaces or simply by surface heating. This ionization mechanism is quite different from the first four and will be considered separately later on in this chapter.

2.1.4. Elastic Scattering and Energy Transfer in Collisions of Charged Particles: Coulomb Collisions

Electron–electron, electron–ion, and ion–ion scattering processes are the so-called **Coulomb collisions**. Their cross sections are quite high with respect to those of collisions with neutral partners, but they are much less frequent in discharges with low degrees of ionization. An important feature of Coulomb collisions is a strong dependence of their cross sections on the kinetic energy of colliding particles. This can be demonstrated by a simple analysis, illustrated in Fig. 2–1, where two particles have the same charge and, for the sake of simplicity, one collision partner is considered at rest. A scattering event takes place if the Coulomb interaction energy ($U \sim q^2/b$, where b is the impact parameter) is about the kinetic energy ε of a moving particle. Then, the impact parameter ($b \sim q^2/\varepsilon$) and the reaction cross section σ can be estimated as πb^2:

$$\sigma(\varepsilon) \approx \pi \frac{q^4}{(4\pi\varepsilon_0 \cdot \varepsilon)^2}. \tag{2–8}$$

Electron–electron scattering cross sections at room temperature are about 1000 times greater than those at an electron temperature of 1 eV, which is typical for electric discharges. Similar considerations for charged particle scattering on neutral molecules having a permanent dipole moment (interaction energy $U \sim 1/r^2$) and induced dipole moment (interaction energy $U \sim 1/r^4$) give, respectively, $\sigma(\varepsilon) \sim 1/\varepsilon$ and $\sigma(\varepsilon) \sim 1/\varepsilon^{1/2}$.

Energy transfer during the elastic collisions is possible only as a transfer of kinetic energy. The average fraction γ of kinetic energy, transferred from one particle of mass m to another one of mass M, is equal to

$$\gamma = \frac{2mM}{(m+M)^2}. \tag{2–9}$$

In an elastic collision of electrons with heavy neutrals or ions, $m \ll M$ and, hence, $\gamma = 2m/M$, which means that the fraction of transferred energy is very small ($\gamma \sim 10^{-4}$). It explains, in particular, why the direct electron impact ionization due to a collision of an incident electron with a valence electron of an atom predominates. Simply stated, only electron–electron collisions allow significant energy transfer. Detailed information regarding the elementary elastic collisions of charged species can be found in Smirnov (2001), McDaniel (1964, 1989), and Massey, Burhop, and Gilbody (1974).

2.1.5. Direct Ionization by Electron Impact: Thomson Formula

Direct ionization is a result of the interaction of an incident electron, having a high energy ε, with a valence electron of a preliminary neutral atom or molecule. Ionization occurs when the energy $\Delta\varepsilon$ transferred to the valence electron exceeds the ionization potential I. Analysis of the elementary process obviously requires a quantum mechanical consideration, but a clear physical understanding can be obtained from the classical Thomson model (1912). The valence electron is assumed to be at rest in this model, and interaction of the two colliding electrons with the rest of the atom is neglected. The differential cross section of the incident electron scattering with energy transfer $\Delta\varepsilon$ to the valence electron can be defined by the Rutherford formula:

$$d\sigma_i = \frac{1}{(4\pi\varepsilon_0)^2} \frac{\pi e^4}{\varepsilon(\Delta\varepsilon)^2} d(\Delta\varepsilon). \tag{2-10}$$

When the transferred energy exceeds the ionization potential, $\Delta\varepsilon \geq I$, direct ionization takes place. Thus, integration of (2–10) over $\Delta\varepsilon \geq I$ gives an expression for the ionization cross section by direct electron impact, known as the **Thomson formula:**

$$\sigma_i = \frac{1}{(4\pi\varepsilon_0)^2} \frac{\pi e^4}{\varepsilon} \left(\frac{1}{I} - \frac{1}{\varepsilon} \right). \tag{2-11}$$

Expression (2–11) should be multiplied, in general, by the number of valence electrons, Z_v. At high electron energies, $\varepsilon \gg I$, the Thomson cross section (2–11) is falling as $\sigma_i \sim 1/\varepsilon$. Quantum-mechanical treatment gives a more accurate but close asymptotic approximation: $\sigma_i \sim \ln \varepsilon/\varepsilon$. When $\varepsilon = 2I$, the Thomson cross section reaches the maximum value:

$$\sigma_i^{max} = \frac{1}{(4\pi\varepsilon_0)^2} \frac{\pi e^4}{4I^2}. \tag{2-12}$$

The Thomson formula can be rewritten taking into account the kinetic energy ε_v of the valence electron (Smirnov, 2001):

$$\sigma_i = \frac{1}{(4\pi\varepsilon_0)^2} \frac{\pi e^4}{\varepsilon} \left(\frac{1}{\varepsilon} - \frac{1}{I} + \frac{2\varepsilon_v}{3} \left(\frac{1}{I^2} - \frac{1}{\varepsilon^2} \right) \right). \tag{2-13}$$

The Thomson formula (2–13) agrees with (2–12), assuming the valence electron is at rest and $\varepsilon_v = 0$. An interesting variation of the Thomson formula (2–13) can be obtained assuming a Coulomb interaction of the valence electron with the rest of the atom and taking $\varepsilon_v = I$:

$$\sigma_i = \frac{1}{(4\pi\varepsilon_0)^2} \frac{\pi e^4}{\varepsilon} \left(\frac{5}{3I} - \frac{1}{\varepsilon} - \frac{2I}{3\varepsilon^2} \right). \tag{2-14}$$

All the modifications of the Thomson formula can be combined using the generalized function $f(\varepsilon/I)$ common for all atoms:

$$\sigma_i = \frac{1}{(4\pi\varepsilon_0)^2} \frac{\pi e^2}{I^2} Z_v f\left(\frac{\varepsilon}{I} \right). \tag{2-15}$$

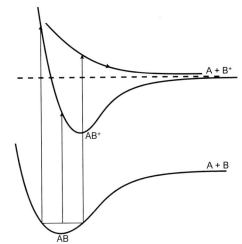

Figure 2–2. Molecular and ionic terms illustrating dissociative ionization of molecules AB.

Here Z_v is the number of valence electrons in an atom. For the simplest Thomson formula (2–11), the generalized function can be expressed as

$$f\left(x = \frac{\varepsilon}{I}\right) = \frac{1}{x} - \frac{1}{x^2}. \qquad (2\text{–}16)$$

The best agreement with experimental data for different atoms and molecules can be achieved when the generalized function $f(\varepsilon/I)$ is taken as follows (Smirnov, 2001):

$$\frac{10(x-1)}{\pi(x+0.5)(x+8)} < f(x) < \frac{10(x-1)}{\pi x(x+8)}. \qquad (2\text{–}17)$$

Useful formulas for estimations of the direct ionization cross sections can be also found in Barnett (1989). The ionization rate coefficient $k_i(T_e)$ can then be calculated by integration of the cross section $\sigma_i(\varepsilon)$ over the electron energy distribution function (see (2–7)). Assuming the Maxwellian EEDF, the direct ionization rate coefficient can be presented as

$$k_i(T_e) = \sqrt{8T_e/\pi m}\,\sigma_0 \exp\left(-\frac{I}{T_e}\right). \qquad (2\text{–}18)$$

In this relation, the cross section $\sigma_0 = Z_v \pi e^4 / I^2 (4\pi\varepsilon_0)^2$ is about the geometric atomic cross section (for molecular nitrogen, $10^{-16}\ \text{cm}^2$, and for argon, $3 \cdot 10^{-16}\ \text{cm}^2$).

2.1.6. Specific Features of Ionization of Molecules by Electron Impact: Frank-Condon Principle and Dissociative Ionization

Non-dissociative ionization of molecules by direct electron impact can be presented for the case of diatomic molecules AB as

$$e + AB \rightarrow AB^+ + e + e. \qquad (2\text{–}19)$$

This process takes place when the electron energy does not greatly exceed the ionization potential. Some peculiarities of ionization of molecules by electron impact can be seen from illustrative potential energy curves for AB and AB^+, shown in Fig. 2–2. The fastest internal motion of atoms inside molecules is molecular vibration. But even molecular vibrations have typical time periods of 10^{-14}–10^{-13} s, which are much longer than the interaction times between plasma electrons and the molecules: $a_0/v_e \sim 10^{-16}$–10^{-15} s (where a_0 is the atomic unit of length and v_e is the mean electron velocity). It means that all kinds of electronic excitation processes under consideration, induced by electron impact (including the molecular ionization; see Fig. 2–1), are much faster than all other atomic motions inside

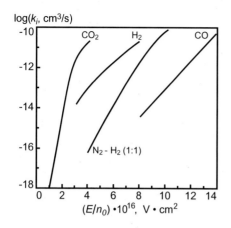

Figure 2–3. Rate coefficients of ionization by direct electron impact in molecular gases (CO_2, H_2, N_2–H_2, CO) as a function of reduced electric field.

the molecules. As a result, all the atoms inside a molecule can be considered as being frozen during the process of electronic transition, stimulated by electron impact. This fact is known as the **Frank-Condon principle.**

According to the Frank-Condon principle, the processes of collisional excitation and ionization of molecules are presented in Fig. 2–2 by vertical lines. The non-dissociative ionization process (2–19) usually results in the formation of a vibrationally excited ion $(AB^+)^*$ and requires a little more energy than corresponding atomic ionization.

When the electron energy is relatively high and substantially exceeds the ionization potential, the dissociative ionization process can take place:

$$e + AB \rightarrow A + B^+ + e + e. \tag{2–20}$$

This ionization process corresponds to electronic excitation into a repulsive state of the ion, $(AB^+)^*$, followed by a decay of this molecular ion. It is also illustrated by a vertical line in Fig. 2–2. One can see from Fig. 2–2 that the energy threshold for the dissociative ionization is essentially greater than that for the non-dissociative one. Data on the electron impact direct ionization for different molecular gases (CO_2, H_2, N_2, etc.) are presented in Fig. 2–3 as a function of reduced electric field E/n_0, which is the ratio of the electric field over the neutral gas concentration.

2.1.7. Stepwise Ionization by Electron Impact

When the plasma density and, therefore, the concentration of excited neutrals are high enough, the energy (I) necessary for ionization can be provided in two different ways. First, like in the case of direct ionization, it could be provided by the energy of plasma electrons. Second, the high energy of preliminary electronic excitation of neutrals can be converted in the ionization act, which is called **stepwise ionization**. If the level of electronic excitation is high enough, stepwise ionization is much faster than direct ionization, because the statistical weight of electronically excited neutrals is greater than that of free plasma electrons. In other words, when $T_e \ll I$, the probability of obtaining high ionization energy I is much lower for free plasma electrons (direct ionization) than for excited atoms and molecules (stepwise ionization).

In contrast to direct ionization, the stepwise process includes several steps to provide the ionization event. At first, electron–neutral collisions prepare highly excited species, and then a final collision with a relatively low-energy electron provides the actual ionization event. In thermodynamic equilibrium, the ionization process $e + A \rightarrow A^+ + e + e$ is inverse to the three-body recombination $A^+ + e + e \rightarrow A^* + e \rightarrow A + e$, which proceeds through a set of excited states. According to the principle of detailed equilibrium, the ionization process $e + A \rightarrow A^+ + e + e$ should go through the set of electronically excited states as well, which means the ionization should be a stepwise process. However, this conclusion can only hold true for quasi-equilibrium or thermal plasmas.

The stepwise ionization rate coefficient k_i^s can be found by the summation of partial rate coefficients $k_i^{s,n}$, corresponding to the nth electronically excited state, over all states of excitation, taking into account their concentrations:

$$k_i^s = \sum_n k_i^{s,n} N_n(\varepsilon_n)/N_0. \tag{2-21}$$

To calculate the maximum stepwise ionization rate (Smirnov, 2001), we can assume that electronically excited neutrals are in quasi-equilibrium with plasma electrons, and the electronically excited states are characterized by the Boltzmann distribution having electron temperature T_e:

$$N_n = \left(\frac{g_n}{g_0}\right) N_0 \exp\left(-\frac{\varepsilon_n}{T_e}\right). \tag{2-22}$$

In this relation, N_n, g_n, and ε_n are number densities, statistical weights, and energies of the electronically excited atoms, radicals, or molecules, respectively; the index n is the principal quantum number. From statistical thermodynamics, the statistical weight of an excited particle $g_n = 2g_i n^2$, where g_i is the statistical weight of an ion; N_0 and g_0 are concentration and statistical weights of ground-state particles, respectively.

The typical energy transfer from a plasma electron to an electron sitting in an excited atomic level is about T_e. This means that excited particles with energy about $\varepsilon_n = I - T_e$ make the major contributions into sum (2–21). Taking into account that $I_n \sim 1/n^2$, the number of states with energy about $\varepsilon_n = I - T_e$ and ionization potential about $I_n = T_e$ has an order of n. Thus, from (2–21) and (2–22), we can derive

$$k_i^s \approx \frac{g_i}{g_0} n^3 \langle \sigma v \rangle \exp\left(-\frac{I}{T_e}\right). \tag{2-23}$$

The cross section σ in (2–23), corresponding to an energy transfer of about T_e between electrons, can be estimated as $e^4/T_e^2(4\pi\varepsilon_0)^2$, velocity $v \sim \sqrt{T_e/m}$, and the principal quantum number can be taken from

$$I_n \approx \frac{1}{(4\pi\varepsilon_0)^2} me^4/\hbar^2 n^2 \approx T_e. \tag{2-24}$$

As a result, the stepwise ionization rate coefficient can be expressed as

$$k_i^s \approx \frac{g_i}{g_0} \frac{1}{(4\pi\varepsilon_0)^5} \left(me^{10}/\hbar^3 T_e^3\right) \exp\left(-\frac{I}{T_e}\right). \tag{2-25}$$

Comparing direct ionization (2–18) with stepwise ionization (2–25), one can see that the second one can be much faster because of the high statistical weight of excited species involved in the stepwise ionization. The ratio of rate coefficients for these two competing mechanisms of ionization can be easily derived from (2–18) and (2–25):

$$\frac{k_i^s(T_e)}{k_i(T_e)} \approx \frac{g_i a_0^2}{g_0 \sigma_0} \left(\frac{1}{(4\pi\varepsilon_0)^2} me^4/\hbar^2 T_e\right)^{\frac{7}{2}} \approx \left(\frac{I}{T_e}\right)^{\frac{7}{2}}. \tag{2-26}$$

In this relation, a_0 is the atomic unit of length; we also took into account the estimations for the geometric collisional cross section, $\sigma_0 \sim a_0^2$, and for the ionization potential, $I \approx \frac{1}{(4\pi\varepsilon_0)^2} me^4/\hbar^2$. For typical discharges with $I/T_e \sim 10$, the stepwise ionization can be 10^3–10^4 times faster than the direct one. Surely, it takes place only in the case of quasi-equilibrium (2–22) between electronically excited species and plasma electrons, which is possible in thermal discharges. A description of the stepwise ionization in non-thermal plasma conditions can be found in Fridman and Kennedy (2004).

2.1.8. Ionization by High-Energy Electrons and Electron Beams: Bethe-Bloch Formula

The electron energy in electron beams applied today usually varies from 50 KeV to 1–2 MeV. Typical energy losses of the beams in atmospheric-pressure air are about 1 MeV per 1 m (≈ 1 keV/mm). Generation of large plasma volumes for this reason requires high electron energies in the beams. The beams with electron energies exceeding 500 KeV are referred to as relativistic electron beams. To describe ionization induced by the electrons with velocities significantly exceeding the velocities of atomic electrons, the Born approximation (Landau, 1997) can be applied. Electron energy losses per unit length, dE/dx, can be evaluated in the **non-relativistic case** by the **Bethe-Bloch formula**:

$$-\frac{dE}{dx} = \frac{2\pi Z e^4}{(4\pi\varepsilon_0)^2 m v^2} n_0 \ln \frac{2m E v^2}{I^2}. \tag{2-27}$$

In this relation, Z is the atomic number of neutral particles, providing the beam stopping; n_0 is their number density; and v is the stopping electron velocity. In the **case of relativistic electron beams** with the electron energies of 0.5–1 MeV, the energy losses can be numerically calculated by the following relation:

$$-\frac{dE}{dx} = 2 \cdot 10^{-22} n_0 Z \ln \frac{183}{Z^{1/3}}, \tag{2-28}$$

where dE/dx is expressed in MeV/cm, and n_0 is the concentration of neutral particles expressed in cm^{-3}. Relation (2–28) can be rewritten in terms of effective ionization rate coefficient k_i^{eff} for relativistic electrons:

$$k_i^{eff} \approx 3 \cdot 10^{-10} (cm^3/s) Z \ln \frac{183}{Z^{1/3}}. \tag{2-29}$$

Numerically, this ionization rate coefficient is about 10^{-8}–10^{-7} cm^3/s. The total rate of ionization by relativistic electron beams can be expressed in this case as a function of the electron beam concentration n_b or the electron beam current density j_b (c is the speed of light):

$$q_e = k_i^{eff} n_b n_0 \approx k_i^{eff} \frac{1}{ec} n_0 j_b. \tag{2-30}$$

2.1.9. Photo-Ionization Processes

Photo-ionization of a neutral particle A with ionization potential I (in electron volts) by a photon $\hbar\omega$ with wavelength λ can be illustrated as

$$\hbar\omega + A \rightarrow A^+ + e, \quad \lambda < \frac{12,400}{I(eV)} \text{Å}. \tag{2-31}$$

To provide the ionization, the photon wavelength should be usually less than 1000 Å, which is ultraviolet radiation. The photo-ionization cross section increases sharply from zero at the threshold energy (2–32) to quite high values – up to the geometrical cross section. The photo-ionization cross sections for some atoms and molecules are presented in Table 2–1.

The cross sections presented in Table 2–1 are quite high. The contribution of the photo-ionization process is usually not very significant because of low concentrations of high-energy photons in most discharge systems. However, sometimes photo-ionization plays a

Table 2–1. Photo-Ionization Cross Sections

Atoms or molecules	Wavelength λ, Å	Cross sections, cm^2	Atoms or molecules	Wavelength λ, Å	Cross sections, cm^2
Ar	787	$3.5 \cdot 10^{-17}$	Ne	575	$0.4 \cdot 10^{-17}$
N_2	798	$2.6 \cdot 10^{-17}$	O	910	$0.3 \cdot 10^{-17}$
N	482	$0.9 \cdot 10^{-17}$	O_2	1020	$0.1 \cdot 10^{-17}$
He	504	$0.7 \cdot 10^{-17}$	Cs	3185	$2.2 \cdot 10^{-19}$
H_2	805	$0.7 \cdot 10^{-17}$	Na	2412	$1.2 \cdot 10^{-19}$
H	912	$0.6 \cdot 10^{-17}$	K	2860	$1.2 \cdot 10^{-20}$

very essential role – mostly by rapidly supplying seed electrons for subsequent ionization by electron impact.

2.1.10. Ionization in Collisions of Heavy Particles: Adiabatic Principle and Massey Parameter

An electron with a kinetic energy only slightly exceeding the ionization potential is often quite effective in producing the ionization event; however, this is not true for ionization by collisions of heavy particles – ions and neutrals. Even when they have enough kinetic energy, they actually cannot provide ionization, because their velocities are much less than those of electrons in atoms. Even if it has enough energy, a heavy particle is very often unable to transfer this energy to an electron inside an atom because the process is far from resonant. Such slow motion may be termed "adiabatic" and is unable to transfer energy to a fast-moving particle. The **adiabatic principle** can be explained in terms of the relationship between low interaction frequency $\omega_{int} = \alpha v$ and high frequency of electron transfers in atoms, $\omega_{tr} = \Delta E / \hbar$. Here $1/\alpha$ is a characteristic size of the interacting neutral particles, v is their velocity, and ΔE is a change of electron energy in the atom during the interaction. Only fast Fourier components of the slow interaction potential between particles with frequencies about $\omega_{tr} = \Delta E / \hbar$ provide the energy transfer between the interacting particles. The relative weight or probability of these fast Fourier components is very low if $\omega_{tr} \gg \omega_{int}$; numerically it is about $\exp(-\omega_{tr}/\omega_{int})$. As a result, the probability P_{EnTr} of energy transfer processes (including the ionization process under consideration) are usually proportional to the so-called Massey parameter:

$$P_{EnTr} \propto \exp\left(-\frac{\omega_{tr}}{\omega_{int}}\right) \propto \exp\left(-\frac{\Delta E}{\hbar \alpha v}\right) = \exp(-P_{Ma}), \qquad (2\text{–}32)$$

where $P_{Ma} = \Delta E / \hbar \alpha v$ is the **adiabatic Massey parameter**. If $P_{Ma} \gg 1$, the process of energy transfer is adiabatic and its probability is exponentially low. It takes place in collisions of heavy neutrals and ions. To get the Massey parameter close to one and eliminate the adiabatic prohibition (2–28) for ionization, the kinetic energy of the colliding heavy particle has to be about 10–100 KeV, which is about three orders of magnitude more than the ionization potential.

2.1.11. Penning Ionization Effect and Associative Ionization

If the electronic excitation energy of a metastable atom A* exceeds the ionization potential of another atom B, their collision can lead to an act of ionization, the so-called **Penning ionization**. The Penning ionization usually proceeds through the intermediate formation of an unstable excited quasi molecule in the state of auto-ionization; cross sections of the process can be very high. Cross sections for the Penning ionization of N_2, CO_2, Xe, and Ar by metastable helium atoms $He(2^3S)$ with an excitation energy of 19.8 eV reach gas-kinetic values of about 10^{-15} cm^2. Similar cross sections can be attained in collisions of metastable neon atoms (excitation energy 16.6 eV) with argon atoms (ionization potential 15.8 eV). An exceptionally high cross section of $1.4 \cdot 10^{-14}$ cm^2 (Smirnov, 1974) can be achieved in Penning ionization of mercury atoms (ionization potential 10.4 eV) by collisions with the metastable helium atoms He(2^3S, 19.8 eV). If the total electronic excitation energy of colliding particles is not sufficient, ionization is still possible when heavy species stick to each other, forming a molecular ion. Such a process is called the **associative ionization;** for example,

$$\text{Hg}(6^3\text{P}_1, E = 4.9\,\text{eV}) + \text{Hg}(6^3\text{P}_0, E = 4.7\,\text{eV}) \rightarrow \text{Hg}_2^+ + e. \qquad (2\text{–}33)$$

The total electronic excitation energy here (9.6 eV) is less than the ionization potential of mercury atoms (10.4 eV) but higher than ionization potential for a Hg_2^- molecule. This is actually the main "trick" of the associative ionization. Cross sections of the associative ionization (similar to the Penning ionization) can be quite high and close to the gas-kinetic value (about 10^{-15} cm^2). Associative ionization $\text{A}^* + \text{B} \rightarrow \text{AB}^+ + e$ takes place effectively when there is a crossing of the electronic energy term of the colliding particles with an electronic energy term of the molecular ion AB^+ and, as a result, the process is non-adiabatic. Such a situation takes place only for a limited number of excited species.

Ionization during collision of vibrationally excited molecules can be very important in non-equilibrium systems. Cross sections of such processes, however, are very low because of the smallness of the Frank-Condon factors. Consider the associative ionization (Rusanov & Fridman, 1984):

$$N_2^*\left(^1\Sigma_g^+, v_1 \approx 32\right) + N_2^*\left(^1\Sigma_g^+, v_2 \approx 32\right) \rightarrow N_4^+ + e. \qquad (2\text{–}34)$$

The nitrogen molecules with 32 vibrational quanta each have enough energy for the associative ionization, but the reaction rate of the process is relatively low: $10^{-15}\exp(-12000\,\text{K/T})$, cm^3/s. More information regarding ionization in collisions of vibrationally excited molecules can be found in Adamovich et al. (1993) and Plonjes et al. (2001).

2.2. ELEMENTARY PLASMA-CHEMICAL REACTIONS OF POSITIVE IONS

Positive ions are obviously major players in plasma-chemical processes. Their exothermic reactions with neutrals usually have no activation energy, which makes their contribution significant in many specific plasma-chemical processes, particularly in plasma catalysis. In addition to high chemical activity, the ions can have significant kinetic energy, which determines their contribution, for example, in reactive ion etching.

2.2.1. Different Mechanisms of Electron–Ion Recombination in Plasma

Electron–ion recombination is a highly exothermic process. Therefore, the process should have a specific channel of accumulation of the energy released during the neutralization of a positive ion and an electron. Most of these channels of recombination energy consumption are related, either to dissociation of molecules or to three-body collisions or to radiation, which determines the following three major groups of mechanisms of electron–ion recombination.

1. The fastest electron neutralization mechanism in molecular gases, or in the presence of molecular ions, is **dissociative electron–ion recombination:**

$$e + \text{AB}^+ \rightarrow (\text{AB})^* \rightarrow \text{A} + \text{B}^*. \qquad (2\text{–}35)$$

2.2. Elementary Plasma-Chemical Reactions of Positive Ions

Recombination energy in these processes goes into dissociation of the intermediately formed molecule ion and to excitation of the dissociation products. These processes are common for molecular gases, but they can also be important in atomic gases because of the formation of molecular ions in the **ion conversion processes**:

$$A^+ + A + A \rightarrow A_2^+ + A. \tag{2-36}$$

2. Electron–ion neutralization is due to a **three-body electron–ion recombination** in atomic gases in the absence of molecular ions:

$$e + e + A^+ \rightarrow A^* + e. \tag{2-37}$$

Energy excess in this case is going into the kinetic energy of a free electron, which participates in the recombination act as "a third-body partner." We must remark that heavy particles (ion and neutrals) are unable to accumulate electron recombination energy fast enough in their kinetic energy and, therefore, are ineffective as the third-body partner.

3. Finally, the recombination energy can be converted into radiation in the process of **radiative electron–ion recombination**:

$$e + A^+ \rightarrow A^* \rightarrow A + \hbar\omega. \tag{2-38}$$

The cross section of this process is relatively low and can compete with the three-body recombination only when the plasma density is not high.

2.2.2. Dissociative Electron–Ion Recombination and Possible Preliminary Stage of Ion Conversion

The recombination mechanism (2–35) is quite fast and plays the major role in molecular gases. Reaction rate coefficients for most of the diatomic and triatomic ions are on the level of 10^{-7} cm^3/s. For some important molecular ions, the kinetic information can be found in Table 2–2. In a group of similar ions, like molecular ions of noble gases, the recombination rate coefficients increase as the number of internal electrons increases: recombination of Kr_2^+ and Xe_2^+ is about 100 times faster than that of helium.

The rate coefficient of dissociative electron–ion recombination decreases with temperature growth (see Table 2–2). The process has no activation energy, so its dependencies on both electron (T_e) and gas (T_0) temperatures are not very strong (Biondi, 1976):

$$k_r^{ei}(T_e, T_0) \propto \frac{1}{T_0 \sqrt{T_e}}. \tag{2-39}$$

If the pressure is high enough, the recombination of atomic ions like Xe^+ proceeds faster, not by means of three-body (2–37) or radiative (2–38) mechanisms but through the preliminary formation of molecular ions, that is, through the so-called **ion conversion reactions** (2–36), such as: $Xe^+ + Xe + Xe \rightarrow Xe_2^+ + Xe$. Then the molecular ion can be quickly neutralized in the rapid process of dissociative recombination (see Table 2–2). The ion conversion reaction rate coefficients are quite high (Smirnov, 1974; Virin et al., 1978); some of them are included in Table 2–3. When the pressure exceeds 10 torr, the ion conversion is usually faster than the following process of dissociative recombination, which becomes a limiting stage in the overall recombination kinetic pathway. An analytical expression for the ion-conversion three-body reaction rate coefficient can be derived just from dimension analysis (Smirnov, 1974):

$$k_{ic} \propto \left(\frac{\beta e^2}{4\pi \varepsilon_0} \right)^{\frac{5}{4}} \frac{1}{M^{1/2} T_0^{3/4}}. \tag{2-40}$$

Table 2–2. Dissociative Electron–Ion Recombination Rate Coefficients k_r^{ei} at Room Gas Temperature $T_0 = 300$ K, and Electron Temperatures $T_e = 300$ K and $T_e = 1$ eV

Recombination process	k_r^{ei}, cm³/s $T_e = 300$ K	k_r^{ei}, cm³/s $T_e = 1$ eV	Recombination process	k_r^{ei}, cm³/s $T_e = 300$ K	k_r^{ei}, cm³/s $T_e = 1$ eV
$e + N_2^+ \rightarrow N + N$	$2\cdot10^{-7}$	$3\cdot10^{-8}$	$e + NO^+ \rightarrow N + O$	$4\cdot10^{-7}$	$6\cdot10^{-8}$
$e + O_2^+ \rightarrow O + O$	$2\cdot10^{-7}$	$3\cdot10^{-8}$	$e + H_2^+ \rightarrow H + H$	$3\cdot10^{-8}$	$5\cdot10^{-9}$
$e + H_3^+ \rightarrow H_2 + H$	$2\cdot10^{-7}$	$3\cdot10^{-8}$	$e + CO^+ \rightarrow C + O$	$5\cdot10^{-7}$	$8\cdot10^{-8}$
$e + CO_2^+ \rightarrow CO + O$	$4\cdot10^{-7}$	$6\cdot10^{-8}$	$e + He_2^+ \rightarrow He + He$	10^{-8}	$2\cdot10^{-9}$
$e + Ne_2^+ \rightarrow Ne + Ne$	$2\cdot10^{-7}$	$3\cdot10^{-8}$	$e + Ar_2^+ \rightarrow Ar + Ar$	$7\cdot10^{-7}$	10^{-7}
$e + Kr_2^+ \rightarrow Kr + Kr$	10^{-6}	$2\cdot10^{-7}$	$e + Xe_2^+ \rightarrow Xe + Xe$	10^{-6}	$2\cdot10^{-7}$
$e + N_4^+ \rightarrow N_2 + N_2$	$2\cdot10^{-6}$	$3\cdot10^{-7}$	$e + O_4^+ \rightarrow O_2 + O_2$	$2\cdot10^{-6}$	$3\cdot10^{-7}$
$e + O_4^+ \rightarrow O + O + O_2$	$7\cdot10^{-6}$	$2\cdot10^{-6}$	$e + H_2O^+ \rightarrow 2H + O$	10^{-6}	$2\cdot10^{-7}$
$e + H_3O^+ \rightarrow H_2 + OH$	10^{-6}	$2\cdot10^{-7}$	$e + (NO)_2^+ \rightarrow 2NO$	$2\cdot10^{-6}$	$3\cdot10^{-7}$
$e + HCO^+ \rightarrow CO + H$	$2\cdot10^{-7}$	$3\cdot10^{-8}$	$e + SiF_3^+ \rightarrow SiF_2 + F$	$2\cdot10^{-7}$	$4\cdot10^{-8}$
$e + SiF_2^+ \rightarrow SiF + F$	$2\cdot10^{-7}$	$4\cdot10^{-8}$	$e + SiF^+ \rightarrow Si + F$	$2\cdot10^{-7}$	$4\cdot10^{-8}$
$e + CF_3^+ \rightarrow CF_2 + F$	$3\cdot10^{-9}$	$5\cdot10^{-10}$	$e + CF_2^+ \rightarrow CF + F$	$3\cdot10^{-9}$	$5\cdot10^{-10}$
$e + CF^+ \rightarrow C + F$	$2\cdot10^{-9}$	$4\cdot10^{-10}$	$e + CO^+ \rightarrow C + O$	$3\cdot10^{-9}$	$5\cdot10^{-10}$

Table 2–3. Ion-Conversion Rate Coefficients at Room Temperature

Ion-conversion process	Reaction rate coefficient	Ion-conversion process	Reaction rate coefficient
$N_2^+ + N_2 + N_2 \rightarrow N_4^+ + N_2$	$8 \cdot 10^{-29}$ cm^6/s	$O_2^+ + O_2 + O_2 \rightarrow O_4^+ + O_2$	$3 \cdot 10^{-30}$ cm^6/s
$H^+ + H_2 + H_2 \rightarrow H_3^+ + H_2$	$4 \cdot 10^{-29}$ cm^6/s	$O_2^+ + O_2 + N_2 \rightarrow O_4^+ + N_2$	$3 \cdot 10^{-30}$ cm^6/s
$N_2^+ + N_2 + N \rightarrow N_4^+ + N$	$8 \cdot 10^{-29}$ cm^6/s	$O_2^+ + O_2 + N \rightarrow O_4^+ + N$	$5 \cdot 10^{-30}$ cm^6/s
$He^+ + He + He \rightarrow He_2^+ + He$	$9 \cdot 10^{-32}$ cm^6/s	$Ne^+ + Ne + Ne \rightarrow Ne_2^+ + Ne$	$6 \cdot 10^{-32}$ cm^6/s
$Ar^+ + Ar + Ar \rightarrow Ar_2^+ + Ar$	$3 \cdot 10^{-31}$ cm^6/s	$Kr^+ + Kr + Kr \rightarrow Kr_2^+ + Kr$	$2 \cdot 10^{-31}$ cm^6/s
$Xe^+ + Xe + Xe \rightarrow Xe_2^+ + Xe$	$4 \cdot 10^{-31}$ cm^6/s	$Cs^+ + Cs + Cs \rightarrow Cs_2^+ + Cs$	$1.5 \cdot 10^{-29}$ cm^6/s

In this relation, M and β are mass and polarization coefficients of colliding atoms, and T_0 is the gas temperature. The ion-conversion effect takes place as a preliminary stage of recombination not only for simple atomic ions but also for some important molecular ions. Polyatomic ions have very high recombination rates (see Table 2–2), often exceeding 10^{-6} cm^3/s at room temperature, which results in an interesting fact: recombination of molecular ions like N_2^+ and O_2^+ at elevated pressures sometimes also goes through intermediate formation of such dimers as N_4^+ and O_4^+.

2.2.3. Three-Body and Radiative Electron–Ion Recombination Mechanisms

The three-body recombination process (2–37) is the most important one in high-density quasi-equilibrium plasmas. Concentrations of molecular ions are very low in this case (because of thermal dissociation) for the fast mechanism of dissociative recombination described earlier, and the three-body reaction dominates. The recombination process starts with the three-body capture of an electron by a positive ion and formation of a highly excited atom with a binding energy of about T_e. This highly excited atom then gradually loses energy in electron impacts. The three-body electron–ion recombination process (2–37) is a reverse one with respect to the stepwise ionization (see Section 2.1.7). For this reason, the rate coefficient of the recombination can be derived from the stepwise ionization rate coefficient k_i^s (2–25) and from the Saha thermodynamic equation for ionization/recombination balance (see Chapter 3):

$$k_r^{eei} = k_i^s \frac{n_0}{n_e n_i} = k_i^s \frac{g_0}{g_e g_i} \left(\frac{2\pi \hbar}{m T_e} \right)^{\frac{3}{2}} \exp \left(\frac{I}{T_e} \right) \approx \frac{e^{10}}{(4\pi \varepsilon_0)^5 \sqrt{m T_e^9}}. \quad (2\text{–}41)$$

In this relation, n_e, n_i, and n_0 are number densities of electrons, ions, and neutrals, respectively; g_e, g_i, and g_0 are their statistical weights; e and m are electron charge and mass; and I is an ionization potential. For practical calculations, relation (2–41) can be presented in numerical form:

$$k_r^{eei}, \frac{cm^6}{s} = \frac{\sigma_0}{I} 10^{-14} \left(\frac{I}{T_e} \right)^{4.5}, \quad (2\text{–}42)$$

where σ_0(cm^2) is the gas-kinetic cross section, and I and T_e are the ionization potential and electron temperature in electron volts. Typical values of k_r^{eei} at room temperature are about 10^{-20} cm^6/s; at $T_e = 1$ eV this rate coefficient is about 10^{-27} cm^6/s. At room temperature, the three-body recombination is able to compete with the dissociative recombination when the electron concentration is quite high and exceeds 10^{13} cm^{-3}. If the electron temperature is about 1 eV, the three-body recombination can compete with the dissociative recombination only in the case of exceptionally high electron densities exceeding 10^{20} cm^{-3}. Excessive energy of recombination goes into the kinetic energy of a free electron acting as a "third body." Heavy particles, both ions and neutrals, are too slow and ineffective in this case as

Table 2–4. Polarizability Coefficients of Different Atoms and Molecules

Atom or molecule	α, 10^{-24} cm^3	Atom or molecule	α, 10^{-24} cm^3	Atom or molecule	α, 10^{-24} cm^3
Ar	1.64	H	0.67	H_2O	1.45
C	1.78	N	1.11	CO_2	2.59
O	0.8	CO	1.95	SF_6	4.44
Cl_2	4.59	O_2	1.57	NH_3	2.19
CCl_4	10.2	CF_4	1.33		

third-body partners. The rate coefficients of such recombination are about 10^8 times lower than in (2–42).

The radiative electron–ion recombination (2–38) is also a relatively slow one, because it requires a photon emission during a short interval of the electron–ion interaction. This type of recombination can play a major role only in the absence of molecular ions, and the plasma density is quite low when three-body mechanisms are suppressed. Cross sections of radiative recombination are usually about 10^{-21} cm^2. Rate coefficients can be estimated (Zeldovich & Raizer, 1966) as a function of electron temperature:

$$k_{\text{rad.rec.}}^{\text{ei}} \approx 3 \cdot 10^{-13} (T_{\text{e}}, eV)^{-3/4} \text{cm}^3/\text{s}. \tag{2–43}$$

Comparison of (2–42) and (2–43) shows that radiative recombination is faster than three-body recombination when the electron concentration is relatively low:

$$n_{\text{e}} < 3 \cdot 10^{13} (T_{\text{e}}, eV)^{3.75}, \text{cm}^3/\text{s}. \tag{2–44}$$

2.2.4. Ion–Molecular Reactions, Ion–Molecular Polarization Collisions, and the Langevin Rate Coefficient

Some ion–molecular reactions were already discussed earlier. Thus, the positive ion conversion $A^+ + B + M \rightarrow AB^+ + M$ was considered in Section 2.2.2 as a preliminary stage of the dissociative electron–ion recombination. Ion–molecular reactions not only make a contribution in the balance of charged particles but also provide plasma-chemical processes by themselves. Ion-cluster growth in dusty SiH_4 plasmas and ion–molecular chain reactions of SO_2 oxidation in air during exhaust gas cleaning are good relevant examples, which will be discussed later in the book.

The ion–molecular processes start with scattering in a polarization potential, leading to the so-called Langevin capture of a charged particle and formation of an intermediate ion–molecular complex. If a neutral particle itself has no permanent dipole moment, the ion–neutral charge-dipole interaction is due to dipole moment P_{m}, induced in the neutral particle by the electric field E of an ion:

$$p_{\text{m}} = \alpha \varepsilon_0 E = \alpha \frac{e}{4\pi r^2}. \tag{2–45}$$

In this relation, r is the distance between the interacting particles, and α is the polarizability of a neutral atom or molecule, which is numerically about the volume of the atom or molecule (see Table 2–4). Typical orbits of relative ion and neutral motion during the polarization scattering are shown in Fig. 2–4. When the impact parameter is high then the orbit is hyperbolic, but when it is low then the scattering leads to the **Langevin polarization capture** and the formation of an ion–molecular complex. The capture occurs when the charge-dipole interaction $p_{\text{m}}E = \alpha \frac{e}{4\pi r^2} \frac{e}{4\pi \varepsilon_0 r^2}$ becomes the order of kinetic energy $\frac{1}{2}Mv^2$,

2.2. Elementary Plasma-Chemical Reactions of Positive Ions

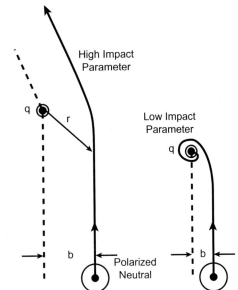

Figure 2–4. Langevin scattering in polarization potential at different impact parameters.

where M is the reduced mass and v is the relative velocity. The **Langevin cross section** of the ion–molecular polarization capture can then be found as $\sigma_L \approx \pi r^2$:

$$\sigma_L = \sqrt{\frac{\pi \alpha e^2}{\varepsilon_0 M v^2}}. \qquad (2\text{--}46)$$

The Langevin capture rate coefficient $k_L \approx \sigma_L v$ in this case does not depend on velocity and, therefore, does not depend on temperature:

$$k_L = \sqrt{\frac{\pi \alpha e^2}{\varepsilon_0 M}}. \qquad (2\text{--}47)$$

Numerically, the Langevin capture rate coefficient can be presented as

$$k_L^{\text{ion/neutral}} = 2.3 \cdot 10^{-9}\ \text{cm}^{-3}/\text{s} \times \sqrt{\frac{\alpha, 10^{-24}\ \text{cm}^3}{M,\ \text{amu}}}. \qquad (2\text{--}48)$$

The typical value of the rate coefficient for ion–molecular reactions is 10^{-9} cm³/s, which is 10 times higher than the gas-kinetic value for neutral particles: $k_0 \approx 10^{-10}$ cm³/s. The preceding relations describe the interaction of a charge with an induced dipole. If an ion interacts with a molecule having a permanent dipole moment μ_D, the Langevin capture cross section becomes larger (Su & Bowers, 1973, 1975):

$$k_L = \sqrt{\frac{\pi e^2}{\varepsilon_0 M}} \left(\sqrt{\alpha} + c \mu_D \sqrt{\frac{2}{\pi T_0}} \right). \qquad (2\text{--}49)$$

In this relation, T_0 is the gas temperature; the parameter $0 < c < 1$ describes the effectiveness of a dipole orientation in the electric field of an ion. For molecules with a high permanent dipole moment, the ratio of the second to the first terms is about $\sqrt{I/T_0}$, where I is an ionization potential. We should note that the Langevin rate coefficient for electron–neutral collisions can be presented as

$$k_L^{\text{electron/neutral}} = 10^{-7}\ \text{cm}^3/\text{s} \times \sqrt{\alpha, 10^{-24}\ \text{cm}^3}. \qquad (2\text{--}50)$$

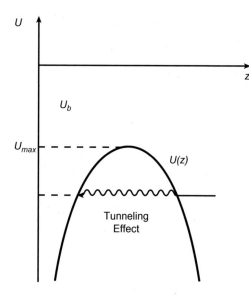

Figure 2–5. Correlation of energy terms for the elementary process of resonant charge exchange.

The numerical value of the Langevin rate coefficient for an electron–neutral collision is about 10^{-7} cm^3/s and also does not depend on temperature.

2.2.5. Ion–Atomic Charge Transfer Processes and Resonant Charge Transfer

The charge exchange processes correspond to an electron transfer from a neutral particle to a positive ion or from negative ion to a neutral particle. If the reaction has no defect of electronic energy, ΔE, it is called a **resonant charge transfer**; otherwise it is referred to as being **non-resonant**. The resonant charge transfer is non-adiabatic and is characterized by a large cross section. Let us consider the charge exchange between a neutral particle B and a positive ion A$^+$ supposing particle B/B$^+$ is at rest:

$$A^+ + B \rightarrow A + B^+. \tag{2–51}$$

The energy scheme of the reaction is illustrated in Fig. 2–5 by the Coulomb potential energy of an electron in the field of A$^+$ and B$^+$:

$$U(z) = -\frac{e^2}{4\pi \varepsilon_0 z} - \frac{e^2}{4\pi \varepsilon_0 |r_{AB} - z|}, \tag{2–52}$$

where r_{AB} is the distance between the centers of A and B. The maximum potential energy is reached at $z = r_{AB}/2$:

$$U_{max} = -\frac{e^2}{\pi \varepsilon_0 r_{AB}}. \tag{2–53}$$

The classical charge transfer (2–51) is possible (see Fig. 2–5) if the maximum of potential energy, U_{max}, is lower than the initial electron energy E_B:

$$E_B = -\frac{I_B}{n^2} - \frac{e^2}{4\pi \varepsilon_0 r_{AB}} \geq U_{max}, \tag{2–54}$$

where I_B is the ionization potential of atom B. The maximum distance between the interacting heavy particles, when the charge transfer is still permitted by classical mechanics, is

$$r_{AB}^{max} = \frac{3e^2 n^2}{4\pi \varepsilon_0 I_B}. \tag{2-55}$$

If the charge exchange is resonant and not energy limited, the classical reaction cross section

$$\sigma_{ch.tr}^{class} = \pi r_{AB}^2 = \frac{9e^4 n^4}{16\pi \varepsilon_0^2 I_B^2}, \tag{2-56}$$

does not depend on the kinetic energy of interacting species and its numerical value for ground-state transfer ($n = 1$) is about equal to the gas-kinetic cross section for collisions of neutrals. The actual cross section of the resonant charge transfer can be much higher than that in (2–56), taking into account the quantum-mechanical effect of electron tunneling from B to A^+ (see Fig. 2–5). This effect can be estimated by calculating the electron tunneling probability P_{tunn} across a potential barrier of height about I_B and width d:

$$P_{tunn} \approx \exp\left(-\frac{2d}{\hbar}\sqrt{2me I_B}\right). \tag{2-57}$$

The frequency of electron oscillations in the ground state of atom B is about $I_B/\hbar$, so the frequency of the tunneling is $I_B P_{tunn}/\hbar$. Taking into account (2–57) we can find the maximum barrier width, d_{max}, when the tunneling frequency still exceeds the reverse ion–neutral collision time: $I_B P_{tunn}/\hbar > v/d$ and as a result the tunneling can take place. Then the cross section πd_{max}^2 of the electron tunneling from B to A^+, leading to the resonant (or energy-permitted) charge exchange, is

$$\sigma_{ch.tr}^{tunn} \approx \frac{1}{I_B}\left(\frac{\pi \hbar^2}{8me}\right)\left(\ln \frac{I_B d}{\hbar} - \ln v\right)^2. \tag{2-58}$$

The numerical relation for the cross section was proposed by Rapp and Francis (1962):

$$\sqrt{\sigma_{ch.tr}^{tunn}, \text{cm}^2} \approx \frac{1}{\sqrt{I_B, eV}}(6.5 \cdot 10^{-7} - 3 \cdot 10^{-8} \ln v, \text{cm/s}). \tag{2-59}$$

The formula can be applied in the velocity range $v = 10^5$–10^8 cm/s; the cross section reaches 10^{-14} cm^2 at 10^5 cm/s. The tunneling cross section decreases to the gas-kinetic value (about $3 \cdot 10^{-16}$ cm^2) at $v > 10^8$ cm/s and follows the classical expression (2–57), which does not depend on energy. In low-energy collisions, when $v \leq 10^5$ cm/s$/\sqrt{M}$, the ion and neutral particle are captured in a complex. In this case, the exchanging electron always has enough time for tunneling and therefore has equal probability of being found on either particle. The resonant (and the non-energy-limited) charge exchange cross section then is about 10^{-14} cm^2 and can be found as being a half of the Langevin capture (2–47) cross section:

$$\sigma_L = \frac{1}{2}\sqrt{\frac{\pi \alpha e^2}{\varepsilon_0 M}\frac{1}{v}}. \tag{2-60}$$

2.2.6. Non-Resonant Charge Transfer Processes and Ion–Molecular Chemical Reactions of Positive and Negative Ions

The O-to-N electron transfer is an example of non-resonant charge exchange with a 1-eV energy defect:

$$N^+ + O \rightarrow N + O^+, \Delta E = -0.9\,eV. \tag{2-61}$$

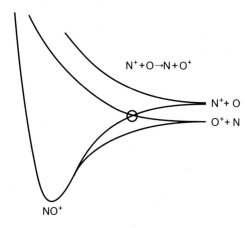

$$N^+ + O \rightarrow N + O^+$$

$$N^+ + O$$
$$O^+ + N$$

$$NO^+$$

Figure 2–6. Correlation of energy terms for elementary process of non-resonant charge exchange.

Illustrative potential curves are shown in Fig. 2–6. The ionization potential of oxygen ($I = 13.6\,\text{eV}$) is lower than that of nitrogen ($I = 14.5\,\text{eV}$), which is why the electron transfer from oxygen to nitrogen is an exothermic process and the separated $N + O^+$ energy level is located 0.9 eV lower than that for $O + N^+$. The reaction starts with N^+ approaching O by the attractive NO^+ energy term. Then, this term crosses with the repulsive NO^+ term and the system undergoes a non-adiabatic transfer, which results in the formation of $O^+ + N$. The cross section of such exothermic charge exchange reactions at low energies is the order of resonant cross sections with tunneling or Langevin capture (2–56)–(2–60). The endothermic reactions for charge exchange, like the reverse process $N + O^+ \rightarrow N^+ + O$, are usually very slow at low energies.

Contributions of the non-resonant charge exchange can be illustrated by the **acidic behavior of non-thermal air plasmas**. Ionization of air in non-thermal discharges primarily leads to large amounts of N_2^+ ions (with respect to other positive ions) because of the high molar fraction of nitrogen in air. The low ionization potential and high dipole moment of water molecules enable the following charge exchange:

$$N_2^+ + H_2O \rightarrow N_2 + H_2O^+, \, k\,(300\,\text{K}) = 2.2 \times 10^{-9}\,\text{cm}^3/\text{s}. \tag{2–62}$$

The whole ionization process in air, as a result, can be significantly focused on the formation of water ions (H_2O^+) even though the molar fraction of water is low. The generated water ions can then react with neutral water molecules via the quite fast ion–molecular reaction:

$$H_2O^+ + H_2O \rightarrow H_3O^+ + OH, \, \Delta H = -12\,\text{kcal/mole}, \, k\,(350\,\text{K}) = 0.5 \times 10^{-9}\,\text{cm}^3/\text{s}. \tag{2–63}$$

Production of H_3O^+ ions and OH radicals determines the acidic behavior of an air plasma, which is in particular a fundamental basis for plasma application for air purification from different pollutants.

The ion–molecular chemical reactions, for example, (2–63), include rearrangement of chemical bonds. These elementary processes can be subdivided into many groups, in particular the following:

$(A)B^+ + C \rightarrow A + (C)B^+$, reactions with an ion transfer (same with a negative ion);
$A(B^+) + C \rightarrow (B^+) + AC$, reactions with a neutral transfer (same with a negative ion);
$(A)B^+ + (C)D \rightarrow (A)D + (C)B^+$, double exchange reactions (same with a negative ion); and
$(A)B^+ + (C)D \rightarrow AC^+ + BD$, reconstruction processes (same with a negative ion).

Most of the exothermic ion–molecular reactions have no activation energy (Talrose, 1952). Quantum-mechanical repulsion between molecules, which provides the activation barrier even in the exothermic reactions of neutrals, can be suppressed by the charge-dipole attraction in the case of ion–molecular reactions. Thus, rate coefficients of the reactions are very high and often correspond to the Langevin relations (2–48)–(2–50), sometimes partially limited by quantum-mechanical factors (Su & Bowers, 1975; Virin et al., 1978). The effect obviously can be applied to both positive and negative ions.

The absence of activation energies in exothermic ion–molecular reactions facilitates organization of chain reactions in ionized media. Thus, the process of SO_2 oxidation into SO_3 and sulfuric acid is highly exothermic but cannot be arranged as an effective chain process without a catalyst, because of essential energy barriers of intermediate elementary reactions between neutral species. The long-length SO_2 chain oxidation, however, becomes possible in ionized media through the negative ion mechanism in water or in droplets in heterogeneous plasma (Daniel & Jacob,1986; Potapkin et al., 1995). This concept is known as **plasma catalysis** (see Section 10.1.1; Potapkin, Rusanov, & Fridman, 1989; Givotov, Potapkin, & Rusanov, 2005) and is useful in organic fuel conversion, hydrogen production, and so forth.

The ion–molecular reactions also make a significant contribution in some nucleation and cluster growth processes related to dusty plasma generation in electric discharges (Bouchoule, 1993, 1999) as well as in soot formation during combustion. Besides the effect of Langevin capture (Section 2.2.4), the **Winchester mechanism** of ion–molecular cluster growth (Bougaenko, Kuzmin, & Polak, 1993) plays an important role in these cases. The Winchester mechanism is based on the thermodynamic advantage of the ion-cluster growth processes, for example, in the sequence of negative ion clusterization:

$$A_1^- \rightarrow A_2^- \rightarrow A_3^- \rightarrow \cdots A_n^- \rightarrow \cdots, \tag{2–64}$$

where electron affinities EA_1, EA_2, EA_3, ..., EA_n, are usually increasing in the sequence to finally reach the value of the work function (electron extraction energy), which is generally higher than the electron affinity for small molecules. As a result each elementary reaction of the cluster growth $A_n^- \rightarrow A_{n+1}^-$ has an a priori tendency to be exothermic. Each elementary step $A_n^- \rightarrow A_{n+1}^-$ includes the cluster rearrangements with an electron usually going to the farthest end of the complex, which explains the term "Winchester." The Winchester mechanism enables positive ion-cluster growth as well:

$$A_1^+ \rightarrow A_2^+ \rightarrow A_3^+ \rightarrow \cdots A_n^+ \rightarrow \cdots, \tag{2–65}$$

where ionization energies $I(A_1)$, $I(A_2)$, $I(A_3)$, ..., $I(A_n)$ are usually decreasing in sequence to finally reach the value of the work function, which is generally lower than the ionization energy for small molecules. It explains the thermodynamic advantage of the ion-cluster growth for positive ions as well.

2.3. ELEMENTARY PLASMA-CHEMICAL REACTIONS INVOLVING NEGATIVE IONS

2.3.1. Dissociative Electron Attachment to Molecules as a Major Mechanism of Negative Ion Formation in Electronegative Molecular Gases

The dissociative attachment is effective when the products have positive electron affinities:

$$e + AB \rightarrow (AB^-)^* \rightarrow A + B^-. \tag{2–66}$$

The mechanism of the process is similar to the dissociative recombination (Section 2.2.2) and proceeds by intermediate formation of an auto-ionization state $(AB^-)^*$. This excited state is unstable and decays, leading either to the reverse reaction of auto-detachment

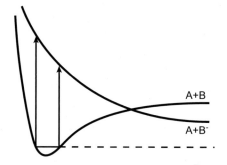

Figure 2–7. Elementary process of dissociative attachment in the case of low electron affinity.

$(AB + e)$ or to the dissociation $(A + B^-)$. An electron is captured during the attachment and is not able to provide the energy balance of the elementary process. Therefore, the dissociative attachment is a resonant reaction requiring a quite definite electron energy. Typical potential energy curves (2–67) are presented in Fig. 2–7. The process starts with a vertical transition from the AB ground state to a repulsive state of AB^- following the Frank-Condon principle (Section 2.1.6). During the repulsion, before $(AB^-)^*$ reaches the intersection point of AB and AB^- electronic terms, the reverse auto-detachment reaction $(AB + e)$ is very possible. But after passing the intersection, the AB potential energy exceeds that of AB^- and further repulsion results in dissociation $(A + B^-)$.

Repulsion time with possible auto-detachment is proportional to the square root of the reduced mass of the AB molecule: $\sqrt{M_A M_B/(M_A + M_B)}$ (Massey, 1976). The characteristic electron transition time is much shorter and is proportional to the square root of its mass m. For this reason (Lieberman & Lichtenberg, 1994), the maximum cross section for dissociative attachment is

$$\sigma_{\text{d.a.}}^{\max} \approx \sigma_0 \sqrt{\frac{m(M_A + M_B)}{M_A M_B}}. \tag{2–67}$$

The maximum cross section is two orders of magnitude less than the gas-kinetic cross section σ_0 and is about 10^{-18} cm^2. Some halogen compounds have an electron affinity of a product exceeding the dissociation energy; corresponding potential energy curves are presented in Fig. 2–8. In this case, in contrast to Fig. 2–7, even very-low-energy electrons provide dissociation. The intersection point of AB and AB^- electronic terms (Fig. 2–8) is actually located in this case inside of the so-called geometric size of the dissociating molecule. As a result, during the repulsion of $(AB^-)^*$, the probability of the reverse auto-detachment reaction $(AB + e)$ is very low and the cross section of the dissociative attachment reaches the gas-kinetic value σ_0 of about 10^{-16} cm^2. Cross sections for the dissociative attachment of several molecular gases are presented in Fig. 2–9.

The dissociative attachment cross section as a function of electron energy $\sigma_a(\varepsilon)$ has a resonant structure. It permits one to estimate the dissociative attachment rate coefficient

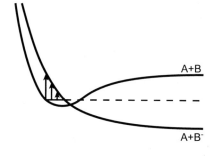

Figure 2–8. Elementary process of dissociative attachment in the case of high electron affinity.

2.3. Elementary Plasma-Chemical Reactions Involving Negative Ions

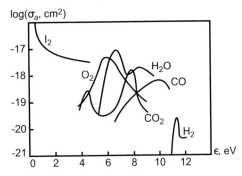

Figure 2–9. Cross sections of elementary processes of dissociative attachment of electrons to different molecules as a function of electron energy.

k_a as a function of electron temperature T_e by integration of $\sigma_a(\varepsilon)$ over the Maxwellian distribution:

$$k_a(T_e) \approx \sigma_{d.a.}^{max}(\varepsilon_{max}) \sqrt{\frac{2\varepsilon_{max}}{m}} \frac{\Delta\varepsilon}{T_e} \exp\left(-\frac{\varepsilon_{max}}{T_e}\right). \tag{2–68}$$

In this relation, a single $\sigma_a(\varepsilon)$ resonance is taken into account; ε_{max} and $\sigma_{d.a.}^{max}$ are the electron energy and maximum cross section, corresponding to the resonance; and $\Delta\varepsilon$ is its energy width. The parameters: ε_{max}, $\sigma_{d.a.}^{max}$, and $\Delta\varepsilon$, required for numerical calculations (2–68), are provided in Table 2–5.

2.3.2. Three-Body Electron Attachment and Other Mechanisms of Formation of Negative Ions

Electron collisions with two heavy particles also can result in the formation of negative ions:

$$e + A + B \rightarrow A^- + B. \tag{2–69}$$

The three-body electron attachment can be a principal channel for electron losses when electron energies are not high enough for the dissociative attachment, and when pressure is elevated (usually more than 0.1 atm) and the third-order kinetic processes are preferable. In contrast to the dissociative attachment, the three-body process is exothermic and its rate coefficient does not depend strongly on electron temperature. Electrons are usually kinetically less effective as a third body B because of a low degree of ionization.

Atmospheric-pressure non-thermal discharges in air are probably the most important systems, where the three-body attachment plays a key role in the balance of charged particles:

$$e + O_2 + M \rightarrow O_2^- + M. \tag{2–70}$$

Table 2–5. Resonance Parameters for Dissociative Attachment of Electrons

Dissociative attachment	ε_{max}, eV	$\sigma_{d.a.}^{max}$, cm^2	$\Delta\varepsilon$, eV	Dissociative attachment	ε_{max}, eV	$\sigma_{d.a.}^{max}$, cm^2	$\Delta\varepsilon$, eV
$e + O_2 \rightarrow O^- + O$	6.7	10^{-18}	1	$e + H_2O \rightarrow H^- + OH$	6.5	$7 \cdot 10^{-18}$	1
$e + H_2 \rightarrow H^- + H$	3.8	10^{-21}	3.6	$e + H_2O \rightarrow O^- + H_2$	8.6	10^{-18}	2.1
$e + NO \rightarrow O^- + N$	8.6	10^{-18}	2.3	$e + H_2O \rightarrow H + OH^-$	5	10^{-19}	2
$e + CO \rightarrow O^- + C$	10.3	$2 \cdot 10^{-19}$	1.4	$e + D_2O \rightarrow D^- + OD$	6.5	$5 \cdot 10^{-18}$	0.8
$e + HCl \rightarrow Cl^- + H$	0.8	$7 \cdot 10^{-18}$	0.3	$e + CO_2 \rightarrow O^- + CO$	4.35	$2 \cdot 10^{-19}$	0.8

Table 2–6. Reaction Rate Coefficients of Electron Attachment to Oxygen Molecules at Room Temperature and Different Third-Body Partners

Three-body attachment	Rate coefficient	Three-body attachment	Rate coefficient
$e + O_2 + Ar \rightarrow O_2^- + Ar$	$3 \cdot 10^{-32}$ cm^6/s	$e + O_2 + Ne \rightarrow O_2^- + Ne$	$3 \cdot 10^{-32}$ cm^6/s
$e + O_2 + N_2 \rightarrow O_2^- + N_2$	$1.6 \cdot 10^{-31}$ cm^6/s	$e + O_2 + H_2 \rightarrow O_2^- + H_2$	$2 \cdot 10^{-31}$ cm^6/s
$e + O_2 + O_2 \rightarrow O_2^- + O_2$	$2.5 \cdot 10^{-30}$ cm^6/s	$e + O_2 + CO_2 \rightarrow O_2^- + CO_2$	$3 \cdot 10^{-30}$ cm^6/s
$e + O_2 + H_2O \rightarrow O_2^- + H_2O$	$1.4 \cdot 10^{-29}$ cm^6/s	$e + O_2 + H_2S \rightarrow O_2^- + H_2S$	10^{-29} cm^6/s
$e + O_2 + NH_3 \rightarrow O_2^- + NH_3$	10^{-29} cm^6/s	$e + O_2 + CH_4 \rightarrow O_2^- + CH_4$	$>10^{-29}$ cm^6/s

This process proceeds by the two-stage **Bloch-Bradbury mechanism** (Bloch & Bradbury, 1935; Alexandrov, 1981) starting with the formation of a negative ion (rate coefficient k_{att}) in an unstable auto-ionization state (τ is the time of collisionless detachment):

$$e + O_2 \xleftrightarrow{k_{att}, \tau} (O_2^-)^*. \tag{2–71}$$

The second stage of the Bloch-Bradbury mechanism includes collision with the third-body particle M (density n_0), leading to relaxation and stabilization of O_2^- or collisional decay of the unstable ion:

$$(O_2^-)^* + M \xrightarrow{k_{st}} O_2^- + M, \tag{2–72}$$

$$(O_2^-)^* + M \xrightarrow{k_{dec}} O_2 + e + M. \tag{2–73}$$

Taking into account the steady-state conditions for number density of the intermediate excited ions $(O_2^-)^*$, the rate coefficient for the total attachment process (2–71) can be expressed as

$$k_{3M} = \frac{k_{att} k_{st}}{\dfrac{1}{\tau} + (k_{st} + k_{dec})n_0}. \tag{2–74}$$

Usually, when the pressure is not too high, $(k_{st} + k_{dec})n_0 \ll \tau^{-1}$, and (2–74) can be simplified:

$$k_{3M} \approx k_{att} k_{st} \tau. \tag{2–75}$$

Thus, the three-body attachment (2–70) has a third kinetic order and depends equally on the rate coefficients of formation and stabilization of negative ions. The latter strongly depends on the type of the third particle: the more complicated molecule a third body (M) is, the easier it stabilizes the $(O_2^-)^*$ and the higher the rate coefficient k_{3M}. Numerical values of the total rate coefficients k_{3M} are presented in Table 2–6; dependence of the total rate coefficients on electron temperature is shown in Fig. 2–10. For simple estimations, when $T_e = 1$ eV and $T_0 = 300$ K, one can take $k_{3M} \approx 10^{-30}$ cm^6/s. The rate of the three-body process is greater than for dissociative attachment (k_a) when the gas number density exceeds a critical value $n_0 > k_a(T_e)/k_{3M}$. Numerically in oxygen, with $T_e = 1$ eV, $T_0 = 300$ K, the rate requires $n_0 > 10^{18}$ cm^{-3}, or in pressure units $p > 30$ torr. Three other mechanisms of formation of negative ions are usually less significant. The first process is **polar dissociation**:

$$e + AB \rightarrow A^+ + B^- + e. \tag{2–76}$$

This process includes both ionization and dissociation; therefore, the threshold energy is quite high. On the other hand, an electron is not captured here, and the process is non-resonant and is effective over a wide range of high electron energies. For example, in

2.3. Elementary Plasma-Chemical Reactions Involving Negative Ions

Figure 2–10. Rate coefficients of electron attachment to different molecules in three-body collisions as a function of electron temperature; molecular gas is assumed to be at room temperature.

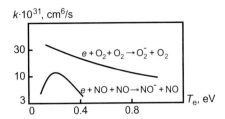

molecular oxygen, the maximum value of the cross section for polar dissociation is about $3 \cdot 10^{-19}$ cm^2 and corresponds to an electron energy of 35 eV.

To stabilize the formation of a negative ion during electron attachment, the excessive energy can be in principal emitted. As a result, the negative ion can be formed in the process of radiative attachment:

$$e + M \rightarrow (M^-)^* \rightarrow M^- + \hbar\omega. \tag{2–77}$$

Such an electron capture can take place at low electron energies, but the probability of the process is very low: 10^{-5}–10^{-7}. Corresponding values of attachment cross sections are about 10^{-21}–10^{-23} cm^2.

Finally, some electronegative polyatomic molecules like SF$_6$ have a negative ion state very close to a ground state (only 0.1 eV in the case of SF$_6^-$). As a result, the lifetime of such metastable negative ions is fairly long. Such a direct one-stage attachment process is resonant and for very low electron energies has maximum cross sections of about 10^{-15} cm^2.

2.3.3. Destruction of Negative Ions: Associative Detachment, Electron Impact Detachment, and Detachment in Collisions with Excited Particles

Different mechanisms of destruction of negative ions releasing an electron are discussed in special books by Massey (1976), McDaniel (1964), and Smirnov (1982). We are going to consider three detachment mechanisms most important in plasma-chemical systems. The first one, which is especially important in non-thermal discharges, is **associative detachment**:

$$A^- + B \rightarrow (AB^-)^* \rightarrow AB + e. \tag{2–78}$$

This is the reverse process with respect to the dissociative attachment (2–66) and therefore it can also be illustrated by Fig. 2–7. The associative detachment is a non-adiabatic process, which occurs via intersection of electronic terms of a complex negative ion A$^-$–B and corresponding molecule AB. Rate coefficients of the non-adiabatic reactions are quite high, typically $k_d = 10^{-10}$–10^{-9} cm^3/s. The kinetic data and enthalpy of some associative detachment processes are presented in Table 2–7.

Another detachment mechanism, **electron impact detachment,** is important for high degrees of ionization and can be described as

$$e + A^- \rightarrow A + e + e. \tag{2–79}$$

This process is somewhat similar to direct ionization by electron impact (Thomson mechanism, see Section 2.1.5). For electron energies of about 10 eV, the cross section of the detachment process can be fairly high, about 10^{-14} cm^2. The cross-sectional dependence on incident electron velocity v_e (Inocuti, Kim, & Platzman, 1967) can be illustrated by electron detachment from a hydrogen ion: $e + H^- \rightarrow H + 2e$:

$$\sigma(v_e) \approx \frac{\sigma_0 e^4}{(4\pi\varepsilon_0)^2 \hbar^2 v_e^2} \left(-7.5 \ln \frac{e^2}{4\pi\varepsilon_0 \hbar v_e} + 25\right). \tag{2–80}$$

Table 2–7. Room-Temperature Rate Coefficients and Enthalpies of Associative Electron Detachment Processes

Associative detachment of electrons	Reaction enthalpy, eV	Rate coefficient, cm³/s	Associative detachment of electrons attachment	Reaction enthalpy, eV	Rate coefficient, cm³/s
$H^- + H \rightarrow H_2 + e$	-3.8	$1.3 \cdot 10^{-9}$	$O^- + O_2(^1\delta_g) \rightarrow O_3 + e$	-0.6	$3 \cdot 10^{-10}$
$H^- + O_2 \rightarrow HO_2 + e$	-1.25	$1.2 \cdot 10^{-9}$	$O^- + N_2 \rightarrow N_2O + e$	-0.15	10^{-11}
$O^- + O \rightarrow O_2 + e$	-3.8	$1.3 \cdot 10^{-9}$	$O^- + NO \rightarrow NO_2 + e$	-1.6	$5 \cdot 10^{-10}$
$O^- + N \rightarrow NO + e$	-5.1	$2 \cdot 10^{-10}$	$O^- + CO \rightarrow CO_2 + e$	-4	$5 \cdot 10^{-10}$
$O^- + O_2 \rightarrow O_3 + e$	0.4 Endo	10^{-12}	$Cl^- + O \rightarrow ClO + e$	0.9 Endo	10^{-11}
$O^- + H_2 \rightarrow H_2O + e$	-3.5	10^{-9}	$O_2^- + N \rightarrow NO_2 + e$	-4.1	$5 \cdot 10^{-10}$
$O^- + CO_2 \rightarrow CO_3 + e$	Endo	10^{-13}	$OH^- + O \rightarrow HO_2 + e$	-1	$2 \cdot 10^{-10}$
$C^- + CO_2 \rightarrow CO + CO + e$	-4.3	$5 \cdot 10^{-11}$	$OH^- + N \rightarrow HNO + e$	-2.4	10^{-11}
$C^- + CO \rightarrow C_2O + e$	-1.1	$4 \cdot 10^{-10}$	$OH^- + H \rightarrow H_2O + e$	-3.2	10^{-9}
$O_2^- + O \rightarrow O_3 + e$	-0.6	$3 \cdot 10^{-10}$	$F^- + CF_3 \rightarrow CF_4 + e$	Exo	$4 \cdot 10^{-10}$
$F^- + CF_2 \rightarrow CF_3 + e$	Exo	$3 \cdot 10^{-10}$	$F^- + CF \rightarrow CF_2 + e$	Exo	$2 \cdot 10^{-10}$
$F^- + C \rightarrow CF + e$	Exo	10^{-10}			

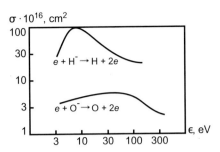

Figure 2–11. Cross sections of negative ions H^- and O^-, destruction by direct electron impact as a function of electron energy.

In this relationship, σ_0 is the geometric atomic cross section. The detachment cross sections as a function of electron energy are presented in Fig. 2–11 for hydrogen and oxygen ions. Maximum values of the cross sections are about 10^{-15}–10^{-14} cm^2 and correspond to electron energies of 10–50 eV.

The third mechanism is **detachment in collisions with excited particles:**

$$A^- + B^* \rightarrow A + B + e. \qquad (2\text{–}81)$$

When particle B is *electronically excited*, the process is similar to Penning ionization (see Section 2.1.11). If the electronic excitation energy of a collision partner B exceeds the electron affinity of particle A, the detachment process can proceed effectively as an electronically non-adiabatic reaction (without essential energy exchange with translational degrees of freedom of the heavy particles). Exothermic detachment of an electron from an oxygen ion in collision with a metastable electronically excited oxygen molecule (excitation energy 0.98 eV) is an example of such processes:

$$O_2^- + O_2(^1\Delta_g) \rightarrow O_2 + O_2 + e, \quad \Delta H = -0.6\,\text{eV} \qquad (2\text{–}82)$$

The rate coefficient of the detachment is very high: $2 \cdot 10^{-10}$ cm^3/s at room temperature. Electron detachment can be also effective in collisions with *vibrationally excited molecules*; for example,

$$O_2^- + O_2^*(v > 3) \rightarrow O_2 + O_2 + e. \qquad (2\text{–}83)$$

Rate coefficients of the detachment process, which are especially important in the quasi-equilibrium thermal systems, are presented in Table 2–8. The kinetics of the detachment process can be described in this case in the conventional manner for all reactions stimulated by the vibrational excitation of molecules. The traditional Arrhenius formula, $k_d \propto \exp(-E_a/T_v)$, is applicable here. The activation energy of the detachment process can be taken in this case to be equal to the electron affinity to oxygen molecules ($E_a \approx 0.44$ eV).

2.3.4. Recombination of Negative and Positive Ions

Actual losses of charged particles in electronegative gases are mostly due to ion–ion recombination, which involves neutralization of positive and negative ions in binary or three-body collisions. Ion–ion recombination can proceed by a variety of different mechanisms, which dominate at different pressure ranges, but all of them are usually characterized by very high rate coefficients.

Table 2–8. Thermal Destruction of Molecular Oxygen Ions

Detachment process	Reaction enthalpy	Rate coefficient, 300 K	Rate coefficient, 600 K
$O_2^- + O_2 \rightarrow O_2 + O_2 + e$	0.44 eV	$2.2 \cdot 10^{-18}$ cm^3/s	$3 \cdot 10^{-14}$ cm^3/s
$O_2^- + N_2 \rightarrow O_2 + N_2 + e$	0.44 eV	$<10^{-20}$ cm^3/s	$1.8 \cdot 10^{-16}$ cm^3/s

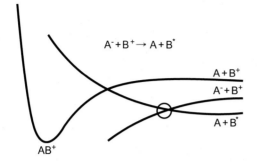

$A^- + B^+ \rightarrow A + B^*$

$A + B^+$

$A^- + B^+$

$A + B^*$

AB^+

Figure 2–12. Correlation of energy terms illustrating the elementary process of ion–ion recombination.

At high pressures (usually above 30 torr), a three-body mechanism dominates the recombination. The recombination rate coefficient in this case reaches a maximum of about $1\text{–}3 \cdot 10^{-6}$ cm^3/s at near-atmospheric pressures for room temperature. Traditionally, the rate coefficient of ion–ion recombination is recalculated with respect to concentrations of positive and negative ions, that is, to the second kinetic order. Because of the tri-molecular nature of the recombination mechanism in this pressure range, the recalculated (to the second kinetic order) recombination rate coefficient depends on pressure. Both increases and decreases of pressure with respect to the optimal pressure (near 1 atm) result in proportional reduction of the three-body ion–ion recombination rate coefficient. At low pressures, obviously, the three-body mechanism becomes relatively slow and binary collisions with transfer of released energy into electronic excitation make a major contribution to the ion–ion recombination.

2.3.5. Ion–Ion Recombination in Binary Collisions

This neutralization process proceeds in a binary collision of a negative and a positive ion, with released energy going into electronic excitation of a neutral product:

$$A^- + B^+ \rightarrow A + B^*. \tag{2–84}$$

This second-order reaction dominates the ion–ion recombination in low-pressure discharges ($p < 10\text{–}30$ torr). Electronic terms, illustrating the recombination (2–85), are presented in Fig. 2–12. The ions A^- and B^+ approach each other following the attractive A^-–B^+ Coulomb potential curve. When the distance between the heavy particles is large, this potential curve lies below the term of A–B^+ (lower value of electron affinity of particle A: EA_A) and above the final A–B^* electronic term on the energy interval

$$\Delta E \approx \frac{I_B}{n^2} - EA_A. \tag{2–85}$$

where I_B is the ionization potential of particle B. If the principal quantum number of particle B after recombination is quite high ($n = 3\text{–}4$), then ΔE is low and the electronic terms of A^-–B^+ and A–B^* become relatively close. When the principal quantum number n is not specifically defined, then electron affinity EA_A can be taken as a reasonable estimation for the energy interval $\Delta E \approx EA_A$. A low value of ΔE results in the possibility of effective transitions between electronic terms (from A^-–B^+ to A–B^*) when the distance R_{ii} between ions is still large. The Coulomb attraction energy, even for the long-distance R_{ii} between the ions, is already enough to compensate for the initial energy gap ΔE between the terms:

$$R_{ii} \approx \frac{e^2}{4\pi \varepsilon_0} \frac{1}{\left(\dfrac{I_B}{n^2} - EA_A \right)} \approx \frac{e^2}{4\pi \varepsilon_0 EA_A}. \tag{2–86}$$

2.3. Elementary Plasma-Chemical Reactions Involving Negative Ions

The high value of R_{ii} results in high recombination cross sections, which can be estimated by taking into account the conservation of angular momentum during the Coulomb collision and taking the maximum kinetic (the same as potential) energy as EA_A. The impact parameter b can be found in this case as a function of the ion kinetic energy in the center-of-mass system:

$$b \approx R_{ii} \frac{\sqrt{EA_A}}{\sqrt{\varepsilon}}. \qquad (2\text{--}87)$$

The cross section of the ion–ion recombination process can then be presented as

$$\sigma_{rec}^{ii} = \pi b^2 \approx \pi \frac{e^4}{(4\pi\varepsilon_0)^2} \frac{1}{EA_A} \frac{1}{\varepsilon}. \qquad (2\text{--}88)$$

Quantum-mechanical calculation of the ion–ion recombination cross section takes into account electron tunneling, which results in a more accurate dependence of the cross section of ion–ion recombination in binary collisions on the ion kinetic energy: $\sigma_{rec}^{ii} \propto \frac{1}{\sqrt{\varepsilon}}$. Relation (2–88) can be rewritten in atomic units: $\sigma_{rec}^{ii} \approx \sigma_0 \frac{I}{EA_A} \frac{I}{\varepsilon}$, where I is a typical value of ionization energy. It shows that the recombination cross section exceeds the gas-kinetic value by several orders of magnitude. The same is true for the reaction rate coefficients, which are usually about 10^{-7} cm^3/s. Some binary recombination rate coefficients are presented in Table 2–9 together with energy released in the process.

2.3.6. Three-Body Ion–Ion Recombination: Thomson's Theory and Langevin Model

The neutralization process at moderate and high pressures ($p > 10$–30 torr) effectively proceeds via a triple collision of a heavy neutral with a negative and a positive ion:

$$A^- + B^+ + M \rightarrow A + B + M. \qquad (2\text{--}89)$$

The three-body reaction (2.5.6) has third kinetic order only in the moderate pressure range, usually less than 1 atm, where it can be described by **Thomson theory** (Thomson, 1924). At higher pressures the process is limited by ion mobility (see the discussion that follows. According to the Thomson theory, the act of recombination takes place if negative and positive ions approach each other closer than the critical distance:

$$b \approx e^2/4\pi\varepsilon_0 T_0, \qquad (2\text{--}90)$$

and when the Coulomb interaction reaches the level of thermal energy (T_0) and can be effectively absorbed by the third body, enabling the act of recombination. The probability P_+ for a positive ion to be closer after the collision than the critical distance b in (2–90) to a negative ion (density n_-) and to be neutralized as a result is

$$P_+ \approx n_- b^3 \approx n_- \frac{e^6}{(4\pi\varepsilon_0)^3 T_0^3}. \qquad (2\text{--}91)$$

The frequency of collisions of a positive ion with neutral particles, resulting in ion–ion recombination, can then be found as $\nu_{ii} = (\sigma v_t n_0) P_+$, where n_0 is the neutral particle density, σ is the cross section of ion-neutral elastic scattering, and v_t is the average velocity of heavy particles. The total three-body ion–ion recombination rate is

$$w_{ii} \approx (\sigma v_t) \frac{e^6}{(4\pi\varepsilon_0)^3 T_0^3} n_0 n_- n_+. \qquad (2\text{--}92)$$

Table 2–9. Rate Coefficients of Ion–Ion Recombination in Binary Collisions at Room Temperature

Recombination process	Energy, $I-EA$, eV	Rate coefficient, cm^3/s	Recombination process	Energy, $I-EA$, eV	Rate coefficient, cm^3/s
$H^- + H^+ \rightarrow H + H$	12.8	$3.9 \cdot 10^{-7}$	$SF_6^- + SF_5^+ \rightarrow SF_6 + SF_5$	15.2	$4 \cdot 10^{-8}$
$O^- + O^+ \rightarrow O + O$	12.1	$2.7 \cdot 10^{-7}$	$NO_2^- + NO^+ \rightarrow NO_2 + NO$	5.7	$3 \cdot 10^{-7}$
$O^- + N^+ \rightarrow O + N$	13.1	$2.6 \cdot 10^{-7}$	$O_2^- + O^+ \rightarrow O_2 + O$	13.2	$2 \cdot 10^{-7}$
$O^- + O_2^+ \rightarrow O + O_2$	11.6	10^{-7}	$O_2^- + O_2^+ \rightarrow O_2 + O_2$	11.6	$4.2 \cdot 10^{-7}$
$O^- + NO^+ \rightarrow O + NO$	7.8	$4.9 \cdot 10^{-7}$	$O_2^- + N_2^+ \rightarrow O_2 + N_2$	15.1	$1.6 \cdot 10^{-7}$
$F^- + Si^+ \rightarrow F + Si$	–	$4 \cdot 10^{-8}$	$F^- + SiF^+ \rightarrow F + SiF$	–	$4 \cdot 10^{-8}$
$F^- + SiF_2^+ \rightarrow F + F + SiF$	–	$4 \cdot 10^{-8}$	$F^- + SiF_3^+ \rightarrow F + F + SiF_2$	–	$4 \cdot 10^{-8}$
$F^- + CF_3^+ \rightarrow F + F + CF_2$	–	$4 \cdot 10^{-8}$	$F^- + CF_2^+ \rightarrow F + CF_2$	–	$4 \cdot 10^{-8}$
$F^- + CF^+ \rightarrow F + F + C$	–	$4 \cdot 10^{-8}$	$F^- + CO^+ \rightarrow F + CO$	–	$4 \cdot 10^{-8}$
$F^- + O^+ \rightarrow F + O$	–	$4 \cdot 10^{-8}$	$F^- + O_2^+ \rightarrow F + O_2$	–	$3 \cdot 10^{-7}$
$F^- + NF_3^+ \rightarrow F + NF_3$	–	10^{-7}	$F^- + N_2^+ \rightarrow F - N_2$	–	10^{-8}
$F^- + F^+ \rightarrow F + F$	–	$3 \cdot 10^{-8}$	$F^- + F_2^+ \rightarrow F + F_2$	–	$4 \cdot 10^{-8}$

Table 2–10. Rate Coefficients of Three-Body Ion–Ion Recombination at Room Temperature and Moderate Pressures

Ion–ion recombination processes	Rate coefficients, third order	Rate coefficients, second order, 1 atm
$O_2^- + O_4^+ + O_2 \rightarrow O_2 + O_2 + O_2 + O_2$	$1.55 \cdot 10^{-25}$ cm^6/s	$4.2 \cdot 10^{-6}$ cm^3/s
$O^- + O_2^+ + O_2 \rightarrow O_3 + O_2$	$3.7 \cdot 10^{-25}$ cm^6/s	10^{-5} cm^3/s
$NO_2^- + NO^+ + O_2 \rightarrow NO_2 + NO + O_2$	$3.4 \cdot 10^{-26}$ cm^6/s	$0.9 \cdot 10^{-6}$ cm^3/s
$NO_2^- + NO^+ + N_2 \rightarrow NO_2 + NO + N_2$	10^{-25} cm^6/s	$2.7 \cdot 10^{-6}$ cm^3/s

Note: In the last column, the coefficients are recalculated to the second kinetic order and pressure of 1 atm, which is multiplied by the concentration of neutrals, $2.7 \cdot 10^{19}$ cm^{-3}.

The recombination here has third kinetic order and rate coefficient k_{r3}^{ii} decreasing with T_0:

$$k_{r3}^{ii} \approx \left(\sigma \sqrt{\frac{1}{m}} \right) \frac{e^6}{(4\pi\varepsilon_0)^3} \frac{1}{T_0^{5/2}}. \tag{2–93}$$

The rate coefficients are presented in Table 2–10 for some specific processes, where they are also recalculated to the second kinetic order: k_{r2}^{ii} ($k_{r2}^{ii} = k_{r3}^{ii} n_0$). Binary and triple collisions (see Tables 2.9 and 2.10) contribute equally to the ion–ion recombination at pressures of about 10–30 torr.

The ion–ion recombination rate coefficients recalculated to the second kinetic order, $k_{r2}^{ii} = k_{r3}^{ii} n_0$, grow linearly with gas density. The growth is limited to moderate pressures by the framework of Thomson's theory, which requires the capture distance b to be less than the ion mean free path ($1/n_0\sigma$):

$$n_0 \sigma b \approx n_0 \frac{e^2}{(4\pi\varepsilon_0)T_0} \sigma < 1. \tag{2–94}$$

At high pressures, $n_0\sigma b > 1$, the recombination is kinetically limited by positive and negative ion drift with respect to each other overcoming multiple collisions with neutrals. **The Langevin model**, developed in 1903, describes the ion–ion recombination in this pressure range. The ions are approaching by drifting in Coulomb field $e/(4\pi\varepsilon_0)r^2$ with velocity determined by positive and negative ion mobilities μ_+ and μ_-:

$$v_d = \frac{e}{(4\pi\varepsilon_0)r^2}(\mu_+ + \mu_-). \tag{2–95}$$

If a sphere with radius r surrounds a positive ion, then a flux of negative ions (concentration n_-) approaching the positive ion and recombination frequency v_{r+} for the positive ion can be expressed as

$$v_{r+} = 4\pi r^2 v_d n_-. \tag{2–96}$$

The ion–ion recombination rate is $w = n_+ v_{r+} = 4\pi r^2 v_d n_- n_+$, and the Langevin expression for the recombination rate coefficient, $k_r^{ii} = w/n_+ n_-$, in the high pressure limit (above 1 atm) is

$$k_r^{ii} = 4\pi e(\mu_+ + \mu_-). \tag{2–97}$$

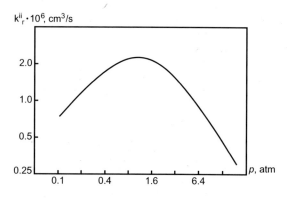

Figure 2–13. Pressure dependence of rate coefficient of ion–ion recombination in air.

Comparison of the Thomson and Langevin models shows that recombination rate coefficients grow with pressure $k_{r2}^{ii} = k_{r3}^{ii} n_0$ at moderate pressures and begin to decrease as $1/p$ together with ion mobility at high pressures (2–97). The highest recombination rate coefficient is achieved at neutral density:

$$n_0 \approx \frac{4\pi \varepsilon_0 T_0}{\sigma e^2}, \tag{2–98}$$

corresponding to atmospheric pressure. The maximum value of the ion–ion recombination rate coefficient is

$$k_{r,max}^{ii} \approx \frac{e^4}{(4\pi \varepsilon_0)^2} \frac{v}{T_0^2}. \tag{2–99}$$

Numerically, the maximum recombination rate coefficient is about $1–3 \cdot 10^{-6}\,\mathrm{cm^3/s}$. This coefficient decreases both with growth and with reduction of pressure from the near-atmospheric value (see Fig. 2–13). A generalized ion–ion recombination model, combining the Thomson and Langevin approaches for moderate and high pressures, was developed by Natanson (1959).

2.4. ELECTRON EMISSION AND HETEROGENEOUS IONIZATION PROCESSES

2.4.1. Thermionic Emission: Sommerfeld Formula and Schottky Effect

Ionization processes considered in Section 2.1 take place in the discharge volume. Here we are going to discuss surface-related ionization processes, which are especially important in supporting electric currents in cathode layers (see Chapter 4) and in the charge balance of aerosol-based plasma-chemical systems. The most important surface ionization process in thermal plasmas is thermionic emission, which is electron emission from a high-temperature metal surface due to the high thermal energy of electrons located in the metal. The emitted electrons can stay in the surface vicinity, creating a negative space charge and preventing further emission. The electric field in the cathode vicinity is sufficient to push the negative space charge out of the electrode and reach the saturation current density, which is the main characteristic of the cathode thermionic emission and can be quantified by using the Sommerfeld formula:

$$j = \frac{4\pi m e}{(2\pi \hbar)^3} T^2 (1 - R) \exp\left(-\frac{W}{T}\right). \tag{2–100}$$

In the Sommerfeld formula, W is the **work function**, which is the minimum energy necessary to extract an electron from a metal surface (numeric values for some cathode materials

2.4. Electron Emission and Heterogeneous Ionization Processes

Table 2–11. Work Function for Different Cathode Materials

Material	C	Cu	Al	Mo
Work function	4.7 eV	4.4 eV	4.25 eV	4.3 eV
Material	W	Pt	Ni	W/ThO$_2$
Work function	4.54 eV	5.32 eV	4.5 eV	2.5 eV

are shown in Table 2–11); $R = 0 - 0.8$ is a quantum-mechanical coefficient describing the reflection of electrons from a potential barrier at the metal surface; the Sommerfeld constant $\frac{4\pi m e}{(2\pi \hbar)^3} = 120 \frac{A}{cm^2 K^2}$; T is the surface temperature; e and m are electron charge and mass; and $\hbar$ is Planck's constant.

The thermionic current grows with electric field until the negative space charge near the cathode is eliminated and saturation is achieved. This saturation, however, is rather relative. A further increase of electric field leads gradually to an increase of the saturation current level, which is related to a reduction of work function due to an electric field known as the **Schottky effect:**

$$W, eV = W_0 - 3.8 \cdot 10^{-4} \cdot \sqrt{E, V/cm}. \qquad (2-101)$$

This decrease of work function W is relatively small at reasonable values of electric field E. The Schottky effect can result, however, in essential change of the thermionic current because of its strong exponential dependence on the work function in accordance with the Sommerfeld formula (2–100). Dependence of the Schottky W decrease and thermionic emission current density on electric field are illustrated in Table 2–12 together with corresponding data on field and thermionic field emission (Raizer, 1997). The 4× change of electric field results in 800× increase of the thermionic current density.

2.4.2. Field Emission of Electrons in Strong Electric Fields: Fowler-Nordheim Formula and Thermionic Field Emission

High electric fields (about $(1 \div 3) \cdot 10^6$ V/cm) are able not only to decrease the work function but also directly extract electrons from cold metal surfaces due to the quantum-mechanical effect of tunneling. A simplified triangular potential energy barrier for electrons inside of metal, taking into account an external electric field E but neglecting the mirror forces,

Table 2–12. Current Densities of Thermionic, Field, and Thermionic Field Emissions at Different Electric Fields E

Electric field, 10^6 V/cm	Schottky decrease of W, V	Thermionic emission, j, A/cm^2	Field emission j, A/cm^2	Thermionic field emission, j, A/cm^2
0	0	$0.13 \cdot 10^3$	0	0
0.8	1.07	$8.2 \cdot 10^3$	$2 \cdot 10^{-20}$	$1.2 \cdot 10^4$
1.7	1.56	$5.2 \cdot 10^4$	$2.2 \cdot 10^{-4}$	$1.0 \cdot 10^5$
2.3	1.81	$1.4 \cdot 10^5$	1.3	$2.1 \cdot 10^5$
2.8	2.01	$3.0 \cdot 10^5$	130	$8 \cdot 10^5$
3.3	2.18	$6.0 \cdot 10^5$	$4.7 \cdot 10^3$	$2.1 \cdot 10^6$

Note: Electrode temperature, work function, Fermi energy, and pre-exponential Sommerfeld factor are $T = 3000$ K, $W = 4$ eV, $\varepsilon_F = 7$ eV, $A_0(1 - R) = 80$ A/cm^2 K^2.

43

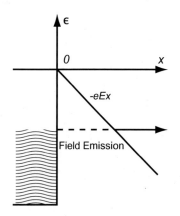

Figure 2–14. Potential energy of electrons on a metal surface in the presence of external electric field.

is presented in Fig. 2–14. Electrons are able to escape from the metal across the barrier due to tunneling, which is called the field emission effect and can be described by the Fowler-Nordheim formula:

$$j = \frac{e^2}{4\pi^2\hbar} \frac{1}{(W_0 + \varepsilon_F)} \sqrt{\frac{\varepsilon_F}{W_0}} \exp\left[-\frac{4\sqrt{2m}\,W_0^{3/2}}{3e\hbar E}\right]. \tag{2–102}$$

where ε_F is the Fermi energy of the metal, and W_0 is the metal's work function not perturbed by an external electric field (2–101). The field emission is very sensitive to small changes of electric field and work function, including those related to the Schottky effect (2–102). Electron tunneling across the potential barrier influenced by the Schottky effect and corresponding field emission are illustrated in Fig. 2–15. A related correction in the Fowler-Nordheim formula can be done by introducing a special numerical factor $\xi(\Delta W / W_0)$ depending on the relative Schottky decrease of the work function:

$$j = 6.2 \cdot 10^{-6} \text{A/cm}^2 \times \frac{1}{(W_0, \text{eV} + \varepsilon_F, \text{eV})} \sqrt{\frac{\varepsilon_F}{W_0}} \exp\left[-\frac{6.85 \cdot 10^7 W_0^{3/2}(\text{eV}) \cdot \xi}{E, \text{V/cm}}\right]. \tag{2–103}$$

The corrective factor $\xi(\Delta W / W_0)$ is shown in Table 2–13 (Granovsky, 1971). Numerical examples of the field emission calculations (2–103) are presented in Table 2–12. The field emission current density dependence on electric field is really sharp in this case: a 4×change of electric field results in an increase of the field emission current density by more than 23 orders of magnitude. According to the corrected Fowler-Nordheim formula and Table 2–12, the field electron emission becomes significant when the electric field exceeds 10^7 V/cm.

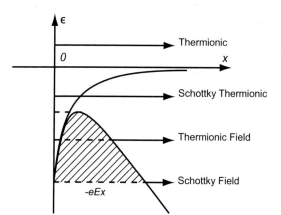

Figure 2–15. Schottky effect in thermionic and field emission.

Table 2–13. Correction Factor $\xi(\Delta W / W_0)$ in the Fowler-Nordheim Formula for Field Emission

Relative Schottky decrease of work function, $\Delta W / W_0$	0	0.2	0.3	0.4	0.5
Fowler-Nordheim correction factor $\xi(\Delta W / W_0)$	1	0.95	0.90	0.85	0.78
Relative Schottky decrease of work function, $\Delta W / W_0$	0.6	0.7	0.8	0.9	1
Fowler-Nordheim correction factor $\xi(\Delta W / W_0)$	0.70	0.60	0.50	0.34	0

In reality, however, the electron field emission makes a significant contribution already at electric fields about $3 \cdot 10^6$ V/cm because of the field enhancement at the microscopic protrusions on metal surfaces.

When the cathode temperature is high in addition to an external electric field, both thermionic and field emission make essential contributions, which is usually referred to as thermionic field emission. To compare the emission mechanisms we can subdivide electrons escaping surfaces into four groups, as illustrated in Fig. 2–15. The first group has energies below the Fermi level, so the electrons are able to escape the metal only through tunneling or, in other words, by the field emission mechanism. Electrons of the fourth group leave the metal via the thermionic emission mechanism without any support from the electric field.

Electrons of the third group overcome the potential energy barrier because of its reduction by the external electric field. This Schottky effect of the applied electric field is obviously a purely classical one. The second group of electrons is able to escape the metal only quantum-mechanically by tunneling similarly to those from the first group. But the potential barrier of tunneling is not so high and large in this case because of the relatively high thermal energy of the second group electrons. These electrons escape the cathode by the mechanism of thermionic field emission.

Because thermionic emission is based on the synergistic effect of temperature and electric field, these two key parameters of electron emission can be just reasonably high enough to provide a significant emission current. Results of calculations of the thermionic field emission are also presented in Table 2–12. The thermionic field emission dominates over other mechanisms at $T = 3000$ K and $E > 8 \cdot 10^6$ V/cm. We should note that at high temperatures but lower electric fields $E < 5 \cdot 10^6$ V/cm, electrons of the third group usually dominate the emission. The Sommerfeld relation in this case includes the work function diminished by the Schottky effect.

2.4.3. Secondary Electron Emission

Mechanisms of electron emission from solids, related to surface bombardment by different particles, are called secondary electron emission. The most important mechanism from this group is **secondary ion-electron emission**. According to the adiabatic principle (see Section 2.10), heavy ions are not efficient for transferring energy to light electrons to provide ionization. This general statement can be referred to the direct electron emission from solid surfaces induced by ion impact. As one can see from Fig. 2–16 (Dobretsov &

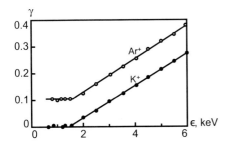

Figure 2–16. Secondary electron emission from tungsten as a function of ion energy.

Table 2–14. Secondary Emission Coefficient γ for the Potential Electron Emission Induced by Collisions with Metastable Atoms

Metastable atom	Surface material	Secondary emission coefficient γ
He(2^3S)	Pt	0.24 electron/atom
He(2^1S)	Pt	0.4 electron/atom
Ar*	Cs	0.4 electron/atom

Gomounova, 1966), the secondary electron emission coefficient γ (electron yield per ion) becomes relatively high only at very high ion energies exceeding 1 keV, when the Massey parameter becomes low. Although the secondary electron emission coefficient γ is low at lower ion energies, it is not negligible and remains almost constant at ion energies below the kilovolt range. It can be explained by the **Penning mechanism of secondary ion-electron emission**, also called the **potential mechanism**. An ion approaching the surface extracts an electron from there because the ionization potential I exceeds the work function W. The defect of energy $I - W$ is usually large enough ($I - W > W$) to enable the escape of more than one electron from the surface. Such a process is non-adiabatic and its probability is not negligible. The secondary ion-electron emission coefficient γ can be estimated using the empirical formula

$$\gamma \approx 0.016(I - 2W). \qquad (2\text{–}104)$$

Another secondary electron emission mechanism is related to surface bombardment by excited metastable atoms with an excitation energy exceeding the surface work function. This so-called **potential electron emission induced by metastable atoms** can have a quite high secondary emission coefficient γ. Some of them are presented in Table 2–14.

Secondary electron emission can also be provided by the photo effect. **The photo-electron emission** is usually characterized by the **quantum yield** $\gamma_{\hbar\omega}$, which gives the number of emitted electrons per one quantum $\hbar\omega$ of radiation. The quantum yields as a function of photon energy $\hbar\omega$ for different metal surfaces (Lozansky & Firsov, 1975) are shown in Fig. 2–17. Visual light and low-energy UV radiation give the quantum yield $\gamma_{\hbar\omega} \approx 10^{-3}$, which is fairly sensitive to the quality of the surface. High-energy UV radiation provides an emission with a quantum yield of about 0.01–0.1, which is in this case less sensitive to the surface characteristics.

Secondary electron–electron emission means electron emission from a solid surface induced by electron impact. This emission mechanism can be important in the high-frequency breakdown of discharge gaps at very low pressures, and also in heterogeneous discharges. The secondary electron–electron emission is usually characterized by the multiplication coefficient γ_e, which gives the number of emitted electrons produced by the initial electron. Dependence of the multiplication coefficient γ_e on electron energy for different metals and dielectrics (Dobretsov & Gomounova, 1966) is shown in Fig. 2–18. Additional information can also be found in books by Modinos (1984) and Komolov (1992).

2.4.4. Photo-Ionization of Aerosols: Monochromatic Radiation

Discharges in aerosols play an important role in plasma-chemical applications. Photo-electron emission often makes an important contribution in the ionization of aerosols, especially if a radiation quantum exceeds the work function of macro-particles but is below the ionization potential of neutral gas species. In this case, the steady-state electron density is determined by photo-ionization of aerosols and electron attachment to the macro-particles.

2.4. Electron Emission and Heterogeneous Ionization Processes

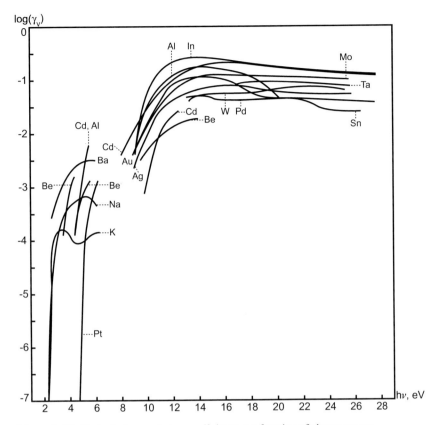

Figure 2–17. Photoelectron emission coefficients as a function of photon energy.

Positive ions in the system are replaced by charged aerosol particulates with some specific distribution of charges (Spitzer, 1941, 1944; Sodha, 1963).

To calculate the steady-state electron density n_e provided by monochromatic ($E = \hbar\omega$) photo-ionization of spherical macro-particles of radius r_a and concentration n_a (Karachev-tsev & Fridman, 1974, 1975), we should first take into account that the **work function of aerosols** depends on their charge:

$$\varphi_m = \varphi_0 + \frac{me^2}{4\pi\varepsilon_0 r_a}, \tag{2–105}$$

which results in upper n_+ and lower n_- limits of the macro-particle charge m. The upper charge limit is due to the fixed value of photon energy, $E = \hbar\omega$, and growth of the work function with a particle charge:

$$n_+ = \frac{4\pi\varepsilon_0(E - \varphi_0)r_a}{e^2}. \tag{2–106}$$

In this relation, φ_0 is the work function of a non-charged particle ($m = 0$), which slightly exceeds work function A_0 for a flat surface (Maksimenko & Tverdokhlebov, 1964; $x_0 \approx 0.2$ nm):

$$\varphi_0 = A_0 \left(1 + \frac{5}{2}\frac{x_0}{r_a}\right). \tag{2–107}$$

47

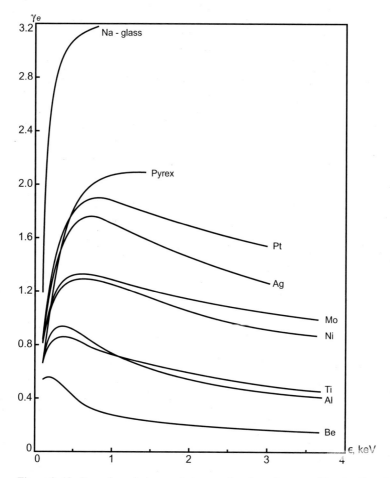

Figure 2–18. Secondary electron emission as a function of energy of bombarding electrons.

The lower limit $n_-e(n_- < 0)$ corresponds to zero work function, when attachment is not effective:

$$n_- = -\frac{4\pi\varepsilon_0\varphi_0 r_a}{e^2}. \qquad (2\text{--}108)$$

Equations describing the concentration of aerosol particles, n_i, with electric charge ie are as follows:

$$\frac{dn_{i=n-}}{dt} = -\gamma p n_{n-} + \alpha n_e n_{n-+1}, \qquad (2\text{--}109)$$

$$\frac{dn_i}{dt} = \gamma p n_{i-1} - \alpha n_e n_i - \gamma p n_i + \alpha n_e n_{i+1}, \quad n_- + 1 \le i \le n_+ - 1, \qquad (2\text{--}110)$$

$$\frac{dn_{i=n+}}{dt} = \gamma p n_{n+-1} - \alpha n_e n_{n+}. \qquad (2\text{--}111)$$

In this system, $\alpha \approx \sigma_a v_e$ is the coefficient of attachment of electrons with thermal velocity v_e to particles of cross section σ_a, which is supposed here to be constant at $i > n_-$; γ is the photo-ionization cross section of the macro-particles, also constant at $i < n_+$; and p is the flux of monochromatic photons. Considering steady-state conditions $d/dt = 0$, and

2.4. Electron Emission and Heterogeneous Ionization Processes

adding equations of the system, we can express the balance equation for charged species in the heterogeneous system under consideration as follows:

$$\alpha n_e n_{i+1} = \gamma p n_i, \quad q = \frac{n_{i+1}}{n_i} = \frac{\gamma p}{\alpha n_e}. \tag{2-112}$$

The factor $q = n_{i+1}/n_i$ describes the distribution of aerosol particles over electric charges. Taking into account both electro-neutrality and total mass balance, the equation for electron density can be presented as

$$\frac{n_e}{n_a} = \frac{\sum\limits_{k=n-}^{k=n+} kq^k}{\sum\limits_{k=n-}^{k=n+} q^k} \approx \frac{\int\limits_{n-}^{n+} xq^x dx}{\int\limits_{n-}^{n+} q^x dx}. \tag{2-113}$$

The integration of (2–113) results in the final equation for electron density due to photoionization:

$$\frac{n_e}{n_a} = \frac{n_+ q^{n+} - n_- q^{n-}}{q^{n+} - q^{n-}} - \frac{1}{\ln q}. \tag{2-114}$$

The chart for the calculation of electron density n_e as a function of photon flux $\log p$ and number density of aerosols $\log(n_a n_+)$ at monochromatic photo-ionization assuming $\alpha/\gamma = 10^8$ cm/s is presented in Fig. 2–19. The chart also shows the relative accuracy of the calculations, which is presented in a form of relative accuracy, $\varsigma = \Delta n_e/n_e$, as well as relative accuracy of calculations of $\ln n_e$ ($\varsigma_1 \approx \frac{\Delta n_e}{2.3 \cdot n_e \log n_e}$).

2.4.5. Photo-Ionization of Aerosols: Continuous-Spectrum Radiation

The particle distribution over charges n_i in the case of the radiation with continuous spectrum can be found from the balance equation

$$\gamma p_i n_i = \alpha n_e n_{i+1}, \tag{2-115}$$

where p_i is the flux of photons with energy sufficient for the $(n + 1) - st$ photo-ionization. Using the energy distribution of photons in the form $\xi_i = p_i/p_0$ ($\xi_0 = 1$), the concentration of macro-particles with charge ke can be presented as

$$n_1 = n_0 \left(\frac{\gamma p_0}{\alpha n_e}\right) \xi_0, \tag{2-116}$$

$$n_2 = n_1 \left(\frac{\gamma p_0}{\alpha n_e}\right) \xi_1 = n_0 \left(\frac{\gamma p_0}{\alpha n_e}\right)^2 \xi_0 \xi_1, \tag{2-117}$$

$$n_k = n_0 \left(\frac{\gamma p_0}{\alpha n_e}\right)^k \prod_{i=0}^{k-1} \xi_i. \tag{2-118}$$

Then electron density can be calculated from the following equation ($q_0 = \gamma p_0/\alpha n_e$):

$$\frac{n_e}{n_a} = \frac{\sum\limits_{k=n-}^{\infty} kn_k}{\sum\limits_{n-}^{\infty} n_k} = \frac{\sum\limits_{k=n-}^{\infty} kq_0^k \prod\limits_{i=0}^{k-1} \xi_i}{\sum\limits_{k=n-}^{\infty} q_0^k \prod\limits_{i=0}^{k-1} \xi_i}. \tag{2-119}$$

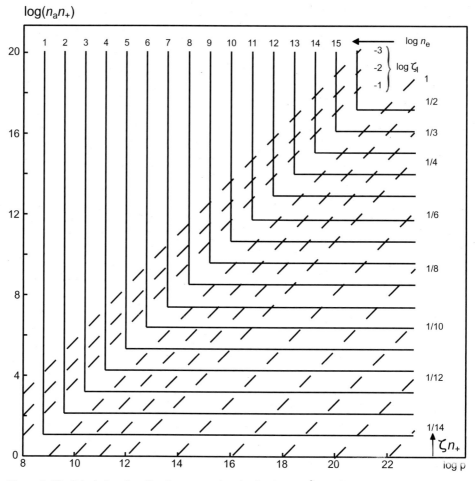

Figure 2–19. Calculation chart for electron number density (n_e, cm^{-3}) provided by photo-ionization (flux p, cm^{-2} s^{-1}) of aerosols (number density n_a, cm^{-3}).

For the specific case of an exponential spectrum $p(E) = p\exp(-\delta E)$, which is relevant for the tail of the thermal radiation, equation (2–119) can be converted to

$$\frac{n_e}{n_a} = \frac{\sum\limits_{-\infty}^{\infty} kq_0^k \exp\left(-\lambda\frac{k(k-1)}{2}\right)}{\sum\limits_{-\infty}^{\infty} q_0^k \exp\left(-\lambda\frac{k(k-1)}{2}\right)}, \qquad (2\text{–}120)$$

where $\lambda = \delta e^2/4\pi\varepsilon_0 r_a$. This equation can be solved, and the electron density can be expressed in differential form using the elliptical functions (Sayasov, 1958; Karachevtsev & Fridman, 1974, 1975):

$$\frac{n_e}{n_a} = y + \frac{\rho}{2\pi}\frac{d}{dy}\ln\theta_3(y, \rho). \qquad (2\text{–}121)$$

Here $\theta_3(y, \rho)$ is the elliptical θ-function of variables: $y = \frac{1}{\lambda}\ln q_0 + \frac{1}{2}$, $\rho = 2\pi/\lambda$. In the extreme case of $\rho \gg 1$,

$$\frac{n_e}{n_a} = \frac{1}{\lambda}\ln q_0 + \frac{1}{2}. \qquad (2\text{–}122)$$

The opposite extreme, $\rho \ll 1$, results in another asymptotic expression for electron density:

$$\frac{n_e}{n_a} = \frac{q_0}{1 + q_0}.$$ (2–123)

2.4.6. Thermal Ionization of Aerosols: Einbinder Formula

Ionization of macro-particles is determined by their work function, which is usually essentially lower than the ionization potential of atoms and molecules. Therefore, thermal ionization of aerosol particles provides high electron density and conductivity at relatively low temperatures, which is applied in particular in magneto-hydrodynamic generators (Kirillin & Sheindlin, 1971) and rocket engine torches (Musin, 1974). Absorption and reflection of radiowaves by the plasma of the rocket engine torches affects and complicates control of the rocket trajectory. Also, the electron density in flames can be high because of thermal ionization of macro-particles (see, for example, Shuler & Weber, 1954).

Simple relations for electron concentration in aerosol systems with an average macro-particle charge less than the unit charge can be based on the Saha equation (Arshinov & Musin, 1958; Samuilov, 1966, 1967). If temperature is relatively high, the average charge of a macro-particle can exceed that of an electron and the electric field of an aerosol particle affects the level of thermal ionization (Einbinder, 1957; Arshinov & Musin, 1958, 1959, 1962). The space distribution of electrons around a thermally ionized particle was analyzed by Samuilov (1968, 1973), as was related electric conductivity (Samuilov, 1966, 1967).

The average distance between macro-particles usually exceeds their radius: r_a: $n_a r_a^3 \ll 1$ (n_a is the concentration of aerosols). If we further assume that the average charge of a particle is less than the elementary charge ($n_e \ll n_a$), the electron density can be found by a relation similar to the Saha equation (see Chapter 3):

$$n_e = K(T) = \frac{2}{(2\pi\hbar)^3}(2\pi m T)^{3/2} \exp\left(-\frac{A_0}{T}\right).$$ (2–124)

A more interesting case corresponds to the average particle charge exceeding the elementary charge. Here the electron density can be expresed by the **Einbinder formula** (Einbinder, 1957; Arshinov & Musin, 1958), taking into account the work function dependence on the macro-particle charge:

$$\frac{n_e}{n_a} = \frac{4\pi\varepsilon_0 r_a T}{e^2} \ln\frac{K(T)}{n_e} + \frac{1}{2}.$$ (2–125)

For numerical calculations over the typical range of parameters, $n_a \approx 10^4 \div 10^8 \, \text{cm}^{-3}$ and work functions about 3 eV, relations (2–124) and (2–125) can be presented as

$$n_e = 5 \cdot 10^{24} \, \text{cm}^{-3} \cdot \exp\left(-\frac{A_0}{T}\right), \quad \text{if} \quad T < 1000 \, \text{K},$$ (2–126)

$$n_e = n_a \frac{4\pi\varepsilon_0 r_a T}{e^2}\left(20 - \frac{A_0}{T}\right), \quad \text{if} \quad T > 1000 \, \text{K}.$$ (2–127)

The Einbinder formula can be generalized to non-equilibrium plasma conditions when the electron temperature exceeds that of aerosol particles ($T_e > T_a$) (Fridman, 1976):

$$\frac{n_e}{n_a} = \frac{4\pi\varepsilon_0 r_a T_e}{e^2} \ln\frac{K(T_a)}{n_e} + \frac{1}{2}.$$ (2–128)

For numerical calculation, this formula can be expressed as

$$n_e = n_a \frac{4\pi \varepsilon_0 r_a T_e}{e^2} \left(20 - \frac{A_0}{T_a} \right).$$

(2–129)

2.4.7. Space Distribution of Electrons and Electric Field Around a Thermally Ionized Macro-Particle

A low aerosol density, $n_a r_a^3 \ll 1$, is conventional in plasma chemistry. In this case, considered by Fridman (1976), the combination of the Poisson equation with the Boltzmann distribution leads to equations for the potential $\varphi(r)$ and electron density $n_e(r)$ around a spherical thermally ionized particle:

$$\frac{d^2\varphi}{dr^2} + \frac{2}{r}\frac{d\varphi}{dr} = \frac{e}{\varepsilon_0} K(T) \exp\left(\frac{e\varphi}{T}\right),$$

(2–130)

$$\frac{d}{dr}\left(\frac{r^2}{n_e}\frac{dn_e}{dr} \right) = \frac{e^2}{\varepsilon_0 T} n_e r^2.$$

(2–131)

Integration of the last equation results in the following equation for the number $N(r)$ of electrons located in the radius interval from $r(r \geq r_a)$ to r_{av} ($r_{av} \approx n_a^{-1/3}$):

$$N(r) = \int_r^{r_{av}} n_e(r) 4\pi r^2 dr = -\frac{4\pi \varepsilon_0 T r^2}{e^2} \frac{1}{n_e(r)} \frac{dn_e(r)}{dr}.$$

(2–132)

The Boltzmann relation $n_e(r) = K(T) \exp[e\varphi(r)/T]$ can be rewritten as

$$E(r) = -\frac{d\varphi}{dr} = -\frac{T}{e}\frac{1}{n_e}\frac{dn_e}{dr}.$$

(2–133)

Combination of (2–132) and (2–133) gives

$$N(r) = \frac{E(r)}{e/4\pi \varepsilon_0 r^2}.$$

(2–134)

In the close vicinity of a macro-particle ($z = r - r_a \ll r_a$), the electron density distribution can be expressed using the **Langmuir relation** (Langmuir, 1961):

$$n_e(z) = \frac{K(T)}{\left(1 + \dfrac{z}{\sqrt{2} r_D^0} \right)^2},$$

(2–135)

where $K(T)$ represents the electron density on the surface, and r_D^0 is the corresponding Debye radius. The total charge of an aerosol particle can then be calculated as

$$Z_a = N(r_a) = \frac{4\pi \varepsilon_0 r_a T}{e^2} \left(\frac{r_a}{r_D^0} \right).$$

(2–136)

Integration of the Langmuir relation (2–135) shows that most of the electrons are located in the thin layer about $r_D^0 \ll r_a$ around the macro-particle. These electrons are confined in the Debye layer by a strong electric field, which according to the preceding relations can be expressed as

$$E(r) = \frac{T}{e\left(\dfrac{r_D^0}{\sqrt{2}} + \dfrac{r - r_a}{2} \right)}.$$

(2–137)

2.4.8. Electric Conductivity of Thermally Ionized Aerosols

Thermally ionized aerosol particles have strong inherent electric fields (2–137) when their charge Z_a is high. As a result, the electric conductivity of aerosols at $n_a r_a^3 \ll 1$ essentially depends on the external electric field E_e, whereas more electrons can be released from trapping by the inherent electric field of a macro-particle at higher E_e. The electric conductivity of free space between macro-particles can be presented as

$$\sigma_0 = e n_a b N(r_e), \qquad (2\text{–}138)$$

where $N(r_e) = \int_{r_e}^{r_{av}} n_e(r) 4\pi r^2 dr$ determines how many electrons per particle ($r_e < r < r_{av} \approx n_a^{-1/3}$) participate in electric conductivity in the external electric field E_e, and b is electron mobility. If the external electric field is low, the density of electrons n_e^* participating in conductivity can be found from the Einbinder formula:

$$N(r_{e1}) = \frac{n_e^*}{n_a} = \frac{4\pi \varepsilon_0 r_a T}{e^2} \ln \frac{K(T)}{n_e^*} \gg 1. \qquad (2\text{–}139)$$

This expression for $N(r_{e1})$ specifies a virtual sphere of radius r_{e1} around a macro-particle. Electrons located outside of this sphere are "not trapped" by the inherent electric field of the aerosol particle and provide the Einbinder value of electron density:

$$r_{e1} - r_a \approx r_a \ln^{-1} \frac{K(T)}{n_e^*} < r_a. \qquad (2\text{–}140)$$

The electric field of a macro-particle on the virtual sphere of radius r_e is

$$E(r_{e1}) = \frac{T}{r_a e} \ln \frac{K(T)}{n_e^*}. \qquad (2\text{–}141)$$

If the external electric field is less than the critical value, $E_e < E(r_{e1})$, then the "free" electron concentration follows the Einbinder formula and does not depend on the external electric field. As a consequence, the electric conductivity of aerosols also does not depend on the electric field in this case. When the external electric field exceeds the critical value (2–141), then the number of electrons per macro-particle participating in electric conductivity $N(r_e)$ depends on the external electric field (Fridman & Kennedy, 2004):

$$N(r_e) = \frac{4\pi \varepsilon_0 E_e r_a^2}{e}. \qquad (2\text{–}142)$$

As a result, the dependence of electric conductivity of thermally ionized aerosols on the electric field is:

$$\sigma_0 = \frac{4\pi \varepsilon_0 r_a T}{e^2} \left(\frac{r_a}{r_D^0} \right) n_a e b \quad \text{if} \quad E_e \geq \frac{T}{e r_D^0}, \qquad (2\text{–}143)$$

$$\sigma_0 = 4\pi \varepsilon_0 E_e r_a^2 n_a b \quad \text{if} \quad \frac{T}{e r_a} \ln \frac{K(T)}{n_e^*} < E_e < \frac{T}{e r_D^0}, \qquad (2\text{–}144)$$

$$\sigma_0 = \frac{4\pi \varepsilon_0 r_a T}{e^2} \left(\ln \frac{K(T)}{n_e^*} \right) n_a e b \quad \text{if} \quad E_e \leq \frac{T}{e r_a} \ln \frac{K(T)}{n_e^*}. \qquad (2\text{–}145)$$

The electric conductivity of aerosols starts depending on the electric field, when $E_e > \frac{T}{e r_a} \ln \frac{K(T)}{n_e^*}$. This means that the electric field should exceed about 200 V/cm, if $T = 1700\,\text{K}, r_a = 10\,\mu\text{m}$, and $A_0 = 3\,\text{eV}$. The maximum number of electrons participate in electric conductivity if $E_e \geq \frac{T}{e r_D^0}$. This corresponds numerically, for the same values of parameters, to an external electric field of about 1000 V/cm. To calculate the total electric conductivity of the thermo-ionized heterogeneous medium, we must also take into account the conductivity σ_a of the macro-particle material (Mezdrikov, 1968):

$$\sigma = \sigma_0 \left(1 - \frac{4}{3} \pi r_a^3 n_a \frac{\sigma_0 - \sigma_a}{2\sigma_0 + \sigma_a} \right). \tag{2-146}$$

2.5. EXCITATION AND DISSOCIATION OF NEUTRAL PARTICLES IN IONIZED GASES

The extremely high chemical activity of plasmas is based on a high and often super-equilibrium concentration of active species. The active species generated in plasma include chemically aggressive atoms and radicals, charged particles (electron and ions), and excited atoms and molecules. Elementary processes of atoms and radicals are traditionally considered in frameworks of chemical kinetics (see, for example, Eyring, Lin, & Lin, 1980; Kondratiev & Nikitin, 1981), elementary processes of charged particles were discussed earlier in Chapter 2, and elementary processes of the excited species are to be considered in the following.

2.5.1. Vibrational Excitation of Molecules by Electron Impact

Vibrational excitation is probably the most important elementary process in non-thermal molecular plasmas. It is responsible for the major part of energy exchange between electrons and molecules, and it makes a significant contribution in kinetics of non-equilibrium chemical processes in plasmas. Elastic collisions of electrons and molecules are not effective in the process of vibrational excitation because of the significant difference in their masses ($m/M \ll 1$). The typical energy transfer from an electron with kinetic energy ε to a molecule in an elastic collision is about $\varepsilon \left(\frac{m}{M} \right)$, whereas a vibrational quantum can be estimated as $\hbar\omega \approx I \sqrt{\frac{m}{M}}$ (Fridman & Kennedy, 2004), which is a much higher energy. For this reason, the classical cross section of the vibrational excitation in an elastic electron–molecular collision is low:

$$\sigma_{\text{vib}}^{\text{elastic}} \approx \sigma_0 \frac{\varepsilon}{I} \sqrt{\frac{m}{M}}. \tag{2-147}$$

In this relation, $\sigma_0 \sim 10^{-16}$ cm^2 is the gas-kinetic cross section, and I is an ionization potential. It gives numerical values of the cross section of about 10^{-19} cm^2 for electron energies about 1 eV. It is much lower than relevant experimental cross sections, which are about the same as atomic values (10^{-16} cm^2). Also experimental cross sections are non-monotonic functions of electron energy and the probability of multi-quantum excitation is not very low. Schultz (1976) presented a detailed review of these cross sections, which indicates that vibrational excitation of a molecule AB from vibrational level v_1 to v_2 is usually not a direct elastic process but a resonant process proceeding through formation of an intermediate non-stable negative ion:

$$\text{AB}(v_1) + e \xrightleftharpoons[]{\Gamma_{1i}, \Gamma_{i1}} \text{AB}^-(v_i) \xrightarrow{\Gamma_{i2}} \text{AB}(v_2) + e. \tag{2-148}$$

In this relation, v_i is the vibrational quantum number of a non-stable negative ion, and $\Gamma_{\alpha\beta}$ (in s^{-1}) are probabilities of transitions between vibrational states. The cross section of the resonant vibrational excitation process (2–148) can be found in the quasi-steady-state approximation using the **Breit-Wigner formula:**

$$\sigma_{12}(v_i, \varepsilon) = \frac{\pi \hbar^2}{2m\varepsilon} \frac{g_{\text{AB}^-}}{g_{\text{AB}} g_e} \frac{\Gamma_{1i} \Gamma_{i2}}{\frac{1}{\hbar^2}(\varepsilon - \Delta E_{1i})^2 + \Gamma_i^2}, \tag{2-149}$$

2.5. Excitation and Dissociation of Neutral Particles in Ionized Gases

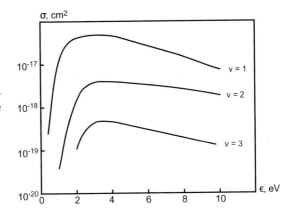

Figure 2–20. Cross sections of vibrational excitation as a function of electron energy in the case of short-lifetime resonances (H_2, $v = 1, 2, 3$).

where ε is the electron energy; δE_{1i} is the energy of transition to the intermediate state $AB(v_1) \rightarrow AB^-(v_i); g^-_{AB}, g_{AB}$, and g_e are statistical weights; and Γ_i is the probability of $AB^-(v_i)$ decay through all channels. Relation (2–149) illustrates the resonance structure of the cross-section dependence on electron energy. The energy width of the resonance pikes is about $\hbar\Gamma_i$, which is related to the lifetime of the non-stable intermediate negative ion $AB^-(v_i)$. The maximum value of the cross section (2–149) is about the same as the atomic value (10^{-16} cm^2).

The energy dependence of the vibrational excitation cross section depends on the lifetime of the intermediate ionic states (the so-called resonances). First let us consider the so-called **short-lifetime resonances** (e.g., H_2, N_2O, H_2O, etc.), where the lifetime of the auto-ionization states $AB^-(v_i)$ is much shorter than the period of oscillation ($\tau \ll 10^{-14}$ s). The energy width of the auto-ionization level $\sim\hbar\Gamma_i$ is very large for the short-lifetime resonances in accordance with the Uncertainty Principle. It results in wide maximum pikes (several electron volts) and no fine energy structure for $\sigma_{12}(\varepsilon)$, which can be seen in Fig. 2–20. Because of the short lifetime of the auto-ionization state $AB^-(v_i)$, displacement of nuclei during the lifetime period is small, which leads mostly to excitation of low vibrational levels. Electron energies that provide the most effective vibrational excitation as well as corresponding maximum cross sections are presented in Table 2–15.

If an intermediate ion lifetime is about a molecular oscillation period ($\sim 10^{-14}$ s), such a resonance is referred to as the **boomerang resonance** (e.g., low energy resonances in N_2, CO, CO_2, etc.). The boomerang model, developed by A. Herzenberg (1968), treats the

Table 2–15. Cross Sections of Vibrational Excitation of Molecules by Electron Impact

Molecule	Most effective electron energy	Maximum cross section	Molecule	Most effective electron energy	Maximum cross section
N_2	1.7–3.5 eV	$3 \cdot 10^{-16}$ cm^2	NO	0–1 eV	10^{-17} cm^2
CO	1.2–3.0 eV	$3.5 \cdot 10^{-16}$ cm^2	NO_2	0–1 eV	–
CO_2	3–5 eV	$2 \cdot 10^{-16}$ cm^2	SO_2	3–4 eV	–
C_2H_4	1.5–2.3 eV	$2 \cdot 10^{-16}$ cm^2	C_6H_6	1.0–1.6 eV	–
H_2	~3 eV	$4 \cdot 10^{-17}$ cm^2	CH_4	Thresh. 0.1 eV	10^{-16} cm^2
N_2O	2–3 eV	10^{-17} cm^2	C_2H_6	Thresh. 0.1 eV	$2 \cdot 10^{-16}$ cm^2
H_2O	5–10 eV	$6 \cdot 10^{-17}$ cm^2	C_3H_8	Thresh. 0.1 eV	$3 \cdot 10^{-16}$ cm^2
H_2S	2–3 eV	–	Cyclopropane	Thresh. 0.1 eV	$2 \cdot 10^{-16}$ cm^2
O_2	0.1–1.5 eV	10^{-17} cm^2	HCl	2–4 eV	10^{-15} cm^2

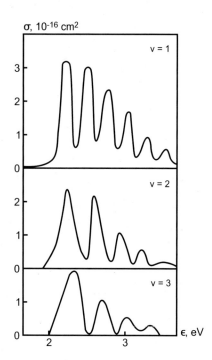

σ, 10^{-16} cm^2

$v = 1$

$v = 2$

$v = 3$

ϵ, eV

Figure 2–21. Cross sections of vibrational excitation as a function of electron energy in the case of boomerang resonances (N_2, $v = 1, 2, 3$).

formation and decay of the negative ion during one oscillation as interference of incoming and reflected waves. The interference of the nuclear wave packages results in an oscillating dependence of the excitation cross section on electron energy with typical peaks of about 0.3 eV (see Fig. 2–21 and Table 2–15). Boomerang resonances require higher electron energies for excitation of higher vibrational levels. For example, the excitation threshold of $N_2(v = 1)$ is 1.9 eV, and that of $N_2(v = 10)$ from the ground state is about 3 eV. Excitation of $CO(v = 1)$ requires a minimum electron energy of 1.6 eV, and the threshold for $CO(v = 10)$ excitation from the ground state is about 2.5 eV. The maximum value of the vibrational excitation cross section decreases here with vibrational quantum number.

The **long-lifetime resonances** (e.g., low energy resonances in O_2, NO, C_6H_6, etc.) correspond to auto-ionization states $AB^-(v_i)$ with much longer lifetime ($\tau = 10^{-14}$–10^{-10} s). The long-lifetime resonances result in quite narrow isolated peaks (about 0.1 eV) in cross-section dependence on electron energy (see Fig. 2–22 and Table 2–15). In contrast to boomerang resonances, here the maximum value of the vibrational excitation cross section remains the same for different vibrational quantum numbers.

2.5.2. Rate Coefficients of Vibrational Excitation by Electron Impact: Semi-Empirical Fridman Approximation

The electron energies most effective in vibrational excitation are 1–3 eV, which usually corresponds to a maximum in the electron energy distribution function. The vibrational excitation rate coefficients, which are results of integration of the cross sections over the electron energy distribution function, are obviously very high in this case and reach 10^{-7} cm^3/s, (see Table 2–16). For molecules such as N_2, CO, and CO_2, almost every electron–molecular collision leads to a vibrational excitation at $T_e = 2$ eV. This explains why a significant fraction of electron energy in non-thermal discharges is going into vibrational excitation at $T_e = 1$–3 eV.

Vibrational excitation by electron impact is preferably a one-quantum process. Nevertheless, multi-quantum-vibrational excitation is also important. Rate coefficients $k_{eV}(v_1, v_2)$

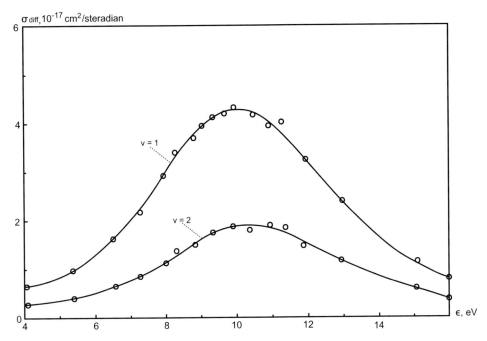

Figure 2–22. Cross sections of vibrational excitation as a function of electron energy in the case of long-lifetime resonances (O_2, $v = 1, 2, 3$).

for excitation of molecules from an initial vibrational level v_1 to a final level v_2 can be found using the semi-empirical **Fridman approximation for multi-quantum-vibrational excitation** (Rusanov, Fridman, & Sholin, 1981; Rusanov & Fridman, 1984). This approach permits one to find the excitation rate coefficients $k_{eV}(v_1, v_2)$ based on much better known vibrational excitation rate coefficient $k_{eV}(0, 1)$, corresponding to excitation from the ground state to the first vibrational level and presented in Table 2–16. According to the Fridman approximation,

$$k_{eV}(v_1, v_2) = k_{eV}(0, 1)\frac{\exp[-\alpha(v_2 - v_1)]}{1 + \beta v_1}. \qquad (2\text{–}150)$$

Parameters α and β of the Fridman approximation are summarized in Table 2–17.

Table 2–16. Rate Coefficients of Vibrational Excitation of Molecules by Electron Impact

Molecule	$T_e = 0.5\,\text{eV}$	$T_e = 1\,\text{eV}$	$T_e = 2\,\text{eV}$
H_2	$2.2 \cdot 10^{-10}\,\text{cm}^3/\text{s}$	$2.5 \cdot 10^{-10}\,\text{cm}^3/\text{s}$	$0.7 \cdot 10^{-9}\,\text{cm}^3/\text{s}$
D_2	–	–	$10^{-9}\,\text{cm}^3/\text{s}$
N_2	$2 \cdot 10^{-11}\,\text{cm}^3/\text{s}$	$4 \cdot 10^{-9}\,\text{cm}^3/\text{s}$	$3 \cdot 10^{-8}\,\text{cm}^3/\text{s}$
O_2	–	–	$10^{-10}\text{–}10^{-9}\,\text{cm}^3/\text{s}$
CO	–	–	$10^{-7}\,\text{cm}^3/\text{s}$
NO	–	$3 \cdot 10^{-10}\,\text{cm}^3/\text{s}$	–
CO_2	$3 \cdot 10^{-9}\,\text{cm}^3/\text{s}$	$10^{-8}\,\text{cm}^3/\text{s}$	$3 \cdot 10^{-8}\,\text{cm}^3/\text{s}$
NO_2	–	–	$10^{-10}\text{–}10^{-9}\,\text{cm}^3/\text{s}$
H_2O	–	–	$10^{-10}\,\text{cm}^3/\text{s}$
C_2H_4	–	$10^{-8}\,\text{cm}^3/\text{s}$	–

Table 2–17. Parameters of Multi-Quantum-Vibrational Excitation by Electron Impact

Molecule	α	β	Molecule	α	β
N_2	0.7	0.05	H_2	3	–
CO	0.6	–	O_2	0.7	–
$CO_2(v3)$	0.5	–	NO	0.7	–

2.5.3. Rotational Excitation of Molecules by Electron Impact

If electron energies exceed ~ 1 eV, rotational excitation can proceed resonantly through the auto-ionization states similarly to the case of vibrational excitation. However, the relative contribution of this multi-stage rotational excitation is small when taking into account the low value of a rotational energy quantum with respect to vibrational quanta. Non-resonant rotational excitation by electron impact can be illustrated using the classical approach. The energy transfer from an electron (kinetic energy ε, mass m) to a molecule (of mass M) in an elastic collision, inducing rotational excitation, is about $\varepsilon(\frac{m}{M})$. The typical spacing between rotational levels (rotational quanta) is about $I(\frac{m}{M})$, where I is the ionization potential. Therefore, cross sections of the non-resonant rotational excitation can be related to $\sigma_0 \sim 10^{-16}$ cm^2 (gas-kinetic collisional cross section) as

$$\sigma_{\text{rotational}}^{\text{elastic}} \approx \sigma_0 \frac{\varepsilon}{I}. \tag{2–151}$$

The cross section of the non-resonant rotational excitation can exceed the non-resonant value by about 100 times. A quantum-mechanical approach leads to similar conclusions. An electron collision with a dipole molecule induces rotational transitions with a change of rotational quantum number $\Delta J = 1$. Quantum-mechanical cross sections for a linear dipole molecule and a low-energy electron are as follows (Crowford, 1967):

$$\sigma(J \rightarrow J+1, \varepsilon) = \frac{d^2}{3\varepsilon_0 a_0 \varepsilon} \frac{J+1}{2J+1} \ln \frac{\sqrt{\varepsilon} + \sqrt{\varepsilon'}}{\sqrt{\varepsilon} - \sqrt{\varepsilon'}}. \tag{2–152}$$

In this relation, d is the dipole moment, a_0 is the Bohr radius, $\varepsilon' = \varepsilon - 2B(J+1)$ is the electron energy after collision, and B is a rotational constant. Numerically, this cross section is about $1.3 \cdot 10^{-16}$ cm^2 at $\varepsilon = 0.1$ eV.

Homonuclear molecules, like N_2 or H_2, have no dipole moment, and rotational excitation is due to electron interaction with their quadruple moment Q. In this case the rotational transition takes place with a change of rotational quantum number, $\Delta J = 2$. The cross section of the rotational excitation by a low-energy electron is lower in this case (about 10^{-17} cm^2 at $\varepsilon = 0.1$ eV; Gerjoy & Stein, 1955):

$$\sigma(J \rightarrow J+2, \varepsilon) = \frac{8\pi Q^2}{15e^2 a_0^2} \frac{(J+1)(J+2)}{(2J+1)(2J+3)} \ln \sqrt{\frac{\varepsilon}{\varepsilon'}}. \tag{2–153}$$

To evaluate "quasi-elastic" energy transfer from an electron gas to neutral molecules, the rotational excitation can be combined with the elastic collisions. The process is then characterized by a gas-kinetic rate coefficient $k_{e0} \approx \sigma_0 \langle v_e \rangle \approx 3 \cdot 10^{-8}$ cm^3/s (where $\langle v_e \rangle$ is the average thermal velocity of electrons), and each collision is considered as a loss of about $\varepsilon \left(\frac{m}{M} \right)$ of electron energy.

2.5.4. Electronic Excitation of Atoms and Molecules by Electron Impact

Electronic excitation by electron impact requires higher electron energies ($\varepsilon > 10\,\text{eV}$) than does vibrational and rotational excitation. The **Born approximation** can be applied to calculate the cross section of these processes when the electron energies are high enough. For excitation of optically permitted transitions from an atomic state i to another state k, the Born approximation gives the following cross section:

$$\sigma_{ik}(\varepsilon) = 4\pi a_0^2 f_{ik} \left(\frac{Ry}{\Delta E_{ik}}\right)^2 \frac{\Delta E_{ik}}{\varepsilon} \ln \frac{\varepsilon}{\Delta E_{ik}}. \tag{2-154}$$

In this relation, Ry is the Rydberg constant, a_0 is the Bohr radius, f_{ik} is force of oscillator for transition $i \rightarrow k$, and ΔE_{ik} is the energy of the transition. The formula is valid for high electron energies ($\varepsilon \gg \Delta E_{ik}$). Semi-empirical formulas can be used for calculations; one was proposed by Drawin (1968, 1969):

$$\sigma_{ik}(\varepsilon) = 4\pi a_0^2 f_{ik} \left(\frac{Ry}{\Delta E_{ik}}\right)^2 \frac{x-1}{x^2} \ln(2.5x). \tag{2-155}$$

Another semi-empirical formula was proposed by Smirnov (1968):

$$\sigma_{ik}(\varepsilon) = 4\pi a_0^2 f_{ik} \left(\frac{Ry}{\Delta E_{ik}}\right)^2 \frac{\ln(0.1x + 0.9)}{x - 0.7}. \tag{2-156}$$

In both relations, $x = \frac{\varepsilon}{\Delta E_{ik}}$; the semi-empirical formulas are valid at $x \gg 1$. Maximum cross sections for the excitation of optically permitted transitions are about the same as the gas-kinetic cross section $\sigma_0 \sim 10^{-16}\,\text{cm}^2$. To reach this cross section, the electron energy should be two to three times greater than the transition energy ΔE_{ik}.

Dependence $\sigma_{ik}(\varepsilon)$ is different for excitation of electronic terms, from which optical transitions (radiation) are forbidden. The maximum cross section, which here is about the same as the atomic value $\sigma_0 \sim 10^{-16}\,\text{cm}^2$, can be reached at much lower electron energies $\frac{\varepsilon}{\Delta E_{ik}} \approx 1.2\text{–}1.6$, which leads to an interesting effect of predominant excitation of the optically forbidden and metastable states by electron impact in non-thermal discharges, where the electron temperature T_e is usually much less than the transition energy ΔE_{ik}. Some cross sections of electronic excitation by electron impact are presented in Fig. 2–23 (Slovetsky, 1980).

Rate coefficients of electronic excitation are calculated by integration of the cross sections $\sigma_{ik}(\varepsilon)$ over the EEDFs. In the simplest case of a Maxwellian EEDF ($T_e \ll \Delta E_{ik}$),

$$k_{\text{el.excit.}} \propto \exp\left(-\frac{\Delta E_{ik}}{T_e}\right). \tag{2-157}$$

A semi-empirical relation for electronic excitation and ionization rate coefficients as a function of reduced electric field E/n_0 was proposed by Kochetov et al. (1979):

$$\log k_{\text{el.excit.}} = -C_1 - \frac{C_2}{E/n_0}. \tag{2-158}$$

The rate coefficient $k_{\text{el.excit.}}$ is expressed in cm^3/s; E is the electric field strength, in V/cm; and n_0 is gas density in $1/\text{cm}^3$. Numerical values of the parameters C_1 and C_2 for different electronically excited states (and ionization) of CO_2 and N_2 are presented in Table 2–18. If the level of vibrational excitation is high, superelastic collisions provide higher electron energies and faster electronic excitation at the same value of E/n_0. This can be taken into account by adding terms related to vibrational temperature T_v(in K):

$$\lg k_{\text{el.excit.}} = -C_1 - \frac{C_2}{E/n_0} + \frac{40z + 13z^2}{[(E/n_0) \cdot 10^{16}]^2} - 0.02 \left(\frac{T_v}{1000}\right)^{\frac{2}{3}}. \tag{2-159}$$

Table 2–18. Parameters of Semi-Empirical Approximation of Rate Coefficients of Electronic Excitation and Ionization of CO_2 and N_2 by Electron Impact

Molecule	Excitation level or Ionization	C_1	$C_2 \cdot 10^{16}$, $V \cdot cm^2$	Molecule	Excitation level or Ionization	C_1	$C_2 \cdot 10^{16}$, $V \cdot cm^2$
N_2	$A^3\Sigma_u^+$	8.04	16.87	N_2	$c^1\Pi_u$	8.85	34.0
N_2	$B^3\Pi_g$	8.00	17.35	N_2	$a^1\Pi_u$	9.65	35.2
N_2	$W^3\Delta_u$	8.21	19.2	N_2	$b'^1\Sigma_u^+$	8.44	33.4
N_2	$B'^3\Sigma_u^-$	8.69	20.1	N_2	$c^3\Pi_u$	8.60	35.4
N_2	$a'^1\Sigma_u^-$	8.65	20.87	N_2	$F^3\Pi_u$	9.30	32.9
N_2	$a^1\Pi_g$	8.29	21.2	N_2	Ionization	8.12	40.6
N_2	$W^1\Delta_u$	8.67	20.85	CO_2	$^3\Sigma_u^+$	8.50	10.7
N_2	$C^3\Pi_u$	8.09	25.5	CO_2	$^1\Delta_u$	8.68	13.2
N_2	$E^3\Sigma_g^+$	9.65	23.53	CO_2	$^1\Pi_g$	8.84	14.8
N_2	$a''^1\Sigma_g^+$	8.88	26.5	CO_2	$^1\Sigma_g^+$	8.23	18.9
N_2	$b^1\Pi_u$	8.50	31.88	CO_2	Other levels	8.34	20.9
N_2	$c'^1\Sigma_u^+$	8.56	35.6	CO_2	Ionization	8.38	25.5

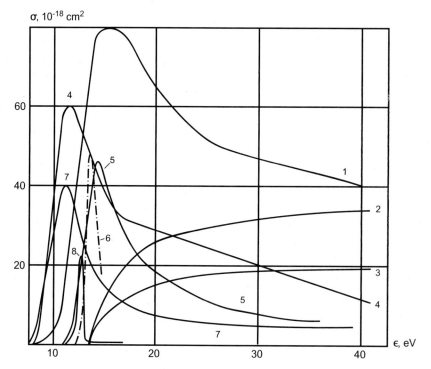

Figure 2–23. Cross section of excitation of different electronic states of $N_2(X^1\Sigma_g^+, v = 0)$ by electron impact: (1) $a^1\Pi_g$, (2) $b^1\Pi_u(v_k = 0\text{–}4)$, (3) transitions 12.96 eV, (4) $B^3\Pi_g$, (5) $C^3\Pi_u$, (6) $a'''^1\Sigma_g^+$, (7) $A^3\Sigma_u^+$, (8) $E^3\Sigma_g^+$.

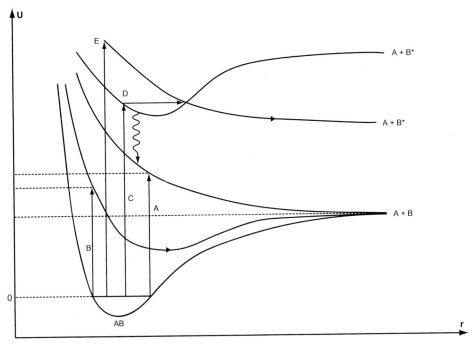

Figure 2–24. Different mechanisms of dissociation of molecules through electronic excitation.

The Boltzmann factor z can be expressed as $z = \exp(-\hbar\omega/T_v)$. Relation (2–159) provides accuracy of about 20% at $5 \cdot 10^{-16}\,\mathrm{V} \cdot \mathrm{cm}^2 < E/n_0 < 30 \cdot 10^{-16}\,\mathrm{V} \cdot \mathrm{cm}^2$ and $T_v < 9000\,\mathrm{K}$.

2.5.5. Dissociation of Molecules by Direct Electron Impact

Electron impact is able to stimulate dissociation of molecules by both vibrational and electronic excitation. Vibrational excitation usually results in the initial formation of molecules with single or few quanta. Dissociation takes place in this case as a non-direct multi-step process, including energy exchange (vibrational–vibrational (VV) relaxation) between molecules to collect vibrational energy sufficient for dissociation. Such processes are effective only for a limited (but very important) group of gases like N_2, CO_2, H_2, and CO. In contrast to this, dissociation through electronic excitation can proceed in just one collision; therefore, it is described as being stimulated by direct electron impact. The elementary process can proceed through different intermediate steps of intramolecular transitions, which is illustrated in Fig. 2–24.

Mechanism A starts with direct electronic excitation from the ground state to a repulsive one with resultant dissociation. In this case the required electron energy can significantly (few electron volts) exceed the dissociation energy. This mechanism generates, therefore, hot (high-energy) neutral fragments, which could, for example, significantly affect surface chemistry in low-pressure non-thermal discharges.

Mechanism B includes direct electronic excitation of a molecule from the ground state to an attractive state with energy exceeding the dissociation threshold. Excitation of the state then results in dissociation. As one can see from Fig. 2–24, the energies of the dissociation fragments are lower in this case.

Mechanism C consists of direct electronic excitation from the ground state to an attractive state corresponding to electronically excited dissociation products. Excitation of this

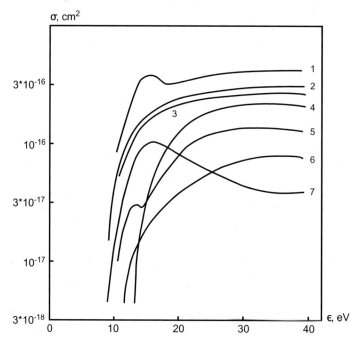

Figure 2–25. Cross sections of dissociation of molecules through electronic excitation as a function of electron energy: (1) CH_4, (2) O_2, (3) NO, (4) N_2, (5) CO_2, (6) CO, (7) H_2.

state can lead to radiative transition to a low-energy repulsive state (see Fig. 2–24) with subsequent dissociation. The energy of the dissociation fragments in this case is similar to those of mechanism A.

Mechanism D (similarly to mechanism C) starts with direct electronic excitation from the ground state to an attractive state corresponding to electronically excited dissociation

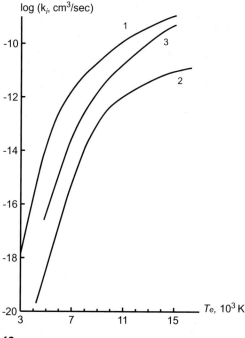

Figure 2–26. (1) Rate coefficients of stepwise N_2 electronic state excitation, (2) direct dissociation of initially non-excited nitrogen molecules, and (3) N_2 dissociation in stepwise electronic excitation sequence.

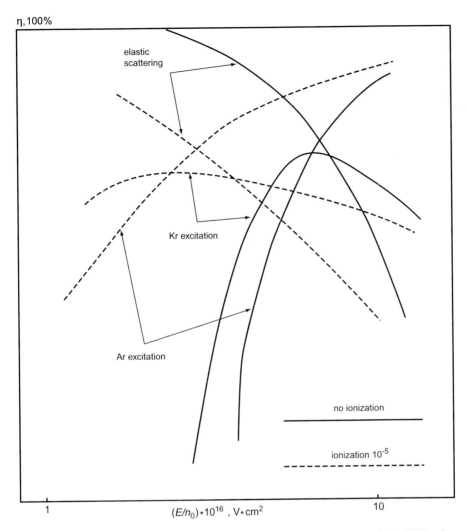

Figure 2–27. Electron energy distribution between excitation channels in Kr (5%)–Ar (95%) mixture.

products. Excitation of this state leads to radiationless transfer to a highly excited repulsive state (Fig. 2–24) with subsequent dissociation into electronically excited fragments. This mechanism is usually referred to as **predissociation**.

Mechanism E is similar to mechanism A and consists of direct electronic excitation from the ground state to a repulsive state, but with the ensuring dissociation into electronically excited fragments. This mechanism requires the highest values of electron energies.

Cross sections of the dissociation by direct electron impact are presented as a function of electron energy in Fig. 2–25. The relevant rate coefficients are given in Fig. 2–26 as a function of electron temperature.

2.5.6. Distribution of Electron Energy in Non-Thermal Discharges Between Different Channels of Excitation and Ionization

Electron gas energy received from the electric field in non-thermal plasma is distributed between elastic energy losses and different channels of excitation and ionization. Such distributions for different atomic and molecular gases are presented in Figs. 2–27–2–34 as a

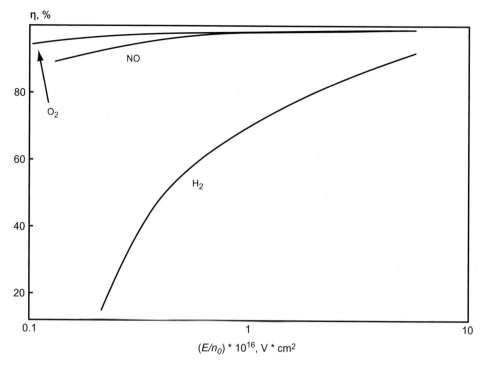

Figure 2–28. Fraction of electron energy spent on vibrational excitation in discharges in molecular gases: O_2, NO, and H_2.

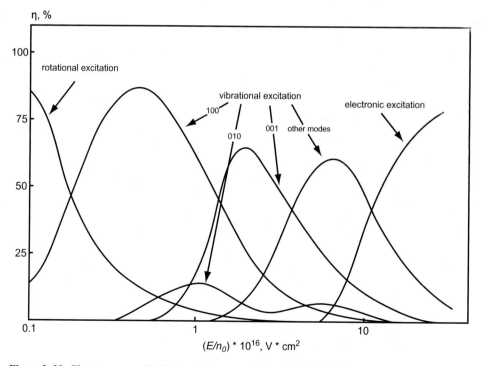

Figure 2–29. Electron energy distribution between excitation channels in CO_2.

2.5. Excitation and Dissociation of Neutral Particles in Ionized Gases

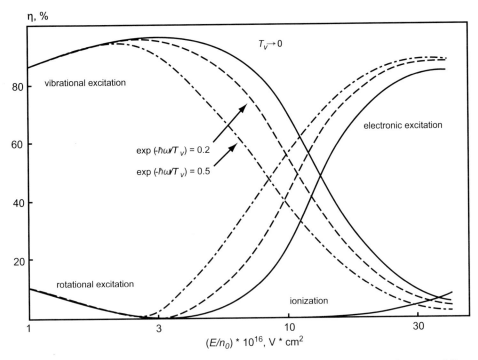

Figure 2–30. Electron energy distribution between excitation channels in molecular nitrogen at different vibrational temperatures.

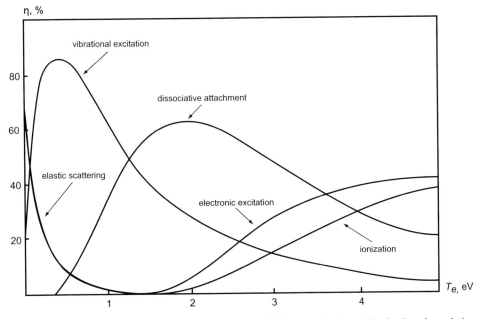

Figure 2–31. Electron energy distribution between different excitation and ionization channels in water vapor.

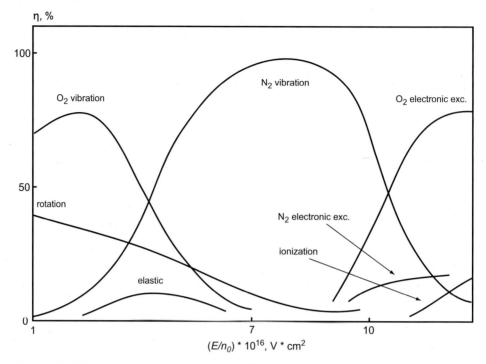

Figure 2–32. Electron energy distribution between different excitation and ionization channels in air.

function of reduced electric field E/n_0. All the energy distributions have quite similar general features. For example, the contribution of rotational excitation of molecules and elastic energy losses are essential only at low E/n_0. It is natural, because these processes are non-resonant and take place at low electron energies ($\ll 1$ eV). At electron temperatures of about of 1 eV, which is conventional for non-thermal discharges, most of the electron energy and,

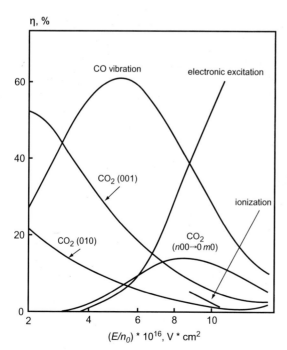

Figure 2–33. Electron energy distribution between different excitation and ionization channels in $CO_2(50\%)$–$CO(50\%)$ mixture.

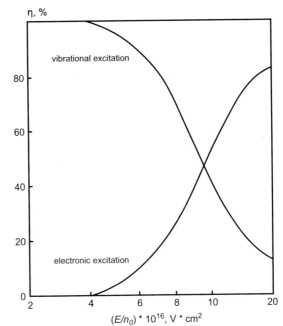

Figure 2–34. Electron energy distribution between different excitation and ionization channels in CO.

therefore, most of the discharge power can be localized within vibrational excitation of molecules. It makes the process of vibrational excitation very important and special in the non-equilibrium plasma chemistry of molecular gases. Obviously, the contribution of electron attachment processes, including dissociative attachment, can effectively compete with vibrational excitation at similar electron temperatures, but only in strongly electronegative gases. Finally, the contribution of electronic excitation and ionization becomes significant at higher electron temperatures because of the high-energy thresholds of these processes.

The electron energy distribution between different excitation channels strongly depends on gas composition. For example, addition of CO to CO_2 changes the distribution quite significantly (compare Figs. 2–29 and Fig. 2–33). The vibrational temperature influence on the electron energy distribution can be illustrated by that of molecular nitrogen (Fig. 2–30). At higher vibrational temperatures, the efficiency of further vibrational excitation is lower, which results in an increase in the fraction of high-energy electrons and, hence, in intensification of electronic excitation and ionization. In general, Figs. 2–27–2–34 demonstrate that, at electron temperatures of about 1 eV, a major fraction of the electron energy can go into the vibrational excitation of molecules, meaning that most of the non-equilibrium cold discharge power can be selectively focused on the single mechanism of vibrational excitation, ensuring chemical reactions. It makes plasma-chemical reactions stimulated by vibrational excitation the most promising for minimizing the energy cost of plasma technologies, based on discharges in molecular gases.

2.6. ELEMENTARY RELAXATION PROCESSES OF ENERGY TRANSFER INVOLVING VIBRATIONALLY, ROTATIONALLY, AND ELECTRONICALLY EXCITED MOLECULES

2.6.1. Vibrational–Translational (VT) Relaxation: Slow Adiabatic Elementary Process

This elementary process of energy transfer between vibrational and translational degrees of freedom is usually called vibrational relaxation or VT relaxation. Qualitative features of vibrational relaxation can be demonstrated by considering the collision of a classical harmonic oscillator with an atom or molecule. The oscillator is considered in this case to be under the influence of an external force $F(t)$, which represents the intermolecular collision. One-dimensional motion of the harmonic oscillator can then be described in the center-of-mass system by the Newton equation:

$$\frac{d^2 y}{dt^2} + \omega^2 y = \frac{1}{\mu_0} F(t). \tag{2-160}$$

In this relation, y is the vibrational coordinate, ω is the oscillator frequency, and μ_0 is its reduced mass. The oscillator is initially ($t \to -\infty$) not excited, which means $y(t \to -\infty) = 0$, $\frac{dy}{dt}(t \to -\infty) = 0$. The vibrational energy transferred to the oscillator during the collision is given by

$$\Delta E_v = \frac{\mu_0}{2} \left[\left(\frac{dy}{dt} \right)^2 + \omega^2 y^2 \right]_{t=\infty}. \tag{2-161}$$

Setting the complex variable $\xi(t) = dy/dt + i\omega y$ gives (Nikitin, 1970; Gordiets, Osipov, & Shelepin, 1980)

$$\Delta E_v = \frac{\mu_0}{2} |\xi(t)|^2_{t=\infty}. \tag{2-162}$$

Dynamic equation (2–160) can be rewritten in this case as a first-order differential equation:

$$\frac{d}{dt}\left(\frac{dy}{dt} + i\omega y \right) - i\omega \left(\frac{dy}{dt} + i\omega y \right) = \frac{1}{\mu_0} F(t); \qquad \frac{d\xi}{dt} - i\omega\xi = \frac{1}{\mu_0} F(t). \tag{2-163}$$

The solution of this equation with initial conditions $y(t \to -\infty) = 0$, $dy/dt(t \to -\infty) = 0$ is

$$\xi(t) = \exp(i\omega t) \int_{-\infty}^{t} \frac{1}{\mu_0} F(t') \exp(-i\omega t') \, dt'. \tag{2-164}$$

As a result, energy transfer during the VT-relaxation process can be presented as

$$\Delta E_v = \frac{1}{2\mu_0} \left| \int_{-\infty}^{+\infty} F(t) \exp(-i\omega t) \, dt \right|^2. \tag{2-165}$$

Only the Fourier component of the perturbation force on the oscillator frequency is effective in collisional excitation (or deactivation) of vibrational degrees of freedom of molecules. Considering time t as a complex variable (Landau, 1997) and assuming $F(t) \to \infty$ in a singularity point $t = \tau + i\tau_{col}$ ($\tau_{col} = 1/\alpha v$ is the collisional time; α is the reverse radius of interaction between molecules, and v is a relative velocity of the colliding particles), integration of (2–165) gives

$$\Delta E_v \propto \exp(-2\omega\tau_{col}). \tag{2-166}$$

This relation (Landau & Teller, 1936) demonstrates the adiabatic behavior of vibrational relaxation. Usually the **Massey parameter** at low gas temperatures is high for molecular vibration $\omega\tau_{col} \gg 1$, which explains the adiabatic behavior and results in the exponentially slow vibrational energy transfer during the VT relaxation. During the adiabatic collision, a molecule has enough time for many vibrations and the oscillator can actually be considered "structureless," which explains such a low level of energy transfer. An exponentially slow adiabatic VT relaxation and intensive vibrational excitation by electron impact result in the unique role of vibrational excitation in plasma chemistry. Molecular vibrations for gases like N_2, CO_2, H_2, and CO are able to "trap" energy of non-thermal discharges; therefore, it is easy to activate them and difficult to deactivate them. Quantitatively, classical VT energy transfer between atom A and molecule BC is

$$\Delta E_v = \frac{8\pi \omega^2 \mu^2 \lambda^2}{\alpha^2 \mu_0^2} \exp\left(-\frac{2\pi \omega}{\alpha v}\right). \qquad (2\text{--}167)$$

In this relation, μ is the reduced mass of A and BC, μ_0 is the reduced mass of the molecule BC, and $\lambda = m_C/(m_B + m_C)$. The classical probability of a single quantum $\hbar\omega$ transfer can be expressed based on (2–167) as

$$P_{01}^{VT}(v) = \frac{\Delta E_v}{\hbar\omega} = \frac{8\pi^2 \omega \mu^2 \lambda^2}{\hbar \alpha^2 \mu_0^2} \exp\left(-\frac{2\pi \omega}{\alpha v}\right). \qquad (2\text{--}168)$$

The quantum-mechanical expression for probability, $P_{mn}^{VT}(v)$, of oscillator transition from an initial state with vibrational quantum number m to a final state n generalizes the classical expression (2–168):

$$P_{mn}^{VT}(v) = \frac{16\pi^2 \mu^2 \omega_{mn}^2 \lambda^2}{\alpha^2 \hbar^2} \langle m \left| y \right| n \rangle^2 \exp\left(-\frac{2\pi \left| \omega_{mn} \right|}{\alpha v}\right). \qquad (2\text{--}169)$$

In this relation, $\hbar\omega_{mn} = E_m - E_n$ is the transition energy; $\langle m \left| y \right| n \rangle$ is a matrix element corresponding to eigenfunctions (m and n) of the non-perturbed Hamiltonian of the oscillator. The matrix elements $\langle m \left| y \right| n \rangle$ are not equal to zero for harmonic oscillators, only for transitions $n = m \pm 1$ (Landau, 1997):

$$\langle m \left| y \right| n \rangle = \sqrt{\frac{\hbar}{2\mu_0 \omega}}(\sqrt{m}\delta_{n,m-1} + \sqrt{m+1}\delta_{n,m+1}). \qquad (2\text{--}170)$$

The symbol $\delta_{ij} = 1$ if $i = j$, and $\delta_{ij} = 0$ if $i \neq j$. Multi-quantum VT-relaxation processes are possible due only to anharmonicity of oscillations and have much lower probability.

2.6.2. Landau–Teller Formula for VT-Relaxation Rate Coefficients

To calculate the rate coefficient for **VT relaxation of anharmonic oscillators**, the relaxation probability as a function of relative velocity v of colliding particles can be presented as

$$P_{n+1,n}^{VT}(v) \propto (n+1)\exp\left(-\frac{2\pi \omega}{\alpha v}\right), \qquad (2\text{--}171)$$

and integrated over the Maxwellian distribution function. The integration results in an expression for the probability of VT relaxation as a function of translational temperature T_0 and vibrational quantum number n:

$$P_{n+1,n}^{VT}(T_0) = (n+1)P_{1,0}^{VT}(T_0). \qquad (2\text{--}172)$$

The temperature dependence of VT-relaxation probability can be expressed after the integration as

$$P_{10}^{VT}(T_0) \propto \exp\left[-3\left(\frac{\hbar^2 \mu \omega^2}{2\alpha^2 T_0}\right)^{\frac{1}{3}}\right]. \qquad (2\text{--}173)$$

The formula can be rewritten for the rate coefficient of VT relaxation of a single vibrational quantum (Landau & Teller, 1936), which is known as the **Landau-Teller formula**:

$$k_{VT}^{10} \propto \exp\left(-\frac{B}{T_0^{1/3}}\right), \qquad B = 3\sqrt{\frac{27\hbar^2 \mu \omega^2}{2\alpha^2 T_0}}. \qquad (2\text{--}174)$$

Table 2–19. Vibrational VT Relaxation Rate Coefficients $k_{VT}^{10}(T_0 = 300\,\text{K})$ for One-Component Gases at Room Temperature

Molecule	$k_{VT}^{10}(T_0 = 300\,\text{K})$, cm^3/s	Molecule	$k_{VT}^{10}(T_0 = 300\,\text{K})$, cm^3/s
O_2	$5 \cdot 10^{-18}$	F_2	$2 \cdot 10^{-15}$
Cl_2	$3 \cdot 10^{-15}$	D_2	$3 \cdot 10^{-17}$
Br_2	10^{-14}	$CO_2(01^10)$	$5 \cdot 10^{-15}$
J_2	$3 \cdot 10^{-14}$	$H_2O(010)$	$3 \cdot 10^{-12}$
N_2	$3 \cdot 10^{-19}$	N_2O	10^{-14}
CO	10^{-18}	COS	$3 \cdot 10^{-14}$
H_2	10^{-16}	CS_2	$5 \cdot 10^{-14}$
HF	$2 \cdot 10^{-12}$	SO_2	$5 \cdot 10^{-14}$
DF	$5 \cdot 10^{-13}$	C_2H_2	10^{-12}
HCl	10^{-14}	CH_2Cl_2	10^{-12}
DCl	$5 \cdot 10^{-15}$	CH_4	10^{-14}
HBr	$2 \cdot 10^{-14}$	CH_3Cl	10^{-13}
DBr	$5 \cdot 10^{-15}$	$CHCl_3$	$5 \cdot 10^{-13}$
HJ	10^{-13}	CCl_4	$5 \cdot 10^{-13}$
HD	10^{-16}	NO	10^{-13}

The Landau-Teller formula was generalized in SSH theory (Schwartz, Slawsky, & Herzfeld, 1952) and, in particular, by Billing (1986). A semi-empirical numerical relation was proposed by Lifshitz (1974):

$$k_{VT}^{10} = 3.03 \cdot 10^6 (\hbar\omega)^{2.66} \mu^{2.06} \exp\left[-0.492(\hbar\omega)^{0.681}\mu^{0.302}T_0^{-1/3}\right]. \quad (2\text{--}175)$$

Here the rate coefficient is in cm^3/mole · sec, T_0 and $\hbar\omega$ are given in degrees Kelvin, and the reduced mass of colliding particles, μ, is in atomic mass units. A semi-empirical formula for VT-relaxation time (in seconds) as a function of pressure (in atmospheres) was proposed by Milliken & White (1963):

$$\ln(p\tau_{VT}) = 1.16 \cdot 10^{-3}\mu^{1/2}(\hbar\omega)^{4/3}\left(T_0^{-1/3} - 0.015\mu^{1/4}\right) - 18.42. \quad (2\text{--}176)$$

Some VT-relaxation rate coefficients at room temperature are shown in Table 2–19. Temperature dependencies of the rate coefficients are given in Table 2–20; most of them follow the Landau-Teller functionality.

A peculiarity of **VT relaxation of anharmonic oscillators** is due to a reduction of transition energy with vibrational quantum number n:

$$\omega_{n,n-1} = \omega(1 - 2x_e n). \quad (2\text{--}177)$$

where x_e is the coefficient of anharmonicity of an excited diatomic molecule. According to (2–171), an increase of vibrational quantum number n leads to reduction of the Massey parameter and makes the VT relaxation less adiabatic and accelerates the process exponentially:

$$P_{n+1,n}^{VT}(T_0) = (n+1)P_{1,0}^{VT}(T_0)\exp(\delta_{VT}n). \quad (2\text{--}178)$$

The temperature dependence for the probability of VT relaxation for anharmonic oscillators (2–178) is similar to that of harmonic oscillators (2–172), (2–173), and corresponds to the Landau-Teller formula. The exponential parameter δ_{VT} can be expressed for anharmonic oscillators as

$$\delta_{VT} = 4\gamma_n^{2/3}x_e, \quad \text{if } \gamma_n \geq 27. \quad (2\text{--}179)$$

$$\delta_{VT} = \frac{4}{3}\gamma_n x_e, \quad \text{if } \gamma_n < 27. \quad (2\text{--}180)$$

Table 2–20. Temperature Dependence of Vibrational VT-Relaxation Rate Coefficients $k_{VT}^{10}(T_0)$ for One-Component Gases

Molecule	Temperature dependence $k_{VT}^{10}(T_0)$, cm^3/s; temperature T_0 in kelvin
O_2	$10^{-10} \exp(-129 \cdot T_0^{-1/3})$
Cl_2	$2 \cdot 10^{-11} \exp(-58 \cdot T_0^{-1/3})$
Br_2	$2 \cdot 10^{-11} \exp(-48 \cdot T_0^{-1/3})$
J_2	$5 \cdot 10^{-12} \exp(-29 \cdot T_0^{-1/3})$
CO	$10^{-12} T_0 \exp(-190 \cdot T_0^{-1/3} + 1410 \cdot T_0^{-1})$
NO	$10^{-12} \exp(-14 \cdot T_0^{-1/3})$
HF	$5 \cdot 10^{-10} T_0^{-1} + 6 \cdot 10^{-20} T_0^{2.26}$
DF	$1.6 \cdot 10^{-5} T_0^{-3} + 3.3 \cdot 10^{-16} T_0$
HCl	$2.6 \cdot 10^{-7} T_0^{-3} + 1.4 \cdot 10^{-19} T_0^2$
F_2	$2 \cdot 10^{-11} \exp(-65 \cdot T_0^{-1/3})$
D_2	$10^{-12} \exp(-67 \cdot T_0^{-1/3})$
$CO_2(01^10)$	$10^{-11} \exp(-72 \cdot T_0^{-1/3})$

The adiabatic factor γ_n is the Massey parameter for vibrational relaxation transition $n + 1 \rightarrow n$:

$$\gamma_n(n + 1 \rightarrow n) = \frac{\pi(E_{n+1} - E_n)}{\hbar\alpha}\sqrt{\frac{\mu}{2T_0}}. \tag{2–181}$$

The Massey parameter γ_n can be calculated as (μ is in a.u., α in A^{-1}, $\hbar\omega$ and T_0 are in K):

$$\gamma_n = \frac{0.32}{\alpha}\sqrt{\frac{\mu}{T_0}}\hbar\omega(1 - 2x_e(n - 1)). \tag{2–182}$$

2.6.3. Fast Non-Adiabatic Mechanisms of VT Relaxation

Vibrational relaxation is slow in adiabatic collisions when there is no chemical interaction between colliding partners. For example, the probability of deactivation of a vibrationally excited N_2 molecule in collision with another N_2 molecule can be as low as 10^{-9} at room temperature. The vibrational relaxation process can be much faster in non-adiabatic collisions, when the colliding partners interact chemically.

   1. *VT relaxation in molecular collisions with atoms and radicals* can be illustrated by relaxation of vibrationally excited N_2 on atomic oxygen (Andreev & Nikitin, 1976; see Fig. 2–35). The interval between degenerated electronic terms grows when an atom approaches the molecule. When the energy interval becomes equal to a vibrational quantum, non-adiabatic relaxation (the so-called **vibronic transition**, Fig. 2–35) can take place. The temperature dependence of this relaxation is not significant, and typical rate coefficients are high: 10^{-13}–10^{-12} cm^3/s. Sometimes, as in the case of relaxation on alkaline atoms, the non-adiabatic VT-relaxation rate coefficients reach those of gas-kinetic collisions, that is, about 10^{-10} cm^3/s.

   2. *VT relaxation through intermediate formation of long-life complexes* takes place, in particular, in collisions such as H_2O^*–H_2O, CO_2^*–H_2O, CH_4^*–CO_2, CH_4^*–H_2O, $C_2H_6^*$–O_2, NO^*–Cl_2, N_2^*–Li, K, and Na. Relaxation rate coefficients also can reach gas-kinetic values in this case. The temperature dependence is not strong here and is usually negative. The probability of one-quantum and multi-quantum transfers of vibrational energy is relatively close in this case.

   3. *VT relaxation in symmetrical exchange reactions* such as

$$A' + (BA'')^*(n = 1) \rightarrow A'' + BA'(n = 0), \tag{2–183}$$

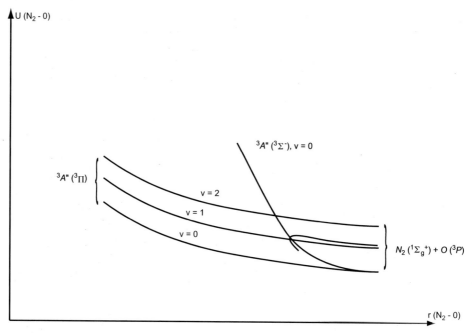

Figure 2–35. Non-adiabatic vibrational VT relaxation of nitrogen molecules (N_2) in collision with O atoms.

are very fast when relevant activation energies are low. An important example here is the barrierless oxygen exchange reaction: $O_2^* + O \rightarrow O_3^* \rightarrow O + O_2$. The relaxation rate coefficient here is about $10^{-11}\,\mathrm{cm^3/s}$ and does not depend on temperature. A similar mechanism takes place in the relaxation of H_2 molecules on H atoms, different halogen molecules on corresponding halogen atoms, and hydrogen halides on the hydrogen atoms.

4). *Heterogeneous VT relaxation* is non-adiabatic and therefore is a fast process usually proceeding through an adsorption phase (Gershenson, Rosenstein, & Umansky, 1977). Relevant losses of vibrational energy are described in terms of **accommodation coefficients**, which describe the probability of the vibrational relaxation calculated with respect to one surface collision (see Table 2–21).

Specific fast mechanisms of VT relaxation take place in collisions of polyatomic molecules and are discussed by Kennedy & Fridman (2004), and in more detail by Rusanov, Fridman, and Sholin (1986).

2.6.4. Vibrational Energy Transfer Between Molecules: Resonant VV Relaxation

The kinetics of plasma-chemical reactions of vibrationally excited molecules is determined not only by their concentration but mostly by the fraction of highly excited molecules able to dissociate or participate in endothermic chemical reactions. The formation of highly vibrationally excited molecules at elevated pressures is due not to direct electron impact but to collisional energy exchange called VV relaxation. Most conventional resonant VV processes usually imply vibrational energy exchange between molecules of the same kind, for example, $N_2^*(v = 1) + N_2(v = 0) \rightarrow N_2(v = 0) + N_2^*(v = 1)$, and are characterized by a probability $q_{mn}^{sl}(v)$ of a transition, when one oscillator changes its vibrational quantum number from s to l, and the other from m to n. Quantum mechanics gives a

2.6. Elementary Relaxation Processes of Energy Transfer

Table 2–21. Accommodation Coefficients for Heterogeneous VT Relaxation

Molecule	Mode	$\hbar\omega$, cm^{-1}	T_0, K	Surface	Accommodation coefficient
CO_2	$v_2 = 1$	667	277–373	Platinum	0.3–0.4
CO_2	$v_2 = 1$	667	297	NaCl	0.22
CO_2	$v_3 = 1$	2349	300–350	Pyrex	0.2–0.4
CO_2	$v_3 = 1$	2349	300	Brass, Teflon, Mylar	0.2
CO_2	$v_3 = 1$	2349	300–1000	Quartz	0.05–0.45
CO_2	$v_3 = 1$	2349	300–560	Molybdenum glass	0.3–0.4
CH_4	$v_4 = 1$	1306	273–373	Platinum	0.5–0.9
H_2	$v = 1$	4160	300	Quartz	$5 \cdot 10^{-4}$
H_2	$v = 1$	4160	300	Molybdenum glass	$1.3 \cdot 10^{-4}$
H_2	$v = 1$	4160	300	Pyrex	10^{-4}
D_2	$v = 1$	2990	77–275	Molybdenum glass	$(0.2–8) \cdot 10^{-4}$
D_2	$v = 1$	2990	300	Quartz	10^{-4}
N_2	$v = 1$	2331	350	Pyrex	$5 \cdot 10^{-4}$
N_2	$v = 1$	2331	282–603	Molybdenum glass	$(1–3) \cdot 10^{-3}$
N_2	$v = 1$	2331	350	Pyrex	$5 \cdot 10^{-4}$
N_2	$v = 1$	2331	300	Steel, aluminum, copper	$3 \cdot 10^{-3}$
N_2	$v = 1$	2331	300	Teflon, alumina	10^{-3}
N_2	$v = 1$	2331	295	Silver	$1.4 \cdot 10^{-2}$
CO	$v = 1$	2143	300	Pyrex	$1.9 \cdot 10^{-2}$
HF	$v = 1$	3962	300	Molybdenum glass	10^{-2}
HCl	$v = 1$	2886	300	Pyrex	0.45
OH	$v = 9$	–	300	Boron acid	1
N_2O	$v_2 = 1$	589	273–373	Platinum, NaCl	0.3
N_2O	$v_3 = 1$	2224	300–350	Pyrex	0.2
N_2O	$v_3 = 1$	2224	300–1000	Quartz	0.05–0.33
N_2O	$v_3 = 1$	2224	300–560	Molybdenum glass	0.01–0.03
CF_3Cl	$v_3 = 1$	732	273–373	Platinum	0.5–0.6

probability $q_{mn}^{sl}(v)$ similar to that for VT relaxation (2–170), by replacing ω_{mn} with a frequency corresponding to a change of vibrational energy during the collision:

$$\omega_{ms,nl} = \frac{1}{\hbar}(E_m + E_s - E_l - E_n) = \frac{\Delta E}{\hbar}. \tag{2–184}$$

The probability also depends on matrix elements of transitions $m \to n$ and $s \to l$:

$$q_{mn}^{sl}(v) = \frac{\langle m|y_1|n\rangle^2 \langle s|y_2|l\rangle^2}{\hbar^2} \left| \int_{-\infty}^{+\infty} F(t)\exp(i\omega_{ms,nl}t)\,dt \right|^2. \tag{2–185}$$

The Fourier component of interaction force $F(t)$ on the transition frequency (2–184) characterizes the level of resonance. Matrix elements for harmonic oscillators $\langle m|y|n\rangle$ are non-zero only for one-quantum transitions, $n = m \pm 1$. The VV relaxation probability of the one-quantum exchange as a function of translational temperature T_0 can be found by averaging the probability $q_{mn}^{sl}(v)$ over Maxwellian distribution:

$$Q_{n+1,n}^{m,m+1}(T_0) = (m+1)(n+1)Q_{10}^{01}(T_0). \tag{2–186}$$

Table 2–22. Resonant One-Quantum VV Relaxation Rate Coefficients at Room Temperature and Their Ratio to Those of Gas-Kinetic Collisions (k_{VV}/k_0)

Molecule	k_{VV}, cm^3/s	k_{VV}/k_0	Molecule	k_{VV}, cm^3/s	k_{VV}/k_0
$CO_2(001)$	$5 \cdot 10^{-10}$	4	HF	$3 \cdot 10^{-11}$	0.5
CO	$3 \cdot 10^{-11}$	0.5	HCl	$2 \cdot 10^{-11}$	0.2
N_2	10^{-13}–10^{-12}	10^{-3}–10^{-2}	HBr	10^{-11}	0.1
H_2	10^{-13}	10^{-3}	DF	$3 \cdot 10^{-11}$	–
$N_2O(001)$	$3 \cdot 10^{-10}$	3	HJ	$2 \cdot 10^{-12}$	–

Higher powers of intermolecular interaction potential permit multi-quantum VV exchange between harmonic oscillators but obviously with lower probability (Nikitin, 1970):

$$Q_{0,k}^{m,m-k} \approx \frac{m!}{(m-k)!k!2^{k-1}}(\Omega\tau_{col})^{2k}.$$

(2–187)

In this relation, τ_{col} is the time of a collision, and Ω is the transition frequency during the collision. The factor $\Omega\tau_{col}$ is small because the vibrational amplitude is small with respect to intermolecular interaction radius. VV relaxation for most of the molecules is due to the exchange interaction (short-distance forces), which results in the short time of collision, τ_{col}, and $\Omega\tau_{col} \approx 0.1$–$0.01$. The probability of a single quantum transfer is $Q_{10}^{01} \approx (\Omega\tau_{col})^2 \approx 10^{-2}$–$10^{-4}$; thus, the probability of multi-quantum exchange in this case is very low (for example, it is about 10^{-9} even for resonant three-quantum exchange). We should note that, for some molecules like CO_2 and N_2O, VV relaxation is due to dipole or multi-pole interaction (long-distance forces). Collision times are much longer in this case, which results in $Q_{10}^{01} \approx (\Omega\tau_{col})^2 \approx 1$ and a higher probability of VV relaxation including multi-quantum exchange.

The temperature dependence of VV-relaxation probability $Q_{10}^{01} \approx (\Omega\tau_{col})^2$ is different for dipole and exchange interactions. The transition frequency Ω is proportional to the average interaction energy; therefore, for short-distance interaction ΩT_0, $\tau_{col} \propto \frac{1}{v} \propto T_0^{1/2}$, and $Q_{10}^{01} \approx (\Omega\tau_{col})^2 \propto T_0$. Thus, the probability of VV relaxation provided by short-distance forces is proportional to the temperature. Conversely, in the case of long-distance dipole and multi-pole interactions, the VV-relaxation probability decreases with T_0. The transition frequency Ω does not depend on temperature in this case and, hence, $Q_{10}^{01} \approx (\Omega\tau_{col})^2 \propto \tau_{col}^2 \propto \frac{1}{T_0}$.

Some resonant VV-relaxation rate coefficients are presented in Table 2–22 at room temperature for one-quantum exchange. Resonant VV exchange is usually much faster at room temperature than VT relaxation, which promotes the population of highly vibrationally excited and reactive molecules.

2.6.5. Non-Resonant VV Exchange: Relaxation of Anharmonic Oscillators and Intermolecular VV Relaxation

VV exchange of anharmonic oscillators is non-resonant, slightly adiabatic, and, as a result, slower than the resonant process:

$$Q_{n+1,n}^{m,m+1} = (m+1)(n+1)Q_{10}^{01}\exp(-\delta_{VV}|n-m|)\left[\frac{2}{3} - \frac{1}{2}\exp(-\delta_{VV}|n-m|)\right].$$

(2–188)

The VV transition probability includes an exponential adiabatic factor (Nikitin, 1970):

$$\delta_{VV} = \frac{4}{3}x_e\gamma_0 = \frac{4}{3}\frac{\pi\omega x_e}{\alpha}\sqrt{\frac{\mu}{2T_0}}.$$

(2–189)

2.6. Elementary Relaxation Processes of Energy Transfer

In one-component gases, $\delta_{VV} = \delta_{VT}$; see relation (2–179). For numerical calculations,

$$\delta_{VV} = \frac{0.427}{\alpha}\sqrt{\frac{\mu}{T_0}}x_e\hbar\omega. \tag{2–190}$$

In this relation, the reduced mass of colliding molecules, μ, is in atomic units, the inverse intermolecular interaction radius α is in Å^{-1}, and the translational gas temperature T_0 and vibrational quantum $\hbar\omega$ are in kelvin.

It is interesting to compare rate coefficients of VV-exchange and VT-relaxation processes taking into account anharmonicity. The effect of anharmonicity on VV exchange is negligible in resonant collisions, $|m - n|\delta_{VV} \ll 1$, but essential in collisions of highly excited molecules with molecules on lower vibrational levels. The rate coefficient of the VV-exchange process, $k_{VV}(n) = k_0 Q_{n+1,n}^{0,1}$, decreases with vibrational quantum number n in contrast to the VT-relaxation rate coefficient, $k_{VT}(n) = k_0 P_{n+1,n}^{VT}$, which increases with n ($k_0 \approx 10^{-10}$ cm^3/s is the rate coefficient of gas-kinetic collisions):

$$\xi(n) = \frac{k_{VT}(n)}{k_{VV}(n)} \approx \frac{P_{10}^{VT}}{Q_{10}^{01}}\exp[(\delta_{VV} + \delta_{VT})n]. \tag{2–191}$$

where $\xi \ll 1$ at low excitation levels: VT relaxation is much slower than VV exchange because $\frac{P_{10}^{VT}}{Q_{10}^{01}} \ll 1$. It means that the population of highly vibrationally excited and chemically active molecular states can be much faster in non-thermal plasma than losses of vibrational energy. The ratio $\xi(n)$, however, grows exponentially with n and VT relaxation catches up with VV exchange ($\xi = 1$) at some critical value of the vibrational quantum number and vibrational energy $E^*(T_0)$ (Rusanov, Fridman, & Sholin, 1979):

$$E^*(T_0) = \hbar\omega\left(\frac{1}{4x_e} - b\sqrt{T_0}\right). \tag{2–192}$$

The parameter b depends on the gas: for CO, $b = 0.90\,\text{K}^{-0.5}$; for N$_2$, $b = 0.90\,\text{K}^{-0.5}$; for HCl, $b = 0.90\,\text{K}^{-0.5}$.

Non-resonant vibrational energy exchange between molecules of a different kind is usually referred to as VV$'$ exchange. Let us first consider the VV$'$ exchange for diatomic molecules A and B with slightly different vibrational quanta $\hbar\omega_A > \hbar\omega_B$. Similar to VV exchange of anharmonic molecules of the same kind, the adiabatic factors here also mainly determine the smallness of the probability. If a molecule A transfers a quantum ($v_A + 1 \to v_A$) to a molecule B ($v_B + 1 \to v_B$),

$$Q_{v_A+1,v_A}^{v_B,v_B+1} = (v_A + 1)(v_B + 1)Q_{10}^{01}(AB)\exp(-|\delta_B v_B - \delta_A v_A + \delta_A p|)\exp(\delta_A p). \tag{2–193}$$

In this equation $Q_{10}^{01}(AB)$ is the probability of a quantum transfer from A to B for the lowest levels. The adiabatic parameters δ_A and δ_B can be found from (2–189) and (2–190), assuming for each molecule a separate coefficient of anharmonicity: x_{eA} and x_{eB}. The parameter $p > 0$ is a vibrational level of the oscillator A, corresponding to the exact resonant transition A($p + 1 \to p$) – B($0 \to 1$) of a quantum from molecule A to B:

$$p = \frac{\hbar(\omega_A - \omega_B)}{2x_{eA}\hbar\omega_A}. \tag{2–194}$$

The VV$'$ probability (2–193) can be expressed through the factor $(\Omega\tau_{col})^2 \approx 10^{-2}$–$10^{-4}$:

$$Q_{v_A+1,v_A}^{v_B,v_B+1} = (v_A + 1)(v_B + 1)(\Omega\tau_{col})^2\exp(-|\delta_B v_B - \delta_A v_A + \delta_A p|). \tag{2–195}$$

Formula (2–195) leads to a simple expression for the probability of a single-quantum VV' exchange,

$$Q_{10}^{01}(AB) = (\Omega \tau_{col})^2 \exp(-\delta p), \qquad (2\text{–}196)$$

and to a related numerical formula (Rapp, 1965) with defect of resonance $\hbar \Delta \omega$ in cm^{-1}, and T_0 in kelvin:

$$Q_{10}^{01}(AB) = 3.7 \cdot 10^{-6} T_0 \, ch^{-2}(0.174 \Delta \omega \hbar / \sqrt{T_0}). \qquad (2\text{–}197)$$

The probability of VV' exchange decreases with $\hbar \Delta \omega$. However, when the defect of resonance is big ($\hbar \Delta \omega \approx \hbar \omega$), multi-quantum resonant VV'-exchange processes become important (Nikitin, 1970):

$$Q_{n,n-s}^{m,m+r} = \frac{1}{r!s!} \frac{n!(m+r)!}{(n-s)!m!} Q_{s0}^{0r}. \qquad (2\text{–}198)$$

The probability of the multi-quantum resonant exchange Q_{s0}^{0r} can be estimated as

$$Q_{s0}^{0r} \propto (\Omega \tau_{col})^{r+s}, \quad Q_{n,n-s}^{m,m+r} = \frac{1}{r!s!} \frac{n!(m+r)!}{(n-s)!m!} (\Omega \tau_{col})^{r+s}. \qquad (2\text{–}199)$$

Some rate coefficients of non-resonant VV' exchange are presented in Table 2–23. VV relaxation of polyatomic molecules is discussed by Kennedy and Fridman (2004) and Rusanov, Fridman, and Sholin (1986).

2.6.6. Rotational Relaxation Processes: Parker Formula

Rotational–rotational (RR) and rotational–translational (RT) energy transfer processes are usually non-adiabatic and fast, because rotational quanta and, therefore, the Massey parameter are small. As a result, the collision of a rotator with an atom or another rotator can be considered a classical collision accompanied by essential energy transfer. The Parker formula for calculation of number of collisions, Z_{rot}, required for RT relaxation was proposed by Parker (1959) and Bray and Jonkman (1970):

$$Z_{rot} = \frac{Z_{rot}^{\infty}}{1 + \left(\frac{\pi}{2}\right)^{\frac{3}{2}} \sqrt{\frac{T^*}{T_0}} + \left(2 + \frac{\pi^2}{4}\right) \frac{T^*}{T_0}} \qquad (2\text{–}200)$$

Where T^* is the energy of molecular interaction; $Z_{rot}^{\infty} = Z_{rot}(T_0 \to \infty)$; and T_0 is the gas temperature. The growth of Z_{rot} with temperature is due to intermolecular attraction, which accelerates the RT energy exchange and becomes less effective at higher temperatures. Numerical values of Z_{rot} are given in Table 2–24 for room temperature and for high temperatures: $Z_{rot}^{\infty} = Z_{rot}(T_0 \to \infty)$.

Factors Z_{rot} are low at room temperature: values of about 3–10 for most molecules except hydrogen and deuterium, where Z_{rot} is about 200–500 due to high rotational quanta and Massey parameters. RT and RR relaxations are fast processes requiring only several collisions, and their kinetic rates are close to those of thermalization. Rotational degrees of freedom, therefore, are often considered in quasi-equilibrium with translational degrees of freedom and characterized by the same temperature T_0 even in strongly non-equilibrium systems.

2.6.7. Relaxation of Electronically Excited Atoms and Molecules

Relaxation of electronically excited atoms and molecules is due to different mechanisms. Superelastic collisions (energy transfer back to plasma electrons) and radiation are essential mostly in thermal plasma. Relaxation in collision with other heavy particles dominates in non-thermal discharges. **Relaxation of electronic excitation into translational degrees**

Table 2–23. Rate Coefficients of Non-Resonant VV' Relaxation at Room Temperature

VV' exchange process	k_{VV}, cm³/s	VV' exchange process	k_{VV}, cm³/s
$H_2(v=0) + HF(v=1) \rightarrow H_2(v=1) + HF(v=0)$	$3 \cdot 10^{-12}$	$O_2(v=0) + CO(v=1) \rightarrow O_2(v=1) + CO(v=0)$	$4 \cdot 10^{-13}$
$DCl(v=0) + N_2(v=1) \rightarrow DCl(v=1) + N_2(v=0)$	$5 \cdot 10^{-14}$	$N_2(v=0) + CO(v=1) \rightarrow N_2(v=1) + CO(v=0)$	10^{-15}
$N_2(v=0) + HCl(v=1) \rightarrow N_2(v=1) + HCl(v=0)$	$3 \cdot 10^{-14}$	$N_2(v=0) + HF(v=1) \rightarrow N_2(v=1) + HF(v=0)$	$5 \cdot 10^{-15}$
$NO(v=0) + CO(v=1) \rightarrow NO(v=1) + CO(v=0)$	$3 \cdot 10^{-14}$	$O_2(v=0) + COl(v=1) \rightarrow O_2(v=1) + CO(v=0)$	10^{-16}
$DCl(v=0) + CO(v=1) \rightarrow DCl(v=1) + CO(v=0)$	10^{-13}	$CO(v=0) + D_2(v=1) \rightarrow CO(v=1) + D_2(v=0)$	$3 \cdot 10^{-14}$
$CO(v=0) + H_2(v=1) \rightarrow CO(v=1) + H_2(v=0)$	10^{-16}	$NO(v=0) + CO(v=1) \rightarrow NO(v=1) + CO(v=0)$	10^{-14}
$HCl(v=0) + CO(v=1) \rightarrow HCl(v=1) + CO(v=0)$	10^{-12}	$HBr(v=0) + CO(v=1) \rightarrow HBr(v=1) + CO(v=0)$	10^{-13}
$HJ(v=0) + CO(v=1) \rightarrow HJ(v=1) + CO(v=0)$	10^{-14}	$HBr(v=0) + HCl(v=1) \rightarrow HBr(v=1) + HCl(v=0)$	10^{-12}
$HJ(v=0) + HCl(v=1) \rightarrow HJ(v=1) + HCl(v=0)$	$2 \cdot 10^{-13}$	$DCl(v=0) + HCl(v=1) \rightarrow DCl(v=1) + HCl(v=0)$	10^{-13}
$O_2(v=0) + CO_2(01^10) \rightarrow O_2(v=1) + CO_2(00^00)$	$3 \cdot 10^{-15}$	$H_2O(000) + N_2(v=1) \rightarrow H_2O(010) + N_2(v=0)$	10^{-15}
$CO_2(00^00) + N_2(v=1) \rightarrow CO_2(00^01) + N_2(v=0)$	10^{-12}	$CO(v=1) + CH_4 \rightarrow CO(v=1) + CH_4^*$	10^{-14}
$CO(v=1) + CF_4 \rightarrow CO(v=1) + CF_4^*$	$2 \cdot 10^{-16}$	$CO(v=1) + SF_6 \rightarrow CO(v=1) + SF_6^*$	10^{-15}
$CO(v=1) + SO_2 \rightarrow CO(v=1) + SO_2^*$	10^{-15}	$CO(v=1) + CO_2 \rightarrow CO(v=1) + CO_2^*$	$3 \cdot 10^{-13}$
$CO_2(00^00) + HCl(v=1) \rightarrow CO_2(00^01) + HCl(v=0)$	$3 \cdot 10^{-13}$	$CO_2(00^00) + HF(v=1) \rightarrow CO_2(00^01) + HF(v=0)$	10^{-12}
$CO_2(00^00) + DF(v=1) \rightarrow CO_2(00^01) + DF(v=0)$	$3 \cdot 10^{-13}$	$CO(v=0) + CO_2(00^01) \rightarrow CO(v=1) + CO_2(00^00)$	$3 \cdot 10^{-15}$
$H_2O^* + CO_2 \rightarrow H_2O + CO_2^*$	10^{-12}	$O_2(v=0) + CO_2(10^00) \rightarrow O_2(v=1) + CO_2(00^00)$	10^{-13}
$CS_2(00^00) + CO(v=1) \rightarrow CS_2(00^01) + CO(v=0)$	$3 \cdot 10^{-13}$	$N_2O(00^00) + CO(v=1) \rightarrow N_2O(00^01) + CO(v=0)$	$3 \cdot 10^{-12}$

Table 2–24. Number of Collisions Z_{rot} Necessary for RT Relaxation in One-Component Gases

Molecule	$Z_{rot}^{\infty}(T_0 \to \infty)$	$Z_{rot}(T_0 = 300\,\text{K})$	Molecule	$Z_{rot}^{\infty}(T_0 \to \infty)$	$Z_{rot}(T_0 = 300\,\text{K})$
Cl_2	47.1	4.9	N_2	15.7	4.0
O_2	14.4	3.45	H_2	–	~500
D_2	–	~200	CH_4	–	15
CD_4	–	12	$C(CH_3)_4$	–	7
SF_6	–	7	CCl_4	–	6
CF_4	–	6	$SiBr_4$	–	5
SiH_4	–	28			

of freedom (several electron volts) is a strongly adiabatic process with very high Massey parameters ($\omega\tau_{col} \sim 100$–1000). The adiabatic relaxation is very slow. For example,

$$Na(3^2P) + Ar \to Na + Ar \qquad (2\text{–}201)$$

is characterized by a cross section $<10^{-19}\,\text{cm}^2$, which corresponds to probability 10^{-9} or less. We should note that some specific relaxation processes of the kind can be fast; for example, $O(^1D)$ relaxation on atoms of noble gases can proceed through an intermediate complex and requires only several collisions to occur.

Electronically excited atoms and molecules transfer energy not only into translational, but also into **vibrational and rotational degrees of freedom**, which is less adiabatic and faster. Fast relaxation takes place by formation of intermediate ionic complexes. For example, electronic energy transfer from excited metal atoms Me^* to vibrational excitation of nitrogen takes place for almost every collision:

$$Me^* + N_2(v = 0) \to Me^+N_2^- \to Me + N_2(v > 0). \qquad (2\text{–}202)$$

Initial and final Me–N_2 energy terms cross an ionic term (Fig. 2–36), which leads to a fast non-adiabatic relaxation transition (Bjerre & Nikitin, 1967). The maximum cross sections of such relaxation processes are presented in Table 2–25.

Electronic excitation energy transfer processes can be effective only very close to a resonance (0.1 eV or less), which limits them to some specific collision partners. The He–Ne gas laser provides examples:

$$He(^1S) + Ne \to He + Ne(5s), \qquad (2\text{–}203)$$

$$He(^3S) + Ne \to He + Ne(4s). \qquad (2\text{–}204)$$

The interaction radius for collisions of electronically excited atoms and molecules is usually high, up to 1 nm. Cross sections, therefore, can obviously be very high if processes are close to resonance. For example, electronic excitation transfer from Hg atoms to sodium has a cross section reaching $10^{-14}\,\text{cm}^2$.

Table 2–25. Maximum Cross Sections of Non-Adiabatic Relaxation of Electronically Excited Sodium Atoms

Relaxation processes	Cross sections, cm^2
$Na^* + N_2(v = 0) \to Na + N_2^*(v > 0)$	$2 \cdot 10^{-14}\,\text{cm}^2$
$Na^* + CO_2(v = 0) \to Na + CO_2^*(v > 0)$	$10^{-14}\,\text{cm}^2$
$Na^* + Br_2(v = 0) \to Na + Br_2^*(v > 0)$	$10^{-13}\,\text{cm}^2$

2.7. Elementary Chemical Reactions of Excited Molecules

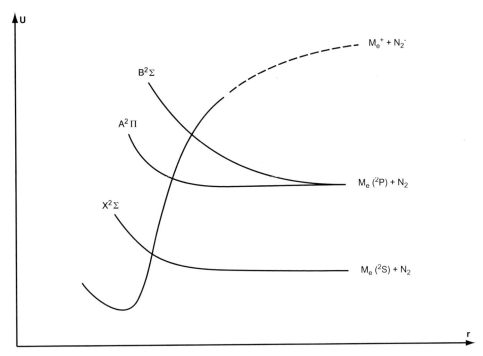

Figure 2–36. Electronic energy relaxation through formation of intermediate ionic complexes; example of relaxation of an alkaline atom $M_e(^2P) \rightarrow M_e(^2S)$ by collision with molecular nitrogen N_2 and vibrational excitation.

When the plasma ionization degree exceeds 10^{-6}, the electronic excitation transfer can be effectively provided via interaction with an electron gas. These relaxation processes are due to superelastic collisions and are conventional for thermal plasma conditions.

2.7. ELEMENTARY CHEMICAL REACTIONS OF EXCITED MOLECULES: FRIDMAN-MACHERET α-MODEL

2.7.1. Rate Coefficient of Reactions of Excited Molecules

The high rates of plasma-chemical reactions are often due to a high concentration of excited atoms and molecules in electric discharges. Vibrational and electronic excitations play the most important role in the stimulation of endothermic processes in plasma. We next focus on reactions of vibrationally excited molecules, which are easier to analyze. The kinetic relations, however, can also be applied to some extent to reactions of electronically excited particles.

A convenient formula for the calculation of rate coefficients of elementary reactions of an excited molecule with vibrational energy E_v at translational gas temperature T_0 can be expressed in the framework of the **theoretical-informational approach** (Levine & Bernstein, 1975):

$$k_R(E_v, T_0) = k_{R0} \exp\left(-\frac{E_a - \alpha E_v}{T_0}\right) \theta(E_a - \alpha E_v). \tag{2–205}$$

In this relation, E_a is the Arrhenius activation energy of an elementary chemical reaction; the coefficient α is the **efficiency of excitation energy in overcoming the activation barrier**; k_{R0} is the pre-exponential factor; and $\theta(x - x_0)$ is the Heaviside function

Table 2–26. Efficiency α of Vibrational Energy in Overcoming Activation Energy Barrier

Reaction	α_{exp}	α_{MF}	Reaction	α_{exp}	α_{MF}
$F + HF^* \rightarrow F_2 + H$	0.98	0.98	$F + DF^* \rightarrow F_2 + D$	0.99	0.98
$Cl + HCl^* \rightarrow H + Cl_2$	0.95	0.96	$Cl + DCl^* \rightarrow D + Cl_2$	0.99	0.96
$Br + HBr^* \rightarrow H + Br_2$	1.0	0.98	$F + HCl^* \rightarrow H + ClF$	0.99	0.96
$Cl + HF^* \rightarrow ClF + H$	1.0	0.98	$Br + HF^* \rightarrow BrF + H$	1.0	0.98
$J + HF^* \rightarrow JF + H$	1.0	0.98	$Br + HCl^* \rightarrow BrCl + H$	0.98	0.98
$Cl + HBr^* \rightarrow BrCl + H$	1.0	0.97	$J + HCl^* \rightarrow JCl + H$	1.0	0.98
$SCl + HCl^* \rightarrow SCl_2 + H$	0.96	0.98	$S_2Cl + HCl^* \rightarrow S_2Cl_2 + H$	0.98	0.97
$SOCl + HCl^* \rightarrow SOCl_2 + H$	0.95	0.96	$SO_2Cl + HCl^* \rightarrow SO_2Cl_2 + H$	1.0	0.96
$NO + HCl^* \rightarrow NOCl + H$	1.0	0.98	$FO + HF^* \rightarrow F_2O + H$	1.0	0.98
$O_2 + OH^* \rightarrow O_3 + H$	1.0	1.0	$NO + OH^* \rightarrow NO_2 + H$	1.0	1.0
$ClO + OH^* \rightarrow ClO_2 + H$	1.0	1.0	$CrO_2Cl + HCl^* \rightarrow CrO_2Cl_2 + H$	0.94	0.9
$PBr_2 + HBr^* \rightarrow PBr_3 + H$	1.0	0.97	$SF_5 + HBr^* \rightarrow SF_5Br + H$	1.0	0.98
$SF_3 + HF^* \rightarrow SF_4 + H$	0.89	0.98	$SF_4 + HF^* \rightarrow SF_5 + H$	0.97	0.99
$H + HF^* \rightarrow H_2 + F$	1.0	0.95	$D + HF^* \rightarrow HD + F$	1.0	0.95
$Cl + HF^* \rightarrow HCl + F$	0.96	0.97	$Br + HF^* \rightarrow HBr + F$	1.0	0.98
$J + HF^* \rightarrow HJ + F$	0.99	0.99	$Br + HCl^* \rightarrow HBr + Cl$	1.0	0.95
$J + HCl^* \rightarrow HJ + Cl$	1.0	0.98	$J + HBr^* \rightarrow HJ + Br$	1.0	0.96
$OH + HF^* \rightarrow H_2O + F$	1.0	0.95	$HS + HF^* \rightarrow H_2S + F$	1.0	0.98
$HS + HCl^* \rightarrow H_2S + Cl$	1.0	1.0	$HO_2 + HF^* \rightarrow H_2O_2 + F$	1.0	0.98
$HO_2 + HF^* \rightarrow H_2O_2 + F$	1.0	0.98	$NH_2 + HF^* \rightarrow NH_3 + F$	1.0	0.97
$SiH_3 + HF^* \rightarrow SiH_4 + F$	1.0	1.0	$GeH_3 + HF^* \rightarrow GeH_4 + F$	1.0	1.0
$N_2H_3 + HF^* \rightarrow N_2H_4 + F$	0.97	0.98	$CH_3 + HF^* \rightarrow CH_4 + F$	0.98	0.97
$CH_2F + HF^* \rightarrow CH_3F + F$	1.0	0.97	$CH_2Cl + HF^* \rightarrow CH_3Cl + F$	1.0	0.97
$CCl_3 + HF^* \rightarrow CHCl_3 + F$	0.98	0.98	$CH_2Br + HF \rightarrow CH_3Br + F$	1.0	0.97
$C_2H_5 + HF^* \rightarrow C_2H_6 + F$	1.0	0.99	$CH_2CF_3 + HF^* \rightarrow CH_3CF_3 + F$	1.0	0.98
$C_2H_5O + HF^* \rightarrow (CH_3)_2O + F$	1.0	1.0	$C_2H_5Hg + HF^* \rightarrow (CH_3)_2Hg + F$	0.99	0.97
$HCO + HF^* \rightarrow H2CO + F$	1.0	0.99	$FCO + HF^* \rightarrow HFCO + F$	1.0	0.99
$C_2H_3 + HF^* \rightarrow C_2H_4 + F$	0.99	0.97	$C_3H_5 + HF^* \rightarrow C_3H_6 + F$	1.0	0.98
$C_6H_5 + HF^* \rightarrow C_6H_6 + F$	1.0	0.97	$CH_3 + JF^* \rightarrow CH_3J + F$	0.81	0.9
$S + CO^* \rightarrow CS + O$	1.0	0.99	$F + CO^* \rightarrow CF + O$	1.0	1.0
$CS + CO^* \rightarrow CS_2 + O$	0.83	0.9	$CS + SO^* \rightarrow CS_2 + O$	0.90	0.95
$SO + CO^* \rightarrow COS + O$	0.96	0.92	$F_2 + CO^* \rightarrow CF_2 + O$	1.0	1.0
$C_2H_4 + CO^* \rightarrow (CH_2)_2C + O$	0.94	0.98	$CH_2 + CO^* \rightarrow C_2H_2 + O$	0.90	0.94
$H_2 + CO^* \rightarrow CH_2 + O$	0.96	–	$OH + OH^* \rightarrow H_2O + O(^1D_2)$	1.0	0.97
$NO + NO^* \rightarrow N_2O + O(^1D_2)$	1.0	–	$CH_3 + OH^* \rightarrow CH_4 + O(^1D_2)$	1.0	1.0
$O + NO^* \rightarrow O2 + N$	0.94	0.86	$H + BaF^* \rightarrow HF + Ba$	0.99	1.0
$O + AlO^* \rightarrow Al + O_2$	0.67	0.7	$O + BaO^* \rightarrow Ba + O_2$	1.0	0.99
$CO + HF^* \rightarrow CHF + O$	1.0	1.0	$CO_2 + HF^* \rightarrow O_2 + CHF$	1.0	–
$CO^* + HF \rightarrow CHF + O$	1.0	1.0	$CF_2O + HF^* \rightarrow CHF_3 + O(^1D_2)$	1.0	1.0
$H_2CO + HF^* \rightarrow CH_3F + O(^1D_2)$	1.0	1.0	$J + HJ^* \rightarrow J_2 + H$	1.0	1.0
$O + N_2 \rightarrow NO + N$ (non-adiabatic)	1.0	1.0	$O + N_2 \rightarrow NO + N$ (adiabatic)	0.6	1.0
$F + ClF^* \rightarrow F_2 + Cl$	1.0	0.93	$O + (CO^+)^* \rightarrow O_2 + C^+$	0.9	1.0
$H + HCl^* \rightarrow H_2 + Cl$	0.3	0.4	$NO + O_2^* \rightarrow NO_2 + O$	0.9	1.0
$O + H_2^* \rightarrow OH + H$	0.3	0.5	$O + HCl^* \rightarrow OH + Cl$	0.60	0.54
$H + HCl^* \rightarrow HCl + H$	0.3	0.5	$H + H_2^* \rightarrow H_2 + H$	0.4	0.5
$NO + O_3^* \rightarrow NO_2(^3B_2) + O_2$	0.5	0.3	$SO + O_3^* \rightarrow SO_2(^1B_1) + O_2$	0.25	0.15
$O^+ + N_2^* \rightarrow NO^+ + N$	0.1	–	$N + O_2^* \rightarrow NO + O$	0.24	0.19
$OH + H_2^* \rightarrow H_2O + H$	0.24	0.22	$F + HCl^* \rightarrow HF + Cl$	0.4	0.1
$H + N_2O^* \rightarrow OH + N_2$	0.4	0.2	$O_3 + OH^* \rightarrow O_2 + O + OH$	0.02	0
$H_2 + OH^* \rightarrow H_2O + H$	0.03	0	$Cl_2 + NO^* \rightarrow ClNO + Cl$	0	0
$O_3 + NO^* \rightarrow NO_2(^2A_1) + O_2$	0	0	$O_3 + OH^* \rightarrow HO_2 + O_2$	0.02	0

Note: α_{exp}, coefficient obtained experimentally, α_{MF}, coefficient calculated from Fridman-Macheret α-model.

($\theta(x - x_0) = 1$ when $x > 0$; and $\theta(x - x_0) = 0$ when $x < 0$). According to relation (2–206), the reaction rate coefficients of vibrationally excited molecules follow the traditional Arrhenius law with an activation energy reduced on the value of vibrational energy taken with efficiency α. If the vibrational temperature exceeds the translational one, $T_v \gg T_0$, and the chemical reaction is mostly determined by the vibrationally excited molecules, then relation (2–206) can be simplified:

$$k_R(E_v) = k_{R0}\, \theta(\alpha v - E_a). \qquad (2\text{–}206)$$

The effective activation energy in this case is E_a/α. Relation (2–205) describes rate coefficients after averaging over the Maxwellian distribution function for translational degrees of freedom. Probabilities for reactions of excited molecules without averaging can be found based on the **Le Roy formula** (Le Roy, 1969), which gives the probability $P_v(E_v, E_t)$ of elementary reaction as a function of vibrational and translational energies (E_v and E_t, respectively):

$$P_v(E_v, E_t) = 0, \quad \text{if} \quad E_t < E_a - \alpha E_v,\ E_a \geq \alpha E_v, \qquad (2\text{–}207)$$

$$P_v(E_v, E_t) = 1 - \frac{E_a - \alpha E_v}{E_t}, \quad \text{if} \quad E_t > E_a - \alpha E_v,\ E_a \geq \alpha E_v, \qquad (2\text{–}208)$$

$$P_v(E_v, E_t) = 1, \quad \text{if} \quad E_a < \alpha E_v, \quad \text{any } E_t \qquad (2\text{–}209)$$

Averaging of $P_v(E_v, E_t)$ over the Maxwellian distribution gives the formula (2–206), which is actually the most important relation for kinetic calculations of elementary reactions of excited particles.

2.7.2. Efficiency α of Vibrational Energy in Overcoming Activation Energy of Chemical Reactions: Numerical Values and Classification Table

The coefficient α is a key parameter describing the influence of plasma excitation of molecules on their chemical reaction rates (2–205). Numeric values of the α-coefficients for some specific chemical reactions are presented in Table 2–26 (Levitsky, Macheret, & Fridman, 1983; Levitsky et al., 1983c; Rusanov & Fridman, 1984).

Table 2–26 permits one to classify chemical reactions into groups with specific probable values of the α-coefficients in each class. Such a classification (Levitsky, Macheret, & Fridman, 1983) is presented in Table 2–27. Reactions are divided in this table into endothermic, exothermic, and thermoneutral categories; and into simple- and double-exchange elementary processes. The classification also separates reactions with breaking bonds into excited or non-excited molecules. This classification table approach is useful for determining the efficiency of vibrational energy α in elementary reactions if it is not known experimentally or from special detailed modeling.

2.7.3. Fridman-Macheret α-Model

This model permits us to calculate the α-coefficient for the efficiency of vibrational energy in elementary chemical processes based mostly on information about the activation energies of the corresponding direct and reverse reactions (Macheret, Rusanov, & Fridman, 1984). The model describes the exchange reaction A + BC → AB + C with profile shown in Fig. 2–37. Vibration of the molecule BC can be taken into account using the approximation of **vibronic terms** (Andreev & Nikitin, 1976; Secrest, 1973), which are shown in Fig. 2–37 by a dash-dotted line. The energy profile, corresponding to the reaction of a vibrationally excited molecule with energy E_v (vibronic term), can be obtained in this approach by a parallel up-shifting of the initial profile A + BC on the value of E_v. Part of the reaction

Table 2–27. Classification of Chemical Reactions for Determination of Vibrational Energy Efficiency Coefficients α in Overcoming Activation Energy Barrier

Reaction type	Simple exchange	Simple exchange	Simple exchange	Double exchange
	Bond break in excited molecule	Bond break in un-excited molecule	Bond break in un-excited molecule	
		Through complex	Direct	
Endothermic	0.9–1.0	0.8	<0.04	0.5–0.9
Exothermic	0.2–0.4	0.2	0	0.1–0.3
Thermo-neutral	0.3–0.6	0.3	0	0.3–0.5

path profile corresponding to products AB + C remains the same, if the products are not excited.

Simple geometry (Fig. 2–37) shows that an effective decrease of activation energy related to vibrational excitation E_v is equal to

$$\Delta E_a = E_v \frac{F_{A+BC}}{F_{A+BC} + F_{AB+C}}. \tag{2–210}$$

In this relation, F_{A+BC} and F_{AB+C} are characteristic slopes of the terms A + BC and AB + C (Fig. 2–37B). If these energy terms depend exponentially on reaction coordinates with decreasing parameters γ_1 and γ_2 (reverse radii of corresponding exchange forces), then

$$\frac{F_{A+BC}}{F_{AB+C}} = \frac{\gamma_1 E_a^{(1)}}{\gamma_2 E_a^{(2)}}. \tag{2–211}$$

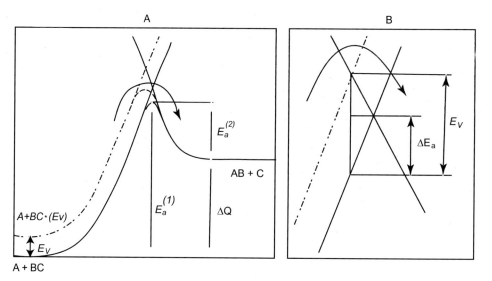

Figure 2–37. Efficiency of vibrational energy in a simple exchange reaction A + BC → AB + C. (A) Solid curve – reaction profile; dashed line represents a vibronic term, corresponding to an atom A interaction with a vibrationally excited molecule BC*(E_v); (B) part of the reaction profile near the summit of the exchange reaction barrier.

2.7. Elementary Chemical Reactions of Excited Molecules

The subscripts 1 and 2 stand for direct (A + BC) and reverse (AB + C) reactions. Formulas (2–210) and (2–211) not only explain the main kinetic relation (2–205) for reactions of vibrationally excited molecules but also determine the value of the coefficient α (which is actually equal to $\alpha = \Delta E_a / E_v$):

$$\alpha = \frac{\gamma_1 E_a^{(1)}}{\gamma_1 E_a^{(1)} + \gamma_2 E_a^{(2)}} = \frac{E_a^{(1)}}{E_a^{(1)} + (\gamma_1/\gamma_2) E_a^{(2)}}. \tag{2–212}$$

Usually the exchange force parameters for direct and reverse reactions γ_1 and γ_2 are close ($\gamma_1/\gamma_2 \approx 1$), which leads to the approximate but very convenient main formula of the Fridman-Macheret α-model:

$$\alpha \approx \frac{E_a^{(1)}}{E_a^{(1)} + E_a^{(2)}}. \tag{2–213}$$

Formula (2–213) is generally in good agreement with experimental data (see Table 2.26) and reflects the three most important tendencies of the α-coefficients:

1. The efficiency α of vibrational energy is the highest – that is, close to 100% – for strongly endothermic reactions with activation energies close to the reaction enthalpy.
2. The efficiency α of vibrational energy is the lowest – that is, close to zero – for exothermic reactions without activation energies.
3. The sum of α-coefficients for direct and reverse reactions is equal to unity ($\alpha^{(1)} + \alpha^{(2)} = 1$).

Formula (2–213) does not include any detailed information on the dynamics of the elementary chemical reaction and type of excitation. For this reason we can expect good results from applying the formula to the wide range of different chemical reactions of different excited species.

2.7.4. Efficiency of Vibrational Energy in Elementary Reactions Proceeding Through Intermediate Complexes: Synthesis of Lithium Hydride

As an example in such plasma-chemical reactions, we can consider using the synthesis of lithium hydride:

$$\text{Li}(^2S) + \text{H}_2^*(^1\Sigma_g^+, v) \rightarrow \text{LiH}(^1\Sigma^+) + \text{H}(^2S). \tag{2–214}$$

The potential energy surface for the reaction with the C_{2v} group of symmetry configuration was calculated using the method of diatomic complexes in molecules (Polishchuk, Rusanov, & Fridman, 1980) and it is presented in Fig. 2–38. The correspondent reaction path profile is shown in Fig. 2–39, where one can see a potential well with an energy depth of about 0.4 eV. The potential well makes the LiH synthesis proceed through an intermediate complex, which can be described using the **statistical theory of chemical processes** (Engelgardt, Felps, & Risk, 1964). The cross section of reaction (2–214) is expressed in the framework of the theory as a product of the cross section of the intermediate complex formation σ_a and the probability of its subsequent decay with the formation of LiH + H. The attachment cross section σ_a at fixed orbital quantum number l for relative motion of the reactants can be presented as

$$\sigma_a^l = \frac{2\pi}{k^2}(2l + 1)\frac{\hbar\Gamma}{\Delta E_k}, \tag{2–215}$$

where ΔE_k is the average spacing between energy levels of the complex; $\hbar\Gamma$ is the level energy width; and k is the wave vector of relative motion of the reactants. Taking into

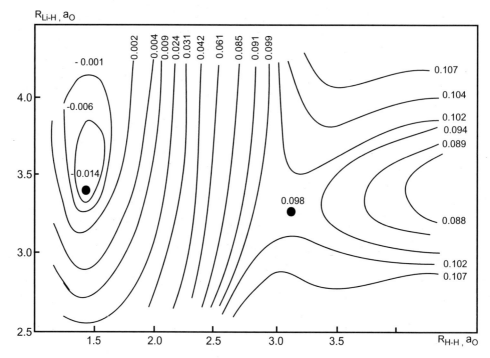

Figure 2–38. Potential energy surface for Li–H$_2$ interaction. Potential energy and interatomic distances are given in atomic units.

account that $\frac{\hbar\Gamma}{\Delta E_k} = \frac{1}{2\pi}$ and considering probabilities of different decay channels, the total cross section of reaction (2–214) can be expressed as

$$\sigma_R(E) = \sum_l \sigma_a^l \frac{k_1(E,l)}{k_1(E,l) + k_{-1}(E,l)} = \sum_l \frac{\pi}{k^2}(2l+1)A_\omega \left(\frac{E - E_1^a}{E}\right)^{s-1} \theta(E - E_1^a)$$

$$(2\text{–}216)$$

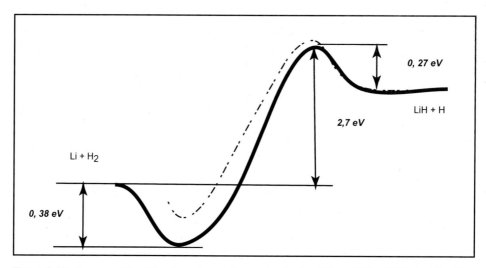

Figure 2–39. Energy profile of the elementary exchange reaction Li + H$_2 \rightarrow$ LiH + H. Dashed line illustrates the effect of conservation of angular momentum.

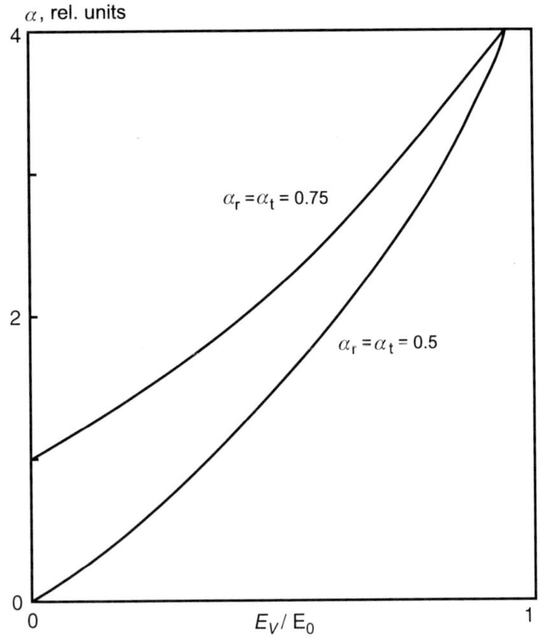

2.7. Elementary Chemical Reactions of Excited Molecules

α, rel. units

Figure 2–40. Cross section of the elementary exchange reaction $Li + H_2 \rightarrow LiH + H$ as a function of H_2 vibrational energy; contribution of different degrees of freedom in the cross section.

where $E_1^a = 2.77\,\mathrm{eV}$ is the activation energy of (2–214); k_1 and k_{-1} are rate coefficients of direct and reverse channels of intermediate complex decay; s is the number of vibrational degrees of freedom of the complex; $A_\omega \approx 3$ is a frequency factor; and $\theta(E - E_1^a)$ is the Heaviside function ($\theta(x - x_0) = 1$ when $x > 0$; $\theta(x - x_0) = 0$ when $x < 0$). If E_v, E_r, and E_t are, respectively, vibrational energy of the hydrogen molecule and rotational and translational energies of the reacting particles, then the total effective energy of the intermediate complex E, calculated with respect to the initial level of potential energy, can be found as the following sum:

$$E = E_v + \alpha_r E_r + \alpha_t E_t. \qquad (2\text{--}217)$$

The efficiencies α_r and α_t of rotational and translational energies in overcoming the activation energy barrier are significantly less than 100% and can be estimated as

$$\alpha_r \approx \alpha_t \approx 1 - \frac{r_e^2}{R_a^2}. \qquad (2\text{--}218)$$

The lower efficiencies of rotational and translational energies are due to conservation of angular momentum. Part of the rotational and translational energies remains in rotational form at the last moment of decay (see Fig. 2–39), when the characteristic distance between products is R_a; r_e is the equilibrium distance between hydrogen atoms in H_2. The cross section of reaction (2–214) is shown in Fig. 2–40 as a function of the fraction E_v/E_0 of vibrational energy E_v in the total energy E_0, assuming a fixed total energy $E_0 = 2E_1^a = $ const. The vibrational energy of hydrogen is more efficient than rotational and translational energies in the stimulation of the reaction, which proceeds via formation of an intermediate complex. According to (2–216) and (2–217), the efficiency of vibrational energy is close to 100% ($\alpha = 1$; see (2.205)).

2.7.5. Dissociation of Molecules in Non-Equilibrium Conditions with Essential Contribution of Translational Energy: Non-Equilibrium Dissociation Factor Z

The vibrational temperature in most non-thermal discharges tends to exceed the translational temperature: $T_v > T_0$. Sometimes, however, this occurs in particular after a shock wave front $T_v < T_0$. In this case actually both vibrational and translational temperatures make a contribution toward the stimulation of dissociation. It is convenient to use in this case a **non-equilibrium dissociation factor Z** (Losev et al., 1996):

$$Z(T_0, T_V) = \frac{k_R(T_0, T_v)}{k^0(T_0, T_0 = T_v)} = Z_h(T_0, T_v) + Z_l(T_0, T_v). \qquad (2\text{–}219)$$

In this formula, $k_R(T_0, T_v)$ is the dissociation rate coefficient; $k_R^0(T_0, T_v = T_0)$ is the corresponding equilibrium rate coefficient for temperature T_0; and the non-equilibrium dissociation factors Z_h and Z_l are related to dissociation from high (h) and low (l) vibrational levels, respectively.

The non-equilibrium dissociation factor Z can be found using the **Macheret-Fridman model** (Macheret et al., 1994). The model is based on the assumption of classical impulsive collisions. Dissociation is considered to occur mostly through an optimum configuration, which is a set of collisional parameters minimizing the energy barrier. The probability of finding the colliding system near the optimum configuration determines the pre-exponential factor of the rate. In the framework of the Macheret-Fridman model, the non-equilibrium dissociation factor related to high vibrational levels can be expressed as

$$Z_h(T_0, T_v) = \left\{ \frac{1 - \exp\left(-\dfrac{\hbar\omega}{T_v}\right)}{1 - \exp\left(-\dfrac{\hbar\omega}{T_0}\right)}(1 - L) \right\} \exp\left[-D\left(\frac{1}{T_v} - \frac{1}{T_0}\right)\right]. \qquad (2\text{–}220)$$

In this relation, D is the dissociation energy and L is the pre-exponential factor related to the geometric configuration of collision. For dissociation from low vibrational levels, the Macheret-Fridman model gives

$$Z_l(T_0, T_v) = L \exp\left[-D\left(\frac{1}{T_{\text{eff}}} - \frac{1}{T_0}\right)\right]. \qquad (2\text{–}221)$$

An effective temperature is introduced here by the following relation (where m is the mass of an atom in the dissociating molecule, and M is the mass of an atom in other colliding partner):

$$T_{\text{eff}} = \alpha T_v + (1 - \alpha)T_0, \quad \alpha = \left(\frac{m}{M + m}\right)^2. \qquad (2\text{–}222)$$

In contrast to the general formula (2–205), showing the efficiency of vibrational energy, relation (2–222) shows efficiencies of both vibrational and translational energies in the dissociation of molecules. Experimental values of the non-equilibrium dissociation factor Z are presented in Fig. 2–41 in comparison with theoretical predictions. The non-equilibrium dissociation factor $Z(T_0)$ at $T_0 > T_v$ has a minimum. When T_0 is close to T_v, the dissociation is mostly due to the contribution of high vibrational levels. While the translational temperature increases, the relative population of high vibrational levels decreases, which results in reduction of the non-equilibrium factor Z. When the translational temperature becomes very high, the dissociation becomes "direct," that is, mostly due to strong collisions with vibrationally non-excited molecules. In this case, the non-equilibrium factor Z increases. Thus, the minimum point of the function $Z(T_0)$ corresponds to a transition to direct dissociation from vibrationally non-excited states.

2.7. Elementary Chemical Reactions of Excited Molecules

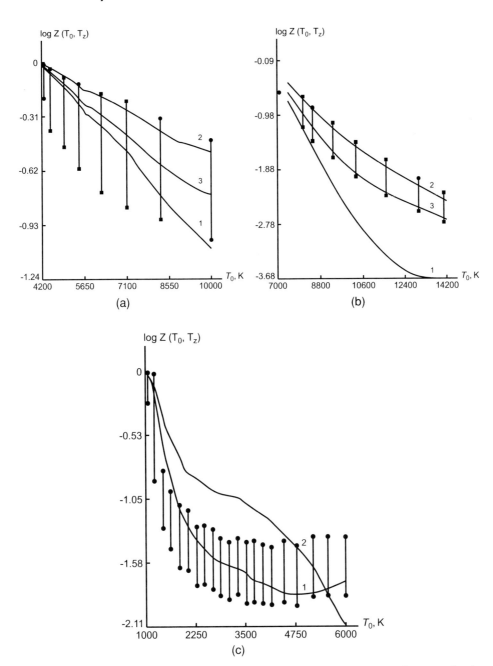

Figure 2–41. Experimental and theoretical values of the non-equilibrium dissociation factor Z as function of translational gas temperature in (a) molecular oxygen, (b) molecular nitrogen, and (c) molecular iodine: (1) Macheret-Fridman model, (2) Kuznetsov model, and (3) Losev model.

2.7.6. Semi-Empirical Models of Non-Equilibrium Dissociation of Molecules Determined by Vibrational and Translational Temperatures

Several semi-empirical models were proposed for the calculation of non-equilibrium dissociation, specifically at $T_0 > T_v$.

Park Model (Park, 1987). This model is widely used in practical calculations because of its simplicity. According to the model, the dissociation rate coefficient can be found by using the conventional Arrhenius formula for equilibrium reactions, by simply replacing temperature with an effective value:

$$T_{\text{eff}} = T_0^s T_v^{1-s}. \tag{2-223}$$

The factor s is a fitting parameter of the Park model, which is recommended to be taken as $s = 0.7$. In initial publications, the effective temperature was taken approximately as $T_{\text{eff}} = \sqrt{T_0 T_v}$. The Macheret-Fridman model permits the Park parameter to be found theoretically:

$$s = 1 - \alpha = 1 - \left(\frac{m}{m + M}\right)^2. \tag{2-224}$$

If two atomic masses are not very different ($m \sim M$), then the parameter $s \approx 0.75$ according to (2–224), which is close to the value $s = 0.7$ recommended by Park.

Losev Model (Losev & Generalov, 1961). According to this model, the dissociation rate coefficient can be found using the conventional Arrhenius formula for equilibrium reactions with vibrational temperature T_v and effective value D_{eff} of dissociation energy. To find D_{eff}, the actual dissociation energy D should be decreased by the value of translational temperature taken with efficiency β:

$$D_{\text{eff}} = D - \beta T_0. \tag{2-225}$$

The coefficient β is a fitting parameter of the Losev model, and for most estimations it can be taken as $\beta \approx 1.0$–1.5. The Macheret-Fridman model permits the Losev semi-empirical parameter to be found theoretically :

$$\beta \approx \frac{D(T_0 - T_v)}{T_0^2}, \quad T_0 > T_v. \tag{2-226}$$

The Losev parameter β can be considered a constant only for a narrow range of temperatures.

Marrone-Treanor Model (Marrone & Treanor, 1963). This model assumes an exponential distribution (with parameter U) of probabilities of dissociation from different vibrational levels. If $U \to \infty$, the probabilities of dissociation from all vibrational levels are equal. The non-equilibrium dissociation factor Z can be found within the framework of the model as

$$Z(T_0, T_v) = \frac{Q(T_0)Q(T_f)}{Q(T_v)Q(-U)}, \quad T_0 > T_v. \tag{2-227}$$

Statistical factors $Q(T_j)$ are determined here for four "temperatures" T_j: translational T_0, vibrational T_v, and two special temperatures, $-U$ and $T_f = \frac{1}{\frac{1}{T_v} - \frac{1}{T_0} - \frac{1}{-U}}$:

$$Q(T_j) = \frac{1 - \exp\left(-\dfrac{D}{T_j}\right)}{1 - \exp\left(-\dfrac{\hbar\omega}{T_j}\right)}. \tag{2-228}$$

The negative temperature $-U$ is a fitting parameter of the model with recommended values $U = 0.6\, T_0 - 3\, T_0$. The Macheret-Fridman model permits us to find theoretically the Marrone-Treanor parameter

$$U = \frac{T_0[\alpha T_0 + (1 - \alpha)T_v]}{(T_0 - T_v)(1 - \alpha)}. \tag{2-229}$$

The Marrone-Treanor parameter U is constant only over a narrow range of temperatures. Details about two-temperature dissociation kinetics for $T_0 > T_v$ can be found in Sergievska, Kovach, & Losev (1995).

PROBLEMS AND CONCEPT QUESTIONS

2–1. Coulomb Collisions. Using integral relation (2–7), try to calculate the rate coefficient for electron–electron Coulomb collisions in plasma with electron temperature $T_e = 1\,eV$. Assume the Maxwellian distribution function (2–1) for plasma electrons and take expression (2–8) for the Coulomb collisions cross section. Analyze the contribution of slow and fast electrons in the Coulomb collision reaction rate.

2–2. Ionization by Direct Electron Impact, Thomson Formula. Compare cross sections of the direct ionization by electron impact calculated using different modifications of the Thomson formula: (2–11), (2–13), and (2–14). Calculate the maximum values of the cross sections and corresponding values of electron energy for each of the Thomson formula modifications.

2–3. Stepwise Ionization by Electron Impact. Estimate a stepwise ionization rate coefficient in Ar at electron temperatures of 0.5 and 1 eV, assuming quasi-equilibrium between plasma electrons and electronic excitation of atoms. Comment on why the stepwise ionization is a preferential one in thermal plasmas and less important in non-thermal discharges.

2–4. Ionization in Collisions of Heavy Particles. Estimate the adiabatic Massey parameter and ionization probability and cross section for ionization of an Ar atom at rest in collision with an Ar^+ ion having kinetic energy twice that of the ionization potential. Compare the ionization probability in this case with the one for the Penning ionization.

2–5. Dissociative Electron–Ion Recombination. Check theoretical relation (2–39) by comparing experimental data presented in Table 2–2 for different temperatures. According to relation (2–39), the rate coefficient of dissociative recombination is reversibly proportional to gas temperature. However, in reality, when the electric field is fixed, this rate coefficient decreases much faster than the reverse proportionality. How can you explain this?

2–6. Langevin Cross Section, Resonant Charge Transfer Processes. Calculate the Langevin cross section for Ar^+ ion collisions with an argon atom and with a water molecule. Compare the contribution of a polarization term and a charge-dipole interaction term in the case of Ar^+ ion collision with a water molecule. Are the two mentioned charge exchange processes limited by the collision energy? Calculate the charge transfer cross section for both collisions as a function of relative velocity of colliding partners.

2–7. Generation of Negative Ions in Oxygen. Calculate the dissociative attachment rate coefficients for molecular oxygen at electron temperatures of 1 and 5 eV. What are the discharge and gas parameters supposed to be if the negative oxygen ions are mostly generated by three-body attachment processes and not by dissociative attachment?

2–8. Thomson and Langevin Models of Three-Body Ion–Ion Recombination. Comparing Thomson's and Langevin's models for the three-body ion–ion recombination, (2–93) and (2–97), find a typical value of pressure when the recombination rate coefficient is the highest. Specifically consider oxygen plasma at room gas temperature.

2–9. Sommerfeld Formula for Thermionic Emission. Using the Sommerfeld formula (2–100), calculate the saturation current densities of thermionic emission for a tungsten cathode at 2500 K. Compare the calculated value of cathode current density with a typical value for the hot thermionic cathodes presented in Table 2–12.

2–10. Secondary Ion–Electron Emission, Penning Mechanism. Explain why the coefficient γ of the secondary electron emission induced by the ion impact is almost constant at relatively low energies when it is provided by the Penning mechanism. Estimate the probability of the electron emission following the Penning mechanism. Explain why the coefficient γ of the secondary electron emission induced by the ion impact grows linearly with ion energy when the ion energy is high enough.

2–11. Work Function Dependence on Radius and Charge of Aerosol Particles. Give your physical interpretation of relation (2–105) for work function dependence on radius and charge of aerosol particles. Consider separately cases of conductive and non-conductive materials of the macro-particles. Estimate the relative change of work function for a metal particle with $\varphi_0 = 4\,\text{eV}$, radius $10\,\mu\text{m}$, and positive charge $100\,e$.

2–12. Space Distribution of Electron Density and Electric Field Around a Thermo-Ionized Aerosol Particle. Consider the thermal ionization of a macro-particle of radius $10\,\mu\text{m}$, work function 3 eV, and temperature 1500 K. Find the electron concentration just near the surface of the aerosol particle and corresponding value of the Debye radius, r_D^0. Calculate the total electric charge of the macro-particle, the electric field on its surface, and the electric field at a distance $3\,\mu\text{m}$ from the surface.

2–13. Multi-Quantum Vibrational Excitation by Electron Impact. Based on the Fridman approximation and taking into account the known basic excitation rate coefficient $k_{\text{eV}}(0, 1)$ as well as parameters α and β of the approximation, calculate the total rate coefficient of the multi-quantum vibrational excitation from an initial state v_1 to all other vibrational levels $v_2 > v_1$. Compare the calculated total vibrational excitation rate coefficient with the basic one, $k_{\text{eV}}(0, 1)$.

2–14. Influence of Vibrational Temperature on Electronic Excitation Rate Coefficients. Calculate the relative increase of the electronic excitation rate coefficient of molecular nitrogen in a non-thermal discharge with reduced electric field, $E/n_0 = 3 \cdot 10^{-16}\,\text{V} \cdot \text{cm}^2$, when the vibrational temperature is increased from room temperature to $T_v = 3000\,\text{K}$.

2–15. Dissociation of Molecules Through Electronic Excitation by Direct Electron Impact. Why does the electron energy threshold of dissociation through electronic excitation almost always exceed the actual dissociation energy in contrast to dissociation, stimulated by vibrational excitation, where the threshold is usually equal to the dissociation energy? Use the dissociation of diatomic molecules as an example.

2–16. Distribution of Electron Energy Between Different Channels of Excitation and Ionization. After analyzing Figs. 2–27–2–34, identify molecular gases with the highest ability to selectively localize discharge energy into vibrational excitation. What is the highest percentage of electron energy that can be selectively transferred to vibrational excitation? Which values of reduced electric field or electron temperatures are optimal to reach this selective regime?

2–17. VT-Relaxation Rate Coefficient as a Function of Vibrational Quantum Number. The VT-relaxation rate grows with the number of quanta for two reasons (see (2–71) or (2–178)). First, the matrix element increases as $n + 1$; second, the vibrational frequency decreases because of anharmonicity. Show numerically which effect dominates the acceleration of VT relaxation with the level of vibrational excitation.

2–18. Resonant Multi-Quantum VV Exchange. Estimate the smallness of the double-quantum resonance VV exchange $A_2^*(v = 2) + A_2(v = 0) \rightarrow A_2(v = 0) + A_2^*(v = 2)$ and triple-quantum resonance VV exchange $A_2^*(v = 3) + A_2(v = 0) \rightarrow A_2(v = 0) + A_2^*(v = 3)$ in diatomic molecules, taking the matrix element factor $(\Omega \tau_{\text{col}})^2 = 10^{-3}$.

2–19. Rotational RT Relaxation. Using the Parker formula and data from Table 2–24, calculate the energy well depth of the intermolecular attraction potential T^* and the probability of RT Relaxation in pure molecular nitrogen at a translational gas temperature of 400 K.

2–20. LeRoy Formula and α-Model. Derive the general relation (2–205), describing the stimulation of chemical reactions by vibrational excitation of molecules, by averaging the LeRoy formula over translational energy E_t. Use the Maxwellian function for the distribution of molecules over translational energies.

2–21. Accuracy of Fridman-Macheret α-Model. Based on Table 2–26 for the α-efficiency coefficients, determine the relative accuracy for application of the Fridman-Macheret α-model to (1) endothermic, (2) exothermic, and (3) thermo-neutral reactions.

2–22. Efficiency of Translational Energy in Elementary Endothermic Reactions. According to (2–218) and basic principles (conservation of momentum and angular momentum), the efficiency of translational energy in promoting endothermic reactions should be less than 100%; this means that the effective dissociation energy should exceed the actual value. On the other hand, the Fridman-Macheret model permits dissociation when the total translational energy is equal to D (Section 2.7.5). Explain this "contradiction."

3

Plasma-Chemical Kinetics,
Thermodynamics, and Electrodynamics

3.1. PLASMA STATISTICS AND THERMODYNAMICS, CHEMICAL AND IONIZATION EQUILIBRIUM, AND THE SAHA EQUATION

3.1.1. Statistical Distributions: Boltzmann Distribution Function

Plasma-chemical reaction rates depend on the probability of relevant elementary processes from a fixed quantum-mechanical state with fixed energy, which was considered in the previous chapter, and the number density of particles with this energy and in this particular quantum-mechanical state, which is to be considered in this chapter. A straightforward determination of the particles distribution in plasma over different energies and different quantum-mechanical states is related to detailed physical kinetics (see Fridman & Kennedy, 2004). However, wherever possible, the application of quasi-equilibrium statistical distributions is the easiest and clearest way to describe the kinetics and thermodynamics of plasma-chemical systems.

Assume an isolated system with energy E that consists of N of particles in different states i:

$$N = \sum_i n_i, \quad E = \sum_i E_i n_i. \qquad (3\text{--}1)$$

The objective of a statistical approach is to find a distribution function of particles over the different states i, taking into account that the probability to find n_i particles in these states is proportional to the number of ways in which the distribution can be arranged. **Thermodynamic probability** $W(n_1, n_2, \ldots, n_i)$ is the probability to have n_1 particles in the state "1," n_2 particles in the state "2," etc.:

$$W(n_1, n_2, \ldots, n_i) = A \frac{N!}{N_1! N_2! \ldots N_i!} = A \frac{N!}{\prod_i n_i!}. \qquad (3\text{--}2)$$

Here A is a normalizing factor. Let us find the most probable numbers of particles, $\overline{n_i}$, in a state i, when probability (3–2) and its logarithm,

$$\ln W(n_1, n_2, \ldots, n_i) = \ln(AN!) - \sum_i \ln n_i! \approx \ln(AN!) - \sum_i \int_0^{n_i} \ln x\, dx, \qquad (3\text{--}3)$$

have a maximum. Maximizing the function $\ln W$ of many variables requires

$$0 = \sum_i \left(\frac{\partial \ln W}{\partial n_i} \right)_{n_i = \overline{n_i}} dn_i = \sum_i \ln \overline{n_i}\, dn_i. \qquad (3\text{--}4)$$

Differentiation of the conservation laws (3–1) results in

$$\sum_i dn_i = 0, \quad \sum_i E_i dn_i = 0. \tag{3–5}$$

Multiplying equations (3–5) by parameters $-\ln C$ and $1/T$, respectively, and adding (3–4) gives

$$\sum_i \left(\ln \overline{n}_i - \ln C + \frac{E_i}{T} \right) dn_i = 0. \tag{3–6}$$

Sum (3–6) equals zero at any independent values of dn_i, which is possible only in the case of the **Boltzmann distribution function:**

$$\overline{n}_i = C \exp\left(-\frac{E_i}{T} \right). \tag{3–7}$$

Here C is the normalizing factor related to the total number of particles, and T is the temperature related to the average particle energy (the same units are assumed for energy and temperature). If level i is degenerated, we should add statistical weight g, showing the number of states with given quantum number,

$$\overline{n_j} = C g_j \exp\left(-\frac{E_j}{T} \right). \tag{3–8}$$

The Boltzmann distribution can then be expressed in terms of number densities N_j and N_0 of particles in j states with statistical weights g_j and the ground state (0) with statistical weight g_0:

$$N_j = N_0 \frac{g_j}{g_0} \exp\left(-\frac{E_j}{T} \right). \tag{3–9}$$

The general Boltzmann distributions (3–7) and (3–9) lead to several specific distribution functions, for example, the Maxwell-Boltzmann (or just the Maxwellian) distribution (2–1). Other specific quasi-equilibrium statistical distributions important in plasma chemistry are to be discussed next.

3.1.2. Equilibrium Statistical Distribution of Diatomic Molecules over Vibrational–Rotational States

Assuming the vibrational energy of diatomic molecules with respect to ground state ($v = 0$) is $E_v = \hbar \omega v$; the number density of molecules with v vibrational quanta according to (3–9) is

$$N_v = N_0 \exp\left(-\frac{\hbar \omega v}{T} \right). \tag{3–10}$$

The total number density of molecules, N, is a sum of densities in different vibrational states (v):

$$N = \sum_{v=0}^{\infty} N_v = \frac{N_0}{1 - \exp\left(-\frac{\hbar \omega}{T} \right)}. \tag{3–11}$$

Distribution (3–10) can then be renormalized with respect to the total number density N:

$$N_v = N \left[1 - \exp\left(-\frac{\hbar \omega}{T} \right) \right] \exp\left(-\frac{\hbar \omega v}{T} \right). \tag{3–12}$$

To find the Boltzmann vibrational–rotational distribution $N_{vJ} (\sum_J N_{vJ} = N_v)$, we can use again general relation (3–9), taking rotational energy as $E_r = B J (J + 1)$ and rotational statistical weight as $2J + 1$, where J is the rotational quantum number and B is the rotational constant:

$$N_{vJ} = N\frac{B}{T}(2J + 1)\left[1 - \exp\left(-\frac{\hbar\omega}{T}\right)\right]\exp\left(-\frac{\hbar\omega v + BJ(J + 1)}{T}\right). \quad (3\text{--}13)$$

3.1.3. Saha Equation for Ionization Equilibrium in Thermal Plasma

The Boltzmann distribution (3–9) can be applied to describe ionization equilibrium $A^+ + e \Leftrightarrow A$ in plasma (Smirnov, 1982a,b; Fridman & Kennedy, 2004):

$$\frac{N_e N_i}{N_a} = \frac{g_e g_i}{g_a}\left(\frac{mT}{2\pi\hbar^2}\right)^{\frac{3}{2}}\exp\left(-\frac{I}{T}\right). \quad (3\text{--}14)$$

In this relation, I is the ionization potential; g_a, g_i, and g_e are the statistical weights of atoms, ions, and electrons; N_a, N_i, and N_e are their number densities; and m is the electron mass. Formula (3–14) is known as the **Saha equation** and is widely used for calculations of ionization degree in thermal plasmas.

According to the Saha formula, the effective statistical weight of the continuum spectrum is very high. As a result, the Saha equation predicts very high values of degree of ionization, N_e/N_a, which can be close to unity even when temperature is still much less than the ionization potential, $T \ll I$.

The Saha formula is commonly used for quasi-equilibrium thermal plasma. It can also be applied, however, for estimations of non-thermal discharges assuming the temperature in (3–14) is that of electrons, T_e. One should take into account in this case that the Saha equation describes ionization equilibrium $A^+ + e \Leftrightarrow A$, which corresponds to a detailed balance of ionization by electron impact and three-body recombination: $e + A \rightarrow e + e + A^+$. As was discussed in Section 2.2, much faster recombination mechanisms (not directly balanced with ionization) take place in non-thermal plasma. For this reason, application of the Saha formula for non-thermal discharges usually significantly overestimates the degree of ionization.

3.1.4. Dissociation Equilibrium in Molecular Gases

The Saha equation (3–14) presented earlier for ionization equilibrium $A^+ + e \Leftrightarrow A$ can be generalized to describe the dissociation equilibrium $X + Y \Leftrightarrow XY$, which is especially important in thermal plasma chemistry. The relation between densities N_X of atoms X, N_Y of atoms Y, and N_{XY} of the molecules XY in the ground state can be written based on (3–14) as follows:

$$\frac{N_X N_Y}{N_{XY}(v = 0, J = 0)} = \frac{g_X g_Y}{g_{XY}}\left(\frac{\mu T}{2\pi\hbar^2}\right)^{\frac{3}{2}}\exp\left(-\frac{D}{T}\right). \quad (3\text{--}15)$$

In this relation, g_X, g_Y, and g_{XY} are relevant statistical weights; μ is the reduced mass of atoms X and Y; and D is the dissociation energy of the molecule XY. Most molecules are not in a ground state but rather in excited states at high thermal plasma temperatures. For this reason it is convenient to replace the ground-state concentration $N_{XY}(v = 0, J = 0)$ in (3–15) by the total N_{XY} concentration, using (3–13):

$$N_{XY}(v = 0, J = 0) = \left[1 - \exp\left(-\frac{\hbar\omega}{T}\right)\right]\frac{B}{T}N_{XY}. \quad (3\text{--}16)$$

Based on (3–15) and (3–16), the statistical relation for equilibrium dissociation in thermal plasma is

$$\frac{N_X N_Y}{N_{XY}} = \frac{g_X g_Y}{g_{XY}} \left(\frac{\mu T}{2\pi\hbar^2}\right)^{\frac{3}{2}} \frac{B}{T} \left[1 - \exp\left(-\frac{\hbar\omega}{T}\right)\right] \exp\left(-\frac{D}{T}\right). \qquad (3\text{–}17)$$

3.1.5. Complete Thermodynamic Equilibrium (CTE) and Local Thermodynamic Equilibrium (LTE) in Plasma

The application of quasi-equilibrium statistics and thermodynamics to plasma-chemical systems requires a clear understanding and distinction between the concepts of **complete thermodynamic equilibrium** (CTE) and **local thermodynamic equilibrium** (LTE). CTE is related to uniform plasma, in which chemical equilibrium and all plasma properties are unambiguous functions of temperature. This temperature is supposed to be homogeneous and the same for all degrees of freedom, all components, and all possible reactions. In particular, the following five equilibrium statistical distributions should take place for the same temperature T:

1. The Maxwell-Boltzmann translational energy distribution (2.1) pertains to all plasma components.
2. The Boltzmann distribution (3–9) describes the population of excited states for all plasma components.
3. The Saha equation (3–14) applies to ionization equilibrium.
4. Dissociation balance (3–17) and other thermodynamic relations describe chemical equilibrium.
5. The Plank distribution and other equilibrium relations pertain to the spectral density of electromagnetic radiation.

Plasma in CTE conditions cannot be practically realized in the laboratory. Nevertheless, thermal plasmas sometimes are modeled this way for simplicity. To imagine CTE plasma, one should consider so large a plasma volume that its central part is homogeneous and not sensitive to boundaries. Electromagnetic plasma radiation can be considered in this case as that of a blackbody with the plasma temperature.

Actually, even thermal plasmas are quite far from these ideal conditions. Most plasmas are optically thin over a wide range of wavelengths, which results in radiation much less intense than that of a blackbody. Plasma non-uniformity leads to irreversible losses related to conduction, convection, and diffusion, which also disturb the CTE. A more realistic approximation is the LTE. Thermal plasma is considered in this case to be optically thin, and radiation is not required to be in equilibrium. Collisional (not radiative) processes are required to be locally in equilibrium similar to that described earlier for CTE with a single temperature T, which can differ from point to point in space and time. Detailed consideration of LTE conditions can be found in Boulos, Fauchais, and Pfender (1994).

3.1.6. Thermodynamic Functions of Quasi-Equilibrium Thermal Plasma Systems

Finding out the thermodynamic plasma properties first requires a statistical calculation of partition functions (Pathria, 1996). The partition function Q of an equilibrium particle system at temperature T can be expressed as a statistical sum over states s of the particle with energies E_s and statistical weights g_s:

$$Q = \sum_s g_s \exp\left(-\frac{E_s}{T}\right). \qquad (3\text{–}18)$$

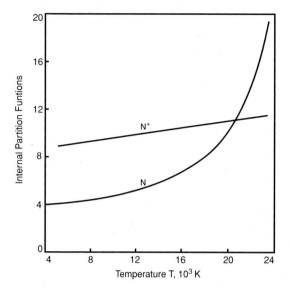

Figure 3–1. Partition functions of nitrogen atoms and nitrogen ions.

Translational and internal degrees of freedom of the particles of a chemical component i can be considered independently. Therefore, their energy can be expressed as a sum $E_s = E^{\text{tr}} + E^{\text{int}}$ and the partition function for plasma volume V as a product:

$$Q_i = Q_i^{\text{tr}} Q_i^{\text{int}} = \left(\frac{m_i T}{\hbar^2} \right)^{\frac{3}{2}} V Q^{\text{int}}. \qquad (3\text{–}19)$$

Translational partition function (3–19) corresponds to continuous-spectrum statistical weight; the partition functions of internal degrees of freedom depend on the system characteristics more specifically (Landau & Lifshitz, 1980). A particular numerical example of the partition functions of nitrogen atoms and ions (Boulos et al., 1994) is presented in Fig. 3–1. Plasma thermodynamic functions can be calculated based on total partition function Q_{tot} of all the particles in the equilibrium system. For example, the Helmholtz free energy F, related to a reference free energy, F_0, can be found as

$$F = F_0 - T \ln Q_{\text{tot}}. \qquad (3\text{–}20)$$

Assuming weak interaction between particles, the total partition function can be expressed as a product of partition functions Q_i of a single particle of a chemical component i:

$$Q_{\text{tot}} = \frac{\prod_i Q_i^{N_i}}{\prod_i N_i!}, \qquad (3\text{–}21)$$

where N_i is the total number of particles of the species i in the system. Taking into account that $\ln N! = N \ln N - N$, (3–30) and (3–21) give the Helmholtz free energy of non-interacting particles:

$$F = F_0 - \sum_i N_i T \ln \frac{Q_i e}{N_i}. \qquad (3\text{–}22)$$

Internal energy U and pressure can be expressed in this case as

$$U = U_0 + T^2 \left(\frac{\partial (F/T)}{\partial T} \right)_{V,N_i} = \sum_i N_i \left[\frac{3}{2} T + T^2 \left(\frac{\partial \ln Q_i^{int}}{\partial V} \right)_{T,N_i} \right], \qquad (3\text{--}23)$$

$$p = - \left(\frac{\partial (F/T)}{\partial V} \right)_{T,N} = \sum_i N_i T \left(\frac{\partial \ln Q_i}{\partial V} \right)_{T,N_i}. \qquad (3\text{--}24)$$

The plasma density increases at higher temperatures, and Coulomb interaction should be added to the thermodynamic functions. According to the Debye model, free energy (3–22) is modified as (Drawin, 1972)

$$F = F_0 - \sum_i N_i T \ln \frac{Q_i e}{N_i} - \frac{TV}{12\pi} \left(\frac{e^2}{\varepsilon_0 T V} \sum_i Z_i^2 N_i \right)^{\frac{3}{2}}. \qquad (3\text{--}25)$$

Here Z_i is the charge of a component i (for electrons $Z_i = -1$). The **Debye correction** of Helmholtz free energy leads to modification of other thermodynamic functions, in particular the Gibbs energy and pressure:

$$G = G_0 - \sum_i N_i T \ln \frac{Q_i}{N_i} - \frac{TV}{8\pi} \left(\frac{e^2}{\varepsilon_0 T V} \sum_i Z_i^2 N_i \right)^{\frac{3}{2}}, \qquad (3\text{--}26)$$

$$p = \frac{NT}{V} - \frac{T}{24\pi} \left(\frac{e^2}{\varepsilon_0 T V} \sum_i Z_i^2 N_i \right)^{\frac{3}{2}}. \qquad (3\text{--}27)$$

The Debye correction becomes significant at temperatures of thermal plasma about 14,000–15,000 K. The thermodynamic functions help to calculate chemical and ionization compositions of thermal plasmas. Examples of such calculations can be found in Boulos et al. (1994), as well as in Chapter 5 and later in this book with regards to the description of the composition and energy efficiency of plasma-chemical processes.

3.1.7. Non-Equilibrium Statistics of Thermal and Non-Thermal Plasmas

A detailed description of non-equilibrium plasma-chemical systems and processes generally requires an application of kinetic models. The application of statistical models leads sometimes to significant errors (Slovetsky, 1980). In some specific systems, however, statistical approaches can be not only simple but also quite successful in describing non-equilibrium plasma, which is discussed next.

The first example is related to thermal discharges with electron temperature deviating from the temperature of heavy particles, which can take place, in particular, in boundary layers separating plasma from electrodes and walls. In this case, the **two-temperature statistics and thermodynamics** can be developed (Boulos et al., 1994). These models assume that partition functions depend on two temperatures. Electron temperature determines the partition functions related to ionization processes, whereas chemical processes are determined by the temperature of heavy particles. The partition functions can then be applied to calculate thermodynamic functions, composition, and properties. An example of such a calculation of composition in two-temperature Ar plasma is given in Fig. 3–2.

A second example is related to strongly non-equilibrium plasma gasification processes,

$$A(\text{solid}) + b B^*(\text{gas}) \rightarrow c C(\text{gas}). \qquad (3\text{--}28)$$

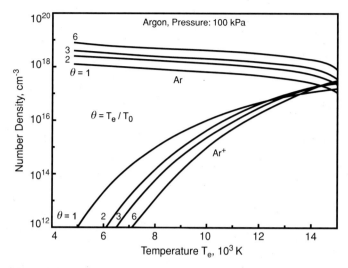

Figure 3–2. Argon atoms and argon ions number densities as functions of electron temperature T_e at atmospheric pressure. Two-temperature plasma calculations with temperature parameters: $\theta = T_e / T_0 = 1, 2, 3, 6$.

stimulated by the selective excitation of particles B (Veprek, 1972a,b; Legasov, Rusanov & Fridman, 1978c). Deviation from the conventional equation for the equilibrium constant,

$$K = \frac{(Q_C)^c}{Q_A(Q_B)^b}, \qquad (3\text{–}29)$$

can be introduced by a generalization of the partition functions, separately taking into account translational rotational, and vibrational temperatures – T_t, T_r, and T_v as well as a non-equilibrium population $f_e^X(\varepsilon_e^X)$ of electronically excited states for each reactant and products $X = A, B, C$:

$$Q_X = \sum_e g_e^X f_e^X\left(\varepsilon_e^X\right) \sum_{(t,r,v)} \prod_{k=t,r,v} g_{ke}^X \exp\left(-\frac{\varepsilon_{ke}^X}{T_k^X}\right). \qquad (3\text{–}30)$$

In this relation, g and ε are statistical weights and energies of corresponding states. A statistical approach can be applied in a consistent way for reactions (3–28) if the non-thermal plasma stimulation of the process is limited to the electronic excitation of a single state (energy E_b). The population of the excited state can be expressed then by δ-function and Boltzmann factor with an effective electronic temperature T^*:

$$f_e^B\left(\varepsilon_e^B\right) = \delta\left(\varepsilon_e^B - E_B\right) \exp\left(-\frac{E_B}{T^*}\right). \qquad (3\text{–}31)$$

All other degrees of freedom can be taken in quasi-equilibrium with a single gas temperature T_0:

$$f_e^{A,C}\left(\varepsilon_e^{A,C}\right) = 1, \quad T_{k=t,r,v}^X = T_0. \qquad (3\text{–}32)$$

The **single excited state approach** leads to the "quasi-equilibrium constant of the non-equilibrium process" (3–28):

$$K \approx \exp\left\{-\frac{1}{T_0}[(\Delta H - \Delta F^A) - bE_B]\right\} = K_0 \exp\frac{bE_B}{T_0}. \qquad (3\text{–}33)$$

In this relationship, ΔH is the reaction enthalpy; ΔF_A is the free energy change corresponding to heating to T_0; and K_0 is an equilibrium constant at temperature T_0. The equilibrium constant can be significantly increased due to the electronic excitation and equilibrium significantly shifted to the reaction products.

3.1.8. Non-Equilibrium Statistics of Vibrationally Excited Molecules: Treanor Distribution

The equilibrium distribution of diatomic molecules over vibrationally excited states follows the Boltzmann formula (3–12). Vibrational excitation in non-thermal plasma can be much faster than vibrational–translation (VT)-relaxation; therefore, the vibrational temperature T_v can significantly exceed the translational temperature T_0. The vibrational temperature in this case is usually defined in accordance with (3–10) as

$$T_v = \frac{\hbar\omega}{\ln\left(\frac{N_0}{N_1}\right)}. \tag{3–34}$$

If diatomic molecules are considered harmonic oscillators, the vibrational distribution function follows the same Boltzmann formula (3–12) even when $T_v > T_0$. Interesting non-equilibrium statistical phenomena take place, however, when we take into account anharmonicity. Then vibrational–vibrational (VV) exchange is not resonant and translational degrees of freedom become involved in the vibrational distribution, which results in a strong deviation from the Boltzmann distribution. Considering vibrational quanta as quasi-particles, and using the Gibbs distribution with a variable number of quasi-particles, v, the relative population of vibrational levels can be expressed (Kuznetzov, 1971) as

$$N_v = N_0 \exp\left(-\frac{\mu v - E_v}{T_0}\right). \tag{3–35}$$

In this relation, the parameter μ is the chemical potential; E_v is the energy of the vibrational level "v" taken with respect to the zero level. Comparing relations (3–35) and (3–10) gives

$$\mu = \hbar\omega\left(1 - \frac{T_0}{T_v}\right). \tag{3–36}$$

The Gibbs distribution (3–35), together with (3–34) and (3–36), leads to a non-equilibrium vibrational distribution of diatomic molecules known as the **Treanor distribution** (Treanor, Rich, & Rehm, 1968):

$$f(v, T_v, T_0) = B \exp\left(-\frac{\hbar\omega v}{T_v} + \frac{x_e \hbar\omega v^2}{T_0}\right). \tag{3–37}$$

Here x_e is the coefficient of anharmonicity and B is the normalizing factor. Comparison of the parabolic-exponential Treanor distribution with the linear-exponential Boltzmann distribution is illustrated in Fig. 3–3. A population of highly vibrationally excited levels at $T_v > T_0$ can be many orders of magnitude higher than that predicted by the Boltzmann distribution even at vibrational temperature. The Treanor distribution results in very high rates and energy efficiencies of chemical reactions stimulated by vibrational excitation in plasma.

The exponentially parabolic Treanor distribution function has a minimum:

$$v_{min}^{Tr} = \frac{1}{2x_e}\frac{T_0}{T_v}. \tag{3–38}$$

Population of the vibrational level, corresponding to the Treanor minimum, is

$$f_{min}^{Tr}(T_v, T_0) = B \exp\left(-\frac{\hbar\omega T_0}{4x_e T_v^2}\right). \tag{3–39}$$

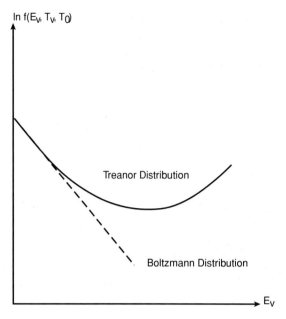

Figure 3–3. Comparison of Treanor and Boltzmann vibrational distribution functions.

This relation can be applied sometimes to describe the exponential quasi-Arrhenius part of plasma-chemical reaction rates (Legasov et al., 1977a,b). The expression does not include activation energy, which will be discussed later when analyzing the regime of strong vibrational excitations in plasma chemistry. We should also note that the population of vibrational states becomes inverse at $v > v_{\min}^{Tr}$ according to the Treanor distribution, which, in particular, plays an important role in CO lasers.

3.2. ELECTRON ENERGY DISTRIBUTION FUNCTIONS (EEDFs) IN NON-THERMAL PLASMA

Electron energy distribution functions (EEDFs) in non-thermal discharges can be very sophisticated and quite different from the quasi-equilibrium statistical Boltzmann distribution discussed earlier, and are more relevant for thermal plasma conditions. EEDFs are usually strongly exponential and significantly influence plasma-chemical reaction rates.

3.2.1. Fokker-Planck Kinetic Equation for Determination of EEDF

EEDFs can obviously be determined by using the Boltzmann kinetic equation (Lieberman & Lichtenberg, 1994). A better physical interpretation of EEDF evolution in plasma can be given, however, using the Fokker-Planck approach. EEDF evolution is considered in this case as an electron diffusion and drift in the space of electron energy (Rusanov & Fridman, 1984; Raizer, 1991). To derive the Fokker-Planck kinetic equation for EEDF $f(\varepsilon)$, we consider the dynamics of energy transfer from an electric field to electrons. An electron has velocity $\vec{v}$ after collision with a neutral and then receives additional velocity during free motion between collisions, corresponding to its drift in electric field $\vec{E}$:

$$\vec{u} = -\frac{e\vec{E}}{m\,\nu_{en}}, \qquad (3\text{–}40)$$

where ν_{en} is frequency of the electron–neutral collisions, and e and m are an electron charge and mass. The corresponding change of the electron kinetic energy between two collisions can be expressed as

$$\Delta\varepsilon = \frac{1}{2}m(\vec{v} + \vec{u})^2 - m\vec{v}^2 = m\vec{v}\vec{u} + \frac{1}{2}m\vec{u}^2. \tag{3–41}$$

The average contribution of the first term $m\vec{v}\vec{u}$ to $\Delta\varepsilon$ is equal to zero, and the average electron energy increase between two collisions is related only to the square of the drift velocity:

$$\langle\Delta\varepsilon\rangle = \frac{1}{2}m\vec{u}^2. \tag{3–42}$$

Thus, electron motion along the energy spectrum (or in energy space) can be considered a diffusion process. An electron with energy $\varepsilon = \frac{1}{2}mv^2$ receives or loses during one collision an energy about mvu, depending on the direction of its motion – along or opposite the electric field. The energy portion mvu can be considered the electron "mean free path along the energy spectrum" (or in the energy space). Taking into account also the possibility of an electron motion across the electric field, we can introduce a corresponding coefficient of electron diffusion along the energy spectrum as

$$D_\varepsilon = \frac{1}{3}(mvu)^2\nu_{en} = \frac{2}{3}mu^2\varepsilon\nu_n. \tag{3–43}$$

Besides the diffusion along the energy spectrum, there is also drift in the energy space related to a permanent average energy gain (3–42) or loss. The average energy loss per collision is mostly due to elastic scattering $\frac{2m}{M}\varepsilon$ and vibrational excitation $P_{eV}(\varepsilon)\hbar\omega$ in the case of molecular gases, where M and m are a neutral particle and electron mass, and $P_{eV}(\varepsilon)$ is the probability of vibrational excitation by electron impact. The drift velocity in energy space can then be expressed as

$$u_\varepsilon = \left[\frac{mu^2}{2} - \frac{2m}{M}\varepsilon - P_{eV}(\varepsilon)\hbar\omega\right]\nu_{en}. \tag{3–44}$$

EEDF $f(\varepsilon)$ can be considered in this approach as the number density of electrons in energy space and can be found from the continuity equation in the energy space, called the **Fokker-Planck kinetic equation:**

$$\frac{\partial f(\varepsilon)}{\partial t} = \frac{\partial}{\partial\varepsilon}\left[D_\varepsilon\frac{\partial f(\varepsilon)}{\partial\varepsilon} - f(\varepsilon)u_\varepsilon\right]. \tag{3–45}$$

3.2.2. Druyvesteyn Distribution, Margenau Distributions, and Other Specific EEDF

The steady-state solution of the Fokker-Planck equation (3–45) for EEDF in non-equilibrium plasma, corresponding to $f(\varepsilon \to \infty) = 0$, $\frac{df}{d\varepsilon}(\varepsilon \to \infty) = 0$, can be presented as

$$f(\varepsilon) = B\exp\left\{\int_0^\varepsilon \frac{u_\varepsilon}{D_\varepsilon}d\varepsilon'\right\}, \tag{3–46}$$

where B is the pre-exponential normalization factor. Using relations (3–43) and (3–44) for the diffusion coefficient and drift velocity in the energy space, EEDFs can be expressed in the following integral form:

$$f(\varepsilon) = B\exp\left[-\int_0^\varepsilon \frac{3m^2}{Me^2E^2}\nu_{en}^2\left(1 + \frac{M}{2m}\frac{\hbar\omega}{\varepsilon'}P_{eV}(\varepsilon)\right)d\varepsilon'\right]. \tag{3–47}$$

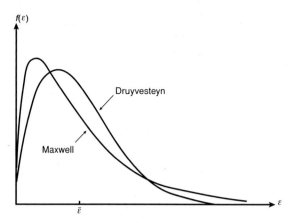

Figure 3–4. Comparison of Maxwell and Druyvesteyn electron energy distribution functions (EEDFs) at the same value of mean electron energy. Statistical weight effect related to the pre-exponential factor B and resulting in $f(0) = 0$ is taken into account.

This distribution function is derived for a constant electric field but can be generalized to the alternating electric fields with frequency ω by replacing the electric field strength E by the effective one:

$$E_{\text{eff}}^2 = E^2 \frac{\nu_{\text{en}}^2}{\omega^2 + \nu_{\text{en}}^2}. \qquad (3\text{–}48)$$

Thus, the electric field is quasi-constant at $T = 300\,\text{K}$ and $p = 1\,\text{Torr}$ if frequencies $\omega \ll 2000\,\text{MHz}$.

3.2.2.1. Maxwellian distribution. If elastic collisions dominate electron energy losses $(P_{\text{eV}} \ll \frac{2m}{M} \frac{\varepsilon}{\hbar\omega})$ and electron–neutral collision frequency is approximated by a constant $\nu_{\text{en}}(\varepsilon) = \text{const}$, integration of (3–47) gives the Maxwellian distribution (2.1) with electron temperature:

$$T_{\text{e}} = \frac{e^2 E^2 M}{3m^2 \nu_{\text{en}}^2}. \qquad (3\text{–}49)$$

3.2.2.2. Druyvesteyn distribution. If, in contrast to the preceding distribution, the electron mean free path $\lambda = \text{const}$, then $\nu_{\text{en}} = \frac{v}{\lambda}$ and integral (3–47) gives the exponential-parabolic Druyvesteyn EEDF, first derived in 1930:

$$f(\varepsilon) = B \exp\left[-\frac{3m}{M} \frac{\varepsilon^2}{(eE\lambda)^2}\right], \qquad (3\text{–}50)$$

the Druyvesteyn distribution decreases with energy much faster than the Maxwellian distribution for the same mean energy (see Fig. 3–4).

3.2.2.3. Margenau distribution. This distribution, first derived in 1946, is a generalization of the Druyvesteyn distribution for the case of alternating electric fields:

$$f(\varepsilon) = B \exp\left[-\frac{3m}{M} \frac{1}{(eE\lambda)^2}(\varepsilon^2 + \varepsilon m \omega^2 \lambda^2)\right]. \qquad (3\text{–}51)$$

3.2.2.4. Distributions controlled by vibrational excitation. Vibrational excitation of molecules by electron impact strongly affects EEDF in molecular gases when $P_{eV} \gg \frac{2m}{M}\frac{\varepsilon}{\hbar\omega}$. Assuming $P_{eV}(\varepsilon) = $ const in the energy interval $\varepsilon_1 < \varepsilon < \varepsilon_2$ and $\lambda = $ const ($\nu_{en} = \frac{v}{\lambda}$), the EEDF in the energy interval $\varepsilon_1 < \varepsilon < \varepsilon_2$ is

$$f(\varepsilon) \propto \exp\left[-\frac{6P_{eV}\hbar\omega}{(eE\lambda)^2}\varepsilon\right]. \tag{3-52}$$

Vibrational excitation takes more electron energy than elastic collisions ($P_{eV}\hbar\omega \gg \frac{m}{M}\varepsilon$) and, therefore, results in a significant EEDF decrease in the energy interval $\varepsilon_1 < \varepsilon < \varepsilon_2$:

$$\alpha_v = \frac{f(\varepsilon_2)}{f(\varepsilon_1)} \approx \exp\left(-\frac{2M}{m}P_{eV}\hbar\omega\frac{\varepsilon_2 - \varepsilon_1}{\langle\varepsilon\rangle^2}\right), \tag{3-53}$$

where $\langle\varepsilon\rangle$ is the average electron energy. The parameter α_v describes the probability for an electron to pass through the energy interval of intensive vibrational excitation; therefore, the ionization and electronic excitation rate coefficients are usually proportional to the parameter α_v.

The parameter α_v can be applied to analyze the influence of vibrational temperature T_v on the ionization and electronic excitation rate coefficients. The probability P_{eV} is proportional to the number density of vibrationally non-excited molecules $N(v = 0) \propto 1 - \exp(-\frac{\hbar\omega}{T_v})$ (3–12), taking into account the possibility of superelastic collisions. As a result, the influence of vibrational temperature T_v on the ionization and electronic excitation rate coefficients can be characterized as the double exponential function

$$\alpha_v \approx \exp\left[-\frac{2M}{m}P_{eV}^0\hbar\omega\frac{\varepsilon_2 - \varepsilon_1}{\langle\varepsilon\rangle^2}\left(1 - \exp\left(-\frac{\hbar\omega}{T_v}\right)\right)\right], \tag{3-54}$$

where P_{eV}^0 is the vibrational excitation probability at $T_v = 0$. Relation (3–54) illustrates the strong increase of EEDF as well as ionization and electronic excitation rates with vibrational temperature.

3.2.3. Effect of Electron–Molecular and Electron–Electron Collisions on EEDF

EEDF in non-equilibrium discharges in noble gases are usually close to the Druyvesteyn distribution if the ionization degree is not high enough for a significant contribution of electron–electron collisions. At the same mean energies, EEDF $f(\varepsilon)$ in molecular gases is closer to the Maxwellian function. Significant deviations of $f(\varepsilon)$ in molecular gases from the Maxwellian distribution take place only for high-energy electrons at relatively low mean energies, about 1.5 eV. Even a small admixture of molecular gas (about 1%) into a noble gas changes EEDF substantially, strongly decreasing the fraction of high-energy electrons. The effect of a small molecular gas admixture is strongest at relatively low reduced electric fields $\frac{E}{n_0}$ where (n_0 is the number density of gas), when the **Ramsauer effect** is essential. For example, the addition of only 10% air into argon makes EEDF look like molecular gas (Slovetsky, 1980).

Electron–electron collisions can affect EEDF and make it Maxwellian, when the ionization degree $\frac{n_e}{n_0}$ is high. The effect can be characterized by the following numerical factor (Kochetov et al., 1979):

$$a = \frac{\nu_{ee}}{\delta\nu_{en}} \approx 10^8\frac{n_e}{n_0} \times \frac{1}{\langle\varepsilon, eV\rangle^2} \times \frac{10^{-16}\,\text{cm}^2}{\sigma_{en},\,\text{cm}^2} \times \frac{10^{-4}}{\delta}. \tag{3-55}$$

In this expression, ν_{ee} and ν_{en} are frequencies of electron–electron and electron–neutral collisions, corresponding to an average electron energy $\langle \varepsilon \rangle$; σ_{en} is the cross section of electron–neutral collisions at the same energy; and δ is the average fraction of electron energy transferred to a neutral particle during collision. The electron–electron collisions create an EEDF Maxwellian at high values of the factor $a \gg 1$. Maxwellization of electrons in plasma of noble gases (smaller δ) requires lower degrees of ionization than in the case of molecular gases (higher δ). For example, effective EEDF Maxwellization in argon plasma at $\frac{E}{n_0} = 10^{-16}$ V $\cdot$ cm^2 begins at $\frac{n_e}{n_0} = 10^{-7} \div 10^{-6}$. In nitrogen at the same $\frac{E}{n_0}$ and ionization degrees up to 10^{-4}, the EEDF perturbation by the electron–electron collisions is still not significant.

3.2.4. Relation Between Electron Temperature and the Reduced Electric Field

The aforementioned EEDF permits one to correlate the reduced electric field and average electron energy, which is related to electron temperature as $\langle \varepsilon \rangle = \frac{3}{2} T_e$, even for non-Maxwellian distributions. Such a relation can be derived, for example, from averaging the electron drift velocity in energy space (3–44):

$$\frac{e^2 E^2}{m \nu_{en}^2} = \delta \cdot \frac{3}{2} T_e. \tag{3–56}$$

The factor δ characterizes the fraction of an electron energy lost in a collision with a neutral particle:

$$\delta \approx \frac{2m}{M} + \langle P_{eV} \rangle \frac{\hbar \omega}{\langle \varepsilon \rangle}. \tag{3–57}$$

Relation (3–57) permits us to find the exact value of the factor $\delta = 2m/M$. In molecular gases, however, this factor is usually considered a semi-empirical one. For example, $\delta \approx 3 \cdot 10^{-3}$ in typical conditions of non-equilibrium discharges in nitrogen. Taking into account that $\nu_{en} = n_0 \langle \sigma_{en} v \rangle$, $\lambda = 1/n_0 \sigma_{en}$, and $\langle v \rangle = \sqrt{8 T_e / \pi m}$, relation (3–56) can be presented as

$$T_e = \frac{e E \lambda}{\sqrt{\delta}} \sqrt{\frac{\pi}{12}}. \tag{3–58}$$

This relation is in good agreement with both Maxwell and Druyvesteyn EEDFs. It is convenient to rewrite (3–58) as a relation between electron temperature and the reduced electric field E/n_0:

$$T_e = \left(\frac{E}{n_0} \right) \frac{e}{\langle \sigma_{en} \rangle} \sqrt{\frac{\pi}{12\delta}}. \tag{3–59}$$

This linear relation between electron temperature and the reduced electric field is obviously only a qualitative one. As one can see from Fig. 3–5, the relation $T_e(\frac{E}{n_0})$ can be more complicated in reality.

3.2.5. Isotropic and Anisotropic Parts of the Electron Distribution Functions: EEDF and Plasma Conductivity

The previously considered EEDFs are related to the isotropic part of the electron velocity distribution $f(\vec{v})$. This distribution $f(\vec{v})$, in general, is anisotropic in an electric field, which determines electric current and plasma conductivity. Electrons receive additional velocity $\vec{u}$ (3–40) during the free motion between collisions. If the anisotropy is not very strong ($u \ll v$), we can assume that the fraction of electrons in the point $\vec{v}$ of real anisotropic

3.2. Electron Energy Distribution Functions (EEDFs) in Non-Thermal Plasma

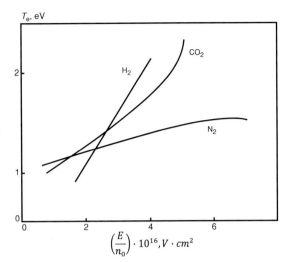

Figure 3–5. Electron temperature as a function of reduced electric field in different molecular gases.

distribution f is directly related to fraction of electrons in the point $\vec{v} - \vec{u}$ of the corresponding isotropic distribution $f^{(0)}$:

$$f(\vec{v}) = f^{(0)}(\vec{v} - \vec{u}) \approx f^{(0)}(\vec{v}) - \vec{u}\frac{\partial}{\partial \vec{v}}f^{(0)}(\vec{v}) = f^{(0)}(\vec{v}) + u\cos\theta\frac{\partial f^{(0)}(v)}{\partial v}, \quad (3\text{–}60)$$

where θ is the angle between directions of velocity $\vec{v}$ electric field $\vec{E}$. Taking into account the azimuthal symmetry of $f(\vec{v})$ and omitting the dependence on θ for the isotropic function $f^{(0)}$,

$$f(v, \theta) = f^{(0)}(v) + \cos\theta \cdot f^{(1)}(v). \quad (3\text{–}61)$$

The function $f^{(1)}(v)$ determines the anisotropy of the electron velocity distribution $f(\vec{v})$:

$$f^{(1)}(v) = u\frac{\partial f^{(0)}(v)}{\partial v} = \frac{eE}{mv_{\text{en}}}\frac{\partial f^{(0)}(v)}{\partial v}. \quad (3\text{–}62)$$

Relation (3–61) can be interpreted as the first two terms of the series expansion of $f(\vec{v})$ using the orthonormalized system of Legendre polynomials. The first term is an isotropic part of the distribution, which is related to the EEDF as

$$f(\varepsilon)\,d\varepsilon = f^{(0)}(v) * 4\pi v^2 dv. \quad (3\text{–}63)$$

The second term of expansion (3–61) is related to the electron current density in plasma:

$$j = -e\int \vec{v}f(\vec{v})\vec{d}v = -\frac{4\pi}{3}e\int_0^\infty v^3 f^{(1)}(v)dv. \quad (3\text{–}64)$$

Relation (3–62) compares the anisotropic and isotropic parts of the distribution function:

$$f^{(1)} = u\frac{\partial f^{(0)}}{\partial v} \sim \frac{u}{\langle v\rangle} * f^{(0)}. \quad (3\text{–}65)$$

The smallness of the anisotropy $f^{(1)} \ll f^{(0)}$ is directly related to the smallness of the electron drift velocity with respect to the thermal velocity of plasma electrons.

Based on relation (3–62) for the anisotropic part of the distribution function $f^{(1)}(v)$ and Ohm's law $j = \sigma E$, we can express the electron conductivity in plasma in terms of an EEDF as

$$\sigma = \frac{4\pi e^2}{3m} \int_0^\infty \frac{v^3}{\nu_{en}(v)} \left[-\frac{\partial f^{(0)}(v)}{\partial v} \right] dv. \qquad (3\text{–}66)$$

3.3. DIFFUSION, ELECTRIC/THERMAL CONDUCTIVITY, AND RADIATION IN PLASMA

3.3.1. Electron Mobility, Plasma Conductivity, and Joule Heating

Assuming $\nu_{en}(v) = \nu_{en} = \text{const}$, and integrating (3–66) by parts, gives the well-known formula for electron conductivity:

$$\sigma = \frac{n_e e^2}{m \nu_{en}}. \qquad (3\text{–}67)$$

This formula can be presented in a convenient numerical form:

$$\sigma = 2.82 \cdot 10^{-4} \frac{n_e (\text{cm}^{-3})}{\nu_{en}(\text{s}^{-1})}, \text{Ohm}^{-1}\,\text{cm}^{-1}. \qquad (3\text{–}68)$$

Plasma conductivity (3–67) is determined by electron density n_e (the contribution of ions will be discussed next) and the frequency of electron–neutral collisions, ν_{en}. The electron density can be found using the Saha equation (3–14) for quasi-equilibrium thermal discharges and from the balance of charged particles in non-equilibrium non-thermal discharges. The frequency of electron–neutral collisions, ν_{en}, is proportional to pressure and can be found numerically for some specific gases from Table 3–1. Relations (3–67) and (3–68) determine the power transferred from the electric field to plasma electrons. This power, calculated per unit volume, is referred to as **Joule heating**:

$$P = \sigma E^2 = \frac{n_e e^2 E^2}{m \nu_{en}}. \qquad (3\text{–}69)$$

Also relations (3–67), (3–68), and (4.3.9) determine **electron mobility** μ_e, which is the coefficient of proportionality between the electron drift velocity v_d and electric field:

$$v_d = \mu_e E, \quad \mu_e = \frac{\sigma_e}{e n_e} = \frac{e}{m \nu_{en}}. \qquad (3\text{–}70)$$

Table 3–1. Similarity Parameters Describing Electron–Neutral Collision Frequency, Electron Mean Free Path, and Electron Mobility and Conductivity at $\frac{E}{p} = 1 \div 30\,\text{V} \cdot \text{cm/Torr}$

Gas	$\lambda \cdot p$, $10^{-2}\,\text{cm} \cdot \text{Torr}$	$\frac{\nu_{en}}{p}$, $10^9 \text{s}^{-1}\,\text{Torr}^{-1}$	$\mu_e * p$, $10^6 \frac{\text{cm}^2\,\text{Torr}}{\text{V·s}}$	$\frac{\sigma * p}{n_e}$, $10^{-13} \frac{\text{Torr cm}^2}{\text{Ohm}}$
Air	3	4	0.45	0.7
N_2	3	4	0.4	0.7
H_2	2	5	0.4	0.6
CO_2	3	2	1	2
CO	2	6	0.3	0.5
Ar	3	5	0.3	0.5
Ne	12	1	1.5	2.4
He	6	2	0.9	1.4

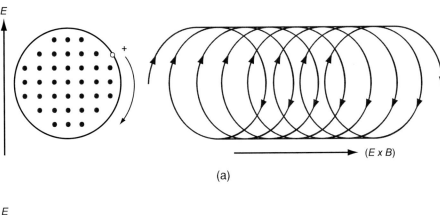

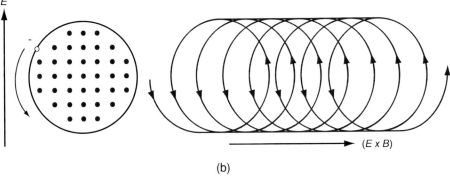

Figure 3–6. Illustration of drift of (a) positively and (b) negatively charged particles in perpendicular electric and magnetic fields.

To calculate the electron mobility, the following numerical relation, similar to (3–68), can be used:

$$\mu_e = \frac{1.76 \cdot 10^{15}}{\nu_{en}(s^{-1})}, \frac{cm^2}{V \cdot s}. \tag{3–71}$$

We should note that electron mobility and plasma conductivity at high ionization degrees $(n_e/n_0 \geq 10^{-3})$ are essentially affected by collisions with charged particles (Fridman & Kennedy, 2004).

Taking into account relations (3–67) and (3–70), it is convenient to construct the so-called **similarity parameters** $\frac{\nu_{en}}{p}$, λp, $\mu_e p$, and $\frac{\sigma p}{n_e}$. These parameters are approximately constant for each gas at room-temperature conditions and $\frac{E}{p} = 1 \div 30 \frac{V \cdot cm}{Torr}$; therefore, they can be applied for the determination of electron–neutral collision frequency ν_{en}, electron mean free path λ, electron mobility μ_e, and conductivity σ. Numerical values of the similarity parameters for several gases are collected in Table 3–1.

3.3.2. Plasma Conductivity in Crossed Electric and Magnetic Fields

A current in the presence of a magnetic field is not collinear with electric field and conductivity should be considered as the tensor σ_{ij}:

$$j_i = \sigma_{ij} E_j. \tag{3–72}$$

The motion of charged particles is quite complex in this case (Uman, 1964). Consider at first the collisionless motion of a particle with charge q in static uniform perpendicular electric and magnetic fields E and B ($E \ll cB$, where c is the speed of light). This motion is illustrated in Fig. 3–6 and described by

$$m\frac{d\vec{v}}{dt} = q(\vec{E} + \vec{v} \times \vec{B}). \tag{3–73}$$

Consider this motion in a reference frame moving with velocity $\vec{v}_{EB}$. The particle velocity in the frame is $\vec{v}' = \vec{v} + \vec{v}_{EB}$. Then equation (3–73) in the new moving reference frame can be rewritten as

$$m\frac{d\vec{v}'}{dt} = q(\vec{E} + \vec{v}_{EB} \times \vec{B}) + q\vec{v}' \times \vec{B}. \tag{3–74}$$

One can find the velocity $\vec{v}_{EB}$, which corresponds to $\vec{E} + \vec{v}_{EB} \times \vec{B} = 0$. It reduces the particle motion to $m\frac{d\vec{v}'}{dt} = q\vec{v}' \times \vec{B}$, which is rotation around the magnetic field lines (Fig. 3–6). The frequency of this rotation is the so-called **cyclotron frequency**:

$$\omega_B = \frac{eB}{m}. \tag{3–75}$$

Multiplying $\vec{E} + \vec{v}_{EB} \times \vec{B} = 0$ by $\vec{B}$, we find the expression for the velocity $\vec{v}_{EB}$. The "spiral" moves (Fig. 3–6) with this velocity in the direction perpendicular to both electric and magnetic fields:

$$\vec{v}_{EB} = \frac{\vec{E} \times \vec{B}}{B^2}. \tag{3–76}$$

This motion of a charged particle is usually referred to as the **drift in crossed electric and magnetic fields**. Direction and velocity of the drift are the same for electrons an ' ' Thus, the entire plasma can move as a whole with the drift velocity as a whole ... is applied in **plasma centrifuges**.

Collisions with neutrals make the charged particles move additionally along the electric field. The conductivity tensor (3–72), describing the combined motion, includes two components: one representing conductivity along the electric field, $\sigma_{\parallel}$, and another corresponding to the current perpendicular to both electric and magnetic fields, $\sigma_{\perp}$:

$$\sigma_{\parallel} = \frac{\sigma_0}{1 + \left(\dfrac{\omega_B}{\nu_{en}}\right)^2}, \quad \sigma_{\perp} = \sigma_0 \frac{\left(\dfrac{\omega_B}{\nu_{en}}\right)}{1 + \left(\dfrac{\omega_B}{\nu_{en}}\right)^2}, \tag{3–77}$$

where σ_0 is conventional conductivity (3–67). When the magnetic field is low and pressure high ($\omega_B \ll \nu_{en}$), the transverse conductivity $\sigma_{\perp}$ can be neglected, and the longitudinal conductivity $\sigma_{\parallel}$ actually coincides with the conventional conductivity, σ_0. At high magnetic fields and low pressures ($\omega_B \gg \nu_{en}$), electrons become "trapped" by the magnetic field and start drifting across the electric and magnetic fields. The longitudinal conductivity can then be neglected, and the transverse conductivity becomes independent of the pressure and mass of a charged particle:

$$\sigma_{\perp} \approx \frac{n_e e^2}{m\omega_B} = \frac{n_e e}{B}. \tag{3–78}$$

This relation is valid when the cyclotron velocity exceeds the electron–neutral collision frequency ($\omega_B \gg \nu_{en}$) but not by too much; ion motion can also become quasi-collisionless at very high magnetic fields, which makes ions and electrons move together (3–76) without any net current.

3.3.3. Ion Energy and Ion Drift in Electric Field

The relationship between the ions' average energy $\langle \varepsilon_i \rangle$, gas temperature T_0, and electric field E can be found from the balance of ion–molecular collisions (Raizer, 1991):

$$\langle \varepsilon_i \rangle = \frac{3}{2} T_0 + \frac{M}{2M_i} \left(1 + \frac{M_i}{M} \right)^3 \frac{e^2 E^2}{M_i v_{in}^2}, \tag{3–79}$$

where M_i and M are ion and neutral mass and v_{in} is the frequency of ion–neutral collisions. If the reduced electric field in the plasma is not too high ($E/p < 10$ V/cm · Torr), the ion energy only slightly exceeds that of neutrals. At high electric fields ($E/p \gg 10$ V/cm · Torr), ion velocity, collision frequency v_{in}, and ion energy increase with electric field (λ is the ion mean free path):

$$\langle \varepsilon_i \rangle \approx \frac{1}{2} \sqrt{\frac{M}{M_i}} \left(1 + \frac{M_i}{M} \right)^{\frac{3}{2}} eE\lambda. \tag{3–80}$$

At weak and moderate electric fields ($E/p < 10$ V/cm · Torr), the ion drift velocity is proportional to the electric field and the ion mobility is constant:

$$\vec{v}_{id} = \frac{e\vec{E}}{v_{in} \cdot MM_1/(M + M_1)}, \quad \mu_i = \frac{e}{v_{in} \cdot MM_1/(M + M_1)}. \tag{3–81}$$

A convenient numerical relation can be used for calculations of the ion mobility:

$$\mu_i = \frac{2.7 \cdot 10^4 \sqrt{1 + M/M_i}}{p(\text{Torr}) \sqrt{A \cdot (\alpha/a_0^3)}}, \tag{3–82}$$

where α is the polarizability of a neutral particle, A is its molecular mass, and a_0 is the Bohr radius. At high electric fields ($E/p \gg 10$ V/cm · Torr), the drift velocity (Fridman & Kennedy, 2004) is

$$v_{id} \approx \sqrt[4]{\frac{M_i}{M}} \left(1 + \frac{M_i}{M} \right) \sqrt{\frac{eE\lambda}{M_i}}. \tag{3–83}$$

3.3.4. Free Diffusion of Electrons and Ions; Continuity Equation; and Einstein Relation Between Diffusion Coefficient, Mobility, and Mean Energy

If electron and ion densities are low, their diffusion in plasma can be considered independent. In this case, the total flux of the charged particles includes drift in the electric field described earlier and diffusion, following Fick's law ("+" and "−" denote charge; "e" and "i" denote electrons or ions, respectively):

$$\vec{\Phi}_{e,i} = \pm n_{e,i} \mu_{e,i} \vec{E} - D_{e,i} \frac{\partial n_{e,i}}{\partial \vec{r}}. \tag{3–84}$$

The continuity equation can be written in this case as

$$\frac{\partial n_{e,i}}{\partial t} + \frac{\partial}{\partial \vec{r}} \left(\pm n_{e,i} \mu_{e,i} \vec{E} - D_{e,i} \frac{\partial n_{e,i}}{\partial \vec{r}} \right) = G - L. \tag{3–85}$$

The source term $G-L$ stands for "generation minus losses." Diffusion coefficients can be estimated as a function of thermal velocity, mean free path, and collisional frequency ($\langle v_{e,i} \rangle$, $\lambda_{e,i}$, $v_{en,in}$):

$$D_{e,i} = \frac{\langle v_{e,i}^2 \rangle}{3 v_{en,in}} = \frac{\lambda_{e,i} \langle v_{e,i} \rangle}{3}. \tag{3–86}$$

Table 3–2. Coefficients of Free Diffusion of Electrons and Ions at Room Temperature

Diffusion	$D \cdot p, \frac{cm^2}{s}$ Torr	Diffusion	$D \cdot p, \frac{cm^2}{s}$ Torr	Diffusion	$D \cdot p, \frac{cm^2}{s}$ Torr
e in He	$2.1 \cdot 10^5$	e in Ne	$2.1 \cdot 10^6$	e in Ar	$6.3 \cdot 10^5$
e in Kr	$4.4 \cdot 10^4$	e in Xe	$1.2 \cdot 10^4$	e in H_2	$1.3 \cdot 10^5$
e in N_2	$2.9 \cdot 10^5$	e in O_2	$1.2 \cdot 10^6$	N_2^+ in N_2	40

Diffusion coefficients for electrons and ions are inversely proportional to pressure; therefore, $D_{e,i} \cdot p$ can be used at room temperature as a similarity parameter (see Section 3.3.1 and Table 3–2).

The ratio of the free electron diffusion coefficient (3–86) and the electron mobility (3–70) is proportional to the average electron energy, which is known as the **Einstein relation**:

$$\frac{D_e}{\mu_e} = \frac{\frac{2}{3}\langle \varepsilon_e \rangle}{e}. \tag{3–87}$$

A similar Einstein relation obviously can be applied to ions as well.

3.3.5. Ambipolar Diffusion and Debye Radius

Electron and ion diffusion cannot be considered "free" and independent at high ionization degrees. Electrons move faster than ions and form the charge separation zone with a strong polarization field equalizing the electron and ion fluxes $\vec{\Phi}_e = \vec{\Phi}_i$:

$$\vec{\Phi}_e = -\mu_e \vec{E} n_e - D_e \frac{\partial n_e}{\partial r}, \quad \vec{\Phi}_i = \mu_i \vec{E} n_i - D_i \frac{\partial n_i}{\partial r}. \tag{3–88}$$

Dividing the first relation by μ_e, the second one by μ_i, and adding up the results, assuming plasma quasi neutrality, obtains a general relation for electron and ion fluxes:

$$\vec{\Phi}_{e,i} = -\frac{D_i \mu_e + D_e \mu_i}{\mu_e + \mu_i} \frac{\partial n_{e,i}}{\partial \vec{r}}. \tag{3–89}$$

This generalized diffusion of electrons and ions, $\vec{\Phi}_e = \vec{\Phi}_i$, is called **ambipolar diffusion**, $\vec{\Phi}_{e,i} = -D_a \frac{\partial n_{e,i}}{\partial \vec{r}}$, and is characterized by the diffusion coefficient

$$D_a = \frac{D_i \mu_e + D_e \mu_i}{\mu_e + \mu_i}. \tag{3–90}$$

Taking into account the Einstein relation and that $\mu_e \gg \mu_i$, $D_e \gg D_i$, we can rewrite (3–90) as

$$D_a \approx D_i + \frac{\mu_i}{e} T_e. \tag{3–91}$$

In quasi-equilibrium plasma, $D_a = 2D_i$; in non-equilibrium plasma ($T_e \gg T_i$), the ambipolar diffusion $D_a = \frac{\mu_i}{e} T_e$ corresponds to the temperature of the fast electrons and mobility of the slow ions. To determine conditions of the ambipolar diffusion with respect to free diffusion, we should estimate the polarization field from equations (3–88):

$$E \approx \frac{D_e}{\mu_e} \frac{1}{n_e} \frac{\partial n_e}{\partial r} = \frac{T_e}{e} \frac{\partial \ln n_e}{\partial r} \propto \frac{T_e}{eR}. \tag{3–92}$$

Here R is the characteristic length of the change in electron concentration. The difference between ion and electron concentrations, $\Delta n = n_i - n_e$, characterizing the space charge, is related to the electric field by the Maxwell equation: $\frac{\partial}{\partial r} \vec{E} = \frac{e\Delta n}{\varepsilon_0}$. We can simplify the

Maxwell equation for estimations as $\frac{E}{R} \propto \frac{e\Delta n}{\varepsilon_0}$. Combining this relation with (4.92), we can estimate deviation from quasi neutrality:

$$\frac{\Delta n}{n_e} \approx \frac{T_e}{e^2 n_e} \frac{1}{R^2} = \left(\frac{r_D}{R}\right)^2, \quad r_D = \sqrt{\frac{T_e}{e^2 n_e}}. \tag{3-93}$$

The **Debye radius** r_D is a plasma parameter characterizing the quasi neutrality. It represents the characteristic size of the charge separation and plasma polarization. If the electron density is high and the Debye radius is small ($r_D \ll R$), then the deviation from quasi neutrality is small, electrons and ions move "together," and diffusion is ambipolar. If, vice versa, electron density is relatively low and the Debye radius is large ($r_D \geq R$), then the plasma is not quasi neutral, and electrons and ions move separately and diffusion is free. For calculations of the Debye radius it is convenient to use the following numerical formula:

$$r_D = 742\sqrt{\frac{T_e, \text{eV}}{n_e, \text{cm}^{-3}}}, \text{cm}. \tag{3-94}$$

A large collection of numerical data concerning diffusion and drift of electrons and ions can be found in the books of McDaniel and Mason (1973) and Huxley and Crompton (1974).

3.3.6. Thermal Conductivity in Plasma

The thermal conductivity equation in a moving one-component gas can be expressed as

$$\frac{\partial}{\partial t}(n_0 \langle \varepsilon \rangle) + \nabla \cdot \vec{q} = 0, \tag{3-95}$$

where n_0 and $\langle \varepsilon \rangle$ are gas number density and average energy of a molecule; $\vec{q}$ is the heat flux. Average energy can be replaced in the equation by gas temperature T_0 and specific heat c_v with respect to one molecule. The heat flux includes two terms related to gas motion with velocity $\vec{u}$ and to thermal conductivity with coefficient κ: $\vec{q} = \langle \varepsilon \rangle n_0 \vec{u} - \kappa \nabla \cdot T_0$, which results in

$$\frac{\partial}{\partial t} T_0 + \vec{u}\nabla \cdot T_0 = \frac{\kappa}{c_v n} \nabla^2 T_0. \tag{3-96}$$

The thermal conductivity coefficient in a one-component gas without dissociation, ionization, or chemical reactions can be estimated as

$$\kappa \approx \frac{1}{3}\lambda \langle v \rangle n_0 c_v \propto \frac{c_v}{\sigma}\sqrt{\frac{T_0}{M}}. \tag{3-97}$$

Here σ is a typical cross section for neutral collisions, and M is the molecular mass. Thermal conductivity growth with temperature in plasma at high temperatures, however, can be much faster than (3–97), because of the influence of dissociation, ionization, and chemical reactions. Consider the effect of dissociation and recombination ($2A \Leftrightarrow A_2$) on the acceleration of thermal conductivity. Molecules are mostly dissociated into atoms in a zone with higher temperature and are much less dissociated in lower-temperature zones. Then the quasi-equilibrium diffusion of the molecules (D_m) to the higher-temperature zone leads to their intensive dissociation, consumption of dissociation energy E_D, and to the related large heat flux:

$$\vec{q}_D = -E_D D_m \nabla n_m = -\left(E_D D_m \frac{\partial n_m}{\partial T_0}\right)\nabla T_0, \tag{3-98}$$

which can be interpreted as acceleration of thermal conductivity. When the concentration of molecules is less than that of atoms ($n_m \ll n_a$) and ($T_0 \ll E_D$), the equilibrium relation

(3–17) gives $\frac{\partial n_m}{\partial T_0} = \frac{E_D}{T_0^2} n_m$. It results in an expression for the coefficient of **thermal conductivity related to the dissociation of molecules**:

$$\kappa_D = D_m \left(\frac{E_D}{T_0} \right)^2 n_m. \tag{3–99}$$

The ratio of the coefficient κ_D to the conventional coefficient in (3–97) can be very high because $T_0 \ll E_D$:

$$\frac{\kappa_D}{\kappa} \approx \frac{1}{c_v} \left(\frac{E_D}{T_0} \right)^2 \frac{n_m}{n_a}. \tag{3–100}$$

Thermal conductivity is an important topic, especially in thermal plasma. Details on the subject can be found in books by Boulos et al. (1994) or Eletsky, Palkina, and Smirnov (1975).

3.3.7. Non-Equilibrium Thermal Conductivity and Treanor Effect in Vibrational Energy Transfer

Cold gas flowing around a high-temperature thermal plasma zone provides the VT non-equilibrium in the area of their contact (Kurochkin, Polak, & Pustogarov, 1978). This effect, in particular, is due to a higher rate of vibrational energy transfer from the quasi-equilibrium high-temperature zone with respect to the rate of translational energy transfer (Dobkin & Son, 1982). An average value of a vibrational quantum is lower because of anharmonicity at higher vibrational temperatures. The fast VV exchange during the transfer of the vibrational quanta prefers transfer from high T_v to lower T_v over transfer in the opposite direction. This Treanor effect in vibrational energy transfer is somewhat similar to the general Treanor effect in plasma statistics (Section 3.1.8) and results in relative domination of the vibrational energy transfer coefficient over that of the translational energy transfer (Liventsov, Rusanov, & Fridman, 1983, 1984a,b; Fridman & Kennedy, 2004):

$$\frac{\Delta D_v}{D_0} \approx 4 Q_{01}^{10} q \frac{1 + 30q + 72q^2}{(1 + 2q)^2}, \quad q = \frac{x_e T_v^2}{T_0 \hbar \omega}. \tag{3–101}$$

The effect is significant for molecules like CO_2, CO, and N_2O with VV exchange provided by long-distance forces, which results in one-quantum transfer probability $Q_{01}^{10} \approx 1$. Also, as always with the Treanor effects, anharmonicity ($x_e \neq 0, q \neq 0$) is necessary for the phenomenon.

3.3.8. Plasma Emission and Absorption of Radiation in Continuous Spectrum and Unsold-Kramers Formula

Radiation is one of the most commonly known plasma properties because of its application in different lighting devices. Radiation also plays an important role in plasma diagnostics, including plasma spectroscopy, in the propagation of some electric discharges, and sometimes even in plasma energy balance. Details on the subject can be found, in particular, in Boulos et al. (1994), Fridman and Kennedy (2004), Modest (1993), Griem (1974), and Cabannes and Chapelle (1971). Here, we only mention some useful general relations regarding emission and absorption of continuous-spectrum radiation in plasma as well as radiation transfer.

The **total spectral density of quasi-equilibrium plasma emission** in continuous spectrum consists of the bremsstrahlung and recombination components, which can be combined in one general expression:

$$J_\omega d\omega = C \frac{n_i n_e}{T^{1/2}} \Psi\left(\frac{\hbar\omega}{T}\right) d\omega. \tag{3-102}$$

In this relation, n_i and n_e are ion and electron densities; $C = 1.08 \cdot 10^{-45}\, \text{W} \cdot \text{cm}^3 \cdot \text{K}^{1/2}$ is a numerical parameter; and $\Psi(x)$ is the dimensionless function:

$$\Psi(x) = 1, \qquad\qquad\qquad\qquad\qquad \text{if } x = \frac{\hbar\omega}{T} < x_g = \frac{|E_g|}{T}, \tag{3-103}$$

$$\Psi(x) = \exp[-(x - x_g)], \qquad\qquad\qquad \text{if } x_g < x < x_1, \tag{3-104}$$

$$\Psi(x) = \exp[-(x - x_g)] + 2x_1 \exp[-(x - x_1)], \quad \text{if } x = \frac{\hbar\omega}{T} > x_1 = \frac{I}{T}, \tag{3-105}$$

where $|E_g|$ is the energy of the first excited state with respect to transition to continuum; I is the ionization potential. Total radiation losses can be found via integration of the spectral density (3–102) as

$$J, \text{kW/cm}^3 = 1.42 \cdot 10^{-37} \sqrt{T, \text{K}}\; n_e n_i (\text{cm}^{-3}) \cdot \left(1 + \frac{|E_g|}{T}\right). \tag{3-106}$$

The **total plasma absorption coefficient in the continuum,** which corresponds to the inverse length of absorption, can be expressed as a function of dimensionless parameter $x = \hbar\omega/T$ as

$$\kappa_\omega = 4.05 \cdot 10^{-23}\, \text{cm}^{-1} \frac{n_e n_i (\text{cm}^{-3})}{(T, \text{K})^{7/2}} \frac{e^x \Psi(x)}{x^3}. \tag{3-107}$$

The product $n_e n_i$ can be replaced by the gas density n_0 using the Saha equation (3–14):

$$\kappa_\omega = \frac{16\pi}{3\sqrt{3}} \frac{e^6 T n_0}{\hbar^4 c \omega^3 (4\pi\varepsilon_0)^3} \frac{g_i}{g_a} \exp\left(\frac{\hbar\omega - I}{T}\right) = 1.95 \cdot 10^{-7}\, \text{cm}^{-1} \frac{n_0(\text{cm}^{-3})}{(T, \text{K})^2} \frac{g_i}{g_a} \frac{e^{-(x_1 - x)}}{x^3}. \tag{3-108}$$

This relation is usually referred to as the **Unsold-Kramers formula,** where g_a and g_i are statistical weights of an atom and an ion, $x = \hbar\omega/T$, $x_1 = I/T$ and I is the ionization potential. It is interesting to note that the Unsold-Kramers formula can be used only in quasi-equilibrium conditions, while the relation (3–107) can be applied for non-equilibrium plasma as well.

3.3.9. Radiation Transfer in Plasma: Optically Thin and Optically Thick Plasmas

The intensity of radiation, I_ω, decreases along its path "s" due to absorption (scattering in plasma can be neglected) and increases because of spontaneous and stimulated emission. The **radiation transfer equation** in quasi-equilibrium plasma can then be presented as

$$\frac{dI_\omega}{ds} = \kappa'_\omega (I_{\omega e} - I_\omega), \quad \kappa'_\omega = \kappa_\omega \left[1 - \exp\left(-\frac{\hbar\omega}{T}\right)\right], \tag{3-109}$$

where $I_{\omega e}$ is the quasi-equilibrium Planck radiation intensity:

$$I_{\omega e} = \frac{\hbar\omega^3}{4\pi^3 c^2} \frac{1}{\exp(\hbar\omega/T) - 1}. \tag{3-110}$$

The optical coordinate ξ is calculated from plasma surface $x = 0$ into the plasma body,

$$\xi = \int_0^x \kappa_\omega'(x)\,dx, \quad d\xi = \kappa_\omega'(x)\,dx, \tag{3-111}$$

and permits one to rewrite the radiation transfer equation in terms of the optical coordinate as

$$\frac{dI_\omega(\xi)}{d\xi} - I_\omega(\xi) = -I_{\omega e}. \tag{3-112}$$

If the plasma thickness is d, the radiation intensity on its surface, $I_{\omega 0}$, according to (3–112) is

$$I_{\omega 0} = \int_0^{\tau_\omega} I_{\omega e}[T(\xi)]\exp(-\xi)\,d\xi, \quad \tau_\omega = \int_0^d \kappa_\omega'\,dx. \tag{3-113}$$

The quasi-equilibrium radiation intensity $I_{\omega e}[T(\xi)]$ is shown here as a function of temperature and, therefore, an indirect function of the optical coordinate. In relation (3–113), τ_ω is the **optical thickness of plasma** referred to a specific ray and specific spectral range. If optical thickness is small ($\tau_\omega \ll 1$), it is usually referred to as the **transparent** or **optically thin plasma**. In this case, the radiation intensity on the plasma surface, $I_{\omega 0}$, (3–313) can be presented as

$$I_{\omega 0} = \int_0^{\tau_\omega} I_{\omega e}[T(\xi)]d\xi = \int_0^{\tau_\omega} I_{\omega e}\kappa_\omega'\,dx = \int_0^d j_\omega\,dx. \tag{3-114}$$

Here the emissivity term $j_\omega = I_{\omega e}\kappa_\omega'$ corresponds to the spontaneous emission. Radiation of the optically thin plasma is a result of the summation of independent emission from different intervals dx along the ray. Radiation generated within the plasma volume is able in this case to escape it. Assuming uniformity of the plasma parameters, the radiation intensity on the plasma surface can be expressed as

$$I_{\omega 0} = j_\omega d = I_{\omega e}\kappa_\omega'd = I_{\omega e}\tau_\omega \ll I_{\omega e}. \tag{3-115}$$

Thus, the radiation intensity of the optically thin plasma is much less ($\tau_\omega \ll 1$) than the equilibrium one.

The opposite case takes place when the optical thickness is high ($\tau_\omega \gg 1$), which is referred to as a **non-transparent** or **optically thick system.** If temperature is constant, equation (3–113) gives for optically thick systems the quasi-equilibrium Planck intensity $I_{\omega 0} = I_{\omega e}(T)$, leading to the Stefan-Boltzmann law of blackbody emission, $J = \sigma T^4$. Plasma, however, is usually optically thin for radiation in the continuous spectrum. As a result, the Stefan-Boltzmann law cannot be applied without an emissivity coefficient ε ($J = \varepsilon \cdot \sigma T^4$), which coincides in optically thin plasma with the optical thickness: $\varepsilon = \tau_\omega$.

3.4. KINETICS OF VIBRATIONALLY AND ELECTRONICALLY EXCITED MOLECULES IN PLASMA: EFFECT OF HOT ATOMS

3.4.1. Fokker-Plank Kinetic Equation for Non-Equilibrium Vibrational Distribution Functions

Vibrationally and electronically excited atoms and molecules, which make a significant contribution to plasma-chemical kinetics, can be described in terms of distribution functions. We already introduced the vibrational distribution functions (population of vibrationally excited states) in Section 3.1.2 in relation to quasi-equilibrium systems, and then we generalized the concept to the case of non-equilibrium statistics (Section 3.1.8). Kinetic equations permit one to describe the evolution of the vibrational and electronic distribution functions by taking into account the variety of relaxation processes and chemical reactions. First consider the kinetics of vibrationally excited molecules. Electrons in non-thermal discharges mostly provide excitation of lower vibrational levels, which determines vibrational temperature. Formation of highly excited and chemically active molecules depends on a variety of different processes in plasma but mostly on the competition of VV exchange processes and vibrational quanta losses in VT relaxation (Nikitin & Osipov, 1977; Likalter & Naidis, 1981; Capitelli, 1986, 2000; Gordiets & Zhdanok, 1986). Similarly to the EEDFs, (see Section 3.2.1), the clearest approach to the vibrational distribution functions is based on the Fokker-Planck equation (Rusanov, Fridman, & Sholin, 1979a,b). In the framework of this approach, the evolution of vibrational distribution functions is considered the diffusion and drift of molecules in the space of vibrational energies. The continuous distribution function $f(E)$ of molecules over vibrational energies E is considered in this case as the density in energy space and is found from the continuity equation along the energy spectrum:

$$\frac{\partial f(E)}{\partial t} + \frac{\partial}{\partial E} J(E) = 0. \tag{3-116}$$

In this continuity equation, $J(E)$ is the flux of molecules in the energy space, which includes all relaxation and reaction processes. We are next going to consider the contribution of VV- and VT-relaxation processes as well as chemical reactions, starting with one-component diatomic molecular gases. The contribution of other major relaxation processes is considered, in particular by Kennedy and Fridman (2004).

3.4.2. VT and VV Fluxes of Excited Molecules in Energy Space

The **VT relaxation of molecules** with vibrational energy E and vibrational quantum $\hbar\omega$, considered diffusion along the vibrational energy spectrum, is characterized by the flux

$$j_{VT} = f(E)k_{VT}(E, E + \hbar\omega)n_0\hbar\omega - f(E + \hbar\omega)k_{VT}(E + \hbar\omega, E)n_0\hbar\omega, \tag{3-117}$$

where n_0 is neutral gas density; $k_{VT}(E + \hbar\omega, E)n_0$ and $k_{VT}(E, E + \hbar\omega)n_0$ are frequencies of direct and reverse VT processes, whose ratio is $\exp\frac{\hbar\omega}{T_0}$. After expanding, $f(E + \hbar\omega) = f(E) + \hbar\omega\frac{\partial f(E)}{\partial E}$, and defining $k_{VT}(E + \hbar\omega, E) \equiv k_{VT}(E)$, we can rewrite (3–117) as

$$j_{VT} = -D_{VT}(E)\left[\frac{\partial f(E)}{\partial E} + \tilde{\beta}_0 f(E)\right]. \tag{3-118}$$

We have just introduced the diffusion coefficient D_{VT} of excited molecules in vibrational energy space:

$$D_{VT}(E) = k_{VT}(E)n_0 \cdot (\hbar\omega)^2, \tag{3-119}$$

which can be interpreted by taking into account that the vibrational quantum $\hbar\omega$ in energy space corresponds to the mean free path in the conventional coordinate space, and the quantum transfer frequency $k_{VT}(E)n_0$ in the energy space corresponds to the frequency of collisions in coordinate space. The temperature parameter $\tilde{\beta}_0$ is

$$\tilde{\beta}_0 = \frac{1 - \exp\left(-\dfrac{\hbar\omega}{T_0}\right)}{\hbar\omega}, \tag{3-120}$$

which corresponds to an inverse temperature $\tilde{\beta}_0 = \beta_0 = 1/T_0$ if $T_0 > \hbar\omega$. The first term in (3–118) can be interpreted as diffusion and the second one as drift in energy space. The ratio of diffusion coefficient and mobility at high temperature is given by $\tilde{\beta}_0 = \beta_0 = 1/T_0$, which is in agreement with the Einstein relation.

When VT relaxation is the dominating process, kinetic equation (3–116) can be presented as

$$\frac{\partial f(E)}{\partial t} = \frac{\partial}{\partial E}\left\{ D_{VT}(E)\left[\frac{\partial f(E)}{\partial E} + \frac{1}{T_0} f(E)\right]\right\}. \qquad (3\text{–}121)$$

At steady-state conditions, the Fokker-Planck kinetic equation (3–121) yields, after integration, the quasi-equilibrium exponential Boltzmann distribution with temperature T_0: $f(E) \propto \exp(-E/T_0)$.

The VV-exchange process involves two excited molecules (energies E, E'), which makes the VV flux non-linear with respect to the vibrational distribution function (Demura, Macheret, & Fridman, 1984):

$$j_{VV}(E) = k_0 n_0 \hbar\omega \int_0^\infty \left[Q_{E+\hbar\omega, E}^{E', E'+\hbar\omega} f(E+\hbar\omega) f(E') - Q_{E, E+\hbar\omega}^{E'+\hbar\omega, E'} f(E) f(E'+\hbar\omega)\right] dE'.$$

$$(3\text{–}122)$$

Taking into account the defect of vibrational energy in the VV-exchange process, $2x_e(E'-E)$, where x_e is the coefficient of anharmonicity and k_0 is the rate coefficient of gas-kinetic collisions, the VV-exchange probabilities Q in integral (3–122) are related to each other by the following detailed equilibrium relation:

$$Q_{E, E+\hbar\omega}^{E'+\hbar\omega, E'} = Q_{E+\hbar\omega, E}^{E', E'+\hbar\omega} \exp\left[-\frac{2x_e(E'-E)}{T_0}\right]. \qquad (3\text{–}123)$$

The Treanor distribution function (see Section 3.1.8) makes the VV flux (3–122) equal to zero. Thus, the Treanor distribution is a steady-state solution of the Fokker-Planck kinetic equation (3–116), if VV exchange is a dominating process and the vibrational temperature T_v exceeds the translational temperature T_0:

$$f(E) = B \exp\left(-\frac{E}{T_v} + \frac{x_e E^2}{T_0 \hbar\omega}\right), \quad \frac{1}{T_v} = \frac{\partial \ln f(E)}{\partial E}\bigg|_{E\to 0}. \qquad (3\text{–}124)$$

The exponentially parabolic Treanor distribution function, which provides a significant overpopulation of the highly vibrationally excited states, was illustrated in Fig. 3–3. To analyze the quite complicated VV flux (3–122), it can be divided into linear ($j^{(0)}$) and non-linear ($j^{(1)}$) components:

$$j_{VV}(E) = j_{VV}^{(0)}(E) + j_{VV}^{(1)}(E). \qquad (3\text{–}125)$$

The **linear VV flux component** $j_{VV}^{(0)}(E)$ corresponds to non-resonant VV exchange between a highly vibrationally excited molecule of energy E with the bulk of lower excited molecules ($0 < E' < T_v$):

$$j_{VV}^{(0)} = -D_{VV}(E)\left[\frac{\partial f(E)}{\partial E} + \left(\frac{1}{T_v} - \frac{2x_e E}{T_0 \hbar\omega}\right) f(E)\right]. \qquad (3\text{–}126)$$

The diffusion coefficient D_{VV} of excited molecules in the vibrational energy space, related to the non-resonant VV exchange of a molecule of energy E with the bulk of low vibrational energy molecules is given by

$$D_{VV}(E) = k_{VV}(E) n_0 (\hbar\omega)^2. \qquad (3\text{–}127)$$

The VV-exchange rate coefficient can be expressed as: $k_{VV}(E) \approx \frac{E T_v}{(\hbar\omega)^2} Q_{01}^{10} k_0$ $\exp(-\delta_{VV} E)$ (see Section 2.6.5), where δ_{VV} is an adiabatic relaxation parameter. Solution of the linear kinetic equation $j_{VV}^{(0)}(E) = 0$ with flux (3–126) gives the Treanor distribution function (3–124).

The **non-linear flux component** $j_{VV}^{(1)}(E)$ corresponds to the resonant VV exchange between two highly vibrationally excited molecules ($|E - E'| \leq \delta_{VV}^{-1}$):

$$j_{VV}^{(1)} = -D_{VV}^{(1)} \frac{\partial}{\partial E} \left[f^2(E) E^2 \left(\frac{2x_e}{T_0} - \hbar\omega \frac{\partial^2 \ln f(E)}{\partial E^2} \right) \right]. \tag{3–128}$$

The diffusion coefficient $D_{VV}^{(1)}$ describing the resonant VV exchange can be expressed as

$$D_{VV}^{(1)} = 3k_0 n_0 Q_{10}^{01} (\delta_{VV} \hbar\omega)^{-3}. \tag{3–129}$$

One solution of the non-linear kinetic equation $j_{VV}^{(1)}(E) = 0$ with flux (3–128) is again the Treanor distribution function (3–124), which is, however, not the only solution of the equation; another solution, which is a plateau-like vibrational distribution, will be discussed in some detail in the next section.

3.4.3. Non-Equilibrium Vibrational Distribution Functions: Regime of Strong Excitation

Vibrational distributions in non-equilibrium plasma are mostly controlled by VV-exchange and VT-relaxation processes, while excitation by electron impact, chemical reactions, radiation, and so on determine averaged energy balance and temperatures. At steady state, the Fokker-Planck kinetic equation (3–116) gives $J(E) = $ const. At $E \to \infty$: $\frac{\partial f(E)}{\partial E} = 0$, $f(E) = 0$; therefore, const$(E) = 0$, which leads to

$$j_{VV}^{(0)}(E) + j_{VV}^{(1)}(E) + j_{VT}(E) = 0. \tag{3–130}$$

With fluxes (3–126), (3–128), and (3–118), the Fokker-Planck equation can be presented as

$$D_{VV}(E) \left(\frac{\partial f(E)}{\partial E} + \frac{1}{T_v} f(E) - \frac{2x_e E}{T_0 \hbar\omega} f(E) \right) + D_{VV}^{(1)} \frac{\partial}{\partial E} \left[f(E)^2 E^2 \left(\frac{2x_e}{T_0} - \hbar\omega \frac{\partial^2 \ln f(E)}{\partial E^2} \right) \right]$$
$$+ D_{VT}(E) \left(\frac{\partial f(E)}{\partial E} + \tilde{\beta}_0 f(E) \right) = 0. \tag{3–131}$$

The first two terms are related to VV relaxation and prevail at low vibrational energies; the third term is related to VT relaxation and dominates at higher vibrational energies. The linear part of equation (3–131), including the first and third terms, is easy to solve, but the second (non-linear) term makes the solution more complicated. It is helpful from this perspective to point out the three cases of **strong, intermediate, and weak excitation,** corresponding to different levels of contribution from the non-linear term and determined by two dimensionless parameters: $\delta_{VV} T_v$ and $\frac{x_e T_v^2}{\hbar\omega T_v}$.

The regime of **strong excitation** takes place at high vibrational temperatures ($\delta_{VV} T_v \geq 1$). The adiabatic factor δ_{VV} (see Section 2.6.5) at room temperature is $\delta_{VV} \approx (0.2 \div 0.5)/\hbar\omega$. Thus, the strong excitation regime requires vibrational temperatures exceeding 5000–10,000 K. In this case the resonant VV exchange between highly excited molecules (the non-linear VV flux) dominates over the non-resonant VV exchange of the excited molecules with the bulk of low excited molecules (the linear VV flux), and kinetic equation (3–131) neglecting VT relaxation can be expressed as

$$E^2 f^2(E) \left(\frac{2x_e}{T_0} - \hbar\omega \frac{\partial^2 \ln f(E)}{\partial E^2} \right) = F. \tag{3–132}$$

Here F is a constant proportional to quantum flux along the vibrational spectrum. In addition to the Treanor distribution, solving kinetic equation (3–132) also gives the **hyperbolic plateau distribution**:

$$E^2 f^2(E) \frac{2x_e}{T_0} = F, \quad f(E) = \frac{C}{E}. \tag{3–133}$$

The "plateau level" C is determined by the power of vibrational excitation per single molecule, P_{eV}:

$$C(P_{eV}) = \frac{1}{\hbar\omega} \sqrt{\frac{P_{eV} T_0 (\delta_{VV} \hbar\omega)^3}{4x_e k_0 n_0 Q_{10}^{01}}}. \tag{3–134}$$

The vibrational distribution is determined by the Treanor function (3–124) at lower energies:

$$E < E_{Tr} - \hbar\omega \sqrt{\frac{T_0}{2x_e \hbar\omega}}, \tag{3–135}$$

where $E_{Tr} = \hbar\omega \cdot v_{min}^{Tr}$ is the Treanor minimum energy (see (3–38)). At the vibrational energies exceeding the Treanor minimum point ($E > E_{Tr}$), the distribution function becomes the hyperbolic plateau:

$$f(E) = B \frac{E_{Tr}}{E} \exp\left(-\frac{T_0 \hbar\omega}{4x_e T_v^2} - \frac{1}{2} \right), \tag{3–136}$$

where B is the Treanor normalization factor. Transition from plateau (3–136) to the quickly decreasing Boltzmann distribution with temperature T_0 occurs at vibrational energies exceeding the critical one:

$$E(\text{plateau} - VT) = \frac{1}{\delta_{VT}} \left[\ln \frac{k_0 Q_{10}^{01}}{k_{VT}(E=0)} \frac{E_{Tr}}{\hbar\omega} - \frac{T_0 \hbar\omega}{4x_e T_v^2} - \frac{1}{2} \right]. \tag{3–137}$$

This plateau–Boltzmann transitional energy corresponds to the equality of probabilities of the VT relaxation and the resonant VV exchange. The vibrational distribution function in the regime of strong excitation in non-thermal plasma is illustrated in Fig. 3–7.

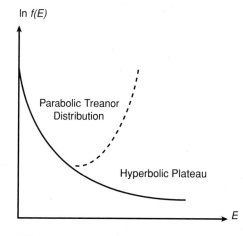

Figure 3–7. Non-equilibrium vibrational energy distribution functions in diatomic molecular gases: strong and intermediate excitation regimes.

Vibrational distribution similar to that presented in Fig. 3–7 also takes place in the **regime of intermediate excitation**, which occurs when $T_v < \delta_{VV}^{-1}$, but the Treanor effect is strong:

$$\frac{x_e T_v^2}{T_0 \hbar \omega} \geq 1. \tag{3–138}$$

The population of vibrationally excited states at the Treanor minimum ($E = E_{Tr}$) is large in this regime, and the non-linear resonant VV exchange dominates and provides a plateau at $E > E_{Tr}$ even though $T_v < \delta_{VV}^{-1}$. At low levels ($E < E_{Tr}$), the linear non-resonant VV exchange dominates over the non-linear one. It does not change the vibrational distribution function, however, because both non-resonant and resonant components of the VV exchange result in the same Treanor distribution at $E < E_{Tr}$.

3.4.4. Non-Equilibrium Vibrational Distribution Functions: Regime of Weak Excitation

Weak excitation means lower vibrational temperature (<5–$10 \cdot 10^3$ K) and a lower Treanor factor:

$$\delta_{VV} T_v < 1, \ \frac{x_e T_v^2}{T_0 \hbar \omega} < 1. \tag{3–139}$$

In this case the non-linear term in the Fokker-Planck kinetic equation (3–131) can be neglected:

$$\frac{\partial f(E)}{\partial E}(1 + \xi(E)) + f(E) \cdot \left(\frac{1}{T_v} - \frac{2x_e E}{T_0 \hbar \omega} + \xi(E)\tilde{\beta}_0\right) = 0, \tag{3–140}$$

where $\xi(E)$ is an exponentially growing ratio (2–191) of VT- and non-resonant VV-relaxation rates. Solution of differential equation (3–140) gives the following distribution (Rusanov, Fridman, & Sholin, 1979a,b):

$$f(E) = B \exp\left[-\frac{E}{T_v} + \frac{x_e E^2}{T_0 \hbar \omega} - \frac{\tilde{\beta}_0 - \frac{1}{T_v}}{2\delta_{VV}} \ln(1 + \xi(E))\right]. \tag{3–141}$$

The continuous function $f(E)$ corresponds to the discrete distribution $f(v)$ over vibrational quantum numbers, known as the **Gordiets vibrational distribution** (Gordiets et al., 1972):

$$f(v) = f_{Tr}(v) \prod_{i=0}^{v} \frac{1 + \xi_i \exp\dfrac{\hbar \omega}{T_0}}{1 + \xi_i}, \tag{3–142}$$

where $f_{Tr}(v)$ is the discrete Treanor distribution (3–124). At low vibrational energies ($E < E^*(T_0)$, see relation (2.192)), $\xi(E) \ll 1$ and the vibrational distribution is close to the Treanor distribution. At higher energies ($E > E^*(T_0)$), $\xi(E) \gg 1$ and the vibrational distribution is exponentially decreasing with temperature T_0 according to the Boltzmann law.

The logarithm of the vibrational distribution ($\ln f(E)$) (3–141) always has an **inflection point** ($\frac{\partial^2 \ln f(E)}{\partial E^2} = 0$) corresponding to the vibrational energy:

$$E_{infl} = E^*(T_0) - \frac{1}{2\delta_{VV}} \ln\frac{\delta_{VV} T_0}{x_e} < E^*(T_0). \tag{3–143}$$

Vibrational distribution function (3–141) can include a **domain of inverse population** ($\frac{\partial \ln f(E)}{\partial E} > 0$; see Fig. 3–8), which occurs if

$$2\delta_{VV}(E^*(T_0) - E_{Tr}) > \ln\frac{x_e}{\delta_{VV} T_0}. \tag{3–144}$$

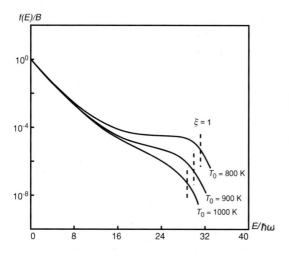

Figure 3–8. Non-equilibrium vibrational distributions in nitrogen at vibrational temperature $T_v = 3000\,\text{K}$ and different translational temperatures; $\xi = 1$ indicates vibrational energy when VV- and VT-relaxation rates are equal.

The vibrational distribution in N_2 at $T_v = 3000$ K and different translational temperatures is shown in Fig. 3–8. Comparison of different theoretical vibrational distribution functions for the same conditions is presented in Fig. 3–9 (dashed lines designate $\xi(E) = 1$). In addition to the difference between Treanor and Boltzmann distributions, Fig. 3–9 also shows (1) the Bray distribution (1968), taking into account VT relaxation only when $\xi(E) \geq 1$ and overestimating the actual $f(E)$; and (2) the completely classical ($\tilde{\beta}_0 = \beta_0$) Brau distribution (1972) which, in contrast, underestimates the actual distribution function $f(E)$.

3.4.5. Kinetics of Population of Electronically Excited States in Plasma

The transfer of electronic excitation energy in collisions of heavy particles is effective only for a limited number of specific electronically excited states (see Section 2.6.7). Even for high levels of electronic excitation, transitions between them are mostly due to collisions with plasma electrons at ionization degrees exceeding 10^{-6}. At relatively low pressures, radiation transitions can be significant as well (Slovetsky, 1980). The population of highly electronically excited states in plasma, $n(E)$, provided by energy exchange with electron gas, can also be described by the Fokker-Planck kinetic equation (Beliaev & Budker,

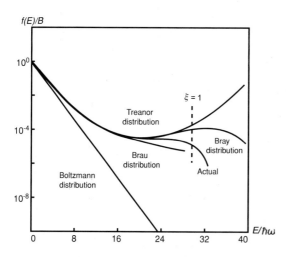

Figure 3–9. Comparison of different models of non-equilibrium vibrational energy distribution functions in nitrogen: $T_v = 3000\,\text{K}$, $T_0 = 800\,\text{K}$.

3.4. Kinetics of Vibrationally and Electronically Excited Molecules in Plasma

1958; Pitaevsky, 1962; Biberman, Gudzenko & Yakovlenko, 1978; Vorobiev, & Yakubov, 1979):

$$\frac{\partial n(E)}{\partial t} = \frac{\partial}{\partial E}\left[D(E)\left(\frac{\partial n(E)}{\partial E} - \frac{\partial \ln n^0}{\partial E}n(E)\right)\right].\qquad(3\text{--}145)$$

In this relation, $n^0(E)$ is a quasi-equilibrium population corresponding to electron temperature T_e:

$$n^0(E) \propto E^{-\frac{5}{2}}\exp\left(-\frac{E_1 - E}{T_e}\right),\qquad(3\text{--}146)$$

where E is the absolute value of the electronic excitation energy (transition to continuum at $E = 0$); E_1 is the ground-state energy ($E_1 \geq E$); Λ is the Coulomb logarithm for the electronically excited state; and the diffusion coefficient in energy space, $D(E)$, can be expressed (Kuznetsov, Raizer & Yu, 1965) as

$$D(E) = \frac{4\sqrt{2\pi}e^4 n_e E}{3\sqrt{mT_e}(4\pi\varepsilon_0)^2}\Lambda.\qquad(3\text{--}147)$$

Taking boundary conditions as $y(E_1) = y_1$, $y(0) = y_e y_i$ (where $y(E) = n(E)/n^0(E)$ is a new variable, y_e and y_i are electron and ion densities in plasma, divided by their equilibrium values), the steady-state solution of kinetic equation (3–145) can be expressed as

$$y(E) = \frac{y_1\chi\left(\dfrac{E}{T_e}\right) + y_e y_i\left[\chi\left(\dfrac{E_1}{T_e}\right) - \chi\left(\dfrac{E}{T_e}\right)\right]}{\chi\left(\dfrac{E_1}{T_e}\right)}.\qquad(3\text{--}148)$$

In this relation, $\chi(x)$ is a special function determined by the integral

$$\chi(x) = \frac{4}{3\sqrt{\pi}}\int_0^x t^{\frac{3}{2}}\exp(-t)dt.\qquad(3\text{--}149)$$

For electronically excited levels close to continuum ($E \ll T_e \ll E_1$), the relative population is

$$y(E) \approx y_e y_i\left[1 - \frac{1}{2\sqrt{\pi}}\left(\frac{E}{T_e}\right)^{\frac{5}{2}}\right] + y_1\frac{1}{2\sqrt{\pi}}\left(\frac{E}{T_e}\right)^{\frac{5}{2}} \to y_e y_i.\qquad(3\text{--}150)$$

This population of electronically excited states decreases exponentially with effective Boltzmann temperature T_e and has an absolute value corresponding to equilibrium with continuum $y(E) \to y_e y_i$. In the opposite case, the population of electronically excited states far from continuum ($E \gg T_e$) can be found as

$$y(E) = y_1 + y_e y_i\frac{4}{3\sqrt{\pi}}\exp\left(-\frac{E}{T_e}\right)\cdot\left(\frac{E}{T_e}\right)^{\frac{3}{2}} \to y_1.\qquad(3\text{--}151)$$

When electronic excitation energy is far from continuum, the population is also exponential with effective temperature T_e, but the absolute value corresponds in this case to equilibrium with the ground state. The Boltzmann distribution of electronically excited states with temperature T_e requires very high ionization degrees ($n_e/n_0 \geq 10^{-3}$; although domination of energy exchange with electron gas requires only $n_e/n_0 \geq 10^{-6}$). The requirement is mostly due to the influence of some resonance transitions and the non-Maxwellian behavior of electron energy distribution functions at lower ionization degrees. The radiative deactivation of electronically excited particles is essential at low pressures, usually when $p < 1$–10 Torr. The contribution of the radiative processes decreases with excitation energy approaching the continuum. It can be explained by both a reduction of the intensity of the radiative processes and an intensification of collisional energy exchange when the electron bonding energy in an atom becomes lower. As a result, we can suggest a numerical formula for the critical value of the electron bonding energy:

$$E_R, eV = \left(\frac{n_e, cm^{-3}}{4.5 \cdot 10^{13} cm^{-3}} \right)^{\frac{1}{4}} *(T_e, eV)^{-\frac{1}{8}}. \tag{3–152}$$

Collisional energy exchange dominates when the excitation level is higher and electron binding energy is lower ($E < E_R$). When the excitation level is not so high and bonding energy is significant ($E > E_R$), then deactivation is mostly due to radiation whereas electronic excitation still occurs collisionally.

3.4.6. Non-Equilibrium Translational Energy Distribution Functions of Heavy Neutrals: Effect of "Hot" Atoms in Fast VT-Relaxation Processes

The relaxation of translational energy of neutral particles requires only a couple of collisions and usually determines the shortest time scale in plasma-chemical systems. Assumption of a local quasi-equilibrium is usually valid for the translational energy subsystem of neutral particles, even in strongly non-equilibrium plasma. Thus, translational energy distributions are usually Maxwellian with one local temperature T_0 for all neutral components. However, this general rule has some important exceptions when high-energy atoms are formed that strongly perturb the conventional Maxwellian distribution. Generation of energetic "hot" atoms can be due to relaxation processes that are faster than Maxwellization (to be considered in the following), and also due to fast exothermic chemical reactions (to be discussed next). One of the fastest non-adiabatic VT-relaxation processes occurs between excited molecules (Mo) and alkaline atoms (Me) through the formation of ionic complexes $[Me^+Mo^-]$ (see Section 2.6.3). For example, VT relaxation of N_2^* on Li, K, and Na atoms takes place at almost every collision, which makes these alkaline atoms "hot" when they are added to non-equilibrium nitrogen plasma ($T_v \gg T_0$).

Consider evolution of the translational energy distribution function $f(E)$ of a small admixture of alkaline atoms in non-equilibrium diatomic molecular gas ($T_v \gg T_0$). The distribution is determined by competition of fast VT-relaxation energy exchange between the alkaline atoms and diatomic molecules and Maxwellization translational–translational (TT) processes in collisions of the same partners. It can be described by the Fokker-Planck kinetic equation for diffusion of the atoms along the translational energy spectrum (Vakar et al., 1981a,b,c,d):

$$D_{VT} \left(\frac{\partial f}{\partial E} + \frac{f}{T_v} \right) + D_{TT} \left(\frac{\partial f}{\partial E} + \frac{f}{T_0} \right) = 0. \tag{3–153}$$

where D_{VT} and D_{TT} are diffusion coefficients of the alkaline atoms Me along the energy spectrum, related to VT relaxation and Maxwellization, respectively; $\mu(E) = D_{TT}(E)/D_{VT}(E)$ is proportional to the ratio of the corresponding relaxation rate coefficients. Integration of linear differential equation (3–153) gives

$$f(E) \propto \exp \left[-\int \frac{\frac{1}{T_v} + \frac{\mu(E)}{T_0}}{1 + \mu(E)} dE \right]. \tag{3–154}$$

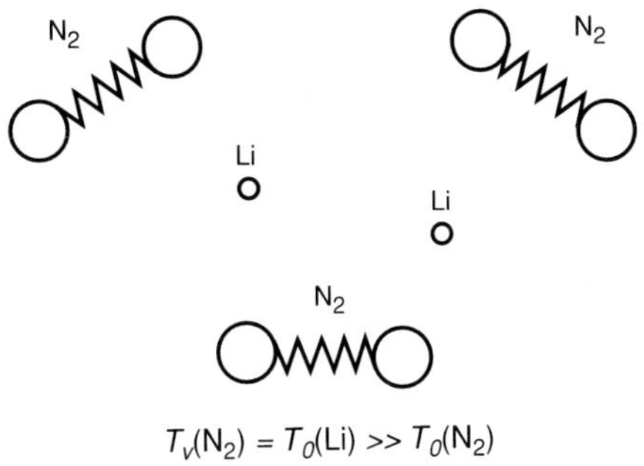

Figure 3–10. Quasi-equilibrium between N_2 vibrational degrees of freedom and Li atoms' translational degrees of freedom.

$$T_v(N_2) = T_0(Li) \gg T_0(N_2)$$

Typically, Maxwellization is much faster than VT relaxation ($\mu(E) \gg 1$), which leads to a Maxwellian distribution with temperature equal to T_0 of molecular gas. It can be different for an admixture of light alkaline atoms (like Li, atomic mass m) with relatively heavy molecular gas (like N_2 or CO_2, molecular mass $M \gg m$). VT- and TT-relaxation frequencies are almost equal in this mixture. Therefore, because of the difference in masses, energy transfer during the Maxwellization can be lower than $\hbar\omega$, which leads to $\mu \ll 1$ at low energies. Thus, according to (3–154), the translational temperature of the light alkaline atoms can be equal to vibrational (T_v) rather than translational (T_0) temperature of the molecular gas (Fig. 3–10).

For further calculation, the factor $\mu(E)$ can be expressed when $m \ll M$ as

$$\mu(E) \approx \frac{E + T_v}{\hbar\omega} \left(\frac{m}{M} \frac{E}{T_v} \right)^2. \tag{3–155}$$

The translational distribution $f(E)$ can be Maxwellian with vibrational temperature T_v only at relatively low translational energies ($E < E^*$). At higher energies ($E > E^*$), $\mu \gg 1$, and the exponential decrease of $f(E)$ always corresponds to temperature T_0. The critical translational energy $E^*(T_v)$ is

$$E^*(T_v) = \frac{M}{m}\sqrt{T_v\hbar\omega}, \quad \text{if } T_v > \hbar\omega\left(\frac{M}{m}\right)^2, \tag{3–156}$$

$$E^*(T_v) = (\hbar\omega)^{\frac{1}{3}} \left(\frac{T_v M}{m} \right)^{\frac{2}{3}}, \quad \text{if } T_v < \hbar\omega\left(\frac{M}{m}\right)^2. \tag{3–157}$$

Thus, spectroscopic Doppler broadening diagnostics of light alkaline atoms can be used to measure the vibrational temperature of molecular gases. It was applied, in particular, to measure T_v of CO_2 in non-equilibrium microwave discharges by adding Li and Na (Givotov, Rusanov, & Fridman, 1985b). The Doppler broadening was observed by Fourier analysis of Li-spectrum (transition $2^2 s_{1/2}$–$2^2 p_{1/2}$, $\lambda = 670.776$ nm) and Na-spectrum lines (transition $3^2 s_{1/2}$–$3^2 p_{1/2}$, $\lambda = 588.995$ nm). Some results are presented in Fig. 3–11 to demonstrate the effect of "hot" atoms related to fast VT relaxation. Although the gas temperature was less than 1000 K, the alkaline temperature was up to 10 times higher. The sodium temperature ($M/m = 1.9$) was always lower than that of lithium ($M/m = 6.3$). Similar measurements in N_2 plasma using the in-cavity laser spectroscopy of Li, Na, and Cs additives were accomplished by Ahmedganov and colleagues (1986).

3.4.7. Generation of "Hot" Atoms in Chemical Reactions

If fast atoms generated in exothermic processes react before Maxwellization, they are able to perturb translational distribution and accelerate the following reactions. A laser-chemical

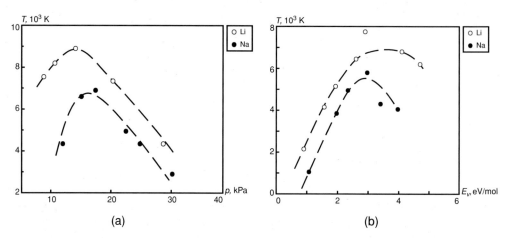

Figure 3–11. Temperatures of Li and Na atoms in non-equilibrium CO_2 microwave discharge as a function of (a) pressure, at specific energy input $E_v = 3\,J/cm^3$; and (b) specific energy input at pressure $p = 15.6\,kPa$.

chain, stimulated in H_2–F_2 mixture by electron beam plasma, is an example of this effect (Piley & Matzen, 1975):

$$H + F_2 \rightarrow HF^* + F, \quad F + H_2 \rightarrow HF^* + H. \tag{3–158}$$

In non-equilibrium plasma, the "hot" atoms can be generated in endothermic chemical reactions as well. For example, vibrationally excited molecules participate in endothermic reactions with some excess of energy, which goes into the translational energy of products (Fridman & Kennedy, 2004):

$$\langle \Delta E_v \rangle = \left| \left[\frac{\partial \ln f}{\partial E}(E = E_a) \right]^{-1} \right|. \tag{3–159}$$

Vibrationally excited molecules are very effective in the stimulation of endothermic chemical reactions. But the exothermic reactions with activation barriers are not stimulated by molecular vibrations (see Section 2.7), which slows down the whole process. In this case the hot atoms can make a difference by accelerating exothermic processes. This effect will be illustrated in Section 6.3.7 in the discussion of NO synthesis in non-equilibrium plasma.

3.5. VIBRATIONAL KINETICS OF GAS MIXTURES, CHEMICAL REACTIONS, AND RELAXATION PROCESSES

3.5.1. Kinetic Equation and Vibrational Distributions in Gas Mixtures: Treanor Isotopic Effect in Vibrational Kinetics

Consider the kinetic equation for vibrational distributions $f_i(E)$ for a two-component gas mixture (Mamedov, 1979; Rich & Bergman, 1986), which is relevant for molecular isotopes:

$$\frac{\partial f_i(E)}{\partial t} + \frac{\partial}{\partial E}\left[j_{VV}^{(i)}(E) + j_{VT}^{(i)}(E) + j_{VV'}^{(i)}(E) \right] = 0. \tag{3–160}$$

Subscripts $i = 1, 2$ correspond to first (lower oscillation frequency) and second (higher frequency) molecular components. VV and VT fluxes, $j_{VV}^{(i)}(E)$ and $j_{VT}^{(i)}(E)$, are similar to those for a one-component gas; the VV′-relaxation flux describing the energy transfer between components, providing the predominant population of a component with lower oscillation frequency ($i = 1$), can be expressed (Macheret et al., 1980a,b) as

$$j_{VV'}^{(i)} = -D_{VV'}^{(i)} \left(\frac{\partial f_i}{\partial E} + \beta_i f_i - 2x_e^{(i)} \beta_0 \frac{E}{\hbar \omega_i} f_i \right). \tag{3–161}$$

In this relation, $x_e^{(i)}$ and ω_i are the anharmonicity coefficient and oscillation frequency for component i; $D_{VV'}^{(i)} = k_{VV'}^{(i)} n_{l \neq i} (\hbar \omega_i)^2$, is the corresponding diffusion coefficient in energy space; and the rate coefficient $k_{VV'}^{(i)}$ corresponds to intercomponent VV' exchange and is related to the rate coefficient $k_{VV}^{(0)}$ of the resonant VV exchange at the low vibrational levels of the component i as

$$k_{VV'}^{(i)} = k_{VV}^{(0)} \exp \left[\frac{\delta_{VV} \hbar (\omega_i - \omega_l)}{2x_e^{(i)}} - \delta_{VV} E \right]. \tag{3–162}$$

The inverse temperature factor β_i in the VV' flux (3–161) can be found from the relation

$$\beta_i = \frac{\omega_l}{\omega_i} \beta_{vl} + \frac{\omega_i - \omega_l}{\omega_i} \beta_0. \tag{3–163}$$

Other inverse temperature parameters in relation (3–163) are $\beta_0 = T_0^{-1}$, $\beta_{vi} = (T_{vi})^{-1}$. Isotope mixtures usually consist of a larger fraction of a light gas component (higher frequency of molecular oscillations, concentration n_2) and only a small fraction of a heavy component (lower oscillation frequency, concentration n_1). In this case, the steady-state solution of the Fokker-Planck kinetic equation (3–160) gives

$$f_{1,2}(E) = B_{1,2} \exp \left[-\int_0^E \frac{\beta_{1,v2} - 2x_e^{(1,2)} \beta_0 \frac{E'}{\hbar \omega_{1,2}} + \tilde{\beta}_0^{(1,2)} \xi_{1,2}}{1 + \xi_{1,2}} dE' \right]. \tag{3–164}$$

In this relation, $B_{1,2}$ are normalization factors; factor ξ_i describes the relative contribution of VT and VV (VV')-relaxation processes (Section 3.4.4): $\xi_i(E) = \xi_i(0) \exp(2\delta_{VV} E)$. Neglecting VT relaxation at low vibrational energies ($\xi_i \ll 1$), integral (3–164) gives, for both gas components ($i = 1, 2$), the Treanor distribution (3–124) with the same T_0, but different vibrational temperatures T_{v1} and T_{v2}:

$$\frac{\omega_1}{T_{v1}} - \frac{\omega_2}{T_{v2}} = \frac{\omega_1 - \omega_2}{T_0}. \tag{3–165}$$

This relation is known as the **Treanor formula for an isotopic mixture**. It indicates that, in non-equilibrium conditions ($T_{v1,v2} > T_0$) of an isotopic mixture, a component with lower oscillation frequency (heavier isotope) has higher vibrational temperature. An experimental illustration of the Treanor effect in an isotopic mixture of $^{12}C^{16}O$ and $^{12}C^{18}O$ is presented in Fig. 3–12 (Bergman et al., 1983).

The Treanor effect (3–165) can be applied for isotope separation in plasma (Belenov et al., 1973). The ratio of rate coefficients of chemical reactions for two different vibrationally excited isotopes ($\kappa = k_R^{(1)}/k_R^{(2)}$), called the **coefficient of selectivity,** can be expressed as

$$\kappa \approx \exp \left[\frac{\Delta \omega}{\omega} E_a \left(\frac{1}{T_0} - \frac{1}{T_{v2}} \right) \right], \tag{3–166}$$

where $\frac{\Delta \omega}{\omega} = \frac{\omega_2 - \omega_1}{\omega_2}$ is the relative defect of resonance. In quasi-equilibrium kinetics, light isotopes react faster because they have higher "zero vibration level" $1/2 \, \hbar \omega$. In non-equilibrium plasma, on the other hand, heavy isotopes react faster due to the Treanor effect.

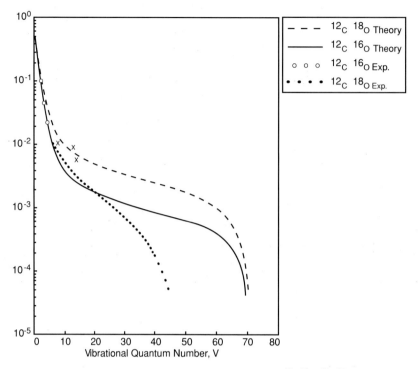

Figure 3–12. Isotopic effect in vibrational population of $^{12}C^{18}O/^{12}C^{16}O$ mixture in strongly non-equilibrium conditions ($T_v \gg T_0$).

The non-equilibrium effect is much stronger than the equilibrium effect. Numerical values of the coefficient of selectivity for different plasma-chemical processes of isotope separation stimulated by vibrational excitation are presented in Fig. 3–13. A detailed consideration of the Treanor-effect isotope separation can be found in Akulintsev, Gorshunov, and Neschimenko (1977, 1983) for nitrogen and carbon monoxide isotopes and in Eletsky and Zaretsky (1981) and Margolin, Mishchenko, and Shmelev (1980) for hydrogen isotopes.

3.5.2. Reverse Isotopic Effect in Plasma-Chemical Kinetics

Taking into account VT relaxation, integration of (3–164) gives the vibrational distribution for isotopes:

$$f_{1,2}(E) = B_{1,2} \exp\left[-\beta_{1,v2}E + \frac{x_e^{(1,2)}\beta_0 E^2}{\hbar\omega_{1,2}} - \frac{\tilde{\beta}_0^{(1,2)}}{2\delta_{VV}}\ln(1+\xi_{1,2}) \right]. \qquad (3\text{--}167)$$

Distributions (3–167) for the mixture of two isotopes are illustrated in Fig. 3–14. The population of lower vibrational levels is larger for the heavier isotope (1, usually small additive), which corresponds to the Treanor effect and relation (3–165). The situation is opposite at higher levels of excitation, where the vibrational population of a relatively light isotope exceeds that of a heavier one. This phenomenon is known as the **reverse isotopic effect** (Macheret et al., 1980a,b).

The vibrational distribution function $f_1(E)$ for the heavier isotope (small additive) is determined by VV' exchange between the isotopes, which is slower than VV exchange because of the defect of resonance. As a result, VT relaxation makes the vibrational distribution $f_1(E)$ decrease at lower energies $E_1(\xi_1 = 1)$ with respect to the distribution

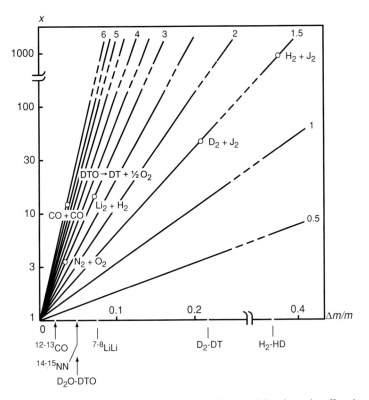

Figure 3–13. Coefficients of selectivity (x) characterizing isotopic effect in non-equilibrium plasma-chemical reactions of different molecules as a function of relative molecular mass difference $\Delta M / M$ of the molecular isotopes. Numbers on curves represent activation energies of specific plasma-chemical reactions (in electron volts); circles represent positions of the specific plasma-chemical reactions of vibrationally excited molecules.

Figure 3–14. Vibrational distribution functions of two di-atomic molecular isotopes (1,2) in non-equilibrium conditions ($T_v \gg T_0$). Dashed lines represent relevant Boltzmann distribution functions for the isotopes; E_1 and E_2 indicate vibrational energies corresponding to a strong exponential decrease of the distribution functions related to VT relaxation from highly excited vibrational levels.

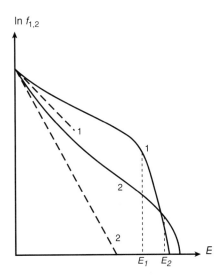

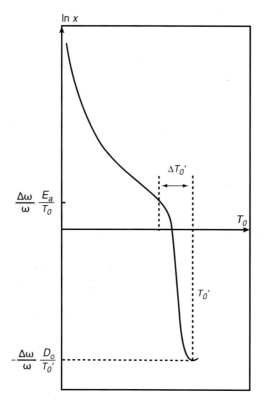

Figure 3–15. The non-equilibrium plasma $(T_v \gg T_0)$ isotopic effect dependence on translational gas temperature T_0; temperature T_0' corresponds to the maximum value of the selectivity coefficient of the inverse isotopic effect.

function of the main isotope, $f_2(E)$, which is determined by VV exchange and starts decreasing at higher vibrational energy $E_2(\xi_2 = 1) > E_1$:

$$E_1 = \frac{1}{2\delta_{VV}} \ln \frac{k_{VV}(0)}{k_{VT}(0)} - \frac{\hbar\Delta\omega}{4x_e}, \quad E_2 = \frac{1}{2\delta_{VV}} \ln \frac{k_{VV}(0)}{k_{VT}(0)}. \tag{3–168}$$

The reverse isotopic effect takes place if the activation energy of the plasma-chemical process appears in the specific interval $E_1 < E_a < E_2$ (see Fig. 3–14). The light isotope is excited much more and reacts much faster in this case than the heavier one. The coefficient of selectivity for the reverse isotopic effect can be calculated from (3–167) and (3–168) as (Macheret et al., 1980):

$$\kappa \approx \exp\left(-\frac{\Delta\omega}{\omega}\frac{D_0}{T_0}\right), \tag{3–169}$$

where D_0 is the dissociation energy of the vibrationally excited diatomic molecule participating in the reaction. The selectivity coefficient (3–169) for the reverse isotopic effect, similarly to (3–166), does not depend on vibrational temperature. Even more interesting, the activation energy $E_a < D_0$ is not explicitly presented in (3–169). Taking into account that $E_a < D_0$, the reverse isotopic effect (3–169) is much stronger than the direct effect (3–166). However, the reverse isotopic effect can be achieved in the pretty narrow range of translational temperatures:

$$\frac{\Delta T_0}{T_0} = 2\frac{\Delta\omega}{\omega}\left(1 - \frac{E_a}{D_0}\right)^{-1}. \tag{3–170}$$

The selectivity coefficient dependence on translational temperature is illustrated in Fig. 3–15. The direct effect takes place at relatively low T_0. By increasing the translational temperature, one can find the narrow temperature range (3–170) where the isotopic effect changes "direction" and becomes much stronger.

3.5.3. Macrokinetics of Chemical Reactions of Vibrationally Excited Molecules

Macrokinetic reaction rates of vibrationally excited molecules are self-consistent with the influence of the reactions on vibrational distribution functions $f(E)$, which can be taken into account by introducing into the Fokker-Planck kinetic equation (3–130) an additional flux related to the reaction:

$$j_R(E) = -\int_E^\infty k_R(E')n_0 f(E')dE' = -J_0 + n_0 \int_0^E k_R(E')f(E')dE', \quad (3\text{–}171)$$

where $J_0 = -j_R(E = 0)$ is the total flux of molecules entering into the chemical reaction (total reaction rate $w_R = n_0 J_0$); $k_R(E)$ is the microscopic reaction rate coefficient. In the weak excitation regime controlled by non-resonant VV and VT relaxation ($j_{VV}^{(0)} + j_{VT} + j_R = 0$), the Fokker-Planck kinetic equation is given as

$$\frac{\partial f(E)}{\partial E}(1 + \xi(E)) + f(E) \cdot \left(\frac{1}{T_v} - \frac{2x_e E}{T_0 \hbar \omega} + \tilde{\beta}_0 \xi(E)\right) = \frac{1}{D_{VV}(E)} j_R(E). \quad (3\text{–}172)$$

The solution $f(E)$ of non-uniform linear equation (3–172) can be presented with respect to solution $f^{(0)}(E)$ of the corresponding uniform equation (see (3–141)):

$$f(E) = f^{(0)}(E)\left[1 - \int_0^E \frac{-j_R(E')dE'}{D_{VV}(E')f(E')(1 + \xi(E'))}\right]. \quad (3\text{–}173)$$

The function $-j_R(E)$ determines the flux of molecules in the energy spectrum, which are going to react when they have enough energy ($E \geq E_a$). At relatively low energies ($E < E_a$), where chemical reaction can be neglected, $-j_R(E) = \int_{E_a}^\infty k_R(E')n_0 f(E')dE' = J_0 = $ const. As a result, perturbation of the vibrational distribution function $f^{(0)}(E)$ at $E < E_a$ by chemical reaction can be presented as

$$f(E) = f^{(0)}(E)\left[1 - J_0 \int_0^E \frac{dE'}{D_{VV}(E')f^{(0)}(E')(1 + \xi(E'))}\right]. \quad (3\text{–}174)$$

At $E \geq E_a$:

$$j_R(E) \approx -k_R(E)n_0 f(E)\hbar\omega, \quad (3\text{–}175)$$

and integral equation (3–173) can be solved:

$$f(E) \propto f^{(0)}(E) \exp\left[-\int_{E_a}^E \frac{k_R(E')\hbar\omega dE'}{D_{VV}(E')(1 + \xi(E'))}\right], \quad (3\text{–}176)$$

which determines the fast exponential decrease of the vibrational distribution at $E \geq E_a$. The total vibrational distribution, taking into account chemical reactions, is illustrated in Fig. 3–16. Reaction rates of vibrationally excited molecules can be calculated in two specific limits of slow and fast reactions:

1. **The fast reaction limit** implies that chemical reaction is fast at $E \geq E_a$:

$$D_{VV}(E = E_a) \ll n_0 \cdot k_R(E + \hbar\omega) \cdot (\hbar\omega)^2, \quad (3\text{–}177)$$

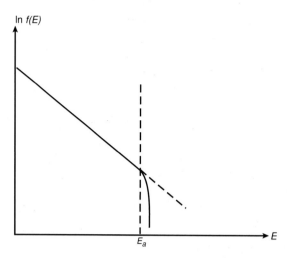

Figure 3–16. Influence of a chemical reaction on the vibrational distribution function in the weak excitation regime of non-equilibrium plasma ($T_v \gg T_0$).

and chemical process is limited by VV diffusion to the threshold $E = E_a$. In this case, the distribution function $f(E)$ falls very fast at $E > E_a$ (3–176), and we can assume $f(E = E_a) = 0$ in (3–173):

$$1 = \int_0^{E_a} \frac{-j_R(E')dE'}{D_{VV}(E')f^{(0)}(E')(1 + \xi(E'))}. \tag{3–178}$$

The preceding equation permits us to find the total chemical process rate for the fast reaction limit, taking into account that $-j_R(E) = J_0 = \text{const}$ at $E < E_a$:

$$w_R = n_0 J_0 = n_0 \left\{ \int_0^{E_a} \frac{dE'}{D_{VV}(E')f^{(0)}(E')(1 + \xi(E'))} \right\}^{-1}. \tag{3–179}$$

The chemical reaction rate in this case is determined by frequency of VV relaxation and by non-perturbed vibrational distribution $f^{(0)}(E)$. The rate (3–179) is not sensitive to the characteristics of elementary chemical reactions. Such a situation in practical plasma chemistry takes place, for example, in CO_2 and H_2O monomolecular dissociation processes (see Chapter 5).

2. **The slow reaction limit** corresponds to an inequality opposite that of (3–177). In this case the population of highly reactive states ($E > E_a$) by VV exchange is faster than the elementary chemical reaction itself. The vibrational distribution function in this case is almost not perturbed by chemical reaction $f(E) \approx f^{(0)}(E)$, and the total macroscopic reaction rate coefficient can be found as

$$k_R^{\text{macro}} = \int_0^{\infty} k_R(E')f(E')dE'. \tag{3–180}$$

In the slow reaction limit, information about the elementary chemical process is explicitly presented in the formula for the rate coefficient (Levitsky, Macheret, & Fridman, 1983a,b).

3.5.4. Vibrational Energy Losses Due to VT Relaxation

Consider losses of average vibrational energy, $\varepsilon_v = \int_0^\infty E f(E)\, dE$. Multiplication of the Fokker-Planck equation (3–116) with the following integration leads to the energy balance equation:

$$\frac{d\varepsilon_v}{dt} = \int\limits_0^\infty j_{VV}(E)dE + \int\limits_0^\infty j_{VT}(E)dE. \qquad (3\text{–}181)$$

Only losses due to VT- and VV-relaxation processes are taken into account here. Consider first those losses related to VT relaxation. Assuming the VT diffusion coefficient in energy space is given as

$$D_{VT}(E) = k_{VT}(E)n_0(\hbar\omega)^2 = k_{VT}^0\left(\frac{E}{\hbar\omega} + 1\right)\exp(\delta_{VT}E)n_0(\hbar\omega)^2, \qquad (3\text{–}182)$$

vibrational energy losses due to VT relaxation can be found from (3–181) as

$$\left(\frac{d\varepsilon_v}{dt}\right)^{VT} = D_{VT}(0)f(0) - \int\limits_0^\infty D_{VT}(0)\left[\frac{(\tilde{\beta}_0 - \delta_{VT})E - 1}{\hbar\omega} - (\tilde{\beta}_0 - \delta_{VT})\right]\exp(\delta_{VT}E)f dE. \qquad (3\text{–}183)$$

The losses can be subdivided into those related to low vibrational levels and prevailing at weak excitation, and those due to high levels and dominating at strong excitation. The **VT losses, related to low levels**, can be described based on (3–182) by the **Losev formula** (Losev, Shatalov, & Yalovik, 1970):

$$\left(\frac{d\varepsilon_v}{dt}\right)_L = -k_{VT}(0)n_0\left[1 - \exp\left(-\frac{\hbar\omega}{T_0}\right)\right] \cdot \left[\frac{1 - \exp\left(-\dfrac{\hbar\omega}{T_v}\right)}{1 - \exp\left(-\dfrac{\hbar\omega}{T_v} + \delta_{VT}\hbar\omega\right)}\right]^2 (\varepsilon_v - \varepsilon_{v0}), \qquad (3\text{–}184)$$

where $\varepsilon_{v0} = \varepsilon_v(T_v = T_0)$. Neglecting anharmonicity ($\delta_{VT} = 0$), the Losev formula can be rewritten as the **Landau-Teller relation**:

$$\left(\frac{d\varepsilon_v}{dt}\right)_L = -k_{VT}(0)n_0\left[1 - \exp\left(-\frac{\hbar\omega}{T_0}\right)\right](\varepsilon_v - \varepsilon_{v0}). \qquad (3\text{–}185)$$

Contribution of high levels into the VT losses is usually related to the highest vibrational levels before the fast decrease in the distribution function due to the VT relaxation or chemical reaction. These losses in the fast reaction limit can be calculated from (3–183) as ($T_0 \ll \hbar\omega$, $\delta_{VT}T \ll 1$):

$$\left(\frac{d\varepsilon_v}{dt}\right)_H^{VT} \approx -D_{VT}(0)f(0) - \left[\frac{(\tilde{\beta}_0 - \delta_{VT})E_a - 1}{\hbar\omega} - (\tilde{\beta}_0 - \delta_{VT})\right]D_{VT}(0)$$
$$\times \exp(\delta_{VT}E_a)f^{(0)}(E_a)\Delta \approx -k_{VT}(E_a)n_0\hbar\omega f^{(0)}(E_a)\Delta, \qquad (3\text{–}186)$$

where $f^{(0)}(E)$ is the distribution not perturbed by reaction, and $\Delta = |\frac{1}{T_v} - \delta_{VT} - \frac{2x_e E_a}{T_0\hbar\omega}|^{-1}$. The losses due to VT relaxation from high levels can be presented per reaction ($\xi(E_a) \ll 1$) as

$$\Delta\varepsilon_{VT} \approx \frac{k_{VT}(E_a)}{k_{VV}(E_a)}\Delta. \qquad (3\text{–}187)$$

Fast reaction and VT relaxation from high levels are related to the same molecules with energies slightly exceeding E_a. The frequencies of the processes are, however, different and proportional, respectively, to k_{VV} and k_{VT}. (More details on the subject can be found in Demura et al., 1981.)

3.5.5. Vibrational Energy Losses Due to Non-Resonance VV Exchange

These vibrational energy losses can be calculated based on the general relation (3–181), but it is more convenient to consider them per stage of chemical reaction. Consider, for example, a diatomic molecule excited by electron impact to the first vibrational level; thus, a quantum comes to the system as a "big one": $\hbar\omega$. Further population of the higher vibrational levels is due to the single-quantum VV exchange. The quanta become smaller during the VV exchange due to anharmonicity. Each step up on the "vibrational ladder" results in the transfer of energy $2x_e\hbar\omega v$ from vibrational to translational degrees of freedom. Thus, the total vibrational energy losses corresponding to excitation of a molecule to the nth level are given by

$$\Delta\varepsilon^{VV}(n) = \sum_{v=0}^{v=n-1} 2x_e\hbar\omega v = x_e\hbar\omega(n-1)n. \qquad (3\text{–}188)$$

In particular, for the process of dissociation of diatomic molecules stimulated by vibrational excitation, $n = n_{max} \approx 1/2x_e$. In this case, the total losses of vibrational energy per dissociation event, associated with the non-resonant nature of VV relaxation, are equal to the dissociation energy of the molecule:

$$\Delta\varepsilon_D^{VV}(n = n_{max}) \approx \frac{\hbar\omega}{4x_e} = D_0. \qquad (3\text{–}189)$$

Thus, the total vibrational energy required for dissociation through VV exchange equals not D_0, but $2D_0$. This amazing fact was first mentioned by Sergeev and Slovetsky (1979) and was then described by Demura et al. (1981). If reaction is stimulated by vibrational excitation and has activation energy $E_a \ll D_0$, the VV losses can be calculated as

$$\Delta\varepsilon_R^{VV}(E_a) \approx \frac{1}{4}D_0\left(\frac{E_a}{D_0}\right)^2 = x_e\frac{E_a^2}{\hbar\omega}. \qquad (3\text{–}190)$$

Consider, as an example, plasma-chemical NO synthesis in air, stimulated by vibrational excitation of N_2 molecules (see Chapter 6). According to the Zeldovich mechanism, this synthesis is limited by

$$O + N_2^* \rightarrow NO + N, \quad E_a = 1.3\,eV, \quad D_0 = 10\,eV. \qquad (3\text{–}191)$$

Energy losses (3–191) in this reaction are equal to 0.28 eV per NO molecule, which results in a 14% decrease of energy efficiency for the total plasma-chemical process (Macheret et al., 1980a,b).

3.6. ENERGY BALANCE AND ENERGY EFFICIENCY OF PLASMA-CHEMICAL PROCESSES

3.6.1. Energy Efficiency of Quasi-Equilibrium and Non-Equilibrium Plasma-Chemical Processes

One of the most important problems of applied plasma chemistry is the minimization of energy consumption. Plasma-chemical processes mostly consume electricity, which is presently a relatively expensive form of energy. It usually results in tough energy-efficiency requirements for the processes to be competitive with other technologies using just heat or chemical sources of energy. The energy efficiency of a plasma-chemical process is closely related to its mechanism. The same processes, organized in different discharge systems or under different conditions (corresponding to different mechanisms), require very different energy. For example, exhaust air cleaning from SO_2 using corona discharge requires 50–70 eV/mol. The same process organized in special plasma conditions provided by relativistic electron beams requires about 1 eV/mol (Baranchicov et al., 1990a,b), almost a hundred times less electric energy.

The energy-efficiency analysis is individualized for each specific plasma-chemical process and will be discussed in the relevant chapters related to specific plasma-chemical processes. Here we are going to discuss some general features of energy efficiency typical for specific plasma-chemical mechanisms. The **energy efficiency** η is the ratio of the thermodynamically minimal energy cost of a plasma-chemical process (which is usually the reaction enthalpy ΔH) to the actual energy consumption in plasma, W_{plasma}:

$$\eta = \Delta H / W_{plasma}. \tag{3–192}$$

The energy efficiency of the quasi-equilibrium plasma-chemical systems performed in thermal discharges is usually relatively low (less than 10–20%), which is due to two major effects:

1. Thermal energy in quasi-equilibrium plasma is distributed over all components and all degrees of freedom, although most of them are useless in the stimulation of a specific plasma-chemical reaction.
2. If a high-temperature gas, generated in thermal plasma and containing the process products, is cooled down slowly, the process products are converted back into initial substances via reverse reactions. Conservation of products of the quasi-equilibrium plasma-chemical processes requires the application of quenching – very fast product cooling – which considerably limits energy efficiency.

The energy efficiency of non-equilibrium plasma-chemical systems can be much higher. The discharge power can be selectively focused on the chemical reaction of interest without heating the whole gas. Also low gas temperature provides product stability with respect to reverse reactions without quenching. The energy efficiency in this case strongly depends on the process mechanism. The most important mechanisms of non-equilibrium plasma processes are compared in the sections that follow with regard to their energy efficiency.

3.6.2. Energy Efficiency of Plasma-Chemical Processes Stimulated by Vibrational Excitation of Molecules

This mechanism can provide the highest energy efficiency of endothermic plasma-chemical reactions in non-equilibrium conditions because of the following four factors:

1. A major fraction (70–95%) of discharge power in several molecular gases (including N_2, H_2, CO, and CO_2) at $T_e \approx 1$ eV can be transferred from plasma electrons to vibrational excitation (see Section 2.5.6).
2. The VT-relaxation rate is usually low at low gas temperatures. Therefore, the optimization of ionization degree and specific energy input permits one to utilize most of the vibrational energy in chemical reactions.
3. Vibrational energy is most effective in the stimulation of endothermic reactions (see Section 2.7).
4. The vibrational energy required for an endothermic reaction usually equals the activation barrier of the reaction and is much lower than the energy threshold of corresponding processes proceeding through electronic excitation. For example, dissociation of H_2 through vibrational excitation requires 4.4 eV; the same process through excitation of an electronically excited state $^3\Sigma_u^+$ requires twice as much: 8.8 eV.

Some processes, stimulated by vibrational excitation, are able to consume most of the discharge energy. The best example here is dissociation of CO_2, which can be arranged in non-thermal plasma with energy efficiency up to 90% (see Section 5.5). Almost all discharged power is spent selectively in this case through vibrational excitation on organization of the chemical process. No other single mechanism can provide energy efficiency as high as a plasma-chemical process.

3.6.3. Energy Efficiency of Plasma-Chemical Processes Stimulated by Electronic Excitation and Dissociative Attachment

Mechanisms of stimulation of plasma-chemical reactions by **electronic excitation** were discussed in Section 2.5.5. No one of the four kinetic factors mentioned in Section 3.6.2 can be applied in this case; therefore, energy efficiency is relatively low, usually below 20–30%. Plasma-chemical processes through electronic excitation can be energy effective if they initiate chain reactions. Such a situation takes place, for example, in NO synthesis, where the Zeldovich mechanism can be effectively initiated by dissociation of molecular oxygen through electronic excitation (see Section 6.1.2).

The energy threshold of *dissociative attachment* is lower than the threshold of dissociation into neutrals (see Section 2.3.1). When the electron affinity of products is high, like for some halogens and their compounds, the dissociative attachment can even be exothermic; this means not only a low energy requirement for the reaction, but also the transfer of a significant part of electron energy into dissociative attachment, which increases the energy efficiency of the dissociative attachment. The energy efficiency is strongly limited, however, by the energy cost of the production of an electron, which is lost during attachment and following ion–ion recombination. The energy cost of an electron, the ionization cost, is usually about 30–300 eV. As an example, the plasma-chemical dissociation of NF_3, which is a strongly electronegative gas, can be accomplished almost completely into atomic fluorine and molecular nitrogen in low-pressure non-thermal discharges (see Section 8.4.3). The process is provided by dissociative attachment with an energy cost of 30 eV/mol, which exactly corresponds to the energy cost of an electron in the high-electron-temperature NF_3 discharge (Conti, Fridman, & Raoux, 1999).

The plasma-chemical process that is based on dissociative attachment can become energy effective only if the same electron is able to participate in a reaction many times. In such a chain reaction, the detachment and liberation of the electron from a negative ion should be faster than the loss of the charged particle in ion–ion recombination. Water vapor dissociation in non-thermal plasma is an example of such a chain process. This plasma-chemical reaction is to be considered in detail in Sections 5.8.4 and 5.8.5. The energy efficiency of the chain process for this reaction can be relatively high: 40–50%.

3.6.4. Energy Balance and Energy Efficiency of Plasma-Chemical Processes Stimulated by Vibrational Excitation of Molecules

The vibrational energy balance in a plasma-chemical process can be illustrated by the following simplified one-component equation taking into account vibrational excitation by electron impact, VT relaxation, and chemical reaction (Rusanov & Fridman, 1984; Fridman & Kennedy, 2004):

$$\frac{d\varepsilon_v(T_v)}{dt} = k_{eV}(T_e)n_e\hbar\omega\theta\left(E_v - \int_0^t k_{eV}n_e\hbar\omega dt\right)$$

$$- k_{VT}^{(0)}(T_0)n_0[\varepsilon_v(T_v) - \varepsilon_{v0}(T_v = T_0)] - k_R(T_V, T_0)n_0\Delta Q_\Sigma. \quad (3\text{–}193)$$

The left-hand side of the equation presents the change in average vibrational energy per molecule, which in one mode approximation can be expressed by the Planck formula:

$$\varepsilon_v(T_v) = \frac{\hbar\omega}{\exp\left(\dfrac{\hbar\omega}{T_v}\right) - 1}. \tag{3-194}$$

The first right-hand-side term in (3–193) describes vibrational excitation of molecules (rate coefficient k_{eV}) by electron impact (density n_e). The Heaviside function $\theta(x) : [\theta = 1,$ if $x \geq 0; \theta = 0,$ if $x < 0]$ restricts the vibrational energy input per single molecule by the value of E_v. If most of the power is going to vibrational excitation, the **specific energy input** E_v per molecule is equal to the ratio of the discharge power and gas flow rate through the discharge. This experimentally convenient discharge parameter is related to energy efficiency (η) and conversion degree (χ) of a plasma-chemical process by the simple formula

$$\eta = \frac{\chi \times \Delta H}{E_v}, \tag{3-195}$$

where ΔH is the enthalpy of formation of a product molecule. The second right-hand-side term in equation (3–193) describes VT relaxation of the lower vibrational levels; n_0 is gas density. The third right-hand-side term is related to vibrational energy losses in the chemical reaction (reaction rate $k_R(T_v, T_0)$) and relaxation channels having rates corresponding to that of the chemical reaction. In this term the total losses of ΔQ_Σ vibrational energy per chemical reaction event can be expressed as

$$\Delta Q_\Sigma = E_a/\alpha + \langle \Delta E_v \rangle + \Delta\varepsilon_{VT} + \Delta\varepsilon_R^{VV}, \tag{3-196}$$

where E_a/α is the vibrational energy necessary to overcome the activation barrier of chemical reaction (α is the efficiency of vibrational energy; see Section 2.7.2); $\langle \Delta E_v \rangle$ characterizes the excess of vibrational energy in chemical reaction (3–159); and $\Delta\varepsilon_{VT}$ and $\Delta\varepsilon_R^{VV}$ are losses of vibrational energy due to respectively VT relaxation from high levels and non-resonance of VV exchange, respectively, expressed with respect to a single chemical reaction event (see Sections 3.5.4 and 3.5.5). The reaction rate coefficient depends on T_v being weighted more strongly than all other terms in balance equation (3–193). In the case of weak excitation, it can be estimated as

$$k_R(T_v, T_0) = A(T_0)\exp\left(-\frac{E_a}{T_v}\right). \tag{3-197}$$

The pre-exponential factor $A(T_0)$ differs here from the conventional Arrhenius factor A_{Arr}. For example, in the case of the Treanor distribution function, $A(T_0) = A_{Arr}\exp(\frac{x_e E_a^2}{T_0\hbar\omega})$. We can point out the critical value of vibrational temperature when the reaction and VT-relaxation rates are equal:

$$T_v^{min} = E_a \ln^{-1}\frac{A(T_0)\Delta Q_\Sigma}{k_{VT}^{(0)}(T_0)\hbar\omega}. \tag{3-198}$$

An increase of vibrational temperature results in a much stronger exponential acceleration of the chemical reaction than vibrational relaxation. Therefore, at high vibrational temperatures ($T_v > T_v^{min}$), almost all vibrational energy is going into chemical reaction. Conversely, at $T_v < T_v^{min}$, almost all vibrational energy can be lost in vibrational relaxation. The critical vibrational temperature T_v^{min} determines the threshold in the dependence $\eta(E_v)$ of energy efficiency on the specific energy input E_v:

$$(E_v)_{threshold} = \varepsilon_v(T_v^{min}) = \frac{\hbar\omega}{\exp\left(\dfrac{\hbar\omega}{T_v^{min}}\right) - 1}. \tag{3-199}$$

If the specific energy input is lower than the threshold value ($E_v < (E_v)_{threshold}$), the vibrational energy cannot reach the critical value ($T_v < T_v^{min}$) and the energy efficiency is very low. Typical threshold values of energy input for reactions stimulated by vibrational excitation are about $0.1 \div 0.2$ eV/mol.

Another important feature of the processes stimulated by vibrational excitation is a maximum in the dependence $\eta(E_v)$. The optimal value of the specific energy input E_v is usually about 1 eV. At larger values of E_v, the translational temperature becomes higher, which accelerates VT relaxation and decreases energy efficiency. Also at higher E_v, a significant part of the discharge energy is spent on the excitation of products. When the conversion degree χ approaches 100% at high E_v, energy efficiency decreases hyperbolically as $\eta \propto \Delta H / E_v$.

If the vibrational temperature exceeds the critical value ($T_v > T_v^{min}$), the relaxation losses in (3–193) can be neglected and almost all discharge energy proceeds through vibrational excitation into chemical reaction. The steady-state value of vibrational temperature in this case can be found as

$$T_v^{st} = E_a \ln^{-1} \left(\frac{A(T_0) \Delta Q_\Sigma}{k_{eV} \hbar \omega} \cdot \frac{n_0}{n_e} \right). \tag{3–200}$$

An energy-effective process is possible when $T_v^{st} > T_v^{min}$, which requires the ionization degree to exceed the critical value determined by the ratio of rate coefficients for VT relaxation and vibrational excitation:

$$n_e/n_0 \gg k_{VT}^{(0)}(T_0)/k_{eV}(T_e). \tag{3–201}$$

3.6.5. Components of Total Energy Efficiency: Excitation, Relaxation, and Chemical Factors

The total energy efficiency of any non-equilibrium plasma-chemical process can be subdivided into three main components: an excitation factor (η_{ex}), a relaxation factor (η_{rel}), and a chemical factor (η_{chem}):

$$\eta = \eta_{ex} \times \eta_{rel} \times \eta_{chem}. \tag{3–202}$$

The **excitation factor** gives the fraction of the discharge energy directed toward the production of the principal agent of the plasma-chemical reaction. For example, in NF_3 dissociation for chemical vapor deposition (CVD) chamber cleaning (Conti, Fridman, & Raoux, 1999), this factor shows the efficiency of the generation of F atoms responsible for cleaning. If processes are stimulated by vibrational excitation, the excitation factor (η_{ex}) shows the fraction of discharge energy going to this excitation channel.

The **relaxation factor** (η_{rel}) is related to conservation of the principal active species (like F atoms or vibrationally excited molecules in the preceding examples) with respect to their losses in relaxation, recombination, and so on. In processes stimulated by vibrational excitation, the factor η_{rel} shows the fraction of vibrational energy that can avoid vibrational relaxation and thus be spent in the chemical reaction of interest. VT relaxation is slow inside of a discharge zone if $T_v > T_v^{min}$; therefore, near the E_v threshold,

$$\eta_{rel} = \frac{E_v - \varepsilon_v(T_v^{min})}{E_v}. \tag{3–203}$$

The preceding relation indicates the threshold of dependence, $\eta(E_v)$, and shows that when the vibrational temperature finally becomes lower than the critical value, most of the vibrational energy goes to VT relaxation.

The **chemical factor** (η_{chem}) shows the efficiency of the principal active discharge species in the chemical reaction of interest. In the process of NF_3 dissociation and F-atom generation for CVD chamber cleaning, the chemical factor (η_{chem}) shows the efficiency of the atoms in the gasification of impurities. In plasma-chemical reactions stimulated by vibrational excitation, η_{chem} can be presented as

$$\eta_{\text{chem}} = \frac{\Delta H}{E_a/\alpha + \langle \Delta E_v \rangle + \Delta\varepsilon_{\text{VT}} + \Delta\varepsilon_R^{\text{VV}}}. \qquad (3\text{--}204)$$

The chemical factor restricts energy efficiency mostly because the activation energy often exceeds the reaction enthalpy: $E_a > \Delta H$. For example, in the plasma-chemical process of NO synthesis, proceeding by the Zeldovich mechanism (see Section 6.3.5), the second reaction is significantly exothermic. This results in a relatively low value of the chemical factor: $\eta_{\text{chem}} \approx 50\%$.

3.6.6. Energy Efficiency of Quasi-Equilibrium Plasma-Chemical Systems: Absolute, Ideal, and Super-Ideal Quenching

Quasi-equilibrium plasma-chemical processes can be subdivided into two phases. In the first phase, reagents are heated to the high temperatures required to shift chemical equilibrium in the direction of products. In the second phase, which is called quenching, the temperature decreases fast enough to protect the products produced on the first high-temperature phase from reverse reactions. Three specific quenching modes can be introduced to analyze efficiency of the quasi-equilibrium plasma-chemical processes.

1. Absolute quenching means that the cooling process is fast enough to save all products formed in the high-temperature zone. In the high-temperature zone, initial reagents are partially converted into the products of the process, but also into some unstable atoms and radicals. In the case of absolute quenching, stable products are saved, but atoms and radicals are converted back into the initial reagents during the cooling process.

2. Ideal quenching means that the cooling process is very effective and able to keep the total conversion degree on the same level as was reached in the high-temperature zone. The ideal quenching not only saves all the products formed in the high-temperature zone but also provides conversion of all the relevant atoms and radicals into the process products. It is interesting that, during the quenching phase, the total conversion degree can not only be saved, it can even be increased. Such quenching is usually referred to as super-ideal quenching.

3. Super-ideal quenching permits one to increase the conversion degree during the cooling stage using the chemical energy of atoms and radicals as well as the excitation energy accumulated in molecules. Super-ideal quenching can be organized, in particular, when gas cooling is faster than VT relaxation, and the VT non-equilibrium ($T_v < T_0$) can be achieved during quenching. In this case, direct endothermic reactions are stimulated by vibrational excitation, whereas reverse exothermic reactions related to translational degrees of freedom proceed more slowly. Such a misbalance between direct and reverse reactions provides additional conversion of initial substances or, in other words, provides super-ideal quenching. The kinetic details of different quenching mechanisms for different specific plasma-chemical processes will be considered later in relevant chapters.

3.6.7. Mass and Energy Transfer Equations in Multi-Component Quasi-Equilibrium Plasma-Chemical Systems

To analyze the influence of transfer phenomena on the energy efficiency of plasma-chemical processes, consider conservation equations, describing enthalpy (*I*) and mass transfer in a multi-component (number of components, N) quasi-equilibrium reacting gas (Frank-Kamenetsky, 1971; Williams, 1985):

$$\rho \frac{dI}{dt} = -\nabla \vec{q} + \frac{dp}{dt} + \Pi_{ik} \frac{\partial v_i}{\partial x_k} + \rho \sum_{\alpha=1}^{N} Y_\alpha \vec{v}_\alpha \vec{f}_\alpha, \tag{3-205}$$

$$\rho \frac{dY_\alpha}{dt} = -\nabla(\rho Y_\alpha \vec{v}_\alpha) + \omega_\alpha. \tag{3-206}$$

The thermal flux $\vec{q}$ can be expressed by neglecting the contribution of the radiation heat transfer:

$$\vec{q} = -\lambda \nabla T + \rho \sum_{\alpha=1}^{N} I_\alpha Y_\alpha \vec{v}_\alpha. \tag{3-207}$$

The diffusion velocity $\vec{v}_\alpha$ for the α-chemical component can be found in this case from the following relation:

$$\nabla x_\alpha = \sum_{\beta=1}^{N} \frac{x_\alpha x_\beta}{D_{\alpha\beta}} (\vec{v}_\beta - \vec{v}_\alpha) + (Y_\alpha - x_\alpha) \frac{\nabla p}{p} + \frac{\rho}{p} \sum_{\beta=1}^{N} Y_\alpha Y_\beta (\vec{f}_\alpha - \vec{f}_\beta)$$

$$+ \sum_{\beta=1}^{N} \left[\frac{x_\alpha x_\beta}{D_{\alpha\beta}} \frac{1}{\rho} \left(\frac{D_{T,\beta}}{Y_\beta} - \frac{D_{T,\alpha}}{Y_\alpha} \right) \right] \frac{\nabla T}{T}, \tag{3-208}$$

where $I = \sum_{\alpha=1}^{N} Y_\alpha I_\alpha$ is the total enthalpy per unit mass of mixture; Π_{ik} is the tensor of viscosity; $\vec{f}_\alpha$ is the external force per unit mass of the component α; ω_α is the mass change rate of component α per unit volume related to chemical reactions; λ, $D_{\alpha\beta}$, and $D_{T,\alpha}$ are coefficients of thermal conductivity, binary diffusion, and thermo-diffusion, respectively; $x_\alpha = n_\alpha/n$, $Y_\alpha = \rho_\alpha/\rho$ are molar and mass fractions of the component α; v_i is the component of hydrodynamic velocity; and p is the pressure of the gas mixture.

Neglecting enthalpy change due to viscosity, assuming the same forces $\vec{f}_\alpha$ for all components, and taking the binary diffusion coefficients as $D_{\alpha\beta} = D(1 + \delta_{\alpha\beta})$, $\delta_{\alpha\beta} < 1$, the energy and mass transfer equations can be rewritten as

$$\rho \frac{dI}{dt} = -\nabla \vec{q} + \frac{dp}{dt}, \tag{3-209}$$

$$\vec{q} = -\frac{\lambda}{c_p} \nabla I + (1 - Le) \sum_{\alpha=1}^{N} I_\alpha D \rho_\alpha \nabla \ln Y_\alpha + \sum_{\alpha=1}^{N} \rho_\alpha I_\alpha \vec{v}_\alpha^c + \sum_{\alpha=1}^{N} \rho_\alpha I_\alpha \vec{v}_\alpha^g, \tag{3-210}$$

$$\vec{v}_\alpha = -D \nabla \ln Y_\alpha + \vec{v}_\alpha^c + \vec{v}_\alpha^g. \tag{3-211}$$

The continuity equation remains in this set of equations in the form (3–206). Here $Le = \lambda/\rho c_p D$ is the Lewis number, and $c_p = \sum_{\alpha=1}^{N} Y_\alpha \frac{\partial I_\alpha}{\partial T}$ is the specific heat of a unit mass of the mixture; diffusion velocities are

$$\vec{v}_\alpha^c = D \sum_{\beta=1}^{N} Y_\beta \left(\frac{\vec{F}_\beta}{x_\beta} - \frac{\vec{F}_\alpha}{x_\alpha} \right). \tag{3-212}$$

$$\vec{v}_\alpha^g = \sum_{\beta=1}^{N} x_\beta \delta_{\alpha\beta} \left(\vec{v}_\beta^{(0)} - \vec{v}_\alpha^{(0)} \right) + \sum_{\gamma=1}^{N} Y_\gamma \sum_{\beta=1}^{N} \delta_{\beta\gamma} \left(\vec{v}_\beta^{(0)} - \vec{v}_\gamma^{(0)} \right). \tag{3-213}$$

The α-component force and the α-component velocities are given by

$$\vec{F}_\alpha = (Y_\alpha - x_\alpha)\frac{\nabla p}{p} + \sum_{\gamma=1}^{N} \frac{x_\alpha x_\gamma}{\rho D_{\alpha\beta}} \left(\frac{D_{T,\gamma}}{Y_\gamma} - \frac{D_{T,\alpha}}{Y_\alpha} \right) \frac{\nabla T}{T}, \tag{3-214}$$

$$\vec{v}_\alpha^{(0)} = -D\nabla \ln Y_\alpha + D\sum_{\beta=1}^{N} Y_\beta \left(\frac{\vec{F}_\beta}{x_\beta} - \frac{\vec{F}_\alpha}{x_\alpha} \right). \tag{3-215}$$

The set of equations describing energy and mass transfer in quasi-equilibrium plasma-chemical systems are analyzed in the next section to determine the influence of transfer phenomena on energy efficiency.

3.6.8. Transfer Phenomena Influence on Energy Efficiency of Plasma-Chemical Processes

Transfer phenomena do not change the energy efficiency limits established by absolute and ideal quenching (see Section 3.6.6) if (1) binary diffusion coefficients are fixed, equal to each other and to reduced coefficient of thermal conductivity (Lewis number $Le = 1, \delta_{\alpha\beta} = 0$); (2) effective pressure is constant, which implies an absence of external forces and large velocity gradients. To explain this rule, consider the enthalpy balance and continuity equations (Section 3.6.7) in the steady-state one-dimensional case:

$$\rho v \frac{\partial}{\partial x} I = \frac{\partial}{\partial x} \left(\rho D \frac{\partial}{\partial x} I \right), \tag{3-216}$$

$$\rho v \frac{\partial}{\partial x} Y_\alpha = \frac{\partial}{\partial x} \left(\rho D \frac{\partial}{\partial x} Y_\alpha \right) + \omega_\alpha. \tag{3-217}$$

Introduce dimensionless functions for reduced enthalpy and reduced mass fractions:

$$\xi = \frac{I - I^r}{I^l - I^r}, \quad \eta = \frac{Y_\alpha - Y_\alpha^r}{Y_\alpha^l - Y_\alpha^r}, \tag{3-218}$$

where Y_α^l, I^l and Y_α^r, I^r are mass fractions and enthalpy on the left (l) and right (r) boundaries of the region under consideration. With these functions, enthalpy conservation and continuity equations become

$$\rho v \frac{\partial \xi}{\partial x} = \frac{\partial}{\partial x} \rho D \frac{\partial \xi}{\partial x}, \tag{3-219}$$

$$\rho v \frac{\partial \eta_\alpha}{\partial x} = \frac{\partial}{\partial x} \rho D \frac{\partial \eta_\alpha}{\partial x} + \omega_\alpha \left(Y_\alpha^l - Y_\alpha^r \right). \tag{3-220}$$

The boundary conditions for these equations on the left and right sides of the region are

$$\xi(x = x_l) = \eta(x = x_l) = 1, \quad \xi(x = x_r) = \eta(x = x_r) = 0. \tag{3-221}$$

Chemical reactions (ω_α) during the diffusion process are generally able to either increase or decrease the concentrations of the reacting components. However, for the endothermic processes under consideration, chemical reactions only decrease the concentration of products during quenching. To find the maximum product yield, the reaction during diffusion can be neglected ($\omega_\alpha = 0$), which makes the equations and boundary conditions for η and ξ completely identical (meaning they are equal), and the gradients of total enthalpy I and mass fraction Y_α are related by the following formula (Potapkin, Rusanov, & Fridman, 1985):

$$\left(Y_\alpha^l - Y_\alpha^r \right)^{-1} \frac{\partial}{\partial x} Y_\alpha = (I^l - I^r)^{-1} \frac{\partial}{\partial x} I. \tag{3-222}$$

The minimum ratio of the enthalpy flux to the flux of products is equal to

$$A = \frac{I^l - I^r}{Y_\alpha^l - Y_\alpha^r},$$

(3–223)

which determines the energy cost of the product and completely correlates with expressions for the energy efficiency of quasi-equilibrium plasma-chemical processes (Section 3.6.6). Thus, neglecting external forces and assuming equality of the diffusion and reduced thermal conductivity coefficients (Lewis number $Le = 1$), the minimum energy cost of the plasma-chemical reaction products is determined by their formation in the quasi-equilibrium high-temperature zone. The minimum energy cost corresponds to the limits of absolute and ideal quenching and can be calculated as

$$\langle A \rangle = \frac{\int y(T)[I(T) - I_0]dT}{\int y(T)\chi(T)dT},$$

(3–224)

where $y(T)$ is the mass fraction of the initial substance heated to temperature T; $I(T)$ and I_0 are total enthalpy of the mixture at temperature T and initial temperature; and $\chi(T)$ is the conversion degree of the initial substance (for ideal quenching) or conversion degree into the final product (for absolute quenching), achieved in the high-temperature zone. Use of thermodynamic relation (3–224) for calculating the energy cost of chemical reactions in thermal plasma under conditions of intense heat and mass transfer is a consequence of the **similarity principle of concentrations and temperature fields**. This principle is valid at the Lewis number $Le = 1$ in the absence of external forces, and it is widely applied in combustion theory (Frank-Kamenetsky, 1971; Williams, 1985).

If $Le = \frac{\lambda}{\rho c_p D} \neq 1$, maximum energy efficiency can differ from that calculated for ideal quenching. For example, if light and fast hydrogen atoms are products of dissociation, the corresponding Lewis number is relatively small ($Le \ll 1$). Diffusion of products is faster in this case than energy transfer, and the energy cost can be lower than at ideal quenching. A significant deviation of energy efficiency from that of ideal quenching can be achieved by the influence of strong external forces. For example, if relatively heavy molecular clusters are produced in a rotating plasma, centrifugal forces make the product flux much more intensive than the heat flux, which results in a significant increase of energy efficiency. This effect will be discussed in Chapter 10 with regard to plasma-chemical H_2S dissociation.

3.7. ELEMENTS OF PLASMA ELECTRODYNAMICS

3.7.1. Ideal and Non-Ideal Plasmas

Plasma electrodynamics is a very important and widespread branch of plasma physics that covers in particular such topics as plasma sheaths, plasma oscillations and waves, propagation of electromagnetic waves in plasma, plasma instabilities, magneto-hydrodynamics of plasma, and collective and non-linear plasma phenomena. Only the most general aspects of plasma electrodynamics relevant to plasma chemistry will be discussed here. Details on the subject can be found, in particular, in books of Kadomtsev (1976), Ginzburg (1960), Boyd and Sanderson (2003), and Fridman and Kennedy (2004).

Consider first the general concepts of plasma ideality. The majority of plasmas are somewhat similar to gases from the point of view that electrons and ions move mostly in straight trajectories between collisions. It means that a potential energy $U \propto e^2/4\pi\varepsilon_0 R$, corresponding to the average distance between electrons and ions ($R \approx n_e^{-1/3}$), is much lower than the electrons' kinetic energy (about T_e):

$$\frac{n_e e^6}{(4\pi\varepsilon_0)^3 T_e^3} \ll 1.$$

(3–225)

3.7. Elements of Plasma Electrodynamics

Plasmas satisfying this condition are called **ideal plasmas.** A non-ideal plasma, corresponding to inverse inequality (3–226) and a very high density of charged particles, is not found in nature. Even the creation of such plasma in the laboratory is problematic. Probably only dusty plasmas can reveal non-ideal behavior.

Interacting particles can be characterized by a **non-ideality parameter,** which is determined as the ratio of the average potential energy of interaction between particles and their neighbors and their average kinetic energy. The Coulomb coupling parameter for plasma electrons and ions is (Fortov & Yakubov, 1994)

$$\Gamma_{e(i)} = \frac{e^2 n_{e(i)}^{1/3}}{4\pi \varepsilon_0 T_{e(i)}}. \tag{3–226}$$

In most plasmas of interest $\Gamma_{e(i)} \ll 1$, which means they are ideal. In a dusty plasma (Fridman & Kennedy, 2004) with a dust particle density n_d, charge $Z_d e$, and temperature T_d, the non-ideality parameter (which is also called the **Coulomb coupling parameter**) is

$$\Gamma_d = \frac{Z_d^2 e^2 n_d^{1/3}}{4\pi \varepsilon_0 T_d}. \tag{3–227}$$

Because particles can be strongly charged ($Z_d \gg 1$), dusty plasmas can be essentially non-ideal.

3.7.2. Plasma Polarization: Debye Shielding of Electric Field in Plasma

External electric fields induce plasma polarization, which prevents penetration of the field inside the plasma. Poisson's equation for the space evolution of potential φ induced by the external field E is

$$div\vec{E} = -\Delta\varphi = \frac{e}{\varepsilon_0}(n_i - n_e). \tag{3–228}$$

Assuming a Boltzmann distribution for plasma electrons and neglecting ion motion – $n_e = n_{e0}\exp(+\frac{e\varphi}{T_e})$, $n_i = n_{e0}\exp(-\frac{e\varphi}{T_i})$ – gives, after expansion in a Taylor series ($e\varphi \ll T_e$),

$$\Delta\varphi = \frac{\varphi}{r_D^2}, \quad r_D = \sqrt{\frac{\varepsilon_0 T_e}{n_{e0}e^2}}. \tag{3–229}$$

where r_D is the **Debye radius,** which was discussed earlier (Section 3.3.5) regarding ambipolar diffusion. In the one-dimensional case, equation (3–229) becomes $d^2\varphi/d^2x = \varphi/r_D^2$ and describes Debye shielding of an external electric field from a plasma boundary E_0 at $x = 0$ along the axis $x(x > 0)$:

$$\vec{E} = -\nabla\varphi = \vec{E}_0 \exp\left(-\frac{x}{r_D}\right). \tag{3–230}$$

The Debye radius gives the characteristic plasma size scale required for the shielding of an external electric field. The same distance is necessary to compensate the electric field of a specified charged particle in plasma. In other words, the Debye radius indicates the scale of plasma quasi-neutrality. There is the correlation between the Debye radius and plasma ideality. The non-ideality parameter Γ is related to the number of plasma particles in the Debye sphere, N^D. For plasma consisting of electrons and positive ions,

$$N_{e(i)}^D = n_{e(i)}\frac{4}{3}\pi r_{De(i)}^3 \propto \frac{1}{\Gamma_{e(i)}^{3/2}}. \tag{3–231}$$

The number of electrons and ions in the Debye sphere is usually large ($N_{e(i)}^D \gg 1$), which confirms the conventional plasma ideality ($\Gamma \ll 1$). In a dusty plasma, it is possible

Table 3–3. Comparison of Debye Radius and Characteristic Size for Different Plasma Systems

Type of plasma	Typical n_e, cm^{-3}	Typical T_e, eV	Debye radius, cm	Typical size, cm
Earth ionosphere	10^5	0.03	0.3	10^6
Flames	10^8	0.2	0.03	10
He-Ne laser	10^{11}	3	0.003	3
Gliding arc	10^{12}	1	$5 \cdot 10^{-4}$	10
Hg lamp	10^{14}	4	$3 \cdot 10^{-5}$	0.3
Solar chromosphere	10^9	10	0.03	10^9
Lightning	10^{17}	3	$3 \cdot 10^{-6}$	100
Micro-plasma	10^{11}	3	0.003	0.003

that $N_d^D \ll 1$. A subsystem of dust particles, however, is not necessarily non-ideal in this case. The distance between dust particles can exceed their Debye radius (r_{Dd}), but interaction between the particles is not necessarily strong because of possible shielding provided by electrons and ions. Taking into account this shielding, the modified non-ideality parameter for the dust particles (Fridman & Kennedy, 2004) can be presented as

$$\Gamma_{ds} = \frac{Z_d^2 e^e n_d^{1/3}}{4\pi \varepsilon_0 T_d} \exp\left(-\frac{1}{n_d^{1/3} r_D}\right), \tag{3–232}$$

where r_D is the Debye radius determined by the electron density. The non-ideality degree of the dust particle subsystem is determined by a combination of two dimensionless parameters: Γ_d (3–227) related to the number of dusty particles in their Debye sphere, and also $K = 1/n_d^{1/3} r_D$ showing the ratio of the interparticle distance to the length of electrostatic shielding by electrons and ions. The factor $\exp(-K)$ in non-ideality parameter (3–232) describes a weakening of the Coulomb interaction between particles due to the shielding.

3.7.3. Plasmas and Sheaths: Physics of DC Sheaths

Plasma is supposed to be quasi neutral ($n_e \approx n_i$) and provide shielding of external electric fields and a field around a specified charged particle as was discussed earlier. Therefore, the characteristic plasma size should exceed the Debye radius, which is illustrated in Table 3–3. Although plasma is quasi-neutral in general ($n_e \approx n_i$), it contacts walls across non-quasi-neutral, positively charged thin layers called **sheaths**. The example of a sheath between plasma and zero-potential surfaces is illustrated in Fig. 3–17. The formation of positively charged sheaths is due to the fact that electrons can move much faster than ions. For example, an electron thermal velocity (about $\sqrt{T_e/m}$) exceeds that of ions (about $\sqrt{T_i/M}$) by about 1000 times. The fast electrons stick to the walls, leaving the region near the walls (the sheath) for positively charged ions alone. The positively charged sheath results in a potential profile illustrated in Fig. 3–17. The bulk of plasma is quasi-neutral and, hence, iso-potential ($\varphi = \text{const}$) according to Poisson equation (3–228). Near discharge walls, the positive potential falls sharply, providing a high electric field, acceleration of ions, and deceleration of electrons. The sheaths play a significant role in various discharge systems engineered for surface treatment (see Chapter 8). Most of these systems operate at low gas temperatures ($T_e \gg T_0, T_i$) and low pressures, when the sheath can be considered collisionless. In this case the basic one-dimensional equation governing the DC sheath potential φ in the direction perpendicular to the wall can be obtained from the Poisson equation, energy conservation for the ions, and the Boltzmann distribution for electrons:

$$\frac{d^2\varphi}{dx^2} = \frac{en_s}{\varepsilon_0}\left[\exp\frac{\varphi}{T_e} - \left(1 - \frac{e\varphi}{E_i}\right)^{-\frac{1}{2}}\right]. \tag{3–233}$$

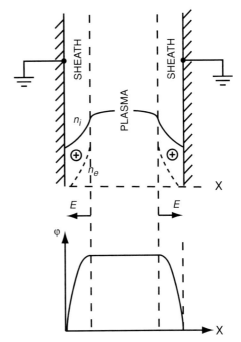

Figure 3–17. Distributions of density of charged particles and potential in plasma and sheaths.

In this equation, n_s is the plasma density at the sheath edge; $E_{is} = \frac{1}{2}Mu_{is}^2$ is the initial energy of an ion entering the sheath (u_{is} is the corresponding velocity); and the potential is assumed to be zero ($\varphi = 0$) at the sheath edge ($x = 0$). Multiplying (3–233) by $d\varphi/dx$ and then integrating the equation assuming boundary conditions ($\varphi = 0$, $d\varphi/dx = 0$ at $x = 0$) permits us to find the electric field in the sheath:

$$\left(\frac{d\varphi}{dx}\right)^2 = \frac{2en_s}{\varepsilon_0}\left[T_e\exp\left(\frac{e\varphi}{T_e}\right) - T_e + 2E_{is}\left(1 - \frac{e\varphi}{E_{is}}\right)^{\frac{1}{2}} - 2E_{is}\right]. \qquad (3\text{–}234)$$

Solution of (3–234) can exist only if its right-hand side is positive. Expanding (3–234) to the second order in a Taylor series shows that the sheath can exist only if the initial ion velocity exceeds the critical one, known as the **Bohm velocity**, u_B:

$$u_{is} \geq u_B = \sqrt{T_e/M}. \qquad (3\text{–}235)$$

Condition (3–235) of a sheath existence is usually referred to as the **Bohm sheath criterion**. To provide ions with the energy required to satisfy the Bohm criterion, there must be a quasi-neutral region (wider than sheath, several Debye radii) with some electric field. This region, illustrated in Fig. 3–18, is called the **presheath**. The minimum presheath potential (between bulk plasma and sheath) should be equal to

$$\varphi_{presheath} \approx \frac{1}{2e}Mu_B^2 = \frac{T_e}{2e}. \qquad (3\text{–}236)$$

Balancing the ion and electron fluxes to the floating wall leads to the expression for the change of potential across the sheath in this case:

$$\Delta\varphi = \frac{1}{e}T_e\ln\sqrt{M/2\pi m}. \qquad (3\text{–}237)$$

The change of potential across the sheath is usually referred to as the **floating potential**. Because the ion-to-electron mass ratio M/m is large, the floating potential exceeds the

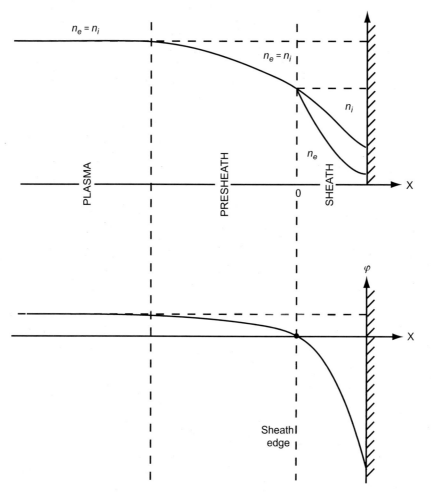

Figure 3–18. Distributions of density of charged particles and potential in sheath and presheath near the wall.

potential across the presheath by 5–8 times. Numerical calculations (3–234) give in the case of floating potential a typical sheath width of about a few Debye radii ($s \approx 3r_D$).

3.7.4. High-Voltage Sheaths: Matrix and Child Law Sheath Models

The floating potential (3–237) exceeds electron temperature by less than 10 times. However, the change of potential across a sheath (sheath voltage V_0) is often driven to be very large compared to the electron temperature. In this case the electron concentration in the sheath can be neglected and only ions should be taken into account. Due to the absence of electrons, the sheath region appears dark when observed visually. A simple model of the high-voltage sheath assumes uniformity of the ion density there. Such a sheath is usually referred to as the **matrix sheath**. In frameworks of the matrix sheath model, the sheath thickness can be expressed in terms of the Debye radius r_D corresponding to a plasma concentration at the sheath edge:

$$s = r_D \sqrt{\frac{2V_0}{T_e}}. \tag{3–238}$$

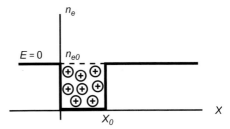

Figure 3–19. Illustration of electrostatic plasma oscillations; distribution of electron density.

Numerically, the matrix sheath thickness is large and exceeds the Debye radius by 10–50 times at high voltages. A more accurate approach, the so-called **Child law sheath,** takes into account a decrease in the ion density due to their acceleration across the sheath. In frameworks of the model, the ion current density $j_0 = n_s e u_B$ is taken from the **Child law of space-charge-limited current** in a plane diode:

$$j_0 = n_s e u_B = \frac{4\varepsilon_0}{9}\sqrt{\frac{2e}{M}}\frac{1}{s^2}V_0^{3/2}. \tag{3–239}$$

The thickness of the Child law sheath "s" can be found from Child law equation (3–239):

$$s = \frac{\sqrt{2}}{3}r_D\left(\frac{2V_0}{T_e}\right)^{\frac{3}{4}}. \tag{3–240}$$

Numerically, the Child law sheath can be of the order of 100 Debye lengths in conditions of typical low-pressure discharges applied for surface treatment. More details regarding sheaths, including collisional sheaths, sheaths in electronegative gases, radiofrequency plasma sheaths, and pulsed potential sheaths can be found, in particular, in the book of Lieberman and Lichtenberg (1994).

3.7.5. Electrostatic Plasma Oscillations: Langmuir or Plasma Frequency
The typical space-size characterizing a plasma is the Debye radius, which is a linear measure of electroneutrality and shielding of external electric fields. The typical plasma time scale and typical time of plasma response to the external fields is determined by the plasma frequency illustrated in Fig. 3–19. Assume in a one-dimensional approach that all electrons at $x > 0$ are initially shifted to the right on the distance x_0, whereas heavy ions are not perturbed and remain at rest. This results in an electric field, which pushes the electrons back. If $E = 0$ at $x < 0$, this electric field acting to restore the plasma quasi-neutrality can be found at $x > x_0$ from the one-dimensional Poisson equation as

$$\frac{dE}{dx} = \frac{e}{\varepsilon_0}(n_i - n_e), \quad E = -\frac{e}{\varepsilon_0}n_{e0}x_0 \text{ (at } x > x_0). \tag{3–241}$$

This electric field makes electrons move back to the left (Fig. 3–19) together with their boundary (at $x = x_0$), which results in electrostatic plasma oscillations:

$$\frac{d^2x_0}{dt^2} = -\omega_p^2 x_0, \quad \omega_p = \sqrt{\frac{e^2 n_e}{\varepsilon_0 m}}, \tag{3–242}$$

where ω_p is the **Langmuir frequency** or **plasma frequency**, which determines the time scale of the plasma response to external electric perturbations. This plasma frequency and the Debye radius are related:

$$\omega_p \times r_D = \sqrt{2T_e/m}. \tag{3–243}$$

The time of plasma reaction to external perturbation $(1/\omega_p)$ corresponds to a time required by a thermal electron (velocity $\sqrt{2T_e/m}$) to travel the distance r_D required to shield the external perturbation. The plasma frequency depends only on plasma density and numerically can be calculated as

$$\omega_p(s^{-1}) = 5.65 \cdot 10^4 \sqrt{n_e(\text{cm}^{-3})}. \tag{3–244}$$

3.7.6. Penetration of Slow-Changing Fields into Plasma: Skin Effect in Plasma

Consider the penetration of a low-frequency electromagnetic field ($\omega < \omega_p$) in plasma. In addition to Ohm's law ($\vec{j} = \sigma \vec{E}$), the Maxwell equation should be taken into account in this case:

$$curl\, \vec{H} = \vec{j} + \varepsilon_0 \frac{\partial \vec{E}}{\partial t}. \tag{3–245}$$

Assuming also that the frequency of the field is low with respect to plasma conductivity and, hence, the displacement current (second current term in (3–245) can be neglected), we can conclude

$$curl\, \vec{H} = \sigma \vec{E}. \tag{3–246}$$

Taking the electric field from equation (3–246) and substituting into another Maxwell equation,

$$curl\, \vec{E} = -\mu_0 \frac{\partial \vec{H}}{\partial t}, \tag{3–247}$$

yields the differential relation for electromagnetic field decrease during penetration in plasma ($\omega < \omega_p$):

$$\frac{\partial \vec{H}}{\partial t} = -\frac{1}{\mu_0 \sigma} curl\, curl\, \vec{H} = -\frac{1}{\mu_0 \sigma} \nabla(div\, \vec{H}) + \frac{1}{\mu_0 \sigma} \Delta \vec{H} = \frac{1}{\mu_0 \sigma} \Delta \vec{H}. \tag{3–248}$$

Equation (3–248) describes the decrease in plasma of the amplitude of the low-frequency electric and magnetic fields with the characteristic space scale:

$$\delta = \sqrt{\frac{2}{\omega \mu_0 \sigma}}. \tag{3–249}$$

If this space scale δ is smaller than the plasma sizes, then the external fields and currents are located only on the plasma surface layer with a penetration depth δ. This effect is known as the **skin effect**. The boundary layer, where the external fields penetrate and where plasma currents are located, is called the **skin layer**. The depth of the skin layer depends on the electromagnetic field frequency ($f = \omega/2\pi$) and plasma conductivity. For calculation of the skin layer depth it is convenient to use the following numeric formula:

$$\delta(\text{cm}) = \frac{5.03}{\sigma^{1/2}(1/\text{Ohm} \cdot \text{cm}) \cdot f^{1/2}(\text{MHz})}. \tag{3–250}$$

3.7.7. Magneto-Hydrodynamics: "Diffusion" of Magnetic Field and Magnetic Field Frozen in Plasma

The motion of high-density plasma in a magnetic field induces electric currents, which together with the magnetic field influence the plasma motion. Such phenomena can be described by the system of **equations of magneto-hydrodynamics** (see Fridman & Kennedy, 2004, for details), including the following.

1. Navier-Stokes equation neglecting viscosity, but taking into account Ampere force acting on plasma current with density $\vec{j}$ (where B is magnetic induction, M is mass of ions, and $n_e = n_i$ is the plasma density):

$$Mn_e \left[\frac{\partial \vec{v}}{\partial t} + (\vec{v}\nabla)\vec{v} \right] + \nabla p = [\vec{j}\vec{B}]. \tag{3–251}$$

2. Continuity equations for electrons and ions, moving as one fluid with macroscopic velocity $\vec{v}$:

$$\frac{\partial n_e}{\partial t} + div\,(n_e\vec{v}) = 0. \tag{3–252}$$

3. Maxwell equations neglecting the displacement current because of relatively low velocities:

$$curl\,\vec{H} = \vec{j}, \quad div\,\vec{B} = 0. \tag{3–253}$$

4. Maxwell equation $curl\,E = -\frac{\partial \vec{B}}{\partial t}$ together with the first equation of (3–253) and Ohm's law $(\vec{j} = \sigma(\vec{E} + [\vec{v}\vec{B}])$ for plasma with conductivity σ give the following relation for a magnetic field:

$$\frac{\partial \vec{B}}{\partial t} = curl\,[\vec{v}\vec{B}] + \frac{1}{\sigma\mu_0}\Delta\vec{B}. \tag{3–254}$$

If the plasma is at rest $(\vec{v} = 0)$, (3–254) can be reduced to the diffusion equation:

$$\frac{\partial \vec{B}}{\partial t} = D_m\Delta\vec{B}, \quad D_m = \frac{1}{\sigma\mu_0}. \tag{3–255}$$

The factor D_m can be interpreted as the coefficient of "diffusion" of a magnetic field in plasma; it is also sometimes called the **magnetic viscosity**. If the characteristic time of the magnetic field change is $\tau = 1/\omega$, then the characteristic length of magnetic field diffusion according to (3–255) is $\delta \approx \sqrt{2D_m\tau} = \sqrt{2/\sigma\omega\mu_0}$, which corresponds to the skin-layer depth (3–249). Relation (3–255) can be also applied to describe damping time τ_m of currents and magnetic fields in a conductor with characteristic size L:

$$\tau_m = \frac{L^2}{D_m} = \mu_0\sigma L^2. \tag{3–256}$$

If the plasma conductivity is high $(\sigma \to \infty)$, the diffusion coefficient of the magnetic field is small $(D_m \to 0)$ and the magnetic field is unable to "move" with respect to plasma. One can say that the magnetic field sticks to the plasma or, in other words, *the magnetic field is frozen in plasma*.

3.7.8. Magnetic Pressure: Plasma Equilibrium in Magnetic Field and Pinch Effect

In steady-state conditions $(d\vec{v}/dt = 0)$, the Navier-Stokes equation (3–251) can be simplified to

$$grad\,p = [\vec{j}\vec{B}]. \tag{3–257}$$

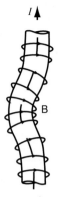

Figure 3–20. Pinch effect, plasma bent Z-pinch instability.

This equation can be interpreted as a balance of hydrostatic pressure p and Ampere force exerted on the plasma. Taking into account the first of equations (3–253), the current can be eliminated from the balance:

$$\nabla p = [\vec{j}\vec{B}] = \mu_0[curl\ \vec{H} \times \vec{H}] = -\frac{\mu_0}{2}\nabla H^2 + \mu_0(\vec{H}\nabla)\vec{H}. \qquad (3\text{–}258)$$

Combination of the gradients leads to the equation for plasma equilibrium in the magnetic field:

$$\nabla\left(p + \frac{\mu_0 H^2}{2}\right) = \mu_0(\vec{H}\nabla)\vec{H} = \frac{\mu_0 H^2}{R}\vec{n}. \qquad (3\text{–}259)$$

where R is the radius of curvature of the magnetic field line and $\vec{n}$ is a unit normal vector to the line. Thus, force $(\mu_0 H^2/R)\vec{n}$ is related to the bending of the magnetic field lines and can be interpreted as the **tension of magnetic lines**. This tension tends to make the magnetic field lines straight. The pressure term $\mu_0 H^2/2$ is called the **magnetic pressure**; the sum of hydrostatic and magnetic pressures, $p + (\mu_0 H^2/2)$, is the total pressure.

The equation (3–259) for plasma equilibrium in a magnetic field can be considered a dynamic balance of the gradient of total pressure and the tension of magnetic lines. If the magnetic field lines are straight and parallel, then $R \to \infty$ and the "tension" of magnetic lines is equal to zero. In this case, equation (3–259) gives the equilibrium criterion:

$$p + \frac{\mu_0 H^2}{2} = \text{const.} \qquad (3\text{–}260)$$

The pinch effect can be presented as an example of magneto-hydrodynamic effects in plasma. The effect consists of the self-compression of plasma in its own magnetic field. Consider the pinch effect in a long cylindrical discharge plasma with an electric current along the axis of the cylinder (Fig. 3–20), which is the so-called Z-pinch (Allis, 1960). The equilibrium of the Z-pinch is determined by the **Bennett relation** (1934):

$$N_L T = \frac{\mu_0}{8\pi}I^2, \qquad (3\text{–}261)$$

where N_L is the linear plasma density and T is temperature. Thus, the plasma temperature should increase proportionally to the square of current. Therefore, to reach thermonuclear temperature (about 100 keV in plasma with a density 10^{15} cm^{-3} and cross section 1 cm^2), the required current should be 100 kA. The current of 100 kA can be achieved in Z-pinch discharges, which in the early 1950s stimulated a large effort in controlled thermonuclear fusion research. The hopes for easily controlled fusion in Z-pinch discharges were shattered, however, because of instabilities related to plasma bending and non-uniform plasma compression, which is illustrated in Fig. 3–20. If the discharge column is bent, the magnetic field and magnetic pressure become higher from the concave side of the plasma, which leads to a breakup of the channel in the "wriggle" instability. If the discharge channel becomes locally thinner, the magnetic field ($B \propto 1/r$) and magnetic pressure increase there, leading to further compression and breaking of the channel.

3.7.9. Two-Fluid Magneto-Hydrodynamics: Generalized Ohm's Law

In the magneto-hydrodynamic approach, discussed earlier, electron and ion velocities ($\vec{v}_e$ and $\vec{v}_i$) were considered equal to each other, which contradicts the current $\vec{j} = en_e(\vec{v}_i - \vec{v}_e)$ in quasi-neutral plasma with density n_e. In two-fluid magneto-hydrodynamics, the Navier-Stokes equation includes the electron's mass m, pressure p_e, and velocity and also takes into account friction between electrons and ions:

$$mn_e\frac{d\vec{v}_e}{dt} + \nabla p_e = -en_e\vec{E} - en_e[\vec{v}_e\vec{B}] - mn_e\nu_e(\vec{v}_e - \vec{v}_i). \qquad (3\text{--}262)$$

where ν_e is the frequency of electron collisions. A similar Navier-Stokes equation for ions includes the friction term with opposite sign. The first term in equation (3–262) related to electron inertia can be neglected because of very low electron mass. Denoting the ion's velocity as $\vec{v}$ and plasma conductivity as $\sigma = n_e e^2/m\nu_e$, equation (3–262) can be presented as the **generalized Ohm's law**:

$$\vec{j} = \sigma(\vec{E} + [\vec{v}\vec{B}]) + \frac{\sigma}{en_e}\nabla p_e - \frac{\sigma}{en_e}[\vec{j}\vec{B}]. \qquad (3\text{--}263)$$

This generalized Ohm's law differs from the conventional one because it takes into account the electron pressure gradient and the $[\vec{j}\vec{B}]$ term related to the **Hall effect**. Solution of equation (3–263) with respect to electric current is complicated because the current is present in two terms. The generalized Ohm's law can be simplified if plasma conductivity is high ($\sigma \to \infty$):

$$\vec{E} + [\vec{v}\vec{B}] + \frac{1}{en_e}\nabla p_e = \frac{1}{en_e}[\vec{j}\vec{B}]. \qquad (3\text{--}264)$$

If electron temperature is uniform, the generalized Ohm's law can be rewritten as

$$\vec{E} = -[\vec{v}_e\vec{B}] - \frac{1}{e}\nabla(T_e \ln n_e). \qquad (3\text{--}265)$$

Substituting (3–265) in the Maxwell equation $curl\, E = -\frac{\partial \vec{B}}{\partial t}$ gives ($\nabla \times \nabla(T_e \ln n_e) \equiv 0$):

$$\frac{\partial \vec{B}}{\partial t} = curl\, [\vec{v}_e\vec{B}]. \qquad (3\text{--}266)$$

Thus, two-fluid magneto-hydrodynamics shows that the magnetic field is frozen into the electron gas.

3.7.10. Plasma Diffusion Across Magnetic Field

The generalized Ohm law (3–263) describes the electron and ion diffusion in the direction perpendicular to a uniform magnetic field (Braginsky, 1963):

$$D_{\perp,e} = \frac{D_e}{1 + \left(\dfrac{\omega_{B,e}}{\nu_e}\right)^2}, \quad D_{\perp,i} = \frac{D_i}{1 + \left(\dfrac{\omega_{B,i}}{\nu_i}\right)^2}, \qquad (3\text{--}267)$$

where D_e and D_i are coefficients of free diffusion of electrons and ions (see Section 3.3.4); ν_e and ν_i are collisional frequencies of electrons and ions; and $\omega_{B,e}$ and $\omega_{B,i}$ are electron and ion cyclotron frequencies (3–75),

$$\omega_{B,e} = \frac{eB}{m}, \quad \omega_{B,i} = \frac{eB}{M}. \tag{3-268}$$

If diffusion is ambipolar, which is usually the case at higher ionization degrees, relation (3–90) can be applied to find D_a in a magnetic field. Obviously, free diffusion coefficients and mobilities for electrons and ions should be replaced by those in the magnetic field (see (3–267) and (3–77)):

$$D_\perp = \frac{D_a}{1 + \dfrac{\omega_{B,i}^2}{\nu_i^2} + \dfrac{\mu_i}{\mu_e}\left(1 + \dfrac{\omega_{B,e}^2}{\nu_e^2}\right)}, \tag{3-269}$$

where D_a is the coefficient of ambipolar diffusion (3–90) in plasma without magnetic field. The expression for ambipolar diffusion perpendicular to a magnetic field can be simplified when electrons are magnetized ($\omega_{B,e}/\nu_e \gg 1$) and ions are not ($\omega_{B,i}/\nu_i \ll 1$):

$$D_\perp = \frac{D_a}{1 + \dfrac{\mu_i}{\mu_e}\dfrac{\omega_{B,e}^2}{\nu_e^2}}. \tag{3-270}$$

In strong magnetic fields ($\frac{\mu_i}{\mu_e}\frac{\omega_{B,e}^2}{\nu_e^2} \gg 1$), the relation can be further simplified:

$$D_\perp \approx D_a \frac{\mu_e}{\mu_i} \frac{\nu_e^2}{\omega_{B,e}^2} \approx D_e \frac{\nu_e^2}{(e/M)^2} \frac{1}{B^2}. \tag{3-271}$$

The value of $D_\perp$ in strong magnetic fields significantly decreases: $D_\perp \propto 1/B^2$, which can be applied to prevent plasma from decaying. Magnetized electrons are trapped by the magnetic field and rotate along the Larmor circles until a collision pushes the electron to another Larmor circle. The **electron Larmor radius**,

$$\rho_L = \frac{\nu_\perp}{\omega_{B,e}} = \frac{1}{eB}\sqrt{2T_e m} \tag{3-272}$$

is the radius of the circular motion of a magnetized electron; $\nu_\perp$ is the component of electron thermal velocity perpendicular to the magnetic field. In the case of diffusion across the strong magnetic field, the Larmor radius plays the same role as the mean free path without magnetic field:

$$D_\perp \approx D_e \frac{\nu_e^2}{\omega_{B,e}^2} \approx \rho_L^2 \nu_e. \tag{3-273}$$

3.7.11. Magneto-Hydrodynamic Behavior of Plasma: Alfven Velocity and Magnetic Reynolds Number

The magneto-hydrodynamic effects are important when fluid dynamics is strongly coupled with the magnetic field. It demands the "diffusion" of the magnetic field to be less than "convection," which means the curl in equation (3–254) should exceed the Laplacian. If the space scale of plasma is L, the following parameters are required:

$$\frac{vB}{L} \gg \frac{1}{\sigma\mu_0}\frac{B}{L^2}, \quad \text{or} \quad v \gg \frac{1}{\sigma\mu_0 L}. \tag{3-274}$$

These requirements of high conductivity and velocity are sufficient to make the magnetic field frozen in plasma (see Section 3.7.7), which is the most important feature of

3.7. Elements of Plasma Electrodynamics

Table 3–4. Magnetic Reynolds Numbers for Different Plasma Systems in Laboratory and Nature

Plasma system	B, Tesla	Space scale, m	ρ, kg/m^3	σ, 1/Ohm $\cdot$ cm	R_m
Ionosphere	10^{-5}	10^5	10^{-5}	0.1	10
Solar atmosphere	10^{-2}	10^7	10^{-6}	10	10^8
Solar corona	10^{-9}	10^9	10^{-17}	10^4	10^{11}
Hot interstellar gas	10^{-10}	10 light yr	10^{-21}	10	10^{15}
Arc discharge plasma	0.1	0.1	10^{-5}	10^3	10^3
Hot confined plasma, $n = 10^{15}$ cm^{-3}, $T = 10^6$ K	0.1	0.1	10^{-6}	10^3	10^4

magneto-hydrodynamics. Using the concept of magnetic viscosity D_m (3–255), the condition (3–274) of magneto-hydrodynamic behavior is

$$Re_m = \frac{vL}{D_m} \gg 1, \qquad (3\text{–}275)$$

where Re_m is the **magnetic Reynolds number**, which is similar to the conventional Reynolds number but with kinematic viscosity replaced by magnetic viscosity. The plasma velocity v in magneto-hydrodynamic systems usually satisfies the approximate balance of dynamic and magnetic pressures:

$$\frac{\rho v^2}{2} \propto \frac{\mu_0 H^2}{2}, \quad \text{or} \quad v \propto v_A = \frac{B}{\sqrt{\rho\mu_0}}, \qquad (3\text{–}276)$$

where $\rho = Mn_e$ is plasma density. The characteristic plasma velocity v_A corresponding to the equality of the dynamic and magnetic pressures is called the **Alfven velocity**. The criterion of magneto-hydrodynamic behavior (3–275) can be presented in terms of the Alfven velocity as

$$Re_m = \frac{v_A L}{D_m} = BL\sigma\sqrt{\frac{\mu_0}{\rho}} \gg 1. \qquad (3\text{–}277)$$

Magneto-hydrodynamic plasma behavior takes place not only at high conductivity and magnetic field but also at large sizes and low densities (Table 3–4; see Chen, 1984; Rutherford & Goldston, 1995).

3.7.12. High-Frequency Plasma Conductivity and Dielectric Permittivity

High-frequency plasma conductivity and dielectric permittivity are important concepts to analyze the propagation of electromagnetic waves in plasma. One-dimensional electron motion in the electric field $E = E_0 \cos \omega t = Re(E_0 e^{i\omega t})$ can be described by the equation:

$$m\frac{du}{dt} = -eE - mu\nu_{en}, \qquad (3\text{–}278)$$

where ν_{en} is the electron–neutral collision frequency, and $u = Re(u_0 e^{i\omega t})$ is the electron velocity. The relation between the amplitudes of electron velocity and electric field is complex and can be found from (3–278) as

$$u_0 = -\frac{e}{m}\frac{1}{\nu_{en} + i\omega}E_0. \qquad (3\text{–}279)$$

The Maxwell equation $curl\ \vec{H} = \varepsilon_0 \frac{\partial \vec{E}}{\partial t} + \vec{j}$ permits one to present the total current density as the sum

$$\vec{j}_t = \varepsilon_0 \frac{\partial \vec{E}}{\partial t} + \vec{j}. \tag{3--280}$$

The first current density component is related to the displacement current, and its amplitude in complex form can be presented as $\varepsilon_0 i \omega E_0$. The second component corresponds to the conductivity current and has an amplitude $-en_e u_0$. Thus, the amplitude of the total current density is

$$j_{t0} = i \omega \varepsilon_0 E_0 - en_e u_0. \tag{3--281}$$

Taking into account the complex electron mobility (3--279), expression (3--281) for the total current density and, hence, the Maxwell equation can be rewritten as

$$j_{t0} = i \omega \varepsilon_0 \left[1 - \frac{\omega_p^2}{\omega(\omega - i\nu_{en})} \right] E_0, \quad curl\ \vec{H}_0 = i \omega \varepsilon_0 \left[1 - \frac{\omega_p^2}{\omega(\omega - i\nu_{en})} \right] \vec{E}_0, \tag{3--282}$$

where ω_p is the electron plasma frequency. Keeping the Maxwell equation (3--282) in the form $curl\ \vec{H}_0 = i \omega \varepsilon_0 \varepsilon E_0$, the complex dielectric permittivity of plasma should be introduced:

$$\varepsilon = 1 - \frac{\omega_p^2}{\omega(\omega - i\nu_{en})}. \tag{3--283}$$

The complex dielectric constant (3--283) can be rewritten to specify the real and imaginary parts:

$$\varepsilon = \varepsilon_\omega - i \frac{\sigma_\omega}{\varepsilon_0 \omega}. \tag{3--284}$$

The real component ε_ω is the high-frequency dielectric constant of the plasma:

$$\varepsilon_\omega = 1 - \frac{\omega_p^2}{\omega^2 + \nu_{en}^2}. \tag{3--285}$$

The imaginary component of (3--283) corresponds to the high-frequency plasma conductivity:

$$\sigma_\omega = \frac{n_e e^2 \nu_{en}}{m(\omega^2 + \nu_{en}^2)}. \tag{3--286}$$

Expressions for the high-frequency dielectric permittivity and conductivity can be simplified in two cases: collisionless plasma and static limit. The **collisionless plasma** limit means $\omega \gg \nu_{en}$. For example, microwave plasma can be considered collisionless at low pressures (about 3 Torr and less). In this case,

$$\sigma_\omega = \frac{n_e e^2 \nu_{en}}{m\omega^2}, \quad \varepsilon_\omega = 1 - \frac{\omega_p^2}{\omega^2}. \tag{3--287}$$

Conductivity in a collisionless plasma is proportional to the electron–neutral collision frequency, and the dielectric constant does not depend on the frequency ν_{en}. The ratio of conduction current to polarization current (which actually corresponds to displacement current) can be estimated as

$$\frac{j_{conduction}}{j_{polarization}} = \frac{\sigma_\omega}{\varepsilon_0 \omega |\varepsilon_\omega - 1|} = \frac{\nu_{en}}{\omega}. \tag{3--288}$$

3.7. Elements of Plasma Electrodynamics

The polarization current in collisionless plasma ($\nu_{en} \ll \omega$) greatly exceeds the conductivity current. In the opposite case for the **static limit** $\nu_{en} \gg \omega$, conductivity and dielectric permittivity are

$$\sigma_\omega = \frac{n_e e^2}{m \nu_{en}}, \quad \varepsilon = 1 - \frac{\omega_p^2}{\nu_{en}^2}. \tag{3-289}$$

Conductivity in the static limit coincides with conventional DC conditions.

3.7.13. Propagation of Electromagnetic Waves in Plasma

Electromagnetic wave propagation in plasma is described by the conventional wave equations:

$$\Delta \vec{E} - \frac{\varepsilon}{c^2} \frac{\partial^2 \vec{E}}{\partial t^2} = 0, \quad \Delta \vec{H} - \frac{\varepsilon}{c^2} \frac{\partial^2 \vec{H}}{\partial t^2} = 0. \tag{3-290}$$

Plasma peculiarities are related to the complex dielectric permittivity ε (3.283, 3.284). The dispersion equation for electromagnetic wave propagation in a dielectric medium (where c is the speed of light),

$$\frac{kc}{\omega} = \sqrt{\varepsilon}, \tag{3-291}$$

is also valid in plasma with the complex dielectric permittivity ε. Assuming the electric and magnetic fields as $\vec{E}, \vec{H} \propto \exp(-i\omega t + i\vec{k}\vec{r})$ with real frequency ω, the wave number k should be complex:

$$k = \frac{\omega}{c}\sqrt{\varepsilon} = \frac{\omega}{c}(n + i\kappa). \tag{3-292}$$

Parameter n is the **refractive index** of the electromagnetic wave; phase velocity is $v = \frac{\omega}{k} = \frac{c}{n}$, and the wavelength is $\lambda = \lambda_0/n$ (where λ_0 corresponds to vacuum). The wave number κ characterizes the **attenuation of electromagnetic wave in plasma**; that is, the wave amplitude decreases e^κ times over the length $\lambda_0/2\pi$. Relations between refractive index and attenuation with high-frequency dielectric permittivity and conductivity are

$$n^2 - \kappa^2 = \varepsilon_\omega, \quad 2n\kappa = \frac{\sigma_\omega}{\varepsilon_0 \omega}. \tag{3-293}$$

Solving this system of equations results in an explicit expression for the attenuation coefficient:

$$\kappa = \sqrt{\frac{1}{2}\left(-\varepsilon_\omega + \sqrt{\varepsilon_\omega^2 + \frac{\sigma_\omega^2}{\varepsilon_0^2 \omega^2}}\right)}. \tag{3-294}$$

Attenuation of the electromagnetic wave is determined by the plasma conductivity: if $\sigma_\omega \ll \varepsilon_\omega \varepsilon_0 \omega$, the electromagnetic field damping can be neglected. The explicit expression for the refractive index is

$$n = \sqrt{\frac{1}{2}\left(\varepsilon_\omega + \sqrt{\varepsilon_\omega^2 + \frac{\sigma_\omega^2}{\varepsilon_0^2 \omega^2}}\right)}. \tag{3-295}$$

If conductivity is negligible, the refractive index $n \approx \sqrt{\varepsilon_\omega}$. Plasma polarization is negative ($\varepsilon_\omega < 1$), which means $n < 1$ at the low conductivity limit. Combining $n \approx \sqrt{\varepsilon_\omega}$ with equations (3–287) at low-conductivity conditions leads to the dispersion equation for electromagnetic waves in a collisionless plasma:

$$\frac{k^2 c^2}{\omega^2} = 1 - \frac{\omega_p^2}{\omega^2}, \quad \omega^2 = \omega_p^2 + k^2 c^2. \tag{3-296}$$

Differentiation of (3–196) relates phase and group velocities of electromagnetic waves in plasma:

$$\frac{\omega}{k} \times \frac{d\omega}{dk} = v_{ph} v_{gr} = c^2. \tag{3–297}$$

3.7.14. Plasma Absorption and Reflection of Electromagnetic Waves: Bouguer Law: Critical Electron Density

The energy flux of electromagnetic waves is determined by the **Pointing vector** $\vec{S} = \varepsilon_0 c^2 [\vec{E} \times \vec{B}]$. Taking into account the relation between electric and magnetic fields ($\varepsilon \varepsilon_0 E^2 = \mu_0 H^2$), damping of the electromagnetic oscillations in plasma can be presented in the form of the **Bouguer law**:

$$\frac{dS}{dx} = -\mu_\omega S, \quad \mu_\omega = \frac{2\kappa\omega}{c} = \frac{\sigma_\omega}{\varepsilon_0 nc}, \tag{3–298}$$

where μ_ω is an absorption coefficient. The energy flux S decreases by a factor e over the length $1/\mu_\omega$. The product $\mu_\omega S$ is the electromagnetic energy dissipated per unit volume of plasma and corresponds to Joule heating:

$$\mu_\omega S = \varepsilon_0 c^2 \langle EB \rangle = \sigma \langle E^2 \rangle. \tag{3–299}$$

If the plasma ionization degree and absorption are relatively low ($n \approx \sqrt{\varepsilon} \approx 1$), the absorption coefficient can be simplified and based on equations (3–298) and (3–286) expressed as

$$\mu_\omega = \frac{n_e e^2 n_e v_{en}}{\varepsilon_0 mc(\omega^2 + v_{en}^2)}. \tag{3–300}$$

This relation for electromagnetic wave absorption can be presented numerically:

$$\mu_\omega, \mathrm{cm}^{-1} = 0.106 n_e \ (\mathrm{cm}^{-3}) \frac{v_{en}(\mathrm{s}^{-1})}{\omega^2(\mathrm{s}^{-1}) + v_{en}^2}. \tag{3–301}$$

At high frequencies ($\omega \gg v_{en}$), the absorption coefficient is proportional to the square of wavelength ($\mu_\omega \propto \omega^{-2} \propto \lambda^2$); therefore, short electromagnetic waves propagate in plasma more readily.

If the plasma conductivity is not high ($\sigma_\omega \ll \omega\varepsilon_0 |\varepsilon|$), the electromagnetic wave propagates in plasma quite easily if the frequency is high enough. When the frequency decreases, however, the dielectric permittivity $\varepsilon_\omega = 1 - (\omega_p^2/\omega^2)$ becomes negative and the electromagnetic wave is unable to propagate. Negative values of dielectric permittivity make the refractive index equal to zero ($n = 0$) and the attenuation coefficient $\kappa \approx \sqrt{|\varepsilon|}$. The penetration depth of the electromagnetic wave in plasma for this case is

$$l = \frac{\lambda_0}{2\pi\sqrt{|\varepsilon_\omega|}} = \frac{\lambda_0}{2\pi} \left| 1 - \frac{\omega_p^2}{\omega^2} \right|^{\frac{-1}{2}} \tag{3–302}$$

and does not depend on conductivity and is not related to energy dissipation. Such a non-dissipative stopping phenomenon is known as the **total electromagnetic wave reflection from plasma**. The electromagnetic wave propagates from an area with low electron density to areas where plasma density is increasing. The electromagnetic wave frequency is fixed, but the plasma frequency increases together with electron density leading to a decrease of ε_ω. At the point when dielectric permittivity, $\varepsilon_\omega = 1 - (\omega_p^2/\omega^2)$, becomes equal to zero, total reflection takes place. The total reflection of electromagnetic waves occurs when the electron density reaches the critical value, which can be found from $\omega = \omega_p$ as

$$n_e^{crit} = \frac{\varepsilon_0 m \omega^2}{e^2}, \quad n_e \ (\text{cm}^{-3}) = 1.24 \cdot 10^4 \cdot [f(\text{MHz})]^2. \qquad (3\text{--}303)$$

PROBLEMS AND CONCEPT QUESTIONS

3–1. Ionization Equilibrium, the Saha Equation. Using the Saha equation, estimate the ionization degree in the thermal Ar plasma at atmospheric pressure and quasi-equilibrium temperature $T = 20,000$ K. Explain why the high ionization degree can be reached at temperatures much less than the ionization potential.

3–2. Debye Correction of Thermodynamic Functions in Plasma. Estimate the Debye correction for Gibbs potential per unit volume of thermal Ar plasma at atmospheric pressure and temperature $T = 20,000$ K. The ionization degree can be calculated using the Saha equation (see Problem 3–1).

3–3. Non-Equilibrium Statistical Treanor Distribution for Vibrationally Excited Molecules. Based on the non-equilibrium Treanor distribution function, find the average value of vibrational energy taking into account only relatively low vibrational levels. Find an application criterion for the result (most of the molecules should be located in the vibrational levels lower than the Treanor minimum).

3–4. Druyvesteyn Electron Energy Distribution Function. Calculate the average electron energy for the Druyvesteyn distribution. Define the effective electron temperature of the distribution and compare it with that of the Maxwellian distribution function.

3–5. Influence of Vibrational Temperature on EEDF. Simplify relation (3–54), describing the influence of vibrational temperature on EEDF, for the case of high vibrational temperatures $(T_v \gg \hbar\omega)$. Estimate the acceleration of the ionization rate coefficient corresponding to a 10% increase of T_v.

3–6. Electron–Electron Collisions and EEDF Maxwellization. Calculate the minimum ionization degree $\frac{n_e}{n_0}$, when the Maxwellization provided by electron–electron collisions becomes essential in establishing EEDF in the case of (1) argon plasma, $T_e = 1$ eV; and (2) nitrogen plasma, $T_e = 1$ eV.

3–7. Plasma Rotation in Crossed Electric and Magnetic Fields, Plasma Centrifuge. Describe plasma motion in the electric field $\vec{E}(r)$ created by a long charged cylinder and uniform magnetic field $\vec{B}$ parallel to the cylinder. Find out the maximum operational pressure of the plasma centrifuge. Is it necessary or not to trap ions in the magnetic field to provide the plasma rotation?

3–8. Ambipolar Diffusion. Estimate the coefficient of ambipolar diffusion in room-temperature atmospheric-pressure nitrogen at electron temperature about 1 eV. Compare the coefficient with that for the free diffusion of electrons and nitrogen molecular ions (take ion temperature as room temperature).

3–9. Thermal Conductivity in Plasma Related to Dissociation and Recombination of Molecules. Based on relations (3–99) and (3–100), find out the temperature range corresponding to a strong influence of the dissociation/recombination on thermal conductivity in plasma.

3–10. Plasma Absorption in Continuum, Unsold-Kramers Formula. Using the Unsold-Kramers formula, discuss the total plasma absorption in a continuum as a function of temperature. Estimate the quasi-equilibrium plasma temperature where the maximum absorption of red light can be achieved.

3–11. Hyperbolic Plateau Distribution of Vibrationally Excited Molecules. Derive a relationship between the hyperbolic plateau coefficient C (3–133, 3–134) for the vibrational energy distribution function and the ionization degree in non-thermal plasma (n_e/n_0). Take into account that, in the strong vibrational excitation regime, the excitation of lower vibrational levels by electron impact is balanced by resonant non-linear VV exchange between higher vibrationally excited molecules.

3–12. Gordiets Vibrational Distribution Function in Non-Thermal Plasma. Compare the discrete Gordiets vibrational distribution with continuous distribution function (3–141). Pay special attention to the exponential decrease of the vibrational distribution functions at high vibrational energies in the case of low translational temperatures ($T_0 < \hbar\omega$).

3–13. Treanor Effect for Isotopic Mixtures. Using the Treanor formula (3–165) and assuming for an isotopic mixture $\frac{\Delta\omega}{\omega} = \frac{1}{2}\frac{\Delta m}{m}$, estimate the difference in vibrational temperatures of nitrogen molecular isotopes at room temperature and the averaged level of vibrational temperature, $T_v \approx 3000$ K.

3–14. Plasma-Chemical Processes, Stimulated by Vibrational Excitation of Molecules. Explain why plasma-chemical reactions can be effectively stimulated by the vibrational excitation of molecules only if the specific energy input exceeds the critical value, whereas reactions related to electronic excitation or dissociative attachment can be equally effective at any levels of specific energy input.

3–15. Ideal and Non-Ideal Plasmas. Based on relation (3–225), calculate the minimum electron density required to reach conditions of the non-ideal plasma (1) at electron temperature of 1 eV and (2) at electron temperature equal to room temperature.

3–16. Floating Potential. Micro-particles or aerosols ($1–10\,\mu m$) are usually negatively charged in a steady-state non-thermal plasma and have floating potential (3–237) with respect to the plasma. Estimate the typical negative charge of such particles as a function of their radius. Assume that micro-particles are spherical and located in non-thermal plasma with electron temperature about 1 eV.

3–17. Matrix and Child Law Sheaths. Calculate the sizes of the matrix and Child law sheaths for non-thermal plasma with electron temperature 3 eV, electron density $10^{12}\,cm^{-3}$, and sheath voltage 300 V. Compare the results obtained for the models of matrix and Child law sheaths.

3–18. Langmuir Plasma Oscillations and Molecular Oscillations. Calculate the plasma density required to reach the resonance condition between plasma oscillations and the vibration of molecules. Is it possible to use such resonance for direct vibrational excitation of molecules in plasma without any electron impacts?

3–19. Magnetic Field Frozen in Plasma. The magnetic field becomes frozen in plasma when the plasma conductivity is high enough. Using the magnetic Reynolds number criterion, find out the minimum value of conductivity required to provide the effect and calculate the corresponding level of plasma density.

3–20. Attenuation of Electromagnetic Waves in Plasma. Based on expression (3–294), estimate attenuation coefficient κ of electromagnetic waves in plasma in the case of low conductivity ($\sigma_\omega \ll \varepsilon_\omega \varepsilon_0 \omega$).

4

Electric Discharges in Plasma Chemistry

4.1. FUNDAMENTALS OF ELECTRIC BREAKDOWN, STREAMER PROCESSES, AND STEADY-STATE REGIMES OF NON-EQUILIBRIUM ELECTRICAL DISCHARGES

4.1.1. Townsend Mechanism of Electric Breakdown and Paschen Curves

Consider breakdown in a plane gap d by DC voltage V corresponding to electric field $E = V/d$ (Fig. 4–1). Occasional primary electrons near a cathode provide low initial current i_0. The primary electrons drift to the anode, ionizing the gas and generating avalanches. The ionization in avalanches is usually described by **Townsend ionization coefficient** α, indicating production of electrons per unit length along the electric field: $dn_e/dx = \alpha n_e$, $n_e(x) = n_{e0} \exp(\alpha x)$. The Townsend ionization coefficient is related to the ionization rate coefficient $k_i(E/n_0)$ (Section 2.1.5) and electron drift velocity v_d as

$$\alpha = \frac{v_i}{v_d} = \frac{1}{v_d} k_i(E/n_0)n_0 = \frac{1}{\mu_e} \frac{k_i(E/n_0)}{E/n_0}, \qquad (4\text{--}1)$$

where v_i is the ionization frequency and μ_e is electron mobility, which is inversely proportional to pressure. The Townsend coefficient α is usually presented as similarity parameter α/p depending on the reduced electric field E/p. Dependences $\alpha/p = f(E/p)$ for different gases can be found in Fridman and Kennedy (2004).

Each primary electron generated near a cathode produces $\exp(\alpha d) - 1$ positive ions moving back to the cathode (Fig. 4–1). The ions lead to extraction of $\gamma^*[\exp(\alpha d) - 1]$ electrons from the cathode due to secondary electron emission characterized by the Townsend coefficient γ (Section 2.4.3). Typical γ-values in discharges are 0.01–0.1. Taking into account the current of primary electrons i_0 and electron current due to the secondary electron emission from the cathode, the total electronic part of the cathode current i_{cath} is

$$i_{cath} = i_0 + \gamma i_{cath}[\exp(\alpha d) - 1]. \qquad (4\text{--}2)$$

Total current in the external circuit is equal to the electronic current at the anode, where the ion current is absent. The total current can be found as $i = i_{cath} \exp(\alpha d)$, which leads to the **Townsend formula**:

$$i = \frac{i_0 \exp(\alpha d)}{1 - \gamma[\exp(\alpha d) - 1]}. \qquad (4\text{--}3)$$

The current in the gap is non-self-sustained as long as the denominator in 4–3 is positive. When the electric field and Townsend coefficient α become high enough, the denominator

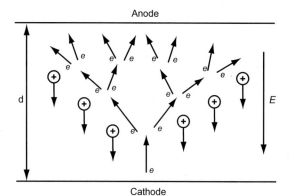

Anode

d

E

Figure 4–1. Townsend breakdown in a gap.

Cathode

in (4–3) goes to zero and transition to self-sustained current takes place, which is called the **Townsend breakdown mechanism**:

$$\gamma[\exp(\alpha d) - 1] = 1, \quad \alpha d = \ln\left(\frac{1}{\gamma} + 1\right). \tag{4-4}$$

The similarity parameters α/p and E/p are related semi-empirically according to (4–1) as

$$\frac{\alpha}{p} = A \exp\left(-\frac{B}{E/p}\right), \tag{4-5}$$

where parameters A and B for different gases at $E/p = 30 - 500$ V/cm · Torr are given in Table 4–1.

Combination of relations (4–4) and (4–5) gives formulas for calculating the breakdown voltage and breakdown reduced electric field as functions of an important similarity parameter pd:

$$V = \frac{B(pd)}{C + \ln(pd)}, \quad \frac{E}{P} = \frac{B}{C + \ln(pd)}. \tag{4-6}$$

Here $C = \ln A - \ln \ln(\frac{1}{\gamma} + 1)$ is almost constant. The breakdown voltage dependence on the similarity parameter pd is usually referred to as the **Paschen curve** (Fig. 4–2; Raizer, 1991). These curves have a minimum corresponding to the easiest breakdown conditions, which can be found from (4–6):

$$V_{min} = \frac{2.72 \cdot B}{A} \ln\left(1 + \frac{1}{\gamma}\right), \quad \left(\frac{E}{P}\right)_{min} = B, \quad (pd)_{min} = \frac{2.72}{A} \ln\left(1 + \frac{1}{\gamma}\right). \tag{4-7}$$

Reduced electric field E/p required for breakdown (4–6) decreases only logarithmically with pd. Breakdown of larger gaps is less sensitive to the secondary electron emission

Table 4–1. Numerical Parameters A and B for Semi-Empirical Calculation of Townsend Coefficient α

Gas	$A, \dfrac{1}{\text{cm} \cdot \text{Torr}}$	$B, \dfrac{\text{V}}{\text{cm} \cdot \text{Torr}}$	Gas	$A, \dfrac{1}{\text{cm} \cdot \text{Torr}}$	$B, \dfrac{\text{V}}{\text{cm} \cdot \text{Torr}}$
Air	15	365	N_2	10	310
CO_2	20	466	H_2O	13	290
H_2	5	130	He	3	34
Ne	4	100	Ar	12	180
Kr	17	240	Xe	26	350

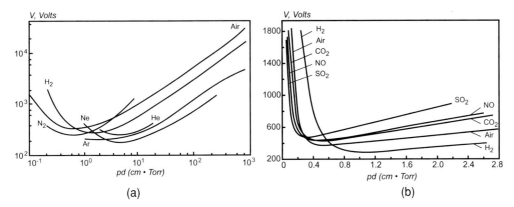

Figure 4–2. Breakdown Paschen curves for different atomic and molecular gases.

and cathode material, which explains the E/p reduction with pd. This reduction of breakdown electric field in electronegative gases is limited by electron attachment processes, characterized by Townsend coefficient β:

$$\beta = \frac{\nu_a}{\nu_d} = \frac{1}{\nu_d} k_a (E/n_0) n_0 = \frac{1}{\mu_e} \frac{k_a(E/n_0)}{E/n_0}. \tag{4–8}$$

In this relation $k_a(E/n_0)$ and ν_a are the attachment rate coefficient and frequency with respect to an electron. The Townsend coefficient β characterizes electron losses due to attachment per unit length:

$$\frac{dn_e}{dx} = (\alpha - \beta)n_e, \quad n_e(x) = n_{e0} \exp[(\alpha - \beta)x]. \tag{4–9}$$

The Townsend coefficient β, similarly to α, is an exponential function of the reduced electric field although not as strong. Therefore, the ionization rate much exceeds attachment at high electric fields, and the coefficient β can be neglected with respect to α in this case (short gaps; see relation (4–6)). When the gaps are relatively large (≥ 1 cm at 1 atm), the Townsend breakdown electric fields in electronegative gases become almost constant and limited by attachment processes. The breakdown electric fields at high pressures and long gaps for electronegative and non-electronegative gases are presented in Table 4–2.

4.1.2. Spark Breakdown Mechanism: Streamer Concept

The Townsend quasi-homogeneous breakdown mechanism can be applied only for relatively low pressures and short gaps ($pd < 4000$, Torr · cm at atmospheric pressure $d < 5$ cm). Another breakdown mechanism, called spark or streamer, takes place in larger gaps at high pressures. The sparks provide breakdown in a local narrow channel, without direct relation to electrode phenomena. Sparks are also primarily related to avalanches, but in large gaps the avalanches cannot be considered independent. The spark breakdown at high pd and

Table 4–2. Electric Fields Required for Townsend Breakdown of Centimeter-Size Gaps in Different Gases at Atmospheric Pressure

Gas	E/p, kV/cm	Gas	E/p, kV/cm	Gas	E/p, kV/cm
Air	32	O_2	30	N_2	35
H_2	20	Cl_2	76	CCl_2F_2	76
CSF_8	150	CCl_4	180	SF_6	89
He	10	Ne	1.4	Ar	2.7

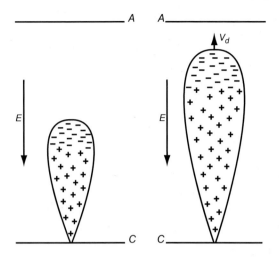

Figure 4–3. Evolution of an avalanche in a gap between cathode and anode.

considerable overvoltage develops much faster than the time necessary for ions to cross the gap and provide the secondary emission. The mechanism of spark breakdown is based on the **streamer** concept. A streamer is a thin ionized channel that rapidly propagates between electrodes along the positively charged trail left by an intensive primary avalanche. This avalanche also generates photons, which in turn initiate numerous secondary avalanches in the vicinity of the primary one. Electrons of the secondary avalanches are pulled by the strong electric field into the positively charged trail of the primary avalanche, creating the rapidly propagating streamer between electrodes (Loeb, 1960; Raether, 1964; Meek & Craggs, 1978).

Avalanche propagation is illustrated in Fig. 4–3. The qualitative change occurs when the charge amplification in the avalanche $\exp(\alpha x)$ becomes large and the created space charge leads to considerable electric field $\vec{E}_a$, which should be added to the external one, $\vec{E}_0$. The electrons are in the head of the avalanche while the positive ions remain behind, creating a dipole with the characteristic length $\frac{1}{\alpha}$ and charge $N_e \approx \exp(\alpha x)$. For the breakdown field, about 30 kV/cm in atmospheric-pressure air, the α-coefficient is about 10 cm^{-1} and the characteristic ionization length can be estimated as $1/\alpha \approx 0.1$ cm. Transverse avalanche size can also be estimated as $1/\alpha \approx 0.1$ cm; thus, the maximum electron density in an avalanche is $10^{12} \div 10^{13}$ cm^{-3} (Fridman & Kennedy, 2004). The external electric field distortion due to the space charge of the dipole is shown in Fig. 4–4. In front of the avalanche head (and behind the avalanche), the external ($\vec{E}_0$) and internal ($\vec{E}_a$) electric fields add up

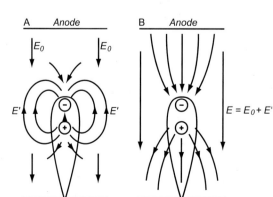

Figure 4–4. Distortion of electric field in an avalanche: (a) external electric field and space charge electric field are shown separately; (b) combination of the external electric field and space charge electric field.

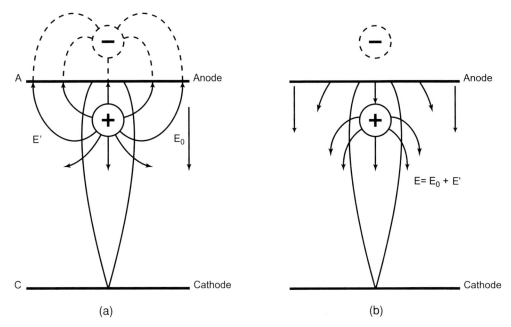

Figure 4–5. Space distribution of electric field when the avalanche reaches the anode: (a) external electric field and electric field of the space charge are shown separately; (b) the fields are shown combined together.

to make a total field stronger, which accelerates ionization. On the contrary, in between the separated charges or "inside the avalanche," the total electric field is lower than the external one, which slows down the ionization. The electric field of the charge $N_e \approx \exp(\alpha x)$ on the distance about the avalanche radius reaches the value of the external field $\vec{E}_0$ at some critical value of αx. Numerically, during the 1-cm-gap breakdown in air, the avalanche radius is about $r_A = 0.02$ cm, and the critical value of αx when the avalanche electric field becomes comparable with E_0 is $\alpha x = 18$. As soon as the avalanche head reaches the anode, electrons flow into the electrode and it is mostly the ionic trail that remains in the gap. Electric field distortion due to the space charge is illustrated in Fig. 4–5. Total electric field is due to the external field, the ionic trail, and also the ionic charge "image" in the anode.

   A strong primary avalanche is able to amplify the external electric field and form a streamer. When the streamer channel connects the electrodes, the current may be significantly increased to form the spark. The avalanche-to-streamer transformation takes place when the internal field of an avalanche becomes comparable with the external one. If the gap is short, the transformation occurs only when the avalanche reaches the anode. Such a streamer that grows from anode to cathode and called the **cathode-directed** or **positive streamer**. If the gap and overvoltage are large, the avalanche-to-streamer transformation can take place far from the anode, and the **anode-directed** or **negative streamer** grows toward both electrodes.

   The mechanism of formation of a cathode-directed streamer is illustrated in Fig. 4–6. High-energy photons emitted from the primary avalanche provide photo-ionization in the vicinity, which initiates the secondary avalanches. Electrons of the secondary avalanches are pulled into the ionic trail of the primary one and create a quasi-neutral plasma channel. The cathode-directed streamer starts near the anode, where the positive charge and electric field of the primary avalanche is the highest. The streamer looks like a thin conductive needle growing from the anode. The electric field at the tip of the "anode needle" is very high, which provides high electron drift and streamer growth velocities, about 10^8 cm/s.

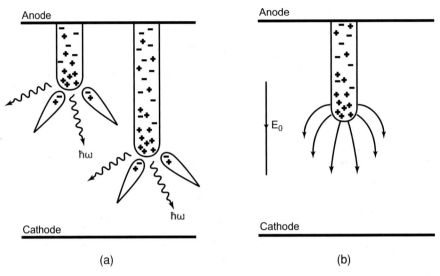

(a)   (b)

Figure 4–6. Illustration of the cathode-directed (positive) streamer: (a) propagation of the positive streamer; (b) electric field near the streamer head.

The diameter of the streamer channel is 0.01–0.1 cm and corresponds to the maximum size of a primary avalanche head, $1/\alpha$. The plasma density in the streamer also corresponds to the maximum electron density in the head of the primary avalanche: $10^{12} \div 10^{13}$ cm^{-3}. The specific energy input in a streamer channel is small during the short period ($\sim$30 ns) of streamer growth between electrodes. In molecular gases it is about 10^{-3} eV/mol, which corresponds to heating of $\sim$10 K.

The anode-directed streamer occurs if the primary avalanche becomes strong before reaching the anode. Such a streamer growing in two directions is illustrated in Fig. 4–7. The mechanism of streamer propagation in the direction of the cathode is the same as for cathode-directed streamers. The mechanism of streamer growth in direction of the anode is similar, but in this case the electrons from the primary avalanche head neutralize the ionic

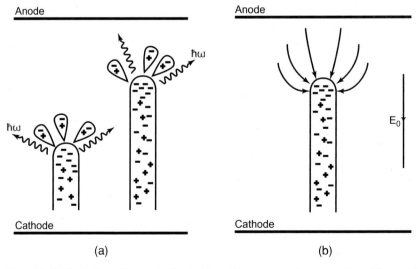

(a)   (b)

Figure 4–7. Illustration of the anode-directed (negative) streamer: (a) propagation of the negative streamer; (b) electric field near the streamer head.

trail of secondary avalanches. The secondary avalanches could be initiated here not only by photons but also by some electrons moving in front of the primary avalanche.

4.1.3. Meek Criterion of Streamer Formation: Streamer Propagation Models

A streamer occurs when the electric field of the space charge in an avalanche, E_a, equals the external field E_0:

$$E_a = \frac{e}{4\pi \varepsilon_0 r_A^2} \exp\left[\alpha\left(\frac{E_0}{p}\right) * d\right] \approx E_0. \qquad (4\text{--}10)$$

Assuming an avalanche head radius of $r_a \approx 1/\alpha$, the formation of a streamer in the gap (d) can be presented as a requirement for the avalanche amplification parameter αd to exceeds the critical value:

$$\alpha\left(\frac{E_0}{p}\right) * d = \ln\frac{4\pi\varepsilon_0 E_0}{e\alpha^2} \approx 20, \qquad N_e = \exp(\alpha d) \approx 3 \cdot 10^8. \qquad (4\text{--}11)$$

This criterion of streamer formation is known as the **Meek breakdown condition** ($\alpha d \geq 20$). Electron attachment mitigates their multiplication in avalanches and increases the electric field required for streamer formation. The ionization coefficient α in the Meek breakdown condition should be replaced in electronegative gases by $\alpha - \beta$. However, when discharge gaps are short (in air $d \leq 15$ cm), electric fields required by the Meek criterion are relatively high; therefore, $\alpha \gg \beta$ and the attachment can be neglected. Increasing distance d between electrodes in electronegative gases does not lead to a gradual decrease (4–11) of the required electric field. The minimal field required for streamer formation is due to ionization attachment balance $\alpha(E_0/p) = \beta(E_0/p)$, which gives 26 kV/cm in air and 117.5 kV/cm in SF$_6$.

Electric field non-uniformity has a strong influence on breakdown conditions and the transformation of an avalanche into a streamer. Voltage applied non-uniformly should provide intensive electron multiplication only near the electrode to initiate a streamer. Once the plasma channel is initiated, it grows mostly due to the high electric field of its own streamer tip. In the case of very long (about a meter and longer) non-uniform systems, the average breakdown electric field can be as low as 2–5 kV/cm. The breakdown threshold for non-uniform electric fields also depends on polarity. The threshold voltage in a long gap between a negatively charged rod and a plane is about twice as high as in the case of a positively charged rod. In the case of a rod anode, the avalanches approach the anode where the electric field becomes stronger, which facilitates the avalanche-to-streamer transition. Also the avalanche electrons easily sink into the anode in this case, leaving the ionic trail near the electrode to enhance the electric field.

There are two streamer-propagating models depending on what assumption is made regarding the conductivity of the streamer channels. The model of **quasi-self-sustained streamers** (Dawson & Winn, 1965; Gallimberti, 1972) assumes low conductivity of the streamer channel, which makes it self-propagating and independent of the anode. Photons initiate an avalanche, which then develops in the self-induced electric field of the positive space charge. To provide continuous and steady propagation of the self-sustained streamer, its positive space charge should be compensated by the negative charge of the avalanche head at the meeting point of the avalanche and streamer. Agreement of Gallimberti's model with an experimental photograph of a streamer corona is presented in Fig. 4–8 (Raizer, 1991). The model correctly describes breakdowns of long non-uniform gaps with high voltage and low average electric fields.

A qualitatively different model (Klingbeil, Tidman, & Fernsler, 1972; Lozansky & Firsov, 1975) assumes a streamer channel as an ideal conductor connected to an anode. The

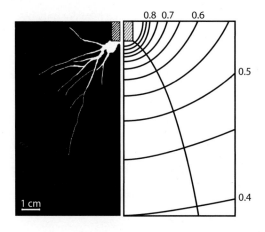

Figure 4–8. Illustration of streamer propagation from a positive 2-cm-diameter rod to a plane located at a distance of 1.5 m; applied voltage 125 kV. Equipotential surfaces are shown as fractions of the total applied voltage.

ideally conducting streamer channel is considered an anode elongation in the direction of external electric field E_0 with the shape of an ellipsoid. The propagation velocity is determined by electron drift in the appropriate electric field E_m on the tip of the streamer with length l and radius r

$$\frac{E_m}{E_0} = 3 + \left(\frac{l}{r}\right)^{0.92}, \quad 10 < \frac{l}{r} < 2000. \tag{4–12}$$

4.1.4. Streamers and Microdischarges

Although streamers are elements of the spark breakdown (Section 4.1.2), their visual observation is often related to DBDs and some corona discharges (see special sections that follow). The DBD gap (from 0.1 mm to 3 cm) usually includes one or more dielectric layers located in the current path between metal electrodes; the typical frequency is 0.05–100 kHz and voltage is about 10 kV at atmospheric pressure. In most cases, DBDs are not uniform and consist of numerous microdischarges built from streamers and distributed in the discharge gap (Fig. 4–9). The image of the filament footprints and electrode is sometimes called Lichtenberg figures.

Electrons in the conducting plasma channel established by the streamers dissipate from the gap in about 40 ns, whereas slowly drifting ions remain there for several microseconds (Table 4–3). Deposition of electrons from the conducting channel onto the anode dielectric barrier results in a charge accumulation and prevents new avalanches and streamers from

Figure 4–9. Storage phosphor image of filaments in DBD gap in air obtained from experimental setup using 10 excitation cycles at frequency 20.9 kHz and discharge gap 0.762 mm; the total discharge area is 5 cm × 5 cm.

Table 4–3. Typical Parameters of a Streamer Microdischarge

Lifetime	10–40 ns	Filament radius	50–100 μm
Electron avalanche duration	10 ns	Electron avalanche transported charge	0.01 nC
Cathode-directed streamer duration	1 ns	Cathode-directed streamer charge transfer	0.1 nC
Plasma channel duration	30 ns	Plasma channel charge transfer	1 nC
Microdischarge remnant duration	1 ms	Microdischarge remnant charge	> 1 nC
Peak current	0.1 A	Current density	0.1–1 kA/cm^2
Electron density	10^{14}–10^{15} cm^{-3}	Electron energy	1–10 eV
Total transported charge	0.1–1 nC	Reduced electric field	$E/n = (1$–$2)(E/n)_{Paschen}$
Total dissipated energy	5 μJ	Gas temperature	Close to average, about 300 K
Overheating	5 K		

forming nearby until the cathode and anode are reversed. Usual DBD operation frequency is around 20 kHz; therefore, the voltage polarity reversal occurs within 25 μs. After the voltage polarity reverses, the deposited negative charge facilitates formation of new avalanches and streamers in the same spot. As a result, a multi-generation family of streamers is formed that is macroscopically observed as a bright spatially localized **filament**.

It is important to distinguish the terms streamer and microdischarge. An initial electron starting from some point in the discharge gap (or from the cathode or dielectric that covers the cathode) produces secondary electrons by direct ionization and develops an electron avalanche. If the avalanche is big enough (Meek condition, see earlier discussion), the cathode-directed streamer is initiated. The streamer bridges the gap in a few nanoseconds and forms a conducting channel of weakly ionized plasma. An intensive electron current flows through this plasma channel until the local electric field collapses. Collapse of the local electric field is caused by the charges accumulated on the dielectric surface and the ionic space charge; ions are too slow to leave the gap for the duration of this current peak. The group of local processes in the discharge gap initiated by avalanche and developed until electron current termination is usually called a **microdischarge**. After electron current termination there is no longer an electron–ion plasma in the main part of the microdischarge channel, but high levels of vibrational and electronic excitation in the channel volume, along with charges deposited on the surface and ionic charges in the volume, allow us to separate this region from the rest of the volume and call it a **microdischarge remnant**.

Positive ions (or positive and negative ions in the case of electronegative gas) of the remnant slowly move to electrodes, resulting in a low and very long (∼10 μs for a 1-mm gap) falling ion current. The microdischarge remnant will facilitate the formation of a new microdischarge in the same spot as the polarity of the applied voltage changes. That is why it is possible to see single filaments in DBD. If microdischarges would form at a new spot each time the polarity changes, the discharge would appear uniform. Thus, the filament in DBD is a group of microdischarges that form on the same spot each time the polarity is changed. The fact that the microdischarge remnant is not fully dissipated before formation of the next microdischarge is called the **memory effect**. Typical characteristics of the DBD microdischarges in a 1-mm gap in atmospheric air are summarized in Table 4–3. A snapshot of the microdischarges in a 0.762-mm DBD air gap photographed through a transparent

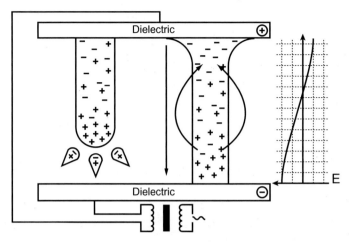

Figure 4–10. Illustration of interaction between the DBD microdischarges (quasi-repulsion) related to an electric field distortion caused by the microdischarge remnant. Streamer formation (left-hand side) and plasma channel (and microdischarge remnant) electric field distortion (right-hand side) are due to space charges. The figure shows the electric field from the microdischarge, the applied external electric field, and their superposition. In the presence of a space positive charge, the electric field is increased at the cathode and decreased at the anode.

electrode is shown in Fig. 4–9; as seen, the microdischarges are spread over the whole DBD zone quite uniformly.

Charge accumulation on the surface of the dielectric barrier reduces the electric field at the location of a microdischarge. It results in current termination within just several nanoseconds after breakdown. The short duration of microdischarges leads to very low overheating of the streamer channel. The principal microdischarge properties for most frequencies do not depend on the characteristics of the external circuit, but only on gas composition, pressure, and the electrode configuration. A power increase leads to the generation of a larger number of microdischarges per unit time, which simplifies DBD scaling. Detailed two-dimensional modeling of the formation and propagation of streamers can be found in numerous publications (see, for example, Kunchardt & Tzeng, 1988; Kulikovsky, 1994; Vitello, Penetrante, & Bardsley, 1994; Djermoune et al., 1995; Babaeba & Naidis, 1996; Fridman, Chirokov, & Gutsol, 2005).

4.1.5. Interaction of Streamers and Microdischarges

The mutual influence of microdischarges in DBD is related to the electric interaction with residual charges left on the dielectric barrier, as well as with the influence of excited species generated in one microdischarge upon formation of another microdischarge (Xu & Kushner, 1998; Chirokov et al., 2004; Chirokov, 2005). The interaction of streamers and microdischarges is responsible for the formation of microdischarge patterns reminiscent of two-dimensional crystals (Fig. 4–9), which may have a significant influence on the performance of plasma-chemical surface treatment, particularly when spatial uniformity is required. Charge distribution associated with streamers and the local electric field in the gap associated with a plasma channel and microdischarge remnant is illustrated in Fig. 4–10. The left-hand side of the figure shows a streamer propagating from anode to cathode while attracting additional avalanches. The resulting plasma channel and microdischarge remnant, shown on the right-hand side, have a net positive charge because electrons leave the gap much faster than ions. The residual positive charge (together with the deposited negative charge in the case of a dielectric surface) influences the formation of nearby families of avalanches and streamers and, therefore, formation of neighboring microdischarges.

The mechanism of interaction of microdischarges is as follows: a positive charge (or dipole field in the case of deposited negative charge) intensifies the electric field in the cathode area of the neighboring microdischarge and decreases the electric field in the anode area. Since the avalanche-to-streamer transition depends mostly on the near-anode electric field (from which new streamers originate), the formation of neighboring microdischarges is actually prevented, and microdischarges effectively repel each other. The **quasi repulsion between microdischarges** leads to the formation of a short-range order that is related to a characteristic repulsion distance between microdischarges. Observation of this cooperative phenomenon depends on several factors, including the number of microdischarges and the operating frequency. For example, when the number of microdischarges is not large enough (when the average distance between microdischarges is larger than the characteristic interaction radius), no significant microdischarge interaction is observed. When the AC frequency is too low to keep the microdischarge remnants from dissipation (low frequency means that the period is longer than the typical lifetime of a microdischarge remnant or "memory effect" lifetime), microdischarge repulsion effects are not observed. DBD cells operating at very high (megahertz) frequencies do not exhibit microdischarge repulsion because the very-high-frequency switching of the voltage interferes with ions still moving to electrodes.

To model the interaction between microdischarges, it can be assumed that avalanche-to-streamer transition depends only on the local value of the electric field and the discharge gap. Once the microdischarge is formed, the electric field of the microdischarge remnant decreases the external applied electric field in the fashion just described. The average effective field near the anode (where the avalanche-to-streamer transition occurs) decreases and, as a consequence, the formation of new streamers at the same location is prevented unless there is an increase in the external applied voltage. Observation of microdischarge filaments in DBD is possible when both electrodes are covered by dielectrics as well as in the case when one electrode is not covered, meaning that no surface discharge can be deposited on this electrode. Thus, repetition of microdischarges at the same spot depends mostly on the volume charge, and surface charge deposition is not critical. When the externally applied field varies quickly with respect to the microdischarge remnant dissipation, the microdischarge stays separated by a distance corresponding to the length scale of the field inhomogeneity. If the applied electric field is high enough, it causes microdischarges to develop in all the unoccupied spaces so that the electrode becomes filled from end to end.

In the unipolar or DC case (before polarity changes), one of the electrodes remains positive and the other is negative. The streamer always moves in one direction so that subsequent streamers and, thus, microdischarges have a small probability of forming in the same place until the microdischarge remnant has dispersed. A different situation appears in the case of alternating voltage. There is no need to wait until the microdischarge remnant dissipates. Instead, the probability of appearance of a streamer in the location of the microdischarge remnant increases when the voltage is switched. After the voltage is switched, the electric field of the microdischarge remnant adds to the strength of the applied electric field, thereby increasing the local field. The increased electric field increases the likelihood for a new streamer to occur at the same place. Thus, if the original streamer is formed just before voltage switching, there is an increased probability of streamers occurring in the same place or nearest vicinity.

4.1.6. Monte Carlo Modeling of Interaction of Streamers and Microdischarges

The cellular automata (CA) scheme, applied for the streamer and microdischarge interaction modeling, consists of a lattice of cells that can have any dimension and size coupled with a set of rules for determining the state of the cells (Chirokov, 2005). At any time a cell can be in only one state. From a physical perspective, each cell represents a volume in the gap located between the electrode surfaces. The upper and lower surfaces of each cell are bounded by dielectric surfaces, and the height of each cell is defined by the gap distance. The CA transformation rules define a new state for a cell after a given time step, using data about the states of all the cells in the CA and additional information such as the driving voltages imposed upon the system as a whole. It is assumed that the probability of streamer occurrence depends only on the local value of the electric field. The position of a streamer strike is determined using a Monte Carlo decision for given probability values in each cell. Once the position of the streamer strike is known, a plasma channel is formed at the same place and the total charge transferred by this microdischarge is assigned to the cell to be used later in electric field calculations. Since the time lag between streamers is random, an additional Monte Carlo simulation is used to decide whether a streamer occurs or not. If a streamer does not strike, a plasma channel will not be formed and there will be no microdischarge.

It is not necessary to specify the charge transferred by a microdischarge. Instead, it can be dynamically calculated during simulation based on the local electric field. The charge transferred by an individual microdischarge decreases the electric field inside the microdischarge channel because it creates a local electric field that opposes the externally applied electric field. Thus, the total charge transferred by the microdischarge is the charge that decreases the local electric field to zero. The probability of a streamer striking is calculated from the local electric field by the following formula (Chirokov, 2005):

$$P(E) = 1 - \frac{1}{1 + \exp\left(S \cdot \dfrac{E - E_0}{E_0}\right)}, \qquad (4\text{--}13)$$

where E is the electric field in the cell, E_0 is the critical electric field necessary for streamer formation given by the Meek condition, and S is a parameter related to the discharge ability to accumulate memory about previous microdischarges. When S is large, the memory effect has a negligible influence on the operation of DBD, and the probability function will be a step function that represents the Meek condition for streamer formation. When S is small, the memory effect significantly affects the probability of streamer formation. The presence of the vibrationally and electronically excited species and negative ions increases the ionization coefficient (first Townsend coefficient) and, thus, a lower electric field is required for avalanche-to-streamer transition. Furthermore, the discharge operating frequency also influences streamer formation so the memory effect is frequency dependent. In light of all the factors that influence streamer formation, the value of S is best determined empirically from experimental data. Typical results of the simulation are presented in Fig. 4–11. The grayscale intensity at any particular cell is proportional to the number of streamers striking the cell. The simulation shows that the occurrence of microdischarges across the simulation lattice is non-uniform: some regions are well covered by microdischarges and some are not treated at all. This non-uniformity of the plasma-chemical surface treatment is the result of interaction between microdischarges.

4.1.7. Self-Organized Pattern of DBD Microdischarges due to Streamer Interaction

Thus, the interaction of streamers in DBD can lead to the formation of an organized structure of microdischarges (similar to Coulomb crystals; see Fridman & Kennedy, 2004), which plays a significant (sometimes positive, sometimes negative) role in plasma-chemical applications. From this perspective it is important to analyze how highly organized the

Figure 4–11. Enlarged central portion of simulated microdischarge pattern in DBD. Simulation conditions are the same as for the experimental image shown in Fig. 4–9. The size of the microdischarge footprint is the same as in Fig. 4–9.

microdischarge structure can be using different image analysis methods, such as the two-dimensional correlation function and the Voronoi polyhedron approach (Chirokov et al., 2004; Chirokov, 2005). Voronoi polyhedra analysis defines polyhedral cells around selected features in an image, and the distribution of polyhedra types in the analysis can be used as a comparative tool. Voronoi polyhedra analysis is a tool for measuring the homogeneity of patterns as well as for comparison of different patterns. Homogeneity can be easily estimated from distributions of the Voronoi cell surface areas. The topology of the pattern can also be compared using a distribution of the number of sides of Voronoi cells. This type of comparison is especially useful as it is invariant to stretching and rotation of the patterns and also invariant to the particular positions of the microdischarge footprints. The Voronoi polyhedra analysis of an experimentally observed microdischarge structure image (Chirokov et al., 2004) and its simulation (see Section 4.1.6, Fig. 4–11) are shown in Figs. 4–12 and 4–13. The Voronoi analysis of a random dot pattern (case without microdischarge interaction) is shown for comparison in Fig. 4–14. The comparison demonstrates the importance of short-range interaction between microdischarges in DBD. One way to numerically express

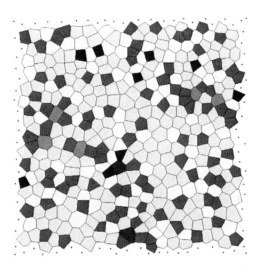

Figure 4–12. Voronoi polyhedra analysis of the experimentally obtained microdischarge locations obtained from the DBD image presented in Fig. 4–9. The polyhedra cells are coded by level of darkness according to the number of angles in each polyhedron. The cells in the image obtained experimentally are mainly six-sided cells and have similar sizes.

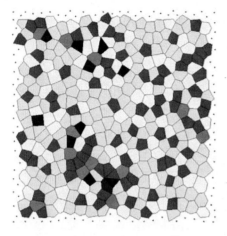

Figure 4–13. Voronoi polyhedra analysis of the simulated microdischarge locations obtained from the DBD image presented in Fig. 4–11. The polyhedra cells are coded by level of darkness according to the number of angles in each polyhedron. The cells in the image obtained in simulation are mainly six-sided cells and have similar sizes.

the difference between the images is to count the number of different-sided polyhedra determined in the Voronoi analysis. Unlike the random dot pattern, most of the polyhedral cells found in the DBD analysis have six interior sides (six angles). This corresponds to the hexagonal lattice and, thus, implies radial symmetry of the interaction. This type of interaction was observed experimentally as well as predicted by the model.

Another approach to the microdischarge structure characterization is the correlation function method, which is widely used for postprocessing in crystallography. The correlation functions of the experimental DBD image and the DBD simulation have been analyzed by Chirokov (2005). The random distribution does not display any periodic oscillations in the correlation function. The correlation function for a completely ordered lattice should show strong oscillations or peaks. Both experimental and simulated DBD images demonstrate non-randomness and a high level of self-organization due to microdischarge interactions.

4.1.8. Steady-State Regimes of Non-Equilibrium Electric Discharges and General Regimes Controlled by Volume and Surface Recombination Processes

Steady-state regimes of non-equilibrium discharges are provided by a balance between generation and loss of charged particles. The generation of electrons and positive ions is mostly due to volume ionization processes (Section 2.1). To sustain the steady-state

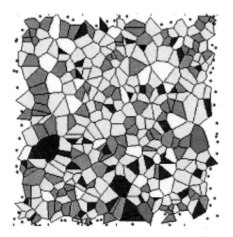

Figure 4–14. Voronoi polyhedra analysis of the random dot pattern for comparison with those presented in Figs. 4–12 and 4–13. The polyhedra cells are coded by level of darkness according to the number of angles in each polyhedron. The areas of different cells vary significantly in this case.

plasma ionization should be quite intensive and usually requires the electron temperature to be at least on the level of one-tenth of the ionization potential (~ 1 eV). Losses of charged particles can also be related to volume processes of recombination or attachment, and they can be provided by the diffusion of charged particles to the walls with further surface recombination. These two charge-loss mechanisms separate two different regimes of sustaining the steady-state discharge: the first is controlled by volume processes and the second is controlled by diffusion to the walls.

If the ionization degree in plasma is relatively high and diffusion can be considered ambipolar, the frequency of charge losses due to diffusion to the walls can be described as

$$\nu_D = \frac{D_a}{\Lambda_D^2},$$ (4–14)

where D_a is the coefficient of ambipolar diffusion and Λ_D is the characteristic diffusion length. Volume-related charge losses dominate and, hence, non-equilibrium discharges are controlled by volume processes when

$$k_i(T_e)n_0 \gg \frac{D_a}{\Lambda_D^2}.$$ (4–15)

In this relation, $k_i(T_e)$ is the ionization rate coefficient and n_0 is the neutral gas density. Criterion 4–15 actually restricts pressure, because $D_a \propto 1/p$ and $n_0 \propto p$. When pressure is low, non-equilibrium discharges are controlled by diffusion to the walls and surface recombination. When pressure exceeds 10–30 Torr (the so-called range of moderate and high pressures), diffusion is relatively slow and the balance of charge particles is due to volume processes:

$$\frac{dn_e}{dt} = k_i n_e n_0 - k_a n_e n_0 + k_d n_0 n_- - k_r^{ei} n_e n_+,$$ (4–16)

$$\frac{dn_+}{dt} = k_i n_e n_0 - k_r^{ei} n_e n_+ - k_r^{ii} n_+ n_-,$$ (4–17)

$$\frac{dn_-}{dt} = k_a n_e n_0 - k_d n_0 n_- - k_r^{ii} n_+ n_-.$$ (4–18)

In this set of equations, n_+, n_- are concentrations of positive and negative ions; n_e, n_0 are concentrations of electrons and neutral species; rate coefficients k_i, k_a, k_d, k_r^{ei}, k_r^{ii} are related to the processes of ionization by electron impact, dissociative or other electron attachment, electron detachment from negative ions, and electron–ion and ion–ion recombination; rate coefficients of processes involving neutral particles (k_i, k_a, k_d) are expressed in the system (4–16)–(4–18) with respect to the total gas density.

If the moderate or high-pressure gas is not electronegative, the volume balance of electrons and positive ions can be reduced to the simple ionization–recombination balance. In electronegative gases, however, two qualitatively different self-sustained regimes can be achieved (at different effectiveness of electron detachment): one controlled by recombination and another controlled by electron attachment.

4.1.9. Discharge Regime Controlled by Electron–Ion Recombination

In some plasma-chemical systems, the destruction of negative ions by, for example, associative electron detachment is faster than ion–ion recombination:

$$k_d n_0 \gg k_r^{ii} n_+.$$ (4–19)

In this case, actual losses of charged particles are also due to electron–ion recombination, in the same way as non-electronegative gases. Such a situation can take place, in particular, in plasma-chemical processes of CO_2 and H_2O dissociation, and NO synthesis in air, when associative detachment processes,

$$O^- + CO \rightarrow CO_2 + e, \quad O^- + NO \rightarrow NO_2 + e, \quad O^- + H_2 \rightarrow H_2O + e, \quad (4\text{--}20)$$

are very fast (about $0.1\,\mu s$ at concentrations of CO, NO, and H_2 molecules of $\sim 10^{17} cm^{-3}$). Electron attachment and detachment (4–18) are in dynamic quasi equilibrium in the recombination regime during time intervals sufficient for electron detachment ($t \gg 1/k_d n_0$). The concentration of negative ions can be considered to be in dynamic quasi equilibrium with the electron concentration:

$$n_- = \frac{k_a}{k_d} n_e = n_e \varsigma. \quad (4\text{--}20)$$

Using the parameter $\varsigma = k_a/k_d$ and presenting quasi neutrality as $n_+ = n_e + n_- = n_e(1 + \varsigma)$, equations (4.5.6)–(4.5.8) can be simplified to a single kinetic equation for electron density:

$$\frac{dn_e}{dt} = \frac{k_i}{1 + \varsigma} n_e n_0 - \left(k_r^{ei} + \varsigma k_r^{ii}\right) n_e^2. \quad (4\text{--}21)$$

The parameter $\varsigma = k_a/k_d$ shows the detachment ability that compensates for electron losses due to attachment. If $\varsigma \ll 1$, the attachment influence is negligible and kinetic equation (4–21) becomes equivalent to one for non-electronegative gases. The kinetic equation includes the effective rate coefficients of ionization, $k_i^{eff} = k_i/1 + \varsigma$, and recombination, $k_r^{eff} = k_r^{ei} + \varsigma k_r^{ii}$. Equation (4–21) describes electron density evolution to the steady-state magnitude of the recombination-controlled regime:

$$\frac{n_e}{n_0} = \frac{k_i^{eff}(T_e)}{k_r^{eff}} = \frac{k_i}{(k_r^{ei} + \varsigma k_r^{ii})(1 + \varsigma)}. \quad (4\text{--}22)$$

4.1.10. Discharge Regime Controlled by Electron Attachment

This regime takes place if the balance of charged particles is due to the volume processes (4–15) and the discharge parameters correspond to the inequality opposite that of (4–19). Negative ions produced by electron attachment go almost instantaneously into ion–ion recombination, and electron losses are mostly due to the attachment process. The steady-state (4–16) solution for the attachment-controlled regime is

$$k_i(T_e) = k_a(T_e) + k_r^{ei} \frac{n_+}{n_0}. \quad (4\text{--}23)$$

In the attachment-controlled regime the electron attachment is usually faster than recombination, and relation (4–13) actually requires $k_i(T_e) \approx k_a(T_e)$. Typical functions of $k_i(T_e)$ and $k_a(T_e)$ are shown in Fig. 4–15 and have a single crossing point T_{st}, which determines the steady-state electron temperature for a non-thermal discharge that is self-sustained in the attachment-controlled regime.

4.1.11. Non-Thermal Discharge Regime Controlled by Charged-Particle Diffusion to the Walls: The Engel-Steenbeck Relation

When pressure is relatively low and the inequality opposite that of (4–15) is valid, the balance of charged particles is provided by competition of ionization in volume and diffusion of

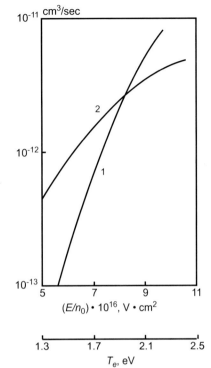

Figure 4–15. Rate coefficients of (1) ionization of CO_2 molecules by direct electron impact; (2) dissociative attachment of electrons to CO_2 molecules.

charged particles to the walls, where they recombine on the surface. The balance of direct ionization by electron impact and ambipolar diffusion to the walls of a long discharge chamber of radius R gives the following relation between electron temperature and pressure (or, better to say, the similarity parameter pR; Fridman & Kennedy, 2004):

$$\left(\frac{T_e}{I}\right)^{\frac{1}{2}} \exp\left(\frac{I}{T_e}\right) = \frac{\sigma_0}{\mu_i p} \left(\frac{8I}{\pi m}\right)^{\frac{1}{2}} \left(\frac{n_0}{p}\right)(2.4)^2 (pR)^2. \qquad (4\text{–}24)$$

This is the **Engel-Steenbeck relation** for the diffusion-controlled regime of non-equilibrium discharges. Here I is the ionization potential, μ_i is the ion mobility, σ_0 is the electron–neutral gas-kinetic cross section, and m is the electron mass. If gas temperature is fixed (for example, at room temperature), the parameters $\mu_i p$ and n_0/p are constant and the Engel-Steenbeck relation can be presented as

$$\sqrt{\frac{T_e}{I}} \exp\left(\frac{I}{T_e}\right) = C(pR)^2. \qquad (4\text{–}25)$$

The constants in the Engel-Steenbeck relation depend on the type of gas and are given in Table 4–4. The universal relation between T_e/I and the similarity parameter cpR for the diffusion-controlled regime is presented in Fig. 4–16. According to the Engel-Steenbeck curve, the electron temperature in the diffusion-controlled regime decreases with growth of the pressure and radius of the discharge tube.

Table 4–4. Engel-Steenbeck Relation Parameters

Gas	C, Torr^{-2} cm^{-2}	c, Torr^{-1} cm^{-1}	Gas	C, Torr^{-2} cm^{-2}	c, Torr^{-1} cm^{-1}
N_2	$2 \cdot 10^4$	$4 \cdot 10^{-2}$	Ar	$2 \cdot 10^4$	$4 \cdot 10^{-2}$
He	$2 \cdot 10^2$	$4 \cdot 10^{-3}$	Ne	$4.5 \cdot 10^2$	$6 \cdot 10^{-3}$
H_2	$1.25 \cdot 10^3$	10^{-2}			

Table 4–5. Parameters of Conventional Low-Pressure Glow Discharge in a Tube

Parameter of a glow discharge	Typical values
Discharge tube radius	0.3–3 cm
Discharge tube length	10–100 cm
Plasma volume	about 100 cm^3
Gas pressure	0.03–30 Torr
Voltage between electrodes	100–1000 V
Electrode current	10^{-4}–0.5 A
Power level	around 100 W
Electron temperature in positive column	1–3 eV
Electron density in positive column	10^9–10^{11} cm^{-3}

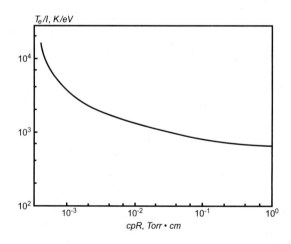

Figure 4–16. Universal relation between electron temperature, pressure, and discharge tube radius in the non-equilibrium discharge regime controlled by diffusion of charged particles.

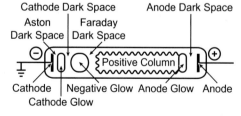

Figure 4–17. Structure of a glow discharge in a long tube.

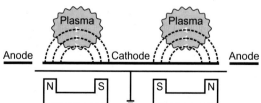

Figure 4–18. Magnetron configuration of glow discharges.

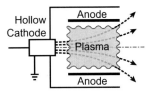

Figure 4–19. Hollow-cathode configuration of glow discharges.

4.2. GLOW DISCHARGES

4.2.1. General Structure and Configurations of Glow Discharges

The glow discharge is the best-known type of non-thermal discharge and has been widely used in plasma chemistry for more than a century. The term "glow" indicates that the plasma of the discharge is luminous in contrast to the relatively low-power dark discharge. The glow discharge can be defined as the self-sustained continuous DC discharge with cold cathode, which emits electrons as a result of secondary emission mostly induced by positive ions. A schematic of the discharge is shown in Fig. 4–17 (Cobine, 1958; Hirsh & Oscam, 1978; Vossen & Kern, 1978, 1991; Sugano, 1985; Boenig, 1988).

A distinctive feature of a glow discharge is the **cathode layer** with positive space charge, strong electric field, and potential drop of about 100–500 V. The thickness of the cathode layer is inversely proportional to gas density. If the distance between the electrodes is large enough, a quasi-neutral plasma with low electric field, the so-called positive column, is formed between the cathode layer and the anode. The **positive column** of glow discharge is the most traditional example of weakly ionized non-equilibrium low-pressure plasma. The positive column is separated from the anode by an anode layer. The **anode layer** is characterized by a negative space charge, slightly elevated electric field, and some potential drop. Typical parameters are given in Table 4–5. This configuration is widely used in fluorescent lamps as a lighting device. Other configurations, applied for thin-film deposition and electron bombardment, are shown in Figs. 4–18 and 4–19. The coplanar magnetron, convenient for sputtering and deposition, includes a magnetic field for plasma confinement (Fig. 4–18). Configurations optimized for electron bombardment (Fig. 4–19) are coaxial and include a hollow cathode ionizer as well as a diverging magnetic field. Glow discharges are applied for gas lasers, and special configurations were developed to increase their power (Fig. 4–20). Usually these are parallel plate discharges with gas flow. The discharge can be transverse with electric current perpendicular to the gas flow (Fig. 4–20a), or longitudinal if they are parallel to each other (Fig. 4–20b). The powerful glow discharges operate at higher currents and voltages, reaching 10–20 A and 30–50 kV.

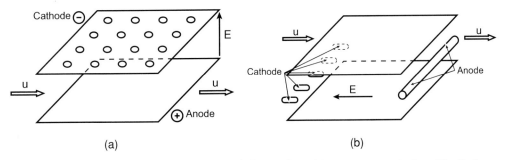

(a)   (b)

Figure 4–20. Organization of glow discharges in fast gas flows: (a) transverse organization of the discharge in flow; (b) longitudinal organization of the discharge in flow.

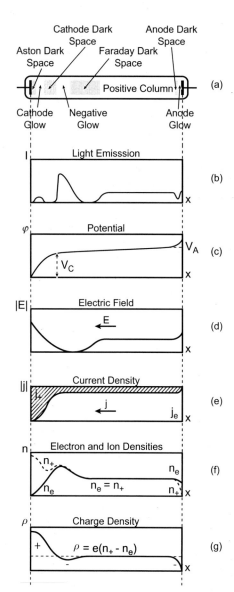

Figure 4–21. Distribution of major physical parameters in a glow discharge.

Consider the pattern of light emission in a classical low-pressure discharge in a tube. One can see along such a discharge tube a sequence of dark and bright luminous layers (Fig. 4–21a). A typical size scale of the structure is proportional to the electron mean free path $\lambda \propto 1/p$ and, hence, inversely proportional to pressure. Therefore, it is easier to observe the glow pattern at low pressures (centimeters at pressures about 0.1 Torr). Special names were given to each layer shown in Fig. 4–21a. Immediately adjacent to the cathode is a dark layer known as the **Aston dark space**. Then there is a relatively thin layer of **the cathode glow** followed by the **cathode dark space**. The next zone is **negative glow**, which is sharply separated from the dark cathode space. The negative glow is gradually less and less bright toward the anode, becoming the **Faraday dark space**. After that the **positive column** begins, which is bright (though not as bright as the negative glow), uniform, and long. Near the anode, the positive column is transferred at first into the **anode dark space**, and finally into a narrow zone of the **anode glow**. The glow pattern can be interpreted based on the distribution of the discharge parameters shown in Fig. 4–21b–g. Electrons are ejected from

4.2. Glow Discharges

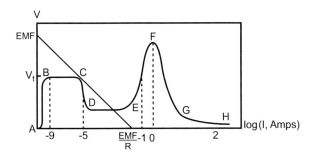

Figure 4–22. Generalized current-voltage characteristic of DC discharges.

the cathode with energy (about 1 eV) insufficient for excitation of atoms, which explains the Aston dark space. Then electrons obtain from the electric field enough energy for electronic excitation, which provides the cathode glow. Further acceleration of electrons in the cathode dark space leads mostly to ionization, not to electronic excitation. This explains the low level of radiation and increase of electron density in the cathode dark space. Slowly moving ions have a high concentration in the cathode layer and provide most of the current.

High electron density at the end of the cathode dark space results in a decrease of the electric field, electron energy, and ionization rate, but leads to intensification of radiation. This explains the transition to the brightest layer, the negative glow. Farther from the cathode, electron energy decreases, resulting in a transition from the negative glow to the Faraday dark space. There plasma density decreases and the electric field grows up, establishing the positive column. The average electron energy in a positive column is about 1–2 eV, which provides light emission. The cathode layer structure remains the same if electrodes are moved closer, whereas the positive column shrinks. The positive column can be extended so it is somewhat long and connecting the electrodes. The anode repels ions and pulls out electrons from the positive column, which creates the negative space charge and leads to an increase of the electric field in the anode layer. A reduction of the electron density explains the anode dark space, whereas the electric field increase explains the anode glow.

4.2.2. Current-Voltage Characteristics of DC Discharges

If the voltage between the electrodes exceeds the critical threshold value V_t necessary for breakdown, a self-sustained discharge can be ignited. The general current-voltage characteristic of such a discharge is illustrated in Fig. 4–22 for a wide range of currents I. The electric circuit also includes an external ohmic resistance R, which results in the Ohm's law usually referred to as the **load line**:

$$EMF = V + RI. \tag{4–26}$$

Here EMF is electromotive force, and V is voltage on the discharge gap. Intersection of the current-voltage characteristic and the load line gives current and voltage in a discharge. If the external ohmic resistance is high and the current in the circuit is low (about 10^{-10}–10^{-5} A), electron and ion densities are negligible and perturbation of the external electric field in plasma can be neglected. Such a discharge is known as the **dark Townsend discharge**. The voltage necessary to sustain this discharge does not depend on current and coincides with the breakdown voltage. The dark Townsend discharge corresponds to the plateau BC in Fig. 4–22. An increase of the EMF or a decrease of the external resistance R leads to a growth of the current and plasma density, which results in significant restructuring of the electric field. It leads to a reduction of voltage with current (interval CD in Fig. 4–22)

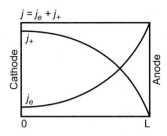

Figure 4–23. Distribution of electron and ion current densities along the axis of a long dark discharge.

and to a transition from dark to glow discharge. This very low current glow discharge is called the subglow discharge. Further EMF increase or R reduction leads to the lower-voltage plateau DE on the current-voltage characteristic, corresponding to the normal glow discharge existing over a large range of currents $10^{-4} \div 0.1$ A.

The current density on the cathode is fixed for normal glow discharges. An increase of the total discharge current is provided by growth of the cathode spot through which the current flows. When the current is so high that no more free surface is left on the cathode, further current growth requires a voltage increase to provide higher values of current density. Such a regime is called the abnormal glow discharge and it corresponds to the growing interval EF on the current-voltage characteristics (Fig. 4–22). Further increase of current and voltage in the abnormal glow regime leads to higher power and to transition to arc discharge. The glow-to-arc transition usually takes place at currents about 1 A.

4.2.3. Dark Discharge and Transition from Townsend Dark to Glow Discharge

A distinctive feature of the dark discharge is the smallness of its current and plasma density, which keeps the external electric field unperturbed and is determined by Townsend breakdown condition (4–4). Relations between electron (j_e), ion (j_+), and total (j) current densities in a dark discharge (distance between electrodes d_0, from cathode x; Townsend coefficient α) (Fridman & Kennedy, 2004) are

$$\frac{j_e}{j} = \exp[-\alpha(d_0 - x)], \quad \frac{j_+}{j} = 1 - \exp[-\alpha(d_0 - x)], \quad \frac{j_+}{j_e} = \exp[\alpha(d_0 - x)] - 1.$$

$$(4-27)$$

According to (4–4) $\alpha d_0 = \ln \frac{\gamma+1}{\gamma} \gg 1$; therefore, the ion current exceeds the electron current over the major part of the discharge gap (Fig. 4–23). The electron and ion currents become equal only near the anode ($j_e = j_+$ at $x = 0.85 d_0$). The difference in concentrations of electrons and ions is even stronger because of an additional big difference in electron and ion mobilities (μ_e, μ_+). Electron and ion concentrations become equal at a point very close to the anode (Fig. 4–24), where

$$1 = \frac{n_+}{n_e} = \frac{\mu_e}{\mu_+} \frac{j_+}{j_e} = \frac{\mu_e}{\mu_+}[\exp \alpha(d_0 - x) - 1].$$

$$(4-28)$$

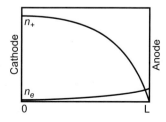

Figure 4–24. Distribution of electron and ion number densities along the axis of a long dark discharge.

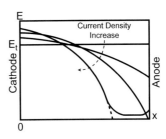

Figure 4–25. Evolution of the electric field distribution along the axis of a long dark discharge with growth of current density. Transition of a dark discharge into a glow discharge.

Assuming $\mu_e/\mu_+ \approx 100$, electron and ion concentrations become equal at $x = 0.998$. Thus, almost the entire gap is charged positively. Dark-to-glow discharge transition at higher currents is due to growth of the positive space charge and distortion of the external electric field, which results in formation of the cathode layer. To describe the transition, the Maxwell equation can be used:

$$\frac{dE}{dx} = \frac{1}{\varepsilon_0} e(n_+ - n_e). \tag{4–29}$$

Here $n_+ \approx j/e\mu_+ E \gg n_e$; therefore, (4–29) gives the following distribution of the electric field;

$$E = E_c \sqrt{1 - \frac{x}{d}}, \quad d = \frac{\varepsilon_0 \mu_+ E_c^2}{2j}, \tag{4–30}$$

where E_c is the electric field at the cathode. The electric field decreases near the anode with respect to the external field and grows in the vicinity of the cathode, as illustrated in Fig. 4–25. Higher current densities lead to greater distortion of the external electric field. The parameter d corresponds to a virtual point where the electric field equals zero. This point is located far beyond the discharge gap ($d \gg d_0$) at the low currents typical for dark discharges. At higher current densities, the point of zero electric field reaches the anode ($d = d_0$). This critical current density is the maximum for the dark discharge and corresponds to formation of the cathode layer and to transition from dark to glow discharge:

$$j_{max} = \frac{\varepsilon_0 \mu_+ E_c^2}{2d_0}. \tag{4–31}$$

In N_2 at pressure 10 Torr, interelectrode distance of 10 cm, electrode area of 100 cm^2, and secondary emission coefficient $\gamma = 10^{-2}$, the maximum dark discharge current is $j_{max} \approx 3 \cdot 10^{-5}$ A.

The dependence of Townsend coefficient α on the electric field is exponential and very strong. It leads to an interesting consequence: the typical voltage of a glow discharge is lower than that of a dark discharge (Fig. 4–22). Growth of the positive space charge during the dark-to-glow transition results in a redistribution of the initially uniform electric field: it becomes stronger near the cathode and lower near the anode. However, the increase of the exponential function $\alpha[E(x)]$ in the cathode side is more significant than its decrease in the anode side. Therefore, the electric field non-uniformity facilitates the breakdown conditions, which explains why the typical voltage of a glow discharge is usually lower than in a dark discharge.

4.2.4. Current-Voltage Characteristics of Cathode Layer: Normal Glow Discharge

The cathode layer is the most distinctive zone of a glow discharge. It provides its self-sustaining behavior and generates enough electrons to balance the plasma current in the positive column. When voltage is applied to a discharge gap, the uniform distribution of the electric field is not optimal; it is easier to sustain the discharge when a sufficiently high potential drop occurs near the cathode. The required electric field non-uniformity is provided by a positive space charge formed near the cathode due to low ion mobility. A theory of the cathode layer has been developed by von Engel and Steenbeck (1934; see also von Engel, 1994). Assuming zero electric field at the end of a cathode layer, neglecting the ion current into a cathode layer from a positive column, and considering the cathode layer as an independent system of length d, the Engel-Steenbeck theory gives the relations between the electric field in the cathode layer, E_c, the cathode potential drop V_c, and the length of cathode layer pd, which are similar to those used to describe the breakdown of a gap:

$$V_c = \frac{B(pd)}{C + \ln(pd)}, \quad \frac{E_c}{p} = \frac{B}{C + \ln(pd)}, \tag{4-32}$$

where $C = \ln A - \ln\ln(\frac{1}{\gamma} + 1)$; A and B are Townsend parameters (see (4–5) and Table 4–1). V_c, E_c, and similarity parameter pd depend on the discharge current density j, which is close to the ion current density because $j_+ \gg j_e$ near the cathode. To find this dependence, first the positive ion density $n_+ \gg n_e$ can be determined from the Maxwell equation:

$$n_+ \approx \frac{\varepsilon_0}{e}\left|\frac{dE(x)}{dx}\right| \approx \frac{\varepsilon_0 E_c}{ed}. \tag{4-33}$$

The total current density in the cathode vicinity is close to the current density of positive ions, which results in the Engel-Steenbeck current-voltage characteristics of the cathode layer:

$$j = en_+\mu_+E \approx \frac{\varepsilon_0\mu_+ E_c^2}{d} \approx \frac{\varepsilon_0\mu_+ V_c^2}{d^3}. \tag{4-34}$$

The cathode potential drop V_c as a function of the similarity parameter pd corresponds to the Paschen curve for breakdown (Section 4.1.1); thus, the function $V_c(pd)$ has a minimum V_n. Taking into account (4–34), the potential drop V_n as a function of current density j also has the same minimum point V_n. Relations between V_c, E_c, pd, and j can be expressed using dimensionless parameters:

$$\tilde{V} = \frac{V_c}{V_n}, \quad \tilde{E} = \frac{E_c/p}{E_n/p}, \quad \tilde{d} = \frac{pd}{(pd)_n}, \quad \tilde{j} = \frac{j}{j_n}, \tag{4-35}$$

where electric field E_n/p and cathode layer length $(pd)_n$ correspond to the minimum point of the cathode voltage drop V_n. The subscript n stands for the "normal" regime of a glow discharge. All three normal parameters – E_n/p, $(pd)_n$, and V_n – can be found from formulas (4–7) originally derived for the electric breakdown as parameters of the Paschen curve. The corresponding value of the normal current density can be found from (4–34) using a numeric formula constructed with similarity parameters:

$$\frac{j_n, A/cm^2}{(p, Torr)^2} = \frac{1}{9 \cdot 10^{11}} \frac{(\mu_+ p), cm^2\, Torr/V s \times (V_n, V)^2}{4\pi[(pd)_n, cm \cdot Torr]^3}. \tag{4-36}$$

Relations between V_c, E_c, j, and the cathode layer length pd can be expressed as

$$\tilde{V} = \frac{\tilde{d}}{1 + \ln\tilde{d}}, \quad \tilde{E} = \frac{1}{1 + \ln\tilde{d}}, \quad \tilde{j} = \frac{1}{\tilde{d}(1 + \ln\tilde{d})^2}. \tag{4-37}$$

Voltage $\tilde{V}$, electric field $\tilde{E}$, and cathode layer length $\tilde{d}$ are presented in Fig. 4–26 as functions of current density, which is called the dimensionless current-voltage characteristic of a cathode layer. According to (4–37) any current densities are possible in a glow discharge. In reality, a cathode layer "prefers" to operate at the only value of current density, the normal one j_n (4–36), which corresponds to a minimum of the cathode potential drop. It can be

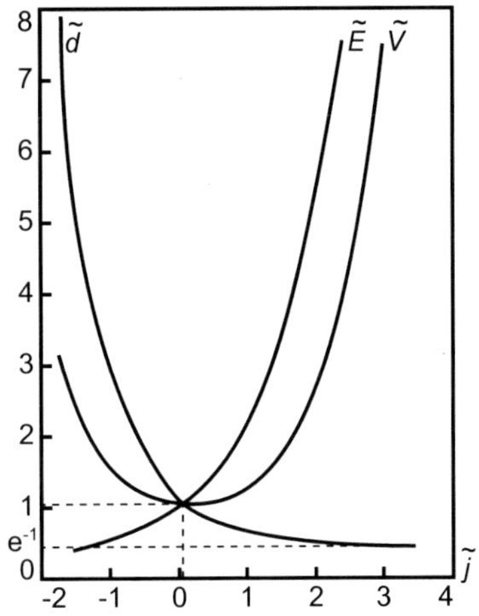

Figure 4–26. Dimensionless current-voltage characteristic and dimensionless parameters of a cathode layer in a glow discharge.

interpreted in terms of the **Steenbeck minimum power principle**; a detailed explanation of the effect can be found, in particular, in Fridman and Kennedy (2004).

The current conducting channel occupies a **cathode spot** with area $A = I/j_n$, which provides the normal current density. Other current densities are unstable, provided the cathode surface is big enough. A glow discharge with the normal cathode current density is referred to as the **normal glow discharge**. Normal glow discharges have fixed current density j_n, corresponding fixed cathode layer thickness $(pd)_n$, and voltage V_n, which depend at room temperature only on the gas composition and cathode material. Normal glow discharge parameters are presented in Table 4–6. A typical normal current density is 100 $\mu A/cm^2$ at pressure of about 1 Torr, the thickness of the normal cathode layer at this pressure is about 0.5 cm, and the normal cathode potential drop is about 200 V and does not depend on pressure and temperature.

4.2.5. Abnormal, Subnormal, and Obstructed Regimes of Glow Discharges

An increase of the current in a normal glow discharge is provided by the growth of the cathode spot at $j = j_n = \text{const}$. As soon as the entire cathode is covered, further current growth results in an increase of current density over the normal value. This discharge is called an **abnormal glow discharge**. The abnormal glow discharge corresponds to the right-hand-side branches ($j > j_n$) in Fig. 4–26. The current-voltage characteristic of the abnormal discharge $\tilde{V}(\tilde{j})$ is growing. It corresponds to the interval EF on the general current-voltage characteristic (Fig. 4–22). According to (4–35), when the current density is growing ($j \to \infty$), the cathode layer thickness decreases asymptotically to a finite value $\tilde{d} = 1/e \approx 0.37$, whereas the cathode potential drop and electric field grow as follows:

$$\tilde{V} = \frac{1}{e^{3/2}}\sqrt{\tilde{j}}, \quad \tilde{E} \approx \frac{1}{e^{1/2}}\sqrt{\tilde{j}}. \tag{4–38}$$

The actual growth of the current and cathode voltage are limited by the cathode overheating. Significant cathode heating at voltages of about 10 kV and current densities of 10–100 A/cm^2 results in a transition of the abnormal glow discharge into an arc discharge.

The normal glow discharge transition to a dark discharge takes place at low currents (about 10^{-5} A) and starts with the **subnormal discharge**, which corresponds to the interval CD on the current-voltage characteristic (Fig. 4–22). The size of the cathode spot at the low currents becomes comparable with the cathode layer thickness, which results in significant electron losses with respect to the normal glow and, therefore, requires higher voltages to sustain the discharge. Another regime, called an **obstructed glow**, occurs at low pressures

Table 4–6. Normal Current Density j_n/p^2, $\mu A/cm^2$ $Torr^2$, Normal Thickness of Cathode Layer $(pd)_n$, cm · Torr, and Normal Cathode Potential Drop V_n, V for Different Gases and Cathode Materials at Room Temperature

Gas	Cathode material	Normal current density	Normal thickness of cathode layer	Normal cathode potential drop
Air	Al	330	0.25	229
Air	Cu	240	0.23	370
Air	Fe	–	0.52	269
Air	Au	570	–	285
Ar	Fe	160	0.33	165
Ar	Mg	20	–	119
Ar	Pt	150	–	131
Ar	Al	–	0.29	100
He	Fe	2.2	1.30	150
He	Mg	3	1.45	125
He	Pt	5	–	165
He	Al	–	1.32	140
Ne	Fe	6	0.72	150
Ne	Mg	5	–	94
Ne	Pt	18	–	152
Ne	Al	–	0.64	120
H_2	Al	90	0.72	170
H_2	Cu	64	0.80	214
H_2	Fe	72	0.90	250
H_2	Pt	90	1.00	276
H_2	C	–	0.90	240
H_2	Ni	–	0.90	211
H_2	Pb	–	0.84	223
H_2	Zn	–	0.80	184
Hg	Al	4	0.33	245
Hg	Cu	15	0.60	447
Hg	Fe	8	0.34	298
N_2	Pt	380	–	216
N_2	Fe	400	0.42	215
N_2	Mg	–	0.35	188
N_2	Al	–	0.31	180
O_2	Pt	550	–	364
O_2	Al	–	0.24	311
O_2	Fe	–	0.31	290
O_2	Mg	–	0.25	310

and narrow gaps, when pd_0 is less than normal $(pd)_n$. The obstructed discharge corresponds to the left-hand branch of the Paschen curve, where voltage exceeds the minimum value V_n. The short interelectrode distance in the obstructed discharge is not sufficient for effective multiplication of electrons, so to sustain the mode, the interelectrode voltage should be greater than the normal one.

4.2.6. Positive Column of Glow Discharge

The positive column can be long, homogeneous, and can release most of the discharge power. Its physical function is simple: to close the electric circuit between the cathode layer and the anode. The balance of charged particles in the positive column follows the general rules described in Sections 4.1.8–4.1.11. Non-equilibrium behavior of the discharge ($T_e \gg T$) is controlled by heat balance, which can be illustrated as

$$w = jE = n_0 c_p (T - T_0) \nu_T, \tag{4–39}$$

where w is the power per unit volume; c_p is the specific heat; T and T_0 are the positive column and room temperatures; and ν_T is the cooling frequency for cylindrical discharge tube of radius R and length d_0:

$$\nu_T = \frac{8}{R^2} \frac{\lambda}{n_0 c_p} + \frac{2u}{d_0}. \tag{4–40}$$

The coefficient of thermal conductivity is λ and u is the gas velocity. The first term in (4–40) is related to thermal conductivity; the second one describes convective heat removal. If heat removal is controlled by thermal conduction, the discharge power causing a doubling of temperature ($T - T_0 = T_0$) is

$$w = jE = \frac{8\lambda T_0}{R^2}. \tag{4–41}$$

The thermal conductivity coefficient λ does not depend on pressure and can be estimated as $\lambda \approx 3 \cdot 10^{-4}$ W/cm $\cdot$ K. Therefore, the specific discharge power (4–41) also does not depend on pressure and can be estimated as 0.7 W/cm^3 for $R = 1$ cm. Higher powers result in higher gas temperatures and glow discharge contraction. The typical current density in the positive column with the heat removal controlled by thermal conduction is inversely proportional to pressure:

$$j = \frac{8\lambda T_0}{R^2} \frac{1}{(E/p)} \frac{1}{p}. \tag{4–42}$$

If $E/p = 3 \div 10$ V/cm $\cdot$ Torr, then j (mA/cm^2) $\approx 100/p$ (Torr). The corresponding electron density in the positive column can be calculated from Ohm's law ($j = \sigma E$):

$$n_e = \frac{w}{E^2} \frac{m\nu_{en}}{e^2} = \frac{w}{(E/p)^2} \frac{m k_{en}}{e^2 T_0} \frac{1}{p}. \tag{4–43}$$

Here ν_{en} and k_{en} are the frequency and rate coefficient of electron–neutral collisions. Numerically, n_e(cm^{-3}) $\approx 3 \cdot 10^{11}/p$ (Torr) in a positive column with conductive heat removal. The reduction of the plasma ionization degree with pressure is significant: $\frac{n_e}{n_0} \propto \frac{1}{p^2}$; therefore, low pressures are generally more favorable for sustaining steady-state homogeneous non-thermal plasma.

4.2.7. Hollow Cathode Glow Discharge

The **hollow cathode discharge** is an electron source based on intensive "non-local" ionization in the negative glow zone (Section 4.2.1), where the electric field is not very high but quite a few electrons are very energetic. These energetic electrons were formed in the cathode vicinity and crossed the cathode layer with only a few inelastic collisions. They provide non-local ionization and lead to electron densities in negative glow exceeding that in the positive column (Fig. 4–21; Gill & Webb, 1977; Boeuf & Marode, 1982; Bronin & Kolobov, 1983). Imagine a glow discharge with a cathode arranged as two parallel plates and the anode to the side. If we gradually decrease the distance between the cathodes, the current at some point grows 100–1000 times without any substantial change of voltage. The effect occurs when the two negative glow regions overlap and accumulate energetic electrons from both cathodes. Strong photo-emission from cathodes also contributes to the

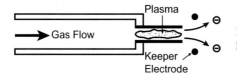

Figure 4–27. General schematic of the Lidsky capillary hollow-cathode discharge.

intensification of ionization (Moskalev, 1969). The hollow cathode can be arranged as a cylinder with anode located farther along the axis. Pressure should be chosen to have the cathode layer thickness comparable with the internal diameter of the hollow cylinder.

The most traditional configuration of the system is the **Lidsky hollow cathode** (Fig. 4–27), which is a narrow capillary-like nozzle operating with axially flowing gas and an anode located about 1 cm downstream. The Lidsky hollow cathode is hard to initiate but it provides electron current densities exceeding the limits of the Child law (Section 3.7.4; see Poeschel et al., 1979; Forrester, 1988).

4.2.8. Other Specific Glow Discharge Plasma Sources

Penning glow discharges. A special feature of this discharge is the quite strong magnetic field up to 0.3 T, which permits the magnetization of both electrons and ions. This discharge has been proposed by Penning (1936, 1937) and was further developed by Roth (1966, 1973a,b, 2000). A classical configuration of the Penning discharge is shown in Fig. 4–28. Pressure in this discharge is low (10^{-6}–10^{-2} Torr) to effectively magnetize charged particles. Two cathodes are grounded and the cylindrical anode has a voltage of about 0.5–5 kV. Although the gas pressure in the Penning glow discharge is low, the plasma density in these systems can be high (up to $6 \cdot 10^{12}$ cm^{-3}) because of low radial electron losses in the strong magnetic field. Electrons are trapped axially in the electrostatic potential well. The electron temperature in the Penning discharge is about 3–10 eV, whereas the ion temperature can be more than 10 times higher (30–300 eV).

The plasma between two cathodes is equipotential and plays the role of an electrode inside of the anode. The electric field inside of the cylindrical anode is radial. Both electrons and ions can be magnetized in the Penning discharge, which leads to azimuthal drift of charged particles in the crossed fields – radial electric, E_r, and axial magnetic, B

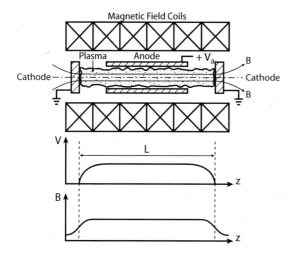

Figure 4–28. Classical Penning discharge with uniform magnetic induction and electrostatic trapping of electrons: general schematic, distribution of electrostatic potential, and magnetic field along the discharge axis.

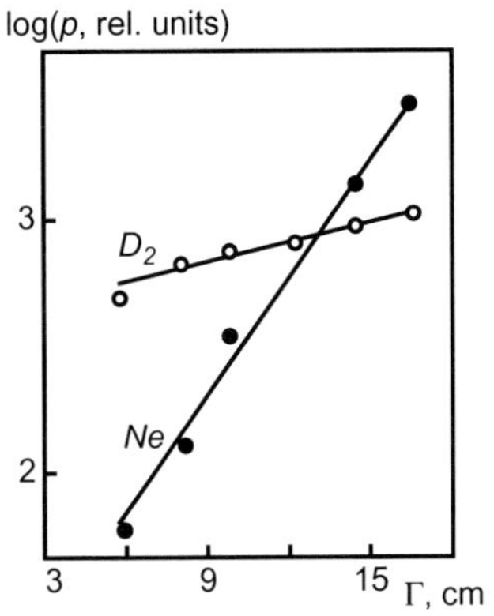

Figure 4–29. Radial distribution of partial pressures of deuterium and neon in their mixture rotating in the plasma centrifuge.

(Section 3.3.2). The tangential velocity v_{EB} of the azimuthal drift is the same for electrons and ions, and their energy $E_K(e, i)$ is proportional to masses $M_{e,i}$,

$$E_K(e, i) = \frac{1}{2} M_{e,i} v_{EB}^2 = \frac{1}{2} M_{e,i} \frac{E_r^2}{B^2}, \tag{4-44}$$

and leads the ion temperature in the Penning discharge to much exceed the electron temperature.

Plasma centrifuge. The effect of azimuthal drift in the crossed electric and magnetic fields (in Penning or similar discharges) and related fast plasma rotation permits the creation of centrifuges, where electrons and ions circulate around the axial magnetic field with large velocities and energies (4–44). The "plasma wind" is able to drag neutral particles with mass M transferring to them the high energies (4–44) limited only by the kinetic energy corresponding to ionization potential I:

$$v_{rA} = \sqrt{2eI/M}. \tag{4-45}$$

The maximum gas rotation velocity v_{rA} is referred to as the **Alfven velocity for plasma centrifuges**. The rotation velocities in plasma centrifuges reach $2–3 \cdot 10^6$ cm/s in the case of light atoms, which exceed those of mechanical centrifuges by about 50-fold. The fast gas rotation v_φ in a plasma centrifuge can be applied for the separation of isotopes with small differences in atomic masses $(M_1 - M_2)$:

$$R = \frac{(n_1/n_2)_{r=r_1}}{(n_1/n_2)_{r=r_2}} = \exp\left[(M_1 - M_2) \int_{r_1}^{r_2} \frac{v_\varphi^2}{T_0} \frac{dr}{r} \right], \tag{4-46}$$

where $(n_1/n_2)_r$ is the density ratio of the binary mixture components at radius r; heavier components move preferentially to the centrifuge periphery. Plasma centrifuges with $n_e/n_0 \approx 10^{-4}$–10^{-2} usually have the parameter $\frac{M v_\varphi^2}{2} / \frac{3}{2} T_0$ of about 3 (Rusanov & Fridman, 1984). The separation coefficient for an He–Xe mixture in such centrifuges exceeds 300; for a ^{235}U–^{238}U mixture the coefficient is about 1.1. The radial distribution of partial pressures of gas components for deuterium–neon separation in the plasma centrifuge is presented in Fig. 4–29. Plasma centrifuges are also able to provide chemical processes with product separation; thus, water vapor can be dissociated with simultaneous separation of hydrogen and oxygen (Poluektov & Efremov, 1998): $H_2O \rightarrow H_2 + \frac{1}{2}O_2$.

Magnetron discharge. This configuration is mostly applied for sputtering of cathode material and film deposition. A schematic of **magnetron discharge with parallel plate**

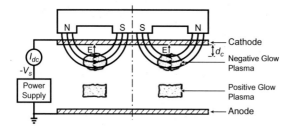

Figure 4–30. Glow discharge plasma generated in the parallel plate magnetron. The negative plasma glow is trapped in the magnetic mirror formed by magnetron magnets.

electrodes is shown in Fig. 4–30. To provide effective film deposition, the mean free path of the sputtered atoms must be large enough and, hence, pressures should be low (10^{-3}–3 Torr). Plasma density on the level of 10^{10} cm^{-3} is achieved in this system, however, because electrons are trapped in the magnetic field by the magnetic mirror. Ions are not supposed to be magnetized in this system to provide sputtering. Typical voltage between electrodes is about several hundred volts, and the magnetic field is about 5–50 mT.

Negative glow electrons are trapped in the magnetron discharge by the **magnetic mirror** effect, which means "reflection" of electrons from area with elevated magnetic field (see Fig. 4–31). The effect is based on the fact that, if spatial gradients of the magnetic field are small, the magnetic moment of a charged particle gyrating around the magnetic lines remains constant. The electric field between the cathode and the negative glow zone is relatively strong, which provides ions with the energy necessary for effective sputtering of the cathode material. The magnetron discharges can be arranged in different configurations. For example, **coplanar configuration of the magnetron discharge** is shown in Fig. 4–32 (Roth, 1995).

4.2.9. Energy Efficiency Peculiarities of Glow Discharge Application for Plasma-Chemical Processes

Glow discharges are widely used as light sources, as an active medium for gas lasers, for the treatment of different surfaces, sputtering, film deposition, and so forth. The application of conventional low-pressure glow discharges for highly energy-effective plasma-chemical processes is limited by three major factors:

1. The specific energy input in the traditional glow discharges controlled by diffusion are of about 100 eV/mol and much exceed the optimal value $E_v \approx 1$ eV/mol (Section 3.6). The energy necessary for one chemical reaction is usually of about 3 eV/mol; therefore, the maximum energy efficiency is about 3% even at 100% conversion. The high specific energy

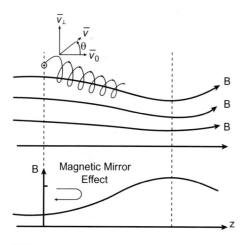

Figure 4–31. Illustration of the magnetic mirror effect.

4.3. Arc Discharges

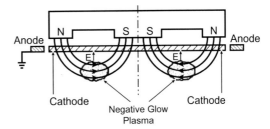

Figure 4–32. Co-planar configuration of the magnetron discharge. Only the negative glow plasma trapped in the magnetic field is normally visible in this configuration of magnetrons.

inputs (about 100 eV/mol) are due to the low gas flows passing though the conventional discharges. When a molecule in an optimal regime receives energy $E_v \approx 1\,\text{eV/mol} \approx 3\,\hbar\omega$, it is supposed to leave the discharge zone, which requires fast gas velocities.

2. Most energy-effective processes stimulated by vibrational excitation require sufficiently high levels of ionization degree (Section 3.6.2). In the discharges under consideration, $n_e/n_0 \propto 1/p^2$ (Section 4.2.6); therefore, the requirement of a high degree of ionization leads to low gas pressures and, hence, to further growth of the specific energy input and decrease of energy efficiency. The pressure reduction also results in an increase in the reduced electric field and electron temperature (Section 4.1.11), which is unfavorable for energy efficiency of the highly effective processes stimulated by vibrational excitation.

3. The specific power of the conventional glow discharges controlled by diffusion does not depend on pressure and is low, about 0.3–0.7 W/cm^3 (Section 4.2.6). Therefore, the specific productivity of relevant plasma-chemical processes is also relatively low. An increase of the specific power and specific productivity can be achieved by increasing the pressure and by applying fast-flow glow discharges with convective cooling.

To summarize, the energy efficiency of plasma-chemical processes in conventional glow discharges is not very high with respect to that in other non-thermal discharges. Lower energy prices can be achieved by applying non-traditional glow discharges in fast flows or those sustained by external sources of ionization (Rusanov & Fridman, 1984).

4.3. ARC DISCHARGES

4.3.1. Classification and Current-Voltage Characteristics of Arc Discharges

Arc discharges have been the major source of thermal plasma applied in illumination devices and in metallurgy for almost for two centuries. Arcs are as a rule self-sustaining DC discharges with relatively low cathode fall voltage of about 10 V. Arc cathodes emit electrons by intensive **thermionic and field emission** (Sections 2.4.1 and 2.4.2). Arc cathodes receive a large amount of Joule heating from the discharge current and are therefore able to reach very high temperatures, which leads to evaporation and erosion of electrodes. The most classical type of arc discharge is the **voltaic arc** (Fig. 4–33), which is the carbon electrode arc in atmospheric air. Arc discharges were first discovered in this form. Cathode and anode layer voltages in the voltaic arc are both about 10 V; the rest of the voltage corresponds to the positive column.

Arcs can be sustained in thermal and non-completely thermal regimes: cathode emission is thermionic in the non-completely thermal regimes, and field emission is thermionic in thermal arcs. The reduced electric field E/p is low in thermal arcs and higher in non-completely thermal arcs. The total arc voltage is low, sometimes only a couple tens of volts. Ranges of plasma parameters are outlined in Table 4–7. The thermal arcs operating at high pressures are much more energy intensive; they have higher currents and current densities, and higher power per unit length. A variety of DC discharges with low cathode fall voltage are usually considered as arc discharges and are classified by principal cathode and positive column mechanisms.

187

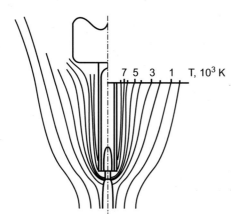

Figure 4–33. Temperature distribution in the carbon arc in air at a current of 200 A.

Hot thermionic cathode arcs. The entire cathode in such arcs has a temperature of 3000 K or higher, which provides a high current due to thermionic emission. The arc is stationary and is connected to the same fixed and quite large cathode spot. The current is distributed over a relatively large cathode area and its density is not so high – about 10^2–10^4 A/cm^2. Only special refractory materials like carbon, tungsten, molybdenum, zirconium, and tantalum can withstand such high temperatures. The hot thermionic cathode can be heated to high temperature not only by the arc current but also by an external heating source. Such cathodes are applied in low-pressure arcs, in thermionic converters in particular. The cathodes in such arc discharges are usually activated to decrease the temperature of thermionic emission.

Arcs with hot cathode spots. If the cathode is made from lower-melting-point metals like copper, iron, silver, or mercury, the high temperature required for emission cannot be sustained permanently. Electric current flows in this case through hot spots that appear, move fast, and disappear on the cathode surface. Current density in the spots is extremely high (10^4–10^7 A/cm^2), which leads to intensive but local and short heating and evaporation of the cathode material while the rest of the cathode actually stays cold. The mechanism of electron emission from the hot spots is thermionic field emission. Cathode spots appear not only on the low-melting-point cathodes but also on refractory metals at low currents and low pressures.

Vacuum arcs. This type of low-pressure arc, operating with cathode spots, is special because the gas-phase working fluid is provided by erosion and evaporation of the electrode material. This type of arc is of importance in high-current electrical equipment, high-current vacuum circuit breakers, and switches.

Table 4–7. Discharge Parameters Typical for Thermal and Non-Completely Thermal Arcs

Discharge plasma parameter	Thermal arc discharge	Non-completely-thermal arc
Gas pressure	0.1–100 atm	10^{-3}–100 Torr
Arc current	30A–30 kA	1–30 A
Cathode current density	10^4–10^7 A/cm^2	10^2–10^4 A/cm^2
Voltage	10–100 V	10–100 V
Power per unit length	>1 kW/cm	<1 kW/cm
Electron density	10^{15}–10^{19} cm^{-3}	10^{14}–10^{15} cm^{-3}
Gas temperature	1–10 eV	300–6000 K
Electron temperature	1–10 eV	0.2–2 eV

High-pressure arc discharges. The positive column of an arc plasma is in quasi equilibrium at pressures exceeding 0.1–0.5 atm. Thermal arcs also operate at very high pressures exceeding 10 atm. The plasma in this case is so dense that most of the discharge power (80–90%) is converted into radiation.

Low-pressure arc discharges. The positive column of arc discharges at low pressures (10^{-3}–1 Torr) consists of non-equilibrium plasma. The ionization degree in the non-thermal arcs is higher than that in glow discharges because arc currents are much larger (see Table 4–7).

The current-voltage characteristic of continuous DC discharges in a wide range of currents was discussed in Section 4.2.2 (Fig. 4–22). The transition from glow to arc corresponds to the interval FG. Current density in the abnormal glow increases, resulting in cathode heating and growth of thermionic emission, which determines the glow-to-arc transition. The glow-to-arc transition is continuous in the case of thermionic cathodes made from refractory metals and takes place at currents of about 10 A. Cathodes made from low-melting-point metals provide the transition at lower currents (0.1–1 A). This transition is sharp, unstable, and accompanied by the formation of hot cathode spots (Finkelburg & Maecker, 1956). When the current grows, the electric field and voltage decrease until a critical point of sharp voltage reduction, followed by an almost horizontal current-voltage characteristic. The transition is accompanied by a specific hissing noise, which is the formation of hot anode spots with intensive evaporation.

4.3.2. Cathode and Anode Layers of Arc Discharges

The function of the cathode layer is to provide the high current necessary for arc operation. Electron emission from the cathode in arcs is due to thermionic and field emissions, which are much more intensive than ion-induced secondary electron emission dominating in the cathode layers of glow discharges (see Sections 2.4.1–2.4.3). In the case of thermionic emission, ion bombardment provides cathode heating, which then leads to the escape of electrons from the surface. Secondary electron emission gives about $\gamma \approx 0.01$ electrons per ion, whereas thermionic emission can generate much more ($\gamma_{\text{eff}} = 2$–9). The fraction of electron current S near the cathode in a glow discharge is very small ($\frac{\gamma}{\gamma+1} \approx 0.01$), but in the cathode layer of an arc,

$$S = \frac{\gamma_{\text{eff}}}{\gamma_{\text{eff}} + 1} \approx 0.7\text{–}0.9. \tag{4–47}$$

This shows that thermionic emission from the cathode provides most of electric current in the arc. Current in the positive column of both arc and glow discharges is almost completely provided by high-mobility electrons. In contrast to a glow, the direct electron-impact ionization in the cathode layer of an arc discharge should provide only a minor fraction of the total discharge current ($1 - S \approx 10$–30%); therefore, the cathode voltage in arcs is relatively low, about or even less than the ionization potential.

Sufficient ion density is generated in the cathode layer to provide the necessary cathode heating in the case of thermionic emission. The gas temperature near the cathode is the same as the cathode surface temperature and a couple of times less than the temperature in the positive column (Fig. 4–34). Therefore, thermal ionization is unable to provide the necessary ionization degree, and additional non-thermal ionization is required. This leads to an elevated electric field near the cathode, which stimulates electron emission by decreasing the work function (Schottky effect) and contributing to the field emission. Intensive ionization in the cathode vicinity leads to a high concentration of ions in the layer and to the formation of a positive space charge, which actually provides the elevated electric field. The distribution of arc parameters – temperature, voltage, and electric field – along the discharge from cathode to anode is illustrated in Fig. 4–34.

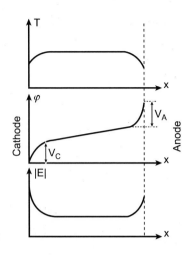

Figure 4–34. Distributions of temperature, electric potential, and electric field from cathode to anode of the arc discharges.

The structure of the cathode layer is illustrated in Fig. 4–35. A large positive space charge, with high electric field and most of the cathode voltage, is located in the narrow layer by the cathode. This layer is shorter than the mean free path and is referred to as the **collisionless zone of the cathode layer.** The longer **quasi-neutral zone of the cathode layer** is located between the narrow collisionless layer and the positive column. The electric field there is not so high, but ionization is intensive because electrons keep the high energy they received in the collisionless layer. Most of the ions carrying current and energy to the cathode are generated here. Electron and ion currents are constant in the collisionless layer, where there are no sources of charge particles. The electron fraction of the current grows from $S \approx 0.7–0.9$ in the cathode layer to about 1 in the positive column. The plasma density in the cathode layer grows in the direction of the positive column.

Electron (J_e) and ion (j_+) components of the total current density j in the collisionless zone are

$$j_e = S \cdot j = n_e e v_e, \quad j_+ = (1 - S)j = n_+ e v_+. \tag{4–48}$$

Electron and ion velocities, v_e, and v_+, can be presented as a function of voltage V, assuming $V = 0$ at the cathode and $V = V_C$ at the end of the collisionless layer (m and M are masses of electrons and ions):

$$v_e = \sqrt{2eV/m}, \quad v_+ = \sqrt{2e(V_C - V)/M}. \tag{4–49}$$

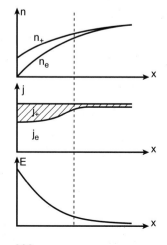

Figure 4–35. Distribution of charge density, electric current, and electric field in the cathode layer of an arc discharge.

4.3. Arc Discharges

Based on (4–48) and (4–49), the Poisson equation for the voltage V in the collisionless layer is

$$-\frac{d^2V}{dx^2} = \frac{e}{\varepsilon_0}(n_+ - n_e) = \frac{j}{\varepsilon_0\sqrt{2e}}\left[\frac{(1-S)\sqrt{M}}{\sqrt{V_c - V}} - \frac{S\sqrt{m}}{\sqrt{V}}\right]. \qquad (4\text{–}50)$$

Taking into account that $\frac{d^2V}{dx^2} = \frac{1}{2}\frac{dE^2}{dV}$, the Poisson equation (4–50) can be integrated assuming the boundary condition $E \approx 0$ at $V = V_C$. This results in a relation between the electric field near the cathode, current density, and the cathode voltage drop (which can be represented by V_C):

$$E_C^2 = \frac{4j}{\varepsilon_0\sqrt{2e}}[(1-S)\sqrt{M} - S\sqrt{m}]\sqrt{V_C}. \qquad (4\text{–}51)$$

This relation can be rewritten by taking into account $S = 0.7\text{–}0.9$ in the numerical form:

$$E_c, \text{V/cm} = 5\cdot10^3\cdot A^{1/4}(1-S)^{1/2}(V_C, \text{V})^{1/4}(j, \text{A/cm}^2)^{1/2}, \qquad (4\text{–}52)$$

where A is the atomic mass of ions in atomic mass units. For example, the arc discharge in nitrogen ($A = 28$) at typical values of current density for hot cathodes ($j = 3\cdot10^3$ A/cm^2), cathode voltage drop ($V_C = 10$ V), and $S = 0.8$ gives, according to (4–52), the electric field near cathode: $E_c = 5.7\cdot10^5$ V/cm. The length of the collisionless zone of the cathode layer can be expressed as

$$\Delta l = 4V_C/3E_C. \qquad (4\text{–}53)$$

Numerically, the length of the collisionless layer is about $\Delta l \approx 2\cdot10^{-5}$ cm.

Arcs can be connected to the anode either by diffuse connection or by anode spots. The diffuse connection occurs on the large-area anodes at current densities of about 100 A/cm^2. The anode spots appear on small and non-homogeneous anodes; current densities in the spots are 10^4–10^5 A/cm^2. The number of spots grows with current and pressure. The anode spots can be arranged and move in patterns, and the anode voltage drop consists of two components. The first one is related to negative space charge near the anode, which repels ions. This small voltage drop (of about ionization potential) stimulates some additional electron generation to compensate for the absence of ion current in the region. The second voltage component is related to the arc discharge geometry. If the anode surface is smaller than the positive column, the current near the electrode should be provided only by electrons. It requires higher electric fields near the electrode and an additional anode voltage drop. Each electron brings to the anode an energy of about 10 eV, so the energy flux to the anode spot at 10^4–10^5 A/cm^2 is about 10^5–10^6 W/cm^2. The temperature at the anode spots of vacuum metal arcs is about 3000 K, and in a carbon arc, about 4000 K.

4.3.3. Cathode Spots in Arc Discharges

The cathodes spots are the localized current centers, which appear on the cathode when significant current should be provided but the cathode cannot be heated enough as a whole. The most typical cause of cathode spots is the application of metals with relatively low melting points. The cathode spots can also be caused by low arc currents, which are only able to provide the necessary electron emission when concentrated to a small area. The cathode spots also appear at low gas pressures (<1 Torr), when metal vapor from the cathode provides atoms to generate positive ions bringing their energy to the cathode to sustain the electron emission. To provide the required evaporation, current is concentrated in spots; at pressures <1 Torr and currents 1–10 A, such spots occur even on refractory metals.

Table 4–8. Typical Characteristics of Cathode Spots of Arc Discharges

Cathode material	Cu	Hg	Fe	W	Ag	Zn
Minimum current through a spot, A	1.6	0.07	1.5	1.6	1.2	0.3
Average current through a spot, A	100	1	80	200	80	10
Current density, A/cm^2	10^4–10^8	10^4–10^6	10^7	10^4–10^6	–	$3 \cdot 10^4$
Cathode voltage drop, V	18	9	18	20	14	10
Specific erosion at 100–200 A, g/C	10^{-4}	–	–	10^{-4}	10^{-4}	–
Vapor jet velocity, 10^5 cm/s	1.5	1	0.9	3	0.9	0.4

The initially formed cathode spots are pretty small (10^{-4}–10^{-2} cm) and move very fast (10^3–10^4 cm/s). These primary spots are non-thermal and related to micro-explosions due to localization of current on tiny protrusions on the cathode surface. After about 10^{-4} s, the small primary spots merge into larger spots (10^{-3}–10^{-2} cm) with temperatures of 3000 K and greater. These spots provide intensive thermal erosion and move much slower (10–100 cm/s). The typical current through an individual spot is 1–300 A. Growth of the current leads to splitting of cathode spots. The minimum current through a single spot is about $I_{min} \approx 0.1$–1 A; the arc as a whole extinguishes at lower currents. This critical minimum current though an individual cathode spot can be calculated as

$$I_{min}, A \approx 2.5 \cdot 10^{-4} \cdot T_{boil}(K) \cdot \sqrt{\lambda, W/cm \cdot K}, \qquad (4\text{–}54)$$

where: $T_{boil}(K)$ is the boiling temperature of the cathode material and $\lambda(W/cm \cdot K)$ is the heat conduction coefficient. The cathode spots are sources of intensive jets of metal vapor. The emission of 10 electrons corresponds to the ablation of about one atom. The metal vapor jet velocities reach 10^5–10^6 cm/s. Data characterizing the cathode spots can be found in Table 4–8; for details see Lafferty (1980) and Lyubimov and Rachovsky (1978). The current density in cathode spots can reach the extremely high level of 10^8 A/cm^2, which can be provided only by field-enhanced thermionic emission. The initial high current densities in a spot can also be due to the **explosive electron emission**, which is related to localization of a strong electric field and follows the explosion of micro-protrusions on the cathode surface (Korolev & Mesiatz, 1982).

Several problems related to the cathode spot phenomenon are not thus far completely solved. For example, the current-voltage characteristics of vacuum arcs are not decreasing but increasing; also there is no complete explanation of the cathode spot motion and splitting. The most intriguing cathode spot paradox is related to the direction of its motion in an

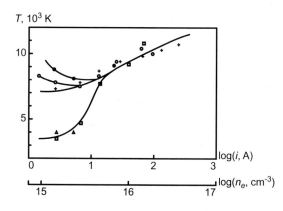

Figure 4–36. Separation of temperatures in the positive arc column in argon (or in argon with an admixture of H$_2$) at pressure $p = 1$ atm as function of the current density or electron density. In the figure: • represents electron temperature T_e; ○ represents effective temperature of the highly electronically excited states T_w; + corresponds to the ion temperature T_i, related to the electron density by the Saha formula; □ corresponds to the effective temperature of the lower electronically excited states T_d; Δ represents the translational gas temperature.

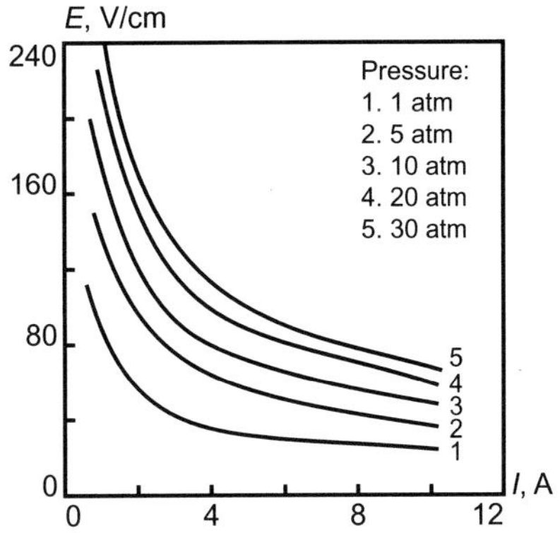

Figure 4–37. Current-voltage characteristics of the positive column of arc discharges in air at different gas pressures.

externally applied magnetic field. If the field is applied along a cathode surface, the cathode spots move in the opposite direction of the magnetic force ($\vec{I} \times \vec{H}$).

4.3.4. Positive Column of High-Pressure Arcs: Elenbaas-Heller Equation

The Joule heating power per unit length of positive column in high-pressure arcs is quite significant (0.2–0.5 kW/cm). Such plasma in molecular gases ($p \geq 1$ atm) is usually in quasi equilibrium at any currents. In inert gases, the electron–neutral energy exchange is less effective and requires high currents and electron densities to reach quasi equilibrium (Granovsky, 1971; Fig. 4–36). Temperatures of electrons and neutrals can differ at low pressures ($p \leq 0.1$ Torr) and currents ($I \approx 1$ A). Current-voltage characteristics of the positive column are illustrated in Fig. 4–37 for different pressures (Granovsky, 1971); electric field E is constant along the positive column and defines the voltage. The current-voltage characteristics are hyperbolic, which indicates that Joule heat per unit length ($w = EI$) does not change significantly with current; I; $w = EI$ grows with pressure due to intensification of heat transfer mostly related to radiation. The contribution of radiation increases proportionally to the square of plasma density and, hence, grows with pressure. The arc radiation losses in atmospheric air are about 1%, but they become significant at pressures above 10 atm and for higher arc power. The highest level of radiation can be reached in Hg, Xe, and Kr, which is applied in mercury and xenon lamps. Empirical formulas for the calculation of plasma radiation (Granovsky, 1971) are presented in Table 4–9. The fraction of power conversion into radiation is high in mercury and xenon even at low values of Joule heat per unit length.

The temperature distribution in a long cylindrical steady-state thermal plasma column stabilized by walls in a tube of radius R is described by the Elenbaas-Heller equation, assuming heat transfer across the positive column provided by heat conduction with the coefficient $\lambda(T)$. According to a Maxwell equation, $curl E = 0$ and the electric field in a long homogeneous arc column is constant across its cross section. Radial distributions

Table 4–9. Radiation Power per Unit Length of Positive Column of Arc Discharges at Different Pressures and Different Values of Joule Heat per Unit Length $w = EI$, W/cm

Gas	Pressure, atm	Radiation power per unit length, W/cm	$w = EI$, W/cm
Hg	≥ 1	$0.72 \cdot (w - 10)$	–
Xe	12	$0.88 \cdot (w - 24)$	>35
Kr	12	$0.72 \cdot (w - 42)$	>70
Ar	1	$0.52 \cdot (w - 95)$	>150

of electric conductivity $\sigma(T)$, current density $j = \sigma(T)E$, and Joule heating density $w = jE = \sigma(T)E^2$ are determined only by radial temperature distribution $T(r)$; therefore,

$$\frac{1}{r}\frac{d}{dr}\left[r\lambda(T)\frac{dT}{dr}\right] + \sigma(T)E^2 = 0. \tag{4–55}$$

This equation is known as the **Elenbaas-Heller equation**. Boundary conditions for (4–55) are $dT/dr = 0$ at $r = 0$, and $T = T_w$ at $r = R$. The experimentally controlled parameter is current:

$$I = E\int_0^R \sigma[T(r)] \cdot 2\pi r\,dr. \tag{4–56}$$

The Elenbaas-Heller equation (4–55) together with (4–56) permits calculations of the function $E(I)$, which is the current-voltage characteristic of the plasma column determined by two material functions: electric conductivity $\sigma(T)$ and thermal conductivity $\lambda(T)$. To reduce the number of material functions, it is convenient to introduce, instead of temperature, the **heat flux potential** $\Theta(T)$:

$$\Theta = \int_0^T \lambda(T)\,dT, \quad \lambda(T)\frac{dT}{dr} = \frac{d}{dr}\Theta. \tag{4–57}$$

Using the heat flux potential, the Elenbaas-Heller equation can be simplified:

$$\frac{1}{r}\frac{d}{dr}\left[r\frac{d\Theta}{dr}\right] + \sigma(\Theta)E^2 = 0, \tag{4–58}$$

but the equation cannot be analytically solved because of the non-linearity of the material function $\sigma(\Theta)$ (Dresvin, 1977).

4.3.5. Steenbeck-Raizer "Channel" Model of Positive Column

The Steenbeck-Raizer channel model (Fig. 4–38) is based on the strong dependence of plasma electric conductivity on temperature (Saha equation, Section 3.1.3). At temperatures below 3000 K, plasma conductivity is low; it grows significantly only when the temperature exceeds 4000 K. The temperature decrease $T(r)$ from the axis to the walls is gradual, whereas the conductivity change with radius $\sigma[T(r)]$ is sharp. Thus, according to the model, arc current is located mostly in a channel of radius r_0. Temperature and electric conductivity are considered constant inside of the arc channel and equal to their maximum value on the discharge axis: T_m and $\sigma(T_m)$. The total arc current can be then expressed as

$$I = E\sigma(T_m) \cdot \pi r_0^2. \tag{4–59}$$

Outside of the channel $r > r_0$; electric conductivity, current, and Joule heating can be neglected; and the Elenbaas-Heller equation can be integrated with the boundary conditions $T = T_m$ at $r = r_0$ and $T = 0$ by the walls at $r = R$. The integration gives the relation between the heat flux potential $\Theta_m(T_m)$ related to the plasma temperature and the discharge power per unit length, $w = EI$:

$$\Theta_m(T_m) = \frac{w}{2\pi}\ln\frac{R}{r_0}. \tag{4–60}$$

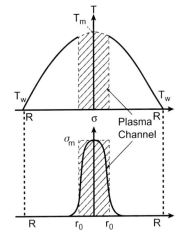

Figure 4–38. Gas temperature and thermal plasma conductivity distributions in arc discharges illustrating the Steenbeck channel model.

The potential $\Theta_m(T_m)$ in the arc channel and the arc power per unit arc length, $w = EI$, are

$$w = \frac{I^2}{\pi r_0^2 \sigma_m(T_m)}, \quad \Theta_m(T_m) = \int_0^{T_m} \lambda(T)\,dT. \tag{4-61}$$

The channel model of an arc includes three parameters to be determined: plasma temperature T_m, arc channel radius r_0, and electric field E. Electric current I and discharge tube radius R are experimentally controlled parameters. To find T_m, r_0, and E, the channel model has only two equations, (4–59) and (4–60). Steenbeck suggested the **principle of minimum power** (see Section 4.2.4) as the third equation to complete the system. The minimum power principle has been proved for arcs by Rozovsky (1972). According to the principle of minimum power, temperature T_m and arc channel radius r_0 should minimize the specific discharge power w and electric field $E = w/I$ at fixed values of current I and discharge tube radius R. The minimization $(\frac{dw}{dr_0})_{I=\text{const}} = 0$ gives the third equation of the model:

$$\left(\frac{d\sigma}{dT}\right)_{T=T_m} = \frac{4\pi \lambda_m(T_m)\sigma_m(T_m)}{w}. \tag{4-62}$$

Raizer (1972a,b) proved that the channel model does not require the minimum power principle to justify the third equation (4–62). It can be derived by analysis of the conduction heat flux J_0 from the arc channel, $w = J_0 \cdot 2\pi r_0$, provided by the temperature difference $\Delta T = T_m - T_0$ across the arc (Fig. 4–38):

$$J_0 \approx \lambda_m(T_m) \cdot \frac{\Delta T}{r_0} = \lambda_m(T_m) \cdot \frac{T_m - T_0}{r_0}. \tag{4-63}$$

Integrating the Elenbaas-Heller equation inside of the arc $0 < r < r_0$ leads to the similar relation:

$$4\pi \Delta\Theta = w \approx 4\pi \lambda_m \Delta T, \quad \Delta\Theta = \Theta_m - \Theta_0. \tag{4-64}$$

The key point of the Raizer modification of the channel model is the definition of an arc channel as a region where electric conductivity decreases not more than "e" times with respect to the maximum value at the axis. It permits one to specify the arc channel radius

r_0 and gives the "third" equation of the channel model. The electric conductivity in the arc channel can be expressed from the Saha equation as

$$\sigma(T) = C \exp\left(-\frac{I_i}{2T}\right),$$ (4–65)

where I_i is ionization potential and C is an almost constant parameter. Electric conductivity in air, nitrogen, and argon at atmospheric pressure and temperatures $T = 8000\text{–}14,000$ K can be expressed as

$$\sigma(T), \text{Ohm}^{-1}\text{cm}^{-1} = 83 \cdot \exp\left(-\frac{36,000}{T, \text{K}}\right),$$ (4–66)

which corresponds to the effective ionization potential $I_{\text{eff}} \approx 6.2$ eV. According to (4–65) and (4–66), the e-times decrease of conductivity corresponds to the following small temperature decrease:

$$\Delta T = T_m - T_0 = \frac{2T_m^2}{I_i}.$$ (4–67)

Combination of (4–63) and (4–64) with (4–67) gives the required "third" equation of the model in the form

$$w = 8\pi \lambda_m(T_m)\frac{T_m^2}{I_i}.$$ (4–68)

The third equation of the Raizer modification of the arc channel model (4–68) coincides with that of the initial Steenbeck model (4–62), which was based on the principle of minimum power. As a reminder, the first two equations of the arc channel model, (4–59) and (4–60), always remain the same.

4.3.6. Steenbeck-Raizer Arc "Channel" Modeling of Plasma Temperature, Specific Power, and Electric Field in Positive Column

Relation (4–68) determines the temperature in the arc as a function of specific discharge power w per unit length (with ionization potential I_i and thermal conductivity coefficient λ_m as parameters):

$$T_m = \sqrt{w \cdot \frac{I_i}{8\pi \lambda_m}}.$$ (4–69)

The temperature does not depend directly on the tube radius, but rather only on specific power w. The principal parameter of an arc discharge is the current. Assuming $\lambda = $ const, $\Theta = \lambda T$, the conductivity in the arc channel is proportional to current (Fridman & Kennedy, 2004):

$$\sigma_m = I \cdot \sqrt{\frac{I_i C}{8\pi^2 R^2 \lambda_m T_m^2}}.$$ (4–70)

The plasma temperature in the arc channel grows with current I, but only logarithmically:

$$T_m = \frac{I_i}{\ln\left(8\pi^2\lambda_m C T_m^2/I_i\right) - 2\ln(I/R)}.$$ (4–71)

4.3. Arc Discharges

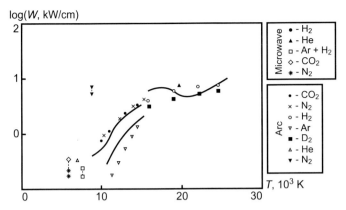

Figure 4–39. Electric power dissipated per unit length of different thermal discharges (arcs and microwave discharges) as function of maximum gas temperature on the discharge axis; solid lines represents results of numerical calculations.

The weak logarithmic growth of temperature in the channel with electric current leads to the similar weak logarithmic dependence on the electric current of the arc discharge power w per unit length:

$$w \approx \frac{\text{const}}{(\text{const} - \ln I)^2}. \tag{4–72}$$

Experimental data demonstrating the dependence are presented in Fig. 4–39. Taking into account that $w = EI$, the decrease of the electric field in the positive column with current I is close to hyperbolic:

$$E = \frac{8\pi \lambda_m T_m^2}{I_i} \cdot \frac{1}{I} \approx \frac{\text{const}}{I \cdot (\text{const} - \ln I)^2}. \tag{4–73}$$

This relation explains the hyperbolic decrease of current-voltage characteristics typical for thermal arc discharges. The radius of an arc discharge may be found in the framework of the channel model:

$$r_0 = R\sqrt{\frac{\sigma_m}{C}} = R\sqrt{\frac{I}{R}} \sqrt[4]{\frac{I_i}{8\pi^2 \lambda_m T_m^2 C}}. \tag{4–74}$$

The arc radius grows as a square root of the discharge current $I (I \propto r_0^2)$; therefore, an increase of current leaves the current density logarithmically fixed similarly to plasma temperature. The channel model always includes current in combination with the discharge tube radius R (the similarity parameter I/R).

4.3.7. Configurations of Arc Discharges Applied in Plasma Chemistry and Plasma Processing

Free-burning linear arcs are the simplest axisymmetric electric arcs burning between two electrodes. They can be arranged in horizontal and vertical configurations (Fig. 4–40). Sir Humphrey Davy first observed such a horizontal arc in the beginning of the nineteenth century. The buoyancy of hot gases in the horizontal arc leads to bowing up or "arching" of the plasma channel, which explains the term "arc." If the free-burning arc is vertical (Fig. 4–40), the cathode is usually placed at the top. Then buoyancy provides more intensive cathode heating. If the arc length is shorter than the diameter (Fig. 4–41), the discharge is

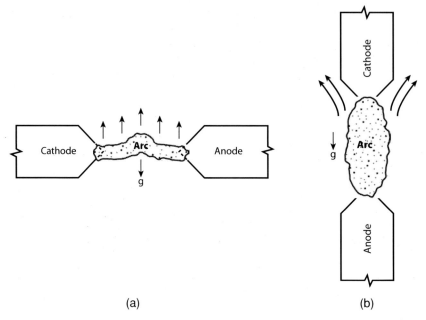

(a)   (b)

Figure 4–40. (a) horizontal and (b) vertical configurations of the free-burning arc discharges.

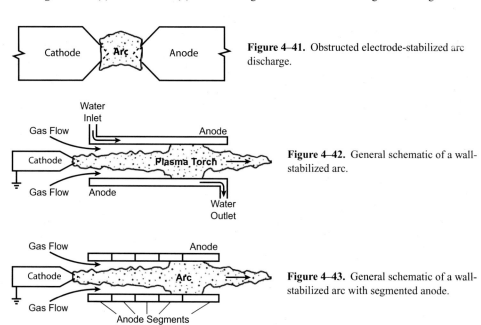

Figure 4–41. Obstructed electrode-stabilized arc discharge.

Figure 4–42. General schematic of a wall-stabilized arc.

Figure 4–43. General schematic of a wall-stabilized arc with segmented anode.

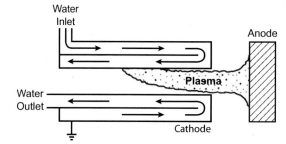

Figure 4–44. General schematic of a transferred arc discharge.

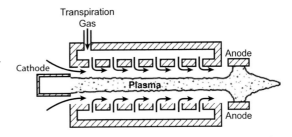

Figure 4-45. General schematic of a transpiration-stabilized arc.

referred to as the **obstructed arc**. Distance between electrodes in such discharges is typically about 1 mm; voltage, nevertheless, exceeds the anode and cathode drops. The obstructed arcs are electrode-stabilized.

Wall-stabilized linear arcs are widely used for gas heating. A simple schematic of the **wall-stabilized arc with a unitary anode** is shown in Fig. 4-42. The cathode is axial in this configuration, and the unitary anode is hollow and coaxial. The arc is axisymmetric and stable with respect to perturbations. If the arc channel is asymmetrically perturbed and approaches a coaxial anode, it leads to intensification of the discharge cooling and to a temperature increase on the axis (according to the Elenbaas-Heller equation). The increase of temperature results in displacement of the arc channel back on the axis of the discharge tube. The arc can attach to the anode at any point along the axis; a better-defined discharge arrangement can be achieved in the so-called **segmented wall-stabilized arc configuration** (see Fig. 4-43). The anode walls in this system are water cooled and electrically segmented and isolated. Such a configuration provides a linear decrease in axial voltage and forces the arc attachment to the wider anode segment – the one farthest from the cathode. The length of the other segments is taken to be small enough to avoid breakdown between them.

Transferred arcs with water-cooled non-consumable cathodes are illustrated in Fig. 4-44. The generation of electrons on the inner walls of the hollow cathodes is provided by field emission, which permits operation of the transferred arcs at the multi-megawatt power for thousands of hours. The electric circuit is completed by transferring the arc to an external anode, which is a conducting material, where the arc is to be applied. The arc root can move over the cathode surface, which further increases its lifetime.

In the **flow-stabilized linear arcs**, the arc channel can be stabilized on the axis of the discharge chamber by radial injection of cooling water or gas. Such a configuration, illustrated in Fig. 4-45, is usually referred to as the **transpiration-stabilized arc**. This discharge is similar to the segmented wall-stabilized arc, but the transpiration of cooling fluid through annular slots between segments increases the lifetime of the interior segments. Another linear arc configuration providing high discharge power is the so-called **coaxial flow-stabilized arc**, illustrated in Fig. 4-46. In this case the anode is located far from the main part of the plasma channel and cannot provide wall stabilization. Instead of a wall, the arc channel is stabilized by a coaxial gas flow moving along the outer surface of the arc. Such a stabilization is effective with essentially no heating of the discharge chamber walls if the coaxial flow is laminar. Similar arc stabilization can be achieved using a flow rotating

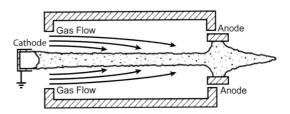

Figure 4-46. General schematic of a flow-stabilized arc.

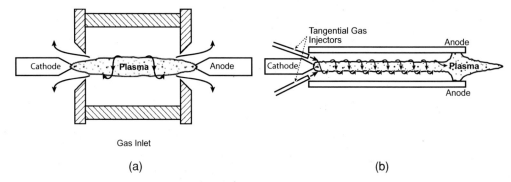

Figure 4–47. General schematics of vortex-stabilized arc discharges.

around the arc column. Different configurations of the **vortex-stabilized arcs** are shown in Fig. 4–47a and b. The arc channel is stabilized by a vortex gas flow, which is introduced from a special tangential injector. The vortex gas flow cools the edges of the arc and keeps the arc column confined to the axis of the discharge chamber.

A non-linear wall-stabilized **non-transferred arc** is shown in Fig. 4–48. It consists of a cylindrical hollow cathode and coaxial hollow anode located in a water-cooled chamber and separated by an insulator. Gas flow blows the arc column out of the anode opening to heat a downstream material, which is supposed to be treated. In contrast to transferred arcs, the treated material is not supposed to operate as an anode. Magnetic $\vec{I} \times \vec{B}$ forces cause the arc roots to rotate around electrodes (Fig. 4–48), which provides longer electrode lifetime. The generation of electrons on the cathode is provided in this case by field emission. An axisymmetric version of the non-transferred arc, usually referred to as the **plasma torch** or the **arc jet**, is illustrated in Fig. 4–49. The arc is generated in a conical gap in the anode and pushed out of this opening by gas flow. The heated gas flow forms a very-high-temperature arc jet, sometimes at supersonic velocities.

Magnetically stabilized rotating arcs are illustrated in Fig. 4–50a and b. An external axial magnetic field provides $\vec{I} \times \vec{B}$ forces, which cause arc rotation and protect the anode from local overheating. Fig. 4–50a shows magnetic stabilization of the wall-stabilized arc; Fig. 4–50b shows a magnetically stabilized plasma torch. Magnetic stabilization is an essential supplement to the wall and aerodynamic stabilizations.

4.3.8. Gliding Arc Discharge
A gliding arc is an auto-oscillating periodic discharge between at least two diverging electrodes submerged in gas flow (Fig. 4–51). Self-initiated in the upstream narrowest gap, the

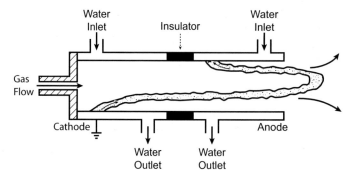

Figure 4–48. General schematic of a non-transferred arc discharge.

4.3. Arc Discharges

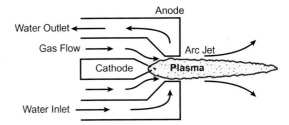

Figure 4–49. General schematic of a plasma torch.

discharge forms the plasma column connecting the electrodes. This column is dragged by the gas flow toward the diverging downstream section. The arc grows with the increase of interelectrode distance until it extinguishes, but it reignites itself at the minimum distance between the electrodes to start a new cycle. Time-space evolution of the discharge is illustrated in Fig. 4–52 by a series of snapshots. The gliding arc plasma can be either thermal or non-thermal depending on the power and flow rate. Along with completely thermal and non-thermal modes, it is possible to operate the arc in the transitional regime when the discharge starts thermal, but during the space-time evolution becomes a non-thermal one. The powerful and energy-efficient transitional discharge combines the benefits of both equilibrium and non-equilibrium discharges.

Generally two very different kinds of plasmas are used for chemical applications. Thermal plasma generators have been designed for many diverse industrial applications covering a wide range of operating power levels from less than 1 kW to over 50 MW. However, in spite of providing sufficient power levels, these appear not to be well adapted to the purposes of plasma chemistry, where the selective treatment of reactants and high energy efficiency are required. An alternative approach to plasma chemistry is the non-thermal approach, which offers good selectivity and energy efficiency of chemical processes but is limited usually to low pressures and powers. In the two general types of discharges it is impossible to simultaneously maintain a high level of non-equilibrium, high electron temperature and high electron density, whereas most prospective plasma-chemical applications require high power for high reactor productivity and a high degree of non-equilibrium to support selective chemical process at the same time. This can be achieved in the transitional regime of the gliding arc, which can be operated at atmospheric pressure or higher, with power at non-equilibrium conditions reaching up to 40 kW per electrode pair.

The gliding arc has been known for more than one hundred years in the form of **Jacob's ladder** and was first used for producing nitrogen-based fertilizers by Naville and Guye (1904). Recent contributions to the development of gliding arcs have been made by A.

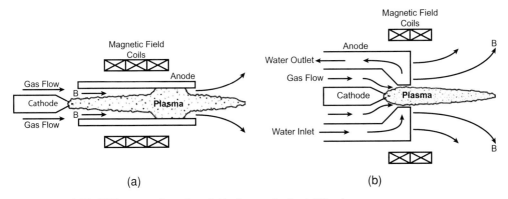

(a)   (b)

Figure 4–50. Different configurations (a,b) of magnetically stabilized arcs.

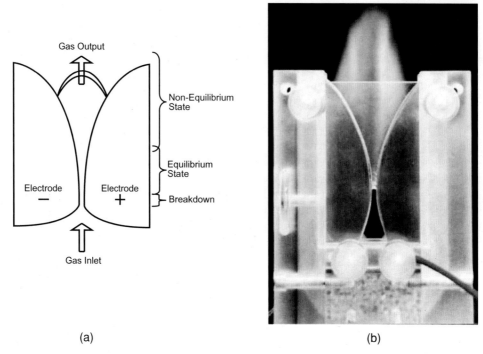

Gas Output

Non-Equilibrium
State

Equilibrium
State

Electrode
−

Electrode
+

Breakdown

Gas Inlet

(a)   (b)

Figure 4–51. Gliding arc discharge: (a) general schematic of flat configuration; (b) photo of the discharge in flat configuration.

Czernichowski and his colleagues (Lesueur, Czernichowski, & Chapelle, 1990; Czerni-chowski, 1994). The non-equilibrium nature of the transitional regime of a gliding arc was first reported by Fridman et al. (1994) and Rusanov et al. (1993). A simple electrical scheme of the DC gliding arc is shown in Fig. 4–53. The high-voltage generator (up to 5 kV) is used in this scheme to ignite the discharge, and the second one is the power generator with

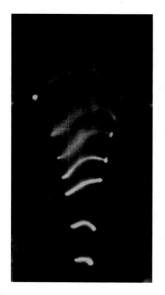

Figure 4–52. Gliding arc evolution with transition from quasi-thermal to non-thermal regime of the discharge; series of consequent snapshots.

4.3. Arc Discharges

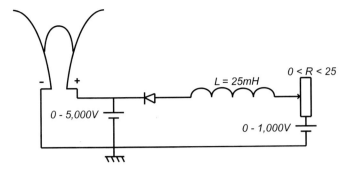

Figure 4–53. Conventional electric circuit schematic for the gliding arc discharge.

voltages up to 1 kV and a total current I up to 60 A. A variable resistor $R = 0$–25 Ω is in series with a self-inductance $L = 25$ mH.

Initial breakdown at the shortest distance (1–2 mm) between the two electrodes (Fig. 4–54) begins the cycle of the gliding arc evolution. For atmospheric air and 1 mm distance between the electrodes, the breakdown voltage V_b is about 3 kV. After about 1 μs, low-resistance plasma is formed and the voltage between the electrodes falls. The **equilibrium stage** occurs after the formation of the plasma channel. The gas flow pushes the plasma column with a velocity of about 10 m/s; their velocities, however, can be slightly different (Richard et al., 1996; Deminsky et al., 1997). The length l of the arc increases together with voltage; the power increases up to the maximum value, P_{max}, provided by the power supply. Electric current increases up to its maximum value of $I_m = V_0/R \approx 40$ A. During the quasi-equilibrium stage, the gas temperature T_0 does not change significantly (remaining around 3000 K; Fridman et al., 1993, 1997, 1999). The **non-equilibrium stage** starts when the length of the gliding arc exceeds its critical value l_{crit}. Heat losses from the plasma column begin to exceed the energy supplied by the source, and it is not possible to sustain the plasma in quasi equilibrium. Then the plasma rapidly cools down to $T_0 = $ 1000–2000 K, while conductivity is sustained by high electron temperature, $T_e = 1$ eV. After decay of the non-equilibrium discharge, a new breakdown takes place at the shortest distance between the electrodes and the cycle repeats.

Figure 4–54. Consequent phases of evolution of a gliding arc discharge: (A) region of gas breakdown; (B) quasi-equilibrium plasma phase; (C) non-equilibrium plasma phase.

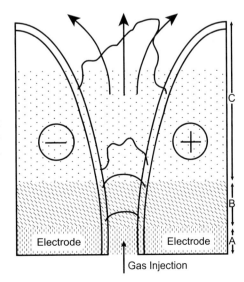

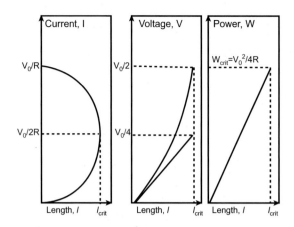

Figure 4–55. Evolution of current, voltage, and power with increasing length of the gliding arc discharge.

4.3.9. Equilibrium Phase of Gliding Arc, Its Critical Parameters, and Fast Equilibrium-to-Non-Equilibrium Transition

Assuming that the specific power w remains constant (Sections 4.3.5 and 4.3.6) permits a description of the evolution of the current, voltage, and power during the gliding arc quasi-equilibrium phase. Neglecting the self-inductance L, Ohm's law can be presented in this case as (Fridman & Kennedy, 2004):

$$V_0 = RI + w \cdot l/I, \qquad (4\text{–}75)$$

where V_0, R, and I are open-circuit voltage of the power supply, external resistance, and current. The arc current can be determined from Ohm's law as a function of the growing arc length l:

$$I = (V_0 \pm (V_0^2 - 4wlR)^{1/2})/2R. \qquad (4\text{–}76)$$

The solution with "+" describes the steady state of the gliding arc column; the solution with $I < V_0/2R$ is unstable and corresponds to negative differential resistance ρ ($\rho = dV/dJ$) of the circuit. The current (4–76) slightly decreases during the quasi-equilibrium period while the arc voltage is growing as W, l/I, and the total arc power $P = w\,l$ increase almost linearly with the length l (see Fig. 4–55). The quasi-equilibrium evolution of a gliding arc is terminated when the arc length approaches the critical value:

$$l_{\text{crit}} = V_0^0 2/(4wR) \qquad (4\text{–}77)$$

and the square root in (4–76) becomes equal to zero. The current decreases at this point to its minimal stable value of $I_{\text{crit}} = V_0/2R$, which is half of the initial value. Plasma voltage, electric field E, and total power at the same critical point approach their maximum values, respectively:

$$V_{\text{crit}} = V_0/2, \quad E_{\text{crit}} = w/I_{\text{crit}}, \quad W_{\text{crit}} = V_0^2/4R. \qquad (4\text{–}78)$$

Plasma resistance at the critical point becomes equal to the external one, and maximum value of the discharge power equals half of the maximum generator power. When the arc length exceeds the critical value ($l > l_{\text{crit}}$), the heat losses wl continue to grow. But the power from the power supply cannot be increased anymore, and the gas temperature rapidly decreases. Plasma conductivity can still be maintained by electron temperatures (≈ 1 eV) and stepwise ionization (Fridman et al., 1999). The fast equilibrium-to-non-equilibrium transition is due to the increase of electric field $E = w/I$ and electron temperature T_e:

$$T_e = T_0(1 + E^2/E_i^2), \qquad (4\text{–}79)$$

where E_i corresponds to the transition from thermal to direct electron impact ionization (Section 2.1).

Equilibrium arcs are stable: a small temperature increase leads to a growth of electric conductivity σ and to a decrease of electric field E (because $j = \sigma E$ is fixed) and the Joule heating power σE^2. In contrast, the direct electron impact ionization, typical for non-equilibrium plasma, is usually not stable. A small temperature increase leads, at fixed pressure and electric field, to a reduction of the gas density n_0 and to an increase of the specific electric field E/n_0 and the ionization rate. This results in an increase in the conductivity σ, Joule heating power σE^2, and finally in the discharge instability. Thus, when the conductivity σ depends mainly on the gas translational temperature, the discharge is stable. When the electric field increases and the conductivity σ starts depending mostly on E/n_0, the ionization becomes unstable. The gliding arc passes such a critical point during its evolution and growth of the electric field:

$$E = 2wR/(V_0 + \left(V_0^2 - 4wlR\right)^{0.5}), \tag{4–80}$$

which results in the fast equilibrium-to-non-equilibrium transition of the gliding arc.

4.3.10. Gliding Arc Stability Analysis and Transitional and Non-Equilibrium Phases of the Discharge

When the electric field is high, the gliding arc conductivity $\sigma(T_0, E)$ depends not only on gas temperature, as in quasi-equilibrium plasma, but also on the field E. The logarithmic sensitivity of the electric conductivity to the gas temperature corresponds to the quasi-equilibrium Saha ionization:

$$\sigma_T = \partial \ln \sigma(T_0, E)/\partial \ln T_0 \approx I_i/2T_0, \tag{4–81}$$

where I_i is ionization potential. The logarithmic sensitivity of $\sigma(T_0, E)$ to the electric field is

$$\sigma_E = \partial \ln \sigma(T_0, E)/\partial \ln E = I_i E^2/E_i^2 T_0. \tag{4–82}$$

The gliding arc stability can be analyzed by linearization of the thermal balance equation:

$$n_0 c_p dT_0/dt = \sigma(T_0, E)E^2 - 8\pi \lambda T_0^2/I_i S, \tag{4–83}$$

where S is arc cross section, and n_0 and c_p are gas density and specific heat. The linearization procedure permits time evolution of a temperature fluctuation ΔT to be described in the exponential form:

$$\Delta T(t) = \Delta T_0 \exp(\Omega t), \tag{4–84}$$

where ΔT_0 is the initial temperature fluctuation and Ω is the instability decrement. A negative decrement ($\Omega < 0$) corresponds to discharge stability, whereas $\Omega > 0$ corresponds to the case of instability (Fridman et al., 1999):

$$\Omega = -\frac{\sigma E^2}{n_0 c_p T_0} \frac{\sigma_T}{1 + \sigma_E} \left(1 - \frac{E}{E_{crit}}\right). \tag{4–85}$$

The discharge remains stable ($\Omega < 0$) in the initial phase of the gliding arc evolution when electric fields are relatively low, $E < E_{crit}$ ($l < l_{crit}$), and becomes unstable ($\Omega > 0$) when the electric field grows stronger. A gliding arc loses its stability even before the critical point (Fridman & Kennedy, 2004):

$$E_i(T_0/I_i)^{0.5} < E < E_{crit}. \tag{4–86}$$

Table 4–10. Critical Gliding Arc Parameters Before Fast Equilibrium-to-Non-Equilibrium Transition as Function of Initial Current I_0

I_0, A	100	80	60	40	20	10	2	1	0.2
I_{crit}, A	50	40	30	20	10	5	1	0.5	0.1
w_{crit}, W/cm	1600	1450	1300	1100	850	650	400	350	250
E_{crit} V/cm	33	37	43	55	85	130	420	700	2400
$(\frac{E}{n_0})\ 10^{-16}$ Vcm²	0.49	0.51	0.57	0.66	0.9	1.3	3.1	4.8	14

Although the electric field is less than the critical one, its influence on $\sigma(T_0, E)$ becomes dominant, which strongly perturbs the stability. In this "quasi-instability" regime, the stability factor Ω (4–85) is still negative, but decreases 10- to 30-fold due to the influence of the logarithmic sensitivity σ_E. During the transitional stage, the electric field and the electron temperature T_e in the gliding arc are slightly increasing, while the translational temperature decreases by about three-fold; electron density falls to about 10^{12} cm^{-3}. The decrease of electron density and conductivity σ leads to a reduction of the specific discharge power σE^2. It cannot be compensated by the electric field growth as in the case of a stable quasi-equilibrium arc because of the strong sensitivity of the conductivity to electric field. This is the main physical reason for the quasi-unstable discharge behavior during the fast equilibrium-to-non-equilibrium transition.

During the transitional phase, specific heat losses w decrease to a smaller value $w_{\text{non-eq}}$. Specific discharge power σE^2 also decreases with the reduction of translational temperature T_0. According to Ohm's law ($E < E_{crit}$), however, the total discharge power increases due to the growth of plasma resistance. This discrepancy between the total power and specific power (per unit of length or volume) results in a possible "explosive" increase in the arc length. This increase of length can be estimated as

$$l_{max}/l_{crit} = w/w_{\text{non-eq}} \approx 3, \tag{4–87}$$

and has been experimentally observed in gliding arcs by Cormier et al. (1993).

After the fast transition, the gliding arc continues to evolve under the non-equilibrium conditions $T_e \gg T_0$ (Rusanov et al., 1993; Kennedy et al., 1997). Up to 70–80% of the total power can be dissipated in the non-equilibrium plasma phase with $T_e \approx 1\,\text{eV}$ and

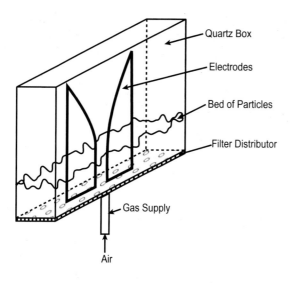

Quartz Box

Electrodes

Bed of Particles

Filter Distributor

Gas Supply

Air

Figure 4–56. General schematic of a gliding arc discharge in a fluidized bed.

4.3. Arc Discharges

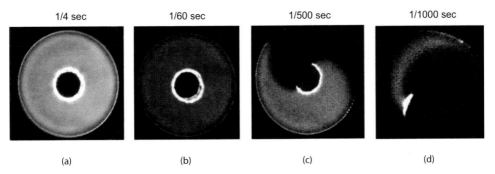

1/4 sec	1/60 sec	1/500 sec	1/1000 sec
(a)	(b)	(c)	(d)

Figure 4–57. Non-equilibrium gliding arc discharge rotating in magnetic field: pictures made with different exposure times.

$T_0 < 1000$–1500 K. Such gliding arc evolution is possible only if the electric field during the transition is high enough and the current is not too high. The critical gliding arc parameters before transition in atmospheric air are shown in Table 4–10 as a function of initial current. During the decay of the thermal arc, even a relatively low field, $(\frac{E}{n_0})_{\text{crit}} \approx (0.5$–$1.0) \times 10^{-16}$ V $\cdot$ cm^2, is sufficient to sustain the non-equilibrium phase due to stepwise and Penning ionization. If $I_0 > 5$–10 A, the arc discharge extinguishes after reaching the critical point; alternatively, for $J_0 < 5$–10 A, electric field $(\frac{E}{n_0})_{\text{crit}}$ is sufficient for maintaining the discharge in the non-equilibrium regime. Three different types of gliding arcs can be observed. At low currents and high flow rates, the gliding arc is non-equilibrium throughout all stages of its development. The opposite occurs at high currents and low flow rates; the discharge is thermal and dies out at the critical point. Only at intermediate values of currents and flow rates does the transitional regime take place (Mutaf-Yardimci et al., 1999). Electron (T_e), vibrational (T_v), and translational (T_0) temperatures in the non-equilibrium regimes of a gliding arc, according to Czernichowski (1994), Dalaine et al. (1998), Pellerin et al. (1996), and Mutaf-Yardimci et al. (1998, 1999), typically are $T_e \approx 10,000$ K, $T_v \approx 2000$–3000 K, and $T_0 \approx 800$–2100 K.

4.3.11. Special Configurations of Gliding Arc Discharges: Gliding Arc Stabilized in Reverse Vortex (Tornado) Flow

The gliding arc discharges can be organized in various ways for specific applications. Of note are gliding arc stabilization in a fluidized bed (Fig. 4–56) and a gliding arc rotating in a magnetic field (Fig. 4–57). An expanding arc configuration is applied in switchgears, as shown in Fig. 4–58. The contacts start out closed. When the contacts open, an expanding arc is formed and glides along the opening of increasing length until it extinguishes. Extinction of the gliding arc can be promoted by using SF$_6$.

Interesting for applications is gliding arc stabilization in the reverse vortex (tornado) flow. This approach is opposite that of the conventional **forward-vortex stabilization** (Fig. 4–59a), where the swirl generator is placed upstream with respect to discharge and the rotating gas provides the walls with protection from the heat flux (Gutsol, 1997). Reverse

Figure 4–58. Switchgears based on gliding arcs.

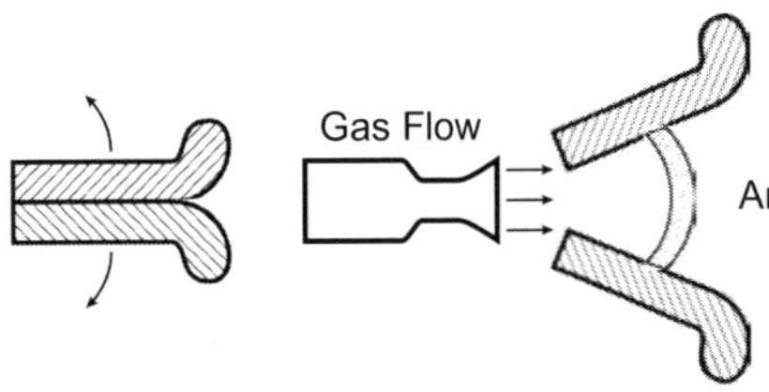

Gas Flow

Arc

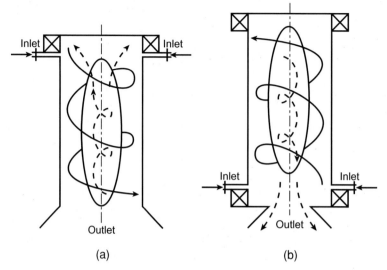

(a) (b)

Figure 4–59. Plasma stabilization by (a) forward vortex and (b) reverse vortex flows.

axial pressure gradient and central reverse flow appears due to fast flow rotation and a strong centrifugal effect near the gas inlet, which becomes a bit slower and weaker downstream. The hot reverse flow mixes with incoming cold gas and increases heat losses to the walls, which makes the discharge walls insulation less effective. More effective wall insulation is achieved by **reverse-vortex stabilization** (Kalinnikov & Gutsol, 1997; Gutsol & Fridman, 2001), illustrated in Fig. 4–59b. In this case, the outlet of the plasma jet is directed along the axis to the swirl generator side. Cold incoming gas moves at first by the walls, providing cooling and insulation, and only after that it goes to the central plasma zone and becomes hot. Thus, in the case of reverse-vortex stabilization, the incoming gas is entering the discharge zone from all directions except the outlet side, which makes it effective for gas heating/conversion efficiency and for discharge walls protection. A picture of the gliding arc discharge trapped in the reverse vortex (tornado) flow is shown in Fig. 4–60. The reverse vortex flow is effective in stabilizing different discharges, including 60-kW thermal ICP

Figure 4–60. Non-equilibrium gliding arc discharge moving along a spiral electrode and stabilized in the reverse vortex (tornado) flow.

discharge in argon (Gutsol, Larjo, & Hernberg, 1999), microwave discharge (Gutsol, 1995), and burners (Kalinnikov & Gutsol, 1999).

4.4. RADIOFREQUENCY AND MICROWAVE DISCHARGES IN PLASMA CHEMISTRY

4.4.1. Generation of Thermal Plasma in Radiofrequency Discharges

Although the arcs are conventional generators of thermal plasma ($0.1–0.5/W), more expensive ($1–5/W) RF discharges are also effective for application, especially when it is important to avoid direct contact between plasma and electrodes. Thermal plasma generation in the **inductively coupled plasma (ICP)** discharges is illustrated in Fig. 4–61. High-frequency electric current passes through a solenoid coil providing an axial magnetic field, which induces the vortex electric field sustaining the RF ICP discharge. The magnetic field in the ICP discharge is determined by the current in the solenoid, while the electric field according to the Maxwell equations is also proportional to the frequency. To achieve sufficient electric fields, RF of 0.1–100 MHz is usually required. A dielectric tube is usually inserted inside the solenoid coil to sustain the thermal plasma in the gas of interest. To avoid interference with radio communication systems, specific frequencies f were assigned for operation of the industrial RF discharges; between those, 13.6 MHz (wavelength 22 m) is used most often. The RF ICP RF discharges are effective in sustaining thermal plasma at atmospheric pressure. However, low electric fields are not sufficient here for discharge ignition, which requires, for example, insertion of an additional rod electrode heated and partially evaporated by the Foucault currents.

RF-ICP discharge plasma should be coupled as a load with an RF generator. Electrical parameters of the plasma load, such as resistance and inductance, influence the operation of the electric circuit as a whole and determine the effectiveness of the coupling. Relevant analysis of the ICP parameters can be accomplished in the framework of the **metallic cylinder model**, where plasma is considered a cylinder with the fixed conductivity σ corresponding to the maximum temperature T_m on the axis. Thermal plasma conductivity is high, and the skin effect prevents deep penetration of the electromagnetic field. Therefore, heat release, related to the inductive currents, is localized in the relatively thin skin layer of the column. Thermal conductivity λ_m inside the cylinder provides the temperature plateau in the central part of the discharge where the inductive heating is negligible. Radial distributions of plasma temperature, conductivity, and Joule heating in the RF-ICP discharge are illustrated in Fig. 4–62. Electromagnetic flux S_0 and power per unit surface of the long cylindrical ICP column can be calculated in the framework of the metallic cylinder model as a function of current I_0 in the solenoid and number of turns per unit length of the coil, n:

$$S_0, \frac{W}{cm^2} = 9.94 \cdot 10^{-2} \cdot \left(I_0 n, \frac{A \cdot turns}{cm} \right)^2 \sqrt{\frac{f, MHz}{\sigma, Ohm^{-1}cm^{-1}}}. \tag{4–88}$$

Figure 4–61. Physical schematic of the generation of inductively coupled plasma (ICP).

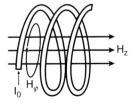

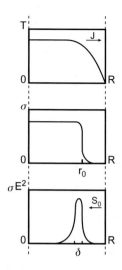

Figure 4–62. Radial distributions of major physical characteristics of the RF-ICP discharge: (a) distribution of gas temperature $T(R)$, R is discharge tube radius, J is heat flux to the periphery; (b) distribution of electric conductivity $\sigma(R)$, r_0 is radius of the high-conductivity plasma zone; (c) distribution of Joule heating $\sigma E^2(r)$, E is electric field, δ is depth of the skin layer, S_0 is flux of electromagnetic energy into plasma.

The plasma temperature in the "metallic cylinder" can be expressed as (I_i is ionization potential)

$$2\sqrt{2}\lambda_m \frac{T_m^2}{I_i}\sigma_m = I_0^2 n^2. \tag{4–89}$$

Conductivity σ_m strongly depends on temperature, whereas T_m and λ_m are changing only slightly; therefore, $\sigma_m \propto (I_0 n)^2$. Then, the Saha equation results in weak logarithmic dependence $T_m(I_0, n)$:

$$T_m = \frac{\text{const}}{\text{const} - \ln(I_0 n)}. \tag{4–90}$$

The dependence of ICP power per unit length on solenoid current and frequency can be shown as

$$w \propto I_0 n \sqrt{\omega}. \tag{4–91}$$

Typical atmospheric pressure thermal RF-ICP discharge ($f = 13.6$ MHz, tube radius 3 cm) has a temperature of about 10,000 K, thermal conductivity $\lambda_m = 1.4 \cdot 10^{-2}$ W/cm · K, electric conductivity $\sigma_m \approx 25$ Ohm^{-1} cm^{-1}, and skin layer $\delta \approx 0.27$ cm. The electromagnetic flux needed to sustain such plasma is $S_0 \approx 250$ W/cm^2; the corresponding solenoid current and number of turns are $I_0 n \approx 60$ A $\frac{\text{turns}}{\text{cm}}$, and magnetic field is $H_0 \approx 6$ kA/m. The electric field on the external plasma boundary is about 12 V/cm, the density of the circular current is 300 A/cm^2, the total current per unit length of the column is about 100 A/cm, the thermal flux potential is about 0.15 kW/cm, the distance between the effective plasma surface and discharge tube is $\Delta r \approx 0.5$ cm, and finally the ICP discharge power per unit length is about 4 kW/cm.

The ICP torches illustrated in Fig. 4–63 are used as industrial plasma sources (Gross, Grycz, & Miklossy, 1969; Boulos, 1985; Dresvin, 1991; Boulos, Fauchais, & Pfender, 1994; Dresvin & Ivanov, 2004, 2006; Dresvin, Shi, & Ivanov, 2006). The heating coil is water-cooled and not in direct contact with plasma, which provides the plasma purity. The ICP torches are often operated in transparent quartz tubes to avoid radiative heat load on the walls. At power exceeding 5 kW, the walls are water-cooled. Plasma stabilization and the wall insulation can be achieved in the fast-rotating gas flows (see Section 4.3.11). The ICP discharge, where the magnetic field is "primary," has lower electric fields than **capacitively coupled plasma** (CCP) discharge, where the electric field is primary. For this reason, the

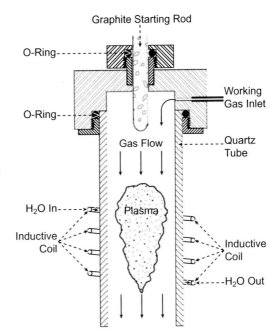

Figure 4–63. Schematic of kilowatt power level inductively coupled radio-frequency (ICP-RF) plasma torch.

ICP discharges at moderate to high pressures usually generate thermal plasma, whereas the RF-CCP discharges in the same conditions can generate non-thermal plasma.

4.4.2. Thermal Plasma Generation in Microwave and Optical Discharges

In contrast to RF discharges, microwave plasma is sustained by the centimeter-range electromagnetic waves interacting with plasma in the quasi-optical way. The optical discharges are sustained by laser radiation with even shorter wavelengths. Thermal plasma generation in microwave and optical discharges is usually related to high-pressure systems. Microwave generators, in particular magnetrons, operating with power exceeding 1 kW in the gigahertz frequency range, are able to maintain the steady-state thermal microwave discharges at atmospheric pressure. Electromagnetic energy in the microwave discharges can be coupled with plasma in different ways. The most typical coupling is provided in waveguides (Fig. 4–64), where the dielectric transparent tube for the electromagnetic waves (usually quartz) crosses the rectangular waveguide. Plasma is ignited and maintained inside the tube. Different modes of electromagnetic waves, formed in the rectangular waveguide, can be used to operate the microwave discharges. The most typical one is H_{01} mode (Fig. 4–65), where the electric field is parallel to the narrow walls of the waveguide and its maximum E_{max} (kV/cm) is related to power P_{MW} (kW) as

$$E_{max}^2 = \frac{1.51 \cdot P_{MW}}{a_w b_w} \left[1 - \left(\frac{\lambda}{\lambda_{crit}} \right)^2 \right]^{-1/2}, \qquad (4\text{–}92)$$

Figure 4–64. Schematic of a microwave discharge organized in a waveguide; S_0 is flux of electromagnetic energy to plasma.

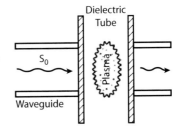

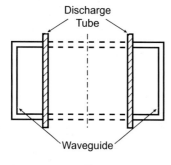

Figure 4–65. Distribution of electric field for H_{01} mode of electromagnetic wave in a microwave waveguide.

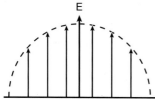

where λ is the wavelength, λ_{crit} is its maximum when propagation is still possible ($\lambda_{\text{crit}} = 2a_w$), and a_w and b_w are lengths of the waveguide walls in centimeters. The H_{01} mode is convenient for microwave plasma generation because the electric field in this case has a maximum in the center of the discharge tube. To provide plasma stabilization, gas flow is usually supplied tangentially. The waveguide dimensions are related to the microwave frequency: if $f = 2.5\,\text{GHz}$ ($\lambda = 12\,\text{cm}$), the wide waveguide wall should be longer than 6 cm, usually 7.2 cm. The narrow waveguide wall is typically 3.4 cm long, the dielectric tube diameter is about 2 cm, and the diameter of the plasma about 1 cm. When microwave power at atmospheric pressure is 1–2 kW, the temperature in molecular gases is about 4000–5000 K, which is lower with respect to RF-ICP discharges. An incident electromagnetic wave interacts with the plasma, which results in partial dissipation and reflection of the electromagnetic wave. Typically about half of the incident microwave power is dissipated in plasma, about one-quarter is transmitted, and another quarter is reflected. To increase the coupling effectiveness to 90–95%, the transmitted wave can be reflected back (Blinov et al., 1969; Batenin et al., 1988). Powerful 1.2-MW microwave discharges for plasma-chemical applications have been developed at the Kurchatov Institute of Atomic Energy (Rusanov & Fridman, 1984; Givotov, Rusanov, & Fridman, 1985).

A powerful atmospheric-pressure-thermal microwave discharge in a resonator has been developed by Kapitsa (1969). A schematic of the resonator with standing microwave mode E_{01} is shown in Fig. 4–66 (microwave power 175 kW, frequency 1.6 GHz). Plasma formed in the resonator looks like a filament located along the axis of the resonator cylinder. The length of the plasma filament is about 10 cm, which corresponds to half of the wavelength; the diameter of the plasma filament is about 1 cm. The plasma temperature in the microwave discharge does not exceed 8000 K due to intensive reflection at higher conductivities.

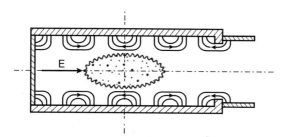

Figure 4–66. Microwave plasma and distribution of electric fields in a resonator-based microwave discharge.

Table 4–11. Characteristics of Atmospheric-Pressure Microwave Discharge Plasma in Air at Frequency $f = 10\,GHz$ ($\lambda = 3\,cm$) as Function of Plasma Temperature

Plasma temperature, T_m	4500 K	5000 K	5500 K	6000 K
Electron density, n_e, $10^{13}\,cm^{-3}$	1.6	4.8	9.3	21
Plasma conductivity, σ_m, $10^{11}\,s^{-1}$	0.33	0.99	1.9	4.1
Thermal conductivity, λ, $10^{-2}\,W/cm \cdot K$	0.95	1.1	1.3	1.55
Refractive index n of plasma surface	1.3	2.1	2.8	4.3
Plasma attenuation coefficient, κ	2.6	4.7	7.3	11
Depth of the microwave absorption layer, $l_\omega = \frac{1}{\mu_\omega}$, $10^{-2}\,cm$	9.1	5.0	3.2	2.2
Energy flux absorbed, S_1, kW/cm^2	0.23	0.35	0.56	1.06
Microwave reflection coefficient, ρ	0.4	0.65	0.76	0.81
Microwave energy flux to sustain plasma, S_0, kW/cm^2	0.38	1.0	2.3	5.6

Some parameters of the thermal microwave discharge at frequency $f = 10\,GHz$ ($\lambda = 3\,cm$) in atmospheric-pressure air are presented in Table 4–11 (Fridman & Kennedy, 2004). The depth of microwave absorption in plasma, l_ω, grows significantly with temperature reduction; at $T = 3500\,K$ it reaches 2 cm. At lower temperatures, microwave plasma becomes transparent for the electromagnetic wave. High microwave power should be provided to sustain such plasma. The minimal temperature of the microwave discharges can be estimated from $l_\omega(T_{min}) = R_f$, where R_f is plasma radius. It is about 4200 K for a filament radius of 3 mm; the corresponding absorbed flux is $S_1 = 0.2\,kW/cm^2$, and the minimum microwave energy flux needed to sustain the thermal plasma is about $S_0 = 0.25\,kW/cm^2$. The maximum temperature is limited by reflection at high plasma conductivity. This effect is seen in Table 4–11: the reflection coefficient becomes very high ($\rho = 81\%$ at $T_m = 6000$ K). For this reason, the quasi-equilibrium temperature of the microwave plasma in atmospheric air usually does not exceed 5000–6000 K.

Waveguide mode converters can be applied for better coupling of microwave radiation with plasma. The scheme of such a plasma torch, developed by Mitsuda, Yoshida, and Akashi (1989), is shown in Fig. 4–67. Microwave power is delivered first by a rectangular waveguide, and then goes through a quartz window into an impedance-matching mode converter, which couples the microwave power into a coaxial waveguide. The coaxial waveguide then operates as an arc jet: the center conductor of the waveguide forms one electrode and the other electrode comprises an annual flange on the outer coaxial electrode. Another configuration

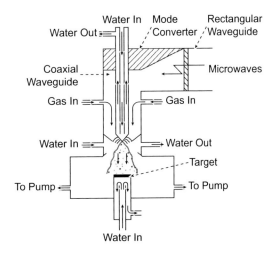

Figure 4–67. Schematic of microwave plasma torch.

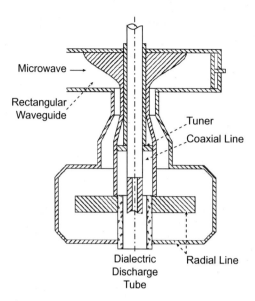

Figure 4–68. Schematic of radial microwave plasmatron, based on rectangular-to-circular microwave mode converter.

of microwave discharge based on conversion of the rectangular waveguide mode into a circular one is illustrated in Fig. 4–68 (MacDonald & Tetenbaum, 1978; Batenin et al., 1988).

Thermal atmospheric-pressure plasma can also be generated by optical radiation of powerful lasers similarly to RF and microwave discharges (Raizer, 1970; Generalov et al., 1970, 1971; see Fig. 4–69). The **optical breakdown** effect has been demonstrated by Maker, Terhune, and Savage (1964) after development of Q-switched lasers, which are able to produce sufficiently powerful light pulses (Raizer, 1974, 1977). The optical breakdown thresholds are presented in Fig. 4–70. A scheme of the continuous optical discharge is shown in Fig. 4–71, where a CO_2 laser beam is focused by lens or mirror to sustain the discharge. The laser power should be high; at least 5 kW for the continuous discharge in atmospheric air. To sustain the high-pressure discharge in Xe, a much lower power of about 150 W is sufficient. Light absorption in plasma significantly decreases at higher

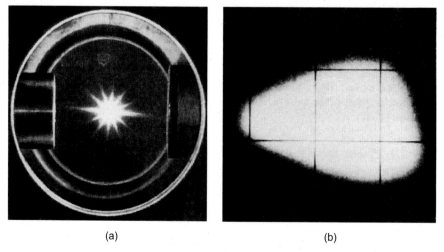

(a)   (b)

Figure 4–69. Continuous optical discharge: (a) general view of the discharge; (b) large image of the discharge with the beam traveling right to left (Generalov et al., 1971).

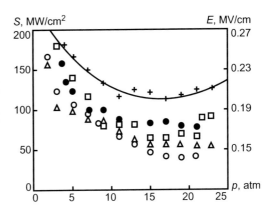

Figure 4–70. Thresholds of breakdown of different inert gases at different pressures by radiation of a CO_2 laser: □ Ar, △ Ne, ○ He (technical helium with possible admixtures), ● He (very-high-purity helium), + Xe.

frequencies; therefore, application of visible radiation requires 100–1000 times higher power than that of the CO_2 laser. Plasma directly sustained by focusing solar radiation would have a temperature less than the surface temperature of the sun (6000 K). Equilibrium plasma density at 6000 K is not sufficient for effective radiation absorption. For effective absorption, the plasma density should be close to complete ionization. Therefore, the plasma temperature in the continuous optical discharges is high (15,000–20,000 K). Fundamentals of the optical discharges can be found in the literature (Raizer, 1974, 1977, 1991).

4.4.3. Non-Thermal Radiofrequency Discharges: Capacitive and Inductive Coupling of Plasma

At low pressures, RF plasma is strongly non-equilibrium and cold; electron–neutral collisions are less frequent while gas cooling by the walls is intensive; therefore, $T_e \gg T_0$ similarly to glow discharges (Section 4.2). The cold electrodeless RF discharges are widely applied, especially in electronics and other technologies involving high-precision surface treatment. The upper frequency limit of the discharges is due to wavelengths close to the system sizes (shorter wavelengths are referred to as microwaves). The lower RF limit is due to frequencies of ionization and ion transfer. Ion density in RF discharge plasmas and sheaths usually can be considered constant during a period of electromagnetic field oscillation. Thus, RFs usually exceed 1 MHz (sometimes smaller); the one used most often in industry is 13.6 MHz. Non-thermal RF discharges can be subdivided into those of moderate and low pressure. In the moderate pressure discharges (1–100 Torr), the electron energy relaxation length is smaller than characteristic system sizes and the electron energy distribution function (EEDF) is determined by the local electric field (Raizer, Shneider, & Yatsenko, 1995). In the low-pressure discharges (p (Torr) $\cdot$ L (cm) < 1 in inert gases), the electron energy relaxation length is comparable to the discharge sizes and EEDF is determined by electric field distribution in the entire discharge (Lieberman & Lichtenberg, 1994).

Non-thermal RF discharges can be either capacitively (CCP) or inductively (ICP) coupled (Fig. 4–72). The CCP discharges provide an electromagnetic field by electrodes located either inside or outside of the chamber (Fig. 4–72a,b). They primarily stimulate an electric field, facilitating ignition. An electromagnetic field in the ICP discharges is induced by the

Figure 4–71. Schematic of continuous optical discharge.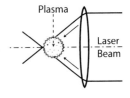

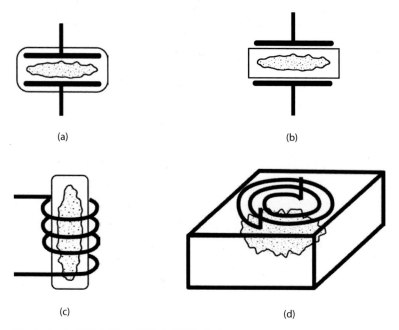

(a)　　　　　　　　　　　　　　(b)

(c)　　　　　　　　　　　　　　(d)

Figure 4–72. CCP (a,b) and ICP (c,d) RF discharges.

inductive coil: the discharge can be located either inside of the coil (Fig. 4–72c) or adjacent to the plane coil (Fig. 4–72d). The ICP discharges primarily stimulate the magnetic field, while the corresponding non-conservative electric field is relatively low. Therefore, the non-thermal ICP is usually organized at lower pressures, when E/p is sufficient for ionization. Coupling between the inductive coil and plasma can be interpreted as a voltage-decreasing transformer: the coil represents the primary multiturn windings, and the plasma represents the secondary single-turn winding. Effective coupling with the RF power supply requires low plasma resistance. As a result, the ICP discharges are convenient to reach high currents, electric conductivities, and electron densities. In contrast, the CCP discharges are more convenient to provide higher electric fields.

RF power supplies for plasma generation typically require an active load of 50 or 75 ohm. To provide effective correlation between resistance of the leading line from the RF generator and the CCP or ICP discharge impedance, the special coupling circuit should be applied (Fig. 4–73). The RF generator should not only be correlated with the RF discharge during its continuous operation but also should provide initially sufficient voltage for breakdown. To provide the ignition, the coupling electric circuit should form the AC resonance in series with the generator and the discharge system during its idle operation. For this reason, the CCP coupling circuit includes inductance in series with the generator and discharge system (Fig. 4–73a). Variable capacitance is there for adjustment, because the design of variable inductances is more complicated. In the case of ICP discharge, a variable capacitance in the electric coupling circuit in series with the generator and idle discharge system provides the necessary effect of initial breakdown.

4.4.4. Non-Thermal RF-CCP Discharges in Moderate Pressure Regimes

The external circuit in CCP discharges provides either fixed voltage or fixed current, which is a typical choice for practical application. The fixed-current regime can be represented simply by current density $j = -j_0 \sin(\omega t)$ between two parallel electrodes. It can be assumed that

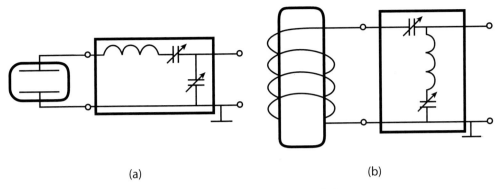

(a)   (b)

Figure 4–73. Coupling circuits for (a) RF-CCP and (b) RF-ICP discharges.

RF plasma density ($n_e = n_i$) is high and the electron conductivity current exceeds the displacement current:

$$\omega \ll \frac{1}{\varepsilon_0}|\sigma_e| \equiv \frac{1}{\tau_e}, \tag{4–93}$$

where τ_e is the Maxwell time for electrons, required for the charged particles to shield the electric field; $\omega \ll \nu_e$, ν_e is the electron collisional frequency, and σ_e is the complex electron conductivity:

$$\sigma_e = \frac{n_e e^2}{m(\nu_e + i\omega)}. \tag{4–94}$$

The inequality opposite to (4–93) is usually valid for ions because of much higher mass ($M \gg m$):

$$\omega \gg \frac{1}{\varepsilon_0}|\sigma_i| \equiv \frac{1}{\tau_i}, \quad \sigma_i = \frac{n_i e^2}{M(\nu_i + i\omega)}. \tag{4–95}$$

Thus, the ion conductivity current can be neglected with respect to the displacement current and the ion drift motion during an oscillation period of electric field can be neglected. Ions are at rest and form the "skeleton" of plasma, while electrons oscillate between the electrodes (Fig. 4–74). Electrons are present in the sheath of width L near an electrode only for a part of the oscillation period called the plasma phase. Another part of the oscillation period (no electrons in the sheath) is called the space charge phase. The oscillating space charge creates an electric field, which forms the displacement current and closes the circuit. The quasi-neutral plasma zone is called the positive column. The electric field of the space charge has oscillating as well as constant components, which are directed from plasma to electrodes. The constant component of the space charge field provides faster ion drift to the electrodes than in the case of ambipolar diffusion. As a result, the ion density in the space charge layers near electrodes is lower than that in plasma.

Moderate-pressure CCP discharges assume energy relaxation lengths smaller than sizes of plasma zone and sheaths $\lambda_\varepsilon < L, L_p$, which corresponds to the pressure interval 1–100 Torr. In this pressure range, $\omega < \delta \nu_e$, where δ is the fraction of electron energy lost per one electron collision and ν_e is the frequency of the electron collisions. In this regime, electrons lose and gain their energy during a time interval shorter than the period of electromagnetic RF oscillations. EEDFs and, hence, ionization and excitation rates are determined in the moderate-pressure discharges by local and instantaneous values of the electric field. In particular, the result is an essential contribution of ionization in sheaths, where electric fields have maximum values and are able to provide maximum electron energies.

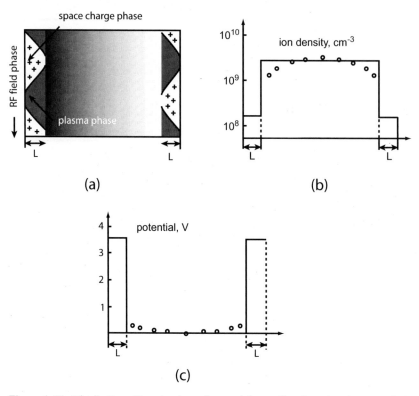

Figure 4–74. Distribution of ion density and potential as well as boundary-layer oscillation in RF-CCP discharge.

Moderate-pressure RF-CCP discharges can be sustained in two forms, referred to as α-discharge and γ-discharge (Levitsky, 1957). The main differences between the α- and γ-regimes are related to current density and luminosity distribution in the discharge gap; see Figs. 4–75 and 4–76 (Yatsenko, 1981; Raizer, Shneider, & Yatsenko, 1995). The α-**discharge** is characterized by low luminosity in the plasma volume. Brighter layers are located closer to electrodes, but layers immediately adjacent to the electrodes are dark. The

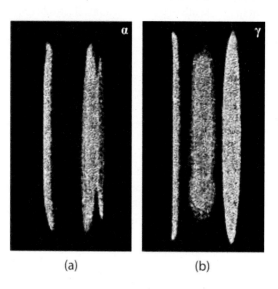

Figure 4–75. RF-CCP discharge in α- and γ-modes: air, pressure 10 Torr, frequency 13.56 MHz, distance between electrodes 2 cm.

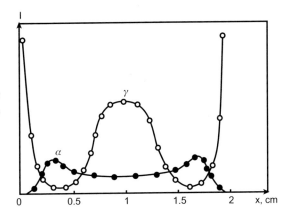

Figure 4–76. Distribution of emission intensity along a discharge gap in α- and γ-regimes: air, 10 Torr, 13.56 MHz, 2 cm.

γ-**discharge** occurs at much higher current density. In this regime, the discharge layers immediately adjacent to the electrodes are very bright but relatively thin. The plasma zone is also luminous and separated from the bright electrode layers by dark layers similar to the Faraday dark space in glow discharges. In the γ-regime, ionization processes occur due to electrons formed on the electrode by the secondary electron emission and are accelerated in a sheath, which is similar to glow discharges. The α- and γ-discharges operate at normal current density similarly to glow discharges: current increase is provided by the growth of the electrode area occupied by the discharge, while the current density remains constant. Normal current density in both regimes is proportional to the frequency of electromagnetic oscillations, and in the case of γ-discharges exceeds current density of α-discharges more than 10-fold.

Current-voltage characteristics of the moderate-pressure CCP discharge are shown in Fig. 4–77 (Raizer et al., 1995). Currents less than 1 A correspond to α-discharges, and higher currents are related to γ-discharges. A normal α-discharge can be observed on the curves 2 and 4 of the figure, where there is no essential change of voltage at low currents. Most of the curves show that α-discharge is abnormal just after breakdown. The current density is so small in this regime that discharge occupies the entire electrode immediately after breakdown, which leads to the voltage growth together with current (abnormal regime). The transition from α-regime to γ-regime occurs when the current density exceeds a critical value, which depends on pressure, frequency, electrode material, and type of gas. The α–γ transition is accompanied by discharge contraction and by a more than 10-fold growth of the current density. Increase of current in the γ-regime does not change voltage; the discharge remains in the normal regime. The RF-CCP discharges of moderate pressure were applied in CO_2 lasers. The first He–Ne and CO_2 lasers were based on these discharges (Javan, Bennett, & Herriot, 1961; Patel, 1964; Raizer et al., 1995).

4.4.5. Low-Pressure Capacitively Coupled RF Discharges

Low-pressure RF discharges (≤ 0.1 Torr), widely used in electronics, are illustrated in Fig. 4–78 (Vossen & Kern, 1978, 1991; Manos & Flamm, 1989; Konuma, 1992; Lieberman & Lichtenberg, 1994). The luminosity pattern is different from that of moderate pressure. The plasma zone is bright and separated from electrodes by dark pre-electrode sheaths. The discharge is usually asymmetric: a sheath located by the electrode where RF voltage is applied is about 1 cm thick, whereas another one is only about 0.3 cm. In contrast to the moderate-pressure discharges, plasma fills out the entire gap, and the normal current density does not occur. Electron energy relaxation length λ_ε in the low-pressure RF-CCP discharges exceeds the typical sizes of the plasma zone and sheaths ($\lambda_\varepsilon > L_p, L$). The EEDF is not local

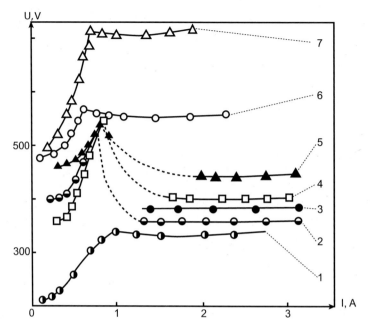

Figure 4–77. Typical current-voltage characteristics of an RF-CCP discharge with frequency 13.56 MHz at different conditions: 1, He ($p = 30$ Torr; $L_0 = 0.9$ cm); 2, air ($p = 30$ Torr; 0.9 cm); 3, air ($p = 30$ Torr; 3 cm); 4, air ($p = 30$ Torr; 0.9 cm, compare to 2); 5, CO_2 ($p = 15$ Torr; 3 cm); 6, air ($p = 7.5$ Torr; 1 cm); 7, air ($p = 7.5$ Torr; 1 cm).

and is determined by the electric fields in the zone of about λ_ε. For He discharges, it requires $p \cdot (L_p, L) \leq 1$ Torr $\cdot$ cm. Also $\omega \gg \delta \cdot \nu_e$ (where δ is the average fraction lost per collision, and ν_e is the frequency of the electron collisions); thus, the time of EEDF formation is longer than the period of RF oscillations and permits EEDF to be considered stationary. The electric field can be divided into constant and oscillating components. The constant one provides a balance of electron and ion fluxes, and quasi-neutrality; it provides current and heating of the plasma electrons. The distribution of plasma density and electric potential φ (corresponding to the constant component) is shown in Fig. 4–74. Electric field in the space charge sheath exceeds that in plasma; therefore, a sharp change of potential

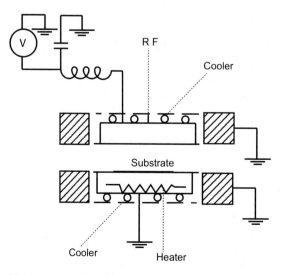

Figure 4–78. General schematic of low-pressure RF-CCP discharge.

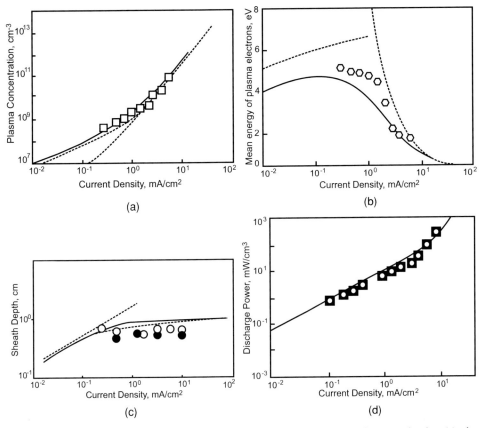

Figure 4–79. Low-pressure RF-CCP discharge parameters as functions of current density: (a) plasma concentration, (b) mean energy of plasma electrons, (c) depth of sheath, (d) discharge power.

occurs on the boundary of the plasma and the sheath. Plasma electrons move between the sharp potential barriers.

Electron heating in RF-CCP discharges of not-very-low pressures is due to electron–neutral collisions ν_e. The energy of the systematic oscillations, received by an electron from the electromagnetic field after a previous collision, can be transferred to chaotic electron motion during the next collision. If $\omega^2 \gg \nu_e^2$, electric conductivity and Joule heating are proportional to ν_e and the discharge power should be low at low pressures. Experimentally, however, the discharge power in such conditions can be significantly higher because of the contribution of the **stochastic heating effect**, which provides heating even in a collisionless regime (Godyak, 1971, 1975, 1986; Lieberman, 1989). For example, when the mean free path of electrons exceeds the sheath size, an electron coming in the sheath is reflected by the space charge potential of the sheath boundary as an elastic ball is reflected from a massive wall. If the sheath boundary moves from the electrode, then the reflected electron receives energy; on the other hand, if the sheath boundary moves to the electrode and a fast electron is "catching up" with the sheath, then the reflected electron loses its kinetic energy. The electron flux to the boundary moving from the electrode exceeds that of the other boundary; therefore, energy is mostly transferred to the fast electrons, which is the stochastic heating effect. The parameters of the low-pressure RF-CCP discharge in argon (Smirnov, 2000) are presented in Fig. 4–79 as functions of current density (see, for example, Godyak, 1986; Godyak & Piejak, 1990; Godyak, Piejak & Alexandrovich, 1991).

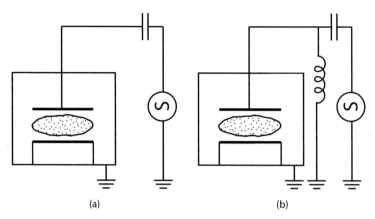

(a)   (b)

Figure 4–80. General schematics of RF-CCP discharges with (a) disconnected and (b) DC-connected electrodes.

The RF-CCP discharges are usually organized in grounded metal chambers: one electrode is connected to the chamber wall and grounded and another one is powered, which makes the discharge asymmetric (Lieberman, 1989). The current between an electrode and the grounded metallic wall is capacitive in moderate-pressure discharges and does not play any important role. In the low-pressure discharges, plasma occupies a bigger volume because of diffusion, and some fraction of the discharge current goes from the loaded electrode to the grounded walls. As a result, current density in the sheath located near the powered electrode exceeds that in the sheath related to the grounded electrode. Also the surface area of the powered electrode is smaller; therefore, current densities there are higher. Lower current density in the sheath corresponds to lower voltage. The constant component of the voltage, which is the plasma potential with respect to the electrode, is also lower at lower current density. Thus, constant potential difference (called **auto-displacement voltage**) occurs between the electrodes if they are covered by a dielectric layer or if capacitance is installed in the electric circuit (Fig. 4–80) to avoid direct current between electrodes.

Low-frequency ($<100\,kHz$) RF-CCP discharges are also applied in plasma chemistry. In particular, nitrogen plasma generated in such discharges is effective in polymer surface treatment that promotes adhesion of silver to polyethylene terephthalate (Spahn & Gerenser, 1994; Gerenser et al., 2000), treatment of polyester web to promote adhesion of gelatin containing layers related to production of photographic film (Grace, 1995), and sputter deposition of metals (Este & Westwood, 1984, 1988; Butterbaugh, Baston, & Sawin, 1990; Schiller et al., 1982; Ridge & Howson, 1982; Affinito & Parson, 1984). These discharges are discussed in detail in Fridman & Kennedy (2004) and Conti et al. (2001).

4.4.6. RF Magnetron Discharges

Sheath voltages in the RF-CCP discharges are relatively high, leading to low ion densities and high ion-bombarding energies, which can be disadvantageous. Also the ion-bombarding energies cannot be varied in these discharges independently of the ion flux. The RF magnetron discharges (called **magnetically enhanced reactive ion etchers [MERIEs]**) were developed to make the relevant improvements. In this case, a relatively weak DC magnetic field (50–$200\,G$) is imposed on the low-pressure RF-CCP discharge parallel to the powered electrode and perpendicular to the RF electric field and current. The RF magnetron permits an increase in the ionization degree at lower RF voltages, and a decrease in the energy of ions bombarding the powered electrode. Also this discharge permits an increase in the ion flux intensifying etching. The RF magnetron discharges can be effectively sustained at much lower pressures (down to 10^{-4} Torr); therefore, a well-directed ion

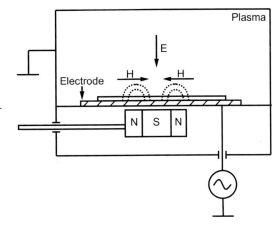

Figure 4–81. General schematic of RF magnetron.

beam is able to penetrate the sheath without any collisions. The ratio of the ion flux to the flux of active neutral species grows, which also leads to higher-quality etching.

A schematic of the RF magnetron used for ion etching (Okano, Yamazaki, & Horiike, 1982) is shown in Fig. 4–81. The RF voltage is applied to the smaller lower electrode, where a sample is located. A rectangular samarium–cobalt magnet is placed under the powered electrode. The part of the electrode where the magnetic field is horizontal does not cover the entire electrode. A scanning device is used to move the magnet and to provide the horizontal magnetic field for the entire electrode. The RF magnetron can also be organized in cylindrical systems with coaxial electrodes (Lin, 1985; Knypers & Hopman, 1988, 1990; Yeom, Thornton, & Kushner, 1989). The magnetic field in these systems is directed along the cylinder axis.

Electrons, oscillating together with the sheath boundary, additionally rotate in the RF magnetrons around the horizontal magnetic field lines with the cyclotron frequency. When the magnetic field and the cyclotron frequency are high enough, the amplitude of the electron oscillations along the RF electric field decreases significantly. The magnetic field "traps electrons." The cyclotron frequency plays the same role as the frequency of electron–neutral collisions: electrons become unable to reach the amplitude of their free oscillations in the RF electric field. The amplitude of electron oscillations determines the thickness of the sheaths. Thus, a decrease in the amplitude of electron oscillations in the magnetic field results in smaller sheaths and lower sheath voltage near the powered electrode, which leads to lower values of the auto-displacement, lower ion energies, and lower voltage necessary to sustain the RF discharge.

The RF magnetron discharge retains major properties of the low-pressure RF-CCP discharges, taking into account peculiarities related to the constant magnetic field (Gurin & Chernova, 1985; Lukyanova, Rahimov, & Suetin, 1990, 1991; Porteous & Graves, 1991). The constant horizontal magnetic field perpendicular to the electric field decreases the electron flux and prevents the electron losses on the powered electrode. The residence time of the electrons increases, promoting ionization. The space charge sheaths are also created in the RF magnetrons near electrodes because electrons leave the discharge gap. In the strong magnetic field, however, the effective electron mobility across the magnetic field may become lower than the ion mobility. The amplitude of ion oscillations exceeds that of the electrons. Electron current to the electrode does not vary a lot during the oscillation period; it is mostly due to diffusion, which is effective in spite of the magnetic field, because electron temperature significantly exceeds the ion temperature. Stochastic heating of the magnetized electrons is also possible at some conditions on the oscillating sheath boundaries (Lieberman, Lichtenberg, & Savas, 1991). The asymmetric magnetron discharge keeps

223

most of the properties typical for the non-magnetized asymmetric low-pressure RF-CCP discharges. However, sheath thickness in the RF magnetrons is much smaller than in non-magnetized discharges. For this reason the auto-displacement in the asymmetric magnetrons is also smaller. The energy spectrum of ions bombarding the electrode in magnetrons is similar to that of the non-magnetized discharges. However, the ion energies depend on the applied magnetic field, which determines the sheath voltage.

4.4.7. Non-Thermal Inductively Coupled RF Discharges in Cylindrical Coil

The electromagnetic field in the RF-ICP discharges is induced by inductive coil (Fig. 4–72), where the magnetic field is primary, and non-conservative electric fields are low. Therefore, the discharges are organized at low pressures (usually less than 50 mTorr) to provide a reduced electric field E/p sufficient for ionization. Coupling between the inductive coil and plasma can be interpreted as a transformer, where the coil is the primary multiturn windings, and plasma is the secondary single turn. The ICP discharge can be considered a voltage-decreasing and current-increasing transformer, which permits it to reach a high current, high electric conductivity, and high electron density at relatively low electric fields and voltages. The low-pressure ICP discharges operate at electron densities of 10^{11}–10^{12} cm^{-3}, exceeding those in CCP more than 10-fold; therefore, they are also referred to as the **high-density plasma (HDP)** discharges. The HDP discharges are widely used in electronics and other high-precision surface treatment technologies.

An important advantage of the discharges for high-precision surface treatment is the RF power coupling to plasma across a dielectric window or wall. Such "non-capacitive" power transfer to plasma permits operation at lower voltages across all sheaths at electrode and wall surfaces. The DC plasma potential and energies of ions accelerated in the sheaths are typically 20–40 V, which is very good for the numerous surface treatment applications. The ion energies can be independently controlled in this case by an additional capacitively coupled RF source called the **RF bias**, which drives the electrode on which the substrate for material treatment is placed (Fig. 4–72). Thus, the ICP discharges are able to provide independent control of the ion and radical fluxes by means of the main ICP source power and the ion-bombarding energies by means of power of the bias electrode (Lieberman & Gottscho, 1994).

Consider the inductive RF discharge in a long cylindrical tube placed inside of the cylindrical coil. The physical properties of such a discharge are similar to those of the planar one which is more convenient for applications. Electric field $E(r)$ induced in the discharge tube located inside of the long coil is

$$E(r) = -\frac{1}{2\pi r}\frac{d\Phi}{dt}, \qquad (4\text{--}96)$$

where Φ is the magnetic flux crossing the loop of radius r perpendicular to the axis of the discharge tube; r is distance from the tube axis. The magnetic field is created in the system by electric current in the coil, $I = I_c e^{i\omega t}$, and in plasma. Assuming constant plasma conductivity σ and expressing plasma current as $j(r) = j_0(r)e^{i\omega t}$, the current density distribution along the plasma radius is determined by

$$\frac{\partial^2 j_0}{\partial r^2} + \frac{1}{r}\frac{\partial j_0}{\partial r} - \frac{1}{r^2}j_0 = i\frac{\sigma\omega}{\varepsilon_0 c^2}j_0. \qquad (4\text{--}97)$$

If pressure is very low and plasma can be considered collisionless, its conductivity is inductive:

$$\sigma = -i\varepsilon_0\frac{\omega_p^2}{\omega}, \qquad (4\text{--}98)$$

where ω_p is plasma frequency. From (4–97) and (4–98), the current density distribution in the ICP plasma is

$$j_0(r) = j_b I_1 \left(\frac{r}{\delta}\right),\tag{4–99}$$

where $I_1(x)$ is the modified Bessel function and δ is the skin layer. The current density on the plasma boundary near the discharge tube is determined by the non-perturbed electric field and plasma conductivity:

$$j_b = \sigma \frac{\omega a N}{2\varepsilon_0 c^2 l} I_c,\tag{4–100}$$

where a is the radius of the discharge tube, N is the number of turns in the coil, l is the length of the coil, c is the speed of light, and I_c is the amplitude of the current in the coil. Plasma conductivity grows with the electric field and the skin layer is smaller than the discharge radius. Then most of the electric current is located in the relatively thin δ-layer on the discharge periphery. The Bessel function (4–99) can be simplified in this case:

$$j_0(r) \approx j_b \exp\left(\frac{r-a}{\delta}\right).\tag{4–101}$$

The current in the coil, I_c, is determined by an external circuit and can be considered a parameter. The electric field in the plasma is related to the voltage on the plasma loop, $E_p = U_p/2\pi a$, and logarithmically depends on other plasma parameters. The electric field in the plasma at high currents is small with respect to that in the idle regime without plasma: $E = \frac{\omega M I_c}{2\pi a}$, where M is the mutual inductance. The voltage drop related to plasma, U_p, can be neglected at high currents. The discharge current can then be expressed as

$$I_d = -\frac{M}{L_d} I_c = -N I_c,\tag{4–102}$$

where $L_d = \mu_0 \pi a^2/l$ is "geometric inductance" attributed to plasma considered as a conducting cylinder. The discharge current flows in the opposite direction with respect to the inductor current, and the value of the plasma current significantly exceeds that in the inductor. Then electron density can be expressed as

$$n_e = j \frac{m\nu_{en}}{e^2 E_p} = \frac{N I_c}{l\delta} \frac{m\nu_{en}}{e^2 E_p}.\tag{4–103}$$

The thickness of the skin layer in the RF-ICP discharge can be calculated as

$$\delta = \frac{2E_p l}{\omega \mu_0 N I_c} \propto \frac{1}{I_c}.\tag{4–104}$$

The skin layer is inversely proportional to the electric current in the inductor coil. Taking into account (4–104), the plasma density is proportional to the square of the current in the inductor coil:

$$n_e = \left(\frac{N I_c}{el E_p}\right)^2 \frac{\omega \mu_0 m\nu_{en}}{2} \propto I_c^2.\tag{4–105}$$

Similarly to the CCP discharges, ICP regimes are different at moderate and low pressures. In the **moderate-pressure regime**, the energy relaxation length is less than the thickness of the skin layer. Heating of electrons is determined by the local electric field and takes place in the skin layer. Therefore, ionization processes as well as plasma luminosity are concentrated in the skin layer, which is illustrated in Fig. 4–82. The internal volume of the discharge tube, located closer to the tube axis with respect to the skin layer, is filled up with plasma only due to the radial inward plasma diffusion from the discharge periphery. If losses of charged particles in the internal volume due to recombination and diffusion

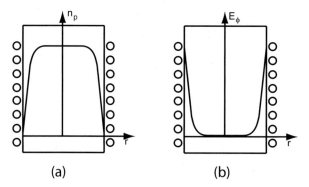

Figure 4–82. Radial distributions of plasma density (n_p) and electric field (E_φ) in moderate-pressure ICP discharge.

along the axis of the discharge tube are significant, the plasma density can be lower in the central part of the discharge than in the periphery. In the **low-pressure regime**, the energy relaxation length exceeds the skin layer. Heating of electrons takes place in the skin layer, but ionization processes are effective in the plasma volume, where the electrons have a maximum value of kinetic energy. Distributions of plasma density, amplitude of the oscillating electric field, and ambipolar potential along the radius of the discharge tube are illustrated in Fig. 4–83. The electron density on the discharge axis can significantly exceed that in the skin layer in this regime.

4.4.8. Planar-Coil and Other Configurations of Non-Thermal Inductively Coupled RF Discharges

A schematic of ICP discharges in the planar configuration, widely used in electronics, is illustrated in Fig. 4–84 (Roth, 1995). It is similar geometrically to the conventional RF-CCP parallel plate reactor, but RF power is applied in this scheme to a flat spiral inductive coil, which is separated from the plasma by quartz or some other dielectric insulating plate. The RF currents in the spiral coil induce the image currents in the upper surface of the plasma corresponding to the skin layer. Thus, this discharge is inductively coupled and is similar to the case of cylindrical geometry of the inductive coils. The analytical relations derived earlier for the low-pressure RF-ICP discharge inside an inductive coil can be applied qualitatively for the planar-coil configuration of the discharge. The planar ICP discharge also includes two important elements: multi-polar permanent magnets and DC wafer bias. The multi-polar permanent magnets are located around the outer circumference of the plasma

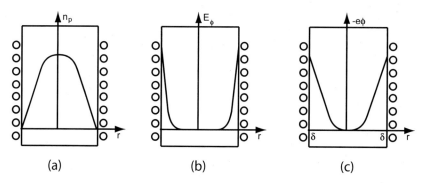

Figure 4–83. Radial distributions of plasma density (n_p), electric field (E_φ), and potential (φ) in low-pressure ICP discharge.

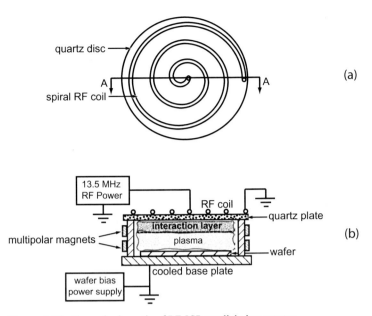

Figure 4–84. General schematic of RF-ICP parallel plate reactor.

to improve plasma uniformity and plasma confinement, and to increase plasma density. The DC wafer bias power supply is used to control the energy of ions impinging on the wafer, which have ranged over 30 to 400 eV in the planar discharge system.

The structure of magnetic field lines in the planar-coil configuration of the ICP discharges is more complicated than in the case of the cylindrical inductive coil. The RF magnetic field lines in the planar coil configuration in the absence of plasma are illustrated in Fig. 4–85a (Wendt & Lieberman, 1993). These magnetic field lines encircle the coil and are symmetric with respect to the plane of the coil. The deformation of the magnetic field in the presence of plasma, formed below the coil, is shown in Fig. 4–85b. In this case an azimuthal electric field and an associated current (in the direction opposite that in the coil) are induced in the plasma skin layer. The total magnetic field is generated by both the multi-turn coil current and "single-turn" induced plasma current. The dominant magnetic field component

Figure 4–85. Distribution of RF magnetic field near planar inductive coil: (a) without nearby plasma, (b) with nearby plasma.

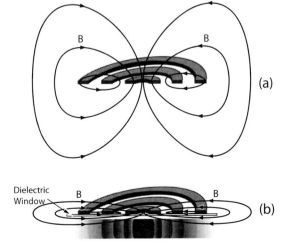

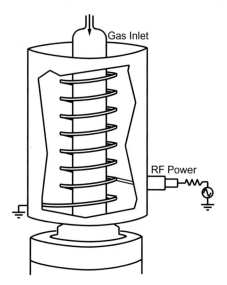

Figure 4–86. General schematic of a helical resonator discharge system.

within the plasma is vertical near the axis of the planar coil and horizontal away from the axis (Hopwood et al., 1993a,b).

The planar RF-ICP discharge, illustrated in Fig. 4–84, is effective in uniform plasma processing of wafers with diameters of at least 20 cm. The power of this reactor is 2 kW, which is an order of magnitude higher than the power of CCP discharges at similar conditions; it results in higher fluxes of ions and other active species, which accelerate the surface treatment. The ion flux in the system is 60 mA/cm^2, which corresponds to an etch rate of 1–2 μm/min for polyimide film. The reactor has been operated at frequencies from 1 to 40 MHz, but the usual operating frequency is 13.56 MHz. Operation pressure is 1–20 mTorr, which is far lower than the typical pressure of RF-CCP discharges (which is about several hundreds of milliTorr). The lower pressures are desirable for chemical vapor deposition and etching because of longer mean free paths and little scattering of ions and active species before they reach the wafer to participate in the surface treatment.

The **helical resonator discharge** is another special type of low-pressure RF-ICP discharge. The helical resonator consists of an inductive coil (helix) located inside the cylindrical conductive screen and can be considered a coaxial line with an internal helical electrode. A schematic of the helical resonator plasma source is shown in Fig. 4–86. An electromagnetic wave propagates in such a coaxial line with phase velocity much lower than the speed of light: $v_{ph} = \omega/k \ll c$, where k is the wavelength and c is the speed of light. This property allows the helical resonator to operate in the megahertz frequency range and to generate low-pressure plasma. The coaxial line of the helical discharge becomes resonant when an integral number of quarter-waves of the RF field fits between the two ends of the system:

$$2\pi r_{\text{h}} N = \frac{\lambda}{4}, \tag{4–106}$$

where r_{h} is the helix radius, N is the number of turns in the coil, and λ is the electromagnetic wavelength in a vacuum. The helical resonator discharges effectively operate at frequencies of 3–30 MHz and do not require a DC magnetic field. The resonators exhibit high Q-values (600–1500 without plasma); therefore, electric fields are high, facilitating the initial breakdown. Also the helical resonator discharges have high impedance and can be operated without a matching network (Cook et al., 1990; Niazi et al., 2004).

4.4.9. Non-Thermal Low-Pressure Microwave and Other Wave-Heated Discharges

Consider three types of low-pressure wave-heated plasmas: electron cyclotron resonance (ECR) discharges, helicon discharges, and surface wave discharges. In ECR discharges, a right circularly polarized wave (usually at microwave frequencies, e.g., 2.45 GHz) propagates along the DC magnetic field (about 850 G) in the ECR conditions, which provides wave energy absorption via collisionless heating. In the helicon discharges, an antenna radiates the whistler wave, which is subsequently absorbed in plasma. The helicon wave-heated discharges are excited at RFs (13.56 MHz), and weak magnetic fields of 20–200 G are required for wave propagation and absorption. In the surface wave discharges, a wave propagates along the plasma surface and is absorbed by collisional heating of the plasma electrons near the surface. The heated electrons then diffuse from the surface into the bulk plasma. The surface wave discharges can be excited by RF or microwave sources and do not require a DC magnetic field. The plasma potential with respect to the walls is low (about $5T_e$, similarly to ICP). It results in the effective generation of high-density plasmas at reasonable absorbed powers, which is attractive for intensive surface treatment. The ECR resonance between electromagnetic wave frequency ω and the electron cyclotron frequency $\omega_{Be} = eB/m$ allows electron heating that is sufficient for ionization at lower electric fields. The electron cyclotron frequency can be calculated as $f_{Be}(\text{MHz}) = 2.8 \cdot B(\text{G})$. Electron heating in the ECR takes place because the gyrating electrons rotate in phase with the right-hand polarized wave, seeing a steady electric field over many gyro-orbits. Pressure in the ECR discharge is supposed to be low to have low electron–neutral collision frequency ($\nu_{en} \ll \omega_{Be}$) and to provide the electron gyration long enough to obtain the energy necessary for ionization. A schematic of the **ECR discharge** with the microwave power injected along the axial magnetic field is shown in Fig. 4–87 (Lieberman & Lichtenberg, 1994). The magnetic field profile is chosen to provide effective propagation of the electromagnetic wave from the quartz window to the zone of the ECR resonance without reflection at high plasma densities. A special magnetic field profile can provide multiple ECR resonance positions as shown in the figure by the dashed line. Low-pressure gas introduced into the discharge chamber forms plasma, which streams and diffuses along the magnetic field toward a wafer (Fig. 4–87). Generated ions and radicals are able to provide a surface treatment effect. A magnetic field coil at the wafer holder can additionally be used to modify the uniformity of etch or deposition. Typical ECR microwave discharge parameters are pressure 0.5–50 mTorr, power 0.1–5 kW, microwave frequency 2.45 MHz, volume 2–50 L, magnetic field 1 kG, plasma density 10^{10}–10^{12} cm^{-3}, ionization degree 10^{-4}–10^{-1}, electron temperature 2–7 eV, ion acceleration energy 20–500 eV, and source diameter 15 cm (Matsuoka & Ono, 1988; Stevens et al., 1992).

Another HDP wave-heated discharge is the **helicon discharge** (Boswell, 1970; Chen, 1991). It is sustained by electromagnetic waves propagating in magnetized plasma in the helicon mode. The frequency is in the RF range 1–50 MHz. The phase velocity of electromagnetic waves in magnetized plasma can be much lower than the speed of light, which permits operation in the wave propagation regime with wavelengths comparable with the system size even at RFs. The magnetic field in the helicon discharges applied for material processing varies from 20 to 200 G, and it is much below the level of the magnetic field applied in ECR microwave discharges. The plasma density in these wave-heated discharges is 10^{11}–10^{12} cm^{-3}. Excitation of the helicon wave is provided by an RF antenna that couples to the transverse mode structure across an insulating chamber wall. The electromagnetic wave mode propagates along the plasma column in the magnetic field and is absorbed by plasma electrons. A schematic of a helicon discharge is illustrated in Fig. 4–88. A material processing chamber is located downstream from the plasma source. Plasma potentials in the helicon discharges are typically low (15–20 V). Advantages of the helicon discharges

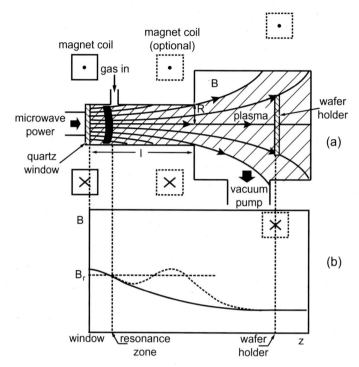

Figure 4–87. General schematic of high-profile electron-cyclotron resonance (ECR) discharge system: (a) configuration of the discharge system; (b) axial variation of magnetic field, showing one or more resonance zones.

with respect to ECR discharges are related to relatively low magnetic fields and applied frequency.

Electromagnetic surface wave discharges are based on the wave propagation along the surface of a plasma column and its absorption by the plasma (Smullin & Chorney,

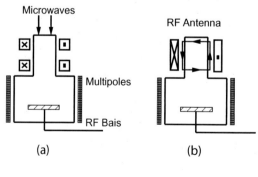

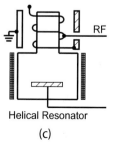

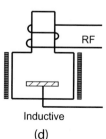

Figure 4–88. General schematics of helicon discharges.

1958; Trivelpiece & Gould, 1959; Moisan & Zakrzewski, 1991). The surface wave discharges can generate HDP with large diameter, about as large as 15 cm. The absorption lengths of the electromagnetic wave surface modes are long in comparison with the ECR discharge. The surface wave discharge typically operates at frequencies in the microwave range of 1–10 GHz without an imposed axial magnetic field. Planar rectangular configurations of this discharge have been developed by Komachi (1992). The electromagnetic surface wave, damping in both directions away from the surface, can be arranged in different configurations. One of them is a planar configuration on the plasma-dielectric interface. In another configuration, plasma is separated from a conducting plane by a dielectric slab. This planar system also admits the propagation of a surface wave that decays into the plasma region. Although this electromagnetic wave does not decay into the dielectric, it is confined within the dielectric layer by the conducting plane. Finally, a surface wave is also able to propagate in the cylindrical discharge geometry. The surface wave propagates in this case on a non-magnetized plasma column confined by a thick dielectric tube.

4.4.10. Non-Equilibrium Plasma-Chemical Microwave Discharges of Moderate Pressure

The highest energy efficiency can be achieved in non-equilibrium conditions with a contribution of vibrationally excited reagents (Section 3.6.2). Such regimes require the generation of plasma with special parameters: electron temperature T_e about 1 eV and higher than the translational temperature (≤ 1000 K); the ionization degree and specific energy input should be sufficiently high ($n_e/n_0 \geq 10^{-6}$, $E_v \approx 1$ eV/mol). Simultaneous achievement of these parameters is difficult. For example, low-pressure non-thermal discharges have too high a specific energy input (30–100 eV/mol), the streamer-based atmospheric pressure discharges have low specific power and average energy input, and powerful steady-state atmospheric pressure discharges usually operate close to quasi equilibrium. Moderate-pressure microwave discharges are able to generate non-equilibrium plasma with optimal parameters. The formation of an overheated filament within the plasma zone does not lead to an electric field decrease because of the skin effect (Vakar et al., 1981). The electrodynamic structure permits microwave discharges to be sustained in non-equilibrium conditions ($T_e > T_v \gg T_0$) at high-specific-energy inputs. For example, a steady-state microwave discharge (Krasheninnikov, 1981) is sustained in CO_2 at frequency 2.4 GHz, power 1.5 kW, pressure 50–200 Torr, and flow rate 0.15–2 sl/s. The specific energy input is 0.2–2 eV/mol and the specific power is up to 500 W/cm^3 (in conventional glow discharges it is about 0.2–3 W/cm^3). Vibrational temperature in the discharge can be on the level of 3000–5000 K and significantly exceeds rotational and translational temperatures, which are about 1000 K (Vakar, Krasheninnikov, & Tischenko, 1984; Givotov et al., 1985b).

Moderate-pressure microwave discharges operate in three major regimes: diffusive, contracted, and combined (Fig. 4–89). They occur at different pressures p and electric fields E (Fig. 4–90). The area above curve 1–1 corresponds to breakdown conditions; microwave discharges are sustained below the curve. Curve 2-2 corresponds to $(E/p)_{max}$ sufficient to sustain ionization on the plasma front even at room temperature. The **diffusive regime** (Fig. 4–89a) takes place at $E/p < (E/p)_{max}$ and relatively low pressures (20–50 Torr). The ratio (E/p) decreases at higher pressures, and temperature on the discharge front should increase to provide the necessary ionization rate. Curve 3-3 determines the minimum reduced electric field $(E/p)_{max}$, when the microwave discharge is still non-thermal. The **combined regime** (Fig. 4–89b) occurs at intermediate pressures (70–200 Torr) and $(E/p)_{min} < E/p < (E/p)_{max}$. Curve 4-4 separates the lower-pressure regime of the homogeneous discharge and higher-pressure regime of the combined discharge. In the

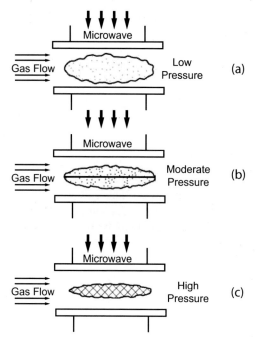

Figure 4–89. Transition from a diffusive microwave discharge (a) to a contracted microwave discharge (c) by increasing pressure; (b) illustrates an intermediate regime of the discharge.

combined regime, a hot thin filament is formed inside of a relatively big surrounding non-thermal plasma. The skin effect prevents penetration of the electromagnetic wave into the filament, and most of the energy is absorbed in the strongly non-equilibrium plasma. The combined regime sustains the non-equilibrium plasma with high values of ionization degree and specific energy input. This feature makes the regime interesting for plasma-chemical applications, where high energy efficiency is the most important factor. At high pressures (to the right of curve 5-5), radiative heat transfer and the discharge front over-heating become significant, and the contracted discharge converts into the **thermal regime** (Fig. 4–89c).

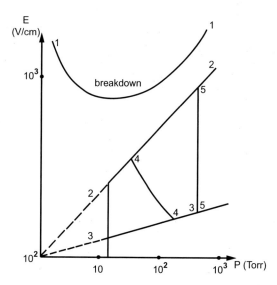

Figure 4–90. Three major regimes of non-equilibrium microwave discharges of moderate pressure.

4.5. NON-THERMAL ATMOSPHERIC PRESSURE DISCHARGES

4.5.1. Corona Discharges

A corona is a weakly luminous discharge, which appears at atmospheric pressure near sharp points, edges, or thin wires where the electric field is sufficiently large. The corona discharges are always non-uniform: a strong electric field, ionization, and luminosity are located in the vicinity of one electrode. Charged particles are dragged by the weak electric fields from one electrode to another to close the electric circuit. At the initial stages of the breakdown, the corona circuit is closed by displacement current rather than charged particle transport (Chyin, 2002). The corona can be observed around high-voltage transmission lines, around lightning rods, and masts of ships, where they are called "Saint Elmo's fire." Here we discuss only the main physical and engineering principles of the corona discharges; more details can be found in Loeb (1965), Goldman and Goldman (1978), Fridman and Kennedy (2004), and A. Fridman et al. (2005).

The polarity of an electrode where the high electric field is located distinguishes between **negative corona** (around the cathode) and **positive corona** (around the anode). Ignition of the negative corona is based on secondary emission from the cathode similarly to the Townsend breakdown (see Section 4.1.1):

$$\int_0^{x_{max}} [\alpha(x) - \beta(x)]dx = \ln\left(1 + \frac{1}{\gamma}\right), \tag{4–107}$$

where x_{max} is the distance from the cathode where $\alpha(x_{max}) = \beta(x_{max})$. The distance $x = x_{max}$ can also be considered to be the visible size of the corona. The ionization processes in a positive corona are related to the cathode-directed streamers (see Section 4.1.2). Ignition can be described by the generalized Meek breakdown criterion:

$$\int_0^{x_{max}} [\alpha(x) - \beta(x)]dx \approx 18\text{–}20. \tag{4–108}$$

Taking into account that $\ln(1/\gamma) \approx 6\text{–}8$, ignition of the positive corona requires slightly higher electric fields (Lowke & D'Alessandro, 2003). The igniting electric field for the coaxial electrodes in air can be calculated using the empirical **Peek formula**:

$$E_{cr}, \frac{kV}{cm} = 31\delta\left(1 + \frac{0.308}{\sqrt{\delta r (cm)}}\right), \tag{4–109}$$

where δ is the ratio of air density to the standard value and r is the radius of the internal electrode. The critical corona-initiating electric field in the case of two parallel wires in air can be calculated using a similar formula:

$$E_{cr}, \frac{kV}{cm} = 30\delta\left(1 + \frac{0.301}{\sqrt{\delta r (cm)}}\right). \tag{4–110}$$

The current-voltage characteristic $I(V)$ of a corona around a thin wire with radius r, length L, and characteristic size R can be expressed as (Fridman & Kennedy, 2004; A. Fridman et al., 2005)

$$I = \frac{4\pi \varepsilon_0 \mu V (V - V_{cr}) L}{R^2 \ln(R/r)}, \tag{4–111}$$

where V_{cr} is the corona ignition voltage and μ is mobility of the charged particles providing conductivity outside the active corona volume. Noting that the mobilities of positive and negative ions are nearly equal, the electric currents in positive and negative corona discharges

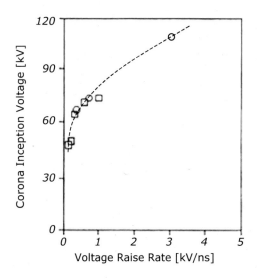

Figure 4–91. Voltage of corona breakdown as a function of the voltage raise rate.

are also close. A negative corona in gases without electron attachment (e.g., noble gases) provides much larger currents because electrons are able to rapidly leave the discharge gap without forming a significant space charge. Parabolic current-voltage characteristics like (4–111) are valid not only for thin wires but also for other corona configurations:

$$I = \text{const} \cdot V(V - V_{cr}). \qquad (4\text{–}112)$$

For example, a relevant expression for corona generated in atmospheric air between a sharp-point cathode with radius $r = 3$–50 μm and a perpendicular flat anode on the distance of $d = 4$–16 mm is

$$I, \mu A = \frac{52}{(d, \text{mm})^2}(V, \text{kV})(V - V_{cr}). \qquad (4\text{–}113)$$

The corona ignition voltage in this case, $V_{cr} = 2.3$ kV, does not depend on d (Goldman & Goldman, 1978). Based on the current-voltage characteristic (4–112), power released in the corona discharge is

$$P = \text{const} \cdot V^2(V - V_{cr}). \qquad (4\text{–}114)$$

For example, corona discharges generated in atmospheric-pressure air around the thin wire ($r = 0.1$ cm, $R = 10$ cm, $V_{cr} = 30$ kV) with voltage 40 kV releases relatively low power of about 0.2 W/cm. An increase in voltage and power of the continuous coronas results in transition to sparks (see Section 4.1).

4.5.2. Pulsed Corona Discharges

Applications of the continuous coronas are limited by low current and power, which results in a low rate of treatment of materials and exhaust streams. Increasing the corona power without transition to sparks becomes possible by using pulse-periodic voltages. The pulsed corona is one of the most promising atmospheric-pressure, non-thermal discharges. The streamer velocity is about 10^8 cm/s and exceeds by a factor of 10 the typical electron drift velocity in an avalanche. If the distance between electrodes is about 1–3 cm, the total time necessary for the development of avalanches, avalanche-to-streamer transition, and streamer propagation between electrodes is about 100–300 ns. Therefore, the voltage pulses of this

Figure 4–92. 10-kW pulsed corona discharge, view from one of six windows.

duration are able to sustain streamers and effective power transfer into non-thermal plasma without streamer transformations into sparks. It should be mentioned that even continuous coronas are characterized by non-steady-state effects, including ignition delay, flashing, and Trichel pulses (see Kennedy & Fridman, 2004).

For the pulsed corona discharges, the key is the development of the pulse power supplies generating sufficiently short voltage pulses with a steep front and short rise times. The nanosecond pulse power supplies generate pulses with duration 100–300 nsec, sufficiently short to avoid the corona-to-spark transition. The power supply should provide a high voltage rise rate (0.5–3 kV/ns), which results in a higher corona ignition voltage and higher power. As an illustration, Fig. 4–91 shows the corona inception voltage as a function of the voltage rise rate (Mattachini, Sani, & Trebbi, 1996). The high-voltage-rise rates also result in better efficiency of several plasma-chemical processes requiring higher electron energies. In such processes (for example, plasma cleansing of gas and liquid steams), high values of mean electron energy are necessary to decrease the fraction of the discharge power going to vibrational excitation of molecules, which stimulates ionization, electronic excitation, and dissociation of molecules. The nanosecond pulse power supplies used for the pulsed coronas include Marx generators, simple and rotating spark gaps, thyratrons, thyristors with possible further magnetic compression of pulses (Pu & Woskov, 1996), and special transistors for the high-voltage pulse generation. The pulsed corona can be relatively powerful and sufficiently luminous. One of the most powerful pulsed corona discharges (10 kW in average power) operates in Philadelphia, in the Drexel Plasma Institute (Fig. 4–92). More information about the physical aspects and applications of the pulsed corona discharge can be found in Penetrante et al. (1998), Korobtsev et al. (1997), Pu and & Woskov (1996), and Fridman and Kennedy (2004).

The most typical configuration of the pulsed and continuous coronas is based on thin wires, which maximize the active discharge volume. One of these configurations is illustrated in Fig. 4–93. Limitations of the wire configuration of the corona are related to the durability of the electrodes and to the non-optimal interaction of the discharge volume with incoming gas flow. From this point of view, it is useful to use another corona configuration (Park et al., 1998) based on multiple stages of pin-to-plate electrodes (see Fig. 4–94). A combination of the pulsed corona with other methods of gas treatment can be practical for applications. The pulsed corona can be combined with catalysis (Penetrante et al., 1998; Sobacchi et al., 2002, 2003) to achieve improved results in the treatment of automotive

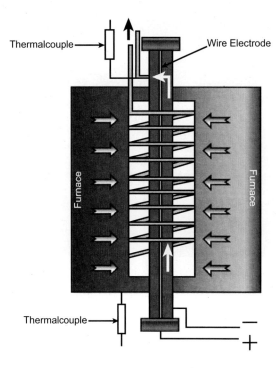

Figure 4–93. General schematic of a pulsed corona discharge in the wire-cylinder configuration with preheating.

exhausts and for hydrogen production from heavy hydrocarbons. Another interesting technological hybrid is related to the pulsed corona coupled with water flow. Such a system can be arranged either in the form of a shower, which is called a spray corona, or with a thin water film on the walls, which is usually referred to as a wet corona (Fig. 4–95).

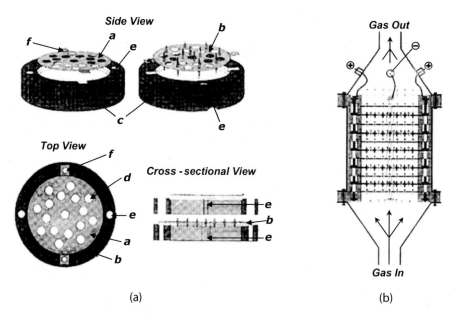

Figure 4–94. Configuration of a pin-to-plate pulsed corona discharge: (a) schematic of the electrodes and mounting blocks; and (b) cross-sectional view of the assembled plasma reactor and gas streams. In the figures, a, anode plates; b, cathode plates; c, mounting block; d, holes for the gas flow; e, holes for a connecting post; f, connection wings.

4.5. Non-Thermal Atmospheric Pressure Discharges

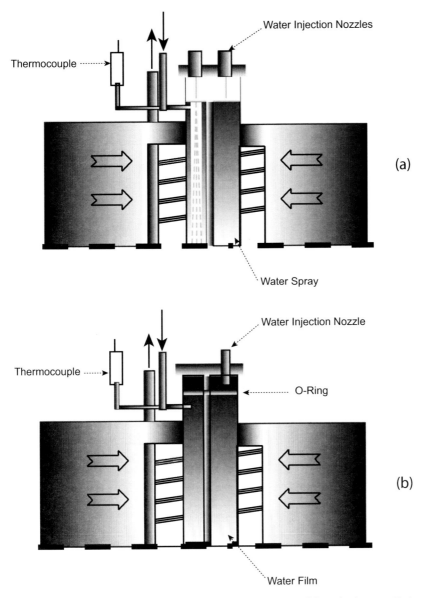

Figure 4–95. Schematics of (a) wet and (b) spray configurations of the pulsed corona discharges.

4.5.3. Dielectric Barrier Discharges

The corona-to-spark transition is prevented in a pulsed corona by employing nanosecond pulse power supplies. Another approach which avoids spark formation in streamer channels is based on the use of a dielectric barrier in the discharge gap that stops current and prevents spark formation. Such a discharge is called the DBD. The presence of a dielectric barrier precludes DC operation of DBD, which usually operates at frequencies of 0.05–500 kHz. Sometimes DBDs are called silent discharges due to the absence of sparks, which are accompanied by local overheating and the generation of local shock waves and noise. DBD has numerous applications because it operates at strongly non-equilibrium conditions at atmospheric pressure of different gases, including air, at reasonably high power levels and (in contrast to the pulsed corona) without using sophisticated pulse power supplies. DBD is

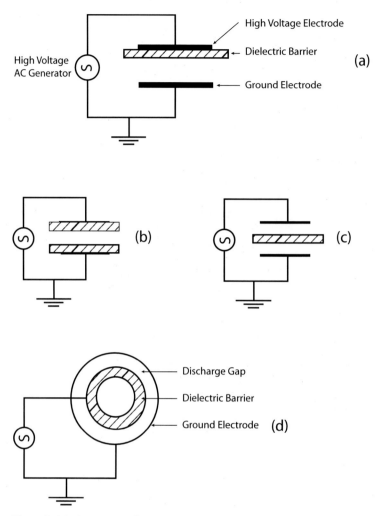

Figure 4–96. Common configurations of the dielectric-barrier discharges (DBDs): planar (a,b,c), cylindrical (d).

widely applied for ozone generation; in UV sources and excimer lamps; in polymer treatment (particularly to promote wettability, printability, and adhesion); for pollution control and exhaust cleaning from CO, NO_x, SO_2, and volatile organic compounds; for biological and medical applications; in CO_2 lasers; in plasma-assisted combustion; and so on. One of the largest expected DBD applications is related to plasma display panels for large-area flat screens, including plasma TVs. Specific DBD applications are discussed in the relevant sections of the book; see also Hood (1980), Uehara (1999), Penetrante et al. (1999), and Rosocha et al. (2004).

Important contributions to the fundamental understanding and industrial applications of DBD were made by Kogelschatz, Eliasson, and their group at ABB (see for example, Kogelschatz, Eliasson, & Egli, 1997). DBD, however, has a long history. It was first introduced by Siemens in 1857 to create ozone, which determined the main direction for DBD for many decades (Siemens, 1857). Important steps in understanding the physical nature of DBD were made by Klemenc, Hinterberger, and Hofer (1937), who showed that DBD occurs in a number of individual tiny breakdown channels, which are now referred to as microdischarges, and intensively investigated their relationship with streamers (see Section 4.1).

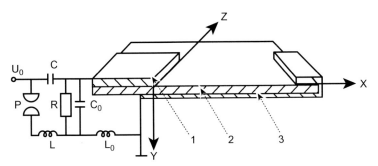

Figure 4–97. General schematic of a pulsed surface discharge: 1, initiating electrode; 2, dielectric; 3, shielding electrode.

The DBD gap includes one or more dielectric layers, which are located in the current path between metal electrodes. Two specific DBD configurations, planar and cylindrical, are illustrated in Fig. 4–96. Typical clearance in the discharge gaps varies from 0.1 mm to several centimeters. The breakdown voltages of these gaps with dielectric barriers are practically the same as those between metal electrodes. If the DBD gap is a few millimeters, the required AC driving voltage with frequency 500 Hz to 500 kHz is typically about 10 kV at atmospheric pressure. The dielectric barrier can be made from glass, quartz, ceramics, or other materials of low dielectric loss and high breakdown strength.

DBD is not uniform and consists of numerous microdischarges distributed in the discharge gap, often moving between and interacting with each other. The formation of the DBD microdischarges, their plasma parameters, and interaction between them resulting in the formation of the microdischarge patterns is an interesting and practically important subject, which has already been discussed (see Sections 4.1.4–4.1.7). In some special cases, particularly in helium, DBD can be uniform without any streamers and microdischarges, which will be discussed in the relevant section to follow. Special DBD modification, operating with one high-voltage electrode and keeping external conductive objects (like a human body, in medical applications) as a second electrode, is called a floating-electrode DBD (FE-DBD) (see Chapter 12).

4.5.4. Special Modifications of DBD: Surface, Packed-Bed, and Ferroelectric Discharges

Closely related to the DBD are **surface discharges**, which are generated at dielectric surfaces imbedded by metal electrodes and supplied by AC or pulsed voltage. The dielectric surface essentially decreases the breakdown voltage because of significant non-uniformities of the electric field and local overvoltage. An effective decrease of the breakdown voltage can be reached in the surface discharge configuration (called the **sliding discharge**; Baselyan & Raizer, 1997) with one electrode located on the dielectric plate and the other one partially wrapped around (see Fig. 4–97). The sliding discharge can be quite uniform on large surfaces with linear sizes over 1 m at voltages not exceeding 20 kV. The component of the electric field normal to the dielectric surface plays an important role in generating the pulse periodic sliding discharge, which does not depend essentially on the distance between electrodes along the dielectric.

Two different modes of surface discharges can be achieved by changing voltages: the complete mode or **sliding surface spark** and the incomplete mode or **sliding surface corona**. The sliding surface corona is ignited at voltages below the critical one and has a low current limited by charging the dielectric capacitance. Active volume and luminosity are localized near the igniting electrode and do not cover all the dielectric. The sliding surface

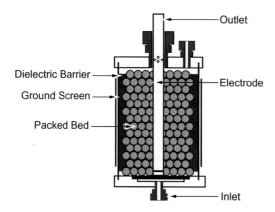

Figure 4–98. General schematic of a packed-bed corona discharge.

spark occurs at voltages exceeding the critical one corresponding to breakdown. The plasma channels actually connect electrodes of the surface discharge gap. At low overvoltages, the breakdown delay is about 1 μs; the multi-step breakdown phenomenon starts with the propagation of a direct ionization wave, which is followed by a possibly more intense reverse wave related to the compensation of charges left on the dielectric surface. At higher overvoltages, the breakdown delay becomes shorter (nanoseconds). The complete mode takes place immediately after the direct ionization wave reaches the opposite electrode. The surface discharge consists of many current channels in this regime. In general, the sliding spark surface discharge is able to generate the luminous current channels of very sophisticated shapes, usually referred to as **Lichtenberg figures**.

Packed-bed discharge is a combination of DBD and sliding surface discharges. High AC voltage (about 15–30 kV) is applied to a packed bed of dielectric pellets and creates a non-equilibrium plasma in the void spaces between the pellets (Birmingham & Moore, 1990; Heath & Birmingham, 1995). The pellets refract the electric field, making it non-uniform and stronger than the externally applied field by a factor of 10 to 250 depending on the shape, porosity, and dielectric constant of the pellets. A schematic of a packed-bed discharge is shown in Fig. 4–98. The inner electrode is connected to a high-voltage AC power supply operated on the level of 15–30 kV at a frequency of 60 Hz. The glass tube serves as a dielectric barrier to inhibit direct charge transfer between electrodes and as a plasma-chemical reaction vessel.

A special mode of DBD (called the **ferroelectric discharge**) uses ferroelectric ceramics with high dielectric permittivity ($\varepsilon > 1000$) as the dielectric barriers (Opalinska & Szymanski, 1996). Ceramics based on $BaTiO_3$ are most commonly employed for the discharges. The ferroelectric materials can have a significant dipole moment in the absence of an external electric field. Electric discharge from contact of gas with a ferroelectric sample was first observed by Robertson and Baily (1965) and analyzed by Kusz (1978). An external AC voltage leads to overpolarization of the ferroelectric material and reveals strong local electric fields on the surface, which can exceed 10^6 V/cm, and stimulates the discharge on the ferroelectric surfaces (Hinazumi, Hosoya, & Mitsui, 1973). The active volume of the ferroelectric discharge is located in the vicinity of the dielectric barrier. The discharge can be arranged, for example, using a packed bed of ferroelectric pellets. Non-equilibrium plasma is created in such a system in the void spaces between the pellets.

4.5.5. Spark Discharges

When a streamer connects the electrodes and neither a pulse power supply nor a dielectric barrier prevents further growth of the current, it opens an opportunity for development of a spark. The initial streamer channel is not very conductive and provides only a low current of about 10 mA. The fast ionization processes lead to a higher ionization degree, to higher current, and to intensification of the spark. When the streamer approaches the cathode, its electric field grows and stimulates intensive formation of electrons in the cathode vicinity. New ionization waves much more intense than the original streamer start propagating along the streamer channel but in the opposite direction (from the cathode to anode) with velocities up to 10^9 cm/s, which is referred to as the **back ionization wave**. The high velocity of the back ionization wave is not directly the velocity of electron motion but rather the phase velocity of the ionization wave. The back wave is accompanied by a front of intensive ionization and the formation of a plasma channel with sufficiently high conductivity to form a channel of intensive spark. The radius of the powerful spark channel grows up to about 1 cm, which corresponds to a spark current of 10^4–10^5 A and current densities of 10^4 A/cm^2. The plasma conductivity becomes high and a cathode spot can be formed on the electrode surface. The voltage between electrodes decreases, and the electric field becomes about 100 V/cm. If the voltage is supplied by a capacitor, the spark current also decreases after reaching a maximum. The detailed theory of the electric sparks has been developed by Drabkina (1951) and Braginsky (1958).

The sparks can be modified by the synergetic application of high voltages with laser pulses (Vasilyak et al., 1994; Asinovsky & Vasilyak, 2001). The laser beams can direct spark discharges not only along straight lines but also along more complicated trajectories. Laser radiation is able to stabilize and direct the spark channel in space due to three major effects: local preheating, local photo-ionization, and optical breakdown of the gas. At the laser radiation density 30 J/cm^2, the breakdown voltage decreases by an order of magnitude. The length of the laser-supported spark in these experiments was as high as 1.5 m. Photo-ionization by laser radiation is able to stabilize and direct coronas by means of local pre-ionization of the discharge channel without significant change of the gas density. UV laser radiation (for example, Nd laser or KrF laser) should be applied in this case. The most intensive laser effect on spark generation can be provided by the optical breakdown of the gases. The length of such a laser spark can exceed 10 m. The laser spark in pure air requires a power density of the Nd laser ($\lambda = 1.06$ μm) exceeding 10^{11} W/cm^2.

4.5.6. Atmospheric Pressure Glow Mode of DBD

DBD can be organized not only in a filamentary mode (Sections 4.5.3, and 4.5.4) but also in a homogeneous or glow mode. The APG mode of DBD with the average power densities comparable to those of filamentary discharges but operating without streamers is of significant interest for applications. It is also important that the glow mode of DBD can be operated at lower voltages (down to hundreds of volts). The streamers can be avoided when electric fields are below the Meek criterion (Section 4.1.3) and discharge operates in the Townsend ionization regime (Section 4.1.1). Secondary electron emission from dielectric surfaces, which sustains the Townsend ionization regime, relies upon adsorbed electrons (with binding energy only about 1 eV) that were deposited during a previous DBD excitation (high-voltage) cycle. If enough electrons "survive" the voltage switching time without recombining, they can trigger a transition to the homogeneous Townsend mode of DBD. The "survival" of electrons and crucial active species between cycles, or the **DBD memory effect**, is critical for organization of APG and depends on properties of the dielectric surface as well as the operating gas (Massines & Gouda, 1998). In electronegative gases, the memory effect is weaker because of attachment losses of electrons. If the memory effect is strong, the transition to Townsend mode can be accomplished, and uniform discharge can be organized without streamers (A. Fridman et al., 2005). A streamer DBD is easy to produce, whereas organization of APG at the same conditions is not always possible. This can be explained by taking into account that the streamer discharge is not sensitive to the secondary electron emission from the dielectric surface, whereas it is critical for operation of APG.

Glow discharges usually undergo contraction with pressure increase due to the thermal instability (Fridman & Kennedy, 2004). The thermal instability in the DBD APG is somewhat suppressed by using alternating voltage; thus, the discharge operates only when the voltage is high enough to satisfy the Townsend criteria, and the rest of the time the discharge is idle, which allows the dissipation of heat and active species. If time between excitation cycles is not enough for the dissipation then an instability will develop and discharge will undergo a transition to filamentary mode. The transition from APG to filamentary mode with increasing frequency has been observed by Croquesel et al. (2000).

The avalanche-to-streamer transition in the APG DBD depends on the level of pre-ionization. The Meek criterion (Section 4.1.3) is related to an isolated avalanche, whereas in the case of intensive pre-ionization, avalanches are produced close to each other and interact. If two avalanches occur close enough, their transition to streamers can be electrostatically prevented and the discharge remains uniform (A. Fridman et al., 2005). A modified Meek criterion of the avalanche-to-streamer transition can be obtained by considering two simultaneously starting avalanches with maximum radius R, separated by the distance L (α is Townsend coefficient, d is distance between electrodes):

$$\alpha d - (R/L)^2 \approx \text{const}, \tag{4–115}$$

$$\alpha d \approx \text{const} + n_e^p R^2 d. \tag{4–116}$$

In (4–116), the distance between avalanches is approximated using the pre-ionization density n_e^p. The constant in (4–115) and (4–116) depends on the gas: in air at 1 atm this constant equals 20. Thus, according to the modified Meek criterion, the avalanche-to-streamer transition can be avoided by increasing the avalanche radius and by providing sufficient pre-ionization. The type of gas is important in the transition to the APG. Helium is relevant for the purpose: it has high-energy electronic excitation levels and no electron energy losses on vibrational excitation, resulting in higher electron temperatures at lower electric fields. Also, fast heat and mass transfer processes prevent contraction and other instabilities at high pressures (Kanazawa et al., 1987, 1988; Lacour & Vannier, 1987; Honda, Tochikubo, & Watanabe, 2001).

4.5.7. APGs: Resistive Barrier Discharge

In the following sections, some special types of the APG characterized by a different level of uniformity will be discussed, starting with the resistive barrier discharge (RBD), proposed by Alexeff et al. (1999; Alexeff, 2001). The RBD can be operated with DC or AC (60 Hz) power supplies and is based on DBD configuration, where the dielectric barrier is replaced by a highly resistive sheet (few MΩ/cm) covering one or both electrodes (Laroussi et al., 2002; Thiyagarajan et al., 2005). The system can consist of a top-wetted high-resistance ceramic electrode and a bottom electrode. The highly resistive sheet plays a role of distributed resistive ballast, which prevents high currents and arcing. If He is used and the gap distance is not too large (5 cm or less), a spatially diffuse discharge can be maintained in the system for several tens of minutes. If 1% of air is added to helium, the discharge forms filaments. Even when driven by a DC voltage, the current signal of RBD is pulsed with the pulse duration of a few microseconds at a repetition rate of a few tens of kilohertz. When the discharge current reaches a certain value, the voltage drop across the resistive layer becomes large to the point where the voltage across the gas became insufficient to sustain the discharge. The discharge extinguishes and the current drops rapidly; then voltage across the gas increases to a value sufficient to reinitiate the RBD.

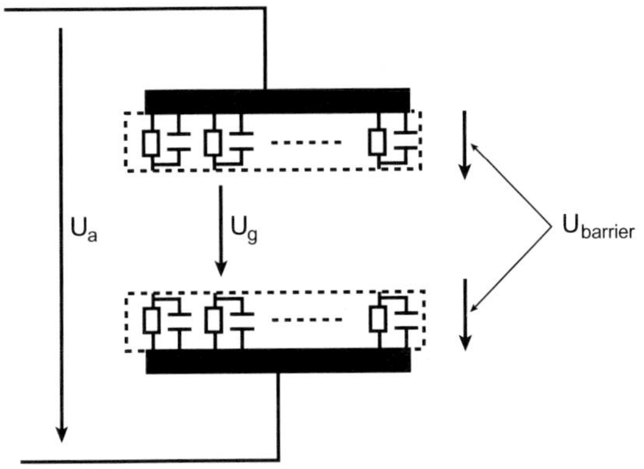

Figure 4–99. Illustrative equivalent circuit for the resistive-barrier discharge (RBD).

Wang et al. (2003) reproduced the pulsed RBD and analyzed the role and necessary parameters of the dielectric, assuming that the resistive layer is equivalent to distributed resistors and capacitors in parallel (Fig. 4–99). When the applied electric field (E_{gas}) reaches the value of the gas breakdown field (E_{on}), the distributed capacitors are rapidly charged, forcing E_{gas} to decrease (to E_{off}) and the discharge to extinguish. Then, the capacitors discharge through the distributed resistors and E_{gas} recovers to E_{on}, leading to reigniting the RBD. Based on the preceding analysis, the development of RBD has been numerically simulated by Wang et al. (2003) with different values of resistivity ρ and permittivity ε of the resistive layer, and taking for He $E_{on} = 300$ V/mm and $E_{off} = 125$ V/mm. Product $\rho\varepsilon_r$ for the resistive layer required for a kilohertz repetition rate of the RBD current pulses using 50 Hz AC voltage was calculated to be in a range of $\rho\varepsilon_r = (10^9 - 10^{11})$ $\Omega \cdot$ cm. The resistive barrier in RBD absorbs significant power to act as resistive ballast, which limits applications of the discharge.

4.5.8. One-Atmosphere Uniform Glow Discharge Plasma as Another Modification of APG

One-atmosphere uniform glow discharge plasma (OAUGDP) is a discharge developed initially at the University of Tennessee by Roth and his colleagues (Spence & Roth, 1994; Roth, 1995, 2001). OAUGDP is similar to a traditional DBD, but it can be much more uniform, which has been interpreted via the ion-trapping mechanism (Roth, 1995; Gardi, 1999). The discharge transition from filamentary to diffuse mode in atmospheric air has been analyzed (Jozef & Sherman, 2005) in the system shown in Fig. 4–100. The transition can lead not only to the diffuse mode but also into another non-homogeneous mode where the filaments are more numerous and less intense.

As discussed in Section 4.5.6, the stability of the filaments and transition to uniformity is related to the "memory effect." In particular, electrons deposited on the dielectric surface promote the formation of new streamers at the same place again and again by adding their own electric fields to the external electric field. According to Roth (2006), the key feature of OAUGDP promoting the transition to uniformity may be hidden in the properties of particular dielectrics that are not stable in plasma and probably become more conductive during a plasma treatment. In particular, plasma can further increase the conductivity of borosilicate glass used as a barrier, for example, by UV radiation. It transforms the discharge into the RBD (see Section 4.5.7) working at a high frequency in a range of 1–15 kHz.

Not only volumetric but also surface conductivity of a dielectric can promote DBD uniformity, if it is in an appropriate range. The memory effect can be suppressed by removing the negative charge spot formed by electrons of a streamer during the half-period of voltage

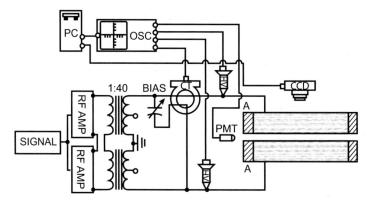

Figure 4–100. OAUGDP experimental setup schematic: A, water electrodes; HV, high-voltage probes; PMT, photomultiplier probe; CT, current transformer; BIAS, parasitic current elimination tool; C_v, variable capacitor; OSC, oscilloscope; PC, computer; SIGNAL, harmonic signal generator; RF AMP, radiofrequency power amplifier; CCD, digital camera.

oscillation (i.e., before polarity changes), and the surface conductivity can help with this. On the other hand, when the surface conductivity is very high, the charge cannot accumulate on the surface during the DBD current pulse of several nanoseconds and cannot stop the filament current. Fig. 4–101 presents the equivalent circuit for one DBD electrode with surface conductivity: for 10 kHz, it should be $0.1\,ms \gg RC \gg 1\,ns$, where resistances and capacitances are determined for the characteristic radius of streamer interaction (about 1 mm, Chirokov et al., 2006). More information can be found in Aldea et al. (2005), Golubovskii et al. (2002), and Fridman, Gutsol, and Cho (2007).

4.5.9. Electronically Stabilized APG Discharges

Electronic stabilization of APG has been demonstrated, in particular, by Aldea et al. (2005, 2006). Uniform plasma has been generated in argon DBD during the first cycles of voltage oscillations with relatively low amplitude (i.e., αd of about 3; see Section 4.5.6). The existence of the Townsend discharge at such low voltage requires an unusually high secondary electron emission coefficient (above 0.1). The high electron emission and the breakdown during the first low-voltage oscillations can be explained, probably, by taking into account the low surface conductivity of most polymers applied as barriers in the system that have very low surface conductivities (Fridman, Gutsol, & Cho, 2007). Surface charges occurring due to cosmic rays can then be easily detached by the applied electric field. A long induction time of the dark discharge is not required in this case, in contrast to the OAUGDP with glass electrodes.

Assuming that the major cause of the DBD filamentation is instability leading to the glow-to-arc transition (Fridman & Kennedy, 2004), it has been suggested by Aldea et al. (2005, 2006) to stabilize the glow mode using an electronic feedback to fast current variations. Fig. 4–102 shows the total current and voltage waveforms for DBD in argon with an active displacement current control achieved in the system. If the plasma is in series

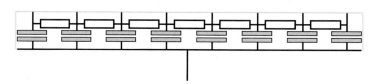

Figure 4–101. Illustrative equivalent circuit for one DBD electrode with surface conductivity.

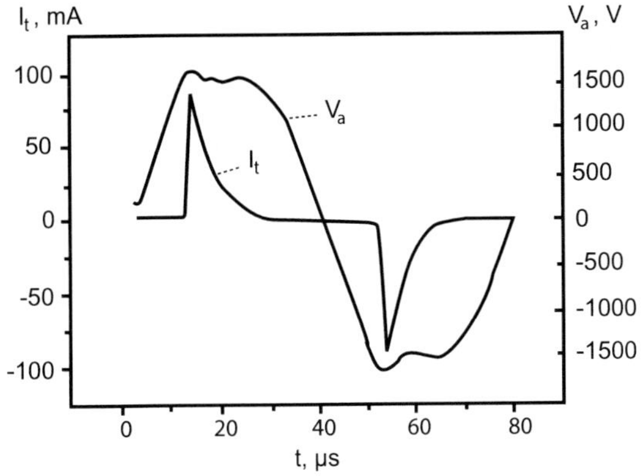

Figure 4–102. Total current and voltage waveforms for DBD in argon with active displacement current control.

with a dielectric, an RC circuit is formed. The filaments are characterized by higher current densities and a smaller RC constant. Therefore, the difference in RC constant can be used to "filter" the filaments because they react differently to a drop of the displacement current (displacement current pulse) of different frequency and amplitude. A simple LC circuit, in which the inductance is saturated during the pulse generation, has been used to generate the displacement current pulses. The method of electronic uniformity stabilization has been used by Aldea et al. (2005, 2006) for relatively high power densities (around 100 W/cm^3) and in a large variety of gases including Ar, N$_2$, O$_2$, and air.

4.5.10. Atmospheric-Pressure Plasma Jets

The RF atmospheric glow discharge or atmospheric-pressure plasma jet (APPJ) (Schütze, et al., 1998; Babayan, 2000; J. Park et al., 2001) is one of the most developed APG systems and has been used in particular for plasma-enhanced chemical vapor deposition (PECVD) of silicon dioxide and silicon nitride thin films. The APPJ can be organized as a planar and coaxial system with discharge gap of 1–1.6 mm, and frequency in the megahertz range (13.56 MHz). Laimer et al. (2005a,b) studied α- and γ-modes of APPJ in contaminated helium and argon. Interesting modifications in the APPJ design have been proposed by Moon, Choe, & Kang (2004). A theoretical model of the APPJ has been developed by Xiaohui and Raja (2003), Chirokov (2005), and Shi and Kong (2005). The APPJ is an RF-CCP discharge (see Section 4.4) that can operate uniformly at atmospheric pressure in noble gases, mostly in helium. In most APPJ configurations, electrodes are placed inside the chamber (Fig. 4–72a) and not covered by any dielectric in contrast to DBD. The discharge in pure helium has limited applications; therefore, various reactive species such as oxygen, nitrogen, nitrogen trifluoride, and so on are added. To achieve higher efficiency and a higher reaction rate, the concentration of the reactive species in the discharge has to be increased. If the concentration of the reactive species exceeds a certain level (which is different for different species but in all cases is on the order of a few percent), the discharge becomes unstable (Selwyn et al., 1999; Yang et al., 2005).

Distance between electrodes in APPJ is usually about 1 mm, which is much smaller than the size of the electrodes (about 10 cm × 10 cm). Therefore, the discharge can be considered one-dimensional, and effects of the boundaries on the discharge can be neglected. The electric current in the discharge is the sum of the current due to the drift of electrons and ions, and the displacement current. Since mobility of the ions is usually 100 times smaller than the electron mobility, the current in the discharge is mostly due to electrons. Considering that the typical ionic drift velocity in APPJ discharge conditions is about 3 × 10^4 cm/s, the time needed for ions to cross the gap is about 3 μs, which corresponds to a frequency of

245

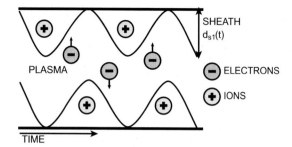

SHEATH
$d_{s1}(t)$

PLASMA

ELECTRONS

IONS

TIME

Figure 4–103. Illustration of the space-time structure of RF-CCP discharge.

0.3 MHz. The frequency of the electric field is much higher and, thus, ions in the discharge do not have enough time to move, whereas electrons move from one electrode to another as the polarity of the applied voltage changes. The typical space-time APPJ structure, showing the two sheaths and the positive column, is shown in Fig. 4–103.

The overall APPJ voltage consists of the voltage on the positive column, V_p (plasma voltage), and the voltage on the sheath, V_s. The voltage on the positive column, V_p, slightly decreases with an increase of the discharge current density (Vitruk, Baker, & Hall, 1992). It happens because a reduced electric field E/N in plasma is almost constant; $E/p \approx 2$ V/(cm · Torr) for helium discharge. If the density of neutral species is constant, the plasma voltage will be constant as well. But the density of neutrals slightly decreases with the electric current density since high currents cause a gas temperature to rise. At higher gas temperatures, a lower voltage is needed to support the discharge and subsequently the plasma voltage decreases. Sheath thickness can be approximated from the amplitude of electron drift oscillations $d_s = 2 \mu E/\omega \approx 0.3$ mm, where μ is an electron mobility, $\omega = 2\pi f$ is the frequency of the applied voltage, and E is an electric field in plasma. Assuming the secondary emission coefficient $\gamma = 0.01$, the critical ion density $n_{p(crit)}$ in helium RF before the α-to-γ transition is about 3×10^{11} cm^{-3}. The corresponding critical sheath voltage is about 300 V.

The typical power density for the APPJ helium discharge is on the level of 10 W/cm^2, which is approximately 10 times higher than that in the DBDs, including its uniform modifications. The power density that can be achieved in the uniform RF discharge is limited by two major instability mechanisms: thermal instability and α-to-γ transition instability (see Fridman & Kennedy, 2004). The critical power density for the thermal instability in APPJ is about 3 W/cm^2. Stable APPJ can be organized, however, with a power density exceeding this threshold. Suppressing of the thermal instability in the APPJ conditions is due to a stabilizing effect of the sheath capacitance (Vitruk et al., 1992), which can be described by the R parameter: the square of the ratio of the plasma voltage to the sheath voltage. The smaller R, creates a more stable discharge with respect to the thermal instability (see Fig. 4–104). For example, if $R = 0.1$, the critical power density with respect to thermal instability is 190 W/cm^2 (Chirokov, 2005). For the helium APPJ with $d_s = 0.3$ mm, $V_s = 300$ V, and $d = 1.524$ mm, the parameter R is $R = (V_p/V_s)^2 = 0.36$, which corresponds to the critical discharge power density 97 W/cm^2 (Chirokov, 2005). Thus, the APPJ discharge remains thermally stable in a wide range of power densities as long as the sheath remains intact. The major instability of the APPJ and loss of its uniformity, therefore, is mostly determined by the α-to-γ transition, or in simple words via breakdown of the sheath, which is to be the focus of later discussion.

The α-to-γ transition in APPJ happens because of the Townsend breakdown (4–4) of the sheath, which occurs when ion density and sheath voltage exceed the critical values ($n_{p(crit)} = 3 \times 10^{11}$ cm^{-3}, $V_s = 300$ V; see Chirokov, 2005; Raizer et al., 1995). Chirokov (2005) described the behavior of the APPJ discharge under the following conditions: in pure

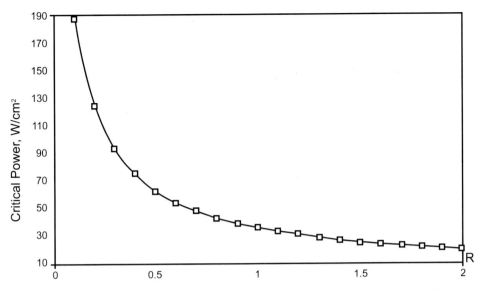

Figure 4–104. Critical power density for the discharge stability as a function of the ratio R. Effect of sheath on the thermal instability of plasma. Area above the curve indicates unstable discharge.

helium and in helium with additions of nitrogen and oxygen at different applied voltages and frequencies, with different coefficients of the secondary electron emission. It has been shown that the main mechanism of the discharge instability is the sheath breakdown that eventually leads to the thermal instability. Therefore, more effective discharge cooling would not solve the stability problem because it did not protect the discharge from the sheath breakdown. Nevertheless, the discharge cooling is important since the sheath breakdown depends on the reduced electric field that increased with temperature. Despite the fact that the thermal stability of helium discharge is better compared to the discharge with oxygen addition, a higher power is achieved with oxygen addition, which prevents the sheath breakdown.

In summary, we should note that it is much easier to generate a uniform APG in helium and argon than in other gases, especially electronegative ones. The effect cannot be explained only by the high thermal conductivity of helium. It is more important for uniformity that noble-gas-based discharges have a significantly lower voltage, and therefore lower power density, which helps to avoid thermal instability. Pure nitrogen provides better conditions for uniformity than does air. The presence of oxygen results in the electron attachment, which causes higher voltage, higher power, and finally leads to thermal instability. More details on the subject can be found in Fridman et al. (2007a) and in Chirokov (2005).

4.6. MICRODISCHARGES

4.6.1. General Features of Microdischarges

The term microdischarge is not clearly defined; sometimes it just indicates the generation of plasma smaller than 1 mm. According to such a definition, even DBD (Section 4.5.3) can be considered a microdischarge: the DBD gaps are often smaller than 1 mm, and the DBD filaments have a typical diameter of 0.1 mm. However, the applied DBD systems are large: the polymer-film ozonizers have a size of several meters. Thus, there is no reason to consider a discharge system as a microdischarge unless scaling down brings some new properties and physics. For example, scaling down with a similarity parameter (pd) should not change significantly the properties of discharges. Taking into account that conventional non-equilibrium discharges are operated in the *pd* range around 10 cm · Torr (Sections 4.1 and 4.2), it is clear that the organization of strongly non-equilibrium discharges at atmospheric pressure should be effective in submillimeter sizes.

Some specific new properties can be achieved by scaling down the plasma size to submillimeters:

1. Size reduction of non-equilibrium plasmas permits an increase of their power density to a level typical for thermal discharges because of intensive heat losses of the tiny systems.
2. At high pressures, volumetric recombination and especially three-body processes can go faster than diffusion losses, which results in significant changes in plasma composition. For example, high-pressure microdischarges can contain a significant amount of molecular ions in noble gases (Kushner, 2004).
3. Sheathes (about 10–30 μm at atmospheric pressure) occupy a significant portion of the plasma volume.
4. Plasma parameters move to the "left" side of the Paschen curve (Section 4.1.1). The Paschen minimum is about *pd* = 3 cm · Torr for some gases; therefore, a 30-μm gap at 1 atm corresponds to the left-hand side of the curve.

The specific properties can lead to positive differential resistance of a microdischarge, which allows the support of many discharges in parallel from a single power supply without using multiple ballast resistors. Another important consequence is the relatively high electron energy in the microplasmas, which are mostly strongly non-equilibrium. A significant development in the fundamentals and applications of microdischarges has been achieved recently by B. Farouk and his group in the Drexel Plasma Institute.

4.6.2. Micro-Glow Discharge

Conventional glow discharges (Section 4.2) operate at low pressure, which permits effective cooling by the walls. At atmospheric pressure and macroscopic sizes, the glow discharges can be stabilized by intensive convective cooling (Akishev et al., 2005; Arkhipenko et al., 2002, 2005; Fridman, Gutsol, & Cho, 2007). The conventional glow discharges can be also cooled down and stabilized at atmospheric pressure like the microdischarges due to their small sizes (see Section 4.6.1), which is discussed next. An atmospheric-pressure DC micro-glow discharge has been generated between a thin cylindrical anode and a flat cathode (Staack et al., 2005a,b; Farouk et al., 2006). The discharge has been studied using an interelectrode gap spacing in the range of 20 μm to 1.5 cm so that one can see the influence of the discharge scale on the plasma properties. Current-voltage characteristics, visualization of the discharge (Fig. 4–105), and estimations of the current density indicate that the discharge operates in the normal glow regime (see Section 4.2.4). Emission spectroscopy and gas-temperature measurements using the second positive band of N_2 indicate that the discharge generates non-equilibrium plasma. For 0.4- and 10-mA discharges, rotational temperatures are 700 and 1550 K, whereas vibrational temperatures are 5000 and 4500 K, respectively. It is possible to distinguish a negative glow in Fig. 4–105, as well as Faraday dark space and positive column regions of the discharge. The radius of the column is about 50 μm and remains relatively constant with changes in the electrode spacing and discharge current. Such a radius permits balance of the heat generation and conductive cooling to help prevent thermal instability and the transition to an arc.

Generally there is no significant change in the current-voltage characteristics of the discharge for different electrode materials or polarity. There are several notable exceptions to this for certain configurations. (a) For a thin upper electrode wire (<100 μm) and high discharge currents, the upper electrode melts. This occurs when the wire is the cathode,

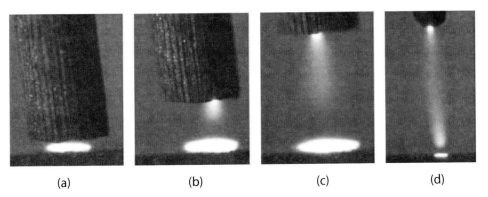

(a)   (b)   (c)   (d)

Figure 4–105. Images of glow discharge in atmospheric-pressure air with electrode spacing: (a) 0.1 mm, (b) 0.5 mm, (c) 1 mm, and (d) 3 mm.

indicating that the heating is due to energetic ions from the cathode sheath and not resistive heating. (b) For a medium-sized wire (~200 μm) as a cathode, the width of the negative glow increases as the current increases until it covers the entire lower surface of the wire. If the current is further increased, the negative glow "spills over" the edge of the wire and begins to cover the side of the wire. This effect is similar to the transition from a normal glow to an abnormal glow in low-pressure glow discharges (Section 4.2). However, there is no increase in the current density since the cathode area is not limited. For sufficiently large electrode wires this effect does not occur. (c) In air discharges with oxidizable cathode materials, the negative glow moves around the cathode electrode leaving a trail of oxide coating behind until there is no clean surface within reach of the discharge and the discharge extinguishes.

Fig. 4–106 shows the characteristics of the discharges corresponding to powers between 50 mW and 5 W (Staack et al., 2005a,b). For a small spacing of electrodes the current-voltage

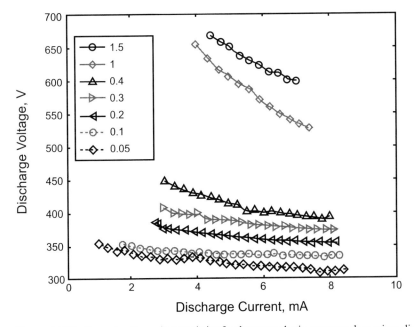

Figure 4–106. Current-voltage characteristics for the atmospheric-pressure glow-micro-discharge in air at several electrode spacings.

Table 4–12. Parameters of a DC Atmospheric-Pressure Micro-Glow Discharge in Air at Currents 0.4 and 10 mA

Microdischarge and micro-plasma parameters	Microdischarge current (mA)	
	0.4	10
Electrode spacing (mm)	0.05	0.5
Microdischarge voltage (V)	340	380
Microdischarge power (W)	0.136	3.8
Diameter of negative glow (μm)	39	470
Positive column diameter (μm)	–	110
Electric field in the positive column (kV $\cdot$ cm^{-1})	5.0	1.4
Translational gas temperature (K)	700	1550
Vibrational gas temperature (K)	5000	4500
Negative glow current density (A $\cdot$ cm^{-2})	33.48	5.8
Positive column current density (A $\cdot$ cm^{-2})	–	105
Reduced electric field E/n_0 (V $\cdot$ cm^2)	4.8×10^{-16}	3×10^{-16}
Electron temperature T_e (eV)	1.4	1.2
Electron density n_e in negative glow (cm^{-3})	3×10^{13}	7.2×10^{12}
Electron density n_e in positive column (cm^{-3})	–	1.3×10^{14}
Ionization degree in negative glow	3×10^{-6}	15×10^{-7}
Ionization degree in positive column	–	3×10^{-5}

characteristics are relatively flat, which is consistent with the normal discharge mode. For a normal glow discharge in air, the potential drop at the normal cathode sheath is around 270 V. A higher voltage drop occurs mostly in the positive column. For larger electrode spacing, the current-voltage characteristics have a negative differential resistance dV/dI. This is because the discharge temperature increases with gap length, resulting in growth of conductivity (Staack et al., 2006). A short discharge loses heat through the thermal conductivity of electrodes. A long discharge cooling is not efficient because the thermal conductivity of the gas is much lower than that of metal electrodes; therefore, the temperature of the long discharge is higher. Such a behavior demonstrates a new property related to the size reduction to micro-scale. Diffusive heat losses can balance the increased power density only at elevated temperatures of a microdischarge and the traditionally cold glow discharge becomes "warm."

Atmospheric-pressure micro-glow discharges in different gases are generally similar to those in air, with some exceptions observed by Staack et al. (2005a,b): (a) Each gas has a distinct discharge color and specific spectral lines. (b) In helium, the scale of the discharge in every dimension is larger than that for a discharge in air with the same current. (c) In helium, the maximum electrode spacing achieved is 75 mm. (d) In hydrogen, the column has standing striations for some conditions (see Fig. 4–107). (e) In both hydrogen and helium, higher discharge currents can be achieved without the transition to an arc or overheating of the electrodes. (f) The discharge in argon is narrower and is prone to transition to an arc at lower currents than in air. Table 4–12 summarizes the micro-glow discharge parameters corresponding to currents of 0.4 and 10 mA (Staack et al., 2005a,b). Thus, the atmospheric-pressure DC microdischarge is a normal glow discharge thermally stabilized by its size and maintaining a high degree of vibrational–translational non-equilibrium. It should be pointed out that the micron-sized precise micro-glow discharges or their arrays can be effectively used for direct micro-scale surface treatment without application of any masks.

4.6. Microdischarges

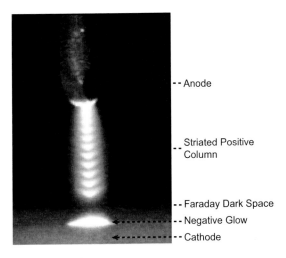

Figure 4–107. Image of the micro-glow discharge in atmospheric-pressure hydrogen. Positive column and negative glow are clearly visible. In addition, standing striations can be seen in the positive column (Staack et al., 2005).

4.6.3. Micro-Hollow-Cathode Discharge

HCDs (see Section 4.2.7) can be effectively organized at the micro-scale (see, for example, White, 1959; Kushner, 2004, 2005; Kim, Iza, & Lee, 2006; Becker et al., 2005). The micro-HCDs, similarly to the conventional HCDs, are interesting for applications because of their ability to generate high-density plasma. While conventional HCDs are organized at low pressures and macro-scale, the micro-HCD can operate at atmospheric pressure in agreement with the pD similarity. The micro-HCD can be effectively arranged in the form of special arrays. If pD is in the range of $0.1–10$ Torr · cm, the discharge develops in stages (see Fig. 4–108). At low currents, a "pre-discharge" is observed, which is a glow discharge with the cathode falling outside the hollow-cathode structure. As the current increases and the glow discharge starts its transformation into the abnormal glow with a positive differential resistance, a positive space charge region moves closer to the hollow-cathode structure and can enter the cavity. After that, the positive space charge in the cavity acts as a virtual anode, resulting in the redistribution of the electric field inside the cavity. At the center of the cavity, a potential well for electrons appears, forming a cathode sheath along the cavity walls. At this transition from the axial pre-discharge to a radial discharge, the sustaining voltage drops (e.g., see 2 mA point in Fig. 4–108). Sometimes this transition is not so sharp, in which case a negative slope in the current-voltage characteristic (i.e., a negative differential resistance) appears, which is traditionally referred to as the "hollow-cathode mode." From this standpoint, the microdischarges developed at Old Dominion University (Schoenbach et al., 1996, 1997; see Fig. 4–109) and Uppsala University (Baránková & Bárdos, 2002; see

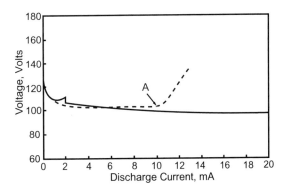

Figure 4–108. Current-voltage characteristics of a 0.75-mm-diameter cavity cathode (solid curve) and plane cathode (dashed curve) having 20 times the area. The voltage discontinuity at the current 2 mA marks the point at which the discharge transfers from the face of the cathode to the cavity (working gas, neon; pressure, 100 Torr).

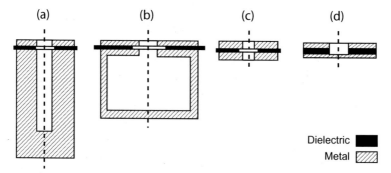

Figure 4–109. Hollow-cathode discharge geometries: system (d) demonstrates hollow-cathode mode when operating in argon with pressure at least as high as 250 Torr.

Fig. 4–110) can be qualified as micro-HCDs. A more detailed review of the micro-HCDs, and interpretations of the micro-HCD effect can also be found in Fridman, Gutsol & Cho (2007).

4.6.4. Arrays of Microdischarges: Microdischarge Self-Organization and Structures

The power of a microdischarge is so small that individual microdischarges have limited applications. Thus, most industrial applications require microdischarge arrays or micro-plasma integrated structures. The plasma TV is an example of such a complex structure. The simplest structure may be the one that consists of multiple identical microdischarges electrically connected in parallel. For a stable operation of such structures, each discharge should have a positive differential resistance (see Fig. 4–110). Most microdischarges have this property as a result of a significant increase in power losses with a current increase. One example is the array consisting of microdischarges with inverted, square pyramidal cathodes (Fig. 4–111). An optical micrograph of the 3×3 array of microdischarges of 50 μm $\times$ 50 μm each, separated (center-to-center) by 75 μm, has been made by Park et al. (2001). All of the microdischarges (700 Torr of Ne) have a common anode and cathode (i.e., the devices are connected in parallel). The ignition voltage and current for the array are 218 V

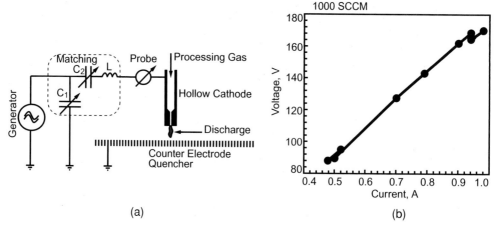

Figure 4–110. Schematic diagram and $V\text{–}I$ characteristics of the atmospheric-pressure cylindrical RF hollow-cathode discharge with cathode diameter 0.4 mm. Neon flow in the system is 1000 sccm.

4.6. Microdischarges

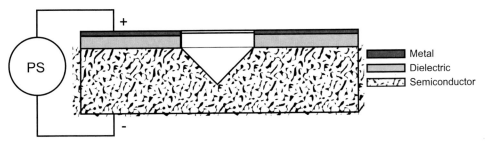

Figure 4–111. Schematic of a single micro-discharge from the arrays described by Park et al. (2001) and numerically simulated by Kushner (2004).

and 0.35 μA. The array has been able to operate at a high-power loading (433 V and 21.4 μA); emission from each discharge is spatially uniform. Another example of microdischarge arrays is the so-called **"fused" hollow cathode** (FHC; Baránková & Bárdos, 2002), which is based on the simultaneous RF generation of HCD plasmas in an integrated open structure with flowing gas. The resulting discharges are stable, homogeneous, luminous, and volume filling without streamers. The power is on the order of one Watt per square centimeter of the electrode structure area. Experiments have been carried out with the system having a total discharge area of 20 cm². The concept of the source is extremely suitable for scaling up for different gas throughputs. In some cases it is beneficial to connect microdischarges in series, for example, to increase a radiant excimer emittance. Such a system can consist of two HCDs with negative differential resistance (Kushner, 2005) and be applied for the excimer laser. Laser devices require a long gain length to achieve the threshold. One of the strategies to produce the long gain length is to alternately stack cathode and anode structures in a single bore. The dynamics of the multi-stage microdischarge devices with 100–200-μm diameter with a current of a few milliamperes and a pressure of several hundreds of Torr has been investigated by Kushner (2005).

The interesting phenomenon of self-organization has been observed in a DC microdischarge of geometry similar to that presented in Fig. 4–109d, but with a larger opening and opposite polarity (hole in the anode); Schoenbach, Moselhy, & Shi (2004). The cathode in the system consists of a molybdenum foil 250 μm thick. A 250-μm-thick layer of alumina with a circular opening is placed on the top of the cathode. The anode, a 100-μm-thick molybdenum foil, is placed on the top of the dielectric with the same size circular opening. The diameter of the dielectric and anode opening varies from 0.75 to 3.5 mm. Xenon is used as the plasma gas at pressures of 75–760 Torr. The glow discharge structure in this electrode configuration has been reduced to only the cathode fall and negative glow, with the negative glow plasma conducting the current radially to the circular anode. Therefore, the discharge is a called a cathode boundary-layer microdischarge (Schoenbach, Moselhy, & Shi, 2004; Becker et al., 2005). Photographs indicate a transition from homogeneous plasma to structured plasma when the current is reduced below a critical value depending on the pressure. The plasma pattern consists of the filamentary structures arranged in concentric circles. The structures (see Fig. 4–112; Schoenbach, Moselhy, & Shi, 2004) are most pronounced at pressures below 200 Torr.

Non-equilibrium microdischarges at atmospheric pressure can also exist at relatively high powers. For example, Dreizin (2004) made a **micro-arc discharge** in a gap of 0.01–0.1 mm with a voltage of 1.5–4.5 V and current of 40–120 A. This arc is similar to some extent to the cathode boundary-layer microdischarge discussed earlier because it exists without an "arc column." The main difference between the two discharges is that, in the micro-arc, up to 95% of the electrical energy is transferred to the anode (similar to an e-beam),

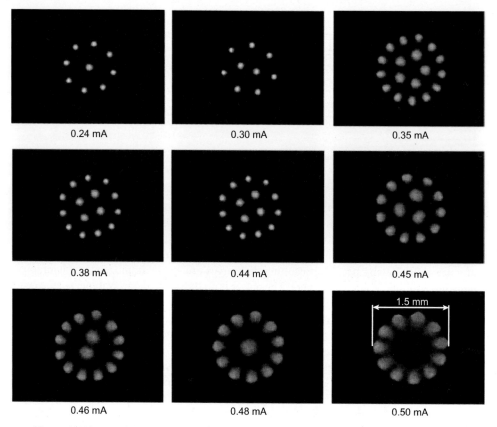

Figure 4–112. Formation of the plasma patterns in a xenon microdischarge at reduced current and pressure of 75 Torr. Diameter of the anode opening in this case is 1.5 mm (Schoenbach, Moselhy, & Shi, 2004).

which is qualitatively different from the case of the cathode boundary-layer discharge. The micro-arc has been used to generate metal droplets and nanopowders as well as for local hardening of metal surfaces.

4.6.5. Kilohertz-Frequency-Range Microdischarges

AC microdischarges can be organized at all possible frequencies. Low- and medium-frequency AC microdischarges are related to DBDs (see Section 4.5.3). A relatively simple large-area plasma source based on the micro-DBD approach has been developed by Sakai, Kishimoto, & Tachibana (2005). An integrated structure called the **coaxial-hollow micro dielectric-barrier discharges (CM-DBD)** has been made by stacking two metal meshes covered with a dielectric layer of alumina with thickness of about 150 μm. The test panel (diameter 50 mm) with hundreds of hollow structures (0.2 mm × 1.7 mm) has been assembled. He or N_2 have been used as the plasma gases at pressures of 20–100 kPa, and voltage below 2 kV even at the maximum pressure. Bipolar square-wave voltage pulses have been applied to one of the mesh electrodes. The pulse duration of both positive and negative voltages varied from 3 to 14 μs; intermittent time was 1 μs and repetition frequency was 10 kHz. In each coaxial hole, the discharge occurs along the inner surface. The intensity of each microdischarge is uniform over the whole area. The extended glow with a length of some millimeters is observed in He but not in N_2. The electron density in He at 100 kPa is about 3×10^{11} cm^{-3}. Thus, the CM-DBD configuration has a rather low operating voltage (typically 1–2 kV); the scaling parameter pd is several tens of Pa · m, corresponding to the

4.6. Microdischarges

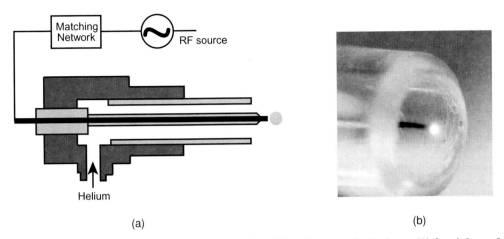

(a)   (b)

Figure 4–113. Schematic (left) and picture (right) of the plasma needle discharge (Kieft, v.d. Laan, & Stoffels, 2004).

Paschen minimum. Plasma in the system is stable over a wide range of external parameters without filamentation or arcing.

Another kilohertz-range microdischarge is the so-called **capillary plasma electrode (CPE)** discharge (Becker et al., 2005; Foest, Schmidt, & Becker, 2006). The CPE discharge uses an electrode design, which employs dielectric capillaries that cover one or both electrodes. Although the CPE discharge looks similar to a conventional DBD, the CPE discharge exhibits a mode of operation that is not observed in DBD: the "capillary jet mode." The capillaries with diameter 0.01–1 mm and a length-to-diameter (L/D) ratio from 10:1 to 1:1 serve as plasma sources and produce jets of high-intensity plasma at high pressure. The jets emerge from the end of the capillary and form a "plasma electrode." The CPE discharge displays two distinct modes of operation when excited by pulsed DC or AC. When the frequency of the applied voltage pulse is increased above a few kilohertz, one first observes a diffuse mode similar to the diffuse DBD as described by Okazaki and Kogoma (1993). When the frequency reaches a critical value (which depends on the L/D value and the plasma gas), the capillaries become "turned on," and bright intense plasma jets emerge from the capillaries. When many capillaries are placed in close proximity to each other, the emerging plasma jets overlap and the discharge appears uniform. This "capillary" mode is the preferred mode of operation of the CPE discharge and is somewhat similar to the FHC. At high frequency even dielectric capillaries can work as hollow cathodes, because for CCP-RF plasma in the gamma mode, a dielectric surface is also a source of secondary emitted electrons similar to metal cathodes in glow or hollow-cathode discharges.

4.6.6. RF Microdischarges

In the RF range (13.56 MHz), the so-called **plasma needle** (Fig. 4–113) attracts interest due to its potential medical applications (Kieft, v.d. Laan, and Stoffels, 2004). This discharge has a single-electrode configuration and is operating in helium. It operates near room temperature, allows the treatment of irregular surfaces, and has a small penetration depth. The plasma needle is capable of bacterial decontamination and localized cell removal without causing necrosis to the neighboring cells. Areas of detached cells can be made with a resolution of 0.1 mm. Radicals and ions from the plasma as well as UV radiation interact with the cell membranes and cell adhesion molecules, causing cell detachment. The plasma needle is confined in a plastic tube through which helium flow is supplied (Fig. 4–113). The discharge is entirely resistive with voltage 140–270 V_{rms}. The electron density is

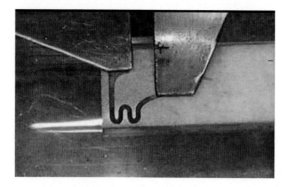

Figure 4–114. Miniaturized ICP jet source.

about 10^{11} cm^{-3}. Optical measurements show substantial UV emission in the range 300–400 nm; radicals O and OH have been detected. At low helium flow rates, densities of molecular species in the plasma are higher.

Conventional RF discharges, both inductively coupled (ICP) and capacitively coupled (CCP), have also been organized in micro-scale at atmospheric pressure (see Fridman, Gutsol & Cho, 2007). These plasmas are non-equilibrium because of their small sizes and effective cooling. Reduction in size requires a reduction in wavelength and increase of frequency. A **miniaturized atmospheric-pressure ICP jet** has been developed for a portable liquid analysis system (Ichiki, Koidesawa, & Horiike, 2003). The plasma device is a planar ICP source (Fig. 4–114), consisting of a ceramic chip with an engraved discharge tube and a planar metallic antenna in a serpentine structure. The chip consists of two dielectric plates with an area of 15 mm × 30 mm. A discharge tube (1 mm × 1 mm × 30 mm [h/w/l]) is mechanically engraved on one side of the dielectric plate. A planar antenna is fabricated on the other side of the plate. The atmospheric-pressure plasma jet with density about 10^{15} cm^{-3} is produced using a compact very-high-frequency (VHF) transmitter at 144 MHz and power 50 W. The electronic excitation temperature in the system is 4000–4500 K.

Ichiki, Taura, and Horiike (2004) also developed a miniaturized VHF-driven ICP jet source for the production of high-temperature and high-density plasmas in a small space and application to localized and ultra-high-rate etchings of silicon wafers. The plasma source consists of a discharge tube of 1 mm diameter with a fine nozzle of 0.1 mm diameter at one end and a three-turn solenoidal antenna around it. The electron density of

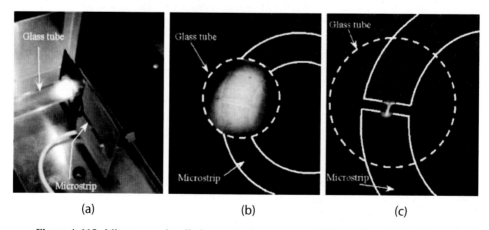

| (a) | (b) | (c) |

Figure 4–115. Microwave microdischarge: Ar-plasma, power 1 W. (a) View of the device operating at 9 Torr; (b) close-up view of the diffuse plasma at 20 Torr; (c) close-up view of the confined plasma at 760 torr (Iza & Hopwood, 2003).

atmospheric-pressure argon plasma jets blowing out from the nozzle is 10^{14}–10^{15} cm^{-3}. By adding halogens into the downstream region of the argon plasma jets, high-speed etching of fine holes of several hundred microns in diameter has been organized. Etching rates of 4000 mm/min and 14 mm/min have been achieved for silicon wafers and fused silica glass wafers, respectively.

4.6.7. Microwave Microdischarges

The low-power microwave micro-plasma source based on a microstrip split-ring 900-MHz resonator (Fig. 4–115) and operating at pressures of 0.05 Torr to 1 atm has been developed by Iza and Hopwood (2003). Argon and air discharges can be self-started in the system with power less than 3 W. An ion density of 1.3×10^{11} cm^{-3} in argon at 400 mTorr can be produced with only 0.5 W of power. Atmospheric discharges can also be sustained in argon with 0.5 W. The low power allows for portable air-cooled operation of the system. This kind of micro-plasma source can be integrated into portable devices for applications such as bio-MEMS sterilization, small-scale material processing, and micro-chemical analysis systems.

We should mention that the highest frequency range discharges are optical ones. The optical discharges or so-called laser sparks are always microdischarges, because they are formed in the focus of a lens that concentrates the laser light (see Section 4.5.5). This chapter reviewed only some of the major discharges applied in plasma chemistry. Much more information on the subject can be found in special publications focused on physics and engineering of gas discharge plasmas; see, for example, Fridman and Kennedy (2004). Some discharges including discharges in the liquid phase, nanosecond pulsed discharges, electron beam discharges, and so on, are to be discussed in sections related to their special plasma-chemical applications.

PROBLEMS AND CONCEPT QUESTIONS

4–1. Townsend Breakdown Mechanism. Using relations (4–4), (4–6), and Table 4–1, compare the Townsend breakdown conditions in molecular gases (take air as an example) and monatomic gas (take argon as an example). Comment on the big difference between these two cases.

4–2. Streamer Propagation Velocity. In the framework of the model of an ideally conducting streamer channel (4–12), estimate the difference between streamer velocity (at l/r ratio of about 100) and electron drift velocity in the external electric field.

4–3. Engel-Steenbeck Model, Diffusion-Controlled Discharges. Calculate the electron temperature in a non-thermal discharge in nitrogen with pressure 1 Torr and radius 1 cm, based on the Engel-Steenbeck model. Estimate the reduced electric field corresponding to the received electron temperature. Compare this reduced electric field with the one required for nitrogen breakdown in similar conditions.

4–4. Maximum Current of Dark Discharge. Using (4–31), show that the maximum dark discharge current is proportional to the square of the gas pressure. Derive the relation between similarity parameters: maximum current j_{max}/p^2, ion mobility $\mu_+ p$, reduced electric field E/p, and interelectrode distance pd_0.

4–5. Normal Cathode Potential Drop, Normal Current Density, and Normal Thickness of Cathode Layer in Glow Discharges. Using (4–32) prove that the normal cathode potential drop does not depend in first approximation either on pressure or on gas temperature. Determine the dependence of normal current density, and normal thickness of the cathode layer on temperature at constant pressure and vice versa.

4–6. Glow Discharge with Hollow Cathode. For the Lidsky configuration of a hollow cathode, explain how the electric field is able to penetrate into the capillary hollow cathode, taking into account that initially there is almost no electric field inside of the hollow metal tube.

4–7. Arc Cathode Spots. The minimum current though a cathode spot for different non-ferromagnetic materials can be found using the empirical formula (4–54). Using this relation, calculate the minimum current density for copper and silver electrodes and compare the results with data presented in Table 4–8.

4–8. Arc Temperature in Framework of the Channel Model. Derive relation (4–71) for plasma temperature in arc discharges based on the principal equations of arc channel model. Explain why the arc temperature is close to a constant with only weak logarithmic dependence on current and radius R.

4–9. Critical Length of Gliding Arc Discharge. Analyze relation (4–76) for electric current in the quasi-equilibrium phase of a gliding arc, and asymptotically simplify it in the vicinity of the critical point when the discharge power approaches its maximum value. Why is this relation unable to describe the current evolution for bigger lengths of the arc?

4–10. Thermal ICP Temperature as a Function of Solenoid Current. Combining equation (4–89) with the formula for plasma conductivity and the Saha equation, derive relation (4–90) for the ICP temperature with explicit expression for constants. Analyze this relation and show that ICP temperature in the case of a strong skin effect does not depend on the frequency of the electromagnetic field.

4–11. Plasma Density in ICP Discharges. The low-pressure ICP discharges operate effectively at electron densities of 10^{11}–$10^{13} \mathrm{cm}^{-3}$, significantly exceeding those of CCP. Analyzing relations (4–103) and (4–105), give your interpretation of this important feature of the ICP discharges. Explain the dependence of the plasma density on the electromagnetic oscillation frequency and electric current in the inductor coil.

4–12. Current-Voltage Characteristics of Corona Discharge. Give your interpretation of the parabolic type of the current-voltage characteristics of the corona discharges. Using formula (4–113), calculate current in an atmospheric-pressure corona discharge from a sharp-point cathode with radius $r = 10\ \mu\mathrm{m}$ to a flat anode separated by $d = 10$ mm.

4–13. Modified Meek Criterion of Avalanche-to-Streamer Transition for Atmospheric-Pressure Glow (APG) mode of DBD. Using equation (4–16), estimate the minimum level of pre-ionization electron density required for a significant change of the Meek criterion of transition from quasi-uniform to filamentary DBD mode.

5

Inorganic Gas-Phase Plasma Decomposition Processes

5.1. CO₂: DISSOCIATION IN PLASMA, THERMAL, AND NON-THERMAL MECHANISMS

5.1.1. Fundamental and Applied Aspects of the CO₂ Plasma Chemistry

The endothermic plasma-chemical process of carbon dioxide decomposition, illustrated in Fig. 5–1, can be presented by the summarizing formula

$$CO_2 \rightarrow CO + \frac{1}{2}O_2, \quad \Delta H = 2.9 \, eV/mol. \tag{5–1}$$

The enthalpy of the process is fairly high and close to that corresponding to hydrogen production from water. The total decomposition process (5–1) starts with and is limited by CO₂ dissociation:

$$CO_2 \rightarrow CO + O, \quad \Delta H = 5.5 \, eV/mol, \tag{5–2}$$

and then ends with O conversion into O_2 by means of either recombination or reaction with another CO_2 molecule. Major research efforts in CO_2 plasma chemistry have been reviewed by Rusanov and Fridman (1984) and Givotov, Rusanov, and Fridman (1982, 1984, 1985b). The experiments were carried out in a wide variety of discharges, including glow, arc, hollow cathode, radiofrequency (both capacitively coupled plasma [CCP] and inductively coupled plasma [ICP]), plasma beam, plasma radiolysis, and non-self-sustained discharges supported by electron beams and ultraviolet (UV) radiation. We can point out four reasons why CO_2 dissociation in plasma attracts the attention of scientists and engineers. First, the dissociation is an important process in CO_2 lasers. Second, it can be stimulated by vibrational excitation with high energy efficiency. Third, CO_2 decomposition can be considered a model of a more complicated process of the reduction of metals from their oxides and halogenides. Finally, the fourth reason is there is a wide range of industrial applications for the decomposition, including the treatment of power plant exhausts, synthesis of new transportation fuels, and even possible fuel production on Mars, where the atmosphere consists predominantly of CO_2. Besides that, carbon monoxide generated in plasma (5–1) can be easily converted into hydrogen without spending additional energy in the thermocatalytic shift reaction:

$$CO + H_2O \rightarrow CO_2 + H_2, \quad \Delta H = -0.4 \, eV/mol. \tag{5–3}$$

Combination of reactions (5–1) and (5–3) together with recirculation of CO_2 results in the two-stage cycle of hydrogen production from water, which is of special interest in the

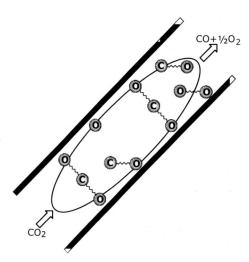

Figure 5–1. Illustration of plasma-chemical process of CO_2 dissociation with production of CO and oxygen.

nuclear/hydrogen energy approach (Legasov et al., 1977, 1978, 1981, 1983; Givotov et al., 1980, 1981, 1983).

5.1.2. Major Experimental Results on CO_2: Dissociation in Different Plasma Systems and Energy Efficiency of the Process

CO_2 dissociation was investigated in numerous thermal and non-thermal plasma systems. The most critical point in the experiments was the maximization of the energy efficiency of the process (5–1), η, determined by the ratio of the dissociation enthalpy $\Delta H = 2.9\,\text{eV/mol}$ to the actual energy cost E_{CO} of one CO molecule produced in a plasma system:

$$\eta = \Delta H / E_{CO}. \tag{5–4}$$

The special interest in the energy efficiency is related to applications in the nuclear/hydrogen energy system. Although the electric energy in this case is supposed to be generated by nuclear power plants at night, the consumption of electric energy is always a key issue in nuclear/hydrogen energetics (Legasov et al., 1977, 1978, 1981, 1983). Thermal plasma experiments in particular were carried out in quasi-equilibrium arc discharges by Polak, Slovetsky, and Butylkin (1977). The highest value of energy efficiency achieved in these experiments was about 15%, which corresponds to a CO energy price of about 20 eV/mol. The theoretical maximum value of the energy efficiency of CO_2 dissociation in thermal plasma is also not very high. It is only about 43–48% according to Levitsky (1978) and Butylkin et al. (1978, 1979). Increased energy efficiency can be achieved in non-thermal plasma systems. Discharge energy in these systems can be selectively focused on effective channels of the process. Low-pressure discharges, in particular glow discharges, are the simplest means to achieve the stationary non-equilibrium conditions for CO_2 dissociation. The dissociation in these systems is mostly controlled by electronic excitation of CO_2 molecules, which is actually not a very energy-effective mechanism (see later discussion). As a result, the energy efficiency of CO_2 dissociation in glow discharges (including those with hollow cathodes) usually does not exceed 8% (Metel & Nastukha, 1977; Slovetsky, 1980). The most energy-efficient CO_2 dissociation by means of electronic excitation is plasma radiolysis, where the process is sustained at atmospheric pressure by high-current relativistic electron beams. Legasov et al. (1978) and Vakar et al. (1978) reported 30% energy efficiency in such experiments. A low-pressure plasma-beam discharge permits one

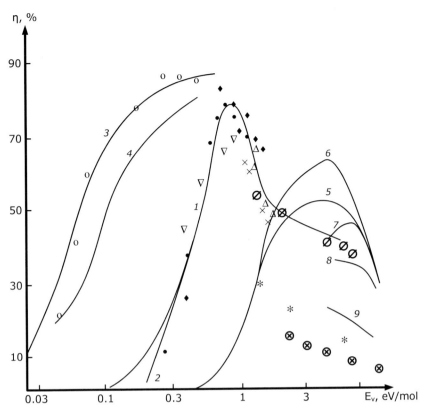

Figure 5–2. Energy efficiency of CO_2 dissociation as a function of specific energy input. (1, 2), non-equilibrium calculations in one- and two-approximations; non-equilibrium calculations for supersonic flows: (3) M = 5; (4) M = 3.5; calculations of thermal dissociation with (5) ideal and (6) super-ideal quenching; (7) thermal dissociation with quenching rates 10^9 K/s, (8) 10^8 K/s, (9) 10^7 K/s. Different experiments in microwave discharges: o, ♦, Δ, ×. Experiments in supersonic microwave discharges: ●. Experiments in different RF-CCP discharges: o,∇. Experiments in RF-ICP discharges: ∅. Experiments in different arc discharges: ⊗, *.

to combine contributions of electronic and vibrational excitation, which leads to higher energy efficiencies, from 20% (Ivanov & Nikiforov, 1978) to 50% (Nikiforov, 1979).

Higher values of energy efficiency of CO_2 dissociation have been achieved experimentally in moderate-pressure non-equilibrium discharges. In particular, experiments with pulsed microwave discharge at 300 Torr in a magnetic field under conditions of electron cyclotron resonance (ECR) give 60% energy efficiency of the CO_2 dissociation (Asisov, Givotov, & Rusanov, 1977). The same 60% energy efficiency was achieved at similar conditions in non-equilibrium radiofrequency (RF) discharge (Butylkin et al., 1981). The highest energy efficiency of CO_2 dissociation was demonstrated in a non-equilibrium microwave plasma at moderate pressures of 50–200 Torr. Performing the process in subsonic flow led to an energy efficiency of 80% (Legasov et al., 1978); in supersonic flow, energy efficiency reaches 90% (Asisov et al., 1981, 1983). Although numerous chemical and relaxation processes take place in CO_2 plasma at the same time, almost all discharge energy (90%) in these experiments can be focused on the dissociation process and CO production. This fact is important in both fundamental and applied aspects. Major data on the energy efficiency of CO_2 dissociation obtained in different thermal and non-thermal discharges under different conditions are presented in Fig. 5–2 together with results of theoretical plasma-chemical modeling.

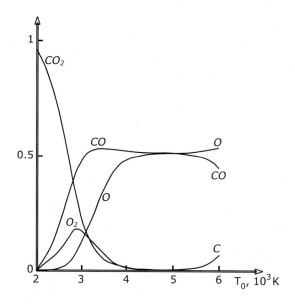

Figure 5–3. Equilibrium molar fraction of products of CO_2 decomposition in thermal plasma as function of temperature in hot discharge zone at fixed gas pressure $p = 0.16$ atm.

5.1.3. Mechanisms of CO_2 Decomposition in Quasi-Equilibrium Thermal Plasma

Thermal plasma systems such as electric arcs or high-pressure RF discharges provide CO_2 dissociation by a high-temperature shift of thermodynamic equilibrium (5–1) in the direction of product formation. It can be seen from Fig. 5–3 that significant CO_2 dissociation requires plasma temperatures of about 2500–3000 K in conditions of thermodynamic quasi equilibrium. Plasma in this case is only a heater – a provider of the required high temperature. The limitations of energy efficiency and conversion degree are related in the quasi-equilibrium systems to two major effects:

1. The products of decomposition (5–1) generated at high temperature (Fig. 5–3), CO and oxygen, can be protected from reverse reactions only by quenching, that is, by fast non-adiabatic cooling. If the products escape from the hot plasma zone too slowly, the thermodynamic quasi equilibrium is continuously sustained during the temperature decrease and products can be converted back to CO_2. The dependence of the dissociation energy efficiency on the quenching rate (cooling rate) is presented in Fig. 5–4 (Polak et al., 1977). As one can see from the figure, the effective protection of the dissociation products from reverse reactions requires very fast cooling rates of about 10^7–10^8 K/s.
2. Absolute quenching of the CO_2 dissociation requires a cooling process sufficient to save all CO formed in the quasi-equilibrium high-temperature zone. But even in the case of absolute quenching, the dissociation degree and energy efficiency of the thermal process are limited by the uniformity of energy distribution over all degrees of freedom of the plasma-chemical system. All degrees of freedom receive an equal energy portion, but only few of them are relevant to dissociation. As a result (see Fig. 5–4), the maximum energy efficiency value of CO_2 dissociation in quasi-equilibrium thermal plasmas is only 43%.

Solving the two aforementioned major problems and achieving higher values of energy efficiency become possible only in essentially non-equilibrium plasma systems. The difference in temperatures related to different molecular degrees of freedom and different subsystems of the non-equilibrium discharges permit selective organization of the CO_2 dissociation, providing energy only into relevant channels of the process and solving the second aforementioned problem. Alternatively, using low gas temperature leads to low

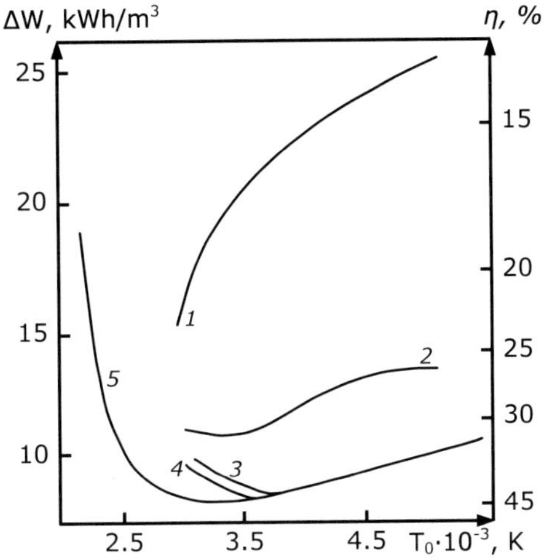

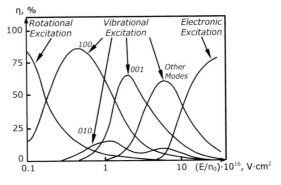

Figure 5–4. Energy efficiency η and energy cost ΔW of CO_2 thermal dissociation as functions of temperature at atmospheric pressure and different quenching rates: (1) 10^6 K/s, (2) 10^7 K/s, (3) 10^8 K/s, (4) 10^9 K/s, (5) instantaneous cooling.

rates of reverse reactions and actually makes quenching unnecessary, which solves the first problem.

The difference between vibrational and translational CO_2 temperatures ($T_v > T_0$) results in a maximum energy efficiency increase to 60% even in the case of the quasi-equilibrium balance of direct and reverse reactions (Evseev, Eletsky, & Palkina, 1979), because direct endothermic reactions are mostly stimulated by molecular vibration, whereas reverse exothermic reactions are mostly stimulated by translational temperature (see the Fridman-Macheret α-model in Chapter 2). This efficiency corresponds to the case of super-ideal non-equilibrium ($T_v > T_0$) quenching of the CO_2 thermal plasma dissociation products (Potapkin et al., 1983).

5.1.4. CO₂ Dissociation in Plasma, Stimulated by Vibrational Excitation of Molecules

This mechanism is the most effective channel for CO_2 dissociation in plasma. First of all, the major portion of the discharge energy is transferred from plasma electrons to CO_2 vibration at electron temperature typical for non-thermal discharges ($T_e \approx 1\,eV$) (see Fig. 5–5). The rate coefficient of CO_2 vibrational excitation by electron impact in this case reaches maximum values of about $k_{eV} = 1$–$3 \times 10^{-8}\,cm^3$/s. Vibrational energy losses through vibrational–translational (VT) relaxation at the same time are mostly related to symmetric vibrational modes and they are relatively slow:

$$k_{VT} \approx 10^{-10} \exp\left(-72/T_0^{1/3}\right),\ cm^3/s,$$

Figure 5–5. Fractions of non-thermal CO_2 discharge energy transferred from plasma electrons to different channels of excitation of the molecule.

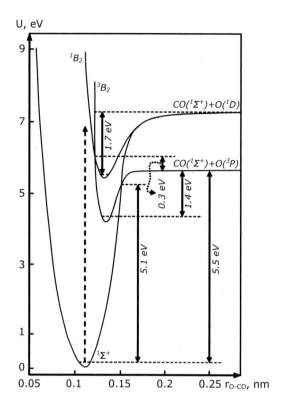

Figure 5–6. Low electronic terms of CO_2.

where T_0 (K) is the translational gas temperature. As a result, sufficiently high values of plasma ionization degrees ($n_e/n_0 \geq 10^{-6}$) permit one to reach significant overequilibrium population of the CO_2 ground electronic state $^1\Sigma^+$ vibrational levels. Plasma electrons mostly provide excitation of low vibrational levels of the ground electronic state $CO_2(^1\Sigma^+)$. Population of highly excited vibrational levels is due to vibrational–vibrational (VV) relaxation processes. When the highly vibrationally excited CO_2 molecules have energy exceeding the dissociation threshold E_D, they are able to dissociate (5–2). The elementary dissociation act can be illustrated based on the scheme of electronic terms of a CO_2 molecule shown in Fig. 5–6. As one can see from the figure, the straightforward adiabatic dissociation of a vibrationally excited CO_2 molecule (with total electron spin conservation),

$$CO_2^*(^1\Sigma^+) \rightarrow CO(^1\Sigma^+) + O(^1D), \qquad (5-5)$$

is related to formation of an O atom in an electronically excited state and requires more than 7 eV. The non-adiabatic transition $^1\Sigma^+ \rightarrow {}^3B_2$ in the point of crossing of the terms (change of spin during the transition, Fig. 5–6) provides a more effective dissociation process stimulated by the step-by-step vibrational excitation:

$$CO_2^*(^1\Sigma^+) \rightarrow CO_2^*(^3B_2) \rightarrow CO(^1\Sigma^+) + O(^3P), \quad E_a = 5.5 \text{ eV/mol}. \qquad (5-6)$$

The step-by-step vibrational excitation, which takes place as a result of VV quantum exchange, determines the second kinetic order of CO_2 dissociation. The dissociation process (5–6) results in the formation of an oxygen atom in the electronically ground state $O(^3P)$ and requires the exact energy of an OC=O bond to be spent (5.5 eV). The non-adiabatic elementary dissociation process (5–6) has much lower activation energy than (5–5) and, therefore, is exponentially faster. The CO_2 dissociation, stimulated by vibrational excitation in plasma, usually proceeds for this reason according to this non-adiabatic mechanism (5–6). Atomic oxygen created in the dissociation (5–6) is able then to participate in a secondary reaction, also with a vibrationally excited CO_2 molecule, to produce a second CO molecule and molecular oxygen:

$$O + CO_2^* \rightarrow CO + O_2, \quad \Delta H = 0.3 \, \text{eV/mol}, \quad E_a \approx 0.5\text{--}1 \, \text{eV/mol}. \qquad (5\text{--}7)$$

This reaction is faster than the three-body recombination of atomic oxygen ($O + O + M \rightarrow O_2 + M$) and permits one to produce a second CO molecule per dissociation event, when the vibrational temperature is not too low ($T_v \geq 0.1$ eV; Rusanov & Fridman, 1978a,b).

The mechanism of CO_2 dissociation stimulated by vibrational excitation in plasma (5–6), (5–7) has the following three essential qualitative advantages in energy efficiency with respect to alternative non-equilibrium plasma-chemical mechanisms of CO_2 dissociation:

1. No less than 95% of the total non-thermal discharge energy at electron temperature $T_e = 1\text{--}2$ eV can be transferred from plasma electrons to vibrational excitation of CO_2 molecules, mostly to their asymmetric vibrational mode (Kochetov et al., 1979, Rusanov et al., 1981; see Fig. 5–5 and also Chapter 2). This fact alone already makes the vibrational dissociation mechanism (5–6), (5–7) a special one.
2. The vibrational energy of CO_2 molecules is the most effective means for stimulation of endothermic reactions related to CO_2 dissociation, in particular reactions (5–6) and (5–7); see the Fridman-Macheret α-model in Chapter 2.
3. The vibrational energy required for CO_2 dissociation (5–6) equals the OC=O bond energy (5.5 eV), which is less than that required for the dissociation by means of electronic excitation in direct electron impact. This effect can be explained using Fig. 5–6, which illustrates the low-energy electronic terms of CO_2 molecules. Direct electron impact provides, due to the Frank-Condon principle, the vertical transition from the ground state $CO_2(^1\Sigma^+)$ to the electronically excited state $CO_2(^1B_2)$ shown in Fig. 5–6, which requires at least 8 eV of electron energy. The dissociation process then takes places by the subsequent transition from $CO_2(^1B_2)$ to $CO_2(^3B_2)$.

Advantages of vibrational excitation in CO_2 dissociation are related not only to direct electron impact but also to any mechanisms of sustaining VT non-equilibrium in plasma, including the preferential diffusion of vibrationally excited molecules from high-temperature discharge zones (see Chapter 3). As a bottom line, the vibrational-translational non-equilibrium of non-thermal discharges permits one to achieve the highest energy efficiency of CO_2 dissociation, which is 80% in subsonic flow (Legasov et al., 1978) and 90% in supersonic flow (Asisov et al., 1981, 1983).

5.1.5. CO₂ Dissociation in Plasma by Means of Electronic Excitation of Molecules

CO_2 dissociation through electronically excited states is mostly related in plasma with direct electron impact excitation of relatively low electronic terms, $CO_2(^1B_2)$ and $CO_2(^3B_2)$, which are illustrated in Fig. 5–6. When electron energies are high, a significant contribution to dissociation can be related also to the product of formation of electronically excited CO:

$$e + CO_2(^1\Sigma^+) \rightarrow CO(a^3\Pi) + O(^3P). \qquad (5\text{--}8)$$

The CO_2 dissociation through electronic excitation can be a dominant mechanism of dissociation in non-thermal plasma with high values of reduced electric fields E/p (usually low-pressure discharges) or when plasma is generated by degradation of very energetic particles (high-energy electron beams or nuclear fission fragments, for example). In both cases, the vibrational excitation of CO_2 molecules is relatively suppressed. Cross sections of CO_2 dissociation through electronic excitation are shown in Fig. 5–7 as a function of electron energy (Corvin & Corrigan, 1969; Slovetsky, 1977, 1980). The rate coefficients of the CO_2 dissociation process can then be found by integrating the cross sections presented

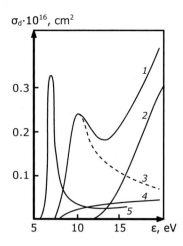

Figure 5–7. Cross sections of $CO_2(^1\Sigma^+)$ dissociation by direct electron impact as functions of electron temperature at room temperature $(T_v = T_0 = 300\,K)$: (1) total dissociation cross sections (sum of partial cross sections 2–4); (2) partial dissociation cross section related to formation of $O(^3P) + CO(a^3\Pi)$; (3) partial cross section of dissociation through excitation of forbidden transitions; (4) partial cross section of dissociation through excitation of permitted transitions $(^1B_2)$; (5) total cross section according to Corvin & Corrigan (1969).

earlier with relevant electron energy distribution functions (EEDFs). Such rate coefficients of the CO_2 dissociation through electronic excitation are presented in Fig. 5–8 as functions of reduced electric field E/n_0 (Slovetsky, 1977, 1980).

Three major effects limit the energy efficiency of the CO_2 dissociation in plasma by means of electronic excitation of the molecules:

1. The electron energy required for the dissociation via electronic excitation exceeds the O=CO bond energy (Fig. 5–6), which obviously leads to electron energy losses.
2. The fraction of discharge energy transferring from plasma electrons to electronic excitation of CO_2 molecules is relatively low at typical electron temperatures ($T_e = 1$–2 eV; see Fig. 5–5).
3. Relatively high electron temperatures, required for a significant contribution of the CO_2 dissociation through electronic excitation, leads to the simultaneous excitation of numerous different states of the molecule, most of which are not relevant or not effective for dissociation.

Results of calculations of energy efficiency of the CO_2 dissociation through electronic excitation of the molecules in non-thermal electric discharges are shown in Fig. 5–9 (Willis, Saryend, & Marlow, 1969). The maximum energy efficiency of the mechanism is not high ($\sim 25\%$), which corresponds to a CO energy cost of ~ 11.5 eV. Thus, the electronic excitation mechanism is able to explain experiments with glow discharge, where the CO energy cost is 40–180 eV/mol (Polak et al., 1975); but obviously it cannot explain experiments with supersonic microwave discharge (Asisov et al., 1981, 1983), where the CO energy cost is

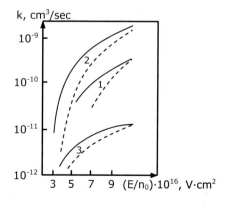

Figure 5–8. Rate coefficients as functions of reduced electric field: (1) CO_2 dissociation by direct electron impact; (2) electronic excitation $CO(X^1\Sigma^+) \rightarrow CO(a^3\Pi)$; (3) dissociative attachment of electrons to CO_2 molecules. Solid lines correspond to calculations based on electron energy distribution functions in CO_2; dashed lines, in CO.

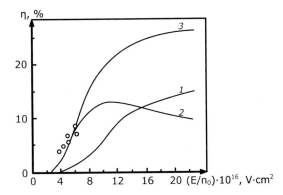

Figure 5–9. Energy efficiency of carbon dioxide dissociation by means of electronic excitation of the molecules as functions of reduced electric field: (1) contribution of singlet states (including $A^1\Sigma$); (2) contribution of triplet states (including $A^3\Pi$); (3) total energy efficiency; dots, experiments of Willis, Saryend, and Marlow (1969).

3.2 eV, which is even lower than the energy threshold of the dissociation by means of electronic excitation.

The energy threshold of dissociation through electronic excitation can be reduced by preliminary vibrational excitation. This effect can be understood from Fig. 5–6, taking into account the Frank-Condon principle of vertical transition between electronic terms induced by electron impact. Such a dissociation mechanism is non-linear with respect to electron density and improves the energy efficiency of the process. The improvement in energy efficiency, however, is not significant, because the aforementioned second and third effects limiting the energy efficiency are still in place. Similar limited improvement can also be expected through another dissociation process which is non-linear with respect to electron density:

$$CO_2(^1\Sigma^+) + CO(a^3\Pi) \rightarrow CO(^1\Sigma^+) + CO(^1\Sigma^+) + O(^3P), \qquad (5\text{–}9)$$

which is based on the preliminary fast electronic excitation of $CO(a^3\Pi)$ (Fig. 5–8).

5.1.6. CO₂ Dissociation in Plasma by Means of Dissociative Attachment of Electrons

The dissociative attachment also makes a contribution in CO_2 decomposition in plasma:

$$e + CO_2 \rightarrow CO + O^-. \qquad (5\text{–}10)$$

The cross section of the process is shown in Fig. 5–10 as a function of electron energy. The energy threshold of (5–10) is lower than that of dissociation through electronic excitation. However, the maximum value of the cross section is not high ($\approx 10^{-18}$ cm²) and the contribution of the dissociative attachment toward the total kinetics is not significant, which is shown in Fig. 5–8. At the same time the dissociative attachment plays a key role in the balance of charged particles and, hence, in sustaining non-thermal discharges in CO_2.

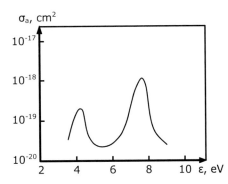

Figure 5–10. Cross section of dissociative electron attachment to CO_2 molecules as a function of electron energy.

The energy efficiency of the CO_2 dissociation through dissociative attachment is limited by the loss of an electron in the process (the energy price of an electron is quite high, usually 30–100 eV). Getting the electron back in the process of associative detachment,

$$O^- + CO \rightarrow CO_2 + e, \tag{5–11}$$

is not efficient because the dissociation product CO is lost in this case. Contribution of an effective associative detachment is possible:

$$O^- + CO_2 \rightarrow CO_3 + e, \tag{5–12}$$

but not significant. Other CO_2 dissociation mechanisms related to losses of charged particles, like dissociative recombination or ion–molecular reactions, have low energy efficiency for the same reason as the dissociative attachment (5–10). Thus, we can conclude that different mechanisms of CO_2 dissociation in plasma lead to very different values of energy efficiency for the process. The highest energy efficiency can be achieved if the process is stimulated in non-thermal plasma by vibrational excitation of the molecules in the ground electronic state $CO_2(^1\Sigma^+)$.

5.2. PHYSICAL KINETICS OF CO_2 DISSOCIATION, STIMULATED BY VIBRATIONAL EXCITATION OF THE MOLECULES IN NON-EQUILIBRIUM PLASMA

5.2.1. Asymmetric and Symmetric CO_2 Vibrational Modes

At $T_e = 1$–2 eV, plasma electrons mostly provide excitation of low vibrational levels $CO_2^*(^1\Sigma^+)$; see Section 5.1.4. Then VV relaxation processes lead to population of highly excited vibrational states with non-adiabatic transition $^1\Sigma^+ \rightarrow {}^3B_2$ and dissociation (5–6); see Fig. 5–6. The VV exchange between vibrationally excited CO_2 molecules is the limiting stage of the dissociation kinetics. It is kinetically a quite complicated process, which first takes place along individual vibrational modes at low energies and then through mixed vibrations (vibrational quasi-continuum) at high energies. The CO_2 molecules are linear and they have three normal vibrational modes like all other molecules of symmetry group $D_{\infty h}$. These three modes are illustrated in Fig. 5–11: asymmetric valence vibration ν_3 (energy quantum $\hbar\omega_3 = 0.30$ eV), symmetric valence vibration ν_1 (energy quantum $\hbar\omega_1 = 0.17$ eV), and a double degenerate symmetric deformation vibration ν_2 (energy quantum $\hbar\omega_2 = 0.085$ eV). The degenerated symmetric deformation vibrations ν_2 are polarized in two perpendicular planes (see Fig. 5–11), which can result after summation in quasi rotation of the linear molecule around its principal axis. Angular momentum of the quasi-rotation is characterized by a special quantum number "l_2", which assumes the values

$$l_2 = v_2, v_2 - 2, v_2 - 4, \ldots, 1 \text{ or } 0, \tag{5–13}$$

where v_2 is the number of quanta on the degenerate mode.

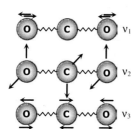

Figure 5–11. Vibrational modes of CO_2 molecules: ν_1, symmetric valence vibrations; ν_2, double degenerate symmetric deformation vibrations; ν_3, asymmetric valence vibrations.

The level of vibrational excitation of the linear triatomic molecule can then be denoted as $CO_2(v_1, v_2^{l_2}, v_3)$, showing the number of quanta on each mode. For example, the molecule CO_2 $(0,2^2,0)$ has only symmetric deformation vibrations excited with effective "circular" polarization. The total vibrational energy of a CO_2 molecule at relatively low excitation levels can then be presented as follows:

$$E_v(v_1, v_2, v_3) = \hbar\omega_1 \left(v_1 + \frac{1}{2}\right) + \hbar\omega_2(v_2 + 1) + \hbar\omega_3 \left(v_3 + \frac{1}{2}\right) + x_{11}\left(v_1 + \frac{1}{2}\right)^2$$

$$+ x_{22}(v_2 + 1)^2 + g_{22}l_2^2 + x_{33}\left(v_3 + \frac{1}{2}\right)^2 + x_{12}\left(v_1 + \frac{1}{2}\right)(v_2 + 1)$$

$$+ x_{13}\left(v_1 + \frac{1}{2}\right)\left(v_3 + \frac{1}{2}\right) + x_{23}(v_2 + 1)\left(v_3 + \frac{1}{2}\right). \tag{5–14}$$

In this relation, $x_{11} = -0.47 \times 10^{-3}$ eV, $x_{22} = -0.08 \times 10^{-3}$ eV, $x_{33} = -1.6 \times 10^{-3}$ eV, $x_{12} = 0.45 \times 10^{-3}$ eV, $x_{13} = -2.4 \times 10^{-3}$ eV, and $x_{23} = -1.6 \times 10^{-3}$ eV are anharmonicity coefficients; the energy coefficient g_{22} corresponds to the quantum number l_2 (5–13), which describes polarization of the double degenerate symmetric deformation vibrations. There happens to be a resonance in CO_2 molecules between the two different types of symmetric vibrations – valence and deformational: $\hbar\omega_1 = 2\hbar\omega_2$. For this reason the symmetric modes are sometimes considered in plasma-chemical calculations for simplicity as one triple degenerate vibration. These generalized symmetric vibrations are characterized by lowest vibrational quantum $\hbar\omega_s = \hbar\omega_2$ and generalized vibrational quantum number $v_s = 2v_1 + v_2$ to fit the following simplified expression for the total value of vibrational energy of all symmetric modes in harmonic approximation:

$$E_{sym} = \hbar\omega_1 v_1 + \hbar\omega_2 v_2 = \hbar\omega_2(2v_1 + v_2) = \hbar\omega_s v_s, \tag{5–15}$$

The total vibrational energy of a CO_2 molecule (5–14) can then be expressed as

$$E_v(v_a, v_s) = \hbar\omega_a v_a + \hbar\omega_s v_s - x_a v_a^2 - x_s v_s^2 - x_{as} v_a v_s, \tag{5–16}$$

where x_a, x_s, x_{as} are generalized coefficients of anharmonicity ($x_{as} = 12 \text{ cm}^{-1}$; compare to (5–14)).

5.2.2. Contribution of Asymmetric and Symmetric CO_2 Vibrational Modes into Plasma-Chemical Dissociation Process

The VV-exchange process proceeds at low levels of vibrational excitation independently along symmetric and asymmetric CO_2 vibrational modes. The symmetric and asymmetric CO_2 vibrational modes can be characterized by individual vibrational temperatures T_{va} and T_{vs}, as well as by individual vibrational energy distribution functions. A major contribution to the population of highly vibrationally excited states of CO_2 and, hence, to dissociation is related to the excitation of the asymmetric vibrational modes (Rusanov et al., 1985b; Rusanov, Fridman, & Sholin, 1986). This important statement is based on the following three effects of CO_2 plasma-chemical kinetics:

1. The asymmetric vibrational mode of CO_2 is 1, which can be predominantly excited by plasma electrons at electron temperatures $T_e = 1$–3 eV (see Fig. 5–5). Discharge energy localization within that vibrational mode becomes stronger when CO_2 is mixed with CO, which is a product of its dissociation (see Fig. 5–12).
2. The VT-relaxation rate from the asymmetric vibrational mode is much slower than that of symmetric vibrations, which helps to accumulate more energy within the asymmetric mode.

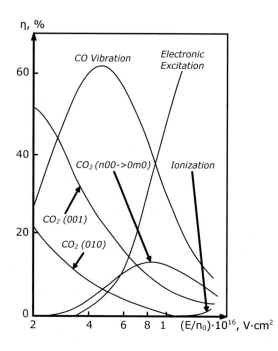

Figure 5–12. Fractions of electron energy transferred in non-thermal discharge conditions from plasma electrons to different channels of molecular excitation and ionization in the gas mixture CO_2 (50%)–CO (50%).

3. VV exchange along the asymmetric mode is several orders of magnitude faster than that along symmetric modes, which intensify the population of highly excited vibrational states in this type of vibration.

Vibrational relaxation rate coefficients for the asymmetric mode are shown in Fig. 5–13 (Rusanov et al., 1979). The VV-exchange process for the asymmetric mode is essentially faster than VT-relaxation and intermode collisional VV′ exchange, which can be explained by taking into account that the actual non-adiabatic CO_2 dissociation energy is less than the adiabatic energy (see Fig. 5–6). The domination of VV-relaxation processes results in a strong Treanor effect (see Chapter 3) and a significant overpopulation of highly excited levels of the asymmetric CO_2 vibrations in non-equilibrium plasma conditions, when the vibrational temperatures T_{va} and T_{vs} exceed the translational temperature T_0 (Fig. 5–14; Rusanov & Fridman, 1984). The Treanor effect, although specific and not so strong, can be observed in symmetric vibrational modes as well. It can be illustrated by experimental

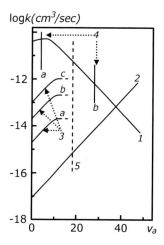

Figure 5–13. Rate coefficients of relaxation processes related to asymmetric mode of CO_2 vibrations at room temperature $T_0 = 300$ K as functions of number of quanta on the mode: (1) VV relaxation; (2) VT relaxation; (3) intermode VV′ relaxation (assuming symmetric vibrational temperature [3a] $T_{vs} = 1000$ K, [3b] $T_{vs} = 2000$ K, [3c] $T_{vs} = 3000$ K); (4) intramolecule VV′ relaxation, transition to vibrational quasi-continuum (assuming [4a] equal excitation of all vibrational modes; [4b] predominant excitation of asymmetric vibrations); (5) CO_2 dissociation energy.

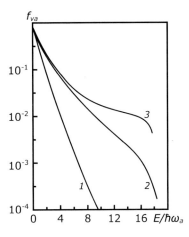

Figure 5–14. Vibrational energy distribution function, population of the asymmetric mode of CO_2 vibrations: (1) $T_{va} = 3000\,K$, $T_{vs} = 2000\,K$, $T_0 = 1000\,K$; (2) $T_{va} = 5000\,K$, $T_{vs} = 3000\,K$, $T_0 = 1000\,K$; (3) $T_{va} = 5000\,K$, $T_{vs} = 3000\,K$, $T_0 = 500\,K$.

measurements of vibrational distribution functions for symmetric valence and symmetric deformation modes of CO_2 at strongly non-equilibrium conditions, presented in Fig. 5–15 (Baronov et al., 1989). The distribution, considered over low and still discrete levels, is on average exponential with vibrational temperature T_v, while groups of quasi-resonant levels are also exponentially distributed but with the rotational–translational temperature T_0. As the level of excitation increases, the vibrations of different types are collisionlessly mixed due to the intermode anharmonicity and the Coriolis interaction, which is also illustrated in Fig. 5–13. The intramolecular VV′ quantum exchange results in the **vibrational quasi continuum** of the highly excited states of CO_2 molecules. One can say that the elementary act of CO_2 dissociation can take place in plasma not from a single mode but from a mixed vibrationally excited state: the vibrational quasi continuum.

5.2.3. Transition of Highly Vibrationally Excited CO₂ Molecules into the Vibrational Quasi Continuum

Kinetics of CO_2 dissociation in plasma through vibrational excitation depends on the lowest energy level when vibrational modes are already mixed. Shchuryak (1976) developed the classical description of the CO_2–like molecule transition to quasi continuum in terms of vibrational beating. The energy of the asymmetric mode changes in the course of vibrational beating corresponding to its interaction with symmetric modes ($\nu_3 \to \nu_1 + \nu_2$), and as a result the effective frequency of this oscillation mode changes as well:

$$\Delta\omega_3 \approx (x_{33}\omega_3)^{\frac{1}{3}}(A_0\omega_3 n_{\text{sym}})^{\frac{2}{3}}, \qquad (5\text{–}17)$$

where $A_0 \approx 0.03$ is a dimensionless characteristic of the interaction between modes ($\nu_3 \to \nu_1 + \nu_2$), and n_{sym} is the total number of quanta on the symmetric modes (ν_1 and ν_2), assuming quasi equilibrium between the modes inside of a CO_2 molecule. As the level of excitation, n_{sym}, increases, the value of $\Delta\omega_3$ grows and at a certain critical number of quanta, $n_{\text{sym}}^{\text{cr}}$, it covers the defect of resonance $\Delta\omega$ of the asymmetric-to-symmetric quantum transition:

$$n_{\text{sym}}^{\text{cr}} \approx \frac{1}{\sqrt{x_{33}A_0^2}}\left(\frac{\Delta\omega_3}{\omega_3}\right)^{\frac{2}{3}}. \qquad (5\text{–}18)$$

When the number of quanta exceeds the critical value (5–18) (the so-called Chirikov stochasticity criterion; Zaslavskii & Chirikov, 1971), the molecular motion becomes quasi random and the modes become mixed in the vibrational quasi continuum. In the general case of polyatomic molecules, the critical excitation level n^{cr} decreases with number N of

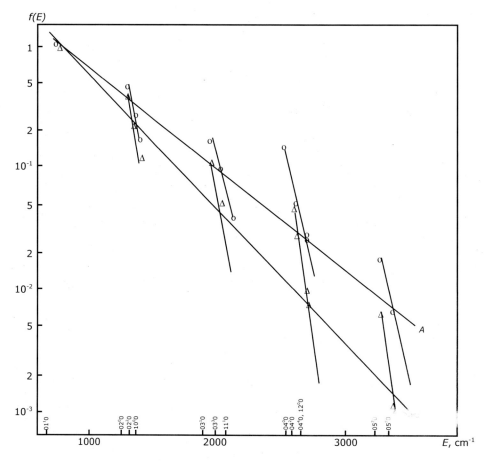

Figure 5–15. Experimental measurements of vibrational distribution function $f(E)$ of CO_2 molecules, including the symmetric valence mode and the deformation modes at different levels of vibrational and rotational (close to translational) temperatures: (A) $T_v = 780 \pm 40$ K, $T_r = 150 \pm 15$ K; (B) $T_v = 550 \pm 40$ K, $T_r = 110 \pm 10$ K.

vibrational modes: $n^{cr} \approx \frac{100}{N^3}$. It means that, for molecules with four or more atoms, the transition to the vibrational quasi continuum takes place at quite low levels of vibrational excitation. CO_2 as a triatomic molecule, however, can maintain the individuality of vibrational modes up to much higher levels of excitation than four-atom molecules. A simplified transition criterion for CO_2 can be obtained based on (5–16) by comparison of the intermode asymmetric–symmetric anharmonicity $x_{as} v_a v_s$ ($x_{as} \approx 12\,\mathrm{cm}^{-1}$, $v_a = v_3$, $v_s = 2v_1 + v_2$) with the defect of resonance $\Delta \omega = \Delta \omega_3 \approx 300\,\mathrm{cm}^{-1}$ for the asymmetric–symmetric transition (Rusanov et al., 1986):

$$x_{as} v_a v_s \geq \Delta \omega. \tag{5–19}$$

Thus, transition to quasi continuum takes place not at some fixed total vibrational energy E^* in general, but according to (5–19) at some special point of critical relation between the number of quanta on asymmetric and symmetric modes. It leads to two different approaches in the description of CO_2 dissociation in non-thermal plasma through vibrational excitation: **one-temperature approximation** and **two-temperature approximation.** If one mode (usually the asymmetric one) is predominantly excited, then the asymmetric vibrational temperature significantly exceeds the symmetric one ($T_{va} \gg T_{vs}$). Transition to quasi continuum takes place in this case from highly excited asymmetric levels: $v_a \gg v_s = \overline{n_s}$ ($\overline{n_s}$ is an average number of vibrational quanta on symmetric modes), and oscillations are not mixed for the most part of relevant vibrational distribution functions (see Fig. 5–14). Dissociation of CO_2 molecules based on excitation of asymmetric and symmetric modes should be considered separately, as should asymmetric and symmetric temperatures for the **two-temperature approximation**. If, conversely, asymmetric and symmetric modes have similar levels of excitation and their vibrational temperatures are close ($T_{va} \approx T_{vs} \equiv T_v$), the transition takes place when v_a and v_s are not too different. In this case, called **one-temperature approximation**, we can introduce the minimal total vibrational energy of a CO_2 molecule corresponding to the transition to quasi continuum:

$$E^* = 2\hbar \sqrt{\frac{\omega_a \omega_s \Delta\omega}{x_{as}}} \approx 6\hbar\omega_a. \qquad (5\text{–}20)$$

This critical transition can be found from the condition for the constant energy straight line,

$$E = \hbar\omega_a v_a + \hbar\omega_s v_s, \qquad (5\text{–}21)$$

to be tangent to the hyperbola (5–19) on a plane of quantum numbers (v_a, v_s). Here $v_s = 2v_1 + v_2$ and $\omega_s = \omega_2$. The transitional total vibrational energy (5–20) is low, so most of the CO_2 molecules relevant to the dissociation process can be considered in this case as being in the vibrational quasi continuum.

5.2.4. One-Temperature Approximation of CO₂ Dissociation Kinetics in Non-Thermal Plasma

This kinetic approximation assumes a single vibrational temperature T_v for CO_2 molecules and, therefore, is sometimes referred to as **quasi equilibrium of vibrational modes**. As one can see from (5–20), most of the vibrationally excited molecules can be considered as being in quasi continuum in this case. Vibrational kinetics of polyatomic molecules in quasi continuum was discussed in Chapter 3. The CO_2 dissociation rate is limited not by elementary dissociation itself, but via energy transfer from a low to high vibrational excitation level of the molecule in the VV-relaxation processes. Such a kinetic situation was referred to in Chapter 3 as the fast reaction limit. The population of highly excited states with vibrational energy E depends in this case on the number of vibrational degrees of freedom "s" and is proportional to the density of the vibrational states $\rho(E) \propto E^{s-1}$. The CO_2 dissociation rate coefficient can be expressed in the quasi equilibrium of vibrational modes as (Potapkin et al., 1980)

$$k_R(T_v) = \frac{k_{VV}^0}{\Gamma(s)} \frac{\hbar\omega}{T_v} \left(\frac{E_a}{T_v}\right)^s \exp\left(-\frac{E_a}{T_v}\right) \sum_{r=0}^{\infty} \frac{(s+r-1)!}{(s-1)!r!} \frac{\gamma(r+1, E_a/T_v)}{(E_a/T_v)^r}. \qquad (5\text{–}22)$$

In this relation, $\Gamma(s)$ is the gamma function; $\gamma(r+1, E_a/T_v)$ is the incomplete gamma function; $E_a = 5.5\,eV$ is the activation energy of the CO_2 dissociation (5–6); k_{VV}^0 and $\hbar\omega$ are the lowest vibrational quantum and corresponding low-energy VV-exchange rate coefficient. The sum in relation (5–20) is not a strong function of E_a/T_v and can be taken for numerical calculations as about 1.1–1.3 if $T_v = 1000$–$4000\,K$. The reaction rate coefficient (5–22) does not include any details related to the elementary act of dissociation itself, which reflects the nature of the fast reaction limit controlling the plasma-chemical CO_2 decomposition stimulated by vibrational excitation.

5.2.5. Two-Temperature Approximation of CO_2 Dissociation Kinetics in Non-Thermal Plasma

If temperatures related to different vibrational modes are different, then transition to quasi continuum takes place along the mode characterized by the highest temperature. Let us specify the asymmetric vibrational temperature as a higher one ($T_{va} \gg T_{vs}$). Transition to the vibrational quasi continuum then takes place during excitation along the asymmetric mode, and the vibrational quantum number for the symmetric modes can be chosen as fixed on its average value $\overline{n_s}$:

$$v_s = \overline{n_s} = \frac{2}{\exp(\hbar\omega_s/T_{vs}) - 1} + \frac{2}{\exp(2\hbar\omega_s/T_{vs}) - 1}. \qquad (5\text{–}23)$$

The vibrational energy on the asymmetric mode necessary for transition to quasi-continuum can be determined from (5–19) taking into account (5–23):

$$E^* = \frac{\Delta\omega}{x_{as}\overline{n_s}} \times \hbar\omega_a. \qquad (5\text{–}24)$$

Numerically, $E^* \approx (15\text{–}20) \times \hbar\omega_a$ at $T_{vs} = 1000\,K$, which means that the transition to quasi continuum actually takes place at very high levels of excitation of the asymmetric vibrational mode, close to the dissociation energy (see Fig. 5–14). Thus, most of the vibrational distribution function relevant to CO_2 dissociation in this case, in contrast to the one-temperature approach, is not continuous but discrete. The discrete distribution function $f(v_a, v_s)$ over vibrational energies (5–16) can be presented analytically according to Licalter (1975a,b, 1976) in the Treanor form:

$$f(v_a, v_s) \propto \exp\left(-\frac{\hbar\omega_a v_a}{T_{va}} - \frac{\hbar\omega_s v_s}{T_{vs}} + \frac{x_a v_a^2 + x_s v_s^2 + x_{as} v_a v_s}{T_0}\right). \qquad (5\text{–}25)$$

The discrete vibrational distribution (5–25) correlates with the Boltzmann distribution function over vibrational energies (5–16) at equilibrium conditions ($T_{va} = T_{vs} = T_0$). It leads, however, to a significantly over-Boltzmann population of highly excited states when the vibrational temperatures T_{va}, T_{vs} exceed the translational temperature T_0 (see Figs. 5–14 and 5–15). To calculate the CO_2 dissociation rate coefficient in general, we should combine the discrete distribution (5–25) for most of the vibrational states with possible continuum-based distribution at high energies exceeding (5–24). Taking into account again the fast reaction limit typical for the kinetics of CO_2 dissociation stimulated by vibrational excitation of the molecules, the combination of the discrete and continuous distributions results in the rate coefficient applied in the following two-temperature approximation (Rusanov et al., 1979a,b, 1986; Fridman, 1986):

$$k_R(T_{va}, T_0) = k_{vv}^0 \frac{E_a}{4\Gamma(s)\hbar\omega_a} \left(\frac{E_a}{E^*}\right)^{s-1} \times \exp\left[-\frac{E_a}{T_{va}} - \frac{1}{\gamma T_0}\ln\frac{1 + \xi(E_a)}{1 + \xi(E^*)}\right.$$

$$\left. + \frac{x_a\hbar\omega_a(v_a^*)^2 + x_s v_a^*\overline{n_s} + x_m(E_a^2 - E^{*2})/\hbar\omega_a}{T_0}\right]. \qquad (5\text{–}26)$$

Most of the symbols in (5–26) are the same as those in expression (5–22) for the dissociation rate coefficient in a one-temperature approximation. Additionally, $v_a^* = \Delta\omega/x_{as}\overline{n_s}$ is a vibrational quantum number corresponding to the transition to quasi continuum along the asymmetric mode (5–24); $\xi(E)$ is the vibrational energy-dependent ratio of rate coefficients of VT and VV relaxation (see Fig. 5–12); $\gamma = \frac{\partial \ln\xi(E)}{\partial E}$ is the logarithmic sensitivity of the VT/VV relaxation ratio to vibrational energy accumulated on the asymmetric mode (see Fig. 5–12); and x_m is the effective CO_2 oscillation anharmonicity in quasi continuum (for details see Chapter 3 or Rusanov et al., 1986). If the transition to vibrational quasi-continuum takes place at energies exceeding the dissociation energy ($E^* > E_a$), then the dissociation rate coefficient (5–26) can be simplified assuming $E^* = E_a$. Such simplification was considered, in particular, by Rusanov, Fridman, and Sholin (1979). An expression similar to (5–26) can also be applied in the opposite case ($T_{vs} \gg T_{va}$), when dissociation is mostly controlled by the vibrational temperature of symmetric modes.

5.2.6. Elementary Reaction Rates of CO$_2$ Decomposition, Stimulated in Plasma by Vibrational Excitation of the Molecules

Thus, the CO$_2$ dissociation rate through vibrational excitation can be described in one- and two-temperature approximations, (5–22) and (5–26). The one- and two-temperature dissociation rate coefficients are not much different numerically at typical values of non-thermal discharge parameters. The contribution of strong Treanor factor, which mostly determines the deviation from Arrhenius law in a two-temperature approximation (5–26), is compensated by a statistical factor $(E_a/T_v)^s$ related to quasi-equilibrium mixing of modes in the one-temperature approximation. The kinetics of dissociation of vibrationally excited CO$_2$ molecules is framed by the so-called fast reaction limit both in one- and two-temperature approximations. In this case, the dissociation rate is mostly controlled not by an elementary act of dissociation of a CO$_2$ molecule already having sufficient energy, but by VV-exchange processes providing the molecule with the necessary vibrational energy. Therefore, the rate coefficients (5–22) and (5–26) are independent of any specific mechanism of dissociation. In particular, the rate coefficients (5–22) and (5–26) can be applied not only to direct dissociation (5–6) but also to formation of a quasi-stable excited molecule CO$_2(^3B_2)$ (see Fig. 5–6 and chemical formula (5–6)).

The quasi-stable excited molecule CO$_2(^3B_2)$ can lead to CO formation not only through direct dissociation but also through formation of CO$_3$ (McEwen & Phillips, 1975):

$$CO_2(^3B_2) + CO_2(^1\Sigma^+) \to CO_3 + CO. \tag{5–27}$$

The formation of the CO$_3$ radical can affect the kinetics of CO$_2$ decomposition due to the essential influence on the secondary reaction (2.7), and also due to reduction of the concentration of atomic oxygen related to fast non-adiabatic CO$_2$ vibrational relaxation.

The CO$_2$ dissociation rate coefficients (5–22) and (5–26) are related to the limiting stage (5–6) of the decomposition process (5–1). The total CO$_2$ decomposition process and total generation of CO are also related to the secondary reaction of atomic oxygen (5–7). This reaction is also slightly endothermic and, therefore, can be effectively stimulated by vibrational excitation of CO$_2$ molecules. However, the efficiency α of vibrational energy for the reaction (5–7) is less than 1 in this case (see Chapter 2), because the activation barrier of a direct reaction does not significantly exceed the activation energy of a reverse one. Reliable information on the kinetics of this elementary reaction in non-equilibrium condition is not available thus far. Even data on activation energies of the reaction (7–3) are quite contradictory and vary from 0.5 to 1.2 eV (Sulzmann, Myer, & Bartle, 1965; Avramenko & Kolesnikov, 1967; Baber & Dean, 1974). An analysis of kinetic pathways of atomic oxygen in vibrationally excited CO$_2$ shows (Potapkin et al., 1980) that the reaction of O atoms with CO$_2$ molecules prevails over the three-body recombination:

$$O + O + M \to O_2 + M, \tag{5–28}$$

even at lower vibrational temperatures ($T_v \geq 0.1$ eV). One act of primary dissociation (5–6) leads in most cases to the formation of two CO molecules.

5.3. VIBRATIONAL KINETICS AND ENERGY BALANCE OF PLASMA-CHEMICAL CO_2 DISSOCIATION

5.3.1. Two-Temperature Approach to Vibrational Kinetics and Energy Balance of CO_2 Dissociation in Non-Equilibrium Plasma: Major Energy Balance and Dynamic Equations

Let us analyze the energy balance of CO_2 dissociation stimulated in plasma by vibrational excitation in the two-temperature approximation, assuming one-dimentional gas motion with density ρ through the plasma in the x-direction with velocity u. Such an energy balance can be illustrated in the framework of the following equations describing major energy transfer, relaxation, and chemical reaction processes separately for different individual vibrational modes in the plasma-chemical system, which includes CO_2 and products of its dissociation (Rusanov & Fridman, 1984):

1. The balance of average vibrational energy, ε_v^a, located on an asymmetric mode of a CO_2 molecule is mostly determined by eV and VV'-relaxation processes as well as dissociation itself:

$$\rho u \frac{d}{dx}\left(\varepsilon_v^a n_{CO_2}/\rho\right) = w_{eV}^a + w_{VV'}^{as} + w_{VV'}^{CO_2-CO} + w_R^a. \qquad (5\text{--}29)$$

2. The balance of average vibrational energy ε_v^s located on symmetric modes of a CO_2 molecule is mostly determined by eV, VV'- and VT-relaxation processes as dissociation itself:

$$\rho u \frac{d}{dx}\left(\varepsilon_v^s n_{CO_2}/\rho\right) = w_{eV}^s + w_{VV'}^{sa} + w_{VT}^s + w_R^s. \qquad (5\text{--}30)$$

3. The balance of average energy ε_v^{CO} located on a CO molecule, produced by CO_2 dissociation, is

$$\rho u \frac{d}{dx}\left(\varepsilon_v^{CO} n_{CO}/\rho\right) = w_{eV}^{CO} + w_{VV'}^{CO-CO_2} + w_{VT}^{CO} + w_R^{CO}. \qquad (5\text{--}31)$$

4. The equation of energy conservation for the total gas flow (n_i, h_i are densities and full enthalpies of an i component of the mixture), taking into account the electron-neutral energy transfer, is

$$\rho u \frac{d}{dx}\left(\frac{\sum n_i h_i}{\rho} + \frac{u^2}{2}\right) = w_{eV}^a + w_{eV}^s + w_{eV}^{CO}. \qquad (5\text{--}32)$$

Equations (5–29)–(5–32) take into account eV processes of vibrational excitation by electron impact (energy transfer rates: w_{eV}^a, w_{eV}^s, w_{eV}^{CO}); VV'-exchange between asymmetric and symmetric CO_2 modes (energy transfer rates: $w_{VV'}^{as}$, $w_{VV'}^{sa}$) as well as between CO_2 and CO (energy transfer rates: $w_{VV'}^{CO_2-CO}$, $w_{VV'}^{CO-CO_2}$); VT relaxation of CO and symmetric deformation mode of CO_2 (energy transfer rates: w_{VT}^{CO}, w_{VT}^s); and chemical reactions; w_R^a, w_R^s, w_R^{CO} (these terms also represent vibrational energy losses related to anharmonicity of VV exchange and VT relaxation from high energy levels, which are usually attributed to one act of chemical reaction; see Chapter 3). In addition to the energy balance equations (5–29)–(5–32), the two-temperature approximation of CO_2 dissociation also includes traditional equations of chemical kinetics and fluid mechanics and requires detailed kinetic relations controlling rates of eV, VV'- and VT-relaxation processes (Potapkin, Rusanov, & Fridman, 1984a,b, 1987). Thus, the chemical kinetic equations can be introduced as

$$\rho u \frac{d}{dx}\left(\frac{n_i}{\rho}\right) = \Phi_i, \qquad (5\text{--}33)$$

where Φ_i is the total rate of density change for component i of the mixture as a result of chemical reactions. The continuity equation can be taken simply as

$$\frac{du}{u} + \frac{d\rho}{\rho} + \frac{dA}{A} = 0, \tag{5-34}$$

where A is the cross-sectional area of the one-dimentional simplified plasma-chemical reactor. Finally, the momentum equation and equation of state for a gas with pressure p and density ρ can be presented as

$$\rho u \frac{du}{dx} = -\frac{\partial p}{\partial x}, \quad p = \sum \frac{n_i}{\rho} \rho T_0, \tag{5-35}$$

where p and ρ are pressure and density of the gas mixture.

5.3.2. Two-Temperature Approach to Vibrational Kinetics and Energy Balance of CO$_2$ Dissociation in Non-Equilibrium Plasma: Additional Vibrational Kinetic Relations

To complete the two-temperature approximation we should introduce vibrational kinetic relation controlling rates of eV, VV'-, and VT-relaxation processes and chemical reactions (Potapkin, Rusanov, & Fridman, 1984a,b, 1987). They have to be considered as functions of the average number of vibrational quanta in CO$_2$ symmetric valence mode (α_1, quantum $\hbar\omega_1$), CO$_2$ symmetric deformation mode (α_2, quantum $\hbar\omega_2$), CO$_2$ asymmetric mode (α_3, quantum $\hbar\omega_3$), and in CO vibrations (α_4, quantum $\hbar\omega_4$). Rates of VV'-relaxation exchange between CO and CO$_2$ molecules (rate coefficient $k_{VV'}^{43}$) and between asymmetric and symmetric modes of CO$_2$ molecules in collisions with any i components of the mixture (rate coefficient $k_{VV',i}^{32}$) can be expressed in this case as

$$w_{VV'}^{CO-CO_2} = -w_{VV'}^{CO_2-CO} \frac{\hbar\omega_4}{\hbar\omega_3} = n_{CO}n_{CO_2}k_{VV'}^{43}\left[\frac{\beta_4}{\beta_3}(1+\alpha_4)\alpha_3 - (1+\alpha_3)\alpha_4\right]\hbar\omega_4, \tag{5-36}$$

$$w_{VV'}^{as} = -w_{VV'}^{sa} \frac{\hbar\omega_3}{3\hbar\omega_2} = \sum_i n_i n_{CO_2}k_{VV',i}^{32} \frac{(1-\beta_2)^3}{8(1-\beta_3)}\left[\frac{\beta_3}{\beta_2^3}\alpha_2^3(1+\alpha_3) - \alpha_3(2+\alpha_2)^3\right]\hbar\omega_3. \tag{5-37}$$

In these relations, $\beta_j = \exp(-\hbar\omega_j/T_0)$ are the Boltzmann factors related to translational gas temperature. Rates of VT relaxation of CO molecules (rate coefficient $k_{VT,CO}^i$) and CO$_2$ molecules from symmetric deformation mode (rate coefficient k_{VT,CO_2}^i) related to collisions with any i components of the mixture can be presented as

$$w_{VT}^{CO} = \sum_i n_i n_{CO}k_{VT,CO}^i(\alpha_4^0 - \alpha_4)\hbar\omega_4, \tag{5-38}$$

$$w_{VT}^s = \sum_i n_i n_{CO_2}k_{VT,CO_2}^i(\alpha_2^0 - \alpha_2)\hbar\omega_2. \tag{5-39}$$

Here the superscript 0 means that the relevant number of vibrational quanta α is calculated for equilibrium conditions ($T_v = T_0$). The vibrational excitation rate of the CO$_2$ asymmetric mode (rate coefficient k_{eV}^a, temperature of the mode T_v^a) by plasma electrons with density n_e and temperature T_e is

$$w_{eV}^a = \left[1 - \exp\left(\frac{\hbar\omega_3}{T_e} - \frac{\hbar\omega_3}{T_v^a}\right)\right]n_{CO_2}n_e k_{eV}^a\hbar\omega_3. \tag{5-40}$$

Table 5–1. Chemical Mechanism of CO_2 Decomposition in Plasma, Elementary Reactions of Neutral Species with their Enthalpies ΔH, Activation Energies E_a, Pre-exponential Factors A, and Efficiencies α of Vibrational Excitation (relation between different units for activation energy: 1 eV/mol = 23 kcal/mol)

#	Elementary chemical reactions	$\Delta H, \dfrac{\text{kcal}}{\text{mole}}$	$A, \dfrac{\text{cm}^3}{\text{s}} \dfrac{\text{cm}^6}{\text{s}}$	$E_a, \dfrac{\text{kcal}}{\text{mol}}$	α
1	$CO_2 + CO_2 \rightarrow CO + O + CO_2$	125.8	$4.39 \cdot 10^{-7}$	128.6	1
2	$CO_2 + CO \rightarrow CO + O + CO$	125.8	$4.39 \cdot 10^{-7}$	128.6	1
3	$CO_2 + O_2 \rightarrow CO + O + O_2$	125.8	$3.72 \cdot 10^{-10}$	119.6	1
4	$CO + O + CO_2 \rightarrow CO_2 + CO_2$	-125.8	$6.54 \cdot 10^{-36}$	4.3	0
5	$CO + O + CO \rightarrow CO_2 + CO$	-125.8	$6.54 \cdot 10^{-36}$	4.3	0
6	$CO + O + O_2 \rightarrow CO_2 + O_2$	-125.8	$6.51 \cdot 10^{-36}$	-3.7	0
7	$O + CO_2 \rightarrow CO + O_2$	7.8	$7.77 \cdot 10^{-12}$	33.0	0.5
8	$CO + O_2 \rightarrow CO_2 + O$	-7.8	$1.23 \cdot 10^{-12}$	25.2	0
9	$O_2 + O_2 \rightarrow O + O + O_2$	118.0	$8.14 \cdot 10^{-9}$	118.6	1
10	$O_2 + O \rightarrow O + O + O$	118.0	$2.0 \cdot 10^{-8}$	114.9	1
11	$O_2 + CO \rightarrow O + O + CO$	118.0	$2.4 \cdot 10^{-9}$	118.0	1
12	$O_2 + CO_2 \rightarrow O + O + CO_2$	118.0	$2.57 \cdot 10^{-9}$	111.5	1
13	$O + O + O_2 \rightarrow O_2 + O_2$	-118.0	$6.8 \cdot 10^{-34}$	0	0
14	$O + O + O \rightarrow O_2 + O$	-118.0	$2.19 \cdot 10^{-33}$	-4.5	0
15	$O + O + CO \rightarrow O_2 + CO$	-118.0	$2.75 \cdot 10^{-34}$	0	0
16	$O + O + CO_2 \rightarrow O_2 + CO_2$	-118.0	$2.75 \cdot 10^{-34}$	0	0

The vibrational excitation rate of CO_2 symmetric modes (rate coefficients k_{eV}^1 and k_{eV}^2; combined temperature of the symmetric modes (1) and (2) is T_v^s) by plasma electrons can be expressed as

$$w_{eV}^s = \left[1 - \exp\left(\frac{\hbar\omega_1}{T_e} - \frac{\hbar\omega_1}{T_v^s}\right)\right] n_{CO_2} n_e k_{eV}^1 \hbar\omega_1 + \left[1 - \exp\left(\frac{\hbar\omega_2}{T_e} - \frac{\hbar\omega_2}{T_v^s}\right)\right] n_{CO_2} n_e k_{eV}^2 \hbar\omega_2.$$

$$(5\text{–}41)$$

Similarly, the vibrational excitation rate of CO molecules (mode 4; the rate coefficient of the excitation is k_{eV}^4; and CO vibrational temperature is T_v^4) can be given as

$$w_{eV}^{CO} = \left[1 - \exp\left(\frac{\hbar\omega_4}{T_e} - \frac{\hbar\omega_4}{T_v^4}\right)\right] n_{CO} n_e k_{eV}^4 \hbar\omega_4.$$

$$(5\text{–}42)$$

Finally, the relations describing the influence of chemical reactions on the average energy of vibrational modes can be presented as

$$w_R^{(a,s,CO)} = \sum k_R^{nm} n_n n_m \Delta Q_{(a,s,CO)}^{nm}.$$

$$(5\text{–}43)$$

Summation (5–43) covers all chemical reactions related to formation or consumption of molecules with a given vibrational mode (a, s, CO). When a CO_2 molecule reacts from the asymmetric mode, energy losses from symmetric modes correspond to their average value, and vice versa. As was mentioned earlier, the reaction terms $\Delta Q_{(a,s,CO)}^{nm}$ also represent vibrational energy losses related to the anharmonicity of VV exchange and VT relaxation from high energy levels, which are usually attributed to a single act of chemical reaction (see Chapter 3). To be solved numerically, aforementioned equations of the two-temperature approximation of CO_2 dissociation in plasma require quite an extensive kinetic database. This database was put together based on numerous publications related to vibrational and chemical kinetics in plasma and lasers: Bekefi (1976); Chernyi et al. (2002); Gordiets, Osipov, and Shelepin (1980); Losev (1977); Eletsky, Palkina, and Smirnov (1975); Ablekov, Denisov, and Lubchenko (1982); Blauer and Nickerson (1974); Achasov and Ragosin

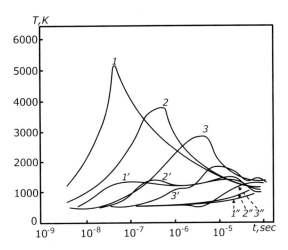

Figure 5–16. Time evolution of vibrational asymmetric T_{va} (curves 1–3), vibrational symmetric T_{vs} (curves 1'–3'), and translational T_0 (curves 1''–3'') temperatures of CO₂ molecules at the specific energy input $E_v = 0.5\,\text{eV/mol}$ and different plasma ionization degrees: $n_e/n_0 = 3 \cdot 10^{-6}$ (curves 1), $n_e/n_0 = 10^{-6}$ (curves 2), $n_e/n_0 = 3 \cdot 10^{-7}$ (curves 3).

(1986); Schultz (1976); Molchanov and Starik (1984); Birukov, Kulagin, and Shelepin (1977); Taylor and Bitterman (1969); Ormonde (1975); Baulch et al. (1976); Kondratiev (1971).

A simplified set of chemical reactions numerically describing neutral chemistry of the CO₂ plasma dissociation is presented in Table 5–1. Elementary chemical reactions are accompanied in the table by corresponding activation energies, pre-exponential factors (given in cm³/s for bimolecular reactions, cm⁶/s for three-body processes), and efficiencies α of vibrational energy in overcoming the activation energy barrier.

5.3.3. Results of CO₂ Dissociation Modeling in the Two-Temperature Approach to Vibrational Kinetics

Numerical solutions of equations (5–29)–(5–35) give the time evolution of major vibrational and translational temperatures of the plasma-chemical system (T_v^a, T_v^s, T_v^{CO}, and T_0) at different ionization degrees n_e/n_0 and different specific energy inputs E_v(eV/mol). The time evolution of CO₂ asymmetric and symmetric vibrational temperatures is shown in Fig. 5–16 together with translational gas temperature at a typical value of specific energy input $E_v = 0.5\,\text{eV/mol}$ and different levels of electron density (Fridman, 1986). The degree of thermal non-equilibrium (difference between vibrational [T_v] and translational [T_0] gas temperatures) grows with an increase of the plasma ionization degree. Also, an interesting effect of energy oscillations between asymmetric and symmetric vibrational modes, related to the non-linearity of interaction between the modes, can be observed in Fig. 5–16 at relatively low degrees of ionization (Novobrantsev & Starostin, 1974). The energy efficiency of CO₂ dissociation in non-equilibrium plasma, calculated in the two-temperature approximation as a function of ionization degree at a fixed value of specific energy input $E_v = 0.5\,\text{eV/mol}$, is shown in Fig. 5–17. This function has a threshold at an ionization degree of about $3 \cdot 10^{-7}$ and a tendency to saturate at higher ionization degrees (see Fig. 5–17). The two-temperature calculations of the CO₂ dissociation energy efficiency dependence on specific energy input $\eta(E_v)$ are presented in Fig. 5–2 in comparison with other calculations and numerous experimental results obtained in different discharges. The ionization degree is assumed fixed in this case ($n_e/n_0 = 3 \cdot 10^{-6}$). As one can see from the figure, the maximum value of the energy efficiency, about 80%, can be achieved in the case of subsonic gas flow through a non-thermal discharge at optimal values of specific energy input $E_v = 0.8$–$1.0\,\text{eV/mol}$. Fig. 5–2 also illustrates an important qualitative feature of the function $\eta(E_v)$ for the CO₂ dissociation process – the threshold in the energy efficiency dependence on specific energy input. This threshold is about 0.2 eV/mol for subsonic completion of the plasma-chemical

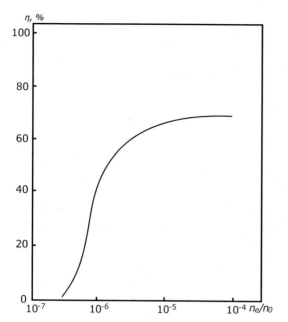

Figure 5–17. Energy efficiency of CO_2 dissociation in plasma at the specific energy input $E_v = 0.5\,\text{eV/mol}$ as a function of plasma ionization degree n_e/n_0.

system. The same dependence $\eta(E_v)$ for CO_2 dissociation in supersonic discharges with an initial gas temperature of $T_{00} = 300\,\text{K}$ (before supersonic nozzle) is also shown in Fig. 5–2. The maximum value of the energy efficiency in this case can reach a record high value of 90% (because of small VT losses at gas temperatures in a discharge zone much below room temperature). The $\eta(E_v)$ threshold position in this case shifts for the same reason to values about ten times lower than the specific energy input.

5.3.4. One-Temperature Approach to Vibrational Kinetics and Energy Balance of CO_2 Dissociation in Non-Equilibrium Plasma: Major Equations

The one-temperature approximation (see Section 5.2.4) permits one to find out all major qualitative features of the plasma-chemical CO_2 dissociation energy balance, including threshold, maximum, and decrease of energy efficiency dependence on specific energy input $\eta(E_v)$, energy efficiency dependence on ionization degree, initial gas temperature, and Mach number in the case of conducting the process in a supersonic flow. The one-temperature approach takes into account a single vibrational temperature T_v and, hence, a single vibrational energy $(\varepsilon_v(T_v))$ of a CO_2 molecule, which is determined by the following balance equation:

$$\frac{d\varepsilon_v}{dt} = k_{eV}n_e\hbar\omega_a \times \theta \left(E_v - \int_0^t k_{eV}n_e\hbar\omega_a dt \right) - k_R(T_v)n_0\Delta Q_\Sigma - k_{VT}^S(T_0)n_0 \left(\varepsilon_v - \varepsilon_v^0 \right).$$

$$(5\text{--}44)$$

The time evolution of translational temperature of the gas (T_0) follows the related kinetic equation describing the translational and rotational energy balance (where c_T is the specific heat corresponding to CO_2 non-vibrational degrees of freedom):

$$c_T \frac{dT_0}{dt} = \beta \cdot k_R(T_v)n_0\Delta Q_\Sigma + k_{VT}^s(T_0)n_0 \left(\varepsilon_v - \varepsilon_v^0 \right). \qquad (5\text{--}45)$$

The first term in (5–44) describes the vibrational excitation of CO_2 molecules by electron impact with the effective rate coefficient, k_{eV}, calculated with respect to the asymmetric vibrational quantum $\hbar\omega_a$. The step function in this term, $\theta(E_v - \int_0^t k_{eV}n_e\hbar\omega_a dt)$, represents the fact that vibrational excitation takes place only in an active discharge zone ($n_e \neq 0$) until each CO_2 molecule receives energy E_v corresponding to the specific energy input (energy losses related to excitation of dissociation products will be considered later in Section 5.3). As always, n_e is electron density and n_0 is neutral gas density. The third term in equation (5–44) and its equivalent in equation (5–45) describe VT relaxation from low vibrational levels of deformation mode (rate coefficient k_{VT}^s), which is significantly faster than that from other modes; ε_v^0 is the vibrational energy of a molecule in quasi-equilibrium conditions, while $T_v = T_0$. The second term in equation (5–44) takes into account the energy consumption in the chemical reaction for CO_2 dissociation proceeding with a rate coefficient $k_R(T_v)$ (see Section 5.2.4 and equation (5–22)). This term also covers (through generalization of ΔQ_Σ) the vibrational energy losses related to VT relaxation from high vibrational levels and to anharmonicity of the VV exchange, which take place with rates proportional to those of the chemical reaction (see Chapter 3). The energy parameter ΔQ_Σ is the total vibrational energy consumed per single event of CO_2 dissociation; the parameter β in the first right-hand term of equation (5–45) shows the portion of the total energy ΔQ_Σ going not to chemical reaction but to translational and rotational degrees of freedom.

5.3.5. Threshold Values of Vibrational Temperature, Specific Energy Input, and Ionization Degree for Effective Stimulation of CO₂ Dissociation by Vibrational Excitation of the Molecules

Consumption of vibrational energy in the plasma-chemical system under consideration takes place in two competitive processes: chemical reaction and VT relaxation, illustrated in (5–44) by the second and third terms. It is important that the chemical reaction rate grows much faster with vibrational temperature than the VT relaxation rate. Taking into account the very strong exponential behavior of the function $k_R(T_v)$, one can determine the **critical vibrational temperature** T_v^{min}: most of the vibrational energy is going to CO_2 dissociation if $T_v > T_v^{min}$, and most of the vibrational energy is lost to VT relaxation into translational degrees of freedom if $T_v < T_v^{min}$. The value of the critical vibrational temperature T_v^{min} can be found by comparison of the second and third terms in (5–44):

$$T_v^{min} = \frac{E_a}{\ln\left[\dfrac{k_{vv}^0\hbar\omega_a\Delta Q_\Sigma}{k_{VT}(T_0)T_v^{min}E_a}\dfrac{1}{4(s-1)!}\left(\dfrac{E_a}{T_v^{min}}\right)^{s+1}\right] + \dfrac{x_m E_a^2}{T_0\hbar\omega_a}}. \tag{5–46}$$

Typical numerical values of the critical vibrational temperature are about 1000 K. The existence of the critical minimal vibrational temperature explains the threshold in the dependence $\eta(E_v)$ for the energy efficiency of CO_2 dissociation on the specific energy input. The specific energy input is the average plasma energy delivered into the gas, calculated per molecule or per unit volume. In stationary systems, the specific energy input can be calculated as a ratio of discharge power and gas flow rate. The threshold value of the specific energy input, $(E_v)_{thr}$, related to the critical vibrational temperature T_v^{min} (5–46) is determined by the average vibrational energy of a CO_2 molecule in the critical conditions when $T_v = T_v^{min}$ and the discharge energy consumption into chemical reaction and to VT relaxation are equal:

$$(E_v)_{thr} = \varepsilon_v(T_v = T_v^{min}) = \sum_i \frac{g_i\hbar\omega_i}{\exp\left(\dfrac{\hbar\omega_i}{T_v^{min}}\right) - 1}. \tag{5–47}$$

The expression for the average vibrational energy (5–47) includes summation over all CO_2 vibrational modes (i) with statistical weights g_i when necessary. When the specific energy input is less than the critical one ($E_v < (E_v)_{thr}$), the vibrational temperature cannot reach the critical value $T_v < T_v^{min}$, and energy losses into VT relaxation are always exponentially faster than energy consumption into CO_2 dissociation, which results in a low energy efficiency. The $\eta(E_v)$ threshold can be observed in experimental and simulation results presented in Fig. 5–2. Numerically, the threshold is about 0.2 eV/mol for subsonic plasma-chemical systems, and it is about 0.03 eV/mol for the supersonic systems. Comparison of the first and third terms in the vibrational energy balance equation (5–44) together with requirement of $T_v > T_v^{min}$ leads to a restriction of ionization degree to obtain efficient CO_2 dissociation. Vibrational excitation by electron impact is faster than VT relaxation, and the CO_2 dissociation process is energy effective if the ionization degree exceeds its critical value:

$$\frac{n_e}{n_0} > \left(\frac{n_e}{n_0}\right)_{thr} = \frac{k_{VT}^s \cdot \varepsilon_v(T_v^{min})}{k_{eV} \cdot \hbar\omega_a}. \tag{5–48}$$

Numerically, the threshold value of the ionization degree n_e/n_0 is about $3 \cdot 10^{-7}$, which is in good agreement with results presented in Fig. 5–17.

5.3.6. Characteristic Time Scales of CO_2 Dissociation in Plasma Stimulated by Vibrational Excitation of the Molecules: VT-Relaxation Time

Effective CO_2 dissociation takes place not only in the active discharge zone (residence time τ_{eV}), where the vibrational temperature is stationary and exceeds the critical one ($T_v > T_v^{min}$), but also in the discharge "after-glow" or in the discharge passive zone (residence time τ_p), where the vibrational temperature decreases (but still $T_v > T_v^{min}$). Therefore, residence times in both the discharge τ_{eV} and in the after-glow τ_p should be compared with the VT-relaxation time τ_{VT}, which characterizes the time interval before the explosive growth of translational temperature and as a consequence loss of the VT non-equilibrium. To describe the explosive growth of translational temperature and find out τ_{VT}, the balance equation (5–45) can be solved analytically by assuming that most heating is due to VT relaxation from low vibrational levels. The time interval necessary for the translational temperature growth from the initial value of T_0^0 to some selected final value T_0^f can be found as

$$t_{VT}\left(T_0^0, T_0^f\right) = \tau_{VT} \cdot \sum_{r=0}^{\infty} \frac{(3+r)!}{6r!} \cdot \frac{\gamma(r+1, \delta)}{(u_0)^r}. \tag{5–49}$$

In this relation, $\delta = u_0 - u_f$, $u_{0,f} = B(T_0^{0,f})^{-1/3}$, B is a constant of the Landau-Teller formula (see Chapter 2), and $\gamma(r+1, \delta)$ is the incomplete gamma-function. The characteristic time τ_{VT} of vibrational relaxation is determined in (5–49) as follows:

$$\tau_{VT} = \frac{c_T}{k_{VT}(T_0^0)n_0\hbar\omega_S} \cdot \frac{T_0^0}{\hat{k}_{VT}}, \tag{5–50}$$

where $\hat{k}_{VT} = \frac{\partial \ln k_{VT}}{\partial \ln T_0}$ is the logarithmic sensitivity of the VT-relaxation rate coefficient for CO_2 molecules to the variation of their translational temperature, which can be found based on the Landau-Teller formula (see Chapter 2). Results of numerical calculations of the function $t_{VT}(T_0^0, T_0^f)$ (5–49) at an initial gas temperature $T_0^0 = 300$ K are shown in Fig. 5–18 and actually represent the time evolution of translational temperature during CO_2 dissociation in a strongly non-equilibrium plasma ($T_v > T_0$). Growth of the translational temperature is clearly "explosive" with a characteristic time τ_{VT} determined by (5–50).

5.3. Vibrational Kinetics and Energy Balance of Plasma-Chemical CO_2 Dissociation

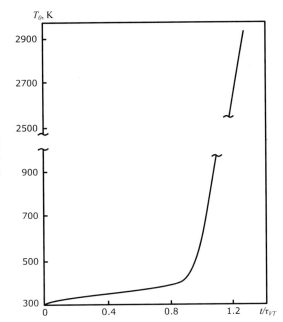

Figure 5–18. Time evolution of translational temperature during VT relaxation; one-temperature approximation of CO_2 dissociation in plasma.

5.3.7. Flow Velocity and Compressibility Effects on Vibrational Relaxation Kinetics During Plasma-Chemical CO_2 Dissociation: Maximum Linear Preheating Temperature

The CO_2 flow velocity and compressibility were not taken into account in the purely kinetic equation (5–45). Taking into account the fluid mechanics effects leads to the following correction of formula (5–50) for characteristic VT-relaxation time (Fridman, 1986):

$$\tau_{VT} = \frac{c_p^*}{k_{VT}n_0\hbar\omega_s} \cdot \frac{T_0^0}{\hat{k}_{VT} + 2/(\gamma M_0^2 - 1)}, \tag{5–51}$$

where M_0 is the Mach number corresponding to the reaction zone inlet; γ is the specific heat ratio; and $c_p^* = c_p \frac{M_0^2 - 1}{\gamma M_0^2 - 1}$ is an effective specific heat corresponding to c_p at low velocities and corresponds to c_v in strongly supersonic conditions ($M_0 \gg 1$). Effects of gas compressibility and flow velocity are significant mostly when the initial velocity is near the speed of sound. Translational temperature growth is close to being linear in the beginning of the relaxation process when $t < \tau_{VT}$ (see Fig. 5–18):

$$T_0 = T_0^0 + \frac{\Delta T_{0,VT}}{\tau_{VT}}t. \tag{5–52}$$

After the time interval exceeds the critical one ($t > \tau_{VT}$), the translational temperature growth becomes "explosive" (see Fig. 5–18). The explosion of T_0 is due to the strong exponential dependence of the VT-relaxation rate on temperature (see Chapter 2, Landau-Teller formula). VT-relaxation determines gas heating and increase of translational temperature, which then leads to an exponential acceleration of the VT relaxation and further and faster gas heating. A linear growth of translational temperature is limited by the characteristic time τ_{VT} and, hence, by the maximum linear preheating before the explosion:

$$\Delta T_{0,VT} = \frac{T_0^0}{\hat{k}_{VT} + 2/(\gamma M_0^2 - 1)}. \tag{5–53}$$

Thus, the translational temperature grows linearly due to VT relaxation from the initial temperature T_0^0 (which is usually room temperature) to the maximum one $T_0^0 + \Delta T_{0,VT}$

before explosion. The maximum linear preheating temperature is less than T_0^0 mostly due to the logarithmic sensitivity $\hat{k}_{VT} = \frac{\partial \ln k_{VT}}{\partial \ln T_0} \gg 1$; in other words, the small increase of translational temperature leads to its following explosive growth. Numerically, the maximum linear preheating temperature is usually less than 300 K, which can be observed in Fig. 5–18. A more detailed kinetic consideration, taking into account VT relaxation from high vibrational levels, heating through non-resonant VV-relaxation chemical reactions, and so on, can be found, for example, in Rusanov and Fridman (1984); and Givotov, Rusanov, and Fridman (1982, 1984).

5.3.8. CO₂ Dissociation in Active and Passive Discharge Zones: Discharge (τ_{eV}) and After-Glow (τ_p) Residence Time

If the ionization degree is sufficient (5–48), the stationary vibrational temperature T_v^{st} is established after a necessary time interval of about $\approx 1/k_{eV}n_e$. The stationary vibrational temperature T_v^{st} in the active discharge zone can be found from the energy balance equation (5–44), taking the CO₂ dissociation rate in the framework of the one-temperature approximation (compare to (5–46)):

$$T_v^{st} = \frac{E_a}{\ln\left[\dfrac{k_{vv}^0 n_0}{k_{eV}n_e 4(s-1)!}\left(\dfrac{E_a}{T_v^{st}}\right)^{s+1}\right] + \dfrac{x_m E_a^2}{T_0 \hbar \omega_a}}. \tag{5–54}$$

Typical numerical values of the stationary vibrational temperature in the active discharge zone are about 3000 K and exceed the critical vibrational temperature T_v^{min} (5–46). The residence time of CO₂ molecules in the active discharge zone can be expressed as a total time interval of vibrational excitation:

$$\tau_{eV} = \frac{E_v}{k_{eV}n_e \hbar \omega_a}. \tag{5–55}$$

Because optimal energy inputs E_v are usually about 1 eV/mol (see Chapter 3 and Fig. 5–2), the total residence time of CO₂ molecules in the active zone, τ_{eV}, is longer than the time interval $\approx 1/k_{eV}n_e$ necessary for establishment of the stationary vibrational temperature (5–54). When gas leaves the active zone, CO₂ dissociation continues in the "after-glow" (passive zone) during the additional time interval τ_p, where vibrational temperature decreases from T_v^{st} to T_v^{min}. Solving the energy equation (5–44), the time interval τ_p can be found as (Rusanov & Fridman, 1984; Givotov et al., 1982, 1984):

$$\tau_p = \frac{c_p^* \Delta T_p}{k_{VT}(T_0^m)n_0 \hbar \omega_s}. \tag{5–56}$$

where superscript m means that the physical parameter is taken at the critical point $T_v = T_v^{min}$; the subscript 0 means that it is taken at the discharge inlet, in the initial moment of the process. ΔT_p is the translational temperature increase during the effective CO₂ dissociation in the after-glow:

$$\Delta T_p = \frac{c_v^m}{c_p^*}\Delta T_{v,r}^0 \left(1 - \frac{c_v^m}{2c_p^*}\cdot\frac{\Delta T_{v,r}^m}{\Delta T_{0,r}^m}\right). \tag{5–57}$$

The sensitivity parameters $\Delta T_{v,r}^m$ and $\Delta T_{v,r}^m$ (similar to $\Delta T_{0,VT}$, see (5–53)) show intervals of vibrational and translational temperatures corresponding to a significant change in the CO₂ dissociation rate:

$$\Delta T_{v,r}^m = \frac{(T_v^{min})^2}{E_a}\left(1 + \frac{T_v^{min}}{E_a}\right), \quad \Delta T_{0,r}^m = \frac{(T_0^m)^2 \hbar \omega_a}{x_m E_a^2}\left(1 + \frac{2T_0 \hbar \omega_a}{x_m E_a^2}\right). \tag{5–58}$$

When the translational temperature corresponding to $T_v = T_v^{min}$ is high enough (numerically $T_0^m \geq 500\,K$), the CO_2 dissociation rate sensitivity to T_v is stronger than to T_0: $c_v^m \Delta T_{v,r}^m < 2c_p^* \Delta T_{0,r}^m$. Therefore, parameter ΔT_p (5–57), characterizing the T_0 increase during the effective CO_2 dissociation in the after-glow, is positive. Comparing relations (5–51) and (5–56), one can see that both τ_{VT} and τ_p are mostly determined by the same vibrational relaxation frequency $k_{VT} n_0$. This results in a clear relation between the chemical reaction time in the after-glow, τ_p, and the time interval τ_{VT} before the explosive growth of translational temperature and loss of the effective non-equilibrium conditions:

$$\frac{\tau_p}{\tau_{VT}} = \frac{c_v^m \Delta T_v^r}{c_p^* \Delta T_{0,VT}^m}. \tag{5–59}$$

Numerically, when the translational temperature is not very high ($T_0 \leq 2000\,K$), the dissociation sensitivity to vibrational temperature is stronger than VT-relaxation sensitivity to translational temperature ($\Delta T_{0,VT}^m > \Delta T_{v,r}^m$). As a result, relation (5–57) leads to an important qualitative conclusion: *reaction in after-glow is always faster than VT relaxation* ($\tau_p < \tau_{VT}$).

5.3.9. Ionization Degree Regimes of the CO₂ Dissociation Process in Non-Thermal Plasma

Analysis of the three characteristic time scales – τ_{eV} (residence time in discharge zone), τ_p (effective reaction time in after-glow), and τ_{VT} (vibrational relaxation time) – shows that only the first is determined by ionization degree n_e/n_0, whereas the two others are connected ($\tau_p < \tau_{VT}$). The analysis determines three ionization degree regimes of the process stimulated by vibrational excitation.

A. High ionization degree. The CO_2 dissociation process including the after-glow reaction in this regime is completed before the thermal explosion:

$$\tau_{eV} + \tau_p < \tau_{VT}. \tag{5–60}$$

The ionization degree required for this, the most energy-effective regime, should satisfy the requirement somewhat stronger than (5–48):

$$\frac{n_e}{n_0} > \frac{k_{VT}(T_0^m)\hbar\omega_s}{k_{eV}\hbar\omega_a} \cdot \frac{E_v}{c_p^* \Delta T_p} \left[\frac{\Delta T_{0,VT}^0}{\Delta T_p} \cdot \frac{k_{VT}(T_0^m)}{k_{VT}(T_0^0)} - 1 \right]^{-1}. \tag{5–61}$$

The energy efficiency of the plasma-chemical processes is usually divided into three factors: excitation factor (η_{ex}), relaxation factor (η_{rel}), and chemical factor (η_{chem}). The most important factor in the case of CO_2 dissociation stimulated by vibrational excitation is the relaxation factor, which shows the fraction of the total vibrational energy going not into VT relaxation but rather to the useful chemical reaction. The relaxation factor is almost equal in this case to the total energy efficiency. Neglecting energy losses related to the excitation of the dissociation products (see next section), the relaxation factor can be expressed as

$$\eta_{rel} = \frac{E_v - (\tau_{eV} + \tau_p)k_{VT}(T_0^0)n_0\hbar\omega_s - (1-\chi)\varepsilon_v(T_v = T_v^{min})}{E_v}. \tag{5–62}$$

Taking into account that the CO_2 conversion degree χ into CO can be expressed as

$$\chi = \frac{E_v \eta_{rel} \eta_{chem}}{\Delta H}, \tag{5–63}$$

the energy efficiency (5–62) can be finally rewritten as

$$\eta_{rel} = \frac{E_v - (\tau_{eV} + \tau_p)k_{VT}(T_0^0)n_0\hbar\omega_s - \varepsilon_v(T_v^{min})}{E_v \cdot [1 - \varepsilon_v(T_v^{min})\eta_{chem}/\Delta H]}. \tag{5–64}$$

This formula shows the threshold in $\eta(E_v)$ and growth of the CO_2 dissociation energy efficiency with increasing ionization degree. If the ionization degree is so high that $\tau_{eV}(n_e/n_0) \gg \tau_p$, then the energy efficiency cannot increase any more with ionization degree and becomes saturated (see Fig. 5–17). The ionization degree necessary for the saturation can be found from the equation (5–64):

$$\frac{n_e}{n_0} > \frac{k_{VT}(T_0^m)\hbar\omega_s}{k_{eV}\hbar\omega_a} \cdot \frac{E_v}{c_p^*\Delta T_p}.$$
(5–65)

B. Low ionization degree. This regime corresponds to electron densities so low that requirement (5–48) is not satisfied:

$$\frac{n_e}{n_0} < \left(\frac{n_e}{n_0}\right)_{thr} = \frac{k_{VT}^s \cdot \varepsilon_v(T_v^{min})}{k_{eV} \cdot \hbar\omega_a}.$$
(5–66)

In this case vibrational excitation is slower than VT relaxation and the vibrational temperature cannot reach the critical value of $T_v = T_v^{min}$. Effective CO_2 dissociation through vibrational excitation of the molecules becomes impossible in such conditions (see Fig. 5–17).

C. Intermediate ionization degree. This regime takes place when requirement (5–48) is satisfied, the critical value of $T_v = T_v^{min}$ can be reached, but the ionization degree is not high enough:

$$\frac{n_e}{n_0} < \frac{k_{VT}(T_0^0)\hbar\omega_s}{k_{eV}\hbar\omega_a} \cdot \frac{E_v}{c_p^*\Delta T_{0,VT}^0},$$
(5–67)

and the thermal explosion of VT relaxation takes place before the effective chemical reaction is completed. In this regime, the effective CO_2 dissociation can take place only in the beginning of the gas residence time in the active discharge zone when $t < \tau_{VT}$. During the rest of the residence time $\tau_{VT} < t < \tau_{eV}$, and most of discharge energy accumulated in molecular vibrations goes to VT relaxation and gas heating. The relaxation factor of energy efficiency can be expressed in this case as

$$\eta_{rel} = \frac{[k_{eV}n_e\hbar\omega_a - k_{VT}(T_0^0)n_0\hbar\omega_s]\tau_{VT}^0 - \varepsilon_v(T_v^{st})}{E_v \cdot [1 - \eta_{chem}\varepsilon_v(T_v^{st})/\Delta H]}.$$
(5–68)

This energy efficiency decreases hyperbolically with the specific energy input ($\eta \propto 1/E_v$), in good agreement with experiments (see Fig. 5–2). This decrease takes place when E_v exceeds the optimal value $E_v \geq E_v^{max}$, where the energy efficiency of CO_2 dissociation reaches a maximum:

$$E_v^{max} = c_p^*\Delta T_{0,VT}^0 \frac{k_{eV}\hbar\omega_a}{k_{VT}(T_0^0)\hbar\omega_s} \frac{n_e}{n_0}.$$
(5–69)

5.3.10. Energy Losses Related to Excitation of CO_2 Dissociation Products: Hyperbolic Behavior of Energy Efficiency Dependence on Specific Energy Input

The CO_2 conversion into CO and O_2 approaches 100% at sufficient values of specific energy input. The high conversion, achieved already in the active discharge zone, leads to significant energy losses related to the excitation of dissociation, products, and especially

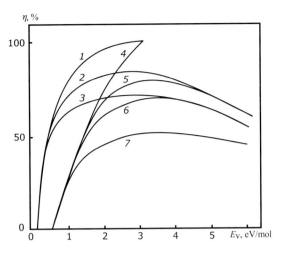

Figure 5–19. One-temperature approximation of energy efficiency of CO$_2$ dissociation as a function of specific energy input, specifically taking into account energy losses related to excitation of the dissociation products: $\varepsilon_{CO} = 0\,\mathrm{eV}$ (curves 1 and 4), $\varepsilon_{CO} = 1\,\mathrm{eV}$ (curves 2 and 5), $\varepsilon_{CO} = 2.8\,\mathrm{eV}$ (curves 3 and 6), $\varepsilon_{CO} = 5\,\mathrm{eV}$ (curve 7); $\varepsilon_v^{CO_2}(T_v^{min}) = 0.2\,\mathrm{eV}$ (curves 1, 2, and 3), $\varepsilon_v^{CO_2}(T_v^{min}) = 0.7\,\mathrm{eV}$ (curves 4, 5, 6, and 7).

CO. Calculations of the relaxation factor of energy efficiency of CO$_2$ dissociation, taking into account the energy losses related to vibrational excitation of CO, were performed by Liventsov, Rusanov, and Fridman (1984a,b):

$$\eta_{rel} = \frac{\Delta H}{E_v}\left\{1 - \exp\left[-\frac{E_v - \varepsilon_v^{CO_2}\left(T_v^{min}\right)}{\varepsilon_{CO}}\cdot\left(1 + \frac{\varepsilon_{CO} + \varepsilon_v^{CO_2}\left(T_v^{min}\right) - \Delta H}{\Delta H - 2\varepsilon_v^{CO_2}\left(T_v^{min}\right) + E_v}\right)\right]\right\}.$$
(5–70)

In this relation, ΔH is the enthalpy of CO$_2$ dissociation, which is not much different from the activation energy E_a; $\varepsilon_v^{CO_2}(T_v^{min})$ is the average total vibrational energy of a CO$_2$ molecule corresponding to the critical temperature T_v^{min} and determining the threshold of the $\eta(E_v)$ dependence (see (5–47)); the energy factor ε_{CO} characterizes the ratio of reaction rates of vibrational excitation by electron impact for CO (k_{eV}^{CO}) and CO$_2$ ($k_{eV}^{CO_2}$) molecules:

$$\varepsilon_{CO} = \Delta H \frac{k_{eV}^{CO}\hbar\omega_{CO}}{k_{eV}^{CO_2}\hbar\omega_a},$$
(5–71)

where $\hbar\omega_{CO}$ is the vibrational quantum of CO, and $\hbar\omega_a$ is that of the asymmetric mode of CO$_2$. Asymptotic presentation of $\eta_{rel}(E_v)$ at high-specific-energy inputs $E_v \to \infty$ according to (5–70) is

$$\eta_{rel}(E_v) = \frac{\Delta H}{E_v} \propto 1/E_v.$$
(5–72)

The asymptotic solution describes the hyperbolic behavior of the function $\eta_{rel}(E_v)$ at high values of the specific energy input. Results of numerical calculations of the relaxation factor of energy efficiency of CO$_2$ dissociation are presented in Fig. 5–19, taking into account the energy losses related to vibrational excitation of CO in plasma for different specific values of the energy factor ε_{CO} (Liventsov, Rusanov, & Fridman, 1984a,b). Reduction of the energy factor ε_{CO} results in an increase of the energy efficiency and in some shift of the curve $\eta_{rel}(E_v)$ to the higher values of specific energy input. The best agreement of the calculations with experimental results can be achieved when $\varepsilon_v(T_v^{min}) = 0.2\,\mathrm{eV}$ and $\varepsilon_{CO} = 2\,\mathrm{eV}$. Most qualitative and quantitative features of kinetics and energy balance of CO$_2$ dissociation in non-thermal plasma can be analyzed based on Fig. 5–2. The most important qualitative features are the threshold and hyperbolic behavior of the $\eta(E_v)$ dependence.

The most important quantitative features are the possibility of getting the conversion level close to 100%, and a uniquely high energy efficiency of the process, which reaches 90% in strongly non-equilibrium conditions. Taking into account that non-thermal discharge energy can be dispersed over thousands of different reaction, radiation, and relaxation channels, the selective localization of 90% discharge energy on the dissociation process is truly impressive.

5.4. ENERGY EFFICIENCY OF CO_2 DISSOCIATION IN QUASI-EQUILIBRIUM PLASMA, AND NON-EQUILIBRIUM EFFECTS OF QUENCHING PRODUCTS OF THERMAL DISSOCIATION

5.4.1. Ideal and Super-Ideal Modes of Quenching Products of CO_2 Dissociation in Thermal Plasma

CO_2 dissociation in thermal plasma includes two phases: heating to temperatures necessary to shift chemical equilibrium in the direction of products (see Fig. 5–3) and fast cooling (quenching, see Fig. 5–4) of the dissociation products to stabilize them from reverse reactions. Major quenching modes (absolute and ideal) were discussed in Chapter 3. The maximum energy efficiency of quasi-equilibrium plasma-chemical systems corresponds to the ideal quenching mode, when the CO_2 conversion degree achieved in the high-temperature (heating) phase is saved during the cooling. Ideal quenching means not only saving all main dissociation products (in this case CO) generated at high temperature, but also conversion during fast cooling of all relevant atoms and radicals such as C, C_2, or C_2O into CO. The CO_2 conversion degree can not only be maintained during the quenching phase but sometimes can even be increased using energy accumulated in atoms, radicals, and internal degrees of freedom of the reagents, first of all in molecular vibrations (Givotov et al., 1984). It can take place when cooling is faster than VT relaxation, and the VT non-equilibrium ($T_v > T_0$) is achieved during quenching. Direct reactions related to dissociation are endothermic and, therefore, significantly stimulated by vibrational energy. Reverse exothermic reactions proceed slower because they cannot be accelerated by elevated vibrational temperature and require mostly translational energy (see the Fridman-Macheret α-model in Chapter 2). This effect is known as super-ideal quenching and makes plasma-chemical CO_2 dissociation a non-equilibrium process even when it is conducted in a thermal discharge with quasi-equilibrium dissociation in the initial high-temperature phase.

5.4.2. Kinetic Evolution of Thermal CO_2 Dissociation Products During Quenching Phase

Possible paths for the time evolution of product composition during the quenching phase can be analyzed using the chemical mechanism for CO_2 dissociation presented in Table 5–1. If the cooling rate is so slow that chemical reactions are able to sustain quasi equilibrium during the "quenching," then products generated at high temperature will be converted back into the initial substance, CO_2. On the other hand, if cooling is very fast, then composition, achieved at high temperature (see Fig. 5–3), remains initially the same after quenching. The major components of the mixture are O, CO, and CO_2. The final composition of the CO_2 dissociation products and, hence, the final conversion degree are determined in this case by the competition of recombination of the oxygen atoms, which doesn't change the conversion degree:

$$O + O + M \rightarrow O_2 + M \text{ (reactions 13–16 in Table 5–1)}, \tag{5–73}$$

and recombination of the oxygen atoms with the dissociation products (CO molecules), which is a reverse reaction and obviously leads to a decrease of the conversion degree:

$$O + CO + M \rightarrow CO_2 + M \text{ (reactions 4–6 in Table 5–1).} \quad (5\text{–}74)$$

As one can see from Table 5–1, the reaction rate coefficients for recombination (5–73) are more than two orders of magnitude higher than those of recombination reactions (5–74). For this reason, recombination of atomic oxygen at low temperatures doesn't significantly affect the concentration of CO molecules. Practically, if the quenching rate is sufficiently high and exceeds 10^8 K/s, atomic oxygen recombines into O_2 without changing the initial CO_2 conversion. This is a kinetic scenario of the ideal quenching of CO_2 dissociation products.

5.4.3. Energy Efficiency of CO_2 Dissociation in Thermal Plasma Under Conditions of Ideal Quenching of Products

The thermal plasma energy, required for heating up CO_2 from room temperature to a necessary high temperature T at constant pressure p, can be expressed by taking into account dissociation:

$$\Delta W(T) = \frac{\sum x_i(T) I_i(T)}{x_{CO_2}(T) + x_{CO}(T)} - I_{CO_2}(T = 300\,\text{K}). \quad (5\text{–}75)$$

In this relation, $x_i(p, T) = n_i/n_0$ is the relative concentration of component i of the mixture of dissociation products; I_i is the total enthalpy of component i; and n_0 is the total density of the quasi-equilibrium gas mixture. Let us assume that each CO_2 molecule gives β CO molecules after quenching; therefore, the parameter β characterizes the conversion degree of the dissociation process. Then the energy efficiency of the quasi-equilibrium dissociation process can be presented as

$$\eta = \frac{\Delta H_{CO_2}}{\Delta W(T)/\beta(T)}, \quad (5\text{–}76)$$

where $\Delta H_{CO_2} = 2.9\,\text{eV/mol}$ is the enthalpy of the CO_2 decomposition into CO and O_2. In the dissociation, one CO_2 molecule gives one CO molecule; therefore, the parameter β for the ideal quenching mode is equal to the fraction of CO_2 molecules that have been decomposed in the high-temperature phase:

$$\beta^0 = \frac{x_{CO}(T)}{x_{CO_2}(T) + x_{CO}(T)}. \quad (5\text{–}77)$$

Combination of equations (5–75)–(5–77) permits one to derive the formula for calculations of the energy efficiency of thermal CO_2 dissociation in plasma in the case of ideal quenching:

$$\eta = \Delta H_{CO_2} \left[\frac{\sum x_i I_i(T)}{x_{CO}} - \frac{x_{CO_2} + x_{CO}}{x_{CO}} I_{CO_2}(T = 300\,\text{K}) \right]^{-1}. \quad (5\text{–}78)$$

The cooling rate limitations required to reach the conditions of ideal quenching can be estimated from Fig. 5–4; one can say that a cooling rate of about 10^8 K/s is practically sufficient at atmospheric pressure to use relation (5–78) for practical calculations. The energy efficiency of thermal CO_2 dissociation with ideal quenching $\eta(T)$, calculated as a function of temperature in the thermal zone using equation (5–78) for pressure ($p = 0.16\,\text{atm}$), is presented in Fig. 5–20. The maximum value of energy efficiency (about 50%) can be achieved at temperature 2900 K.

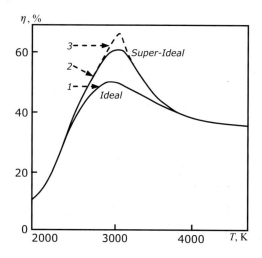

Figure 5–20. CO_2 thermal dissociation energy efficiency as a function of heating temperature in plasma; gas pressure $p = 0.16\,atm$: (1) ideal quenching, (2) super-ideal quenching, (3) upper limit of super-ideal quenching related to energy balance.

5.4.4. Vibrational–Translational Non-Equilibrium Effects of Quenching Products of Thermal CO_2 Dissociation in Plasma: Super-Ideal Quenching Mode

The quenching of CO_2 dissociation products is related to the fast cooling of the gas mixture and, therefore, can lead to the essential separation of vibrational and translational temperatures. The fast cooling indeed means an initially fast decrease of only the translational temperature, followed by a secondary reduction of vibrational temperature through VT relaxation, similarly to the case of conventional gasdynamic lasers (see, for example, Losev, 1977). So if the process of cooling of the dissociation products proceeds faster than VT relaxation, a notable difference between vibrational and translational temperature

$$T_v > T_0 \tag{5–79}$$

can be achieved during the quenching. In the case of an essentially non-equilibrium cooling process ($T_v > T_0$), the conversion degree of CO_2 into CO and the energy efficiency of the process can exceed the corresponding values described in Section 5.4.3 for the case of ideal quenching. This effect of additional CO_2 decomposition during the quenching phase is usually referred to as super-ideal quenching. The effect becomes possible because $T_v > T_0$, and direct endothermic reactions are stimulated by vibrational excitation, whereas reverse exothermic reactions are determined mostly by the translational gas temperature (see the Fridman-Macheret α-model in Chapter 2, and Table 5–1). Saving the dissociation products in the ideal quenching mode is kinetically related to the competition of the three-body chemical reactions (5–73) and (5–74), from which the recombination of oxygen atoms is faster. A kinetic scenario of the additional CO_2 decomposition during quenching is mostly related to the competition of the direct endothermic reaction of atomic oxygen with CO_2:

$$O + CO_2 \rightarrow CO + O_2, \tag{5–80}$$

which can be significantly accelerated by vibrational energy (see Table 5–1), and the corresponding exothermic reverse reaction of molecular oxygen and CO, which is characterized by low sensitivity to vibrational energy (see Table 5–1):

$$O_2 + CO \rightarrow CO_2 + O. \tag{5–81}$$

Super-ideal quenching takes place when $T_v > T_0$ during the fast product cooling phase, reaction (5–81) stimulated by translation temperature proceeds relatively slowly, and reaction (5–80) effectively consumes the vibrational energy and determines the additional CO_2 decomposition and CO production during the quenching.

5.4.5. Maximum Value of Energy Efficiency of CO₂ Dissociation in Thermal Plasma with Super-Ideal Quenching of the Dissociation Products

To calculate the upper limit of the energy efficiency of thermal CO_2 dissociation in the super-ideal quenching mode let us make the following kinetic assumptions. Cooling is so fast during super-ideal quenching that neither the chemical composition nor the vibrational temperature of molecules is changed until the translational gas temperature approaches room temperature. Also it can be assumed that atomic oxygen mostly reacts with CO_2 (5–81) during the conditions of VT non-equilibrium ($T_v > T_0$); and when the quasi-equilibrium is finally restored ($T_v = T_0$), oxygen atoms mostly recombine (5–73), without decreasing the product concentration (5–74), similarly to the case of ideal quenching. The maximum amount of CO, additionally generated through reaction (5–80) during super-ideal quenching, can be taken as the amount of atomic oxygen in the high-temperature phase if $x_O(T) < x_{CO_2}(T)$, and vice versa as the amount of carbon dioxide in the high-temperature phase if $x_O(T) > x_{CO_2}(T)$. The corresponding maximum energy efficiency of CO_2 dissociation with super-ideal quenching can be expressed as

$$\eta(T) = \frac{\Delta H_{CO_2}}{\Delta W(T)} \cdot \frac{x_{CO}(T) + x_O(T)}{x_{CO_2}(T) + x_{CO}(T)}, \quad \text{if } x_{CO_2} > x_O, \tag{5–82}$$

$$\eta(T) = \Delta H / \Delta W(T), \quad \text{if } x_{CO_2} > x_O. \tag{5–83}$$

In these relations, $\Delta W(T)$ is the heating enthalpy of the reactive mixture characterizing the high-temperature phase of the thermal dissociation process. Results of numerical calculations for the energy efficiency of the plasma-chemical process, which are based on relations (5–82) and (5–83), are presented in Fig. 5–20. The maximum energy efficiency of CO_2 dissociation with super-ideal quenching can reach a quite high value of about 60% for gas temperature in the high-temperature phase, about 3000 K.

5.4.6. Kinetic Calculations of Energy Efficiency of CO₂ Dissociation in Thermal Plasma with Super-Ideal Quenching

Relations (5–82) and (5–83) only determine the maximum value of the energy efficiency without taking into account kinetic limitations related to requirements of non-equilibrium ($T_v > T_0$). Reaction (5–80), which plays the key role in additional CO_2 decomposition during the cooling phase of the plasma-chemical process, dominates the quenching process only before restoration of the VT equilibrium ($\tau_{VT} \approx 1/k_{VT}n_0$). Taking into account that kinetic limitation, the additional number of CO molecules produced during the non-equilibrium quenching per single initial CO_2 molecule can be found (Givotov et al., 1984):

$$\beta_1^1 = \frac{x_{CO_2}}{x_{CO} + x_{CO_2}} \cdot \frac{\exp\left[\frac{k_R}{k_{VT}}(x_O - x_{CO_2})\right] - 1}{\exp\left[\frac{k_R}{k_{VT}}(x_O - x_{CO_2})\right] - \frac{x_{CO_2}}{x_O}}. \tag{5–84}$$

In this relation, k_R is the rate coefficient of reaction (5–80) generating super-ideal quenching CO; both reaction and relaxation rate coefficients (k_R, k_{VT}) as well as relative concentrations of all major gas components (see (5–75); x_O, x_{CO}, x_{CO_2}) are assumed in equation (5–84) corresponding to the initial high-temperature value in the plasma. Relation (5–84) reflects only the kinetic limitations of super-ideal quenching; obviously, the energy balance of the quenching process should also be taken into account to calculate the additional CO_2 dissociation. Generation of the additional amount of CO, predicted by formula (5–84), becomes possible only if the vibrational energy accumulated by CO_2 molecules in the high-temperature plasma zone is sufficient to do that. Consumption of vibrational energy per single elementary reaction (5–80) is determined by the activation energy of the reaction divided by the efficiency of vibrational energy in the reaction: $E_a^{(2)}/\alpha$ (see Table 5–1). Taking into account the limitation related to vibrational energy balance, the additional number of CO molecules produced during non-equilibrium quenching per single initial CO_2 molecule cannot exceed the following value (Givotov et al., 1984):

$$\beta_1^2 = \frac{x_{CO}}{x_{CO} + x_{CO_2}} \cdot \frac{\varepsilon_v^{CO_2} + (x_{CO}/x_{CO_2})\varepsilon_v^{CO}}{E_a^{(2)}/\alpha}. \tag{5–85}$$

Vibrational energies $\varepsilon_v^{CO_2}$, ε_v^{CO} are also taken in (5–85) at vibrational temperature corresponding to the high-temperature plasma zone. Taking into account kinetic (5–84) and energy balance (5–85) limitations, the energy efficiency of thermal CO_2 dissociation with super-ideal quenching is

$$\eta = \frac{\Delta H_{CO_2}}{\Delta W}(\beta_0 + \beta_1^1), \quad \text{if } \beta_1^1 < \beta_1^2, \tag{5–86}$$

$$\eta = \frac{\Delta H_{CO_2}}{\Delta W}(\beta_0 + \beta_1^2), \quad \text{if } \beta_1^1 > \beta_1^2. \tag{5–87}$$

The parameter β^0 characterizes CO generation per CO_2 molecule in the high-temperature plasma zone for the case of ideal quenching (5–77). Results of calculations (based on relations (5–86) and (5–87)) of the energy efficiency for thermal CO_2 dissociation at different temperatures and pressure $p = 0.16$ atm are also presented in Fig. 5–20 (curve 2). The non-equilibrium ($T_v > T_0$) organization of the quenching permits more than 10% improvement of energy efficiency with respect to the case of ideal quenching.

5.4.7. Comparison of Thermal and Non-Thermal Plasma Approaches to CO_2 Dissociation: Comments on Products ($CO-O_2$) Oxidation and Explosion

Fig. 5–20 presents the energy efficiency of CO_2 dissociation in thermal plasma, $\eta(T)$, as a function of gas temperature in a discharge hot zone. To compare this energy efficiency with that related to non-thermal plasma, it is convenient to recalculate it as a function of specific energy input in the discharge $\eta(E_v)$. Such recalculation is possible using equation (4–75), which relates the temperature in a discharge hot zone with the energy necessary to heat up the gas mixture to this temperature, $\Delta W(T)$. This energy of the quasi-equilibrium gas heating (4–75) is equal to the specific energy input in the discharge, assuming its quasi-uniformity. Recalculated this way, the energy efficiency of CO_2 dissociation in thermal plasma with different modes of quenching is presented in Fig. 5–2. The maximum value of the CO_2 dissociation energy efficiency in non-thermal plasma (90%) is significantly higher than that in thermal plasma (60%) even in the case of super-ideal quenching. Optimal values of specific energy input are also different: in thermal plasma they are 5–10 times higher than in non-thermal plasma and are about 3–5 eV/mol. The cooling rates required to avoid significant contribution of reverse reactions and to reach a sufficiently high level of conversion and energy efficiency are about 10^8–10^9 K/s. Non-equilibrium "cold" plasma systems in contrast to thermal discharges don't require quenching of products at all, because their translational temperature is low, whereas vibrational temperature is not effective in the stimulation of reverse reactions, which are exothermic (Evseev et al., 1979).

Explosion in the $CO-O_2$ mixture has a branching-chain nature and, once initiated, results in exponential growth of active species even before the system heats up. The branching-chain explosion is based on the following elementary reactions:

$$CO + O \rightarrow CO_2^*, \quad CO_2^* + O_2 \rightarrow CO_2 + O + O. \tag{5–88}$$

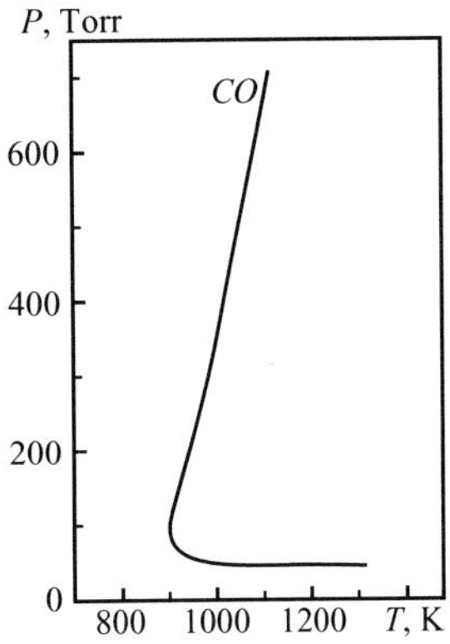

Figure 5–21. Stability limits of CO_2 dissociation products (CO and O_2) outside of discharge zone with respect to chain explosion.

If the plasma-chemical conversion of CO_2 into $CO–O_2$ is relatively low and the explosive mixture is diluted by CO_2, then the excited molecule CO_2^* mostly loses its excitation energy in relaxation collisions instead of dissociating O_2 (5–88), which stabilizes the mixture. The threshold explosion curve of the $CO–O_2$ mixture (known among some students as "Semenov's nose") is presented in Fig. 5–21 (see details, for example, in Lewis & von Elbe, 1987). Reverse conversion of the products ($CO–O_2$) back into CO_2 can happen not only through explosion but also through combustion and slow oxidation, which is usually catalytically sustained by water or other hydrogen-containing substances. These compounds generate a small concentration of atomic hydrogen, which stimulates the chain oxidation of carbon oxide:

$$H + O_2 + M \rightarrow HO_2 + M, \quad HO_2 + CO \rightarrow CO_2 + OH, \quad OH + CO \rightarrow CO_2 + H.$$
$$(5–89)$$

This chain oxidation converts CO back into CO_2 at quite low gas temperatures. Plasma-chemical dissociation in "dry" CO_2 helps to protect products from reverse reactions.

5.5. EXPERIMENTAL INVESTIGATIONS OF CO_2 DISSOCIATION IN DIFFERENT DISCHARGE SYSTEMS

5.5.1. Experiments with Non-Equilibrium Microwave Discharges of Moderate Pressure, Discharges in Waveguide Perpendicular to Gas Flow Direction, and Microwave Plasma Parameters in CO_2

The CO_2 discharge, where the highest energy efficiency (for subsonic flows) was achieved, was organized in a quartz discharge tube crossing a rectangular waveguide perpendicular to its wide wall. The radius of the discharge tube was smaller than the size of the wide waveguide wall and smaller than the skin layer; therefore, electric field in the discharge tube was quite uniform (for more details see Chapter 4 or Fridman & Kennedy, 2004). A schematic of the plasma-chemical microwave reactor applied for CO_2 dissociation is shown in Fig. 5–22 (Butylkin et al., 1981; Krasheninnikov, 1981; Vakar et al., 1981a,b). The microwave frequency in the experiment was 2.4 GHz, the typical power level was 1.5 kW, the diameter of the quartz tube was 28 mm, and the waveguide cross section was 72 mm × 34 mm. Discharge stabilization around the axis of the quartz tube and thermal insulation of the walls are achieved by tangential injection of gas and, therefore, rotation of the gas flow. The gasdynamic parameters of the rotating gas flow, namely radial distributions of rotation velocity and gas density at different distances from the discharge inlet, are

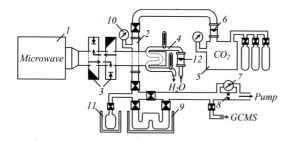

Figure 5–22. General schematic of plasma-chemical microwave reactor: (1) magnetron, (2) quartz tube, plasma-chemical reactor, (3) complex microwave tuning system, (4) calorimetric load, (5) CO_2 gas inlet, (6) and (12) gas flow meters, (7) and (10) manometers, (8) gas flow control, (9) liquid nitrogen trapping system, (11) gas sampling system.

presented in Figs. 5–23 and 5–24. The gas density near the axis of the discharge tube is relatively low along the entire microwave discharge zone. As a result, the reduced electric field E/n_0 near the axis is the highest, which determines the position stabilization of the microwave discharge. The microwave generator in this case is based on a magnetron source. Typical absorption of microwave energy in the discharge without special effort is about 70–85%; the rest of the energy is reflected. Optimization of the microwave generator coupling with the plasma load results in a much more effective microwave absorption of 99% (Alexeeva, Bobrov, & Lysov, 1977, 1979; Bobrov, Kudrevtova, & Lysov, 1979a,b). The experimental system sustains stable microwave discharge in CO_2 in the pressure range 50–760 Torr. At relatively low pressures (50–100 Torr), the microwave discharge is uniform, with characteristic plasma diameter of about 20 mm. A gradual transition to the quasi-contracted mode, characterized by the formation of a relatively hot filament with diameter of about 6 mm on the discharge axis, starts at about 120–150 Torr. The microwave discharge remains strongly non-equilibrium ($T_e > T_v > T_0$) until the pressure exceeds 150–200 Torr. Variation of CO_2 pressure in the non-equilibrium range, from 50 to 250 Torr, leads to the following changes of the microwave discharge parameters:

- Electron density grows from $8 \cdot 10^{11}$ to $8 \cdot 10^{13}$ cm^{-3}, which corresponds to an increase of ionization degree from 10^{-6} to $8 \cdot 10^{-5}$;
- Electron–neutral collision frequency grows from $4 \cdot 10^{10}$ to $6 \cdot 10^{11}$ s^{-1};
- Electron temperature decreases from 1.2 to 0.5 eV; and
- Vibrational temperature reaches a maximum value of about $T_v = 4000$ K at electron temperature $T_e = 0.8$ eV; this electron temperature corresponds to the most effective excitation of the asymmetric vibrational mode of CO_2 molecules.

Variation of the gas (CO_2) flow rate in these experiments was in the range 0.15–2 L/s. Taking into account the aforementioned microwave power absorption in the discharge, the specific energy input in the experiments was in the range $E_v = 0.1$–1.4 eV/mol. A characterization of the discharge, where the highest energy efficiency (for subsonic flows)

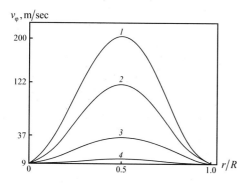

Figure 5–23. Radial distribution of gas rotation velocity $v_\varphi(r)$ at different distances from inlet of the plasma-chemical reactor with radius $R = 1.65$ cm. Distances from reactor inlet: (1) 1 cm, (2) 20 cm, (3) 40 cm, (4) 49 cm.

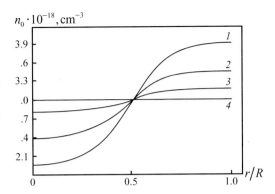

$n_0 \cdot 10^{-18}, \text{cm}^{-3}$

Figure 5–24. Radial distribution of CO_2 gas density $n_0(r)$ at different distances from inlet of the plasma-chemical reactor with radius $R = 1.65$ cm. Distances from reactor inlet: (1) 1 cm, (2) 15 cm, (3) 30 cm, (4) 49 cm.

was achieved, is described by Givotov, Kalachev, and Krasheninnikov (1983), and Givotov et al. (1985b).

5.5.2. Plasma-Chemical Experiments with Dissociation of CO₂ in Non-Equilibrium Microwave Discharges of Moderate Pressure

The results summarizing measurements of conversion degree α and corresponding values of energy efficiency η are presented in Fig. 5–25 as a function of specific energy input E_v at fixed pressure $p = 120$ Torr (Givotov et al., 1980; Butylkin et al., 1981; Vakar et al., 1981, Rusanov & Fridman, 1984). The energy efficiency of CO_2 dissociation was determined conventionally as $\eta = \alpha \cdot \Delta H / E_v$. As one can see from Fig. 5–25, the energy efficiency reaches a maximum value of 80%. An average energy efficiency is about 75% at specific energy inputs in the relatively wide range from 2.5 to 4.5 J/cm³. Such high energy efficiency values can be provided only through selective non-equilibrium stimulation of the dissociation process by vibrational excitation of the asymmetric mode of CO_2 molecules by electron impact (see Sections 5.2 and 5.3). The energy efficiency of CO_2 dissociation as a function of specific energy input $\eta(E_v)$ in the non-equilibrium moderate-pressure microwave discharges is also shown in Fig. 5–2 in comparison with other experimental discharge systems and different modeling approaches.

The experimental dependence of the energy efficiency on gas pressure is presented in Fig. 5–26. The dependence shows that the optimal pressure for CO_2 dissociation in the

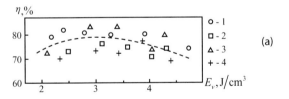

(a)

Figure 5–25. (a) Energy efficiency η and (b) conversion α of CO_2 dissociation in microwave plasma as a function of specific energy input: (1 and 2) results of different mass-spectral measurements, (3) manometric measurements, (4) flow-control-based measurements.

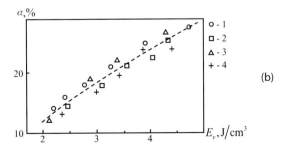

(b)

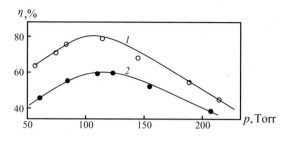

Figure 5–26. Energy efficiency of CO_2 dissociation in microwave plasma as function of pressure at different values of specific energy input: (1) $E_v = 3$ J/cm³, (2) $E_v = 2$ J/cm³.

microwave discharges under consideration is about 120 Torr. A detailed explanation of the effect can be found, for example, in Fridman & Kennedy (2004), in connection with a discussion of moderate-pressure microwave discharges. In simple words, it can be explained as follows. The microwave discharge actually becomes thermal at higher pressures, whereas at low pressures it is non-equilibrium but not optimal for vibrational excitation of the asymmetric mode of CO_2 molecules. It is interesting to note that the optimal value of pressure, 120 Torr, remains almost the same in the microwave discharges at different values of specific energy input if they are close enough to the optimal range 2.5 to 4.5 J/cm³.

An important qualitative feature of the CO_2 dissociation energy efficiency dependence on specific energy input is a threshold in the function $\eta(E_v)$. Results of experimental investigations of the threshold in the function $\eta(E_v)$ for the non-equilibrium moderate-pressure microwave discharges are presented in Fig. 5–27. Variations of values of the specific energy input in the discharge, E_v, were provided in these experiments in two different ways. Curve 1 in Fig. 5–27 corresponds to variations of CO_2 gas flow rate at a constant level of microwave power and fixed pressure $p = 120$ Torr. Curve 2 in Fig. 5–27 corresponds to variations of the microwave power and a fixed value of CO_2 gas flow rate; the fixed pressure in this case is $p = 50$ Torr, which is lower than the optimal value discussed earlier. The threshold value of specific energy input, which can be clearly observed in Fig. 5–27, is about $E_v = 0.8$ eV/mol.

5.5.3. Experimental Diagnostics of Plasma-Chemical Non-Equilibrium Microwave Discharges in Moderate-Pressure CO₂: Plasma Measurements

Experimental diagnostics of the plasma-chemical CO_2 microwave discharge were reviewed by Givotov, Rusanov, and Fridman (1985b). Space-resolved two-dimensional measurements of plasma density and vibrational and translational temperatures in the discharge systems can be found in Golubev, Krasheninnikov, and Tishchenko (1981). Characterization of the plasma ionization degree $n_e/n_0(p)$ and reduced electric field $E/n_0(p)$ as functions of pressure were performed by means of microwave diagnostics at wavelength $\lambda = 1.7$ mm. The microwave amplitude and phase measurements were used to directly determine the radial distributions of electron density n_e and electron–neutral collision frequency ν_{en}.

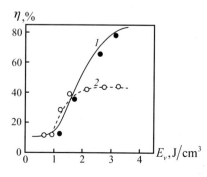

Figure 5–27. Energy efficiency of CO_2 dissociation in microwave plasma near the threshold value of specific energy input: (1) varying CO_2 gas flow rate at fixed power, pressure 120 Torr; (2) varying microwave power at fixed gas flow rate, pressure 50 Torr; white dots, mass-spectral measurements; black dots, flow-control-based measurements.

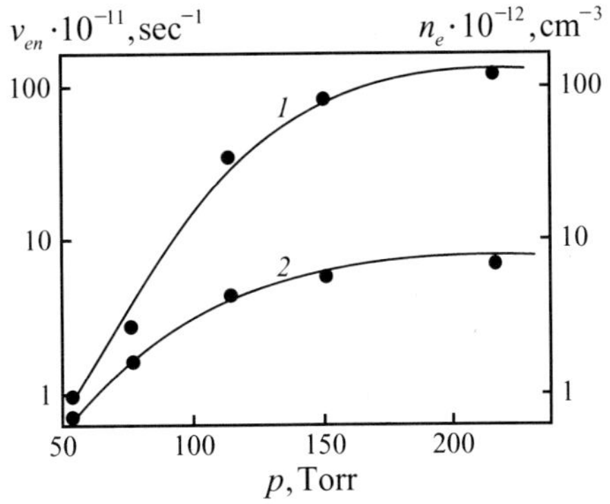

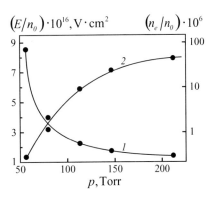

Figure 5–28. (1) Electron density n_e and (2) frequency of electron–neutral collisions ν_{en} as functions of pressure in microwave discharge.

The corresponding space-averaged values of the electron density n_e and electron-neutral collision frequency ν_{en} are presented in Fig. 5–28. Variation of pressure corresponding to the change in the discharge mode from quasi-uniform (at lower pressures) to filamentary (at higher pressures, see Section 5.5.1) leads to a two-order-of-magnitude increase in the electron concentration.

Experimental pressure dependence of ionization degree $n_e/n_0(p)$ is compared with that of the reduced electric field $E/n_0(p)$ in Fig. 5–29. The low-pressure regime of the microwave discharge is characterized by a relatively low ionization degree ($n_e/n_0 \approx 3 \cdot 10^{-6}$) and relatively high values of reduced electric field E/n_0, which correspond to a significant electronic excitation of CO_2 molecules (see Section 5.1.3 and Fig. 5–5). At relatively higher values of gas pressure ($p = 120\,\text{Torr}$, optimum for CO_2 dissociation; see Section 5.5.2), the plasma ionization degree is significantly larger ($n_e/n_0 \approx 2 \cdot 10^{-5}$) and the reduced electric field is somewhat lower ($E/n_0 = 2 \cdot 10^{-16}\,\text{V} \cdot \text{cm}^2$). This exactly corresponds to an excitation maximum for the asymmetric CO_2 vibrational mode, the most effective in CO_2 dissociation. Further increase of the CO_2 gas pressure and, hence, further decrease of reduced electric field E/n_0 results in a more intensive excitation of the symmetric CO_2 vibrational modes; it leads to a higher VT-relaxation rate, faster gas heating, a decrease in the degree of VT non-equilibrium, and finally to lower values of energy efficiency for the CO_2 dissociation.

5.5.4. Experimental Diagnostics of Plasma-Chemical Non-Equilibrium Microwave Discharges in Moderate-Pressure CO₂: Temperature Measurements

CO_2 vibrational temperature in the non-equilibrium microwave discharge is presented in Fig. 5–30 as a function of pressure and in Fig. 5–31 as a function of specific energy input. The

Figure 5–29. (1) Reduced electric field E/n_0 and (2) ionization degree n_e/n_0 as functions of pressure in microwave discharge.

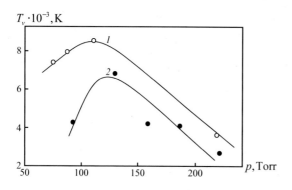

Figure 5–30. Vibrational temperature of CO_2 molecules in plasma-chemical microwave discharge as function of gas pressure. Diagnostic measurements based on Doppler broadening of spectral lines of (1) lithium and (2) sodium.

vibrational temperature measurements were performed based on the analysis of Doppler broadening of spectral lines of alkaline metals (mostly Li) introduced as a small additive to CO_2. This interesting method of diagnostic is related to the phenomenon of quasi equilibrium between the vibrational temperature of molecular gases and the translational temperature of light alkaline atoms such as Li. The method was developed by Vakar et al. (1981), and Ahmedganov et al. (1986), and reviewed by Fridman and Kennedy (2004). The experimental dependences $T_v(p)$ and $T_v(E_v)$, shown in Figs. 5–31 and 5–32, illustrate a good correlation between the level of vibrational excitation of CO_2 molecules and energy efficiency of the dissociation process. In particular, the maximum value of vibrational temperature T_v corresponds to the microwave discharge regime $E_v \approx 3$ J/cm^3, $p = 120$ Torr, when the energy efficiency also has the maximum value. On the other hand, a decrease of the vibrational temperature corresponds to a threshold value of specific energy input, $\eta(E_v)$. To distinguish between the temperatures of different vibrational modes of CO_2 molecules, an experimental emission spectrum in the range 2.5–3.1 μm was compared with relevant theoretically simulated spectra (Givotov et al., 1985b). In the plasma-chemical regime ($E_v \approx 4.2$ J/cm^3, $p = 120$ Torr), close to the optimal regime, it was found that the vibrational temperature of the asymmetric mode was $T_{va} \approx 3000$ K, and that of symmetric modes was $T_{vs} \approx 2000$ K, whereas the translational temperature was only about $T_0 = 800$ K. Non-equilibrium conditions in such plasma-chemical discharges were also characterized by Bragin and Matukhin (1979), Kutlin, Larionov, and Yarosh (1980), and Krasheninnikov (1981). Experiments with CO_2 dissociation in non-equilibrium microwave discharges at a much higher power level (up to 100 kW) were carried out by Asisov (1980) and Legasov et al. (1981a,b). Some measurements related to these high-power microwave experiments are presented in Fig. 5–32.

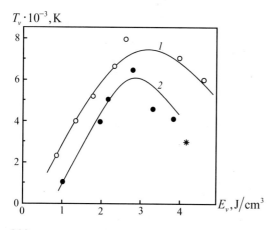

Figure 5–31. Vibrational temperature of CO_2 molecules in plasma-chemical microwave discharge as function of pressure. Measurements based on Doppler broadening of spectral lines of (1) lithium and (2) sodium; stars indicate infra-red spectral measurement.

5.5. Experimental Investigations of CO_2 Dissociation in Different Discharge Systems

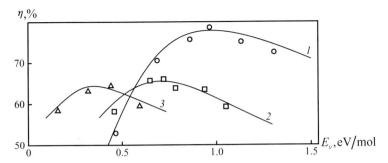

Figure 5–32. Energy efficiency of CO_2 dissociation as a function of specific energy input in microwave discharges of different power. Power: (1) 1 kW; (2) 100 kW, discharge tube diameter, 3.5 cm; (3) 100 kW, discharge tube diameter, 5 cm.

5.5.5. CO_2 Dissociation in Non-Equilibrium Radiofrequency Discharges: Experiments with Inductively Coupled Plasma

Non-equilibrium CO_2 dissociation was widely investigated in RF discharges, both with inductive and capacitive coupling (see, for example, Capezzutto, Cramarossa, & D'Agostino, 1976; Patrushev, Rykunov, & Spector, 1977; Kolesnichenko et al., 1979; Fridman & Rusanov, 1994). The energy efficiency of the dissociation in these discharges is high (up to 60%), although not as high as in the previously considered microwave discharges. The interpretation of the major dissociation mechanism in RF discharges is also not as clear as in microwave discharges. However, the vibrational excitation of CO_2 molecules is usually considered the most important mechanism in experiments where the highest energy efficiency values of dissociation were achieved. Experiments with RF-ICP discharges, described by Patrushev, et al. (1977) and Legasov et al. (1978), were performed in a quartz tube with 5 cm diameter, at CO_2 pressures of 0.1–2 Torr, and CO_2 flow rate of 2–50 cm^3/s. Inductive coupling at 0.5–1.5 kW was provided by a ring of frequency 21 MHz. The plasma density was in the range $1–4 \cdot 10^{12}$ cm^{-3}, and the electron temperature ($T_e \approx 1$ eV) was much higher than the gas temperature; thus, the dissociation was strongly a non-equilibrium one. The CO_2 conversion degree α is presented in Fig. 5–33 as a function of gas flow rate. It approaches 100% at low flow rate and, hence, relatively high values of specific energy input. Energy efficiency, however, is not high (10–20%). Experimental results for CO_2 conversion degree and energy efficiency of CO_2 dissociation in RF discharges are compared with the results of theoretical modeling in 5–34.

Figure 5–33. CO_2 conversion degree in RF-ICP discharge as function of flow rate at different values of power: (1) 0.5 kW, (2) 1 kW, (3) 1.5 kW.

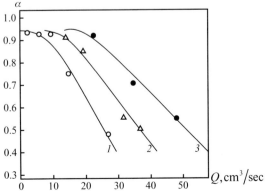

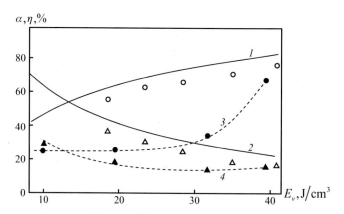

Figure 5–34. CO$_2$ (1, 3) conversion degree and (2, 4) energy efficiency in RF discharges as functions of specific energy input: (1, 2) ICP discharge, (3, 4) CCP discharge; dots, experimental results, curves, results of theoretical simulations.

5.5.6. CO$_2$ Dissociation in Non-Equilibrium Radiofrequency Discharges: Experiments with Capacitively Coupled Plasma

Electric fields are generally higher in CCP discharges than in ICP discharges. For this reason, non-equilibrium CCP discharges can be effectively performed at pressures of 100 Torr and even higher, whereas non-equilibrium ICP is usually limited to pressures not exceeding several Torr to provide the reduced electric field necessary for ionization. The higher pressures help to avoid energy losses to the walls, increase the gas flow rate, reach optimal values of specific energy input, and achieve energy efficiency comparable with that of microwave discharges. CO$_2$ dissociation experiments with non-equilibrium CCP discharges (Givotov et al., 1980a,b) were carried out between cylindrical electrodes in a quartz chamber, at frequencies of 5, 20, and 60 MHz, with power 1–30 kW. The diameter of the discharge zone was about 1 cm, and the length was about 10 cm. Maximum values of energy efficiency in these experiments (40–60%) were achieved at a gas pressure of 60 Torr, CO$_2$ flow rate of 300 cm^3/s, and specific energy inputs of 0.7–1 eV/mol. Some of these results are presented in Fig. 5–2. The CO$_2$ conversion degree in these experiments was only about 15–20% at the conditions optimal for highest energy efficiency.

Experiments of Capezzutto, Cramarossa, and D'Agostino (1976) were carried out in 10 kW CCP discharge with RF 35 MHz, pressure 5–40 Torr, flow rate 3–17 L/min, specific power 4–40 W/cm^3, specific energy input 5–70 J/cm^3, and translational gas temperature 500–1200 K. The diameter of the discharge zone was 3.5 cm, and the distance between ring electrodes (discharge length) was 18 cm. The distribution of the CO$_2$ conversion degree along the discharge axis, observed in the experiments, is shown in Fig. 5–35. Values of reduced electric field in the CCP discharge at different pressures are presented in Fig. 5–36. Finally, the experimentally measured CO$_2$ dissociation rate coefficients (assuming the first

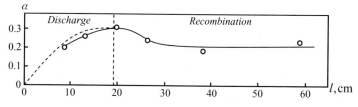

Figure 5–35. Evolution of CO$_2$ conversion degree along axis of RF-CCP discharge; conversion is shown in discharge zone and in recombination zone.

5.5. Experimental Investigations of CO₂ Dissociation in Different Discharge Systems

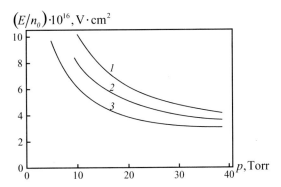

Figure 5–36. Reduced electric field E/n_0 in RF-CCP discharge as function of pressure at different values of specific power: (1) 16 W/cm³, (2) 12 W/cm³, (3) 7 W/cm³.

kinetic order with respect to densities of electrons and neutrals) are shown in Fig. 5–37 as a function of the reduced electric field for different pressures. They are compared in Fig. 5–37 with theoretical calculations of the CO₂ dissociation rate coefficients (assuming dissociation through electronic excitation) and experimental results obtained in other discharge systems. Figure 5–37 (together with Fig. 5–36) shows that an increase of gas pressure leads to a change in the major CO₂ dissociation mechanism from electronic excitation to vibrational excitation. Experimental results obtained at gas pressure $p = 10$ Torr $(E/n_0 \approx 8 \cdot 10^{-16}\,V \cdot cm^2)$ can be explained by electronic excitation (good agreement in Fig. 5–37). The experimental value of the dissociation rate coefficient at $p = 40$ Torr $(E/n_0 \approx 4 \cdot 10^{-16}\,V \cdot cm^2)$ is shown in Fig. 5–37 and is already almost 10-fold higher than the theoretical value that assumes dissociation through electronic excitation.

Interesting results for CO₂ dissociation in non-equilibrium RF-CCP discharges were published by Kolesnichenko et al. (1979) and Bragin and Matukhin (1979). The discharge in these experiments was arranged between perpendicular crossed copper tubes (diameter 2 mm) covered with quartz (0.5 mm) and separated by a 1–8-cm gap. The discharge power

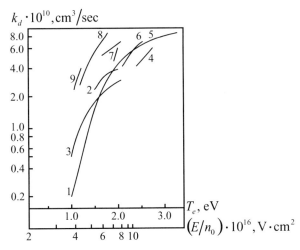

Figure 5–37. Rate coefficients of CO₂ dissociation in different discharge systems as functions of reduced electric field and electron temperature, assuming first kinetic order of the dissociation with respect to electron density. (1) theoretical simulations assuming dissociation through electronic excitation (Capezzutto, Cramarossa, & D'Agostino, 1976); (2) glow discharge experiments at pressure 2 Torr (Ivanov, Polak, & Slovetsky, 1971); (3) experiments at pressures 0.3–3 Torr (Corvin & Corrigan, 1969). Non-self-sustained discharge experiments (Kutlin, Larionov, & Yarosh, 1980), pressures: (4) 160 Torr, (5) 220 Torr, (6) 300 Torr. Experiments with RF-CCP discharge (Capezzutto, Cramarossa, & D'Agostino, 1976), pressures: (7) 10 Torr, (8) 20 Torr, (9) 40 Torr.

varied between 100 W and 2 kW, gas pressure was 30–200 Torr, and the CO_2 flow rate was up to 3 L/s. The maximum energy efficiency achieved in these experiments was quite high (50–65%); the corresponding value of CO_2 conversion was 25%. Interesting experiments on CO_2 dissociation in the RF-CCP were carried out in relation to oxygen extraction from the Martian atmosphere (Vuskovic et al., 1997). The Martian atmosphere consists mostly of CO_2, so its dissociation with production of CO and O_2 can be applied for life support and production of propellants in Mars conditions (Ash, Dowler, & Varsi, 1978). The experiments were performed at a pressure of about 5–7 Torr, corresponding to the Martian conditions. Energy efficiency was on the level of 25%. It is important that atomic oxygen generated in the plasma-chemical system was effectively separated from the gas mixture before recombination using a special silver membrane (Wu, Outlaw, & Ash, 1993).

5.5.7. CO_2 Dissociation in Non-Self-Sustained Atmospheric-Pressure Discharges Supported by High-Energy Electron Beams or UV Radiation

The non-self-sustained discharges, supported in particular by relativistic electron beams, permit one to make non-equilibrium CO_2 plasma uniformly in a large volume, and at high pressures of 1 atm and higher (Basov et al., 1973). The plasma-chemical experiments on CO_2 dissociation in such discharges were carried out by Vakar, Denisenko, and Rusanov (1977). Non-equilibrium atmospheric-pressure plasma was sustained in these experiments by pulsed electron beams with electron energy of 300 keV, pulse duration about 10^{-6} s, and electron beam current density up to 100 A/cm^2. The plasma density was about 10^{14} cm^{-3}, and the electron temperature $T_e \approx 1$ eV. Experiments with the non-self-sustained discharges clearly illustrate the threshold in energy efficiency dependence on specific energy input $\eta(E_v)$. Values of the specific energy input in these uniform and high-volume discharge systems are much lower than those of RF and microwave discharges; they are essentially limited by thermal instability. At specific energy inputs below 0.25 eV/mol, CO_2 dissociation is almost negligible. When E_v exceeds the threshold value, the CO_2 conversion degree and energy efficiency grow to a maximum value of about $\eta = 14\%$, achieved in these experiments at $E_v = 0.6$–0.7 eV/mol. Further increase of specific energy input and, hence, conversion degree and energy efficiency are limited by thermal instability. Dissociation of CO_2 in a non-self-sustained pulse discharge supported by UV radiation is reported by Kutlin et al. (1980). The pulse duration in these experiments was $3 \cdot 10^{-7}$ s, discharge volume was 40 cm^3, voltage 20–25 kV, pressure 160–300 Torr, electron density $5 \cdot 10^{12}$ cm^{-3}, and electron temperature 1.7–2.2 eV. CO_2 dissociation was analyzed at different values of reduced electric field ($E/n_0 = 0.5$–$1.3 \cdot 10^{-15}$ V $\cdot$ cm^2). The CO_2 dissociation rate coefficient grew in the range 2–8 $\cdot 10^{-10}$ cm^3/s with an increase of pressure and decrease of the reduced electric field. It indicates the essential contribution of vibrational excitation in the dissociation process in these experiments, at least at relatively high gas pressures.

5.5.8. CO_2 Dissociation in Different Types of Glow Discharges

The major mechanism of CO_2 dissociation in **low-pressure glow discharges** is electronic excitation of the molecules (Slovetsky, 1980; Volchenok et al., 1980). In particular, the cross section of the CO_2 dissociation was experimentally measured in a glow discharge at pressure 0.3–0.4 Torr and current 0.1 mA (Corvin & Corrigan, 1969). The cross section has a threshold corresponding to an energy of electronic excitation of $CO_2(A^3\Pi)$. The EEDF in a CO_2 glow discharge was measured at 2 Torr, at current densities of $j = 0.6$–12 mA/cm^2, at electric currents of $I = 5$–100 A, electron densities of $n_e = 0.3$–$6 \cdot 10^9$ cm^{-3}, and translational gas temperatures of $T_0 = 370$–780 K (Ivanov, Polak, & Slovetsky, 1971). The EEDF is shown in Fig. 5–38 and is an essentially non-Maxwellian distribution. Increase of current density leads to a growing number of high-energy electrons and, hence, to an increase of CO_2 conversion degree, which is illustrated in Fig. 5–39. The CO_2 dissociation

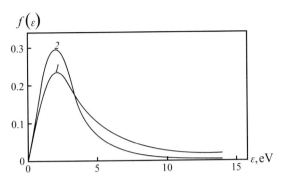

Figure 5–38. Electron energy distribution function (EEDF) in glow discharge at different currents: (1) 70 mA, (2) 30 mA.

rate coefficient (assuming first kinetic order with respect to the densities of electrons and neutrals), calculated in the described experiments with the glow discharge, is in the range $1.5-3.7 \cdot 10^{-10}$ cm³/s. The theoretical value of the CO_2 dissociation rate coefficient, calculated based on the experimentally measured EEDF (See Fig. 5–38) and cross section of electronic excitation, is in good agreement with that measured experimentally (Ivanov, Polak, & Slovetsky, 1971). The experimentally measured value of the reduced electric field in the glow discharge, $E/n_0 = 5-7 \cdot 10^{-16}$ V · cm², also confirms the conclusion that the major mechanism of CO_2 dissociation in the systems is electronic excitation (see Fig. 5–5).

Experiments with **pulsed glow discharge** (Willis et al., 1969) were carried out at pressures of 200–800 Torr (even slightly above atmospheric value), voltages of 10–22 kV, pulse duration of 1 μs, and reduced electric field $E/n_0 = 4-20 \cdot 10^{-16}$ V · cm². Pressure decrease and growth of the reduced electric field led to a reduction of the dissociation energy cost, which in these experiments was about 30 eV/mol CO. Calculations of energy efficiency, assuming electronic excitation of $CO_2(A^1\Sigma)$ and $CO_2(A^3\Pi)$ (see Fig. 5–9), are in good agreement with the experiments. The calculations also permit one to find the minimum energy cost for CO_2 dissociation through electronic excitation of the molecules as 11.5 eV/mol CO.

Glow discharge with a hollow cathode was applied for CO_2 dissociation with the following conditions: pressure 1–8 Torr, current density 10–150 A/cm², voltage 500 V, and discharge chamber size 0.3 cm × 0.3 cm × 1.6 cm (Metel & Nastukha, 1977). The dissociation rate in these experiments is significantly higher than in the previously considered glow discharges. A CO_2 conversion degree of 80% can be achieved at the specific energy input of 2 eV/mol during the very short time interval of 1 ms in these systems. It is 100–1000 times faster than in conventional glow discharges at comparable values of pressure and current density. The CO_2 dissociation rate coefficient was determined as 10^{-8} cm³/s in the discharge with hollow cathode at a specific energy input of 0.4 eV/mol, which is

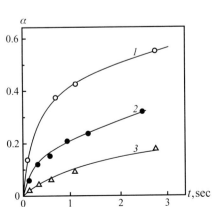

Figure 5–39. Time evolution of CO_2 conversion degree in glow discharge at different currents: (1) 70 mA, (2) 30 mA, (3) 10 mA.

much higher than in other configurations of glow discharges. The fast dissociation rates in the discharge with hollow cathode are due to the large fraction of very energetic electrons related to high values of the specific electric field (10^{-15}–10^{-14} V · cm^2), which correspond to $E = 200$–300 V/cm and $p = 1$–10 Torr.

5.5.9. CO$_2$ Dissociation in Other Non-Thermal and Thermal Discharges: Contribution of Vibrational and Electronic Excitation Mechanisms

Experiments with CO$_2$ dissociation in quasi-equilibrium plasmas of **arc discharges** were carried out, in particular, at voltage 200 V, current 5–10 A, and atmospheric pressure (Butylkin et al., 1979). An optimal dissociation regime was achieved at gas flow rates of 150–700 L/s and specific energy inputs of 5–50 J/cm^3. The degree of CO$_2$ conversion was in this case up to 5–10%, and the dissociation energy cost reached about 16 eV/mol CO, which corresponds to an energy efficiency of about 18%. Obviously, such conversion and energy efficiency were achieved with special performance of quenching with a cooling rate of 10^7 K/s. Conversion degree without special quenching was about two-fold less.

CO$_2$ conversion in non-equilibrium plasma conditions can significantly exceed conversion at thermodynamic quasi equilibrium, which is, for example, equal to $\alpha = 1.6\%$ and $\alpha = 3.6\%$ at $T_0 = 2000$ K and pressures $p = 2$ atm and $p = 50$ Torr, respectively (Maltsev, Eremin, & Ivanter, 1967). To compare, in addition to the non-thermal discharges considered earlier, the CO$_2$ conversion degree in **corona discharge** reaches $\alpha = 28\%$ at a gas temperature T_0 not much above room temperature.

The CO$_2$ conversion degree in another essentially non-equilibrium plasma system – in dielectric barrier discharge (DBD) – reaches about 30% at an electron temperature of $T_e = 1$–3 eV and gas temperature of $T_0 = 400$–500 K (Andreev, 1964). This level of conversion degree was achieved in the DBD at a power level of 100 W and gas pressure 500 Torr after 20 min of gas circulation. An increase of the specific energy input from 18 to 800 J/cm^3 at power 100 W and flow rates up to 80 L/h leads to growth of the CO$_2$ conversion degree from 3 to 30%. The energy efficiency of the CO$_2$ dissociation was falling at the same time, from 1 to 0.4%. A comparison of experimental results for CO$_2$ dissociation rate coefficients obtained in different non-equilibrium discharges with simulation results calculated assuming dissociation through electronic excitation is presented in Fig. 5–37. The theoretical simulations are in good agreement with experiments carried out in glow discharges at pressures of about 1 Torr (Corvin & Corrigan, 1969; Ivanov, Polak, & Slovetsky, 1971) and in a pulsed non-self-sustained discharge at pressures of 160–220 Torr (Kutlin et al., 1980). However, at higher pressures (300 Torr), experimental dissociation rates exceed the simulated rates, which confirms the conclusion that higher pressures and lower values of reduced electric field make the vibrational excitation mechanism more favorable than that of electronic excitation.

5.6. CO$_2$ DISSOCIATION IN SPECIAL EXPERIMENTAL SYSTEMS, INCLUDING SUPERSONIC STIMULATION AND PLASMA RADIOLYSIS

5.6.1. Dissociation of CO$_2$ in Supersonic Non-Equilibrium Discharges: Advantages and Gasdynamic Characteristics

The highest energy efficiency of CO$_2$ dissociation in plasma can be achieved through its stimulation by vibrational excitation. This mechanism permits, in particular, 80% energy efficiency to be reached in non-thermal microwave discharges of moderate pressure (see Section 5.5.2). The main cause of energy losses in these systems is VT relaxation from low vibrational levels of CO$_2$ molecules, which decreases exponentially with reduction of translational gas temperature. A significant decrease of translational gas temperatures (to

5.6. CO_2 Dissociation in Special Experimental Systems

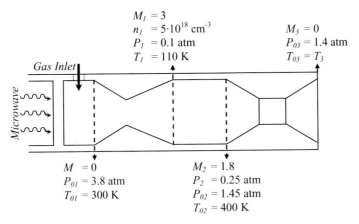

$M_1 = 3$
$n_1 = 5 \cdot 10^{18}$ cm^{-3}
$P_1 = 0.1$ atm
$T_1 = 110$ K

$M_3 = 0$
$P_{03} = 1.4$ atm
$T_{03} = T_3$

$M = 0$
$P_{01} = 3.8$ atm
$T_{01} = 300$ K

$M_2 = 1.8$
$P_2 = 0.25$ atm
$P_{02} = 1.45$ atm
$T_{02} = 400$ K

Figure 5–40. Typical parameters of a nozzle system in supersonic plasma-chemical experiments. Subscripts 1 and 2 are related to inlet and exit of a discharge zone; subscript 3 is related to exit from the nozzle system; subscript 0 is related to stagnation pressure and temperature.

the level of about 100 K; see Fig. 5–40) and, hence, a further increase of energy efficiency (to 90%!) can be achieved by carrying out plasma-chemical CO_2 dissociation in a supersonic gas flow. Another reason for supersonic organization of CO_2 dissociation in plasma is related to limitations of productivity in subsonic discharges. The optimal energy input in energy-effective plasma-chemical systems as about 1 eV/mol at moderate and high pressures. This determines the proportionality between discharge power and gas flow rate in the optimal regimes: high discharge power requires high flow rates through the discharge. The gas flow rates through the discharge zone, however, are limited (although the limits are high). Instabilities restrict space sizes and pressures of the steady-state discharge systems, and velocities are usually limited by the speed of sound. As a result, flow rate and, hence, power are restricted by some critical value in subsonic plasma systems, which is about 100 kW for non-equilibrium microwave discharges (Ageeva et al., 1986). Also the high flow velocities make this discharge more stable with respect to contraction at higher pressures and energy inputs (Provorov & Chebotaev, 1977). For example, homogeneous glow discharges can be performed in atmospheric-pressure fast flows at an energy input of 0.5 kJ/g (Gibbs & McLeary, 1971; Hill, 1971).

Higher powers of the steady-state non-equilibrium discharges and, hence, higher process productivity can be reached in supersonic gas flows. The maximum power increase of a steady-state non-equilibrium supersonic microwave discharge is up to 1 MW (Legasov et al., 1983a). A simplified scheme and flow parameters of such a discharge system (power 500 kW, Mach number $M = 3$) is presented in Fig. 5–40. Some gasdynamic characteristics of the supersonic discharges are presented in Fig. 5–41 (Potapkin et al., 1983a; Zyrichev, Kulish, & Rusanov, 1984). Ignition of the non-equilibrium discharge takes place in this case after a supersonic nozzle in a relatively low-pressure zone, which is optimal for non-equilibrium discharges. Pressure in the after-discharge zone is restored in a diffuser, so initial and final pressures are above atmospheric value, which is again an important advantage of the supersonic performance of the plasma-chemical process. It requires an effective supersonic diffuser to restore pressure back to high values after the discharge zone. Characteristics of pressure restoration after the supersonic discharge with different Mach numbers are presented in Fig. 5–42. The initial pressure before the supersonic nozzle in this case was 3.8 atm at room temperature, and the heat release was about half of the critical value. The pressure restoration can be quite decent (up to 1.5–2.5 atm), although heating and this method of stopping the supersonic flow are not optimal for pressure restoration.

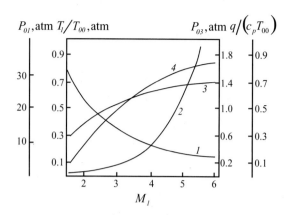

Figure 5–41. Gasdynamic characteristics of a plasma-chemical discharge in supersonic flow: (1) discharge inlet temperature T_1; (2) initial tank pressure p_{01}; (3) exit pressure p_{03} in the conditions of critical heat release; (4) critical heat release q. All the parameters (1–4) are shown as functions of Mach number M_1 in front of the discharge. Initial gas tank temperature $T_{00} = 300\,\text{K}$; static pressure in front of discharge $p_1 = 0.1\,\text{atm}$.

The energy efficiency of the plasma-chemical process of CO_2 dissociation in a supersonic discharge is presented in Fig. 5–43, taking into account the energy spent on gas compression for non-converted gas recirculation (Zyrichev et al., 1984). Comparison of these results with Fig. 5–2 (pure plasma-chemical efficiency of the process) shows that the energy spent on compression in the supersonic system is about 10% of the total energy cost. Also, as one can see from Fig. 5–43, the energy cost of compression makes the whole plasma-chemical process less effective at high Mach numbers ($M > 3$). This effect takes place although plasma-chemical efficiency is increasing with Mach number because of the reduction of vibrational relaxation losses.

5.6.2. Kinetics and Energy Balance of Non-Equilibrium Plasma-Chemical CO_2 Dissociation in Supersonic Flow

The major peculiarity of CO_2 dissociation kinetics in supersonic plasma flow is related to a very low translational temperature (about 100 K) and, hence, a very high degree of VT non-equilibrium ($T_v/T_0 \geq 30$). The population of highly excited vibrational levels of the CO_2 asymmetric mode can be essentially affected by non-linear resonant VV exchange (strong excitation case, see Chapter 3; for details, see Fridman & Kennedy, 2004). The critical vibrational temperature T_v^{min}, corresponding to equal rates of dissociation and VT relaxation (see Section 5.3.5), can be estimated in this case as (Asisov et al., 1983; compare with (5–46)):

$$\left(T_v^{\text{min}}\right)^2 = T_0 \hbar\omega \left[4x_e \ln\left(Z_{\text{VT}}\frac{E_a}{\hbar\omega}\right)\right]^{-1}. \qquad (5\text{–}90)$$

In this relation, x_e and $\hbar\omega$ are the coefficient of anharmonicity and vibrational frequency related to the asymmetric mode of CO_2; T_0 is the translational gas temperature at the discharge inlet; E_a is the CO_2 dissociation energy (see (5–6)); Z_{VT} is the ratio of the rate coefficients related to resonant VV exchange at high vibrational energies and to

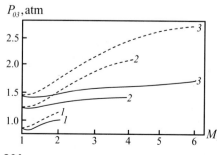

Figure 5–42. Pressure restoration in a diffuser (p_{03}). M_1 and M_2 are Mach numbers in the discharge inlet and exit. (1) $M_1 = 2$, (2) $M_1 = 4$, (3) $M_1 = 6$. Dashed lines correspond to an ideal diffuser; solid lines take into account non-ideal shock waves; static pressure at the discharge inlet is 0.1 atm.

5.6. CO$_2$ Dissociation in Special Experimental Systems

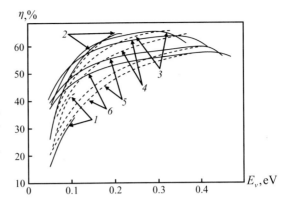

Figure 5–43. Energy efficiency of plasma-chemical CO$_2$ dissociation in supersonic flow, taking into account energy cost of compression, as a function of specific energy input at different Mach numbers: (1) $M = 2.5$; (2) $M = 3$; (3) $M = 4$; (4) $M = 5$; (5) $M = 7$; (6) $M = 8$. Solid lines correspond to pressure restoration in a diffuser, calculated based on Oswatich theory; dashed curves correspond to diffuser consideration as a normal shock.

VT relaxation from low vibrational levels of the asymmetric mode. A comparison of relations (5–90) and (5–46) shows that the critical vibrational temperature $T_v^{\min}$ and, hence, the threshold in $\eta(E_v)$ are essentially lower and strongly depend on initial translational temperature T_0, when the plasma-chemical CO$_2$ dissociation is organized in supersonic flow. Experimental confirmation of this effect is seen in Fig. 5–2. Taking into account the dependence of the initial translational temperature at the discharge inlet on the Mach number in front of the discharge, the energy efficiency of CO$_2$ dissociation (relaxation factor, see Section 5.3) in a supersonic non-thermal plasma flow can be calculated as (Asisov et al., 1981a,b; Rusanov, Fridman, & Sholin, 1982):

$$\eta_{\mathrm{rel}} = 1 - \frac{1 - \chi}{E_v}\varepsilon_v \left(T_v = \sqrt{\frac{\alpha_{\mathrm{VT}} T_{00}\hbar\omega}{1 + \frac{\gamma - 1}{2}M^2}} \right). \tag{5–91}$$

In this relation, $\alpha_{\mathrm{VT}}(T_0)$ is a slowly varying VT function ($\alpha(T_0) \approx 5$ at $T_0 = 100\,\mathrm{K}$); T_{00} is the gas temperature in the inlet tank before the supersonic nozzle (usually $T_{00} = 300\,\mathrm{K}$); M is the Mach number after the supersonic nozzle in front of plasma; χ is the CO$_2$ conversion degree; and γ is the specific heat ratio. Numerically, relation (5–91) gives the relaxation factor $\eta_{\mathrm{rel}} \approx 85\%$ for subsonic plasma-chemical systems, where $M = 0$, $T_{00} = 300\,\mathrm{K}$, and $E_v = 0.8\,\mathrm{eV}$. Contribution of other "non-relaxation related" energy losses takes out an additional 5% of energy efficiency and reduces it to the well known value of 80%. In the supersonic plasma-chemical systems with $M = 3\text{–}5$ and other parameters the same as those just given, the relaxation factor of CO$_2$ dissociation energy efficiency according to relation (5–91) reaches an extremely high level of about 98%. Such high values of the relaxation factor (98%) in supersonic plasma shift attention to other energy losses, which are often neglected in subsonic systems. The most important of those are losses related to the non-resonant VV exchange and to the excess of activation energy of the secondary reaction (5–7) over its enthalpy. These losses, which are usually referred to as the so-called "chemical factor" (see Chapter 3, or for more details Fridman & Kennedy, 2004), lead to an energy efficiency decrease of about 5–10% (Asisov et al., 1981).

The total energy efficiency of CO$_2$ dissociation at different Mach numbers, which takes into account the aforementioned effects, is presented in Fig. 5–44 as a function of specific energy input ($n_e/n_0 = 3 \cdot 10^{-6}$, $n_0 = 3 \cdot 10^{18}\,\mathrm{cm}^{-3}$). Two qualitative effects related to increasing Mach number and a decrease of energy losses in VT relaxation can be observed (Fig. 5–44) from the dependence $\eta(E_v, M)$:

- maximum energy efficiency grows, while optimal specific energy input decreases;
- threshold value of the specific energy input decreases.

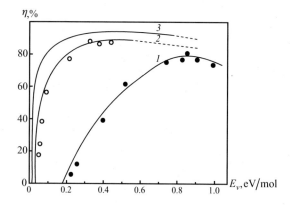

Figure 5–44. Energy efficiency of plasma-chemical CO_2 dissociation (without cost of compression) as a function of specific energy input at different Mach numbers: (1) $M \ll 1$; (2) $M = 3$; (3) $M = 5$. Solid lines correspond to results of theoretical modeling; black dots, experiments with subsonic microwave plasma; clear dots, experiments with supersonic microwave discharge. Dashed lines correspond to conditions of non-stationary perturbations of the discharge in supersonic flow.

A decrease of energy efficiency at higher values of the specific energy input is due to overheating and as a result of acceleration of VT relaxation. Serious restrictions of the specific energy input in supersonic plasma systems are also related to the critical heat release and choking of the flow.

5.6.3. Limitations of Specific Energy Input and CO_2 Conversion Degree in Supersonic Plasma Related to Critical Heat Release and Choking the Flow

The specific energy input and conversion degree are limited in supersonic plasma reactors by the **critical heat release**, which corresponds to a drop of the initial Mach number from $M > 1$ before the discharge to $M = 1$ afterward and leads to choking of the supersonic flow. The critical heat release for the supersonic reactor with constant cross section is equal to

$$q_{cr} = c_p T_{00} \left[\frac{(1 + \gamma M^2)^2}{2(\gamma + 1) M^2 \left(1 + \frac{\gamma - 1}{2} M^2\right)} - 1 \right]. \quad (5-92)$$

Here T_{00} is the initial gas temperature in the tank before the supersonic nozzle. If the initial Mach number is not very close to unity, the critical heat release can be estimated as $q_{cr} \approx c_p T_{00}$. Numerical values of the critical heat release at different initial Mach numbers for the supersonic CO_2 flow can also be found in Fig. 5–41. Further increase of the heat release in plasma over the critical value leads to the formation of non-steady-state flow perturbations like shock waves, which do no good to a non-equilibrium plasma-chemical system. Even taking into account the high energy efficiency of chemical reactions in supersonic flows, the critical heat release seriously restricts the specific energy input:

$$E_v (1 - \eta) < q_{cr}. \quad (5-93)$$

The restriction of specific energy input E_v limits the conversion degree of the process. As a result, the maximum conversion degree of CO_2 dissociation in supersonic microwave discharge ($T_{00} = 300 \, \text{K} = 300 \, \text{K}$ and $M = 3$) does not exceed 15–20%, even at the extremely high energy efficiency of about 90% (Zyrichev et al., 1984). We should note that special profiling of a supersonic discharge chamber makes the restrictions not so strong (see Potapkin, 1986; Fridman & Kennedy, 2004).

5.6.4. Experiments with Dissociation of CO_2 in Non-Equilibrium Supersonic Microwave Discharges

Experiments with a supersonic microwave discharge, operating at a frequency of 915 MHz at a power level of 10–100 kW, confirm the possibility of achieving 90% energy efficiency

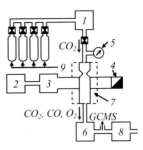

Figure 5–45. Schematic of an experiment with supersonic plasma-chemical microwave discharge: (1) CO_2 gas inlet system; (2) power supply; (3) microwave generator; (4) tuning system; (5) manometer; (6) heat exchanger; (7) microwave plasma reactor; (8) vacuum pump; (9) CO_2 gas tanks.

of CO_2 dissociation in non-equilibrium plasma (maximum power of the discharge is about 1 MW; Asisov et al., 1981a,b, 1983). A schematic of the experiments is shown in Fig. 5–45. The plasma-chemical reactor is arranged in a quartz tube (after a supersonic nozzle) with a diameter of 3.5 cm and effective length of 30 cm, crossing a waveguide. The critical cross-sectional diameter of the supersonic nozzle is 0.8 cm, and the expansion degree is about 20. The initial pressure in the tank before the nozzle varies from 1.8 to 7 atm, which corresponds to static pressure in the plasma-chemical reactor from 0.05 atm (40 Torr) to 0.2 atm (150 Torr) and to flow rates of 5–50 L/s. This pressure range is optimal for process stimulation by vibrational excitation. Distributions of static pressure along the axis of the nozzle and the microwave discharge chamber with and without discharge are shown in Fig. 5–46. These distributions as well as the values of flow rate measurements prove that the flow remains supersonic in the presence of microwave plasma. The experimental Mach number value is in the range $M \approx 2$–3, in good agreement with calculations based on the nozzle geometry. The electron density, measured in the discharge chamber by means of microwave interferometry on wavelength $\lambda = 0.3$–0.8 cm, varies in the range 3–$8 \cdot 10^{12} \text{cm}^{-3}$; the electron temperature is about 1 eV. The vibrational temperature was determined in these experiments by measurements of the Doppler broadening of spectral lines of alkaline atoms, which were in quasi equilibrium with vibrational degrees of freedom of CO_2 molecules (Givotov et al., 1985b). The vibrational temperature in the optimal regime was found to be $T_v \approx 3500 \text{ K}$, while the translational gas temperature, $T_0 \approx 160 \text{ K}$ (!), remained below room temperature. Gas composition was measured by chromatography and by gas analyzers. The energy efficiency of CO_2 dissociation corresponding to these measurements is shown in Fig. 5–44. The minimum energy cost of the CO_2 dissociation process, measured in these experiments, was 3.2 eV/mol, which corresponds to the record high energy efficiency of 90%.

5.6.5. Gasdynamic Stimulation of CO₂ Dissociation in Supersonic Flow: "Plasma Chemistry Without Electricity"

Stimulation of CO_2 dissociation can be achieved not only by means of an increase of gas temperature (better the vibrational one) but also by means of cooling the gas(!). It

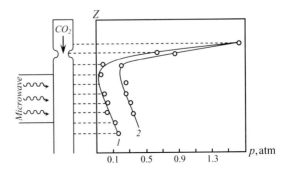

Figure 5–46. Evolution of static gas pressure along axis of supersonic plasma-chemical system: (1) without discharge; (2) in presence of microwave discharge.

sounds non-realistic at first, but it is correct theoretically and proven experimentally. The explanation is simple: the dissociation rate in non-equilibrium systems depends not only on vibrational temperature (T_v) of CO_2 molecules but also strongly on the degree of VT non-equilibrium (T_v / T_0) (see, in particular, Section 5.2.5 and relation (5–26)). If the level of VT non-equilibrium (T_v / T_0) is high, the Treanor effect leads to a significant overpopulation of highly vibrationally excited states even at relatively low values of vibrational temperature. In other words, gas cooling can accelerate the endothermic process almost as effectively as heating. A significant decrease of translational gas temperature T_0 ("gas cooling") without essential reduction of vibrational temperature T_v of CO_2 molecules (and as a result, strong VT non-equilibrium $T_v \gg T_0$, strong Treanor effect, and effective CO_2 dissociation) can be done in a supersonic nozzle, similarly to the case of supersonic plasma considered in Sections 5.6.1–5.6.4. Most of the energy accumulated in CO_2 vibrational degrees of freedom before a supersonic nozzle can be localized through the Treanor effect in highly excited levels after the nozzle and, finally, transferred into CO_2 dissociation.

This interesting effect of **gasdynamic stimulation of CO_2 dissociation in supersonic flow** is similar to plasma-chemical CO_2 dissociation but doesn't require direct consumption of electricity to sustain the plasma. For this reason the effect sometimes is referred to as **"plasma chemistry without electricity."** The effect of gasdynamic stimulation of dissociation is somewhat similar to gasdynamic lasers (Losev, 1977), but non-equilibrium vibrational energy is transferred in this case into a chemical reaction instead of radiation. The phase "plasma chemistry without electricity" is due to first experimental observation of the effect. The electricity used to sustain microwave plasma was turned off in a supersonic plasma-chemical experiment, but dissociation of CO_2 in the supersonic flow remained at a significant level even though the initial gas temperature was 300 K(!). The effect can be observed not only in CO_2 dissociation but also in other endothermic processes stimulated by vibrational excitation (Rusanov et al., 1982; Zhdanok & Soloukhin, 1982; Asisov et al., 1986). One can say that the gasdynamic stimulation effectively increases the "kinetic temperature" of the gas. It means that, due to the Treanor effect even at room temperature, the CO_2 dissociation rate is equivalent to that at more than 2000 K in quasi equilibrium. The conversion degree, however, corresponds to the total energy accumulated in vibrational degrees of freedom at actual CO_2 temperature. Obviously, the gasdynamic stimulation effect is not limited to CO_2 dissociation and can be applied to other gases where the Treanor effect is strong enough, for example CO, N_2, and H_2. Experimental results on the gasdynamic stimulation of CO_2 dissociation in supersonic flow with Mach number $M \approx 6$ (Asisov et al., 1986) are shown in Fig. 5–47 in comparison with kinetic simulations at Mach numbers ($M = 4.5, 5.5, 6.5, 9.5$) corresponding to $T_v / T_0 = 5, 7, 10, 20$. The figure presents CO_2 conversion degree α as a function of initial gas temperature before the supersonic nozzle. As one can see, the conversion degree can reach the level of 1–3% without direct use of electricity and without significant heating of the gas. At room temperature (that is without heating at all!), the CO_2 conversion degree is not high (about 0.2%) but is quite visible.

5.6.6. Plasma Radiolysis of CO_2 Provided by High-Current Relativistic Electron Beams

Dissociation of CO_2 can be effectively stimulated by relativistic electron beams (Legasov et al., 1978; Vakar et al., 1978). The dominating mechanism of CO_2 dissociation is usually related in this case to electronic excitation of the molecules. The major difference between plasma radiolysis and conventional radiolysis is related to the contribution of collective non-linear effects, or in other words synergetic effects, in the dissociation process. While conventional radiolysis is based on CO_2 dissociation provided by individual high-energy

5.6. CO_2 Dissociation in Special Experimental Systems

Figure 5–47. CO_2 conversion degree into CO and O_2 as function of preheating temperature in supersonic system of gas-dynamic stimulation of CO_2 dissociation. Mach numbers: (1) $M = 4.5$ (corresponds to $T_v/T_0 = 5$); (2) $M = 5.5$ (corresponds to $T_v/T_0 = 7$); (3) $M = 6.5$ (correspond to $T_v/T_0 = 10$); (4) $M = 9.5$ (correspond to $T_v/T_0 = 20$). Solid lines correspond to theoretical calculations, dots correspond to experiments with $M = 6$.

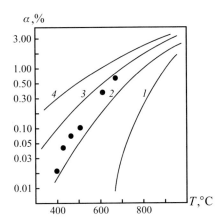

particles (like relativistic electrons) and the conversion degree is linear with respect to the flux of the energetic particles, the plasma radiolysis is essentially affected by such collective synergetic effects as three-body recombination of oxygen atoms (preventing the reverse reaction $O + CO + M \rightarrow CO_2 + M$), as well as Maxwellization of generated plasma electrons, possible vibrational excitation, and so on. The effects of plasma radiolysis lead to a non-linear dependence of CO_2 conversion on the flux of energetic particles and to an increase of energy efficiency of CO_2 dissociation. The CO_2 plasma radiolysis experiments were carried out using the high-current relativistic electron beam with the following parameters: electron energy 400 keV, current density 80 A/cm^2, gas pressure 0.5–2 atm, pulse duration 1 μs, and energy in each pulse of 150 J. The energy cost of CO_2 dissociation in these experiments with relativistic electron beams was about 10–12 eV/mol. Energy efficiency is usually expressed in radiolysis by means of the **G-factor**, which represents the number of produced molecules per 100 eV of absorbed radiation. So, in these terms, the plasma-radiolytic process can be characterized by the quite high G-factor: $G = 10$–12 eV/mol CO. Modeling of the plasma radiolysis produced by a relativistic electron beam was accomplished by Legasov et al. (1978) using the kinetic equation for the degradation spectrum (Sholin, Nikerov, & Rusanov, 1980; Nikerov & Sholin, 1978, 1985). The simulated value of the dissociation energy cost was about 10 eV/mol, corresponding to an energy efficiency of about 30%.

5.6.7. Plasma Radiolysis of CO_2 in Tracks of Nuclear Fission Fragments

Consider CO_2 plasma generation based on non-electric energy sources particularly in the stopping process of nuclear fission fragments (Fridman & Kennedy, 2004). The nuclear fission process can be illustrated by the reaction of a slow neutron with uranium-235:

$$_{92}U^{235} + {}_0n^1 \rightarrow {}_{57}La^{147} + {}_{35}Ba^{87} + 2{}_0n^1. \tag{5–94}$$

The total fission energy is 195 MeV; most of it (162 MeV) goes to the kinetic energy of the fission fragments – barium and lanthanum (see, for example, Glasstone & Edlung, 1952). Plasma-chemical effects can be induced by the nuclear fission fragment if they are transmitted into the gas phase before thermalization. The fission fragment appears in gas as multiply charged ions with $Z = 20$–22 and an initial energy of about $E_0 \approx 100$ MeV (Hassan & Deese, 1976). Degradation of these fission-fragment ions in gas is related to the formation of so-called δ-electrons moving perpendicularly to the fission fragment with a kinetic energy of about $\varepsilon_0 \approx 100$ eV. These δ-electrons form a radial high-energy electron beam, which then generates plasma in the long cylinder surrounding the trajectory of the fission fragment. This plasma cylinder is usually referred to as the **track of the nuclear fission fragments.** If the plasma density in the track of nuclear fission fragments is relatively low, then degradation of each δ-electron in the radial beam is quasi independent. Plasma radiolysis can be observed in the tracks of the nuclear fission fragments if plasma density in the tracks of nuclear fission fragments is relatively high and the collective effects of electron–electron interaction, vibrational excitation, and so forth take place (Belousov et al., 1979).

The radiation yield of CO_2 dissociation includes a conventional part, G_{rad}, typical for radiation chemistry, and a special plasma-radiolytic part related to the contribution of electrons under the threshold of electronic excitation. The energy efficiency of radiolysis is usually characterized by the so-called G-factor, which shows the number of chemical processes of interest (in this case, the number of dissociated CO_2 molecules) per 100 eV of radiation energy spent. The first conventional radiolytic part is due to electronic excitation and dissociative recombination and can reach at maximum $G_{rad} = 8 \frac{mol}{100\,eV}$ (Aulsloos, 1968). The contribution of low-energy electrons (under the threshold of electronic excitation), which determines the plasma radiolysis in CO_2 is mostly due to vibrational excitation of CO_2 molecules. This effect is proportional to the ionization degree n_e/n_0 in the track of the nuclear fission fragment:

$$\Delta T_v \approx \frac{n_e}{n_0} \left[\frac{\int_0^I \varepsilon Z(\varepsilon)\,d\varepsilon}{c_v \int_0^I Z(\varepsilon)\,d\varepsilon} \right]. \tag{5–95}$$

In this relation, c_v is the specific heat related to vibrational degrees of freedom of CO_2 molecules; the factor in parentheses is determined by the degradation spectrum $Z(\varepsilon)$ of δ-electrons in tracks (Nikerov & Sholin, 1978, 1985) and numerically is about several electron volts (Belousov et al., 1979). The plasma effect in CO_2 radiolysis is mostly related to stimulation of the reaction

$$O + CO_2 \rightarrow CO + O_2, \tag{5–96}$$

by vibrational excitation of CO_2 molecules. This reaction prevails over trimolecular recombination of oxygen atoms,

$$O + O + M \rightarrow O_2 + M, \tag{5–97}$$

at vibrational temperatures corresponding in accordance with relation (5–95) to ionization degrees $n_e/n_0 > 10^{-4}$–10^{-3}. The radiation yield of CO_2 dissociation in this case becomes twice as high as in conventional radiolysis and reaches $G = 16 \frac{mol}{100\,eV}$, which corresponds to an energy efficiency of about 50%. The radiation yield of CO in conventional radiolysis is essentially suppressed by reverse reactions:

$$O + CO + M \rightarrow CO_2 + M, \tag{5–98}$$

which are much faster at standard conditions than recombination (5–97), because the CO concentration usually greatly exceeds the concentration of atomic oxygen. As a result, the radiation yield of conventional CO_2 dissociation, $G_{rad} = 8 \frac{mol}{100\,eV}$, leads to a radiation yield of CO production of only $G = 0.1 \frac{mol}{100\,eV}$ (Hartech & Doodes, 1957). Only high-intensity radiolysis, which provides a high concentration of oxygen atoms, permits one to suppress the reverse reactions (5–98) in favor of (5–97) and increase the radiation yield (Kummler, 1977).

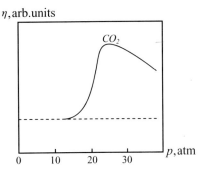

Figure 5–48. Energy efficiency of CO_2 dissociation (shown in arbitrary units) during plasma radiolysis in tracks of nuclear fission fragments as a function of pressure.

5.6.8. Ionization Degree in Tracks of Nuclear Fission Fragments, Energy Efficiency of Plasma Radiolysis of CO₂, and Plasma-Assisted Chemonuclear Reactors

The energy efficiency of CO_2 dissociation can be increased about two-fold, when the ionization degree in tracks of nuclear fission fragments exceeds some critical value $n_e/n_0 \geq 10^{-4}$–10^{-3}. To find the conditions required to reach such a high ionization degree, one should analyze the degradation of δ-electrons with kinetic energy $\varepsilon_\delta \approx 100$ eV formed by the fragments. Theory of the degradation (see Fridman & Kennedy, 2004) gives the ionization degree in the tracks of nuclear fission fragments:

$$\frac{n_e}{n_0} = \left(\frac{3Z_0 e^4 \sigma_i^2}{4\pi \varepsilon_0^2 m \varepsilon_\delta} \right) \cdot \left(\frac{Z}{v} \right)^2 n_0^2. \tag{5–99}$$

In this relation, σ_i is the cross section of ionization for δ-electrons; Z_0 is the sum of charge numbers of atoms forming CO_2 ($Z_0 = 44$); Z and v are the charge number and velocity of a nuclear fission fragment; and ε_δ is the energy of a δ-electron. The ionization degree in the plasma channel of a nuclear track is proportional to the square of neutral gas density and pressure; numerically, at atmospheric pressure, it is $n_e/n_0 \approx 3 \cdot 10^{-6}$ in CO_2.

Thus, the necessary value of ionization degrees for significant plasma effects in radiolysis (10^{-3}) cannot be achieved in atmospheric-pressure CO_2. However, taking into account that $n_e/n_0 \propto n_0^2$, a sufficient ionization degree can be achieved by increasing the pressure to 10–20 atm. A further increase of pressure and ionization degree is limited by energy transfer from δ-electrons to plasma ions. This ineffective ion heating and energy transfer to thermal plasma in tracks takes place in carbon dioxide at pressures about 30 atm. The calculated dependence of energy efficiency for CO_2 decomposition in tracks of nuclear fission fragments is shown in Fig. 5–48, which is in qualitative agreement with experiments described by Hartech and Doodes (1957) and Belousov et al. (1979). The highest values of energy efficiency for H_2O and CO_2 decomposition in tracks of nuclear fission fragments can be achieved, in principal, if δ-electrons transfer their energy not directly to neutral gas but through an intermediate transfer of their energy into plasma formed in tracks. The major portion of the energy of nuclear fragments can be transferred to plasma electrons by mechanisms similar to those in plasma-beam discharges described in Chapter 4 (see also Ivanov & Soboleva, 1978). Then the energy of plasma electrons can be transferred to the effective channels of stimulation of plasma-chemical reactions, in particular to vibrational excitation. Thus, nuclear fission fragments can transfer their energy in such systems primarily not into heating but into the formation of chemical products (in particular, in hydrogen). The chemonuclear reactor based on such principles can actually be "cold" because only a small portion of its energy generation is related to heating. One such concept, the **plasma-assisted chemonuclear reactor**, was considered in detail by Bochin et al. (1983).

5.7. COMPLETE CO₂ DISSOCIATION IN PLASMA WITH PRODUCTION OF CARBON AND OXYGEN

5.7.1. Complete Plasma-Chemical Dissociation of CO₂: Specifics of the Process and Elementary Reaction Mechanism

Plasma parameters can be chosen to provide a predominantly complete dissociation of CO_2 with the production of solid carbon and oxygen:

$$CO_2 \rightarrow C(s) + O_2, \quad \Delta H = 11.5\,\text{eV/mol}. \qquad (5\text{--}100)$$

This plasma-chemical process not only attracts attention itself for application in chemical technology but it also can be applied as an engineering model for the production of metals from their oxides and halogenides (Tsvetkov & Panfilov, 1980). The plasma-chemical mechanism for the complete dissociation process (5–100) includes all reactions leading to the formation of CO (see, in particular, Table 5–1). First of all, it includes elementary reactions (5–6) and (5–7), which can be effectively stimulated by vibrational excitation under non-equilibrium plasma conditions. For complete dissociation of carbon dioxide, the conversion of CO into elementary carbon should also take place. This process is usually kinetically limited (Legasov et al., 1977a,b) because of the record high value (11 eV!) of bonding energy in a CO molecule. Conversion of CO into carbon can be effectively performed by means of the reaction of disproportioning, which can be, in particular, significantly accelerated by the vibrational excitation of CO molecules:

$$CO + CO \rightarrow CO_2 + C, \quad \Delta H = 5.5\,\text{eV/mol}, \quad E_a = 6\,\text{eV/mol}. \qquad (5\text{--}101)$$

The activation energy of disproportioning reaction (5–101) is still high, but it is almost two-fold less than the CO bonding energy mentioned earlier (Walker, Rusinko, & Austin, 1963). Elementary reaction (5–101) leads to the formation of free carbon atoms in the gas phase. The kinetics of stimulation of the elementary process of disproportioning (5–101) by vibrational excitation of CO molecules will be discussed in the next section. The disproportioning (5–101) can also be accelerated via electronic excitation:

$$CO(a^3\Pi) + CO \rightarrow CO_2 + C. \qquad (5\text{--}102)$$

This elementary process dominates, however, at relatively high values of electron temperature in plasma (Volchenok & Egorov, 1976). In general, the disproportioning of CO with the production of elementary carbon (5–101) can be easily stimulated in different plasma systems (Maximov et al., 1979; Legasov et al., 1981). It is much more difficult to provide reduction of elementary carbon not from CO, but directly from CO_2. The major complication in this case is due to the fast reverse reaction of carbon atoms:

$$C + CO_2 \rightarrow CO + CO, \quad E_a = 0.5\,\text{eV/mol}. \qquad (5\text{--}103)$$

Effective suppression of the reverse reaction (5–103) becomes possible only by heterogeneous stabilization of atomic carbon produced in (5–101) on discharge walls or by carbon clusterization in the discharge volume. The kinetics of heterogeneous carbon stabilization and clusterization is considered in detail in Legasov et al. (1978c).

5.7.2. Kinetics of CO Disproportioning Stimulated in Non-Equilibrium Plasma by Vibrational Excitation of Molecules

The efficiency of vibrational energy in overcoming the activation energy barriers of strongly endothermic reactions is usually close to 100% ($\alpha = 1$; see Chapter 2). Usually only one molecule retains the vibrational energy, as in the case of CO_2 dissociation (see Section 5.1.4, reactions (5–6) and (5–7)). In the case of CO disproportioning, two molecules involved in reaction can be strongly excited:

$$CO^*(v_1) + CO^*(v_2) \rightarrow CO_2 + C. \quad (5\text{--}104)$$

This elementary reaction was studied by Nester, Demura, and Fridman (1983) using the quantum-mechanical approach of vibronic terms. It was shown that the high activation energy of the process ($E_a = 6$ eV/mol; see (5–101)) can be reduced by the vibrational energy of reagents. Both excited molecules contribute their energy to decrease the activation energy in this case:

$$\Delta E_a = \alpha_1 E_{v1} + \alpha_2 E_{v2}. \quad (5\text{--}105)$$

It is assumed that the rest of the activation barrier should be overcome by energy in translational and rotational degrees of freedom of the CO molecules. In relation (5–105), E_{v1} and E_{v2} are the vibrational energies of the two excited molecules (subscript 1 corresponds to the molecule losing an atom and subscript 2 is related to another molecule accepting the atom); α_1 and α_2 are coefficients of efficiency for using vibrational energy from these two sources. The calculations show that $\alpha_1 = 1$ and $\alpha_2 \approx 0.2$. It reflects the fact that the vibrational energy of the donating molecule is much more efficient in stimulation of endothermic exchange reactions than that of the accepting molecule. The rate coefficient for disproportioning reaction (5–104) was calculated by Rusanov, Fridman, and Sholin (1977), assuming that an effective microscopic reaction $k_R(E', E'')$ takes place, when the sum of vibrational energies of two partners exceeds the activation energy: $E' + E'' \geq E_a$ (compare with (5–105)). The total reaction rate ($w_R = n_0 J_0$) can be found based on the vibrational distribution function $f^{(0)}(E)$ not perturbed by the chemical process (slow reaction regime; see Chapter 2):

$$J_0 = \iint_{E' + E'' \geq E_a} f^0(E') f(E'') k_R(E', E'') n_0 \, dE' dE''. \quad (5\text{--}106)$$

If $E_a > 2E^*(T_0)$, the process stimulation by vibrational excitation is ineffective, because the population of excited states is low due to VT relaxation; here $E^*(T_0)$ is the critical vibrational energy, when rates of VT and VV relaxation become equal (see Chapter 2). If $E_a < E^*(T_0)$, the Treanor effect is stronger when one molecule has all the vibrational energy, and the reaction of two excited molecules actually effectively uses only one of them. Indeed, when the total vibrational energy is fixed ($E' + E'' = $ const), then $(E')^2 + (E'')^2 = (E' + E'')^2 - 2E'E''$ and the Treanor effect is the strongest if $E'' = 0$. As the result, the non-trivial kinetics (5–106) takes place only if $E^*(T_0) < E_a < 2E^*(T_0)$. In this case, integration (5–106) determines the optimal vibrational energies of reaction partners, E_{opt}, and $E_a - E_{opt}$, which make the biggest contribution to the total disproportioning reaction rate:

$$E_{opt} = E^*(T_0) - \frac{1}{2\delta_{vv}} \ln \frac{T_0}{2x_e[2E^*(T_0) - E_a]}. \quad (5\text{--}107)$$

In this relation, x_e is the CO anharmonicity coefficient, $\delta_{vv} \approx (0.2\text{--}0.5)/\hbar\omega$ is the adiabatic parameter of VV relaxation, and $\hbar\omega$ is the CO vibrational quantum. The macroscopic rate coefficient for the disproportioning reaction of two vibrationally excited CO molecules can then be presented as

$$k_R^{macro}(T_v, T_0) \propto \exp\left[-\frac{E_a}{T_v} + \frac{x_e}{T_0\hbar\omega}\left(E_{opt}^2 + (E_a - E_{opt})^2\right) - \frac{x_e}{\delta_{vv}T_0^2}(2E^*(T_0) - E_a)\right]. \quad (5\text{--}108)$$

Kinetic analysis of (5–108) and the energy balance of CO disproportioning by itself and as a part of the plasma-chemical process of complete CO₂ dissociation to elementary

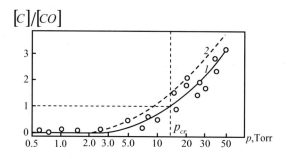

Figure 5–49. Concentration ratio [C]/[CO] as a function of pressure, measured by means of (1) pressure defect and (2) by mass spectrometry, in experiments on complete CO_2 dissociation in microwave discharge operating in conditions of electron cyclotron resonance (ECR).

carbon can be found in Rusanov, Fridman, and Sholin (1977) and Rusanov and Fridman (1984).

5.7.3. Experiments with Complete CO_2 Dissociation in Microwave Discharges Operating in Conditions of Electron Cyclotron Resonance

The microwave discharge operating in a magnetic field under ECR conditions, where the experiments on complete CO_2 dissociation were performed (Asisov, 1980), will be discussed in Chapter 6 regarding NO synthesis in air. It was a pulse-periodic discharge, sustained by microwave pulses with wavelength 8 mm, power in each pulse of 30 kW, pulse duration 0.3 μs, and frequency 0.5 kHz. The reactor was organized as part of a round waveguide with diameter 1.7 cm and length 10 cm. Experiments were carried out in CO_2 at pressures of 0.5–100 Torr. Two alternative channels of CO_2 dissociation (complete and incomplete) are characterized by different changes in the number of moles in the gas phase. Decomposition (5–1) 1.5 times increases the number of moles ($CO_2 \rightarrow CO + 0.5O_2$), whereas complete dissociation (5–100) doesn't change the number of moles in the gas phase ($CO_2 \rightarrow C(s) + O_2$). Based on that, the C/CO ratio can be calculated from pressure change in the constant-volume reactor maintained without gas flow. This ratio at a CO_2 pressure of 50 Torr reaches quite a high value of C/CO $\approx$ 4 (see Fig. 5–49), which means that complete dissociation can essentially dominate over incomplete dissociation. When the CO_2 temperature was preliminarily decreased below 300 K and the reverse reaction (5–103) was suppressed, the C/CO ratio grew significantly in the experiments, providing, almost exclusively, complete CO_2 dissociation. A stoichiometric amount of soot was collected on the reactor walls in the experiments. Results of mass-spectroscopic measurements of the dissociation products are also presented in Fig. 5–49 and confirmed the domination of complete dissociation. At lower pressures (0.5–2 Torr), the major product of CO_2 dissociation was CO. The translational temperature increase ΔT_0 during the plasma-chemical process and non-equilibrium degree T_e/T_0 are shown in Fig. 5–50. Both parameters were spectroscopically measured in the discharge and theoretically simulated, showing good agreement with experiments. Good agreement was also achieved between the modeling and diagnostic measurements of electron density ($n_e \approx 10^{13}$ cm^{-3}), accomplished using the special microwave interferometry at wavelength

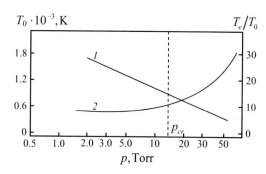

Figure 5–50. Translational temperature increase ΔT_0 during (1) the discharge and (2) the discharge non-equilibrium degree T_e/T_0 as function of pressure in experiments with complete CO_2 dissociation in microwave plasma in conditions of electron cyclotron resonance (ECR).

5.7. Complete CO_2 Dissociation in Plasma with Production of Carbon and Oxygen

Figure 5–51. CO_2 dissociation in plasma-beam discharge. Partial pressures of the dissociation products as function of electron beam power: (1) CO_2; (2) CO; (3) O_2.

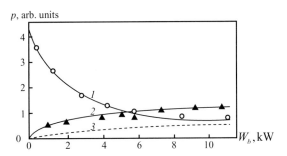

$\lambda = 4\,\text{mm}$. The non-equilibrium degree in the microwave discharge operating in a magnetic field under ECR conditions was quite high ($T_e / T_0 = 10\text{–}30$) and continued growing with pressure increase.

Analysis of the experimental results presented in Figs. 5–49 and 5–50 leads to some critical value of the gas pressure, $p_{cr} \approx 15\,\text{Torr}$. At lower pressures ($p < p_{cr}$), gas temperature is elevated and intensive reverse reaction (5–103) results in the domination of incomplete CO_2 dissociation. At higher pressures ($p > p_{cr}$), gas temperature remains low, and the degree of non-equilibrium (T_e / T_0) high, which leads to domination of complete CO_2 dissociation and elementary carbon production in agreement with the concept of **heterogeneous stabilization of products of plasma-chemical reactions** (Legasov et al., 1978c). Effective CO_2 dissociation in the discharge was achieved when vibrational temperature exceeded the critical value ($T_v^{\min} = 0.2\,\text{eV}$). Energy efficiency strongly depends in this case on magnetic field and, at the ECR conditions, reaches 60% with respect to CO production. We should note that, in this pulse-periodic microwave discharge, it is not necessary to exceed the critical value of vibrational temperature ($T_v^{\min}$) just after one microwave pulse. The required vibrational temperature can be collected in several pulses (see description of the ECR discharge in Chapter 6).

5.7.4. Experiments with Complete CO_2 Dissociation in Stationary Plasma-Beam Discharge

In these experiments (Ivanov & Nikiforov, 1978; Atamanov, Ivanov, & Nikiforov, 1979a,b), the electron energy in the beam was 30 keV and the beam current was up to 2 A. During the electron beam transportation through the plasma, 70% of the beam energy was transferred to plasma electrons. The gas pressure in the experiments was about 10 mTorr to provide effective beam–plasma interaction; the gas flow rate was about 40 cm^3/s. The power varies from 0.5 to 10 kW, which corresponds to rather high values of specific energy input ($E_v = 13\text{–}270\,\text{J/cm}^3$). Plasma parameters were experimentally determined as $T_e = 1\text{–}2\,\text{eV}$, $n_e / n_0 = 10^{-3}\text{–}10^{-2}$. Results of mass-spectrometric measurements of CO_2, CO, and O_2 partial pressures after treatment of CO_2 in the plasma-beam discharge are presented in Fig. 5–51 as functions of electron beam power. At electron beam powers of 8–10 kW, most of the CO_2 is dissociated. A relatively low partial pressure of CO indicates the domination of complete CO_2 decomposition with the formation of elementary carbon (some fraction of oxygen was lost to the oxidation of discharge walls). The formation of elementary carbon was also confirmed in these experiments by means of special spectral diagnostics. Figure 5–52 shows that the conversion degree of CO_2 in the plasma-beam discharge reaches 85%. At the same time, formed CO molecules are further decomposed to form elementary carbon and oxygen with a conversion degree of up to 90%. Dependence of the conversion degree on electron beam power (Fig. 5–52) is interpreted by Atamanov, Ivanov, and Nikiforov (1979a,b) by taking into account that dissociation is stimulated by vibrational excitation at powers of about 1.5 kW, and by electronic excitation at powers exceeding 2 kW, when electron energy

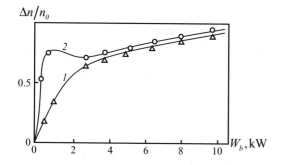

Figure 5–52. CO_2 dissociation in plasma-beam discharge. Conversion degree of (1) CO_2 and (2) CO as function of electron beam power.

in the beam is higher. We should note at this point that the specific energy input is $E_v = 30$–40 J/cm^3 when the beam power is about 1.5 kW, which is far from the optimal E_v values (see Section 5.3). The energy efficiency of CO_2 dissociation was calculated in the experiments with plasma-beam discharge using the following formula, which takes into account both incomplete and complete decomposition:

$$\eta = \frac{(\alpha_{CO_2} \Delta H_1 + \alpha_{CO_2} \alpha_{CO} \Delta H_2) \cdot Q}{W \cdot \eta_{beam}}. \qquad (5\text{–}109)$$

In this relation, α_{CO_2} is the conversion degree related to the incomplete dissociation process $CO_2 \rightarrow CO + 0.5O_2$, $\Delta H_1 = 2.9$ eV/mol; α_{CO} is the conversion degree related to the CO conversion into elementary carbon, $CO \rightarrow C(s) + 0.5O_2$, $\Delta H_2 = 8.6$ eV/mol; Q is the total gas flow rate through the discharge; W is the electron beam power; and η_{beam} is the efficiency of energy transfer from the electron beam to plasma electrons. The energy efficiency of CO_2 dissociation in the plasma-beam discharge, calculated using relation (5–109), reaches the maximum value of 60–70% at specific energy inputs $E_v = 15$–30 J/cm^3. The energy efficiency of CO_2 dissociation decreases to about 30% at higher values of specific energy input in the discharge.

5.8. DISSOCIATION OF WATER VAPOR AND HYDROGEN PRODUCTION IN PLASMA-CHEMICAL SYSTEMS

5.8.1. Fundamental and Applied Aspects of H_2O Plasma Chemistry

The direct decomposition of water vapor in plasma is obviously the most natural and straight-forward approach to plasma-chemical hydrogen production:

$$H_2O \rightarrow H_2 + \frac{1}{2}O_2, \quad \Delta H = 2.6 \text{ eV/mol}. \qquad (5\text{–}110)$$

The plasma-chemical method of water decomposition with production of hydrogen and oxygen is able to compete successfully with more conventional approaches to hydrogen production from water, such as electrolysis, thermochemical and thermocatalytic cycles, and radiolysis. The principal advantage of the plasma-chemical approach is very high specific productivity (in other words, very high hydrogen productivity per unit volume or per unit mass of the industrial device; Legasov et al., 1980). Because of its volumetric nature, the plasma-chemical process can produce up to 1000 times more hydrogen than electrolytic or thermocatalytic systems of the same size. However, the energy efficiency of plasma-chemical methods can be about the same as that of electrolytic and thermocatalytic systems (Asisov et al., 1985; Gutsol et al., 1985; Melik-Aslanov et al., 1985; Rusanov & Fridman, 1978). The high specific productivity and energy efficiency obviously call attention to the plasma-chemical methods of hydrogen production from water, especially if there is a relatively inexpensive source of electricity such as, for example, a nuclear reactor during

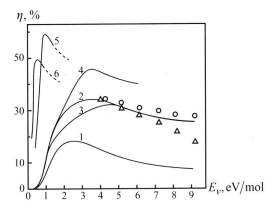

Figure 5–53. Energy efficiency of water vapor dissociation as function of specific energy input: (1, 2) thermal dissociation with absolute and ideal quenching; (3, 4) super-absolute and super-ideal quenching mechanisms; (5) non-equilibrium dissociation stimulated by vibrational excitation; (6) non-equilibrium decomposition due to dissociative attachment. Circles and triangles represent different experiments with microwave discharges.

periods of low electricity consumption (Alexandrov, 1978; Legasov, 1978, 1988; Gutsol et al., 1985; Legasov, et al., 1988). The cumulative dependence of energy efficiency of H_2O dissociation on energy input (see Fig. 5–53) is calculated for different specific mechanisms of the process in comparison with experimental results.

5.8.2. Kinetics of Dissociation of Water Vapor Stimulated in Non-Thermal Plasma by Vibrational Excitation of Water Molecules

Similarly to the CO_2 dissociation, a significant contribution of vibrational excitation in plasma-chemical H_2O dissociation kinetics (Bochin et al., 1977, 1979) is due to the possibility of transfering most of the discharge energy there (more than 80% at $T_e \approx 1\,eV$; see Fig. 5–54, Givotov et al., 1981). The dissociation of H_2O molecules in plasma, stimulated by vibrational excitation, follows three major kinetic steps:

1. The excitation of lower vibrational levels of H_2O molecules is provided by electron impact. In contrast to CO_2 molecules, it is difficult to point out here which vibrational mode is mostly excited by electron impact (Massey, 1969).
2. The excitation of higher vibrational levels of H_2O is due to VV relaxation. Relaxation rates of the intermode vibrational exchange $\nu_1(A_1) \Leftrightarrow 2\nu_2(A_1)$ and of the intermode Fermi-resonant vibration exchange $\nu_3(B_1) \Leftrightarrow \nu_1(A_1)$ are close to conventional intramode VV-relaxation rates (Birukov, Gordiets, and Shelepin, 1968). For this reason, vibrationally excited H_2O molecules can be characterized by a single vibrational temperature T_v. The fastest VT relaxation takes place from the vibrational mode $\nu_2(A_1)$.
3. Decomposition of highly vibrationally excited H_2O molecules proceeds as a chain reaction. The total chemical process (5–110) is initiated by the dissociation of vibrationally excited molecules:

$$H_2O^* + H_2O \rightarrow H + OH + H_2O. \tag{5–111}$$

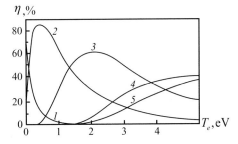

Figure 5–54. Electron energy distribution between excitation, dissociation, and ionization channels in water vapor: (1) elastic scattering; (2) vibrational excitation; (3) dissociative attachment; (4) electronic excitation; (5) ionization.

After chain initiation (5–111), chain propagation takes place through the following reactions of vibrationally excited molecules (Bochin et al., 1979; Gutsol et al., 1984):

$$H + H_2O^* \rightarrow H_2 + OH, \quad \Delta H_1 = 0.6\,eV/mol, \quad E_{a2} = 0.9\,eV/mol, \quad (5\text{–}112)$$

$$OH + H_2O^* \rightarrow H_2O_2 + H, \quad \Delta H_2 = 2.6\,eV/mol, \quad E_{a2} = 3.0\,eV/mol. \quad (5\text{–}113)$$

Primary products of the chain reaction are molecular hydrogen (H_2) and hydrogen peroxide (H_2O_2), which is usually unstable at typical plasma conditions and decays quite fast into water (H_2O) and oxygen (O_2). H_2O_2 can be stabilized in the products of water dissociation at low temperatures, in particular if the process is carried out in a supersonic flow. Such a process is not only very fundamentally interesting, but also permits us to effectively separate the dissociation products and protect them from reverse reactions. The chain termination, finally, is related to different three-body recombination processes, and first of all to the following reaction:

$$H + OH + H_2O \rightarrow H_2O + H_2O, \quad k_a = 3 \cdot 10^{-33}\,cm^6/s. \quad (5\text{–}114)$$

To describe the kinetics of the chain mechanism of H_2O dissociation (5–111)–(5–114) stimulated by vibrational excitation, we present the first rate coefficient of the chain initiation process (5–111):

$$k_R(T_v) = k_{vv}^0 \left(\frac{E_D}{T_v} \right)^s \exp\left(-\frac{E_D}{T_v} \right). \quad (5\text{–}115)$$

In this relation, k_{vv}^0 is the rate coefficient for resonant VV exchange of lower vibrational levels (the rate coefficient is close to the gaskinetic one, which is about $10^{-10}\,cm^3/s$), and $E_D \approx 5.1\,eV/mol$ is the H_2O dissociation energy. The total chain reaction rate obviously depends on the chain length, which can be numerically estimated as $\nu \approx 100$ at vibrational temperature $T_v = 0.5\,eV$ and water vapor density $[H_2O] = 3 \cdot 10^{18}\,cm^{-3}$. The total chain reaction rate can be calculated by taking into account both the chain initiation rate (5–115) and the chain length as follows:

$$w_R \approx k_0 [H_2O]^2 \sqrt{\frac{k_0}{k_a[H_2O]}} \cdot \exp\left(-\frac{E_D + E_{a1} + E_{a2}}{2T_v} \right). \quad (5\text{–}116)$$

In addition to mechanism (5–112), (5–113), there is another chain propagation path, which is able to provide plasma-chemical H_2O dissociation stimulated by vibrational excitation:

$$OH + H_2O^* \rightarrow H_2 + HO_2, \quad \Delta H = 2.1\,eV/mol, \quad E_a = 3.3\,eV/mol, \quad (5\text{–}117)$$

$$HO_2 + H_2O^* \rightarrow H_2O_2 + OH, \quad \Delta H = 1.1\,eV/mol, \quad E_a = 1.3\,eV/mol. \quad (5\text{–}118)$$

These alternative chain propagation reactions are essentially slower, however, than reactions (5–112) and (5–113) because their activation energies are higher.

5.8.3. Energy Efficiency of Dissociation of Water Vapor Stimulated in Non-Thermal Plasma by Vibrational Excitation

The total energy efficiency of the considered mechanism of H_2O dissociation includes three major factors: the excitation factor η_{ex}, related to the discharge energy fraction going to vibrational excitation; the relaxation factor η_{rel}, showing the effectiveness of the dissociation with respect to VT relaxation; and finally the chemical factor η_{chem}, which is related to energy losses in exothermic stages and reverse reactions.

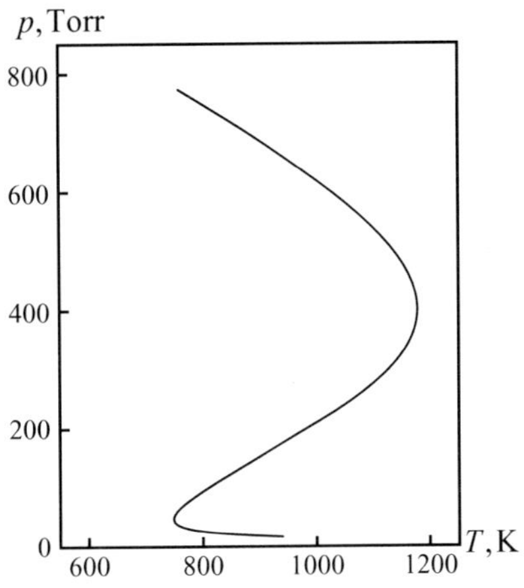

5.8. Dissociation of Water Vapor and Hydrogen Production

Figure 5–55. Explosion limits of the products of H_2O dissociation (H_2, O_2); explosion takes place to the right of the curve; stability to the left.

A. The excitation factor. This factor depends on reduced electric field and electron temperature. It can be found directly from Fig. 5–54, which shows the distribution of electron energy transfer between different channels of excitation and ionization of molecular gas. This factor can exceed 80% at $T_e \approx 1$ eV.

B. The chemical factor. This factor depends on the chain length ν. The energy efficiency is higher, when the reaction chain is long and the contribution of chain initiation stage (5–111) is relatively low:

$$\eta_{chem} = \frac{\Delta H_1 + \Delta H_2}{\frac{1}{\nu}(E_D - E_{a2}) + E_{a1} + E_{a2}}. \tag{5–119}$$

Numerically, $\eta_{chem} = 85\%$ for long chains ($\nu = 100$), while the chemical factor is relatively low ($\eta_{chem} = 50\%$) in the absence of the chain. Low values of the chemical factor and a decrease in total energy efficiency can also be due to the contribution of reverse reactions, predominantly

$$OH + H_2 \rightarrow H_2O + H, \quad E_{(-)} = 0.5 \text{ eV/mol}. \tag{5–120}$$

This reaction is important because the high activation energy of (5–113) leads to a high concentration of OH radicals:

$$\frac{[OH]^2}{[H_2O]^2} = \frac{k_0}{k_a[H_2O]} \exp\left(\frac{E_{a2} - E_{a1} - E_D}{T_v}\right). \tag{5–121}$$

In this relation, k_a is the rate coefficient of recombination (5–114), and $k_0 = 10^{-10}$ cm^3/s is the rate coefficient for gas-kinetic collisions. The influence of the reverse reaction (5–120) results in a limitation of the hydrogen concentration in products:

$$\frac{[H_2]}{[H_2O]} \ll \exp\left(-\frac{E_{(-)}T_v - E_{a2}T_0}{T_v T_0}\right). \tag{5–122}$$

The dissociation yield and temperature limitation (5–122) is usually stronger than the restriction related to explosion of the hydrogen–oxygen mixture (Fig. 5–55), which is mitigated because of essential vapor fraction, and hence dilution, in the produced mixture (H_2–O_2–H_2O). The domain of vibrational and translational temperatures of water vapor, where the product stability with respect to explosion and reverse reactions can be achieved, are shown in Fig. 5–56.

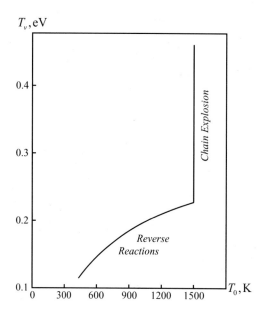

Figure 5–56. Temperature restrictions for stability of the products of CO_2 dissociation in plasma stimulated by vibrational excitation. Explosion and reverse reactions take place to the right and down from the curve; stability is to the left and up.

C. The relaxation factor. Energy losses related to vibrational relaxation take place in both active and passive discharge zones. It is especially serious in the case of plasma-chemical water dissociation because of the relatively high value of the VT-relaxation rate coefficient for H_2O molecules ($k_{VT} \approx 10^{-12}$ cm^3/s). To overcome vibrational relaxation in the active discharge zone, the ionization degree in plasma should be fairly high:

$$\frac{n_e}{[H_2O]} > \frac{k_{VT}(T_0)}{k_{eV}(T_e)}. \tag{5–123}$$

In this relation, $k_{eV}(T_e, \approx 10^{-8}$cm^3/s is the rate coefficient of the vibrational excitation of H_2O molecules by direct electron impact. Numerically, requirement (5–123) means $\frac{n_e}{[H_2O]} > 10^{-4}$. When the ionization degree is sufficiently high (5–123), the vibrational temperature of water molecules exceeds the critical value T_v^{min}:

$$T_v^{min} = \frac{E_D + E_{a1} + E_{a2}}{2\ln\left[\dfrac{k_0}{k_{VT}} \cdot \dfrac{(E_{a1} + E_{a2})}{\hbar\omega}\sqrt{\dfrac{k_0}{k_a[H_2O]}}\right]}. \tag{5–124}$$

The critical vibrational temperature corresponds to equality between the rates of vibrational excitation by electron impact and vibrational relaxation (see Chapter 3, and Section 5.3 for a similar case involving plasma-chemical CO_2 dissociation). Stimulation of plasma-chemical processes by vibrational excitation becomes effective when $T_v > T_v^{min}$. The total energy efficiency of H_2O dissociation stimulated by vibrational excitation (including the three factors considered earlier) is shown in Fig. 5–53 as a function of specific energy input. Qualitatively, the dependences $\eta(E_v)$ look similar for H_2O and CO_2 dissociation stimulated by vibrational excitation (compare Figs. 5–53 and 5–2). The maximum value of $\eta(E_v) \approx 60\%$, however, is lower for H_2O because of faster VT relaxation.

5.8.4. Contribution of Dissociative Attachment of Electrons into Decomposition of Water Vapor in Non-Thermal Plasma

The contribution of the dissociative attachment in plasma-chemical H_2O dissociation kinetics (Bochin et al., 1977, 1978) can be essential, because more than 60% of discharge electron

5.8. Dissociation of Water Vapor and Hydrogen Production

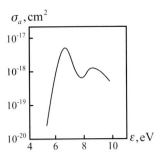

Figure 5–57. Cross sections of dissociative attachment of electrons to water molecules as a function of electron energy.

energy can be transferred into dissociative attachment at slightly higher electron temperatures ($T_e \geq 1.5\,\text{eV}$; see Fig. 5–54, Givotov et al., 1981). The dissociation process proceeds in this case mostly as

$$e + H_2O \rightarrow H^- + OH. \tag{5–125}$$

Cross sections of the dissociative attachment (5–125) are presented in Fig. 5–57 as a function of electron energy. More details regarding the kinetics of the elementary process (5–125) can be found, in particular, in Fridman and Kennedy (2004). The energy efficiency of the H_2O dissociation based on (5–125) is limited by the loss of an electron in the process, and the energy price of an electron is usually very high (not less than 30 eV). Negative ions, formed as a result of dissociative attachment, are subjects of very fast processes of ion–ion recombination resulting in charge neutralization. The only way for the dissociative attachment (5–125) to become energy effective is via recuperation of electrons from negative ions by means of detachment processes, in particular, through electron detachment by electron impact:

$$H^- + e \rightarrow H + e + e. \tag{5–126}$$

Electron detachment (5–126) together with dissociative attachment (5–125) creates a chain reaction of water decomposition. One electron is able to participate in the H_2O dissociation process many times in this case, which makes the whole kinetic mechanism energy effective. Cross sections of the electron detachment by electron impact (5–126) are shown in Fig. 5–58; a typical value of rate coefficient for the detachment process is very high ($k_d = 10^{-6}\,\text{cm}^3/\text{s}$; Smirnov & Chibisov, 1965; Tisone & Branscome, 1968; Inocuti et al., 1967). The chain mechanism, (5–125) and (5–126), is initiated by ionization of H_2O molecules. The chain termination is related to fast ion–ion recombination,

$$H^- + H_2O^+ \rightarrow H_2 + OH, \quad k_r^{ii} = 10^{-7}\,\text{cm}^3/\text{s}, \tag{5–127}$$

and also to the fast ion–molecular reaction,

$$H^- + H_2O \rightarrow H_2 + OH^-, \quad k_{i0} = 3 \cdot 10^{-9}\,\text{cm}^3/\text{s}, \tag{5–128}$$

which leads to fast ion–ion recombination (of OH^-) as well.

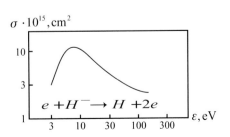

Figure 5–58. Cross sections of electron detachment from H^- negative ions by electron impact as a function of electron energy.

5.8.5. Kinetic Analysis of the Chain Reaction of H_2O Dissociation via Dissociative Attachment/Detachment Mechanism

The energy efficiency of plasma-chemical H_2O dissociation through the dissociative attachment/detachment chain mechanism (5–125) and (5–126) depends on the length of the chain μ characterizing how many times one electron can participate in the dissociative attachment. Let us start with H_2O dissociation provided by high-current relativistic electron beams (Rusanov & Fridman, 1984). The factor μ in this case can be calculated using the degradation cascade method (Fridman, 1976a,b; Fridman & Kennedy, 2004):

$$\mu = \frac{\int_0^I \varepsilon \cdot Z(\varepsilon) \cdot d\varepsilon}{\varepsilon_a \int_0^I Z(\varepsilon) \cdot d\varepsilon}. \qquad (5\text{--}129)$$

In this relation, I is the ionization potential; ε_a is the average energy required for dissociative attachment (5–125) (see Fig. 5–57); ε is electron energy; $Z(\varepsilon)$ is the degradation spectrum of a relativistic electron in a molecular gas; $Z(\varepsilon) \cdot d\varepsilon$ is proportional to the average number of electrons, that used to have energy ε during the degradation (stopping) process (Alkhasov, 1971). The degradation spectrum can be calculated under the ionization threshold ($\varepsilon \leq I$) as follows (Nikerov & Sholin, 1985):

$$Z(\varepsilon) = 2 \left(1 + \frac{\varepsilon}{I}\right)^{-4} \ln\left(2 + \frac{\varepsilon}{I}\right) + \left(1 + \frac{\varepsilon}{I}\right)^{-2}. \qquad (5\text{--}130)$$

The combination of (5–129) and (5–130) permits us to find $\mu = 1$, which is actually limited by the energy of the cascade electrons below the ionization threshold. Although the factor $\mu = 1$ for the relativistic electron beams is not large, it is significantly higher than that for conventional radiolysis of water vapor, $\nu = 0.1$. It means that the H_2O dissociation yield in the case of high-current relativistic electron beams is higher than that for radiolysis. The effect of larger μ-factors is due to the intensive Maxwellization of low-energy electrons when their density is high enough. The critical value of plasma density required for the effective Maxwellization and higher dissociation yield can be found from the analysis of the kinetic equation for the electron velocity distribution function f_e:

$$L\left[f_e^2 - \frac{\partial^2 f_e}{\partial v_i \partial v_k} \frac{\partial^2 \Psi_e}{\partial v_i \partial v_k}\right] - f_e \sigma_a v [H_2O] + Z_d k_d n_e [H^-] + Z(v) q_e - f_e \sigma_r^{ei} v n_i = 0. \qquad (5\text{--}131)$$

In this relation, $L = \lambda(\frac{e^2}{\varepsilon_0 m})$; λ is the Coulomb logarithm; k_d and Z_d are the rate coefficient and velocity distribution related to detachment (5–126); σ_a and σ_r^{ei} are cross sections of dissociative attachment and electron–ion recombination; q_e is the generation rate of plasma electrons by the electron beam and $Z(v)$ is their velocity distribution function related to the degradation spectrum (5–130); n_e, n_i, $[H^-]$, and $[H_2O]$ are densities of electrons, positive ions, hydrogen negative ions, and water molecules; ε_0 is the electric permittivity of free space; and the function $\Psi(\vec{v})$ is determined as

$$\Psi_e(\vec{v}) = \frac{1}{8\pi} \int |\vec{v} - \vec{v}'| f_e(\vec{v}') \, d\vec{v}'. \qquad (5\text{--}132)$$

Analysis of kinetic equation (5–131) shows the effective Maxwellization of plasma electrons, generated by a high-current relativistic electron beam, and as a result $\mu \approx 1$ becomes possible when ionization degrees are sufficiently high:

$$\frac{n_e}{[H_2O]} \gg \frac{4\varepsilon_0^2 \sigma_a^{max} T_e^2}{e^4 \lambda}. \qquad (5\text{--}133)$$

In this relation, $\sigma_a^{max} = 6 \cdot 10^{-18} \text{ cm}^2$ is the maximum value of the cross section of dissociative attachment of electrons to H_2O molecules. Numerically, criterion (5–133) means $\frac{n_e}{[H_2O]} \gg 10^{-6}$ for electron temperature $T_e = 2\,\text{eV}$. Further increase of the factor μ above 1 (the long chain (5–125), (5–126), and the actual multiple use of one electron) becomes possible only in electric discharges, where electrons are able to receive energy from an electric field. The chain length μ of H_2O dissociation is limited in this case by chain termination due to ion–ion recombination (5–127) and ion–molecular reactions (5–128):

$$\mu = \frac{k_d + k_r^{\text{ii}} + k_{i0}\frac{[H_2O]}{n_e}}{k_r^{\text{ii}} + k_{i0}\frac{[H_2O]}{n_e}}. \tag{5–134}$$

According to (5–134), to permit the long-chain dissociative attachment process, the ionization degree should be very high:

$$\frac{n_e}{[H_2O]} > \frac{k_{i0}}{k_d}. \tag{5–135}$$

Criterion (5–135) requires that electron detachment from H^- (5–126) should be faster than the ion–molecular reaction (5–128) leading to fast ion–ion recombination. Numerically, the requirement (5–135) means $\frac{n_e}{[H_2O]} > 3 \cdot 10^{-4}$. The energy efficiency of H_2O dissociation provided by the dissociative attachment chain mechanism (Bochin et al., 1977, 1978) at high ionization degree (5–135) is shown in Fig. 5–53 in comparison with other mechanisms and with experimental results. The maximum energy efficiency in this case is about 50%; the threshold value of specific energy input is lower than for process stimulation by vibrational excitation. Comparing the $\eta(E_v)$ functions (Fig. 5–53) and n_e/n_0 requirements (5–123) and (5–135) for H_2O dissociation mechanisms related to vibrational excitation and dissociative attachment leads to the following general conclusion. The energy efficiency of H_2O dissociation in non-thermal plasma can be quite high (50–60%), but it requires very high values of ionization degree $\frac{n_e}{[H_2O]} > 1–3 \cdot 10^{-4}$, which is difficult to achieve in most non-equilibrium discharges.

5.8.6. H_2O Dissociation in Thermal Plasma and Quenching of the Dissociation Products: Absolute and Ideal Quenching Modes

The high-efficiency regimes of water dissociation in plasma are related to high ionization degrees. At moderate and high pressures, dissociation leads to thermal and quasi-thermal discharge regimes (see Chapter 4). H_2O dissociation in thermal plasma, similarly to the case of CO_2 (Section 5.4), essentially depends on the quenching mode. Consider the equilibrium composition of H_2O thermal dissociation products shown in Fig. 5–59 (Potapkin, Rusanov, & Fridman, 1984a,b). The main products of water vapor dissociation (5–110) are H_2 and O_2. In contrast to the case of CO_2, thermal dissociation of water results in a variety of atoms and radicals. Concentrations of O, H, and OH are fairly significant in the process. Even when H_2 and O_2, initially formed in the high-temperature plasma zone, are saved from reverse reactions, the active species O, H, and OH can be converted either into products (H_2 and O_2) increasing the overall conversion or back into H_2O. In the case of absolute quenching, when the active species are converted back to H_2O, the energy efficiency of the plasma-chemical process (5.7.8) for hydrogen production is

$$\eta = \frac{\Delta H_{H_2O}}{(\Delta W_{H_2O}/\beta_{H_2}^0)}, \tag{5–136}$$

where $\beta_{H_2}^0(T)$ is the quasi-equilibrium conversion of water into hydrogen or, in other words, the number of hydrogen molecules formed in the quasi-equilibrium phase per single initial H_2O molecule:

$$\beta_{H_2}^0 = \frac{x_{H_2}}{x_{H_2O} + x_O + 2x_{O_2} + x_{OH}}. \tag{5–137}$$

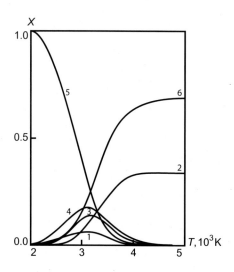

Figure 5–59. Water vapor dissociation in thermal plasma; equilibrium gas composition in hot thermal discharge zone as function of its temperature at atmospheric-pressure conditions: (1) O_2, (2) O, (3) OH, (4) H_2, (5) H_2O, (6) H.

The discharge energy per single initial H_2O molecule to heat up the dissociating water at constant pressure from room temperature to temperature T can be calculated similarly to (5–75) as

$$\Delta W_{H_2O} = \frac{\sum x_i I_i(T)}{x_{H_2O} + x_O + 2x_{O_2} + x_{OH}} - I_{H_2O}(T = 300\,\text{K}). \qquad (5\text{–}138)$$

In this relation, $x_i(p, T) = n_i/n_0$ is the relative concentration (molar fraction for ideal gases) of the i component of the mixture of the H_2O dissociation products; I_i is the total enthalpy of component i; n_0 is the total density of the quasi-equilibrium gas mixture. Based on (5–136)–(5–138), the final formula for the calculation of the energy efficiency of water dissociation and hydrogen production in thermal plasma with absolute quenching of the process products can be presented as

$$\eta = \frac{\Delta H_{H_2O} x_{H_2}}{\sum x_i I_i(T) - (x_{H_2O} + x_O + 2x_{O_2} + x_{OH}) I_{H_2O}(T = 300\,\text{K})}. \qquad (5\text{–}139)$$

Energy efficiency in the case of ideal quenching can be calculated taking into account the additional conversion of active species (H, OH, and O) into major products of the process (H_2 and O_2):

$$\eta = \Delta H_{H_2O} \left[\frac{\sum x_i I_i(T)}{x_{H_2} + 1/2(x_H + x_{OH})} - \frac{x_{H_2O} + x_O + 2x_{O_2} + x_{OH}}{x_{H_2} + 1/2(x_H + x_{OH})} I_{H_2O}(T = 300\,\text{K}) \right]^{-1}.$$
$$(5\text{–}140)$$

The energy efficiencies of water dissociation in thermal plasma for cases of absolute and ideal quenching are presented in Fig. 5–60. Complete usage of atoms and radicals to form additional products (ideal quenching) permits an increase of energy efficiency by almost a factor of 2.

5.8.7. Cooling Rate Influence on Kinetics of H_2O Dissociation Products in Thermal Plasma: Super-Ideal Quenching Effect

To clarify what cooling rate corresponds to which specific quenching mode, we should analyze the major reactions of the quenching kinetics. Slow cooling shifts the equilibrium

5.8. Dissociation of Water Vapor and Hydrogen Production

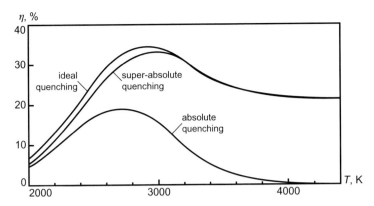

Figure 5–60. Energy efficiency of water vapor dissociation and hydrogen production in thermal plasma with different quenching modes as function of hot plasma zone temperature at fixed gas pressure $p = 0.05$ atm.

between direct and reverse reactions in the exothermic direction of the destruction of molecular hydrogen:

$$O + H_2 \rightarrow OH + H, \tag{5–141}$$

$$OH + H_2 \rightarrow H_2O + H. \tag{5–142}$$

Following a three-body recombination completes the reverse process to water dissociation:

$$H + OH + M \rightarrow H_2O + M. \tag{5–143}$$

The rate coefficient of the reverse reaction (5–143) is about 30-fold higher than the rate coefficient of the alternative (not reverse) three-body recombination leading to the formation of molecular hydrogen:

$$H + H + M \rightarrow H_2 + M. \tag{5–144}$$

Thus, reactions (5–141)–(5–144) explain the destruction of hydrogen during the slow cooling of dissociation products. If the cooling rate is high enough ($>10^7$ K/s), the reactions of O and OH with saturated molecules like (5–141) and (5–142) become less effective. This provides conditions for absolute quenching, because the major dissociation products generated in the high-temperature plasma zone – hydrogen and oxygen – are protected from reverse reactions. Instead of participating in processes (5–141) and (5–142), the active species OH and O react at high cooling rates with each other to form mostly atomic hydrogen in the following elementary processes (Malkov, 1981):

$$OH + OH \rightarrow O + H_2O, \tag{5–145}$$

$$O + OH \rightarrow O_2 + H. \tag{5–146}$$

The atomic hydrogen (being in excess with respect to OH) recombines (5–144), providing, at higher cooling rates ($>2 \cdot 10^7$ K/s), the additional conversion of radicals into stable process products. It means that super-absolute (almost ideal, see Fig. 5–60) quenching can be achieved at these cooling rates. The energy efficiency of hydrogen production as a function of the cooling rate is presented in Fig. 5–61 (Givotov et al., 1983). VT non-equilibrium during product cooling can lead to super-ideal quenching similarly to that in CO_2 dissociation

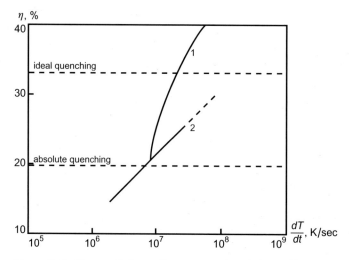

Figure 5–61. Energy efficiency of water vapor dissociation and hydrogen production in thermal plasma as function of quenching rate. Initial temperature in a hot plasma zone 2800 K, pressure $p = 0.05$ atm: (1) quenching in conditions of VT non-equilibrium ($T_v > T_0$); (2) quenching in conditions of VT equilibrium ($T_v = T_0$).

(Sections 5.4.4–5.4.6). Super-ideal quenching can be related in this case to the shift of the quasi equilibrium (at $T_v > T_0$) of the reactions:

$$H + H_2O \Leftrightarrow H_2 + OH, \quad \Delta H = 0.6 \, eV/mol. \quad (5\text{–}147)$$

The direct endothermic reaction of hydrogen production is stimulated here by vibrational excitation, whereas the reverse reaction remains relatively slow. The energy efficiency of water dissociation in such non-equilibrium quenching conditions is presented in Fig. 5–61 as a function of cooling rate (Givotov et al., 1983). Super-ideal quenching requires cooling rates exceeding $5 \cdot 10^7$ K/s and provides energy efficiency up to about 45%. Both super-absolute and super-ideal quenching modes are presented in Fig. 5–53 in comparison with other H_2O dissociation mechanisms and experimental results.

5.8.8. Water Dissociation and H_2 Production in Plasma-Chemical System CO_2–H_2O

The effectiveness of plasma-chemical H_2O dissociation is essentially restricted by require-ments of high ionization degree and by reverse reactions of OH radicals. Both limitations can be mitigated by the addition of CO_2 (and its dissociation product CO) to water vapor. The presence of CO decreases the concentration of OH radicals in the fast exothermic reaction $OH + CO \rightarrow CO_2 + H$. On the other hand, CO_2 and CO have larger cross sections of vibrational excitation and much lower probabilities of VT relaxation, which makes the ionization degree requirements less strong. CO_2 molecules can play the role of a "physical catalyst" of water decomposition (Bochin et al., 1978). The mechanism of hydrogen pro-duction in the non-thermal CO_2–H_2O plasma includes reactions (5–6) and (5–7) of CO_2 dissociation with CO production, followed by exothermic shift reaction (5–3), converting CO into hydrogen. Plasma stimulation of the shift (5–3) is related to the following chain mechanism (Bochin et al., 1978a,b). Most of the oxygen atoms participate in reaction (5–7) with vibrationally excited CO_2 molecules (activation energy $E_x = 0.5$–1 eV/mol), but some of them react with vibrationally excited water molecules:

$$O + H_2O^* \rightarrow H_2 + OH, \quad E_y = 1 \, eV/mol. \quad (5\text{–}148)$$

5.8. Dissociation of Water Vapor and Hydrogen Production

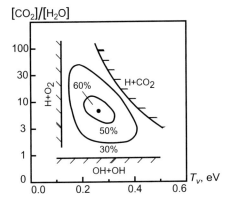

Figure 5–62. Energy efficiency of hydrogen production and restrictions of H_2O–CO_2 mixture composition for effective water dissociation at different values of vibrational temperature in non-equilibrium plasma conditions: closed curves represent lines of constant value of energy efficiency shown as percent.

This process can initiate the chain mechanism of shift (5–3) via the chain propagation:

$$OH + CO \rightarrow H + CO_2, \tag{5–149}$$

$$H + H_2O^* \rightarrow H_2 + OH. \tag{5–150}$$

The chain length ν can be quite long and can be calculated in the framework of the one-temperature approximation (see Sections 5.2.4 and 5.3.4), assuming that the plasma system is characterized by translational temperature T_0 and a single vibrational temperature T_v for both CO_2 and water vapor:

$$\nu = \frac{[CO_2]}{[H_2O]} \exp\left(\frac{E_x - E_y}{T_v}\right). \tag{5–151}$$

The vibrational excitation in the CO_2–H_2O mixture can lead not only to hydrogen production, but also to production of CO and other products. Selective formation of hydrogen is mostly limited by following three alternative reactions: $OH + OH \rightarrow H_2O + O$, $H + CO_2^* \rightarrow OH + CO$, and $H + O_2 \rightarrow OH + O$. Restrictions of the CO_2–H_2O mixture composition and the vibrational temperature, required for selective hydrogen production, are shown in Fig. 5–62. Requirements for high vibrational temperature lead to a limitation of the ionization degree, as illustrated in Fig. 5–63. This limitation is less strong when CO_2 dominates in the CO_2–H_2O mixture. In addition to the chain mechanism of hydrogen production (5–149), (5–150), related to the vibrational excitation of H_2O molecules, the shift process (5–3) can also be provided by the following chain of fast ion–molecular reactions:

$$OH^- + CO \rightarrow H^- + CO_2, \tag{5–152}$$

$$H^- + H_2O \rightarrow H_2 + OH^-. \tag{5–153}$$

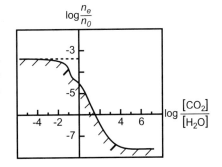

Figure 5–63. Limitations of ionization degree for high-energy-efficiency water dissociation and hydrogen production in non-equilibrium plasma-chemical process organized in H_2O–CO_2 mixture of different composition.

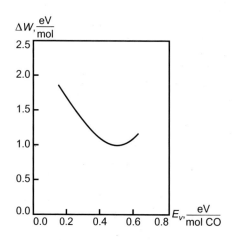

Figure 5–64. Energy cost of hydrogen production by means of non-equilibrium plasma-chemical process related to mixture $CO-O_2-H_2O$.

This ion–molecular chain mechanism doesn't involve relaxing the vibrationally excited H_2O molecules, which makes it favorable at relatively low degrees of ionization. On the other hand, this mechanism suffers from fast ion–ion recombination processes, making the length of the chain (5–152), (5–153) relatively short at high ionization degrees. The total energy efficiency of hydrogen production in the non-equilibrium CO_2-H_2O plasma-chemical mixture is shown in Fig. 5–62 as a function of mixture composition ($[CO_2]/[H_2O]$) and vibrational temperature T_v (assuming the electron density is high enough; see Fig. 5–63). The energy efficiency in optimal conditions (essentially an excess of CO_2) can be close to 60%, which corresponds to the process energy cost of 4.3 eV/mol H_2.

5.8.9. CO-to-H_2 Shift Reaction: Plasma Chemistry of CO–O_2–H_2O Mixture

The CO-to-H_2 shift reaction (5–3) in plasma imposes strong limitations on mixture composition (see Fig. 5–62). Nevertheless, the plasma-chemical shift reaction (5–3) in the presence of O_2 – that is, CO–O_2–H_2O plasma chemistry – has been analyzed (Bochin et al., 1978a,b; Legasov et al., 1980). Chemical processes in the CO–O_2–H_2O mixture can be plasma-stimulated in three different regimes. The first regime, which does not produce hydrogen, corresponds to the branched-chain explosion of CO–O_2 (see Fig. 5–62 and, for example, Benson, 1960), which takes place without any significant involvement of water molecules:

$$O + CO \rightarrow CO_2^*, \quad CO_2^* + O_2 \rightarrow CO_2 + O + O. \qquad (5-154)$$

The second regime in the CO–O_2–H_2O mixture is related not to a branched chain, but to chain reaction of CO oxidation, which is catalytically supported by hydrogen atoms (Lewis & Von Elbe, 1987):

$$H + O_2 + M \rightarrow HO_2 + M, \quad HO_2 + CO \rightarrow CO_2 + OH, \quad OH + CO \rightarrow CO_2 + H. \qquad (5-155)$$

Although water molecules are involved in the mechanism as sources of hydrogen atoms, they just accelerate CO oxidation into CO_2. The shift process (5–3) is suppressed in this case, and hydrogen cannot be produced as well in this regime. Hydrogen can be produced only through stimulation of the chain mechanism (5–149)–(5–153). Organization of the H_2-to-CO shift under such conditions is quite complicated. As analyzed by Bochin et al. (1978); calculated values of the energy cost of hydrogen production are shown in Fig. 5–64 as a function of specific energy input in the mixture. The hydrogen energy cost in optimal conditions is about 1 eV/mol; hydrogen yield, however, is relatively low, on the level of 5%.

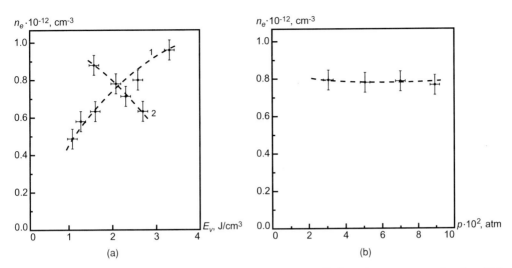

Figure 5–65. Electron density n_e in non-equilibrium microwave discharge in water vapor as function of (a) specific energy input E_v and (b) pressure p. (a)(1) fixed flow rate $Q_{H_2O} = 0.45$ g/s, pressure $p = 0.05$ atm; (2) fixed power $W = 1.1$ kW; (b) fixed flow rate $Q_{H_2O} = 0.45$ g/s and fixed power $W = 1.1$ kW.

5.9. EXPERIMENTAL INVESTIGATIONS OF H₂O DISSOCIATION IN DIFFERENT DISCHARGE SYSTEMS

5.9.1. Microwave Discharge in Water Vapor

The highest value of the energy efficiency of direct water dissociation into molecular hydrogen and oxygen (35–40%) was experimentally achieved in moderate-pressure microwave discharges (Malkov, 1981; Givotov et al., 1983; Gutsol et al., 1985). The microwave frequency for these experiments was 2.4 GHz, and discharge power was about 1.5 kW; the plasma-chemical reactor consisted of a quartz tube with an internal diameter of 28 mm crossing a rectangular 72×34 mm² waveguide. A schematic of the discharge setup is shown in Fig. 5–22 and discussed in Section 5.5.1. The water flow rate in the microwave experiments was 0.05–0.6 g/s, the vapor temperature T_0 at the reactor inlet was 450–500 K, and pressure in the reactor was 30–50 Torr. A small addition of potassium atoms was permitted to intensify ionization and increase the pressure to 250 Torr. The discharge diameter is almost independent of specific energy input and decreases from 22 to 15 mm with a pressure increase. The plasma is mostly uniform in the radial direction, although it is brighter from the side of incoming microwave radiation on the size of the skin layer (Maximov, Sizov, & Chebotarenko, 1970). Measurements of electron density n_e by microwave interferometry are presented in Fig. 5–65 as functions of specific energy input (Fig. 5–65a) and pressure (Fig. 5–65b). The electron density always exceeded the critical value ($n_{e,cr} = 6.7 \cdot 10^{10}$ cm^{-3}) corresponding to the frequency of the applied microwave radiation ($\nu = 2.4$ GHz). Figure 5–65 shows that the microwave plasma ionization degree increases with power at fixed flow rate, increase with flow rate at fixed power, and increases a little bit with pressure decrease. The maximum ionization degree is $n_e/n_0 = 1$–$3 \cdot 10^{-5}$ (Baltin & Batenin, 1970). Stark broadening measurements also give high values of electron density with a maximum of $n_e = 3 \cdot 10^{12}$. The diagnostic measurements determine the microwave plasma ionization degree ($n_e/n_0 = 1$–$3 \cdot 10^{-5}$), electron temperature ($T_e = 2$ eV), and a reduced electric field ($E/n_0 = 10^{15}$ V $\cdot$ cm²). High ionization degrees at moderate and elevated pressures result in relatively high gas temperatures in the microwave discharge. Doppler measurements of the translational gas temperature, presented in Fig. 5–66 for different values of pressure and specific energy input, give in this case $T_0 \geq 3000$ K.

331

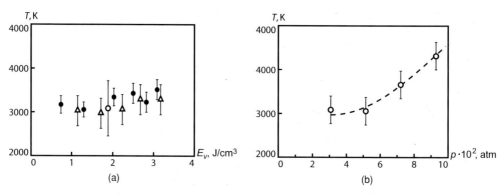

Figure 5–66. Translational gas temperature T in non-equilibrium microwave discharge in water vapor as function of (a) specific energy input E_v and (b) pressure p. (a) fixed flow rate $Q_{H_2O} = 0.45$ g/s, pressure $p = 0.05$ atm; (b) fixed flow rate $Q_{H_2O} = 0.45$ g/s and fixed power $W = 1$ kW.

Radial measurements of Doppler broadening show that the gas temperature is almost uniform in the entire discharge, which is mostly due to the high thermal conductivity of the dissociated water vapor, which is about 10 W/m · K at 3000 K and pressure around 10 Torr (Budko et al., 1981). A weak dependence of gas temperature T_0 on the specific energy input is due to a significant growth of water enthalpy in the temperature range 2500–4000 K. Some temperature increase with pressure growth can be explained by a decrease of the microwave plasma diameter with pressure.

5.9.2. Plasma-Chemical Experiments with Microwave Discharge in Water Vapor

The experiments were carried out in the waveguide microwave discharge described earlier. Products were measured by mass spectrometry and chromatography. Simultaneous diagnostics of the discharge power, the flow rate of products (H_2, O_2), and of the water vapor through the reactor permit the energy efficiency (η) and conversion of water vapor into hydrogen (α) to be determined. Results are presented in Fig. 5–67 as functions of specific energy input and in Fig. 5–68 as a function of pressure (Givotov et al., 1981). The maximum energy efficiency exceeds that corresponding to absolute quenching of products of thermal dissociation by a factor of 2 (Fig. 5–53). On the other hand, the maximum experimental energy efficiency is 1.5–2 times lower than that corresponding to non-equilibrium mechanisms related to vibrational excitation and dissociative attachment (see Sections 5.8.2–5.8.5, and Fig. 5–53). This is due to a high but not sufficient value of ionization degree ($n_e/n_0 = 1$–$3 \cdot 10^{-5}$). These results can be interpreted by the super-absolute and super-ideal quenching modes (Fig. 5–53). Realization of these non-equilibrium quenching modes is due to high cooling rates, which exceeded 10^7 K/s in these experiments (see Fig. 5–61). The plasma-chemical water dissociation experiments of Maximov, Sizov, and Chebotarenko (1970) in a microwave discharge at lower pressures (0.2–4 Torr) and higher electron temperatures (6 eV) were characterized by lower energy efficiency of hydrogen production.

5.9.3. Dissociation of Water Vapor in Glow Discharges

Glow discharge in water vapor was investigated in detail by Dubrovin, Maximov, and their co-authors (Dubrovin et al., 1979; Dubrovin, 1983; Dubrovin & Maximov, 1980, 1981, 1982; Dubrovin, Maximov, & Menagarishvili, 1980). It was shown that characteristics of the discharge are determined by the electronegativity of H_2O and dissociation products. The

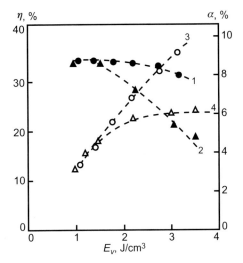

Figure 5–67. (1, 2) Energy efficiency η and (3, 4) conversion degree α of water vapor dissociation and hydrogen production in non-equilibrium microwave discharge as functions of specific energy input E_v. Pressure 0.05 atm; (1, 3) fixed flow rate $Q_{H_2O} = 0.42$ g/s; (2, 4) fixed power $W = 0.5$ kW.

major negative ion is OH^-, which is formed by dissociative attachment and ion–molecular conversion of H^- into OH^-:

$$e + H_2O \rightarrow H^- + OH, \quad H^- + H_2O \rightarrow H_2 + OH^-. \qquad (5\text{--}156)$$

The energy efficiency of H_2O dissociation is relatively low in glow discharges. According to Bugaev and Kukhta (1979), its maximum value is about 12%. According to Dubrovin, Maximov, and their co-authors (Dubrovin et al., 1979; Dubrovin, 1983; Dubrovin & Maximov, 1980, 1981, 1982; Dubrovin, Maximov, & Menagarishvili, 1980), the maximum energy efficiency is slightly higher (20%), but still lower than in the previously considered microwave discharges. In the glow discharge with hollow cathode (see Chapter 4), the dissociation degree reaches 80%. The energy efficiency of dissociation is quite low in this case, only about 2%. The hollow-cathode glow discharge in water vapor, where these results were achieved, operates at a power level of 1–300 W, gas flow rates of 0.1–10 L/h, and a water vapor pressure of 0.4–4 Torr.

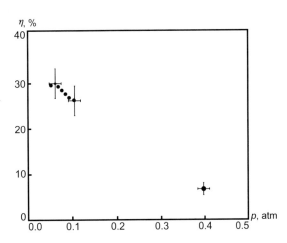

Figure 5–68. Energy efficiency η of water vapor dissociation and hydrogen production in non-equilibrium microwave discharge as function of pressure p at fixed flow rate $Q_{H_2O} = 0.42$ g/s and fixed power $W = 1.7$ kW. Experiments at gas pressures below 0.1 atm were carried out in pure water vapor; those at pressure 0.4 atm were done with Kr admixture.

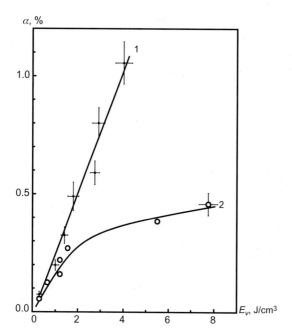

Figure 5–69. Conversion degree of water vapor α into (1) hydrogen and (2) hydrogen peroxide in a non-equilibrium supersonic microwave discharge as function of specific energy input E_v.

5.9.4. Dissociation of H_2O with Production of H_2 and H_2O_2 in Supersonic Microwave Discharges

The effective H_2O dissociation mechanisms are related not to the initial formation of H_2 and O_2 but to the initial formation of hydrogen and hydrogen peroxide (5.112), (5.113). Because of H_2O_2 instability at elevated plasma temperatures, final H_2O dissociation products usually contain oxygen instead of H_2O_2. Water dissociation in plasma with production of H_2 and H_2O_2, based on highly effective mechanisms (5.112), (5.113) and (5.117), (5.118) or just on simple OH recombination, can be summarized as follows:

$$H_2O + H_2O \rightarrow H_2 + H_2O_2, \quad \Delta H = 3.2 \, \text{eV/mol}. \quad (5–157)$$

This process requires more enthalpy for H_2 production but provides better product stabilization with respect to reverse reactions, and effective separation of the dissociation products. Realization of the process (5–157) in plasma requires stabilization of hydrogen peroxide to prevent decomposition into water and oxygen. Stabilization can be achieved at low temperatures, in particular, if the plasma process is performed in a supersonic flow where the gas temperature after the nozzle can be very low (see Figs. 5–40 and 5–41). Kinetic modeling (Melik-Aslanova et al., 1985) shows that the process (5–157) can be performed in supersonic flow with an energy efficiency as high as 45% and a conversion degree up to 20%. According to the simulations, most of the hydrogen peroxide can be stabilized under this condition with respect to decomposition into oxygen and water vapor. Experiments with microwave discharge in supersonic steam flow with the production of hydrogen and stabilization of hydrogen peroxide were carried out by Gutsol et al. (1990, 1992). The microwave power was 2 kW and frequency was 2.45 GHz. A typical Mach number was about $M = 2$ and initial vapor temperature was 400 K. At higher specific energy inputs ($E_v \approx 2.5 \, \text{J/cm}^3$), the Mach number fell to $M \approx 1.5$. The gas temperatures at all those Mach numbers are low and sufficient for H_2O_2 stabilization. Water condensation at those low temperatures is thermodynamically favorable but kinetically slow to have a significant effect on the process yield. Water conversion into hydrogen and hydrogen peroxide achieved in these experiments is shown in Fig. 5–69 as a function of specific

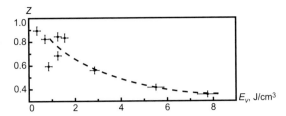

Figure 5–70. Relative efficiency Z of hydrogen peroxide formation with respect to formation of oxygen in a non-equilibrium supersonic microwave discharge as function of specific energy input E_v.

energy input. The conversion degree is about 2%, which is not very high. Production of hydrogen peroxide is lower than that of hydrogen at higher values of specific energy inputs (see (5–157)) when the temperature is higher. On the other hand, at lower energy inputs, the formation of H_2 and H_2O_2 is close to stoichiometic (5–157), which means that peroxide is well stabilized in water dissociation products. Fig. 5–70 presents the dependence of relative efficiency Z of H_2O_2 formation on specific energy input, which is defined as the ratio of H_2O molecules reacting (5–157) to form peroxide to the total number of dissociated water molecules. The relative efficiency of peroxide production reaches 85% at energy inputs of $E_v \leq 1.5\,\text{J/cm}^3$ and remains quite high ($Z \geq 50\%$), up to $E_v \approx 5\,\text{J/cm}^3$. The energy efficiency of the process (5–157) is about 6% in this regime.

5.9.5. Plasma Radiolysis of Water Vapor in Tracks of Nuclear Fission Fragments

The effect of plasma radiolysis in the tracks of nuclear fission fragments was introduced in Section 6.5.7 with regard to the CO_2 dissociation process. Consideration of the plasma radiolysis of water vapor requires a preliminary analysis of conventional H_2O radiolysis by individual high-energy electrons without collective (plasma) effects. Radiolysis is usually characterized by the G-factor, which is the number of dissociations per 100 eV (see Section 6.5.7). The value of the G-factor for H_2O radiolysis by individual high-energy electrons can be calculated according to Pikaev (1965):

$$\frac{G}{100} = \left[\frac{1}{U_i(H_2O)} + \frac{U_i(H_2O) - I_i(H_2O)}{U_i(H_2O) \cdot I_{ex}(H_2O)} \right] + \frac{\mu}{U_i(H_2O)}. \qquad (5\text{–}158)$$

In this relation, $U_i(H_2O) \approx 30\,\text{eV}$ is the total ionization energy cost for water molecules (Fridman & Kennedy, 2004); $I_i(H_2O) = 12.6\,\text{eV}$, $I_{ex}(H_2O) = 7.5\,\text{eV}$ are energy thresholds of ionization and electronic excitation of water molecules, respectively. The factor μ shows how many times one electron with energy below the threshold of electronic excitation ($\varepsilon < I_{ex}(H_2O)$) can be used in the water molecule decomposition process. The first two terms in relation (5–158) are mostly related to dissociation through electronic excitation and recombination of charged particles. They give the radiation yield $G \approx 12\,\frac{\text{mol}}{100\,\text{eV}}$ (Firestone, 1957), which describes H_2O dissociation in experiments with β-radiation and low-current-density electron beams. The third term in (5–158) describes the contribution of the dissociative attachment with the following ion–molecular reactions:

$$e + H_2O \rightarrow H^- + OH, \quad H^- + H_2O \rightarrow H_2 + OH^-. \qquad (5\text{–}159)$$

The contribution of mechanism (5–159) for conventional radiolysis is relatively low ($\mu = 0.1$, Bochin et al., 1978) because of the resonance character of the dissociative attachment cross-sectional dependence on energy and low ionization degree. The factor μ is significantly higher at higher plasma ionization degree because of the Maxwellization of plasma electrons with energies below the threshold of electronic excitation ($\varepsilon < I_{ex}(H_2O)$). Effective Maxwellization takes place when the ionization degree is sufficiently high: $n_e/[H_2O] \gg 10^{-6}$ (Section 5.8.5). In this case the μ-factor, which shows how many times

335

one electron with energy $\varepsilon < I_{ex}(H_2O)$ can be used for H_2O dissociation (5–158), increases from about $\mu = 0.1$ (typical for conventional radiolysis) to $\mu \approx 1$, determined according to (5–129) and (5–130) from the degradation spectrum $Z(\varepsilon)$. Thus, the first two terms in formula (5–158), which are related to H_2O dissociation through electronic excitation and recombination of charged particles, give the radiation yield $G \approx 12\frac{mol}{100\,eV}$. The third term, specific for plasma radiolysis, contributes an additional $G \approx 7\frac{mol}{100\,eV}$ due to dissociative attachment; it brings the total radiation yield of water dissociation in plasma radiolysis to $G \approx 19\frac{mol}{100\,eV}$, which corresponds to an energy efficiency of about 55%.

5.9.6. Effect of Plasma Radiolysis on Radiation Yield of Hydrogen Production in Tracks of Nuclear Fission Fragments

The main final saturated products of water radiolysis are hydrogen and oxygen. However, the radiation yields of water dissociation do not directly correspond to the radiation yield of hydrogen production because of the contribution of reverse reactions. If the radiation yield of conventional water radiolysis reaches $G \approx 12\frac{mol}{100\,eV}$ without plasma effects, the corresponding radiation yield of hydrogen production is only $G \approx 6\frac{mol}{100\,eV}$ (Aulsloos, 1968). Plasma radiolysis in tracks of nuclear fission fragments can bring the total radiation yield of water dissociation to $G \approx 19\frac{mol}{100\,eV}$; the corresponding radiation yield of hydrogen production grows in this case only to $G \approx 13\frac{mol}{100\,eV}$, which in terms of energy efficiency means about 40% (Belousov et al., 1979). The energy efficiency of hydrogen production in water vapor radiolysis depends strongly on gas temperature and intensity of the process (Dzantiev, Ermakov, & Popov, 1979). The primary products of water dissociation related to the mechanisms of electronic excitation and electron–ion recombination are atomic hydrogen (H) and hydroxyl radical (OH). The formation of molecular hydrogen is then related to the following endothermic bimolecular reactions:

$$H + H_2O \rightarrow H_2 + OH, \quad E_a = 0.6\,eV/mol. \tag{5–160}$$

The reverse reaction leading back to the formation of water molecules is mostly recombination:

$$H + OH + M \rightarrow H_2O + M, \tag{5–161}$$

which is about 100 times faster than molecular hydrogen formation in the recombination

$$H + H + M \rightarrow H_2 + M. \tag{5–162}$$

The high intensity of radiolysis leads to a higher concentration of radicals and, therefore, to a bigger contribution of the trimolecular processes and lower yield of H_2 production. The yield can be significantly increased by a temperature increase, which intensifies the endothermic reaction (5–160).

5.10. INORGANIC GAS-PHASE PLASMA-CHEMICAL PROCESSES OF DECOMPOSITION OF TRIATOMIC MOLECULES: NH₃, SO₂, AND N₂O

5.10.1. Gas-Phase Plasma Decomposition Reactions in Multi-Phase Technologies

Several inorganic plasma-chemical dissociation processes, such as the dissociation of gas-phase metal compounds during production of metals in plasma metallurgy, the dissociation of fluorine compounds (C_3F_8, NF_3, and others) with generation of fluorine atoms for etching and chamber cleaning in electronics, the dissociation of H_2S for hydrogen and sulfur production, and many others, include an important gas-phase stage in integrated heterogeneous multi-phase processes. For these processes, gas-phase dissociation of inorganic compounds is one part of a more complicated technology. These gas-phase processes are to be considered in sections related to their special applications. For example, the dissociation of gas-phase metal compounds during production of metals will be considered as part of plasma metallurgy (Chapter 7), the dissociation of fluorine compounds (C_3F_8, NF_3) during generation of F atoms for etching and chamber cleaning will be considered part of electronics (Chapter 8), and the dissociation of H_2S for hydrogen and sulfur production will be considered part of fuel conversion and hydrogen production technologies (Chapter 10). When considering important gas-phase plasma the decomposition processes, we should especially mention the decomposition of chlorine–fluorine–carbon compounds, particularly, CF_2Cl_2 and CHF_2Cl. These compounds are usually referred to in industry as freons. Application of these plasma-chemical processes is related to environmental control; therefore, they are discussed in Chapter 12.

5.10.2. Dissociation of Ammonia in Non-Equilibrium Plasma: Mechanism of the Process in Glow Discharge

The dissociation of ammonia with formation of N_2 and H_2 is a slightly endothermic process:

$$NH_3 \rightarrow \frac{1}{2}N_2 + \frac{3}{2}H_2, \quad \Delta H = 0.46 \, \text{eV/mol.} \tag{5-163}$$

The plasma-chemical process for ammonia dissociation was analyzed, in particular, in glow discharge at relatively low pressures (1–3 Torr; Urbas, 1978a,b). In these experiments, the current varied in the range 5–60 mA, the gas temperature was kept below 470 K, the reduced electric field was $E/n_0 = 4$–$8 \cdot 10^{-16}$ V $\cdot$ cm^2, and the electron density varied in the range $n_e = 10^9$–$2 \cdot 10^{10}$ cm^{-3}. The dissociation degree achieved for NH_3 in these experiments was quite high. Along with ammonia dissociation (5–163), hydrazine synthesis was analyzed in this plasma-chemical system. The characterization of plasma and process parameters was provided based on spectral, mass-spectrometric, and electric probe measurements. In particular it was experimentally proved that a major contribution is made to ammonia dissociation in glow discharge by the following seven primary elementary plasma-chemical reactions.

- Three elementary reactions are related to NH_3 dissociation by direct electron impact:

$$e + NH_3 \rightarrow NH_2 + H + e, \tag{5-164}$$

$$e + NH_3 \rightarrow NH + H_2 + e, \tag{5-165}$$

$$e + NH_3 \rightarrow NH + H + H + e. \tag{5-166}$$

- Two elementary reactions are related to dissociative attachments of electrons to NH_3 molecules:

$$e + NH_3 \rightarrow NH_2^- + H, \tag{5-167}$$

$$e + NH_3 \rightarrow NH_2 + H^-. \tag{5-168}$$

Rate coefficients of the NH_3 dissociation (5–164)–(5–168) vary at conditions typical of low-pressure glow discharges according to Slovetsky (1980) in the wide range 10^{-15}–10^{-10} cm^3/s.

- Two elementary reactions are provided by ammonia collisions with highly vibrationally excited nitrogen molecules in electronically ground state:

$$NH_3 + N_2^*(X, v \geq 20) \rightarrow NH_2 + H + N_2, \tag{5-169}$$

$$NH_3 + N_2^*(X, v \geq 20) \rightarrow NH + H_2 + N_2. \tag{5-170}$$

The total rate coefficient of the two ammonia dissociation processes stimulated by vibrationally excited nitrogen molecules is quite high and is equal to $k = 6 \cdot 10^{-13}$ cm^3/s at room temperature (Slovetsky, 1980). It is interesting to note that the ammonia dissociation processes (5–169) and (5–170) are stimulated in non-equilibrium plasma by the vibrational excitation of molecular nitrogen, which is a product of the NH$_3$ decomposition. Also we should point out that 20 vibrational quanta of nitrogen, which correspond to the dissociation energy of NH$_3$ molecules, are sufficient for ammonia dissociation; the efficiency α of vibrational energy in overcoming activation barriers of reactions (5–169) and (5–170) is quite high. It is interesting because the dissociation process is related to multi-quantum energy transfer between molecules N$_2 \rightarrow$ NH$_3$ (see Chapter 2).

5.10.3. Mechanism of Formation of Molecular Nitrogen and Hydrogen in Non-Equilibrium Plasma-Chemical Process of Ammonia Dissociation

Elementary processes (5–164)–(5–170) are responsible for the formation of primary ammonia dissociation products NH$_2$, NH, and H. The formation of molecular hydrogen and nitrogen, which are major stable products of ammonia dissociation (5–163), proceeds according to Slovetsky (1980) in the following fast exothermic reactions:

$$\text{NH} + \text{NH}_2 \rightarrow \text{N}_2 + \text{H}_2 + \text{H}, \quad \Delta H = -3.16 \, \text{eV/mol}, \tag{5–171}$$

$$\text{NH} + \text{NH} \rightarrow \text{N}_2 + \text{H}_2, \quad \Delta H = -7.05 \, \text{eV/mol}, \tag{5–172}$$

$$\text{NH} + \text{NH} \rightarrow \text{N}_2 + \text{H} + \text{H}, \quad \Delta H = -2.55 \, \text{eV/mol}, \tag{5–173}$$

$$\text{NH} + \text{H} \rightarrow \text{N} + \text{H}_2, \quad \Delta H = -0.88 \, \text{eV/mol}, \tag{5–174}$$

$$\text{NH}_2 + \text{N} \rightarrow \text{N}_2 + \text{H}_2, \quad \Delta H = -6.76 \, \text{eV/mol}. \tag{5–175}$$

The rate coefficients of the exothermic reactions (5–171)–(5–175) leading to H$_2$ and N$_2$ formation are very high and close to the gas-kinetic value $k_0 \approx 10^{-10}$ cm^3/s. As a result, the lifetime of the NH$_2$ and NH radicals in the low-pressure glow discharge experiments (see the previous section; Urbas, 1978a,b) was very short (less than 10 ms). For this reason, the concentration of these radicals (NH$_2$ and NH) in the discharge zone was very low. For the same reason, most H$_2$ and N$_2$ formation was observed in these experiments in the active discharge zone where residence time was about 0.3 s. Atomic hydrogen was actually regenerated in reactions (5–171)–(5–175), which results in its longer lifetime and higher concentration in the discharge.

5.10.4. Plasma Dissociation of Sulfur Dioxide

The process of dissociation of sulfur oxide was analyzed in both thermal and non-thermal plasma conditions. Non-thermal plasma experiments (Meschi & Meyers, 1956; Ruehrwein, 1960; Longfield, 1969) demonstrated the possibility of dissociating SO$_2$ with the production of elementary sulfur and oxygen, as well as with production of lower sulfur oxides (SO and S$_2$O). The energy efficiency achieved in the non-equilibrium plasma systems was not as high as that reported earlier in the chapter regarding CO$_2$ dissociation. The lower energy efficiency of the plasma-chemical process is probably due to the wide variety of SO$_2$ dissociation products and, hence, lower process selectivity. The low energy efficiency in this case is also due to the low contribution of vibrational excitation mechanism in SO$_2$ decomposition. Thermal plasma dissociation of SO$_2$ was considered with the objective of producing elementary sulfur:

$$\text{SO}_2 \rightarrow \text{S}(s) + \text{O}_2, \quad \Delta H = 3.1 \, \text{eV/mol}. \tag{5–176}$$

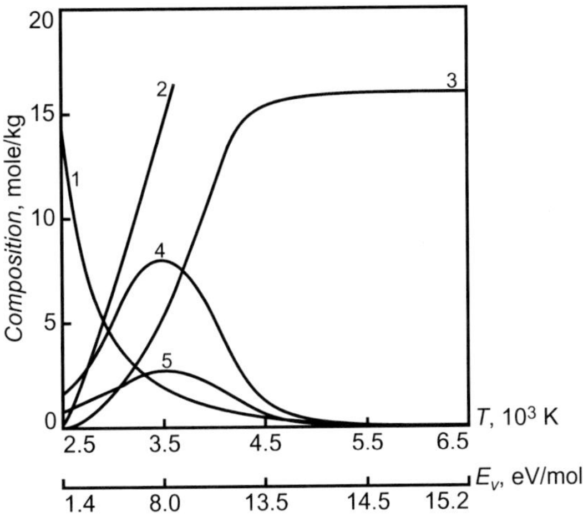

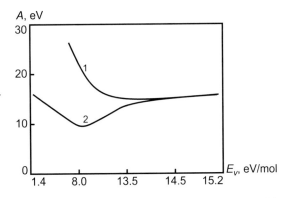

Figure 5–71. SO$_2$ dissociation in atmospheric-pressure thermal plasma. Composition of products: (1) SO$_2$, (2) O, (3) S, (4) SO, (5) O$_2$.

The composition of products of thermal plasma SO$_2$ dissociation is shown in Fig. 5–71 as a function of temperature in a hot thermal plasma zone and as a function of specific energy input in plasma (Nester et al., 1988). The process is at atmospheric pressure, the composition is presented in moles per kilogram of mixture, and the initial value of the parameter is 15.7 mol/kg. The energy cost and conversion degree of the process (5–176) are presented in Figs. 5–72 and 5–73 also as functions of specific energy input in thermal plasma for different modes of product quenching. The minimal energy cost of SO$_2$ dissociation for the production of elemental sulfur and molecular oxygen (5–176) in thermal plasma conditions is equal to 14 eV/S atom in the case of absolute quenching and 9.1 eV/S atom in the case of ideal quenching. Thus, the energy efficiency of the thermal plasma dissociation process is not very high and does not exceed 34% even at optimal plasma conditions. The energy cost of SO$_2$ decomposition in thermal plasma with the production of sulfur and oxygen can be significantly decreased by adding hydrogen and making the whole plasma-chemical process slightly exothermic (Nester et al., 1988):

$$\frac{1}{2}SO_2 + H_2 \rightarrow \frac{1}{2}S + H_2O, \quad \Delta H = -0.97 \, \text{eV/mol}. \tag{5–177}$$

The composition of the thermal plasma-chemical products is shown in Fig. 5–74 as a function of temperature in a hot thermal plasma zone and as a function of specific energy input in the plasma. The plasma-chemical process takes place at atmospheric pressure. The composition of products is presented in the figure in moles per kilogram of mixture. The initial number of moles per unit mass of the mixture in this case is equal to 14.7 mol/kg.

Figure 5–72. SO$_2$ dissociation in atmospheric-pressure thermal plasma. Energy cost required for production of a sulfur atom as function of specific energy input E_v in the case of (1) absolute quenching; (2) ideal quenching.

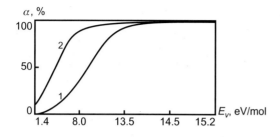

Figure 5–73. SO$_2$ dissociation in atmospheric-pressure thermal plasma. Conversion degree of sulfur dioxide into molecular sulfur as function of specific energy input E_v in the case of (1) absolute quenching; (2) ideal quenching.

5.10.5. Destruction and Conversion of Nitrous Oxide in Non-Equilibrium Plasma

The practical importance of the process is related, in particular, to the reduction of atmospheric emission of N$_2$O, which is a greenhouse gas and also contributes to stratospheric ozone depletion. One of the most important anthropogenic sources of N$_2$O emission is adipic acid production by nitric oxidation of cyclohexanol and/or cyclohexanon. Since the concentration of N$_2$O in adipic-acid-production off-gases is high, the most effective way of reducing N$_2$O emission, according to the Rhone-Poulenc approach, is conversion back to NO$_x$:

$$N_2O + \frac{1}{2}O_2 \rightarrow NO + NO, \quad \Delta H = 1\,\text{eV/mol}. \tag{5–178}$$

The nitrogen oxides generated in this way from N$_2$O can be converted into nitric acid and recycled back to the adipic acid production, which makes this approach attractive for industrial application.

Heating of N$_2$O in conventional chemical or thermal plasma-chemical systems leads mostly to N$_2$O dissociation and reduction of molecular nitrogen. The selective oxidation of N$_2$O (5–178) becomes possible in non-equilibrium plasma conditions by means of oxygen dissociation and by the following exothermic reactions of NO formation:

$$O + N_2O \rightarrow NO + NO, \quad \Delta H = -1.6\,\text{eV/mol}, \quad E_a \approx 0.7\,\text{eV/mol}. \tag{5–179}$$

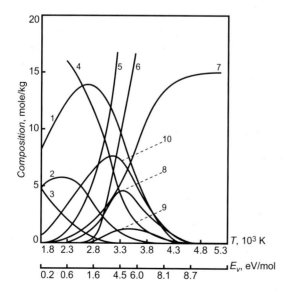

Figure 5–74. SO$_2$ dissociation in atmospheric-pressure thermal plasma in presence of molecular hydrogen $\frac{1}{2}$SO$_2$ + H$_2$. Composition of products as function of temperature and specific energy input: (1) H$_2$, (2) SO$_2$, (3) S$_2$, (4) H$_2$O, (5) H, (6) O, (7) S, (8) OH, (9) O$_2$, (10) SO.

We should keep in mind, however, that even in non-equilibrium plasma conditions with intensive oxygen dissociation, the NO yield from N_2O is limited by competing exothermic reactions:

$$O + N_2O \rightarrow N_2 + O_2, \quad \Delta H = -3.4\,\text{eV/mol}, \quad E_a \approx 1\,\text{eV/mol}. \quad (5\text{--}180)$$

The plasma-chemical performance of the non-equilibrium N_2O-to-NO_x conversion (5–178) was experimentally demonstrated and analyzed in a non-thermal gliding arc discharge (Czernichowski, 1994; Fridman et al., 1995). Experiments were carried out in an N_2O–N_2–O_2–NO_2 mixture at flow rates from 22 to 82 slm with a molar fraction of N_2O in the range 31.6–50.6%. The efficiency of N_2O oxidation to NO_x achieved in the experiments was about 70%. In the absence of an oxidizer, dissociation of nitrous oxide in a non-thermal plasma leads mostly to the formation of molecular nitrogen and oxygen. Such an N_2O dissociation process in a silent discharge at pressures of 200–800 Torr is described by Shekhter (1935). In particular, the NO_2 dissociation degree reached 90% at 200 Torr. A more detailed consideration of nitrous oxide dissociation kinetics in plasma can be found in Gilles and Clump (1969).

5.11. NON-THERMAL AND THERMAL PLASMA DISSOCIATION OF DIATOMIC MOLECULES

5.11.1. Plasma-Chemical Decomposition of Hydrogen Halides: Example of HBr Dissociation with Formation of Hydrogen and Bromine

Dissociation of hydrogen halides in plasma with production of hydrogen and halogens can successfully compete with electrolysis and in some cases can be very attractive for practical applications. The experimental data related to these processes, however, are not sufficient today for clear conclusions to be drawn about their detailed mechanism, kinetics, and energy efficiency. The dissociative attachment of electrons to hydrogen halides is usually very fast (see Chapter 2) because of the high electron affinity of halogen atoms:

$$e + HHal \rightarrow H + Hal^-. \quad (5\text{--}181)$$

Although the HHal dissociation can be very intensive in non-equilibrium plasma, its energy efficiency is limited in general by electron losses in the dissociative attachment process. The production of an electron required for dissociation (5–181) is characterized by an energy cost of at least 30–50 eV, and a charged particle is usually lost after (5–181) in the fast process of ion–ion recombination. The restoration of an electron from a negative ion (Hal^-) has low probability at low concentration of atomic hydrogen, when the associative detachment reverse to (5–181) is ineffective. The energy efficiency of hydrogen halides dissociation can be higher in thermal plasma systems; however, it requires special consideration for product quenching. This can be illustrated by the plasma-chemical dissociation of hydrogen bromide:

$$HBr \rightarrow \frac{1}{2}H_2 + \frac{1}{2}Br_2, \quad \Delta H = 0.38\,\text{eV/mol}. \quad (5\text{--}182)$$

This endothermic plasma-chemical process was considered, in particular, to be an important step in the thermochemical calcium–bromine–water splitting cycle for hydrogen production (Doctor, 2000). The plasma-chemical HBr decomposition (5–182) assumes in this case effective quenching and separation of products by fast rotation of quasi-thermal plasma (10.7.5), (10.7.6). The composition of products of HBr dissociation in thermal plasma is shown in Fig. 5–75 as a function of temperature in the hot thermal plasma zone and as a function of specific energy input in the plasma (Nester et al., 1988). The process being

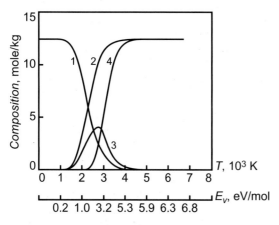

Figure 5–75. HBr dissociation in atmospheric-pressure thermal plasma. Composition of products: (1) HBr, (2) Br, (3) H_2, (4) H.

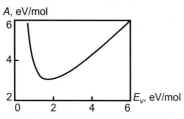

Figure 5–76. HBr dissociation in atmospheric-pressure thermal plasma. Energy cost of the dissociation as a function of specific energy input in the case of absolute quenching.

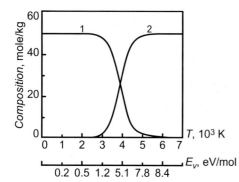

Figure 5–77. HF dissociation in atmospheric-pressure thermal plasma. Composition of products: (1) HF, (2) H, F.

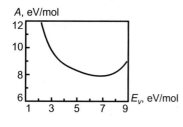

Figure 5–78. HF dissociation in atmospheric-pressure thermal plasma. Energy cost of the dissociation as a function of specific energy input in the case of absolute quenching.

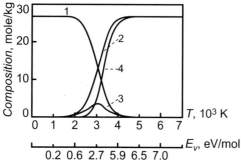

Figure 5–79. HCl dissociation in atmospheric-pressure thermal plasma. Composition of products: (1) HCl, (2) Cl, (3) H_2, (4) H.

Figure 5–80. HCl dissociation in atmospheric-pressure thermal plasma. Energy cost of the dissociation as a function of specific energy input in the case of absolute quenching.

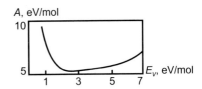

considered is at atmospheric pressure, the composition is presented in moles per kilogram of mixture, and the initial value of the parameter is 12.36 mol/kg. The energy cost of HBr decomposition (5–182) is presented in Fig. 5–76 for the case of the absolute quenching mode also as a function of specific energy input in the thermal plasma. The minimal energy cost of HBr dissociation with formation of molecular hydrogen and bromine (5–182) in thermal plasma is 3 eV/mol.

5.11.2. Dissociation of HF, HCl, and HI in Plasma

Dissociation of HF, in contrast to HBr (5–182), leads to the formation of two gas-phase products:

$$HF \rightarrow \frac{1}{2}H_2 + \frac{1}{2}F_2, \quad \Delta H = 2.83\,eV/mol. \tag{5–183}$$

It makes the centrifugal product separation ineffective and decreases the energy efficiency of the thermal plasma process. The energy cost of the dissociation is also higher than that of HBr simply because of the higher value of HF dissociation enthalpy. The composition of products of HF dissociation in thermal plasma is shown in Fig. 5–77 as a function of temperature in the hot thermal plasma zone and as a function of specific energy input (Nester et al., 1988). The process considered is at atmospheric pressure, the composition is presented in moles per kilogram of mixture, and the initial value of the parameter is 49.98 mol/kg. In contrast to dissociation of HBr (Fig. 5–75), that of HF (Fig. 5–77) requires much higher temperatures when molecular hydrogen is almost completely dissociated (0.5 mol/kg at 3500 K). It explains why concentrations of atomic hydrogen and bromine are actually equal in Fig. 5–77. The energy cost of HF decomposition (5–183) is presented in Fig. 5–78 for the absolute quenching mode. The minimal energy cost of HF dissociation with formation of molecular hydrogen and fluorine (5–183) in thermal plasma is 7.9 eV/mol.

The plasma dissociation of HCl is somewhat similar to that of HF but is less endothermic:

$$HCl \rightarrow \frac{1}{2}H_2 + \frac{1}{2}Cl_2, \quad \Delta H = 0.96\,eV/mol. \tag{5–184}$$

The composition of products (initial molar content 27.43 mol/kg) and the energy cost for the process of HCl dissociation in atmospheric-pressure thermal plasma are shown in Figs. 5–79 and 5–80 (Nester et al., 1988). Dissociation does not require as high a temperature; therefore, H and Cl atoms can coexist in this case with molecular hydrogen (compare with Fig. 5–77). The minimum dissociation energy cost, which can be achieved through absolute quenching, is 5.1 eV/mol.

Dissociation of HI is the only exothermic process in the set of hydrogen halide decompositions:

$$HI \rightarrow \frac{1}{2}H_2 + \frac{1}{2}I_2, \quad \Delta H = -0.27\,eV/mol. \tag{5–185}$$

The initial molar content of the mixture is 7.82 mol/kg. The composition of products for the HI dissociation in atmospheric-pressure plasma is shown in Fig. 5–81 as a function

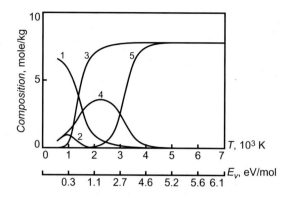

Figure 5–81. HI dissociation in atmospheric-pressure thermal plasma. Composition of products: (1) HI, (2) I_2, (3) I, (4) H_2, (5) H.

of temperature in the hot thermal plasma zone and as a function of specific energy input in the discharge (Nester et al., 1988).

5.11.3. Non-Thermal and Thermal Dissociation of Molecular Fluorine

The bonding energy of fluorine molecules is very low, which makes the F_2 dissociation process quite easy both in thermal and non-thermal systems:

$$F_2 \rightarrow F + F, \quad \Delta H = 1.64 \, \text{eV/mol}. \tag{5–186}$$

The electron affinity of a fluorine atom ($EA = 3.4 \, \text{eV/atom}$) exceeds the bonding energy (5–180); therefore, the dissociative attachment of electrons to fluorine molecules is a very fast exothermic process proceeding without an energy barrier:

$$e + F_2 \rightarrow F + F^-, \quad \Delta H = -1.8 \, \text{eV/mol}. \tag{5–187}$$

The rate coefficient of the dissociative attachment (5–181) is very high ($k = 3 \cdot 10^{-}$ cm^3/s), and it actually has no activation energy. This process dominates the fluorine dissociation in non-thermal plasma and makes it fast, but it is not very energy effective because of an electron loss in the reaction (see Table 5–2, and explanation of a similar phenomenon in Section 5.10.5). Lower values of the dissociation energy cost and, hence, higher energy

Table 5–2. Energy Cost of Dissociation of Molecular Fluorine in Different Non-Thermal and Thermal Experimental Systems

Dissociation method	Dissociation cost, (eV/mol, eV/2 atoms)	Reference
Glow discharge	15	Parker & Stephens, 1973
Non-equilibrium microwave Discharge (argon diluted)	120	Fehsenfeld, Evenson & Broida, 1965
Low-pressure glow discharge	10	Zaitsev, 1978
Non-equilibrium microwave Discharge (no dilution)	6	Lebedev, Murashov & Prusakov, 1972
Low-energy electron beam (40–50 keV)	20	Gross & Wesner, 1973
High-energy electron beam (1 MeV)	10	Whittier, 1976
Hot wall dissociation	6	Legasov, Smirnov & Chaivanov, 1982

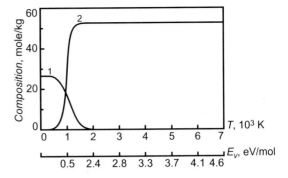

Figure 5–82. F_2 dissociation in atmospheric-pressure thermal plasma. Composition of products: (1) F_2, (2) F.

efficiency can be achieved in plasma systems with higher electron energies when the dissociative attachments compete with fluorine dissociation through electronic excitation without the direct formation of negative ions and loss of an electron (Zaitsev, 1983):

$$e + F_2(^1\Sigma_g^+) \rightarrow F_2^*(^1\Pi_u) \rightarrow F + F + e, \qquad (5\text{–}188)$$

$$e + F_2(^1\Sigma_g^+) \rightarrow F_2^*(^3\Pi_{0u}) \rightarrow F + F + e. \qquad (5\text{–}189)$$

This effect can be seen by comparing different discharge systems presented in Table 5–2, particularly by comparing high-energy and low-energy electron beams. The dissociation energy cost sometimes becomes lower than the energy required for formation of an electron in the discharges.

Lower values of dissociation energy cost can be achieved in thermal dissociation processes, where the dissociation is not related to the electron losses (operation with high-temperature fluorine requires special precautions). The composition of products at high temperature (initial molar content 26.32 mol/kg) and the energy cost for the process of F_2 dissociation in atmospheric-pressure thermal plasma are shown in Figs. 5–82 and 5–83 (Nester et al., 1988). Because of low bonding energy (5–186), dissociation in this case requires relatively low temperatures. The minimal dissociation energy cost, which can be achieved through absolute quenching, is 2.2 eV/mol. The thermal dissociation of molecular fluorine can also be effectively stimulated by the Ni surface at relatively low temperature of 900 K (Goetschel, 1969; Artukhov, 1976; Bezmelnitsyn, Siniansky, & Chaivanov, 1979).

5.11.4. Dissociation of Molecular Hydrogen in Non-Thermal and Thermal Plasma Systems

The dissociation of H_2 in non-equilibrium plasma has been analyzed in detail in relation to its numerous applications, including diamond film deposition, negative and positive ion

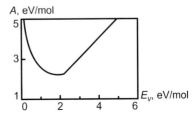

Figure 5–83. F_2 dissociation in atmospheric-pressure thermal plasma. Energy cost of the dissociation as a function of specific energy input in the case of absolute quenching.

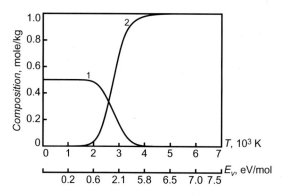

Figure 5–84. H_2 dissociation in atmospheric-pressure thermal plasma. Composition of products: (1) H_2, (2) H.

beams, fusion, and so on (see, for example, Capitelli, 2001; Hassouni et al., 1999b; Hassouni, Gicquel, & Capitelli, 1998, 1999a):

$$H_2 \rightarrow H + H, \quad \Delta H = 4.5 \, \text{eV/mol}. \tag{5–190}$$

The major contribution to H_2 dissociation in low-pressure non-equilibrium discharges is provided by excitation of the electronically excited states $H_2(b^3\Sigma_u)$ with the following decomposition (Khare, 1966, 1967):

$$e + H_2 \rightarrow H_2(b^3\Sigma_u) + e \rightarrow H + H + e. \tag{5–191}$$

Dissociative attachment also can make a significant contribution (Cacciatore et al., 1986):

$$e + H_2 \rightarrow H + H^-. \tag{5–192}$$

The vibrational excitation of molecular hydrogen by electron impact is very effective at electron temperatures around $T_e = 1-2 \, \text{eV}$. Therefore, the vibrational temperature in non-thermal H_2 discharges can be very high and can significantly affect the H_2 dissociation rate (Nelson, Saunder, & Harwey, 1971; Capezzuto, Cramarossa, & d'Agostino, 1975; Capitelli & Dilonardo, 1977; Shaub, Nibler, & Harvey, 1977). A major contribution of vibrational excitation into H_2 dissociation in non-equilibrium discharges is related not directly to the dissociation of vibrationally excited molecules (as in the case of CO_2 dissociation; see Section 5.2), but to acceleration of elementary processes (5–191) and (5–192) when an electron hits a vibrationally excited molecule. An increase of the H_2 vibrational quantum number decreases the threshold of the processes (5–191) and (5–192), increasing their magnitude (Cacciatore & Capitelli, 1981; Gauyacq, 1985). In particular, the dissociative attachment cross section drastically increases (by five orders of magnitude) with the vibrational quantum number passing from $v = 0$ to $v = 5$ (Wadehra, 1986).

Molecular hydrogen can also be quite effectively dissociated in thermal plasma conditions. The composition of products of H_2 dissociation in thermal plasma is shown in Fig. 5–84 as a function of temperature in the hot thermal plasma zone and as a function of specific energy input (Nester et al., 1988). The process considered is at atmospheric pressure, the composition is presented in moles per kilogram of mixture, and the initial value of the parameter is 49.6 mol/kg. The energy cost of the hydrogen dissociation process (5–190) is presented in Fig. 5–85 for the case of the absolute quenching mode as a function of specific energy input. The minimal energy cost of H_2 dissociation in thermal plasma is 6.2 eV/mol.

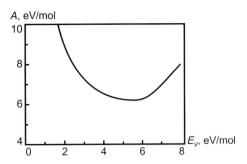

Figure 5–85. H$_2$ dissociation in atmospheric-pressure thermal plasma. Energy cost of the dissociation as a function of specific energy input in the case of absolute quenching.

5.11.5. Dissociation of Molecular Nitrogen in Non-Thermal and Thermal Plasma Systems

Dissociation of molecular nitrogen is a strongly endothermic process because of the very high bonding energy of the molecules:

$$N_2 \rightarrow N + N, \quad \Delta H = 9.8 \, \text{eV/mol.} \quad (5\text{--}193)$$

Vibrational excitation is very effective in stimulating the process (5–193). In low-pressure glow discharges ($p < 1.5$ Torr), the dissociation is dominated by electronic excitation. However, at higher pressures ($p \geq 1.5$ Torr), the vibrational excitation of N$_2$ molecules leads to dissociation, not in combination with electronic excitation but directly through VV exchange and formation of molecules with vibrational energy sufficient for dissociation (Slovetsky & Todesaite, 1973a,b; Polak, Slovetsky, & Todesaite, 1975b). In order to reach high values of N$_2$ vibrational temperature (T_v), the electron density should be sufficiently high (similar to the case of CO$_2$ dissociation; see Section 5.2). The energy cost of formation of atomic nitrogen is high in non-thermal plasma, but it decreases significantly with an increase of vibrational temperature (Alexandrov, Konchakov, & Son, 1978a,b). This effect is clearly demonstrated in Table 5–3. Dissociation through electronic excitation dominates at $T_v = 300$ K and requires 370 eV per atom, whereas dissociation through vibrational excitation ($T_v = 300$ K) requires almost four-fold less energy.

The energy cost of N$_2$ dissociation in thermal plasma is much less, which was demonstrated in particular by Legasov, Smirnov, and Chaivanov (1982). Table 5–4 illustrates the energy cost of formation of a nitrogen atom in thermal plasma at different pressures and temperatures. The dissociation is more effective at lower pressures. At $p = 0.1$–10 Torr, the minimum energy cost value of N-atoms is 6.3 eV/atom, which requires heating the gas to 6400–6600 K as well as absolute quenching.

5.11.6. Thermal Plasma Dissociation of Other Diatomic Molecules (O$_2$, Cl$_2$, Br$_2$)

A. Dissociation of molecular oxygen. This process is mostly investigated in strongly non-equilibrium atmospheric-pressure plasma conditions in relation to ozone synthesis (Section 6.5). Here we discuss the characteristics of thermal plasma decomposition of molecular

Table 5–3. Energy Cost of Formation of a Nitrogen Atom by Means of N$_2$ Dissociation in Non-Thermal Plasma at Room Temperature and Reduced Electric Field ($E/n_0 = 6 \cdot 10^{-16}$ V $\cdot$ cm^2)

Vibrational temperature, T_v	300 K	2100 K	4900 K
Energy cost of a nitrogen atom	370 eV	220 eV	100 eV

Table 5–4. Energy Cost of Formation of Nitrogen Atoms (electron volts) and Their Relative Concentration (%) in Thermal Plasma N_2 Dissociation Process with Absolute Quenching at Different Discharge Temperatures and Pressures

Pressure, p, Torr	$T =$ 5200 K	$T =$ 5400 K	$T =$ 5600 K	$T =$ 5800 K	$T =$ 6000 K	$T =$ 6200 K	$T =$ 6400 K	$T =$ 6600 K	$T =$ 6800 K
0.1	620 eV	120 eV	27 eV	9.6 eV	6.4 eV	6.2 eV	6.3 eV	6.3 eV	6.4 eV
	0.32%	1.8%	8.5%	32%	67%	97%	100%	100%	100%
0.3	1100 eV	210 eV	45 eV	13 eV	6.9 eV	6.3 eV	6.3 eV	6.3 eV	6.4 eV
	0.19%	1.0%	5.0%	20%	58%	92%	99%	100%	100%
1.0	1900 eV	370 eV	80 eV	21 eV	8.5 eV	6.4 eV	6.3 eV	6.3 eV	6.4 eV
	0.10%	0.57%	2.8%	11%	38%	80%	97%	100%	100%
3.0	3400 eV	640 eV	140 eV	35 eV	10.6 eV	6.7 eV	6.3 eV	6.3 eV	6.4 eV
	0.06%	0.33%	1.6%	6.8%	25%	63%	93%	99%	100%
10	6100 eV	1200 eV	250 eV	60 eV	18 eV	8.1 eV	6.4 eV	6.3 eV	6.4 eV
	0.03%	0.18%	0.88%	3.8%	14%	43%	82%	97%	100%

oxygen, when the formation of ozone is almost impossible and atomic oxygen is actually the only product:

$$O_2 \rightarrow O + O, \quad \Delta H = 5.2 \, \text{eV/mol.} \tag{5–194}$$

The composition of products of O_2 dissociation in atmospheric pressure thermal plasma is shown in Fig. 5–86 (Nester et al., 1988); it is presented in moles per kilogram of mixture, and the initial value of the parameter is 31.25 mol/kg. The maximum concentration of ozone in the thermal plasma is about 10^{-4} mol/kg and is reached at $T = 3000$ K. The concentration of ozone required for O_2 dissociation during quenching from high temperatures is also low. When the plasma temperature is reduced to values sufficient for O_3 stabilization (which are not much above room temperature), the concentration of atomic oxygen is already significantly decreased by recombination. The energy cost of the oxygen dissociation (5–194) is presented in Fig. 5–87 for the case of the absolute quenching. The minimal energy cost of O_2 dissociation in atmospheric-pressure thermal plasma is 6.9 eV/mol. As a result, the minimum value of discharge energy required to form an O atom at atmospheric pressure is 3.45 eV/atom. At lower pressures, the energy cost for formation of atomic oxygen is lower (Legasov et al., 1982). This effect is illustrated in Table 5–5. At pressures of $p = 0.1$–10 Torr, the minimum O-atom energy cost is 3.3 eV/atom, which requires heating molecular oxygen to temperatures of about 3400–3600 K.

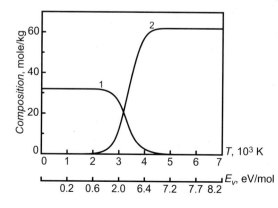

Figure 5–86. O_2 dissociation in atmospheric-pressure thermal plasma. Composition of products: (1) O_2, (2) O.

Table 5–5. Energy Cost of Formation of Oxygen Atoms (electron volts) and Their Relative Concentration (%) in Thermal Plasma O_2 Dissociation Process with Absolute Quenching at Different Discharge Temperatures and Pressures

Pressure, p, Torr	$T = 2600$ K	$T = 2800$ K	$T = 3000$ K	$T = 3200$ K	$T = 3400$ K	$T = 3600$ K	$T = 3800$ K
0.1	7500 eV	250 eV	14 eV	3.4 eV	3.3 eV	3.3 eV	3.4 eV
	0.013%	0.43%	8.6%	70%	99%	100%	100%
0.3	13 keV	430 eV	24 eV	3.8 eV	3.3 eV	3.3 eV	3.4 eV
	0.007%	0.25%	5.0%	52%	98%	100%	100%
1.0	24 keV	800 eV	42 eV	4.9 eV	3.3 eV	3.3 eV	3.4 eV
	0.004%	0.14%	2.8%	33%	95%	100%	100%
3.0	41 keV	1.4 keV	72 eV	7.0 eV	3.3 eV	3.3 eV	3.4 eV
	0.002%	0.08%	1.6%	21%	88%	100%	100%
10	75 keV	2,5 keV	130 eV	11 eV	3.4 eV	3.3 eV	3.4 eV
	0.01 %	0.04%	0.89%	12%	72%	99%	100%

Dissociation of molecular chlorine. The bonding energy of chlorine molecules is low:

$$Cl_2 \rightarrow Cl + Cl, \quad \Delta H = 2.51 \text{ eV/mol}. \tag{5–195}$$

The relatively low enthalpy results in the low temperatures required for the chlorine dissociation. Temperatures well below 2000 K are sufficient to achieve a significant degree of dissociation. The composition of products of Cl_2 dissociation in atmospheric-pressure thermal plasma is shown in Fig. 5–88 as a function of temperature in the hot thermal plasma zone and as a function of specific energy input in plasma (Nester et al., 1988). The composition is presented in moles per kilogram of mixture, and the initial value of the parameter is 14.1 mol/kg. Atomic chlorine is the only product of the thermal plasma decomposition process. The energy cost of the dissociation (5–195) is presented in Fig. 5–89 for the case of absolute quenching as a function of specific energy input. The minimum energy cost of Cl_2 dissociation in atmospheric-pressure thermal plasma is 3.5 eV/mol and can be achieved at a specific energy input of about 2.7 eV/mol.

Dissociation of molecular bromine. This slightly endothermic thermal plasma process is similar to that for chlorine (5–195):

$$Br_2 \rightarrow Br + Br, \quad \Delta H = 2.32 \text{ eV/mol}. \tag{5–196}$$

The composition of products and energy cost of Br_2 dissociation in atmospheric-pressure thermal plasma are shown in Figs. 5–90 and 5–91 (Nester et al., 1988). The initial value of moles per kilogram for the case of bromine decomposition is 6.26. Atomic chlorine is the only product of the thermal plasma decomposition. The minimal energy cost of Cl_2 dissociation in atmospheric-pressure thermal plasma is 3.1 eV/mol.

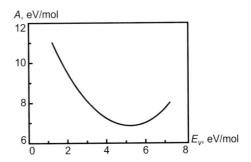

Figure 5–87. O_2 dissociation in atmospheric-pressure thermal plasma. Energy cost of the dissociation as a function of specific energy input in the case of absolute quenching.

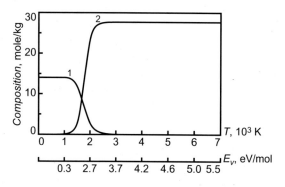

Figure 5–88. Cl_2 dissociation in atmospheric-pressure thermal plasma. Composition of products: (1) Cl_2, (2) Cl.

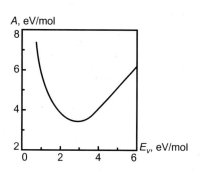

Figure 5–89. Cl_2 dissociation in atmospheric-pressure thermal plasma. Energy cost of the dissociation as a function of specific energy input in the case of absolute quenching.

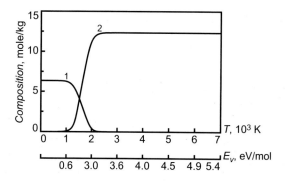

Figure 5–90. Br_2 dissociation in atmospheric-pressure thermal plasma. Composition of products: (1) Br_2, (2) Br.

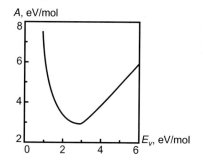

Figure 5–91. Br_2 dissociation in atmospheric-pressure thermal plasma. Energy cost of the dissociation as a function of specific energy input in the case of absolute quenching.

PROBLEMS AND CONCEPT QUESTIONS

5–1. CO_2 Dissociation by Means of Electronic Excitation. Using Fig. 5–6, calculate the decrease in energy threshold of the CO_2 dissociation through electronic excitation related to preliminary vibrational excitation of the molecules. What is the kinetic order of the process with respect to the electron density?

5–2. CO_2 Dissociation by Means of Dissociative Attachment of Electrons. Estimate the energy cost of CO_2 dissociation in non-thermal plasma related to the dissociative attachment of electrons. Compare the energy cost with that due to electronic excitation. How does the energy cost related to dissociative attachment depend on electron temperature, and on electron density?

5–3. Anharmonicity of CO_2 Vibrations, Generalization of Symmetric Modes. Compare general and simplified expressions for energy of a CO_2 molecule as an anharmonic oscillator (5–14) and (5–16). Establish a relationship between the simplified coefficient of anharmonicity, x_{as}, generalizing the symmetric modes with the conventional ones.

5–4. Treanor Effect for Low Discrete Levels of CO_2 Symmetric Oscillations. Explain the vibrational distribution function related to low discrete levels of CO_2 symmetric modes, presented in Fig. 5–15. Why does this vibrational distribution clearly combine two exponential functions with two separate temperatures – a vibrational one and a rotational–translational one?

5–5. Transition of Highly Vibrationally Excited CO_2 Molecules into Vibrational Quasi Continuum. Explain the mechanism for mixing different vibrational modes of CO_2 molecules when the oscillation amplitudes are relatively high. Consider the influence of vibrational beating on energy transfer between modes and the transition of a polyatomic molecule to a quasi continuum of vibrational states. Analyze how the critical maximum number of quanta on an individual mode depends on the number of vibrational modes (see (5–18) and below).

5–6. One-Temperature Approximation of CO_2 Dissociation Kinetics. Explain why the pre-exponential factor for the rate coefficient of CO_2 dissociation in one-temperature approximation (5–22) is proportional to $(E_a/T_v)^s$. How does it reflect the fact that CO_2 dissociation kinetics through vibrational excitation in plasma is limited by VV exchange?

5–7. Average Number of Quanta on Combined Symmetric CO_2 Vibrational Modes. Derive the Planck relation (5–23) for the average number of quanta on combined symmetric valence and double degenerate deformation vibrational modes of CO_2 molecules. Take into account that the deformation mode quantum is considered to be 1 for the combined symmetric CO_2 vibrational modes.

5–8. Two-Temperature Approximation of CO_2 Dissociation Kinetics. Estimate the Treanor factor in the rate coefficient for CO_2 dissociation in the two-temperature approximation (5–26) at typical non-thermal plasma conditions (take, for example, $T_v = 3000\,K$, $T_0 = 500\,K$; also assume for simplicity that $E_a = E^*$). Considering the estimated Treanor factor as a dimensionless part of the pre-exponential factor of dissociation rate coefficient, compare it numerically with the corresponding pre-exponential factor $(E_a/T_v)^s$ appearing in the rate coefficient (5–22) in the one-temperature approximation.

5–9. Contribution of a Secondary Reaction of Atomic Oxygen with Vibrationally Excited CO_2. Assuming a simple Arrhenius expression for the rate coefficient of the secondary reaction (5–7), compare the reaction rate with the three-body recombination of O atoms. Estimate the vibrational temperature at which the contributions of these two channels are comparable.

5–10. Two-Temperature Approach to Vibrational Kinetics and CO_2 Dissociation in Plasma. Explain the energy oscillations between asymmetric and symmetric vibrational modes, related to the non-linearity of their interaction, which takes place at relatively low degrees of ionization (see Fig. 5–16).

5–11. Critical Vibrational Temperature for Effective CO_2 Dissociation. Calculate the critical vibrational temperature for CO_2 dissociation in non-thermal plasma in the framework of a one-temperature approximation using relation (5–46). Take into account that T_v^{min} appears not only on the left-hand side of the equation but also under the logarithm.

5–12. Threshold Value of Specific Energy Input in $\eta(E_v)$ – Comparison of One- and Two-Temperature Approximations. Using Fig. 5–2, compare the threshold of specific energy input, maximum, and other qualitative features of the CO_2 dissociation energy efficiency $\eta(E_v)$, calculated in one- and two-temperature approximations.

5–13. Time of Explosive Growth of Translational Temperature During CO_2 Dissociation Through Non-Equilibrium Vibrational Excitation. Derive relation (5–49), describing the explosive growth of translational temperature by solving non-linear differential equation (5–45). Assume that most of the heating is due to VT relaxation from low vibrational levels, and neglect the relaxation rate dependence on vibrational temperature.

5–14. Flow Velocity and Compressibility Effects on CO_2 Vibrational Relaxation Kinetics. Analyze the characteristic VT-relaxation time (5–51) and the maximum linear preheating temperature (5–53) during CO_2 dissociation in plasma stimulated by vibrational excitation, taking into account the gas compressibility. Describe the qualitative difference of the systems performed in subsonic flows ($M \ll 1$), supersonic flows ($M \gg 1$), and near the speed of sound.

5–15. Hyperbolic Behavior of Energy Efficiency Dependence on Specific Energy Input. Taking into account that CO_2 conversion at high-specific-energy inputs is close to 100%, derive relation (5–72), which explains hyperbolic behavior of $\eta_{rel}(E_v)$ at high values of specific energy input.

5–16. Kinetic Scenario of Ideal Quenching of Products of CO_2 Dissociation in Thermal Plasma. Comparing rate coefficients of recombination processes (5–73) and (7–74), explain the kinetic scenario of the ideal quenching mode. How does a significant difference in the concentrations of O and CO affect the mechanism of ideal quenching? Estimate the difference in conversion degree and in the energy efficiency corresponding to ideal and absolute modes of quenching.

5–17. Super-Ideal Quenching of Products of CO_2 Dissociation in Thermal Plasma. Analyze a kinetic scenario for performing super-ideal quenching. Calculate the possible separation of vibrational and translational temperatures of CO_2 molecules during cool down of the hot gas mixture from the temperatures optimal for thermal CO_2 dissociation. Which mechanism of VT relaxation is the most important during the quenching (take composition of the mixture from Fig. 5–3).

5–18. Critical Vibrational Temperature T_v^{min} for CO_2 Dissociation in Non-Thermal Supersonic Discharges. Explain why the critical vibrational temperature T_v^{min} and, hence, the threshold in dependence $\eta(E_v)$ are essentially lower and strongly depend on initial translational temperature T_0 when the plasma-chemical CO_2 dissociation is done in supersonic flow. Use relations (5–46) and (5–90) for estimations; compare the results with those shown in Fig. 5–2.

5–19. Limitations of Specific Energy Input and CO_2 Conversion Degree in Supersonic Discharges. Using relation (5–93) and assuming that major energy losses are related to the non-resonant VV exchange and to the excess of activation energy of reaction (5–7) over its enthalpy (numerically about 5–10%), estimate the maximum the specific energy input E_v and CO_2 conversion degree in supersonic plasma. Take the gas temperature in the tank before the supersonic nozzle to be $T_{00} = 300\,K$.

5–20. Plasma Radiolysis of CO_2 in Tracks of Nuclear Fission Fragments. Why is the ionization degree in tracks of nuclear fission fragments proportional to the square of gas pressure (relation (5–99))? Estimate the pressure sufficient for complete ionization in the nuclear tracks in CO_2. Why does the energy efficiency of CO_2 dissociation decreases at high pressures (see Fig. 5–48)?

5–21. Plasma-Stimulated Disproportioning of CO, and Complete Dissociation of CO_2 with Production of Elementary Carbon. C atoms generated in CO disproportioning in plasma can be lost in fast reverse reactions of the atoms with CO_2. Estimate the requirements for gas temperature and pressure to provide effective heterogeneous stabilization of the elementary carbon. Which temperature is more important for the stabilization: vibrational or translational?

5–22. Chain Mechanism of Plasma-Chemical H_2O Dissociation Stimulated by Vibrational Excitation. Derive a formula for calculating of the H_2O dissociation chain length, related to mechanism (5–112) and (5–113), as a function of vibrational temperature and density of water vapor. Compare your result with the ratio of the total H_2O dissociation rate (5–116) to the rate of chain initiation (5–115).

5–23. Reverse Reactions and Explosion of Products of Plasma Dissociation of Water Vapor. Interpret the explosion limitation of translational gas temperature, as shown in Fig. 5–56. Why does the restriction not depend on vibrational temperature in contrast to the limitation related to reverse reactions?

5–24. Contribution of Dissociative Attachment to H_2O Dissociation in Non-Thermal Plasma. Explain the $\eta(E_v)$ curve for H_2O dissociation stimulated by dissociative attachment presented in Fig. 5–53. What is the physical cause of the threshold? Estimate the E_v threshold in this case. What is the physical cause of the decrease in energy efficiency at high values of specific energy input?

5–25. Different Modes of Quenching of Products of H_2O Dissociation in Plasma. Explain the difference between ideal and super-absolute quenching of H_2O dissociation products. Both quenching modes assume no losses of H_2 molecules and assume additional production of hydrogen from atoms and radicals during cooling stage. Take into account that reactions (5–145) and (5–146) have related fast reverse processes, establishing their own detailed balance shifted in the direction of conversion of OH radicals into H atoms.

5–26. Ion–Molecular Chain Mechanism of Shift Reaction $CO + H_2O \rightarrow H_2 + CO_2$; Plasma Chemistry of CO_2–H_2O Mixture. Calculate the length of the ion–molecular chain (5–152) and (5–153), providing the CO-to-hydrogen shift reaction as a function of ionization degree and degree of initial dissociation of CO_2 into CO in the plasma-chemical CO_2–H_2O mixture. Estimate the energy cost of hydrogen production in the shift process as a function of ionization degree.

5–27. H_2O Dissociation in Non-Equilibrium Supersonic Plasma with Formation of H_2 and Stabilization of Peroxide in Products. Based on data on supersonic microwave discharge in water vapor (Section 5.9.4), calculate the translational gas temperature after the supersonic nozzle before plasma heat release. Analyze the stability of H_2O_2 at these conditions, and comment why there is no significant water vapor condensation in the active discharge zone at such low temperatures.

5–28. Dissociation of NH_3 in Non-Equilibrium Plasma Stimulated by Vibrational Excitation of Nitrogen. Dissociation of NH_3 molecules from (5–168) and (5–169) can be stimulated by vibrational excitation of nitrogen, which is a product of the dissociation. It is known that 20 vibrational quanta of nitrogen are sufficient for ammonia dissociation. Calculate the efficiency α of vibrational energy in overcoming the activation barriers of reactions (5–169) and (5–170). Comment on the relatively high α-value for elementary processes related to multi-quantum energy transfer between molecules.

6

Gas-Phase Inorganic Synthesis in Plasma

6.1. PLASMA-CHEMICAL SYNTHESIS OF NITROGEN OXIDES FROM AIR AND NITROGEN–OXYGEN MIXTURES: THERMAL AND NON-THERMAL MECHANISMS

Plasmas have been applied both for NO production from air and, conversely, for cleaning air of nitrogen oxides produced in combustion systems. Both NO production and NO destruction are of great industrial interest: the first process is the basis for the chemical technology of fertilizers and explosives, and the second process is a key to the treatment of exhaust gases of power plants, automotive systems, and so on. It is not a surprise that plasma is able to stimulate the process in two opposite directions – plasma behavior can be so much different under different conditions. This chapter deals with NO synthesis in plasma; plasma application for cleaning NO from air will be considered in Chapter 11.

6.1.1. Fundamental and Applied Aspects of NO Synthesis in Air Plasma

The synthesis of nitrogen oxides in air plasma is one of the "old timers" in plasma technology. This endothermic plasma-chemical process is usually presented simply as

$$\frac{1}{2}N_2 + \frac{1}{2}O_2 \rightarrow NO, \quad \Delta H \approx 1\,eV/mol. \tag{6-1}$$

Henry Cavendish and Joseph Priestly were the first scientists who investigated this process, in the eighteenth century. Industrial implementation of the plasma technology was performed in 1900 by Kristian Birkeland and Samuel Eyde. Birkeland and Eyde developed special thermal arc furnaces for air conversion that produced about 1–2% of nitrogen oxides. The energy price for NO production in these systems was about 25 eV/mol, which corresponds in accordance with (6–1) to a relatively low energy efficiency – about 4%. The energy efficiency is calculated as the ratio of the enthalpy of the process (in this case, 1 eV/mol) to the actual energy cost of the product. An alternative ammonia production technology developed shortly after by Fritz Haber and Karl Bosch appeared to be less energy intensive. Competition with the Haber-Bosch method motivated an almost century-long research and development effort to increase the yield and energy efficiency of plasma-chemical NO synthesis from air. Modernization of thermal plasma systems, including operation at pressures of 20–30 atm and temperatures in the range of 3000–3500 K, permitted a decrease in energy cost to 9, eV/mol, which corresponds to an energy efficiency of 11% (Polak & Shchipachev, 1965). To reach such energy efficiencies in thermal plasma, a very high quenching rate of about 10^8 K/s is required (Guliaev, Kozlov, & Polak, 1965). At slightly lower quenching rates, the energy cost of thermal plasma NO synthesis is about 20 eV/mol (energy efficiency 5%), and the conversion degree is not higher than 4–5% (Timmins & Amman, 1970; Zyrichev et al., 1979).

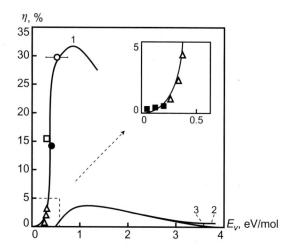

Figure 6–1. Energy efficiency of NO synthesis in air plasma as a function of energy input: (1) non-equilibrium process stimulated by vibrational excitation; (2,3) thermal synthesis with (2) ideal and (3) absolute quenching. Experiments with microwave discharges: ○, □; with discharges sustained by electron beams: ●, ■, △.

Better efficiencies were achieved in non-thermal plasma, where the discharge energy was selectively focused on the most productive channels of NO synthesis and fast quenching was not required. It was first tried in glow discharges in the 1930s, but the energy cost remained high (Shekhter, 1935; Eremin, Vasiliev, & Kobosev, 1936; Eremin & Maltsev, 1956; Krykhtina, Maltsev, & Eremin, 1966). In non-self-sustained discharges stimulated by relativistic electron beams, the energy efficiency was better at 14% (energy cost of 7 eV/mol). The threshold for the energy efficiency dependence on specific energy input $\eta(E_v)$ was observed at $E_v = 0.1$–0.2 eV/mol, which indicated a contribution of vibrationally excited molecules in the NO synthesis (Vakar, Denisenko, & Rusanov, 1977; Panteleev, 1978). Experiments with a non-thermal pulsed microwave discharge (pulse duration 50–100 ns, repetition frequency 1000 Hz, pulse power 0.5–1.5 MW; Polak et al., 1975a) gave similar results. The energy cost of NO synthesis was 6 eV/mol, an energy efficiency of 16%. The highest energy efficiency (about 30%) and lowest energy cost (about 3 eV/mol) were achieved in a non-thermal microwave discharge operating in a magnetic field under electron cyclotron resonance (ECR; Asisov et al., 1980). The most energy effective regimes of NO synthesis are related to process stimulation by vibrational excitation of N_2. Plasma conversion of air into NO is limited by the products' reactions with N atoms. The highest conversion degree (about 20%) was achieved in experiments using a low-pressure plasma-beam discharge (Ivanov, 1975; Ivanov & Nikiforov, 1978; Nikiforov, 1979; Krasheninnikov, 1980; Ivanov, Starykh, & Filiushkin, 1979). The energy efficiency for NO synthesis in plasma as a function of specific energy input $\eta(E_v)$ is presented in Fig. 6–1 for different experimental systems in comparison with modeling results.

6.1.2. Mechanisms of NO Synthesis Provided in Non-Thermal Plasma by Excitation of Neutral Molecules: Zeldovich Mechanism

The process of NO synthesis (6–1) is produced in plasma by numerous elementary processes. The limiting stage is related to the breaking of the strong bond (10 eV) in the N_2 molecule. Such breaking of the N–N chemical bond takes place very efficiently in reactions of vibrationally excited nitrogen molecules:

$$O(^3P) + N_2^*(^1\Sigma_g^+, v) \rightarrow NO(^2\Pi) + N(^4S), \quad E_a \approx \Delta H \approx 3 \text{ eV/mol}. \quad (6\text{–}2)$$

The formation of highly vibrationally excited N_2 molecules takes place in vibrational–vibrational (VV) exchange between the molecules at the lower vibrational states, excited

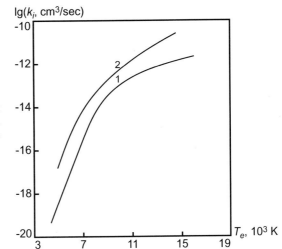

Figure 6–2. Rate coefficients of (1) direct dissociation of initially non-excited nitrogen molecules and of (2) N_2 dissociation in stepwise electronic excitation sequence.

by electron impact with reaction rate coefficient $k_{eV} \approx 10^{-8}$ cm³/s at $T_e = 1$ eV. Reaction (6–2) is followed by a secondary exothermic process:

$$N + O_2 \xrightarrow{k_2} NO + O, \quad E_a \approx 0.3 \,\text{eV/mol}, \quad \Delta H \approx -1 \,\text{eV/mol}. \quad (6\text{–}3)$$

This reaction leads to the formation of atomic oxygen, which can react with a vibrationally excited nitrogen molecule, closing the chain. The chain mechanism, (6–2), and (6–3), is well known in chemical kinetics as the **Zeldovich mechanism**. A sum of the two reactions of chain propagation, (6–2) and (6–3), gives the total NO synthesis (6–1). In contrast to the traditional Zeldovich mechanism, which is attributed to high-temperature oxidation of nitrogen in combustion and explosions (Zeldovich, 1947), the chain (6–2) and (6–3) can take place in non-equilibrium plasma at low temperatures. It is an example of a strongly endothermic chain reaction stimulated by non-equilibrium vibrational excitation of nitrogen molecules. The energy required to overcome the high activation barrier of the limiting reaction (6–2) is provided from nitrogen molecular vibration. Part of this energy can be provided in non-equilibrium plasma conditions (especially at low pressures) by electronic excitation of atomic oxygen to $O(^1D)$. The formation of NO proceeds in this case in a fast electronically adiabatic (see Section 6–2) reaction (Young, Black, & Slanger, 1968):

$$O(^1D) + N_2^*(^1\Sigma_g^+, v) \rightarrow NO(^2\Pi) + N(^2D). \quad (6\text{–}4)$$

The contribution of this electronically adiabatic reaction (see Section 6.2) in moderate- and high-pressure plasma systems is limited by the fast relaxation of electronically excited oxygen atoms $O(^1D)$ (Andreev & Nikitin, 1976). In general, NO synthesis through electronic excitation in plasma is mostly due to the dissociation of molecular nitrogen by direct electron impact:

$$e + N_2(^1\Sigma_g^+) \rightarrow N_2(a^1\Pi_g, B^3\Pi_g, b^1\Pi_u) + e \rightarrow N + N + e. \quad (6\text{–}5)$$

N_2 dissociation in non-thermal plasma has been discussed in Section 5.11.4. The rate coefficients of the process (6–5) are shown in Fig. 6–2 as a function of electron temperature for the cases of dissociation by direct electron impact from the ground state as well as stepwise dissociation through the electronic excitation sequence. The dissociation rate coefficient can reach $k_d = 10^{-11}$ cm³/s at $T_e = 1$ eV. After dissociation of N_2, NO synthesis takes place in the exothermic reaction (6–3). The energy efficiency of NO synthesis through electronic excitation is limited by the high energy cost of atomic nitrogen (6–5) and is much lower than the efficiency of the Zeldovich mechanism, (6–2) and (6–3).

When NO is already formed, it can be further oxidized in plasma to NO_2 and other nitrogen oxides, which is especially interesting, not for NO production but for cleaning air of nitrogen oxides. The conversion of NO into NO_2 is important for removal of nitrogen oxides from exhaust gases, because NO_2 is easier to neutralize or to reduce to nitrogen using catalysis or other conventional chemical approaches.

6.1.3. Mechanisms of NO Synthesis Provided in Non-Thermal Plasma by Charged Particles

Dissociative recombination makes a contribution to the formation of N atoms, and NO synthesis:

$$e + N_2^+ \rightarrow N + N. \tag{6–6}$$

The rate coefficient of the recombination (6–6) is $k_r^{ei} \approx 10^{-8}\ cm^3/s$ at electron temperature $T_e = 1\ eV$. After formation of atomic nitrogen, NO synthesis takes place in the exothermic reaction (6–3). N_2^+ ions in non-thermal plasmas at elevated pressures have a tendency to form complex ions $N_2^+ \cdot N_2$ (or N_4^+; see Fridman & Kennedy, 2004), which decreases the yield of N atoms in the recombination.

A significant contribution to NO synthesis in non-thermal air plasma can be provided by ion–molecular reactions, especially those involving positive atomic oxygen ions:

$$O^+ + N_2 \rightarrow NO^+ + N, \quad k_{io} = 10^{-12}\text{–}10^{-11}\ cm^3/s. \tag{6–7}$$

This reaction proceeds fast without any energy barrier but requires the formation of an expensive ion, O^+. An additional NO molecule can be produced in this case through exothermic reaction (6–3). NO synthesis can be also provided by ion–molecular reactions of molecular ions:

$$O_2^+ + N_2 \rightarrow NO^+ + NO, \tag{6–8}$$

$$N_2^+ + O_2 \rightarrow NO^+ + NO. \tag{6–9}$$

Concentration of the ions N_2^+ and O_2^+ is much higher than that of atomic ions, but the contribution of reactions (6–8) and (6–9) is less significant because of low rate coefficients ($\approx 10^{-16}\ cm^3/s$).

6.1.4. NO Synthesis in Thermal Plasma Systems

NO synthesis in thermal air plasmas is somewhat similar to that in combustion and follows the quasi-equilibrium chain Zeldovich mechanism (Williams, 1985), including reactions (6–2) and (6–3). The vibrational excitation of N_2 molecules is mostly responsible for the limiting endothermic reaction of NO synthesis ($O + N_2 \rightarrow NO + N$) in both quasi-equilibrium and strongly non-equilibrium systems. The difference is only in the selectivity and non-selectivity in the distribution of discharge energy over different reaction channels. Non-equilibrium discharges at electron temperatures of about 1–3 eV are able to transfer most of their energy selectively into N_2 vibrational excitation and, hence, into the limiting NO synthesis reaction (6–2). Thermal discharges are not so selective and distribute their energy uniformly over all degrees of freedom. The composition of products in a high-temperature plasma zone is shown in Fig. 6–3 for the case of atmospheric-pressure stoichiometric (molar 1:1) N_2–O_2 mixture. Maximum NO production requires a high plasma temperature of about 3300 K. The molar fraction of nitrogen oxides in thermal plasma is not high even at these temperatures. Nitrogen oxides are already unstable at the temperatures

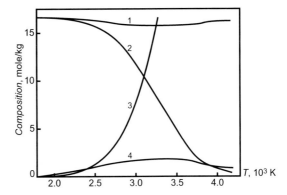

Figure 6–3. NO synthesis in atmospheric-pressure thermal $N_2 + O_2$ plasma. Composition of products: (1) N_2, (2) O_2, (3) O, (4) NO.

required for N_2 dissociation, because the bonding energy of NO molecules is not as high as that of N_2.

The conversion degree of the stoichiometric N_2–O_2 mixture into NO is presented in Fig. 6–4 as a function of specific energy input for absolute and ideal quenching. The maximum conversion only slightly exceeds 5% in both cases. Achieving even this conversion in the thermal plasma requires a fast product cooling of about 10^7–10^8 K/s (Guliaev et al., 1965). The conversion degree achieved in non-equilibrium plasma is also shown in Fig. 6–4 for comparison. The conversion degree can be much higher in the non-equilibrium systems, where NO production is due to non-thermal mechanisms (mainly the contribution of vibrationally excited N_2), and already-produced NO is stable in the cold discharges. The energy efficiency of NO synthesis in thermal plasma is shown for the absolute and ideal quenching modes in Fig. 3–1. The energy efficiency of non-equilibrium NO synthesis (although not very high by itself) can be almost 10-fold greater than that in quasi-equilibrium systems.

6.1.5. Energy Efficiency of Different Mechanisms of NO Synthesis in Thermal and Non-Thermal Discharge Systems

The energy cost of NO generation varies between different plasma-chemical mechanisms. The energy efficiency of NO synthesis in thermal plasma is not high and is limited by three major factors:

1. A low conversion degree is required because at the high temperatures required for breaking the strong N–N chemical bonds, NO is not stable.

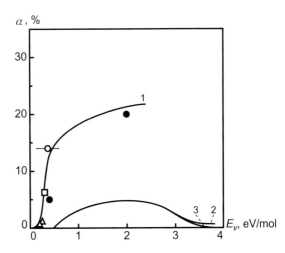

Figure 6–4. Conversion degree α of N_2–O_2 mixture into nitrogen oxides as functions of specific energy input E_v: (1) simulation of the process stimulated by vibrational excitation of nitrogen molecules; (2) simulation of the thermal process assuming ideal quenching of products; (3) simulation of the thermal process assuming absolute quenching of products. Experiments with non-equilibrium microwave discharges, ○, □; with discharges sustained by electron beams, ●, △; with non-equilibrium plasma-beam discharge, ■.

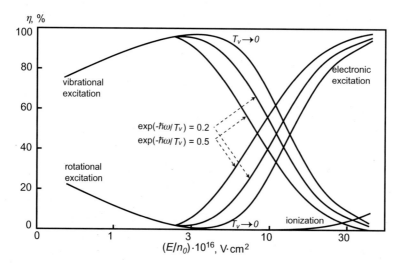

Figure 6–5. Distribution of electron energy in non-thermal discharges between different excitation and ionization channels in nitrogen as functions of reduced electric field at different levels of vibrational temperature.

2. The thermal NO synthesis from air follows the Zeldovich mechanism, which is stimulated by vibrational excitation of N_2 molecules even at quasi-equilibrium, and the energy in thermal systems is distributed over all degrees of freedom, including those not effective in the synthesis.

3. Preserving the NO generated in a high-temperature plasma zone requires very fast product cooling.

As a result, the energy efficiency of the NO synthesis in thermal plasma is almost 10-fold lower than that in strongly non-equilibrium discharges (see Fig. 3–1). The energy efficiency of NO synthesis in non-equilibrium systems is different for different mechanisms. The energy efficiency of mechanisms involving charged species (Section 6.1.3) and electronically excited particles (Section 6.1.2) is limited by the high energy cost of formation of charged and electronically excited particles. The energy efficiency of non-equilibrium NO synthesis through electronic excitation, ion–molecular reactions, and electron–ion recombination usually does not exceed 3% (Rusanov & Fridman, 1976; Macheret et al., 1980a,b).

The most energy-effective mechanism for plasma-chemical NO synthesis is provided by the vibrational excitation of molecules, because most of the energy of non-equilibrium discharges is transferred from electrons to vibrational excitation of N_2 molecules at $T_e = 1$–3 eV, which is seen from Fig. 6–5. The mechanism of synthesis through vibrational excitation of N_2 is not only the most energy effective but is often the fastest and therefore the most dominating mechanism. The vibrational excitation dominates in NO synthesis over electronic excitation and ionic mechanisms when (Macheret et al., 1980a,b)

$$k_d(T_e) \ll k_{eV}\frac{\hbar\omega}{\Delta H}, \quad \frac{n_e}{n_0} \ll \frac{k_{eV}\hbar\omega}{k_r^{ei}\Delta H}, \quad \frac{[O^+]}{n_e} \ll \frac{k_{eV}\hbar\omega}{k_{io}\Delta H}, \qquad (6\text{–}10)$$

where $k_d(T_e)$, k_{eV}, k_r^{ei}, and k_{io} are the rate coefficients of N_2 dissociation through electronic excitation by direct electron impact (6–5), of N_2 vibrational excitation, of electron–ion recombination (6–6), and of ion–molecular reactions (6–7); ΔH is the enthalpy of synthesis;

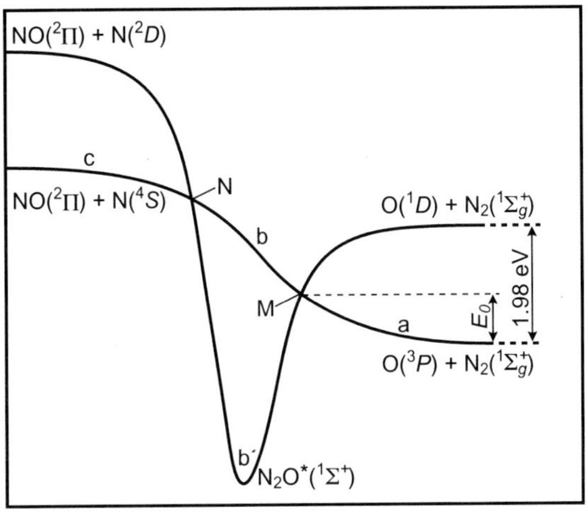

Figure 6–6. Reaction path profile (potential energy curve) for the elementary process of NO synthesis: $O + N_2 \rightarrow NO + N$, showing adiabatic and non-adiabatic reaction channels.

$\hbar\omega \approx 0.2$ eV is a vibrational quantum of N_2; and n_e, $[O^+]$ are number densities of electrons and positive atomic oxygen ions.

6.2. ELEMENTARY REACTION OF NO SYNTHESIS STIMULATED BY VIBRATIONAL EXCITATION OF MOLECULAR NITROGEN

6.2.1. Limiting Elementary Reaction of Zeldovich Mechanism: Adiabatic and Non-Adiabatic Channels of NO Synthesis

The most energy-effective mechanism of NO synthesis in plasma is related to stimulation of the process under non-equilibrium conditions by vibrational excitation of N_2 molecules. The kinetics of this process is controlled by the Zeldovich mechanism (see Section 6.1.2) and is limited by the elementary endothermic reaction (6–2) of a vibrationally excited N_2 molecule. Thus, elementary reaction (6–2) plays a key role in the entire plasma-chemical NO synthesis. This elementary reaction is limited not by VV relaxation and formation of molecules with sufficient energy (as in the case of CO_2 dissociation; see Section 5.3), but by the elementary process of the chemical reaction itself. That is why the elementary process (6–2) should be considered to describe the Zeldovich kinetics of NO synthesis in non-equilibrium plasma.

The elementary process (6–2) is illustrated in Fig. 6–6 by its reaction path profile. The process can proceed in two ways: through the electronically adiabatic channel aMbNc or through the electronically non-adiabatic channel aMb′Nc with transitions localized in points M and N. The adiabatic channels correspond to direct transfer of a nitrogen atom from N_2 to atomic oxygen, whereas the non-adiabatic channel proceeds through the formation of an intermediate long-lifetime vibrationally excited state $N_2O^*(^1\Sigma^+)$. Although the probability of the non-adiabatic channel is limited by small factors related to the transition between different electronic states, it can be kinetically faster than the adiabatic channel in non-equilibrium plasma conditions. The kinetic advantages of the non-adiabatic channel are due to the high efficiency of vibrational energy in reactions proceeding through the formation of stable intermediate molecular complexes.

6.2.2. Electronically Adiabatic Channel of NO Synthesis $O + N_2 \rightarrow NO + N$ Stimulated by Vibrational Excitation of Molecular Nitrogen

The probability of the electronically adiabatic channel aMbNc (Fig. 6–6) of reaction (6–2), $w_a(v)$, can be found as a function of the vibrational quantum number v of N_2 based on the principle of detailed equilibrium (Nikitin, 1970; Landau & Lifshitz, 1981a,b). The reverse elementary process $N + NO \rightarrow N_2(v) + O$ can be characterized by the distribution function of its products over their translational energy, $F_v(\varepsilon_T, \varepsilon_T')$. Thus, the adiabatic channel probability is (Macheret et al., 1980a,b):

$$w_a(v) = \iint\limits_{0,\infty} F_v(\varepsilon_T, \varepsilon_T') \cdot \delta(\varepsilon_T' + E_a - \hbar\omega v - \varepsilon_T) \cdot \theta(\varepsilon_T') \cdot \frac{\sqrt{\varepsilon_T} \cdot \exp(-\varepsilon_T)/T_0}{\Gamma(3/2) \cdot T_0^{3/2}} \cdot d\varepsilon_T \cdot d\varepsilon_T'.$$

(6–11)

In this relation, ε_T, ε_T' are translational energies of particles before and after elementary reaction (6–2); $E_a \approx 3\,\text{eV/mol}$ is the activation energy of reaction (6–2); $\delta(\varepsilon), \theta(\varepsilon), \Gamma(n)$ are the Dirac δ-function, the Heaviside step function, and the gamma-function, respectively; and $\hbar\omega$ and v are the vibrational quantum and vibrational quantum number of N_2. The distribution function $F_v(\varepsilon_T, \varepsilon_T')$ of products of the reverse elementary process $N + NO \rightarrow N_2(v) + O$ over their translational energy can be presented as the prior-expectation distribution function of the theoretical–informational approach developed by Ben-Shaul, Levine, & Bernstein (1974; Levine & Bernstein, 1975):

$$F_v(\varepsilon_T, \varepsilon_T') = \frac{2\hbar\omega\varepsilon_T}{(E_a + \varepsilon_T')^2} \cdot \theta(E_a + \varepsilon_T' - \varepsilon_T).$$

(6–12)

Based on distribution function (6–12), the probability of the electronically adiabatic channel aMbNc (Fig. 6–6) of elementary reaction (6–2) can be found after integration of (6–11):

$$w_a(v) = \frac{1}{v^2} \exp\left[-\frac{E_a - \hbar\omega v}{T_0} \cdot \theta(E_a - \hbar\omega v)\right].$$

(6–13)

The small factor in reaction probability (6–13) is a pre-exponential one, whereas the α-factor characterizing the efficiency of vibrational energy in the elementary process high and close to unity ($\alpha = 1$). A detailed calculation of the adiabatic elementary reaction in the framework of the method of classical trajectories gives $\alpha \approx 0.5$ (Levitsky & Polak, 1980). To take this result into account, the distribution function $F_v(\varepsilon_T, \varepsilon_T')$ can be chosen in more deterministic Dirac form (Macheret et al., 1980a,b):

$$F_v(\varepsilon_T, \varepsilon_T') = \delta\left\{\frac{1}{\hbar\omega}[\varepsilon_T - \xi(E_a + \varepsilon_T')]\right\},$$

(6–14)

where $\xi \approx 0.3$ is a fraction of energy released in the reverse reaction $N + NO \rightarrow N_2(v) + O$ going into translational and rotational degrees of freedom (in the framework of the method of classical trajectories). Integration (6–11) gives the probability of an electronically adiabatic channel:

$$w_a(v) = \exp\left[-\frac{\hbar\omega v}{T_0} \cdot \frac{\xi}{1 - \xi}\right] \cdot \left[\theta\left(\frac{\hbar\omega v}{1 - \xi} - E_a\right)\right].$$

(6–15)

Expressions (6–13) and (6–15) determine the upper and lower limits of the probability of the electronically adiabatic channel aMbNc of NO synthesis (Fig. 6–6). This probability is relatively small ($w_a \approx 10^{-3} – 10^{-5}$ for $\hbar\omega v \geq E_a$ and $T_0 = 1000\,\text{K}$) and exponentially decreases at lower translational temperatures (see (6–15)). These estimations of the adiabatic probability are in good agreement with numerical results obtained in the framework of

6.2. Elementary Reaction of NO Synthesis Stimulated by Vibrational Excitation

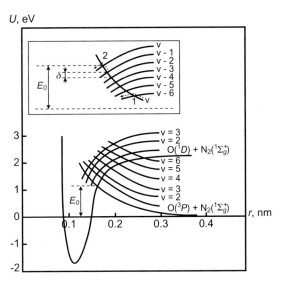

Figure 6–7. System of electronic–vibrational (vibronic) terms for linear configuration of N_2–O interaction. Arrows (1,2) point out the most probable transitions leading to formation of the N_2O complex: (1) transition in the case of lower temperatures, $T_0/\delta \ln(1/\gamma^2 v) < 1$; (2) transition in the case of higher temperatures, $T_0/\delta \ln(1/\gamma^2 v) > 1$.

the theoretical–informational approximation (Dmitrieva & Zenevich, 1983, 1984) and the method of classical trajectories (Levitsky, 1978a,b; Levitsky & Polak, 1980).

6.2.3. Electronically Non-Adiabatic Channel of NO Synthesis ($O + N_2 \rightarrow NO + N$): Stages of the Elementary Process and Method of Vibronic Terms

The electronically non-adiabatic channel aMb′Nc of NO synthesis (Fig. 6–6) proceeds through the formation and decay of an intermediate long-lifetime vibrationally excited complex:

$$O(^3P) + N_2^*(^1\Sigma_g^+, v) \Leftrightarrow N_2O^*(^1\Sigma^+) \rightarrow NO(^2\Pi) + N(^4S). \tag{6–16}$$

This non-adiabatic channel of the elementary reaction limiting the Zeldovich mechanism is kinetically more preferential than the electronically adiabatic one at low temperatures that is typical for strongly non-equilibrium plasmas. The total probability of the kinetic channel (6–16) depends on the probability of formation of the intermediate complex $N_2O^*(^1\Sigma^+)$ as well as the probability of its decay with formation of required products (NO + N). To illustrate the interaction of atomic oxygen with an excited nitrogen molecule (O–NN), let us consider the cross section of the potential energy surface O–N–N corresponding to the angle $\gamma = 180°$, the axis of the complex between N–N and N–O. The dependence of interaction energy between the O atom and N_2 molecule on the distance r between them can be presented by a set of intersecting electronically vibrational or so-called **vibronic** terms (Nikitin, 1984) shown in Fig. 6–7. The vibronic terms are formed by a vertical $\hbar\omega v$ displacement of electronic terms of linear N_2–O configuration. The potential energy difference related to intersection of the $v_1 + 1$ and v_1 singlet electronic terms with the v triplet term is equal to $\delta = 0.15\,\text{eV}$, and it is independent of v and v_1 in the harmonic approximation of N_2 vibrations. The correlation diagram of electronic terms of N_2–O is shown in Fig. 6–8 (Andreev & Nikitin, 1976). Formation of the $N_2O(^1\Sigma^+)$ intermediate complex takes place as a result of non-adiabatic transitions between vibronic terms corresponding to adiabatic electronic terms $^3\Pi$ and $^1\Sigma^+$ (see Fig. 6–8). Note that transitions, where $^3\Pi \rightarrow {}^1\Sigma^+$ are preceded by $^3\Pi \rightarrow {}^3\Sigma^-$ and $3A' \rightarrow 3A''$ (both related to vibrational relaxation), are characterized by a low probability of about 10^{-3} (Nikitin & Umansky, 1971).

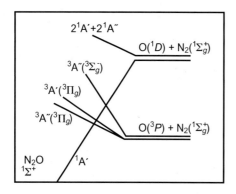

Figure 6–8. Correlation of electronic terms for the atomic–molecular interaction N_2–O: electronic terms $^3A'(^3\Pi)$, $^3A''(^3\Pi)$, and $^3A''(^3\Sigma^-)$ appear from electronic terms $^3\Pi$ and $^3\Sigma^-$ as a result of deviation from linear configuration.

6.2.4. Transition Probability Between Vibronic Terms Corresponding to Formation of Intermediate $N_2O^*(^1\Sigma^+)$ Complex

Thus, the formation of the intermediate complex $N_2O^*(^1\Sigma^+)$ takes place through the non-adiabatic transitions between the vibronic terms (see Fig. 6–7). The probability of transition in an intersection can be calculated assuming separation of electronic and nuclear motion in the molecular system. Perturbation matrix elements can be factorized into electronic and vibrational parts; the transition probability between electronic ($n \rightarrow n'$) and vibrational ($v_1 \rightarrow v_2$) states can be presented as

$$P_{v_1 v_2}^{nn'} = \left| S_{v_1 v_2}^{nn'} \right|^2 P^{nn'}, \qquad (6\text{–}17)$$

where $P^{nn'}$ is the probability of an electronic transition between the vibronic terms at the fixed distance between nitrogen atoms in N_2 and $S_{v_1 v_2}^{nn'}$ is the Frank-Condon factor for overlapping the vibrational ψ-functions corresponding to the initial and final state of N_2 (Landau & Lifshitz, 1981a,b):

$$S_{v_1 v_2}^{nn'} = \int \psi_{v_1}^* \left(r - r_e^n \right) \cdot \psi_{v_2} \left(r - r_e^{n'} \right) \cdot dr. \qquad (6\text{–}18)$$

Integral (6–18) is calculated for the intersection point of the vibronic terms. In integral (6–18), n and n' correspond to the terms $^3\Pi$ and $^1\Sigma^+$; r_e^n and $r_e^{n'}$ are relevant equilibrium interatomic N–N distances in these states. The Frank-Condon integral (6–18) can be calculated using the harmonic oscillator ψ-functions (Frank-Kamentsky & Lukashin, 1975):

$$S_{v, v-\Delta v} = \exp(-\gamma^2) \cdot \frac{(v - \Delta v)!}{v!} \cdot (\gamma \sqrt{2})^{\Delta v} \cdot L_{v-\Delta v}^{\Delta v}(2\gamma^2), \qquad (6\text{–}19)$$

where $L_{v-\Delta v}^{\Delta v}(x)$ is a generalized Laguerre polynomial; v and $v - \Delta v$ are initial and final vibrational quantum numbers; the factor $\gamma = \Delta r_e \sqrt{\frac{M\omega}{\hbar}}$ is the ratio of displacement Δr_e of equilibrium N–N distance during the non-adiabatic transition to the amplitude $\sqrt{\hbar/M\omega}$ of N–N zero vibrations; M is the nitrogen atomic mass; and ω is the N_2 vibration frequency. To finalize determination of the transition probability between relevant electronic ($n \rightarrow n'$) and vibrational ($v_1 \rightarrow v_2$) states (6–17), the electronic transition probability $p^{nn'}$ can be found based on the **Landau-Zener formula** (Landau & Lifshitz, 1981a,b):

$$p^{nn'} = \frac{4\pi V_{nn'}^2}{\hbar u \Delta F}, \qquad (6\text{–}20)$$

where $V_{nn'} \approx 80 \, \text{cm}^{-1}$ is the matrix element of spin–orbital interaction between the electronic terms (Yamanouchi & Horie, 1952), ΔF is the difference of slopes of the electronic terms in the point of their quasi intersection, and u is the relative velocity of O–N_2 interaction.

6.2.5. Probability of Formation of Intermediate $N_2O^*(^1\Sigma^+)$ Complex in Electronically Non-Adiabatic Channel of NO Synthesis

The probability of formation of the intermediate complex $N_2O^*(^1\Sigma^+)$ in the non-adiabatic channel of NO synthesis can be found by summation of independent transitions from an initial triplet term (quantum number v) to any singlet $^1\Sigma^+$ terms (Figs. 6–7, and 6–8; Osherov, 1965; Macheret et al., 1978):

$$w(v, \varepsilon_T) = \sum_{\Delta v = \Delta v_{min}}^{6} p^{nn'} \cdot \left| S_{v, v-\Delta v}^{nn'} \right|^2 = \sum_{\Delta v = \Delta v_{min}}^{6} \frac{4\pi V_{nn'}^2}{\hbar u \Delta F} \cdot \frac{(2\gamma^2)^{\Delta v} \cdot v!}{\Delta v! \cdot (v - \Delta v)!}. \quad (6\text{–}21)$$

Here the lower summation limit $\Delta v_{min} = |(E_0 - \varepsilon_T)/\delta| = |6 - \frac{\varepsilon_T}{\delta}|$; E_0 is the activation barrier of formation of the complex $N_2O^*(^1\Sigma^+)$ without changing vibrational quantum number v (see Fig. 6–7); and ε_T is the kinetic energy of the relative O–N_2 motion. The Landau-Zener factors don't change a lot with Δv, whereas the Frank-Condon factors abruptly decrease at higher differences Δv in vibrational quantum numbers during transition. Summation (6–21) gives the formation probability of $N_2O^*(^1\Sigma^+)$:

$$w(v, \varepsilon_T) = \frac{4\pi V_{nn'}^2}{\hbar u \Delta F} \left[\frac{2e}{\Delta v_{min}} \gamma^2 v \right]^{\Delta v_{min}}. \quad (6\text{–}22)$$

Probability (6–22) depends on the vibrational and translational (ε_T) energy of reagents. It can be integrated over the Maxwellian distribution for kinetic energies of relative O–N_2 motion to get the probability as a function of N_2 vibrational quantum number and translational gas temperature:

$$w_1(v, T_0) = \frac{\pi V_{nn'}^2}{\hbar u \Delta F} \sqrt{\frac{\mu}{T_0}} (\gamma^2 v)^{\frac{E_0}{\delta}} \frac{1 - \exp\left[-\frac{E_0}{T_0} \left(1 - \frac{T_0}{\delta} \ln \frac{1}{\gamma^2 v} \right) \right]}{1 - \frac{T_0}{\delta} \ln \frac{1}{\gamma^2 v}}, \quad (6\text{–}23)$$

where μ is the reduced mass of reagents (O–N_2). The probability of formation of the intermediate complex $N_2O^*(^1\Sigma^+)$ can be simplified at different extremes of the temperature parameter $\frac{T_0}{\delta} \ln \frac{1}{\gamma^2 v}$:

1. At relatively high translational gas temperatures ($\frac{T_0}{\delta} \ln \frac{1}{\gamma^2 v} \gg 1$), overcoming the activation barrier E_0 for formation of an intermediate complex is mostly due to the translational temperature of reagents and is limited by the Maxwellian factor $\exp(-E_0/T_0)$:

$$w_1(v, T_0) = \frac{\pi V_{nn'}^2}{\hbar \Delta F} \sqrt{\frac{\mu}{T_0}} \cdot \left[\frac{T_0}{\delta} \ln \frac{1}{\gamma^2 v} \right]^{-1} \exp\left(-\frac{E_0}{T_0} \right). \quad (6\text{–}24)$$

2. At low and intermediate gas temperatures ($\frac{T_0}{\delta} \ln \frac{1}{\gamma^2 v} \ll 1$ and $\frac{T_0}{\delta} \ln \frac{1}{\gamma^2 v} \approx 1$), conventional for non-thermal plasmas, overcoming the activation barrier E_0 is mostly due to the vibrational energy of the N_2 molecule. The process proceeds as a multi-quanta vibrational transition and is limited by the factor $(\gamma^2 v)^{E_0/\delta}$. The probability at low temperatures ($\frac{T_0}{\delta} \ln \frac{1}{\gamma^2 v} \ll 1$) can be expressed as

$$w_1(v, T_0) = \frac{\pi V_{nn'}^2}{\hbar \Delta F} \sqrt{\frac{\mu}{T_0}} \cdot (\gamma^2 v)^{\frac{E_0}{\delta}}. \quad (6\text{–}25)$$

The probability of a complex formation at intermediate temperatures ($\frac{T_0}{\delta} \ln \frac{1}{\gamma^2 v} \approx 1$) is

$$w_1(v, T_0) = \frac{\pi V_{nn'}^2}{\hbar \Delta F} \sqrt{\frac{\mu}{T_0}} \frac{E_0}{T_0} (\gamma^2 v)^{\frac{E_0}{\delta}}. \quad (6\text{–}26)$$

6.2.6. Decay of Intermediate Complex $N_2O^*(^1\Sigma^+)$: Second Stage of Electronically Non-Adiabatic Channel of NO Synthesis

Non-adiabatic NO synthesis proceeds through the formation and decay of the intermediate complex (see Fig. 6–6). The probability of complex formation is described by (6–24)–(6–26); decay of the complex will now be considered using the statistical theory of monomolecular reactions (Nikitin, 1970). Such an approach is acceptable because the characteristic time of vibrational energy exchange inside $N_2O^*(^1\Sigma^+)$ is much shorter than the lifetime of the complex, and vibrational–translational (VT) relaxation of $N_2O^*(^1\Sigma^+)$ is much longer than the lifetime of the complex (Rusanov, Fridman, & Sholin, 1978). According to the statistical theory, the direct decomposition frequency (marked "+") of the complex $N_2O \rightarrow NO + N$, and reverse decomposition frequency (marked "−") $N_2O \rightarrow N_2 + O$ can be expressed as

$$k_{\pm}(E) = \frac{\Omega^{\otimes}\left(E - E_a^{(\pm)}\right)}{2\pi\hbar \cdot \rho(E)} \cdot w_{LZ}^{(\pm)}, \tag{6-27}$$

where $w_{LZ}^{(\pm)} \approx 0.1$ is the Landau-Zener probability for non-adiabatic transitions in points M and N (see Fig. 6–6) during the direct and reverse N_2O decay with $\Delta v = 0$. $\rho(E)$ is the density of vibrational levels in an N_2O molecule with vibrational energy E (the Marcus-Rice approximation):

$$\rho(E) = \frac{(E + E_z)^{s-1}}{(s-1)! \prod_i \hbar\omega_i}\left(1 + \frac{E}{E_D}\right), \tag{6-28}$$

where $s = 3$ is the number of active vibrational degrees of freedom of an N_2O molecule; E_z is the energy of zero vibrations; E_D is the energy of the N–N bond in the complex; ω_i are vibrational frequencies of an active molecule; $\Omega(E)$ is the phase volume in vibrational space, related to the density $\rho(E)$ as

$$\rho(E) = \frac{d\Omega(E)}{dE}; \tag{6-29}$$

$E_a^{(+)}$ and E_a^- are activation energies of the N_2O decay in direct ($N_2O \rightarrow NO + N$) and reverse ($N_2O \rightarrow N_2 + O$) directions; and the symbol $\otimes$ attributes the value to the activated state of the intermediate complex. Calculating the phase volume $\Omega(E)$ from (6–28) and (6–29), and attributing relevant vibrational frequencies to the activated state of the intermediate complex, the direct and reverse decomposition frequencies of the intermediate complex can be presented as

$$k_{(\pm)}(E) = \frac{\omega_1\omega_2\omega_3}{\omega_1^{\otimes(\pm)}\omega_2^{\otimes(\pm)}\omega_3^{\otimes(\pm)}}w_{LZ}^{(\pm)}\left[\frac{E + E_z - E_a^{(\pm)}}{E + E_z}\right]^{s-1}\left(1 + \frac{x_e E}{\hbar\omega(s-1)}\right), \tag{6-30}$$

where $\hbar\omega$ and x_e are the N_2 vibrational quantum and anharmonicity. Frequencies (6–30) describe different decay channels of the complex $N_2O^*(^1\Sigma^+)$; they give the probability of direct decomposition ($N_2O \rightarrow NO + N$) as a function of vibrational ($\hbar\omega$), translational (ε_T), and rotational (ε_r) energies:

$$w_2(v, \varepsilon_T, \varepsilon_r) = \frac{\omega_1^{\otimes(-)}\omega_2^{\otimes(-)}\omega_3^{\otimes(-)}w_{LZ}^{(+)}}{\omega_1^{\otimes(+)}\omega_2^{\otimes(+)}\omega_3^{\otimes(+)}w_{LZ}^{(-)}}\varsigma\left(\frac{\hbar\omega v + \alpha_T\varepsilon_T + \alpha_r\varepsilon_r - E_a}{\hbar\omega v + \alpha_T\varepsilon_T + \alpha_r\varepsilon_r - E_0}\right)^{s-1}. \tag{6-31}$$

This relation takes into account that translational (ε_T) and rotational (ε_r) energies contribute to the N_2O vibrational energy with efficiencies $\alpha_T, \alpha_r < 1$ due to conservation of angular momentum; the factor $\varsigma < 1$ is related to the adiabatic decay of the complex into $N_2(^1\Sigma_g^+) + O(^1D)$. The direct decay probability (6–31) should be integrated over the

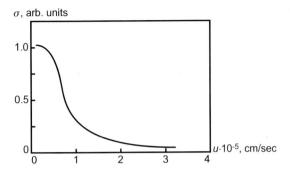

Figure 6–9. Cross section of the non-adiabatic channel of NO synthesis, $O + N_2 \rightarrow NO + N$, as function of relative velocity u of atomic oxygen and molecular nitrogen assuming fixed values of their total energy, including internal degrees of freedom.

Maxwellian distribution to have it as a function of vibrational quantum number and translational temperature $w_2(v, T_0)$ similar to (6–23).

6.2.7. Total Probability of Electronically Non-Adiabatic Channel of NO Synthesis ($O + N_2 \rightarrow NO + N$)

The total probability of the non-adiabatic NO synthesis (6–16) is a product of probabilities (6–23) and (6–31), related to formation and direct decay of an intermediate complex. The cross section of the synthesis (6–16) is shown in Fig. 6–9 as a function of relative velocity of reagents (N_2–O), assuming their total energy is fixed ($\hbar\omega v + \varepsilon_T + \varepsilon_r = \text{const}$). The analytical calculations are in agreement with molecular dynamic modeling (method of classical trajectories; Basov, Danilychev, & Lobanov, 1978a), and demonstrate that the vibrational energy of N_2 is the most effective in stimulating the reaction. The total probability of non-adiabatic NO synthesis (6–16) at strongly non-equilibrium conditions ($T_v \gg T_0$) and translational temperatures not exceeding about 1000 K can be simplified and expressed as (Macheret et al., 1978)

$$w_n(v) = \frac{\pi V_{nn'}^2}{\hbar \Delta F} \sqrt{\frac{\mu}{T_0}} \cdot (\gamma^2 v)^{\frac{E_0}{\delta}} \varsigma \left(\frac{\hbar\omega v - E_a}{\hbar\omega v - E_0} \right)^{s-1}. \tag{6–32}$$

The probability of the non-adiabatic channel of synthesis (6–32) is about 10^{-2}–10^{-4} and doesn't essentially depend on translational temperature, in contrast to that of adiabatic channel (6–15). The difference in $w_a(v)$ and $w_n(v)$ dependences on T_0 permits one to specify the temperature range where each mechanism dominates. When the vibrational energy slightly exceeds the activation energy, the non-adiabatic mechanism dominates at translational temperatures $T_0 < 1000$ K; the adiabatic channel dominates at higher temperatures. Thus, non-adiabatic channel (6–16) proceeding through an intermediate complex is the major process in conventional non-equilibrium plasma conditions.

6.3. KINETICS AND ENERGY BALANCE OF PLASMA-CHEMICAL NO SYNTHESIS STIMULATED IN AIR AND O_2–N_2 MIXTURES BY VIBRATIONAL EXCITATION

6.3.1. Rate Coefficient of Reaction $O + N_2 \rightarrow NO + N$ Stimulated in Non-Equilibrium Plasma by Vibrational Excitation of Nitrogen Molecules

The reaction $O + N_2 \rightarrow NO + N$ (6–2), limiting the Zeldovich mechanism of NO synthesis in air plasma, proceeds in non-equilibrium conditions mostly through the non-adiabatic channel (Fig. 6–6) with probability (6–32). To find out the rate coefficient of the reaction under non-equilibrium conditions ($T_v > T_0$) as a function of vibrational (T_v) and translational (T_0) temperatures, probability (6–32) should be averaged over the vibrational distribution function $f(v)$. In contrast to CO_2 dissociation (see Section 5.2), the reaction $O + N_2 \rightarrow NO + N$ is classified in vibrational kinetics as a slow reaction, which means that the reaction is slower than VV relaxation and, therefore, does not perturb the vibrational distribution function. Slow reaction speed is due to low concentration of atomic oxygen and low probability of the elementary chemical process with respect to that of VV exchange between N_2 molecules:

$$\left(\frac{T_v}{E_a}\right)^{s-1} w_{LZ} \left(\gamma^2 \frac{E_a}{\hbar\omega}\right)^{E_0/\delta} \frac{[O]}{n_0} \ll \frac{k_{VV}(E_a)}{k_0}, \qquad (6\text{--}33)$$

where $k_{VV}(E_a)$ is the rate coefficient for the non-resonant VV relaxation of N_2 molecules at vibrational energies corresponding to the activation energy of reaction (6–2); $k_0 \approx 10^{-10}$ cm^3/s is the rate coefficient of gas-kinetic collisions; $[O]/n_0$ is the relative concentration of atomic oxygen with respect to the total density of neutrals; and for other symbols (see Section 6.2). Assuming the Treanor distribution function, the rate coefficient of $O + N_2 \rightarrow NO + N$ (6–2) is (Macheret et al., 1980a,b)

$$k_R(T_v, T_0) \approx \left[k_0 \frac{\pi V_{nn'}^2}{\hbar \Delta F} \sqrt{\frac{\mu}{T_0}} \left(\frac{T_v}{E_a}\right)^{s-1} \left(\gamma^2 \frac{E_a}{\hbar\omega}\right)^{E_0/\delta} \right] \cdot \exp\left(-\frac{E_a}{T_v} + \frac{x_e E_a^2}{T_0 \hbar\omega}\right). \qquad (6\text{--}34)$$

The formula can be applied in the case of $2x_e E_a T_v/\hbar\omega T_0 < 1$, where $\hbar\omega$ and x_e are the vibrational quantum and anharmonicity coefficient of N_2 molecules. Compare the rate coefficient (6–34) for NO synthesis stimulated by vibrational excitation with that of CO_2 dissociation (see Section 5.2). Because NO synthesis represents the slow reaction, the rate coefficient (6–34) includes detailed characteristics of the elementary process in contrast to the case of fast reaction in plasma-chemical CO_2 dissociation. For calculations of the rate coefficient using equation (6–34) at typical non-equilibrium plasma conditions ($T_v \approx 3000$ K, $T_0 \le 1000$ K), one can take the pre-exponential factor $A_0 = 10^{-12}\text{--}10^{-13}$ cm^3/s.

6.3.2. Energy Balance of Plasma-Chemical NO Synthesis: Zeldovich Mechanism Stimulated by Vibrational Excitation

The equation of N_2 vibrational energy balance can be applied to illustrate the qualitative features of non-equilibrium plasma-chemical NO synthesis stimulated by vibrational excitation:

$$\frac{d\varepsilon_v}{dt} = k_{eV} n_e \hbar\omega - k_R(T_v, T_0)[O]E_a - (k_{VT}n_0 + k'_{VT}[O])(\varepsilon_v - \varepsilon_{v0}), \qquad (6\text{--}35)$$

where ε_v is the average vibrational energy of N_2 molecules related through the Planck formula to vibrational temperature T_v; $\varepsilon_{v0} = \varepsilon_v(T_v = T_0)$ is the average vibrational energy corresponding to the translational temperature T_0; k_{VT} and k'_{VT} are rate coefficients of VT relaxation of N_2 molecules on neutrals (mostly N_2 and O_2) and on atomic oxygen; $k_{eV} = 1\text{--}3 \cdot 10^{-8}$ cm^3/s is the rate coefficient of N_2 vibrational excitation by electron impact; and $k_R(T_v, T_0)$ is the NO synthesis reaction rate (6–34). Although the concentration of O atoms is not very high, its non-adiabatic relaxation is fast ($k'_{VT} \approx 3 \cdot 10^{-13}$ cm^3/s) and the contribution into VT relaxation is significant. The rates of both chemical reaction (6–2) and the VT relaxation of nitrogen depend on the density of O atoms formed mostly by means of dissociative attachment (rate coefficient $k_a \approx 10^{-12}$ cm^3/s at $T_e = 1$eV):

$$e + O_2 \rightarrow O + O^-. \qquad (6\text{--}36)$$

Comparable formation of O atoms is due to dissociation through electronic excitation:

$$e + O_2 \rightarrow O + O + e. \qquad (6\text{--}37)$$

Losses of atomic oxygen are mostly related to trimolecular reactions of ozone formation:

$$O + O_2 + M \rightarrow O_3 + M, \quad k_3 = 10^{-33}\,cm^6/s. \tag{6-38}$$

When the NO yield is high, trimolecular reactions of NO oxidation to NO_2 also make a significant contribution in the consumption of atomic oxygen. We should point out that reactions (6–36)–(6–38) can be considered as chain initiation and chain termination processes for the Zeldovich mechanism. Reactions (6–2) and (6–3) are the Zeldovich chain propagation processes. Stationary conditions for the Zeldovich mechanism give the following steady-state density of atomic oxygen:

$$[O]_0 = \frac{n_e}{n_0} \cdot \frac{k_a}{k_3}. \tag{6-39}$$

Thus, the absolute density of atomic oxygen is proportional to the ionization degree; specifically $[O]_0 \approx 10^{16}\,cm^{-3}$ at electron temperature $T_e \approx 1\,eV$ and ionization degree $n_e/n_0 = 10^{-5}$.

6.3.3. Macro-Kinetics of Plasma-Chemical NO Synthesis: Time Evolution of Vibrational Temperature

NO synthesis is characterized by a critical value of vibrational temperature T_v^{min}, corresponding to the equality of rates of chemical reaction and VT relaxation (see Chapter 3):

$$T_v^{min} = E_a \ln^{-1} \frac{A_0 [O]_0 E_a \exp\left(\frac{x_e E_a^2}{T_0 \hbar \omega}\right)}{(k_{VT} n_0 + k'_{VT} [O]_0)\hbar\omega}, \tag{6-40}$$

where A_0 is a pre-exponential factor of the rate coefficient (6–34), and E_a is the activation energy of reaction (6–2). For NO synthesis, typically $T_v^{min} = 0.2$–$0.25\,eV$. At lower vibrational temperatures ($T_v < T_v^{min}$), VT relaxation is faster than chemical reaction, and synthesis is ineffective. For this reason, the critical vibrational temperature T_v^{min} determines (through the Planck formula) the threshold $\varepsilon_v(T_v = T_v^{min})$ of the dependence of energy efficiency of the NO synthesis on specific energy input. Typically, $\varepsilon_v(T_v^{min}) \approx 0.5$–$0.8\,J/cm^3$ (see Fig. 6–1). The stationary vibrational temperature in an active plasma zone, T_v^{st}, can be estimated (if $T_v^{st} > T_v^{min}$) from the energy balance (6–35) as

$$T_v^{st} = E_a \ln^{-1} \frac{[O]_0 E_a A_0 \exp\left(\frac{x_e E_a^2}{T_0 \hbar \omega}\right)}{k_{eV} n_e \hbar\omega}. \tag{6-41}$$

A typical stationary vibrational temperature is $T_v = 3000$–$5000\,K$. VT relaxation can be neglected with respect to chemical reactions when $T_v^{st} > T_v^{min}$, which requires high ionization degree:

$$\frac{n_e}{n_0} \gg \frac{k_{VT}}{k_{eV}} + \frac{k'_{VT}}{k_{eV}} \cdot \frac{[O]_0}{n_0}. \tag{6-42}$$

The restriction (6.42) of ionization degree to reach high-efficiency NO synthesis means $\frac{n_e}{n_0} \gg 10^{-8}$, which is less serious than that for CO_2 and H_2O dissociation due to low N_2 relaxation rate. At high ionization degrees (6–42), the total reaction rate of NO synthesis in an active discharge zone is

$$w_R \approx k_{eV} n_e n_0 \frac{\hbar\omega}{\Delta H}, \tag{6-43}$$

where $\Delta H \approx 1\,eV/mol$ is the enthalpy of NO synthesis (6–1). NO synthesis continues outside of the active discharge zone in the so-called passive zone (where $n_e = 0$), using vibrational energy accumulated by nitrogen. The effective synthesis continues in the passive zone until the vibrational temperature decreases from initial value T_v^{st} to the critical value T_v^{min}. The vibrational energy balance in the passive discharge zone, taking into account losses of atomic oxygen (for example, in ozone formation (6–38)), can be presented as

$$\frac{d\varepsilon_v}{dt} = -k_R(T_v, T_0)[O]_0 \exp(-k_3 n_0[O_2]) \cdot E_a - (k_{VT} n_0 + k'_{VT}[O])(\varepsilon_v - \varepsilon_{v0}), \quad (6\text{–}44)$$

where $[O_2]$ is the oxygen concentration. Integration of (6–44), assuming $t < 1/k_3 n_0[O_2]$ and $T_v > T_v^{min}$, gives the time evolution of vibrational temperature in the passive zone:

$$T_v(t) = T_v^{st}\left\{1 + \frac{T_v^{st}}{E_a}\ln\left[1 + \exp\left(-\frac{E_a}{T_v^{st}} + \frac{x_e E_a^2}{T_0 \hbar\omega}\right) \cdot A_0[O]_0 \left(\frac{E_a}{\hbar\omega}\right)^2 \cdot t\right]\right\}^{-1}, \quad (6\text{–}45)$$

The duration of the effective chemical reaction in the passive discharge zone ($T_v^{min} \leq T_v \leq T_v^{st}$) is

$$\tau_p \approx \frac{1}{k_{VT} n_0} \cdot \frac{\hbar\omega}{E_a}. \quad (6\text{–}46)$$

Taking into account that $\tau_{VT} = 1/k_{VT} n_0$ is the time of VT relaxation, the effective reaction time in the passive zone is much shorter than the relaxation time ($\tau_p \ll \tau_{VT}$). The strong non-equilibrium ($T_v > T_0$) can be sustained during the entire period of the effective synthesis in the passive zone ($T_v^{min} \leq T_v \leq T_v^{st}$).

6.3.4. Energy Efficiency of Plasma-Chemical NO Synthesis: Excitation and Relaxation Factors

Total energy efficiency (η) of the non-equilibrium plasma synthesis of NO from air or N_2–O_2 mixtures, stimulated by vibrational excitation, can be subdivided into three factors: the excitation factor (η_{ex}), the relaxation factor (η_{rel}), and the chemical factor (η_{chem}, to be considered in the next section):

$$\eta = \eta_{ex} \cdot \eta_{rel} \cdot \eta_{chem}. \quad (6\text{–}47)$$

The excitation factor (η_{ex}) is due to energy losses related to the excitation of low productive degrees of freedom. For the aforementioned mechanism, the percentage of discharge energy going to vibrational excitation of N_2 is shown. Thus, at $T_e \approx 1$ eV, $\eta_{ex} = 0.8$–0.9; see Fig. 6–5.

If the ionization degree is high (6–42), the relaxation factor (η_{rel}) is due to losses outside of the active discharge zone. When T_v decreases below the critical value T_v^{min}, almost all the rest of the vibrational energy goes to gas heating through VT relaxation. From the energy balance (6–44), the energy losses from lower vibrational levels outside of active discharge zone are (Macheret et al., 1980a,b)

$$E_{VT} \approx \int_{T_v^{st}}^{T_0} \frac{(\hbar\omega)^2}{\left(\exp\frac{\hbar\omega}{T_v} - 1\right)^2}\left[1 + \exp\left(\frac{E_a}{T_v^{min}} - \frac{E_a}{T_v}\right)\right]^{-1} d\left(\frac{1}{T_v}\right). \quad (6\text{–}48)$$

With an accuracy of about $(\hbar\omega/E_a)^2$, the total VT-relaxation energy losses (6–48) can be approximated by the Planck vibrational energy corresponding to the critical vibrational temperature: $E_{VT} \approx \varepsilon_v(T_v^{min})$. The relaxation factor can then be expressed as a function of specific energy input E_v:

$$\eta_{rel} \approx \frac{E_v - \varepsilon_v\left(T_v^{min}\right)}{E_v}. \quad (6\text{–}49)$$

The relaxation factor determines the threshold of the dependence $\eta(E_v)$ for NO synthesis stimulated by vibrational excitation: $E_v = \varepsilon_v(T_v^{min}) \approx 0.5\text{--}0.8 \text{ J/cm}^3$ (see Fig. 6–1).

6.3.5. Energy Efficiency of Plasma-Chemical NO Synthesis: Chemical Factor

The chemical factor (η_{chem}) of energy efficiency is due to losses in exothermic reactions; see, for example, (6–3). Vibrational energy losses, attributed to a single act of chemical reaction (like non-resonance VV exchange and VT relaxation from highly vibrationally excited levels), can also be included in the chemical factor. The chemical factor, in contrast to the relaxation factor, doesn't depend on energy input but is sensitive to chemical mechanism. Reaction (6–3) of the Zeldovich chain $N + O_2 \rightarrow NO + N$ leads to losses of about 1 eV per single NO molecule formed, and to a 33% decrease in energy efficiency of NO synthesis in non-thermal plasma. Other losses are due to the fact that $N_2^*(^1\Sigma_g^+, v)$ participates in reaction (6–2), not with minimal required energy ($E_a \approx 3 \text{ eV/mol}$) but on average with a little higher energy ($E_a + T_v'$). The excess energy T_v' is determined by the logarithmic slope of the vibrational distribution function $f(E)$ near the activation energy:

$$T_v' = \left| \left(\left. \frac{\partial \ln f(E)}{\partial E} \right|_{E=E_a} \right)^{-1} \right|. \tag{6–50}$$

The losses in (6–50) are about 0.3–0.5 eV per single NO molecule produced. Vibrational energy losses related to non-resonance VV exchange and VT relaxation from highly vibrationally excited levels can also be attributed to a single reaction event and are about 0.2–0.3 eV per single NO molecule. In total, the chemical factor for NO synthesis at a sufficiently long Zeldovich chain is about $\eta_{chem} = 50\%$.

Major energy losses in plasma-chemical NO synthesis are due to exothermic reaction (6–3) producing vibrationally excited NO molecules: $N + O_2 \rightarrow NO^* + N$. In most cases this energy is lost in fast non-adiabatic VT relaxation of NO^* ($k_{VT}^{NO} \approx 10^{-13} \text{cm}^3/\text{s}$ at $T_0 \approx 1000 \text{ K}$), but for some conditions this energy can be used in further NO oxidation:

$$NO^* + O_2 \rightarrow NO_2 + O. \tag{6–51}$$

This reaction provides branching of the Zeldovich chain mechanism (6–2), (6–3), which is unique for endothermic processes. The branching of (6–51) doesn't lead to explosive synthesis acceleration, which is anyway limited by vibrational excitation rate (6–43), but results in a significant increase in atomic oxygen concentration (up to $[O]/[O_2] = 0.1$) and, therefore, of ozone. NO is mostly converted in this case into NO_2. An increase of atomic oxygen density results in a decrease of vibrational temperature, while the total reaction rate and total energy efficiency remain almost the same. The branching of the Zeldovich chain mechanism was experimentally observed in a strongly non-equilibrium microwave discharge with magnetic field, operating under ECR conditions (Asisov et al., 1980). The total energy efficiency of the plasma-chemical NO synthesis, taking into account all the factors (6–47), is shown in Fig. 6–1 in comparison with experimental results achieved in different discharge systems.

6.3.6. Stability of Products of Plasma-Chemical Synthesis to Reverse Reactions in Active Zone of Non-Thermal Plasma

Protection of products from active species in the active plasma zone seriously restricts the yield of non-thermal plasma-chemical NO synthesis. The most important fast barrierless reverse reaction, leading to destruction of NO inside the active discharge zone, is

$$N + NO \rightarrow N_2 + O, \quad k_N = 10^{-11} \text{ cm}^3/\text{s}. \tag{6–52}$$

Reverse reaction (6–52) competes with the propagation (6–3) of the Zeldovich chain, and it is able to terminate the chain when the concentration of NO becomes high in the active discharge zone:

$$\frac{[\text{NO}]}{[\text{O}_2]} > \frac{k_2}{k_\text{N}}, \tag{6–53}$$

where $k_2 \approx 10^{-10} \exp(-4000/T_0, \text{K})$, cm³/s is the rate coefficient of reaction (6–3). To estimate the influence of reverse reaction (6–52) in the passive zone, we should take into account the quasi-stationary evolution of atomic nitrogen concentration outside of the active discharge zone:

$$N(t) = [\text{O}]_0 \frac{k_\text{R}(T_\text{v}, T_0)}{k_2} \exp(-k_3 n_0 [\text{O}_2] t), \tag{6–54}$$

where $[\text{O}]_0$ is the stationary concentration of O atoms in the active discharge zone (6–39); $k_\text{R}(T_\text{v}, T_0)$ is the rate coefficient (6–34) of the principal reaction of NO synthesis; k_3 is the rate coefficient of ozone formation (6–38); and the evolution of vibrational temperature T_v follows (6–45) in the passive zone. Then NO destruction due to reverse reaction (6–52) in the passive zone can be described by the kinetic equation

$$\frac{d[\text{NO}]}{dt} = -k_\text{N}[\text{NO}] \cdot [\text{N}](t). \tag{6–55}$$

Integration of equation (6–55) gives a characteristic time τ_chem of NO destruction in the passive zone with respect to total duration τ_p (6–46) of the effective synthesis outside of the active zone:

$$\frac{\tau_\text{p}}{\tau_\text{chem}} = \frac{k_\text{eV} n_\text{e}}{k_\text{VT} n_0} \exp\left[-\frac{k_2}{k_\text{N}} \left(\frac{E_\text{a}}{\hbar\omega} \right)^2 \right]. \tag{6–56}$$

The produced nitrogen oxides are stabilized in the passive zone if $\tau_\text{p}/\tau_\text{chem} \ll 1$. To stabilize NO in the active and passive zones ((6–53), (6–56)), it is necessary to have $k_\text{N}/k_2 \ll 1$. To suppress reverse reaction (6–52), atomic nitrogen should preferentially react not with NO but with O_2, propagating the Zeldovich chain. Reverse reaction (6–52) is fast and has no activation barrier whereas the direct reaction (6–3) has an activation barrier of about 0.3 eV. Therefore, requirements (6–53) usually limit NO yield at temperatures $T_0 < 1000\,\text{K}$ to about 3–5%. Chain propagation reaction (6–3) is exothermic and characterized by a low efficiency of vibrational energy in overcoming activation barriers ($\alpha \approx 0.2$). Thus, reaction (6–3) cannot be effectively accelerated by vibrational excitation of oxygen until the vibrational temperature exceeds 5000 K. Even at these high vibrational O_2 temperatures, the chemical factor of energy efficiency for NO synthesis alone decreases to less than 40%. Higher NO yields can be reached without loss of energy efficiency under conditions of the effect of "hot atoms" (Grigorieva et al., 1984).

6.3.7. Effect of "Hot Nitrogen Atoms" on Yield of NO Synthesis in Non-Equilibrium Plasma in Air and Nitrogen–Oxygen Mixtures

The generation of hot atoms has been discussed in Sections 3.4.6 and 3.4.7. When energetic atoms generated in exothermic reactions are able to participate in the following reactions before Maxwellization, they can significantly perturb the translational distribution function of the atoms. The "hot atoms" are able to accelerate the exothermic chemical reactions, because it is the translational (not vibrational) energy of reagents that stimulates overcoming the activation barriers of these reactions. This effect is able to accelerate exothermic reaction (6–3) and mitigate the restriction of NO yield. Vibrationally excited $\text{N}_2^*(^1\Sigma_\text{g}^+, v)$ participates in reaction (6–2) not exactly with the minimal required energy ($E_\text{a} \approx 3\,\text{eV/mol}$), but on

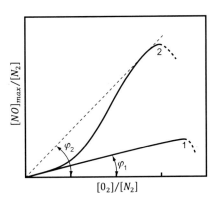

Figure 6–10. Maximum relative nitrogen oxide yield $[NO]_{max}/[N_2]$ as function of initial mixture composition ($[O_2]/[N_2]$ ratio) at relatively low translational temperature ($T_0 < 1500\,K$) (1) without taking into account the effect of "hot atoms," $\tan\varphi_1 = k_2/k_N \ll 1$; (2) taking into account the effect of "hot atoms," $\tan\varphi_2 = k_2/k_N \approx 1$.

average with a higher energy ($E_a + T_v'$). The excess energy (T_v' determined by (6–50) is about 0.3–0.5 eV. An essential portion of this energy goes into the kinetic energy of N atoms, stimulating the following exothermic process (6–3). The N-atom distribution over translational energy $f_N(E)$ can be found from the kinetic equation taking into account their formation in endothermic reactions (6–2), Maxwellization in translational–translational (TT) relaxation processes, and losses in chemical processes (Grigorieva et al., 1984):

$$f_N(E) \approx f_N^{(0)}(E) + \frac{\nu_R}{\nu_{max}} f_{N_2}^{vib}\left(E_a + \frac{E}{\beta}\right). \tag{6–57}$$

Here $f_N^{(0)}(E)$ is the distribution not perturbed by the hot atoms; ν_R/ν_{max} is the frequency ratio of reaction (6–3) and the TT Maxwellization of N atoms; $f_{N_2}^{vib}(E)$ is the vibrational distribution of N_2; E_a is the activation energy of (6–2); and the coefficient β represents a fraction of the vibrational energy released in the reaction that goes into translational energy of N atoms. Hot atoms are represented in the distribution by the second term and mirror the vibrational distribution of N_2 after the activation barrier with scaling factor β. The hot atoms increase the rate coefficient k_2 of N-atom reaction (6–3) with O_2 (6–53, 6–56) and, therefore, increase the maximum conversion of air into NO. This effect is demonstrated in Fig. 6–10 (Grigorieva et al., 1984) and is stronger at higher oxygen fractions in the N_2–O_2 mixtures.

6.3.8. Stability of Products of Plasma-Chemical NO Synthesis to Reverse Reactions Outside of the Discharge Zone

Outside of the discharge, NO can be converted back into N_2 and O_2 via the following reverse reaction:

$$NO + NO \rightarrow N_2 + O_2, \quad E_a \approx 3\,eV/mol. \tag{6–58}$$

The activation energy of (6–58) is high; therefore, its contribution is significant at high translational temperatures ($T_0 \geq 1500\,K$). When $T_0 \geq 1500\,K$, and the vibrational temperature of N_2 is higher ($T_v > T_0$), the two-temperature quasi-equilibrium state of the plasma-chemical system can be established:

$$N_2^*\left(^1\Sigma_g^+, T_v\right) + O_2 \Leftrightarrow NO + NO. \tag{6–59}$$

The direct endothermic reaction is determined by the vibrational temperature of N_2 ($T_v > T_0$), while the reverse exothermic reaction is stimulated by the translational temperature T_0 (Section 2.7.3). The two-temperature quasi-equilibrium (6–59) can be described in terms of the quasi-equilibrium constant:

$$K_q(T_v, T_0) = \frac{[NO]^2}{[N_2^*(T_v)] \cdot [O_2]}. \qquad (6\text{–}60)$$

The constant $K_q(T_v, T_0)$ can be expressed through the thermodynamic equilibrium constant for reaction (6–59) not perturbed by vibrational excitation $K_e(T_0)$ and through vibrational distribution functions related to two-temperature ($T_v > T_0$) and one-temperature ($T_v = T_0$) situations (Macheret et al., 1980a,b):

$$K_q(T_v, T_0) = K_e(T_0) \cdot \frac{f_q(E_a, T_v, T_0)}{f_e(E_a, T_0)}, \qquad (6\text{–}61)$$

where $f_q(E_a, T_v, T_0)$ and $f_e(E_a, T_0)$ are the two-temperature ($T_v > T_0$) and one-temperature ($T_v = T_0$) vibrational distribution functions of N_2 at a vibrational energy equal to the activation energy E_a of limiting reaction (6–2). The two-temperature quasi-equilibrium (6–59) doesn't lead immediately to a significant decrease in the energy efficiency of NO synthesis. Only when VT relaxation establishes conventional one-temperature equilibrium is most of the NO generated in the active plasma zone lost in the reverse reaction (6–58). The pair of two-temperature quasi-equilibrium reactions (6–59), similarly to VT relaxation, also leads to energy transfer from vibrational to translational degrees of freedom. In general, the gas temperature increase to 1500 K makes non-equilibrium NO synthesis much less efficient. Thus, this temperature is usually considered to be an upper limit of overheating of the relevant non-thermal plasma systems. This temperature determines the upper limit of specific energy input, $E_v^{max} \approx 5 \, \text{J/cm}^3$. Optimal specific energy inputs at moderate and high pressures are $E_v = 2\text{–}3 \, \text{J/cm}^3$. The corresponding energy efficiency for NO synthesis is about 35%, and the maximum yield of nitrogen oxides is about 10–15%.

6.4. EXPERIMENTAL INVESTIGATIONS OF NO SYNTHESIS FROM AIR AND N_2–O_2 MIXTURES IN DIFFERENT DISCHARGES

6.4.1. Non-Equilibrium Microwave Discharge in Magnetic Field Operating in Conditions of Electron Cyclotron Resonance

The highest energy efficiency of plasma-chemical NO synthesis from air and N_2–O_2 mixtures is achieved in a microwave discharge operating in a magnetic field under ECR conditions. A schematic of the experimental setup is shown in Fig. 6–11 (Asisov et al., 1980). Microwave power with an electromagnetic wavelength of 8 mm is supplied by a stationary klystron (power 30 W; K in the figure) and pulse-periodic magnetron (M in the figure; pulse power 30 KW, pulse duration 300 ns, pulse repetition frequency 1000 Hz). These power supplies sustain the strongly non-equilibrium microwave discharge, both in stationary and in pulse-periodic regimes. Electromagnetic radiation H_{10} initially propagates in a rectangular waveguide, and then the microwave is converted into the H_{01} mode in a round waveguide with maximum field at half of the waveguide radius (for details see Fridman & Kennedy, 2004). The discharge has a ring shape and is separated from the walls. The reactor is organized in a section of the round waveguide that is 10 cm long with a diameter of 17 mm. Cross-sectional reactor windows are made from sapphire ($\tan \delta < 0.0015$) and tightened by indium. Microwave receivers measure reflected and transmitted radiation power, allowing one to find the energy input in the discharge. Cooling is provided if necessary by liquid nitrogen. The resonance value of the magnetic field required for ECR conditions is 1.2 Tesla. To achieve the ECR conditions, the reactor is located on the axis of a solenoid providing up to 1.5 Tesla.

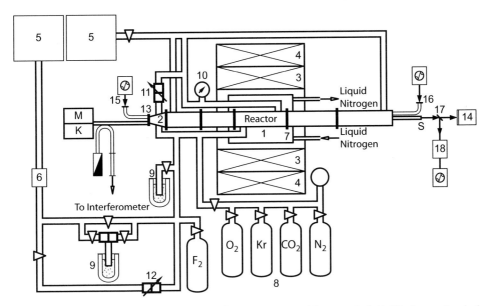

Figure 6–11. Schematic of plasma-chemical microwave system with magnetic field: (1) plasma-chemical reactor; (2) converter of type of electromagnetic wave; (3, 4) solenoids; (5) vacuum pump; (6) liquid nitrogen trap; (7) refrigerator; (8) gas tanks; (9) control volumes; (10) vacuum-meter; (11, 12) differential manometers; (13) waveguide branching system; (14) spectrograph; (15, 16) microwave detectors; (17) semi-transparent mirror; (18) photo-electronic amplifier; (M) magnetron microwave source; (K) klystron microwave source; (S) window for diagnostics.

Microwave interferometry has been applied for the characterization of plasma parameters (Givotov et al., 1985b). The results of electron density measurements in a stationary discharge regime are shown in Fig. 6–12a as a function of pressure. Electron densities in the pulsed regime are about 10-fold higher. Fig. 6–12b presents the dependence of absorbed electromagnetic power in the stationary microwave plasma as a function of pressure. The non-equilibrium microwave discharge under ECR conditions (magnetic field of 1.2 Tesla) is stable at pressures up to one atmosphere. The absorption of electromagnetic power obviously decreases outside of the resonance conditions to only about 3–5% in the pulse discharge regime. Oscilloscope curves showing electromagnetic power transmitted through the discharge at different magnetic fields that demonstrate the resonance effect are presented in Fig. 6–13. Relatively low absorption in the beginning of the electromagnetic pulse even under resonance conditions is due to the low electron density in plasma after less than 0.1 μs from the pulse front. The microwave resonance absorption at 100 Torr is shown in Fig. 6–14. Absorption slightly decreases at higher pressures because the electron–neutral collision

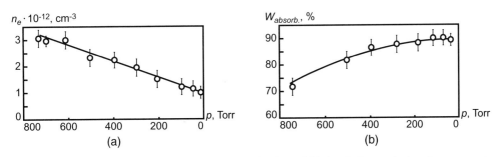

Figure 6–12. Pressure dependence of (a) electron density and (b) absorbed power in the microwave system with magnetic field $H = 12\,\text{kOe}$.

375

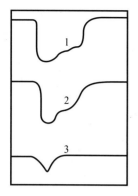

Figure 6–13. Oscillograms of electromagnetic wave power transmitted through non-equilibrium microwave discharge plasma in magnetic field at different values of magnetic field strength: (1) 3 kOe; (2) 8 kOe; (3) 12 kOe.

frequency ν_{en} exceeds the resonance frequency ω_B in this case. Plasma electrons are able to receive a significant portion of electromagnetic energy before they leave the Larmor circle orbit even in quasi-collisional conditions ($\omega_B \leq \nu_{en}$), therefore, the resonance plasma absorption of 70% microwave power is observed in the system even at atmospheric pressure.

6.4.2. Evolution of Vibrational Temperature of Nitrogen Molecules in Non-Equilibrium ECR: Microwave Discharge During Plasma-Chemical NO Synthesis

First specify the non-equilibrium ECR microwave discharge parameters in an N_2–O_2 mixture at 50 Torr in stationary and pulse regimes. In the stationary regime, the electron density is $n_e \approx 2 \cdot 10^{11}$ cm^{-3}, the electron temperature is $T_e = 1.5$ eV, and the gas temperature is $T_0 = 0.12$ eV. At the same pressure in the pulse regime, the electron density reaches $n_e \approx 8 \cdot 10^{13}$ cm^{-3} at electron temperature $T_e = 1.5$ eV after a short time period of 10^{-7} s. Evolution of the plasma in the pulse-periodic regime is sophisticated. When an electromagnetic pulse is terminated, the electron temperature and density decrease during a relaxation period of $1.5 \cdot 10^{-6}$ s. The gas temperature slightly increases during a pulse; cooling down after a pulse requires about $3 \cdot 10^{-6}$ s. Whereas the gas temperature decreases each time between pulses to the initial gas temperature, the vibrational temperature does not decrease much between the pulses (see Fig. 6–15). Thus, there is an accumulation of vibrational energy in the pulse regime. The accumulation of vibrational energy continues until the consumption of vibrational energy into NO synthesis stabilizes the vibrational temperature growth. The vibrational temperature reaches $T_v = 0.3$ eV and essentially exceeds the gas

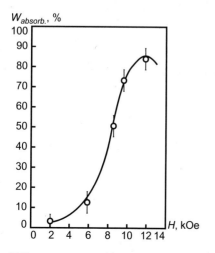

Figure 6–14. Resonant curve of absorption of electromagnetic wave power in non-equilibrium microwave discharge plasma.

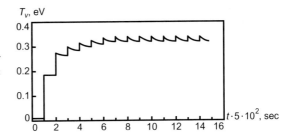

Figure 6–15. Time evolution of vibrational temperature of nitrogen molecules in pulsed N_2–O_2 microwave discharge in magnetic field.

temperature, providing energy-effective NO synthesis stimulated by vibrationally excited N_2 molecules.

6.4.3. NO Synthesis in the Non-Equilibrium ECR Microwave Discharge

Plasma-chemical experiments with NO synthesis in the non-equilibrium ECR microwave discharge (see Sections 6.4.1 and 6.4.2) resulted in observation of the highest energy efficiency of the process ($\eta \approx 30\%$), which corresponds to an NO energy cost close to 0.3 eV/mol (Asisov et al., 1980). The experiments were carried out in mixture $[N_2] : [O_2] = 1$ at pressures of 10–100 Torr. The best results were achieved in a pulse-periodic regime with pulse power 30 kW, pulse duration 0.3 μs, and repetition frequency up to 1 kHz. The 100 cm^3 reactor was cooled down with liquid nitrogen. ECR permitted not only effective absorption of microwave radiation ($>90\%$; see Section 6.4.1) but also control and optimization of the electron temperature and other plasma parameters: electron temperature $T_e = 1$–2 eV, electron density $n_e = 10^{12}$–10^{14} cm^{-3}, gas temperature $T_0 = 600$–1300 K.

Kinetic curves of NO production in the discharge are shown in Fig. 6–16. The NO concentration is presented as a function of residence time and energy input in the discharge. The NO energy cost in the experiments is low (≈ 3 eV/mol), which corresponds to an energy efficiency of about 30% (see Fig. 6–1). At higher pressures ($p > 20$ Torr), 80–90% of the total product yield is NO; the rest is NO_2 and N_2O_4 (a fraction of O_3 is negligible). At lower pressures ($p < 20$ Torr), the energy efficiency remains high, but fractions of NO_2 and O_3 grow significantly, which corresponds to branching of the Zeldovich chain (see Section 6.3.5). The efficiency of NO synthesis in non-equilibrium microwave discharges arranged without a magnetic field was not as high as that under ECR conditions (Polak et al., 1975a; Tarras et al., 1981).

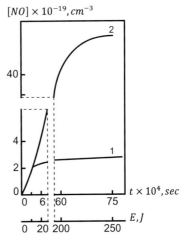

Figure 6–16. Dependence of NO yield on time and discharge energy input at different pressures: (1) $p = 20$–100 Torr; (2) $p < 20$ Torr.

6.4.4. NO Synthesis in Non-Self-Sustained Discharges Supported by Relativistic Electron Beams

These experiments clearly demonstrated a threshold in the dependence of the process energy efficiency on specific energy input $\eta(E_v)$ and a significant contribution of vibrationally excited nitrogen molecules in plasma-chemical NO synthesis from N_2–O_2 mixtures (Basov, Danilychev, & Panteleev, 1977b, 1978b; Basov et al., 1978a). Strongly non-equilibrium atmospheric-pressure non-self-sustained discharge was supported by a pulse-periodic relativistic electron beam (EB) with current density 0.5–2.5 A/cm^2, electron energy 600 keV, EB pulse duration 2.5 μs, and EB pulse repetition frequency 1 Hz. Experiments were also carried out with shorter EB pulses for current density 300–500 A/cm^2, electron energy 300 keV, EB pulse duration 25 ns, and pulse repetition frequency 1 Hz. The relativistic EB provided the ionization necessary for the discharge, whereas the main energy input was directed to vibrational excitation of N_2 and supplied by a relatively low electric field. A discharge chamber with electrodes was located 7–10 cm away from the accelerator exit window. A 0.5 μF battery of capacitors charged to 15 kV was connected to the electrodes to provide principal energy input in the discharge at the reduced electric field of $E/p \approx 10$ V/cm $\cdot$ Torr. The gas flow through the discharge chamber with an active volume of 5 cm^3 provided a residence time of about 1 ms. After the uniform ionization zone, restricted by metallic grids, a special system enabled the mixing of plasma-excited N_2 with oxygen. Products were analyzed using photo-calorimetry and chromatography.

NO synthesis was investigated in these experiments in two different ways. In one approach, only pure nitrogen was excited in the discharge chamber and after that mixed with oxygen. In the other approach, the discharge provided excitation of the N_2–O_2 mixture. Experiments with longer EB pulses (2.5 μs), reduced electric field $E/p = 8$–10 V/cm $\cdot$ Torr, and excitation of pure nitrogen before mixing with oxygen were characterized by specific energy input from the electric field up to $E_v = 1$ J/cm^3. The regime of short EB pulses (25 ns) provided specific energy input from the electric field up to $E_v = 0.6$ J/cm^3 in the case of plasma excitation of pure nitrogen before mixing with oxygen; the energy input from the relativistic EB itself was about $E_v = 0.06$ J/cm^3. The dependence of NO yield on specific energy input in the discharge from electric field is shown in Fig. 6–17. To clarify NO synthesis specifically provided by the non-self-sustained discharge, NO production related only to the radiolytic effect of the relativistic EB was subtracted from the total NO yield. A comparison of regimes with longer and shorter EB pulses shows that the product yield and energy efficiency don't depend separately on EB current density and pulse duration; they depend on specific energy input E_v in the discharge. Fig. 6–17 demonstrates

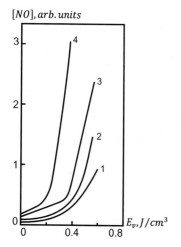

Figure 6–17. Dependence of NO yield on specific energy input into non-equilibrium microwave discharge with magnetic field in premixed N_2–O_2 system at different nitrogen-to-oxygen concentration ratios: (1) $[N_2] : [O_2] = 8 : 1$; (2) $[N_2] : [O_2] = 6 : 1$; (3) $[N_2] : [O_2] = 4 : 1$; (4) (argon dilution) $[N_2] : [O_2] : [Ar] = 4 : 1 : 5$.

6.4. Experimental Investigations of NO Synthesis from Air and N_2–O_2 Mixtures

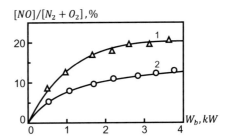

$[NO]/[N_2 + O_2], \%$

Figure 6–18. Dependence of NO yield in non-equilibrium stationary plasma-beam discharge on electron beam power at gas flow rate 120 L/s and different levels of gas pressure: (1) $p = 5 \cdot 10^{-2}$ Torr; (2) $p = 0.2$ Torr.

the threshold in the yield and energy efficiency dependences on specific energy input E_v, which demonstrates stimulation of the process by vibrational excitation of N_2 molecules. The threshold in $\eta(E_v)$ for the experiments can also be seen in Fig. 6–1. The yield of NO and energy efficiency of the plasma-chemical process are equal at equal values of specific energy input, no matter if nitrogen was excited in plasma before mixing with oxygen or directly in the N_2–O_2 mixture. However, the maximum value of the specific energy input in the N_2–O_2 mixture ($E_v = 0.3 \, \text{J/cm}^3$) was lower than that for pure nitrogen ($E_v = 1 \, \text{J/cm}^3$). Therefore, higher product yields and higher energy efficiencies were achieved when N_2 was excited in plasma before mixing with oxygen. Similar experiments with NO synthesis in non-self-sustained discharges were also carried out by Vakar, Denisenko, & Rusanov (1977), and by Fominsky (1978; see Fig. 6–1). Most of the experiments of Fominsky (1978) were carried out at specific energy inputs lower than the critical one, which significantly decreased the contribution of vibrational excitation and, therefore, energy efficiency of the synthesis.

6.4.5. Experiments with NO Synthesis from Air in Stationary Non-Equilibrium Plasma-Beam Discharge

A high rate of air conversion into NO has been achieved using this strongly non-equilibrium discharge (Ivanov & Nikiforov, 1978; Nikiforov, 1979; Krasheninnikov, 1980; Ivanov, 1975; Ivanov et al., 1979). Physical principles of plasma-beam discharges can be found in Fridman & Kennedy (2004). Data collected in these discharge at different air pressures, flow rates, and powers are presented in Figs. 6–18 to 6–21. The experiments with plasma-beam discharges lead to the highest (20%) plasma-chemical air conversion into nitrogen oxides. Energy efficiency of the NO synthesis was, however, not so high, with a maximum of about 10% (energy cost about 10 eV/mol). Typical specific energy inputs in the discharge were relatively high (around $E_v = 2 \, \text{eV/mol}$), which created some problems with system overheating and stabilization of the generated products. Kinetic analysis of the strongly non-equilibrium discharge proved that NO synthesis was provided through the Zeldovich chain mechanism (6–2), (6–3), stimulated by vibrational excitation of nitrogen molecules (Alexeev, Ivanov, & Starykh, 1979).

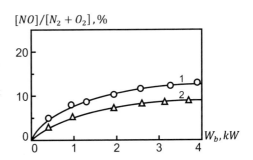

$[NO]/[N_2 + O_2], \%$

Figure 6–19. Dependence of NO yield in non-equilibrium stationary plasma-beam discharge on electron beam power at gas flow rate 1500 L/s and different levels of gas pressure: (1) $p = 0.1$ Torr; (2) $p = 0.2$ Torr.

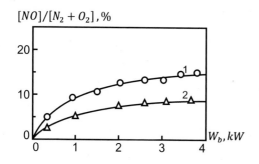

Figure 6–20. Dependence of NO yield in non-equilibrium stationary plasma-beam discharge on electron beam power at gas pressure $p = 0.2$ Torr and different levels of gas flow rate: (1) 120 L/s; (2) 1500 L/s.

6.4.6. Experiments with NO Synthesis from N_2 and O_2 in Thermal Plasma of Arc Discharges

Numerous experiments as well as pilot and industrial-scale tests were performed to optimize NO synthesis from air and N_2–O_2 mixtures in the thermal arcs. Thus, experiments of Polak and Schipachev (1965), Timmins and Amman (1970), and Guliaev, Kozlov, and Polak (1965) were carried out using a nitrogen arc jet with injection of oxygen containing gases (including water vapor) at pressures from 1 to 10 atm and power level up to 30 kW. Data on the NO yield at different pressures and temperatures, collected in the laboratory-scale experiments, are summarized in Fig. 6–22 in comparison with kinetic modeling. The experimental agreement with modeling is quite good at temperatures below 4000 K. The kinetics of NO generation and quenching in plasma are in good agreement with the **Zeldovich equation** for thermal NO synthesis in stoichiometric mixture at atmospheric pressure (Zeldovich, Sadovnikov, & Frank-Kamenetsky, 1947):

$$\frac{d[NO]}{dt}, \frac{mol}{cm^3 \cdot s} = 1.5 \cdot 10^3 \cdot \exp\left(-\frac{43000}{T, K}\right) \frac{1}{\sqrt{[O_2]}} \{[NO]_{eq}^2(T) - [NO]^2\}, \quad (6\text{--}62)$$

where [NO] and $[O_2]$ are the concentrations of NO and O_2 in moles per cubic centimeter and $[NO]_{eq}(T)$ is the equilibrium NO concentration. In a stoichiometric mixture at atmospheric pressure and temperature of 3500 K, generation of the quasi-equilibrium NO concentration (7.8 vol %) requires $2 \cdot 10^{-4}$ s. A quenching rate of 10^8 K/s saves 95% of the products generated at 3500 K. Quenching takes place at temperatures of about 2500 K. Lower quenching rates, 10^7 K/s and 10^6 K/s, save 85 and 63% of the initially produced nitrogen oxides, respectively. The final concentration of nitrogen oxides strongly depends on the process temperature before the start of the quenching. The optimal heating temperature for an atmospheric-pressure stoichiometric N_2–O_2 mixture is about 3500 K. Heating the mixture to the higher temperature of 7000 K decreases the NO concentration in products after fast quenching to only 2 vol %.

NO synthesis from air in a high-temperature H_2O plasma jet was demonstrated by Parkhomenko, Rudenko, and Ganz (1969). The NO concentration after quenching and condensation of water reached a relatively high value of 14%. The energy cost of NO in

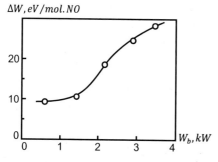

Figure 6–21. Dependence of energy cost of generation of nitrogen oxides in non-equilibrium stationary plasma-beam discharge on electron beam power at gas pressure $p = 0.1$ Torr and gas flow rate 1500 L/s.

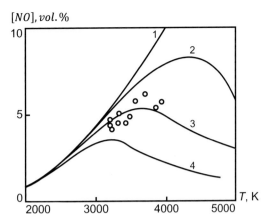

Figure 6–22. Quasi-equilibrium NO concentration produced from initial N$_2$–O$_2$ = 1:4 mixture as function of temperature at different pressures: (1) infinitely high pressure; (2) 10 atm, (3) 1 atm, (4) 0.1 atm; circles, experimental data corresponding to pressure 5–10 atm.

such system, however, is very high. The experiments with arcs at 30 kW or less are usually referred to as laboratory-scale experiments. Industrial-scale tests of plasma-chemical NO synthesis at 2500 kW are described by Zyrichev (1968), Lytkin, Zyrichev, and Pendrakovsky (1970), and Lytkin, Pendrakovsky, and Varapaeva (1969). NO synthesis takes place there directly in a high-power arc in air. Results obtained in this industrial-scale air-plasma system are presented in Fig. 6–23. They are in good agreement with laboratory experiments at the average enthalpy up to 1200 cal/g, which corresponds to a plasma temperature of 3100 K.

6.4.7. General Schematic and Parameters of Industrial Plasma-Chemical Technology of NO Synthesis from Air

The industrial technology of NO synthesis in thermal arcs is described by Polak and Shchipachev (1965). Air enters the plasma-chemical reactor after preheating to 1600–1800 K in a heat exchanger using the high enthalpy of N$_2$-containing products after a plasmatron. Preheated air is heated in the plasmatron to a high operational temperature of 3000–3300 K, where NO synthesis takes place. Quenching of products is arranged in two stages. First, fast quenching from the initial temperature of 3000–3300 K to an intermediate value of 1800–2000 K is provided by special mixing of hot product gases with already

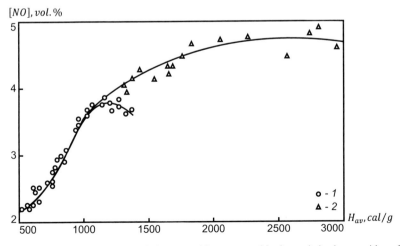

Figure 6–23. Concentration of nitrogen oxides generated in thermal air plasma with surface quenching as a function of flow enthalpy. Gas pressure 1 atm; triangles correspond to experimental installation, circles to industrial installation.

cold product gases with a cooling rate of about 10^8 K/s. Quenching from the intermediate temperature (1800–2000 K) to low temperatures requires slower cooling ($2–5 \cdot 10^5$ K/s) in a conventional heat exchanger, where energy is transferred to incoming air. Optimal conditions for industrial plasma technology of NO synthesis are plasma temperature 3000–3500 K and gas pressure 20–30 atm. The thermal efficiency of the plasmatron with the reactor is 80%, and the energy efficiency of a turbo-compressor is 30%. The energy cost of generating of nitrogen oxides is 8 kWh/kg NO (10 eV/mol NO), which corresponds to a total energy efficiency of about 10%. Although the energy efficiency and NO yield in strongly non-equilibrium plasmas are higher than in the quasi-equilibrium plasmas, non-thermal discharges have not been tested for industrial NO production mostly because of their relatively low total power and productivity.

6.5. PLASMA-CHEMICAL OZONE GENERATION: MECHANISMS AND KINETICS

6.5.1. Ozone Production as a Large-Scale Industrial Application of Non-Thermal Atmospheric-Pressure Plasma

Ozone generation from oxygen or air is one of the oldest plasma-chemical processes, with a 150-year history of fundamental research and large-scale industrial application. Synthesis of ozone (O_3) from oxygen is an endothermic process:

$$\frac{3}{2} O_2 \rightarrow O_3, \quad \Delta H \approx 1.5 \, \text{eV/mol}. \tag{6–63}$$

The reverse process is obviously exothermic. Diluted ozone mixtures are relatively stable at low temperatures. However, even a relatively small heating leads to ozone decomposition, which can be explosive if the O_3 concentration is sufficiently high. Thus, effective O_3 synthesis should be performed at low temperatures, typically at room temperature. Carrying out the highly endothermic process (6–63) at room temperature requires the application of strongly non-equilibrium methods of chemical synthesis. Non-equilibrium plasma of a dielectric barrier, pulsed corona, and other non-thermal atmospheric-pressure discharges are some of the most efficient methods for practical ozone production.

The history of ozone generation and application has always been closely related to the history of plasma chemistry. It was Martinus van Marum who first observed in 1785 the formation of ozone in electric sparks generated by his electrostatic machine. Ozone was first identified as a new chemical compound in 1839 by Schonbein (1840). Soon after that, Werner Siemens (1857) made a breakthrough step and proposed a reliable method of ozone generation by passing air or oxygen through an alternating-current discharge bounded by at least one dielectric barrier. This breakthrough was actually an invention of a dielectric barrier discharge (DBD), which is today one of the most useful discharges not only for ozone production but also for the entirety of non-thermal atmospheric-pressure plasma chemistry (Kogelschatz, 2003). The application of DBD in air or oxygen at pressures of 1–3 atm first proposed by Werner Siemens remains today the major approach to large-scale ozone generation.

The practical application of ozone is related to its very high ability to oxidize organic compounds (first of all in the liquid phase) and to its ability to react with double bonds. Different applications of ozone include treatment of drinking water, water in pools, and other reservoirs; cleaning of waste water; bleaching processes; and oxidation steps in numerous chemical technologies. The largest scale ozone production is related to water cleaning. The first large-scale plasma installations in drinking water plants were built in Paris (1897), Nice (1904), and St. Petersburg (1910). Modern plasma-chemical ozone generators are more effective and powerful. The power of DBD ozone generators extends up to several

Table 6–1. Comparison of Different Types of Industrial Ozone Generators

Ozone generator type; feedstock gas	Dew point, °C	Cooling fluid	Discharge gap, mm	Voltage, kV	Frequency, Hz	Dielectric barrier, mm	Energy cost, kWh/kg
Otto (flat, glass dielectric); air	−40	water	3.1	7.5–20	50–500	3–5	10.2
Tubes; air	−60	water	2.5	15–19	60	2.5	7.5–10
Tubes; oxygen	−60	water	2.5	15–19	60	2.5	3.75–5.0
Lowther (flat, ceramic dielectric); air	−60	air	1.25	8–10	2000	0.5	6.3–8.8
Lowther (flat, ceramic dielectric); oxygen	−60	air	1.25	8–10	2000	0.5	2.5–5.5

megawatts, which results in ozone generating capacities of 100–500 kg/h (Razumovsky & Zaikov, 1974).

6.5.2. Energy Cost and Energy Efficiency of Plasma-Chemical Production of Ozone in Some Experimental and Industrial Systems

While industrial application of ozone is widespread, it is still limited by the energy cost of its generation. The thermodynamic minimum of energy cost for O_3 production (6–63) is 0.8 kWh/kg. However, conventional generators require about 15–20 kWh/kg for ozone synthesis from oxygen (energy efficiency 4–6%) and about 30–40 kWh/kg for ozone synthesis from air (energy efficiency 2–3%). Such low energy efficiencies are due to the high energy cost of O_2 dissociation, as well as to losses related to chemical admixtures and temperature effects. Careful performance of the plasma-chemical process permits one to decrease the energy cost. Thus, Funahashi (1953) decreased it to 8–9 kWh/kg, which corresponds to an energy efficiency of up to 10%. Experiments with a glow discharge (Brewer & Westhaver, 1930) permitted a decrease in energy cost to 6 kWh/kg (energy efficiency 13%). Some O_3 energy cost data achieved in different industrial ozone generators (Razumovsky & Zaikov, 1974) are presented in Table 6–1. Especially low energy costs of ozone production were achieved in experiments with O_3 quenching by cooling to temperatures sufficient for its condensation. Thus, the energy cost in experiments with O_2 plasma (Briner & Ricca, 1955a,b) was decreased to 3.3 kWh/kg (energy efficiency 25%) by cooling the plasma-chemical system to cryogenic temperature (−183°C).

The lowest energy cost of O_3 synthesis, 2.5 kWh/kg (energy efficiency about 30%), was demonstrated in experiments with non-uniform DBD in O_2 (Kozlov et al., 1979), where the reactor walls were cooled for O_3 stabilization by liquid nitrogen. The economics of cryogenic O_3 stabilization is questionable, although deep cooling of the feed gas is also necessary to reach the low humidity required for effective synthesis (see Section 6.5.7). Very low energy costs for ozone generation, about 2.5 kWh/kg (energy efficiency about 30%), were also achieved in experiments with a pulsed corona discharge not at cryogenic but at room temperature (Korobtsev et al., 1997).

6.5.3. Plasma-Chemical Ozone Formation in Oxygen

Ozone generation in oxygen plasma requires about twice less energy than that in air plasma, which is interesting for applications even when taking into account the price of oxygen.

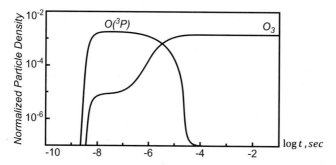

Figure 6–24. Kinetic curves related to ozone formation due to one DBD microdischarge in pure oxygen at atmospheric pressure and room temperature.

Also plasma-chemical kinetics in pure oxygen is not as complicated as in air (see Eliasson, Hirth, & Kogelschatz, 1987; Eliasson & Kogelschatz, 1991). When oxygen plasma is produced in a DBD, the strength of microdischarges significantly influences the kinetics and efficiency of ozone generation. The microdischarges in atmospheric-pressure oxygen last for only a few nanoseconds and transport electric charge less than 1 nC. The microdischarge current density in an optimal DBD regime is about 100 A/cm^2, the local electron density is of the order 10^{14} cm^{-3}, the specific energy input is 10 mJ/cm^3, and gas heating due to an individual microdischarge is negligible.

The plasma-chemical kinetics of ozone formation in pure oxygen is determined by O$_2$ dissociation (see Section 5.11.5). Electronic and vibrational excitation, dissociative attachment, and ion-related processes contribute to oxygen dissociation followed by ozone synthesis in strongly non-equilibrium plasmas (Gibalov, Samoilovich, & Filippov, 1981; Yoshida & Tagashira, 1987; Peyrous, Pignolet, & Held, 1989). The O$_2$ dissociation and ozone formation in strongly non-equilibrium discharges is mostly due to an intermediate excitation of an excited state $A^3\Sigma_u^+$ with energy threshold of about 6 eV,

$$e + O_2 \rightarrow O_2^*\left(A^3\Sigma_u^+\right) + e, \rightarrow O(^3P) + O(^3P) + e, \tag{6–64}$$

and to intermediate excitation of an electronically excited state $B^3\Sigma_u^-$ with energy threshold 8.4 eV:

$$e + O_2 \rightarrow O_2^*\left(B^3\Sigma_u^-\right) + e \rightarrow O(^3P) + O(^1D) + e. \tag{6–65}$$

The O atoms formed in reactions (6–64) and (6–65) then recombine with O$_2$ in the three-body process:

$$O + O_2 + O_2 \rightarrow O_3 + O_2, \quad k(\text{cm}^6/\text{s}) = 6.9 \cdot 10^{-34} \cdot (300/T, \text{K})^{1.25}. \tag{6–66}$$

Taking into account the rate coefficient (6–66), it is clear that, even though electrons disappear from the DBD gap in oxygen (from microdischarges) in less than 10 ns, ozone formation requires much more time, about a few microseconds. The third body in recombination process (6–66) can be represented not only by molecular oxygen but by any other neutrals (O + O$_2$ + M → O$_3$ + M). The kinetics of such recombination processes was analyzed by Kossyi et al. (1992). When ozone is generated in a DBD consisting of numerous microdischarges, a large ozone concentration is built up by a large number of the microdischarges. Kinetic curves in Fig. 6–24 represent the evolution of concentrations of some major chemical species produced due to one microdischarge in DBD, calculated by Eliasson & Kogelschatz (1991). The efficiency of ozone generation in a DBD depends on the strength of the microdischarges.

6.5.4. Optimum DBD Microdischarge Strength and Maximization of Energy Efficiency of Ozone Production in Oxygen Plasma

A strong DBD microdischarge provides a relatively high specific energy input and hence a relatively high concentration of atomic oxygen and ozone. It leads, however, to the undesired side reactions of three-body recombination of atomic oxygen (the third body is not necessarily O_2):

$$O + O + O_2 \rightarrow O_2 + O_2, \quad k = 2.45 \cdot 10^{-31} \cdot (T, K)^{-0.63}, \mathrm{cm^3/s}, \quad (6\text{--}67)$$

as well as to reverse reactions of destruction of ozone molecules in collisions with atomic oxygen,

$$O + O_3 \rightarrow O_2 + O_2 \quad k = 2 \cdot 10^{-11} \exp(-2300\,\mathrm{K}/T_0), \mathrm{cm^3/s}. \quad (6\text{--}68)$$

Reactions (6–67) and (6–68) compete with the ozone formation (6–66). The effectiveness of ozone synthesis decreases due to reactions (6–67) and (6–68) when the DBD microdischarges are so strong that relative concentration of atomic oxygen $[O]/[O_2]$ exceeds 10^{-4}. Significant losses of energy efficiency occur when the relative concentration of atomic oxygen $[O]/[O_2]$ exceeds 0.3–0.5%. Thus, the upper limit of strength of non-thermal atmospheric-pressure discharges is limited by the chemical kinetics of ozone production. The lower limit of microdischarge strength in DBD is due to losses related to ions, which dissipate energy significant to support the discharge even at low specific energy inputs. An optimal relative concentration of O atoms is about $[O]/[O_2] = 2 \cdot 10^{-3}$ (Eliasson, Hirth, & Kogelschatz, 1987). In some sense, the optimal microdischarge strength and optimal concentration of atomic oxygen can be considered a compromise of the chemical kinetics limitation of O_3 concentration from above and the discharge physics limitation of the O_3 concentration from below. Anyway, the 0.2% relative concentration of atomic oxygen in a microdischarge, which then leads to the 0.2% relative concentration of ozone, is a reasonable estimation for optimal parameters of plasma-chemical O_3 synthesis.

The kinetic modeling of Eliasson, Hirth, and Kogelschatz (1987) predicts a minimum energy cost of ozone production on the level of 2.5 kWh/kg (energy efficiency about 30%), which corresponds to the experimental results of Kozlov et al. (1979) mentioned in the previous section. The lowest O_3 energy cost in experiments with DBDs is about 4 kWh/kg (energy efficiency 20%). Taking into account this energy efficiency, at least 70–80% of the total discharge energy is dissipated into heat. This heat should be quickly removed, because ozone molecules are very unstable at elevated temperature. Effective heat removal is therefore an important aspect of ozone generators.

Why is the energy efficiency of O_3 generation even in optimal regimes and even in pure oxygen so low (20–30%) with respect to CO_2 dissociation (see Sections 5.1–5.6) where the energy efficiency reaches 90%? The effect is related to the vibrational excitation of molecules, which initially collect most of the discharge energy in CO_2, O_2, and many other molecular gases. CO_2 dissociation can use most of the energy collected in molecular vibration before VT relaxation, whereas O_2 molecules lose their vibrational energy before dissociation via vibrational excitation in symmetrical exchange reactions:

$$O_2^* + O \rightarrow O_3^* \rightarrow O + O_2, \quad k \approx 10^{-11}\,\mathrm{cm^3/s}\ (T = 300\,\mathrm{K}). \quad (6\text{--}69)$$

Losses of vibrational energy are mostly due to atomic oxygen, which is inevitable during ozone generation. An effective contribution of vibrational excitation and, therefore, high energy efficiency of O_3 production would only be possible in very unrealistic systems where the fraction of vibrationally excited O_2 molecules is high (high vibrational temperature) while the atomic oxygen concentration is low.

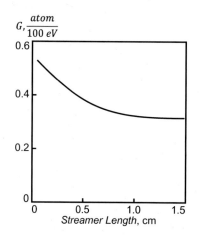

Figure 6–25. Dependence of the G-factor value for nitrogen-atom production in atmospheric-pressure air on the streamer length.

6.5.5. Plasma-Chemical Ozone Generation in Air

Non-thermal electric discharges in air distribute energy between nitrogen and oxygen, which makes them less energy effective in O_3 synthesis than pure oxygen discharges. Discharge energy initially localized in N_2 is not completely lost and can be partially used for O_3 synthesis. Therefore, ozone production from air requires only about twice as much energy as needed from oxygen. Taking into account the price of pure oxygen, smaller ozone generators are easier to make based on discharges in dry air. The mechanism of ozone synthesis in pure oxygen, (6–64)–(6–66), is still used and makes a significant contribution in case of O_3 generation in air. The third-body particle in ozone synthesis (6–66) can be effectively represented in this case by all air components: $O + O_2 + M \rightarrow O_3 + M$ (Kossyi et al., 1992). In addition to the pure oxygen mechanism (6–64)–(6–66), a significant contribution is made to the synthesis by molecular nitrogen (Yagi & Tanaka, 1979; Eliasson, Kogelschatz, & Baessler, 1984; Braun et al., 1988, 1991, 1992; Kossyi et al., 1992; Kitayama & Kuzumoto, 1999; Herron, 1999). Kinetic modeling and analysis of the plasma-chemical process was accomplished in numerous works, including Eliasson & Kogelschatz (1986) and Peyrous (1990). The contribution of N_2 for O_3 synthesis is mostly due to atomic nitrogen and electronically excited nitrogen molecules. Atomic nitrogen leads to ozone synthesis (in recombination like (6–66)) by means of the intermediate formation of oxygen atoms in exothermic reactions:

$$N + O_2 \rightarrow NO + O, \quad E_a \approx 0.3\,\text{eV/mol}, \quad \Delta H \approx -1\,\text{eV/mol}, \qquad (6\text{–}70)$$

$$N + NO \rightarrow N_2 + O, \quad k = 10^{-11}\,\text{cm}^3/\text{s}, \quad \Delta H \approx -3\,\text{eV/mol}. \qquad (6\text{–}71)$$

These processes are limited by the strongly endothermic formation of N atoms (see Section 5.11.4). The energy cost of N atoms in DBD streamers in air is presented in Fig. 6–25 in the form of G-factors, which characterize number of N atoms formed per 100 eV of streamer energy (Naidis, 1997). For comparison, G-factors for formation of O atoms are shown in Fig. 6–26. The contribution of N atoms can add about 10% to the total production of atomic oxygen and, hence, ozone. More significant contribution to production of O atoms and ozone is related to reactions of electronically excited nitrogen molecules:

$$N_2\left(A^3\Sigma_u^+\right) + O_2 \rightarrow N_2O + O, \qquad (6\text{–}72)$$

$$N_2\left(A^3\Sigma_u^+, B^3\Pi_g\right) + O_2 \rightarrow N_2 + O + O. \qquad (6\text{–}73)$$

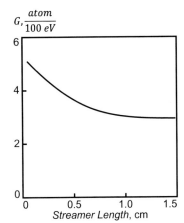

Figure 6–26. Dependence of the G-factor value for oxygen-atom production in atmospheric-pressure air on the streamer length.

In total, about 50% of O_3 generated in the non-thermal atmospheric-pressure air discharges is attributed to the indirect processes (6–70)–(6–73) related to nitrogen. The characteristic time of ozone formation in atmospheric-pressure oxygen plasma is about $10\,\mu s$ (see Fig. 6–24). The nitrogen-related indirect reactions (6–70)–(6–73) are significantly longer than direct processes of ozone synthesis from oxygen (6–64)–(6–66). As a result, ozone formation in air takes longer, about $100\,\mu s$. This effect is clearly seen from kinetic simulations of Eliasson and Kogelschatz (1991), illustrated in Fig. 6–27.

6.5.6. Discharge Poisoning Effect During Ozone Generation in Air Plasma

An increase of specific energy input E_v in non-thermal discharges, due to higher power or lower air flow rate, leads to an increase in the concentrations of NO and NO_2. A high concentration of nitrogen oxides (NO/NO_2), usually referred to as NO_x, completely stops the generation of O_3. This phenomenon is known as the **discharge poisoning effect** and is related to fast reactions of O atoms with NO_x:

$$O + NO + M \rightarrow NO_2 + M, \quad k = 6 \cdot 10^{-32}\,cm^6/s, \qquad (6\text{–}74)$$

$$O + NO_2 \rightarrow NO + O_2, \quad k = 10^{-11} \cdot (T_0/1000\,K)^{0.18}\,cm^3/s. \qquad (6\text{–}75)$$

When NO_x concentration exceeds the threshold value of about 0.1%, the atomic oxygen reactions with nitrogen oxides, (6–74) and (6–75), become faster than their reactions with molecular oxygen (6–66) to form ozone. That actually explains the discharge poisoning effect, which was first observed long ago by Andrews and Tait (1860). Even ozone molecules

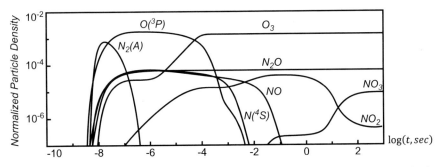

Figure 6–27. Kinetic curves describing ozone formation due to one DBD microdischarge in air at atmospheric pressure and room temperature.

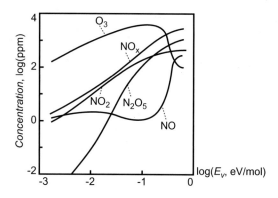

Figure 6–28. Concentrations of ozone and nitrogen oxides generated by a series of DBD microdischarges in room-temperature atmospheric-pressure air as a function of specific energy input: energy density per microdischarge is 5.4 mJ/cm³; time interval between microdischarges is 3 ms.

already formed in the non-thermal atmospheric-pressure discharge under conditions of discharge poisoning are converted back to oxygen in the fast low-barrier reactions with nitrogen oxides (Kogelschatz, 1992):

$$O_3 + NO \rightarrow NO_2 + O_2, \quad k = 4.3 \cdot 10^{-12} \cdot \exp(-1560\,K/T_0)\,cm^3/s. \quad (6\text{–}76)$$

Nitrogen oxides are not generally consumed in the reactions of discharge poisoning; therefore, they can be interpreted as the catalytic recombination of oxygen atoms and catalytic destruction of ozone in the presence of NO_x. The discharge poisoning effect can be illustrated by results of kinetic simulations of DBD with multiple microdischarges in atmospheric air (Eliasson & Kogelschatz, 1991), shown in Fig. 6–28. The simulations assume the energy density per microdischarge is 5.4 mJ/cm³ and the time interval between microdischarges is 3 ms. The maximum ozone concentration is about 0.3% and it starts decreasing when the specific energy input exceeds 0.1 eV/mol and the concentration of nitrogen oxides (NO_x) exceeds 0.03%. A similar effect takes place in the stratosphere, where nitrogen oxides at a level of 0.1% significantly decrease the ozone concentration (Crutzen, 1970; Johnston, 1992). In general, ozone and nitrogen oxides don't coexist well in plasma. Non-thermal atmospheric-pressure discharges produce nitrogen oxides at elevated specific energy inputs and temperatures without any ozone formation (see next section). Conversely, at low specific energy inputs and temperatures close to room temperature, ozone is effectively produced but not NO_x, and the discharges can even be used in air purification of NO_x (see Section 11.7). Taking into account that ozone and nitrogen oxides have different smells, one can actually "smell" the temperature of the atmospheric-pressure discharges in air based on this effect.

6.5.7. Temperature Effect on Plasma-Chemical Generation and Stability of Ozone

Elevated temperatures significantly reduce ozone production in plasma. Thus, experiments with ozone generators operated with air at 800°C result in the production of nitrogen oxides only (Schultz & Wulf, 1940). On the one hand, it is related to the previously discussed effect of discharge poisoning: higher temperatures are related to higher specific energy inputs and higher concentrations of nitrogen oxides (see Fig. 6–28), which suppresses ozone synthesis. On the other hand, ozone itself even diluted in air is unstable at elevated temperatures. Decomposition of ozone is an exothermic process reverse to the synthesis (6–63), which proceeds through the following kinetic mechanism (Intesarova & Kondratiev, 1967):

$$O_3 + M \Leftrightarrow O_2 + O + M, \quad (6\text{–}77)$$

$$O + O_3 \rightarrow O_2 + O_2. \quad (6\text{–}78)$$

Rate coefficients of the reverse reaction of O_3 formation (6–77) and O_3 destruction (6–78) can be taken from (6–66) and (6–68). The rate coefficient of O_3 dissociation (6–77) is (Razumovsky & Zaikov, 1974)

$$k_{O_3} = 1.3 \cdot 10^{-6} \cdot \exp(-11750\,\text{K}/T_0)\,\text{cm}^3/\text{s}. \qquad (6\text{–}79)$$

The stability of ozone with respect to dissociation depends exponentially on temperature T_0. According to (6–79), ozone decomposition time in pure dry air (without catalytic admixtures, see following sections) at room temperature is quite long, about an hour. However, already at $100°C$, the ozone decomposition time is only 1 s. When O_3 dissociates and creates atomic oxygen, the oxygen atom is able almost immediately to stick to molecular oxygen and form ozone again. In other words, the reverse reaction of ozone formation (6–66) is very fast and is able to balance the decomposition (6–79) if the ozone concentration is low and reaction (6–78) can be neglected. A detailed balance of the direct and reverse reactions (6–77) in pure dry air results in the relative concentration of atomic oxygen with respect to ozone:

$$[O]/[O_3] \approx 3 \cdot 10^8 \cdot \exp(-11750\,\text{K}/T_0). \qquad (6\text{–}80)$$

At a gas temperature of $100°C$, when ozone decomposition is already intensive (the characteristic time of O_3 dissociation process (6–77) is about 1 s), the reverse process of ozone formation (6–77) is still significantly faster and the stationary atomic oxygen concentration is negligible: $[O]/[O_3] \approx 10^{-5}$. According to relation (6–80), the stationary concentrations of ozone and atomic oxygen become comparable in pure dry air conditions only at a gas temperature of about $300°C$. These data are valid only outside of plasma, where atomic oxygen is produced not from O_2 but from ozone dissociation (6–77). Actual losses of ozone related to the thermal decomposition mechanism (6–77) and (6–78), leading to O_3 conversion into O_2, take place in the fast reaction of ozone with atomic oxygen (6–78). The characteristic time of ozone thermal decomposition (6–78), assuming a stationary $[O]/[O_3]$ ratio, can be calculated based on (6–68) and (6–80) as

$$\tau_{O_3} = \frac{5 \cdot 10^{-18}\,\text{s}}{\alpha_{O_3}} \cdot \exp\left(\frac{14050\,\text{K}}{T_0}\right). \qquad (6\text{–}81)$$

Thus, the O_3 decomposition time depends not only exponentially on temperature but also strongly on the initial concentration of ozone, $\alpha_{O_3} = [O_3]/n_0$; n_0 is the total gas density. Diluted ozone is stable at low temperatures. For an ozone relative concentration of 0.2%, the ozone decomposition time is about 1 s for a gas temperature of about $145°C$. In comparison, when the ozone relative concentration is 1%, the ozone decomposition time is about 1 s at a gas temperature of about $125°C$. If the temperature is higher and the ozone concentration is higher, the O_3 decomposition process becomes very fast. Taking into account that ozone decomposition (6–63) is a highly exothermic process, the fast decomposition leads to an intensive overheating, further acceleration of decomposition, and finally to thermal explosion.

6.5.8. Negative Effect of Water Vapor on Plasma-Chemical Ozone Synthesis

Humidity is a strong negative factor that significantly suppresses ozone synthesis in non-thermal atmospheric-pressure plasma, specifically for DBDs. To provide high efficiency for O_3 synthesis, ozone generators are usually operated in dry air or dry oxygen conditions. The feed gases, air or oxygen, are normally supplied with a dew point below $-60°C$, which corresponds to a water vapor content of about a few parts per million or less. The negative effect of water vapor on ozone synthesis is due to two major factors. The first, specifically related to DBD, is that it increases the surface conductivity of dielectrics at high humidity. The dielectric barrier with increased surface conductivity leads to stronger

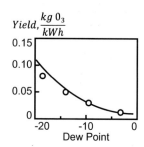

Figure 6–29. Influence of humidity (water vapor) on ozone yield in a DBD in room-temperature oxygen.

and more powerful individual microdischarges, which decreases the efficiency of ozone generation because of too high a concentration of atomic oxygen, possible temperature effects, and the generation of nitrogen oxides poisoning the discharge (see Sections 6.5.4, 6.5.6, and 6.5.7). The negative kinetic effect of water vapor on plasma-chemical O_3 synthesis is due to the formation of OH and HO_2 radicals in plasma, leading to catalytic ozone destruction in the chain mechanism (Heidt & Forbes, 1934):

$$OH + O_3 \rightarrow HO_2 + O_2, \quad k = 5 \cdot 10^{-13} \, cm^3/s, \tag{6–82}$$

$$HO_2 + O_3 \rightarrow OH + O_2 + O_2. \tag{6–83}$$

The chain length can be pretty long in this case, because of the relative slowness of chain termination in pure air or oxygen. Losses of OH radicals are mainly due to gas impurities, in particular NO_x and hydrocarbons. NO and NO_2 in the presence of OH radicals are rapidly converted to HNO_2 and HNO_3. In general, the typical dependence of ozone yield (kgO_3/kWh) in a DBD in oxygen on the gas humidity (dew point) is shown in Fig. 6–29 (Razumovsky & Zaikov, 1974). The curve is still far from saturation at a dew point of about $-20°C$, and efficiency of ozone synthesis is still growing with humidity reduction.

6.5.9. Effect of Hydrogen, Hydrocarbons, and Other Admixtures on Plasma-Chemical Ozone Synthesis

Small additions of molecular hydrogen have a negative effect on ozone synthesis in non-thermal atmospheric-pressure discharges. Molecular hydrogen present in oxygen or air, similarly to water vapor, leads to the formation in plasma of OH and HO_2 radicals, which catalytically decompose ozone by the chain mechanism (6–82) and (6–83) (Inoue & Sugino, 1959; Cromwell & Manley, 1959; Kogelschatz, 1999). The dependence of ozone yield (kgO_3/kWh) in DBD oxygen plasma on the volume fraction of hydrogen in the gas is shown in Fig. 6–30 (Razumovsky & Zaikov, 1974). The figure demonstrates a significant decrease in ozone production efficiency when the volume fraction of hydrogen exceeds 0.03% (300 ppm).

The negative effect of the admixture of hydrocarbons on ozone synthesis is also partially due to the formation of OH and HO_2 radicals with the chain mechanism (6–82), (6–83). The major effect of hydrocarbons, however, is related to trapping of oxygen atoms by the hydrocarbon molecules:

$$O + RH + M \rightarrow products + M, \tag{6–84}$$

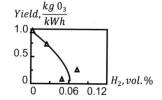

Figure 6–30. Influence of admixture of molecular hydrogen on ozone yield in a DBD in room-temperature oxygen.

6.5. Plasma-Chemical Ozone Generation: Mechanisms and Kinetics

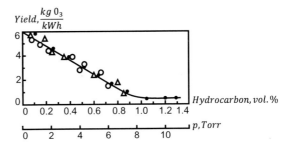

Figure 6–31. Influence of admixtures of different hydrocarbons on ozone yield in a DBD in room-temperature oxygen: black circles correspond to cyclohexane, white circles correspond to heptane, and triangles correspond to hexane.

which competes with O_3 production from atomic oxygen (6–66). Comparing the reactions of atomic oxygen, (6–66) and (6–84), explains the linear dependence of ozone yield on the volume fraction of hydrocarbons presented in Fig. 6–31 (Razumovsky & Zaikov, 1974). The influence of different hydrocarbons of similar size on the yield of ozone synthesis is almost identical. An increase in hydrocarbon volume fraction to about 1% suppresses ozone production almost completely.

Even small admixtures of halogens, like bromine and chlorine, also have a significant negative effect on production and stability of ozone (Benson & Axworthy, 1957). Such admixtures sometimes lead to ozone explosions. The halogens stimulate catalytic thermal decomposition of ozone via the fast chain mechanism, which in the case of chlorine, for example, starts with the formation of atomic chlorine and ClO radicals in the chain initiation reactions (Schumacher, 1957):

$$Cl_2 + O_3 \rightarrow ClO + Cl + O_2, \tag{6–85}$$

$$Cl_2 + O_3 \rightarrow ClO + ClO_2. \tag{6–86}$$

Cl and ClO participate in chain propagation, which determines the catalytic effect of the halogens:

$$OCl + O_3 \rightarrow Cl + O_2 + O_2, \tag{6–87}$$

$$Cl + O_3 \rightarrow ClO + O_2. \tag{6–88}$$

The chain termination is mostly related to the recombination of Cl atoms and ClO radicals:

$$ClO + ClO \rightarrow Cl_2 + O_2, \tag{6–89}$$

$$Cl + Cl + M \rightarrow Cl_2 + M. \tag{6–90}$$

A qualitatively similar chain process of O_3 decomposition occurs in the case of admixtures of Br_2. In contrast to the admixtures with a negative effect on ozone synthesis (water vapor, nitrogen oxides, hydrogen, hydrocarbons), some additives can stimulate O_3 generation in plasma. Specifically nitrogen and carbon oxide provide such enhancement, which is illustrated in Fig. 6–32 (Razumovsky & Zaikov, 1974). Small N_2 additions increase ozone generation efficiency and maximum ozone concentration with respect to pure oxygen. The positive effect of the small N_2 additions to oxygen is significant at high values of specific energy input (Kogelschatz, 1999). For this reason, high-productivity industrial ozone generators, operating with oxygen, often have an additional 1% of nitrogen added to the feed gas. The positive nitrogen effect is due to the additional channels of oxygen dissociation related to nitrogen (see Section 6.5.5); large additions of nitrogen and carbon oxide obviously lead to decreased efficiency of O_3 synthesis.

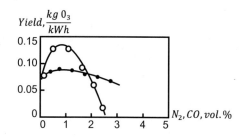

Figure 6–32. Influence of admixtures of N_2 and CO on ozone yield in a DBD in room-temperature oxygen: black circles correspond to molecular nitrogen, white circles correspond to carbon monoxide.

6.6. EXPERIMENTAL AND INDUSTRIAL PLASMA-CHEMICAL OZONE GENERATORS

Although O_3 synthesis can also be performed in photochemical and electrolytic systems, most ozone generation is done using plasmas. Effective ozone generation requires product stability and, therefore, low gas temperatures (surely below $100°C$; see Section 6.5.7). It also requires elevated gas pressures to provide effective three-body attachment of oxygen atoms to molecular oxygen (6–66). These requirements make non-thermal atmospheric-pressure discharges, and specifically DBDs and pulsed corona discharges, the most interesting practical systems for ozone production.

6.6.1. Synthesis of Ozone in Dielectric Barrier Discharges as the Oldest and Still Most Successful Approach to Ozone Generation

Werner Siemens invented the DBD in 1857, which immediately became the champion plasma source for ozone synthesis. The DBD systems were installed in the first large-scale ozone generators in water purification plants, which were built a century ago in Paris (1897), Nice (1904), and St. Petersburg (1910). Nowadays, 150 years after the Siemens invention, DBD is still a champion in large-scale ozone production. Recent developments of more effective DBD ozone generators are due to the fundamental understanding of O_3 synthesis in DBD, and first of all to knowledge of streamer processes in DBD microdischarges (Kogelschatz & Eliasson, 1995). It was demonstrated (Starke, 1923) that the energy efficiency of O_3 formation in DBD does not depend on applied frequency in the range between 10 and 50 kHz. Such frequency increase leads to a larger number of DBD microdischarges, to an increase of total power, and total ozone production, keeping the O_3 energy cost on the same level. The use of higher frequencies, therefore, permits the same total discharge power to be sustained at lower operating voltages. The lower operating voltages result, in particular, in a reduced strain on the dielectrics. The higher applied frequencies, as well as other modern DBD modifications, also lead to a substantial increase of power density in the discharge. The higher power density results in higher ozone production rates and higher ozone concentrations, which can be produced in much more compact ozone generators (Kogelschatz, 1988). Starting from the first industrial DBD ozone generators to the present, the most popular configurations use dielectric plates and tubular dielectrics in the form of glass tubes.

6.6.2. Tubular DBD Ozone Generators and Large Ozone Production Installations

A majority of industrial ozone generators are tubular (see Fig. 6–33; Emelianov & Filippov, 1960; Filippov & Vendillo, 1961; Samoilovich & Filippov, 1964; Filippov, Emelianov, & Semiokhin, 1968). The tubular ozone generators look somewhat like heat exchangers. They consist of 200–300 or more cylindrical discharge tubes of about 2–5 cm diameter and 1–3 m length, as illustrated in Fig. 6–34 (Aliev, Emelianov, & Babayan, 1971; Ashurly et al., 1971; Emelianov, 1971; Filippov & Popovich, 1971; Kogelschatz, 2003). Inner dielectric

Figure 6–33. Illustration of a small industrial ozone generator.

glass tubes, closed at one side in most ozone generators, are inserted into outer metallic (steel) tubes. The dielectric tube material is usually borosilicate (Pyrex [Duran]) glass. The two tubes form an annual discharge gap of about 0.5–1 mm radial width. Ozone generator performance critically depends on the accuracy of the gap, which affects microdischarge properties, flow rate, electric power deposition, and heat flux to the cooled metal (steel) electrode. Precise and extremely narrow discharge gaps (0.1 mm), considered by Kuzumoto, Tabata, and Yagi (1996), lead to especially high performance of ozone synthesis. The high-voltage electrode is constructed as a thin metal film (from aluminum or copper) on the inner wall of the glass tube, which is contacted to the power supply by metal brushes (see Fig. 6–34). Modern high-performance ozone generators use non-glass dielectric coatings on the steel tubes instead of glass dielectric tubes, which makes the tubes less fragile. Individual DBD tubes are protected by high-voltage fuses (see Fig. 6–34), which permit other tubes and the whole system to stay operational in the case of dielectric failure in one DBD tube. This is especially important when taking into account that large DBD-based

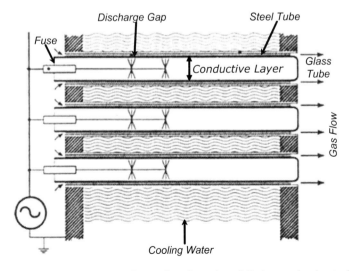

Figure 6–34. Illustration of general configuration of discharge tubes in a technical ozone generator.

Figure 6–35. Large-scale ozone generator at Los Angeles Aqueduct Filtration Plant.

ozone generators use up to a thousand individual DBD tubes inside one large steel tank. Taking into account the high sensitivity of ozone synthesis to gas temperature, cooling of the high-power ozone generators is an important issue. To provide effective cooling the outer steel tubes are welded to two parallel end flanges, forming a sealed compartment in conventional water-cooled heat exchanger configuration. Incoming cold water first cools the exit sections of the DBD tubes, where the ozone concentration and the total specific energy input are the highest and cooling is the most critical.

Modern large-scale industrial ozone generators operate in a frequency range between 0.5 and 5 kHz. Thyristor- or transistor-controlled frequency converters impress square-wave currents or specially formed pulse trains to effectively operate the DBDs. This in particular, allows for the reduction of operating voltages to about 5 kV, dramatically reducing the failure of dielectric barriers. Typical power densities in this case can reach $0.1–1$ W/cm^2, which is fairly high for DBD. Large ozone production installations have not only high power densities but also large total electrode surface areas. This allows operation of one production unit at a power level of several megawatts to produce about 100 kg of ozone per hour (see Fig. 6–35). Modern industrial systems produce ozone with concentrations up to 5 mass % using air as a feed gas, and up to 18 mass % using oxygen. Water treatment plants utilize ozone with a mass fraction of up to 12%. Related DBD energy efficiencies and ozone energy costs were discussed in Section 6.5.2.

6.6.3. Planar and Surface Discharge Configurations of DBD Ozone Generators

Flat or planar configurations of DBD-based ozone generators were also widely applied, particularly, in France. Some characteristics of such ozone generators (Otto and Lowther types) are presented in Table 6–1. A schematic of an element of the planar ozone generator is shown in Fig. 6–36. The dielectric barrier is arranged in this configuration in the form of a ceramic layer. Using ceramics allows for the use of thinner dielectric barriers and ceramics provide higher dielectric permittivity than glass. Masuda et al. (1988) proposed the application of surface discharges for ozone generation. This approach to ozone synthesis is simple and practical. The discharge is maintained between thin parallel electrode strips

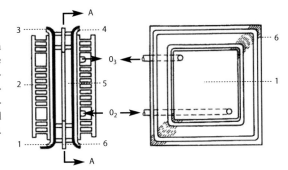

Figure 6–36. Schematic of ozone generator in planar configuration: (A-A) cross section of the device is shown to the right; (1) ceramic dielectric; (2) aluminum structure for effective cooling of the dielectric barrier discharge; (3) high-voltage steel electrode; (4) low-voltage steel electrode; (5) discharge gap; (6) glass separators.

deposited on ceramic tubes. This very practical device is applied mainly in small ozone generators; a slightly modified version of the discharge is used in plasma displays today.

6.6.4. Synthesis of Ozone in Pulsed Corona Discharges

Recently pulsed corona discharges, and especially positive streamer coronas, have been investigated as an alternative to ozone generation in DBD (Gordeyenya & Matveyev, 1994; Hegeler & Akyiama, 1997; Korobtsev et al., 1997; Samaranayake et al., 2000; Simek & Clupek, 2002). Numerical simulation of the plasma-chemical process in positive streamers together with the experiments demonstrate an important advantage of the pulsed coronas (Kulikovsky, 1997a,b, 2001; Knizhnik et al., 1999a,b) although they require pulse power supplies much more sophisticated than those applied in DBD.

The conventional configuration and parameters of a pulsed corona discharge applied for ozone synthesis are investigated by Korobtsev et al. (1997). The record low value of O_3 energy cost was achieved in this discharge system. The discharge chamber in the pulsed corona system is configured in a coaxial "wire in cylinder" geometry. The outer water-cooled metal electrode is 30 cm long with an inner diameter of either 26 or 50 mm; its temperature is kept at 20°C. The inner electrode is a nickel wire with 1 mm diameter. High-voltage pulses with 25–40 kV amplitude and duration of about 150 ns are applied to the nickel wire electrode. The pulse repetition frequency is varied from 50 Hz to 2 kHz. Dry air and oxygen with a dew point of about $-50°C$ are used as feed gases in the system with operating pressures of either 1 or 2 atm. A detailed analysis of heat exchange processes in the pulsed corona discharge shows that they are mostly due to the contribution of convective flows. Conduction contributes only about 10% to the total heat transfer in the pulsed corona, in contrast to the typical situation with DBD which is characterized by very narrow discharge gaps (Horvath, Bilitzky, & Hattner, 1985). Intensive convection leads to effective heat exchange between the gas and walls, which leads to a near proportionality of average gas temperature (with respect to the walls ΔT_0) and discharge power. For example, a discharge power increase from 3 to 10 W in the previously described corona configuration results in a relative gas temperature increase from $\Delta T_0 \approx 5\,K$ to $\Delta T_0 \approx 15\,K$.

The energy cost for ozone synthesis in the pulsed corona discharge is shown in Fig. 6–37 as a function of the produced ozone concentration (which is related to specific energy input) at different levels of specific power. The dependence is presented for different feed gases and different discharge polarities (negative and positive coronas). The negative pulsed corona is less effective in ozone synthesis than a positive one; the efficiency of ozone generation from oxygen is about twice as high as that from air; increase of specific power and, hence, gas temperature (see earlier discussion) at the fixed level of produced O_3 concentration leads to a decrease of ozone energy cost. The most interesting conclusion of the experiments is the possibility of ozone production with a very low energy cost of about 2.5 kWh/kg at room temperature.

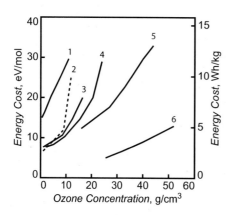

Figure 6–37. Energy cost of ozone production in pulsed corona discharge as function of concentration of produced ozone in different conditions (positive/negative corona, air/oxygen) and different discharge power (Watt per liter): (1) 40 W/L, air, negative corona; (2) 8 W/L, air, positive corona; (3) 20 W/L, air, positive corona; (4) 45 W/L, air, positive corona; (5) 70 W/L, air, positive corona; (6) 50 W/L, oxygen, positive corona.

6.6.5. Peculiarities of Ozone Synthesis in Pulsed Corona with Respect to DBD

Features of the pulsed corona as an ozone generator are quite different from those of DBDs. Probably the most important difference is related to the discharge zone sizes. DBDs, operating in a quasi-uniform electric field, usually require very narrow discharge gaps (see Sections 6.6.1–6.6.3). The pulsed corona operates in essentially non-uniform electric fields, which results in possibly sustaining the discharge in much larger volumes. Thus, characteristic volumes of corona may be much larger than those of DBD, which leads to lower specific power of the discharge and lower temperature and ozone concentration. To achieve higher concentrations of ozone in the corona, the residence time of gas in the discharge zone should be much longer. A large pulsed corona volume also leads to effective convective gas mixing in the discharge. The convection intensifies heat transfer in the corona discharge system by about 10-fold. Together with lower specific discharge powers and related lower gas temperature, the intensive heat transfer from the walls of the discharge chamber results in no overheating of the corona and high stability of the synthesized ozone. Convective gas mixing also leads to ozone decomposition on the walls of the discharge chamber. This negative effect is important because achievement of higher O_3 concentration requires longer residence time and, hence, longer gas contact with the walls. In general, most of the destruction of ozone produced in pulsed coronas is due not to volumetric but to surface processes, which is in contrast to the DBD characterized by higher temperatures, which more strongly affect O_3 decomposition in volume (Korobtsev et al., 1997). This effect explains why an increase of specific power (and therefore gas temperature), at the fixed level of produced O_3 concentration, leads not to an increase but to a decrease of the ozone energy cost (see Fig. 6–37). An increase of gas temperature first of all intensifies the ozone decomposition in the discharge volume, which is less important in a corona. It is more important that the power increase at a fixed ozone concentration (and specific energy input) leads to a decrease of gas residence time and less contact of the gas with the walls, where most of the O_3 decomposition takes place. Another peculiarity of ozone generation in pulsed coronas is the effect of the corona polarity on the effectiveness of the synthesis. As one can see from Fig. 6–37, the O_3 energy cost in a negative corona is two to three times higher than in a positive corona. It is mostly due to the higher electron temperatures in the pulsed coronas with positive polarity, where the discharge is completely based on the formation of streamers. Higher electron temperatures increase the fraction of the discharge energy going into electronic excitation of molecular oxygen leading to ozone synthesis (Samoilovich, Gibalov, & Kozlov, 1989).

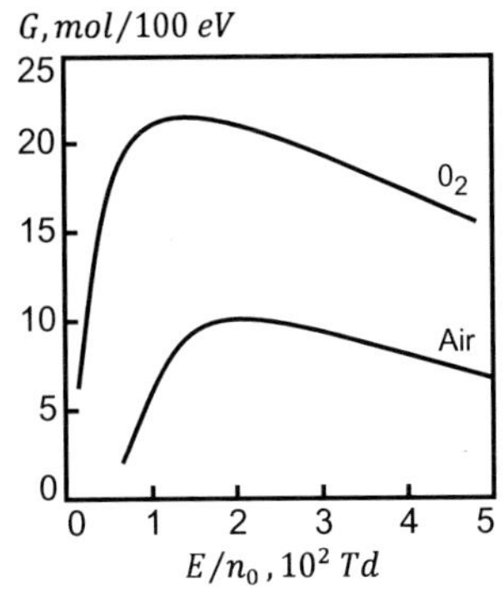

Figure 6–38. G-factors of ozone production as a function of reduced electric field E/n_0 in room-temperature atmospheric-pressure discharges in air and pure oxygen.

6.6.6. Possible Specific Contribution of Vibrational Excitation of Molecules to Ozone Synthesis in Pulsed Corona Discharges

The high energy efficiency of ozone synthesis achieved in pulsed corona discharges in air and in oxygen (Abolentsev et al., 1992; Korobtsev et al., 1997; see two previous sections) challenged scientists to explain the phenomenon. The most consistent explanation was proposed by Knizhnik et al. (1999a,b) and is related to the contribution of vibrationally excited molecules to ozone synthesis. Kinetic calculations of the G-factors (number of O_3 molecules synthesized per 100 eV) corresponding to the mechanism of ozone synthesis, described in Sections 6.5.3 and 6.5.5, were performed by Mnatsakanyan and Naidis (1992), Kogelschatz, Eliasson, and Hirth (1987), and Braun and Pietsch (1991). The results can be summarized as the G-factor dependences on reduced electric field and are presented in Fig. 6–38 for the cases of O_3 generation from air and oxygen. The unit of the reduced electric field used in the figure, E/n_0, is 1 Td $= 10^{-17}$ V $\cdot$ cm^2; n_0 is gas density. The maximum G-factors for air and oxygen are, respectively, $G \approx 11$ mol/100 eV and $G \approx 22$ mol/100 eV, which correspond to minimum energy costs of about 5 kWh/kg for air and 2.5 kWh/kg for oxygen. In principle, these values of minimum O_3 energy cost are in agreement with the experimental results presented in Fig. 6–37. However, the values of the reduced electric field, required to reach the minimum O_3 energy cost, are in discrepancy with the E/n_0 values in major regions of streamers.

Ozone synthesis, both in the DBDs and pulsed corona discharges, takes place in streamers. The streamers consist of the streamer heads with high electric fields (140 kV/cm, or about 600 Td) and streamer channels with relatively low reduced electric fields of about 100 Td and less (Kulikovsky, 1997a,b). According to Fig. 6–38, the O_3 energy cost in the streamer head for the case of air is about 10 kWh/kg because the electric field there is too high, and the energy cost in the streamer channel is more than 15 kWh/kg because the electric field there is too low. In both regions of a streamer, the actual energy cost is higher than the experimental one, which was about 5 kWh/kg in the case of a pulsed corona discharge in air. This discrepancy between kinetic modeling and experiments with pulsed corona was explained by Knizhnik et al. (1999a,b) as the contribution of vibrationally excited molecules to ozone synthesis.

Streamers are relatively short in DBD; therefore, the major contribution to O_3 synthesis is related to streamer heads with relatively high electric fields and the aforementioned limit of minimum energy cost. Pulsed corona discharges, which operate with much larger gaps, consist of longer streamers, where the contribution of the streamer channels to O_3 synthesis can be much more significant. Relatively low electric fields typical for the streamer channels are actually optimal for vibrational excitation of both nitrogen and oxygen molecules. For this reason, vibrational excitation plays an important role in ozone synthesis, specifically

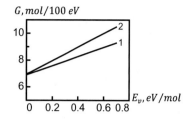

$G, mol/100\ eV$

Figure 6–39. Effect of vibrational excitation of molecules on G-factor of ozone synthesis in atmospheric-pressure pulsed corona discharges in air as function of specific energy input: (1) taking into account the contribution of the effect of saturation of vibrational excitation; (2) total contribution of vibrational excitation including the cross-sectional effect.

in pulsed corona discharges. We can subdivide the contribution of vibrational excitation of molecular nitrogen and oxygen into ozone synthesis in streamer channels into three kinetic effects.

1. Effect of saturation of vibrational excitation. The reduced electric field in a streamer channel (about 100 Td and less) corresponds to a regime when most of the discharge energy is transferred from electrons to vibrational excitation of N_2 (see Fig. 6–5). Ozone synthesis (Sections 6.5.3 and 6.5.5) is mostly provided by electronic excitation of molecular nitrogen and oxygen; therefore, vibrational excitation of the molecules is often considered to be an energy loss. Furthermore, vibrational excitation absorbs electron energy at $E/n_0 \leq 100$ Td so intensively that it results in a significant "drop" of the electron energy distribution function (EEDF) near the threshold of N_2 vibrational excitation. This kinetic effect exponentially slows down the electronic excitation of both nitrogen and oxygen molecules and, therefore, significantly decreases the energy efficiency of ozone production at the low reduced electric fields E/n_0 (see Fig. 6–38). However, the situation changes at higher specific energy inputs and higher levels of vibrational excitation. When most of the molecules are already vibrationally excited, they start to transfer energy back to electrons in the so-called superelastic collisions, which mitigates the EEDF "drop," exponentially accelerates electronic excitation, and finally decreases the energy cost of O_3 production; it is the effect of saturation of vibrational excitation on the electronic excitation rate and energy cost of O_3 synthesis. The results of kinetic modeling of the effect for O_3 synthesis from air, performed by Knizhnik et al. (1999a,b), are presented in Fig. 6–39. Thus, the G-factor of ozone production at $E/n_0 = 100$ Td increases by about 30% due to the saturation of vibrational excitation at the specific energy input to molecular vibration 0.75 eV/mol.

2. Evolution of electronic excitation cross section with vibrational excitation. The vibrational excitation of molecules also accelerates the electronic excitation due to changes in the elementary process cross sections. First, the process is related to an increase in the maximum of the cross sections, but what is more important is that it is due to a shift of the threshold of electronic excitation for the vibrationally excited molecules to lower electron energies. The G-factor of ozone production from air (number of ozone molecules produced per 100 eV) at $E/n_0 = 100$ Td increases by about 20% due to the cross-sectional effect at the specific energy input to molecular vibration, 0.75 eV/mol. The total effect of vibrational excitation including the saturation and cross-sectional effects is also presented in Fig. 6–39; the G-factor at $E/n_0 = 100$ Td and energy input 0.75 eV/mol increases by about 50% total due to vibrational excitation. The total effect of vibrational excitation on ozone synthesis in the long streamer channels can, in principle, explain the experimental data obtained in pulsed corona discharges (Fig. 6–37).

3. Effect of energy transfer from vibrational excitation to electronically excited molecular states and directly to ozone synthesis. Additional ozone synthesis, specifically in the streamer channels, can take place via the direct use of vibrational energy. In the case of pulsed corona discharges in air, it can be first of all related to the so-called vibrational–electronic (VE) relaxation processes, which corresponds to energy transfer from vibrationally excited nitrogen molecules to those that are electronically excited:

$$N_2^*(v_1) + N_2^*(v_2) \rightarrow N_2\left(A^3\Sigma_u^+, B^3\Pi_g\right) + N_2. \tag{6-91}$$

Nitrogen molecules electronically excited through VE relaxation can produce atomic oxygen and ozone by reactions (6–72) and (6–73). The VE-relaxation processes are adiabatic, related to energy transfer from heavy particles to electrons, and therefore their probability is relatively low. The kinetics of these processes is similar to that of associative ionization of vibrationally excited molecules (2–34). Another mechanism of direct energy transfer from vibrational excitation to ozone production is related to disproportioning in the collision of vibrationally excited oxygen molecules:

$$O_2^*(v_1) + O_2^*(v_2) \rightarrow O_3 + O.$$

The contribution of such processes of ozone generation in streamer channels is usually low because of fast VT-relation rate of vibrationally excited O_2 molecules in the presence of O atoms.

6.7. SYNTHESIS OF KrF₂ AND OTHER AGGRESSIVE FLUORINE OXIDIZERS

6.7.1. Plasma-Chemical Gas-Phase Synthesis of KrF₂ and Mechanism of Surface Stabilization of Reaction Products

The production of noble gas compounds is quite challenging because of their low stability and (even more importantly) the low stability of intermediate compounds for their synthesis. Non-thermal plasma is effective in such synthetic processes, because high electron temperature intensifies the synthesis, and low gas and wall temperatures stabilize the products. The synthesis of noble gas compounds in plasma can be represented by the low endothermic process of synthesis of krypton fluoride (KrF₂):

$$\text{Kr} + \text{F}_2 \rightarrow \text{KrF}_2, \quad \Delta H = 0.6\,\text{eV/mol}. \tag{6-92}$$

This process is important for utilization of radioactive ^{85}Kr and for the production of highly reactive fluorine oxidizers (Makeev, Sokolov, & Chaivanov, 1975). Problems of gas-phase synthesis of KrF₂ are related to a low bonding energy (about 0.1 eV) of an intermediate compound, KrF, which dissociates easily in thermal collisions. Synthesis of KrF₂ becomes possible in strongly non-equilibrium conditions by means of the intermediate formation of a stable electronically excited molecular compound, KrF* (Mangano & Jacob, 1975), which can be formed in reactions of molecular fluorine with electronically excited krypton:

$$\text{Kr}^* + \text{F}_2 \rightarrow \text{KrF}^* + \text{F}. \tag{6-93}$$

Electronically excited KrF* can then be converted into KrF₂ in reactions with F₂,

$$\text{KrF}^* + \text{F}_2 \rightarrow \text{KrF}_2 + \text{F}, \tag{6-94}$$

or it can dissociate back into atomic krypton and fluorine:

$$\text{KrF}^* + \text{F}_2 \rightarrow \text{Kr} + \text{F} + \text{F}_2 + \hbar\omega. \tag{6-95}$$

The rate coefficient of dissociation (6–95) is higher than that of synthesis (6–94), which explains the rather low efficiency of the gas-phase mechanism. Gas-phase synthesis of KrF₂ also takes place in triatomic collisions of fluorine with electronically excited krypton:

$$\text{Kr}^* + \text{F} + \text{F} \rightarrow \text{KrF}_2. \tag{6-96}$$

The rate of the trimolecular reaction (6–96) is also not very high because of the possible relaxation of the electronically excited krypton before formation of KrF_2. A much higher efficiency of KrF_2 production can be achieved if the intermediate product of the synthesis (KrF) is stabilized in a crystal structure of krypton (Legasov, Rusanov, & Fridman, 1978c). The mechanism of synthesis includes four major steps:

1. Reactor walls are cooled down to liquid nitrogen temperature, which leads to condensation of Kr.
2. Atomic fluorine is generated from F_2 via plasma (or some other method) and fluorine atoms are transported across the gas phase to the surface with the condensed krypton matrix.
3. Interaction of the F atoms with the Kr film leads to the formation of the intermediate compound KrF with low bonding energy (about 0.1 eV) and its stabilization in the Kr crystal structure.
4. Reaction between two KrF groups is stabilized in the Kr matrix, leading to the formation of KrF_2 molecules.

Diffusion of F atoms in the Kr crystal structure proceeds with formation of the Kr–F bonds. When the distance between two fluorine atoms diffusing in the crystal matrix becomes less than a critical value r_0 (which is about the characteristic size of a unit cell), the groups KrF are able to react, forming KrF_2:

$$KrF + KrF \rightarrow KrF_2 + Kr. \tag{6–97}$$

An alternative outcome of the reaction between two intermediate KrF groups is recombination:

$$KrF + KrF \rightarrow F_2 + Kr + Kr. \tag{6–98}$$

The relative probability of the productive channel of KrF_2 synthesis (6–97) with respect to recombination is about $\alpha \approx 0.02$ (Artukhov, 1976).

6.7.2. Physical Kinetics of KrF_2 Synthesis in Krypton Matrix

Thus, plasma-chemical synthesis of krypton fluoride includes only the generation of F atoms in the plasma zone, whereas formation of KrF_2 can take place at the discharge walls in the Kr crystal structure. The reaction rate of KrF_2 synthesis in the krypton matrix (6–97) is determined by diffusion of the reactive species to the reaction zone (West, 1999), which at the concentration n_F of F atoms can be expressed as

$$v_R = 8\pi r_0 D_F \alpha n_F^2 \left[1 + \frac{r_0}{\sqrt{2\pi D_F t}} \right], \tag{6–99}$$

where r_0 is the critical reactive distance between F atoms in the matrix, $\alpha \approx 0.02$ is the probability of KrF_2 synthesis (6–97), and D_F is the diffusion coefficient of F atoms in the Kr crystal structure (Manning, 1971):

$$D_F = \frac{1}{6} \lambda^2 v_0 f \exp\left(-\frac{E_a^D}{T_0} \right). \tag{6–100}$$

In this equation, λ is displacement of fluorine atoms during transition over the potential barrier ("jumping length"), $v_0 = (1/2\pi)\sqrt{K/M}$ is the "jumping" frequency at the high-temperature limit, M is the atomic mass, K is the force constant for displacement of an atom in the crystal structure, f is a numeric parameter, and E_a^D is the activation barrier of F-atom diffusion in the krypton matrix. Assuming a steady-state flux Φ of F atoms into the krypton film, the concentration of atomic fluorine, n_F, across the film (x is the distance from the surface) is determined by the equation of diffusion with chemical reaction:

$$D_F \frac{\partial^2}{\partial x^2} n_F - 8\pi r_0 D_F n_F^2 = 0. \tag{6–101}$$

Boundary conditions on the surface ($x = 0$) and deep in the krypton film ($x \to \infty$) are

$$-D_F \frac{\partial}{\partial x} n_F \Big|_{x=0} = \Phi, \quad \frac{\partial}{\partial x} n_F \Big|_{x=\infty} = n_F \Big|_{x=\infty} = 0. \tag{6–102}$$

Solution of the differential equation of diffusion (6–101) with boundary conditions (6–102) gives a depth concentration profile of fluorine groups (KrF) within the krypton film:

$$n_F(x) = \left[\left(\frac{16\pi r_0 D_F^2}{3\Phi^2} \right)^{\frac{1}{6}} + x \cdot \left(\frac{4\pi r_0}{3} \right)^{\frac{1}{2}} \right]^{-2}. \tag{6–103}$$

The depth of the layer, δ, where effective synthesis of KrF$_2$ takes place can then be expressed as

$$\delta = \sqrt[3]{\frac{D_F}{\Phi r_0}}. \tag{6–104}$$

Concentration of the intermediate fluorine groups (KrF) in the δ-layer (6–104) is

$$n_{F0} = \sqrt[3]{\frac{\Phi^2}{r_0 D_F^2}}. \tag{6–105}$$

The energy efficiency of plasma-chemical KrF$_2$ synthesis with surface stabilization of products is here determined mostly by the energy cost of F$_2$ dissociation and by the effectiveness of F-atom transportation to the surface of the krypton film.

6.7.3. Synthesis of KrF$_2$ in Glow Discharges, Barrier Discharges, and Photo-Chemical Systems

The synthesis of KrF$_2$ with surface stabilization of products was first performed photo-chemically with low energy efficiency (0.02%) and productivity (Artukhov, Legasov, & Makeev, 1977). Better efficiencies have been achieved in plasma-chemical systems with surface stabilization of products. The first experiments were carried out in a contracted glow discharge with walls cooled by liquid nitrogen or liquid oxygen (Klimov, Legasov, & Chaivanov, 1968). The KrF$_2$ yield in this system at pressure of 40 Torr was 7–10 g/kWh, which corresponds to an energy efficiency of 0.12%. A pressure increase in the glow discharge above 40 Torr leads to a decrease in product yield because of the low efficiency of the glow discharge at such pressures, and because of a decrease of flux of F atoms to the Kr surface.

Higher energy efficiencies for KrF$_2$ synthesis have been achieved in DBD (Isaev, Legasov, & Rakhimbabaev, 1975). The small size of the discharge gap permits a pressure increase to atmospheric pressure, keeping diffusion time short. A high surface-to-volume ratio intensifies KrF$_2$ surface synthesis and product accumulation in DBD. The KrF$_2$ yield in this system is higher (50 g/kWh), which corresponds to an energy efficiency of 0.6% with respect to the process enthalpy (6–92). The small discharge gaps typical for DBD, which are in general effective for KrF$_2$ synthesis with surface stabilization, have a disadvantage related to direct contact of the krypton surface with the DBD plasma. Such contact leads to intensive reverse reactions and partial decomposition of synthesized KrF$_2$. The most

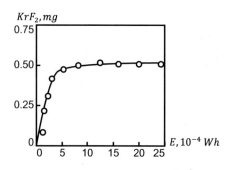

Figure 6–40. KrF$_2$ yield as a function of energy input in non-equilibrium microwave discharge with magnetic field.

effective KrF$_2$ synthesis can be performed in systems where the F$_2$ dissociation zone is separated from the zone of surface stabilization of products in the solid matrix. Such systems are built based on F$_2$ surface dissociation (Bezmelnizyn, Legasov, & Chaivanov, 1977) on non-equilibrium microwave discharges with magnetic fields.

6.7.4. Synthesis of KrF$_2$ in Non-Equilibrium Microwave Discharge in Magnetic Field

High energy efficiency of KrF$_2$ synthesis with surface product stabilization was achieved in experiments with non-equilibrium microwave discharge in a magnetic field (Asisov, 1980). The characteristics of the discharge were considered in Section 6.4.1. The microwave discharge was arranged in a reactor with a metallic wall cooled with liquid nitrogen. A layer of solid krypton was frozen on the wall. The saturated Kr vapor pressure is 2 Torr at the wall temperature (77 K), and the partial pressure of fluorine in the discharge was 10 Torr. The experiments were carried out in the pulsed regime of the system (see Section 6.4.1) with pulse frequency from 40 to 500 Hz. The yield of KrF$_2$ is presented in Fig. 6–40 as a function of total energy input. Saturation of the dependence of KrF$_2$ output on the energy input is due to the effect of reverse reactions. The energy cost of KrF$_2$ synthesis in this plasma-chemical system is a record low (1.2 kWh/kg), which corresponds to a yield of about 800 g/kWh and to an energy efficiency of about 10%. This energy efficiency is very high, because the theoretical maximum of energy efficiency for KrF$_2$ production through preliminary F$_2$ dissociation is only 20%. Dissociation of F$_2$ through electronically excited states $^1\Pi_u$ and $^3\Pi_{0u}$ requires at least 3 eV/mol; it results in the maximum energy efficiency of the slightly endothermic process (6–92): 0.6/3 = 20%. Comparison of the actual energy efficiency (10%) with a theoretical maximum of 20% shows that the efficiency of F-atom transportation to the walls and the efficiency of KrF$_2$ synthesis at the walls is very high in these experiments (not less than 50%).

6.7.5. Plasma F$_2$ Dissociation as the First Step in Synthesis of Aggressive Fluorine Oxidizers

Atomic fluorine is a major starting agent in the synthesis of many fluorides, especially aggressive fluorine oxidizers. Plasma chemistry, therefore, plays an important role in these synthetic processes because application of non-thermal discharges is the most developed approach to F$_2$ dissociation and generation of fluorine atoms. Generation of fluorine atoms from CF$_4$, C$_2$F$_6$, C$_3$F$_8$, NF$_3$ and others is an important part of plasma-chemical material prossessing in electronics and will be especially considered in Chapter 8. The basics of F$_2$ dissociation in non-thermal and thermal plasma systems, general kinetics, and the energy efficiency of the process were discussed in Section 5.11.3 (see, in particular, Figs. 5–82 and 5–83). Experiments with F$_2$ dissociation for synthetic purposes were carried out in glow discharge by Maeg and Geshi (1971). Their typical experimental conditions were fluorine pressure of 10 Torr, gas temperature 500 K, and electron density 10^8 cm^{-3}. The concentration

Table 6–2. Generation of F Atoms by Dissociation of Molecular Fluorine in Mixture with Helium Using High-Energy Electron Beams

Electron beam energy	Electron beam current density (Flux), $cm^{-2} s^{-1}$	Concentration of F atoms, cm^{-3}	F atom production per one EB/F₂ collision
53 keV	$5.4 \cdot 10^{13}$	$5.2 \cdot 10^{15}$	4
43 keV	$2.9 \cdot 10^{13}$	$2.4 \cdot 10^{15}$	3
39 keV	$1.8 \cdot 10^{13}$	$2.1 \cdot 10^{15}$	7

of produced atomic fluorine was $5 \cdot 10^{15}$ cm^{-3}, the flux of F atoms from the generation zone was 10^{18} $cm^{-2} s^{-1}$, and the energy cost of a produced F atom was 10 eV/atom. The generation of fluorine atoms in the glow discharge is of second kinetic order, the dissociation rate is proportional to the densities of electrons and fluorine molecules, and the reaction rate coefficient can be presented empirically as the following function of translational gas temperature, T_0:

$$k_F(T_0), \frac{cm^3}{s} = 0.52 \cdot T_0(K)^{-2.155} \cdot \exp\left(-\frac{1910\,K}{T_0}\right). \qquad (6\text{–}106)$$

Experiments involving F₂ dissociation in a microwave discharge for synthetic purposes were carried out by Fehsenfeld, Evenson, and Broida (1965). These experiments utilized microwave power up to 100 W, frequency 3 GHz, gas pressure 10^{-3}–700 Torr, gas velocity 0.1–100 m/s, and argon–fluorine mixture composition Ar:F₂ = 10:1. The degree of fluorine dissociation was 90% at a gas velocity of 10 m/s, and almost constant in the power range 50–100 W. The maximum production rate of fluorine atoms was 10^{19} s^{-1}.

We should especially draw attention to experiments involving F₂ dissociation and generation of fluorine atoms by high-energy EBs (Wilson, Hao-Line, & Fyfe, 1973). These experiments were actually related to the development of chemical HF lasers, where the generated F atoms are used to initiate a chain reaction with molecular hydrogen. A typical gas mixture in these experiments was F₂ 3–6%, HCl 1%, CO₂ 1%, and the remainder helium. The concentration of generated fluorine atoms and the number of atoms created per single collision of an electron from the beam with an F₂ molecule are presented in Table 6–2 for different EB energies and currents. Several (4–7) F atoms can be created per single collision of an electron from the beam with an F₂ molecule. This effect is due to the production of multiple secondary electrons by a high-energy electron from the beam, and each secondary plasma electron is still able to dissociate a fluorine molecule. The principal advantage of the EB systems is related to the possibility of operation at high gas pressures and in large reactor volumes. The application of F atoms generated by high-energy electron beams from F₂ for further chemical synthesis of fluorine oxidizers is, however, limited by the high energy cost of the atoms (about 12 eV/atom) in these systems.

6.7.6. Plasma-Chemical Synthesis of O₂F₂ and Other Oxygen Fluorides

The effective generation of atomic fluorine in non-thermal plasma permits one to perform the following synthesis of aggressive fluorine oxidizers, in particular oxygen fluorides. Numerous relevant experiments were carried out in glow discharges at gas pressures of 10–40 Torr and at cryogenic temperatures (Nikitin & Rosolovsky, 1970; Prusakov & Sokolov, 1971). The mechanism of plasma-chemical synthesis of oxygen fluorides is related to the formation of the intermediate active radical O₂F, which is stable in the gas phase even at room temperature (Nikitin & Rosolovsky, 1971). Plasma-chemical synthesis of oxygen fluorides starts with the generation of F atoms from fluorine molecules by means of electronic excitation and dissociative attachment as described in Section 5.11.3. Atomic fluorine then sticks to molecular oxygen, forming the O_2F radical in the following fast barrierless trimolecular reaction:

$$F + O_2 + M \rightarrow O_2F + M, \quad M = F_2, O_2, \text{wall}. \tag{6-107}$$

A major product of the synthesis, oxygen fluoride (O_2F_2), is formed in the reactions:

$$O_2F + F + M \rightarrow O_2F_2 + M, \tag{6-108}$$

$$O_2F + O_2F \rightarrow O_2F_2 + O_2. \tag{6-109}$$

An alternative channel of the reaction between O_2F radicals leads to restoration of O_2 and F_2:

$$O_2F + O_2F \rightarrow O_2 + O_2 + F_2. \tag{6-110}$$

It is interesting that the generation of O_2F_2 according to the mechanism (6–107)–(6–109) does not require dissociation of oxygen molecules. Dissociation of oxygen can indeed be neglected with respect to dissociation of fluorine at relatively low electron temperatures in glow discharges, because F_2 bonding energy is significantly lower than that of O_2 molecules. This effect was actually observed in the experiments with glow discharges, where the low rate of O_2 dissociation resulted in the absence of OF_2 molecules in reaction products. In the relevant experiments with glow discharges, the initial mixture composition was $O_2:F_2 = 1:1$, the discharge current was 30–80 mA, the temperature of the discharge walls was $-196°C$, and the yield of the product O_2F_2 was relatively high, about 50%. Similar glow discharges with higher oxygen concentrations (initial mixture composition $O_2:F_2 = 2:1$) and lower temperature of the discharge walls ($-203°C$) permits one to produce other aggressive oxygen fluorides (O_4F_2) with a yield 0.18 g/h. The formation of O_4F_2 is attributed to recombination of O_2F radicals:

$$O_2F + O_2F + M \rightarrow O_4F_2 + M. \tag{6-111}$$

The oxyfluoride O_4F_2 can be considered a precursor of the production of polyoxyfluorides (Nikitin & Rosolovsky, 1970).

6.7.7. Plasma-Chemical Synthesis of NF_3 and Other Nitrogen Fluorides

Synthesis of NF_3 from N_2 and F_2 was investigated in glow discharges with currents of 4–40 mA, at pressures of 10–40 Torr, and wall temperatures from $-196°$ to $-150°C$ (Nikitin & Rosolovsky, 1970; Prusakov & Sokolov, 1971). The typical initial gas composition was $N_2:O_2 = 3:1$ and gas flow rate was 0.09–1.2 L/h. Although NF_3 synthesis is essentially limited by intensive reverse reactions, the product yield in these experiments reached 30–70%. A glow discharge in NF_3 mostly results in dissociation with formation of molecular nitrogen and atomic fluorine (see Chapter 8). However, N_2F_4 can be synthesized from NF_3 with high yield (50–65%) in some special discharge regimes, as well as N_2F_2 with a yield of 12–16%. Conversely, the glow discharge in N_2F_4 not only leads to dissociation of the molecules with formation of N_2 and F_2, but also results in the formation of NF_3. Glow discharge in a mixture of NF_3 and O_2 (1:1) with current 30–50 mA, pressure 10–15 Torr, gas flow rate 1.7–3.4 L/h, and wall temperature $-196°C$ provides an effective synthesis of nitrogen oxyfluoride (NF_3O). The yield of aggressive fluorine oxidizer – the nitrogen oxyfluoride – reaches 10–15%.

Especially interesting is plasma-chemical synthesis in the mixture NF_3–AsF_5–F_2 (composition 1:1:2), which results in the formation of NF_4AsF_6. This fluorine oxidizer is a derivative from NF_5, which is not stably synthesized by itself. The mechanism of synthesis of NF$_4$ArF$_6$ (Makeev et al., 1975) includes the following two stages after conventional formation of F atoms (Section 5.11.3):

$$F + NF_3 + M \rightarrow NF_4 + M, \tag{6–112}$$

$$F + NF_4 + AsF_5 \rightarrow NF_4AsF_6. \tag{6–113}$$

Guertin, Christe, and Pavlath (1966) considered another mechanism of NF$_4$ArF$_6$ synthesis in plasma through a set of ion–molecular reactions. The mechanism starts with the formation of positive ions (NF$_3^+$ and F$_2^+$ through ionization, and the formation of negative ions (F$^-$) through dissociative electron attachment to NF$_3$, AsF$_5$, or F$_2$ molecules. The initial positive ions lead to the formation of an ion radical NF$_4^+$:

$$NF_3^+ + F_2 \rightarrow NF_4^+ + F, \tag{6–114}$$

$$F_2^+ + NF_3 \rightarrow NF_4^+ + F. \tag{6–115}$$

F^- reacts with AsF$_5$ in three-body collision and forms the complex negative ion AsF$_6^-$:

$$F^- + AsF_5 + M \rightarrow AsF_6^- + M. \tag{6–116}$$

Finally, fast ion–ion recombination of NF$_4^+$ and AsF$_6^-$ leads to formation of the product – the strong N$^{(V)}$ fluorine oxidizer NF$_4$AsF$_6$:

$$NF_4^+ + AsF_6^- + M \rightarrow NF_4AsF_6 + M. \tag{6–117}$$

The experiments with NF$_4$AsF$_6$ synthesis in glow discharge in an NF$_3$–AsF$_5$–F$_2$ mixture (1:1:2) were carried out in a plasma-chemical reactor with gas recirculation at relatively high pressures (80 Torr) and wall temperature $-78°$C. The yield of the product NF$_4$AsF$_6$ in the continuous process was 0.025 g/h. Synthesis of nitrogen fluorides was also investigated in different kinds of arc discharges including those rotating in a magnetic field. Conversion degrees in these experiments were not higher than 1%.

6.7.8. Plasma-Chemical Synthesis of Xenon Fluorides and Other Fluorine Oxidizers

Plasma-chemical synthesis of xenon fluorides (XeF$_2$, XeF$_4$, and XeF$_6$) is similar to that of KrF$_2$, considered in Sections 6.7.1–6.7.4. Experiments with glow discharges ((Nikitin & Rosolovsky, 1970; Prusakov & Sokolov, 1971)) in Xe–F$_2$ (1:1) and Xe–CF$_4$ (1:1) mixtures were focused on the synthesis of XeF$_2$. The glow discharge current in these experiments was 30–90 mA, wall temperature was $-78°$C, and gas flow rate was 1–1.5 L/h. The yield of the product XeF$_2$ was 3.5 g/h in the experiments with Xe–F$_2$ (1:1) mixture, and 0.18 g/h in the experiments with Xe–CF$_4$ (1:1) mixture. XeF$_2$ synthesis can take place completely in the gas phase via the intermediate formation of XeF radicals:

$$F + Xe + M \rightarrow XeF + M. \tag{6–118}$$

The XeF radicals then can be converted into XeF$_2$ by recombination or disproportioning reactions:

$$F + XeF + M \rightarrow XeF + M, \tag{6–119}$$

$$XeF + XeF \rightarrow XeF_2 + Xe. \tag{6–120}$$

The plasma-chemical production of xenon fluorides containing more F atoms (XeF$_4$ and XeF$_6$) is more challenging. Synthesis of XeF$_4$ was performed in a Xe–F$_2$ mixture (1:2) treated in a glow discharge with currents of 10–30 mA. The flow rate was 0.14 L/h and yield of XeF$_4$ was 0.43 g/h. Similar experiments but with a Xe–F$_2$ mixture (1:2) and lower flow rate of 0.016 L/h results in effective generation of XeF$_6$.

Among other gas-phase synthetic processes performed in glow discharges and related to the production of fluorine oxidizers, we can mention the generation of ClF_5 from a $Cl_2–F_2$ mixture (1:10). The pressure in these experiments was 30 Torr, the wall temperature was $-78°C$, and the gas flow rate was 1.5 L/h. The process proceeds through formation of an intermediate active radical, ClF_2, and stable compound, ClF_3. The most challenging phase of the process is actually synthesis of the final product ClF_5 from ClF_3, which takes place in the sequence of three-body recombination processes (Krieger, Gatti, & Schumacher, 1966; Makeev et al., 1975),

$$ClF_3 + F + M \rightarrow ClF_4 + M, \qquad (6\text{--}121)$$

$$ClF_4 + F + M \rightarrow ClF_5 + M, \qquad (6\text{--}122)$$

and also through disproportioning of the intermediate radical ClF_4:

$$ClF_4 + ClF_4 \rightarrow ClF_5 + ClF_3. \qquad (6\text{--}123)$$

Thermal synthesis of ClF_5 requires high temperatures and pressures of 30 atm and higher.

Finally, it is interesting to point out the synthesis of O_2BF_4 from a mixture of $O_2–F_2–BF_3$. The process yield at temperature $-196°C$ is high and characterized by the G-factor $G = 26 \, mol/100 \, eV$; energy cost of the synthesis is relatively low, only about 4 eV per molecule.

6.8. PLASMA-CHEMICAL SYNTHESIS OF HYDRAZINE (N_2H_4), AMMONIA (NH_3), NITRIDES OF PHOSPHORUS, AND SOME OTHER INORGANIC COMPOUNDS

6.8.1. Direct Plasma-Chemical Hydrazine (N_2H_4) Synthesis from Nitrogen and Hydrogen in Non-Equilibrium Discharges

Direct synthesis of hydrazine (N_2H_4) from nitrogen and hydrogen is attractive for chemical technology. It is a slightly endothermic process similar to the synthesis of ozone and krypton fluoride:

$$N_2 + 2H_2 \rightarrow N_2H_4, \quad \Delta H = 1.1 \, eV/mol. \qquad (6\text{--}124)$$

Application of synthetic process (6–124) is limited by the low stability of anhydrous hydrazine. Electric sparking of hydrazine vapor at $100°C$ results in explosive decomposition of N_2H_4 with the formation of ammonia, nitrogen, and hydrogen. Thus, somewhat similarly to the synthesis of ozone, compounds of noble gases, and aggressive fluorine oxidizers, the application of non-thermal plasma for direct N_2H_4 production is promising because it stimulates the process through high electron temperature and, at the same time, stabilizes products due to low gas temperature. An important feature of plasma-chemical process (6–124) is related to the fact that both molecules participating in the synthesis (nitrogen and hydrogen) are very effective for vibrational excitation in non-equilibrium discharges (see Chapters 2 and 3; for more details see Fridman & Kennedy, 2004). For this reason, the most detailed experiments for direct N_2H_4 synthesis were carried out in discharges that were especially effective in selective vibrational excitation of molecular nitrogen and hydrogen. A good example of such a non-equilibrium discharge with effective vibrational excitation is non-self-sustained discharge supported by EBs.

6.8. Plasma-Chemical Synthesis of Hydrazine (N₂H₄), Ammonia (NH₃)

Figure 6–41. Hydrazine yield as function of pressure: (1) discharge in N₂–H₂ mixture; (2) discharge in nitrogen, followed by mixing with hydrogen. Gas velocity in discharge zone is 10 m/s.

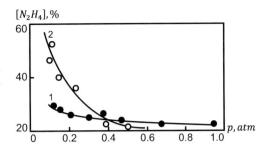

6.8.2. Hydrazine (N₂H₄) Synthesis from N₂–H₂ Mixture in Non-Self-Sustained Stationary Discharge Supported by Electron Beam

Hydrazine synthesis from N_2 and H_2 in a non-self-sustained EB-supported discharge was investigated by Koval, Petrishchev, and Khmelevoy (1979). In these experiments, the EB current density was $1–40\ \mu\text{A/cm}^2$, the electron beam diameter was about 1 cm, and the reactor chamber volume was about 3 cm³. The voltage applied to the discharge was about 6 kV, the gas pressure varied in the range 0.05–1 atm, the gas velocity was in the range 0.5–40 m/s, and the non-self-sustained discharge current density was $1–100\ \text{mA/cm}^2$. Products generated in the plasma-chemical process were N_2H_4 as well as ammonia (NH_3). The specific energy input in the discharge at atmospheric pressure was up to 0.1 J/cm³. The reduced electric field was relatively low ($E/p \leq 3$ V/cm · Torr). A significant increase of discharge current and reduced electric field E/p occurs at lower gas pressures. The dependence of hydrazine yield on gas pressure at different gas velocity and for different means of supplying hydrogen and nitrogen into the reactor is presented in Fig. 6–41. The hydrazine yield always increases with decreasing pressure. Increased hydrazine yield is related to the growth of the reduced electric field E/p and the specific energy input in the discharge at lower gas pressures. The hydrazine yield increases significantly when nitrogen alone is excited in the discharge and then subsequently mixed with hydrogen. Taking into account the high efficiency of vibrational excitation of pure nitrogen, the aforementioned effect confirms the essential contribution of vibrationally excited nitrogen molecules for hydrazine synthesis in the non-self-sustained discharge.

6.8.3. Kinetics of Hydrazine (N₂H₄) Synthesis from N₂–H₂ Mixture in Non-Thermal Plasma Conditions

The kinetic modeling of N_2H_4 synthesis from N_2–H_2 mixture in the non-self-sustained discharge supported by an EB was accomplished by Diskina, Koval, and Petrishchev (1979). Stimulation of the synthesis within the discharge is mostly due to effective vibrational excitation of nitrogen and hydrogen molecules, leading to their dissociation. Hydrazine synthesis was considered in two cases: directly in the N_2–H_2 discharge, and through N_2 vibrational excitation and dissociation in the non-equilibrium discharge with a subsequent admixture of molecular hydrogen (see Figs. 6–42 and 6–43). The case when the process starts with plasma dissociation of molecular nitrogen only is illustrated in Fig. 6–42. As one can see from the figure, atomic nitrogen is quickly converted in this case into NH_2 radicals after mixing with molecular hydrogen. When the NH_2 radicals are sufficiently accumulated, their recombination and disproportioning lead to the formation of NH_3 and N_2H_4 with comparable production rates:

$$NH_2 + NH_2 \rightarrow NH_3 + NH, \qquad (6–125)$$

$$NH_2 + NH_2 + M \rightarrow N_2H_4 + M. \qquad (6–126)$$

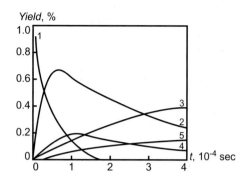

Figure 6–42. Kinetic curves of plasma-chemical synthesis in N_2–H_2 mixture (1:1) at atmospheric pressure. Concentrations: (1) atomic nitrogen; (2) NH_2 radicals; (3) hydrazine (N_2H_4); (4) NH radicals ([NH] $\times$ 10); (5) NH_3 (ammonia).

In the case of the discharge in the N_2–H_2 mixture, the effective direct dissociation of hydrogen essentially affects the plasma-chemical synthesis. When the initial dissociation degree of hydrogen in the discharge is 1%, the yield of ammonia exceeds the yield of hydrazine by about 10-fold. The effect is mostly due to recombination of NH_2 radicals with atomic hydrogen:

$$NH_2 + H + M \rightarrow NH_3 + M. \tag{6–127}$$

Kinetic competition of the three-body recombination processes (6–126) and (6–127) requires about 10-fold excess nitrogen over hydrogen atoms for effective hydrazine synthesis. This effect can be clearly seen from the kinetic curves presented in Fig. 6–43. Atomic hydrogen also shifts the yield of the process in the direction of ammonia because of the reaction sequence:

$$H + N_2H_4 \rightarrow N_2H_3 + H_2, \tag{6–128}$$

$$N_2H_3 + N_2H_3 \rightarrow NH_3 + NH_3 + N_2. \tag{6–129}$$

The contribution of reaction (6–128) is especially important when generation of H atoms in the discharge is delayed with respect to the formation of N atoms. Such a kinetic delay can be due to a more effective N_2 dissociation through vibrational excitation and contribution of dissociative attachment and electronic excitation to dissociation of hydrogen molecules.

6.8.4. Synthesis of Ammonia in DBD and Glow Discharges

Experiments with the plasma-chemical synthesis of ammonia from nitrogen and hydrogen in DBD and glow discharges are described by Shekhter (1935). Experiments with glow discharge were carried out at pressures of 1–4 Torr; experiments with a DBD were carried out at atmospheric pressure. Ammonia was a major product of the synthetic processes, and it was collected in these experiments by freezing the discharge walls using liquid air. A very high conversion degree of nitrogen and hydrogen into ammonia (close to 100%)

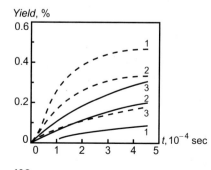

Figure 6–43. Product yield of plasma-chemical process in N_2–H_2 mixture at different initial concentrations of atomic hydrogen: (1) $[H]_0 = 1\%$ ($[N]_0 = [H]_0$); (2) $[H]_0 = 0.5\%$, ($[N]_0 = 2[H]_0$); (3) $[H]_0 = 0.1\%$ ($[N]_0 = 10[H]_0$). Solid curves, N_2H_4 (hydrazine); dashed curves, ammonia (NH_3).

was achieved in the experiments. The energy cost of the produced ammonia was, however, very high (100 kWh/g). Analysis of the mechanism and kinetics of the ammonia synthesis in DBD and glow discharges shows that the major contribution in the synthesis is from ion–molecular reactions, which explain the low energy efficiency of the plasma-chemical process. The energy cost of the plasma synthesis is about 100-fold higher than that of conventional catalytic technology, which makes the approach unattractive for large-scale application.

6.8.5. Plasma-Chemical Synthesis of Nitrides of Phosphorus

The synthesis of nitrides of phosphorus can be effectively conducted in non-thermal plasma by surface reactions of vibrationally excited nitrogen molecules and nitrogen atoms. Experiments with direct PN synthesis from nitrogen and phosphorus in glow discharge were carried out by Miller, Polak, and Raitsis (1980) and Raitsis, Raitsis, and Miller (1976). A major contribution toward production of the PN-molecular groups on the surface of phosphorus in N_2 glow discharge is due to the reaction

$$N + P_4 \rightarrow PN + P_3, \quad \Delta H = 0.92 \, \text{eV/mol}. \tag{6–130}$$

A significant contribution to the plasma-chemical synthesis of nitrides of phosphorus is provided by direct surface interaction of vibrationally excited N_2 molecules with P_4:

$$N_2\left(X^1\Sigma_g^+, v\right) + P_4 \rightarrow PN + P_3 + N, \quad \Delta E_v \geq 8 \, \text{eV/mol}, \tag{6–131}$$

$$N_2\left(X^1\Sigma_g^+, v\right) + P_4 \rightarrow PN + PN + P_2, \quad \Delta E_v \geq 2.9 \, \text{eV/mol}. \tag{6–132}$$

The vibrational energy ΔE_v required for reactions (6–131) and (6–132) is lower than that necessary for dissociation and direct formation of atomic nitrogen. The energy cost of PN synthesis in glow discharge is relatively high mostly because of the low efficiency of vibrational excitation and N_2 dissociation in these systems. Also the efficiency of the synthesis is limited in glow discharge systems due to process auto-inhibition provided by a monolayer of produced nitrides.

A high energy efficiency of phosphorus nitrides synthesis was achieved in experiments with non-self-sustained discharges supported by pulsed EBs (Kurenkov et al., 1979). The experiments were carried out at atmospheric pressure, gas temperature 418°C, specific energy input up to $0.4 \, \text{J/cm}^3$, reduced electric field $E/p = 5$–$7 \, \text{V/cm} \cdot \text{Torr}$, pulse duration $10 \, \mu\text{s}$, and plasma-chemical reactor volume $700 \, \text{cm}^3$. The experimental conditions were chosen to optimize the vibrational excitation of molecular nitrogen and their reactions with P_4. As a result, the energy cost of the nitrides PN in these experiments was relatively low: 6 eV per one PN group. Products of the plasma-chemical synthesis can be identified as polyphosphorus nitrides (P_6N_6). The plasma-generated polyphosphorus nitrides are characterized by high chemical and thermal stability, which permits their application for protective coating. The most important practical use of plasma-generated polyphosphorus nitrides is their application as "balastless" chemical fertilizers (Kurenkov et al., 1979).

6.8.6. Sulfur Gasification by Carbon Dioxide in Non-Thermal and Thermal Plasmas

Sulfur gasification by CO_2 leads to the formation of SO_2 and CO in the endothermic process

$$S_{(\text{solid})} + 2CO_2 \rightarrow 2CO + SO_2, \quad \Delta H = 1.4 \, \text{eV/CO mol}. \tag{6–133}$$

The energy efficiency of this plasma-chemical process can be high, because it can be effectively stimulated in non-equilibrium discharges by vibrational excitation of CO_2 molecules. A relatively high process efficiency was achieved in experiments with non-thermal microwave discharge at moderate pressures (Garsky, Givotov, & Ivanov, 1981).

Sulfur was supplied either as a powder or a vapor (in this case CO_2 was preheated to $450°C$), the gas pressure was 60–120 Torr, the CO_2 flow rate was 0.1–0.5 L/s, the sulfur supply rate was up to 0.2 g/s, and the power was about 1 kW. In the case of initially stoichiometric $S–CO_2$ mixture, the product mixture was also stoichiometric: $[SO_2]:[CO] = 1:2$. Molar fractions of oxygen and sulfur-organic compounds in the reaction products were negligible. A minimal energy cost of the process (6–133) with sulfur supplied as a vapor was 4.5 eV/CO mol. The optimal regime was achieved at a pressure of 100 Torr and a specific energy input of $6\,J/cm^3$ (1.2 eV/mol). The CO_2 dissociation degree in this regime was 24%. Lower energy costs of the process (6–133) were achieved when sulfur was supplied in the microwave discharge as powder. The minimal energy cost in this case was 3.4 eV/CO mol at pressure 80 Torr, and specific energy input in the discharge was 0.6 eV/mol.

A detailed kinetic analysis of the plasma-chemical process (6–133) (Levitsky et al., 1983) shows that it can be provided by dissociation of vibrationally excited CO_2 molecules (see Section 5.2):

$$CO_2^* + CO_2 \rightarrow CO + O + CO_2. \qquad (6–134)$$

Atomic oxygen formed in (6–134) then reacts with sulfur in the form of S_2:

$$O + S_2 \rightarrow SO + S. \qquad (6–135)$$

This reaction is exothermic and fairly fast, which determines its domination over the reaction of atomic oxygen with carbon dioxide ($O + CO_2 \rightarrow CO + O_2$) and explains the absence of molecular oxygen in products. Atomic sulfur generated in (6–135) then reacts with a vibrationally excited CO_2 molecule:

$$S + CO_2^* \rightarrow SO + CO. \qquad (6–136)$$

Reactions (6–134)–(6–136) determine the mechanism for formation of CO and SO from S_2 and CO_2. Formation of SO_2 (a final product of (6–133)) is due to fast exothermic disproportioning of SO:

$$SO + SO \rightarrow SO_2 + S. \qquad (6–137)$$

Atomic sulfur formed in this reaction obviously continues reacting with vibrationally excited CO_2 molecules (6–136), which leads to the additional production of CO, SO, and SO_2. The total chemical formula describing non-equilibrium mechanism (6–134)–(6–137) can be summarized as

$$CO_2 + \frac{1}{4}S_2 \rightarrow CO + \frac{1}{2}SO_2, \quad \Delta H \approx 1\,eV/mol. \qquad (6–138)$$

The total process (6–138) implies double repetition of reaction (6–37) and triple repetition of reaction (6–136). An alternative mechanism to (6–134)–(6–137) that also can be

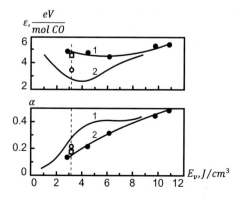

Figure 6–44. Energy cost (ε) and conversion degree (α) as functions of specific energy input for plasma-chemical process $S + CO_2 \rightarrow CO + SO_2$: (1) computer simulations; (2) experiments with gas-phase sulfur; $\circ$, experiment with sulfur powder; $\square$, CO_2 dissociation without sulfur.

6.8. Plasma-Chemical Synthesis of Hydrazine (N₂H₄), Ammonia (NH₃)

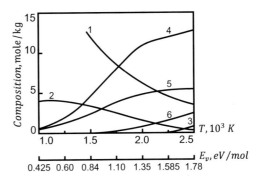

Figure 6–45. Plasma-chemical process in $S + CO_2$ mixture in atmospheric-pressure thermal plasma. Composition of products: (1) CO_2; (2) S_2; (3) S; (4) CO; (5) SO_2; (6) SO.

effectively stimulated by vibrational excitation of CO_2 molecules is related to the intermediate formation of S_2O radicals (Givotov et al., 1985a). The enthalpy of the total process (6–138) starting at higher temperatures with gaseous sulfur (S_2) is about 0.4 eV/mol less than that of (6–133) because of the enthalpy required for formation of S_2 molecules. While the enthalpy of process (6–138) is $\Delta H \approx 1$ eV/mol, the total vibrational energy required for the process is about 1.4 eV/mol, mostly because of exothermic losses in reactions (6–135) and (6–137). As a result, the chemical factor of the energy efficiency for process (6–138) of CO and SO_2 production is 72%. Kinetic modeling of non-equilibrium process (6–133) stimulated by vibrational excitation of CO_2 molecules (Levitsky et al., 1983) is in good agreement with experiments in non-thermal microwave plasma (Garsky et al., 1981), which is illustrated in Fig. 6–44.

Compare the results obtained in non-thermal discharges with those related to thermal plasma with different regimes of product quenching. The composition of products of process (6–133) carried out in thermal plasma is shown in Fig. 6–45 as a function of temperature in a high-temperature zone and with specific energy input (Nester et al., 1988). Next Fig. 6–46 shows the energy cost for production of one CO molecule in different regimes of the process organization. Curve 1 in the figure represents the thermal plasma process with ideal quenching of products. The regime of super-ideal quenching (curve 2) permits the additional use of vibrational energy accumulated in CO_2 molecules during quenching. Finally, for comparison, curve 3 represents the non-thermal performance of process (6–133) with its stimulation by vibrational excitation of CO_2 molecules (Levitsky, Macheret, & Fridman, 1983a,b; Levitsky et al., 1983c; Givotov et al., 1985a). The conversion degree of initial substances into products of process (6–133) is shown in Fig. 6–47 as a function of specific energy input. The thermal plasma process is presented with either absolute or ideal quenching, as well as with non-equilibrium quenching stimulated by vibrational energy of molecules.

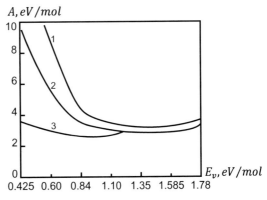

Figure 6–46. Plasma-chemical process in $S + CO_2$ mixture at atmospheric pressure. Energy cost of CO production as function of specific energy input: (1) ideal quenching in thermal plasma conditions; (2) super-ideal quenching in thermal plasma conditions; (3) non-equilibrium process stimulated by vibrational excitation of CO_2 molecules.

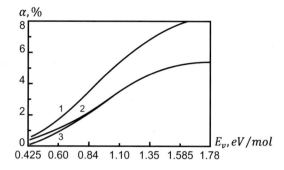

Figure 6–47. Plasma-chemical process in S + CO_2 mixture in atmospheric-pressure thermal plasma. Conversion degree of initial substances into CO as function of specific energy input: (1) absolute quenching; (2) ideal quenching; (3) super-ideal quenching.

6.8.7. CN and NO Synthesis in CO–N$_2$ Plasma

This endothermic plasma-chemical synthesis in CO–N$_2$ mixture can be presented as

$$CO + N_2 \rightarrow CN + NO, \quad \Delta H = 6 \, eV/mol. \tag{6–139}$$

The composition of products of thermal plasma process (6–134) organized in thermal plasma is shown in Fig. 6–48 as a function of temperature (Nester et al., 1988). Both initial molecules, CO and N_2, have high bonding energies (11 and 10 eV), which explains the high temperature (about 5000 K, see Fig. 6–48) required for their dissociation and initiation of synthesis (6–139). As a result, the quasi-equilibrium concentration of products is low in the high-temperature plasma zone and most of the CN and NO production takes place during the quenching phase. The energy costs of products and conversion degree in thermal plasma are shown in Figs. 6–49 and 6–50 for absolute, ideal, and super-ideal quenching modes. All the data are normalized to the production of a single CN molecule. Absolute quenching assumes conservation of CN molecules generated in the high-temperature zone; the minimal energy cost in this case is quite high (640 eV per CN molecule). Ideal quenching assumes additional CN formation due to recombination of carbon and nitrogen atoms during the cooling phase; the minimal energy cost in this case is lower: 27.6 eV per CN molecule. Super-ideal quenching takes place in the case of VT non-equilibrium during the cooling phase and permits the additional transfer of vibrational energy into the formation of CN molecules; the minimal energy cost in this case is the lowest: 12.9 eV per CN molecule.

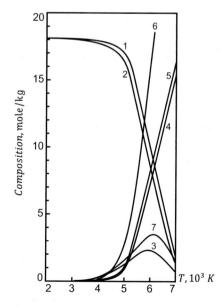

Figure 6–48. Synthesis of CN and NO in atmospheric-pressure thermal plasma. Composition of products: (1) CO; (2) N_2; (3) [CN] $\times$ 10; (4) C; (5) O; (6) N; (7) [NO] $\times$ 100.

6.8. Plasma-Chemical Synthesis of Hydrazine (N$_2$H$_4$), Ammonia (NH$_3$)

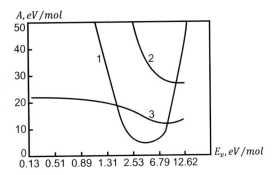

Figure 6–49. Synthesis of CN and NO in atmospheric-pressure thermal plasma. Energy cost of CN production as function of specific energy input: (1) absolute quenching ($A \times 100$); (2) ideal quenching; (3) super-ideal quenching.

6.8.8. Gas-Phase Synthesis Related to Plasma-Chemical Oxidation of HCl and SO$_2$

Oxidation of HCl is practically important for the production of molecular chlorine and has been investigated in both thermal and non-thermal plasma systems:

$$2\,HCl + \frac{1}{2}O_2 \rightarrow H_2O + Cl_2. \tag{6–140}$$

The maximum yield of Cl$_2$ achieved in atmospheric-pressure thermal plasma systems (air plasma jet) is 85% at temperatures of about 4000 K and the stoichiometric mixture composition of HCl to O$_2$ is 4:1. The energy cost of the product in this regime is 5.5 kWh/kgCl$_2$. At higher plasma temperature (4200 K), the yield reaches 96%; however, the energy cost becomes 14 kWh/kg Cl$_2$. Conversely, at lower plasma temperature (3000 K), the yield becomes lower (45%); however, the energy cost becomes much better – 3.4 kWh/kgCl$_2$ (Enduskin et al., 1969). Experiments with a water vapor thermal plasma jet (Margeloff, 1966) at lower plasma temperatures (1500–3000 K) result in a relatively low product yield of 45%, and the energy cost is lower than that for an air plasma jet: 3.2 kWh/kgCl$_2$.

The plasma-chemical process (6–140) was also investigated in DBDs and microwave discharges. Experiments with DBD (VanDrumpt, 1972) were carried out at atmospheric pressure and different HCl–O$_2$ mixture compositions. The maximum conversion degree in these experiments was 18% and was achieved for a stoichiometric mixture composition. The energy cost of the product in this regime was 10 kWh/kgCl$_2$. The conversion degree in the experiments with DBD increased linearly with increase of reduced electric field E/n_0 and electron temperature. The increase can be interpreted by assuming an essential contribution of electronic excitation into the Cl$_2$ synthesis. Plasma-chemical synthesis of chlorine by HCl oxidation was investigated in a non-equilibrium microwave discharge by Baddur and Dandus (1970). The experiments were carried out in stoichiometric HCl–O$_2$ mixture at pressures of 5–40 Torr, and the microwave discharge power was 200–350 W. An optimal regime for Cl$_2$ synthesis has been achieved both for a discharge power of 200 W at pressure 15 Torr and for a discharge power 340 W at pressure 20 Torr. This optimal regime permits

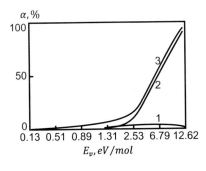

Figure 6–50. Synthesis of CN and NO in atmospheric-pressure thermal plasma. Conversion degree of initial substances into CN as function of specific energy input: (1) absolute quenching; (2) ideal quenching; (3) super-ideal quenching.

a conversion degree of 50% to be achieved and the energy cost of the products is only 20 kWh/kgCl$_2$, which is relatively high. The mechanism of the non-equilibrium process is mostly based on reactions of atomic oxygen and atomic hydrogen. Both the conversion degree of HCl into chlorine and the energy efficiency of the non-thermal process (6–140) are lower than those in thermal plasma.

Non-equilibrium plasma is effective in stimulating the exothermic process of SO$_2$ oxidation to SO$_3$:

$$SO_2 + \frac{1}{2}O_2 \rightarrow SO_3. \qquad (6\text{--}141)$$

This process in stoichiometric SO$_2$–O$_2$ mixture was investigated in strongly non-equilibrium plasma conditions of a microwave discharge (Baddur and Dandus, 1970), similar to the one mentioned earlier regarding Cl$_2$ synthesis. Optimal SO$_3$ synthesis was achieved in these experiments at a pressure of 10 Torr and a microwave discharge power of 250 W. The high efficiency of the process (6–141) was achieved in atmospheric-pressure plasmas where SO$_2$ was only a small admixture to air (see Section 11.7).

PROBLEMS AND CONCEPT QUESTIONS

6–1. NO Synthesis in Non-Thermal Plasma Provided by Positive Ions and Electronically Excited Molecules. Estimate the energy cost of NO molecule formation in air at non-thermal plasma conditions produced by each of the three following processes: (1) dissociation of molecular nitrogen through its electronic excitation by direct electron impact; (2) electron–ion recombination of positive molecular nitrogen ion; and (3) ion–molecular reactions of O$^+$, O$_2^+$, and N$_2^+$.

6–2. Kinetics of Different Mechanisms of NO Synthesis in Non-Equilibrium Plasma Conditions. Based on (6–10) estimate the electron temperatures and plasma ionization degrees required for the Zeldovich mechanism of NO synthesis, (6–2) and (6–3), to be kinetically faster than alternative mechanisms related to the contribution of positive ions and electronically excited atoms and molecules.

6–3. Adiabatic Channel of Elementary Reaction of Vibrationally Excited Nitrogen with Atomic Oxygen: Limiting Zeldovich Mechanism of NO Synthesis. At $\hbar\omega v \geq E_a$ and $T_0 = 1000$ K, compare stochastic and deterministic formulas (6–13) and (6–15) for the probability of the adiabatic channel of reaction (6–2) limiting the Zeldovich mechanism of NO synthesis in plasma. Explain the difference in temperature dependences of the probabilities for the stochastic and deterministic approaches.

6–4. Formation of Intermediate N$_2$O*($^1\Sigma^+$) Complex in Electronically Non-Adiabatic Channel of NO Synthesis. Estimate the maximum gas temperature T_0, when the formation of the intermediate N$_2$O*($^1\Sigma^+$) complex proceeds as a multi-quantum transition and is determined by the vibrational energy of N$_2$ molecules. Use relations (6–25) and (6–26) and estimate the probability of the multi-quantum transition. Explain why this mechanism is a preferred one in cold plasma-chemical systems.

6.5. Non-Adiabatic Channel of Elementary Reaction of Vibrationally Excited Nitrogen with Atomic Oxygen: Limiting Zeldovich Mechanism of NO Synthesis. Compare probabilities (6–32) and (6–15) of non-adiabatic and adiabatic channels of the reaction O + N$_2 \rightarrow$ NO + N, stimulated by vibrational excitation. Calculate the critical gas temperature T_0 when the contributions of the channels in kinetics of NO synthesis are similar. How does the critical temperature depend on N$_2$ vibrational energy?

6–6. Rate Coefficient of Reaction $O + N_2 \rightarrow NO + N$ Stimulated by Vibrational Excitation. Compare the logarithmic sensitivity of the rate coefficient (6–34) to the vibrational $(\partial \ln k_R / \partial \ln T_v)$ and translational $(\partial \ln k_R / \partial \ln T_0)$ temperatures. Find out the range of parameters of the plasma system when the reaction stimulated by vibrational temperature is more sensitive to translational temperature.

6–7. Zeldovich Chain Mechanism of Plasma-Chemical NO Synthesis. Based on expressions (6–39) for the stationary concentration of atomic oxygen and (6–34) for the rate coefficient of limiting reaction of chain propagation, derive a formula for calculating the chain length for the Zeldovich mechanism of NO synthesis. Analyze the dependence of the chain length on electron temperature T_e, vibrational temperature T_v, and translational temperature.

6–8. Plasma-Chemical NO Synthesis from Air Outside of the Active Discharge Zone (in Passive Zone). NO synthesis stimulated by vibrational excitation continues outside of the active discharge zone, using previously accumulated vibrational energy. The effective chemical process continues in the passive zone during a time interval τ_p until the vibrational temperature from an initial value T_v^{st} decreases to the critical value T_v^{min}. Based on (6–45), derive relation (6–46) for τ_p.

6–9. Chemical Factor of Energy Efficiency of NO Synthesis in Non-Thermal Air Plasma. Calculate the vibrational energy losses related to non-resonance VV exchange and VT relaxation from highly vibrationally excited N_2 molecules during NO synthesis stimulated by vibrational excitation of nitrogen molecules in non-thermal air plasma. Demonstrate that these losses can be attributed to a single act of chemical reaction of synthesis and, therefore, can be included in the chemical factor.

6–10. Two-Temperature Quasi-Equilibrium Regime of Plasma-Chemical NO Synthesis. Direct reaction (6–59) consumes vibrational energy of molecular nitrogen, whereas the reverse reaction consumes translational energy. For this reason the pair of two-temperature quasi-equilibrium reactions (6–59), similarly to VT relaxation, leads also to intensive energy transfer from vibrational to translational degrees of freedom. Compare the rates of vibrational to translational energy transfer in these two mechanisms at conventional conditions of plasma-chemical NO synthesis.

6.11. Zeldovich Equation for NO Synthesis in Quasi-Equilibrium Thermal Systems. Using the Zeldovich equation (6–62), calculate the time required for reaching quasi-equilibrium concentration of nitrogen oxides (NO_x) in stoichiometric N_2–O_2 atmospheric-pressure mixture at different thermal plasma temperatures: 3000, 3500, and 4000 K. Analyzing activation energies in the Zeldovich equation, find out the limiting reaction responsible for establishing of the quasi-equilibrium concentration of NO_x.

6–12. Plasma-Chemical Ozone Generation from Oxygen. Comparing the rate coefficients of ozone generation (6–66) and atomic oxygen recombination (6–67), calculate the upper limit of relative concentration of atomic oxygen $[O]/[O_2]$ acceptable for effective ozone synthesis. Analyze the effect of gas temperature and influence of the reverse reaction (6–68) on the derived limit of discharge strength.

6–13. Plasma-Chemical Ozone Generation from Air. Calculate the contribution of nitrogen atoms and electronically excited N_2 molecules in the energy efficiency of ozone synthesis, assuming that all nitrogen atoms are converted to oxygen atoms in reactions (6–70) and (6–71), and that half of the O_3 is generated via indirect processes (6–70)–(6–73). Use typical G-factor values for production of oxygen and nitrogen atoms presented in Figs. 6–25 and 6–26.

6–14. Discharge Poisoning Effect. Comparing the rate of oxygen recombination (6–66) leading to the formation of O_3 with rates of oxygen reactions with NO and NO_2, (6–74) and

(6–75), estimate the concentration of nitrogen oxides, which results in the discharge poisoning effect, that is in the inhibition of ozone generation. Which nitrogen oxide is more effective in suppression of ozone synthesis in non-thermal plasma?

6–15. Temperature Effect on Ozone Stability. Based on enthalpy released during the ozone decomposition (6–63), calculate the relative O_3 concentration in air that can lead to a doubling of initial room temperature of the air (heating to a temperature of about 600 K). Estimate the characteristic time of ozone decomposition (6–81) at this relative ozone concentration in air, and at this temperature (about 600 K).

6–16. Negative Effect of Humidity on Ozone Generation in DBD. Explain the negative effect of humidity on O_3 synthesis related to water enhancement of surface conductivity of dielectric barriers. Why does humidity influence surface conductivity? Why does the elevated surface conductivity make microdischarges stronger? Why does making the microdischarges stronger lead to a reduction of O_3 synthesis efficiency? Compare the humidity-related inhibition of O_3 synthesis through an increase of the barrier surface conductivity with kinetic inhibition due to OH and HO_2 radicals.

6–17. Positive Effect of N_2 and CO Admixtures to Oxygen on Plasma-Chemical Ozone Synthesis. Taking into account the mechanism of CO oxidation chain reaction with branching (5–88), explain the positive effect of CO addition on ozone synthesis from oxygen. Why do small N_2 and CO admixtures have a positive effect on ozone generation, whereas a more significant addition of these molecular gases leads to a decrease in efficiency of the synthesis?

6–18. Plasma-Chemical KrF_2 Synthesis with Product Stabilization in a Krypton Matrix. Solve the diffusion equation with chemical reaction (6–101) assuming the boundary conditions (6–102) and determine (1) the space distribution of fluorine groups KrF across the krypton film (6–103), (2) the depth of the synthesized KrF_2 layer (6–104), and (3) the concentration of intermediate KrF in the treated layer (6–105).

6–19. Synthesis of Ammonia and Hydrazine from N_2–H_2 Mixtures in Non-Thermal Plasma. Why does plasma-chemical synthesis in N_2–H_2 mixtures mostly lead to the formation of ammonia and not hydrazine? What is the role of atomic hydrogen in the competition of two channels of synthesis in N_2–H_2 mixtures? Which plasma-chemical technology is the most effective for fixing nitrogen: the synthesis of nitrogen hydrides or nitrogen oxides?

7

Plasma Synthesis, Treatment, and Processing of Inorganic Materials, and Plasma Metallurgy

7.1. PLASMA REDUCTION OF OXIDES OF METALS AND OTHER ELEMENTS

Metallurgical processes usually require high temperatures, which attracts attention to the application of plasma. High plasma temperatures permit one to intensify metallurgical processes and make metallurgical equipment smaller. An important advantage of plasma metallurgy is the possibility of conducting one-stage or direct processes, which is especially important for reduction of metals from their oxides, because conventional approaches include multiple stages. Plasma reduction of oxides is usually carried out in thermal discharges; however, non-thermal plasmas are also the focus of investigation.

7.1.1. Thermal Plasma Reduction of Iron Ore, Iron Production from Oxides Using Hydrogen and Hydrocarbons, and Plasma-Chemical Steel Manufacturing

The conventional industrial blast furnace process of iron production from its oxides (Fe_2O_3, Fe_3O_4) with the manufacturing of steel includes multiple stages and uses large-scale equipment, which is illustrated in Fig. 7–1. Iron oxides are supplied for reduction in a blast furnace in the form of ore, which requires preliminary agglomeration. Coke is burned as a fuel in a coke furnace to heat the blast furnace and produce CO, which reacts in the blast furnace with metal oxides to produce iron in the following high-temperature reaction:

$$Fe_2O_3 + 3CO \rightarrow 3CO_2 + 2Fe, \quad \Delta H = -0.28\,eV/mol. \tag{7–1}$$

Disproportioning of carbon monoxide (see, for example, Section 5.7.2) leads to the formation of carbon mixed with the iron (7–1). Iron produced in this way in the blast furnace is called pig-iron and requires decarbonization in a special converter to become steel. The application of thermal plasma discharges permits the whole steel manufacturing process to be accomplished directly and in one stage, which is illustrated in Fig. 7–2. Reduction of metal oxides can be provided in plasma by hydrogen or natural gas. High plasma temperatures and the application of hydrogen intensify the process and help to avoid carbon in products, which leads to one-stage steel production (Tsvetkov & Panfilov, 1980):

$$Fe_2O_3 + 3H_2 \rightarrow 2Fe + 3H_2O. \tag{7–2}$$

One-stage production of pure iron from oxides by a hydrogen-based reduction process in helium thermal discharge plasma was investigated by Gold et al. (1975). A detailed analysis of the reduction of iron ore with hydrogen in a direct-current plasma jet was carried out by Gilles and Clump (1970). It was shown there that heat transfer to oxide particulates is a key factor determining the kinetics of the reduction process. Plasma-metallurgical reduction of

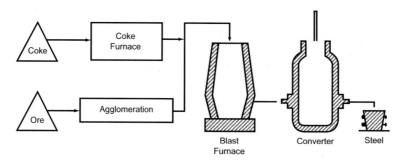

Figure 7–1. Schematic of blast furnace technology of iron production by reduction of its oxides.

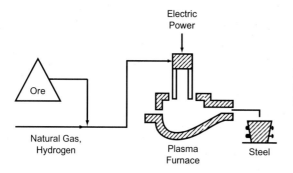

Figure 7–2. Schematic of plasma-metallurgical process of iron production by reduction of its oxides.

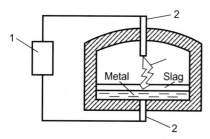

Figure 7–3. Plasma furnace for iron reduction from oxides. Reduction in slag zone by thermal arc electrically attached to the melt: (1) plasma generator, (2) electrodes, (3) arc.

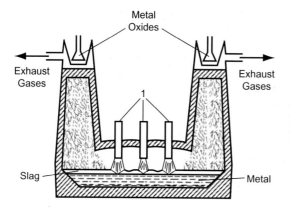

Figure 7–4. Plasma furnace for iron reduction from oxides. Treatment of the oxide melt by gas energy carrier generated by the three-phase thermal arc plasmatron: (1) three-phase thermal arc plasmatron.

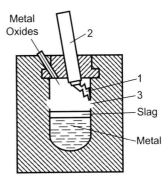

Figure 7–5. Iron reduction from oxides in a plasma furnace with arc rotating between cathode and stationary circular anode: (1) arc, (2) cathode, (3) circular anode.

iron ore can be organized in different configurations. Figure 7–3 illustrates the process of iron production from oxides (Daith & Marcio, 1967), where the thermal arc directly contacts the melt and reduction is provided by hydrogen or hydrocarbons. Another scheme (Fig. 7–4) uses iron ore preheating by the reduction process exhaust gases. In this case, the ore reduction process is stimulated in slag by reduction gases heated up in a three-phase arc plasmatron. An arc, illustrated in Fig. 7–5, is produced between a cathode and a stationary ring anode. The arc moves along the ring anode and forms a plasma cone. Particulates of iron oxide and reduction gases cross the plasma cone, where ore reduction takes place (Tylko, 1973, 1974). Another arc discharge, shown in Fig. 7–6, is produced between two cylindrical electrodes. The arc rotates between the cylindrical electrodes, and hydrogen is supplied into the reactor, also between the electrodes, and pushes the rotating arc inside of the reactor. Oxides pass through the reactor along the cylinders with sufficient time for ore reduction and metal production. Most such thermal plasma systems applied for ore reduction operate at atmospheric pressure.

7.1.2. Productivity and Energy Efficiency of Thermal Plasma Reduction of Iron Ore

Electric power and, therefore, productivity of plasma-metallurgical furnaces can be high, about 1 MW (Gold et al., 1975). Arc discharges in this case are usually equipped with copper anodes and tungsten or copper cathodes, and the arcs are stabilized by magnetic forces or by special gas flows (Fey & Kemeny, 1973). Pure hydrogen, natural gas, or their mixtures are used as reduction agents and heat carriers. The thermal efficiency of the plasma furnaces with respect to heat losses can be high – up to 90%. The theoretical minimum energy cost for iron production from magnetite (Fe_3O_4) and hematite (Fe_2O_3) concentrates is 2.21 kWh/kg Fe. Experiments with 100 kW arc, using a hydrogen – natural gas (80%–20%) mixture as a heat carrier and reduction agent, result in single-stage pure iron production with an energy cost of 4.8–5.9 kWh/kg Fe. Similar experiments with a 10-fold more powerful arc (1 MW) result in a decrease of the process energy cost under the same conditions to 4.2–5.2 kWh/kg Fe. The best energy efficiency in these experiments was achieved by using a reduction mixture with higher molar fraction of natural gas ($H_2:CH_4 = 2:1$). The energy cost of pure iron production in this case was 3.3 kWh/kg Fe. Higher energy efficiency in

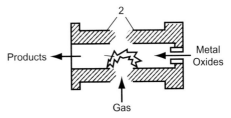

Figure 7–6. Plasma furnace for iron reduction from oxides by injecting them into rotating and expanding DC arc: (1) arc, (2) electrodes.

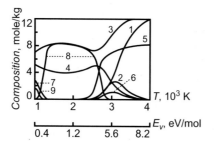

Figure 7–7. Iron reduction from Fe_2O_3 with CO in atmospheric-pressure thermal plasma. Composition of products: (1) O; (2) O_2; (3) CO; (4) CO_2; (5) Fe; (6) FeO; (7) Fe (condensed); (8) FeO (condensed); (9) F_3O_4 (condensed).

the experiments with more powerful arcs is mostly due to an increase of thermal efficiency of the reactor (in these experiments, from 72 to 84%) and better efficiency of mixing and decreased sizes of particulates of magnetite (Fe_3O_4) and hematite (Fe_2O_3). The significant influence of particle sizes of the hematite (Fe_2O_3) concentrate on the energy efficiency of iron production can be illustrated by a simple example. An increase of the fraction of hematite particles smaller than 37 μm from 2 to 45% results in decrease of energy cost from 4.2 to 2.9 kWh/kg Fe. A high-efficiency plasma reduction of iron ore was demonstrated by Gold et al. (1975) using a powerful 1 MW thermal arc discharge and hydrogen–natural gas mixture (H_2:CH_4 = 2:1). The productivity of the plasma furnace was quite high (318 kg/h of pure iron); the energy cost was 2.65 kWh/kg Fe, which corresponds to an energy efficiency of 83%. Iron produced in the described plasma furnaces from low sulfur and phosphorus hematite–magnetite concentrate is quite pure and contains only 50 ppm S (sulfur), 10 ppm P (phosphorus), 60 ppm C (carbon), 60 ppm Si (silicon), and 70 ppm Cu (copper).

The aforementioned iron ore reduction in plasma implies the application of hydrogen and natural gas as reducing agents to minimize carbon admixtures in the product. In principal, carbon monoxide can be used for plasma reduction (7–1) as well. A quasi-equilibrium product composition in the high-temperature zone of thermal plasma discharge at atmospheric pressure and energy cost of iron production in such a case is shown in Figs. 7–7 and 7–8 as functions of temperature and specific energy input in thermal discharge, respectively (Nester et al., 1988). The initial composition in thermal plasma process (7–1) is close to being stoichiometric: [Fe_2O_3] = 4.11 mol/kg, [CO] = 12.29 mol/kg. The minimal energy cost of iron production (7–1) in the case of absolute quenching is 13.2 eV/mol Fe and can be achieved at a specific energy input of 5.6 eV/mol and conversion degree of 85%. A more than two-fold-lower energy cost of iron production (7–1) can be achieved in the case of super-ideal quenching (see Fig. 7–8). Super-ideal quenching requires centrifugal separation of the products in this process (see Section 10.7.5).

7.1.3. Hydrogen Reduction of Refractory Metal Oxides in Thermal Plasma and Plasma Metallurgy of Tungsten and Molybdenum

Hydrogen reduction of tungsten oxide (as well as oxides of other refractory metals) proceeds in thermal plasma at high-temperatures when molecular products of the reduction process are not stable. Thermodynamic and kinetic analysis shows that the high-temperature thermal plasma reduction processes follow the so-called **dissociative reduction mechanism**

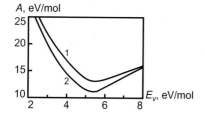

Figure 7–8. Iron reduction from Fe_2O_3 with CO in atmospheric-pressure thermal plasma. Energy cost of iron production as function of specific energy input: (1) absolute quenching, (2) super-ideal quenching.

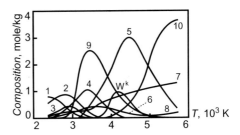

Figure 7–9. Tungsten reduction from WO_3 by direct decomposition in atmospheric-pressure thermal plasma. Composition of products: (1) W_3O_9; (2) W_2O_6; (3) W_3O_8; (4) WO_3; (5) WO; (6) O_2; (7) $[O] \times 10$; (8) W^+; (9) WO_2; (10) W.

(Chizhikov, Tsvetkov, & Tagirov, 1970). According to the dissociative reduction mechanism, the reduction process starts with dissociation or atomization of the oxides with the formation of metal atoms and oxygen atoms. Then, during the quenching stage, oxygen atoms recombine with reduction agents, saving metal atoms from reverse reactions and providing effective metal production. The effective interaction of hydrogen with metal oxides is a serious engineering challenge in plasma-metallurgical systems. Some plasma-metallurgical processes can be performed using systems similar to those considered earlier regarding plasma reduction of iron ore. In general, plasma reduction of metal oxides with a melting temperature below the boiling temperature of metal can be performed in so-called melting furnaces (Stewart, 1967; Rains & Kadlek, 1970). Some reduction processes, in particular tungsten production, can be done by treating oxides in plasma jets to produce metallic powder, and in other configurations of a plasma reactor (Tsvetkov & Panfilov, 1980; Tumanov, 1981). The reduction of tungsten oxide with hydrogen to produce metallic tungsten can be represented as

$$WO_3 + 3H_2 \rightarrow W + 3H_2O. \tag{7-3}$$

The quasi-equilibrium composition of products of thermal WO_3 dissociation at atmospheric pressure is shown in Fig. 7–9 as a function of temperature. The dissociative reduction mechanism means in this case that highly energy intensive dissociation of tungsten oxide (WO_3) and formation of atomic tungsten (W) ($WO_3 \rightarrow W + (3/2)O_2$, $\Delta H = 8.73$ eV/mol) takes place at high temperatures (about 5000 K) where oxygen and water are completely atomized. So water formation (7–3) takes place only at the quenching stage of the plasma-metallurgical process. The energy cost of tungsten production from tungsten oxide WO_3 in thermal plasma is shown in Fig. 7–10 for atmospheric pressure as a function of specific energy input for different quenching modes. The minimal energy cost in the absolute quenching mode is 35 eV per W atom, which can be achieved at a specific energy input of 30.5 eV/mol. Ideal quenching assumes in this process that all intermediate tungsten oxides participate in disproportioning reactions during the fast cooling phase, forming atomic tungsten (W) and WO_3. The minimal energy cost of metal production in the case of ideal quenching is 30.6 eV per W atom. Super-ideal quenching in this process requires centrifugal separation of products (see Section 10.7.5). The centrifugal effect permits one to decrease the energy cost to 19 eV per W atom at a plasma temperature of 4500 K (Nester et al., 1988).

Figure 7–10. Tungsten reduction from WO_3 by direct decomposition in atmospheric-pressure thermal plasma. Energy cost of tungsten production as function of specific energy input: (1) absolute quenching; (2) ideal quenching; (3) super-ideal quenching.

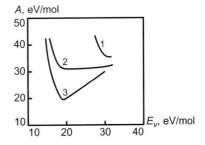

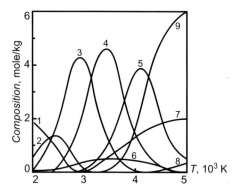

Figure 7–11. Molybdenum reduction from MoO_3 by direct decomposition in atmospheric-pressure thermal plasma. Composition of products: (1) Mo_3O_9; (2) Mo_2O_6; (3) MoO_3; (4) MoO_2; (6) O_2; (7) [O] × 0.1; (8) Mo^+; (9) Mo.

Between the different experimental arc discharge systems applied for effective hydrogen reduction of WO_3 and production of tungsten powder, we can mention one developed by Tsvetkov, Panfilov, & Kliachko (1976). The thermal arc discharges were arranged in these experiments as hydrogen plasma jets with powers of 150 and 250 kW. The reduction degree from WO_3 to tungsten powder using hydrogen as a reduction agent was 95%; using methane gave a value of 60% (Stokes, 1971). The tungsten powder produced was characterized by a specific surface area of 6–8 m^2/g. A similar process developed for thermal plasma production of metallic powders of titanium (Ti), vanadium (V), and zirconium (Zr) by reduction of their oxides in a hydrogen arc jet are described by Chizhikov et al. (1968).

Another example of H_2 plasma reduction of oxides of refractory metals is the production of Mo:

$$MoO_3 + 3H_2 \rightarrow Mo + 3H_2O. \qquad (7\text{–}4)$$

The characteristics of molybdenum production are similar to those of tungsten (7–3). The quasi-equilibrium composition of products of thermal MoO_3 dissociation at atmospheric pressure is shown in Fig. 7–11 as a function of plasma temperature. MoO_3 is characterized by a high enthalpy of decomposition into atomic metal and molecular oxygen: $MoO_3 \rightarrow Mo + (3/2)O_2$, $\Delta H = 7.7$ eV/mol. The dissociative reduction mechanism in this case means that dissociation of MoO_3 and formation of Mo atoms requires a high temperature exceeding 4500 K when oxygen and water are completely atomized. The energy cost of Mo production from MoO_3 in thermal plasma at atmospheric pressure is shown in Fig. 7–12 as a function of specific energy input. The minimal energy cost in the absolute quenching mode is 28 eV per Mo atom at a specific energy input of 25 eV/mol. Ideal quenching assumes that all intermediate oxides disproportionate during the quenching into Mo and MoO_3. The minimal energy cost in this case is 25.4 eV per Mo atom and can be achieved at a specific energy input of 22.8 eV/mol. Super-ideal quenching assumes a centrifugal separation of the products, which permits one to decrease the energy cost to 17.3 eV per Mo atom; this can be achieved at a plasma temperature of 3900 K (Nester et al., 1988). Effective plasma-metallurgical

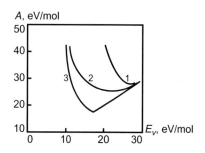

Figure 7–12. Molybdenum reduction from MoO_3 by direct decomposition in atmospheric-pressure thermal plasma. Energy cost of molybdenum production as function of specific energy input: (1) absolute quenching; (2) ideal quenching; (3) super-ideal quenching.

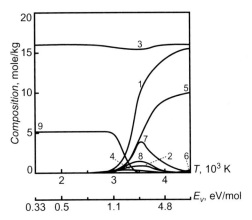

Figure 7–13. Aluminum reduction from Al_2O_3 with CO in atmospheric-pressure thermal plasma. Composition of products: (1) O; (2) O_2; (3) CO; (4) CO_2; (5) Al; (6) Al^+; (7) AlO; (8) Al_2O; (9) Al_2O_3 (condensed).

production of molybdenum (Mo) from molybdenum oxide (MoO_3) was experimentally demonstrated by Bolotov, Isikov, & Filkov (1976b). The experiments were carried out in a high-temperature fluidized-bed reactor using hydrogen as a reduction agent and energy carrier. The hydrogen was heated up in arc to temperatures required for effective oxide reduction. Reduction of tungsten and molybdenum can be also organized in thermal plasma of nitrogen with admixtures of hydrogen and different hydrocarbons (Kornilov, Lamikhov, & Barashkina, 1976).

7.1.4. Thermal Plasma Reduction of Oxides of Aluminum and Other Inorganic Elements

Aluminum can be produced from alumina (Al_2O_3) using CO, CH_4, or hydrogen as reduction agents. Plasma jets are arranged based on atmospheric-pressure arcs (Stokes, 1971) as well as thermal atmospheric-pressure radiofrequency (RF) discharges (Rykalin & Sorokin, 1987). The reduction gaseous agents (CO, CH_4, or H_2) are in this case diluted by argon. The comparison of aluminum production from alumina simply by Al_2O_3 decomposition in pure Ar plasma with the actual reduction process of using plasma of argon mixed with the gaseous reduction agents (CO, CH_4, or H_2) will be analyzed in Section 7.2. The plasma-metallurgical process for reduction of alumina with carbon monoxide is a strongly endothermic process:

$$Al_2O_3 + 3CO \rightarrow 2Al + 3CO_2, \quad \Delta H = 8.57 \, eV/mol. \qquad (7\text{–}5)$$

The quasi-equilibrium composition of products of the atmospheric-pressure process is shown in Fig. 7–13 as a function of plasma temperature and specific energy input in the thermal discharge. The initial composition of the reaction mixture is $[Al_2O_3] = 5.38$ mol/kg and $[CO] = 16.12$ mole/kg. Alumina reduction takes place at high temperatures, but lower than those required for reduction of tungsten. In contrast to the dissociative reduction mechanism, aluminum atoms are produced at temperatures when CO is not yet dissociated. The energy cost of aluminum production from Al_2O_3 in atmospheric-pressure thermal plasma is shown in Fig. 7–14 as a function of specific energy input. The minimal

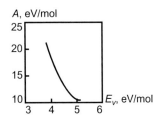

Figure 7–14. Aluminum reduction from Al_2O_3 with CO in atmospheric-pressure thermal plasma. Energy cost of aluminum production as function of specific energy input in the case of absolute quenching.

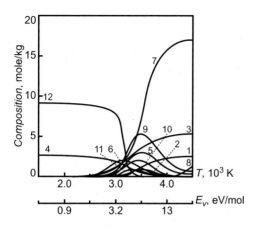

Figure 7–15. Aluminum reduction from Al_2O_3 with molecular hydrogen in atmospheric-pressure thermal plasma. Composition of products: (1) [O] × 0.1; (2) O_2; (3) [H] × 0.1; (4) [H_2] × 0.1; (5) OH; (6) H_2O; (7) Al; (8) Al^+; (9) AlO; (10) Al_2O; (11) Al_2O_3 (condensed).

energy cost of the process in absolute quenching mode is 10.6 eV per Al atom. This energy cost can be achieved at a specific energy input of 5.2 eV/mol with a conversion degree of 97%.

The plasma-metallurgical process of reduction of alumina with hydrogen requires a little less energy:

$$Al_2O_3 + 3H_2 \rightarrow 2Al + 3H_2O, \quad \Delta H = 7.51 \text{ eV/mol}. \tag{7–6}$$

The quasi-equilibrium composition of products is shown in Fig. 7–15 for the case of atmospheric pressure as a function of plasma temperature and specific energy input. The initial composition of the reaction mixture is [Al_2O_3] = 9.26 mol/kg and [H_2] = 16.12 mol/kg. Alumina reduction takes place at temperatures of about 3500 K. The energy cost in aluminum production by Al_2O_3 reduction in thermal plasma is shown in Fig. 7–16 as a function of specific energy input in different quenching modes. The minimal energy cost in the absolute quenching mode is 29.4 eV per Al atom, which can be achieved at a specific energy input of 13.9 eV/mol with a conversion degree of 94%. Ideal quenching assumes disproportioning of AlO, AlO_2, and Al_2O during the fast cooling phase with formation of Al_2O_3 and Al. The minimal energy cost of aluminum production from alumina in the absolute quenching mode is 27.9 eV per Al atom.

Thermal plasmas have been applied for oxide reduction of many other metals and non-metallic inorganic elements. Different reduction agents have been used in these processes. In experiments of Frolova, Minkin, and Frolov (1976), dispersed oxides of chromium (Cr), nickel (Ni), and silicon (Si) were reduced to produce the elements via the oxide treatment with reduction gases heated up in a thermal arc discharge. The process of Si reduction from oxides in thermal plasma was also considered by Kornilov, Lamikhov, and Glukhov (1976b) and Yukhimchuk and Shinka (1976). Plasma reduction of Ni, cobalt (Co), and tin (Sn) from their oxides using a hydrogen–nitrogen mixture heated up in thermal arc discharge was investigated by Bolotov et al. (1976) and Bobylev, Ivashov, and Krutiansky (1976). The

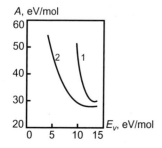

Figure 7–16. Aluminum reduction from Al_2O_3 with molecular hydrogen in atmospheric-pressure thermal plasma. Energy cost of aluminum production as function of specific energy: (1) absolute quenching; (2) ideal quenching.

plasma-metallurgical process of reducing tantalum (Ta) from its pentoxide (Ta_2O_5) was investigated in a plasma jet of helium with additions of hydrogen (Stokes, 1971). The conversion degree of tantalum pentoxide into pure metallic tantalum was about 50% in this system. Specific features of reduction of transition metal oxides in hydrogen plasma are also considered by Wertheimer and Bailon (1977).

7.1.5. Reduction of Metal Oxides and Production of Metals Using Non-Thermal Hydrogen Plasma

Although most experiments for hydrogen reduction of metal oxides were carried out in thermal plasmas, strongly non-equilibrium discharges were also investigated, both theoretically and experimentally. A major advantage of non-equilibrium plasma is the high energy efficiency for generation of atomic hydrogen, excited hydrogen molecules, and other active species stimulating the reduction process at low temperatures of gas and solid oxides. A major challenge of non-thermal plasma metallurgy is related to the kinetics of secondary surface reactions and the efficiency of desorption of gas-phase products (in particular water vapor) to provide an excess of active plasma species deeper in the solid oxide. Another challenge for non-thermal plasma metallurgy is discharge power and, hence, productivity. Atomic hydrogen generated in non-thermal plasma reaches the surface of a solid metal oxide, stimulating the "cold" reduction process:

$$Me_mO_k + 2k\,H \rightarrow m\,Me + k\,H_2O. \tag{7–7}$$

The process can be stimulated by vibrationally excited H_2 molecules through their surface dissociation and diffusion of H atoms into the crystal structure. The process can be considered to be gasification of oxygen from the crystal structure by vibrationally excited hydrogen (Legasov et al., 1978c). Metal reduction starts from the surface. The front of metal formation propagates from the surface into the solid body. The depth of the reduction layer is limited by the recombination of H atoms:

$$\delta = \left(\frac{D_0}{\Phi r_0}\right)^{\frac{1}{3}} \exp\left(-\frac{E_a^D}{3T_0}\right), \tag{7–8}$$

where D_0 (usually three orders of magnitude lower than in the gas phase) and E_a^D (about 1 eV/mol) are the pre-exponential factor and activation energy for diffusion of atomic hydrogen in the crystal structure of the reduced metal, respectively; Φ is the flux of atomic hydrogen from plasma into the treated surface; and r_0 is a critical distance between H atoms in the metal crystal structure sufficient for their recombination. The thickness of the reduction layer δ (7–8) grows exponentially with surface temperature. The surface temperature in non-equilibrium discharges ($T_e \gg T_0$) may be much higher than the gas temperature, which is an effect of **non-equilibrium surface heating**. This effect can stimulate the surface oxide reduction processes, keeping the gas temperature low and therefore the efficiency of H_2 dissociation in plasma high. Reduction of metal oxides by atomic hydrogen was first experimentally demonstrated almost a century ago by Bonhoeffer (1924), using oxides of copper, bismuth, lead, silver, tin, and mercury as examples. Some metals, specifically titanium and vanadium, are not completely reduced under non-equilibrium conditions but only decrease their level of oxidation (McTaggart, 1967). Non-equilibrium microwave discharges have been applied to generate atomic hydrogen, which is then used to provide treatment and partial reduction of powders and ceramic samples of oxides of zirconium (ZrO_2) and hafnium (HfO_2) (McTaggart, 1963). Surface layers of the oxides (about 10 μm) changed color from white to golden and became conductive (specific resistance 10^{-3} Ohm · cm). In the case of zirconium oxide, the composition of the partially reduced film can be characterized as $H_{0.2}ZrO_{1.7}$. Atomic hydrogen generated in the non-equilibrium microwave discharge has also been applied for the reduction of TiO_2. Titanium dioxide was reduced to monoxide (TiO); further reduction was accompanied by saturating the products with hydrogen. Black titanium dioxide changed color to gold and became conductive; the typical product composition was $H_{0.2}TiO_{1.3}$. The plasma-generated, partially reduced oxides were called "hydrogen bronzes." Effective formation of the hydrogen bronzes took place at the microwave discharge power exceeding 100 W.

In contrast to Zr, Hf, and Ti, reduction of thorium, beryllium, and aluminum oxides in the non-equilibrium microwave plasma was not effective. A non-equilibrium RF hydrogen discharge was less effective than a microwave discharge with comparable parameters. Reduction of silica (SiO_2) with H atoms was effective but was accompanied by formation of silicon hydrides (McTaggart, 1964). The color change during the partial reduction of metal oxides (MoO_3 and WO_3) in thin surface layers of a sensor can be applied for the detection of H atoms. In the case of molybdenum trioxide in particular, atomic hydrogen has been detected due to color change of a sensor from light yellow for MoO_3 to dark blue for lower oxides.

7.1.6. Non-Equilibrium Surface Heating and Evaporation Effect in Heterogeneous Plasma-Chemical Processes in Non-Thermal Discharges

Heterogeneous non-thermal plasma processes ($T_e \gg T_0$; such as the hydrogen reduction of oxides) include the production of active species in plasma (controlled by T_e and usually very efficient) and following surface reactions (controlled by surface temperature and limiting the total kinetics). If molecular plasma is characterized by vibrational–translational (VT) non-equilibrium ($T_e > T_v \gg T_0$), the surface temperature (T_S) can significantly exceed the translational gas temperature (T_0), because surface VT relaxation is much faster than that in the gas phase. This overheating ($T_e > T_v > T_S \gg T_0$) can be accompanied by surface evaporation, desorption of chemisorbed complexes, and passivating layers. It is called the **non-equilibrium surface heating and evaporation effect**. This effect can be significant for plasma interaction with powders (see following discussion) and flat surfaces (see the next section). The preferential macro-particle surface heating without essential gas heating becomes possible in molecular gases, when the total surface area of the macro-particles is so high that VT relaxation on the surface dominates over VT relaxation due to intermolecular collisions in the plasma volume:

$$4\pi r_a^2 n_a \geq \frac{k_{VT} n_0}{P_{VT}^S v_T}. \tag{7–9}$$

In this relation, n_a and r_a are the density and average radius of macro-particles; $k_{VT}(T_0)$ is the rate coefficient of gas-phase VT relaxation; n_0 is the gas density; v_T is the average thermal velocity of molecules; P_{VT}^S is the probability of VT relaxation on the macro-particle surface, called the accommodation coefficient. We can take for a numerical example the following parameters, $P_{VT}^S = 3 \cdot 10^{-3}$; $k_{VT} = 10^{-17}$ cm^3/s; $n_0 = 3 \cdot 10^{19}$ cm^{-3}, which correspond to atmospheric pressure and normal conditions; $v_T = 10^5$ cm/s. In this case, according to (7–9) the amount of the dispersed phase required for the non-equilibrium heating effect should satisfy the criterion $n_a r_a^2 > 0.1$ cm^{-1}.

On the other hand, vibrational energy accumulated in a gas is sufficient to evaporate a surface layer of the macro-particles with depth $r_S < r_a$ if the total volume fraction of the macro-particles is small:

$$n_a r_a^3 \leq \frac{n_0 T_v}{n_k(T_S + r_S \varepsilon_k / r_a)}, \tag{7–10}$$

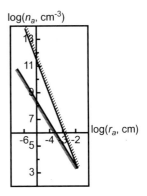

Figure 7–17. Parameters of heterogeneous system (concentration of aerosols, n_a, and radius of particles, r_a) required for non-equilibrium heating of dispersed phase. Probability of surface relaxation is taken as $P_{VT}^S = 3 \cdot 10^{-3}$; gas density is $n_0 = 3 \cdot 10^{19}$ cm^{-3}. The effective values of parameters are located between the two lines.

where T_v is the vibrational temperature, n_k is number density of atoms in a macro-particle, and ε_k is the energy required for evaporation of an atom. If the thickness of the reduced layer is relatively low,

$$r_S < r_a \frac{T_S}{\varepsilon_k}; \tag{7-11}$$

then the major portion of vibrational energy input in (7–10) is related to heating the macro-particles up to temperature T_S. Numerically in this case the limitation (7–10) means $n_a r_a^3 \leq 10^{-3}$ for the conventional conditions of the heterogeneous non-equilibrium plasma system: $n_0 = 3 \cdot 10^{19}$ cm^{-3}, $n_k = 3 \cdot 10^{22}$ cm^{-3}, $T_v = 0.3$ eV, $T_S = 0.3$ eV. Limitations of the volume fraction of particulates for effective non-equilibrium heating are determined by relations (7–9) and (7–10) and are illustrated in Fig. 7–17. Typical parameters of a heterogeneous plasma system relevant for non-equilibrium surface heating and evaporation effect are $r_a = 10^{-4}$ cm (1 μm) and $n_a = 10^8$ cm^{-3} (Friedlander, 2000).

The non-equilibrium surface heating and evaporation effect in plasma treatment of powders can play a significant role not only in metal reduction from oxides but in other technologies related to powder treatment as well (Mosse & Burov, 1980). One of those non-thermal plasma technologies is the reduction of $CaSO_4$ for the following sulfuric acid production (Givotov et al., 1982a). Heating of $CaSO_4$ to temperatures of about 1400 K results in decomposition:

$$CaSO_4 \rightarrow CaO + SO_2 + \frac{1}{2}O_2. \tag{7-12}$$

SO_2 produced in the decomposition process can be further oxidized to SO_3 in conventional sulfuric acid production technology. An effective performance of the heterogeneous process (7–12) under strongly non-equilibrium microwave plasma conditions in nitrogen was demonstrated by Givotov, Pakhomov, & Raddatis (1982a). Complete conversion of the $CaSO_4$ powder has been achieved at low gas temperature but high N_2 vibrational temperature, proving the non-equilibrium heating of the 1 μm particles.

7.1.7. Non-Equilibrium Surface Heating and Evaporation in Plasma Treatment of Thin Layers of Flat Surfaces: Effect of Short Pulses

Hydrogen reduction of oxides as well as gasification reactions, solid-state synthesis, and surface treatment processes can be focused on a thin layer of a flat surface and stimulated by non-thermal plasma ($T_e \gg T_0$). Similar to the powder-related cold plasma processes, thin-layer processes in cold gas can also be limited by surface temperature. Again, VT non-equilibrium ($T_e > T_v \gg T_0$) in this case can provide surface temperatures (T_S) significantly exceeding the gas temperature (T_0), because surface VT relaxation is much faster than that in the gas phase. The non-equilibrium heating of a thin layer can be accomplished by using short non-thermal discharge pulses. Taking into account the low temperature of the surrounding gas, the short heating pulses can provide significant local temperature increase in a thin surface layer without heating up the whole system. To describe the transient process, we can assume that heat flux Q to the surface, related to VT relaxation, starts at initial moment $t = 0$:

$$Q_0 = n_0 P_{VT}^S v_T \hbar \omega. \tag{7-13}$$

The symbols used in this formula are the same as in the previous section, the vibrational temperature of molecular gas is assumed to be high ($T_v > \hbar \omega$). Solution of the heat transfer equation with the flux (7–13) from the surface ($x = 0$) taken as a boundary condition is (Bochin, Rusanov, & Lukianchuk, 1976),

$$T(x, t) = T_0 + \int_0^t \frac{q_0}{\sqrt{\pi(t - \tau)\chi}} \exp\left[-\frac{x^2}{4\chi(t - \tau)}\right] d\tau, \tag{7-14}$$

where $q_0 = Q_0/\rho c$ is the reduced heat flux; ρ and c are density and specific heat of the surface layer of treated solid material, respectively; χ is the reduced heat transfer coefficient; and T_0 is the initial surface temperature. The surface temperature evolution ($x = 0$) can be found from (7–14):

$$T(0, t) = T_0 + \int_0^t \frac{q_0 d\tau}{\sqrt{\pi(t - \tau)\chi}} = T_0 + \frac{2q_0\sqrt{t}}{\sqrt{\pi \chi}}. \tag{7-15}$$

To describe the temperature profile in the plasma-treated layer, it is convenient to rewrite expression (7–14) to introduce a dimensionless special function $\psi(a)$:

$$T(x, t) = T_0 + \frac{2q_0\sqrt{t}}{\sqrt{\pi \chi}} \cdot \psi(a). \tag{7-16}$$

The dimensionless special function $\psi(a)$ determines the temperature profile in the plasma-treated layer as a function of dimensionless depth a of the layer:

$$\psi(a) = \int_a^\infty \frac{1}{u^2} \exp(-u^2) du. \tag{7-17}$$

The dimensionless depth a of the layer is defined through the thermal conductivity relation:

$$a = \frac{x}{\sqrt{4\chi t}}. \tag{7-18}$$

Relations (7–16)–(7–18) demonstrate that heating of a thin layer by short non-thermal discharge pulses provides significant local temperature increase in a thin surface layer without heating up the whole system. The gas temperature also remains low because of the application of strongly non-equilibrium discharges ($T_e > T_v \gg T_0$). Based on relations (7–16)–(7–18) the heat flux Q_0 (7–13) required to provide a temperature increase ΔT in a layer with thickness h can be found as

$$Q_0 \approx 0.6\sqrt{\pi} \frac{\rho c \chi \Delta T}{h}. \tag{7-19}$$

The effective non-equilibrium thin-layer heating can be easily achieved in materials with low thermal conductivity; for metals it requires relatively high discharge powers.

7.2. PRODUCTION OF METALS AND OTHER ELEMENTS BY CARBOTHERMIC REDUCTION AND DIRECT DECOMPOSITION OF THEIR OXIDES IN THERMAL PLASMA

7.2.1. Carbothermic Reduction of Elements from Their Oxides

Carbothermic reduction of elements (El) is a chemical reaction between reactants in the condensed phase with formation of the pure element also in condensed phase and carbon oxides:

$$(El_xO_y)_{cond} + y(C)_{cond} \rightarrow x(El)_{cond} + (CO)_{gas}. \qquad (7-20)$$

The carbothermic reduction processes are usually strongly endothermic and require high temperatures. For example, carbothermic reduction of uranium (U), boron (B), zirconium (Zr), niobium (Nb), and tantalum (Ta) from their oxides requires 2000–4000 K and, therefore, application of thermal plasma. In most plasma-chemical carbothermic reduction processes, an arc electrode is prepared from well-mixed and pressed oxide and carbon particles. The arc provides heating of the mixture, stimulating the reduction process on the electrode. Carbon oxides leave the electrode, finalizing the reduction process.

7.2.2. Production of Pure Metallic Uranium by Carbothermic Plasma-Chemical Reduction of Uranium Oxides

This thermal plasma process is an advanced technology of uranium metallurgy (Tumanov, 1981). The conventional approach starts with uranium oxide (UO_2) and includes two technological stages. The first one is exothermic conversion of the oxide into fluoride with HF:

$$(UO_2)_{solid} + 4(HF)_{gas} \rightarrow (UF_4)_{solid} + 2(H_2O)_{gas}, \quad \Delta H = -41.4\,\text{kcal/mol.} \quad (7-21)$$

The reaction enthalpy can be converted to electron volts per mole (1 eV/mol $\approx$ 23 kcal/mol). The second stage is solid-state exothermic UF_4 reduction with calcium, producing uranium and fluorite:

$$(UF_4)_{solid} + 2(Ca)_{solid} \rightarrow (U)_{solid} + 2(CaF_2)_{solid}, \quad \Delta H = -44.2\,\text{kcal/mol.} \quad (7-22)$$

In contrast to the conventional two-stage technology of (7–21) and (7–22), the carbothermic approach to the reduction process (Gibson & Weidman, 1963; Wilhelm & MacClusky, 1969) is the one-stage technology, which can be described by the only (but strongly endothermic) reaction:

$$(UO_2)_{solid} + 2(C)_{solid} \rightarrow (U)_{solid} + 2(CO)_{gas}, \quad \Delta H = 206.4\,\text{kcal/mol.} \quad (7-23)$$

The carbothermic process provides metallic uranium reduction not only from dioxide (UO_2) but also from a much more convenient and accessible oxide, U_3O_8 (Wilhelm & MacClusky, 1969):

$$(U_3O_8)_{solid} + (x+y)(C)_{solid} \rightarrow 3(U)_{solid} + x(CO_2)_{gas} + y(CO)_{gas}. \qquad (7-24)$$

A schematic of the carbothermic process of metallic uranium production by reduction of either UO_2 or U_3O_8 is shown in Fig. 7–18. The anode of an arc discharge was prepared in the experiments of Gibson and Weidman (1963) from well-mixed and pressed uranium oxide and carbon particles. An arc provided heating of the solid UO_2–C or U_3O_8–C mixtures on the electrode. The thermal arc discharge temperature in these experiments was in the range 4000–5000 K. Carbon was supplied either as graphite or soot. The anode was consumed during the experiments, and carbon oxides were pumped out of the discharge zone. Melted products of the process pour from the arc discharge zone into a special collector underneath the reactor. Admixtures to melted uranium that have boiling temperatures lower than that

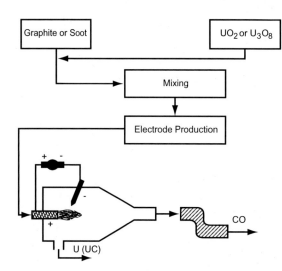

Figure 7–18. Schematic of the carbothermic plasma reduction of uranium from oxides.

of the uranium are easily separated from the product. The described system also can be applied to produce uranium carbides.

7.2.3. Production of Niobium by Carbothermic Plasma-Chemical Reduction of Niobium Oxides

Carbothermic reduction of Nb_2O_5 is an endothermic process:

$$(Nb_2O_5)_{solid} + 5(C)_{solid} \rightarrow 2(Nb)_{solid} + 5(CO)_{gas}. \tag{7–25}$$

Experiments with carbothermic reduction of Nb_2O_5 in an arc with production of pure niobium were carried out by Akashi, Ishizuka, and Egami (1972). The anode of the arc was a water-cooled copper container filled with well-mixed and pressed niobium pentoxide and either carbon or niobium carbide. The arc was sustained in an Ar–H_2 mixture (molar fraction of hydrogen 5%) and provided heating of the condensed-phase Nb_2O_5–C or Nb_2O_5–NbC mixtures to temperatures exceeding 2900 K. Reduction of niobium from its pentoxide was quite fast in these conditions with formation of carbon monoxide and oxygen in the gas phase. After only a few minutes of interaction with carbon, melted Nb_2O_5 was 90% converted into niobium; a conversion degree of 99% was reached not much longer after that. The best results were achieved using niobium carbide as a reducing agent. The highest purity of produced metal can be achieved by additional application of the electron beam treatment of the melted metal in vacuum.

7.2.4. Double-Stage Carbothermic Thermal Plasma Reduction of Rare and Refractory Metals from Their Oxides

Two-stage plasma technology of rare and refractory metal reduction was developed by Komkov et al. (1976). The first stage of carbothermic reduction is conducted in a plasma furnace and results in production of so-called black metal, which can still contain some carbon or oxygen. The second stage is conducted in a high-current vacuum discharge with high-temperature hollow cathode, called an electron-plasma furnace, to provide refining and produce pure metals. A schematic of this double-stage plasma-metallurgic technology is shown in Fig. 7–19. The plasma furnace for the first stage contains an intermediate melting volume for direct carbothermic reduction of metal oxides. Heating is provided by arcs of reverse polarity with an internal plasma anode. Hydrogen used as an energy carrier is heated in the furnace to 5000 K. The reverse polarity prevents formation of a double arc, because electric breakdown of relatively cold gas in the boundary layer inside of the exit

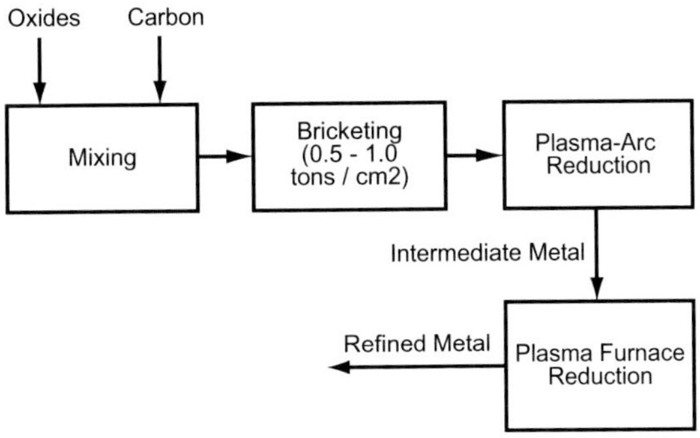

Figure 7–19. Schematic of the carbothermic plasma reduction of metals from their oxides following by metal refining.

electrode is due to ionic compounding of electric current. Low ion mobility mitigates the electric breakdown in the reverse polarity system, which leads to stability of the discharge with length of internal arc column up to 20 mm; it also results in compression of the arc channel inside the exit electrode and in increased hydrogen temperature to 5000–6000 K. The current-voltage characteristic of the discharge grows at operational conditions. The effective lifetime of the 150 kW reverse polarity plasmatron exceeds 200 h. Several reverse polarity discharges, located in the lower part of the plasma furnace, provide heating of the metal oxide–carbon mixture, which is supplied in the furnace in counterflow with respect to the gaseous energy carrier. The MeO–C mixture is melted in a horizontal part of the system. Carbothermic reduction takes place in the melt, which flows into the crystallizer. The first technological stage of the metallurgical process produces black metal, which can contain carbon and oxygen admixtures depending on reduction conditions.

The second technological stage, metal refining, can be conventionally performed in an electron beam vacuum furnace. However, if the black metal contains 3–4% of admixtures, pumping gas product from the vacuum furnace becomes ineffective. For this reason, the refining stage of the double-stage technology under consideration is based on the application of a special high-current vacuum discharge with high-temperature hollow cathode. The system is called an **electron-plasma furnace** (Kruchinin, Orlov, & Komkov, 1976). The electron-plasma furnace operates at pressures much higher (≈ 10 Pa) than those in conventional electron-beam furnaces. The higher pressure in the plasma-metallurgical refining system permits one to increase the efficiency of vacuum pumping. The furnace contains six cathodes with 200 kW power each. The operational electric current is 3000 A and voltage is 65 V. The total power of the electron-plasma refining furnace is 1.2 MW. The heating temperature of the black metal in the furnace is 2300–2500 K. The pumping speed of the vacuum system is 4000 m^3/h.

7.2.5. Carbothermic Reduction of Iron from Iron Titanium Oxide Concentrates in a Thermal Plasma Fluidized Bed

Plasma reduction of metals from their oxides can also be carried out in fluidized beds. Such technology was applied, in particular, for the carbothermic reduction of iron from $FeO \cdot TiO_2$ (Borodulia, 1973):

$$FeO \cdot TiO_2 + C(H_2) \rightarrow Fe + TiO_2 + CO(H_2O). \tag{7–26}$$

A schematic of the plasma-metallurgical reactor based on the application of a plasma fluidized bed is shown in Fig. 7–20. The reduction process is due to both carbon mixed with $FeO \cdot TiO_2$ and hydrogen used as a gas energy carrier in the discharge. Quasi-thermal plasma is generated in the reduction reactor by means of a spark discharge between two electrodes.

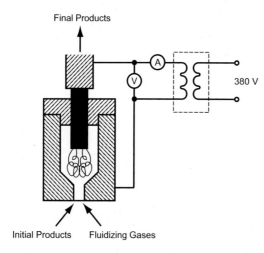

Figure 7–20. Schematic of the thermal plasma fluidized bed.

The discharge electrodes also guide the hydrogen gas flow, which provides fluidization of the FeO · TiO$_2$ concentrate and carbon powders.

7.2.6. Production of Silicon Monoxide by SiO$_2$ Decomposition in Thermal Plasma

Silicon monoxide is used in electronics as a dielectric material for manufacturing capacitors, triodes, and other micro-film elements. Thermal plasma decomposition of SiO$_2$ at 1 atm and temperatures of 3500–5000 K leads to the conversion of 45–100% of dioxide into SiO. The monoxide disproportionates at temperatures exceeding 670 K:

$$SiO + SiO \rightarrow SiO_2 + Si. \tag{7–27}$$

Nevertheless, solid silicon monoxide can be produced by fast quenching of gaseous SiO to room temperature (Vurzel et al., 1965). Electronics requires the production of pure silicon monoxide. For this reason, plasma reduction of SiO$_2$ to SiO should be done without using any special reduction agents, which can contaminate the product. Silicon itself, however, can be used to intensify SiO$_2$ decomposition:

$$SiO_2 + Si \rightarrow SiO + SiO. \tag{7–28}$$

The Si-supported SiO$_2$ decomposition (7–28), reverse to the disproportioning (7–27), is the most effective plasma-chemical approach to production of pure silicon monoxide (Kulagin, Lubimov, & Marin, 1967). The process can be carried out at temperatures not exceeding the silicon melting point (1420°C). However, the most attractive results were achieved in thermal plasma discharges at much higher temperatures (3000–6000 K), when all initial substances are gases. Intensive mixing of the gas flows as well as the high temperature of reagents significantly accelerate the process. The synthesis of silicon monoxide from gas-phase reagents also permits continuity of the process, which stabilizes the properties of the produced monoxide. It is important to point out that even at such high temperatures (3000–6000 K) the gas-phase SiO remains stable, which is illustrated by Table 7–1. The table demonstrates that the SiO fraction in silicon-containing products at 4000 K is 98.5%, and at 3000 K exceeds 99.9%. Effective quenching by fast gas cooling from 3000 K to temperatures below 670 K, which is a limit of SiO disproportioning (7–27), permits one to obtain pure silicon monoxide with a quality required for electronic applications.

Table 7–1. Partial Pressures (atm) of Atomic and Molecular Components During Thermal Plasma Production of Silicon Monoxide from Silicon Dioxide at Different Temperatures in High-Temperature Plasma Zone

Components	$T = 3000$ K	$T = 4000$ K	$T = 5000$ K	$T = 6000$ K
O	$1.02 \cdot 10^{-4}$	$1.36 \cdot 10^{-2}$	$1.44 \cdot 10^{-1}$	$4.04 \cdot 10^{-1}$
Si	$4.72 \cdot 10^{-4}$	$1.42 \cdot 10^{-2}$	$1.44 \cdot 10^{-1}$	$4.05 \cdot 10^{-1}$
SiO	$9.99 \cdot 10^{-1}$	$9.7 \cdot 10^{-1}$	$7.12 \cdot 10^{-1}$	$1.90 \cdot 10^{-1}$
SiO$_2$	$3.66 \cdot 10^{-4}$	$4.67 \cdot 10^{-4}$	$2.7 \cdot 10^{-4}$	$2.67 \cdot 10^{-5}$
O$_2$	$8.34 \cdot 10^{-7}$	$8.4 \cdot 10^{-5}$	$4.34 \cdot 10^{-4}$	$4.23 \cdot 10^{-4}$
SiO molar fraction	99.9%	98.5%	83.1%	31.5%

7.2.7. Experiments with SiO$_2$ Reduction to Pure Silicon Monoxide in High-Temperature Radiofrequency ICP Discharges

Arc discharges have been applied as thermal plasma sources in many reduction technologies. High product purity requirements restrict the use of arcs because of erosion of electrode materials, which can result in contamination of the products. For this reason, experiments with SO$_2$ reduction to produce pure silicon monoxide for use in electronics were carried out in RF inductively coupled plasma (ICP) discharges (Rykalin & Sorokin, 1987), which are thermal but electrodeless. Thermal plasmas have been generated in an RF-ICP discharge in atmospheric-pressure argon at a frequency of 20 MHz. A schematic of the installation is shown in Fig. 7–21. Ignition of the discharge was provided by voltage increase on the inductor without introducing any igniting inserts, to avoid contamination of the reaction

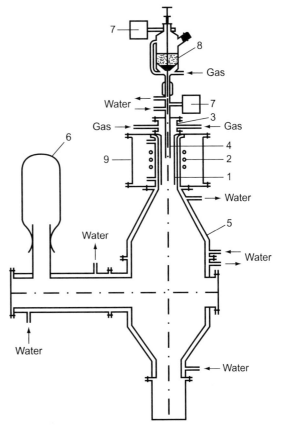

Figure 7–21. Schematic of RF-ICP discharge system for production of silicon monoxide: (1) discharge chamber; (2) inductor; (3) tangential gas inlet; (4) water-cooled feeding system for initial product injection; (5) chemical reactor; (6) filter; (7) vibrator; (8) tank with initial product; (9) safety box for the RF-ICP discharge system.

Table 7–2. Comparison of Compositions of Admixtures (in mass %) to Initial Reagents (Si/SiO$_2$) and to Final Product (SiO) in Plasma-Chemical Process of Silicon Dioxide Reduction to Pure Silicon Monoxide in RF-ICP Discharges

Admixture →	Cu, mass %	Fe, mass %	Al, mass %	Cr, mass %	Ni, mass %
Si/SiO$_2$, initial	$6 \cdot 10^{-4}$	$4 \cdot 10^{-4}$	$2 \cdot 10^{-2}$	$5 \cdot 10^{-4}$	$5 \cdot 10^{-4}$
SiO, product	$4 \cdot 10^{-4}$	$2 \cdot 10^{-3}$	$8 \cdot 10^{-3}$	$2 \cdot 10^{-4}$	$2 \cdot 10^{-4}$
SiO, requirement	$3 \cdot 10^{-3}$	$3 \cdot 10^{-3}$	$1 \cdot 10^{-2}$	–	–

volume. The internal diameter of the water-cooled discharge chamber was 40 mm; argon was supplied in the discharge chamber tangentially; the stoichiometric mixture of silicon and silicon oxide powder (7–28) was supplied through a special water-cooled injection system and controlled by feeder and vibrator systems. The size of both the silicon macro-particles and those of silicon dioxide was about 40 μm. The powder was swept in the injection system by an argon gas carrier with typical flow rate in the range 120–350 L/h. Heating and interaction of the reagents took place in the RF discharge inside the inductor as well as downstream in the plasma jet located in the reactor area. A typical flow rate of argon tangentially supplied in the discharge was 2.7 m³/h, the discharge power was 7.1 kW, and the throughput was 320 g/h. The product of the process, pure silicon monoxide, was collected in the form of fine powder, which is used in particular in the manufacturing of thin-film capacitors (Sakharov, Marin, & Lubimov, 1968; Rykalin & Sorokin, 1987). Plasma technology doesn't contaminate the silicon monoxide. Furthermore, the contamination level of the plasma-produced SiO is even lower than that of the initial Si–SiO$_2$ mixture, which is illustrated in Table 7–2. The purification effect of plasma is due to evaporation of all initial substances.

7.2.8. Reduction of Aluminum by Direct Thermal Plasma Decomposition of Alumina

The production of metallic aluminum by reduction of Al$_2$O$_3$ in H$_2$, CO, and CH$_4$ plasmas was discussed in Section 7.1.4. Consider the production of aluminum by Al$_2$O$_3$ decomposition without any reduction agents in RF-ICP discharges (Rains & Kadlek, 1970; Rykalin & Sorokin, 1987). Such technology permits the production of higher-purity metal for the same reasons as those discussed regarding the production of SiO. The RF-ICP discharge is arranged in this case in a flow of pure argon. Conversion of alumina in Ar plasma starts with the formation of monoxide:

$$Al_2O_3 \rightarrow 2AlO + O. \tag{7–29}$$

The optimal temperature of dissociation is about 4000 K. At higher temperatures closer to 10^4 K, which are reached in the Ar plasma, AlO is unstable and dissociates (Rains & Kadlek, 1970):

$$AlO \rightarrow Al + O. \tag{7–30}$$

In the presence of gaseous reduction agents (H$_2$, CO, and CH$_4$), Al$_2$O$_3$ decomposition starts with

$$Al_2O_3 \rightarrow Al_2O + 2O. \tag{7–31}$$

The plasma process is then completed by Al formation from Al$_2$O (Rains & Kadlek, 1970):

$$Al_2O \rightarrow 2Al + O. \tag{7–32}$$

7.2. Production of Metals and Other Elements by Carbothermic Reduction

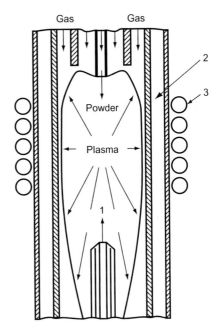

Figure 7–22. Schematic of thermal plasma reactor for aluminum reduction from its oxides: (1) quenching zone; (2) water-cooling system; (3) inductor.

The schematic of the experimental RF-ICP installation is shown in Fig. 7–22. The quartz-made reactor is surrounded by an inductor and water-cooling system. Most of the argon is supplied tangentially, but some is also supplied axially to intensify the interaction of alumina powder with plasma and to provide transportation for the powder. Product quenching is especially important in the absence of reduction agents; it was carried out using a water-cooled probe for fast Al condensation, and was supported by additional cold Ar injected in a direction opposite to the plasma flow. The main characteristics of the process are presented in Figs. 7–23 and 7–24. Figure 7–23 illustrates the increase of reduction degree

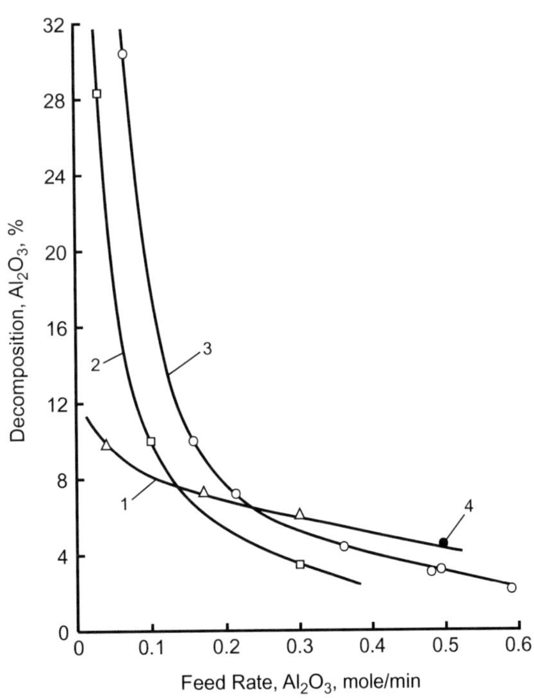

Figure 7–23. Decomposition of Al_2O_3 at argon plasma power of 5 kW as function of its feed rate and particles sizes: (1) 45 μm; (2) 37 μm; (3) 26 μm; (4) 44 μm.

435

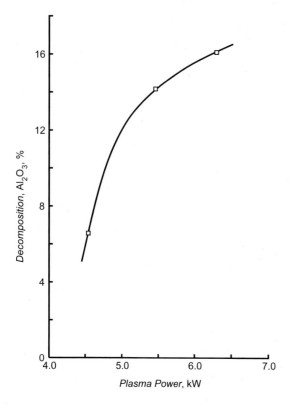

Figure 7–24. Decomposition of Al_2O_3 thermal argon plasma as a function of the discharge power: particle size 26 μm, alumina feeding rate 0.23 g/min.

of Al_2O_3 into aluminum at lower feeding rate of alumina, and by using Al_2O_3 powder with smaller sizes. Figure 7–22 demonstrates that the alumina reduction degree increases with plasma power. Analyzing the influence of adding H_2, CO, and CH_4 to argon, the major effect is related to carbon, which has a higher affinity for oxygen than aluminum and, therefore, intensifies the reduction. Hydrogen over a wide temperature range has a lower affinity for oxygen than aluminum, which makes its contribution less effective. In the experiments with the RF-ICP discharge, it was possible to add the reduction admixtures in the tangential Ar flow. The experiments confirmed the small influence of H_2 on reduction of Al_2O_3 into aluminum, which is illustrated in Fig. 7–25. For comparison, addition of CO doubles the reduction of Al_2O_3, and addition of methane increases the reduction degree four-fold with respect to operating in pure argon without any reduction admixtures.

7.2.9. Reduction of Vanadium by Direct Plasma Decomposition of Its Oxides, V_2O_5 and V_2O_3

Vanadium can be produced by direct thermal decomposition of its oxides: V_2O_5, which has a melting temperature of 953 K and dissociates starting from about 1000 K, as well as V_2O_3, which has a much higher melting temperature of 2243 K and a boiling temperature of 3270 K. Decomposition of V_2O_5 is a strongly endothermic process:

$$V_2O_5 \rightarrow 2V + \frac{5}{2}O_2, \quad \Delta H = 16.1\,\text{eV/mol}. \tag{7–33}$$

The composition of vanadium pentoxide dissociation products in thermal plasma is shown in Fig. 7–26 as a function of plasma temperature and specific energy input (Nester et al., 1988). The process is considered at atmospheric pressure; the composition is presented in moles per kilogram of mixture (initially 6.67 mol/kg). Decomposition with formation of VO_2 starts at relatively low temperatures; however, formation of vanadium atoms requires

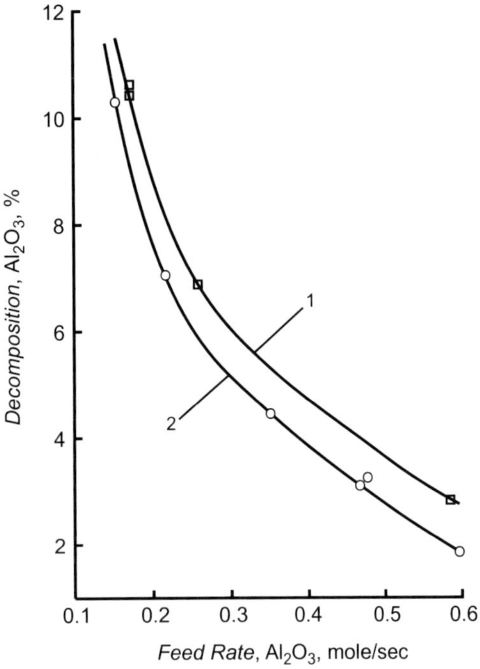

Figure 7–25. Comparison of Al_2O_3 decomposition degrees in thermal argon plasma (1) with and (2) without addition of hydrogen presented as functions of alumina feeding rate. The thermal plasma power is 5 kW, particles size is 26 μm.

much higher temperatures, exceeding 3500 K. The energy cost of vanadium production from V_2O_5 is shown in Fig. 7–27 as a function of specific energy input for absolute, ideal, and super-ideal quenching modes. The minimal energy cost for absolute quenching is 27.1 eV/atom and is reached at a specific energy input of 44.8 eV/mol and conversion degree of about 83%. An ideal quenching mechanism assumes additional production of vanadium due to disproportioning of lower oxides during fast cooling of the products. The minimal energy cost in the case of ideal quenching is 23.9 eV/atom, which can be reached at the same specific energy input of 44.8 eV/mol and conversion degree of about 94%. Super-ideal quenching can be reached by centrifugal separation of the products. The minimal energy cost of vanadium in this regime is 13.7 eV/atom and requires a plasma temperature of about 3600 K.

Direct thermal plasma decomposition of V_2O_3 is also a strongly endothermic process:

$$V_2O_3 \rightarrow 2V + \frac{3}{2}O_2, \quad \Delta H = 13.4\,\text{eV/mol}. \tag{7-34}$$

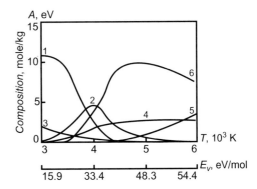

Figure 7–26. Vanadium production by direct decomposition of pentoxide (V_2O_5) in atmospheric-pressure thermal plasma. Composition of products: (1) VO_2; (2) VO; (3) O_2; (4) [O] × 0.1; (5) V^+; (6) V.

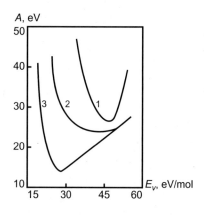

Figure 7–27. Vanadium production by direct decomposition of pentoxide (V_2O_5) in atmospheric-pressure thermal plasma. Energy cost of vanadium production as function of specific energy input: (1) absolute quenching; (2) ideal quenching; (3) super-ideal quenching.

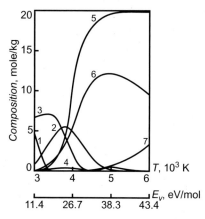

Figure 7–28. Vanadium production by direct decomposition of trioxide (V_2O_3) in atmospheric-pressure thermal plasma. Composition of products: (1) VO (condensed phase); (2) VO; (3) VO_2; (4) O_2; (5) O; (6) V; (7) V^+.

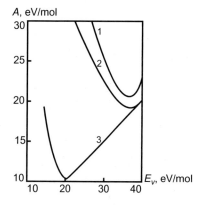

Figure 7–29. Vanadium production by direct decomposition of trioxide (V_2O_3) in atmospheric-pressure thermal plasma. Energy cost of vanadium production as function of specific energy input: (1) absolute quenching; (2) ideal quenching; (3) super-ideal quenching.

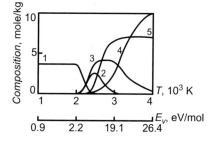

Figure 7–30. Indium production by direct decomposition of its oxide (In_2O_3) in atmospheric-pressure thermal plasma. Composition of products: (1) In_2O_3 (condensed phase); (2) In_2O; (3) O_2; (4) O; (5) In.

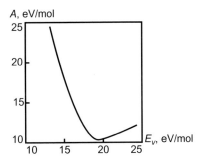

Figure 7–31. Indium production by direct decomposition of its oxide (In_2O_3) in atmospheric-pressure thermal plasma. Energy cost of indium production as function of specific energy input in the case of absolute quenching.

The composition of products of the V_2O_3 dissociation in plasma at atmospheric pressure is shown in Fig. 7–28 as a function of plasma temperature and specific energy input (Nester et al., 1988). The composition is presented in moles per kilogram of mixture (initially 6.67 mol/kg). Vanadium formation takes place at temperatures of about 3500 K. The energy cost of vanadium production from V_2O_3 is shown in Fig. 7–29 for absolute, ideal, and super-ideal quenching. The minimal energy cost for absolute quenching is 20.8 eV/atom, which is lower than that for V_2O_5 decomposition. It can be achieved at a specific energy input 35.5 eV/mol and conversion is about 86%. The minimal energy cost for ideal quenching is 19.4 eV/atom, which can be reached at the same specific energy input of 44.8 eV/mol and conversion degree of about 91%. Super-ideal quenching can be reached by centrifugal separation of products. The minimal energy cost of metallic vanadium for super-ideal quenching is 13.7 eV/atom.

7.2.10. Reduction of Indium and Germanium by Direct Plasma Decomposition of Their Oxides

The direct plasma reduction of indium and germanium from their oxides is similar to that of vanadium. The melting temperature of indium oxide (In_2O_3) is 2183 K. Decomposition of In_2O_3 is also endothermic, although energy consumption here is less than in the case of the direct reduction of vanadium:

$$In_2O_3 \rightarrow 2V + \frac{3}{2}O_2, \quad \Delta H = 9.6 \, eV/mol. \tag{7–35}$$

The composition of In_2O_3 dissociation products in thermal plasma at atmospheric pressure is shown in Fig. 7–30 as a function of plasma temperature and specific energy input (Nester et al., 1988). The initial concentration of In_2O_3 is 3.6 mol/kg. Formation of In takes place at temperatures not so high, about 2300 K, as a result of decomposition of the intermediate oxide (In_2O). The energy cost of In production is shown in Fig. 7–31 as a function of specific energy input for absolute quenching. The minimal energy cost is 10.3 eV/atom and can be achieved at a specific energy input of 19.1 eV/mol and conversion 93%.

Germanium oxide (GeO_2) is characterized by a melting temperature of 1359 K, which is lower than that of indium oxide. It is characterized by a relatively low melting enthalpy: $\Delta H_{melt} = 0.22 \, eV/mol$. Direct thermal plasma decomposition of GeO_2 can be described as

$$GeO_2 \rightarrow Ge + O_2, \quad \Delta H = 6.0 \, eV/mol. \tag{7–36}$$

The composition of products of germanium oxide dissociation at atmospheric pressure in thermal plasma is shown in Fig. 7–32 as a function of plasma temperature and specific energy input (Nester et al., 1988). The initial concentration of GeO_2 is 9.56 mol/kg. Germanium formation takes place at relatively high temperatures, exceeding 4000 K, as a result of decomposition of the intermediate oxide (GeO). The energy cost of germanium production

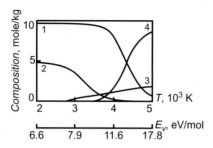

Figure 7–32. Germanium production by direct decomposition of its dioxide (GeO_2) in atmospheric-pressure thermal plasma. Composition of products: (1) GeO; (2) O_2; (3) [O] $\times$ 0.1; (4) Ge.

from GeO_2 is shown in Fig. 7–33 for the case of absolute quenching. The minimal energy cost is 19 eV/atom and can be achieved at a specific energy input of 18 eV/mol and conversion degree of about 95%. In general, by analyzing the direct plasma-chemical processes of reduction of metals by decomposition of their oxides in thermal discharges we can conclude that the processes are feasible but essentially limited by quenching requirements. The strong quenching restrictions to save products generated in the high-temperature zone and the non-selectivity of the quasi-equilibrium processes lead, in most cases, to high-energy-cost technologies (Tumanov, 1981).

7.3. HYDROGEN PLASMA REDUCTION OF METALS AND OTHER ELEMENTS FROM THEIR HALIDES

7.3.1. Using Halides for Production of Metals and Other Elements from Their Compounds

Reduction of pure metals and other elements from their oxides (Sections 7.1 and 7.2) is applied only if the oxides are sufficiently pure. If this is not the case, the oxides can be initially converted into halides (usually fluorides and chlorides), which are mostly volatile, even at relatively low temperatures, and therefore easier to purify. Metals or other elements should then be derived from their halides, where plasma can be successfully applied. The plasma reduction takes place in this case completely in the gas phase. Reduction from halides (both fluorides and chlorides) can be organized with or without a special reduction agent, which is usually H_2. Avoiding hydrogen or other reduction additives permits one to produce uncontaminated elements but makes quenching more challenging. Thermal and non-thermal discharges can be applied for the halide reduction. Reduction of oxides is mostly carried out in thermal plasmas more relevant to the treatment of solids. In contrast, treatment of gaseous halides can be more effective in cold plasmas. First, we will consider

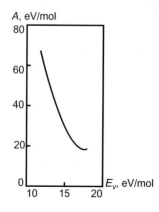

Figure 7–33. Germanium production by direct decomposition of its dioxide (GeO_2) in atmospheric-pressure thermal plasma. Energy cost of germanium production as a function of specific energy input in the case of absolute quenching.

the thermal plasma reduction of metals and other elements from halides using hydrogen as a reducing agent to bind halogens formed during the process. Non-equilibrium plasma processes of hydrogen reduction of halides will be discussed at the end of Section 7.3. Direct plasma decomposition of halides of metals and other elements without using hydrogen in both thermal and non-thermal discharges will be considered in Section 7.4.

7.3.2. Plasma-Chemical Production of Boron: Thermal Plasma Reduction of BCl₃ with Hydrogen

The production of boron from BCl_3 using H_2 as a reducing agent is a slightly endothermic process:

$$(BCl_3)_{gas} + \frac{3}{2}H_2 \rightarrow (B)_{solid} + 3(HCl)_{gas}, \quad \Delta H = 1.35 \, eV/mol. \quad (7\text{--}37)$$

Kinetic analysis of the plasma-chemical process in thermal plasma conditions shows that it follows a complicated mechanism through formation of intermediate compounds: BCl, BCl_2, $HBCl_2$, and H_2BCl. The process starts with dissociation of BCl_3 and formation of a BCl_2 radical:

$$BCl_3 + M \rightarrow BCl_2 + Cl + M, \quad \Delta H \approx E_a = 5.6 \, eV/mol. \quad (7\text{--}38)$$

The Cl atom formed in the dissociation reacts with H_2 in the following fast barrierless exothermic reaction:

$$Cl + H_2 \rightarrow HCl + H, \quad k \approx 10^{-10} \, cm^3/s. \quad (7\text{--}39)$$

Subsequent reactions of atomic hydrogen lead to further reduction of boron chlorides as well as to the formation of boron hydrochlorides. This total boron reduction process can be performed on a hot wire or using hot walls of a pipe reactor (Carleton et al., 1970). Boron formation takes place in this approach at temperatures of 1100–1700 K in the form of dense deposits on the walls, which is not convenient for further use. The productivity and efficiency of such a thermal process are also low. Performing the process in thermal plasma of RF-ICP discharges results in more effective production of boron in the form of fine powder (Hamblym, Reuben, & Thompson, 1972; Cuelleron & Crusiat, 1974).

A schematic of the atmospheric-pressure RF-ICP discharge for H_2 reduction of boron from its trichloride is shown in Fig. 7–34. The water-cooled plasma reactor is made from quartz and surrounded by an inductor of an RF generator. Argon and BCl_3 are introduced in the reactor tangentially, and H_2 is supplied in the opposite direction. The quenching chamber is located downstream from the reactor, with a trap for boron powder and a filter for boron particulates bigger than 1 μm. The residual gas flow, containing the rest of the BCl_3 and H_2 as well as Ar and HCl, goes to a condenser for BCl_3 collection at 195 K, then to an HCl trapping system, and then to an exhaust. The power of the RF generator varied from 2 to 30 kW at 2 MHz frequency. Energy efficiency and conversion achieved in the reactor are fairly high. Injection of BCl_3 in plasma leads to some complications related to plasma instabilities and also in some regimes to boron deposition on the reactor walls. For this reason, another RF-ICP discharge scheme was proposed by Murdoch & Hamblyn (1972), where plasma is generated in argon or Ar–H_2 and the reacting mixture BCl_3–H_2 is injected after the inductor downstream from the discharge. Then the purity of the produced boron is high (99.5%) without its deposition in the discharge area, but the conversion of BCl_3 into boron is only 22–31%, because of H_2 deficiency, low power, and also dilution of hydrogen with argon. A similar approach to boron reduction from BCl_3 was applied by Diana, Russo, and de Marino (1974) with an important difference: the Ar–H_2 mixture was

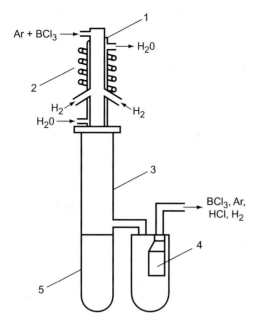

Figure 7–34. Schematic of plasma reduction of boron from BCl_3: (1) quartz plasmatron, (2) RF inductor, (3) quenching zone, (4) filter, (5) powder trapping system.

excited in these experiments not in RF ICP but in an arc discharge. The characteristics of the reduction process using an arc are close to those achieved in ICP.

7.3.3. Hydrogen Reduction of Niobium from Its Pentachloride ($NiCl_5$) in Thermal Plasma

Production of niobium by H_2 plasma reduction of its pentachloride can be described as

$$(NbCl_5)_{gas} + \frac{5}{2}H_2 \rightarrow (Nb)_{solid} + 5(HCl)_{gas}. \tag{7–40}$$

The process was analyzed by Ivanov and Matytsin (1973) using an RF-ICP thermal discharge. The degree of niobium reduction (7–40) at atmospheric pressure approaches 100% when there is about five-fold excess hydrogen over stoichiometry and the temperature in the reaction zone is about 1200 K. Dilution of hydrogen with argon at a fixed temperature and a lower hydrogen excess leads to an increased degree of conversion. If the hydrogen dilution is H_2:$Ar = 1$:4 and $T = 1200$ K, the niobium reduction (7–40) at atmospheric pressure approaches 100% when hydrogen excess in the reactive mixture is only twice the stoichiometric level. Typical parameters of the RF-ICP experiments are RF generator power 63 kW, frequency 5.28 MHz, flow rate of argon 7.2 m^3/h, and hydrogen flow rate 0.4–2 m^3/h. The Ar–H_2 mixture was heated up in RF plasma to temperatures in the range 3000–5100 K. The flow rate of NbF_5 is in the range 0.7–3.3 kg/h, and the residence time of reactants in the system is 10^{-4}–10^{-3} s. The maximum degree of $NiCl_5$ conversion into niobium achieved in the RF-ICP experiments was 98%.

7.3.4. Hydrogen Reduction of Uranium from Its Hexafluoride (UF_6) in Thermal Plasma

Production of metallic uranium from its hexafluoride in a conventional nuclear fuel technology is a two-stage process. The first stage is an exothermic reduction of hexafluoride to tetrafluoride:

$$(UF_6)_{gas} + (H_2)_{gas} \rightarrow (UF_4)_{solid} + 2(HF)_{gas}, \quad \Delta H = -3\,eV/mol. \tag{7–41}$$

Not only hydrogen but other reduction agents like chlorocarbons can also be applied at that stage. The second stage is a complicated solid-state exothermic reaction of UF_4 reduction with calcium:

$$(UF_4)_{solid} + 2(Ca)_{solid} \rightarrow (U)_{solid} + 2(CaF_2)_{solid}, \quad \Delta H = -1.9 \, eV/mol. \quad (7\text{--}42)$$

Fluorine, used initially to produce UF_6, is converted in the two-stage approach, (7–41) and (7–42), into HF (one-third) and into CaF_2 fluorite (two-thirds). Fluorine fixed in the solid fluorite is lost for further conversion. Problems with significant losses of fluorine draw attention to the single-stage uranium reduction:

$$(UF_6)_{gas} + 3(H_2)_{gas} \rightarrow (U)_{solid} + 6(HF)_{gas}, \quad \Delta H = 5.5 \, eV/mol.$$

Such a process has only one stage; and fluorine in this case is fixed in the form of gaseous HF, from where it can be easily produced back. The single-stage uranium reduction from UF_6 is a strongly endothermic process and can be organized only in plasma. This process proceeds through step-by-step reduction from UF_6 to UF_5, from UF_5 to UF_4, from UF_4 to UF_3, and so forth, until it ends with pure uranium. From all the intermediate fluorides, only one, UF_4, is actually a stable product. It makes one-stage reduction from hexafluoride to pure metal quite a challenging process, because UF_4 always has a tendency to remain in the final products. Thermodynamic parameters characterizing each reduction step from UF_k to UF_{k-1} at temperatures between 1000 and 5000 K are shown in Table 7–3 (Tumanov & Galkin, 1972, 1974). The most effective path of the high-temperature process is the formation of UF_4 from hexafluoride. Thermodynamic probabilities of the relevant reactions are significant both at high and relatively low temperatures. In contrast, the formation of fluorides below the stable tetrafluoride (UF_3, UF_2, and UF) and pure uranium require temperatures of about 5000 K and higher. Hydrogen actually loses its reduction ability at such high temperatures because of HF instability at temperatures exceeding 3000 K. Experiments of Tumanov and Galkin (1974) and Sham, Spenceley, and Teetzel (1975) confirm the thermodynamic predictions. Although pure uranium can be formed directly from UF_6, it requires very high temperatures (about 6000 K) and extremely fast quenching. Generated products consist of a mixture of uranium and different fluorides, which can be characterized by the averaged empirical formula $UF_{3.6}$. Taking into account that the most stable intermediate fluoride is UF_4, the generated mixture $UF_{3.6}$ (or $UF_{k<3.6}$) can recombine with HF in an exothermic reverse reaction: $UF_{k-1} + HF \rightarrow UF_k + (1/2)H_2$. As a result, a generated mixture UF_k ($0 < k < 4$) usually ends up in the form of stable solid UF_4, which is a practically important product of UF_6 reduction in thermal plasma.

7.3.5. Hydrogen Reduction of Tantalum (Ta), Molybdenum (Mo), Tungsten (W), Zirconium (Zr), and Hafnium (Hf) from Their Chlorides in Thermal Plasma

Data on reduction of these metals (Ta, Mo, W, Zr, Hf) from chlorides in arc discharges in hydrogen (Case, 1967; Chizhikov et al., 1976) are summarized in Table 7–4. The table proves that hydrogen conversion of chlorides into pure metals can be quite high in arc discharges, up to 98%. Metals produced in these experiments are in powder form. Specific surface areas of the powder can be different – from 0.6 to 13 m^2/g – which is due to differences in methods of separation of gas and the powder. When the separation is performed at temperatures exceeding 600–700 K, agglomeration of particulates takes place and the specific surface area of the powder decreases.

Table 7–3. Thermodynamic Parameters of Atmospheric-Pressure Gas-Phase Reactions Characterizing Step-by-Step Hydrogen Reduction of Uranium Hexafluoride to Uranium

Reduction step	$UF_6 + (1/2)H_2 \rightarrow UF_5 + HF$			Exothermic, $\Delta H = -0.1$ eV/mol		
$T = 1000$ K		$T = 1800$ K	$T = 2600$ K	$T = 3400$ K	$T = 4200$ K	$T = 5000$ K
K_p	x, %	K_p x, %	K_p x, %	K_p x, %	K_p x, %	K_p x, %
$9.1 \cdot 10^4$	100	$2.9 \cdot 10^4$ 100	$1.3 \cdot 10^4$ 100	– 100	– 100	– 100

Reduction step	$UF_6 + H \rightarrow UF_5 + HF$			Exothermic, $\Delta H = -2.3$ eV/mol		
$T = 1000$ K		$T = 1800$ K	$T = 2600$ K	$T = 3400$ K	$T = 4200$ K	$T = 5000$ K
K_p	x, %	K_p x, %	K_p x, %	K_p x, %	K_p x, %	K_p x, %
$2 \cdot 10^{13}$	100	$4 \cdot 10^7$ 100	$1.7 \cdot 10^5$ 100	9600 99	1400 98	44 96

Reduction step	$UF_5 + (1/2H_2) \rightarrow UF_4 + HF$			Endothermic, $\Delta H = 0.4$ eV/mol		
$T = 1000$ K		$T = 1800$ K	$T = 2600$ K	$T = 3400$ K	$T = 4200$ K	$T = 5000$ K
K_p	x, %	K_p x, %	K_p x, %	K_p x, %	K_p x, %	K_p x, %
460	98	260 99.5	5000 100	– 100	– 100	– 100

Reduction step	$UF_5 + H \rightarrow UF_4 + HF$			Exothermic, $\Delta H = -1.8$ eV/mol		
$T = 1000$ K		$T = 1800$ K	$T = 2600$ K	$T = 3400$ K	$T = 4200$ K	$T = 5000$ K
K_p	x, %	K_p x, %	K_p x, %	K_p x, %	K_p x, %	K_p x, %
$6 \cdot 10^{10}$	100	$4 \cdot 10^6$ 100	$7 \cdot 10^4$ 100	8100 99	2100 98.6	780 97

Reduction step	$UF_4 + (1/2H_2) \rightarrow UF_3 + HF$			Endothermic, $\Delta H = 2.3$ eV/mol		
$T = 1000$ K		$T = 1800$ K	$T = 2600$ K	$T = 3400$ K	$T = 4200$ K	$T = 5000$ K
K_p	x, %	K_p x, %	K_p x, %	K_p x, %	K_p x, %	K_p x, %
$2 \cdot 10^{-7}$	0.1	0.02 10	1.7 62	15 84.5	– –	– –

Reduction step	$UF_4 + H \rightarrow UF_3 + HF$			Endothermic, $\Delta H = 0.1$ eV/mol		
$T = 1000$ K		$T = 1800$ K	$T = 2600$ K	$T = 3400$ K	$T = 4200$ K	$T = 5000$ K
K_p	x, %	K_p x, %	K_p x, %	K_p x, %	K_p x, %	K_p x, %
58	83	480 88	36 85.5	30 84	23 83	20 82

Reduction step	$UF_3 + (1/2H_2) \rightarrow UF_2 + HF$			Endothermic, $\Delta H = 2.3$ eV/mol		
$T = 1000$ K		$T = 1800$ K	$T = 2600$ K	$T = 3400$ K	$T = 4200$ K	$T = 5000$ K
K_p	x, %	K_p x, %	K_p x, %	K_p x, %	K_p x, %	K_p x, %
10^{-9}	–	10^{-3} 4	0.1 24	0.82 51	2.9 70	– –

Reduction step	$UF_3 + H \rightarrow UF_2 + HF$			Thermoneutral		
$T = 1000$ K		$T = 1800$ K	$T = 2600$ K	$T = 3400$ K	$T = 4200$ K	$T = 5000$ K
K_p	x, %	K_p x, %	K_p x, %	K_p x, %	K_p x, %	K_p x, %
11	77	4.5 68	2.6 61	1.8 56	1.3 53	1.1 51

Reduction step	$UF_2 + (1/2H_2) \rightarrow UF + HF$			Endothermic, $\Delta H = 0.25$ eV/mol		
$T = 1000$ K		$T = 1800$ K	$T = 2600$ K	$T = 3400$ K	$T = 4200$ K	$T = 5000$ K
K_p	x, %	K_p x, %	K_p x, %	K_p x, %	K_p x, %	K_p x, %
10^{-7}	0.1	0.04 17	5.1 74	60 94	280 97.5	770 99

Table 7–3 (*cont.*)

Reduction step		$UF_2 + H \rightarrow UF + HF$			Endothermic, $\Delta H = 0.2\,\text{eV/mol}$				
$T = 1000\,\text{K}$		$T = 1800\,\text{K}$		$T = 2600\,\text{K}$		$T = 3400\,\text{K}$	$T = 4200\,\text{K}$	$T = 5000\,\text{K}$	
K_p	$x, \%$	K_p	$x, \%$	K_p	$x, \%$	K_p	$x, \%$	K_p	$x, \%$	K_p	$x, \%$
0.56	42	0.8	47	0.84	47	0.82	47	0.75	46	0.7	45

Reduction step		$UF + (1/2H_2) \rightarrow U + HF$			Endothermic, $\Delta H = 0.35\,\text{eV/mol}$				
$T = 1000\,\text{K}$		$T = 1800\,\text{K}$		$T = 2600\,\text{K}$		$T = 3400\,\text{K}$	$T = 4200\,\text{K}$	$T = 5000\,\text{K}$	
K_p	$x, \%$	K_p	$x, \%$	K_p	$x, \%$	K_p	$x, \%$	K_p	$x, \%$	K_p	$x, \%$
–	–	10^{-7}	0.1	$2 \cdot 10^{-4}$	0.3	$8 \cdot 10^{-3}$	8	10	24	0.66	47

Reduction step		$UF + H \rightarrow U + HF$			Endothermic, $\Delta H = 1.3\,\text{eV/mol}$				
$T = 1000\,\text{K}$		$T = 1800\,\text{K}$		$T = 2600\,\text{K}$		$T = 3400\,\text{K}$	$T = 4200\,\text{K}$	$T = 5000\,\text{K}$	
K_p	$x, \%$	K_p	$x, \%$	K_p	$x, \%$	K_p	$x, \%$	K_p	$x, \%$	K_p	$x, \%$
$5 \cdot 10^{-7}$	–	$3 \cdot 10^{-4}$	1.6	$4 \cdot 10^{-3}$	6	0.02	13	0.05	19	0.1	24

Note: K_p is equilibrium constant of a specific reaction; x is conversion degree of an initial fluoride UF_k into UF_{k-1} at stoichiometric composition for the specified reaction.

7.3.6. Hydrogen Reduction of Titanium (Ti), Germanium (Ge), and Silicon (Si) from Their Tetrachlorides in Thermal Plasma

Plasma reduction of tetrachlorides was analyzed by Vurzel et al. (1965) and Krapukhin and Korolev (1968) using atmospheric-pressure plasma jets of arcs and RF-ICP discharges. The formation of Ge,

$$GeCl_4 + 2H_2 \rightarrow Ge + 4HCl, \tag{7–43}$$

starts at 2300 K. The conversion approaches 100% at temperatures of about 4000 K. The formation of silicon,

$$SiCl_4 + 2H_2 \rightarrow Si + 4HCl, \tag{7–44}$$

starts at 3000 K. The conversion approaches 100% at temperatures of about 4500 K. The formation of titanium,

$$TiCl_4 + 2H_2 \rightarrow Ti + 4HCl, \tag{7–45}$$

starts at 3500 K. The conversion approaches 100% at higher temperatures, about 5000 K. At lower pressure (0.1 atm), decomposition of the chlorides requires plasma temperatures about 1000 K lower. The effect of hydrogen on the thermal decomposition of the tetrachlorides is illustrated in Fig. 7–35, using $SiCl_4$ as an example. Using hydrogen decreases the plasma temperatures required for reduction of the chlorides by about 300 K. The reduction effect of hydrogen is limited by the stability of HCl at high temperatures; however, it doubtless simplifies quenching of the products by binding chlorine into HCl at lower temperatures. The kinetics of the reduction processes (7–43)–(7–45) has been investigated by Krapukhin and Korolev (1968) and Vurzel, Polak, and Shchipachev (1967) at temperatures of 3000–6000 K. The reduction proceeds gradually: $ElCl_4 \rightarrow ElCl_3 \rightarrow ElCl_2 \rightarrow ElCl \rightarrow El$. Between the intermediate chlorides, one compound, $ElCl_2$, is especially stable; therefore, the reduction kinetics can be presented as double-stage reaction kinetics: $ElCl_4 \rightarrow ElCl_2 \rightarrow El$. The rate coefficients of both stages are summarized in Table 7–5 for reduction of germanium, silicon, and titanium from their tetrachlorides. The reaction rate coefficients are presented

445

Table 7–4. Hydrogen Reduction of Metals from Their Chlorides in Thermal Plasma

Reduction process		$TaCl_5 + (5/2)H_2 \rightarrow Ta + 5HCl$				
Initial gas flow rate		Arc discharge regime		Metal yield	O_2 content	Specific
Chloride, g/min	Hydrogen, L/min	Current, Amp	Voltage, V	%	%	Surface, m²/g
120	74	200	120	98	0.8	2.8

Reduction process		$MoCl_5 + (5/2)H_2 \rightarrow Mo + 5HCl$				
Initial gas flow rate		Arc discharge regime		Metal yield	O_2 content	Specific
Chloride, g/min	Hydrogen, L/min	Current, Amp	Voltage, V	%	%	Surface, m²/g
100	74	200	120	90	0.6	0.6

Reduction process		$WCl_6 + 3H_2 \rightarrow W + 6HCl$				
Initial gas flow rate		Arc discharge regime		Metal yield	O_2 content	Specific
Chloride, g/min	Hydrogen, L/min	Current, Amp	Voltage, V	%	%	Surface, m²/g
150	74	200	120	94	0.7	6.1

Reduction process		$ZrCl_4 + 2H_2 \rightarrow Zr + 4HCl$				
Initial gas flow rate		Arc discharge regime		Metal yield	O_2 content	Specific
Chloride, g/min	Hydrogen, L/min	Current, Amp	Voltage, V	%	%	Surface, m²/g
30	29	115	98	65	1.8	13.0

Reduction process		$HfCl_4 + 2H_2 \rightarrow Hf + 4HCl$				
Initial gas flow rate		Arc discharge regime		Metal yield	O_2 content	Specific
Chloride, g/min	Hydrogen, L/min	Current, Amp	Voltage, V	%	%	Surface, m²/g
35	29	115	98	70	1.3	6.5

assuming first-order kinetics of the reactions $ElCl_4 \rightarrow ElCl_2$ and $ElCl_2 \rightarrow El$, as well as Arrhenius temperature dependence of the rate coefficients in the range 3000–6000 K. Reduction of all tetrachlorides is limited by the final stage of production of an element from $ElCl_2$. The results for hydrogen reduction of $GeCl_4$, $SiCl_4$, and $TiCl_4$ using plasma jets of arc and RF-ICP discharges are close to the thermodynamic predictions (see Fig. 7–36 for the silicon case). The purity of elements produced in RF-ICP discharges from their chlorides is not less than the purity of initial substances.

7.3.7. Thermal Plasma Reduction of Some Other Halides with Hydrogen: Plasma Production of Intermetallic Compounds

Thermal plasma processes of hydrogen reduction of halides of vanadium (V) and thorium (Th) are described by Tumanov and Galkin (1974). A typical composition of the initial mixture is 5–10 mol hydrogen per 1 mol of halide. Pure metallic powder was generated in this

Table 7–5. Two-Stage Kinetics of Reduction of Germanium, Silicon, and Titanium from Their Tetrachlorides

Element	First kinetic order reaction rate coefficient, k, s^{-1}	
Stage	First Reduction Stage, $ElCl_4 \rightarrow ElCl_2$	Final Reduction Stage, $ElCl_2 \rightarrow El$
Germanium, Ge	$10^9 \cdot \exp\left(-\dfrac{41000 \pm 2000\,K}{T}\right)$	$7 \cdot 10^7 \exp\left(-\dfrac{56000 \pm 5000\,K}{T}\right)$
Silicon, Si	$5 \cdot 10^8 \exp\left(-\dfrac{44000 \pm 2000\,K}{T}\right)$	$5 \cdot 10^7 \exp\left(-\dfrac{63000 \pm 5000\,K}{T}\right)$
Titanium, Ti	$4.5 \cdot 10^9 \exp\left(-\dfrac{41500 \pm 2000\,K}{T}\right)$	$10^8 \cdot \exp\left(-\dfrac{72000 \pm 7000\,K}{T}\right)$

case; sizes of the produced particulates were 0.03–0.1 μm, and the particulates had mostly cubic, octahedral, or spherical shapes. The oxygen content in the produced powder is less than 3 mg per 1 m² of their surface. Reduction of VCl₄ in a hydrogen plasma jet of an RF-ICP discharge is considered by Bashkirov and Medvedev (1968). When VCl₄ is injected into an argon plasma jet, the major reduction product is an intermediate chloride (VCl₂). Injection of VCl₄ into hydrogen plasma permits one to achieve complete reduction and produce pure metallic powder.

Hydrogen reduction of chlorides in thermal plasma can be applied not only to produce powder of individual metals but also to produce different intermetallic compounds. The composition of the generated crystals copies the composition of the chloride mixture in the gas phase. This effect permits the production of the intermetallic compounds with the required composition. Refractory intermetallic compounds based on niobium and vanadium were synthesized from their chlorides (conversion close to 100%) in a hydrogen plasma jet generated by RF-ICP discharge (Bashkirov & Medvedev, 1968). The reaction products were crystallized on a substrate sustained at a temperature of about 1000°C. The plasma-metallurgical technology is proven to be effective in the synthesis of vanadium–silicon

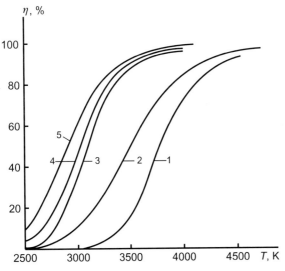

Figure 7–35. Temperature dependence of silicon yield in the process of decomposition of silicon chloride in neutral and reducing atmospheres: (1) $p = 1.0$ atm, H:Cl = 0; (2) $p = 1.0$ atm, H:Cl = 1:1; (3) $p = 0.1$ atm, H:Cl = 1:2; (4) $p = 0.1$ atm, H:Cl = 1:1; (5) $p = 0.1$ atm, H:Cl = 2:1.

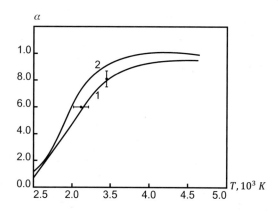

Figure 7–36. Temperature dependence of silicon yield in the process of its hydrogen reduction from silicon tetrachloride: (1) experimental curve; (2) modeling curve.

and vanadium–germanium compounds of different composition, as well as the specific compounds V_3Ge, Nb_3Al, and Nb_3Ge, $Nb_3Al_{0.8}Ge_{0.2}$. Additional information on the subject can be found in reviews of Vurzel (1970) and Mosse and Pechkovsky (1973).

7.3.8. Hydrogen Reduction of Halides in Non-Thermal Plasma

Although most investigations of hydrogen reduction of halides are focused on the application of thermal plasma, some interesting results were obtained in experiments with non-thermal plasma of strongly non-equilibrium discharges. In particular, hydrogen reduction of $TiCl_4$ in a low-pressure glow discharge was considered by Ishizuka (1958). The reduction products collected in the glow discharge system consist of 90% metallic titanium and 10% lower chlorides, $TiCl_2$ and $TiCl$. Vacuum heating of the mixture results in the production of metallic titanium with relatively high purity (99.6%). Hydrogen reduction of BCl_3 with production of pure boron was investigated by Markovsky, Lvova, and Kondrashev (1958) in a glow discharge with moderate pressures (30–200 Torr). The optimal gas composition in the experiments was $BCl_3:H_2 = 1:5$. The reduction process was very effective, with the production of boron at 99.9% purity. The glow discharge experiments with hydrogen reduction of zirconium halides ($ZrCl_4$, $ZrBr_4$, ZrI_4, and ZrF_4) resulted in the production of lower halides ($ZrCl_3$, $ZrBr_3$, ZrI_3, and ZrF_2) but not zirconium. The glow discharge experiments for zirconium halides were carried out at low pressures, 3–4 Torr (Newman & Watts, 1960). Effective hydrogen reduction of halides was also achieved in non-equilibrium microwave discharges. In particular, H_2 reduction of aluminum and scandium from different halides ($AlCl_3$, AlF_3, AlI_3, $ScCl_3$, and ScF_3) was observed in non-equilibrium microwave discharges by McTaggart (1965a,b, 1967). The plasma-metallurgical processes lead to the formation of pure aluminum and scandium in the system. Complete hydrogen reduction of elemental arsenic, tin, and antimony from their chlorides ($AsCl_3$, $SnCl_4$, and $SbCl_5$) was also achieved in non-equilibrium plasma conditions of a non-thermal discharge (Gutmann, 1955).

7.4. DIRECT DECOMPOSITION OF HALIDES IN THERMAL AND NON-THERMAL PLASMA

7.4.1. Direct Decomposition of Halides and Production of Metals in Plasma

The production of metals from their halides can be done in plasma directly without using hydrogen or any other reducing agent. This approach is attractive because the produced metal is not contaminated with hydrogen. On the other hand, the absence of hydrogen makes product quenching more challenging, because aggressive halogen atoms and molecules are not bound with hydrogen and are supposed to coexist with produced metals. The direct decomposition of iodides for refining metals and other elements is used for the production

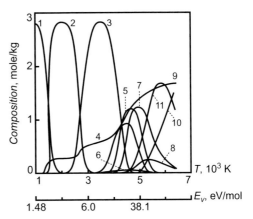

Figure 7–37. Direct decomposition of uranium hexafluoride (UF_6) in atmospheric-pressure thermal plasma. Composition of products: (1) UF_6; (2) UF_5; (3) UF_4; (4) UF_3; (5) UF_2; (6) UF_2^+; (7) UF; (8) UF^+; (9) $[F] \times 0.1$; (10) U^+; (11) U.

of pure zirconium (from ZrI_4) and uranium (from UI_4). Chlorides and fluorides have higher binding energies and are thermally stable at temperatures not exceeding 1000 K. High plasma temperatures make possible the direct decomposition of metal chlorides and fluorides. It is a necessary but not a sufficient requirement for effective production of pure metals in such approaches. Production of pure metals by direct decomposition in thermal plasma is essentially controlled by the efficiency of quenching.

Non-equilibrium plasma processes are much less sensitive to reverse reactions, because of low temperature, and often don't require product quenching. For this reason, several direct halide decomposition processes have been effectively carried out in non-thermal discharges.

7.4.2. Direct UF_6 Decomposition in Thermal Plasma: Requirements for Effective Product Quenching

The decomposition of UF_6 with formation of metallic uranium is a strongly endothermic process:

$$(UF_6)_{gas} \rightarrow U_{solid} + 3(F_2)_{gas}, \quad \Delta H = 22.7\,eV/mol. \quad (7\text{–}46)$$

The composition of UF_6 dissociation products in atmospheric-pressure plasma is shown in Fig. 7–37 as a function of plasma temperature and specific energy input (Nester et al., 1988). The initial concentration of UF_6 is 2.84 mol/kg. Uranium formation takes place at temperatures exceeding 5000 K. The energy cost of uranium production from UF_6 is shown in Fig. 7–38 as a function of specific energy input for the cases of absolute and ideal quenching. The minimal energy cost in the absolute quenching mode is 49 eV/atom and can be achieved at a specific energy input of 46.7 eV/mol and a conversion degree of about

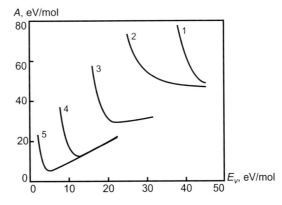

Figure 7–38. Direct decomposition of uranium fluorides in atmospheric-pressure thermal plasma. Energy cost of the process at different process and quenching modes: (1) absolute quenching of uranium production from hexafluoride (UF_6); (2) ideal quenching of uranium production from hexafluoride (UF_6); (3) ideal quenching of (UF_3) production from tetrafluoride (UF_4); (4) ideal quenching of uranium tetrafluoride (UF_4) production from (UF_5); (5) ideal quenching of (UF_5) production from hexafluoride (UF_6).

449

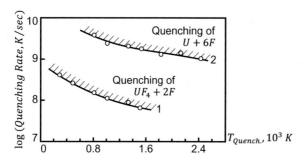

Figure 7–39. Threshold values of the quenching rates required for avoiding product recombination in the uranium fluorine plasma presented as functions of temperature in the quenching zone: (1) composition UF$_4$ + 2F, initial temperature 3200 K; (2) composition U + 6F, initial temperature 6200 K.

95%. Ideal quenching assumes disproportioning of lower fluorides into uranium and its tetrafluoride, UF$_4$. The minimal energy cost in the absolute quenching mode is 47 eV/atom. Figure 7–38 also shows (for comparison) the energy cost of dissociation of UF$_6$, UF$_5$, and UF$_4$ (in ideal quenching mode): the minimal energies required per single act of the dissociation processes are 5.6, 12.7, and 30 eV, respectively. Without special intensification of quenching, the products collected after the thermal plasma reactor can be characterized as being a heterogeneous mixture: UF$_n$ + (3 − n/2)F$_2$, n = 4–5 (Tumanov, 1981). To produce lower fluorides and metallic uranium, the cooling rate and uranium condensation should be faster than the volumetric reverse reactions with fluorine. The characteristic time of reverse reactions of fluorine with an intermediate fluoride (UF$_4$) at atmospheric pressure is about 10^{-5} s. The volumetric reverse reactions of uranium oxidation are even faster at 10^{-6} s (Tumanov, 1975, 1976). The condensation rate can be calculated as (Stachorska, 1965; Frenkel, 1975)

$$w_{\text{cond}} = n_0 \frac{2p}{kT\rho} \sqrt{\frac{\sigma M}{2\pi}} \exp\left(-\frac{4\pi r_{\text{cr}}^2 \sigma}{3kT}\right), \qquad (7\text{–}47)$$

where n_0 is the density of the condensing atoms; M is the mass of the atoms; p and T are pressure and temperature; k is the Boltzmann constant (if temperature is expressed in energy units, the constant is equal to 1); σ is the surface tension for a condensed-phase precursor; and r_{cr} is the critical radius of the precursor, which correspond to maximum thermodynamic potential for the system supersaturated vapor–condensed phase. Formation of a condensed phase during product quenching of UF$_6$ dissociation is faster than the reverse reaction. Condensation with formation of the intermediate fluoride, UF$_4$, requires only about $3 \cdot 10^{-7}$s, whereas condensation of metallic uranium requires about 10^{-7}s. The quenching rate should be fast enough to reach necessary temperatures for condensation before the volumetric reverse reactions with fluorine destroy products generated at high temperatures.

Requirements for the cooling rates to quench products of direct plasma decomposition of UF$_6$ are presented in Fig. 7–39 as a function of temperature in the quenching zone. One of the curves shows the minimal cooling rate for quenching the UF$_4$–F$_2$ mixture with initial temperature of 3200 K. Another curve shows the minimal cooling rates for quenching a uranium–fluorine mixture starting from a plasma temperature of 6200 K, sufficient to produce pure uranium. Effective condensation of UF$_4$ from the high-temperature reactive mixture requires the cooling rate to exceed ($7 \cdot 10^7$–$4 \cdot 10^8$ K/s). To quench metallic uranium in products, the cooling rate should be extremely high (10^9–$4 \cdot 10^9$ K/s). The highest quenching rates achieved today approach 10^9 K/s. Tubular heat exchangers permit cooling rates of about 10^7 K/s, cooling rates achieved using gas or liquid jets are about 10^8 K/s, and using fluidized-bed and Laval nozzles permits slightly higher quenching rates (Ambrazevicius,

7.4. Direct Decomposition of Halides in Thermal and Non-Thermal Plasma

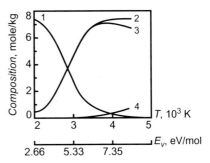

Figure 7–40. Lithium production by direct decomposition of its iodide (LiI) in atmospheric-pressure thermal plasma. Composition of products: (1) LiI; (2) I; (3) Li; (4) Li$^+$.

1983). All these factors explain why quenching of products of direct UF$_6$ decomposition in thermal plasma is such a challenging task. The condensed products should be separated from fluorine before heterogeneous reverse reactions take place. For example, the heterogeneous reverse reaction of tetrafluoride gasification, UF$_4$ + F$_2$ → UF$_6$, decreases the mass of a UF$_4$ particulate in accordance with the following kinetic equation:

$$\frac{dm}{dt} = -K \cdot 4\pi r_0^2 \left(\frac{m}{m_0}\right)^{\frac{2}{3}}, \tag{7-48}$$

where K is the kinetic coefficient constant at fixed temperature and partial pressure of fluorine, and r_0 and m_0 are the initial radius and mass of a particulate, respectively. Integration of kinetic equation (7–47) results in the following time dependence for the mass of a UF$_4$ particle (Galkin, 1961):

$$m = m_0(1 - K^*t)^3, \tag{7-49}$$

where $K^* = 3.4 \cdot 10^{-2} \, 1/\min$ at temperature 593 K. The mass ratio is $m/m_0 = 0.9$ only after 62 s, which means that separation of fluorine and condensed products can be effective. High quenching efficiency of UF$_4$ has been achieved using vibrating heat exchangers (Tumanov, 1981).

7.4.3. Direct Decomposition of Halides of Some Alkali and Alkaline Earth Metals in Thermal Plasma

Halides of lithium (LiF, LiCl, LiBr, and LiI) have relatively high boiling temperatures, especially fluoride ($T_{boil} = 1950$ K); iodide has the lowest boiling temperature. However, all the halides of lithium can be vaporized and directly decomposed in thermal plasma conditions with the production of lithium (Nester et al., 1988). Consider the direct endothermic plasma decomposition of iodide ($T_{boil} = 1440$ K):

$$\text{LiI} \rightarrow \text{Li} + 1/2 \, \text{I}_2, \quad \Delta H = 2.8 \, \text{eV/mol}. \tag{7-50}$$

Composition of LiI dissociation products in atmospheric-pressure thermal plasma is shown in Fig. 7–40. The initial concentration of LiI is 7.47 mol/kg. Lithium formation takes place at temperatures exceeding 2500 K. The energy cost of lithium production from iodide is shown in Fig. 7–41. The minimal energy cost is 7.44 eV/atom in the case of absolute quenching, and 7.35 eV/atom in the case of ideal quenching, which can be achieved at a specific energy input of 6.7 eV/mol.

The situation with sodium halides is somewhat similar. Consider as an example the direct endothermic process for plasma decomposition of chloride ($T_{boil} = 1740$ K):

$$\text{NaCl} \rightarrow \text{Na} + 1/2 \, \text{Cl}_2, \quad \Delta H = 4.26 \, \text{eV/mol}. \tag{7-51}$$

451

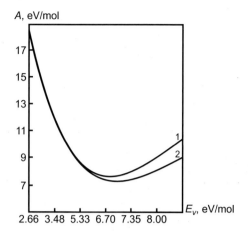

Figure 7–41. Lithium production by direct decomposition of its iodide (LiI) in atmospheric-pressure thermal plasma. Energy cost of the process at different quenching modes: (1) absolute quenching; (2) ideal quenching.

The composition of NaCl dissociation products in atmospheric-pressure thermal plasma is shown in Fig. 7–42. The initial concentration of NaCl is 17.1 mol/kg. Sodium formation takes place at temperatures exceeding 3000 K. The energy cost of sodium production from chloride is shown in Fig. 7–43. The minimal energy cost is 9.4 eV/atom in the case of absolute quenching, and 8.8 eV/atom in the case of ideal quenching, which can be achieved at a specific energy input of 8.2 eV/mol.

Potassium halide thermal decomposition can be represented by that of its iodide also having the lowest boiling temperature with respect to other halides ($T_{boil} = 1590$ K):

$$KI \rightarrow K + 1/2 I_2, \quad \Delta H = 3.4 \, eV/mol. \tag{7–52}$$

The composition of KI dissociation products in atmospheric-pressure thermal plasma is shown in Fig. 7–44. The initial concentration of KI is 6.0 mol/kg. Potassium formation occurs at temperatures exceeding 2500 K. The energy cost of potassium production from iodide is shown in Fig. 7–45. The minimal energy cost is 7.9 eV/atom in the case of absolute quenching, and 7.3 eV/atom in the case of ideal quenching, which can be achieved at a specific energy input of 6.8 eV/mol. Data on the decomposition of halides of rubidium (Rb) and cesium (Cs) in thermal plasma can be found in Nester et al. (1988).

Beryllium halides have lower boiling temperatures and can be more easily introduced into the gas phase for plasma treatment. Consider the direct endothermic plasma decomposition of $BeCl_2$ ($T_{boil} = 793$ K):

$$BeCl_2 \rightarrow Be + Cl_2, \quad \Delta H = 5.1 \, eV/mol. \tag{7–53}$$

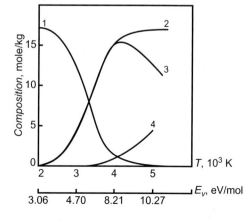

Figure 7–42. Sodium production by direct decomposition of its chloride (NaCl) in atmospheric-pressure thermal plasma. Composition of products: (1) NaCl; (2) Cl; (3) Na; (4) Na^+.

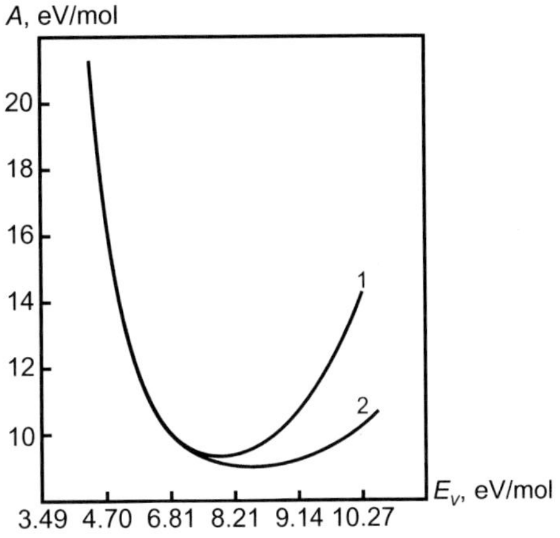

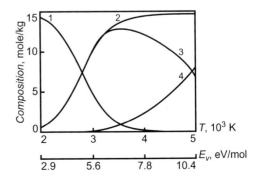

Figure 7–43. Sodium production by direct decomposition of its chloride (NaCl) in atmospheric-pressure thermal plasma. Energy cost of the process at different quenching modes: (1) absolute quenching; (2) ideal quenching.

The composition of $BeCl_2$ dissociation products in atmospheric-pressure thermal plasma is shown in Fig. 7–46. The initial concentration of $BeCl_2$ is 12.5 mol/kg. Formation of beryllium takes place at temperatures exceeding 3500 K. The energy cost of beryllium production from chloride is shown in Fig. 7–47. The minimal energy cost is 14 eV/atom in the case of absolute quenching, which can be achieved at a specific energy input of 13.1 eV/mol, and 12.6 eV/atom in the case of ideal quenching, which corresponds to a specific energy input of 10.4 eV/mol.

Plasma decomposition of magnesium halides can be illustrated by that of MgI_2 (T_{boil} = 923 K):

$$MgI_2 \rightarrow Mg + I_2, \quad \Delta H = 3.7 \, eV/mol. \tag{7-54}$$

The composition of MgI_2 dissociation products in atmospheric-pressure thermal plasma is shown in Fig. 7–48. The initial concentration of MgI_2 is 3.6 mol/kg. Formation of magnesium takes place at temperatures exceeding 2000 K. The energy cost of magnesium production from the iodide is shown in Fig. 7–49. The minimal energy cost is 9.2 eV/atom in the case of absolute quenching, which can be achieved at a specific energy input of 8.8 eV/mol, and 8.9 eV/atom in the case of ideal quenching, which corresponds to a specific energy input of 8.75 eV/mol.

Evaporation of calcium halides requires slightly higher temperatures. Their thermal decomposition can be represented by the dissociation of the bromide, $CaBr_2$ (T_{boil} = 1080 K):

$$CaBr_2 \rightarrow Ca + Br_2, \quad \Delta H = 7 \, eV/mol. \tag{7-55}$$

Figure 7–44. Potassium production by direct decomposition of its iodide (KI) in atmospheric-pressure thermal plasma. Composition of products: (1) KI; (2) I; (3) K; (4) K^+.

453

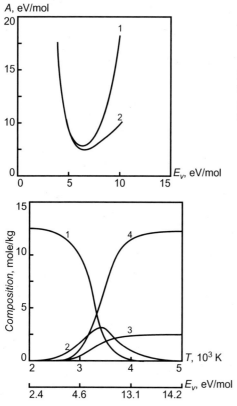

Figure 7–45. Potassium production by direct decomposition of its iodide (KI) in atmospheric-pressure thermal plasma. Energy cost of the process at different quenching modes: (1) absolute quenching; (2) ideal quenching.

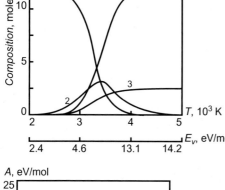

Figure 7–46. Beryllium production by direct decomposition of its dichloride ($BeCl_2$) in atmospheric-pressure thermal plasma. Composition of products: (1) $BeCl_2$; (2) $BeCl$; (3) $Cl \times 0.1$; (4) Be.

Figure 7–47. Beryllium production by direct decomposition of its dichloride ($BeCl_2$) in atmospheric-pressure thermal plasma. Energy cost of the process at different quenching modes: (1) absolute quenching; (2) ideal quenching.

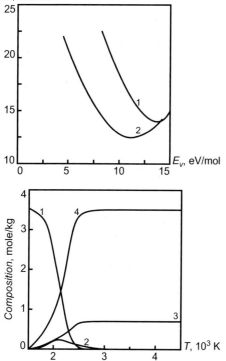

Figure 7–48. Magnesium production by direct decomposition of its iodide (MgI_2) in atmospheric-pressure thermal plasma. Composition of products: (1) MgI_2; (2) MgI; (3) $I \times 0.1$; (4) Mg.

7.4. Direct Decomposition of Halides in Thermal and Non-Thermal Plasma

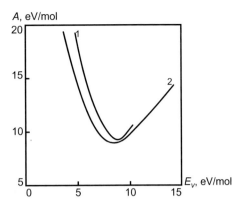

Figure 7–49. Magnesium production by direct decomposition of its iodide (MgI_2) in atmospheric-pressure thermal plasma. Energy cost of the process at different quenching modes: (1) absolute quenching; (2) ideal quenching.

The composition of $CaBr_2$ dissociation products in atmospheric-pressure thermal plasma is shown in Fig. 7–50. The initial concentration of $CaBr_2$ is 5 mol/kg. Calcium formation takes place at temperatures exceeding 3500 K. The energy cost of calcium production from its bromide is shown in Fig. 7–51. The minimal energy cost is 14.9 eV/atom in the case of absolute quenching, which can be achieved at a specific energy input of 13.8 eV/mol, and 13.4 eV/atom in the case of ideal quenching, which corresponds to a specific energy input of 12.2 eV/mol.

Barium halides are characterized by high boiling temperatures, but thermal plasma temperatures are obviously sufficient to vaporize and dissociate them. Their thermal decomposition can be represented by the dissociation of $BaCl_2$ ($T_{boil} = 1830$ K):

$$BaCl_2 \rightarrow Ba + Cl_2, \quad \Delta H = 8.9 \, eV/mol. \tag{7–56}$$

The composition of $BaCl_2$ dissociation products in atmospheric-pressure plasma is shown in Fig. 7–52. The initial concentration of $BaCl_2$ is 4.8 mol/kg. Calcium formation takes place at temperatures exceeding 3500 K. The energy cost of barium production from its chloride is shown in Fig. 7–53. The minimal energy cost is 19.7 eV/atom in the case of absolute quenching, which can be achieved at a specific energy input 17.4 eV/mol, and 18.8 eV/atom in the case of ideal quenching. Data regarding decomposition of halides of strontium (Sr) in thermal plasma can be found in Nester et al. (1988).

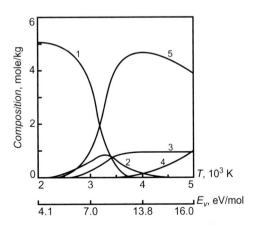

Figure 7–50. Calcium production by direct decomposition of its bromide ($CaBr_2$) in atmospheric-pressure thermal plasma. Composition of products: (1) $CaBr_2$; (2) CaBr; (3) Br $\times$ 0.1; (4) Ca^+, (5) Ca.

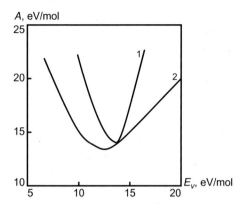

Figure 7–51. Calcium production by direct decomposition of its bromide ($CaBr_2$) in atmospheric-pressure thermal plasma. Energy cost of the process at different quenching modes: (1) absolute quenching; (2) ideal quenching.

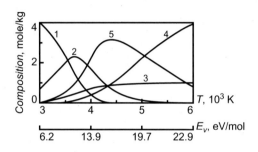

Figure 7–52. Barium production by direct decomposition of its dichloride ($BaCl_2$) in atmospheric-pressure thermal plasma. Composition of products: (1) $BaCl_2$; (2) $BaCl$; (3) $Cl \times 0.1$; (4) Ba^+, (5) Ba.

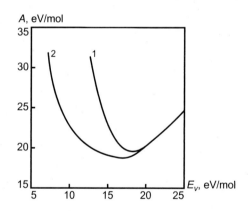

Figure 7–53. Barium production by direct decomposition of its dichloride ($BaCl_2$) in atmospheric-pressure thermal plasma. Energy cost of the process at different quenching modes: (1) absolute quenching; (2) ideal quenching.

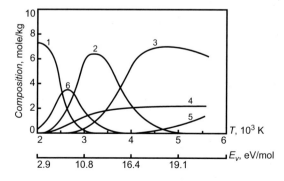

Figure 7–54. Aluminum production by direct decomposition of its trichloride ($AlCl_3$) in atmospheric-pressure thermal plasma. Composition of products: (1) $AlCl_3$; (2) $AlCl$; (3) Al; (4) $Cl \times 0.1$, (5) Al^+; (6) $AlCl_2$.

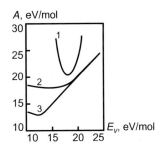

Figure 7–55. Aluminum production by direct decomposition of its trichloride (AlCl$_3$) in atmospheric-pressure thermal plasma. Energy cost of the process at different quenching modes: (1) absolute quenching; (2) ideal quenching; (3) super-ideal quenching.

7.4.4. Direct Thermal Plasma Decomposition of Halides of Aluminum, Silicon, Arsenic, and Some Other Elements of Groups 3, 4, and 5

Between different aluminum halides, AlCl$_3$ has the lowest temperature of transition to the gas phase (atmospheric-pressure sublimation at 453 K), which simplifies the plasma-chemical process:

$$AlCl_3 \rightarrow Al + 3/2Cl_2, \quad \Delta H = 7.3 \, \text{eV/mol}. \tag{7–57}$$

The composition of AlCl$_3$ dissociation products in atmospheric-pressure plasma is shown in Fig. 7–54. The initial concentration of AlCl$_3$ is 7.5 mol/kg. Caluminum formation takes place at temperatures exceeding 3500 K. The energy cost of aluminum production from chloride is shown in Fig. 7–55. The minimal energy cost is 20 eV/atom at absolute quenching, which can be achieved at an energy input of 18.1 eV/mol and conversion degree of 90%, and 17.9 eV/atom in the case of ideal quenching, which corresponds to a specific energy input of 11 eV/mol. Super-ideal quenching can be achieved due to the centrifugal effect with minimal energy cost of 11 eV/atom at a plasma temperature of 2750 K.

Direct decomposition of boron halides can be represented by dissociation of gaseous BF$_3$:

$$BF_3 \rightarrow B + 3/2F_2, \quad \Delta H = 11.8 \, \text{eV/mol}. \tag{7–58}$$

The composition of BF$_3$ dissociation products in atmospheric-pressure plasma is shown in Fig. 7–56. The initial concentration of BF$_3$ is 14.75 mol/kg. Boron formation takes place at temperatures exceeding 4500 K. The energy cost of boron production from fluoride is shown in Fig. 7–57. The minimal energy cost is 19 eV/atom in the case of absolute quenching, which can be achieved at a specific energy input of 18 eV/mol and conversion degree 95%. Super-ideal quenching due to intensive centrifugal effect results in a minimal energy cost of 10.7 eV/atom at plasma temperature 4000 K.

Gallium chloride (GaCl$_3$) also can be vaporized easily ($T_{boil} = 474$ K) and dissociated in plasma:

$$GaCl_3 \rightarrow Ga + 3/2 Cl_2, \quad \Delta H = 5.44 \, \text{eV/mol}. \tag{7–59}$$

The composition of GaCl$_3$ dissociation products in atmospheric-pressure plasma is shown in Fig. 7–58. The initial concentration of GaCl$_3$ is 5.7 mol/kg. Gallium formation

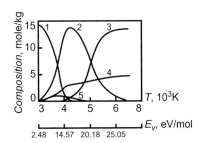

Figure 7–56. Boron production by direct decomposition of its trifluoride (BF$_3$) in atmospheric-pressure thermal plasma. Composition of products: (1) BF$_3$; (2) BF; (3) B; (4) F $\times$ 0.1; (5) BF$_2$.

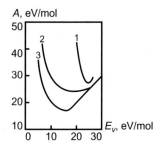

Figure 7–57. Boron production by direct decomposition of its trifluoride (BF$_3$) in atmospheric-pressure thermal plasma. Energy cost of the process at different quenching modes: (1) absolute quenching; (2) ideal quenching; (3) super-ideal quenching.

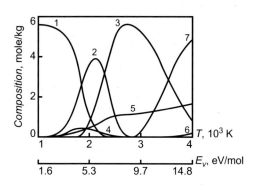

Figure 7–58. Galium production by direct decomposition of its trichloride (GaCl$_3$) in atmospheric-pressure thermal plasma. Composition of products: (1) GaCl$_3$; (2) GaCl$_2$; (3) GaCl; (4) Cl$_2$; (5) Cl × 0.1; (6) Ga$^+$; (7) Ga.

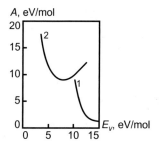

Figure 7–59. Galium production by direct decomposition of its trichloride (GaCl$_3$) in atmospheric-pressure thermal plasma. Energy cost of the process at different quenching modes: (1) Ga production from GaCl$_3$, absolute quenching (A × 0.1); (2) GaCl production from GaCl$_3$, ideal quenching.

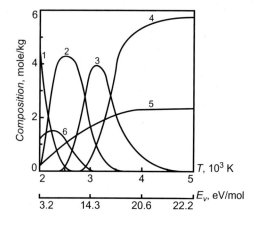

Figure 7–60. Silicon production by direct decomposition of its tetrachloride (SiCl$_4$) in atmospheric-pressure thermal plasma. Composition of products: (1) SiCl$_4$; (2) SiCl$_2$; (3) SiCl; (4) Si; (5) Cl × 0.1; (6) SiCl$_3$.

7.4. Direct Decomposition of Halides in Thermal and Non-Thermal Plasma

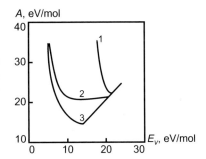

Figure 7–61. Silicon production by direct decomposition of its tetrachloride (SiCl$_4$) in atmospheric-pressure thermal plasma. Energy cost of the process at different quenching modes: (1) absolute quenching; (2) ideal quenching; (3) super-ideal quenching.

takes place at temperatures exceeding 3500 K. The energy cost of gallium production from chloride is shown in Fig. 7–59. The minimal energy cost is 16.5 eV/atom in the case of absolute quenching, which can be achieved at a specific energy input of 14.9 eV/mol and conversion degree of 90%, and 9.3 eV/atom in the case of ideal quenching, which corresponds to a specific energy input of 8.6 eV/mol. Information regarding direct thermal plasma decomposition of different halides of other elements of groups, indium (In) and thallium (Tl), can be found in Nester et al. (1988).

Silicon chloride (SiCl$_4$) is a gas at temperatures above 320 K and can be dissociated in plasma:

$$SiCl_4 \rightarrow Si + 2Cl_2, \quad \Delta H = 4\,eV/mol. \quad (7\text{--}60)$$

The composition of SiCl$_4$ dissociation products in atmospheric-pressure plasma is shown in Fig. 7–60. The initial concentration of SiCl$_4$ is 5.9 mol/kg. Formation of Si occurs at temperatures exceeding 3500 K. The energy cost of Si production from chloride is shown in Fig. 7–61. The minimal energy cost is 22 eV/atom for absolute quenching, which can be achieved at an energy input of 21.5 eV/mol and conversion degree 98%, and 9.3 eV/atom in the case of ideal quenching. Super-ideal quenching due to an intensive centrifugal effect results in a minimal energy cost of 14 eV/atom at plasma temperature 3500 K.

GeF$_4$ is a gas at normal conditions so its direct plasma decomposition can be effectively carried out:

$$GeF_4 \rightarrow Ge + 2F_2, \quad \Delta H = 11.3\,eV/mol. \quad (7\text{--}61)$$

The composition of GeF$_4$ dissociation products in atmospheric-pressure plasma is shown in Fig. 7–62. The initial concentration of GeF$_4$ is 6.7 mol/kg. Formation of germanium takes place at temperatures exceeding 4000 K. The energy cost of germanium production from fluoride is presented in Fig. 7–63. The minimal energy cost is 26 eV/atom in the case of absolute quenching, which can be achieved at a specific energy input of 25 eV/mol and conversion degree 95%, and 23.7 eV/atom in the case of ideal quenching, which can be achieved at a specific energy input of 23.5 eV/mol and conversion degree 99%. Information

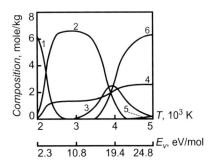

Figure 7–62. Germanium production by direct decomposition of its tetrafluoride (GeF$_4$) in atmospheric-pressure thermal plasma. Composition of products: (1) GeF$_4$; (2) GeF$_2$; (3) GeF; (4) F $\times$ 0.1; (5) Ge$^+$; (6) Ge.

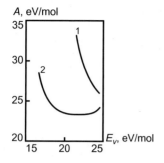

Figure 7–63. Germanium production by direct decomposition of its tetrafluoride (GeF_4) in atmospheric-pressure thermal plasma. Energy cost of the process at different quenching modes: (1) absolute quenching; (2) ideal quenching.

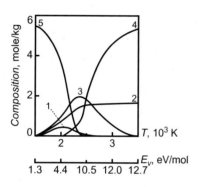

Figure 7–64. Arsenic production by direct decomposition of its trichloride ($AsCl_3$) in atmospheric-pressure thermal plasma. Composition of products: (1) Cl_2; (2) $Cl \times 0.1$; (3) As_2; (4) As; (5) $AsCl_3$.

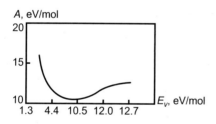

Figure 7–65. Arsenic production by direct decomposition of its trichloride ($AsCl_3$) in atmospheric-pressure thermal plasma. Energy cost of the process at absolute quenching.

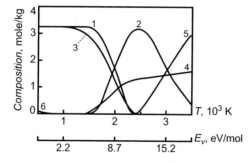

Figure 7–66. Antimony production by direct decomposition of its pentachloride ($SbCl_5$) in atmospheric-pressure thermal plasma. Composition of products: (1) $SbCl_3$; (2) SbCl; (3) Cl_2; (4) $Cl \times 0.1$; (5) Sb; (6) $SbCl_5$.

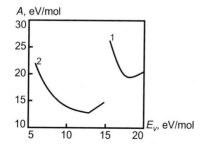

Figure 7–67. Antimony production by direct decomposition of its pentachloride ($SbCl_5$) in atmospheric-pressure thermal plasma. Energy cost of the process at different quenching modes: (1) Sb production from $SbCl_5$, absolute quenching; (2) SbCl production from $SbCl_5$, ideal quenching.

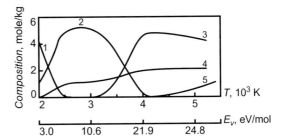

Figure 7–68. Titanium production by direct decomposition of its tetrachloride ($TiCl_4$) in atmospheric-pressure thermal plasma. Composition of products: (1) $TiCl_4$; (2) $TiCl_2$; (3) Ti; (4) $Cl \times 0.1$; (5) Ti^+.

regarding direct thermal plasma decomposition of different inorganic halides of other elements of group 4, tin (Sn) and lead, can be found in Nester et al. (1988).

Halides of elements of group 5 can be represented by $AsCl_3$ and $SbCl_5$. First, arsenic:

$$AsCl_3 \rightarrow As + 3/2Cl_2, \quad \Delta H = 3.2\,\text{eV/mol.} \quad (7\text{--}62)$$

The composition of $AsCl_3$ dissociation products in atmospheric-pressure plasma is shown in Fig. 7–64. The initial concentration of $AsCl_3$ is 5.5 mol/kg. Arsenic formation occurs at temperatures exceeding 2500 K. The energy cost of As production is presented in Fig. 7–65. The minimal energy cost in the absolute quenching mode is 26 eV/atom.

Antimony pentachloride ($SbCl_5$) is a liquid at normal conditions, and its decomposition into gas-phase tetrachloride starts at 379 K. Thermal plasma decomposition of the pentachloride can be presented as

$$SbCl_5 \rightarrow Sb + 5/2Cl_2, \quad \Delta H = 4.5\,\text{eV/mol.} \quad (7\text{--}63)$$

The composition of $SbCl_5$ dissociation products in atmospheric-pressure plasma is shown in Fig. 7–66. The initial concentration of $SbCl_5$ is 3.3 mol/kg. Antimony formation occurs at temperatures exceeding 2700 K. The energy cost of Sb production from pentachloride is shown in Fig. 7–67. The minimal energy cost is 19 eV/atom for absolute quenching, which can be achieved at an energy input of 18.5 eV/mol and conversion degree 98%. The minimal energy cost is 12.8 eV/atom in the case of ideal quenching.

7.4.5. Direct Thermal Plasma Decomposition of Halides of Titanium (Ti), Zirconium (Zr), Hafnium (Hf), Vanadium (V), and Niobium (Nb)

Titanium chloride ($TiCl_4$) has the lowest boiling temperature among titanium halides ($T_{boil} = 410\,\text{K}$). Its decomposition in thermal plasma can be described by the following chemical equation:

$$TiCl_4 \rightarrow Ti + 2Cl_2, \quad \Delta H = 8.3\,\text{eV/mol.} \quad (7\text{--}64)$$

Composition of $TiCl_4$ dissociation products in atmospheric-pressure plasma is shown in Fig. 7–68. The initial concentration of $TiCl_4$ is 5.3 mol/kg. Titanium formation takes place at temperatures exceeding 3500 K. The energy cost of Ti production is shown in Fig. 7–69.

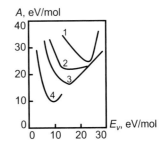

Figure 7–69. Titanium production by direct decomposition of its tetrachloride ($TiCl_4$) in atmospheric-pressure thermal plasma. Energy cost of the process at different quenching modes: (1) absolute quenching; (2) ideal quenching; (3) super-ideal quenching; (4) ideal quenching related to $TiCl_2$ production from $TiCl_4$.

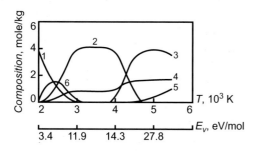

Figure 7–70. Zirconium production by direct decomposition of its tetrachloride ($ZrCl_4$) in atmospheric-pressure thermal plasma. Composition of products: (1) $ZrCl_4$; (2) $ZrCl_2$; (3) Zr; (4) Cl × 0.1; (5) Zr^+.

The minimal energy cost is 24.9 eV/atom in the case of absolute quenching, which can be achieved at a specific energy input of 21.9 eV/mol and conversion degree 88%. The minimal energy cost is 22.2 eV/atom for ideal quenching, which corresponds to an energy input of 14 eV/mol. The minimal energy cost in the super-ideal quenching mode, assuming effective centrifugal effect, is 16.2 eV/mol at plasma temperature 3500 K. Also Fig. 7–69 presents the energy cost of formation of $TiCl_2$ from $TiCl_4$ in the thermal decomposition process $TiCl_4 \rightarrow TiCl_2 + Cl_2$ in the ideal quenching mode. The minimal energy cost of formation of $TiCl_2$ in this case is 9.5 eV/mol, which corresponds to a specific energy input of 9.8 eV/mol.

Direct decomposition of zirconium tetrachloride (sublimation temperature 606 K) is

$$ZrCl_4 \rightarrow Zr + 2Cl_2, \quad \Delta H = 10.15 \, \text{eV/mol}. \quad (7\text{–}65)$$

The composition of $ZrCl_4$ dissociation products in atmospheric-pressure plasma is shown in Fig. 7–70. The initial concentration of $ZrCl_4$ is 4.3 mol/kg. Zirconium formation takes place at temperatures exceeding 4200 K. The energy cost of Zr production is shown in Fig. 7–71. The minimal energy cost is 31 eV/atom for the absolute quenching. The minimal energy cost in the ideal quenching mode is 26.3 eV/atom. The minimal energy cost in the super-ideal quenching mode (centrifugal effect) is 18.9 eV/mol at plasma temperature 4500 K. Figure 7–71 also shows the energy cost of formation of dichloride ($ZrCl_2$) from tetrachloride in the thermal plasma decomposition process, $ZrCl_4 \rightarrow ZrCl_2 + Cl_2$, in the ideal quenching mode. The minimal energy cost of formation of $TiCl_2$ in this case is 10.7 eV/mol.

Direct plasma decomposition of hafnium tetrachloride (sublimation temperature 588 K) is

$$HfCl_4 \rightarrow Hf + 2Cl_2, \quad \Delta H = 10.3 \, \text{eV/mol}. \quad (7\text{–}66)$$

The composition of $HfCl_4$ dissociation products in atmospheric-pressure plasma is shown in Fig. 7–72. The initial concentration of $HfCl_4$ is 3.1 mol/kg. Hafnium formation takes place at temperatures exceeding 4200 K. The energy cost of hafnium production from tetrachloride is shown in Fig. 7–73. The minimal energy cost is 30 eV/atom in

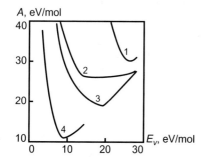

Figure 7–71. Zirconium production by direct decomposition of its tetrachloride ($ZrCl_4$) in atmospheric-pressure thermal plasma. Energy cost of the process at different quenching modes: (1) absolute quenching; (2) ideal quenching; (3) super-ideal quenching; (4) ideal quenching related to $ZrCl_2$ production from $TiCl_4$.

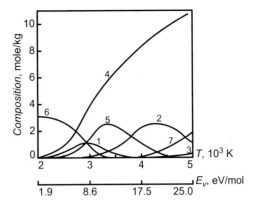

Figure 7–72. Hafnium production by direct decomposition of its tetrachloride ($HfCl_4$) in atmospheric-pressure thermal plasma. Composition of products: (1) $HfCl_3$; (2) HfCl; (3) Hf^+; (4) Cl; (5) $HfCl_2$; (6) $HfCl_4$; (7) Hf.

the case of absolute quenching. The minimal energy cost in the ideal quenching mode is 26.5 eV/atom, and the minimal energy cost in the super-ideal quenching mode (centrifugal effect) is 19 eV/mol at plasma temperature 4500 K. Figure 7–73 also shows the energy cost of formation of dichloride ($HfCl_2$) from tetrachloride in the decomposition process $HfCl_4 \rightarrow HfCl_2 + Cl_2$ in the ideal quenching mode. The minimal energy cost of formation of $HfCl_2$ in this case is 11.3 eV/mol.

Direct plasma decomposition of vanadium tetrachloride (boiling temperature 588 K) is

$$VCl_4 \rightarrow V + 2Cl_2, \quad \Delta H = 5.9 \, \text{eV/mol}. \quad (7–67)$$

The composition of VCl_4 dissociation products in atmospheric-pressure plasma is shown in Fig. 7–74. The initial concentration of VCl_4 is 5.2 mol/kg. Vanadium formation occurs at temperatures exceeding 4500 K. The energy cost of V production is shown in Fig. 7–75. The minimal energy cost is 26.9 eV/atom in the case of absolute quenching, which can be achieved at a specific energy input of 25.4 eV/mol and conversion degree 94%. The minimal energy cost in the ideal quenching mode is 12.6 eV/atom, which can be achieved at a specific energy input of 12.5 eV/mol and conversion degree 99%.

Direct thermal plasma decomposition of niobium pentachloride (boiling temperature 520 K) can be presented by chemical equation

$$NbCl_5 \rightarrow Nb + 5/2 Cl_2, \quad \Delta H = 8.3 \, \text{eV/mol}. \quad (7–68)$$

The composition of $NbCl_5$ dissociation products in atmospheric-pressure plasma is shown in Fig. 7–76. The initial concentration of $NbCl_5$ is 3.7 mol/kg. Niobium formation occurs at temperatures exceeding 3700 K. The energy cost of niobium production is shown in Fig. 7–77. The minimal energy cost is 28.3 eV/atom in the case of absolute quenching, which can be achieved at a specific energy input of 28.1 eV/mol and conversion degree of 99%.

Figure 7–73. Hafnium production by direct decomposition of its tetrachloride ($HfCl_4$) in atmospheric-pressure thermal plasma. Energy cost of the process at different quenching modes: (1) absolute quenching; (2) ideal quenching; (3) super-ideal quenching; (4) ideal quenching related to $HfCl_2$ production from $HfCl_4$.

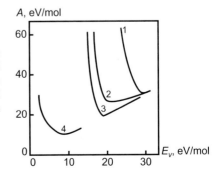

463

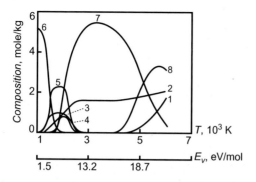

Figure 7–74. Vanadium production by direct decomposition of its tetrachloride (VaCl$_4$) in atmospheric-pressure thermal plasma. Composition of products: (1) V$^+$; (2) Cl $\times$ 0.1; (3) Cl$_2$; (4) VCl$_2$; (5) VCl$_3$; (6) VCl$_4$; (7) VCl; (8) V.

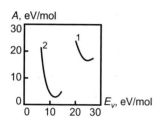

Figure 7–75. Vanadium production by direct decomposition of its tetrachloride (VCl$_4$) in atmospheric-pressure thermal plasma. Energy cost of the process at different quenching modes: (1) V production from VCl$_4$, absolute quenching; (2) VCl production from VCl$_4$, ideal quenching.

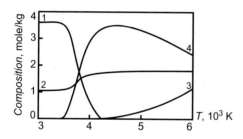

Figure 7–76. Niobium production by direct decomposition of its pentachloride (NbCl$_5$) in atmospheric-pressure thermal plasma. Composition of products: (1) NbCl2; (2) Cl $\times$ 0.1; (3) Nb$^+$; (4) Nb.

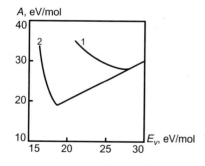

Figure 7–77. Niobium production by direct decomposition of its pentachloride (NbCl$_5$) in atmospheric-pressure thermal plasma. Energy cost of the process at different quenching modes: (1) absolute quenching; (2) super-ideal quenching.

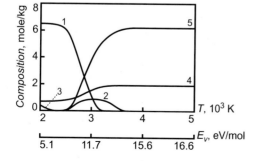

Figure 7–78. Iron production by direct decomposition of its trichloride (FeCl$_3$) in atmospheric-pressure thermal plasma. Composition of products: (1) FeCl$_2$; (2) FeCl; (3) FeCl$_3$; (4) Cl $\times$ 0.1; (5) Fe.

7.4. Direct Decomposition of Halides in Thermal and Non-Thermal Plasma

Figure 7–79. Iron production by direct decomposition of its trichloride ($FeCl_3$) in atmospheric-pressure thermal plasma. Energy cost of the process at different quenching modes: (1) absolute quenching; (2) ideal quenching; (3) super-ideal quenching.

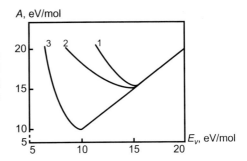

The minimal energy cost in the super-ideal quenching mode, assuming effective centrifugal effect, is 18.7 eV/mol at a plasma temperature of 4000 K.

7.4.6. Direct Decomposition of Halides of Iron (Fe), Cobalt (Co), Nickel (Ni), and Other Transition Metals in Thermal Plasma

Direct thermal plasma decomposition of $FeCl_3$ (boiling temperature 588 K) can be presented as

$$FeCl_3 \rightarrow Fe + 3/2Cl_2, \quad \Delta H = 4.14 \, eV/mol. \quad (7\text{--}69)$$

The composition of $FeCl_3$ dissociation products in atmospheric-pressure plasma is shown in Fig. 7–78. The initial concentration of $FeCl_3$ is 6.2 mol/kg. Iron formation occurs at temperatures exceeding 2700 K. The energy cost of Fe production is shown in Fig. 7–79. The minimal energy cost is 15.5 eV/atom for absolute quenching, which can be achieved at a specific energy input of 15.2 eV/mol. The minimal energy cost in the ideal quenching mode is 15.1 eV/atom. The minimal energy cost in the super-ideal quenching mode (centrifugal effect) is 9.7 eV/mol at a plasma temperature of 3000 K.

Halides of cobalt have essentially higher boiling temperatures, which makes their treatment in plasma more complicated. Direct thermal plasma decomposition of $CoCl_2$ ($T_{boil} = 1320 K$) is

$$CoCl_2 \rightarrow Co + Cl_2, \quad \Delta H = 3.4 \, eV/mol. \quad (7\text{--}70)$$

Composition of $CoCl_2$ dissociation products in atmospheric-pressure plasma is shown in Fig. 7–80. The initial concentration of $CoCl_2$ is 7.7 mol/kg. Cobalt formation occurs at temperatures exceeding 2500 K. The energy cost of Co production is shown in Fig. 7–81. The minimal energy cost is 13 eV/atom for absolute quenching at a specific energy input of 12.5 eV/mol and conversion degree of 98%. The minimal energy cost in the ideal quenching mode is 12.7 eV/atom. The minimal energy cost in the super-ideal quenching mode, assuming effective centrifugal effect, is 7.1 eV/mol at a plasma temperature of 3000 K.

Figure 7–80. Cobalt production by direct decomposition of its dichloride ($CoCl_2$) in atmospheric-pressure thermal plasma. Composition of products: (1) $CoCl_2$; (2) CoCl; (3) $Cl \times 0.1$; (4) Co.

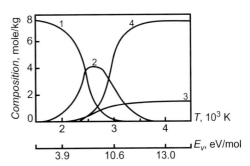

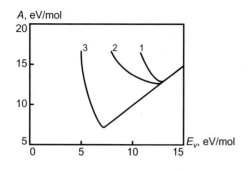

Figure 7–81. Cobalt production by direct decomposition of its dichloride (CoCl$_2$) in atmospheric-pressure thermal plasma. Energy cost of the process at different quenching modes: (1) absolute quenching; (2) ideal quenching; (3) super-ideal quenching.

Halides of nickel also require high temperatures for vaporization. Thermal plasma decomposition of nickel chloride NiCl$_2$ (sublimation temperature 1240 K) can be presented as

$$NiCl_2 \rightarrow Ni + Cl_2, \quad \Delta H = 3.15\,eV/mol. \tag{7–71}$$

The composition of NiCl$_2$ dissociation products in atmospheric-pressure plasma is shown in Fig. 7–82. The initial concentration of NiCl$_2$ is 7.7 mol/kg. Nickel formation takes place at temperatures exceeding 2500 K. The energy cost of nickel production from the chloride is shown in Fig. 7–83. The minimal energy cost is 12.6 eV/atom in the case of absolute quenching, which can be achieved at a specific energy input of 12.2 eV/mol and conversion degree of 97%. The minimal energy cost in the ideal quenching mode is 11.3 eV/atom, which can be achieved at a specific energy input of 11.1 eV/mol and conversion degree of 98%. The minimal energy cost in the super-ideal quenching mode, assuming effective centrifugal effect, is 6.9 eV/mol at a temperature in the plasma zone of 3100 K.

Direct plasma decomposition of CrCl$_3$ (sublimation temperature 1240 K) can be presented as

$$CrCl_3 \rightarrow Cr + 3/2Cl_2, \quad \Delta H = 5.9\,eV/mol. \tag{7–72}$$

The composition of CrCl$_3$ dissociation products in atmospheric-pressure plasma is shown in Fig. 7–84. The initial concentration of CrCl$_3$ is 6.3 mol/kg. Chromium formation occurs at temperatures exceeding 3000 K. The energy cost of Cr production is shown in Fig. 7–85. The minimal energy cost is 17.9 eV/atom for absolute quenching, which can be achieved at a specific energy input of 17.2 eV/mol and conversion degree of 96%. The minimal energy cost in the ideal quenching mode is 16.4 eV/atom at a specific energy input of 16.1 eV/mol and conversion degree of 99%. The minimal energy cost in the super-ideal quenching mode (centrifugal effect) is 11.5 eV/mol at a plasma temperature of 2900 K. Figure 7–84 also shows the energy cost of formation of CrCl$_2$ from CrCl$_3$ in the process CrCl$_3 \rightarrow$ CrCl$_2$ + 1/2Cl$_2$ in the ideal quenching mode. The minimal energy cost of

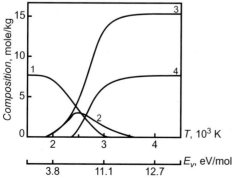

Figure 7–82. Nickel production by direct decomposition of its dichloride (NiCl$_2$) in atmospheric-pressure thermal plasma. Composition of products: (1) NiCl$_2$; (2) NiCl; (3) Cl; (4) Ni.

7.4. Direct Decomposition of Halides in Thermal and Non-Thermal Plasma

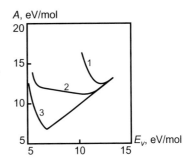

Figure 7–83. Nickel production by direct decomposition of its dichloride ($NiCl_2$) in atmospheric-pressure thermal plasma. Energy cost of the process at different quenching modes: (1) absolute quenching; (2) ideal quenching; (3) super-ideal quenching.

formation of $CrCl_2$ is 8.3 eV/mol, which corresponds to a specific energy input of 7.5 eV/mol and conversion degree of 93%.

Molybdenum fluoride (MoF_6) is a gas at temperatures not much exceeding room temperature ($T_{boil} = 307$ K), which simplifies its treatment in electric discharges. Direct plasma decomposition of molybdenum fluoride is a strongly endothermic process, which can be presented by the following chemical equation:

$$MoF_6 \rightarrow Mo + 3F_2, \quad \Delta H = 16.4 \text{ eV/mol}. \tag{7–73}$$

The composition of MoF_6 dissociation products in atmospheric-pressure plasma is shown in Fig. 7–86. The initial concentration of MoF_6 is 4.8 mol/kg. Molybdenum formation occurs at temperatures exceeding 3500 K. The energy cost of Mo production is shown in Fig. 7–87. The minimal energy cost is 30 eV/atom in the case of absolute quenching, which can be achieved at specific energy input 26.4 eV/mol and conversion degree of 88%. The ideal quenching mode in this process is not much different from the absolute one. The minimal energy cost in the super-ideal quenching mode, assuming effective centrifugal effect, is 25.6 eV/mol. Figure 7–87 also shows the energy cost of formation of MoF_4 from hexafluoride in the thermal plasma decomposition process $MoF_6 \rightarrow MoF_4 + F_2$ in the ideal quenching mode. The minimal energy cost of formation of MoF_4 in this case is 10.5 eV/mol, which corresponds to a specific energy input of 9.8 eV/mol and conversion degree of 93%.

Tungsten fluoride (WF_6) is a gas at normal conditions ($T_{boil} = 290$ K). Direct thermal plasma decomposition of WF_6 is a strongly endothermic process, which can be presented as

$$WF_6 \rightarrow W + 3F_2, \quad \Delta H = 17.8 \text{ eV/mol}. \tag{7–74}$$

The composition of WF_6 dissociation products in atmospheric-pressure plasma is presented in Fig. 7–88. The initial concentration of WF_6 is 3.36 mol/kg. Tungsten formation takes place at high temperatures exceeding 4500 K. The energy cost of W production is shown in Fig. 7–89. The minimal energy cost is 43.2 eV/atom for absolute quenching, which

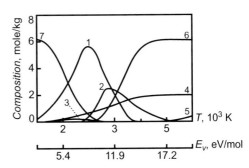

Figure 7–84. Chromium production by direct decomposition of its trichloride ($CrCl_3$) in atmospheric-pressure thermal plasma. Composition of products: (1) $CrCl_2$; (2) $CrCl$; (3) Cl_2; (4) $Cl \times 0.1$; (5) Cr^+; (6) Cr; (7) $CrCl_3$.

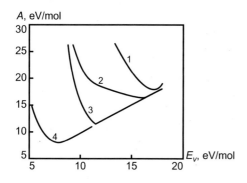

Figure 7–85. Chromium production by direct decomposition of its trichloride ($CrCl_3$) in atmospheric-pressure thermal plasma. Energy cost of the process at different quenching modes: (1) absolute quenching; (2) ideal quenching; (3) super-ideal quenching; (4) ideal quenching related to production of $CrCl_2$ from $CrCl_3$.

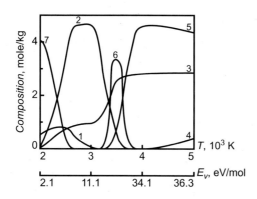

Figure 7–86. Molybdenum production by direct decomposition of its hexafluoride (MoF_6) in atmospheric-pressure thermal plasma. Composition of products: (1) MoF_5; (2) MoF_4; (3) $F \times 0.1$; (4) Mo^+; (5) Mo; (6) Mo (condensed); (7) MoF_6.

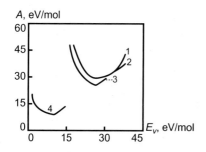

Figure 7–87. Molybdenum production by direct decomposition of its hexafluoride (MoF_6) in atmospheric-pressure thermal plasma. Energy cost of the process at different quenching modes: (1) absolute quenching; (2) ideal quenching; (3) super-ideal quenching; (4) ideal quenching related to production of MoF_4 from MoF_6.

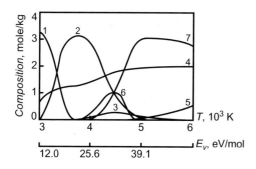

Figure 7–88. Tungsten production by direct decomposition of its hexafluoride (WF_6) in atmospheric-pressure thermal plasma. Composition of products: (1) WF_4; (2) WF_2; (3) WF; (4) $F \times 0.1$; (5) W^+; (6) W (condensed); (7) W.

7.4. Direct Decomposition of Halides in Thermal and Non-Thermal Plasma

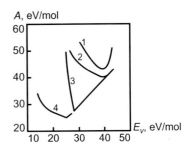

Figure 7–89. Tungsten production by direct decomposition of its hexafluoride (WF_6) in atmospheric-pressure thermal plasma. Energy cost of the process at different quenching modes: (1) absolute quenching; (2) ideal quenching; (3) ideal quenching related to production of WF_2 from WF_6.

can be achieved at a specific energy input of 39.1 eV/mol and conversion degree of 91%. The minimal energy cost in the ideal quenching mode is 40 eV/atom at a specific energy input of 39.1 eV/mol and conversion degree of 98%. The minimal energy cost in the super-ideal quenching mode (centrifugal effect) is 28 eV/mol at plasma temperature 4500 K. Figure 7–89 also presents the energy cost of formation of tungsten fluoride (WF_2) from hexafluoride in the plasma decomposition process $WF_6 \rightarrow WF_2 + 2F_2$ in the ideal quenching mode. The minimal energy cost of formation of WF_2 in this case is 25.6 eV/mol. All thermal plasma processes of direct reduction of elements from their halides require special fast product cooling and stabilization to reach the absolute and ideal quenching conditions (Tumanov, 1981; Ambrazevicius, 1983). Requirements for the cooling and stabilization, however, are not necessarily as strong as those discussed in Section 7.4.2 regarding reduction of uranium fluoride.

7.4.7. Direct Decomposition of Halides and Reduction of Metals in Non-Thermal Plasma

Direct decomposition of halides of alkali metals in a strongly non-equilibrium microwave discharge was investigated by McTaggart (1965a,b). The discharge was maintained in a quartz tube crossing a waveguide resonator; metal halides were diluted by inert gases or molecular nitrogen. Reduced metals were collected on the quartz walls of the plasma-chemical reactor in the form of a mirror-like deposit; the gas-phase products were halogens. The microwave plasma experiments show that most intensive dissociation takes place for lithium and sodium halides. The dissociation degree of halides of potassium and cesium is lower at similar discharge parameters. Specifically, the non-thermal plasma treatment of lithium iodide vapor (LiI) results in the production of metal with a conversion degree of 70%. The dissociation of NaCl vapor is characterized in this system by a 30% conversion degree. Similar results were obtained in experiments with a direct plasma decomposition of halides of alkaline earth metals in non-thermal microwave discharges. Products in this case were mostly monohalides of the metals. Interesting results were achieved in non-thermal plasma decomposition of higher metal halides, particularly ZrI_4 and $TiCl_4$. Application of non-thermal plasma provides effective stabilization of pure metals (specifically zirconium and titanium) without using hydrogen and, therefore, without product contamination with hydrogen. Both processes are of significant practical interest, especially the dissociation of ZrI_4, which is a crucial step in zirconium refinement in the nuclear industry. Experiments with dissociation of ZrI_4 were carried out in non-equilibrium microwave and non-equilibrium RF discharges at moderate pressure (Moukhametshina et al., 1986a,b); they are summarized in Fig. 7–90. Complete conversion of the iodide and pure metal production without adding hydrogen has been achieved at an energy cost of about 40 kWh/kg of zirconium in the microwave discharge, and at an energy cost of about 200 kWh/kg of zirconium in the RF discharge. Experimental results are compared with modeling results, which predict higher energy efficiency at lower energy inputs (about 10 kWh/m³ of gaseous ZrI_4).

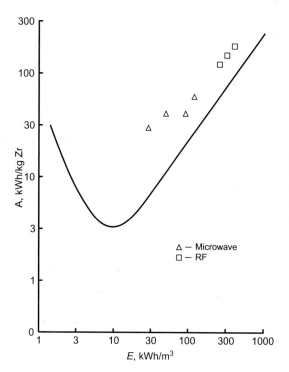

Figure 7–90. Energy cost of direct decomposition of zirconium tetraiodide as function of specific energy input: triangles, experiments with microwave discharge; squares, experiments with radiofrequency discharge; solid line, modeling.

Complete dissociation of chlorides, which have higher bonding energies, is more challenging than that of iodides. The effective one-step dissociation of $TiCl_4$ and production of metallic titanium without adding hydrogen has been achieved in a strongly non-equilibrium stationary plasma-beam discharge (Atamanov, Ivanov, & Nikiforov, 1979a,b). Gaseous titanium tetrachloride passes through an active zone of the plasma-beam discharge and dissociates there; metallic titanium is collected on a special quenching substrate. Some amount of produced chlorine is adsorbed on the metallic surface, which leads to characterization of the product composition as $TiCl_{k<0.2}$. Non-thermal plasma dissociation of SF_6 was observed in continuous corona and spark discharges without using hydrogen as a reducing agent (Edelson, 1953). The main dissociation product in these atmospheric-pressure non-thermal plasma experiments was S_2F_2.

7.4.8. Kinetics of Dissociation of Metal Halides in Non-Thermal Plasma: Distribution of Halides over Oxidation Degrees

The high efficiency of one-step halide decomposition and metal production in non-thermal plasma, without using hydrogen as a reducing agent, is due to the stability of the solid product to halogen atoms at low temperatures. Such low-temperature heterogeneous product stabilization is possible at gas pressures sufficient for volume recombination of halogens but still permits access of metal atoms to a quenching surface. Volume recombination of halogens (Ha) is due to the following process:

$$Ha + MeHa_k \rightarrow MeHa_{k+1}, \quad 1 \leq k \leq n - 1, \quad (7\text{–}75)$$

where n is the maximum number of halogen atoms in the initial halide $MeHa_n$. Volume losses of metal atoms (Me) are mostly due to reactions of disproportioning:

$$Me + MeHa_k \rightarrow MeHa_{k-1} + MeHa, \quad 2 \leq k \leq n. \quad (7\text{–}76)$$

7.4. Direct Decomposition of Halides in Thermal and Non-Thermal Plasma

The heterogeneous stabilization of metals can be achieved when reaction (7–75) is fast enough and occurs before the halogen atoms attack the surface, whereas reaction (7–76) is slow enough and metal atoms are able to reach the surface. Calculation of the rates of reactions (7–75) and (7–76) requires finding the distribution function $f(k)$ of the halides $MeHa_{k \leq n}$ over the oxidation degrees ($k \leq n$). Halogen detachment from the metal halides is essentially due to dissociative attachment processes:

$$e + MeHa_k \rightarrow MeHa_{k-1}^- + Ha, \tag{7–77}$$

$$e + MeHa_k \rightarrow MeHa_{k-1} + Ha^-. \tag{7–78}$$

Distribution $f(k)$ over oxidation degrees is established due to fast disproportioning:

$$MeHa_k + MeHa_l \rightarrow MeHa_{k-1} + MeHa_{l+1}. \tag{7–79}$$

Assuming that the disproportioning rates are not changing significantly with the oxidation degree k, the number density of the halides with oxidation degree k is (Legasov et al., 1978c)

$$n_{k=n-m} = n_0 \cdot f(m) = An_0 \exp(-\alpha m), \tag{7–80}$$

where n_k is the density of $MeHa_k$; n_0 is the density of all compounds $MeHa_k (0 \leq k \leq n)$; m is the number of detached halogen atoms ($m = n - k$); and A is the normalization factor:

$$A = \frac{1 - \exp(-\alpha)}{1 - \exp[-\alpha(n + 1)]}. \tag{7–81}$$

The exponential factor α is determined by the average number $\langle m \rangle$ of detached halogen atoms:

$$\langle m \rangle = \sum_{m=0}^{n} m A \exp(-\alpha m) = \frac{1}{1 - \exp[-\alpha(n + 1)]}$$
$$\times \left\{ \frac{1 - \exp(-\alpha n)}{\exp \alpha - 1} - n \exp[-\alpha(n + 1)] \right\}. \tag{7–82}$$

The asymptotic solution of equation (7–82) with respect to the parameter α results in the general expression for the distribution function of halides over their oxidation degrees, $f(m = n - k)$:

$$f(m = n - k) \approx A(n, \langle m \rangle) \times \left(1 + \frac{1}{\langle m \rangle} \right)^{-m\{1 - \frac{n/\langle m \rangle}{\exp[(n+1)/\langle m \rangle]-1}\}}. \tag{7–83}$$

The normalization factor $A(n, \langle m \rangle)$ can be found in this case from summation:

$$A(n, \langle m \rangle) = \frac{\left(1 + \frac{1}{\langle m \rangle}\right)^{-\{1 - \frac{n/\langle m \rangle}{\exp((n+1)/\langle m \rangle)-1}\}} - 1}{\left(1 + \frac{1}{\langle m \rangle}\right)^{-(n+1)\{1 - \frac{n/\langle m \rangle}{\exp((n+1)/\langle m \rangle)-1}\}} - 1}. \tag{7–84}$$

In the simplest case, when the number of detached halogen atoms is relatively small with respect to the total number of atoms in the halide $1 < (\langle m \rangle) \ll n$, the distribution function can be expressed as

$$f(k) = \frac{1}{\langle m \rangle} \exp\left(-\frac{n - k}{\langle m \rangle} \right). \tag{7–85}$$

7.4.9. Heterogeneous Stabilization of Products During Direct Decomposition of Metal Halides in Non-Thermal Plasma: Application of Plasma Centrifuges for Product Quenching

The heterogeneous stabilization of metals, produced by non-thermal plasma dissociation of halides, can be achieved when the mean free path of halogen atoms, λ_{Ha}, related to recombination (7–75) is shorter than the mean free path of metal atoms, λ_{Me}, related to disproportioning (7–76). Assuming the distribution function of halides $MeHa_k$ over their oxidation degrees $f(k)$ in exponential form (7–85), the ratio of the mean free paths of halogen and metal atoms can be expressed as

$$\gamma = \frac{\lambda_{Ha}}{\lambda_{Me}} = \frac{\frac{1}{\langle m \rangle} \sum_{k=2}^{n} \exp\left(\frac{k-n}{\langle m \rangle}\right) \times \left(\frac{T_0}{E_a^k}\right)^{s(k)} \times \exp\left(-\frac{E_a^k}{T_0}\right)}{1 - \frac{1}{\langle m \rangle} \exp\left(-\frac{n}{\langle m \rangle}\right)}, \qquad (7-86)$$

where E_a^k are activation barriers of reactions (7–76), $s(k)$ is the effective number of vibrational degrees of freedom of molecules $MeHa_k$ and n is the maximum value of k ($MeHa_n$ is an initial halide). Expression (7–86) can be simplified assuming $E_a^k = E_a$ and $s(k) \propto k$, which leads to the formula:

$$\gamma = \frac{\lambda_{Ha}}{\lambda_{Me}} \approx \frac{1}{\langle m \rangle} \exp\left(-\frac{E_a}{T_0}\right) \times \left(\frac{T_0}{E_a}\right)^2 \left[\exp\left(-\frac{n}{\langle m \rangle}\right) - \left(\frac{T_0}{E_a}\right)^n\right], \qquad (7-87)$$

where $\gamma \ll 1$ in non-equilibrium plasma when gas temperature is low ($T_0 \ll E_a$). Thus, if the pressure is high enough and $\lambda_{Ha} \leq d$ (where d is the characteristic size of the discharge chamber), active halogen atoms do not attack the metal surface, which provides effective quenching. On the other hand, if $\gamma \ll 1$ and $\lambda_{Me} = (1/\gamma)\lambda_{Ha} \gg d$, metal atoms are able to safely reach the surface. The effect of chemisorption of intermediate halides on the product quality is discussed by Legasov, Rusanov, and Fridman (1978c).

The heterogeneous stabilization of products is mostly due to the following kinetic effect. Disproportioning reactions (7–75), which eliminate metal atoms, have activation barriers, whereas recombination reactions consuming halogen atoms (7–76) do not. A similar effect provides surface stabilization of carbon produced by complete dissociation of CO_2 in strongly non-equilibrium microwave discharge (see Section 5.7). The quenching effect in non-thermal plasma dissociation of halides can also be achieved by product separation due to the centrifugal effect (Rusanov & Fridman, 1984). Especially interesting is the application of plasma centrifuges, where plasma rotates in crossed electric and magnetic fields with extremely high rotation rates (Fridman & Kennedy, 2004). Such systems provide simultaneous dissociation of molecules and centrifugal product separation, which results in effective quenching (Korobtsev, 1982; Korobtsev et al., 1982). The application of such systems for decomposition of $TiCl_4$ and production of metallic titanium has been demonstrated by Babaritsky, Ivanov, and Shapkin (1977).

7.5. PLASMA-CHEMICAL SYNTHESIS OF NITRIDES AND CARBIDES OF INORGANIC MATERIALS

7.5.1. Plasma-Chemical Synthesis of Metal Nitrides from Elements: Gas-Phase and Heterogeneous Reaction Mechanisms

The production of metal nitrides with exact stoichiometry is made effective by preliminary evaporation of metals in conjunction with gas-phase synthesis. Synthesis of nitrides by evaporation of such refractory metals as titanium (Ti), zirconium (Zr), hafnium (Hf), niobium (Nb), and tantalum (Ta) and subsequent gas-phase reactions with nitrogen require plasma temperatures. Such gas-phase synthesis of nitrides of Ti, Zr, Hf, Al, Nb, and Ta in thermal plasma of RF-ICP discharges has been demonstrated by Grabis et al. (1975) and Polak et al. (1975a). Metal powder in these systems is radially injected for evaporation into an RF-ICP N_2 plasma flow after the RF inductor. Exothermic synthesis of nitrides takes place after the discharge zone. Metal powder particles of 20–80 μm reach boiling temperature after about 3–8 μs, which results in the gas-phase synthesis of nitrides. Heterogeneous synthesis was not observed in this case. Heterogeneous plasma synthesis of metal nitrides was done in arc discharges by Matsumoto (1973), Matsumoto and Abe (1975) and Matsumoto and Hayakawa (1966). The anode spot of the thermal arc in nitrogen, thereby in these systems is attached to the metallic tablet, which melts and reacts with nitrogen, thereby generating the nitrides. The process is determined by the diffusion of atomic and molecular nitrogen into the melted metal. Metal also evaporates from the tablet; therefore, gas-phase synthesis also takes place. The arc parameters are voltage 65–75 V, current 70–120 A, and nitrogen flow rate 1.5 L/min. According to Matsumoto and Hayakawa (1966), Ar plasma interaction with Ta surface leads to the formation of tantalum nitride powder. Sublimation of the powder finally results in tantalum nitride recondensation on a water-cooled copper spiral. The tantalum nitride yield increases with current density and N_2 flow rate, with saturation occurring at certain flow rates. The rate of metal nitride synthesis in the heterogeneous systems is limited by nitrogen diffusion into the molten metals. The diffusion into the melt is characterized specifically for titanium and zirconium by a diffusion coefficient of 10^{-4}–10^{-3} cm^2/s, exceeding that of solid metals by a factor of 10^8. Many heterogeneous processes for nitride synthesis from elements, including Ti, Zr, Al, Ti, B, and Mg, were performed by interaction of an N_2 plasma jet with powders (Vurzel, 1970; Mosse & Pechkovsky, 1973). Nitride film is formed on the surface of the powder particles for a period of about 10^{-3} s. Further synthesis is limited by atomic nitrogen diffusion across the film (Grabis et al., 1971; Miller, 1971a,b).

7.5.2. Synthesis of Nitrides of Titanium and Other Elements by Plasma-Chemical Conversion of Their Chlorides

Nitrides of elements, including titanium (Ti), silicon (Si), aluminum (Al), boron (B), zirconium (Zr), hafnium (Hf), niobium (Nb), and tantalum (Ta) can be produced by conversion of their chlorides in plasma in the presence of hydrogen and nitrogen, or ammonia NH_3 (Neuenschwander, Schnet, & Scheller, 1965; Murdoch & Hamblyn, 1967; DeVink, 1970, 1971a,b; Dale, 1974). This process takes place in the gas phase and, therefore, is quite effective. As an example, the yield of Si_3N_4 produced in this way is 80%.

Synthesis of titanium nitride from $TiCl_4$ interacting with H_2–N_2 is a slightly endothermic reaction:

$$(TiCl_4)_{gas} + 2(H_2)_{gas} + \frac{1}{2}(N_2)_{gas} \rightarrow (TiN)_{solid} + 4(HCl)_{gas}, \quad \Delta H = 0.6\,eV/mol.$$
$$(7\text{–}88)$$

Thermodynamic analysis of the system Ti–N–H–Cl, characterizing the plasma synthesis of TiN from $TiCl_4$, is presented in Fig. 7–91 (Koriagin et al., 1973). The maximum concentration of titanium nitride can be achieved at temperatures of 1000–1700 K. The yield of titanium nitride depends on $TiCl_4$ dilution with nitrogen and hydrogen. The major effect on $TiCl_4$ conversion is related to hydrogen dilution. Increasing the ratios H/Cl and N/Ti from 1 to 5 and from 1 to 100 results in a titanium nitride yield growth from 40 to 100%. TiN synthesis has been demonstrated in microwave nitrogen plasma with injection of $TiCl_4$–H_2 mixture (Troitsky et al., 1971a,b). The conversion degree of $TiCl_4$ to nitride was up to 100%. The nitride was synthesized in the form of a fine powder with a typical composition of $TiN_{0.8}$. The specific surface area of the powder was 45 m^2/g and particle size was 10–50 nm. The energy cost of titanium nitride was 18 kWh/kg, which is close to

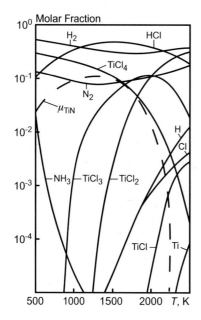

Figure 7–91. Quasi-equilibrium composition of heterogeneous atmospheric-pressure plasma system Ti–H–N–Cl with composition H:Cl = 1, Ti:N = 1, Ti:Cl = 1:4.

the thermodynamic prediction of 15 kWh/kg. Injection of a $TiCl_4$–BCl_3–H_2 mixture into the microwave N_2 plasma results in the formation of such powders as TiN, TiN + TiB_2, and TiN + TiB_2 + B, depending on the composition of the initial gas mixture (Troitsky, Grebtsov, & Aivasov, 1973).

7.5.3. Synthesis of Silicon Nitride (Si_3N_4) and Oxynitrides by Non-Thermal Plasma Conversion of Silane (SiH_4)

Silicon nitride (Si_3N_4), an important material in micro-electronics, is produced by film deposition in the gas-phase reaction of silane and ammonia:

$$(SiH_4)_{gas} + \frac{4}{3}(NH_3)_{gas} \rightarrow \frac{1}{3}(Si_3N_4)_{solid} + 4(H_2)_{gas}. \qquad (7\text{–}89)$$

The quasi-equilibrium performance process requires temperatures of 970–1170 K. Non-thermal plasma requires much lower temperatures (Kirk, 1974). It is carried out at low pressure in a quartz reactor located in an inductor of an RF generator. Reacting gases SiH_4 and NH_3 are injected into the non-equilibrium RF-ICP reactor near a substrate heated by an infrared lamp. Silicon nitride film deposition on the substrate takes place at substrate temperatures of 370–770 K, and at molar compositions of the mixture in the range $\frac{NH_3}{SiH_4} = \frac{9}{1} - \frac{3}{7}$. The deposition rate of Si_3N_4 in the RF-ICP discharges doesn't depend on substrate temperature but strongly depends on the discharge power and amount of silane in the reactive mixture. A stoichiometric composition of silicon nitride (N:Si = 1.33) is achieved in the deposited film when the mixture composition is NH_3:SiH_4 = 3:1 (Yoyce, Sterling, & Alexander, 1968). When the molar fraction of ammonia is lower (NH_3:SiH_4 < 3), the film includes some amount of elemental silicon mixed with the nitride Si_3N_4. Addition of NO into the SiH_4–NH_3 mixture leads to synthesis of silicon oxynitrides in the strongly non-equilibrium RF-ICP discharges. This deposition occurs on the heated substrate at temperatures similar to those mentioned earlier. The plasma-chemical process can be described as

$$x\,SiH_4 + y\,NO + z\,NH_3 \rightarrow Si_xO_yN_z + \left(2x + \frac{3}{2}z\right)H_2 + \frac{y}{2}N_2. \qquad (7\text{–}90)$$

The non-thermal RF-ICP discharge also permits deposition of SiO_2 film by the gas-phase reaction

$$SiH_4 + 2N_2O \rightarrow SiO_2 + 2N_2 + 2H_2, \qquad (7\text{--}91)$$

and by the non-equilibrium plasma-chemical reaction of gaseous tetraethylortosilicate with oxygen.

7.5.4. Production of Metal Carbides by Solid-Phase Synthesis in Thermal Plasma of Inert Gases

A synthesis of metal carbides in plasma can be provided by heating dispersed or bricked mixture of the metal oxide and carbon in the form of graphite or soot and using thermal discharges in inert gases. The synthesis of metal carbides takes place in such systems in the solid phase and proceeds as follows:

$$(Me_xO_y)_{solid} + zC_{solid} \rightarrow (Me_xC_{z-y})_{solid} + y(CO)_{gas}. \qquad (7\text{--}92)$$

This method was applied to produce carbides of titanium, zirconium, lead, and bismuth (Barcicki & Myrdzik, 1974). Applications of the method are limited because of insufficient contact between solid particles in the presence of plasma flow if the solid mixture is not bricked or sintered. The efficiency of the process arranged in bricks or sintered solid mixture is limited by radiation heat losses and insufficient heat transfer inside of the brick, especially taking into account the decrease of density due to CO formation (7–92). The most effective condensed-phase synthesis of carbides (7–92) is that from melt containing carbon compounds (Tumanov, 1981). A relevant example is the synthesis of carbides of uranium and plutonium from a melt containing their nitrites and carbon compounds (Coppinger & Johnson, 1969).

7.5.5. Synthesis of Metal Carbides by Reaction of Solid Metal Oxides with Gaseous Hydrocarbons in Thermal Plasma

The advantage of metal carbide synthesis via reaction of solid metal oxides with gaseous hydrocarbons is because one reactant is in the gas phase. In contrast to (7–92), a solid oxide reacts in this method with a gas-phase hydrocarbon. Consider, as an example, the synthesis of titanium carbide from titanium oxide (Hancock, 1973; Walder & Hancock, 1973):

$$TiO_2 + CH_4 \rightarrow TiC + 2H_2O. \qquad (7\text{--}93)$$

Titanium nitride synthesis (7–93) is conducted by injection of solid TiO_2 with a CH_4 gas carrier into argon thermal plasma generated by RF-ICP discharge. The production of carbide takes place after the inductor at temperatures of 2000–4000 K. A problem with the technology is the possible deposition of conductive products (carbide or carbon) on the dielectric reactor walls, which prevents effective coupling of the RF inductor with plasma. The deposition can be suppressed by applying the porous internal walls of the reactor with special argon or hydrogen flow, preventing contact of carbide or carbon products with the walls (Evans, Wynne, & Marinowsky, 1968). Plasma-chemical reaction (7–93) occurs in this case outside of plasma. Ar is heated up in the RF-ICP discharge as an energy carrier and stimulates the synthesis of titanium carbide. Conversion of oxides into carbides inside of plasma will be discussed in Section 7.5.7.

7.5.6. Gas-Phase Synthesis of Carbides in Plasma-Chemical Reactions of Halides with Hydrocarbons

The plasma-chemical synthesis of carbide of an element El can be presented as

$$(El_xHal_y)_{gas} + (C_mH_n)_{gas} \rightarrow (El_xC_m)_{solid} + y(HHal)_{gas} + \frac{n-y}{2}(H_2)_{gas}. \qquad (7\text{--}94)$$

475

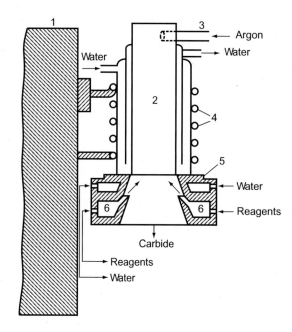

Figure 7–92. Thermal ICP reactor for boron carbide production by BCl_3 conversion in hydrocarbon plasma: (1) RF generator, (2) discharge chamber, (3) argon injection system, (4) RF inductor, (5) water-cooling system, (6) distributor of reagents.

Carbides of metals and other elements have been produced by this approach, including carbides of boron, silicon, titanium, zirconium, hafnium, vanadium, niobium, molybdenum, tungsten, tantalum, and thorium (Funke, Klementiev, & Kosukhin, 1969; Sheppard & Wilson, 1972; MacKinnon & Wickens, 1973; Chase, 1974; Steiger & Wilson, 1974; Swaney, 1974; MacKinnon & Reuben, 1975). The produced carbide particles are very small; their diameter is usually about 20–200 nm. Halides are mixed with hydrocarbons, usually in a ratio $H_2 : Me = 2$–30. The gas mixture is heated up in plasma to temperatures of 1300–4000 K; time of synthesis exceeds 50 ms. As an example, production of submicron boron carbide powder from gaseous boron trichloride and methane occurs in the strongly endothermic process:

$$4(BCl_3)_{gas} + (CH_4)_{gas} \rightarrow (B_4C)_{solid} + 4(HCl)_{gas} + 4(Cl_2)_{gas}, \quad \Delta H = 13.3 \, eV/mol. \tag{7–95}$$

The process is effectively stimulated by argon heated up inside of the RF-ICP discharge and takes place downstream from the thermal discharge, where reactive gases are injected into argon (Tumanov, 1981; see Fig. 7–92). The yield achieved by the process is 93–95%, the composition of the boron chloride is close to a stoichiometric value $B_{3.9}C$, the size of the produced particles is 200–300 nm, and the particles' color varies from gray to black depending on the fraction of carbon in the carbide.

Plasma-chemical processes (7–94) can be applied not only for production of nanopowders of carbides but also for deposition of thin carbide films. Specifically, we mention here the deposition of a boron carbide film based on the gas-phase process similar to (7–95) but with an excess of hydrogen:

$$4BCl_3 + CH_4 + 4H_2 \rightarrow (B_4C)_{deposit} + 12HCl. \tag{7–96}$$

The excess of hydrogen prevents the formation of precursors of the condensed phase and therefore prevents the formation of fine powder, which results in domination by thin-film deposition (Amberger, Druminsky, & Dietze, 1968; Amberger, Druminsky & Ploog, 1970).

7.5.7. Conversion of Solid Oxides into Carbides Using Gaseous Hydrocarbons inside of RF-ICP Thermal Plasma Discharge and Some Other Plasma Technologies for Carbide Synthesis

Most of the plasma conversion processes are conducted after the discharge, where only the energy-carrier gas (often argon) is heated up (see Section 7.5.5). Conversion of a solid oxide into a carbide can be also performed inside of the RF-ICP discharge. As an example, consider the plasma synthesis of silicon carbide from silica in the presence of CH_4 and H_2 (Evans etal., 1968; Tumanov, 1981):

$$CH_4 + SiO_2 + Ar + H_2 \rightarrow SiC + H_2O + Ar + H_2. \qquad (7\text{--}97)$$

A schematic of the plasma reactor is shown in Fig. 7–93. A portion of hydrogen is supplied through the porous wall of the discharge to avoid deposition of carbon there. Such precautions are important because deposition of conductive products on the discharge chamber walls prevents normal RF coupling of the plasma with the inductor. Silicon carbide powder is quenched in a cooling zone below the discharge chamber and is separated from the water vapor, soot, and initial chemicals that are not converted in plasma. Typical parameters of the atmospheric-pressure process are as follows: RF-ICP discharge power 9.2 kW, SiO_2 feeding rate 45 g/min, Ar flow rate 17 L/min, CH_4 flow rate 1.1 L/min, and H_2 flow rate 17 L/min. The conversion degree of silicon oxide into carbide is close to 100%, and the size of produced particles is 40–150 μm. Carbides have also been synthesized in other thermal plasma systems. Synthesis of carbides of uranium and calcium in thermal plasma has been considered by Vurzel (1970) and Mosse and Pechkovsky (1973). Metal carbides were produced via hydrogen plasma reduction of chlorides in the presence of carbon halides: CCl_4 (Murdoch & Hamblyn, 1967) or $C_2H_4Cl_2$ (Swaney, 1968a,b). Titanium carbide produced by this technology from tetrachloride is pure and black. Sizes of synthesized carbide particles are 10–500 nm. Silicon carbide has been produced in thermal hydrogen plasma from gaseous silicon–organic compounds mixed with hydrocarbons (Silbiger & Schnell, 1968). Mixtures of CH_3SiCl_3 or CH_3SiHCl_2 with methane were used in these experiments as initial reagents. Carbides of tungsten and tantalum were produced in a helium plasma jet by injection of a metal or oxide powder in methane flow (Stokes, 1971). The conversion degree into tungsten carbides (W_2C, WC) was as high as 40%, the conversion degree of metallic tantalum into carbide was 70%, and the conversion degree of tantalum oxide into carbide was about 20%.

7.6. PLASMA-CHEMICAL PRODUCTION OF INORGANIC OXIDES BY THERMAL DECOMPOSITION OF MINERALS, AQUEOUS SOLUTIONS, AND CONVERSION PROCESSES

7.6.1. Plasma Production of Zirconia (ZrO₂) by Decomposition of Zircon Sand (ZrSiO₄)

Inorganic oxides with very high melting temperatures can be produced in thermal plasma by decomposition of complex minerals. Zirconia production from zircon sand is an example of such industrial plasma technologies (Thorpe, 1971; Thorpe & Wilks, 1971; Wilks, Ravinder, & Grant, 1972; Wilks, 1973a,b, 1975). Decomposition of zircon sand requires temperatures exceeding 2050 K:

$$(ZrSiO_4)_{solid} \rightarrow (ZrO_2)_{solid} + (SiO_2)_{solid}, \qquad (7\text{--}98)$$

which results in the formation of a mixture of zirconia and silica. Heating of zircon sand to the required temperature is provided by an energy-carrier gas, which is supplied to the reactor through arc discharges. A solid mixture of oxides produced in plasma is then treated

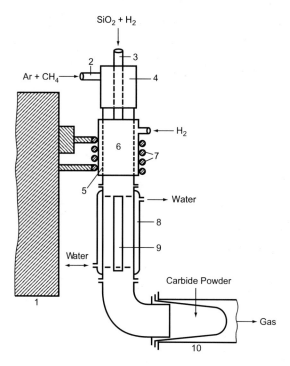

Figure 7–93. Schematic of reactor for silicon carbide production in discharge zone of hydrocarbon plasma: (1) RF generator; (2, 3) inlet of reagents; (4) distributor of initial products; (5) porous wall; (6) discharge zone; (7) inductor; (8) quenching zone; (9) heat exchanger; (10) filter.

by sodium hydroxide (NaOH), which leads to the formation of a soluble sodium salt taking silica SiO_2 into solution:

$$SiO_2 + 2NaOH \rightarrow Na_2SiO_3 + H_2O. \qquad (7\text{–}99)$$

Zirconia remains solid after this treatment and can be easily separated as a product. A general schematic of the technology is shown in Fig. 7–94. Zirconia with purity of 99% is produced as hollow spheres with sizes of 100–200 nm; the energy cost of the zirconia is 1.32 kWh/kg. Thermal decomposition of zircon sand particles starts from their surface and then propagates to the inside of the solid particles. If the residence time of the particles in the high-temperature zone is not sufficient, the central part of the $ZrSiO_4$ particles remains partially converted. Fast plasma heating of the particles results in breakup of the crystals with formation of zirconia (ZrO_2) micro-crystals and amorphous silica (SiO_2). The large-scale plasma technology of zirconia production from zircon sand was first demonstrated in a pilot unit with an arc power of 350 kW and productivity 45.4 t/year, and then later in a bigger unit with an arc power of 1 MW and productivity 450 t/year. Industrial ZrO_2 production by this technology is characterized by about 4550 t/year output and requires about 10 MW of arc discharge power (Tumanov, 1981).

7.6.2. Plasma Production of Manganese Oxide (MnO) by Decomposition of Rhodonite (MnSiO₃)

Thermal decomposition of rhodonite in plasma is similar to that of zircon sand. The mineral rhodonite, which is a manganese silicate ($MnSiO_3$), has a high melting temperature but can be effectively dissociated in thermal plasma with the formation of manganese oxide (MnO) and silica (SiO_2):

$$(MnSiO_3)_{solid} \rightarrow (MnO)_{solid} + (SiO_2)_{solid}. \qquad (7\text{–}100)$$

7.6. Plasma-Chemical Production of Inorganic Oxides

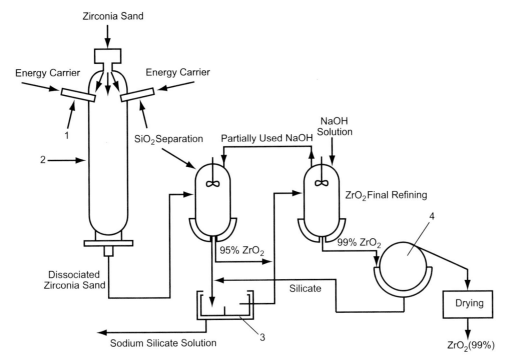

Figure 7–94. Zirconia (ZrO_2) production from zirconia sand ($ZrSiO_4$): (1) plasmatron; (2) plasma reactor; (3) centrifuge; (4) filter.

The decomposition of rhodonite takes place at temperatures exceeding 1560 K. At a little higher temperature (1973 K), the MnO-enriched mixture becomes liquid and can be effectively separated from the silica by conventional approaches, which results in extraction of a final product – pure manganese oxide. The process was done in a thermal RF-ICP discharge by Thursfield and Davies (1974). A schematic of the RF plasma reactor is shown in Fig. 7–95. The mineral rhodonite was supplied to the plasma zone as powder with argon as a gas carrier. Parameters of the atmospheric-pressure system were as follows: RF generator power 30 kW; frequency 4.4–5 MHz; diameter of the quartz plasma reactor, 37–50 mm; flow rate of argon as a plasma gas, 0.3–1 L/s; argon flow rate for mineral powder transportation, 0.01–05 L/s; enthalpy of argon plasma flow, 270–310 kJ/mol; and the plasma flow enthalpy reaches 630 kJ/mol when nitrogen or oxygen is added. The parameters of the MnO production technology (relative and absolute product yield) are presented in Figs.7–96–7–98 as functions of mineral feed rate and specific plasma enthalpy. The conversion of rhodonite into manganese oxide increases by using smaller sizes of initial particles of the mineral and by increasing energy input in the thermal plasma discharge. An increase of the rhodonite feed rate leads to lower values of relative yields of MnO, but the absolute yield of the manganese oxide grows at the same time. The decomposition of rhodonite in an arc discharge was investigated by Harris, Holmgren, and Korman (1959). The anode was made from rhodonite and carbon pressed together, while the cathode was made from carbon. The high temperature of the arc discharge results in rhodonite decomposition in the anode. Further treatment of the anode in HCl leads to extraction of 90% manganese. To compare, HCl treatment of rhodonite without high-temperature decomposition permits a maximum extraction of 25% manganese.

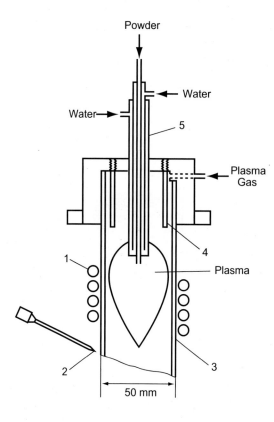

Figure 7–95. Electrodeless RF plasma reactor for manganese oxide (MnO) production from rhodonite (MnSiO₃): (1) inductor of the RF generator; (2) Tesla coil; (3) quartz reactor walls; (4) brass pipe; (5) water-cooled pipe.

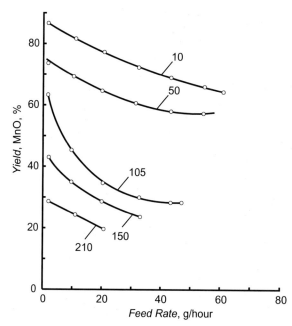

Figure 7–96. Manganese oxide (MnO) yield as function of rhodonite (MnSiO₃) powder feed rate at different particle sizes; torch diameter 3.7 cm, generator power 10 kW, argon flow rate 0.5 L/s; two upper curves correspond to plasma in the quenching zone. Numbers near the curves designate sizes of particles, presented in micrometers.

7.6. Plasma-Chemical Production of Inorganic Oxides

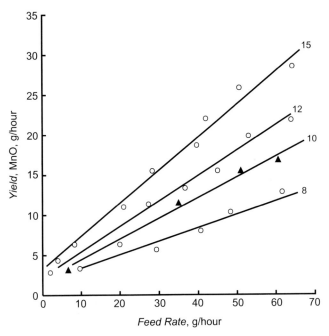

Figure 7–97. Manganese oxide (MnO) yield as function of rhodonite ($MnSiO_3$) powder feed rate and discharge power; torch diameter 3.7 cm, size of particles 50 μm. Numbers near the curves designate discharge power in kilowatts.

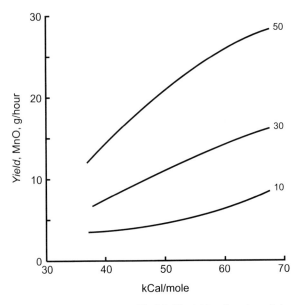

Figure 7–98. Manganese oxide (MnO) yield as function of plasma enthalpy (kCal/mol) at different levels of the rhodonite feed rate; size of the rhodonite particles, 50 μm. Numbers near the curves designate the rhodonite feed rate in grams per hour.

7.6.3. Plasma-Chemical Extraction of Nickel from Serpentine Minerals

Serpentine is a widely spread mineral, $(Mg_xFe_yNi_z)_3Si_2O_5(OH)_4$, that contains low nickel concentration (about 0.25%) but can be effectively used for production of the metal by plasma technology starting from thermal decomposition (Meubus, 1973):

$$(Mg_xFe_yNi_z)_3Si_2O_5(OH)_4 \rightarrow 3x \cdot MnO + 3y \cdot FeO + 3z \cdot NiO + 2SiO_2 + 2H_2.$$

$$(7-101)$$

Nickel can be conventionally extracted from the produced mixture by the intermediate formation of a volatile nickel carbonyl. Also an essential portion of nickel can be extracted from the mixture in the form of a magnetic phase, $MeO(Fe_2O_3)$, where Me is mostly Fe (II) but also nickel. The concentration of nickel in the magnetic phase is 10 times higher than in the rest of the thermal decomposition products. Plasma-chemical serpentine decomposition was investigated in an RF-ICP discharge (Meubus, 1973), which is similar to that shown in Fig. 7–95. The discharge power was 15 kW, frequency 5 MHz, and the reactor chamber was made from quartz. Argon or argon–hydrogen mixture was used as the plasma gas, which was heated up to 3800–6000 K. Some fraction of the gas was supplied as a carrier for the serpentine powder. When serpentine decomposition is performed in neutral atmosphere (pure argon), then the nickel concentration in the magnetic phase, $MeO(Fe_2O_3)$, as well as the fraction of the nickel-rich magnetic phase $MeO(Fe_2O_3)$ increases with temperature. The temperature dependence is reversed when the decomposition is performed in a reducing atmosphere (argon with 9% hydrogen). Small carbon additions to serpentine ($\geq 1.5\%$) lead to further increase in the nickel concentration in the nickel-rich magnetic phase, $MeO(Fe_2O_3)$.

7.6.4. Production of Uranium Oxide (U_3O_8) by Thermal Plasma Decomposition of Uranyl Nitrate ($UO_2(NO_3)_2$) Aqueous Solutions

Consider plasma decomposition of aqueous solutions or other liquids injected in discharges. The first example is thermal decomposition of uranyl nitrate solution, which is important in nuclear fuel processing. An important step in the production and regeneration of nuclear fuel is the conversion of uranyl nitrate solutions, produced during extractive separation of uranium and plutonium, into uranium oxides (Gromov et al., 1971). U_3O_8 can be produced directly by plasma treatment of aqueous $UO_2(NO_3)_2$ (Tumanov, 1989):

$$[UO_2(NO_3)_2]_{solution} \rightarrow \frac{1}{3}(U_3O_8)_{solid} + (NO)_{gas} + (NO_2)_{gas} + \frac{7}{6}(O_2)_{gas} + x \cdot (H_2O)_{gas}.$$

$$(7-102)$$

The aqueous solution is injected into an atmospheric air arc. Decomposition of droplets of the solution proceeds in three major phases. The first phase includes heating the droplets to the boiling temperature of the solution and partial evaporation of water. Complete evaporation of water occurs in the second phase. Further heating and decomposition of the residual salt takes place in the third phase. A special separator provides extraction of the solid product (U_3O_8); gases proceed to a condenser/absorber, where nitric acid is generated as a by-product. The effectiveness of the technology was demonstrated in a pilot unit based on a 300 kW arc in atmospheric-pressure air, which converts 80 kg/h of uranyl nitrate solution into U_3O_8 (Tumanov, 1989; Ivanov, 2000). The system parameters are as follows: reactor diameter is 0.3 m; plasma gas (air) flow rate is 16 g/s; plasma gas temperature is 4000 K; solution concentration is 0.4 kg of uranium per kilogram of solution; initial solution temperature is 293 K; initial diameter of solution droplets is 150 µm; velocity of the solution droplets entering the reactor is 30 m/s; the reactor length required for efficient conversion is about 2 m; heat losses to the reactor walls are about 20%; and temperature at the reactor exit is about 1500 K. The efficiency of the plasma-chemical process is high. Conversion of uranyl nitrate into U_3O_8 is 99.8% in terms of uranium balance. The energy cost of the process is 2.2 kWh/kg of initial solution.

7.6.5. Production of Magnesium Oxide (MgO) by Thermal Plasma Decomposition of Aqueous Solution or Melt of Magnesium Nitrate (Mg(NO$_3$)$_2$)

Another example is plasma decomposition of aqueous solutions of magnesium nitrate, $Mg(NO_3)_2$. Magnesium oxide production from caustic magnesite (mostly $MgCO_3$) proceeds in two stages. Caustic magnesite is first treated in nitric acid to produce magnesium nitrate, which can be then decomposed:

$$Mg(NO_3)_2 \rightarrow MgO + NO + NO_2 + O_2. \tag{7--103}$$

The thermal plasma process (7–103) can be carried out with magnesium nitrate in the form of either melt or aqueous solution. Nitrogen oxides produced as process by-products are going to nitric acid solution in the presence of water, which is recycled to the first stage of treatment of caustic magnesite. The process (7–103) can be applied not only for the conversion of caustic magnesite but also in other technologies requiring decomposition of magnesium nitrates into oxides. The industrial plasma-chemical system for magnesium nitrate solution decomposition with production of magnesium oxide and nitric acid is based on four 100 kW arc discharges (Tumanov, 1989; Ivanov, 2000). The major parameters of the system are as follows: nitrate solution feeding rate, 80 L/h; concentration of magnesium oxide (MgO) in the solution, 150 g/L; flow rate of plasma gas (air), 100 m^3/h; and average temperature of the plasma gas (air), 3000–4000 K. Magnesium oxide is produced as a powder with particle sizes below 3 μm (95%) and 3–5 μm (5%); the average density of the powder is 0.15–0.30 g/cm^3. The effectiveness of the plasma-chemical process for MgO production from nitrate solution or melt is high. The conversion of MgO from initial nitrate is 98.5%. The energy cost of the process mostly depends on the concentration of the initial nitrate solution. When the initial MgO concentration in the solution is 150 g/L, the energy cost of production of high-quality MgO (lemon acid number 70) is 20 kWh/kg. An increase of the initial concentration to 263 g/L results in a decrease of the energy cost to 12 kWh/kg. When magnesium nitrate is supplied in the form of melt, $Mg(NO_3)_2 \cdot 6H_2O$, the energy cost of production of high-quality magnesium oxide (lemon acid number 70) becomes 10 kWh/kg. Plasma can be applied not only for decomposition of complex oxides into simple ones but also for the synthesis of complex oxides from simple ones. As an example, we can mention the thermal plasma production of $MgAl_2O_4$ from MgO and Al_2O_3 (Mayer, 1962; Vishnevska et al., 1969).

7.6.6. Plasma-Chemical Production of Oxide Powders for Synthesis of High-Temperature Superconducting Composites

As was demonstrated earlier with examples for production of U_3O_8 and MgO, inorganic oxides can be effectively produced in plasma via thermal decomposition of aqueous nitrate solutions. The nitrate solutions are preliminarily produced by treatment of relevant compounds with nitric acid. This approach can be applied for producing a wide variety of metal oxides with very high quality:

$$[Me_a(NO_3)_b]_{aq.\,solution} \rightarrow Me_aO_c + x \cdot NO_2 + y \cdot NO + z \cdot O_2 + H_2O. \tag{7--104}$$

The plasma-chemical method produces nitric acid as a by-product, which can be reused to form more of the nitrate solution for further plasma decomposition (7–104). The high-quality ceramics produced in plasma are of special interest for the synthesis of high-temperature superconducting composites. Such processes were demonstrated using arc discharges with power 100 and 300 kW, which were able to treat, respectively, 25 and 60 L/h of relevant nitrate solution (Ivanov, 2000). The method was applied for production of such high-temperature superconducting composites as $YBa_2Cu_3O_x$, $YBa_2Cu_4O_x$, $Bi_2Sr_2Ca_2Cu_3O_x$, and $Bi_{1.7}Pb_{0.3}Sr_2Ca_2Cu_3O_x$. The Y–Ba–Cu oxide composites, produced in plasma, are characterized by an average powder density of 0.5–0.7 g/cm^3, particle diameter of 1–3 μm, specific surface area of 1–5 m^2/g, residual water content of 0.5–0.8%, and carbon content of 0.1–0.2%. The bismuth-based oxide composites are characterized by an average powder density of 0.7–1.0 g/cm^3 and by particle diameter 3–6 μm. The plasma-chemically produced superconducting composites demonstrate excellent electrophysical properties. The Y–Ba–Cu oxide composites are characterized by superconductivity transition temperature 90–95 K with a transition width less than 3 K; critical current density in the zero field is 300 A/cm^2. The bismuth-based oxide composites are characterized by a superconducting transition temperature of 108–109 K with a transition width less than 6 K; critical current density in the zero field is 300–450 A/cm^2. Plasma-chemical production of inorganic oxides by thermal decomposition of nitrate solutions was also applied to produce oxides of zirconium (ZrO_2), aluminum (Al_2O_3), chromium (Cr_2O_3), and zinc (ZnO). The average particle size is 0.1–5 μm, and the specific surface area of the powder is 3–40 m^2/g.

7.6.7. Production of Uranium Oxide (U_3O_8) by Thermal Plasma Conversion of Uranium Hexafluoride (UF_6) with Water Vapor

Consider the production of inorganic oxides by steam conversion of halides in plasma. Conversion of gaseous UF_6 into UO_2, which is an important step in the nuclear fuel cycle, can be done in arc discharges by injection of UF_6 in water vapor plasma (Tumanov, 1989; Ivanov, 2000):

$$UF_6 + 3H_2O \rightarrow \frac{1}{3}U_3O_8 + 6HF + \frac{1}{6}O_2. \qquad (7\text{--}105)$$

The process temperatures 1050–1400 K at pressure 50 kPa are sufficient for almost complete conversion of uranium from UF_6 into U_3O_8, and fluorine into HF. At higher temperatures exceeding 1400 K, the production of uranium oxides is partially suppressed by the formation of gaseous uranyl fluoride (UO_2F_2). The plasma process (7–105) is conducted using a 200 kW arc discharge heating water vapor to the required temperatures; 150 kg/h of gaseous hexafluoride is supplied into the plasma reactor after the discharge. The molar ratio in the reactor is close to stoichiometric: UF_6:H_2O = 1:3. Uranium oxide is produced as powder with an average density of 4.5–5.7 g/cm^3; the specific surface area of the powder is 0.037–0.138 m^2/g. The conversion degree of hexafluoride into oxide is very high (99%); the energy cost is 1.3–1.4 kWh/kg UF_6. The energy cost of uranium oxide decreases at higher arc powers. The higher powers lead to increased energy efficiency of the plasma-chemical reactor and power supply, and larger reactor volumes lead to a decrease of heat losses to the reactor walls.

7.6.8. Conversion of Silicon Tetrafluoride (SiF_4) with Water Vapor into Silica (SiO_2) and HF in Thermal Plasma

Gaseous silicon tetrafluoride can be converted in plasma into silica:

$$SiF_4 + 2H_2O \rightarrow SiO_2 + 4HF. \qquad (7\text{--}106)$$

The process can be carried out by heating water vapor in an arc and injecing silicon tetrafluoride into the plasma. A dedicated industrial pilot plant was built to treat exhaust gases of HF production, which contain a significant amount (15–30%) of SiF_4 (Tumanov, 1989;

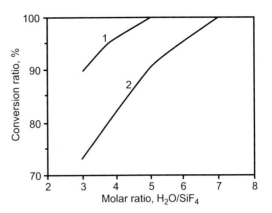

Figure 7–99. Conversion degree of (SiF$_4$) as function of molar ratio H$_2$O:SiF$_4$ at different temperatures: (1) $T = 3000$–3100 K; (2) $T = 2600$–2700 K.

Ivanov, 2000). The parameters of the pilot unit are as follows: discharge power, 120–140 kW; the plasma gas is water vapor at atmospheric pressure; exhaust gas treatment capacity, 30–35 m^3/h (SiF$_4$ concentration in the exhaust gas up to 30%); and HF mass fraction in products, 50–55%. The energy cost of the plasma-chemical treatment is 2.0–2.5 kWh/m^3 of exhaust gases. The conversion of SiF$_4$ into SiO$_2$ is shown in Figs. 7–99 and 7–100 as a function of molar ratio H$_2$O:SiF$_4$ and process temperature. Increase of the molar ratio H$_2$O:SiF$_4$ at fixed reaction temperature leads to growth of the tetrafluoride conversion degree into silica, which reaches 99.5–99.9% at $T = 3000$ K and H$_2$O:SiF$_4$ = 6–8. Such high molar ratio of H$_2$O:SiF$_4$, however, results in a decrease of HF mass fraction in products. A major factor providing both high conversion degree and high HF mass fraction in products is reaction temperature. For example, keeping a reaction temperature of 2500 K at a molar ratio of H$_2$O:SiF$_4$ = 5 results in a high degree of tetrafluoride conversion into silica (95%) and reasonably high HF mass fraction in the products (46%). The silica (SiO$_2$) is produced in the plasma-chemical process in form of γ-tridimite powder. The γ-tridimite is characterized by an average density of 0.05–0.07 g/cm^3 and specific surface area of 120–130 m^2/g.

7.6.9. Production of Pigment Titanium Dioxide (TiO$_2$) by Thermal Plasma Conversion of Titanium Tetrachloride (TiCl$_4$) in Oxygen

The production of pigment titanium dioxide by conversion of gaseous tetrachloride (TiCl$_4$) is performed on an industrial scale (Farmer & Bogdan, 1969; Dundas & Thorpe, 1974; Sakharov & Lukianychev, 1974):

$$(\text{TiCl}_4)_{\text{gas}} + (\text{O}_2)_{\text{gas}} \rightarrow (\text{TiO}_2)_{\text{solid}} + 2(\text{Cl}_2)_{\text{gas}}. \qquad (7\text{–}107)$$

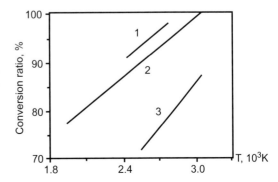

Figure 7–100. Conversion degree of SiF$_4$ as function of temperature in the reaction zone at different mixture compositions: (1) H$_2$O:SiF$_4$ = 6.5; (2) H$_2$O:SiF$_4$ = 5.0; (3) H$_2$O:SiF$_4$ = 3.5.

The thermodynamics equilibrium (7–107) is shifted in the direction of products even at room temperature. The reaction, however, is kinetically limited and therefore requires high temperatures. Reaction rates increase significantly only at temperatures of 1000–1500 K, when process time decreases to 1–10 ms. Oxygen or oxygen-containing gases are heated up to the required temperatures in thermal arc or RF-ICP discharges (Rykalin, Ogurtsov, & Kulagin, 1975). $TiCl_4$ is injected into the oxygen plasma in the reactor chamber located after the discharge. Product quenching and separation of the TiO_2 powder takes place at the reactor chamber exit. Finally, chlorine is separated from the gaseous products.

Technological parameters of the method can be presented for a plasma system (Milko, Gukov, & Isay, 1973) in which air is heated up by three arc jets, forming a swirl. Liquid $TiCl_4$ is injected into the high-temperature swirl, where it is converted into products (7–107). Heterogeneous flow containing TiO_2, Cl_2, N_2, and O_2 passes a quenching nozzle with a velocity of 700 m/s to the TiO_2 powder separation system. The air temperature after the arc discharge heating is 3000–3200 K; the temperature in the reaction zone is 2200–2300 K. The residence time of $TiCl_4$ in the high-temperature zone is 2–20 ms. Pigment titanium dioxide is produced in the form of a powder; most of the particles are spherical with diameters of 100–400 nm. Exhaust gas contains about 25% of Cl_2. The energy cost of the TiO_2 powder is 2–3 kWh/kg.

The energy cost of the process (7–107) can be reduced to 2 kWh/kg TiO_2 if enriched air (for example, 30% air and 70% oxygen) is applied as an oxygen-containing gas (Sakharov & Lukianychev, 1974). The chlorine concentration in the exhaust gases in such systems can be increased to 70%, which means that the exhaust gas can be applied directly for Cl_2 treatment of initial titanium-containing materials. When the excess of oxygen is 40–60%, the average oxygen temperature after discharge is 2000–6000 K; the initial $TiCl_4$ temperature is 720 K. The TiO_2 particle size decreases at shorter $TiCl_4$ residence time in the reactor. For example, production of 95% powder with particle sizes less than 1 μm requires a residence time of 7 ms at temperature 3000 K. The required quenching rate to keep the submicron TiO_2 particle sizes in this regime is 1000 K/s. Not only O_2 but also some oxides, in particular CO_2, can be used for plasma-chemical conversion of $TiCl_4$ into TiO_2 (Wilson, Strain, & Hockie, 1965; Polak et al., 1975a):

$$(TiCl_4)_{gas} + 2(CO_2)_{gas} \rightarrow (TiO_2)_{solid} + 2(Cl_2)_{gas} + 2(CO)_{gas}. \qquad (7–108)$$

7.6.10. Thermal Plasma Conversion of Halides in Production of Individual and Mixed Oxides of Chromium, Aluminum, and Titanium

Fine powders of mixed oxides (catalysts and pigments) can be produced from metal halides using the plasma method similar to conversion of $TiCl_4$ into TiO_2. The conversion of gaseous chromium and aluminum halides into mixed oxides was done in atmospheric-pressure RF-ICP by Barry, Bayliss, and Lay (1968). CrO_2Cl_2 and $AlCl_3$ were used as initial substances; the mixed oxides powder was separated using an electrostatic precipitator. Conversion of $AlCl_3$ alone in Ar–O_2 plasma leads to the formation of the δ-phase of alumina (Al_2O_3) with some admixture of the ϑ-phase (both phases are metastable). Condensation of the Al_2O_3 droplets occurs at 3000 K. Conversion of CrO_2Cl_2 alone in Ar–O_2 plasma leads mostly to the formation of α-phase chromium chloride (Cr_2O_3). Condensation takes place at 1900–2100 K. When a mixture of $AlCl_3$ and CrO_2Cl_2 are injected into Ar–O_2 plasma, a solid oxide solution is formed together with a simple mixture of corresponding oxides. The solid oxide solution can be characterized as mostly solution of Cr_2O_3 in the δ-phase of alumina (Al_2O_3); the fraction of the solid solution in total mixed oxide product is 5–6%. A plasma conversion of mixed halides of chromium and titanium was also demonstrated by Barry, Bayliss, and Lay (1968). Conversion of $TiCl_4$ alone in Ar–O_2 plasma leads to the formation of a metastable phase of titanium oxide (TiO_2). The sizes of Cr_2O_3 and TiO_2

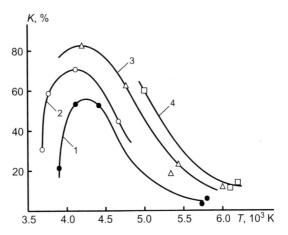

Figure 7–101. Decomposition degree of phosphates, K, as function of temperature at different values of energy cost per 1 kg of initial material: (1) 11 kWh/kg; (2) 16.5 kWh/kg; (3) 29 kWh/kg; (4) 39 kWh/kg.

particles are 10–150 μm. The concentration of Cr_2O_3 in TiO_2 (solid solution) is significantly higher in this case in comparison with the previously discussed Cr_2O_3 –Al_2O_3 system. In all processes considered here, the condensation of oxides takes place at very high temperatures. According to the Frenkel theory (Frenkel, 1975), the crystallization rate during cooling is slower than the condensation rate, which results in the formation of metastable crystal modification of the oxides. The effect of formation of metastable phases becomes even more significant because of the requirements of fast product cooling to achieve a higher quenching efficiency (Shevchenko, Zverev, & Pavlov, 1974).

7.6.11. Thermal Plasma Treatment of Phosphates: Tricalcium Phosphate ($Ca_3(PO_4)_2$) and Fluoroapatite ($Ca_5F(PO_4)_3$)

Major phosphate compounds are tricalcium phosphate, $Ca_3(PO_4)_2$, and fluoroapatite, $Ca_5F(PO_4)_3$. Both compounds can be converted in plasma at temperatures around 3500 K (Mosse & Pechkovsky, 1973). Thermal decomposition of tricalcium phosphate, $Ca_3(PO_4)_2$, can be expressed as

$$Ca_3(PO_4)_2 \rightarrow 3CaO + P_2O_5. \tag{7–109}$$

The yields of different phosphorus oxides depend on temperature. The process was investigated in atmospheric-pressure arc discharges in nitrogen and air at powers up to 50 kW (Wolfkovich & Asiev, 1965; Burov et al., 1970, 1973). The dissociation degrees are presented in Fig. 7–101 as a function of temperature in and energy cost. High dissociation degree (up to 85%) leads to a higher energy cost of the process. Tricalcium phosphate can be converted in plasma not only to phosphorus oxides but also to elementary phosphorus if reduction agents (like carbon) are added (Goldberger, 1966a,b; Longfield, 1969):

$$Ca_3(PO_4)_2 + 3SiO_2 + 5C \rightarrow CaSiO_3 + \frac{1}{2}P_4 + 5CO. \tag{7–110}$$

The conversion from tricalcium phosphate to elementary phosphorus was close to 100%. The process was performed in a plasma fluidized bed supported by argon flow at 1030–1200°C. Mosse and Pechkovsky (1973) applied CH_4 as a plasma gas and reduction agent for processing tricalcium phosphate. The reaction was conducted in an atmospheric-pressure CH_4 plasma jet at 3000–4000 K, the residence time was 0.01–0.05 s, and the phosphate-to-CH_4 ratio was 1:10. The conversion degree from $Ca_3(PO_4)_2$ to elementary phosphorus was up to 75%. Gas-phase by-products included acetylene and other hydrocarbons, PH_3, CO, and CO_2; solid-phase by-products included mostly calcium carbide and phosphide. Another major compound of natural phosphates is fluoroapatite, $Ca_5F(PO_4)_3$. Practical applications of fluoroapatite usually require its preliminary defluorination, which can be done by plasma decomposition of fluoroapatite with formation of tricalcium phosphate and fluorite CaF_2 (Mosse & Pechkovsky, 1973):

$$2Ca_5F(PO_4)_3 \rightarrow 3Ca_3(PO_4)_2 + CaF_2. \tag{7–111}$$

Defluorination of fluoroapatite in arc discharges without additives can exceed 90% at temperatures above 2000 K. The decomposition (7–111) leads to formation of solid products. Defluorination can be easier to do when fluorine is fixed in products in the gas phase. Such processes are conducted in the presence of SiO_2. Plasma defluorination leads to the formation of tricalcium phosphate with fluorine in gaseous SiF_4,

$$Ca_5F(PO_4)_3 + 0.25\,SiO_2 \rightarrow Ca_3(PO_4)_2 + 0.5\,Ca_4P_2O_9 + 0.25\,SiF_4, \tag{7–112}$$

or to formation of more valuable phosphorus pentoxide (P_2O_5), again with fluorine in gaseous form (SiF_4):

$$Ca_5F(PO_4)_3 + 5.25\,SiO_2 \rightarrow 5CaSiO_3 + 1.5\,P_2O_5 + 0.25\,SiF_4. \tag{7–113}$$

The plasma defluorination is 99% at optimal admixtures of SiO_2 (Mosse & Pechkovsky, 1973).

7.6.12. Oxidation of Phosphorus and Production of Phosphorus Oxides in Air Plasma

Phosphorus can be oxidized in air plasma to produce a mixture of phosphorus and nitrogen oxides (fertilizers). Such a plasma-chemical process is carried out in arc discharges (Mosse & Pechkovsky, 1973):

$$P_4 + 6O_2 + N_2 \rightarrow 2P_2O_5 + 2NO. \tag{7–114}$$

NO and oxides of phosphorus (P_2O_5, PO, P_2O_3) are present in plasma at $T = 1000$–4000 K. P_2O_5 dominates at temperatures below 2100 K; mixture P_2O_3 and P_2O_5 dominates at temperatures 2100–3250 K; a mixture of P_2O_3 and PO dominates at temperatures of 3250–4100 K; and P_2O_5 dominates at very high temperatures, above 4100 K. The experiments were carried out in an atmospheric-pressure arc discharge at 5–12 kW power, air flow rates of 1.4–1.9 g/s, and red phosphorus feeding rates of 2–10 g/min. The injection of phosphorus in a flow of air plasma significantly increases the concentration of nitrogen oxides. Typically, the concentration of phosphorus pentoxide was 0.3 g/L, and the concentration of nitrogen oxides was about 3%. The quenching was based on mixing with cold water and air flows and was not efficient enough.

7.7. PLASMA-CHEMICAL PRODUCTION OF HYDRIDES, BORIDES, CARBONYLS, AND OTHER COMPOUNDS OF INORGANIC MATERIALS

7.7.1. Production of Hydrides in Thermal and Non-Thermal Plasma

Thermal plasma can be applied for direct synthesis from elements of metal hydrides (Kubanek, Chevalier & Cauvin, 1968; Shmakin & Marusin, 1970), in particular, hydrides of tin (SnH_4) and lead (PbH_4). The process was performed by heating H_2 in an atmospheric-pressure RF-ICP discharge, and injection of tin and lead powders into the plasma. Another example is the synthesis of H_2S, which was performed in an He plasma jet by injection of sulfur powder with H_2 flow (Stokes, 1971). The conversion degree of sulfur into H_2S and the energy cost of the process are presented in Fig. 7–102 as a function of specific energy input. Taking into account the problems with hydride stability in thermal plasmas required quenching; direct synthesis of hydrides was also carried out in non-thermal discharges.

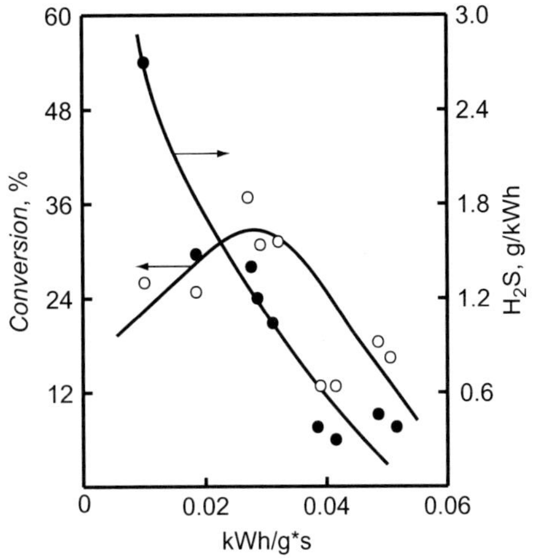

7.7. Plasma-Chemical Production of Hydrides, Borides, Carbonyls

Figure 7–102. Conversion degree and energy yield of H_2S per 1 kWh as function of energy consumption per 1 g of sulfur.

Hydrides of phosphorus, arsenic, and sulfur were produced in a glow discharge in hydrogen, when the elements were deposited on walls of the discharge tube (McTaggart, 1967). Different boron hydrides (B_2H_6, $B_{10}H_{16}$, $B_{20}H_{16}$) were also synthesized in a glow discharge in the mixture H_2:BCl_3 = 12:1 at a pressure of 20 Torr. The conversion degree of boron from chloride into hydride was as high as 40% (McTaggart, 1967). Hydride synthesis in dielectric barrier discharges (DBDs) at higher pressures was considered by Shahin (1971). Saturated germanium hydrides (Ge_nH_{2n+2}) were synthesized from GeH_4 at 300 Torr and discharge wall temperature of 78°C. Hydrides of silicon were produced in DBD from silane (SiH_4) at lower pressures. Major products of the synthesis were saturated compounds: Si_2H_6 (66%), Si_3H_8 (23%), Si_4H_{10} (11%), and even Si_8H_{18}. DBD in a GeH_4–PH_3 mixture leads to the formation of GeH_3PH_2 and Ge_2PH_7. DBD treatment of a GeH_4–AsH_3 mixture results in the production of GeH_3AsH_2. DBD in a GeH_4–SiH_4 mixture results in high conversion degree into GeH_3SiH_3. DBD in a SiH_4–PH_3 mixture leads to the production of $Si_2H_5PH_2$ and $(SiH_3)_2PH$. DBD plasma processes are also carried out in mixtures of SiH_4 and AsH_3. We should also mention early experiments involving the formation of thin films of metal hydrides in hydrogen plasma carried out by Pietsh (1933). Hydrides of silver, beryllium, gallium, indium, and tantalum were synthesized in these experiments.

7.7.2. Non-Thermal Plasma Mechanisms of Hydride Formation by Hydrogen Gasification of Elements and by Hydrogenation of Thin Films

The vibrational excitation of H_2 in non-equilibrium plasma can effectively stimulate gasification of elements with the direct production of hydrides (Legasov et al., 1978c):

$$(Si)_{solid} + H_2^*(X^1\Sigma_g^+, v) \rightarrow (SiH_4)_{gas}, \tag{7–115}$$

$$(Ge)_{solid} + H_2^*(X^1\Sigma_g^+, v) \rightarrow (GeH_4)_{gas}, \tag{7–116}$$

$$(B)_{solid} + 1.5 \cdot H_2^*(X^1\Sigma_g^+, v) \rightarrow 0.5 \cdot (B_2H_6)_{gas}. \tag{7–117}$$

Gasification stimulated by non-thermal plasma excitation and subsequent dissociation of H_2 was also observed in experiments with production of hydrides of sulfur, arsenic, germanium, tin, tellurium, and selenium (Bonhoeffer, 1924; Pearson, Robinson, & Stoddart, 1933; Radford, 1964). The mechanism of hydride production includes in this case volume and surface dissociation of excited hydrogen molecules, chemisorption of atomic

489

hydrogen, surface reactions, and desorption of produced volatile hydrides. Non-thermal plasma-stimulated synthesis of metal hydrides can also take place in thin surface films. Such solid-state processes can be analyzed by taking as an example the synthesis of aluminum hydride (AlH_3) from the elements (Legasov, et al., 1978c):

$$(Al)_{solid} + 1.5 \cdot H_2^*(X^1\Sigma_g^+, v) \rightarrow (Al)_{solid} + 3H \rightarrow (AlH_3)_{solid}. \qquad (7\text{--}118)$$

This process proceeds through the intermediate formation of atomic hydrogen but requires low temperatures, because AlH_3 is stable only at temperatures below 400 K. The formation of atomic hydrogen stimulated by vibrational excitation takes place not only in the plasma volume (where dissociation through vibrational excitation requires 4.4 eV) but also on the surface of aluminum, where it requires only about 2 eV/mol. The gradual hydrogenation of aluminum in a thin film is determined by competition between solid-state reactions of hydrogen attachment and hydrogen exchange:

$$H + AlH_k \rightarrow AlH_{k+1}, \quad k = 0, 1, 2, \qquad (7\text{--}119)$$

$$H + AlH_k \rightarrow AlH_{k-1} + H_2, \quad k = 1, 2, 3. \qquad (7\text{--}120)$$

The average quasi-stationary concentration of AlH_3 in the film can be estimated as

$$\gamma \approx 1 - \exp(-\tilde{E}_a/T_0), \qquad (7\text{--}121)$$

where $\tilde{E}_a$ is a characteristic activation barrier of exchange reactions (7–120), and T_0 is the solid-state temperature, assumed equal to the gas temperature. The formation of aluminum hydrides takes place in a layer whose thickness is determined by losses of atomic hydrogen in solid-state recombination reactions,

$$H + H + AlH_k \rightarrow H_2 + AlH_k, \qquad (7\text{--}122)$$

as well as in reactions (7–119) and (7–120). The depth δ_1 of reactions (7–119) and (7–120) can be found from the one-dimensional diffusion equation for H-atom density [H] in a solid matrix of aluminum during the synthesis

$$D_H \frac{\partial^2}{\partial x^2}[H] - v_0 \exp(-\tilde{E}_a/T_0) = 0, \qquad (7\text{--}123)$$

where D_H is the coefficient of diffusion of H atoms in a crystal structure (see (6–100)), and v_0 is the "jumping" frequency at the high-temperature limit (see Section 6.7.2). Based on equations (7–123) and (6–100), the effective depth of synthesis, δ_1, of aluminum hydride on an aluminum surface in H_2 plasma is

$$\delta_1 \approx \lambda \sqrt{f} \exp \frac{\tilde{E}_a - E_a^D}{2T_0}. \qquad (7\text{--}124)$$

The depth of the layer exponentially depends on the potential barrier $\tilde{E}_a$, which determines the stability of AlH_k in the matrix; $\delta_1 \rightarrow \infty$ in the extreme case if $\tilde{E}_a \rightarrow \infty$. We should mention that atomic hydrogen also provides effective removal of surface oxides before hydrogenation.

7.7.3. Synthesis of Metal Carbonyls in Non-Thermal Plasma: Effect of Vibrational Excitation of CO Molecules on Carbonyl Synthesis

Metal carbonyls can be directly produced by reactions with CO molecules excited in non-thermal plasma (Legasov, et al., 1978c), which is especially important for the synthesis of carbonyls of chromium, molybdenum, manganese, and tungsten, which cannot be alternatively produced without solvents (Calderazzo, 1970):

$$(Cr)_{solid} + 6(CO)_{gas} \rightarrow [Cr(CO)_6]_{gas}, \qquad (7\text{–}125)$$

$$(Mo)_{solid} + 6(CO)_{gas} \rightarrow [Mo(CO)_6]_{gas}, \qquad (7\text{–}126)$$

$$(Mn)_{solid} + 5(CO)_{gas} \rightarrow 1/2[Mn_2(CO)_{10}]_{gas}, \qquad (7\text{–}127)$$

$$(W)_{solid} + 6(CO)_{gas} \rightarrow [W(CO)_6]_{gas}. \qquad (7\text{–}128)$$

The gaseous metal carbonyls are unstable at temperatures exceeding 600 K, which leads to restrictions in performing non-thermal discharges. The mechanism for plasma-chemical formation of metal carbonyls can be considered by taking as an example the synthesis of $Cr(CO)_6$ (7–125) from chromium and CO vibrationally excited in a non-thermal discharge. The process includes the following four stages:

1. Chemisorption of plasma-excited CO molecules on a chromium surface:

$$(Cr)_{solid} + k \cdot (CO)_g \Leftrightarrow Cr[k \cdot CO]_{chemisorbed}. \qquad (7\text{–}129)$$

The chromium crystal structure is characterized by a cubic lattice with valence 6 and coordination number 8. Therefore, each chromium surface atom is able to attach $k \leq 3$ carbonyl CO groups without destruction of the lattice (Darken & Gurry, 1960; Atkins, 1998).

2. Formation of higher carbonyls ($k > 3$) requires taking a chromium atom with an attached CO group from the crystal lattice to a chemisorbed state:

$$Cr[m \cdot CO]_{chemisorbed} + (Cr)_{solid}CO \Leftrightarrow Cr[Cr(CO)_{m+1}]_{chemisorbed}. \qquad (7\text{–}130)$$

The characteristic time for removing the chromium atom from the crystal lattice τ_{to} to form the chemisorbed carbonyl complex ($T_v \gg T_0$) can be estimated as

$$\tau_{to} \approx \frac{1}{a\Phi} \exp\left(\frac{E_{to}}{T_v}\right), \qquad (7\text{–}131)$$

where, E_{to} is the activation barrier of (7–130), Φ is the flux of CO molecules to the chromium surface, a is the surface area related to a chemisorbed complex, and T_v is the vibrational temperature of CO. The kinetics of surface dissociation of the chemisorbed complex $Cr[Cr(CO)_{m+1}]_{chemisorbed}$ (reverse reaction (7–131)) is determined by the chromium surface temperature, T_S.

3. Formation of chromium hexacarbonyl $Cr(CO)_6$ (the product of (7–125) still adsorbed on the surface) takes place in the surface reactions of intermediate chemisorbed chromium carbonyl complexes:

$$Cr[Cr(CO)_l]_{chemisorbed} + (6 - l) \cdot Cr[CO]_{chemisorbed} \Leftrightarrow Cr[Cr(CO)_6]_{adsorbed}. \qquad (7\text{–}132)$$

The characteristic time of the surface reactions with activation barriers $E_{sr,l}$ is (Roginsky, 1948)

$$\tau_{sr,l} \approx \frac{2\pi \hbar b_0}{T_S \cdot [CO]} \exp\left(\frac{E_{sr,l}}{T_S}\right), \qquad (7\text{–}133)$$

where b_0 is the reverse adsorption coefficient at the surface coverage $\xi = 0.5$.

4. The final stage is desorption of volatile chromium hexacarbonyl $Cr(CO)_6$ into gas phase:

$$Cr[Cr(CO)_6]_{adsorbed} \rightarrow (Cr)_{solid} + [Cr(CO_6)]_{gas}. \qquad (7\text{–}134)$$

Table 7–6. Formation of Borides from Metals and Boron Hydrides

Product of synthesis	Process temperature, K	Conversion degree, %
VB_2	2700–3600	11–100
TiB_2	2900–3500	17–100
ZrB_2	2900–4000	19–100
CrB_2	2600–3000	8–100

In the case of $T_v \gg T_0$, the desorption rate can be calculated as

$$w_{des} \approx \xi \left[\frac{\omega_L}{a} \exp\left(-\frac{E_a^{des}}{T_S} \right) + \Phi \exp\left(-\frac{E_a^{des}}{T_v} \right) \right] = \xi \cdot k_{des}(T_v, T_S), \quad (7\text{–}135)$$

where ξ is the surface fraction covered by complexes, ω_L is the characteristic lattice frequency, E_a^{des} is the activation energy of the desorption process, and $k_{des}(T_v, T_S)$ is the desorption rate coefficient.

Thus, acceleration of the desorption process can be due to the high vibrational temperature of CO molecules as well as effective non-equilibrium heating of the chromium surface (T_S; see Section 7.1.7). Summarizing the four stages for synthesis of chromium hexacarbonyl, we can note that conducting the process in non-thermal plasma is supported by an interesting effect: intermediate carbonyls $Cr(CO)_m$ are stabilized on the surface due to chemisorption while vibrationally excited CO molecules stimulate the extraction of Cr atoms from the lattice (7–131). The final product, $Cr(CO)_m$, is not so strongly bound to the surface and can be effectively desorbed also by using energy accumulated in vibrational degrees of freedom of CO molecules.

7.7.4. Plasma-Chemical Synthesis of Borides of Inorganic Materials

The plasma-chemical synthesis of borides can be accomplished using different initial chemicals, and not only in thermal but in non-thermal plasma as well. Borides of refractory metals were effectively produced in thermal $Ar–H_2$ RF-ICP discharges by the simultaneous reduction of metal chlorides and BCl_3 (Murdoch & Hamblyn, 1967; DeVink, 1970, 1971a,b; Dale, 1974):

$$BCl_3 + MeCl_k + H_2 \rightarrow Me_lB_m + HCl. \quad (7\text{–}136)$$

Borides of vanadium, titanium, zirconium, and chromium were produced in thermal plasma by interaction of their melts with boron hydrides (BH_3, B_2H_6):

$$(V)_{melt} + (BH_3)_{gas} \rightarrow (VB_2)_{solid} + (H_2)_{gas}, \quad (7\text{–}137)$$

$$(Ti)_{melt} + (BH_3)_{gas} \rightarrow (TiB_2)_{solid} + (H_2)_{gas}, \quad (7\text{–}138)$$

$$(Zr)_{melt} + (BH_3)_{gas} \rightarrow (ZrB_2)_{solid} + (H_2)_{gas}, \quad (7\text{–}139)$$

$$(Cr)_{melt} + (BH_3)_{gas} \rightarrow (CrB_2)_{solid} + (H_2)_{gas}. \quad (7\text{–}140)$$

Temperatures in a discharge zone, where the formation of borides in reactions (7–137)–(7–140) takes place, are presented in Table 7–6. The corresponding degrees of conversion of metals into borides are also summarized in the table (Fomin & Cherepanov, 1995). Different boron hydrides (B_2H_6, $B_{10}H_{16}$, $B_{20}H_{16}$) were synthesized in non-thermal plasma of a glow

Table 7–7. Properties of Some Intermetallic Compounds Produced in Thermal Plasma

Intermetallic compound	Melting temperature, °C	Density, g/cm^3	Oxidation resistivity, mg/cm^2
ZrBe$_3$	1930	2.72	0.062
CrBe$_2$	1840	4.34	0.023
LaBe$_2$	1810	2.57	0.020
TaBe$_2$	1830	9.79	0.039
MoBe$_2$	1715	3.03	0.026
NbBe$_2$	1580	3.04	0.013
WBe$_5$	1800	6.91	0.051
TaAl$_3$	1690	6.92	0.064
NbAl$_3$	1660	4.57	0.022
Sn$_3$Zr$_5$	1985	6.61	–
NiAl	1640	4.82	–

discharge at a pressure of 20 Torr. The process was done under strongly non-equilibrium conditions in the mixture ratio H$_2$:BCl$_3$ = 12:1 (McTaggart, 1967).

7.7.5. Synthesis of Intermetallic Compounds in Thermal Plasma

Numerous refractory and highly oxidation-resistive intermetallic compounds have been effectively synthesized from individual metals in thermal plasma (Usov & Borisenko, 1965; Fauchais, 2004). Melting of intermetallic compounds takes place at temperatures exceeding those of individual metals. Synthesis of intermetallic compounds is usually combined with plasma coating (see Chapter 8). An initial mixture of metals is injected into the plasma jet, where exothermic synthesis takes place. Because of the homogeneous heat release during the exothermic synthesis, intermetallic compounds are liquid during the coating process. This effect provides high-quality, plasma-produced, dense intermetallic coatings. Some properties of the coatings are summarized in Table 7–7 (Usov & Borisenko, 1965). ZrBe$_3$ (1930°C) and Sn$_3$Zr$_5$ (1985°C) have the highest melting temperature, whereas NbBe$_2$ has the lowest melting temperature (1580°C) in the list of plasma-produced intermetallic compounds presented in the table. Oxidation resistance is shown in this table by an increase of coating weight (per 1 cm^2 of surface area) after 100 h heating in air at a temperature of 1250°C.

7.8. PLASMA CUTTING, WELDING, MELTING, AND OTHER HIGH-TEMPERATURE INORGANIC MATERIAL PROCESSING TECHNOLOGIES

These traditional thermal plasma applications, which are widely used in modern industry, will be only briefly outlined here. For more details, see Storer and Haynes (1994), Fauchais (1997), Dembovsky (1985), Heberlein (2002), and Krasnov, Zilberberg, and Sharivker (1970).

7.8.1. Plasma Cutting Technology

The principle of plasma cutting involves the constriction of an arc formed between an electrode (cathode) and a workpiece by a fine-bore copper nozzle. A schematic of the plasma cutting device is shown in Fig. 7–103. Restricting the opening through which plasma gas passes increases the jet velocity and heat flux to the workpiece. Plasma temperatures in

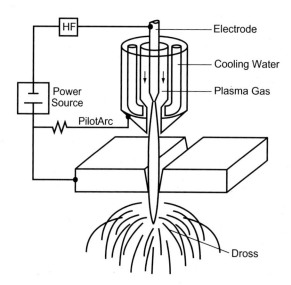

Figure 7–103. General schematic of plasma cutting system.

the system can exceed 20,000 K, providing fast melting. Gas velocities can approach the speed of sound, providing deep penetration of the plasma jet into a workpiece and removal of molten material (Nemchisky, 1996, 1997, 1998; Nemchinsky, Severance, & Showalter, 1999). Plasma cutting is an alternative to the oxyfuel process, where the exothermic oxidation of metals provides heat for melting. Oxyfuel technology can't be applied to cut metals such as stainless steel, aluminum, cast iron, and non-ferrous alloys, which form refractory oxides. Plasma cutters can be applied in this case because their heat source is electric and not related to the formation of refractory metal oxides. A typical operating arc voltage is 50–60 V; the typical current is 50–80 A. The initiating voltage is as high as 400 V to start the arc between the cathode and the nozzle; then the arc is "transferred" by fast gas flow from the nozzle to the workpiece. Conventional plasma cutters use a tungsten electrode (cathode) and a non-oxidizing gas flow of argon, argon–hydrogen mixture, or nitrogen. To increase arc constriction and make "blowing away" of the "dross" more effective, a secondary gas shield can be introduced around the nozzle. The secondary gas is usually selected according to the metal being cut: air, O_2, or N_2 for cutting steel; N_2, Ar–H_2, or CO_2 for cutting stainless steel and aluminum. Additional arc constriction can be achieved by rotating the gas flow and applying a magnetic field surrounding the arc discharge. Instead of secondary gas, water can be injected radially into the arc around the nozzle. The injection of water additionally constricts the jet and considerably increases the temperature to about 30,000 K. The plasma system can also be operated with a water shroud, or even with a workpiece submerged 5–7.5 cm below the water surface. Water acts as a barrier, providing reduction of fume and noise and improving nozzle life. Conventionally applied inert or unreactive plasma gases, argon or nitrogen, can be replaced with air, which can make the technology less expensive. The application of air requires a special electrode of hafnium or zirconium mounted in a copper holder. Cathode erosion in high-current arcs applied for cutting is a serious problem. The erosion puts a limit on the arc current and, thus, limits cutting speed, cut quality, and maximum plate thickness (Nemchinsky, 2002, 2003; Nemchinsky & Showalter, 2003).

7.8.2. Plasma Welding Technology

The arc discharge for plasma welding is formed between a pointed tungsten electrode and a workpiece. A schematic of the plasma welding is shown in Fig. 7–104. By positioning

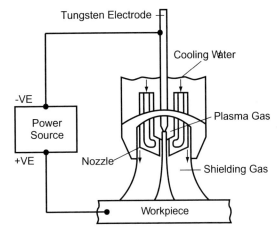

Figure 7–104. General schematic of plasma welding system.

the electrode within the body of the torch, the arc can be separated from the shielding gas envelope. Plasma is then forced through a fine-bore copper nozzle, which constricts the arc. The electrode used for plasma welding is usually tungsten with 2% thoria; the plasma nozzle is made from copper. The normal combination of gases is argon for the plasma gas, with argon plus 2–5% hydrogen for the shielding gas. By varying the bore diameter and plasma gas flow rate, we can distinguish three operating modes of plasma welding technology. The **microplasma welding mode** corresponds to very low discharge currents of 0.1–15 A. This arc is stable even when the arc length is varied up to 20 mm. The microplasma mode is traditionally used for welding thin sheets (down to 0.1 mm thickness), as well as for wire and mesh sections. The **medium current welding mode** corresponds to higher currents (from 15 to 100 A) and is similar to what is known as tungsten inert gas conventional arc welding. The **keyhole plasma welding mode** corresponds to currents over 100 A and therefore to high gas flows. A very powerful plasma beam is created in this mode, which can achieve full penetration in a material, as in laser and electron beam welding. During welding, the hole progressively cuts through the metal with the molten weld pool flowing behind to form the weld bead under surface-tension forces. This mode is characterized by high welding speed and can be used to weld thicker material (up to 10 mm of stainless steel) in a single pass.

7.8.3. About Plasma Melting and Remelting of Metals

Plasma is used in these technologies as an effective source of process heat. Plasma melting makes use of the anode heat transfer characteristics of an arc between a cathode and the metal (Heberlein, 2002). The relatively long characteristic process times (from 0.1 s to minutes) reduce the importance of instability effects in this case. Use of recycled scrap materials has increased lately: 35% of steel was produced from scrap in 2000. As a result, metal melting and remelting applications, in particular those involving thermal plasma, have expanded considerably (Neuschutz, 2000). New approaches to plasma melting and remelting are concentrated mainly on the development of more efficient overall operation of the installation (Johnson, Dowson, & Ward, 1998; see also Section 7.8).

7.8.4. Plasma Spheroidization and Densification of Powders

Plasma technologies for spheroidization and densification of different powders based on different thermal discharges are discussed in detail by Krasnov, Zilberberg, and Sharivker

Table 7–8. Characteristics of Dielectric Powders Before and After Plasma Treatment

Material	Specific surface before treatment, m^2/g	Specific surface after treatment, m^2/g	Particles size after treatment, μm
$Si + SiO_2 \rightarrow SiO$	0.1	75	0.04
Al_2O_3	0.7	25	0.06
SiO_2	0.1	65	0.03

(1970), Boulos, Fauchais, and Pfender (1994), Solonenko (1996), Rykalin et al. (1973), and Waldie (1973). A powder densification and spheroidization process is actually simple with respect to other plasma processes. It takes place in thermal plasma by means of in-flight heating and melting of feed material while passing through the discharge in the form of sintered or crushed powders. Molten spherical droplets formed in this way are gradually cooled down under free-fall conditions. The particles' time of flight must be sufficient for complete solidification before reaching the bottom of reactor chamber. Finer particles are entrained by plasma gases and recovered in a filter. Different atmospheric-pressure thermal plasmas are applied for the treatment of powders. In particular, effective powder densification and spheroidization have been demonstrated in atmospheric-pressure RF-ICP discharge by Tekna Plasma Systems Inc. Their technique has the ability to melt relatively large particles and operates with a wide range of gases. Specifically, we mention silica powder spheroidization performed in an open-air 100 kW induction plasma system. Some characteristics of dielectric powders before and after treatment in thermal plasma systems are summarized in Table 7–8 (Polak et al., 1975a). The table shows that thermal plasma treatment is able to produce dielectric powders with nanolevel sizes and to increase the specific surface area of the powder almost 1000-fold. Thermal plasma treatment of powders results not only in their spheroidization and densification but also in physical and chemical transformations, such as

- decreased powder porosity due to particle melting;
- surface oxidation or nitrification of particles;
- purification and refining of powders due to selective or reactive vaporization of specific impurities (for example, Si and Zn impurities in tungsten powder can be decreased about 3-fold, Mn and Pb impurities in tungsten powder can be decreased 10-fold); and
- phase and chemical transitions (for example, alumina powder after plasma evaporation and fast quenching becomes γ–Al_2O_3; tungsten carbide (WC) powder loses carbon with increase of discharge power and plasma temperature forming new compounds: $WC \rightarrow W_2C \rightarrow W$).

PROBLEMS AND CONCEPT QUESTIONS

7–1. Depth of Metal Oxide Reduction Layer in Non-Thermal Hydrogen Plasma. Estimate the depth of metal oxide reduction layer using relation (7–8) at surface temperatures 300, 700, and 1100 K. Assume the interaction radius of hydrogen atoms in a metal structure is $r_0 \approx 0.4$ nm. To calculate the flux Φ of atomic hydrogen from a discharge, assume a surface power of 1 W/cm^2, energy cost of atomic hydrogen in the discharge of 5 eV per atom, and probability for a hydrogen atom to reach the oxide surface, 30%.

7–2. Non-Equilibrium Surface Heating in Plasma Treatment of Thin Layers. Based on (7–16)–(7–18), derive formula (7–19) for the heat flux Q_0 required to provide a temperature increase ΔT in a layer with thickness h. Calculate the discharge power per unit area required for $\Delta T = 700\,\text{K}$ in a layer with thickness $h = 0.3\,\text{cm}$. For treated materials, consider iron and iron oxide (Fe_2O_3).

7–3. Carbothermic Plasma-Chemical Reduction of Uranium Oxide (U_3O_8). Analyze the stoichiometry of the carbothermic reduction of U_3O_8 (7–24). Explain why the ratio of molar fractions of CO_2 and CO in products of the process is not fixed. Find out the relation between the molar fractions of CO_2 and CO in the products as a function of initial composition of the solid mixture U_3O_8–C. Explain why the carbothermic reduction process of UO_2 (7–23) assumes only CO in products, while that of U_3O_8 (7–24) expects formation of CO and CO_2.

7–4. Hydrogen Reduction of Titanium, Germanium, and Silicon from Tetrachlorides in Thermal Plasma. Using the kinetic data presented in Table 7–5, calculate the residence time required for hydrogen reduction of the tetrachlorides. Take into account that optimal temperatures for the reduction processes are 3500 K for reduction of germanium and 4500 K for plasma reduction of silicon and titanium. Considering titanium as an example, calculate the percent total residence time required for the final stage of the metal reduction, $TiCl_2 \rightarrow Ti$.

7–5. Quenching of Products of Direct UF_6 Decomposition in Thermal Plasma. Integrating kinetic equation (7–48), derive formula (7–49) for the mass decrease of uranium-containing product particles due to their gasification by fluorine. Analyze the effect of initial particulate size on the rate and characteristic time of the heterogeneous reverse reactions. Compare the efficiencies of quenching of UF_4 and metallic uranium.

7–6. Distribution Function of Metal Halides over Oxidation Degrees During Their Dissociation in Non-Thermal Plasma. Derive the simplified distribution function over halide oxidation degrees (7–85) based on the detailed function (7–83) and (7–84) and assume that average number of detached halogen atoms is much less than the total number in the initial molecule.

7–7. Surface Quenching of Metals During Direct Decomposition of Halides in Non-Thermal Plasma. Calculate the ratio γ of the mean free paths of halogen and metal atoms (equation (7–87)), which determines the effect of heterogeneous product stabilization in non-thermal plasma. Take as an example the non-equilibrium dissociation of $TiCl_4$, and assume an average number of detached Cl atoms $\langle m \rangle = 2$, where the gas temperature equals room temperature; activation barriers of disproportioning are $E_a = 0.5\,\text{eV/mol}$.

7–8. Production of Uranium Oxides by Decomposition of Uranyl Nitrate ($UO_2(NO_3)_2$) Aqueous Solutions. Based on data presented in Section 7.6.4, calculate the energy cost of production of uranium from uranyl nitrate in the plasma-chemical process (7–102). Compare the result with the energy cost of uranium production by direct plasma dissociation of uranium hexafluoride (see Section 7.4.2).

7–9. Plasma Conversion of Silicon Tetrafluoride (SiF_4) with Water Vapor into Silica (SiO_2) and HF. Based on data presented in Section 7.6.8, calculate the energy cost for the plasma-chemical conversion of silicon tetrafluoride with respect to production of 1 kg of SiO_2. Calculate the energy cost for production of 1 kg of HF, which is a major product of the plasma process. Explain the effect of temperature and molar ratio H_2O/SiF_4 on the conversion of silicon tetrafluoride into silica (see Figs. 7–99 and Fig. 7–100).

7–10. Plasma Production of AlH_3 from Elements. Using relation (7–124), analyze the temperature dependence for the depth of the AlH_3 layer formed during the interaction of aluminum with molecular hydrogen excited in a non-thermal discharge. To make your conclusion, compare the activation energies of hydrogen-atom diffusion in aluminum, E_a^D, and the reverse solid-state reactions (7–120), $\tilde{E}_a$.

7–11. Synthesis of Metal Carbonyls in Non-Thermal Plasma. Based on relation (7–135), compare the contribution of vibrational temperature of CO molecules and chromium surface temperature on the rate of desorption of $Cr(CO)_6$ during its synthesis in non-equilibrium plasma. Make reasonable assumptions about values of T_v and T_S (see Section 7.1.7).

8

Plasma-Surface Processing of Inorganic Materials: Micro- and Nano-Technologies

8.1. THERMAL PLASMA SPRAYING

8.1.1. Plasma Spraying as a Thermal Spray Technology

Plasma spraying is a thermal spray technology in which finely ground metallic and non-metallic materials are deposited on a substrate in a molten or semimolten state (Kudinov et al., 1990; Zhukov & Solonenko, 1990; Pawlowski, 1995; Vurzel & Nazarov, 2000; Fauchais, Vardelle, & Dussoubs, 2001; Fauchais, 2004; Knight, 1991, 1998, 2005; Yoshida, 2005). Two other technologies are combustion spray and wire-arc spray. The thermal plasma heat source is usually based on direct-current (DC)-arc or radiofrequency inductively coupled plasma (RF-ICP) discharge. It provides very high temperatures, over 8000 K at atmospheric pressure, which allows melting of any material. This is actually a major distinctive feature of plasma spraying. Although thermal plasma enables extremely high temperatures, the operational melting temperature in plasma spraying is usually kept at least 300 K lower than the vaporization or decomposition temperature to avoid losses in energy efficiency. Ground materials are either injected into plasma (in the case of RF discharges) or in the plasma jet (in the case of DC arcs), where they are heated, melted or softened, accelerated, and directed toward the surface or substrate being coated. Particles or droplets making impact with the substrate rapidly cool, solidify, cool further, contract, and build up incrementally to form a deposit. The basic coating building blocks, known as "splats," typically undergo cooling rates in excess of 10^6 K/s in the case of metals (Jackson et al., 1981). Individual splats are typically quite thin, about 1–20 μm; coating thicknesses range from 50 to 700 μm, which is thicker than most physical vapor deposits (PVDs) and chemical vapor deposits (CVDs).

The first industrial plasma spray torches based on DC-arc discharges appeared in the 1960s and later, in the 1980s, were based on thermal RF discharges. Today, thermal spray processes are able to process and consolidate virtually any material that exhibits a stable molten phase to produce coatings or deposits characterized by relatively homogeneous, fine-grained microstructures. Thermal spray is very versatile and is being used for several new applications, including rapid part prototyping, production of fiber-reinforced intermetallic matrix composites, coating of biomedical implants, and the production of "functionally gradient" materials (Knight, 2005). Plasma spraying, with the highest processing temperatures, is the most flexible in the thermal spray "family" of technologies with respect to materials and processing conditions. Plasma spray processes are able to operate in environments ranging from air to inert gases to under water, and they are commonly used for consolidating oxidation-sensitive and reactive materials.

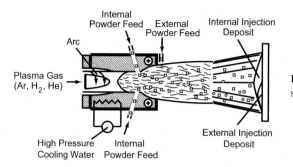

Figure 8–1. Schematic of DC plasma thermal spray process.

8.1.2. DC-Arc Plasma Spray: Air Plasma Spray

A schematic of the DC-arc plasma spray is shown in Fig. 8–1. Plasma spray processes can be broken down into three major categories:

1. Air plasma spray (APS);
2. "Vacuum" plasma spray (VPS) or low-pressure plasma spray (LPPS); and
3. Controlled atmosphere plasma spray, which also includes inert gas shrouded plasma spray (SPS) jets, and underwater plasma spray (UPS).

These distinctions are based on the level of interaction between the process jet and the materials being sprayed with the surrounding atmosphere and, thus, control the microstructure and properties of the sprayed materials. Thermal plasma spray processes utilize high-pressure vortex or axial gas-flow-stabilized non-transferred DC electric arcs. Arc currents generally range from 300 to 2000 A, and power levels from 20 to 200 kW. These produce plasma jets with quasi-equilibrium temperatures of 3000 to 15,000 K. Such temperatures enable processing of any material that exhibits a stable melting point. A typical DC plasma spray torch, or gun, consists of a cylindrical, water-cooled, 1–2% thoriated tungsten cathode, which emits electrons thermionically when heated by an electric arc, located concentrically and coaxially inside and upstream of a cylindrical water-cooled copper nozzle. Electrons emitted from the cathode flow to the nozzle anode wall under the influence of the applied electric field, typically 30 to 80 V (150–300 V for an open circuit). Non-oxidizing gases injected into this interelectrode region are heated, dissociated, ionized, and subsequently expanded to atmosphere through a constricting nozzle as a sub- or supersonic plasma jet, depending on the nozzle design and pressure ratio. The gas velocity may, in some cases, exceed 1500 m/s, depending on the power input, nozzle design, gas composition, and flow rate. Corresponding particle velocities are as high as 500 m/s.

A vortex flow is commonly used to spatially stabilize the arc and associated downstream plasma jet. Powders are injected into the plasma jet as fluidized streams of "carrier" gas and powder, either externally, via a tube directing the stream of powder particles radially into the hot gas jet immediately after it exits the nozzle, or internally, through powder feed ports through the nozzle wall itself. Internal injection produces longer particle heating or "dwell" times and improves melting efficiency, enabling higher-melting-point materials to be sprayed. Ar and N_2 are the most widely used carrier gases. Commercial plasma torch designs typically use different anode/cathode combinations depending on the melting point and particle size of the material being sprayed. Variables include the anode throat diameter and the angle of the powder feed ports relative to the torch axis. The use of upstream or downstream powder injection can also have a significant influence on the melting behavior of injected materials. Convergent/divergent Laval nozzles are used to generate supersonic flows. Plasma spray devices generally use pure Ar or mixtures of Ar plus H_2, He, or sometimes N_2 to generate the hot plasma jets used for spraying. The plasma gas composition and flow rate, as well as the distance between the torch nozzle exit and substrate and the power level, are commonly varied to control the enthalpy of the jet, with the goal of optimizing the heating/melting of the injected particulate feedstock materials. Plasma spraying uses feedstock powders with fairly narrow size distributions to provide uniform particle temperatures on impact at the substrate. Particles are generally fully melted during plasma spraying. Fluctuations in the plasma jet caused by motion and "restriking" of the anode attachment along the inside of the nozzle (Wutzke, Pfender, & Eckert, 1984) and turbulent eddy mixing of the downstream plasma jet with the surrounding atmosphere can entrain "pockets" of cold gas into the jet, resulting in non-uniform heating and oxidation of materials being sprayed.

8.1.3. DC-Arc Plasma Spray: VPS, LPPS, CAPS, SPS, UPS, and Other Specific Spray Approaches

Oxidation during plasma spray is not a problem when spraying ceramics (Al_2O_3, Cr_2O_3, TiO_2, etc.); however, refractory metals such as W, Ta, Re, and "reactive" metals such as Ti and Al are readily oxidized during APS, leading to inferior coating properties. Various approaches have been developed to overcome material oxidation during plasma spraying. Spraying at low pressures (using VPS or low-pressure plasma spray), in inert atmospheres, in enclosed pressure chambers, or using inert gas shrouds, can produce coatings with very low oxidation and porosity levels. This effectively eliminates interaction between the molten particles and oxygen entrained into the jet from the surrounding air. Operation of high-power non-transferred arc plasma jets in low-pressure (50–200 Torr) environments also leads to significant extension of the jet. Spray distances during VPS are typically longer than in APS at the same power level, and may be as high as 1 m. VPS processing also enables a process known as "reverse transferred arc" (RTA) to be used to "clean" the surfaces of metallic substrates prior to spraying. In RTA (Steffens & Hohle, 1980) an auxiliary low-power (about 20 kW) DC power supply is connected between the torch nozzle (anode) and the substrate such that a second, series arc is established between the torch and substrate, with the latter as the cathode. The presence of lower-work-function oxide films on the substrate surface favors preferential arc attachments, which quickly consume the oxide, leaving a metallurgically clean and active surface that promotes improved coating adhesion. In many VPS systems with RTA capability, the polarity of the RTA power supply can also be reversed, making the workpiece the anode with respect to the nozzle of the torch. This configuration can be used to preheat the substrate prior to spraying, if required. VPS coatings generally exhibit low oxide contents, densities up to 99%, and bond strengths up to 70 MPa, such that in many cases VPS-sprayed materials have the same, or better, properties than cast materials. This characteristic has enabled VPS processes to be used to spray-form net shapes of "difficult-to-process" materials including ceramics, refractory metals, intermetallics, and composites. Inert-gas SPS has the potential of achieving VPS-quality coatings under APS conditions, thereby lowering the total processing cost. Several effective inert-gas shroud designs have been developed and reported, including porous metal nozzle inserts (Chen et al., 1996) and discrete, multi-jet, parallel flow designs mounted downstream of the nozzle anode on conventional plasma spray torches (Mohanty et al., 1996). UPS has also been developed to overcome the oxidation issues associated with spraying in air without the need for costly and complex vacuum chambers, loadlocks, and so on. UPS effectively operates in its own "bubble" of inert gas and, rather like VPS, requires additional electrical insulation around the electrical connections at the rear of the torch. This approach has also been used for the plasma densification of agglomerated zirconia powders (Lugscheider & Rass, 1992).

Major limitations of thermal plasma spray torches are related to the non-uniformity of temperature and velocity distributions, and relatively low deposition rates (about 50 g/min) and deposition efficiencies (50–80%). Various novel plasma torch designs have appeared to overcome these limitations, including the shroud attachment (Sahoo & Lewis, 1992), plasma spray torches with multiple cathodes and/or axial powder injection, and torches with rotating nozzles for coating the internal surfaces of cylindrical substrates (Lugscheider et al., 1995). Water-stabilized Gerdien arcs (Chraska & Hrabovsky, 1992) utilize water stabilization, a consumable graphite cathode, and an external rotating anode to achieve power levels of 160 kW at arc voltages and currents of 310–340 V and 460–490 A, respectively. The primary application is reported to be the high-deposition-rate (50 kg/h) spraying of oxide ceramics such as Al_2O_3 at up to 0.3 mm per pass (Chraska et al., 1992). The design of conventional non-transferred DC-arc plasma spraying torches (see Fig. 8–1) does not favor the injection of feedstock materials directly into, or through, the plasma arc region. Excessive "loading" of constricted-arc plasma jets can lead to physical and electrical instabilities in the arc and even arc extinction. Several novel "center-feed" DC torches have been developed for plasma spray coating applications (Delcea & Herrington, 1995; Moreau et al., 1995), which enable the increase of deposition efficiency to 60–80%. Modeling of DC-arc plasma spraying is helpful to optimize the technology but is very complicated. Modeling complexity is due to arc root fluctuations, particle carrier-gas injection orthogonally to the jet, and mixing of entrained cold gas and plasma, which requires application of three-dimensional transient models taking into account two-fluid mixture and turbulency (Huang, Heberlein, & Pfender, 1995; Kolman et al., 1996; Delluc et al., 2003).

8.1.4. Radiofrequency Plasma Spray

RF plasma torches (Boulos et al., 1994), based on RF-ICP discharges, are electrodeless, in contrast to DC designs, allowing the use of a much wider range of plasma-forming gases, including air, O_2, H_2, and even hydrocarbons such as methane along with the more conventional Ar, Ar–H_2, Ar–He, and Ar–N_2. Continuous operation of RF discharges and torches is also unaffected by the electrode erosion and current carrying/cooling limitations of DC designs. RF torch and power supply efficiencies are, however, generally lower than those of DC plasma systems, primarily due to the lower energy transfer efficiency of induction heating and the lower efficiency of the high-frequency power supplies required. RF systems operating in the megahertz frequency range often utilize vacuum tube–driven tuned oscillators, which have inherent efficiencies of less than 50%. More efficient solid-state SCR/GTO-based power supplies operating at frequencies up to ~450 kHz are now being used with some of the newer high-power (>100 kW) RF torch designs. Future improvements in both RF torch and power-supply design will likely improve the overall efficiency of RF systems. Typical RF induction plasma spraying systems operate at pressures ranging from 50 Torr to atmospheric, power levels up to 150 kW, and frequencies of 2–4 MHz. Powders are injected axially via water-cooled probes introduced into the center of the toroidal discharge region between the 4–5 turn coils generally used. RF induction plasma torches generate larger plasma volumes with gas velocities (10–100 m/s) an order of magnitude lower than in DC torches, resulting in residence times of 10–25 ms versus the ~1 ms typical for DC plasma spray. Coarser (>150 μm) powders can also be sprayed owing to the larger-volume plasma region. A schematic of an RF plasma spray system is shown in Fig. 8–2. It comprises the torch plus its associated systems (power and gas supplies, powder feeder, etc.), a vacuum chamber, and some form of substrate support/manipulator. RF plasmas commonly "self-ignite" at reduced pressures. Ignition at atmospheric pressure, however, can also be accomplished by using a high-voltage Tesla coil or some other source of pre-ionization. Spraying is carried out at power levels of 25–50 kW, pressures of 50–400 Torr, and spray distances of 150–300 mm, the latter somewhat greater than those normally used in DC plasma spray processes. In RF plasma systems, "spray distance" is defined as the distance between the tip of the injection probe positioned in the center of the plasma region and the substrate.

8.1. Thermal Plasma Spraying

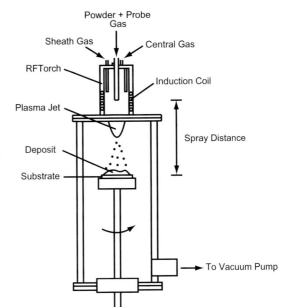

Figure 8–2. Schematic of RF plasma spray coating system.

Historically one of the major applications of RF plasma torches has been spheroidization of particulate materials (Muller et al., 1995), owing to the larger, less-constricted volume of plasma as compared to DC systems and the relatively low gas velocities (100 m/s), which ensure efficient melting of injected materials. RF VPS of various materials has also been reported, including fiber-reinforced metal matrix composite "monotapes" (Boulos, 1992) and tungsten (Jiang et al., 1993). More recently the RF plasma spraying of near-net-shape structured Cr bipolar plates for solid oxide fuel cells (Henne et al., 1999) and carbon fiber-reinforced aluminum coatings (Fleischer et al., 1999) have been reported. A recent process development in this area, likely to further improve the density of RF plasma sprayed deposits, is the incorporation of a supersonic nozzle (Boulos, 1999) into an RF torch. Modeling of thermal plasma spraying using RF torches is a much better developed technique than using DC plasma torches for two major reasons. First of all, the RF discharge plasma is fully axisymmetric, including the axial injection of particles with carrier gas, which permits the application of two-dimensional models. Also, the absence of significant fluctuations permits application of stationary models. Useful models of RF plasma spraying are described, in particular, by Xue, Proulx, and Boulos (2001, 2002, 2003).

8.1.5. Thermal Plasma Spraying of Monolithic Materials

Three types of deposits can be thermally sprayed:

1. Monolithic materials, including metals, alloys, intermetallics, ceramics, and polymers;
2. Composite particles, such as cermets (WC/Co, Cr_3C_2/NiCr, WC/CoCr, NiCrAlY/ Al_2O_3, etc.), reinforced metals (NiCr + TiC), and reinforced polymers; and
3. Layered or functionally gradient materials (FGMs).

Thermal plasma spraying of monolithic materials will be discussed, starting with metals.

1. Spraying metals. Tungsten, molybdenum, rhenium, niobium, titanium, superalloys (nickel, iron, and cobalt base), zinc, aluminum, bronze, cast iron, mild and stainless steels, NiCr and NiCrAl alloys, cobalt-based Stellites, Co/Ni-based Tribaloys, and NiCrBSi "self-fluxing" Colmonoy materials have all been thermally spray consolidated, either as coatings

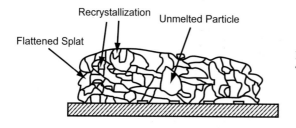

Recrystallization
Unmelted Particle
Flattened Splat

Figure 8–3. Schematic microstructure of an "as-sprayed" metallic deposit.

or free-standing deposits. Tribolite (FeCrNiBSi) and AmaCor amorphous alloys have also been developed for spraying and these exhibit excellent wear and corrosion resistance (Smith et al., 1992). A key feature of thermally sprayed deposits is their fine-grained structures and micro-columnar orientation. Plasma sprayed metals have reported grain sizes of less than 1 μm prior to postdeposition heat treatment and associated grain growth/recrystallization. Grain structure across an individual splat thus normally ranges from 10 to 50 μm, with typical grain diameters of 0.25 to 0.5 μm, due to the high cooling rates (10^6 K/s). Such rapid cooling rates, known to form fine-grained martensitic microstructures in steels, contribute to the high strengths exhibited by thermally sprayed materials. The "as-sprayed" microstructure of a typical metallic coating is shown in Fig. 8–3. Recent developments include the spraying of nanocrystalline metallic coatings, for which improved mechanical properties have been reported (Lugscheider, Born, & Wagner, 1991). Some of the challenges here are feeding of nanosized powders, which tend to agglomerate, and suppression of grain growth after deposition.

2. Spraying ceramics. Oxide ceramics such as Al_2O_3, ZrO_2 (stabilized with MgO, CeO, or Y_2O_3), TiO_2, Cr_2O_3, MgO, and hydroxy-apatite; carbides such as Cr_3C_2, TiC, Mo_2C, SiC (generally with a supporting metal matrix), and diamond; nitrides such as TiN and Si_3N_4; and spinels or perovskites such as mullite and superconducting oxides have all been deposited by plasma spray. Ceramics (with the exception of SiC, which sublimes), diamond, and SiC or Si_3N_4 (which must be incorporated in a metallic binder) are particularly suited to plasma spray consolidation due to high jet temperatures. Processing and materials flexibility as well as high temperatures give plasma spraying a leading role in spraying ceramic thermal barrier coatings (TBCs). APS is most suited to depositing ceramics and has found applications in TBCs, wear coatings on printing rolls (Cr_2O_3), and for electrical insulators (Al_2O_3). The microstructures of sprayed ceramics are similar to those of metals, with two important exceptions: grain orientation and microcracking. Rapid solidification of small droplets (<50 μm) results in very fine grains, as in metals; however, owing to the low thermal conductivity of ceramics, multiple grains may exist though the thickness of a splat, whereas in metals a single grain may cross an entire splat and even grow into an adjacent one. Inter- and intrasplat microcracking is widespread in plasma sprayed ceramic coatings, resulting from the accumulation of highly localized, residual, quenching stresses. Splats normally exhibit through-thickness cracking because of the very low ductility of most ceramics, but these cracks do not usually link up through the entire deposit thickness, at least not until an external stress is applied. Microcracking of splats is a major contributor to the effectiveness of TBC, even under high-temperature gradients and moderate strains – conditions under which conventionally formed bulk ceramics would fail.

3. Spraying intermetallics. Thermal spray consolidation and forming of intermetallic powders can be very effective. Thermal spray's high heating and cooling rates reduce the segregation and residual stresses that ordinarily limit the formability of these brittle materials. Thermal spray processes are also able to deposit materials onto mandrels, building

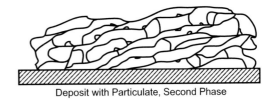

Deposit with Particulate, Second Phase

Figure 8–4. Schematic microstructures of possible thermal-spray-formed composites.

Deposit with a Whisker, Second Phase

Deposition on a Continuous Fiber, Second Phase

up thin layers of material and forming near-net shapes and providing the opportunity for "engineered microstructures" and functionally graded structures. Plasma spraying of TiAl, Ti_3Al, Ni_3Al, NiAl, and $MoSi_2$ was demonstrated with excellent deposit characteristics. Improved ductility has been obtained, with tensile strengths equal to, or better than, those of materials consolidated by other powder processing techniques. Plasma spray in controlled atmospheres is an effective method for the production of bulk intermetallics. Most intermetallics are reactive at high temperatures and sensitive to oxidation, which explains the preference for using inert-atmosphere plasma spray. Plasma spray formation is also well suited for the net-shape forming of brittle intermetallics; it can also increase the ductility of intermetallics such as $MoSi_2$ (Castro, 1992) and $NiAl/Ni_3Al$ (Sampath, Herman, & Sangaswamy, 1987). These improvements were linked to a decrease in grain boundary and/or splat interface contaminants, which limit localized plastic flow under an applied strain and lead to early crack linking, thus lowering the overall plastic flow (measured as ductility). Intermetallics are more porous than sprayed metals because of their limited ductility and low plasticity, which translates into a narrower "processing window" of plasma spray parameters for the production of high-density intermetallics, requiring tighter process control than for spraying metals. Higher-velocity plasma spray is preferred since the increased particle velocities result in more complete deformation of individual splats and denser deposits.

8.1.6. Thermal Plasma Spraying of Composite Materials

Difficult-to-process composites can be readily produced by thermal spray forming, and plasma spray is the process of choice for the most reactive matrix materials. Particulate-, fiber-, and whisker-reinforced composites have all been produced and used in various applications. Particulate-reinforced wear-resistant coatings such as WC/Co, $Cr_3C_2/NiCr$, and TiC/NiCr are the most common applications. Figure 8–4 illustrates the diverse forms of composites that can be thermally spray formed. Whisker particulates can be incorporated into sprayed deposits using so-called engineered powders, mechanical blending, or by co-injecting dissimilar materials into a single spray jet. Mechanical blends and co-injection, although useful, have been found to result in segregation of the reinforcing phase, due to size and/or density differences and degradation of the second-phase whiskers. Thermal spray composite materials can have reinforcing phase contents ranging from 10 to 90% by volume, where the metal matrix acts as a tough binder, supporting the reinforcing phase. The ability to consolidate such fine-grained, high-reinforcing-phase-content materials is a major advantage of thermal spray over other methods used to produce composites.

Thermal spraying of composite materials with discontinuous reinforcements, such as particulates or short fibers, is accomplished by spraying powders or powder blends. Techniques for the production of continuous fiber-reinforced materials which overcome the "line-of-sight" limitations of thermal spray processes were developed. This includes "monotape" fabrication techniques, where continuous fibers are prewrapped around a mandrel and a thin layer of a metal, ceramic, or intermetallic matrix material is sprayed (Westfall, 1987). The fibers are thus encapsulated within thin monolayer tapes which are subsequently removed from the mandrel for consolidation to full density by hot pressing with preferred fiber orientations, producing continuously reinforced bulk composites. Discontinuously reinforced composites produced using thermal spray utilize either composite powders or direct reactive synthesis. Powders can be produced mechanically, chemically, thermomechanically, or by high-temperature synthesis. "Engineered" powders are defined as those in which different phases are incorporated to produce a "microcomposition" of the final desired structure. Blended powders are produced by mixing the required proportions of a binder phase (a metal, intermetallic, or ceramic) together with a reinforcing phase. The matrix and reinforcing phases may segregate over time during shipping, handling, and feeding into the thermal spray process and often yield poor "as-sprayed" deposit uniformities. Many mechanical blends can, however, be "agglomerated" and fixed using an organic binder, which reduces their sensitivity to handling. Spherical agglomerates have been shown to flow and feed better, and spray drying has been used to produce such materials. Agglomerated powders have enabled thermally sprayed deposits with improved uniformity to be produced, although the high viscosity and shear and thermal forces acting on injected agglomerated particles tend to break the agglomerates apart, leading to segregation and direct exposure of the second phase to the thermal spray jet. The mechanical stability of agglomerated composite particles can be improved by agglomeration and sintering, which is a solid or rapid "matrix" melting and cooling process. Metal matrix/ceramic hard-phase reinforced powders (for example, WC/Co, $Cr_3C_2/NiCr$, $NiCrAlY/Al_2O_3$, etc.) can be sintered in the solid state by heating in a protective atmosphere furnace, sometimes using a fluidized bed, followed by a gentle milling to break any weak interagglomerate bonds. Retained porosity, although not a source of phase segregation in thermally sprayed deposits, can lead to higher as-sprayed porosity and increases the spray jet melting enthalpies required to melt the powders because the thermal conductivity of the individual particles is lowered by the porosity. Melting of metallic binding phases and retention of spherical particle morphologies can be achieved by processing the powders through a thermal plasma jet. Known as "plasma densification," this process produces essentially spherical powders exhibiting near 100% particle densities and uniform distributions of reinforcing phases. Plasma-densified powders produce the most uniform thermally spray consolidated deposits, but at significantly increased cost owing to the additional handling and processing steps and lower process yields.

An interesting approach to composite powder production, which is finding increasing application in thermal spray forming, is **self-propagating high-temperature synthesis** (SHS); Borisov, De Luca, & Merzhanov, 2002). The SHS process exploits the high heats of formation released from exothermic reactants which, when mixed together and ignited, produce a self-sustaining combustion-type reaction which propagates until all the reactants are consumed. SHS processing uses mixtures of reactant powders placed in a reactor vessel, generally in an argon atmosphere. The reactant mixture is then ignited using an incandescent filament or laser beam. SHS has now been developed to the point where it is capable of producing composite materials with well-controlled distributions of carbides or other reacted phases in metallic or intermetallic matrices (Munir & Anselmi-Tamburini, 1989; Sampath, 1993; Castro, Kung, & Stanek, 1994). SHS produces low-density porous "ingots" which must be further processed by crushing, milling, and screening to yield powders in size distributions compatible with thermal spraying. SHS-produced $MoSi_2$, SiC-reinforced $MoSi_2$, FeCr/TiC, Ti_3SiC_2, and NiCr/TiC materials have been successfully thermally sprayed.

8.1.7. Thermal Spray Technologies: Reactive Plasma Spray Forming

The thermally sprayed coating deposits considered earlier are usually less than 500 μm thick, more typically ∼250 μm. Thermal spray forming, however, can yield deposit thicknesses in excess of 25 mm. Free-standing shapes can be produced by spraying onto sacrificial mandrels, which are then mechanically or chemically removed after spraying. The incremental nature of thermal spray processes, using particulate feedstocks and with deposited layers typically ∼15–25 μm thick, enables graded and laminated structural materials to be formed. Also, because of the high processing temperatures and high particle kinetic energies at impact, traditionally difficult-to-form materials can be processed into near-net-shape components. Virtually all common and refractory metals, many intermetallics, ceramics, and combinations of these have been spray-formed as "composite" materials. The current major advantage of thermal spray forming techniques is in the forming of refractory metals, both in graded or layered structures and in composites. Plasma spray forming, a rapid particulate consolidation, has an inherent suitability for chemical synthesis. It is a strong candidate for synthesizing multi-phase, advanced materials, which is usually referred to as reactive plasma spray forming (Smith, 1993). Controlled-atmosphere plasma spraying, adapted for reactive plasma spraying, has been developed to combine controlled dissociation and reactions in thermal plasma jets, for in situ formation of new materials, or to produce new phases in sprayed deposits. Reactive plasma spray forming is emerging as a viable method for producing a wide range of advanced materials. The process, a logical evolution from conventional plasma spraying, allows "reactive" precursors to be injected into the particulate and/or hot gas streams. These reactive precursors may be liquids, gases, or mixtures of solid reactants which, on contact with the high-temperature plasma jet, decompose or dissociate to form highly reactive and ionic species which then react with other heated materials within the plasma jet to form new compounds.

Reactive plasma spray applications include the synthesis of composite materials, shaped brittle intermetallic alloys, reinforced or toughened ceramics, and tribological coatings with hard or lubricating phases formed in situ. Reactive plasma spray forming enables a wide range of materials to be produced, for example, Al with AlN, Al_2O_3, or SiC reinforcements; NiCrTi alloys with TiC or TiN; intermetallics such as TiAl, Ti_3Al, $MoSi_2$, and other ceramics with oxides, nitrides, borides, and/or carbides. All of these have been produced in situ in reactive thermal plasma jets. Recent investigations have demonstrated that reactive plasma spray forming utilizing incomplete or non-equilibrium reactions in plasma jets to form mixtures of phases can produce functional composites with the reactant products embedded in metallic, ceramic, and even intermetallic matrices. These materials have been shown to be very hard and have potential application as wear-resistant coatings (Smith, 1993). These reactively plasma sprayed coatings are, in practice, metal matrix composites with in situ reacted hard phases formed when the reactive gas mixtures reacted with molten metal particles. Fine TiC, WC, TaC, and other refractory carbides, oxides, silicides, or nitrides have been spray-formed in situ as particulate phases in near-net-shape metallic, intermetallic, and ceramic matrices, thus eliminating many difficult postforming operations. Lightweight structural materials such as Ti_3Al, TiAl, NiAl, and $MoSi_2$ have also been

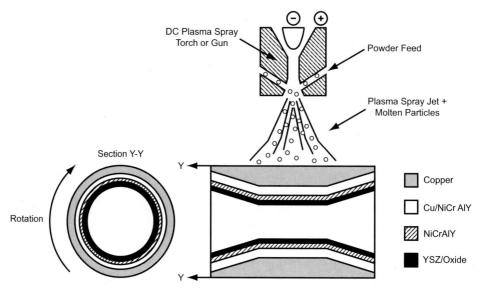

Figure 8–5. Schematic of a plasma sprayed functionally gradient material (FGM) for burner nozzle application.

produced by reactive spray forming. In these cases aluminum metal powders were injected into $TiCl_4$ plasmas to produce TiAl. Mo particles have also been injected into plasma jets seeded with disilane (Si_2H_6) to form $MoSi_2$. Deposition and synthesis of high-T_c ceramic superconductors were demonstrated as an example of reactive spray forming using DC and RF plasma jets (Zhu, Lau, & Pfender, 1991; Tsai et al., 1992). Dense, thick deposits have been produced using pre-alloyed 1-2-3 $YBaCu_xO_y$ powders. One more example is related to the synthesis and deposition of diamonds using either DC or RF thermal plasma jets (Müller & Lugscheider, 1992; Tsai et al., 1992; Girshik, 1993). CH_4 and H_2 were injected into the plasma jets as precursors, which are going to be discussed in a special section related to deposition of diamond films.

8.1.8. Thermal Plasma Spraying of Functionally Gradient Materials

FGMs promise production of improved materials and devices for use in applications subject to large thermal gradients, lower-cost clad materials for combinations of corrosion and strength or wear resistance, and perhaps improved electronic material structures for batteries, fuel cells, and thermoelectric energy conversion devices (Knight, 2005). The most immediate application for FGMs is in thermally protective claddings, where large thermal stresses could be minimized and component lifetimes improved by "tailoring" the coefficients of thermal expansion, thermal conductivity, and oxidation resistance. These FGM TBCs are finding use in turbine components, rocket nozzles, chemical reactor tubes, incinerator burner nozzles, or other critical furnace components. Figure 8.5 illustrates an example of a plasma sprayed FGM for the protection of copper using a layered ceramic FGM structure (Bartlett et al., 1996). Thermally protective FGMs may also be used in applications where reduced weight is required, such as in aircraft or other transportation systems; or for the thermal protection of lightweight polymeric materials by insulating materials. Production of graded metallic, oxide, and intermetallic structures may also enable advanced batteries, solid oxide fuel cells, and thermoelectric devices consisting of alternating layers of semiconducting materials $(FeSi_2)$ to be produced (Fendler, Henne, & Lang, 1995). Thermally sprayed graded materials are now also being considered for the production of oxide/metal/air-type

8.1. Thermal Plasma Spraying

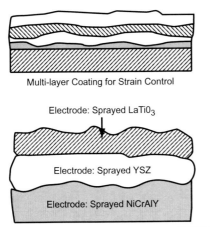

Multi-layer Coating for Strain Control

Electrode: Sprayed LaTiO₃

Electrode: Sprayed YSZ

Electrode: Sprayed NiCrAlY

Sprayed Electrodes for Solid Oxide Fuel Cells (SOFCs)

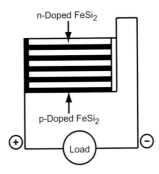

n-Doped $FeSi_2$

p-Doped $FeSi_2$

Load

Sprayed Alternating Layers for Thermoelectric Devices

Figure 8–6. Schematic examples of functionally gradient material (FGM), thermally sprayed using plasma system.

electrode/electrolyte systems. This has been proven on solid oxide fuel cells and electrode developments (Henne et al., 1992) using thermal spray methods. Spray formed battery cells promise to be lightweight and significantly more environmentally acceptable than current lead/acid and Ni/Cd batteries, thus opening up large potential markets in consumer, computer, and automotive electronics. These products would all utilize FGM, which promises to increase performance and lower cost. The forming of composite gun barrels by spraying tubular laminates and superalloy composites (Westfall, 1987), intermetallics, and refractory metals is one example of completed work. Thick thermally sprayed coatings with graded structures are being developed for several commercial applications, including ceramic outer air seals in aircraft gas turbines; thick, multi-layer, TBCs for heavy-duty diesel engine pistons; high-performance dielectric coatings for electronic devices; wear-resistant coatings for diesel engine piston rings; and oxidation-resistant coatings for high-temperature conditions.

Thermal spray forming can produce continuous gradations of metals, ceramics, and intermetallics, either in alternating stepwise layers or as microlaminates. Materials could be sprayed to replicate the mechanical behavior of a material, for example, by making tungsten shear locally in high-impact applications. Normally, under high impact, tungsten and its alloys behave in a ductile fashion, exhibiting significant plastic deformation, or "mushrooming." When localized shear behavior is required, however, very thin, alternating layers of intermetallic materials can be deposited concurrently while spray forming the bulk tungsten alloy. Thermal spray forming thus enables microlaminate structures to be formed, which can be used to locally modify the overall mechanical behavior of the system. The same approach could also be used to modify the electrical or thermal properties of materials. Thermally sprayed gradient structures have also been proposed as strain-controlled coatings and/or structures and for electrical devices, as shown schematically in Fig. 8–6 (Knight, 2005). The powder production route for materials permits oxides and/or brittle intermetallics with unique electrical and thermal properties to be produced. Figure 8–6 shows how oxide-based fuel cells can be spray formed by depositing alternating layers of oxide and metallic electrodes, and oxide electrolytes. The figure also shows the spraying of alternating layers of thermoelectric materials ($FeSi_2$) and the concurrent deposition of metallic electrodes. This processing combines the grading capability of thermal spray together with its ability to spray materials with widely varying melting points, thus truly realizing the advantages of plasma spray forming.

8.1.9. Thermal Plasma Spray Modeling

Thermal plasma spray modeling is focused on the description of RF and DC plasma torches, on plasma-particle interactions, as well as on the behavior of particles at impact and the coating development. Both two- and three-dimensional models are applied to calculate velocity, enthalpy, and temperature distributions at the nozzle exit, matching with the plasma gas flow rate and enthalpy. The properties are presented as follows (Dussoubs, 1998; Freton, Gonzalez, & Gleizes, 2000; Wan et al., 2002):

$$\frac{\phi(r) - \phi_{\mathrm{w}}}{\phi_{\mathrm{c}} - \phi_{\mathrm{w}}} = 1 - \left(\frac{r}{R}\right)^{n}. \tag{8-1}$$

Here $\phi(r)$ represents temperature, flow velocity, or enthalpy at the distance r from the torch axis; ϕ_{c} and ϕ_{w} are values of the variables on the torch axis and at the wall; R is the torch radius; and n is a fitting parameter. The most recent developments of plasma spray modeling are three-dimensional (Vardelle et al., 1996; Ahmed & Bergman, 1999; Williamson, Fincke, & Chang, 2000; Zhang, Gawne & Lin, 2000; Li & Chen, 2001; Vardelle, 2005). The physics of single particle interaction with plasma is summarized by Pfender and Lee (1985); Boulos et al. (1993); Essoltani et al. (1991); Sobolev, Guilmany, and Mart (1999); and Dong-Yan, Xin-Can, and Xi (2002). Tremendous temperature gradients in the boundary layer surrounding a particle as well as a thermal buffer related to the "particle's vapor" traveling with the particle determine specifics of the interaction. It was found that chemical reactions of particles with medium (oxidation, for example) are usually controlled by diffusion (Volenik et al., 1999; Kolman & Volenik, 2002). Models of particles at impact and especially single particle flattening were developed by Mostaghimi and his group at the University of Toronto. These three-dimensional models take into account flow on a rough substrate, cooling and solidification of the flattening particle, and flattening/splashing processes based on the Rayleigh-Taylor instability (Pasandideh-Fard et al., 1998, 2002; Bussmann, Mostaghimi, & Chandra, 1999).

8.2. PLASMA-CHEMICAL ETCHING: MECHANISMS AND KINETICS

8.2.1. Main Principles of Plasma Etching as Part of Integrated Circuit Fabrication Technology

Etching is a non-equilibrium process performed in non-thermal plasma. Plasma etching is one of the key processes applied in microelectronic circuit fabrication. The objective of the etching is to remove material from surfaces, which can be chemically selective and anisotropic. Chemically selective etching means removing one material and leaving another one unaffected. Anisotropic etching involves removing material at the bottom of a trench and leaving the sidewalls unaffected; see an example in Fig. 8–7. Today's plasma etching can create 0.2-μm-wide, 5-μm-deep trenches in silicon films or substrates, which are hundreds of times thinner than a human hair. Application of plasma is the only viable approach today for anisotropic etching, which makes it a unique part of modern integrated circuit (IC) fabrication technology. To analyze plasma applications for **integrated circuit fabrication**, consider as an example the following six-step process for the creation of a metal film patterned with submicron features on a wafer substrate (Fig. 8–8):

1. Film deposition on a substrate;
2. Photoresist layer deposition over the film;
3. Selective exposition of the photoresist to light through a pattern;
4. Development of the photoresist and removal of the exposed resist regions (this leaves behind a patterned resist mask);

Figure 8–7. Example of deep silicon etching accomplished by plasma processing.

5. Transferring the pattern into the film via the etching process (the mask protects the underlying film from being etched); and
6. Removing of the remaining resist mask.

Plasma processing is effectively used for film deposition (step 1) and etching (step 5). Plasma processing can also be used for photoresist development (step 4) and removal (step 6). The etching in step 5 is anisotropic, which is a unique feature of plasma processing that makes it so important in IC fabrication. **Isotropic etching** can be illustrated by the etching of a silicon substrate. The process starts with the production of chemically aggressive atoms (for example, F atoms) by dissociation of its non-active compounds (for example CF_4) in a non-thermal, usually low-pressure RF discharge. The F atoms then react with the silicon substrate, yielding the volatile etch product SiF_4. **Anisotropic etching**, shown in Fig. 8–8e, is due to substrate bombardment by high-energy ions, which mostly bombard the bottom of the trench, leaving the sidewalls untreated. The sidewalls of the trench can be additionally protected by special passivating film to achieve a stronger anisotropic effect from the plasma etching. Taking into account the significant practical importance of the plasma etching and deposition processes, numerous reviews and books have been published on the subject, including those of Hollahan and Bell (1974), Vossen and Kern (1978), Powel (1984), Sugano (1985), Morgan (1985), Manos and Flamm (1989), Liebermann and Lichtenberg (1994), Roth (2001), and Chen and Chang (2003). In the following sections we focus on plasma-chemical aspects of etching technology, in particular applications for IC fabrication.

8.2.2. Etch Rate, Anisotropy, Selectivity, and Other Plasma Etch Requirements

Etch rate requirements can be analyzed by keeping in mind the conventional set of films shown in Fig. 8–9 and including the following: 1500 nm of resist, over 300 nm of polysilicon,

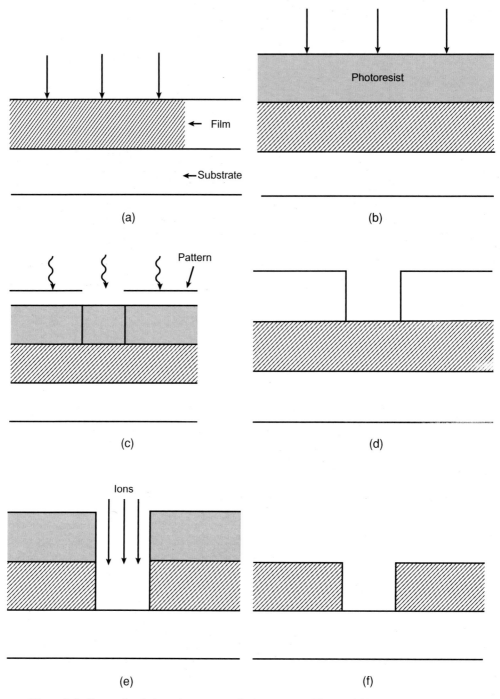

Figure 8–8. Conventional six major steps required to create a thin metal film patterned with submicron features on a large-area wafer substrate: (a) metal film deposition; (b) deposition of photoresist layer; (c) selective optical exposure of the photoresist through a pattern; (d) photoresist development, removing of the exposed resist regions leaving behind a patterned resist mask; (e) anisotropic plasma etch, pattern is transferred into the metal film; (f) removal of the remaining photoresist mask.

8.2. Plasma-Chemical Etching: Mechanisms and Kinetics

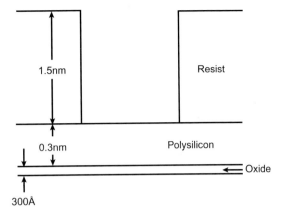

Figure 8–9. Illustration of a typical set of films in plasma etching.

and over 30 nm of gate oxide on an epitaxial silicon wafer. Each of these films should be etched for 2–3 min for a single wafer process. Therefore, etch rate requirements are as follows:

- For a photoresist layer, $w_{pr} = 500$ nm/min (minimum rate).
- For a gate oxide layer, $w_{ox} = 10$ nm/min (minimum rate).
- For a polysilicon layer, $w_{poly} = 500$ nm/min (intermediate requirement).

Selectivity requirements can be analyzed by comparing the etching rates of polysilicon and photoresist used as a mask. To complete the polysilicon etch before eroding the resist, the required etching selectivity should satisfy the following criterion:

$$S = \frac{w_{poly}}{w_{pr}} \gg \frac{300 \text{ nm}}{1500 \text{ nm}} = 0.2. \tag{8–2}$$

Due to non-uniformity of the polysilicon layer across the wafer, twice as much time may be required to clear polysilicon from all unmasked regions. To restrict destruction of the thin oxide layer during the **over-etching**, another selectivity criterion should be satisfied:

$$S = \frac{w_{poly}}{w_{ox}} \gg \frac{300 \text{ nm}}{30 \text{ nm}} = 10. \tag{8–3}$$

Thus, effectiveness of the etching process also imposes requirements on system non-uniformity.

Anisotropy requirements are usually characterized by the anisotropy coefficient, which is the etching rate ratio in vertical and horizontal directions:

$$a_h = w_v/w_h. \tag{8–4}$$

Figure 8–10 illustrates the anisotropic etching of a deep trench of width l into a film of thickness d. Assuming that the mask is not eroded and horizontal overetching under the mask is limited by δ, relation (8–4) gives the anisotropy requirement:

$$a_h \geq d/\delta. \tag{8–5}$$

If the mask width is l_m, the maximum width of the trench is (see Fig. 8–10)

$$l = l_m + 2\delta. \tag{8–6}$$

Based on relations (8–5) and (8–6), the anisotropy requirement can be presented as

$$a_h \geq \frac{2d}{l - l_m}. \tag{8–7}$$

For conventional sizes ($l = 1$ μm, $l_m = 0.5$ μm, $d = 2$ μm), the required anisotropy coefficient should be at least 10. Small trench width can obviously be achieved by using a

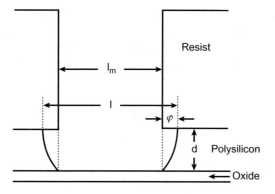

Figure 8–10. Anisotropy in polysilicon plasma etching.

mask width that is as thin as permitted by, for example, the limitations of lithography. The minimal trench width can be estimated from (8–7) as

$$l_{\min} \approx 2d/a_{\mathrm{h}}. \tag{8–8}$$

Fabrication of deep trenches ($d/l \gg 1$) requires significant anisotropy of the etching process.

Process and system non-uniformity (8–3) can affect selectivity requirements. Similarly, the non-uniformity significantly affects anisotropy requirements as well. In general, there are essential trade-offs among anisotropy, selectivity, and uniformity of plasma etching processes, which are discussed, in particular, by Flamm and Herb (1989) and Liebermann and Lichtenberg (1994).

8.2.3. Basic Plasma Etch Processes: Sputtering

We can distinguish four basic types of plasma etching processes applied to remove materials from surfaces: sputtering, pure chemical etching, ion energy-driven etching, and ion inhibitor etching. The first process to discuss is sputtering (see Fig. 8–11), which involves the ejection of atoms from surfaces due to energetic ion bombardment. To provide an effective sputtering process, a low-pressure non-thermal discharge supplies energetic ions to the treated surface. The ion energies in practical sputtering systems are typically above a few hundred electron volts. Ions with energies exceeding 20–30 eV are able to sputter atoms from a surface. The sputtering yield γ_{sput}, which is the number of atoms sputtered per incident ion, rapidly increases with the ion energies up to a few hundred electron volts, where the yield becomes sufficient for practical applications. Usual projectiles for practical physical sputtering are argon ions with energies of 500–1000 eV. In this energy range, a bombarding ion transfers energy to a large group of surface atoms. After redistribution of energy in the group, some atoms can overcome the surface binding energy, E_{s}, and escape the surface, enabling sputtering. The sputtering yield γ_{sput} is independent of the ion energy over a broad range of the energies and surface atom density and depends mostly on the surface binding

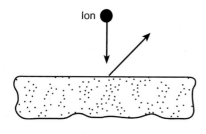

Figure 8–11. Plasma sputtering.

Table 8–1. Sputtering Yields for Different Materials Provided by Ar + Ions with Energy 600 eV

Material	γ_{sput}	Material	γ_{sput}	Material	γ_{sput}
Al	0.83	Cu	2.00	SiO_2	1.34
Si	0.54	Ge	0.82	GaAs	0.9
Fe	0.97	W	0.32	SiC	1.8
Co	0.99	Au	1.18	SnO_2	0.96
Ni	1.34	Al_2O_3	0.18		

energy E_s, and masses M_i and M_s of incident ions and surface atoms (Feldman & Mayer, 1986):

$$\gamma_{\text{sput}} \propto \frac{1}{E_s} \cdot \frac{M_i}{M_i + M_a}. \tag{8–9}$$

Some experimentally measured sputtering yields (γ_{sput}) for 600 eV argon-ion bombardment of different materials (Konuma, 1992) are presented in Table 8–1. The sputtering yields are close to unity and not much different for different materials. Sputtering rates are generally low for this reason: about one atom per one incident ion. Because the ion fluxes from conventional low-pressure etching discharges are relatively small, the sputtering rates are generally not sufficient for commercially significant material removal. Also taking into account that the surface binding energy E_s (few electron volts) is much less than the incident ion energy (500–1000 eV), the energy efficiency of sputtering is only about 0.2–0.5%. Sputtering is an anisotropic process and strongly sensitive to the angle of incidence of the ion, θ ($\theta = 0$ is normal incidence; $\theta = \pi/2$ is grazing incidence). The sputtering yield has a maximum at some specific angle about $\theta = \pi/4$–$\pi/3$, where sputtering is about twice as intensive as at normal incidence ($\theta = 0$). There is no sputtering at grazing incidence ($\theta = \pi/2$); therefore, sidewall removal by ions normally incident on a substrate is negligible. A distinctive feature of sputtering is the possibility of removing involatile product from the surface during the etching. We should also mention that sputtering is generally carried out at low pressures, when redeposition of the sputtered atoms can be neglected.

8.2.4. Basic Plasma Etch Processes: Pure Chemical Etching

Pure chemical etching involves a plasma discharge that supplies gas-phase etchant atoms or molecules, which are able to react with the surface to form gas-phase products and remove the surface material. This type of etching is simply illustrated in Fig. 8–12. As examples, we can mention etching of silicon using fluorine atoms generated in plasma with formation of gaseous SiF_4, or etching of photoresist (which consists of hydrocarbons) using plasma-generated oxygen atoms with the formation of gaseous CO_2 and H_2O. Products in this type of plasma etching should be always volatile to be removed. Pure chemical etching is an isotropic process, because the gas-phase neutral etchant atoms or molecules approach the substrate with near uniform angular distribution. While not anisotropic, pure chemical

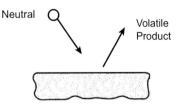

Figure 8–12. Plasma process of pure chemical etching.

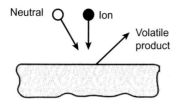

Figure 8–13. Plasma process of ion energy-driven etching

etching is characterized by quite large etching rates. The large etch rates in this case are due to the possibility of having much higher fluxes of active chemicals from the etching plasma discharges than fluxes of ions.

8.2.5. Basic Plasma Etch Processes: Ion Energy-Driven Etching

This etching process, based on the synergetic effect of neutral chemically active etchants (for example, F atoms) and energetic ions, is illustrated in Fig. 8–13. The synergetic etching effect occurs when neutral etchants and energetic ions together are able to provide much higher surface material removal rates than corresponding pure chemical etching alone or sputtering alone. For example, etch rates of silicon by XeF_2 etchant gas alone, as well as by Ar^+ ion beam sputtering alone, were less than 0.5 nm/min (Coburn & Winters, 1979). Ion energy-driven etching, which combines the effects of the XeF_2 etchant gas along with Ar^+ ion beam sputtering, results in an etch rate exceeding 6 nm/min. In the framework of ion energy-driven etching of silicon with a high incident flux of F atoms, a single 1000 eV Ar^+ ion initiates removal of about 25 silicon atoms from the surface in the form of SiF_4. To compare, Ar^+ ion sputtering processes result in this case in the removal of a single silicon atom from the surface (see Section 8.2.3). Thus, ion energy-driven etching is chemical in nature but with a reaction rate determined by energetic ion bombardment. In contrast to sputtering, the etch rate increases in this case with growth of ion energy. Similarly to pure chemical etching, ion energy-driven etching requires products to be volatile. Energetic ions have highly directional angular distribution; therefore, ion energy-driven etching can be highly anisotropic. Etching selectivity in this case might not be as good as that in pure chemical etching. The trade-off between anisotropy and selectivity of surface atom removal can be effectively controlled in this type of etching.

8.2.6. Basic Plasma Etch Processes: Ion-Enhanced Inhibitor Etching

Ion-enhanced inhibitor etching, illustrated in Fig. 8–14, involves the application of inhibitor species. Non-thermal discharge supplies not only neutral chemically active etchants and energetic ions, as in ion energy-driven etching, but also inhibitor precursor radicals and molecules. These inhibitor precursors can be absorbed or deposited anisotropically on the substrate to form a protective layer of polymer film, preventing etching. Anisotropic ion bombardment prevents the formation of the inhibitor layer and, therefore, exposes the surface to chemical etching. Where there is no ion flux, the inhibitor is effectively formed and protects the surface from the chemical etchant. In particular, the inhibitor precursor radicals are CF_2 and CF_3, which are formed by plasma-chemical decomposition of, for example, CF_4, or CCl_2 and CCl_3, generated by plasma-chemical decomposition of CCl_4. These radicals, deposited on the substrate, are able to form fluoro- and chlorocarbon polymer

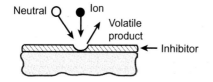

Figure 8–14. Plasma process of ion-enhanced inhibitor etching.

films, protecting the substrate from etching. A classic example of the ion-enhanced inhibitor etching process is anisotropic etching of aluminum trenches in non-thermal, low-pressure discharges in CCl_4–Cl_2 or $CHCl_3$–Cl_2 mixtures. In these systems, both atomic and molecular chlorine (Cl and Cl_2) chemically etch aluminum – intensively, but isotropically. Addition of carbon compounds to the feed gas leads to the formation of a protective chlorocarbon film on the surface. Ion bombardment normal to the substrate eliminates the protective film from the trench bottom, allowing the etch process to proceed there anisotropically. The chlorocarbon film created on the sidewalls of the trench protects them from chemical etching by Cl and Cl_2. Optimization of the process results in highly anisotropic etching with formation of vertical sidewalls. Ion-enhanced inhibitor etching is characterized by features somewhat similar to those of the ion energy-driven etching. The etching process is not as selective as pure chemical etching, and volatility of the etching products is required. Also we should remember that the protective layer can contaminate the substrate and should be removed. In Sections 8.2.3–8.2.6 we distinguished four types of plasma etch processes. In practical applications, however, parallel or serial combinations of the different etching processes are often required.

8.2.7. Surface Kinetics of Etching Processes; Kinetics of Ion Energy-Driven Etching

The kinetics of etching processes includes surface kinetics, which determines the etch rate and anisotropy based on ion and neutral fluxes from a discharge, as well as the discharge kinetics, which describes the generation of etchant atoms and bombarding ions. Consider first the surface etching kinetics, taking as an example the ion energy-driven etching of carbon substrate by atomic oxygen. A simplified mechanism of the etching process starts with chemisorption of O atoms on the carbon substrate:

$$O(gas) + C(solid) \rightarrow C{:}O(chemisorbed), \quad K_a. \tag{8–10}$$

The chemisorbed complexes C:O can then be either simply desorbed from the surface,

$$C{:}O(chemisorbed) \rightarrow CO(gas), \quad K_d, \tag{8–11}$$

or desorbed by ion impact, characterized by kinetic coefficients Y_i and K_i:

$$C{:}O(chemisorbed) + ion \rightarrow CO(gas). \tag{8–12}$$

In the framework of the Langmuir adsorption–desorption model, all O atoms approaching the fraction $1 - \theta$ of the carbon surface not covered by C:O complexes immediately react, forming the complexes. The kinetic equation for the fraction θ can be written as (Lieberman & Lichtenberg, 1994)

$$\frac{d\theta}{dt} = K_a n_O (1 - \theta) - K_d \theta - Y_i K_i n_i \theta. \tag{8–13}$$

A limiting step of the etching is CO desorption into the gas phase, and so steady-state carbon surface coverage with the C:O groups can be found by assuming $d\theta/dt = 0$. Here, n_O and n_i are O and ion densities near the carbon surface (at the plasma-sheath edge), and K_a is the O adsorption rate coefficient,

$$K_a = \frac{1}{4} \frac{v_{tO}}{n_S} = \frac{1}{4n_S} \sqrt{\frac{8T_0}{\pi M_O}}; \tag{8–14}$$

v_{tO} is the velocity of O atoms; n_S is the surface density of the sites; M_O is mass of an oxygen atom; K_d is the rate coefficient of the thermal desorption 8–11; Y_i is the ion energy-driven yield of CO molecules desorbed per ion incident on a fully covered surface, which is greater than unity and can be estimated as

$$Y_i = \eta E_i / E_b. \tag{8-15}$$

In the preceding equation E_i is ion energy; E_b is the C:O group-surface binding energy ($E_i \gg E_b$); $\eta \leq 1$ is the efficiency of ion energy in desorption of the C:O complex from the carbon; K_i is the ion flux coefficient,

$$K_i = \frac{u_B}{n_S} = \frac{1}{n_S} \sqrt{T_e / M_i}; \tag{8-16}$$

u_B is the Bohm velocity (see Chapter 3); T_e is the electron temperature; and M_i is an ion mass.

The steady-state carbon surface coverage with the C:O groups, θ, can be found from (8–13) as

$$\theta = \frac{K_a n_O}{K_a n_O + K_d + Y_i K_i n_i}. \tag{8-17}$$

The flux of CO molecules leaving the surface as a result of the etching can then be expressed as

$$\Gamma_{CO} = (K_d + Y_i K_i n_i)\theta n_S = \frac{(K_d + Y_i K_i n_i)n_S K_a n_O}{K_a n_O + K_d + Y_i K_i n_i}. \tag{8-18}$$

The vertical etch rate is determined by the etching flux of CO (Γ_{CO}) and the C-atom density, n_C:

$$w_v = \Gamma_{CO} / n_c, \tag{8-19}$$

which can finally be rewritten, taking into account the detailed kinetic expression (8–18):

$$w_v = \frac{n_S}{n_C} \cdot \frac{1}{\dfrac{1}{K_d + Y_i K_i n_i} + \dfrac{1}{K_a n_O}}. \tag{8-20}$$

To calculate the horizontal etch rate, it is reasonable to assume that bombarding ions strike the substrate surface at normal incidence. The ion flux incident on a vertical trench sidewall in this case is equal to zero. Then expression (8–20) can be rewritten for the horizontal etch rate as

$$w_h = \frac{n_S}{n_C} \cdot \frac{1}{\dfrac{1}{K_d} + \dfrac{1}{K_a n_O}}. \tag{8-21}$$

The etch anisotropy coefficient in a practically interesting case, $K_a n_O (Y_i K_i n_i \gg K_d)$ can be expressed based on (8–20) and (8–21) as

$$a_h = \frac{w_v}{w_h} = \frac{Y_i K_i n_i}{K_d} \cdot \frac{1}{1 + \dfrac{Y_i K_i n_i}{K_a n_O}}. \tag{8-22}$$

The maximum value of the anisotropy coefficient for the ion energy-driven etching is equal to

$$(a_h)_{max} = \frac{w_v}{w_h} = \frac{Y_i K_i n_i}{K_d}. \tag{8-23}$$

The maximum anisotropy can be achieved when the ion flux is not too high ($Y_i K_i n_i \ll K_a n_O$) and the surface is covered by chemisorbed complexes ($\theta \to 1$) (8–17). In this ion flux-limited regime, the etching anisotropy increases with an increase of ion energies and fluxes (relatively large values of Y_i (8–15) and n_i), as well as with a decrease of substrate temperature, which reduces the desorption, K_d.

In the opposite regime of limited flux of active neutral etching species, $K_d \ll K_a n_O \ll Y_i K_i n_i$, the anisotropy coefficient becomes independent of ion energy and ion flux:

$$a_h = K_a n_O / K_d. \tag{8–24}$$

A similar kinetic analysis clarifies the other specific etching mechanisms and can be applied for sputtering, as well as for ion energy-driven etching limited by the formation and desorption of etch product, which takes place in the case of silicon etching by F atoms (Lieberman & Lichtenberg, 1994).

8.2.8. Discharges Applied for Plasma Etching: RF-CCP Sources, RF Diodes and Triodes, and MERIEs

The **etching discharges** are mostly low-pressure, strongly non-equilibrium discharges. Most of them are RF based; however, some are supplied by microwave radiation or in some different manner. Low translational temperature in the discharges permits one to avoid damaging the treated surfaces and to maintain high selectivity in the plasma processing. Low pressures help to sustain the discharge uniformity and non-equilibrium in sufficient volumes; Low pressures also provide effective contact of active species generated in the plasma volume with the treated surface, without product redeposition. Specifically, the most used etching discharges are RF capacitively coupled plasma (CCP) discharges, called **RF diodes**, and high-density plasma (HDP) discharges, in particular RF-ICP discharges. The RF diodes used for anisotropic etching typically operate at pressures of 10–100 mTorr, at power densities of $0.1–1 \, W/cm^2$, and at a driving frequency of 13.56 MHz. Multiple wafer treatment systems are common. Typical RF driving voltages are 100–1000 V, and plate separation is 2–10 cm. When a wafer is mounted for etching purposes on the powered electrode, the system is usually called a **reactive ion etcher (RIE)**. The term RIE, although widely used, is not very accurate because etching is a chemical process enhanced by energetic ions – not a removal process due to the "reactive ions" alone.

Typical plasma densities used for etching in the RF diodes are relatively low (10^9–$10^{11} \, cm^{-3}$); electron temperatures are about 3 eV. Sheath voltages, which determine ion acceleration energies, are relatively high and exceed 200 V. Dissociation degrees of molecules vary from 0.1% to almost 100%. Typical discharge volumes are 1–10 L, and typical cross-sectional areas are $300–2000 \, cm^2$. A limiting feature of the RF diode application is that the ion bombarding flux in these discharges cannot be varied independently of the ion bombarding energy. Sheath voltages and hence bombarding ion energies should be high to achieve reasonable (although still not high) ion fluxes and sufficient dissociation degrees of the feed gas. Values that are too high can result in damages and losses of linewidth control for wafers placed on the driven electrode. In general, relatively low ion fluxes and high ion energies result in narrow process windows. Also, low process rates related to limited ion fluxes often lead to multi-wafer or batch processing, which negatively affects wafer-to-wafer reproducibility. Control over the energy of ion bombardment can be achieved to some extent by using **RF triodes**. In these discharges, the wafer is located on the undriven electrode independently biased by a second RF source. Although the RF triodes are able to control the ion energy, processing rates and sputtering contamination are still issues in these discharges. Different magnetically enhanced modifications of the RF-CCP discharges have been developed to improve practical performance of the RF diodes and triodes. In particular, we should point out **magnetically enhanced reactive ion etchers (MERIEs)**. A magnetic field of 50–300 G is applied in these discharge systems parallel to the powered electrode, where the wafer is located. The magnetic field increases the efficiency of power transfer to the plasma and enhances plasma confinement, which results in reduction of sheath voltage and an increase of plasma density. MERIE plasma is non-uniform both radially and azimuthally. To increase process uniformity, the magnetic field in the MERIE discharges can be slowly rotated in the wafer plane with frequency of about 0.5 Hz.

8.2.9. Discharges Applied for Plasma Etching: High-Density Plasma Sources

The key feature of low-pressure HDP etching discharges, which are able to overcome the problems of RF diodes and triodes, is the coupling of RF or microwave power to the plasma across a dielectric window (in contrast to RF diodes and triodes, where electrodes have direct contact with the plasma). The non-capacitive coupling and power transfer allow operation at low voltages across sheaths. The ion acceleration energies are typically about 20–30 eV at all surfaces in this case. The electrode with wafer is biased by an independently driven capacitively coupled RF source, which permits independent control of ion energy. Thus, using HDP plasma sources makes possible the generation of plasma with a high density of charged and chemically active species, which leads to high processing rates, with independent control of the ion-bombarding energy. HDP sources significantly differ in the means by which power is coupled to the plasma. It can be an electron cyclotron resonance (ECR) plasma source supported by microwave radiation; it can be a helicon plasma source applying an axial magnetic field with an RF-driven antenna; or it can be a helical resonator plasma source with the external helix and conducting cylinder surrounding the dielectric discharge chamber. HDP etching discharges can also be produced with ICP sources. In the ICP or transformer discharges, plasma acts as a single-turn lossy conductor, which is coupled to a multi-turn non-resonant RF coil across the dielectric discharge chamber. RF power is inductively coupled to the plasma, as in a transformer. Because plasma acts in a single turn, the transformer decreases voltage and increases current, which explains the generation of HDP in the discharges. Although HDP sources are quite different, all of them have RF or microwave power coupled to the plasma across a dielectric window. HDP etching discharges are characterized in general by pressures of 0.5–50 mTorr, which is lower than that in RF diodes and triodes. HDP discharge power is in the range 0.1–5 kW, which is generally higher than that of RF diodes and triodes. Typical electron temperatures are 2–7 eV, which is higher than that in capacitively coupled discharges. In accordance with the term HDP, the plasma density in these discharges is 10^{10}–10^{12} cm^{-3}, exceeding that of capacitively coupled discharges by about 10-fold. The low-pressure discharges described in the previous two sections are conventionally used not only for plasma etching but also for deposition and other material processing applications. More details about these discharges can be found in Chapter 4 and in books especially dedicated to the discharge subject, including Lieberman and Lichtenberg (1994), Roth (1995, 2001), and Fridman and Kennedy (2004).

8.2.10. Discharge Kinetics in Etching Processes: Ion Density and Ion Flux

The surface kinetics of etching (Section 8.2.7) is controlled by concentrations of ions and active neutrals near the surface. Determination of these parameters requires a detailed consideration of etching discharges (Sections 8.2.8 and 8.2.9). Some useful relations, however, can be derived from general kinetics of the low-pressure discharges applied for etching. In this section, we make such estimations for the concentration and flux of ions; concentration and flux of neutral chemically active etchants will be estimated in the next section. A balance of charged particles in plasma between electrodes with area A (characteristic radius R) and narrow gap l between them ($l \ll R$), controlled by ionization and losses to the electrodes, can be estimated as

$$2n_i u_B A \approx k_{ion}(T_e) n_0 n_e l A, \qquad (8-25)$$

where n_e, n_i, and n_g are the number densities of bulk plasma, ions at the plasma-sheath edge, and neutrals; $u_B = \sqrt{T_e/M_i}$ is the Bohm velocity; T_e is the electron temperature; M_i

is an ion mass; and $k_{ion}(T_e)$ is the rate coefficient of ionization by direct electron impact, which strongly depends on electron temperature. The relation between the ion density n_i near the surface at the plasma-sheath edge and the bulk plasma density n_e can be expressed in the low-pressure case as (Lieberman & Lichtenberg, 1994)

$$n_i = h_l \cdot n_e, \quad h_l = 0.86 \cdot \left(3 + \frac{l}{2\lambda_i}\right)^{-1/2}, \tag{8-26}$$

where $\lambda_i = 1/n_0\sigma_{i0}$ is the ion–neutral mean free path; $\sigma_{i0} \approx 10^{-14}\,\text{cm}^2$ is a typical value of the ion–neutral collision cross section in the system. Thus, based on (8–25), electron temperature in the discharge can be determined as a function of $n_0 l$:

$$k_{ion}(T_e) \approx \frac{2u_B h_l}{n_0 l} \approx 1.72 \cdot \frac{\sqrt{T_e/M_i}}{\sqrt{3 + (n_0 l)\sigma_{i0}/2}} \cdot \frac{1}{n_0 l}. \tag{8-27}$$

The total discharge power can be expressed through the generation rate of charged particles and total discharge energy cost corresponding to the creation of a single electron–ion pair, E_{ei} (which depends on electron temperature, and numerically is about 100–300 eV):

$$P \approx 2AE_{ei}n_i u_B. \tag{8-28}$$

Based on equation (8–28), we obtain a formula for calculating the near-surface ion density at the plasma-sheath edge, which, in surface kinetics, is given by (see Section 8.2.7)

$$n_i = \frac{P}{2AE_{ei}u_B} = \frac{P}{2AE_{ei}}\sqrt{M_i/T_e}. \tag{8-29}$$

Finally, the ion flux at the etching surface, which is also required for etching surface kinetics calculations, can also be expressed as a function of the total power of the non-equilibrium low-pressure discharge (see Section 8.2.7):

$$\Gamma_i = n_i u_B = \frac{P}{2AE_{ei}}. \tag{8-30}$$

As an example of applying expressions (8–29) and (8–30), we can derive a formula based on (8–23) for maximum anisotropy of the ion energy-driven etching as a function of the total discharge power:

$$(a_h)_{max} = \frac{w_v}{w_h} = P \cdot \frac{Y_i}{2AE_{ei}n_S K_d}. \tag{8-31}$$

8.2.11. Discharge Kinetics in Etching Processes: Density and Flux of Neutral Etchants

To estimate the concentration and flux of neutral etchant to the surface in the discharge described in the previous section, consider as an example the dissociation of oxygen: $e + O_2 \rightarrow O + O + e$. It corresponds, in particular, to the ion energy-driven carbon etching with oxygen considered in Section 8.2.7. The rate coefficient of the dissociation stimulated by direct electron impact, $k_{diss}(T_e)$, depends on electron temperature and can be expressed in Arrhenius form as

$$k_{diss}(T_e) = k_{diss}^0 \exp\left(-\frac{E_{diss}}{T_e}\right), \tag{8-32}$$

where k_{diss}^0 is the pre-exponential Arrhenius factor and E_{diss} is the dissociation energy. Assuming that the ionization rate coefficient (8–27) also depends on electron temperature in Arrhenius form,

$$k_{ion}(T_e) = k_{ion}^0 \exp\left(-\frac{I}{T_e}\right), \tag{8-33}$$

we can establish a relation between rate coefficients of dissociation and ionization as

$$k_{\text{diss}} = C \cdot (k_{\text{ion}})^{E_{\text{diss}}/I}, \tag{8–34}$$

where I is the ionization potential and $C = (k_{\text{diss}}^0 / k_{\text{ion}}^0)^{E_{\text{diss}}/I}$ is a constant parameter. Taking into account relation (8–27) or the ionization rate coefficient, we can express the dissociation rate coefficient as

$$k_{\text{diss}} = C \left(\frac{2h_l u_B}{n_0 l} \right)^{E_{\text{diss}}/I}. \tag{8–35}$$

The ratio E_{diss}/I is about 0.3–0.5; therefore, the direct dependence of the dissociation rate coefficient on electron temperature is not significant. Assuming that atom formation is due to the dissociation of molecules and atom losses are due to vacuum pumping with speed S_p (the loading effect is discussed next), the kinetic balance equation for atomic oxygen density n_O can be presented as

$$Al \frac{dn_O}{dt} = 2Alk_{\text{diss}} n_e n_0 - S_p n_O. \tag{8–36}$$

In the steady-state conditions ($dn_O/dt = 0$), the O-atom density in the etching discharge is

$$n_O = \frac{2l\,An_0}{S_p} k_{\text{diss}} n_e. \tag{8–37}$$

Taking into account (8–26), (8–29), and (8–35), the atomic oxygen density can be expressed as

$$n_O = \frac{2P}{E_{\text{ei}} S_p} C \left(\frac{n_0 l}{2h_l u_B} \right)^{1-E_{\text{diss}}/I}. \tag{8–38}$$

The atomic oxygen flux incident on an electrode can then be presented based on (8–38) as

$$\Gamma_O = \frac{1}{4} n_O v_O = \frac{P}{2E_{\text{ei}} S_p} C \left(\frac{n_0 l}{2h_l u_B} \right)^{1-E_{\text{diss}}/I} \sqrt{\frac{8T_0}{\pi M_O}}, \tag{8–39}$$

where v_O is the thermal velocity of O atoms, M_O is their mass, and T_0 is the gas temperature. It is important in some regimes to take into account atomic oxygen losses at the electrodes during the etching of N_w wafers each having area A_w, which can be reflected by a loss term in the balance equation (8–36)

$$Al \frac{dn_O}{dt} = 2Alk_{\text{diss}} n_e n_0 - S_p n_O - N_w A_w \cdot \frac{1}{4} v_O n_O. \tag{8–40}$$

which leads to a reduction of atomic oxygen density and to a decrease of etching rate, which is called the **loading effect**. Modification of (8–37), taking into account the loading effect, can be presented as

$$n_O = \frac{2l\,An_0}{S_p + N_w A_w v_O/4} \cdot k_{\text{diss}} n_e. \tag{8–41}$$

This equation can also be rewritten in a form that relates atomic oxygen density n_O, taking into account the loading effect, and atomic oxygen density in the absence of wafers, n_O^0:

$$\frac{1}{n_O} = \frac{1}{n_O^0} + \frac{N_w A_w v_O}{8l\,Ak_{\text{diss}} n_e n_0}. \tag{8–42}$$

Relations (8–38), (8–39), (8–41), and (8–42) for the neutral etchant density and flux can be effectively used in surface etching kinetics, particularly, in equations (8–20)–(8–22).

8.3. SPECIFIC PLASMA-CHEMICAL ETCHING PROCESSES

8.3.1. Gas Composition in Plasma Etching Processes: Etchants-to-Unsaturates Flux Ratio

Consider the role of different chemical compounds applied in etching. Gases used for plasma etching can have, in general, quite complex compositions, because they essentially affect the balance between etch rate, etching selectivity, and anisotropy (see Section 8.2.2). The gas-phase compounds participating in the etching can be classified into the following six groups (Flamm, 1989):

1. **Saturates** like CF_4, CCl_4, CF_3Cl, COF_4, SF_6, and NF_3. The saturates are chemically not very active and unable themselves either to etch or to form a surface film. They dissociate in plasma through collisions with electrons, producing chemically aggressive etchants and unsaturates.
2. **Unsaturates** like radicals CF_3, CF_2, and CCl_3, and molecules such as C_2F_4, C_2F_6, and C_3F_8. The unsaturates can react with substrates, forming surface films. Sometimes, however, the unsaturates are able to etch a substrate (for example, SiO_2), producing volatile products.
3. **Etchants** like atoms F, Cl, Br, and O (applied for resist) and molecules such as F_2, Cl_2, and Br_2. Etchants are obviously the major players in etching processes; they are very aggressive, reacting with substrates and producing volatile products.
4. **Oxidants** like O, O_2, and so forth. The oxidants added to feed gas are able to react with unsaturates, converting them into etchants and volatile products.
5. **Reductants** like H, H_2, and others. The reductants added to the feed gas are able to react with etchants and neutralize those producing passive volatile products.
6. **Non-reactive gases** like N_2, Ar, He, and so forth. The non-reactive additives are sometimes used to control electrical properties of etching discharges and thermal properties of a substrate.

An important parameter of etching is the etchants-to-unsaturates flux ratio at the substrate. When the ratio is high, pure chemical etching can dominate, resulting in mostly isotropic etching. On the other hand, a low etchants-to-unsaturates ratio leads to the formation of surface films, which increases the contribution of the ion energy-driven etching and makes the etching anisotropic. There can also be an intermediate etchants-to-unsaturates flux ratio, when the inhibitor film can be deposited on sidewalls but is cleared from the bottom of trenches by ion bombardment. This regime with the intermediate etchants-to-unsaturates flux ratio corresponds to the mechanism of anisotropic ion-enhanced inhibitor etching (Section 8.2.6). The etchants-to-unsaturates flux ratios can be increased (for more isotropic etching) by adding to the feed gas such etchants as Cl_2 or Br_2, which do not produce unsaturates. Another way to increase the ratio is by adding oxidants, which are able to convert unsaturates into etchants ($CF_3 + O \rightarrow COF_2 + F$). To decrease the etchants-to-unsaturates flux ratio and increase the sidewalls' protection (for more anisotropic etching), unsaturated gases like C_3F_8 or C_2F_4 can be added to the feed gas. The effect can also be achieved by adding H_2 or other reductants to the feed gas, which react with etchants and neutralize them.

8.3.2. Pure Chemical F-Atom Etching of Silicon: Flamm Formulas and Doping Effect

Etching of silicon by fluorine atoms is the most practically important and the best characterized surface etch process (Flamm, 1989, 1990; Winters & Coburn, 1992; Lieberman & Lichtenberg, 1994). We consider first the pure chemical mechanism of the process, leading to isotropic silicon etching; then we discuss ion energy-enhanced anisotropic etching.

F atoms intensively react with silicon, producing volatile silicon fluorides and, hence, providing effective surface etching. The reaction is of first kinetic order with respect to gaseous atomic fluorine (up to the near-surface density of fluorine atoms, $n_F = 5 \cdot 10^{15}$ cm^{-3}). The etching reaction rate follows the Arrhenius exponential temperature dependence over a wide range of substrate temperatures, T. For undoped silicon and for thermally grown silicon dioxide, the F-atom etching rates can be expressed by the **Flamm formulas**:

$$w_{Si}(\text{nm/min}) = 2.86 \cdot 10^{-13} \cdot n_F\sqrt{T(\text{K})}\exp(-1248\,\text{K}/T), \tag{8–43}$$

$$w_{SiO_2}(\text{nm/min}) = 0.61 \cdot 10^{-13} \cdot n_F\sqrt{T(\text{K})}\exp(-1892\,\text{K}/T). \tag{8–44}$$

The Flamm formulas lead to the expression for the silicon-to-silicon dioxide etching selectivity:

$$S = w_{Si}/w_{SiO_2} = 4.66 \cdot \exp(644\,\text{K}/T). \tag{8–45}$$

In accordance with (8–45), the silicon-to-silicon dioxide etching selectivity at room temperature is quite good ($S \approx 40$). The F-atom silicon etching also has a good selectivity over Si_3N_4 etching and reasonable selectivity over resist etching. The mechanism of pure chemical F-atom etching of silicon includes the formation of two to five intermediate silicon fluoride monolayers with a composition corresponding to SiF_3 on the surface. The thickness of the intermediate silicon fluoride layer becomes smaller when the etch rate is higher, and vice versa. The main silicon etching product at room temperature is gaseous SiF_4 (about 65%); the rest is composed of other volatile silicon fluorides, Si_2F_6 and Si_3F_8. The etching process can be described by a simple model assuming that diffusion of F atoms (or negative ions) into the surface is balanced by surface erosion due to the etching. It is convenient to consider the diffusion in the frame $x' = x - w_{Si}t$ moving with the etched front at $x' = 0$. The flux of fluorine atoms vanishes in this frame:

$$\Gamma'_F = -D_F\frac{dn_F}{dx'} - n_F w_{Si} = 0, \tag{8–46}$$

where D_F is the diffusion coefficient of fluorine atoms in the solid surface film and n_F is the density of F atoms diffusing into the surface. Equation (8–46) has an exponential solution,

$$n_F = n_F(x' = 0) \cdot \exp\left(-\frac{w_{Si}x'}{D}\right), \tag{8–47}$$

which gives an expression for the thickness of the layer of intermediate silicon fluorides on the Si surface:

$$L_{SiF_x} = D/w_{Si}. \tag{8–48}$$

This model explains why the thickness of the layer decreases at high rates of etching. The halogen-atom etch rate depends on silicon doping: n-type silicon etching is faster than p-type, which is known as the **doping effect**. The doping effect in the case of F-atom etching gives only a factor-of-2 difference in the etch rates, whereas in the case of Cl-atom etching, the difference in etch rates can reach many orders of magnitude. The doping effect is related to the formation of negative halogen ions on the surface and their contribution in the etching (Winters & Haarer, 1987, 1988; Winters & Coburn, 1992).

8.3.3. Ion Energy-Driven F-Atom Etching Process: Main Etching Mechanisms

High fluxes of energetic ions increase the etch rates of silicon five- to ten-fold with respect to the pure chemical etching described by Flamm formulas (8–43) and (8–44). A single 1 KeV Ar$^+$ ion is able to cause the removal of 25 Si atoms and 100 F atoms (to compare, such an ion removes only one atom in the case of sputtering; see Section 8.2.3). Although anisotropy of Si etching can be about five to ten, pure chemical etching on the trench sidewalls still remains a limiting aspect of fluorine-based anisotropic silicon etching. The following five main mechanisms contribute to the ion energy-driven F-atom etching of silicon:

1. Ion bombardment causes chemical reactions that produce etch products, which can be effectively desorbed from the surface. This mechanism is usually referred to as **chemical sputtering**. Energetic ions destroy a large number of Si–Si and Si–F bonds inside of the SiF_x surface layer and replace them with such molecular groups as SiF_4 and SiF_2. These groups are weakly bound to the surface and can be thermally desorbed during or after the collision cascade initiated by the energetic bombarding ions. The contribution of the chemical sputtering into ion energy-driven F-atom etching of silicon can be significant, which was especially demonstrated in molecular dynamics simulations of Barone and Graves (1995).

2. Ion bombardment reduces the thickness of the layer of intermediate silicon fluorides, which accelerates etching in accordance with (8–46)–(8–48). This effect is an additional one with respect to chemical sputtering and is called **enhanced chemical sputtering** (or enhanced chemical etching). Acceleration of the ion energy-driven F-atom etching of silicon due to the enhanced chemical sputtering effect, however, does not exceed a factor of 2 for typical conditions of etching discharges.

3. Ion bombardment leads to the formation of **damaged surface regions** which are especially reactive with fluorine. Although the surface damage effect, particularly lattice damage, makes a significant contribution toward other etching processes (for example, in ion energy-driven XeF_2 etching of tungsten), its role in ion energy-driven F-atom etching of silicon is limited.

4. Ion bombardment and accompanying reactions result in an **increase of surface temperature**, which leads to some acceleration of etching. The temperature rise, however, is not significant to make an essential contribution to the ion energy-driven F-atom etching of silicon.

5. Binding energies in the SiF_x surface layer can be lower (for example, Si–SiF_4) than those in pure silicon, which results in higher values of the sputtering yield γ_{sput} (see Section 8.2.3). This effect is called **chemically enhanced physical sputtering**. In contrast to the aforementioned chemical sputtering effect, removal of the weakly bound molecular groups in this case is not due to thermal desorption but to physical sputtering. At the neutral–ion flux ratios conventional for etching discharges, however, the contribution of chemically enhanced physical sputtering into ion energy-driven F-atom etching of silicon is not a major one, which was also demonstrated in molecular dynamics simulations of Barone and Graves (1995).

8.3.4. Plasma Etching of Silicon in CF_4 Discharges: Kinetics of Fluorine Atoms

Plasma-generated F atoms are widely used for etching processes. Molecular fluorine (F_2) is not applied, however, as a feed gas or source of F atoms, because F_2 itself etches silicon, leaving a rough and pitted surface as a result. Major feed gases for plasma generation of F atoms are CF_4, SF_6, and NF_3 (obviously, unsaturates and other special admixtures can be added to the feed gases; see Section 8.3.1). The best studied plasma etching feed gas is CF_4; therefore, we are going to discuss its plasma chemistry. Plasma etching of silicon in CF_4 discharges can be described as

$$Si_{solid} + 4 \cdot (CF_4)_{gas} \rightarrow 2 \cdot (C_2F_6)_{gas} + (SiF_4)_{gas}. \qquad (8\text{–}49)$$

Table 8–2. Surface Recombination Coefficients for F Atoms at Substrate Temperature 300 K

Surface material	Recombination probability	Surface material	Recombination probability
Al_2O_3	10^{-4}–10^{-3}	SiO_2	10^{-4}–10^{-3}
Pyrex	10^{-4}–10^{-3}	Teflon	10^{-4}–10^{-3}
Stainless Steel	10^{-4}–10^{-3}	Mo	10^{-4}–10^{-3}
Ni	10^{-4}–10^{-3}	Al + 0.1% Cu	10^{-4}–10^{-3}
Cu	$\geq 10^{-2}$	Brass	≥ 0.05
Zn	≥ 0.2		

The etching process is determined by F atoms created by dissociative attachment,

$$e + CF_4 \rightarrow CF_3 + F^-, CF_3^- + F, \quad k(cm^3/s) = 4.6 \cdot 10^{-9}[T_e(eV)]^{-3/2} \exp(-7\,eV/T_e), \tag{8–50}$$

and by CF_4 molecule dissociation (with formation of CF_3 or CF_2) through electronic excitation in direct electron impact collisions (Plumb & Ryan, 1986; Hayashi, 1987):

$$e + CF_4 \rightarrow CF_3 + F + e, \quad k(cm^3/s) = 2 \cdot 10^{-9} \exp(-13\,eV/T_e), \tag{8–51}$$

$$e + CF_4 \rightarrow CF_2 + F + F + e, \quad k(cm^3/s) = 5 \cdot 10^{-9} \exp(-13\,eV/T_e). \tag{8–52}$$

Formation of F atoms also takes place due to dissociative ionization of CF_4 molecules:

$$e + CF_4 \rightarrow CF_3^+ + F + e + e, \quad k(cm^3/s) = 1.5 \cdot 10^{-8} \exp(-16\,eV/T_e). \tag{8–53}$$

Losses of F atoms are due to volume and surface recombination. Volume recombination processes depend on pressure and proceed following either second kinetic order (at lower pressures, rate coefficients k_2, cm^3/s) or third kinetic order (at higher pressures, rate coefficients k_3, cm^6/s). The most important volume recombination reactions for fluorine atoms in the CF_4 discharges are

$$F + CF_3(+M) \rightarrow CF_4(+M), \quad k_2 = 2 \cdot 10^{-11}\,cm^3/s, \quad k_3 = 7.7 \cdot 10^{-27}\,cm^6/s, \tag{8–54}$$

$$F + CF_2(+M) \rightarrow CF_3(+M), \quad k_2 = 1.3 \cdot 10^{-11}\,cm^3/s, \quad k_3 = 3 \cdot 10^{-29}\,cm^6/s, \tag{8–55}$$

$$F + CF(+M) \rightarrow CF_2(+M), \quad k_2 = 10^{-11}\,cm^3/s, \quad k_3 = 3.2 \cdot 10^{-31}\,cm^6/s. \tag{8–56}$$

The kinetics of surface recombination of F atoms can be characterized by the recombination probability per incident atom, P_{rec}, and depends on the surface material. These probabilities, with some exceptions, are small at substrate temperatures and atomic fluxes typical for the etching discharges. Numerical values of the F-atom surface recombination coefficients are presented in Table 8–2. Surface recombination coefficients for Cl and O atoms are larger and reach values of about 0.1 for many surfaces. It is also important to notice that the probability of an etching reaction of one F atom per one silicon surface collision is $P_r \approx 1.7 \cdot 10^{-3}$ and is negligible for most of other surfaces. The reaction probability, however, can be very high and even reach $P_r \approx 1$ in some specific cases, for example, in F etching of a BN surface.

8.3.5. Plasma Etching of Silicon in CF_4 Discharges: Kinetics of CF_x Radicals and Competition Between Etching and Carbon Deposition

The CF_x radicals play an important role in anisotropic etching, forming protective films on the treated surfaces. Formation of CF_x radicals in CF_4 discharges, similar to the formation of

F atoms, is due to direct electron impact processes (8–50)–(8–52). The volume losses of the radicals are due to recombination with atomic fluorine (8–54)–(8–56), and recombination in collisions between themselves, in particular,

$$CF_3 + CF_3(+M) \rightarrow C_2F_6(+M), \quad k_2 = 8.3 \cdot 10^{-12} \text{ cm}^3/\text{s}, \quad k_3 = 2.8 \cdot 10^{-23} \text{ cm}^6/\text{s},$$
$$(8–57)$$

$$CF_3 + CF_2(+M) \rightarrow C_2F_5(+M), \quad k_2 = 10^{-12} \text{ cm}^3/\text{s}, \quad k_3 = 2.3 \cdot 10^{-26} \text{ cm}^6/\text{s}.$$
$$(8–58)$$

The volume recombination processes, similarly to (8–54)–(8–56), can proceed following either second kinetic order (at lower pressures, rate coefficients k_2, cm^3/s) or third kinetic order (at higher pressures, rate coefficients k_3, cm^6/s). An important surface kinetic feature of the CF$_x$ radicals is their ability to dissociate on active surfaces, including silicon and the layer of intermediate silicon fluorides, SiF$_x$. The surface dissociation with formation of chemisorbed carbon atoms leads to a buildup of carbon or polymer film on the surface, which prevents etching and is unlikely to be removed without ion bombardment. To characterize the reduction of the silicon etch rate due to the flux of CF$_x$ radicals ($x < 4$), the balance of fluorine on the surface can be presented as a balance of adsorbed and desorbed fluxes. The balance of the fluxes of CF$_x$ radicals (Γ_{CF_x}) and F atoms (Γ_F) absorbed on the surface and the fluxes of volatile molecular gases CF$_4$ (Γ_{CF_4}) and SiF$_4$ (Γ_{SiF_4}) desorbed from the surface can be written as

$$x \cdot \Gamma_{CF_x} + \Gamma_F = 4 \cdot \Gamma_{CF_4} + 4 \cdot \Gamma_{SiF_4}. \qquad (8–59)$$

Steady-state conditions for atomic carbon on the treated surface also require a balance of the adsorption–desorption fluxes: $\Gamma_{CF_4} = \Gamma_{CF_x}$, which permits the rewriting of the fluorine balance equation as follows:

$$\Gamma_F = (4 - x) \cdot \Gamma_{CF_x} + 4 \cdot \Gamma_{SiF_4}. \qquad (8–60)$$

Taking into account that the rate of silicon removal from the surface is related to the desorption of volatile tetrafluoride, $w_{Si} = \Gamma_{SiF_4}/n_{Si}$ (where n_{Si} is silicon number density), the etch rate can be presented based on (8–60) and clearly showing competition between fluxes of F atoms and CF$_x$ radicals:

$$w_{Si} = \frac{\Gamma_{SiF_4}}{n_{Si}} = \frac{\Gamma_F - (4 - x) \cdot \Gamma_{CF_x}}{4n_{Si}}. \qquad (8–61)$$

Assuming $x = 3$, etching is possible only if $\Gamma_F > \Gamma_{CF_3}$. In the opposite case, carbon deposition dominates and the etch rate is zero. Ion bombardment can shift the balance in the direction of etching, which is surely used for anisotropic etching. The effect of ion bombardment is related to an increase of the ratio of the net fluxes absorbed, Γ_F/Γ_{CF_x}, and the effective physical sputtering of CF$_y$ ($y < 4$). When sputtering of the CF$_y$ groups dominates the desorption process, expression (8–61) for the etch rate can be modified for the case of ion energy-driven etching:

$$w_{Si} = \frac{\Gamma_{SiF_4}}{n_{Si}} = \frac{\Gamma_F - (y - x) \cdot \Gamma_{CF_x}}{4n_{Si}}. \qquad (8–62)$$

Si etching dominates over polymerization in this regime if $y \leq x$. Increase of bias applied to the surface leads to higher ion bombardment energies, lower values of "y," and further shifting of the treatment in the direction of etching. The **F/C ratio** of fluorine atoms to CF$_x$ radicals in the discharge strongly affects the etching deposition balance (see (8–61) and (8–62)). When the F/C ratio is high (F/C > 3), etching dominates at any applied bias voltages and both trench sidewalls and bottoms are etched. When the F/C ratio is low (F/C < 2), film deposition dominates, stopping etching of both trench sidewalls and bottoms. Highly

Table 8–3. Arrhenius Kinetic Parameters for Cl-Atom Etching of n-Type Silicon

Crystallographic orientation	$A, \dfrac{nm \cdot cm^{3+3\gamma}}{min \cdot K^{1/2}}$	B	γ
Polysilicon	$4 \cdot 10^{-18}$	2365 K	0.39
$\langle 100 \rangle$	$1.1 \cdot 10^{-17}$	2139 K	0.29
$\langle 111 \rangle$	$1.6 \cdot 10^{-31}$	2084 K	1.03

anisotropic etching takes place in the intermediate case ($2 < F/C < 3$), when the trench bottom is etched at sufficient bias voltages (about 200 V) but the sidewalls are protected by the deposited film. Correction of the F/C ratio can be achieved in the CF_4 discharges, in particular by additions of unsaturates, O_2 and H_2 (see Section 8.3.1). Addition of oxygen up to about 16% converts CF_3 and other unsaturates into F atoms, removes carbon from the surface (by O atoms), increases the F/C ratio, and promotes etching (Mogab, Adams, & Flamm, 1979). Addition of hydrogen and unsaturates decreases the F/C ratio and stimulates polymerization.

8.3.6. Plasma Etching of Silicon by Cl Atoms

Silicon etching by Cl atoms has many features in common with the aforementioned etching provided by F atoms. Between major differences in pure chemical etching, the most important are related to strong crystallographic and doping effects in the case of Cl-atom etching. The silicon etch rate by Cl atoms can be described in Arrhenius form (Flamm, 1990; Ogryzlo et al., 1990):

$$w_{Si} = A n_D^{\gamma} n_{Cl} \sqrt{T} \exp(-B/T), \qquad (8\text{--}63)$$

where n_{Cl} is the near-surface density of Cl atoms in the etching discharge; n_D is the number density of donor atoms in the treated material, which determines the contribution of the doping effect (etching of n-type silicon is faster than that of p-type; see Section 8.3.2); T is the substrate temperature; and A, B, and γ are kinetic parameters given in Table 8–3 for etching of n-type silicon with different crystallographic orientations. According to (8–63), pure chemical etch rates are very small for Cl atoms on undoped silicon but can be substantial for heavily n-doped silicon. Etching rates of Si by Cl_2 are not significant at room temperature. For this reason, molecular chlorine (in contrast to fluorine) can be used as a feed gas for Cl-atom plasma etching. Thus, plasma etching of silicon in Cl_2 discharges can be described as

$$Si_{solid} + 2 \cdot (Cl_2)_{gas} \rightarrow (SiCl_4)_{gas}. \qquad (8\text{--}64)$$

Cl_2 dissociates in plasma and forms F atoms, which then provide etching (8–64). Etching proceeds through formation of an $SiCl_x$ intermediate chloride layer, which is thinner than that formed by F atoms and is only a few monolayers thick. When Cl_2 is used as a feed gas, carbon atoms are not involved and there is no formation of an inhibitor film as occurs in the case of CF_4 feed gas (Sections 8.3.4 and 8.3.5).

Ion-assisted etching of silicon can be performed with Cl_2, even without plasma dissociation of Cl_2 and generation of Cl atoms in volume. For example, 1 keV Ar^+ ions with an adequate flux of Cl_2 provide the removal of three to five silicon atoms per incident ion (etching yield 3–5). Dissociation of Cl_2 does not result in a significant increase of the etching yield. To compare, dissociation of F_2 in a similar situation results in a five- to ten-fold increase in yield of ion energy-driven silicon etching. To conclude our considerations of halogen-atom etching of silicon (Sections 8.3.2–8.3.6), we can make some remarks regarding **bromine abilities in etching**. Br atoms are in general less reactive than chlorine atoms, and much less reactive than F atoms. Pure chemical etching of even heavily n-doped silicon is negligible at room temperature. Pure chemical etching by Br atoms is observed at higher temperatures with a very strong doping effect.

8.3.7. Plasma Etching of SiO_2 by F Atoms and CF_x Radicals

The pure chemical etching rate of SiO_2 by F atoms is determined by the Flamm formula (8–44). For a typical density of F atoms ($3 \cdot 10^{14}$ cm^{-3}), the etch rate is 0.59 nm/min at room temperature. The etch rate of silicon dioxide is low, about 40 times slower than that of silicon; CF_x radicals are even less effective in etching SiO_2. In contrast to the case of silicon, CF_x radicals are unable even to dissociate on the SiO_2 surface. Thus, effective SiO_2 etching in fluorocarbon plasmas is usually ion energy driven, which will be discussed in the next section. When ion energies are 500 eV and higher, F and CF_x ion energy-driven etching of SiO_2 is characterized by high rates exceeding 200 nm/min. The etching is obviously anisotropic: the etch rates are correlated with ion bombarding energy and are independent of substrate temperature. The loading effect is much smaller than in the case of silicon etching. F atoms have no etching selectivity for SiO_2 over silicon, whereas CF_x radicals do; therefore, unsaturated fluorocarbon feedstocks, or CH_4–H_2 mixtures, are used to achieve selective SiO_2/Si etching. In these systems, the SiO_2 etch rate exceeds that of Si about 15-fold. The combined effect of ion bombardment and flux of CF_x radicals leads to the formation of an intermediate $SiC_xF_yO_z$ layer on the SiO_2 surface. The thickness of the layer at high CF_x radical fluxes reaches 2 nm and becomes thinner at lower radical fluxes. The collision cascade in the layer accompanying ion bombardment stimulates production of such molecules as SiF_4, SiF_2, CO, CO_2, and COF_2, which can be easily desorbed from the surface. While F atoms from CF_x remove silicon, the carbon atoms of CF_x provide an effective removal of O atoms from the intermediate $SiC_xF_yO_z$ layer and, finally, from SiO_2.

8.3.8. Plasma Etching of Silicon Nitride (Si_3N_4)

The importance of silicon nitride etching is due to application of the material as a mask for patterned oxidation of silicon and as a dielectric. Pure chemical etching of Si_3N_4 by F atoms is quite effective, is isotropic, and is characterized by Arrhenius kinetics with an activation energy of about 0.17 eV. Silicon nitride etching has a selectivity value of 5–10 over that of etching of SiO_2. Pure chemical etching of Si_3N_4 by F atoms is not selective, however, over the etching of silicon. There are two kinds of Si_3N_4 materials, which differ with respect to etching. One type is produced by CVD at high temperatures, and another type produced by plasma-enhanced chemical vapor deposition (PECVD) at low temperatures ($\leq 400°C$). Plasma etch rates of PECVD-produced silicon nitride are usually higher compared to those for Si_3N_4 produced by high-temperature CVD. Anisotropic etching of silicon nitride can be performed, similar to the aforementioned plasma systems, by ion enhancement of the etching process. Ion energy-driven anisotropic Si_3N_4 etching is usually carried out using unsaturated fluorocarbon feed gases. The ion energy-driven Si_3N_4 etching process is characterized by a low selectivity over silicon dioxide (SiO_2), but high selectivity over silicon and resist.

8.3.9. Plasma Etching of Aluminum

The importance of aluminum etching is mostly due to its application as an interconnect material in ICs. Fluorine atoms cannot be applied for aluminum etching, because the process product AlF_3 is not volatile. For this reason, other halogens (Cl_2 and Br_2) are used as feed gases for plasma etching of aluminum. In the absence of ion bombardment, chlorine and bromine provide isotropic etching. Molecular chlorine is able to etch clean aluminum even without plasma. Etching of aluminum in Cl_2 discharges at low temperatures ($T \leq 200°C$) can be described as

$$Al_{solid} + 3/2 \cdot (Cl_2)_{gas} \rightarrow (AlCl_3)_{gas}. \qquad (8\text{–}65)$$

At higher substrate temperatures, the main product of pure chemical etching is Al_2Cl_6, which is also volatile. To provide anisotropic etching of aluminum, ion energy-driven performance of the processes with an inhibitor is required. In this case, special additives such as CCl_4, $CHCl_3$, $SiCl_4$, and BCl_3 are added to the main feed gas, which is molecular chlorine (Cl_2). Water vapor negatively interferes with aluminum etching and should be excluded from the system. Additives such as BCl_3 and $SiCl_4$ are effective in eliminating water. An important issue of aluminum etching is related to breaking through the native oxide film (about 3 nm thick) to access the pure metal. The problem is related to the fact that neither molecular (Cl_2) nor atomic chlorine (Cl) are able to etch the alumina (Al_2O_3) film, even in the presence of ion bombardment. Film breakthrough can be provided by ion bombardment physical sputtering and additives such as CCl_4, $SiCl_4$, and BCl_3. In particular, carbon atoms of the CCl_4 admixture promote oxygen removal from the film by formation of stable and volatile carbon oxides.

8.3.10. Plasma Etching of Photoresist

Photoresist masks play an important role in manufacturing ICs for electronics. Photoresist plasma etching is key for two processes: the first is isotropic etching of the resist mask materials from wafers (called stripping), and the second is anisotropic pattern transfer into the resist (in the surface-imaged dry development schemes). The photoresist mask materials are usually long-chain organic hydrocarbon polymers, so oxygen atoms can be effectively used as etchants for both of these applications. Thus, O_2 discharges are usually used for plasma etching of the photoresist mask materials. Pure chemical etching of photoresists by O atoms is isotropic and highly selective over silicon and SiO_2. When the kinetics of the process is presented in Arrhenius form, the activation energy is 0.2–0.6 eV. In some cases, pure chemical etching of the resists is characterized by two activation energies, depending if the resist temperature is above or below the glass transition temperature of the polymer material. Pure chemical etching of photoresists by O atoms can be significantly enhanced by adding a few percent of C_2F_6 or CF_4 to the main feed gas, which is molecular oxygen (O_2) in this case. The contribution of fluorocarbon additives is by their reactions with hydrogen atoms from the resists to form HF and to open active sites on the polymer surface for subsequent reactions of oxygen atoms. Also, intensive chemisorption of F atoms on the reactor walls reduces surface recombination of oxygen atoms, which also results in acceleration of pure chemical etching of the photoresist materials in O_2 plasma. For dry development of surface-imaged photoresists, ion energy-driven anisotropic etching in O_2 plasma is used. An accurate balance between ion bombardment and flux of O atoms is required in this application. The O-atom flux should be high enough to oxidize the silylated layer to form the SiO_x mask, but the ion bombardment must be weak enough at the same time to avoid physical sputtering of the mask during etching of the unsilylated areas.

8.3.11. Plasma Etching of Refractory Metals and Semiconductors

Plasma can also be effectively applied for etching layers of silicon doped with refractory or rare metals, semiconductors, or via interconnects made with rare or refractory metals. Some of those etching reactions are illustrated in Table 8–4. These processes normally use fluorine-containing gases as a feedstock and have hexafluorides as volatile etching products. However, chlorine is preferred in some specific cases. Only the most common plasma etching systems are discussed here; much more detail on the subject can be found, particularly in the reviews of Flamm (1989) and Orlikovsky (2000).

Table 8–4. General Chemical Characteristics of Plasma Etching of Refractory Metals and Some Other Compounds

Material to be etched	Feedstock gas	Major etching atom	Volatile etching product
W	CF_4, SF_6	F	WF_6
Ta	NF_3, F_2	F	TaF_6
Nb	NF_3, F_2	F	NbF_6
Mo	NF_3, F_2	F	MoF_6
GaAs	BCl_3, Cl_2	Cl	$GaCl_3$, $AsCl_5$
$MoSi_2$	$Cl_2 + O_2$	Cl	$SiCl_4$

8.4. PLASMA CLEANING OF CVD AND ETCHING REACTORS IN MICRO-ELECTRONICS AND OTHER PLASMA CLEANING PROCESSES

8.4.1. In Situ Plasma Cleaning in Micro-Electronics and Related Environmental Issues

Plasma etching considered in Sections 8.2 and 8.3 was directly focused on IC fabrication. Besides direct treatment of wafers, this technology also requires chamber cleaning, where plasma is effectively applied as well. CVD or etching chamber cleaning from deposited silicon materials is actually an isotropic F-atom etching process, which has, however, several individual special features (to be discussed later). In the production of ICs, the cleaning of treatment chambers is a very time-consuming operation because deposits of silicon oxides are difficult to remove from surfaces of the treatment chamber. Cleaning is usually achieved by etching chamber surfaces with active particles, among which atomic fluorine is the most effective. Atomic fluorine can be conveniently produced from stock gases such as NF_3, CF_4, C_2F_6, and SF_6 in low-temperature discharge plasmas. Plasma cleaning does not require ion bombardment and has fewer limitations related to process anisotropy and selectivity (see Section 8.2.2). As a result, operational discharge pressures can be essentially increased with respect to etching discharges to the level of several Torr (compare with etching discharges described in Sections 8.2.8 and 8.2.9). Higher pressures applied in this pure chemical etching permit significant acceleration of the cleaning process (see Flamm formulas (8–43) and (8–44). To avoid ion-related chamber damages during elevated pressure plasma chamber cleaning, **remote plasma sources (RPSs)** are applied, where F atoms are generated in a special discharge and then delivered to the chamber through a special transport tube. The high cleaning rates can lead to significant emissions of **perfluoro compounds (PFCs),** which are products of conventional cleaning and create environmental control problems in semiconductor fabrication plants. PFC emission restrictions stimulated changes in feed gases applied as a source of F atoms for plasma chamber cleaning, leading to a wider application of NF_3 as a feed gas. Plasma dissociation of NF_3 to N_2 and F is effective and not accompanied by PFC emission. However, high NF_3 prices dictate strong requirements for utilization efficiency of F atoms, especially the efficiency of F-atom transportation from the RPS to the chamber. All those factors determine features of in situ plasma cleaning, distinguishing this process from conventional etching.

The in situ plasma cleaning procedure performed after CVD of dielectric thin films is one of the major emitters of PFCs in semiconductor manufacturing; it can represent 50–70% of the total PFC emission in a semiconductor fabrication plant. To clean the process chambers of deposited by-products, conventional cleaning methods use CF_4 or C_2F_6 gases activated by RF CCP (usually 13.56 MHz) inside the process chamber. These PFC gases achieve a relatively low degree of dissociation and unreacted molecules are emitted in the process exhaust. Significant PFC emission reductions have been achieved through optimization of CVD chamber cleans. Over the past years, the semiconductor industry has continually reduced the million metric ton carbon equivalent of its in situ CVD chamber cleaning (Raoux et al., 1999). It was demonstrated that the overall gas flow can be reduced, while the chamber cleaning time is shortened. These improvements resulted from gains in source gas utilization obtained by adjustment of the operating power, pressure, and the number of steps required to achieve complete residue removal. Optimization of the CVD chamber design is also of critical importance in improving cleaning efficiency. The chamber can be designed with reduced chamber volume and surface area to limit the quantity of deposition residues to be cleaned. The use of ceramic materials (Al_2O_3, AlN) for the chamber components (liners, heater, electrostatic chuck, dome) is also preferable because the recombination rate of the reactive species (F radicals) injected in the chamber is much lower for ceramics than metals. Moreover, these ceramic components present better resistance to fluorine corrosion compared to conventional materials. Although these advances have been considerable, they have not achieved the goal of near-complete destruction of PFC gases. For example, efforts to continue increasing RF power with fluorocarbon chemistries have resulted in the generation of other PFCs (for example, CF_4 from C_2F_6 decomposition). Furthermore, increased in situ plasma power density can lead to severe corrosion of the chamber components and can induce process drifts and particulate contamination. To overcome these limitations, a new plasma cleaning technology is applied, which uses a remote RF-ICP source to completely break down NF_3 gas into an effective cleaning chemical. This remote cleaning plasma technology is discussed in the next section.

8.4.2. Remote Plasma Cleaning Technology in Microelectronics: Choice of Cleaning Feedstock Gas

Remote plasma cleaning technology can be simply explained as follows. A fluorine-containing gas (in particular, NF_3) is introduced in a remote discharge chamber, where plasma is sustained by application of microwave or RF energy. In the plasma, the cleaning feed gas is dissociated into charged and neutral species (F, F_2, N, N_2, NF_x, electron, ions, and excited species). Because the plasma is confined inside the applicator, and since ions have a very short lifetime, mainly neutral species are injected through a transport tube and a special shower head into the main deposition chamber for cleaning purposes. The fluorine radicals react in the main chamber to be cleaned with the deposition residues (SiO_2, Si_3N_4, etc.) to form non-global-warming volatile by-products (SiF_4, HF, F_2, N_2, O_2). The volatile by-products are pumped through the exhaust and can be removed from the stream using conventional scrubbing technologies. Due to the high efficiency of the microwave/RF excitation, NF_3 gas utilization removal efficiency can be as high as 99% in standard operating conditions (Raoux et al., 1999). This ensures an efficient source of fluorine while eliminating global-warming emissions. With this "remote" technique, no plasma is sustained in the main deposition chamber and the cleaning is much "softer" on the chamber components, compared to a traditional in situ plasma cleaning technology.

Consider the choice of cleaning feed gas. PFC molecules have very long atmospheric lifetimes (see Table 8–5); this is a direct measure of their chemical stability. It is not surprising then that very high plasma power levels are required to achieve near-complete destruction efficiencies of these gases. Among the commonly used fluorinated gas sources, nitrogen trifluoride (NF_3) has been determined to be the best source gas. NF_3 has similar global warming potential (GWP) compared to most other PFCs (with a 100-year integrated time horizon). However, when considering the GWP of NF_3 over the life of the molecule, it is much lower than that of CF_4, C_2F_6, and C_3F_8 due to a shorter lifetime (740 years vs. 50,000 for CF_4). This should be taken into consideration for estimating the long-term impacts of PFC gas usage. One other reason why NF_3 is well suited to this application

Table 8–5. Lifetime and Global Warming Potential of Different Gases

Gas	Lifetime, years	GWP, 100 years ITH	GWP, ∞ ITH
CO_2	100	1	1
CF_4	50,000	6,500	850,000
C_2F_6	10,000	9,200	230,000
SF_6	3,200	23,900	230,000
C_3F_8	7,000	7,000	130,000
CHF_3	250	11,700	11,000
NF_3	740	8,000	18,000

is the weaker nitrogen–fluorine bond as compared to the carbon–fluorine bonds in CF_4 or C_2F_6 (see Table 8–6). Actually the relative ease of destruction of NF_3 results in a high usage efficiency of the source gas for plasma cleaning. Another advantage of NF_3 plasma chemistry is that it is a non-corrosive carbon-free source of fluorine. The use of fluorocarbon molecules such as CF_4, C_2F_6, and so forth requires dilution with an oxidizer (O_2, N_2O, etc.) to prevent formation of polymeric residues during the cleaning process (Raoux et al., 1997). Dilution of NF_3 by oxygen enhances the etch rate, but this solution was not chosen because of the formation of NO_x by-products (another global warming molecule and a hazardous air pollutant). All those factors show the advantages of the application of remote NF_3 discharges to achieve high etch rates at distances farther downstream from the source, allowing for faster and more complete chamber cleaning.

8.4.3. Kinetics of F-Atom Generation from NF_3, CF_4, and C_2F_6 in Remote Plasma Sources

The RPS kinetics of F-atom generation from NF_3, CF_4, and C_2F_6 can be analyzed for typical conditions of RF-ICP (transformer) discharges conventionally applied for remote plasma cleaning technology (Conti et al., 1999; Iskenderova, 2003). The typical RPS parameters are gas pressure of several Torr, plasma density up to 10^{12} cm^{-3}, electron temperature about 3 eV, and power 1–3 kW. The NF_3 feedstock gas flow rate is usually chosen to provide a high level of dissociation, which corresponds to a specific energy input about 30 kW per standard cubic meter. The plasma dissociation kinetics of NF_3 is illustrated in Fig. 8–15. Dissociation of NF_3 in the discharge approaches 100%, resulting in higher F-atom concentrations and higher etch rates in comparison with fluorocarbon gases. Reaction rate data regarding NF_3 plasma chemistry can be found in Meeks et al. (1997) and detailed kinetics of NF_3 dissociation is described by Iskenderova (2003). Addition of O_2 in the discharge slightly changes the dissociation kinetics (Fig. 8–16), leading in particular to NO production (Kastenmeier et al., 1998). In general, oxygen admixtures accelerate NF_3 decomposition but reduce SiO_2 etching rates because of dilution and shifting etching/oxidation balance in the direction of oxidation:

$$SiO_2 + 4F \Leftrightarrow SiF_4 + O_2. \tag{8–66}$$

Table 8–6. Comparison of Nitrogen Fluoride and Carbon Fluoride Bond Energies

Gas	Specific bond	Bond strength, kcal/mol
NF_3	NF_2–F	59
CF_4	CF_3–F	130
C_2F_6	C_2F_5–F	127

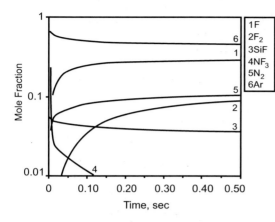

Figure 8–15. Numerical simulation of NF_3 dissociation in a non-thermal discharge. Time dependence of the main species concentration. Initial mixture composition: Ar = 64%, $N_2 = 0.4\%$, $NF_3 = 35\%$.

The fluorocarbon feed gases (in particular, CF_4 and C_2F_6) are usually mixed with oxygen in RPSs to avoid formation of polymeric residues and to stimulate generation of F atoms. The plasma kinetics of dissociation of CF_4 and C_2F_6 in the presence of oxygen is illustrated in Figs. 8–17 and 8–18. The fluorocarbon discharges produce a significant amount of CF_x species, which do not contribute to etching and deposition (see Sections 8.3.1 and 8.3.5). In this case, addition of O_2 increases the atomic fluorine concentration due to the oxidation of CF_x molecules. Dissociation of C_2F_6 is less dependent on the O_2 concentration and more on power consumption. Also, the CF_4 molecule dissociates faster than C_2F_6. Such a tendency is observed in systems with high specific energy inputs (about 200 J/cm^3); Kastenmeier et al., 2000. The dissociation degree of CF_4 molecules reaches 97%, and the dissociation degree of C_2F_6 molecules reaches 80%, which is lower than in the case of NF_3 feedstock. In general, only the NF_3-based RPS chamber cleaning provides near-complete destruction (> 99%) of

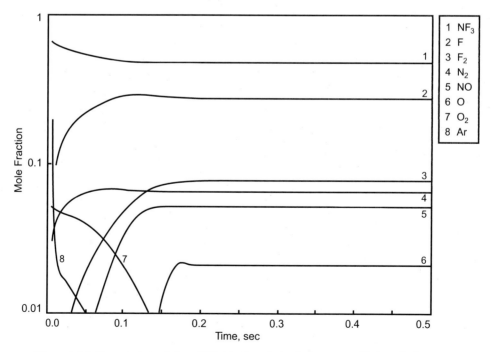

Figure 8–16. Numerical simulation of NF_3/O_2 dissociation in the non-equilibrium discharge. Time dependence of the main species concentration. Initial mixture composition: Ar = 67%, $N_2 = 3\%$, $NF_3 = 20\%$, $O_2 = 5.3\%$.

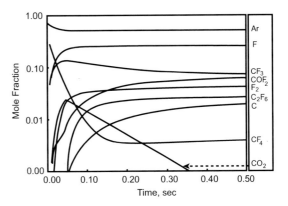

Figure 8–17. Kinetic curves for major species during CF_4/O_2 dissociation in non-equilibrium plasma. Initial mixture composition: Ar = 67%, N_2 = 0.4%, O_2 = 4.2%, CF_4 = 28.3%.

the fluorine-containing source gas. Thus, use of NF_3 as a feedstock for plasma cleaning results in low emission of global warming compounds, higher F-atom concentrations, and higher etch yields as compared to fluorocarbon gases.

8.4.4. Surface and Volume Recombination of F Atoms in Transport Tube

Generation of F atoms in RPSs is effective, especially if NF_3 is used as a feed gas. A more challenging problem is related to transportation of the F atoms from the RPS into the chamber to be cleaned, which we discuss next. Two mechanisms are responsible for the recombination of F atoms in the transport tube: surface recombination, which dominates at low pressures (usually below 2–5 Torr), and volume recombination, which dominates at higher pressures. The **surface recombination of F atoms** can be described in the framework of the **Langmuir-Rideal mechanism** (Gelb & Kim, 1971), which includes two steps that are illustrated in Fig. 8–19. A gas-phase F atom must first collide with the surface and stick at an empty surface site(s) (Fig. 8–19a). Another gas-phase F atom may strike the adsorbed F atom and recombine (Fig. 8–19b), forming F_2 and leaving the site empty again (Fig. 8–19c). The two-step mechanism may be written as follows.

First reaction (adsorption):

$$F + (s) \rightarrow F(s), \qquad (8\text{--}67)$$

Second reaction (recombination):

$$F + F(s) \rightarrow F_2 + (s). \qquad (8\text{--}68)$$

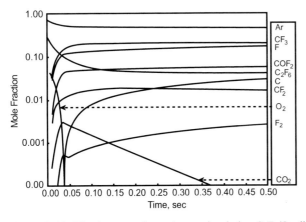

Figure 8–18. Kinetic curves for major species during C_2F_6/O_2 dissociation in non-equilibrium plasma. Initial mixture composition: Ar = 67%, N_2 = 0.4%, O_2 = 4.2%, C_2F_6 = 28.3%.

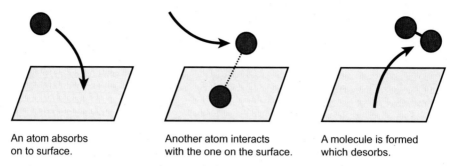

An atom absorbs on to surface.

Another atom interacts with the one on the surface.

A molecule is formed which desorbs.

Figure 8–19. Major kinetic stages of the Langmuir-Rideal heterogeneous recombination mechanism.

The F-atom thermal desorption reverse to (8–67) cannot be neglected. The adsorption rate (8–67), that is, the number of F atoms adsorbed per unit time and area, is (Christman et al., 1974)

$$w_{\mathrm{ad}} = S_0(1 - \theta_{\mathrm{T}}) \cdot \frac{1}{4} n_{\mathrm{F}} \sqrt{\frac{3T}{M_{\mathrm{F}}}}, \qquad (8\text{–}69)$$

where S_0 is a sticking coefficient, which characterizes the probability that any given collision of an F atom with the clean surface at temperature T would cause the particle to be trapped by the surface; θ_{T} is total fraction of the surface sites(s) already occupied; n_{F} and M_{F} are number density and mass of F atoms; and T is surface temperature. The thermal desorption rate for any adsorbed species can be written as

$$w_{\mathrm{des}} = \theta \cdot C_{\mathrm{A}} \frac{T}{2\pi\hbar} \exp\left(-\frac{E_{\mathrm{D}}}{T}\right), \qquad (8\text{–}70)$$

where θ is the fraction of the surface sites occupied by the species; C_{A} is the number of surface sites per unit area; $\hbar$ is the Planck constant; for calculations, $C_{\mathrm{A}}/2\pi\hbar \approx 3.4 \cdot 10^{25}\,\mathrm{K}^{-1}\,\mathrm{s}^{-1}\,\mathrm{cm}^{-2}$; and E_{D} is the desorption energy. The reaction rate of the recombination–desorption process (8–68) can be presented as

$$w_{\mathrm{rec/des}} = P_{\mathrm{S}} \cdot \theta_{\mathrm{F}} \cdot \frac{1}{4} n_{\mathrm{F}} \sqrt{\frac{3T}{M_{\mathrm{F}}}}, \qquad (8\text{–}71)$$

where P_{S} is a steric factor characterizing the probability of recombination–desorption (8–68) during the collision of an incident and an adsorbed fluorine atom (for a nickel surface, for example, $P_{\mathrm{S}} = 0.018$); θ_{F} is the fraction of the surface sites occupied by F atoms. Based on relations (8–69)–(8–71), the surface recombination kinetics for F atoms can be described if desorption energies E_{D} and sticking coefficients S_0 are known for all the species to be absorbed. It should be taken into account that physical adsorption and chemisorption must be considered separately. Characteristic values of the desorption energies are presented in Table 8–7. The sticking coefficients for F atoms, F_2 molecules, and Ar atoms (as an example of inactive species) are shown in Figs. 8–20–8–22 as a function of surface temperature. Cases of physical sorption and chemisorption are presented separately. Both desorption energies and sticking coefficients are given for recombination processes on a nickel surface. Experimental kinetic data regarding surface recombination of F atoms can be found in Butterbaugh, Gray, and Sawin (1991).

Volume recombination of F atoms in the transport tube is due to the three-body reaction

$$\mathrm{F} + \mathrm{F} + M \rightarrow \mathrm{F}_2 + M. \qquad (8\text{–}72)$$

8.4. Plasma Cleaning of CVD and Etching Reactors in Micro-Electronics

Table 8–7. Desorption Energies Related to Surface Recombination Kinetics

Adsorbed particle	Type of sorption	Desorption energy, eV
F	Physical adsorption	0.1
F	Chemisorption	0.8
F_2	Physical sorption	0.08
F_2	Chemisorption	0.76
Ar	Physical sorption	0.09

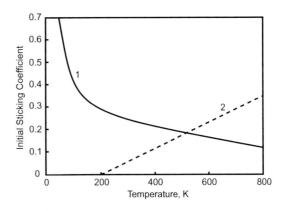

Figure 8–20. Initial sticking coefficient for F atoms on Ni surface as a function of temperature; simulated values of the coefficient corresponding to the cases of (1) physical and (2) chemical bonding.

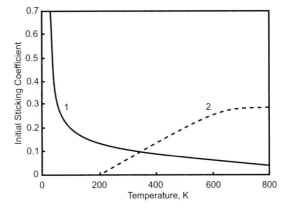

Figure 8–21. Initial sticking coefficient for F_2 molecules on Ni surface as a function of temperature; simulated values of the coefficient corresponding to the cases of (1) physical and (2) chemical bonding.

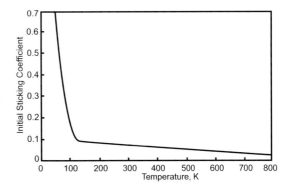

Figure 8–22. Initial sticking coefficient for Ar on Ni as a function of temperature (physical bonding only).

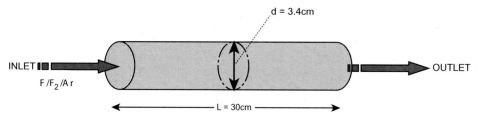

Figure 8–23. Illustrative geometry of the transport tube.

The rate coefficient for the volume recombination is $k = 2.21 \cdot 10^{14}$ cm^6/mol^{2}s (Ultee, 1977) at 300 K and pressures from 1 to 35 Torr. The reverse reaction is not relevant because of low temperature in the tube. Dissociation of F_2 takes place at temperatures above 1300 K.

8.4.5. Effectiveness of F Atom Transportation from Remote Plasma Source

To demonstrate the effectiveness of transportation of F atoms from RPSs into the chamber to be cleaned, consider an example of the transport tube connecting the RPS and CVD chamber, shown in Fig. 8–23. Gas with molar composition F 60%, F_2 10%, and Ar 30% flows into the tube at a rate of 250 sccm. The temperature of the tube walls is room temperature (300 K) and the gas pressure varies from 1 to 8 Torr. Adsorption rates of F atoms along the transport tube are shown in Fig. 8–24 for pressures of 3, 5, and 8 Torr. Adsorption of F atoms is faster near the inlet at higher pressures. Figure 8–25 shows losses of F atoms along the tube at different pressures. Losses of F atoms are more significant at higher pressure, which indicates a contribution of volume recombination. The ratio of F-atom recombination losses attributed to surface and volume processes integrated over the whole tube, $f(p)$, is shown in Fig. 8–26. Volume recombination (rate proportional to p^3) obviously dominates at

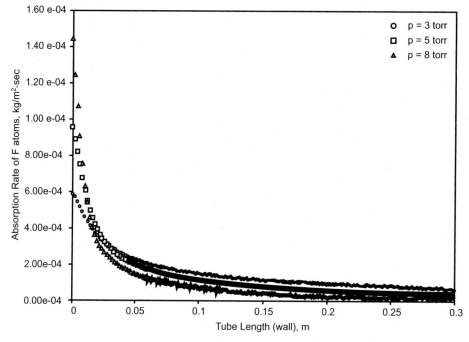

Figure 8–24. Adsorption rate of atomic fluorine along the transport tube length. The curves show computed values for three different pressures: 3, 5, and 8 Torr.

8.4. Plasma Cleaning of CVD and Etching Reactors in Micro-Electronics

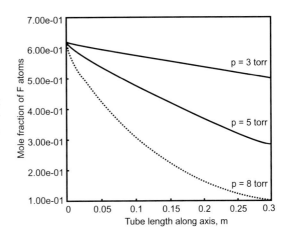

Figure 8–25. Mole fraction of F atoms along the transport tube length. The curves show computed values for three different pressures: 3, 5, and 8 Torr.

higher pressures; contributions of volume and surface mechanisms are equal at a pressure of 4.5 Torr in the specific conditions under consideration (Iskenderova, 2003). A related question involves the optimization of pressure in the transport tube to maximize the F-atom density at the tube exit. If the pressure is too low, the initial density of fluorine atoms is low; if the pressure is too high, recombination losses are significant. The F-atom concentration at the exit of the transport tube is shown in Fig. 8–27 as a function of pressure. The optimal pressure at the conditions under consideration is about 2 Torr (Iskenderova, 2003; Stueber et al., 2003).

8.4.6. Other Plasma Cleaning Processes: Passive Plasma Cleaning

Plasmas are able to remove adherent monolayers of hydrocarbons, thin layers of chemical contaminations like surface oxides, radioactive contaminations, and more. Plasma cleaning systems can be subdivided into two classes: passive and active plasma cleaning. This classification, based on differences in the collection of electric current on the workpiece to be cleaned, is rather artificial but nevertheless differentiates between types of non-equilibrium discharges applied for cleaning. Both types of plasma cleaning discharges are to be briefly discussed in the following two sections. **Passive plasma cleaning** is provided by plasma-generated active species but does not introduce the treated material as a current-collecting electrode; hence, the treated workpiece does not draw any "real" electric current from the plasma. For example, effective passive plasma cleaning of stainless steel samples to provide high lap shear strength for adhesive bonding was demonstrated by Tira (1987) using intermediate-pressure argon and dry-air discharges. Plasma cleaning in this system reduces or removes oxides, contaminants, or hydrocarbons that been on the surface initially and

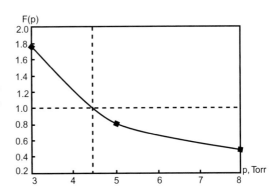

Figure 8–26. Dependence of the function f on pressure. Simulated results for three pressures: 3, 5, and 8 Torr. Critical pressure is 4.5 Torr.

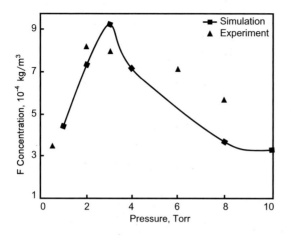

Figure 8–27. F-atom concentration after the transport zone as function of pressure.

interfered with adhesive bonding. An achieved value of the lap shear strength was 36.7 MPa after 30 min of plasma treatment at a power of 75 W.

Another example of effective passive plasma cleaning is related to removal of thin layers (≤ 4 µm) of hydrocarbon vacuum pump oils and poly-a-olefins from the surface of stainless steel and other metals. Korzec et al. (1994) performed the cleaning process using RF discharge (power around 300 W, frequency 13.56 MHz) in oxygen at gas pressures of 15–400 mTorr. The removal rate of the hydrocarbon oil film in this experiments was in the range 1.2–2.0 µm/ min. Surface cleaning results were achieved without drawing any essential electric current from the RF plasma, so the process can be qualified as passive plasma cleaning. The surface temperature of the treated samples remained low – on the level of about 50°C. Effective passive plasma cleaning at atmospheric pressure was also demonstrated by Ben Gardi et al. (2000) and Carr (1997), and was summarized by Roth (2001). One-atmosphere uniform glow discharge plasma (Roth, 1995) was applied to clean samples of automotive steel and aluminum foil contaminated with machining oils. Characteristic parameters of the discharge were frequency around 5 kHz, voltage around 7 kV, and plasma power about 200 W. Atmospheric air was used as a working gas in these experiments. The cleaning rate and cleaning efficiency in the atmospheric-pressure plasma achieved in the experiments are very similar to those mentioned earlier for low-pressure systems.

8.4.7. Other Plasma Cleaning Processes: Active Plasma Cleaning

Active plasma cleaning, in contrast to passive cleaning, is characterized by applying electric current directly to the treated material. The workpiece to be cleaned is placed on an electrode, in particular on the cathode of an abnormal glow discharge. Active plasma cleaning can also be carried out by placing the workpiece on a powered RF electrode that is DC biased to draw current, or on a hollow cathode electrode. Although active plasma cleaning provides a flux of energetic charged particles to the surface in the form of a "real" electric current, the cleaning rate can sometimes be even lower than that of passive plasma cleaning (Roth, 2001). Active plasma cleaning can be represented by cleaning Al surfaces of an ion-beam diode of carbon and oxygen impurities by using a DC 60 Hz abnormal glow discharge (Struckman & Kusse, 1993). The abnormal glow discharge was performed in Ar at a pressure of 100 mTorr, a voltage of 300 V, a current density of 0.5 mA/cm². The rate of removal of the impurities is 1–10 monolayers/min; cleaning time was 300 s. Another example of active plasma cleaning involves the application of DC glow discharge to clean the interior of stainless steel accelerator beam tubes (Hseuh, Chou, & Christianson, 1985). The glow discharge was performed in argon with 10% oxygen at pressure 35 mTorr. The interior surface of the tube (area 1.5 m²) served as the cathode; the voltage was 350 V, the cathode current density was $20\,\mu\text{A/cm}^2$, and the surface temperature was $225\,^\circ\text{C}$. The impurity removal rate was 1.5–2.3 monolayers/min and the cleaning duration was 4–6 h. A simple comparison of active and passive plasma cleaning rates in the aforementioned specific examples shows that the passive cleaning approach can be faster (micrometers vs. monolayers per minute). Cleaning rates in passive non-equilibrium discharge systems can be up to 1000 times higher than in active systems in the process of removing oil films with oxygen-containing plasmas.

8.4.8. Wettability Improvement of Metallic Surfaces by Active and Passive Plasma Cleaning

Wettability is one the most important surface characteristics related to surface energy and can be significantly improved by plasma treatment. Wettability is directly related to the effectiveness of printing, dyeing, and so forth on surfaces, which determines its significant practical importance. Wettability can be determined as the solid surfaces ability to adsorb liquid or to absorb the liquid in the bulk of fibrous materials. Wettability implies a high surface energy on the level of 50–70 dyn/cm and low contact angles. Plasma is very effective in increasing the surface energy, not only by chemical conversion of the surface but often simply through plasma cleaning. Thus, plasma cleaning that increases surface energy can significantly improve the wettability of the surface. Whereas plasma treatment of polymers surfaces and other organic materials often leads to an increase in surface energy and wettability by means of special chemical activation (for example, opening or transformation of chemical bonds; see Chapter 9), improvements in the wettability of metals is usually due to simple plasma surface cleaning. Both active and passive plasma cleaning can be effectively applied to metallic surfaces. Plasma cleaning can make these surfaces so clean that their energies approach 70 dyn/cm, which is sufficient for good wettability.

8.5. PLASMA DEPOSITION PROCESSES: PLASMA-ENHANCED CHEMICAL VAPOR DEPOSITION AND SPUTTERING DEPOSITION

Plasma plays an important role in the production and modification of thin films by means of plasma deposition, including PECVD and sputtering deposition, implantation, and surface modification processes. Some of these processes, like the production of thin films of oxides, nitrides, carbides, hydrides, and so on, using both thermal and non-thermal plasma, were already discussed in Chapter 7 and Section 8.1. Some of these processes, which are related to thin films of organic polymers and other organic or bio-organic materials, will be considered in Chapter 9. In Sections 8.5 and 8.6, we focus on the production of inorganic thin films by means of strongly non-equilibrium plasma processes of deposition (PECVD and sputtering) and implantation.

8.5.1. Plasma-Enhanced Chemical Vapor Deposition: General Principles

Production of a thin film from a feed gas through a set of gas-phase and surface chemical reactions is called chemical vapor deposition (CVD). If the gas-phase and surface reactions, starting from dissociation of the feed gas, are stimulated by plasma, the CVD process is called plasma-enhanced chemical vapor deposition (PECVD). Similar to plasma etching, PECVD also plays a key role in microelectronics for fabrication of ICs. As an example, PECVD is applied for deposition of the final insulating silicon nitride (Si_3N_4) layer in many electronic devices. This PECVD requires a surface temperature near $300\,^\circ\text{C}$, whereas the non-plasma CVD activated thermally requires $900\,^\circ\text{C}$, which is unacceptable. PECVD as a physical and chemical effect has been known for more than a century. It was observed in the late nineteenth century that operating a glow discharge in reactive chemicals can lead to deposition of thin films inside of vacuum systems, which was obviously unwanted. The application of such films to microelectronic IC fabrication started in the late 1960s (Denaro, Owens, & Crawshaw, 1968).

The PECVD often requires reactions between gas-phase precursor components; therefore, gas pressure in the PECVD discharges cannot be too low. Usually the operational pressures in these systems are 0.1–10 Torr, which is considerably higher than the pressure in etching discharges (see Sections 8.2.8 and 8.2.9). The typical plasma density in PECVD discharges is 10^9–10^{11} cm^{-3}, and ionization degrees are 10^{-7}–10^{-4}. Deposition rates are usually not sensitive to the temperature of the substrate. The substrate temperature does, however, determine such properties of the deposited film as morphology, composition, stress, and so on. A critical issue of PECVD is uniformity of the deposited thin film. Relatively high gas pressures and flow rates in the discharge make the uniformity requirements quite challenging. The uniformity requirements also dictate careful control of discharge power deposition per unit area. For this reason, RF-CCP discharges with a parallel-plate configuration (see Chapter 4) are especially relevant for PECVD applications. We should note, however, that several effective plasma deposition processes have been performed in cylindrical HDP discharges, including ECR, helicons, and RF-ICP transformer discharges. Consider characteristic features of some specific PECVD processes. More details regarding the basics and practice of PECVD can be found in Hollahan and Bell (1974), Veprek and Venugopalan (1980), Mort and Jansen (1986), Lieberman and Lichtenberg (1994), Smith (1995), and Roth (2001).

8.5.2. PECVD of Thin Films of Amorphous Silicon

PECVD-produced thin films of amorphous silicon have a wide range of applications, including thin-film transistors for flat-panel displays, solar cells, and exposure drums for xerography. PECVD amorphous silicon produced from SiH_4 usually incorporates 5–20% of H atoms in lattice. Its density is 2.2 g/cm^3 – lower than that of epitaxial crystalline silicon (2.33 g/cm^3). PECVD amorphous silicon is inexpensive and can be deposited over large areas of different substrates such as glasses, metals, polymers, and ceramics. Silane (SiH_4) is a typical feed gas for the process in RF discharges at pressures of 0.2–10 Torr. At somewhat higher pressures, H_2 and Ar can be admixed to the silane. To grow p-type material, B_2H_6 is usually added; PH_3 addition is used to produce n-type material. Deposition rates of the process are typically 5–50 nm/min, which requires an RF discharge power of 0.01–0.1 W/cm^2. The substrate temperatures for the PECVD process are 25–400°C depending on the application.

The kinetics of the PECVD discharge in silane (SiH_4) and SiH_4–H_2–Ar mixtures was described by Kushner (1988). SiH_4^+ positive ions are unstable; therefore, ionization of silane is dissociative:

$$e + SiH_4 \rightarrow SiH_3^+ + H + e + e, \quad k = 3.3 \cdot 10^{-9} \exp(-12 \, eV/T_e), \, cm^3/s, \quad (8\text{–}73)$$

$$e + SiH_4 \rightarrow SiH_2^+ + H_2 + e + e, \quad k = 4.7 \cdot 10^{-9} \exp(-12 \, eV/T_e), \, cm^3/s. \quad (8\text{–}74)$$

Surface bombardment by positive ions (particularly SiH_3^+) plays a critical role in the growth of amorphous silicon films. The silane radicals SiH_3 and SiH_2 have a positive electron affinity; therefore, the silane discharge is essentially electronegative and dissociative attachment processes make a significant contribution in the balance of charged particles and production of negative silane ions:

$$e + SiH_4 \rightarrow SiH_3^- + H, \quad k = 1.5 \cdot 10^{-11} \exp(-9 \, eV/T_e), \, cm^3/s, \quad (8\text{–}75)$$

$$e + SiH_4 \rightarrow SiH_2^- + H_2, \quad k = 9 \cdot 10^{-12} \exp(-9 \, eV/T_e), \, cm^3/s. \quad (8\text{–}76)$$

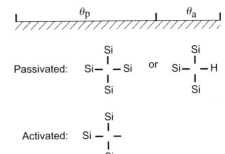

Figure 8–28. Surface coverage during amorphous silicon deposition in plasma: θ_a and θ_p represent the active and passive surface fractions.

The most important precursors for amorphous silicon thin-film growth are silane radicals, SiH_3 and SiH_2, produced by direct electron impact dissociation of the silane molecules (SiH_4):

$$e + SiH_4 \rightarrow SiH_3 + H + e, \quad k = 1.5 \cdot 10^{-8} \exp(-10\,eV/T_e), \, cm^3/s, \quad (8\text{–}77)$$

$$e + SiH_4 \rightarrow SiH_2 + H_2 + e, \quad k = 1.8 \cdot 10^{-9} \exp(-10\,eV/T_e), \, cm^3/s. \quad (8\text{–}78)$$

The non-dissociated SiH_4 molecules also participate in the surface reactions, but mostly as deactivating chemical agents. Detailed kinetics of the gas-phase processes in the silane discharges is much more complicated (Kushner, 1988). In particular, the kinetics of silicon clusterization in the silane discharges will be discussed later in this chapter in connection with the formation of nanoparticles.

8.5.3. Kinetics of Amorphous Silicon Film Deposition in Silane (SiH_4) Discharges

Amorphous silicon deposition rates from silane in the PECVD process are typically in the range 5–50 nm/min. To analyze the kinetics of the deposition process (McCaughey & Kushner, 1989; Lieberman & Lichtenberg, 1994), consider a simple surface model illustrated in Fig. 8–28. The surface of amorphous silicon consists of a combination of active (a) and passive (p) sites in the framework of the simplified model. The active sites contain at least one open free bond; the fraction of the surface covered by active sites is θ_a. In the passive sites, all bonds are occupied by silicon or hydrogen atoms (see Fig. 8–28); the fraction of the surface covered by passive sites is θ_p ($\theta_a + \theta_p = 1$). Creation of new active sites is due to ion bombardment, which also removes hydrogen atoms from the surface with the yield Y_i (per one ion):

$$SiH_3^+ + \theta_p \rightarrow \theta_a + Y_i \cdot H(g), \quad K_i = u_B/n_S, \quad (8\text{–}79)$$

where K_i is the rate coefficient of the process (see relation 8–16), u_B is Bohm velocity, and n_S is the area density of the surface sites. Collisions of SiH_2 radicals with both active and passive sites of the surface lead to their insertion into the lattice. The probability of their insertion into the passive silicon surface is lower than into the active one. It can be assumed that the SiH_2 radicals' insertion does not change the type of the surface sites; that is, the passive sites remain passive and the active sites remain active:

$$SiH_2 + \theta_a \rightarrow \theta_a, \quad K_2 = \frac{1}{4}\frac{s_2 \overline{v_2}}{n_S}, \quad (8\text{–}80)$$

$$SiH_2 + \theta_p \rightarrow \theta_p, \quad K_{2p} = \frac{1}{4}\frac{s_{2p} \overline{v_2}}{n_S}. \quad (8\text{–}81)$$

Here K_2 and K_{2p} are rate coefficients of the surface processes on the active and passive sites, respectively; s_2 and s_{2p} are sticking coefficients for an SiH_2 radical on the active and passive sites, respectively; and $\overline{v_2}$ is the thermal velocity of the radical. Film growth related to the insertion of SiH_2 radicals can be qualified as PVD, which is characterized by poor quality, voids, surface roughness, and other surface defects. Growth of a smooth high-quality film is provided by SiH_3 radicals, which are adsorbed on the surface, diffuse along the surface, and finally can be inserted into the lattice selectively at the active sites only. Insertion of SiH_3 radicals into the active lattice sites converts them into passive sites, fills in the surface roughness, and results in growth of a smooth, high-quality film:

$$SiH_3 + \theta_a \rightarrow \theta_p, \quad K_3 = \frac{1}{4}\overline{M}\frac{s_3\overline{v_3}}{n_S}, \tag{8–82}$$

where K_3 is the rate coefficient of the surface process, s_3 is the sticking coefficient for an SiH_3 radical on the active surface, $\overline{v_3}$ is the thermal velocity of the radical, and factor $\overline{M}$ characterizes the mean number of sites visited by a surface-diffusing adsorbed SiH_3 radical before desorption. Non-dissociated silane molecules (SiH_4) can also be adsorbed upon direct impact at active sites. The saturated SiH_4 molecule can lose a hydrogen atom in this case, passivate the site, and release an SiH_3 radical back to the gas phase:

$$SiH_4 + \theta_a \rightarrow \theta_p + SiH_3(g), \quad K_4 = \frac{1}{4}\frac{s_4\overline{v_4}}{n_S}, \tag{8–83}$$

where K_4 is the rate coefficient of the surface process, s_4 is the sticking coefficient for an SiH_4 molecule on the active surface site, and $\overline{v_3}$ is the thermal velocity of the silane molecule.

The balance equation for the fraction of the active surface sites is (to compare, see Section 8.2.7)

$$\frac{d\theta_a}{dt} = Y_i K_i n_i (1 - \theta_a) - K_3 n_3 \theta_a - K_4 n_4 \theta_a, \tag{8–84}$$

where n_i is the ion density at the plasma-sheath edge, and n_3, n_4 are near the surface number densities of SiH_3 radicals and SiH_4 molecules, respectively. In steady-state conditions ($d/dt = 0$), the fraction of the active surface sites, θ_a, can be found from the balance equation (8–84) as

$$\theta_a = \frac{Y_i K_i n_i}{Y_i K_i n_i + K_3 n_3 + K_4 n_4}. \tag{8–85}$$

The amorphous Si-atom deposition rate is mostly determined by (8–80) and (8–82):

$$w_{D,Si} \approx (K_3 n_3 \theta_a + K_2 n_2)\frac{n_S}{n_{Si}} = \left(\frac{K_3 n_3 \cdot Y_i K_i n_i}{Y_i K_i n_i + K_3 n_3 + K_4 n_4} + K_2 n_2\right)\frac{n_S}{n_{Si}}, \tag{8–86}$$

where n_{Si} is the density of Si atoms in the lattice. For typical PECVD conditions, deactivation of the active surface sites by SiH_4 dominates the kinetics ($K_4 n_4 \gg Y_i K_i n_i + K_3 n_3$), and relation (8–85) becomes

$$\theta_a = \frac{Y_i K_i n_i}{K_4 n_4}, \tag{8–87}$$

which results in the following rate of amorphous silicon film deposition from SiH_4 in the PECVD process:

$$w_{D,Si} \approx (K_3 n_3 \theta_a + K_2 n_2)\frac{n_S}{n_{Si}} = \left(\frac{K_3 n_3 \cdot Y_i K_i n_i}{K_4 n_4} + K_2 n_2\right)\frac{n_S}{n_{Si}}. \tag{8–88}$$

The following process and discharge parameters can be used for calculations: $Y_i = 5$–10, $\overline{M} \approx 10$, $n_i/n_4 \approx 10^{-4}$, and all sticking coefficients s are of order unity ($s \approx 1$). Then, based on relation (8–87), the fraction of the active surface sites is $\theta_a \approx 10^{-2}$. Thus, most of the surface remains passive during the PECVD process, and only about 1% of sites are active for the growth of silicon film. Compare the first and second terms of expressions (8–86) and (8–88), which represent high-quality (related to SiH_3) and low-quality (related to SiH_2) deposition mechanisms. The ratio of these terms, ξ, representing the film quality, is

$$\xi = \frac{K_3 n_3 \theta_a}{K_2 n_2} \approx \overline{M} \theta_a \frac{n_3}{n_2}. \tag{8–89}$$

Under typical SiH_4 PECVD conditions, the concentration of SiH_3 exceeds that of SiH_2: $n_3/n_2 \approx 100$. It results in $\xi \approx 10$, which means that the contribution of a high-quality film deposition mechanism usually dominates the process. Improvement in film quality generally requires higher ion fluxes (that is, higher n_i) and energies (that is, higher Y_i), higher SiH_3/SiH_2 ratios, and finally better SiH_3 surface diffusion (that is, higher $\overline{M}$).

8.5.4. Plasma Processes of Silicon Oxide (SiO$_2$) Film Growth: Direct Silicon Oxidation

Production of SiO_2 thin films is of considerable importance in microelectronics. It can be performed using many approaches, including direct oxidation of silicon and different types of CVD. All these processes can be stimulated by using non-thermal plasma. First consider SiO_2 film growth by direct silicon oxidation in O_2 or H_2O. The most common method of oxidation is the thermal method, which requires 850–1100°C. The process can be enhanced by application of non-thermal plasma, which accelerates the deposition and decreases the required substrate temperature (Moruzzi, Kiermasz, & Eccleston, 1982). High-quality thin SiO_2 films have been grown, in particular, on single-crystal silicon at substrate temperatures of 250–400°C using a non-thermal O_2 discharge (Carl et al., 1991). This direct plasma oxidation process is usually referred to as **plasma anodization**. The substrate is biased positively with respect to plasma in the anodization approach, which draws a net DC current through the film while it grows. The mechanism of direct oxidation is related to the drift of negative ions (O^-) from the surface across the Si–SiO_2 interface into silicon. The O^- transport limits the kinetics of SiO_2 film growth. More effective, from this perspective, are deposition processes, which don't require oxygen transportation through the interface to build up the SiO_2 film; these will be considered in the next two sections. Sputtering contamination during plasma anodization film growth is a serious problem with the method. Therefore, application requires the use of discharges with relatively low sheath voltages. Microwave and other high-density (HDP) discharges can be effectively used in this case.

8.5.5. Plasma Processes of Silicon Oxide (SiO$_2$) Film Growth: PECVD from Silane–Oxygen Feedstock Mixtures and Conformal and Non-Conformal Deposition Within Trenches

PECVD growth of SiO_2 films can be conducted at 100–300°C using either silane–oxygen mixtures, or a special tetraethoxysilane (TEOS)–oxygen mixture as feed gases. To compare, the thermal CVD approach in this case requires much higher temperatures (600–800°C). Use of different feed gases leads to different conformality characteristics of the PECVD on topographical features like trenches. Consideration of the PECVD processes in these different feed-gas mixtures (SiH_4–O_2 and TEOS–O_2), therefore, will be divided, respectively, into two parts between this section and the following. Silane–oxygen PECVD of SiO_2 films can be performed in such gas mixtures as SiH_4–Ar–N_2O, SiH_4–Ar–NO, and SiH_4–Ar–O_2. The molecular gases N_2O, NO, and O_2 are used here as a source of atomic oxygen, which determines the SiO_2 film growth. The most conventional source of oxygen, for this purpose, is N_2O. The SiO_2 deposition precursors are SH_3, SiH_2 radicals, and O atoms, which are created by plasma dissociation of feed gases. Initial surface reactions of the film

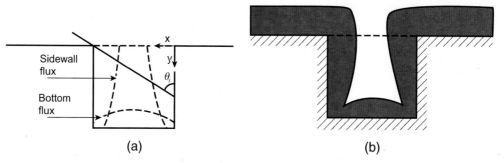

Figure 8–29. Illustration of a non-conformal plasma deposition within a trench: (a) trench before deposition (the dashed lines present the deposition flux incident on the sidewall and bottom); (b) intermediate moment of the deposition.

growth include surface formation of such intermediate molecular oxides as $(SiH_3)_2O$ and SiH_3OH:

$$SiH_3 + SiH_3 + O(surf) \rightarrow (SiH_3)O, \tag{8–90}$$

$$SiH_3 + SiH_3 + O(surf) \rightarrow SiH_3OH + SiH_2. \tag{8–91}$$

Further oxidation removes hydrogen from the surface in the form of water vapor, usually leaving about 2–9% of hydrogen atoms in the final SiO_2 film. The SiO_2 film deposition rate in the described silane–oxygen discharges is quite high, up to 200 nm/min. The high deposition rates are, in particular, due to high sticking coefficients of SiH_3 and SiH_2 radicals (the sticking probabilities are about $s \approx 0.35$). On the other hand, the high sticking coefficients of SiH_3 and SiH_2 radicals result in a non-conformal deposition of SiO_2 films within the trenches. This means that deposition rates at different points of the trench are different, which definitely limits the application of silane–oxygen discharges for PECVD process.

Deposition conformality is an important issue to discuss. Consider a simple model of SiO_2 deposition on the sidewalls and bottom of a trench, which is provided by a uniform isotropic source of precursors at the top of the trench (see Fig. 8–29a). Assume the sticking coefficient equals unity, and the precursor's mean free path is much greater than the characteristic sizes of the trench. Then the deposition flux of SiO_2 on the sidewalls, Γ_{SiO_2}, can be expressed as:

$$\Gamma_{SiO_2} \propto 1 - \cos\theta_s, \tag{8–92}$$

where θ_s is the angle subtended by the trench opening as seen at a position along the sidewall (see Fig. 8–29a). Thus, the maximum deposition rate value corresponds to near the top area of the trench, which can lead to the formation of undesirable keyhole non-conformality within the trench (see Fig. 8–29b).

The two following effects help to make the PECVD process within the trenches uniform and conformal, to avoid formation of the "keyholes," and to permit complete filling of the trench with insulating dielectric material (SiO_2):

1. Low sticking probabilities of active precursors – numerous reflections within the trench before the final random sticking to the surface lead to uniform deposition within a trench.
2. High surface diffusion rates of the adsorbed active precursors in the trench also make the deposition uniform and conformal.

The silane-containing feed gases have high sticking coefficients (they are also characterized by insignificant rates of surface diffusion), which provide high deposition rates on one hand, but poor deposition conformality on the other hand. When deposition conformality is an important issue, another feed-gas mixture, TEOS–O_2, is usually applied for PECVD of silicon dioxide films.

8.5.6. Plasma Processes of Silicon Oxide (SiO$_2$) Film Growth: PECVD from TEOS–O$_2$ Feed-Gas Mixtures

TEOS ($Si(OC_2H_5)_4$) is widely used as an SiO_2 deposition precursor in PECVD processes, especially when conformal deposition is crucial. TEOS produces deposition precursors with low sticking probabilities, which actually permits excellent conformality of the process. It is also important for practical application that TEOS is a relatively inert liquid at room temperature, in contrast to silane (SiH_4), which is an explosive gas at room temperature. Nitrogen and argon can be used as gas carriers of the TEOS vapor applied as a PECVD feedstock. A feedstock mixture, which is typically used for conformal PECVD deposition, is 1% TEOS and 99% O_2. The discharge kinetics is dominated in this case by molecular oxygen. The significant excess of O_2 in the mixture provides effective burning out of carbon and hydrogen from the TEOS-based precursors forming gaseous CO_2 and water vapor. It permits one to avoid the contamination of SiO_2 film with C and H atoms.

SiO_2 deposition from the TEOS–O_2 mixture is usually done in non-thermal plasma at pressures of 0.2–0.5 Torr and at substrate temperatures of 200–300°C. Deposition rates in this case are usually not higher than 50 nm/min (to compare, those in SiH_4-based mixtures are considerably higher, up to 200 nm/min). However, the sticking coefficients for the TEOS precursor deposition ($s \approx 0.045$) are about 10-fold lower than those for the SiH_4 precursor deposition. This results in good deposition conformality and indicates the effectiveness of TEOS application for PECVD on topographical features like trenches. Plasma-chemical reactions in the TEOS–O_2 mixture lead to formation of such SiO_2 growth precursors as molecules ($Si(OC_2H_5)_k(OH)_{4-k}$) and radicals ($Si(OC_2H_5)_kO_{4-k}$) ($k = 0, 1, 2, 3$). In both cases, the OC_2H_5 groups are partially replaced by either OH radicals or O atoms. Formation of the precursors can be attributed in particular to direct electron impact dissociation processes:

$$e + Si(OC_2H_5)_k(OH)_{4-k} \rightarrow Si(OC_2H_5)_{k-1}(OH)_{5-k} + C_2H_4 + e, \quad k = 1, 2, 3, 4.$$
$$(8\text{–}93)$$

Other channels of precursor formation in plasma are related to reactions with atomic oxygen:

$$O + Si(OC_2H_5)_k(OH)_{4-k} \rightarrow Si(OC_2H_5)_{k-1}(OH)_{5-k} + C_2H_4O, \quad k = 1, 2, 3, 4.$$
$$(8\text{–}94)$$

Mechanism (8–94) dominates when the TEOS–O_2 mixture is strongly diluted. While being adsorbed on the surface, the precursors are further oxidized with releasing CO_2 and H_2O into the gas phase and formation of the SiO_2 film. Ion bombardment of the substrate anisotropically stimulates the vertical deposition rate of SiO_2, providing directionality to the PECVD process. Details regarding the surface kinetics of the PECVD process can be found in Stout and Kushner (1993) and Cale et al. (1992).

8.5.7. PECVD Process of Silicon Nitride (Si$_3$N$_4$)

Amorphous silicon nitride films, which are resistant to water vapor, salts, and other chemicals and, therefore, are applied as a final encapsulating layer for ICs, are effectively produced using PECVD. A typical feed-gas mixture for PECVD is SiH_4–NH_3. The process is performed in plasma at pressures of 0.25–3 Torr; conventional substrate temperatures are in the range 250–500°C. Deposition rates of silicon nitride films under such conditions are about 20–50 nm/min. The plasma deposited silicon nitride film can usually be characterized as $SiN_{1-1.2}H_{0.2-0.6}$. Major film precursors are SiH_3, SiH_2, and NH radicals, generated in plasma by dissociation of a feed gas by direct electron impact. Hydrogen admixtures are not desired in silicon nitride films; therefore, thermal and ion-induced desorption of hydrogen from the growing film is an important step of the PECVD process. Lower hydrogen content in the film can be achieved by increasing the substrate temperature and RF power flux. The source of most of the hydrogen atoms in the film is not silane, but ammonia. Therefore, replacing NH_3 with molecular nitrogen significantly decreases the hydrogen content in the produced film. Application of the $SiH_4–N_2$ feed-gas mixture, however, decreases the deposition rate, deposition conformality, and thin-film quality in general.

8.5.8. Sputter Deposition Processes: General Principles
The plasma deposition process can be performed in two ways: as PECVD, considered earlier, and sputter deposition, which is discussed next. General features of sputtering, as an approach to etching, were discussed in Section 8.2.3. Sputtering is a process of ejection of atoms from surfaces due to energetic ion bombardment. The sputtering process can be applied not only for material removal from surfaces, which is called etching, but also as a source of atoms for following deposition. This method is usually referred to as **sputter deposition**. Sputter deposition can be divided into two approaches: physical sputtering and reactive sputtering. **Physical sputtering** implies that atoms sputtered from the target material are directly transported without any special conversion to a substrate, where they are deposited. An important issue of physical sputtering is process uniformity. The sputtered atoms' energy distribution is also very important, because it determines mixing of the sputtered atoms with substrate atoms and influences the properties of the deposited film. In the case of **reactive sputtering**, a special feed gas is also supplied to the plasma deposition system. Molecules of the feed gas dissociate in plasma, producing chemically active species, which then react with the target during ion bombardment and sputtering. As a result, the film deposited on a substrate includes not only material of the ion-bombarded target but also compounds produced due to the reactive gases. Chemical reactions between the target material and reactive gas species take place in this type of sputter deposition process at both target and substrate surfaces. Non-thermal plasma sources used for the sputtering deposition of thin conductive films are generally DC discharges. More specifically, DC planar magnetrons are usually applied in this case. RF-CCP discharges or RF-driven planar magnetrons are usually used in the case of sputtering deposition of insulating films.

8.5.9. Physical Sputter Deposition
Physical sputter deposition implies ion bombardment of a target, which results in sputtering of the target atoms. The sputtered atoms are then ballistically flowing to the substrate and finally are deposited on the substrate. The ion bombardment is typically provided by Ar^+ ions with energies in the range 0.5–1 keV. Sputter yields for different target materials were presented in Table 8.1; most of them are of order unity. Therefore, a wide variety of materials, including metals, alloys, and insulators, can be deposited this way over large areas with reasonable deposition rates, excellent uniformity, and good surface smoothness and adhesion. The sputter deposition is not conformal; the conformality can be improved in this case only by ion bombardment of the deposited film and redeposition. If a multi-component target is sputtered, the deposited film can have the same composition as the bulk target material in the steady state. This can be achieved if the sticking coefficients of all components on the substrate are about the same. In this way, alloy targets can be deposited without changing composition. When some components have lower sticking coefficients and form gas on the substrate (like oxygen atoms), then physical sputter deposition becomes ineffective. Thus, targets such as ceramics and oxides usually cannot be deposited by physical sputtering.

RF- or DC-driven planar magnetron discharges are applied for sputter deposition. The pressures in the discharges are usually quite low, in this case 1–10 mTorr. The pressure is limited in the physical sputter deposition system by the requirement that the sputtered atoms' mean free path be larger than the separation between the sputtered target and substrate where the atoms should be deposited. The physical sputter deposition rate can be estimated by assuming that all sputtered material is deposited on the substrate:

$$w_{\text{sput.dep.}} = \frac{\gamma_{\text{sput}} \Gamma_i}{n_f} \cdot \frac{A_{\text{tar}}}{A_{\text{sub}}}, \tag{8–95}$$

where γ_{sput} is the sputtering yield from the target (see Section 8.2.3 and Table 8.1), Γ_i is the incident ion flux (if Ar^+ ions have energy 1 keV and current 1 mA/cm^2, then $\Gamma_i = 6.3 \cdot 10^{15}$ cm^{-2} s^{-1}), n_f is the density of the deposited film (typical value $n_f = 5 \cdot 10^{22}$ cm^{-3}), A_t is the area of the sputtered target, and A_{sub} is the substrate-area where the film is deposited. A typical physical sputter deposition rate is 75 nm/min. Properties of the deposited film are sensitive to the sputtered-atom energy distribution. It can be explained taking into account that the energy of the atoms striking the substrate determines mixing and diffusion processes between the incoming sputtered atoms and substrate material. These processes are crucial for good bonding and adhesion, resulting in better quality of the sputter-deposited film. The sputtered atoms' energy distribution function can be presented as (Winters & Coburn, 1992)

$$f(\varepsilon) \propto \frac{E_s \varepsilon}{(E + \varepsilon)^3}, \tag{8–96}$$

where ε is the energy of a sputtered atom, and E_s is the surface binding energy ($E_s = 1$– 4 eV; see Section 8.2.3). Distribution function (8–96) determines the average energy of the sputtered atoms as $\bar{\varepsilon} = E_s/2$, which quantitatively means $\bar{\varepsilon} = 0.5$–2 eV. Thus, the sputtered atoms' energy is sufficiently high to provide their intensive mixing with the material. The morphology of the produced film is considerably influenced by the substrate temperature and the deposition pressure. More information regarding morphology, uniformity, and other properties of thin films produced by means of physical sputter deposition can be found in Thornton (1986), Konuma (1992), and Lieberman and Lichtenberg (1994).

8.5.10. Reactive Sputter Deposition Processes

Reactive sputter deposition uses active species from the gas phase for chemical reactions with the target material during ion bombardment. The chemical reactions take place at both the target and the substrate. The gas-phase chemically active species are produced by dissociation of the special feed gas introduced into the discharge. Thus, the deposited film in this case is a compound formed from the target material and the feed gas, which is important for applications, particularly, in microelectronics. The sputtering deposition of SiO_2 explains the advantages of reactive sputter deposition. After SiO_2 target physical sputtering, Si and O atoms approach the substrate. Silicon atoms stick to the substrate surface, while O atoms partially form molecular oxygen, leaving the deposited film with a deficiency of oxygen. To restore the SiO_2 stoichiometry of the deposited film, O_2 is added as a feed gas to the discharge, which leads to the additional formation of O atoms and their incorporation into the growing film. This is the main idea of the reactive sputter deposition approach. The feed gas is able to control oxygen concentration in the film; therefore, even pure silicon can be used for reactive sputter deposition of SiO_2 films in this case. Taking into account the advantages of the approach when one component has a high vapor pressure, reactive sputter deposition is widely applied in the production of thin films of oxides, nitrides, carbides, and silicides. Feed gases commonly used as a source of O atoms are O_2 and H_2O; N_2 and NH_3 are sources of N atoms; CH_4 and C_2H_2 are sources of C atoms; and silane (SiH_4) is a source of Si atoms. Reactive sputter deposition is used for production of thin films of different metal compounds like titanium nitride (TiN). The deposition processes with a metal target and a compound film can be arranged in two modes: "metallic" mode and "covered" mode. The **covered mode** takes place when the ion flux is low and the gas flux is high; hence, the target is covered by the compound during the deposition process. The **metallic mode**, on the other hand, takes place when the ion flux is high and the gas flux is low; hence, the target remains metallic during the deposition. Deposition rates are generally higher in the metallic mode. If the ion flux is fixed, increased gas flow leads to a transition from metallic to covered mode. The transition exhibits hysteresis: the "increasing" transitional gas flux corresponding to the metallic-to-covered-mode transition is larger than the "decreasing" transitional gas flux corresponding to the covered-to-metallic-mode transition. This can be interpreted in the framework of the following kinetic model.

8.5.11. Kinetics of Reactive Sputter Deposition: Hysteresis Effect

The kinetics of reactive sputter deposition, especially for deposition with a metal target and a metal compound film, is described in a model developed by Berg et al. (1989). The model is based on the balance of the compound covered fraction (θ_t) of the total target area (A_{tar}) during the reactive sputter deposition,

$$n_t \frac{d\theta_t}{dt} = i\Gamma_g s_g (1 - \theta_t) - \Gamma_i \gamma_c \theta_t, \tag{8-97}$$

as well as the balance of the compound covered fraction (θ_s) of the total substrate area (A_{sub}),

$$n_s \frac{d\theta_s}{dt} = i\Gamma_g s_g (1 - \theta_s) + \Gamma_i \gamma_c \theta_t \frac{A_{tar}}{A_{sub}} (1 - \theta_s) - \Gamma_i \gamma_m (1 - \theta_t) \frac{A_{tar}}{A_{sub}} \theta_s, \tag{8-98}$$

where n_t and n_s are the target and substrate number densities; γ_m and γ_c are yields for sputtering the metal and the compound from the target; Γ_i and Γ_g are incident fluxes of ions and reactive gas species; s_g is the sticking coefficient of a reactive gas species on the metal part of the target; and i is the number of atoms in the reactive gas species ($i = 2$ for O_2, for example). Solving the system of balance equations (8–97) and (8–98) in steady-state conditions ($d/dt = 0$) gives the surface coverage of the compound on the target as

$$\theta_t = \frac{Y}{1 + Y}, \quad Y = \frac{2\Gamma_g s_g}{\Gamma_i \gamma_c}, \tag{8-99}$$

and the surface coverage of the compound on the substrate as

$$\theta_s = \frac{Y^2 + 2Y}{Y^2 + 2Y + a}, \quad a = \frac{\gamma_m}{\gamma_c}. \tag{8-100}$$

The surface coverages of the compound, θ_t and θ_s, are presented as functions of normalized parameters Y and a; the parameter Y characterizes the relative contribution of the ion and reactive gas fluxes. The total number of reactive gas molecules consumed to form the compound deposited on the substrate, which actually characterizes the reactive gas flow, can be expressed as

$$\frac{dN_g}{dt} = \Gamma_g s_g [(1 - \theta_t) A_{tar} + (1 - \theta_s) A_{sub}]. \tag{8-101}$$

Taking into account the steady-state solutions (8–99) and (8–100), the reactive gas flow can be presented as a function of the parameter Y (characterizing the contribution of the ion and reactive gas fluxes):

$$\frac{dN_g}{dt} \propto Y \left(\frac{1}{1 + Y} + \frac{a}{Y^2 + 2Y + a} \right). \tag{8-102}$$

8.6. Ion Implantation Processes: Ion-Beam Implantation

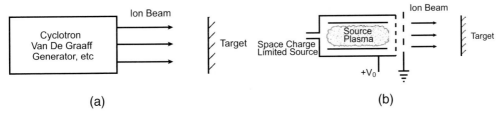

Figure 8–30. General schematics of ion-beam implantation systems: (a) schematic of a high-energy (above 300 keV) implantation system; (b) schematic of an intermediate-energy (between 10 and 300 keV) implantation system.

The total target sputtering flux, including sputtering of metal and compound, related to the ion bombardment flux can be expressed in this case as

$$\Gamma_{\text{sput}} = \Gamma_i[\gamma_m(1 - \theta_t) + \gamma_c\theta_t]. \tag{8–103}$$

Again taking into account the steady-state solutions (8–99) and (8–100), the sputtering flux can be presented as a function of the parameter Y:

$$\Gamma_{\text{sput}} \propto \frac{a + Y}{1 + Y}. \tag{8–104}$$

Based on the dependences $\frac{dN_g}{dt}(Y)$ and $\Gamma_{\text{sput}}(Y)$ described by relations (8–102) and (8–104), it is easy to figure out that the function $\frac{dN_g}{dt}(\Gamma_{\text{sput}})$ is not monotonic and has maximum and minimum points if $a \gg 1$. As a result, the reverse function $\Gamma_{\text{sput}}(\frac{dN_g}{dt})$ has a reverse S-shape and exhibits the hysteresis behavior – increase and decrease of the reactive gas flow rate result in changing the reactive sputter deposition modes (from metallic to covered mode and back) at different flow rates.

8.6. ION IMPLANTATION PROCESSES: ION-BEAM IMPLANTATION AND PLASMA-IMMERSION ION IMPLANTATION

8.6.1. Ion-Beam Implantation

Ion implantation technology is based on directing high-energy ions onto surfaces and penetration of the energetic ions into the atomic structure of a material over many atomic layers. Ion energies typically used for the implantation processes are in the range 10–300 keV and sometimes higher. The penetration depth of the energetic ions can be up to a micron, which is already sufficient for many applications. Ion implantation is an important process for semiconductor doping but also has other practical applications, especially in surface hardening of materials, first of all metals. Silicon doping can be provided, in particular, by ion implantation of boron, phosphorus, and arsenic. Surface hardening of metals can be provided by ion implantation of nitrogen or carbon. The ion implantation technology can be divided into two approaches: conventional ion-beam technology, which is to be briefly considered in this section, and plasma ion implantation, which we discuss in more detail in the following sections. General schematics of the ion-beam implantation systems are shown in Fig. 8–30. Ion-beam implantation can be accomplished by two types of systems, depending on energy of the ions. The first type (Fig. 8–30b) applies lower-energy ions (10–300 keV) and is mostly used in microelectronics for wafer doping (Brown, 1989). Current densities of these ion sources are limited by space charge. The second type (Fig. 8–30b) uses higher-energy ions (above 300 kV), and it is used less. Acceleration of ions is provided in this case by cyclotrons, Van De Graaff generators, or by other accelerators, which makes the method more expensive (Wilson & Brewer, 1973). The second type of ion-beam implantation makes it possible

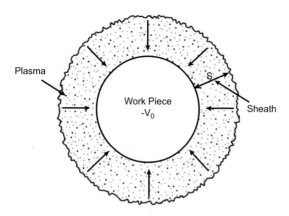

Figure 8–31. Illustration of the plasma ion implantation of a workpiece by negative biasing and the attraction of ions from the plasma.

to implant atoms to greater depths, which determines its application niches. Advantages of ion-beam implantation are related to possibilities of effectively adjusting ion energy and therefore the implantation depth, and adjusting implantation dose by controlling the exposure time. The biggest issue of ion-beam implantation is that the technology is expensive and capital-intensive. As a result, ion-beam implantation is mostly used for high-value products in microelectronics. An important disadvantage of conventional ion-beam implantation is the requirement of a normal incidence of ions to the workpiece, which limits its application for three-dimensional objects. Implantation on three-dimensional objects requires the application of masking and movable fixturing, which makes the method quite complicated and expensive. In other words, between the initial step of ion generation and the final step of ion implantation, three intermediate costly steps are required: beam extraction, beam focusing, and beam scanning. Relatively low pressure in the system requires vacuum or batch processing, which is also an issue for implantation technology. Plasma-immersion ion implantation (PIII), considered in the next section, is a relatively new technology that allows one to overcome at least two of the aforementioned disadvantages by decreasing equipment cost and avoiding masking and fixturing of the workpiece.

8.6.2. Plasma-Immersion Ion Implantation: General Principles

In PIII technology, the target is immersed in plasma. Positive ions are extracted directly from the plasma and accelerated toward the target for implantation by a series of negative high-voltage pulses applied to the target. Comparing PIII with conventional ion-beam implantation, described earlier, we can say that PIII skips the intermediate stages of ion-beam implantation: beam extraction, beam focusing, and beam scanning. PIII technology has been demonstrated for two major applications of ion implantation: for doping of semiconductors in microelectronics (Cheung, 1991) and for surface hardening of metallurgical components (Conrad et al., 1990). Also we should mention that PIII has been applied for hardening of medically implantable hip joints. A schematic of the PIII system is shown in Fig. 8–31. Uniform plasma can be generated by microwave, DC, or RF discharges. As an example, we mention a 3 kW microwave discharge at frequency 2.45 GHz and gas pressures 20–100 mTorr. Plasma volumes in PIII systems are typically large – hundreds of liters or more. To improve efficiency of plasma sources and to increase plasma densities, a magnetic field can be applied to magnetize plasma electrons. Typical electron densities vary over quite a wide range: 10^8–10^{11} cm^{-3}. Electrically conducting workpieces are positioned in the plasma for implantation. Pulsed negative DC voltage (10–200 kV) is applied to the workpieces by a fast switching circuit to attract and accelerate ions. Depending on the plasma density and the workpiece area, current pulses are 1–500 A during the implantation. To control the implantation dose and avoid surface overheating, the pulse repetition rates are varied by the switching circuit in the range 200–500 Hz (pulse duration 10–30 μs); the

8.6. Ion Implantation Processes: Ion-Beam Implantation

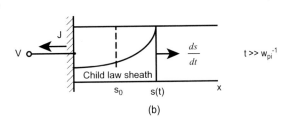

Figure 8–32. Illustration of the planar PIII geometry: (a) density distribution just after formation of the matrix sheath; (b) density distribution after evolution of the quasi-static Child law sheath.

pulse rise time is about 1 μs. Descriptions of different specific PIII systems can be found in Conrad (1988), Bernius and Chutjian (1989), Keebler et al. (1989), Keebler (1990), Spence et al. (1991), Matossian and Goebel (1992), Anders et al. (1994), and Matossian (1994).

8.6.3. Dynamics of Sheath Evolution in Plasma-Immersion Ion Implantation: From Matrix Sheath to Child Law Sheath

Sheath evolution in the PIII process is illustrated in Fig. 8–32 for a planar target in plasma with density n_p. It is assumed that a negative voltage pulse with amplitude $-V_0$ and duration t_p is applied to the target. After a short period of time (about the inverse electron plasma frequency, $1/\omega_{pe}$), electrons are driven away from the target surface and create the matrix sheath, which consists of uniformly distributed positive ions (Fig. 8–32a). The matrix sheath thickness s_0 is determined by the applied voltage and plasma density. After a longer period of time (about inverse ion plasma frequency, $1/\omega_{pi}$), ions are accelerated in the sheath and implanted, and the sheath edge recedes, exposing new ions to be extracted. The process can be interpreted as sheath propagation. The steady-state Child law sheath is finally developed (Fig. 8–32b; see Chapter 3), which is about $(eV_0/T_e)^{1/4}$ times larger than the matrix sheath. The dynamics of sheath evolution determines the implantation current and energy distribution of the ions to be implanted. We discuss collisionless sheath evolution shortly; additional details can be found in Conrad (1987), Donelly and Watterson (1989), Lieberman (1989a,b), Scheuer, Shamim, and Conrad (1990), Vahedi et al. (1991), and Stewart and Lieberman (1991). The ion matrix sheath evolves into the Child law sheath (Fig. 8–32), which has time-dependent current density and sheath thickness. When gas pressure is sufficiently low and ion collisions within the sheath can be neglected, the relation between the current density J_c and the sheath thickness s corresponds to the Child law:

$$J_{Ch} = \frac{4}{9}\varepsilon_0 \left(\frac{2e}{M}\right)^{1/2} \frac{V_0^{3/2}}{s^2}, \qquad (8\text{–}105)$$

where e and M are ion charge and mass; ε_0 is the free-space permittivity. The applied voltage is high ($V_0 \gg T_e$); therefore, the Debye length r_D is much shorter than the matrix sheath size s_0 ($r_D \ll s_0$) and the sheath edge at $s(t)$ is abrupt (see Fig. 8–32). The current demanded by the sheath is supplied by the uncovering of ions at the moving sheath edge, $s(t)$, and by the drift of ions to the target with the ion sound speed, which equals the Bohm velocity, $u_B = \sqrt{T_e/M}$. Thus, balancing the Child law current density (8–105) to the value

553

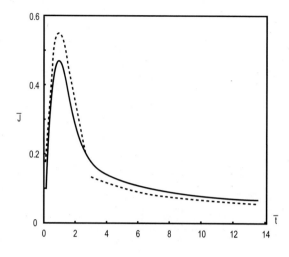

Figure 8–33. Time evolution of the normalized implantation current density, $\bar{J} = J/(en_0u_0)$, presented as a function of normalized time, $\bar{t} = \omega_{pi}t$. The dashed lines present analytical solution for $\bar{t} < 2.7$ and $\bar{t} > 3.0$; the solid line presents the numerical solution of the fluid equations.

of the charge crossing the moving sheath boundary per unit time leads to the following equation:

$$en_p\left(\frac{ds}{dt} + u_B\right) = J_{Ch}.$$ (8–106)

The sheath propagation velocity can then be found as a function of its position ($s(t)$, Fig. 8–32):

$$\frac{ds}{dt} = \frac{2}{9}\frac{s_0^2 u_0}{s^2} - u_B,$$ (8–107)

where s_0 is the thickness of the matrix sheath (see Fig. 8–33), which can be expressed as

$$s_0 = \sqrt{\frac{2\varepsilon_0 V_0}{en_p}},$$ (8–108)

and u_0 is the characteristic ion velocity in the sheath, determined as

$$u_0 = \sqrt{\frac{eV_0}{M}}.$$ (8–109)

Time evolution of the sheath, $s(t)$, can be found by integration of equation (8–107):

$$\tanh^{-1}\frac{s}{s_{Ch}} - \frac{s}{s_{Ch}} = \frac{u_B t}{s_{Ch}} + \tanh^{-1}\frac{s_0}{s_{Ch}} - \frac{s_0}{s_{Ch}},$$ (8–110)

where s_{Ch} is the steady-state Child law sheath thickness to be finally established, which is related to the matrix sheath thickness s_0 (8–108) as follows:

$$s_{Ch} = s_0\sqrt{\frac{u_0}{u_B}}.$$ (8–111)

Because the applied voltage is much higher than the electron temperature ($T_e \gg eV_0$), the steady-state Child law sheath is much larger than the matrix sheath: $s_{Ch}/s_0 \approx (eV_0/T_e)^{1/4} \gg 1$. Expansion of equation (8–110) for $s_0 < s < s_{Ch}$ gives the time evolution of the sheath thickness:

$$\frac{s}{s_0} = \sqrt[3]{1 + \frac{2}{3}\omega_{pi}t},$$ (8–112)

where $\omega_{pi} = (n_p e^2 / \varepsilon_0 M) = u_0 / s_0$ is the plasma ion frequency in the matrix sheath. Relations (8–111) and (8–112) determine the time required to establish the steady-state Child law sheath $(s(t) = s_{Ch})$ as

$$t_{Ch} \approx \frac{\sqrt{2}}{9} \frac{1}{\omega_{pi}} \left(\frac{V_0}{T_e} \right)^{3/4}. \qquad (8\text{–}113)$$

The time (8–113) for formation of the sheath usually exceeds the voltage pulse period $(t_p \ll t_{Ch})$.

8.6.4. Time Evolution of Implantation Current in PIII Systems

Consider first the implantation current evolution during the sheath matrix phase, in the framework of the collisionless sheath model applied in the previous section (Lieberman & Lichtenberg, 1994). This initial charge density in the matrix sheath is uniform; therefore, the initial electric field varies linearly with distance x from the target (see Fig. 8–32):

$$E = \frac{M}{e} \omega_{pi}^2 (x - s). \qquad (8\text{–}114)$$

Assuming, for the matrix sheath phase, $s(t) = s_0 + \left(\frac{ds}{dt} \right)_0 t$ and taking $\left(\frac{ds}{dt} \right)_0$ from (8–107) $(s = s_0)$, the dynamic equation for an ion coordinate x can be presented as

$$\frac{d^2 x}{dt^2} = \omega_{pi}^2 (x - s_0) - \frac{2}{9} u_0 \omega_{pi}^2 t. \qquad (8\text{–}115)$$

Initial conditions $(t = 0)$ for an ion $(x = x_0, dx/dt = 0)$ lead to the solution of (8–115):

$$x = s_0 + (x_0 - s_0) \cosh \omega_{pi} t - \frac{2}{9} \sinh \omega_{pi} t + \frac{2}{9} u_0 t. \qquad (8\text{–}116)$$

Letting $x = 0$ in equation (8–116), we can calculate an ion flight time t_{if} from the position x_0 to the target in the form of function $x_0(t_{if})$. Ions from the interval between x_0 and $x_0 + dx_0$ are implanted in a time interval between t_{if} and $t_{if} + dt_{if}$. Therefore, differentiation of the function $x_0(t_{if})$ determines the implantation current density as $J = en_p(dx_0/dt_{if})$ (see Problem 8.16). As a result, the time evolution of the implantation current density from the matrix sheath can be presented as

$$J = en_p u_0 \left(\frac{\sinh \omega_{pi} t}{\cosh^2 \omega_{pi} t} + \frac{2}{9} \frac{1 + \omega_{pi} t \sinh \omega_{pi} t - \cosh \omega_{pi} t}{\cosh^2 \omega_{pi} t} \right). \qquad (8\text{–}117)$$

This expression describes the implantation of ions from the initial matrix sheath $(0 \leq x_0 \leq s_0)$. Letting $x = 0$ and $x_0 = s_0$ in equation (8–116), we can calculate an ion flight time across the initial matrix sheath: $t \approx 2.7(1/\omega_{pi})$. All ions from the initial matrix sheath $(0 \leq x_0 \leq s_0)$ are implanted during this time interval. The implantation current density, $J/en_p u_0$, is shown in Fig. 8–33 as a function of time $(\omega_{pi} t)$. The left-hand branch of the graph (dashed curve) represents current from the matrix sheath $(t \leq 2.7(1/\omega_{pi}))$. The maximum current value, $J \approx 0.5 en_p u_0$, is achieved at $t \approx 0.95(1/\omega_{pi})$. Now consider implantation from the Child law sheath, that is, implantation of ions initially located at $x_0 > s_0$. Their implantation current density as a function of time is represented by the right-hand branch of the graph in Fig. 8–33. Analytically this monotonically decreasing function can be expressed as

$$J = en_p u_0 \frac{1}{\frac{9}{2} \left(\frac{x_0^2}{s_0^2} \right) + 3}, \qquad (8\text{–}118)$$

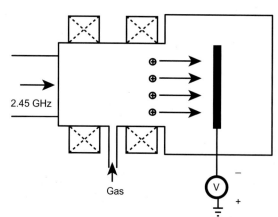

Figure 8–34. Schematic of the diode configuration of PIII for semiconductor implantation.

taking into account that the ions, initially located at the distance $x_0 > s_0$ from the target, reach the target after time interval t:

$$\omega_{\text{pi}} t = \frac{3}{2} \frac{x_0^3}{s_0^3} - \frac{3}{2} + 3\frac{x_0}{s_0}. \tag{8–119}$$

Combination of relations (8–118) and (8–119) actually gives the monotonically decreasing right-hand branch of the $J(t)$ graph in Fig. 8–33. When $t \to \infty$, the ions initially located at the edge of the Child law sheath are coming to the target ($x_0 \to s_{\text{Ch}}$); the implantation current density in this case is determined by the Bohm velocity, $J \to en_p u_{\text{B}}$. When the gas pressure is higher, ion–neutral collisions within the sheath start making a considerable contribution to the PIII process. The ion–neutral collisions reduce implantation energies and make the angular distribution quite broad, which complicates the ion implantation over topography, particularly, within trenches. A detailed analysis of the collisional sheaths in the PIII processes can be found in Vahedi et al. (1991) and Vahedi, Stewart, and Lieberman (1993).

8.6.5. PIII Applications for Processing Semiconductor Materials

Microwave discharges (2.45 GHz) operating under ECR conditions are used for processing of semiconductor materials. These discharges can operate at pressures as low as 0.2 mTorr, providing the relatively high ion densities (10^{10}–10^{11} cm^{-3}) required for sufficiently high implantation current densities. To accelerate the ions, the substrate is biased by pulsed negative voltages (amplitude 2–30 kV, pulse duration 1–3 μs). Two different PIII system configurations are applied for the semiconductor processing applications: diode configurations and triode configurations. The **diode configuration** is shown in Fig. 8–34. Gases like argon, nitrogen, oxygen, water vapor, and BF$_3$ are introduced into the diode systems to be converted into ions for direct implantation. The diode configuration can be applied when the dopant gaseous sources are available. In particular, the PIII diode configuration is successfully applied for such semiconductor doping applications as shallow junction formation and conformal doping of non-planar device structures. The **triode configuration** involves adding an additional negatively biased target controlled by a separate power supply. Atoms from the target are sputtered into the plasma by the carrier-gas plasma ions. Some of the atoms emitted from the intermediate target are ionized in the plasma and, therefore, can be implanted into the major substrate. A main advantage of the triode system is the possibility of implanting components when dopant gaseous sources are not available (Fig. 8–35).

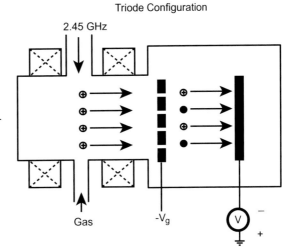

Figure 8–35. Schematic of the triode configuration of PIII for semiconductor implantation.

Between different specific PIII applications for processing of semiconductor materials, we can point out the application of boron implantation in formation of sub-100-nm p^+/n junctions. PIII technology, characterized by high ion fluxes and low implantation energies, is very suitable to form ultrashallow junctions. Prior to the 2 kV BF_3 implantation, the silicon is pre-amorphized with 4 kV SiF_4 implantation. The technology permits one to fabricate the junctions with a total leakage current density lower than 30 nA/cm² at a reverse bias of –5 V (Pico, Lieberman, & Cheung, 1992). Selective metal plating is another PIII application in IC fabrication. PIII has been used, in particular, for selective and planarized plating of copper interconnects using palladium seeding. The process is conducted in the triode configuration. A palladium sputtering target is immersed in plasma and independently negatively biased. The sputtered palladium atoms then provide deposition, while Ar^+ and Pd^+ ions stimulate penetration of the deposited palladium into the substrate. The third specific PIII application in semiconductor processes that we are going to mention is related to conformal doping of silicon trenches. Conformal BF_3 doping of high-aspect-ratio silicon trenches has been achieved using angular divergence of implanting ions at relatively elevated pressures. Silicon trenches (1 μm wide, 5 μm deep) were implanted at a pressure of 5 mTorr and voltage of 10 kV. A p^+/n junction with relatively uniform depth on the top, bottom, and sidewalls of the trench was produced using this approach.

8.6.6. PIII Applications for Modifying Metallurgical Surfaces: Plasma Source Ion Implantation

Plasma-immersion ion implantation (PIII) technology, applied to the treatment of metallurgical surfaces, is usually referred to as plasma source ion implantation (PSII). The PSII objective is to improve wear, hardness, and corrosion resistance of metal surfaces. As with PIII, the PSII approach can be used to treat non-planar targets such as tools and dies (Conrad et al., 1990; Redsten et al., 1992). In contrast to the PIII systems in microelectronics, metallurgical PSII technology can use hot-filament plasma sources because contamination is not as serious an issue in this case. A typical PSII process is done by immersion of a treated target in nitrogen plasma with a density of about $5 \cdot 10^9$ cm^{-3}. A series of negative voltage pulses (50 kV, duration 10 μs, pulse frequency 100 kHz) are applied to the target under treatment. The total PSII treatment time of a target varies depending on the application, from minutes to hours. PSII for metallurgy can be much less sophisticated than that for semiconductor processing applications. Plasma can be generated by a hot filament, in particular by a hot tungsten filament. The hot filament is inserted into a chamber with background gas at a pressure of about 0.1 mTorr and negatively biased at −(100–300) V. The hot filament emits electrons, which are accelerated in the filament's sheath and then ionize the background gas, forming plasma. Multipole magnets are usually required on the surface of the implantation chamber to confine the primary electrons. Dynamics of the hot-filament plasma sources applied for ion implantation (PSII) can be found in Leung et al. (1976).

8.7. MICROARC (ELECTROLYTIC-SPARK) OXIDATION COATING AND OTHER MICRODISCHARGE SURFACE PROCESSING SYSTEMS

8.7.1. Microarc (Electrolytic-Spark) Oxidation Coating: General Features

The microarc oxidation coating is related to microdischarge plasma chemistry not in the gas phase but in the liquid phase, or to be more exact – in an electrolyte. That is why the coating process is often called electrolytic-spark oxidation. The protective coatings can be arranged in this way on articles of certain metals like Al, Ti, Ta, Nb, Zr, Be, and their alloys. The main idea of electrolytic-spark oxidation is to sustain numerous electric sparks (or microarcs) at the surface of a workpiece placed in an electrolyte, which results in buildup of a protective oxide layer. Oxide coating deposition is provided by plasma-chemical processes in the microarc discharges. Generation of the microdischarges at the electrolyte–anode interface is induced by applying pulsed (or sometimes DC) voltage exceeding 100 V to the electrolytic system (Kharitonov, Gogish-Klushin, & Novikov, 1987; Kharitonov et al., 1988; Lukiyanchuk et al., 2002; Yerokhin et al., 2003; Meyer, Gorges, & Kreisel, 2004; Long et al., 2005). The method allows coatings to be formed with a set of properties that cannot be obtained by any other technique: uniform protective characteristics of oxide films with thickness 10–100 μm, high lifetime corrosion and wear resistance, adhesion over 5000 N/cm^2, and high thermal stability. Microarc oxidation coatings were initially designed especially for applications in nuclear reactors. Besides that the coatings can be applied in systems for radioactive waste reprocessing and burial, in dielectric substrates for high-voltage elements, in ceramic coatings on medical implants, in heat- and radiation-resistant coatings on ultralight beryllium alloys in aerospace systems, and so forth. Consider an example of alumina (Al_2O_3) protective-coating formation on the surface of aluminum placed as an anode in a concentrated sulfuric acid electrolytic system with pulsed overvoltage exceeding 300 V (Kharitonov et al., 1987, 1988). Deposition of the Al_2O_3 oxide film on the surface of the aluminum anode takes place in a series of microdischarges (also called microarcs) at the anode–electrolyte interface.

8.7.2. Major Characteristics of the Microarc (Electrolytic-Spark) Oxidation Process

Major characteristics of the process of microarc oxidation coating of aluminum placed as an anode in a concentrated sulfuric acid electrolytic system (Kharitonov et al., 1987, 1988) can be summarized in the following ten statements:

1. Light emission from the electrolyte–anode interface follows each voltage pulse (above 300 V in the experiments under consideration) with some delay and is well correlated with the current pulse.
2. The duration of the light emission pulses and related current pulses (1–3 mA) are 3–5 μs in the experiments under consideration.
3. When the amplitude of current pulses is stabilized, the required voltage V (in volts) grows with the film thickness d (in microns) in accordance with the following empirical formula:

$$V = 160 + \sqrt{1.51 \cdot 10^3 \cdot d}.$$ (8–120)

4. Specific resistance of the deposited coating ρ (in Mohm $\cdot$ m) grows linearly with the film thickness d (in microns) in accordance with the following empirical formula:

$$\rho = 18.6 + 0.24 \cdot d. \qquad (8-121)$$

5. The total electric charge crossing the oxide film corresponds to the electrochemical oxidation:

$$3O^{-2} + 2Al^{+3} \rightarrow Al_2O_5. \qquad (8-122)$$

6. The amount and composition of gas produced on the anode do not depend on the film thickness. The amount of the produced gas is proportional to the total transferred charge. The gas produced on the anode contains 99% oxygen. For each oxygen atom used to form the Al_2O_3 coating, 2.6–2.8 oxygen atoms are converted into molecular oxygen and released to exhaust anode gases.

7. Formation of the oxide film on the anode is accompanied by formation of passive colloid sulfur in the electrolyte.

8. Each current pulse corresponds to the formation of a bright microdischarge (microarc) at the anode–electrolyte interface. The microdischarge has a semispherical shape with radius about 100 µm. Each microdischarge semisphere has a much smaller core or nucleus with much higher brightness.

9. Spectral measurements of the microdischarge radiation indicate that the radiation can be interpreted as the superposition of two blackbody spectra: one corresponding to a relatively "cold" temperature of 1600–2200 K (the semispherical microdischarge), and another corresponding to relatively "hot" temperature, 6800–9500 K (the hot nucleus). The effective radiative surface area of the hot nucleus is $2 \cdot 10^4$–$5 \cdot 10^5$ times smaller than that of the relatively "cold" microdischarge.

10. Technological characteristics of the oxide films, in particular adhesive, protective, and dielectric properties, are determined by the porosity of the films. Decreasing porosity and complete porosity elimination, which improves film characteristics, can be achieved by optimization of the amplitude, duration, and frequency of the voltage pulses, as well as the current density.

8.7.3. Mechanism of Microarc (Electrolytic-Spark) Oxidation Coating of Aluminum in Sulfuric Acid

Consider the qualitative steps of the quite sophisticated mechanism of microarc oxidation. When a voltage pulse is applied, it induces current and a redistribution of electric field in the electrolyte. A dielectric film already deposited on the anode leads to the formation of an electric double layer near the anode surface. Formation of the electric double layer results in localization of the major portion of the applied electric field in the anode vicinity. At some point in time, the electric field near the anode becomes high enough for thermal breakdown of the dielectric film and formation of a narrow conductive channel in the dielectric film. The conductive channel is then heated up by electric current to temperatures of about 10,000 K. Thermal plasma generated in the channel is rapidly expanded and ejected from the narrow conductive channel into the electrolyte, forming a "plasma bubble." This plasma bubble is characterized by a very high temperature gradient and electric field. The energy required to form and sustain this bubble is provided by Joule heating in the narrow conductive channel in the dielectric film. High current density in the channel is limited by conductivity in the relatively cold plasma bubble, where the temperature is about 2000 K. Electric conductivity in the bubble is determined not by electrons but by negative oxygen-containing ions. Further expansion of the plasma bubble leads to its cooling and to a decrease in the density of charged

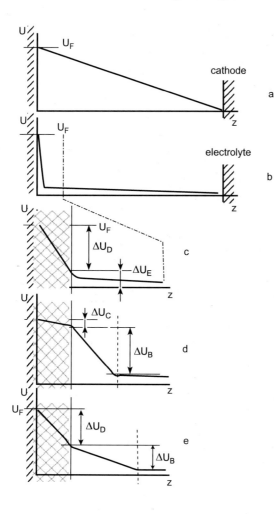

Figure 8–36. Redistribution of voltage in a cell on different phases of an elementary discharge: (a) at the initial moment of applying electric field; (b,c) at the end of the pre-breakdown stage (ΔU_d is the voltage drop across the dielectric film; ΔU_e is the voltage drop across the electrolyte); (d) at the moment of the developed discharge (ΔU_c is the voltage drop in the channel; ΔU_c is the voltage drop in the bubble); (e) at the final stage of the discharge and condensation of reaction products.

particles; it also results in a decrease of electric current and in a significant reduction of Joule heating in the conductive channel in the dielectric film, and finally in a cooling down of the channel itself. Products of the plasma-chemical process (γ-Al_2O_3) are condensed on the bottom and sidewalls of the channel. The oxide film grows mostly inside the substrate material. The described sequence of events takes place during a single voltage pulse. When the next voltage pulse is applied to the electrolytic system, new thermal breakdown and new microarcs occur at another surface spot where the dielectric film is less developed. This effect provides uniformity of the microarc oxidation coating.

8.7.4. Breakdown of Oxide Film and Starting Microarc Discharge

A voltage pulse applied to the electrolyte (concentrated sulfuric acid) stimulates the re-arrangement of electric charges there. A layer of adsorbed HSO_4^- ions is formed on the Al_2O_3 dielectric barrier previously formed on the aluminum anode. Rearrangement of electric charges results in a redistribution of the electric field in the electrolyte, which is illustrated in Fig. 8–36. Most of the voltage change and the highest electric field become concentrated at the oxide film. The voltage drop in the electrolyte decreases to about 30 V, which is related to the electrolyte resistance and leakage current across the thin film. Before breakdown of the thin dielectric film, the system can be represented by an equivalent electric circuit, shown in Fig. 8–37. The total current in the electrolytic cell includes a leakage current through the film as well as current charging of the electric double layer related to the adsorption of

8.7. Microarc (Electrolytic-Spark) Oxidation Coating

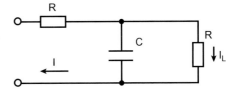

Figure 8–37. Equivalent electric circuit for the cell at the pre-breakdown phase.

HSO_4^- ions on the anode. The evolution of current and voltage in the pre-breakdown stage of the process is shown in Fig. 8–38. Although the total voltage remains almost constant, the electric field in the thin film grows rapidly toward breakdown conditions. Thermal breakdown of the dielectric film occurs at electric fields of 10^5–10^6 V/cm (Tareev, 1978; Poplavko, 1980; Kharitonov et al., 1988). The physical nature of the thermal breakdown is related to the thermal instability of the local leakage currents through the dielectric film with respect to Joule overheating. If the leakage current is slightly higher at one point, Joule heating and, hence, temperature increases. The temperature increase results in a growth of local conductivity and the leakage current closing the loop of thermal instability. An exponential temperature growth to several thousand degrees at a local point leads to the formation of a narrow plasma channel in the dielectric, which determines the breakdown. The phenomenon is usually referred to as **thermal breakdown of dielectrics** (Skanavi, 1958). The thermal breakdown is a critical phenomenon that takes place at applied voltages exceeding some threshold ($V \geq V_{\text{th.br.}}$), when heat release in the conductive channel cannot be compensated by heat transfer. The critical thermal breakdown voltage can be calculated according to Skanavi (1958) and Kharitonov et al. (1988) as the function of initial temperature (T) and thickness of the dielectric coating:

$$V_{\text{th.br.}} = A(T) \cdot \sqrt{\rho \cdot \Phi(c)}. \tag{8-123}$$

In this relation, ρ is the specific electric resistance of the film at room temperature, and $A(T)$ is a function describing the temperature dependence of the critical thermal breakdown voltage:

$$A(T) = \sqrt{\frac{33.6 \cdot \kappa}{\alpha}} \exp\left(-\frac{\alpha T}{2}\right); \tag{8-124}$$

κ is the thermal conductivity of the coating material; and the coefficient α determines the exponential growth of the electric conductivity of the coating material with temperature. The initial temperature of the coating before the breakdown is constant, and the breakdown voltage (8–123) grows due to its dependence on the thickness of the coating, d. The dependence is given in (8–123) by a special function $\Phi(c)$ (Skanavi, 1958; Kharitonov et al.,

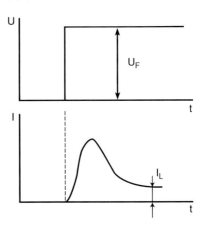

Figure 8–38. Evolution of current and voltage at the pre-breakdown phase.

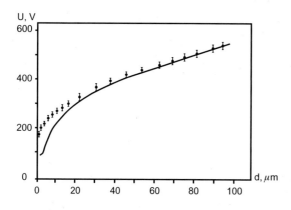

Figure 8–39. Voltage required for the thermal breakdown of coating (solid curve) and experimental values of the breakdown voltage (points with error bars) as a function of the film thickness.

1988), where the thickness of the dielectric oxide coating, d, is represented by the parameter $c = d(\lambda/\kappa)$ (λ is the coefficient of heat transfer from the coating):

$$\Phi(c) = \frac{\beta_0(c)}{\cosh \beta_0(c)} \exp\left(-\frac{\beta_0(c)\tanh\beta_0(c)}{c}\right). \tag{8–125}$$

The parameter $\beta_0(c)$ characterizes the thickness of the dielectric film, d, and is related to c as follows:

$$c = \frac{\beta_0^2 + \beta_0 \sinh\beta_0 \cosh\beta_0}{(1 - \beta_0 \tanh\beta_0)\cosh\beta_0}. \tag{8–126}$$

The thermal breakdown voltage dependence on the coating thickness is presented in Fig. 8–39. The voltage pulse duration is comparable with the duration of a microdischarge (in the range of microseconds), which explains some discrepancy between modeling and experimental results when the coating is very thin. Thus, while coating thickness increases, the required breakdown voltage grows as well. Each new breakdown occurs in the dielectrically weakest spot of the coating where electric resistance is the lowest, which is usually related to defects in the coating structure or morphology. It is interesting that this effect leads to a "cure" of the defects, which in general results in high-quality electrolytic-spark coatings.

8.7.5. Microarc Discharge Plasma Chemistry of Oxide Coating Deposition on Aluminum in Concentrated Sulfuric Acid Electrolyte

Breakdown of the Al_2O_3 film in the microarc aluminum oxidation is a fast process requiring about 0.1 μs. The breakdown leads to the formation of a narrow plasma channel with a temperature of about 10,000 K in the coating material. The high temperatures result in intensive evaporation of sidewalls (Al_2O_3) and bottom (Al) of the channel. The metal evaporation from the bottom of the channel is much more significant. Pressure in the channel grows to thousands of atmospheres, which leads to ejection of the hot plasma into the electrolyte, forming a "plasma bubble" with a temperature of 1000–2000 K (Kharitonov et al., 1988). High electric conductivity in the plasma channel leads to shifting of the high electric fields from the channel to the "bubble" (see Fig. 8–36). These electric fields provide drift of negatively charged particles from the bubble into the channel. Taking into account that temperature in the plasma bubble is not sufficient for thermal ionization and that the electric field there ($2–5 \cdot 10^4$ V/cm) is not sufficient for direct electron impact ionization, the negative oxygen-containing ions from the electrolyte make a major contribution in negative charge transfer from the bubble into the channel. Molar fractions of the major neutral components of the "bubble" plasma are H_2O, 40.7%; SO_2, 36.7%; O_2, 18.3%; SO_3, 4%; H_2SO_4, 0.12%; and OH, 0.08%. The ionization degree in the bubble plasma

8.7. Microarc (Electrolytic-Spark) Oxidation Coating

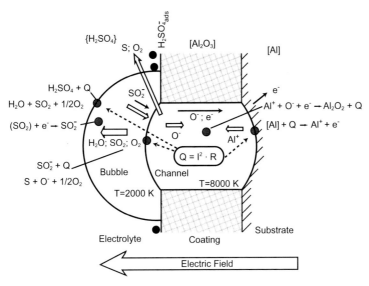

Figure 8–40. Major processes during the electrolytic-spark microdischarges: dashed arrows, thermal fluxes; hollow arrows, mass transfer; solid arrows, and charge transfer.

is quite low: 1–$2 \cdot 10^{-8}$. The major negative ion is SO_2^-; its density exceeds that of free electrons by about 10^4. Major components in the thin and hot plasma channel are O atoms, 60.4%; Al^+ ions, 24.6%; and S atoms, 14.2%. Mass, electric charge, and energy transfer processes taking place in bubble and plasma microchannels are illustrated in Fig. 8–40. Expansion of the plasma bubble leads to its cooling down and to a decrease of the flux of charged particles into the microchannel, which finally results in a temperature decrease in the channel. The bubble shrinks, reaction products move into the plasma channel, and alumina condenses at the bottom and walls of the channel. Because of fast quenching of products in the microchannel, the alumina is produced in the form of a metastable-phase γ-Al_2O_3. While γ-Al_2O_3 fills out the microchannel, stable gases O_2, SO_2, and SO_3 are formed near the electrode.

8.7.6. Direct Micropatterning and Microfabrication in Atmospheric-Pressure Microdischarges

The most widely used technique for creating microstructures in microelectromechanical systems (MEMS) and ICS is photolithography (Madou, 2002). Photolithography involves the application, masking, exposure, and development of photoresists to make polymer micropatterns. Then chemical, physical, or plasma deposition and etching techniques are applied to form metal, crystal, or ceramic microstructures. The application of non-equilibrium plasma in photolithography for creating microstructures in MEMS and ICS was mentioned. The photolithography-based approach is multi-stepped, and the relevant facilities are costly. The resolution of such patterns depends on the facilities used and varies from about 80 nm in high-end IC manufacture to about 10 μm in low-end IC and MEMS applications. Therefore, attention is focused on the application of non-thermal microplasmas for creating microstructures. Inherently small sizes of microdischarges (about 10 μm and below) permit micropatterning with high resolution determined by these sizes. Micropatterning is direct in this case and doesn't require the multi-step procedures that are conventional for photolithography. The microdischarges can be arranged in a special array for simultaneous treatment of large surfaces. An important advantage of microdischarges applied for micropatterning

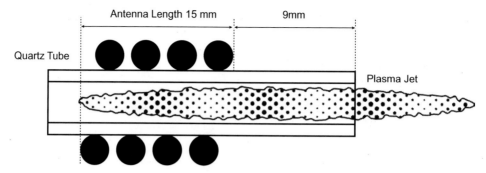

Figure 8–41. General schematic of the atmospheric-pressure microplasma jet etcher.

and microfabrication is related to the possibility of their effective operation at atmospheric pressure (see Section 4.6), which is due to intensive cooling of the micron-size discharges. Advantages of the microplasma technique for micropatterning and microfabrication over conventional widely used photolithography are that it is a single-step, direct-write, and lower-cost atmospheric-pressure process. The use of microplasma for patterning is especially suitable for rapid prototyping of structures.

Micropatterning and microfabrication can be performed using different types of atmospheric-pressure microdischarges. The most straightforward example for the treatment of conductive substrates is DC normal microglow discharge (Staack et al., 2005; see Section 4.6). The use of microwave microplasmas is described by Hopwood and Iza (2004), Iza and Hopwood (2004), and Bilgic (2000). A miniaturized ICP source is described, in particular, by Minayeva and Hopwood (2003), Iza and Hopwood (2003), and Hopwood and Iza (2004). Other microdischarges, developed for micropatterning, microfabrication, and other applications, are presented, in particular, by Okumura and Saitoh (2004), Okumura et al. (2004), Karanassios (2005), Gianchandani and Wilson (2004), and Oehrlein, Hua, and Stolz (2005). All plasma processing technologies, considered earlier regarding treatment of large surfaces (etching, deposition, and surface modification), can be done on a microscale using relevant microdischarge systems.

8.7.7. Microetching, Microdeposition, and Microsurface Modification by Atmospheric-Pressure Microplasma Discharges

Effective **microetching** of Si wafers using a microscale RF-ICP discharge was demonstrated by Ichiki, Taura, and Horiike (2004). The miniaturized inductively coupled atmospheric-pressure thermal plasma was generated at very high frequency (100 MHz) in a 1-mm-diameter discharge tube. The tube was provided by a fine nozzle (100 μm) at one end and a three-turn solenoidal antenna wound around it. A schematic of the microetcher is shown in

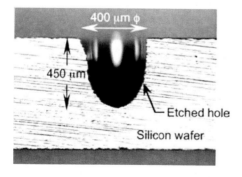

Figure 8–42. Silicon wafer etched using the microplasma etcher for 1 min. VHF (very-high-frequency) power, substrate bias power, and Ar–SF$_6$ flow rates are: 150 W, 4 W, 200 sccm, and 100 sccm, respectively.

8.7. Microarc (Electrolytic-Spark) Oxidation Coating

Figure 8–43. Microdischarge-stimulated oxidation of silicon: the oxide patterns are formed in the same manner as the changes in wettability (motion of the substrate); the difference is that to form the oxide the currents are larger and the speed of the substrate motion is slower.

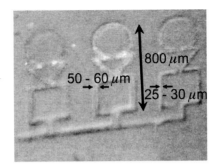

Fig. 8–41. The electron density in the microplasma jet in Ar is 10^{14}–10^{15} cm^{-3}. Different halogen-containing gases are added into the downstream portion of the argon plasma jet for microetching. Achieved etch rates are high in the microetcher: Si wafer etching is characterized by etch rates up to 4000 μm/min; etch rates of fused silica glass wafers are up to 14 μm/min. An Si wafer after 1 min of etching in the microdischarge is shown in Fig. 8–42. The microplasma power is 150 W, the substrate bias power is 4 W, the Ar flow rate is 200 cm^3/min, and the SF$_6$ flow rate is 100 cm^3/min. The diameter of the hole directly microetched in the silicon wafer shown in the figure is 400 μm.

Microdeposition can be illustrated by two processes in the DC normal microglow discharge at atmospheric pressure (Staack et al., 2005a,b, see Section 4.6). First, a micropattern was formed on a silicon substrate (see Fig. 8–43; Staack, 2008) by using microplasma in ambient air, which results in oxidation of the silicon. The figure shows a microscope image of the pattern created by suspending a thin-wire (10 μm) tungsten electrode about 20 μm over the silicon substrate. The microplasma (about 20 μm in diameter) is ignited between the substrate and the wire electrode. The substrate is moved using computer-controlled stepper motors in a pattern dictated by an AutoCAD file, which results in the creation of the pattern shown in Fig. 8–43. The resolution of the microplasma-generated patterns is approximately the size of the discharge.

The same DC normal microglow discharge (Staack et al., 2005a,b) was applied for PECVD of diamond-like carbon (DLC) coatings. In contrast to conventional plasma techniques of PECVD of DLC, application of a microplasma permits deposition at atmospheric pressure and allows deposition rates as fast as microns per minute. The process is done in a microglow discharge in atmospheric-pressure hydrogen with a 2% admixture of methane. An image of the discharge is shown in Fig. 8–44 (Staack, 2008). Higher yields of DLC are observed at higher substrate temperatures and lower methane concentrations in the mixture.

Figure 8–44. DC microdischarge applied for microdeposition.

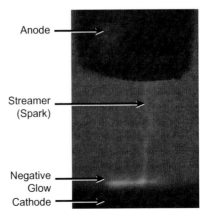

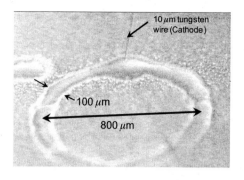

Figure 8–45. Microdischarge-stimulated wettability control: droplets are condensed on the silicon substrate surface from the breath; pattern is made by the substrate motion with X–Y positioning resolution about 10 μm.

The DC normal microglow discharge (Staack et al., 2005a,b) was also used to demonstrate application of microplasma for the **microscale differential wettability patterning**. Because non-thermal plasma is effective in improving of wettability of different surfaces (see Section 8.4.8, as well as Chapter 9), microplasma can be applied for differential wettability patterning at the microscale. The plasma-stimulated differential wettability patterning permits further microdeposition using conventional liquid-phase techniques without any masking and application of photolithography. Fig. 8–45 (Staack, 2008) illustrates this approach. Silicon substrates were also treated in microplasma in atmospheric-pressure air. The substrate was moved by computer-controlled stepper motors significantly faster in this case; therefore, a major change in the silicon surface was related to improved wettability. Spraying microscale water droplets around the created differential wettability pattern led to the development of the water deposition pattern shown in Fig. 8–45. The resolution of the microplasma-generated patterns in this case is larger than the size of the microdischarge. This is probably related to diffusion of the active plasma species, which improve wettability, from the microdischarge zone into the surrounding areas.

8.8. PLASMA NANOTECHNOLOGIES: NANOPARTICLES AND DUSTY PLASMAS

8.8.1. Nanoparticles in Plasma: Kinetics of Dusty Plasma Formation in Low-Pressure Silane Discharges

Plasma is one of the most effective ways to synthesize ultrafine particles with nanoscale dimensions of many different substances, especially advanced ceramics (Girshick, 2005). Some of these processes were discussed in Chapter 7. On the other hand, starting around 1990, interest began to grow in the problem of unwanted nucleation of contaminant particles in low-pressure semiconductor processing plasmas, and attention was directed to the formation and growth of molecular clusters and nano- and microscale particles in plasma. Plasma physics and chemistry are essentially coupled in such dusty plasma systems (Watanabe et al., 1988; Bouchoule, 1993, 1999; Garscadden, 1994). Nanoparticle formation and the behavior of nanoparticles in plasma has been mostly investigated with regard to the contamination of plasma reactors used for etching, sputtering, and PECVD. Therefore, formation of nanoparticles in the low-pressure glow or RF discharges in silane (SiH$_4$, or SiH$_4$ – Ar mixture) has been one of the processes studied most (Spears, Kampf, & Robinson, 1988; Selwyn, Heidenreich, & Haller, 1990; Bouchoule et al., 1991, 1992, 1993; Howling, Hollenstein, & Paris, 1991; Jellum, Graves, & Daugherty, 1991). Detection and dynamics of nanoparticles of 10 nm or larger in silane discharge were the focus of intensive research. Such particles are electrostatically trapped in the plasma bulk and significantly affect the discharge behavior (Sommerer et al., 1991; Barnes et al., 1992; Boeuf, 1992; Daugherty et al., 1992, 1993a,b). Fundamental phenomena take place in the beginning of the process, starting from the initial dust-free silane discharge. Small deviations in plasma parameters in this period (temperature, pressure, electron density) could essentially change cluster growth and subsequent discharge behavior (Bohm & Perrin, 1991; Perrin, 1991; Bouchoule & Boufendi, 1993). Experimental data concerning super-small (2 nm) particle growth and negative ion kinetics (up to 500 or 1000 amu) demonstrated the main role of negative ion clusters in particle growth (Choi & Kushner, 1993; Hollenstein et al., 1993; Howling, Dorier, & Hollenstein, 1993a; Howling et al., 1993b; Perrin, 1993; Perrin et al., 1993; Boufendi & Bouchoule, 1994). Thus, the initial phase of formation and growth of the dust particles in a low-pressure silane discharge is a homogeneous process, stimulated by fast silane molecules and radical reactions with negative ion clusters.

The first particle generation appears monodispersed with a crystallite size of about 2 nm. Due to a selective trapping effect, the concentration of these crystallites increases up to a critical value, where a fast coagulation process (like a phase transition) takes place, leading to the formation of larger (50 nm) dust particles. During the coagulation, when aggregate size exceeds about 6 nm, another critical phenomenon, the so-called α–γ transition, takes place with a strong decrease in electron concentration and significant increase in their energy. A small increase of the gas temperature prolongs the induction period required to achieve the critical phenomena of agglomeration and the α–γ transition. The low-pressure RF discharge for observation of dusty plasma formation has been investigated in the Boufendi-Bouchoule experiments in a grounded cylindrical box (13 cm inner diameter) equipped with a shower-type RF-powered electrode. A grid was used as the bottom of the chamber to allow a vertical laminar flow in the discharge box. Typically, the Ar flow was 30 sccm; the silane flow was 1.2 sccm; the total pressure was 117 mTorr (so the gas density, $4 \cdot 10^{15}$ cm^{-3}; silane, $1.6 \cdot 10^{14}$ cm^{-3}); the gas residence time in the discharge was 150 ms; and RF power was 10 W. The discharge has been surrounded by a cylindrical oven to vary the gas temperature from ambient up to 200°C. The experiments show that the first particle size distribution is monodispersed with a diameter about 2 nm, which does not depend on temperature. The appearance time of the first-generation particles is less than 5 ms. The particle growth proceeds through the successive steps of fast super-small 2-nm-particle formation and the growth of their concentration up to the critical value of about 10^{10}–10^{11} cm^{-3}, when new particle formation terminates and the formation of aggregates with diameters of up to 50 nm begins through coagulation. During the initial discharge phase (until the α–γ transition: 0.5 s for room temperature and several seconds for 400 K), the electron temperature remains at about 2 eV, the electron concentration is about $3 \cdot 10^9$ cm^{-3}, the positive ion concentration is $4 \cdot 10^9$ cm^{-3}, and the negative ion concentration is 10^9 cm^{-3}. After the α–γ transition, the electron temperature increases up to 8 eV while the electron concentration decreases 10-fold and the positive ion concentration increases by 2-fold (Boufendi, 1994; Boufendi et al., 1994). The Boufendi-Bouchoule experiments show that the critical density of supersmall particles before coagulation does not depend on temperature. The induction time before coagulation is a highly sensitive function of temperature: from 150 ms at 300 K, it increases more than 10-fold when heated to only 400 K. For a temperature of 400 K, the time required to increase the supersmall neutral particle density is much longer (more than 10-fold) than the gas residence time in the discharge, which can be explained by the neutral particle trapping phenomenon. Particle density during the coagulation decreases, while the average particle radius increases. The total mass of dust in the plasma remains almost constant during the coagulation.

8.8.2. Formation of Nanoparticles in Silane: Plasma Chemistry of Birth and Catastrophic Evolution

Dust particle formation and growth in SiH$_4$/Ar low-pressure discharge can be divided into four steps: growth of supersmall particles from molecular species, and three successive

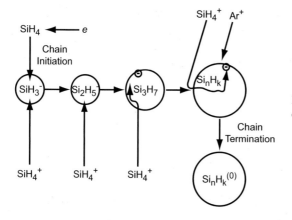

Figure 8–46. Mechanism of the first generation of supersmall particle growth.

catastrophic events – selective trapping, fast coagulation, and finally α–γ transition (Fridman et al., 1996). A scheme for the formation of the **first supersmall particle generation** is presented in Fig. 8–46. It begins with SiH_3^- negative ion formation by dissociative attachment. Non-dissociative three-body attachment to SiH_3 radicals is a complementary way ($e + SiH_3 = (SiH_3^-)^*$, $(SiH_3^-)^* + M = SiH_3^- + M$). Subsequent cluster growth is due to ion–molecular reactions: $SiH_3^- + SiH_4 = Si_2H_5^- + H_2$, $Si2H_5^- + SiH_4 = Si_3H_7^- + H_2$. The chain of reactions can be accelerated by vibrational excitation of molecules. The typical reaction time is about 0.1 ms and is much faster than ion–ion recombination (1–3 ms), which determines the termination of the chain. As the negative cluster size increases, the probability of reactions with the vibrationally excited molecules decreases because of an effect of vibrational–translational (VT) relaxation on the cluster surface. When the particle size reaches a critical value (about 2 nm at room temperature) the chain reaction of cluster growth becomes much slower and is finally stopped by the ion–ion recombination process. The typical time for 2 nm particle formation by this mechanism is about 1 ms at room temperature. A critical temperature effect on particle growth is partially due to VT relaxation, which depends exponentially on translational gas temperature according to the Landau-Teller effect (Section 2.6.2). Even a small increase of gas temperature results in a reduction of the vibrational excitation level and decelerates the cluster growth.

The **selective trapping effect** of neutral particles is illustrated in Fig. 8–47. Trapping of negatively charged particles is due to repelling forces in the sheaths when the particles reach the plasma boundary. For supersmall particles, however, the electron attachment time (100 ms) is two orders of magnitude longer than the ion–ion recombination; therefore, most

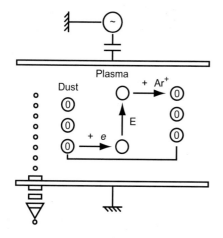

Figure 8–47. Physical scheme of electric trapping of "neutral" particles in plasma.

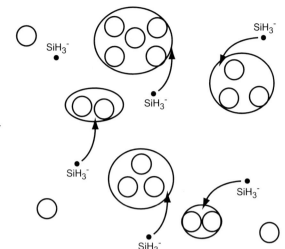

Figure 8–48. Mechanism of critical phenomenon of particle coagulation.

of the particles are neutral. Thus, "trapping of neutral particles in the electric field" should take place to allow the particle density to reach the critical value sufficient for effective coagulation. The trapping of neutral particles can be explained by taking into account that the electron attachment time for 2-nm particles is shorter than residence time. Therefore, each neutral particle is charged at least once and, hence, quickly trapped by the strong electric field before recombination. The rate coefficient of electron attachment increases strongly with particle size. Particles smaller than 2 nm have an attachment time much longer than their residence time; hence, they are not charged even once and cannot be trapped. Only "big particles" (> 2 nm) are selectively trapped and survive. This first "catastrophe" associated with small particles explains why the first-generation particles appear with the well-defined size of crystallites; it also explains the strong effect of temperature on dust production. A small temperature increase results in a significant reduction of the cluster growth velocity and initial cluster size. It leads to losses of the initial neutral particles with gas flow and determines the long delay of coagulation, $\alpha-\gamma$ transition, and dust production process in general.

The **fast coagulation phenomenon** occurs when the density of monodispersed 2-nm particles reaches a critical value of 10^{10}–10^{11} cm^{-3} (see Fig. 8–48). At such densities, attachment of small negative ions like SiH_3^- to 2-nm particles becomes faster than their chain reaction to generate new particles. New chains of dust formation become suppressed and the total particle mass remains constant. During this fast coagulation, mass increase by "surface deposition" is negligible. When such a particle density is reached, the probability of multi-body interaction increases. The aggregate formation rate constant increases drastically, and the coagulation looks like a critical (catastrophic) phenomenon of phase transition. The induction time before coagulation is 200 ms for room temperature and much longer for 400 K. It could be explained by taking into account the selective trapping effect. The trapped particle production rate for 400 K is much slower than that for room temperature, but the critical particle density remains the same.

The **critical phenomenon of fast changing of discharge parameters ($\alpha-\gamma$ transition)** occurs during coagulation when the particle size increases and density decreases. Before this critical moment, the electron temperature and other plasma parameters are mainly determined by the balance of volume ionization and electron losses at the walls. The $\alpha-\gamma$ transition occurs when the electron losses on the particle surfaces become greater than on the reactor walls. The electron temperature increases to support the plasma balance and

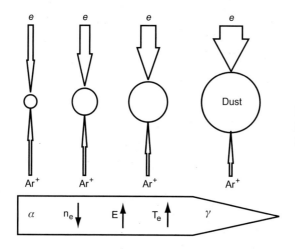

Figure 8–49. Physical scheme of the critical phenomenon of α–γ transition during the dust particle growth.

the electron concentration diminishes dramatically. This fourth step of particle and dusty plasma evolution is the "plasma electron catastrophe" (Fig. 8–49).

The total mass of the particles remains almost constant during coagulation, so the overall particle surface decreases. Thus, the influence of particle surface on plasma parameters is more significant when the specific surface decreases. The effect can be explained by taking into account that the probability of electron attachment to the particles grows exponentially with the particle size. When the particle size exceeds a critical value of about 6 nm, an essential change occurs in the heterogeneous discharge behavior resulting in a significant reduction of electron concentration. Taking into account the acceleration of the electron attachment to these growing particles, it becomes clear that most of them become negatively charged soon after the α–γ transition. The induction time before the α–γ transition about 500 ms for room temperature and more than an order of magnitude longer for 400 K. This strong temperature effect is due to the threshold character of the α–γ transition, which takes place when the particle size exceeds the critical value and, hence, is determined by the strongly temperature-dependent time of the beginning of coagulation. Additional information regarding the formation, growth, and coagulation of particles in different dusty plasmas can be found in Bouchoule (1999); Watanabe, Shiratani, and Koga (2001); Watanabe et al. (2000); Koga et al. (2000); Girshick et al. (1999, 2000); Kim et al. (2002); and Suh et al. (2003).

8.8.3. Critical Phenomena in Dusty Plasma Kinetics: Nucleation of Nanoparticles, Winchester Mechanism, and Growth of First Generation of Negative Ion Clusters

Cluster formation in low-pressure RF silane plasma starts from the formation of negative ions SiH_3^- and their derivatives. For the discharge parameters described earlier, the density of the negative ions can be found from the balance of dissociative attachment and ion–ion recombination:

$$\frac{d[SiH_3^-]}{dt} = k_{ad}n_e[SiH_4] - k_r^{ii}n_i[SiH_3^-], \qquad (8\text{–}127)$$

where the rate coefficient of dissociative attachment is $k_{ad} = 10^{-12}$ cm^3/s for $T_e = 2$ eV, the ion–ion recombination rate coefficient is $k_r^{ii} = 2 \cdot 10^{-7}$ cm^3/s, n_e and n_i are the electron and positive ion densities. Before the α–γ transition, the electron and positive ion densities are close, and the concentration of SiH_3^- (and its derivatives) is about 10^9 cm^{-3}. The steady-state ion balance is established after about 1 ms. Other mechanisms of initial SiH_3^--production, including non-dissociative electron attachment to SiH_3, are discussed by Perrin et al. (1993). In the SiH_4–Ar discharge under consideration, this mechanism probably is not the principal one. Indeed, SiH_3 radical production in this case is due to dissociation by direct electron impact with the rate constant $k_d = 10^{-11}$ cm^{-3}, and losses are due to diffusion to the walls (diffusion coefficient, $D = 3 \cdot 10^3$ cm^2/s; characteristic distance to the walls, $R = 3$ cm):

$$\frac{d[SiH_3]}{dt} = k_D n_e [SiH_4] - \frac{D}{R^2}[SiH_3]. \tag{8–128}$$

The steady state is established after about 3 ms, and the radical density is less than 0.1% from SiH_4. Thus, the dissociative attachment is the main initial source of negative ions in the system.

After initial SiH_3^- formation, the main pathway of cluster growth is the chain of ion–molecular processes, called the **Winchester mechanism** (Section 2.2.6), and is based on the thermodynamic advantage of the ion-cluster growth (2–64) ($A_1^- \rightarrow A_2^- \rightarrow A_3^- \rightarrow \cdots \rightarrow A_n^- \rightarrow$). Electron affinities $EA_1, EA_2, EA_3, \ldots, EA_n$ are generally increasing in sequence to finally reach the work function, which determines tendency for the elementary reactions of cluster growth $A_n^- \rightarrow A_{n+1}^-$ to be exothermic. Exothermic ion–molecular reactions have no activation barrier and are usually very fast; thus, the Winchester mechanism explains effective cluster growth based on ion–molecular processes. Thermal effects of the first four reactions of the silicon cluster growth demonstrate the Winchester mechanism in the case of silane plasma:

$$SiH_3^- + SiH_4 \rightarrow Si_2H_5^- + H_2 - 0.07\,eV, \tag{8–129}$$

$$Si_2H_5^- + SiH_4 \rightarrow Si_3H_7^- + H_2 + 0.07\,eV, \tag{8–130}$$

$$Si_3H_7^- + SiH_4 \rightarrow Si_4H_9^- + H_2 + 0.07\,eV, \tag{8–131}$$

$$Si_4H_9^- + SiH_4 \rightarrow Si_5H_{11}^- + H_2 \pm 0.00\,eV. \tag{8–132}$$

In the case of silane cluster growth, the Winchester mechanism is not so fast compared to ion–ion recombination. For example, the rate coefficient of (8–129) is 10^{-12} cm^3/s and is due to the thermoneutral and even endothermic character of some reactions from the Winchester chain (8–129)–(8–132), which results in an intermolecular energy barrier (Raghavachari, 1992; Reents & Mandich, 1992) for such reactions, even in a bottleneck effect in their kinetics (Mandich & Reents, 1992). The Winchester mechanism can be accelerated by the vibrational energy of the polyatomic molecules (Talrose, Vinogradov, & Larin, 1979; Baronov et al., 1989; Veprek & Veprek-Heijman, 1990, 1991). To estimate the influence of vibrational excitation on cluster growth, consider the vibrational energy balance with SiH_4 vibrational excitation by electron impact ($k_{ev} = 10^{-7}$ cm^3/s), and VT relaxation on molecules ($k_{VT,silane}$), atoms ($k_{VT,Ar}$), and walls (accommodation coefficient P_{VT}; Chesnokov & Panfilov, 1981):

$$k_{ev} n_e [SiH_4] \cdot \hbar\omega = \left(k_{VT,silane}[SiH_4]^2 + k_{VT,Ar} n_0 [SiH_4] + P_{VT}[SiH_4]\frac{D}{R^2} \right)$$

$$\times \left[\frac{\hbar\omega}{\exp(\hbar\omega/T_v) - 1} - \frac{\hbar\omega}{\exp(\hbar\omega/T_0) - 1} \right], \tag{8–133}$$

where n_e and n_0 are electron and neutral gas densities; T_v and T_0 are vibrational and translational temperatures, and $\hbar\omega$ is a vibrational quantum in the approximation of single-mode excitation. Vibrational relaxation on the cluster surface will be especially considered next. Taking into account the Landau–Teller formula for VT relaxation (Section 2.6.2) and activation energy $E_a(N)$ as a function of number of silicon atoms in a cluster, N, the growth rates (8–129)–(8–132) can be expressed as

$$k_{i0}^{\text{initial}}(T_0, N) = k_0 N^{2/3} \exp\left(-\frac{\Lambda + B\dfrac{\Delta T}{3T_{00}^{4/3}}}{\hbar\omega} E_a(N)\right), \qquad (8\text{--}134)$$

where $T_{00} = 300\,\text{K}$ is room temperature, $\Delta T = T_0 - T_{00}$ is the temperature increase, k_0 is the gas-kinetic rate coefficient, B is the Landau-Teller parameter, and Λ is the slightly changing VT factor:

$$\Lambda = \ln \frac{k_{\text{VT,silane}}[\text{SiH}_4] + k_{\text{VT,Ar}} n_0 + P_{\text{VT}} D/R^2}{k_{\text{eV}} n_e}. \qquad (8\text{--}135)$$

The initial cluster growth rate coefficients (8–134) increase with cluster size. When the number of atoms, N, exceeds about 300, the relaxation on the cluster surface (probability $P_{\text{VT}} \approx 0.01$) becomes significant and the rate of negative cluster growth decreases. The number of relaxation-active spots on the cluster surface is $s = (N^{2/3})^2 = N^{2/3}$; therefore, the probability of chemical reaction on the cluster surface must be multiplied as follows (Legasov, Rusanov, & Fridman, 1978):

$$(1 - P_{\text{VT}})^s = \exp(-P_{\text{VT}} N^{2/3}). \qquad (8\text{--}136)$$

The negatively charged cluster growth rate coefficient based on (8–134) and (8–136) can be expressed as

$$k_{i0}(T_0, N) = k_0 N^{2/3} \exp\left[-\frac{\Lambda + \dfrac{B\Delta T}{3T_{00}^{4/3}}}{\hbar\omega} E_a(N) - P_{\text{VT}} N^{2/3}\right]. \qquad (8\text{--}137)$$

This formula demonstrates the strong temperature dependence of the cluster growth rate, as well as the growth limitation for larger clusters $N > P_{\text{VT}}^{-2/3}$. Numerically, the critical N is about 1000 at room temperature (cluster radius about 1 nm); at higher temperatures, the critical size decreases.

8.8.4. Critical Size of Primary Nanoparticles in Silane Plasma

Neglecting ion–ion recombination, the particle size growth in the sequence of ion–molecular reactions (8–129)–(8–132) can be described based on (8–137) by the following equation:

$$\frac{dN}{dt} = k_{i0}^*[\text{SiH}_4] N^{2/3} \exp(-P_{\text{VT}} N^{2/3}), \qquad (8\text{--}138)$$

where $k_{i0}^* = k_{i0}(T_0, N = 1)$ is the rate coefficient of the ion–molecular reactions stimulated by vibrational excitation. Solution of this equation can be expressed using the critical particle size, $N_{\text{cr}} = P_{\text{VT}}^{-3/2}$, and characteristic reaction time $\tau = 1/k_{i0}^*[\text{SiH}_4]$, which is 0.1–0.3 ms at room temperature:

$$t = \tau N_{\text{cr}}^{1/3} \frac{\exp(N/N_{\text{cr}})^{2/3} - 1}{(N/N_{\text{cr}})^{1/3}}. \qquad (8\text{--}139)$$

Initially, when the number of atoms, N, is less than a critical value of 1000 atoms (size about 1 nm), the function $N^{1/3}$ increases linearly with time. This means that the particle

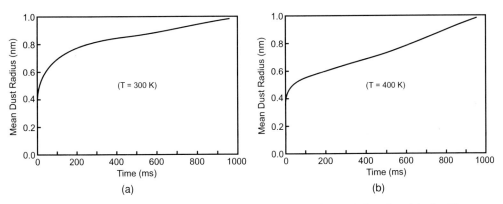

Figure 8–50. Kinetic modeling of initial stage of particle growth; time evolution of particle size. First curve corresponds to temperature 300 K, second curve corresponds to temperature 400 K.

radius, not particle mass and volume, grows linearly with time, and time of formation of a 1000-atom cluster is only 10 times longer than the characteristic time of the first ion–molecular reaction. The particle growth is limited by the critical size of 2 nm ($N_{cr} = 1000$). The formation time of such particles is about 1–3 ms at room temperature. The growth is suppressed then by fast ion–ion recombination. Small changes of temperature accelerate vibrational relaxation and diminish the level of vibrational excitation, which makes the growth process much slower. Figure 8–50 presents calculation results demonstrating the saturation of supersmall particle growth and the strong effect of temperature on particle growth time (Porteous et al., 1994; Fridman et al., 1996).

8.8.5. Critical Phenomenon of Neutral-Particle Trapping in Silane Plasma

The supersmall cluster growth described in the previous section leads to a continuous production of nanoparticles (2 nm for room temperature) with a reaction rate corresponding to that of the initial formation of SiH_3^- by dissociative attachment. Most of these particles are neutral according to the Boufendi-Bouchoule experiments. Ion–ion recombination for such particles ($k_r^{ii} = 3 \cdot 10^{-7}$ cm^3/s) is much faster than electron attachment ($k_a = 3 \cdot 10^{-9}$ cm^3/s). Therefore, the percentage of negatively charged particles is about $k_a / k_r^{ii} = 1\%$. The density of the neutral particles grows during a period longer than the neutral gas residence time in the plasma volume. This means that the trapping effect, well known for negatively charged particles repelled from the walls, takes place here for nanometer-size neutral particles. The neutral particle lifetime in plasma is determined by particle diffusion and drift with gas flow to the walls. The drift lifetime (residence time) does not depend on the particle size and is about 150 ms in the Boufendi-Bouchoule experiments. On the contrary, the diffusion time R^2/D is about 3 ms for molecules and radicals and increases with the number of atoms in a particle proportionally to $N^{7/6}$. The two processes become comparable for N of about 60 and, thus, the main losses of the 2-nm particle are due to gas flow.

Neutral particle trapping is explained by taking into account that the electron attachment time for 2-nm particles ($1/k_a n_e$, about 100 ms) becomes shorter than the residence time (150 ms). Therefore, each neutral particle is charged at least once during the residence time. Even a short time interval of being negatively charged before recombination (which happens after about 1 ms) is sufficient for particle trapping due to repulsion from the walls back into the plasma volume. The trapping of neutral particles is size-selective because the rate of electron attachment to a particle grows strongly with the particle size (r). When clusters are smaller than 2 nm, the attachment time is longer than the residence time. Chances for such clusters to be charged even once are low; hence, they cannot be trapped. The electron attachment probability is low for small particles due to the adiabatic nature of the energy transfer from electrons to the cluster vibrations ($\hbar\omega$; Fridman & Kennedy, 2004):

$$k_\mathrm{a} = \pi r^2 \sqrt{\frac{8T_\mathrm{e}}{\pi m}} \exp\left(-\frac{e^2}{4\pi\varepsilon_0\hbar\omega r}\right). \tag{8–140}$$

For a 5-nm particle, $k_\mathrm{a} = 10^{-7}\,\mathrm{cm^3/s}$; for a 2-nm particle, $k_\mathrm{a} = 3\,\,10^{-9}\,\mathrm{cm^3/s}$. Neutral particle trapping requires at least one electron attachment during the residence time τ_R:

$$k_\mathrm{a}(r)n_\mathrm{e}\tau_\mathrm{R} = n_\mathrm{e}\tau_\mathrm{R}\pi r^2 \sqrt{\frac{8T_\mathrm{e}}{\pi m}} \exp\left(-\frac{e^2}{4\pi\varepsilon_0\hbar\omega r}\right) = 1. \tag{8–141}$$

The critical radius of trapping, R_cr, is about 1 nm; particle growth at room temperature is limited by approximately the same size. Thus, on one hand, most of the particles initially produced at room temperature are trapped; on the other hand, their size distribution is monodispersed due to losses of smaller clusters with gas flow. The losses of small particles with gas flow (the selective trapping effect) explain the strong temperature effect on dust production. A small temperature increase results in a significant reduction of cluster growth velocity (8–137). Therefore, the initial cluster size before recombination, R_in, is reduced. According to (8–141), this leads to losses of the initial neutral particles with gas flow, and to a delay of coagulation, α–γ transition, and dust production in general at higher temperatures.

The nanoparticles with size limited by $R_\mathrm{in}(T)$ are conceived by dissociative attachment (8–127) and lost due to the gas flow, which permits estimation of their density as

$$n(R_\mathrm{in}) = k_\mathrm{ad}n_\mathrm{e}[\mathrm{SiH_4}]\tau_\mathrm{R}. \tag{8–142}$$

Formation of the 2-nm particles effectively trapped in plasma is determined at $T > 300\,\mathrm{K}$ by the slow process of electron attachment to the neutral particles with R_in less than R_cr:

$$\begin{aligned} W_\mathrm{p} &= k_\mathrm{a}(R_\mathrm{in})n_\mathrm{e}n(R_\mathrm{in}) = k_\mathrm{a}(R_\mathrm{in})n_\mathrm{e}k_\mathrm{ad}n_\mathrm{e}[\mathrm{SiH_4}]\tau_\mathrm{R} \\ &= [k_\mathrm{a}(R_\mathrm{cr})n_\mathrm{e}\tau_\mathrm{R}]k_\mathrm{ad}n_\mathrm{e}[\mathrm{SiH_4}][k_\mathrm{a}(R_\mathrm{in})/k_\mathrm{a}(R_\mathrm{cr})]. \end{aligned} \tag{8–143}$$

It can be presented, taking into account (8–140) and (8–141), as the relation between the rate of particle production, W_p, and the initial negative ion $\mathrm{SiH_3^-}$ production by dissociative attachment, W_ad:

$$W_\mathrm{p} = W_\mathrm{ad}\exp\left[\frac{e^2}{4\pi\varepsilon_0\hbar\omega}\left(\frac{1}{R_\mathrm{cr}} - \frac{1}{R_\mathrm{in}}\right)\right]. \tag{8–144}$$

This formula illustrates the radius-dependent catastrophe of nanoparticle production in $\mathrm{SiH_4}$ plasma. If the rate of particle growth is high, and the particle size during the initial period (so-called first generation) is sufficiently large for trapping ($R_\mathrm{in} = R_\mathrm{cr}$), the rate of particle production is close to the rate of $\mathrm{SiH_3^-}$ production, which takes place at room temperature. For higher temperatures, the initial radius (R_in) for first-generation particles is less than the critical value (R_cr) for selective trapping, and the particle production becomes slower than the dissociative attachment. The nanoparticle production rate (8–144) can be presented by taking (8–137) and (8–139) as a function of temperature (Fridman et al., 1996):

$$W_\mathrm{p} = W_\mathrm{ad}\exp\left\{\frac{e^2}{4\pi\varepsilon_0\hbar\omega R_\mathrm{cr}}\left[\exp\left(-\frac{B\Delta T}{3T_{00}^{4/3}}\right) - 1\right]\right\}, \tag{8–145}$$

which explains the extremely strong temperature dependence of the dust production. Discharge parameters effective for nanoparticle production in low-pressure silane plasma are shown in Fig. 8–51. Dust production can be stimulated by increasing the electron

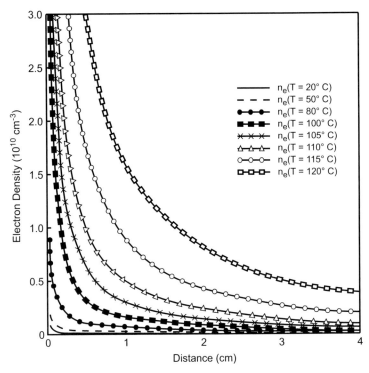

Figure 8–51. Critical conditions of formation of nanoparticles in plasma. Relation between gas temperature, electron density, and interelectrode distance. Gas temperature is expressed in Celsius; gas velocity in the discharge is 20 cm/s.

density and the residence time (for example, by increasing the distance between electrodes) and can be significantly reduced by even small gas heating.

8.8.6. Critical Phenomenon of Super-Small Nanoparticle Coagulation

When an increasing concentration of monodispersed 2-nm particles in silane plasma reaches the critical value of about 10^{10}–10^{11} cm^{-3}, the fast coagulation process begins. The critical value of particle density required for coagulation in contrast to conventional phase transitions doesn't depend on temperature. When the cluster density reaches the critical value of 10^{10}–10^{11} cm^{-3}, attachment of small negative ions (like SiH_3^-) to 2-nm neutral particles becomes faster than the chain reaction of new particle growth. Therefore, the total particle mass remains almost constant during the coagulation (Fridman et al., 1996). Consider the probability of two-, three-, and multi-particle interaction, assuming that the mean radius R of particle interaction is proportional to their physical radius and all direct collisions result in aggregation. The reaction rate and rate coefficient for a binary particle collision are

$$W(2) = \sigma v N^2, \quad K(2) = \sigma v, \tag{8-146}$$

where N is the particle density, and σv is the product of their cross section and velocity. A stationary density of two-particle complexes, $N(2)$, can be estimated by taking into account the characteristic lifetime of the particles, R/v, as

$$N(2) = (\sigma v N^2)\frac{R}{v} = N(\sigma R N). \tag{8-147}$$

The reaction rate and rate coefficient of the agglomerative three-particle collision are

$$W(3) = \sigma v \cdot (\sigma R N)N^2, \quad K(3) = \sigma v(\sigma R N). \tag{8-148}$$

575

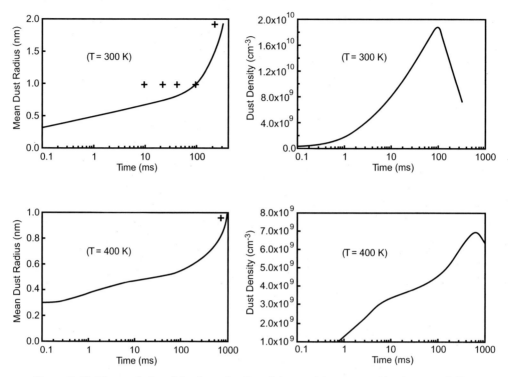

Figure 8–52. Time evolution of density and radius of dust particles generated in plasma at different gas temperatures: solid curve corresponds to kinetic modeling, "plus" corresponds to results of relevant plasma experiments.

Repeating this procedure (Nikitin, 1970), and taking into account that σR is proportional to particle mass, the rate coefficient of a $(k + 2)$-body collision can be presented as

$$K(k) = \sigma v \cdot (\sigma RN)^k \cdot k!. \qquad (8\text{–}149)$$

The total rate constant of the coagulation process can be calculated as the sum of the partial rates (8–149), which can be asymptotically calculated by a special integration procedure (Migdal, 1981):

$$K_c = \sum \sigma v \cdot (\sigma RN)^k \cdot k! = \sigma v \left[1 + \frac{1}{\ln(N_{cr}/N)} \right]. \qquad (8\text{–}150)$$

When the initial particle density N is less than the critical value N_{cr} (which is proportional to $1/\sigma R$, and numerically is about 10^{10}–10^{11} cm^{-3}), the coagulation rate coefficient is the conventional one related to binary collisions. But when the initial particle concentration N approaches the critical value N_{cr}, the coagulation rate sharply increases as if during a phase transition. Because of the very low probability of aggregate decomposition, the critical particle density does not depend on temperature, which is the key difference between the critical phenomenon under consideration and conventional gas condensation. The time evolution of density and mean radius of dust particles in silane plasma at different temperatures is presented in Fig. 8–52 (Fridman et al., 1996). Typical induction times before coagulation are 100–200 ms for room temperature and much longer for 400 K. This temperature effect is due to selective trapping. The particle production rate at 400 K is slower than that at room temperature, but the critical particle concentration remains almost the same; therefore, the induction period becomes much longer.

8.8.7. Critical Change of Plasma Parameters due to Formation of Nano-Particles: α–γ Transition

When the particle size reaches a second critical value during the coagulation process, a significant change occurs in the silane plasma parameters: the electron temperature increases and the electron concentration decreases dramatically. Consider the balance equation of electrons (n_e), positive ions (n_i), and negatively charged particles (N_n). Major processes to be taken into account are ionization ($k_i(T_e)$), electron and positive ion losses at the walls by ambipolar diffusion ($n_e D_a/R^2$), electron attachment to neutral particles ($k_a(r_a)$), and "ion–ion" recombination k_r of positive ions and negatively charged particles. Before the α–γ transition, negatively charged particles have only one electron; therefore,

$$\frac{dn_e}{dt} = k_i(T_e)n_e n_0 - n_e \frac{D_a}{R^2} - k_a(r)n_e N, \tag{8–151}$$

$$\frac{dn_i}{dt} = k_i(T_e)n_e n_0 - n_i \frac{D_a}{R^2} - k_r N_n n_i, \tag{8–152}$$

$$\frac{dN_n}{dt} = k_a(r)n_e N - k_r n_i N_n. \tag{8–153}$$

where N and n_0 are neutral particle and gas densities. For steady-state discharge, the derivatives on the left-hand side can be neglected. The threshold of the α–γ transition can be determined from the electron balance (8–151), assuming ionization coefficient $k_i(T_e) = k_{0i} \exp(-I/T_e)$ and electron attachment coefficient (8–140) $k_a(r) = k_{0a}(\frac{r}{R_0})^2 \exp(-\frac{e^2}{4\pi\varepsilon_0\hbar\omega r})$:

$$k_{0i} \exp\left(-\frac{I}{T_e}\right) n_0 = \frac{D_a}{R^2} + N k_{0a} \left(\frac{r}{R_0}\right)^2 \exp\left(-\frac{e^2}{4\pi\varepsilon_0\hbar\omega r}\right). \tag{8–154}$$

Before the critical moment of α–γ transition, when particles are relatively small and electron attachment is not effective, the electron balance is determined by ionization and electron losses to the walls. The α–γ transition is the moment when the electron attachment to the particles exceeds the electron losses to the walls. The electron temperature increases to support the plasma balance (Belenguer et al., 1992). The total mass and volume of the particles remain almost constant during coagulation; therefore, the specific surface of the particles decreases with the growth of mean particle radius. Hence, the influence of the particle surface becomes more significant, when the specific surface area decreases. Relation (8–154) explains the phenomenon: the exponential part of the electron attachment dependence on particle radius is much more important than the pre-exponential factor. Comparison of the first and second terms on the right-hand side of (8–154) gives a critical particle size R_c required for the α–γ transition:

$$R_c = \frac{e^2}{4\pi\varepsilon_0\hbar\omega \ln(k_{0a}N R^2/D_a)}. \tag{8–155}$$

The critical particle radius is about 3 nm and does not depend significantly on temperature. Equation (8–154) also describes the electron temperature evolution as a function of aggregates radius r during the α–γ transition. There are two exponents in (8–154), which play a major role in the electron balance. Therefore, neglecting the ambipolar diffusion term,

$$\frac{1}{T_e} - \frac{1}{T_{e0}} = \frac{e^2}{4\pi\varepsilon_0\hbar\omega}\left(\frac{1}{r} - \frac{1}{R_c}\right). \tag{8–156}$$

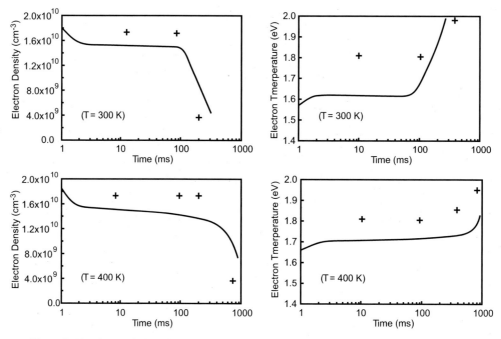

Figure 8–53. Time evolution of electron density and electron temperature in dusty plasma at different gas temperatures, including moment of the $\alpha-\gamma$ transition: solid curve corresponds to kinetic modeling, "plus" corresponds to results of relevant plasma experiments.

where T_{e0} (about 2 eV) is the electron temperature just before the $\alpha-\gamma$ transition, when the aggregate radius is near its critical value ($r = R_c$). During the particle growth after the $\alpha-\gamma$ transition, the electron temperature increases with the saturation on the level of $e^2/4\pi\varepsilon_0\hbar\omega R_c$, which is numerically about 5–7 eV. Thus, as a result of the $\alpha-\gamma$ transition, the electron temperature increases by as much as three-fold. The electron temperature increase as a function of relative difference of positive ion and electron densities, $\Delta = (n_i - n_e)/n_i$, is derived from the positive ion balance equation (8–152):

$$k_{0i}\exp\left(-\frac{I}{T_e}\right)n_0 = \frac{D_a}{R^2} + k_r n_i \frac{\Delta}{1-\Delta}. \qquad (8\text{–}157)$$

During the $\alpha-\gamma$ transition, the electron temperature grows, the electron density decreases, and the factor Δ becomes close to 1. The electron density before the $\alpha-\gamma$ transition is almost constant on the level of about $n_{eo} = 3 \cdot 10^9 \text{ cm}^{-3}$. In the experimental conditions, current density ($j_e = n_e e b_e E$) is proportional to $n_e T_e^2$ and remains fixed during the $\alpha-\gamma$ transition. Therefore, electron and positive ion densities during this period can be derived from the balance equation as

$$n_i = \sqrt{\frac{k_{0i} n_{e0} T_{e0}^2 n_0 \exp(-I/T_e)}{k_r T_e^2}}, \qquad n_e = n_{e0}\frac{T_{e0}^2}{T_e^2}. \qquad (8\text{–}158)$$

The electron density decreases during the $\alpha-\gamma$ transition by approximately 10-fold, while the positive ion density slightly increases. The time evolution of electron density and temperature during the $\alpha-\gamma$ transition in silane plasma are shown in Fig. 8–53 (Fridman et al., 1996).

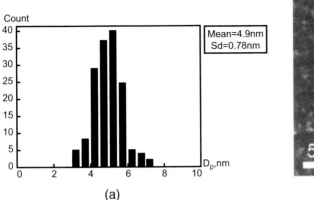
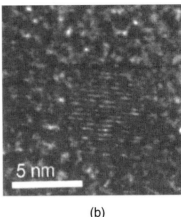

(a)   (b)

Figure 8–54. (a) Size distribution of primary particles; (b) high-magnification image of a 5-nm silicon particle showing the lattice fringes of the (111) plane.

8.8.8. Other Processes of Plasma Production of Nanoparticles: Synthesis of Aluminum Nanopowder and Luminescent Silicon Quantum Dots

Plasma can be effectively applied to the production of nanoparticles from different materials. Unipolar charging of nanoparticles in plasma suppresses their agglomeration, which has attracted interest in plasma-generated nanopowders. Consider two examples of plasma nanotechnologies interesting for special applications. The first technology is related to the synthesis of aluminum nanoparticles using DC thermal plasma (Zhang et al., 2005). Such particles are highly pyrophoric and light. If they are properly surface–passivated, the aluminum nanoparticles are of significant interest for solid fuel propulsion. The particles are effectively synthesized using Ar–H_2 thermal DC-arc plasma at near atmospheric pressure. The Ar–H_2 plasma is expanded in a subsonic flow through a ceramic nozzle to a pressure of 53 kPa. $AlCl_3$ vapor is injected into the plasma at the upstream end of the nozzle. A cold argon counterflow opposes the plasma jet exiting the nozzle to provide dilution and cooling, which prevents coagulation and coalescence and maintains small particles sizes. At $AlCl_3$ flow rates of 20 sccm and carrier gas of 200 sccm, most of the produced aluminum particles have sizes of about 10 nm.

The second nanotechnology is related to high-yield synthesis of luminescent silicon quantum dots in non-thermal RF discharge (Mangolini, Thimsen, & Kortshagen, 2005). The nanosize quantum dots are of interest for a variety of applications, from solid-state lighting and opto-electronic devices to the use of fluorescent tagging agents. Conventional liquid-phase synthesis of the silicon quantum dots is characterized by limited process yields. Gas-phase approaches can provide higher yields; however, they are afflicted with problems of particle agglomeration. Non-thermal plasma synthesis offers an effective and fast gas-to-particle conversion common to gas-phase approaches and strongly reduces agglomeration at the same time since the particles in plasma are unipolarly negatively charged. Silicon quantum dots have been produced in a continuous-flow Ar–SiH_4 non-thermal plasma reactor with a 27 MHz RF-powered ring electrode at a pressure of 1.5 Torr. A typical flow rate is 40 sccm for Ar–SiH_4 (95:5) mixture. The residence time of particles in the plasma region is short (less than 5 ms), and the size of the produced particles is small (about 5 nm). The size distribution of the nanoparticles is shown in Fig. 8–54, together with a higher-magnification image of a 5-nm particle. The transmission electron microscopy (TEM) micrograph shows the lattice fringes of the (111) plane. The plasma-produced particles are collected on filters and washed with methanol. After growth of a native oxide layer, the particles show bright

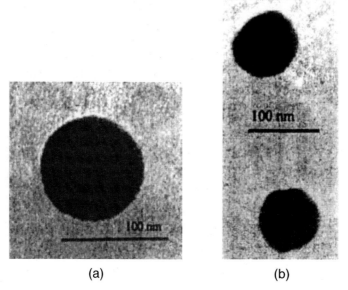

(a)   (b)

Figure 8–55. TEM images of single SiO$_2$ nanoparticles coated in the RF-CCP discharge by a nanolayer of polymer-like C:H film: (a) discharge in methane; (b) discharge in ethane.

photo-luminescence in the red–orange region. The production rate of the nanoscale silicon quantum dots in these experiments is a few tens of milligrams per hour.

8.8.9. Plasma Synthesis of Nanocomposite Particles

Plasma is effective in the fabrication of nanocomposites, such as nanoparticles composed of one material and covered by a nanolayer of another. Relevant examples include, in particular, carbon-coated magnetic metal particles produced in thermal arc plasma (McHenry et al., 1994), as well as polymer-coated nanoparticles with improved adhesion, corrosion resistance, and surface passivation produced in RF plasma in a fluidized bed (Shi et al., 2001). Consider two other specific technologies of the synthesis of nanocomposite particles. The first is related to polymer-like C:H nanocoating of nanopowders of amorphous SiO$_2$ in non-thermal low-pressure RF-CCP discharge (Kouprine et al., 2003). Nanoparticles of SiO$_2$ with diameter less than 100 nm are treated in 500–800 W discharge at 1–5 kPa in pure methane and ethane. Although non-saturated hydrocarbons are partially synthesized, their presence does not lead to soot nucleation. Instead, the SiO$_2$ particles passing through the non-thermal low-pressure RF-CCP discharge become individually coated by a polymer-like C:H film with thickness of about 10 nm. TEM images of such nanocomposites produced in CH$_4$ and C$_2$H$_6$ plasmas are shown in Fig. 8–55. The experiments with methane demonstrate that thermal plasma processes proceeding through generation of CH$_2$ and CH radicals as well as acetylene result in an intensive production of soot. In contrast to that, low-temperature non-thermal plasma processes generate mostly CH$_3$ radicals and H atoms, which is not sufficient for soot production but very effective for deposition of thin films on the surface of already existing particles. The second nanotechnology involves the synthesis of ferromagnetic nanopowders from iron pentacarbonyl, Fe(CO)$_5$ (Kouprine et al., 2005). Decomposition of Fe(CO)$_5$ leads at first to formation of iron particles and then, because of disproportioning of CO (Section 5.7.2), to carbon deposition on the surface of the particles. Magnetic nanoparticles encapsulated in carbon protect them from oxidation as well as reduce interparticle magnetic coupling (Tomita & Hikita, 2000). This process is also

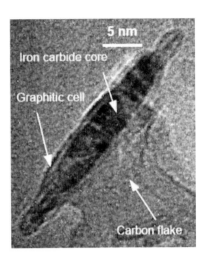

Figure 8–56. TEM of a single iron "nanocigar" in a graphitic cell synthesized in the RF-CCP discharge at pressure 110 Pa (iron carbonyl $Fe(CO)_5$ is a precursor of ferromagnetic iron core; decomposition of carbon monoxide leads to individual encapsulation of the ferromagnetic iron core in a carbon/graphite matrix).

performed in non-thermal low-pressure RF-CCP discharges. This discharge is operated at pressures of 100–300 Pa and power densities of 4–6 W/cm^3. Saturated $Fe(CO)_5$ vapor (4000 Pa) is diluted with argon ($Fe(CO)_5$:Ar = 1:20) and supplied to the plasma at flow rates of 0.015–0.02 slm. The process leads to the production of particles, which are presented in Fig. 8–56 and look like "nanocigars." The core of the "nanocigars" contains iron in the form of Fe_2C. The surrounding layered cells are composed of a graphite-like structure. Graphite planes in the cells are structurally similar to the multi-wall carbon nanotubes.

8.9. PLASMA NANOTECHNOLOGIES: SYNTHESIS OF FULLERENES AND CARBON NANOTUBES

8.9.1. Highly Organized Carbon Nanostructures: Fullerenes and Carbon Nanotubes

The production of non-highly structured nanoparticles was described in Section 8.8. In addition to that, plasma is also effective in the production of highly organized nanostructures. The best examples of such plasma-synthesized structures are fullerenes and carbon nanotubes (Eletsky & Smirnov, 1995; Dresselhaus, Dresselhaus, & Eklund, 1996; Eletsky, 1997). Fullerenes are closed-cage carbon molecules such as C_{36}, C_{60}, C_{70}, C_{76}, C_{84}, and so on in which carbon atoms are located in corners of regular pentagons or hexagons covering the surface of a sphere or a spheroid (see Fig. 8–57). The C_{60} molecule has the highest symmetry and, as a result, is the most stable. It is composed of 20 regular hexagons and 12 regular pentagons with a characteristic distance between carbon atoms of 0.142 nm. For comparison, carbon layers in graphite consist of regular hexagons with a distance between atoms of 0.142 nm. To some extent, one can say that fullerenes are spheroidal modifications of graphite-like sheets. Fullerenes have very specific physical and chemical properties. In particular, they are characterized by high oxidative potential and are able to simultaneously attach up to six electrons. Insertion of alkaline atoms in the crystal structure of C_{60} makes the crystal superconductive, with a transition to superconductivity at about 40 K. High-pressure and high-temperature modification of crystallite fullerenes leads to the formation

Figure 8–57. Structure of fullerenes constructed from pentagonal and hexagonal carbon rings.

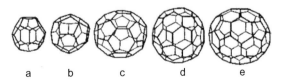

a   b   c   d   e

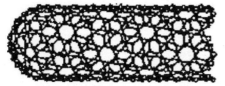

Figure 8–58. Idealized model of a single-wall nanotube.

of a new phase of carbon, which has a hardness similar to that of diamonds. The graphite-like sheets can form not only spheroidal nanosize configurations (fullerenes) but also special high-aspect-ratio nanocylindrical structures called nanotubes (Harris, 2002; Reich, Thomsen, & Maultzsch, 2004). The diameter of nanotubes can be as small as one or a few nanometers, while their length can exceed tens of microns. Depending on the number of graphite-like sheets forming the walls of the nanotubes, they are usually divided into two classes: single-wall and multi-wall nanotubes.

Single-wall carbon nanotubes are formed by rolling the graphite sheet composed of regular carbon hexagons into a high-aspect-ratio cylinder. The properties of the nanotube, especially electrical conductivity, depend on the **chirality** of the structure, which is determined by the angle of orientation of the graphite sheet, α, with respect to the nanotube axis. Two specific chiralities permit rolling a graphite sheet into a nanotube without perturbation of the structure. One such chirality corresponds to the orientation angle $\alpha = 0$, which is called the **armchair configuration**. In armchair single-wall nanotubes, two C–C bonds of each regular hexagon are parallel to the axis of the nanotube ($\alpha = 0$). The armchair nanotubes have high metallic electric conductivity and high chemical stability. Another chirality, not perturbing the graphite sheet, corresponds to the orientation angle $\alpha = \pi/6$. Nanotubes with such chirality are referred to as having a **zigzag orientation**. An idealized model of a single-wall nanotube is shown in Fig. 8–58. The idealized nanotube has no seams related to rolling of a graphite sheet and is also confined at both ends by hemispheres composed of regular hexagons and six regular pentagons. The ideal nanotube can be considered an extension of a fullerene to the extreme of very high aspect ratio. Actual single-wall nanotubes are different from an idealized nanotube, especially regarding the ends, which are in reality far from a spherical shape. **Multi-wall carbon nanotubes** can be formed in a larger variety of configurations. Some of the configurations are shown in Fig. 8–59. Coaxial single-wall nanotubes inserted one into another form configuration called a **Russian doll** (Fig. 8–59a). Another variation of the Russian doll, shown in Fig. 8–59b, is composed of coaxial prisms. The last configuration of multi-wall nanotubes is shown in Fig. 8–58c and looks like a **scroll**. The distance between the graphite sheets in all three configurations is about 0.34 nm, which corresponds to the interlayer distance in conventional crystalline graphite. Carbon nanotubes have a wide range of potential applications in different areas. In particular, carbon nanotubes are effective electron emitters at relatively low voltages, which can be applied in super-high-resolution displays. Single-wall nanotubes with surface defects are a nanoscale heterostructure to be applied in future miniature ICs. The mechanical properties of nanotubes are sufficient for their use as a probe in atomic force microscopes,

Figure 8–59. Idealized models of multi-wall nanotubes: (a) Russian doll; (b) hexagonal prism; (c) scroll.

a   b   c

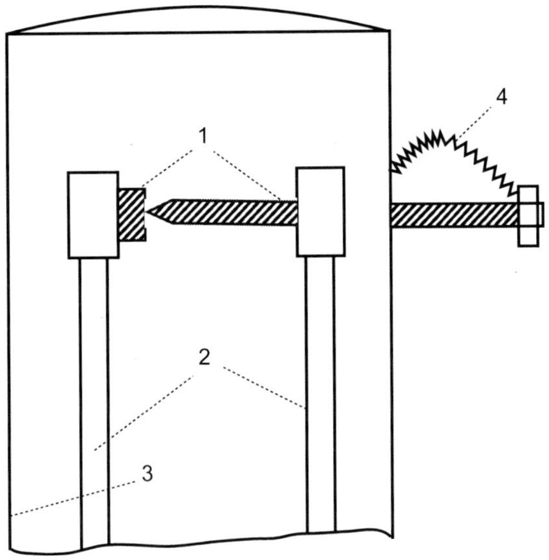

Figure 8–60. General schematic of a system of production of fullerene-containing soot: (1) graphite electrodes; (2) water-cooled copper element; (3) water-cooled surface for condensation of carbon condensate; (4) springs.

as well as for enforcement of composite materials. Carbon nanotubes can also be used for storage of hydrogen and other gases.

8.9.2. Plasma Synthesis of Fullerenes

A conventional approach to plasma synthesis of fullerenes uses thermal arcs (Farhat, Hinkov, & Scott, 2002; Saidane et al., 2004). The process starts with production of fullerene-containing soot (see Fig. 8–60). An AC arc is sustained between two graphite electrodes, with typical parameters: frequency 60 Hz, current 100–200 A, and voltage 10–20 V. Modern fullerene production processes are also based on a DC thermal arc with similar parameters. A plasma-chemical reactor is filled with He at a pressure of about 100 Torr. The evaporation rate of a graphite rode is up to 10 g/h. As a result of the process, walls of the water-cooled reactor are covered with soot that contains up to 15% of fullerenes C_{60} and C_{70} (mass ratio 10:1). The productivity of the plasma system is high, about 1 kg of the fullerene-containing soot per day. Extraction of fullerenes from the plasma-produced soot is based on higher fullerene solubility in toluene, benzene, dimethyl-benzene (xylol), dichlorine-benzene, CS_2, and others. The solubility of fullerenes in such solvents is several grams per liter. The extraction of fullerenes results in the production of polycrystalline black powder, which mostly consists of C_{60} and C_{70} with some admixture (less than 1%) of higher fullerenes (C_{76}, C_{84}, etc.). Further separation of fullerenes can be achieved using liquid chromatography.

8.9.3. Plasma Synthesis of Endohedral Fullerenes

Thermal arc systems, similar to those described earlier (Fig. 8–60), can be applied for the production of endohedral fullerenes (Eletsky, 2000; Lange et al., 2002). The endohedral fullerenes, or simply **endohedrals** ($M_m@C_n$), consist of a conventional fullerene shell C_n with an atom or molecule M_m encapsulated inside. Injection of metal vapor into the arc discharge leads to the formation of endohedrals with a relative concentration not exceeding a few percent from the total amount of fullerenes in the produced soot. The simplest method of metal injection into the thermal arc is based on the use of a composite electrode (anode), which is made of graphite with some small admixture of powder of the metal or its compound, oxide or carbide. Consider, for example, the formation of an endohedral fullerene La@C82 in the thermal arc discharge. The anode is made from a graphite rod (length 300 mm, diameter 15 mm) with a cylindrical hole (length 270 mm, diameter 10 mm). The cavity is filled with a mixture of La_2O_3 powder with graphite powder. The atomic fraction of lanthanum in the anode material is 1.6%. A pure graphite rod is used as a cathode. The thermal arc discharge is sustained in helium flow at a pressure of 500 Torr, and the DC current is 250 A. Produced soot is transported by the helium flow and trapped in a filter in the absence of oxygen. Extraction of endohedral fullerenes from soot is done by using solvents such as toluene or CS_2. Metal injection in a thermal arc can be provided not only by adding metal powder to the electrode material but also by admixing gaseous compounds of the metal in the He flow. This approach was applied in plasma synthesis of $Fe@C_{60}$. Pentacarbonyl $(Fe(CO)_5)$ was used as a gaseous admixture to helium in this case. Another approach to the production of endohedrals is based on ion bombardment of hollow fullerenes by ions of elements to be encapsulated. This approach is especially effective in the production of endohedrals containing highly chemically active atoms, specifically alkaline atoms, as well as $N@C_{60}$.

8.9.4. Plasma Synthesis of Carbon Nanotubes by Dispersion of Thermal Arc Electrodes

The most effective plasma method for the production of carbon nanotubes is based on dispersion of graphite electrodes in thermal arc plasma in helium. Typical voltages are 15–25 V, typical currents are several tens of amperes, and the distance between electrodes is a few millimeters. Formation of nanotubes is due to dispersion of the graphite anode, which is a process characterized by temperature higher than that of the cathode. Carbon nanotubes are deposited on the discharge chamber walls and on the cathode. In contrast to the production of fullerenes, where the typical He pressure is 100–150 Torr, the production of nanotubes is optimal at higher pressures, around 500 Torr. Also, cathodes of larger diameters (exceeding 10 mm) are favorable for higher yields of nanotubes. A schematic of the arc discharge system especially designed for the production of nanotubes is shown in Fig. 8–61 (Eletsky, 2000a,b). The system is able to produce grams of nanotubes, with their concentration in the cathode deposit exceeding 60%. Maximum nanotube yield is achieved at minimal currents sufficient for the arc stability. The current increase leads to conversion of the nanotubes into graphite. Nanotubes with length up to 40 μm grow from the cathode perpendicular to its surface. To separate the nanotubes from the rest of deposit, ultrasonic dispersion can be effectively applied to the deposit preliminarily dissolved in methanol or other solvent. Thermal arc discharges with graphite electrodes produce mostly multi-wall nanotubes with mixed chirality. The production of single-wall nanotubes usually requires application of metallic catalysts of the platinum group (Ru, Rh, Pd, Os, Ir, Pt) and iron group (Co, Fe, Ni), as well as such metals as Mn, Sc, La, V, Ce, Zr, Y, or Ti. The presence of catalysts leads to the formation of a large variety of different nanostructures, which are determined by the type of the catalyst and by the parameters of the arc discharge. The iron group metals selectively promote the formation of single-wall nanotubes. Furthermore, the highest synthesis efficiency of single-wall nanotubes is achieved by application of mixed iron-group–metal catalysts, specifically Fe–Ni and Co–Ni.

8.9.5. Plasma Synthesis of Carbon Nanotubes by Dissociation of Carbon Compounds

Carbon nanotubes can also be produced in thermal plasma by dissociation of carbon compounds, for example by tetrachloroethylene (TCE; Harbec et al., 2005a,b). The process has been conducted by TCE injection into the nozzle of a 100 kW DC non-transferred thermal plasma torch. Metal catalysts stimulate the process of formation of nanotubes. In particular, erosion of tungsten electrodes provides nanoparticles of the metal as precursors for the carbon nanotubes (Harbec et al., 2005a,b). A catalytic stimulation of the process can also be achieved by using electrodes made of a nickel alloy and by using ferrocene dissolved in the

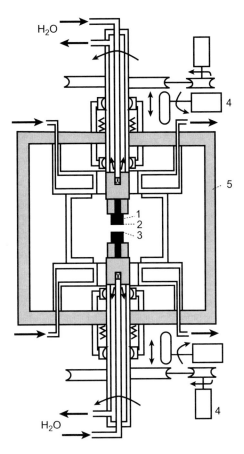

Figure 8–61. Schematic of an electric arc installation for nanotube production (in grams): (1) graphite anode; (2) deposit containing nanotubes; (3) graphite cathode; (4) automatic system maintaining fixed interelectrode distance; (5) chamber wall. Arrows show flow direction of cooling water.

TCE supply (Harbec et al., 2005a,b). The arc plasma erodes the electrodes, which creates metal vapors transported toward the torch exit. Fast quenching in the torch nozzle leads to metal vapor nucleation into metal nanoparticles, which are catalysts for the formation of carbon nanotubes based on carbon atomized from TCE in the thermal plasma. Thermal arc electrodes made of tungsten, by means of generation of metallic tungsten particles, stimulate production of multi-wall nanotubes having 10 concentric rolled-up graphite sheets. External and internal diameters of the multi-wall nanotubes produced in this system are, respectively, 10–30 nm and about 0.8 nm. Nickel catalysts in this case promote the formation of single-wall carbon nanotubes. When thermal plasma electrodes are made from the nickel-containing Inconel, single-wall carbon nanotubes are created and deposited on the reactor walls. Dissolution of the nickel-containing ferrocene in the TCE supply also results in the formation of single-wall nanotubes.

8.9.6. Surface Modification of Carbon Nanotubes by RF Plasma

The surface of carbon nanotubes is non-reactive, which reduces their compatibility with other materials, especially organic solvents and polymer matrices. Surface modification of carbon nanotubes for this reason can significantly increase their applicability (Robertson, 2004). In particular, functionalization of the nanotubes with different gases, including O_2, NH_3, and CF_4, can be especially interesting. Such a functionalization process was effectively accomplished using RF-ICP plasma at a pressure of 0.1 Torr and power of 30 W (Felten et al., 2005). Functionalization in RF oxygen plasma leads to the formation of C–O, C=O, and COO groups on the nanotube surface. The atomic fraction of these groups varies with

Table 8–8. Formation of Different Surface Groups by Functionalization of Carbon Nanotubes in Low-Pressure RF-ICP Discharge in Oxygen

Plasma treatment time, s	Atomic O, at. %	Group C–O, %	Group C=O, %	Group COO, %
5	11	38.5	38.5	23
15	9	25	65.5	9.5
30	12	26	60.5	13.5
120	16	16	55.5	28.5
300	19	10	62	28
600	19	9	59.5	31.5
1200	20	6	82.5	11.5

plasma parameters and with treatment time, which is illustrated in Table 8–8 (Felten et al., 2005). The sum of the atomic fractions of C–O, C=O, and COO in the table gives 100%. The relative amount of the non-incorporated atomic oxygen at the carbon surface is also presented in the table. Treatment in oxygen plasma also results in a significant change in the hydrophilicity of the carbon nanotubes, which is demonstrated by special contact-angle measurements. Treatment in CF_4 plasma, on the contrary, results in a significant change in the hydrophobicity of carbon nanotubes. Another approach to surface modification of carbon nanotubes is related to plasma polymerization of monomers on their surfaces, which was also demonstrated in RF-ICP at a pressure of 0.1 Torr and power of 30 W (Felten et al., 2005). Specifically acrylic acid was used as a monomer in these experiments with the objective of producing water-soluble carbon nanotubes. Plasma treatment leads to the deposition of thin polyacrylic acid film, and indeed to an improvement of solubility. The uncoated nanotubes are sedimenting in water within a few hours, whereas coated ones remains suspended for at least several days.

PROBLEMS AND CONCEPT QUESTIONS

8–1. Thermal Plasma Spray. Analyzing equation (8–1), determine the effect of the exponential factor n on the uniformity of velocity, enthalpy, and temperature distributions. Calculate the relative radius corresponding to a decrease of the parameters twice with respect to those at the torch axis. How does thermal plasma uniformity affect the quality and characteristics of the thermal plasma spray process?

8–2. Anisotropy Requirements of Plasma Etching. Using relation (8–7), make a graph demonstrating the effect of mask feature size l_m on requirements for etching anisotropy. Discuss the minimization of the mask feature size, taking into account the size limitations related to lithography.

8–3. Sputtering. Explain why the sputtering yield (8–9) is low when the mass of an incident atom is much less than that of surface atoms. Illustrate the effect by using the numeric values of the sputtering yield from Table 8–1. Why does the sputtering yield have its maximum not at normal incidence but at some specific angle of incidence about $\theta = \pi/4-\pi/3$.

8–4. Etching Anisotropy Analysis in the Framework of Surface Kinetics of Plasma Etching. Based on relation (8–22), analyze the substrate temperature effect on etching anisotropy. Which step of the ion energy-driven etching is most affected by substrate temperature? Compare effects of translational gas temperature and electron temperature on the plasma etching process.

Problems and Concept Questions

8–5. Etching Anisotropy as a Function of Etching Discharge Power. Analyze the dependence of the etching anisotropy coefficient on discharge power based on formula (8–31). Estimate the anisotropy coefficient for carbon etching in oxygen assuming a discharge power 100 W, pressure of 30 mTorr, cross-sectional area of $400 \, \text{cm}^2$, and $E_{ei} = 300$ eV. Assume the substrate temperature to be $T_0 = 400$ K, and make an assumption regarding the desorption rate coefficient K_D. Which effect limits the growth of the etching anisotropy with the discharge power?

8–6. Loading Effect in Etching Kinetics. Using relation (8–41) describing the loading effect of wafers on the concentration of neutral etchants and therefore on the etching rate, derive a formula for the minimum value of the vacuum pumping velocity, when the loading effect can still be neglected. Does the contribution of the loading effect depend on gas pressure?

8–7. Flamm Formulas for F-Atom Silicon Etching. Using the Flamm formulas (8–43) and (8–44), calculate the F-atom pure chemical etching rates of undoped silicon and thermally grown silicon dioxide for room-temperature substrates. Assume a typical F-atom density in the etching discharges, $3 \cdot 10^{14} \, \text{cm}^{-3}$. Based on the found reaction rates, determine the Si-to-SiO_2 etching selectivity at room temperature.

8–8. Competition Between Silicon Etching and Carbon Film Deposition in CF_4 Discharges. Generalize relation (8–61) for the silicon etch rate in CF_4 discharges additionally including physical sputtering of CF_y groups from the surface by ion bombardment. To derive the generalized relation, modify equations for surface balance of F atoms (8–59) and surface balance of carbon atoms (see (8–60)), taking into account Γ_{CF_y}, the desorption flux provided by sputtering. Compare the derived generalized relation with the simplified one (8–62) assuming the sputter-dominated desorption.

8–9. Langmuir-Rideal Mechanism of Surface Recombination of F Atoms During Their Transportation from Remote Plasma Source for Plasma Cleaning. Analyzing relations (8–69)–(8–71) for F-atom adsorption, thermal desorption, and recombination/desorption, find out the limiting stage of the F-atom surface recombination kinetics in the framework of the Langmuir-Rideal recombination mechanism. Assume conventional transport tube and gas flow conditions in the RPS chamber cleaning technology.

8–10. Rate of Amorphous Silicon Film Deposition in Silane (SiH_4) Discharges. Amorphous silicon deposition rates from silane in the PECVD process are typically in the range 5–50 nm/min. Compare these practical deposition rates with theoretical ones, which you can calculate using relations (8–86) or (8–88). For the calculations, take the process and discharge parameters from Section 8.5.2.

8–11. Non-Conformal Deposition of SiO_2 Within Trenches During PECVD Process in SiH_4–O_2 Mixture. Calculate the non-uniform (non-conformal) deposition flux Γ_{SiO_2} on the sidewalls within a trench as a function of the angle θ_s subtended by the trench opening as seen at a position along the sidewall (see Fig. 8–29). Assume a uniform isotropic source of precursors at the top of the trench, a sticking coefficient equal to unity, and precursor mean free path that is much greater than the trench. Compare your result with proportionality relation (8–92). Derive a similar proportionality relationship for the non-uniform (non-conformal) deposition flux Γ_{SiO_2} on the bottom of the trench.

8–12. Physical Sputter Deposition Rate. Calculate the physical sputter deposition rate assuming a sputtering coefficient $\gamma_{sput} = 1$, an incident ion flux $\Gamma_i = 10^{16} \, \text{cm}^{-2} \, \text{s}^{-1}$, and density of the deposited film $n_f = 5 \cdot 10^{22} \, \text{cm}^{-3}$. Also assume that all sputtered material is deposited on the substrate, and the area of the sputtered target is equal to the substrate area where the film is deposited ($A_t = A_{sub}$).

8–13. Sputtered Atom Energy Distribution Function in Film Deposition. Analyze the sputtered atom energy distribution function (8–96) and calculate the normalization factor for the distribution. Using the normalization factor, calculate the fraction of the sputtered atoms having energies at least twice as large as the average energy of the atoms and, therefore, exceeding the surface binding energy.

8–14. Hysteresis in Transition Between Metallic and Covered Modes of Reactive Sputter Deposition. Illustrate graphically the dependences $\frac{dN_g}{dt}(Y)$ and $\Gamma_{sput}(Y)$, described by relations (8–102), and (8–104) for the asymptotic case ($a \gg 1$). Based on the graphs, illustrate the function $\Gamma_{sput}(\frac{dN_g}{dt})$ and demonstrate the hysteresis behavior in the sputtering flux dependence on reactive gas flow rate.

8–15. Plasma-Immersion Ion Implantation and Sheath Evolution from the Matrix Sheath to the Child Law Sheath. Solving differential equation (8–107), derive the implicit relation (8–110) for the sheath propagation. Expand relation (8–110) using as a small parameter s/s_{Ch} and derive the explicit relation (8–112) for the sheath propagation. By making relevant graphics, compare the accurate and simplified $s(t)$ functions given by relations (8–110) and (8–112).

8–16. Implantation Current from Matrix Sheath During Plasma-Immersion Ion Implantation. Derive relation (8–117) for time evolution of the implantation current from the matrix sheath, using the following steps. Let $x = 0$ in equation (8–116); calculate an ion flight time t_{if} from the position x_0 to the target in the form of a function $x_0(t_{if})$. Ions from the interval between x_0 and $x_0 + dx_0$ are implanted in a time interval between t_{if} and $t_{if} + dt_{if}$. Differentiation of the function $x_0(t_{if})$, therefore, permits you to find the implantation current density as $J = en_p(dx_0/dt_{if})$.

8–17. Trapping of Neutral Nanoparticles in Low-Pressure Silane Plasma. Using relation (8–140), show that the electron attachment time ($1/k_a\, n_e$) for 2-nm particles becomes shorter than the residence time. Each neutral particle in this case can be charged at least once during the residence time. Show that even a short time of being negatively charged before recombination (about 1 ms) is sufficient for particle trapping to happen.

8–18. The α–γ Transition During Coagulation of Nanoparticles in Silane Plasma. The α–γ transition, including a sharp growth of electron temperature and decrease of electron density, is related to a significant contribution of electron attachment to the particle surfaces and takes place during coagulation. It is interesting that the total surface area of particles decreases during the coagulation process, because their diameters grow but total mass remains almost fixed. Give your interpretation of the phenomenon.

9

Organic and Polymer Plasma Chemistry

9.1. THERMAL PLASMA PYROLYSIS OF METHANE AND OTHER HYDROCARBONS: PRODUCTION OF ACETYLENE AND ETHYLENE

Among numerous plasma processes related to the conversion of hydrocarbons, special attention is traditionally focused on the endothermic process of methane conversion into acetylene:

$$CH_4 + CH_4 \rightarrow C_2H_2 + 3H_2, \quad \Delta H = 3.8 \, eV/mol. \tag{9–1}$$

This process has been applied at a large industrial scale using thermal arcs with an optimal energy cost of 8.3 eV/mol (energy efficiency 46%). An alternative carbide method of acetylene production requires 10–11 eV/mol (energy efficiency 35–40%, product yield 90–100 g C_2H_2 per kWh), which explains the advantage of the plasma-chemical approach with respect to applying calcium carbide. This and similar processes will be considered in the following two sections. Before considering specific conversion processes, however, the general kinetic aspects of plasma pyrolysis should be discussed.

9.1.1. Kinetics of Thermal Plasma Pyrolysis of Methane and Other Hydrocarbons: The Kassel Mechanism

Plasma pyrolysis of hydrocarbons proceeds through a long sequence of chemical processes that generally have condensed-phase carbon and hydrogen as final products. The formation of condensed-phase products at 1100–1300 K is mostly due to the contribution of radicals; soot formation at temperatures exceeding 1300 K is mostly due to molecular processes on the carbon surfaces (Valibekov et al., 1975). Thermal decomposition of methane generally follows the phenomenological (reduced) scheme suggested by Kassel (1932). According to the Kassel mechanism, the thermal pyrolysis of methane is determined by the initial formation of CH_2 radicals:

$$CH_4 \rightarrow CH_2 + H_2. \tag{9–2}$$

Other radicals, in particular CH_3, make a significant contribution in the kinetics of plasma pyrolysis. The Kassel mechanism is reduced and focuses only on CH_2, which is sufficient for qualitative analysis. The CH_2 radicals, according to the Kassel mechanism, lead to methane conversion into ethane:

$$CH_4 + CH_2 \rightarrow C_2H_6. \tag{9–3}$$

Further dehydrogenization results in a gradual conversion of ethane into ethylene, ethylene into acetylene, and finally acetylene into soot:

$$C_2H_6 \rightarrow C_2H_4 + H_2, \quad C_2H_4 \rightarrow C_2H_2 + H_2, \quad C_2H_2 \rightarrow 2C_{(s)} + H_2. \tag{9–4}$$

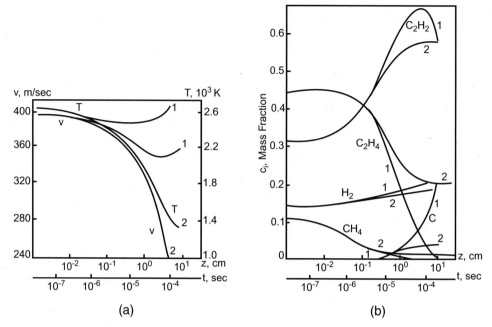

Figure 9–1. Kinetics of the plasma-chemical pyrolysis of gas condensate: (a) without cooling losses in the reactor; (b) with cooling losses in the reactor.

Although acetylene conversion into soot actually includes numerous stages, this exothermic reaction proceeds relatively quickly with respect to other steps of the Kassel mechanism. For this reason, the major intermediate components of methane pyrolysis are ethylene and acetylene. Fixation of a significant amount of higher hydrocarbons preceding soot formation has low probability. During plasma pyrolysis of methane, ethylene is usually formed after 10^{-6}–10^{-5} s, and acetylene after 10^{-4}–10^{-3} s. Then fast quenching should be performed to avoid significant soot formation. Some kinetic details of the Kassel mechanism of methane conversion into acetylene are to be discussed in Section 9.2.3 with regard to the energy efficiency limitations of thermal plasma conversion. Actual hydrocarbon mixtures used for pyrolysis, like gasoline/crude oil compounds and gas condensates, usually consist of not only alkanes but also significant amounts of aromatic compounds. A typical composition of a gas condensate, for example, is normal alkanes (represented as C_8H_{18}), 21%; naphtheno-isoparaphenes (represented as C_7H_{14}), 58%; aromatic compounds (represented by C_9H_{14}), 10%; and multi-nuclei aromatics (represented by $C_{12}H_{10}$), 11%. The kinetics of plasma-chemical pyrolysis of such complicated mixtures is discussed, in particular, by Polak et al. (1975a). Kinetic curves illustrating the plasma pyrolysis of such gas condensates are shown in Fig. 9–1 (those for gasoline are similar). The kinetic curves are not much different qualitatively from those for methane. However, the time required to reach maximum concentrations of ethylene and acetylene is shorter in the case of gas condensate pyrolysis. The ethylene concentration reaches a maximum in this case after 10^{-6} s, and acetylene after 10^{-5}–10^{-4} s. This effect is due to a lower enthalpy of the conversion reaction in the case of gas condensate and gasoline than in the case of methane (9–1). The plasma jet temperature decreases during the initial 10^{-4} s due to endothermic processes of the CH_4 and C_2H_4 decomposition. Then the temperature increases during exothermic C_2H_2 conversion into soot (see Fig. 9–1). To save acetylene and other valuable intermediate products, quenching should start after no later than 10^{-4} s with a cooling rate of $5 \cdot 10^6$ K/s. The cooling processes significantly affect the kinetic curves shown in Fig. 9–1 starting from about 10^{-5} s: the acetylene concentration decreases, while the ethylene concentration increases with respect to those values without cooling. Starting from about $3 \cdot 10^{-5}$ s, the concentrations of products stop changing almost completely. Thus, cooling rates of 10^6–10^7 K/s provides effective quenching for products of pyrolysis.

9.1.2. Kinetics of Double-Step Plasma Pyrolysis of Hydrocarbons

The fixation of valuable products of plasma pyrolysis of hydrocarbons requires quenching after about 10^{-4} s (starting with a temperature of about 1800 K) with cooling rates of about $5 \cdot 10^6$ K/s. The fast cooling process for products of plasma pyrolysis can be provided, in principle, by injection of an additional portion of hydrocarbons. In this case quenching is combined with additional pyrolysis, which increases the effectiveness of the total technology. The approach is usually referred to as double-step plasma pyrolysis. The kinetics of double-step plasma pyrolysis was analyzed in detail, in particular by Gershuni et al. (1975) and Vurzel, Polak, and Epstein (1975). Kinetic curves characterizing double-step plasma pyrolysis are shown in Fig. 9–2. The first step of the process in this case is plasma pyrolysis of methane in an H_2 plasma jet. In the second phase, room-temperature propane (C_3H_8) is injected with some delay (three specific delay intervals are compared in Fig. 9–2). The kinetic results presented in the figure correspond to the molar fraction of the second phase flow rate of hydrocarbons (C_3H_8:$CH_4 = 0.5$). Double-step pyrolysis leads to a larger production of ethylene. Changing the starting moment of the propane injection affects the acetylene-to-ethylene ratio in products of pyrolysis. Earlier injection results in a larger amount of ethylene in products. Double-step plasma pyrolysis permits a significant increase of energy efficiency for the process. Typically, the energy cost of co-production of ethylene and acetylene decreases in this case two- to three-fold (Gershuni et al., 1975; Vurzel et al., 1975).

9.1.3. Electric Cracking of Natural Gas with Production of Acetylene–Hydrogen or Acetylene–Ethylene–Hydrogen Mixtures

The conversion of natural gas into acetylene, ethylene, and hydrogen is the most developed industrial plasma process of hydrocarbon pyrolysis. The process takes place directly in thermal plasma discharges and, therefore, is called electric cracking. Industrial electric cracking units usually operate at power levels of 5–10 MW; the length of the thermal arc is about 1 m. The optimized characteristics of industrial electric cracking include the following (Gladisch, 1962, 1969): total conversion degree of natural gas, 70%; conversion degree into acetylene, 50%; volume fraction of acetylene in products, 14.5%; and consumption of electric energy, 13.6 kWh per 1 m^3 of natural gas. A drawback of the plasma-chemical process is the considerable amount of soot in the reaction products. Westinghouse applied thermal arcs with magnetic stabilization for natural gas pyrolysis (Maniero, Kienast, & Hirayama, 1966). Arcs are performed in a conical shape in this configuration of electric cracking systems. The Westinghouse process is conducted in two steps (see Section 9.1.2). The first step is the electric cracking of natural gas, and the second step is the product quenching by heavy hydrocarbons with their simultaneous pyrolysis. This approach results in high product yields: the conversion of methane into acetylene reaches 80%, the volume fraction of acetylene in products is 20%, and the consumption of electric energy is 12.5–13.3 kWh per 1 kg of acetylene. These data characterize large-scale industrial plasma systems producing 25,000 t of acetylene per year.

Industrial-scale electric cracking can be used not only for the production of acetylene but also for production or co-production of ethylene, which was demonstrated in particular by Huls, Chem. W., Germany. The relative yield of acetylene and ethylene depends on temperature and duration of the process, the type of quenching, the composition of initial products, and so on. The maximum production of acetylene usually corresponds to temperatures of 1800–2000 K (for CH_4 and C_3H_8) and temperatures of 1600–1880 K (for heavier

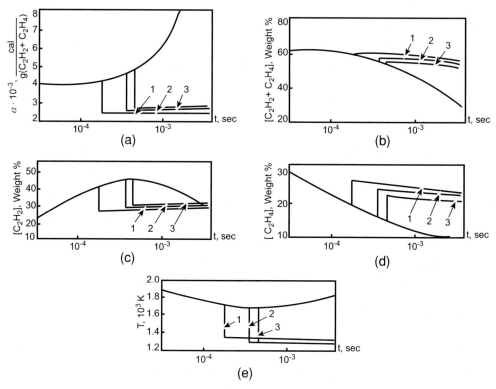

Figure 9–2. Kinetics of the double-step pyrolysis of hydrocarbons in plasma-chemical jet reactor. Step 1, methane pyrolysis in plasma jet of hydrogen; Step 2, propane pyrolysis, injection of propane is delayed with respect to start of the process: (1) 0.17 ms, (2) 0.37 ms, (3) 0.73 ms. Propane injection temperature 293 K; propane-to-methane flow rate ratio 1:2.

hydrocarbons); the process duration is about 10^{-5}–10^{-4} s. In such regimes, the combined yield of acetylene and ethylene can be 65–79 mass %. The energy cost of the co-production is 4.6 kWh/kg in the case of pyrolysis of gasoline or crude oil, and 9.5 kWh/kg in the case of pyrolysis of methane. At lower temperatures (1300–1600 K) and similar process duration (10^{-5}–10^{-4} s), pyrolysis of hydrocarbons with more than three carbon atoms results in much higher concentration of ethylene. Pyrolysis of gasoline, for example, leads in these conditions to the production of 13 vol % of ethylene in products and corresponds to equal production of acetylene and ethylene in the process (C_2H_4:C_2H_2 = 1:1).

9.1.4. Other Processes and Regimes of Hydrocarbon Conversion in Thermal Plasma

Pyrolysis of numerous hydrocarbons has been investigated in thermal plasma systems, mostly in arc jets over a wide range of power, from 10–30 kW to 10 MW. Typical energy carriers in plasma jets in this case are hydrogen or H_2–CH_4. Hydrocarbons are then injected into the plasma jet for pyrolysis. The variety of such processes includes pyrolysis of methane (Guliaev & Polak, 1965; Kozlov, Khudiakov, & Kobsev, 1967, 1974; Souris & Shorin, 1969; Liakhovich, 1974), natural gas of different composition (Gladisch, 1962, 1967; Guliaev & Polak, 1965), propane (Gladisch, 1962, 1967; Vurzel & Polak, 1965; Souris & Shorin, 1969), propylene (Gladisch, 1962, 1967), normal heptane (Il'in & Eremin, 1962, 1965, 1967, 1968; Vurzel & Polak, 1965), butadiene and cyclohexane (Sennewald, Schallus, & Pohl, 1963), different kinds of gasoline (Il'in & Eremin, 1962, 1965, 1967, 1968; Sennewald, Schallus,

Table 9–1. Characteristics of Plasma-Chemical Pyrolysis of Methane, Propane, and Benzene

Process characteristics	Pyrolysis of methane	Pyrolysis of propane	Pyrolysis of benzene
C_2H_2, vol % in products	15.5	13.7	18.1
C_3H_4, vol % in products	0.17	0.23	0.14
C_4H_2, vol % in products	0.36	0.25	0.54
C_4H_4, vol % in products	0.09	0.32	0.35
C_2H_4, vol % in products	0.5	5.83	0.96
C_3H_6, vol % in products	–	0.74	0.09
Allene, vol % in products	–	0.09	0.06
C_4H_8, vol % in products	–	0.13	0.04
C_4H_6, vol % in products	–	0.23	0.08
C_5H_6, vol % in products	–	0.11	0.84
C_6H_6, vol % in products	–	0.17	0.68
CH_4, vol % in products	5.8	8.38	2.96
C_2H_6, vol % in products	–	0.36	–
C_3H_8, vol % in products	–	2.87	0.01
C_4H_{10}, vol % in products	–	0.05	–
H_2, vol % in products	76.0	66.4	75.9
N_2, vol % in products	–	0.23	–
Yield of $C_2H_2 + C_2H_4$, mass %	78.5	70.0	65.0
Yield of soot, g/kWh	–	2.0	40.0
Energy cost, kWh/kg $C_2H_2 + C_2H_4$	9.5	6.6	8.5
Arc discharge power in the pyrolysis	30 kW	8 MW	8 MW

& Pohl, 1963; Vurzel & Polak, 1965; Volodin et al., 1970a,b; Polak, 1971); gas condensate (Vurzel et al., 1973), gasoil (Volodin et al., 1970a,b, 1971a,b), and different kinds of crude oils (Gladisch, 1962, 1967; Volodin et al., 1970a,b, 1971a,b; Herman & Schmidt, 1971; Mosse, Zabrodin & Krylova, 1971, 1973; Polak, 1971).

The characteristics of thermal plasma pyrolysis of different hydrocarbons (Polak et al., 1975a) are summarized in Tables 9–1, 9–2, and 9–3. The tables are organized in a way that simplifies the comparison of the characteristics of pyrolysis for different feedstock hydrocarbons. These tables can be used not only to compare the effectiveness of different hydrocarbon feedstocks in the production of acetylene and ethylene but also to analyze the possibilities of production of other valuable compounds based on thermal plasma pyrolysis. Table 9–1 describes the plasma pyrolysis of methane, propane, and benzene, presenting the volume fraction of gas products, the combined yield of acetylene and ethylene ($C_2H_2 + C_2H_4$, mass percent), the yield of soot in grams per kilowatt-hours and the energy cost of co-production of acetylene and ethylene (kWh/kg$C_2H_2 + C_2H_4$). The table also shows the power of the thermal arc discharge system, where these characteristics were achieved; the power can be recalculated to give the productivity of pyrolysis. Table 9–2 presents the same process characteristics with regard to plasma pyrolysis of gasoline, gas condensate, and gasoil. The table separately shows data related to light gasoline, and gasoline with an upper boiling temperature of 165°C. Finally, Table 9–3 shows again the same characteristics for the case of plasma pyrolysis of different kinds of crude oil. Specifically, three feedstock oils are presented in the table: crude oil, light crude oil, and the Ishimby field (Russia) crude oil. Product yields shown in the tables are partially related to details of initial composition and quenching specifics. These data nevertheless give a good picture for comparison of the effectiveness of different feedstocks in the production of acetylene and ethylene, as well as for analysis of production of other valuable compounds using plasma pyrolysis.

Table 9–2. Characteristics of Plasma-Chemical Pyrolysis of Gasoline, Gasoil, and Gas Condensate

Process characteristics	Pyrolysis of light gasoline	Pyrolysis of gasoline (boiling 165°C)	Pyrolysis of gas condensate	Pyrolysis of gasoil
C_2H_2, vol % in products	14.8	17.0	16.5	16.2
C_3H_4, vol % in products	0.34	0.30	0.56	0.27
C_4H_2, vol % in products	0.25	0.16	0.25	0.10
C_4H_4, vol % in products	0.36	–	–	–
C_2H_4, vol % in products	7.47	7.03	7.05	6.52
C_3H_6, vol % in products	1.13	0.40	0.10	0.69
Allene, vol % in products	0.13	0.15	0.15	0.10
C_4H_8, vol % in products	0.24	0.13	–	0.12
C_4H_6, vol % in products	0.32	–	–	–
C_6H_6, vol % in products	0.11	–	–	–
CH_4, vol % in products	7.92	7.50	5.37	6.11
C_2H_6, vol % in products	0.32	–	–	–
C_3H_8, vol % in products	0.03	–	–	–
H_2, vol % in products	64.10	67.31	69.72	69.89
O_2, vol % in products	0.03	–	–	–
N_2, vol % in products	0.33	0.04	–	–
Hydrocarbons C_5, vol %	2.12	–	–	–
Yield of $C_2H_2 + C_2H_4$, mass %	70.0	72.0	77.0	74.5
Yield of soot, g/kWh	6.0	–	–	–
Energy cost, kWh/kg $C_2H_2 + C_2H_4$	6.4	4.6	4.7	4.9
Arc discharge power in the pyrolysis	8 MW	200 kW	30 kW	30 kW

In the case of pyrolysis of hydrocarbons with more than three carbon atoms (C > 3), the composition of the initial products (alkanes) does not much affect the process yield. However, an increase in the fraction of hydrocarbons with higher boiling temperatures leads to a decrease of acetylene and ethylene yield (Mitscuio & Yasio, 1968, 1970). Also, aromatic compounds in the mixture undergoing plasma pyrolysis lead to higher yields of soot. An increase of gas pressure affects the plasma pyrolysis of hydrocarbons in a negative manner (Kozlov et al., 1967; Polak et al., 1969; Kobsev & Khudiakov, 1971). A pressure increase from 1 to 6 atm in particular results in lower yields of acetylene and higher energy costs of its production. An increase of temperature from 1300 K to 1700 K leads at first to an increase in the yields of vinyl acetylene from 0.18 to 0.30 vol % at 1500 K; then the yield decreases to 0.24 vol %. The concentration of diacetylene in this case does not exceed 0.1 vol %; the concentration of other compounds decreases from 1.44 to 0.4 vol %. Plasma pyrolysis can be conducted with a relatively low yield of products (60–80%) in the gas phase, and the rest are formed mostly in the liquid phase (Polak et al., 1975a). The liquid products in this case include some unconverted initial hydrocarbons and a significant amount of aromatic compounds. For example, the paraphene/naphthene fraction of gas condensate does not initially include any aromatic compounds. Pyrolysis of the fraction can result in the formation of liquid products containing 20% aromatic compounds, specifically toluene, benzene, and xylols. Pyrolysis of terpenes in the plasma jet of nitrogen is considered by Stokes (1971). Terpenes are injected in the nitrogen plasma jet in a liquid form; quenching is done in a water-cooled chamber. The conversion degree has been increased with the discharge power increase – the conversion of β-pinene into myrcene reaches 26% in this process. Plasma pyrolysis of α-terpinene leads to the formation of α-pinene and para-cymene with a conversion degree of about 4%.

Table 9–3. Characteristics of Plasma-Chemical Pyrolysis of Crude Oils

Process characteristics	Pyrolysis of crude oil	Pyrolysis of light crude oil	Pyrolysis of crude oil, Ishimby field
C_2H_2, vol % in products	14.5	15.00	15.95
C_3H_4, vol % in products	0.31	0.35	0.28
C_4H_2, vol % in products	0.30	0.39	0.16
C_4H_4, vol % in products	0.38	0.41	–
C_2H_4, vol % in products	6.51	6.43	7.00
C_3H_6, vol % in products	1.12	1.00	0.70
Allene, vol % in products	0.17	0.18	0.10
C_4H_8, vol % in products	0.23	0.22	0.22
C_4H_6, vol % in products	0.33	0.026	–
C_5H_6, vol % in products	0.20	0.16	–
C_6H_6, vol % in products	0.38	0.25	–
CH_4, vol % in products	6.04	6.26	5.82
C_2H_6, vol % in products	0.26	0.23	–
C_3H_8, vol % in products	0.12	0.05	–
C_4H_{10}, vol % in products	0.22	0.02	–
C_5H_{12}, vol % in products	0.26	0.15	–
H_2, vol % in products	68.5	68.20	69.37
CO, vol % in products	0.15	0.33	0.26
H_2S, vol % in products	–	–	0.14
HCN, vol % in products	–	–	0.06
Yield of $C_2H_2 + C_2H_4$, mass %	–	71.0	74.5
Yield of soot, g/kWh	–	–	9.0
Energy cost, kWh/kg $C_2H_2 + C_2H_4$	–	7.0	4.8
Arc discharge power in the pyrolysis	8.5 MW	8 MW	30 kW

9.1.5. Some Chemical Engineering Aspects of Plasma Pyrolysis of Hydrocarbons

Pyrolysis of different hydrocarbons has been performed in a hydrogen plasma jet. Hydrogen is chosen as an energy carrier for three reasons. First, H_2 is one of the major products of pyrolysis. Second, H_2 is effective as an energy carrier for pyrolysis: when it is heated in plasma to 3700 K and the reaction proceeds at 1300 K, then 90% of its energy is used for the chemical process. Third, H_2 prevents the formation of soot and diacetylene. A mixture of 80–90% H_2 with 10–20% CH_4 can also be effectively applied as an energy carrier for pyrolysis. This mixture is actually a residual fraction of the products of pyrolysis after extraction of acetylene and ethylene. The yield of C_2H_2/C_2H_4 is about 5–10% higher in this case with respect to the use of pure hydrogen as an energy carrier. The application of natural gas as an energy carrier for pyrolysis of other hydrocarbons was investigated by Kozlov, Khudiakov, and Kobsev (1967, 1974). The yield of C_2H_2/C_2H_4 can be increased by using water vapor as an energy carrier; however, the energy cost of the process also increases in this modification of plasma pyrolysis (Il'in & Eremin, 1962, 1965, 1967, 1968). For example, when application of H_2 as an energy carrier provides pyrolysis of gasoline with formation of 5.6 vol % ethylene and 9.4 vol % acetylene, replacement of hydrogen by water vapor leads to formation of 11 vol % ethylene and 19 vol % acetylene. Thus, the yield of co-production of acetylene and ethylene increases by a factor of 2 in this case. The energy cost of the C_2H_2/C_2H_4 co-production increases 25% at the same time. Quenching of products by injection of initial compounds as well as by preheating of the initial compounds makes

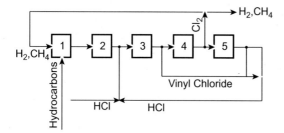

Figure 9–3. Technological scheme of production of the vinyl chloride monomer: (1) plasma-chemical pyrolysis; (2) cleaning from higher unsaturated hydrocarbons; (3) hydrochlorination of acetylene; (4) chlorination of ethylene; (5) thermal pyrolysis of dichloroethane.

plasma pyrolysis more effective (Gladisch, 1962, 1969; Kobsev & Khudiakov, 1971). For example, preheating of natural gas to 500°C increases the acetylene yield and decreases the energy cost of the product by about 25%. The result has been achieved by using hydrogen as well as methane as energy carriers. Recuperation of the heat of plasma pyrolysis gases obviously improves the energy balance of the whole technology. The heat recuperation in the system can be conducted in a wide variety of methods. It can be specifically mentioned that 60–70% of the energy carried by the plasma pyrolysis gases was effectively used for generating technological vapor in the plasma pyrolysis approach developed by Sennewald, Schallus, and Pohl (1963).

9.1.6. Production of Vinyl Chloride as an Example of Technology Based on Thermal Plasma Pyrolysis of Hydrocarbons

Acetylene and ethylene produced by plasma pyrolysis of hydrocarbons can be used for synthesis of different monomers, including vinyl chloride, acryl-nitryl, and chloroprene. The introduction of plasma stages leads in this case to a fundamental reorganization of the entire technology. Consider as an example the industrial process for production of vinyl chloride, which is a monomer applied for large-scale production of polyvinyl chloride (PVC). The technology is based on plasma pyrolysis of hydrocarbons with production of acetylene and ethylene. Vinyl chloride is produced by hydrochlorination of acetylene and by chlorination of ethylene with a subsequent pyrolysis of dichloroethane. Hydrochlorination of acetylene and chlorination of ethylene are completed without extracting them from the plasma pyrolysis gases (Kureha, 1958; Gomi, 1964; Volodin et al., 1969, 1971a,b). A schematic of the vinyl chloride production technology is illustrated in Fig. 9–3. The *first step* of the technology is pyrolysis of hydrocarbons in the plasma jet containing 80–90% hydrogen and 10–20% methane (see Section 9.1.5). Quenching is accomplished by using either water or gasoline jets. The gas composition after separation of the quenching liquid is hydrogen (H_2) 60.8–66.9 vol %, acetylene (C_2H_2) 12.1–20.0 vol %, ethylene (C_2H_4) 5.0–12.9 vol %, C_2H_6 0–0.3 vol %, CH_4 7.5–10.5 vol %, C_3H_6 0.1–1.9 vol %, and higher non-saturated compounds (vinyl acetylene, methyl acetylene, diacetylene, propadiene, and butadiene) 0.5–1.5 vol %. The *second step* is separation of the higher non-saturated hydrocarbons (vinyl acetylene, methyl acetylene, diacetylene, propadiene, and butadiene) by kerosene absorption (temperature −15°C, pressure 6–10 atm). Hydrogen chloride (HCl) is injected in the plasma pyrolysis gases in stoichiometry with acetylene for further hydrochlorination (step 3). Hydrogen chloride applied in this stage is recycled from the dichloroethane pyrolysis (step 4). Catalytic hydrochlorination of acetylene and production of vinyl chloride takes place in the *third step* without C_2H_2 extraction from the pyrolysis gases:

$$C_2H_2 + HCl \rightarrow C_2H_3Cl \qquad (9-5)$$

The gas composition after this stage is hydrogen (H_2) 63.1–67.3 vol %, vinyl chloride (C_2H_3Cl) 12.5–20 vol %, ethylene (C_2H_4) 5.0–13.3 vol %, C_2H_6 0–0.3 vol %, and CH_4 7.6–10.8 vol %. Vinyl chloride is extracted from the mixture by dichloroethane absorption at 10 atm. The *fourth step* of the technology is chlorination of ethylene in pyrolysis gases by injection of chlorine (Cl_2)

$$C_2H_4 + Cl_2 \rightarrow C_2H_2Cl_2. \tag{9–6}$$

The chlorination of ethylene proceeds in liquid dichloroethane in the presence of a catalyst (iron chloride) at a pressure of 5 atm without separation of the pyrolysis gas mixture. The composition of the residual plasma pyrolysis gas mixture after the liquid-phase chlorination is hydrogen (H_2) 85.5–90 vol % and methane (CH_4) 10–14.5 vol %. This mixture is then applied as an energy carrier gas for plasma pyrolysis of hydrocarbons in the first stage. Dichloroethane formed in liquid phase then undergoes thermal pyrolysis in the *fifth step* of the technology. The pyrolysis of dichloroethane is performed at temperatures of 450–550°C and pressures of 7 atm and results in production of vinyl chloride and hydrogen chloride:

$$C_2H_4Cl_2 \rightarrow C_2H_3Cl + HCl. \tag{9–7}$$

Vinyl chloride produced at this stage is a product of the entire technology together with the vinyl chloride produced in the third step (9–6). Hydrogen chloride (HCl) produced in the fifth step (9–7) is recycled back to the third step (9–5) for hydrochlorination of acetylene. The economics of the plasma-chemical technologies based on thermal plasma pyrolysis of hydrocarbons with initial production of acetylene and ethylene is discussed by Rabkina, Borisov, and Ponomareva 1971a,b.

9.1.7. Plasma Pyrolysis of Hydrocarbons with Production of Soot and Hydrogen

When plasma pyrolysis is conducted without special quenching, the hydrocarbons can be mostly converted into soot and hydrogen. The quality of the soot is better than that of the soot generated in furnaces. The soot produced in plasma is highly dispersed and characterized by a developed secondary structure. The mechanism and kinetics of soot formation in plasma can be found, in particular, in Rykalin et al. (1970). The kinetic analysis shows that the most intensive soot production takes place during pyrolysis of aromatic compounds. The process has been performed in the arc discharges and thermal radiofrequency (RF) inductively coupled plasma (ICP) discharges directly, as well as in plasma jets of argon, nitrogen, and hydrogen at temperatures from 1500 to 5000–6000 K and residence times of 10^{-3}–10^{-2} s. Different hydrocarbons were used as a feedstock, including methane, natural gas, propane, toluene, naphthalene, xylol, gasoline, and special mixtures of aromatic compounds (Sheer, 1961; Orbach, 1966; Ryan, 1966; Baddur, 1967; Latham, 1967; Bjornson, 1968; Artukhov et al., 1969). The maximum yield of soot in these plasma systems (45–70 mass %) is achieved by pyrolysis of aromatic compounds. The energy cost of soot is 7–13 kWh/kg in this case. Thermal plasma pyrolysis of paraphenes and naphthenes leads to a lower yield of soot (18 mass %) and a higher energy cost (70 kWh/kg). The structure of soot is usually characterized by the so-called oil number, which is usually not higher than 2.4 cm^3/g for conventional soot. The oil number for plasma produced soot reaches 4.0 cm^3/g, and the specific surface exceeds 150 m^2/g. High-quality soot with fixed crystal structure can be produced by thermal plasma pyrolysis of low-quality soot or other cheap carbon-containing materials (Cabot Corp., 1967). Initial products in this case are injected into a plasma jet, evaporated, and condensed after the quenching process to form special high-quality soot and carbon structures. Plasma pyrolysis of hydrocarbons leads to the formation of a condensed phase not only in the volume but also on the walls of the reactor. The reaction of condensed-phase carbon with hydrocarbons usually results in the formation of pyro-graphite on the reactor walls (Tesner & Altshuller, 1969a,b; Tesner, 1972). The process takes place in thermal plasma conditions at temperatures of 1800–2100 K and atmospheric pressure. We should note that a similar process can also take place under non-thermal plasma conditions (Kiyshy, 1967).

9.1.8. Thermal Plasma Production of Acetylene by Carbon Vapor Reaction with Hydrogen or Methane

High-intensity arc discharges operating at atmospheric pressure and temperatures sufficient for sublimation of graphite have been applied to produce acetylene by reactions of the carbon vapor with hydrogen or methane (Clark, 1970). Reaction with hydrogen provides the formation of acetylene (19.3 vol % of C_2H_2) in products. Using a plasma jet in a mixture of helium and hydrogen ($He:H_2 = 2:1$), the concentration of acetylene in products is 24 vol %. This value exactly corresponds to the quasi-equilibrium thermodynamic calculations. Replacement of hydrogen by methane as the gas energy carrier in plasma jets leads to an increase of acetylene concentration in products reaching 52 vol %, again in good agreement with quasi-equilibrium thermodynamics.

9.2. CONVERSION OF METHANE INTO ACETYLENE AND OTHER PROCESSES OF GAS-PHASE CONVERSION OF HYDROCARBONS IN NON-THERMAL PLASMAS

9.2.1. Energy Efficiency of CH_4 Conversion into Acetylene in Thermal and Non-Thermal Plasmas

One of the most crucial requirements for the plasma conversion of methane into acetylene is improvement of its energy efficiency or, in other words, reduction of the energy cost for acetylene production. The lowest energy cost achieved in thermal plasma systems is 8.5 eV/mol C_2H_2 (Polak, 1965; Kozlov et al., 1974). Comparison with the acetylene from methane production (9–1) enthalpy (3.8 eV/mol C_2H_2) shows that the highest energy efficiency achieved in thermal plasma systems is less than 45%. Lower energy costs can be achieved in non-thermal plasma conditions if the process is effectively stimulated by vibrational excitation of CH_4 molecules. Conducting the process in a non-equilibrium microwave discharge in conditions optimal for vibrational excitation permits an energy cost of acetylene production of 6 eV/mol C_2H_2 (energy efficiency 63%), and a methane-to-acetylene conversion degree of 80% (Babaritsky et al., 1991). The CH_4 conversion into C_2H_2 in a non-equilibrium glow discharge at pressures of 60–80 Torr was done with an energy cost of 9.6 eV/mol C_2H_2 and energy efficiency of 40%. The methane-to-acetylene conversion degree in the system was only 8% (Peters & Pranschke, 1930). In the experiments with non-thermal microwave discharges focused on the higher yields (McCarthy, 1954), the methane-to-acetylene conversion degree reaches 95%. The energy cost of acetylene production, however, was high in these experiments (30 eV/mol C_2H_2; energy efficiency 13%).

9.2.2. High-Efficiency CH_4 Conversion into C_2H_2 in Non-Thermal Moderate-Pressure Microwave Discharges

The highest energy efficiency and the lowest energy cost of methane conversion into acetylene is achieved, as mentioned earlier, in non-equilibrium microwave discharges (Babaritsky et al., 1991) at moderate pressures (10–80 Torr) in the regime providing essential separation of vibrational and translational temperatures ($T_v > T_0$; see Section 3.6.2). A schematic of the plasma-chemical methane conversion system is shown in Fig. 9–4. A magnetron is used as a source of microwave radiation. Microwave radiation with a frequency of 2.45 GHz and power between 100 W and 1.5 kW is guided toward the methane conversion reactor by rectangular waveguides. Methane is injected directly into the discharge with flow rates corresponding to specific energy inputs E_v (electric power absorbed in plasma divided by methane flow rate) in the range $E_v = 0.5–3$ eV/mol. These values of the specific energy input

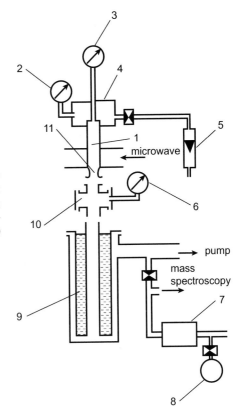

Figure 9–4. General schematic of a microwave plasma system of methane conversion: (1) plasma-chemical reactor; (2, 3, 6) vacuum meters; (4) receiver; (5) flow rotameter; (7) pump; (8) cell for characterization of chemical composition; (9) liquid nitrogen trap; (10) receiving cell; (11) supersonic nozzle.

are optimal for process stimulation by vibrational excitation (see Section 3.6.4). Measurements of vibrational and translational temperatures in the non-thermal microwave discharge are described in Section 5.5.4 and demonstrate the essential vibrational–translational (VT) non-equilibrium ($T_v > T_0$) of the system.

Translational temperature T_0 in the non-equilibrium plasma system was not sufficient for effective methane conversion, whereas vibrational temperature was able to stimulate the process. The conversion degree of methane into acetylene achieved in the plasma system is presented in Fig. 9–5 as a function of the specific energy input at three different pressures: 10, 40, and 80 Torr. When the pressure increases from 10 to 80 Torr, the methane-to-acetylene conversion degree rises and reaches a maximum of 80% at $E_v = 2.6$ eV/mol CH_4. The gas-product composition in this regime is H_2, 73 vol %; CH_4, 5 vol %; and C_2H_2, 22 vol %. The concentration of ethylene, ethane, and higher hydrocarbons is less than 1%; soot production

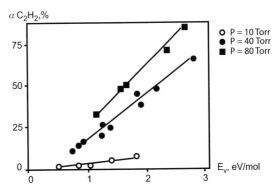

Figure 9–5. Conversion degree of methane into acetylene as a function of specific energy input at three different pressures (10, 40, and 80 Torr).

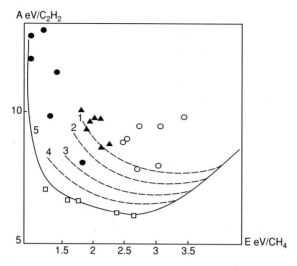

Figure 9–6. Energy cost of acetylene production from methane in plasma: •, Babaritsky's microwave plasma experiments, $p = 40$ Torr; □, Babaritsky's microwave plasma experiments, $p = 80$ Torr; ○, ▲, experiments with thermal arc discharges. Curves represent calculations: (1) quasi-equilibrium plasma modeling; (2–5) non-equilibrium kinetic modeling at different values of the non-equilibrium factor $\gamma = (T_v - T_0)/T_0 = 0.1, 0.2, 0.3, 0.5$.

is not significant. The energy cost of acetylene production in the non-equilibrium plasma system is shown in Fig. 9–6 as a function of the specific energy input E_v at two different pressures, 40 and 80 Torr. The experimental dependence first of all demonstrates a reduction of energy cost (increase of energy efficiency) with an increase of E_v, which is typical for plasma-chemical reactions stimulated by vibrational excitation (see ~~~~~
energy efficiency at a pressure of 80 Torr is higher than that corresponding to a pressure of 40 Torr, which also reflects the general tendency for processes stimulated by vibrational excitation of molecules (see Fig. 5.26 to compare). Most important, the energy cost of acetylene production from methane can be reduced in the microwave discharges to the record low value of 6 eV/mol (energy efficiency 63%), which is interesting to analyze from the points of view of quasi- and non-equilibrium plasma-chemical kinetics (see next sections). Non-equilibrium plasma conversion of methane into acetylene does not require special quenching, because the translational gas temperature remains relatively low. To prove the effect, it has been demonstrated in the aforementioned experiments that introduction of special fast quenching of the conversion products does not affect the yield of the non-equilibrium plasma-chemical process.

9.2.3. Limits of Quasi-Equilibrium Kassel Kinetics for Plasma Conversion of CH$_4$ into C$_2$H$_2$

To determine the upper limits of energy efficiency of this quasi-equilibrium process, kinetic specifics of the Kassel mechanism should be taken into account. The total endothermic process (9–1) of acetylene production from methane (CH$_4$ → 1/2C$_2$H$_2$ + 3/2H$_2$) under quasi-equilibrium conditions corresponds to the first kinetic order and can be characterized by the reaction rate coefficient

$$k = 10^{12} \exp(-39000 \, \text{K/T}), \, 1/\text{s}. \tag{9–8}$$

The detailed Kassel kinetics can be represented as the sequence of the quasi-equilibrium transformations of methane to ethane (endothermic), ethane to ethylene (endothermic),

Figure 9–7. Temperature dependence of the reaction rate coefficients: (1) methane conversion into ethane $(CH_4 \rightarrow 1/2\,C_2H_6 + 1/2H_2)$ at different values of the non-equilibrium factor $\gamma = (T_v - T_0)/T_0$; (2) ethane conversion into ethylene $(C_2H_6 \rightarrow C_2H_4 + H_2)$; (3) ethylene conversion into acetylene $(C_2H_4 \rightarrow C_2H_2 + H_2)$; (4) acetylene conversion into soot $(C_2H_2 \rightarrow 2C_{cond} + H_2)$.

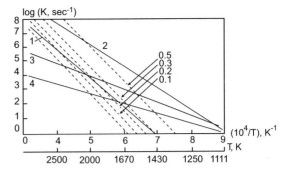

ethylene to acetylene (endothermic), and acetylene to soot (exothermic). The reaction rate coefficients of the Kassel mechanism are all of the first kinetic order and are specified by Gorov (1989; see also Fig. 9–7):

$$2\,CH_4 \rightarrow C_2H_6 + H_2, \quad \Delta H = 0.7\,eV/mol, \quad k = 4.5 \cdot 10^{13} \exp(-46000\,K/T),\ 1/s; \tag{9–9}$$

$$C_2H_6 \rightarrow C_2H_4 + H_2, \quad \Delta H = 1.4\,eV/mol, \quad k = 9.0 \cdot 10^{13} \exp(-35000\,K/T),\ 1/s; \tag{9–10}$$

$$C_2H_4 \rightarrow C_2H_2 + H_2, \quad \Delta H = 1.8\,eV/mol, \quad k = 2.6 \cdot 10^{8} \exp(-20500\,K/T),\ 1/s; \tag{9–11}$$

$$C_2H_2 \rightarrow 2C_s + H_2, \quad \Delta H = -2.34\,eV/mol, \quad k = 1.7 \cdot 10^{6} \exp(-15500\,K/T),\ 1/s. \tag{9–12}$$

To produce acetylene, the sequence of reactions (9–9)–(9–12) should be interrupted after stage (9–11) of C_2H_2 formation is completed and before the fast exothermic reaction (9–12) converts a significant amount of acetylene into soot. It can be achieved when the gas residence time in the discharge is 10^{-3} s and the quenching rate is 10^6 K/s. Results of kinetic calculations of the C_2H_2 production energy cost as a function of the specific energy input E_v in such quasi-equilibrium conditions (Babaritsky et al., 1991) are presented in Fig. 9–6. These quasi-equilibrium kinetic calculations are in a good agreement with experimental results achieved in thermal arc discharges (Polak, 1965; Kozlov et al., 1974), see Fig. 9–6. According to Babaritsky et al. (1991), the quasi-equilibrium kinetic calculations actually determine the lower limit of the energy cost for acetylene production from methane in thermal plasma. The kinetic calculations are in good agreement with experimental results (see Fig. 9–6). The minimum energy cost for a quasi-equilibrium thermal plasma system is 8.5 eV/mol C_2H_2, which can be achieved at a specific energy input $E_v = 2.5$–3 eV/mol. The energy costs achieved in the experiments with non-thermal microwave discharge at pressure 80 Torr are significantly below the limit of the quasi-equilibrium systems and reach in the minimum 6 eV/mol C_2H_2. Interpretation of these results requires taking the contribution of vibrational excitation of CH_4 molecules into account.

9.2.4. Contribution of Vibrational Excitation to Methane Conversion into Acetylene in Non-Equilibrium Discharge Conditions

VT non-equilibrium $(T_v > T_0)$ can increase the energy efficiency of methane conversion in accordance with general principles discussed in Sections 3.6.1 and 3.6.2. Specifically for the case of methane conversion, the first three steps in the Kassel scheme, (9–9)–(9–12), are endothermic and are therefore effectively stimulated by vibrational excitation, while acetylene conversion into soot (9–12) is exothermic and therefore is controlled by translational temperature. Thus, the $T_v > T_0$ non-equilibrium promotes the formation of acetylene without significant soot production. A limiting reaction of the methane conversion mechanism is initial dissociation of CH_4 molecules, which is a part of step (9–9) in the Kassel mechanism and has relatively high activation energy (Isbarn, Ederer, & Ebert, 1981):

$$CH_4 + M \rightarrow CH_3 + H + M, \quad E_a = 4.5\,eV/mol. \tag{9–13}$$

This elementary reaction proceeds mostly through vibrational excitation of CH_4 molecules even in quasi-equilibrium systems. Therefore, providing the non-equilibrium discharge energy selectively to the vibrational degrees of freedom of CH_4 molecules permits a significantly lower energy cost. In contrast to CO_2 dissociation, however, the level of the VT non-equilibrium in CH_4 plasma of moderate pressure is not so high. It is mostly due to fast VT relaxation of CH_4 molecules in collisions with multiple intermediate products from the conversion process, which explains why the maximum energy efficiency of methane conversion into acetylene achieved in non-equilibrium conditions (63%) is lower than that of CO_2 dissociation (90%). However, The vibrational excitation of methane still improves the energy efficiency with respect to quasi-equilibrium systems, even when the parameter of non-equilibrium, $\gamma = (T_v - T_0)/T_0$, is not very high (usually $\gamma = 0.1$–0.5).

9.2.5. Non-Equilibrium Kinetics of Methane Conversion into Acetylene Stimulated by Vibrational Excitation

The kinetics of dissociation of polyatomic molecules, in particular CH_4, stimulated by vibrational excitation in conditions of not very high non-equilibrium parameters $\gamma = (T_v - T_0)/T_0$ has been analyzed by Kuznetsov (1971). The CH_4 dissociation (9–13) proceeds through vibrational excitation of CH_4 molecules at any parameters $\gamma = (T_v - T_0)/T_0$. The vibrational energy distribution is an essentially non-Boltzmann distribution in this case, and it is characterized by both temperatures, T_v and T_0, even when $T_v > T_0$. The rate coefficient $k_R(T_0, T_v)$ of the methane dissociation (9–13) in non-equilibrium conditions ($T_v > T_0$) can be expressed as follows (Kuznetsov, 1971):

$$k_R(T_0, T_v) = k_{R0}(T_0)\frac{Q(T_0)}{Q(T_v)} \exp\left[-E^*(T_0) \cdot \left(\frac{1}{T_v} - \frac{1}{T_0}\right)\right], \tag{9–14}$$

where $k_{R0}(T_0)$ is the rate coefficient of methane dissociation (9–13) in quasi-equilibrium conditions; $Q(T_v)$ and $Q(T_0)$ are relevant partition functions; $E^*(T_0)$ is the vibrational energy of a CH_4 molecule when the VT relaxation rate becomes equal to that of vibrational–vibrational exchange (see Section 3.4.4). For calculations at low values of the parameter $\gamma = (T_v - T_0)/T_0$, expression (9–14) can be simplified:

$$\ln k_R(T_0, T_v) = \ln k_{R0}(T_0) + A \cdot \gamma, \tag{9–15}$$

where the parameter $A = E^*(T_0)/T_v$ is slowly changing with temperature. The CH_4 dissociation rate coefficient recalculated to the first kinetic order is shown in Fig. 9–7 as a function of translational temperature and the non-equilibrium parameter $\gamma = (T_v - T_0)/T_0$ (Babaritsky et al., 1991). To compare, the quasi-equilibrium reaction rates of the Kassel scheme (9–9)–(9–12) are also presented in the figure. Results of the non-equilibrium kinetic calculations of the energy cost of C_2H_2 production from methane as function of the specific energy input E_v and the non-equilibrium parameter $\gamma = (T_v - T_0)/T_0$ (Babaritsky et al., 1991) are also presented in Fig. 9–6 for the case of gas pressure of 80 Torr. These calculations take into account the acceleration by vibrational excitation of the only reaction of CH_4 dissociation, (9–13) other reactions are considered as in Section 9.2.3. These non-equilibrium kinetic calculations are in a good agreement with experimental results achieved in the non-equilibrium microwave discharge (Section 9.2.2, especially assuming the non-equilibrium parameter $\gamma = (T_v - T_0)/T_0 = 0.5$. As seen from Fig. 9–6, the minimum energy cost value

Table 9–4. Isomerization of Hydrocarbons in Non-Thermal Plasma

Initial reagent	Conversion degree	Reaction product	Molar fraction from total products
Trans-stilbene	20%	*Cis*-stilbene	95%
Anisole	67%	Orto-cresol	48%
		Para-cresol	25%
Phenylethanol	30%	Orto-ethyl-phenol	41%
		Para-ethyl-phenol	29%
n-Propyl-phenyl-ester	30%	Orto-propyl-phenol	38%
		Para-propyl-phenol	19%
		Orto-ethyl-phenol	1%
		Para-ethyl-phenol	0.5%
		Orto-cresol	4%
		Para-cresol	7%
1-Naphthyl-methyl-ester	13%	2-Methyl-1-naphthol	48%
2-Naphthyl-methyl-ester	88%	1-Methyl-2-naphthol	45%
Di-Phenyl-Ester	40%	2-Phenyl-phenol	36%
		4-Phenyl-phenol	18%
		Dibenzofuran	9%
N,N-Dimethylaniline	15%	N-Methyl-*o*-toluidine	28%
		N-Methyl-*n*-toluidine	15%
n-Methylaniline	6%	Orto-toluidine	28%
		Para-toluidine	5%
Cyclooctatetraene	80%	Styrene	40%
Pyrol	6%	*Cis*-, *Trans*-Croton-nitryl	57%

of acetylene production in the calculations is $6\,eV/mol\,C_2H_2$, exactly the same as in the microwave experiments.

9.2.6. Other Processes of Decomposition, Elimination, and Isomerization of Hydrocarbons in Non-Equilibrium Plasma: Plasma Catalysis

The decomposition of numerous hydrocarbons has been investigated in non-equilibrium glow, microwave, and RF discharges mostly at low pressures (below 10 Torr). Some non-equilibrium hydrocarbon decomposition processes were analyzed at higher pressures however (including atmospheric pressure), using corona and dielectric barrier discharges (DBDs). Specifically, we can mention non-thermal plasma decomposition of linear alkanes such as CH_4, C_2H_6, n-C_4H_{10}, n-C_5H_{12}, n-C_6H_{14}, and n-C_8H_{18}; cyclic alkanes and their derivatives, such as cyclopentane, cyclohexane, and methyl cyclohexane; alkenes such as C_2H_4 and C_4H_8; and aromatic compounds such as C_6H_6, $C_6H_5CH_3$, and $C_6H_4(CH_3)_2$ (see Slovetsky, 1981). Non-equilibrium plasma application for isomerization of hydrocarbons is illustrated in Table 9–4 (Suhr, 1973). The conversion degree of isomerization can be very high, reaching 80% in the treatment of 2-naphtyl-methyl-ester. The selectivity of the reactions can also be very high. For example, from all conversion products of *trans*-stylbene, 95% is *cis*-stylbene.

Similar characteristics are achieved in non-equilibrium plasma stimulation of elimination reactions (Suhr, 1973). In the elimination processes, plasma provides selective cutting of some specific functional groups from initial hydrocarbon molecules, which is illustrated by Table 9–5. For example, the non-thermal plasma treatment of phthalic anhydride is characterized by a high conversion degree (70%) and selectivity, which means that almost 90% of the products are di- and triphenylene. Such conversion and selectivity characteristics of the non-equilibrium plasma systems allow them to be used in preparative organic chemistry.

Table 9–5. Non-Thermal Plasma-Chemical Reactions of Elimination in Hydrocarbons

Initial reagent	Conversion degree	Reaction product	Molar fraction from total products
Benzen ealdehyde	63%	Benzene	81%
		Diphenyl	16%
2-Pyridine-formaldehyde	25%	Pyridine	70%
		2,2-Dipyridyl	20%
Benzophenol	63%	Diphenyl	70%
Dibenzoyl	16%	Diphenyl	98%
Cyclohexanone	8%	Cyclopentane	5%
1-Naphthol	20%	Indene	76%
2-Naphthol	24%	Indene	93%
Phtalic anhydride	70%	Diphenylene	60%
		Triphenylene	27%
Tetralene	50%	Naphthalene	60%

The combination of plasma with catalysts, usually referred to as plasma catalysis, can be helpful in improving yield characteristics of some processes of plasma conversion of hydrocarbons. The term "plasma catalysis" is applied to two very different situations: combination of plasma chemistry with conventional catalysts, and replacing of conventional catalysts by plasma. Here we consider the simple combination of plasma-chemical conversion of methane with application of catalysts to increase the process yield. As an example, the combination of atmospheric-pressure pulsed corona discharge with Ce/CaO–Al$_2$O$_3$ catalyst results at room temperature in a 14% yield of acetylene from methane with selectivity of C$_2$H$_2$ production exceeding 70% (Gong et al., 1997). A combination of atmospheric-pressure DBD with Ni catalyst increases the energy efficiency of the process by a factor of 40 (Nozaki, 2002; Kado et al., 2003).

9.3. PLASMA SYNTHESIS AND CONVERSION OF ORGANIC NITROGEN COMPOUNDS

9.3.1. Synthesis of Dicyanogen (C$_2$N$_2$) from Carbon and Nitrogen in Thermal Plasma

Dicyanogen (C$_2$N$_2$) can be effectively formed in thermal N$_2$ discharges by the plasma-chemical reaction $2C + N_2 \rightarrow C_2N_2$. Carbon can be introduced into the system either from evaporation of a carbon electrode or by special soot injection into the thermal plasma jet (Leutner, 1962, 1965; Ganz, Parkhomenko, & Krasnokutsky, 1969a,b; Polak et al., 1973). According to the thermodynamics of the C–N mixture (composition C:N = 1:1, pressure 1 atm), the maximum concentration of the dicyanogen (about 0.5 vol %) is formed at 3500 K. The CN concentration reaches 40–55% at 3500–5000 K. Assuming complete CN recombination into dicyanogen results in an energy cost of 10.2–11.2 kWh per 1 kg of C$_2$N$_2$. Kinetic curves of the plasma process for the C:N = 1:1 mixture at atmospheric pressure and initial temperature 7000 K are shown in Fig. 9–8 (Polak et al., 1973). A stationary concentration of CN radicals is achieved after about $5 \cdot 10^{-4}$ s. Formation of the dicyanogen takes place during the quenching stage and reaches 12%. Experiments were carried out at atmospheric pressure based on both evaporation of carbon electrodes and injection of soot in the nitrogen jet at a temperature of 4000 K. The conversion of carbon in the experiments was at least 15%, and the concentration of C$_2$N$_2$ in the products was as high as 18 vol %.

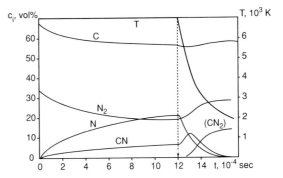

Figure 9–8. Kinetics of plasma-chemical formation of $(CN)_2$.

9.3.2. Co-Production of Hydrogen Cyanide (HCN) and Acetylene (C_2H_2) from Methane and Nitrogen in Thermal Plasma Systems

Methane injection into a nitrogen plasma jet leads to the following reaction:

$$a \cdot CH_4 + N_2 \rightarrow 2HCN + \left(\frac{a}{2} - 1\right) \cdot C_2H_2 + \frac{3a}{2} \cdot H_2. \tag{9–16}$$

Taking into account that H:C = 1:1 in both C_2H_2 and HCN, the conversion of methane in N_2 plasma is always accompanied by hydrogen production. According to the high-temperature thermodynamics, optimal conditions of C_2H_2 and HCN co-production from the mixture $CH_4:N_2 = 1:1$ correspond to a temperature of about 3000 K at atmospheric pressure. The concentration of HCN in the products is 5–6 vol % in this case; the concentration of acetylene is about 1%. Better characteristics can be achieved when the H-to-C ratio is less than 4. Such composition can be achieved by taking into account the contribution of carbon evaporated from the arc discharge electrodes. For example, an optimal synthesis in atmospheric-pressure mixtures C:H:N = 1:2:2 and C:H:N = 1:1:4 takes place at lower temperatures (1500–2500 K) and leads to the formation of 22.1 vol % of C_2H_2 and 20.6 vol % of HCN for the first mixture, and 13.9 vol % of C_2H_2 and 12 vol % of HCN for the second mixture. The energy cost of the C_2H_2–HCN co-production in the case of the second mixture is 4.8–5.3 kWh/kg (Ganz & Krasnokutsky, 1971). Kinetic analysis of the thermal plasma-chemical process shows that production of acetylene in the system proceeds mostly via the Kassel mechanism (see Sections 9.1.1 and 9.2.3), whereas a significant amount of HCN is formed in reactions of atomic nitrogen with acetylene, ethylene, and methyl radicals:

$$N + C_2H_4 \rightarrow HCN + CH_3, \tag{9–17}$$

$$N + C_2H_2 \rightarrow HCN + CH, \tag{9–18}$$

$$N + CH_3 \rightarrow HCN + H_2. \tag{9–19}$$

The formation of atomic nitrogen is due to dissociation of N_2. The production of hydrogen cyanide is also due to preliminary formation of CN radicals followed by reactions with molecular and atomic hydrogen:

$$CN + H_2 \rightarrow HCN + H, \tag{9–20}$$

$$CN + H + M \rightarrow HCN + M. \tag{9–21}$$

Results of the kinetic calculation of the process starting from nitrogen plasma at 6500 K and corresponding to a mixture composition of $CH_4:N_2 = 1:3$ are presented in Fig. 9–9

605

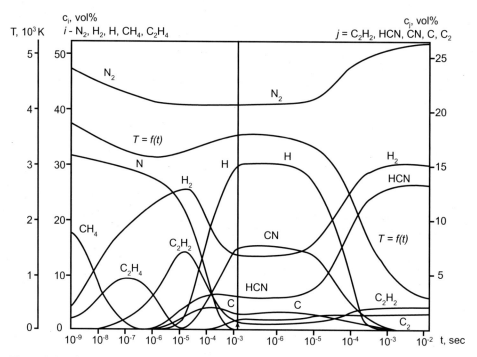

Figure 9–9. Kinetics of plasma-chemical formation of HCN (pointer and vertical line indicate starting moment of quenching).

(Polak et al., 1973). Quenching starts in this kinetic modeling at the moment of maximum HCN concentration; the cooling rate is supposed to be equal to 10^6 K/s. Experiments with thermal plasma co-production of hydrogen cyanide and acetylene were carried out via pyrolysis of hydrocarbons in an N_2 plasma jet, as well as by reaction of carbon evaporated from a cathode with H_2 or NH_3 in the N_2 plasma jet (Leutner, 1962, 1963; Schallus & Gotz, 1962; Neumann, 1964; Ganz, Parkhomenko, & Krasnokutsky, 1969). The second approach results in conversion of 50% of the evaporated carbon into HCN, but the reaction rate in this case is low, however, and determined by the carbon evaporation rate. The reaction rate of HCN and C_2H_2 formation is much faster when methane is injected in the N_2 thermal plasma jet; 91% of the methane is converted in this case into valuable products. The typical composition of products is HCN, 10.7 vol %; C_2H_2, 12.9%; H_2, 74% (Yasui & Akitani, 1965; Hirayama & Maniero, 1967; Ganz, Parkhomenko, & Krasnokutsky, 1969). Process (9–16) has been applied at the industrial scale with an energy cost of about 10 eV/mol HCN (Heilmann, 1963; Ganz & Parkhomenko, 1976).

9.3.3. Hydrogen Cyanide (HCN) Production from Methane and Nitrogen in Non-Thermal Plasma

The production of hydrogen cyanide from methane and nitrogen can be presented as process (9–16) assuming $a = 2$. Plasma-chemical HCN production is an endothermic process:

$$CH_4 + 1/2\,N_2 \rightarrow HCN + 3/2\,H_2, \quad \Delta H = 2.6\,\text{eV/mol} \tag{9–22}$$

Hydrogen cyanide production in thermal plasma requires about 10 eV/mol HCN (Heilmann, 1963; Ganz & Parkhomenko, 1976), which corresponds to an energy efficiency of 26%. Low energy cost and higher energy efficiency can be achieved by conducting HCN synthesis (9–22) in non-equilibrium plasma with the process stimulated through vibrational excitation of N_2 molecules (Panteleev, 1978; Rusanov & Fridman, 1984). HCN synthesis stimulated by vibrational excitation of molecular nitrogen, $N_2^*(^1\Sigma_g^+, v)$, proceeds through the chain mechanism:

$$H + N_2^*(^1\Sigma_g^+, v) \rightarrow NH + N, \quad E_a \approx \Delta H = 7\,eV/mol, \qquad (9\text{--}23)$$

$$N + CH_4 \rightarrow HCN + H_2 + H, \quad \Delta H = -0.5\,eV/mol, \quad E_a = 0.5\,eV/mol. \quad (9\text{--}24)$$

The NH radicals are mostly lost in the fast radical–radical reaction of hydrogen production:

$$NH + NH \rightarrow N_2 + H_2, \quad \Delta H = -7.8\,eV/mol, E_a \approx 0. \qquad (9\text{--}25)$$

The limiting stage (9–23) of the chain process is effectively stimulated by vibrational excitation of nitrogen $N_2^*(^1\Sigma_g^+, v)$. The energy efficiency of the process is limited by losses in the exothermic reactions (9–24) and (9–25) and reaches a maximum of 40%. It corresponds to an energy cost of about 6.5 eV/mol HCN, which is 1.5 times better than that of HCN production in thermal plasma systems. The non-equilibrium plasma experiments with HCN synthesis stimulated by vibrational excitation were carried out using the non-self-sustained discharge supported by a relativistic electron beam (Panteleev, 1978). The plasma process has been performed directly in the methane–nitrogen mixture at atmospheric pressure, as well as by preliminary N_2 excitation in the atmospheric-pressure discharge followed by mixing with CH_4. Mechanism (9–23)–(9–25) of HCN production does not include direct plasma dissociation of CH_4 molecules; it is due to much faster excitation of N_2 to the highly excited and highly reactive vibrational levels (Fridman & Kennedy, 2004). As a result, although CH_4 bonding energies are weaker with respect to nitrogen, vibrational excitation leads preferentially to effective dissociation and reaction of N_2 molecules.

Non-equilibrium plasma systems with higher electron temperatures, especially low-pressure discharges, are characterized by a lower contribution of vibrational excitation and preferential electronic excitation and direct dissociation of reacting molecules (see Section 2.5.6). In such plasma systems, dissociation and reactions of CH_4 molecules mostly through electronic excitation dominate over mechanism (9–23)–(9–25) based on the vibrational excitation of nitrogen. These regimes are characterized by preferential production of acetylene (C_2H_2) from methane, while the hydrogen cyanide (HCN) production is relatively suppressed. The relevant example is the process (9–16) for HCN/C_2H_2 co-production, which has been carried out in the non-thermal plasma of a glow discharge in a CH_4–N_2 mixture at pressures of 10–15 Torr (Shekhter, 1935; Molinary, 1974). The maximum methane conversion into HCN and C_2H_2 (about 80%) has been achieved when the molar fraction of methane in the initial mixture was 15%. The HCN concentration in the product mixture was up to 10%. The energy cost of the co-production process in low-pressure glow discharges is higher than that in the atmospheric-pressure plasma processes stimulated by vibrational excitation. It is even higher than the energy cost achieved in the thermal plasma systems. Details about the experiments with HCN/C_2H_2 co-production in the CH_4–N_2 glow discharges at pressures of 10–15 Torr can be found in Kuster (1931), Peters and Kuster (1931), and Molinary (1974). Conversion degrees of methane into HCN and C_2H_2 are shown in Figs. 9–10 and 9–11 as functions of the glow discharge power and initial gas composition (molar fraction of methane in the CH_4–N_2 mixture), respectively. The maximum HCN yield can be achieved in these systems when the molar fraction of methane is relatively low (below 13%) and the glow discharge power is relatively high (above 1.6 kW). The non-equilibrium reactions in the CH_4–N_2 mixture can also be performed with selective production of ammonia (NH_3) and liquid fuel at ambient pressure and temperature without catalysts (Bai et al., 2005). The process has been demonstrated in a DBD. The conversion degree of CH_4 reaches 50%, yield of liquid fuel is 18%, and concentration of NH_3 in products is up to 1%.

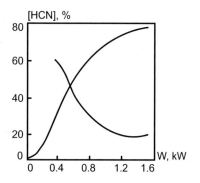

Figure 9–10. Plasma conversion of methane–nitrogen mixture (initial CH₄, 13%) into HCN (curve 1) and acetylene (curve 2).

9.3.4. Production of HCN and H₂ in CH₄–NH₃ Mixture in Thermal and Non-Thermal Plasmas

Plasma production of HCN and H_2 is an endothermic process:

$$CH_4 + NH_3 \rightarrow HCN + 3H_2, \quad \Delta H = 2.6 \, \text{eV/mol.} \qquad (9\text{--}26)$$

The product composition of the atmospheric-pressure thermal plasma process in the temperature range 1000–3500 K, which corresponds to a specific energy input range of 0.95–10.5 eV/mol, is presented in Fig. 9–12 (Nester et al., 1988). At temperatures below 2000 K, dissociation of ammonia mostly leads to the formation of molecular nitrogen. The maximum yield of HCN corresponds to a temperature of 3200 K. That temperature is also optimal for direct production of acetylene. The energy cost of HCN production is presented in Fig. 9–13 as a function of specific energy input for the cases of (1) absolute and (2) ideal quenching mechanisms (Nester et al., 1988). The minimum energy cost in the absolute quenching mode is quite high: about 50 eV/mol HCN (energy efficiency only about 5%). A much lower energy cost (about 3 eV/mol HCN) can be achieved in the ideal quenching mode, when radical–radical reactions during the cooling stage make a significant contribution to HCN production. Production of HCN and H_2 from CH₄ and NH₃ (9–26) can also be accomplished in non-thermal plasma. Conducting the process in a glow discharge results in a high conversion (70%) of methane to hydrogen (Peters & Kuster, 1931). The energy efficiency of the process in non-equilibrium conditions is low with respect to that in the thermal plasma.

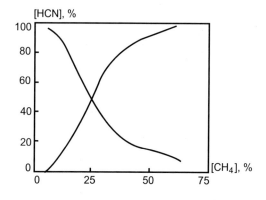

Figure 9–11. Effect of initial methane–nitrogen mixture composition (molar fraction of methane) on formation of HCN; discharge power 1.6 kW. Conversion degrees: (1) yield of HCN; (2) yield of high-boiling-point hydrocarbons.

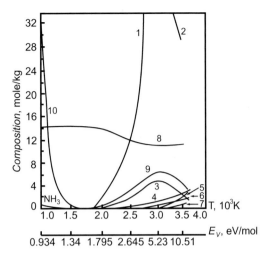

Figure 9–12. Thermodynamic quasi-equilibrium composition of the mixture CH_4–NH_3 at different temperatures and different values of the specific energy input: (1) H; (2) H_2; (3) C_2H_2; (4) C_2H; (5) C; (6) CN; (7) C_2; (8) N_2; (9) HCN; (10) CH_4.

9.3.5. Thermal and Non-Thermal Plasma Conversion Processes in CO–N_2 Mixture

Co-production of NO and CN can be accomplished in the CO–N_2 mixture in both thermal and non-thermal plasma conditions. The endothermic plasma-chemical process can be presented as

$$CO + N_2 \rightarrow CN + NO, \quad \Delta H = 6\,eV/mol. \quad (9\text{–}27)$$

This process has been investigated in atmospheric-pressure thermal plasma by Nester et al. (1988). The composition of products in the temperature range 2000–7000 K is presented in Fig. 9–14. Intensive conversion starts at relatively high temperatures, exceeding 4500 K, due to the very large binding energies of both reagents. The energy efficiency is relatively low in thermal plasma conditions. The energy cost of production of a CN radical is presented in Fig. 9–15 as a function of specific energy input for different quenching modes. The minimum energy cost of CN production in the absolute quenching mode is very high (640 eV/mol) because of the low concentration of the CN radicals at the high temperatures required for the process. Even ideal quenching and super-ideal quenching modes assuming VT non-equilibrium during the fast cooling, decrease the minimum energy cost only to 28 eV CN and 13 eV CN, respectively. Higher energy efficiency of the process (9–27) can be achieved in non-equilibrium plasma with the process stimulation through vibrational excitation because of the excellent characteristics of vibrational excitation for both N_2 and

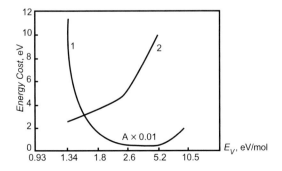

Figure 9–13. Energy cost A (eV/mol) of production of HCN in the process $CH_4 + NH_3 \rightarrow$ HCN + $3H_2$ at different quenching modes: (1) absolute quenching, $A \times 10^{-2}$, $A_{min} = 50$ eV/mol; (2) ideal quenching.

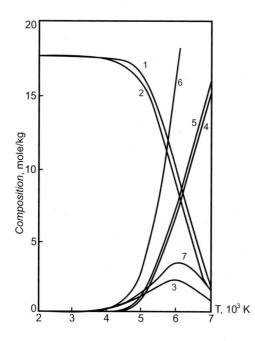

Figure 9–14. Thermodynamic quasi-equilibrium composition of the mixture CO–N_2 at different temperatures: (1) CO; (2) N_2; (3) [CN] × 10; (4) C; (5) O; (6) N; (7) [NO] × 100.

CO in non-equilibrium discharges. Plasma co-production of CN and NO proceeds in this case by the following chain-reaction mechanism:

$$O + N_2^* \rightarrow NO + N, \quad \Delta H \approx E_a = 3\,\text{eV}, \tag{9–28}$$

$$N + CO^* \rightarrow CN + O, \quad \Delta H \approx E_a = 3\,\text{eV}. \tag{9–29}$$

The first reaction in the chain is the limiting reaction of the Zeldovich mechanism and its stimulation by vibrational excitation of N_2 molecules, discussed in Section 6.2. The second reaction is stimulated by vibrational excitation of CO molecules due to the relatively high energy barrier of the reaction and the absence of activation energy of the reverse reaction (see Sections 2.7.2 and 2.7.3).

9.3.6. Other Non-Equilibrium Plasma Processes of Organic Nitrogen Compounds Synthesis

Complex organic synthesis of nitrogen compounds has been performed in a low-pressure glow discharge by Fisher and Peters (1931):

$$N_2 + CO + 3H_2 \rightarrow NH_3 + HCN + H_2O. \tag{9–30}$$

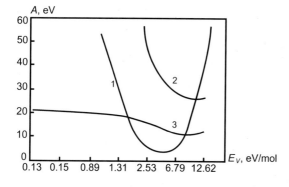

Figure 9–15. Energy cost A (eV/mol) of production of a cyan molecule in the thermal plasma process $CO + N_2 \rightarrow CN + NO$ at different quenching modes: (1) absolute quenching, $A \times 100$, $A_{min} = 640$ eV/mol; (2) ideal quenching, $A_{min} = 27.6$ eV/mol; (3) super-ideal quenching, $A_{min} = 12.9$ eV/mol.

This process leads to the formation of two nitrogen compounds and can be accomplished at specific discharge parameters with relatively high selectivity. The conversion degree in the synthetic process reaches 80–90%. Complex plasma-chemical organic synthesis based on conversion of ethylene has been investigated by Peters and Pranschke (1930). The plasma-chemical reaction of ethylene with hydrogen cyanide results in the formation of ethyl cyanide with relatively high selectivity and conversion degree:

$$C_2H_4 + HCN \rightarrow C_2H_5CN. \qquad (9\text{--}31)$$

The plasma-chemical reaction of ethylene with ammonia results in the formation of ethylamine, also with relatively high selectivity and conversion degree:

$$C_2H_4 + NH_3 \rightarrow C_2H_5NH_2. \qquad (9\text{--}32)$$

Although synthetic processes (9–31) and (9–32) look like simple recombination, they proceed in non-thermal plasma by a sequence of several elementary reactions mostly stimulated by electronic excitation. The energy efficiency of synthetic processes (9–31) and (9–32) in the low-pressure glow discharges is low. Organic synthesis of nitrogen compounds based on conversion of aromatic compounds has been investigated by Gerasimov, Gorogankin, and Gusev (1979):

$$C_6H_6 + NH_3 \rightarrow C_6H_5NH_2 + H_2. \qquad (9\text{--}33)$$

The synthesis of aniline ($C_6H_5NH_2$) from benzene and ammonia differs from the aforementioned processes because it cannot be considered a simple recombination.

9.4. ORGANIC PLASMA CHEMISTRY OF CHLORINE AND FLUORINE COMPOUNDS

9.4.1. Thermal Plasma Synthesis of Reactive Mixtures for Production of Vinyl Chloride

Vinyl chloride (C_2H_3Cl) is a monomer for further production of PVC, which is a widely used polymer material. Vinyl chloride is usually produced by catalytic hydrochlorination of acetylene ($C_2H_2 + HCl \rightarrow C_2H_3Cl$; see Section 9.1.6). The stoichiometric acetylene–hydrogen chloride mixture (C_2H_2:HCl = 1:1), which is applied for vinyl chloride production, can be effectively formed in thermal plasmas either by pyrolysis of hydrocarbons in plasma jets containing H_2, HCl, Cl_2, or CH_4, or by pyrolysis in an H_2 plasma jet of the organic chlorine industrial wastes (Volodin et al., 1969, 1970a,b, 1971a,b). In particular, the stoichiometric C_2H_2–HCl mixture has been produced by pyrolysis of gasoline (with final evaporation temperature 165°C) in atmospheric-pressure plasma jets containing H_2, HCl, Cl_2, and CH_4 in different proportions (Volodin et al., 1969, 1971a,b). The process is done at 1300–1700 K and effective contact time 10^{-5}–10^{-4} s. A temperature increase leads to acetylene concentration growth from 10 to 20 vol %. The maximum concentration of ethylene is 14 vol % and can be reached at 1400 K. Equal concentrations of C_2H_2 and HCl (17–20 vol %), required for the production of vinyl chloride, are achieved in the temperature range 1550–1700 K at effective contact times of 0.5–$1.5 \cdot 10^{-4}$ s. The energy cost of the C_2H_2 and HCl co-production process in this regime is 6.3 kWh/kgC_2H_2. The energy cost is lower when the plasma jet contains Cl_2, which is due to heat release in the reaction of HCl production. The use of gasoline for quenching also decreases the energy cost by about 15–20%; the optimal ratio of gasoline used for pyrolysis with respect to gasoline used for quenching is 1:0.7.

Table 9–6. Thermal Plasma Pyrolysis of Organic Chlorine Compounds

Products and characteristics of plasma pyrolysis	Initial organic chlorine compounds			
	Dichloroethane, rate: 8 kg/h	Butyl chloride, rate: 5.4 kg/h	Hexachlorane	Organic chlorine mixture, rate: 7.2 kg/h
Hydrogen, vol %	42–47	56	56	52–53
Methane, vol %	0.6–0.7	6.5–7.8	1.5	5.2–6.7
Hydrogen chloride (HCl)	34–39	13–14	28.5	19
Acetylene (C_2H_2), vol %	12–17	13–21	12	15–20
Ethylene (C_2H_4), vol %	1.2–2.7	2.0–2.3	1.0	2.1–3.3
Propylene, vol %	–	0.4	–	0.1–0.4
Allene, vol %	–	0.2–1.6	–	0.1–0.5
Methylacetylene, vol %	0.1	0.2–1.8	–	0.2–0.6
Vinylacetylene, vol %	0.1–0.2	0.23	–	0.21
Isobutylene, vol %	–	0.1–1.5	–	–
Vinylchloride (C_2H_3Cl), vol %	0.6–2.6	0.15	–	1.0
Butylchloride, vol %	–	0.1–2.2	–	–
Butadiene, vol %	–	–	–	0.4
Ethylchloride, vol %	–	–	–	0.3
Carbon monoxide (CO), vol %	–	–	0.4	–
Process characteristics ($\downarrow$): Discharge Power, kW	7–10	10–13	10	12–14
H_2 flow rate, st. m^3/h	4	4	4	3.5
CH_4 flow rate, st. m^3/h	–	–	–	0.4–0.6
C_2H_4 flow rate, st. m^3/h	–	–	–	0.14
Reaction temperature, K	1400–1700	1400–1650	–	1400–1600
Products flow rate, st. m^3/h	7.8–9.1	8.6–9.5	–	10.2–11.2
Conversion, %	85–99	85–99	–	90–99
Relative yield of C_2H_2, mass %	67–88	46–75	–	64–88
Energy cost of acetylene, kWh/kg C_2H_2	5.4–6.1	6.0–7.2	–	5.7–6.6
Energy cost of HCl/C_2H_2 Co-Production, kWh/kg $C_2H_2 + HCl$	1.26–1.30	3.0–3.1	–	2.3

9.4.2. Thermal Plasma Pyrolysis of Dichloroethane, Butyl Chloride, Hexachlorane, and Other Organic Chlorine Compounds for Further Synthesis of Vinyl Chloride

Plasma pyrolysis of organic chlorine industrial wastes is interesting for the co-production of acetylene and hydrogen chloride for further synthesis of vinyl chloride and PVC (Volodin et al., 1970a,b). The process is carried out by interaction of thermal plasma jets of hydrogen (or H_2–CH_4, or H_2–C_2H_4 mixtures) with the organic chlorine compounds, which can initially be in different phases. The composition of products and major characteristics of plasma pyrolysis of dichloroethane, butyl chloride, hexachlorane, and organic chlorine industrial waste mixture (Volodin et al., 1970a,b) are presented in Table 9–6. Products of the pyrolysis can be effectively applied for the synthesis of vinyl chloride. The initial composition of the organic chlorine industrial waste mixture used as a feedstock for the pyrolysis is trichloroethane, 15 vol %; dichloroethane, 15 vol %; butyl chloride, 15 vol %; ethyl chloride, 10 vol %; dichloroisobutane, 5 vol %; trichloroisobutylene, 5 vol %; monochlorobutane, 5 vol %; and other organic compounds, 30 vol %. The pyrolysis was carried out in the plasma jet of hydrogen (or H_2–CH_4, or H_2–C_2H_4 mixtures) at 1400–1700 K and at an effective contact time of about 10^{-4} s. The power of the thermal discharges applied for plasma pyrolysis is also presented in the table together with the feeding rates of the initial organic chlorine compounds. In the case of thermal pyrolysis of hexachlorane, the plasma jet has been submerged in a tank containing the dry initial product. Temperature increase of the plasma pyrolysis leads to an increase in the concentration of acetylene (C_2H_2) in the products and to a decrease of concentrations of ethylene (C_2H_4), C_3H_6, iso-C_4H_8, and vinyl chloride (C_2H_3Cl). Quenching of the plasma pyrolysis products can be provided by injection of gasoline or the mixture of organic chlorine compounds, which increases the energy efficiency of the process. Typical pyrolysis-to-quenching feeding rate ratios of the organic chlorine compounds are about 1:2. Kinetic analysis of the thermal plasma pyrolysis of the organic chlorine compounds at temperatures of 1600–2000 K shows that almost all chlorine from the initial products is converted into hydrogen chloride (HCl). The composition of hydrocarbons in the products of pyrolysis of the organic chlorine compounds is similar to that of pyrolysis of pure hydrocarbons (see Section 9.1) at the same C:H ratio and the same temperature. It can be assumed that the organic chlorine compounds are quickly decomposed (about 10^{-6} s), forming HCl, CH_4, C_2H_4, H_2, and also acetylene if aromatic groups are present in the initial products. The ensuring pyrolysis of the hydrocarbons proceeds in accordance with the standard Kassel mechanism (see Sections 9.1.1 and 9.2.3).

9.4.3. Thermal Plasma Pyrolysis of Organic Fluorine Compounds

High-intensity arc discharges with carbon electrodes have been applied for conversion of tetrafluoromethane (CF_4) into tetrafluoroethylene (C_2F_4; Bronfine, 1970):

$$CF_4 + CF_4 \rightarrow C_2F_4 + 2F_2. \tag{9–34}$$

The yield of C_2F_4 increases at higher discharge powers and lower pressures. A maximum yield of C_2F_4 (75 vol %) is achieved at a power of 20 kW and pressure of 0.1 atm. In addition to CF_4 and C_2F_4, products of the plasma-chemical process also contain C_2F_6 and C_3F_8. Quasi-equilibrium heating of organic fluorine compounds to temperatures of 2000–4000 K leads mostly to formation of C_2F_2 and different kinds of carbon clusters. During the quenching phase, significant amounts of C_2F_4 and soot can be generated. The formation of C_2F_4 during the quenching phase is due to attachment of atomic fluorine to C_2F_2 molecules (Bronfine, 1970):

$$C_2F_2 + F \rightarrow C_2F_3, \tag{9–35}$$

$$C_2F_3 + F \rightarrow C_2F_4. \tag{9–36}$$

The additional formation of C_2F_4 from C_2F_2 during cooling is accompanied by soot formation:

$$C_2F_2 + C_2F_2 \rightarrow C_2F_4 + 2C_{(s)}. \tag{9–37}$$

Plasma decomposition of CF_4 can also be optimized for production of soot and fluorine in the following strongly endothermic process:

$$CF_4 \rightarrow C_{(s)} + 2F_2, \quad \Delta H = 9.5 \, \text{eV/mol}. \tag{9–38}$$

The energy cost of the process in atmospheric-pressure thermal plasma is shown in Fig. 9–16 as a function of specific energy input for different quenching modes (Nester et al., 1988). The minimum energy cost in the case of absolute quenching is 25.8 eV/C atom, which can be achieved at the specific energy input of 25.1 eV/mol and CF_4 conversion degree of 97%. The minimal energy cost in the ideal quenching mode is 21.6 eV/C atom and that in the super-ideal quenching mode is 17.4 eV/C atom. Only the thermal pyrolysis

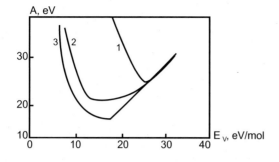

Figure 9–16. Energy cost A (eV/mol) of production of a carbon molecule in the thermal plasma process of dissociation of CF_4 at different quenching modes: (1) absolute quenching; (2) ideal quenching; (3) super-ideal quenching.

of CF_4 has been discussed. Some data related to plasma pyrolysis of C_3F_8, C_3F_6, C_8F_{16}, and other organic fluorine compounds (Bronfine, 1970) are summarized in Table 9–7.

9.4.4. Pyrolysis of Organic Fluorine Compounds in Thermal Plasma of Nitrogen: Synthesis of Nitrogen-Containing Fluorocarbons

The pyrolysis of CF_4, CHF_3, and other organic fluorine compounds in a nitrogen plasma jet has been investigated by Bjornson (1966a,b) and Polak et al. (1975). In the case of CF_4 pyrolysis in an N_2 plasma jet with a quenching rate of 10^6 K/s, the major nitrogen-containing products are NF_3 and CF_3NH_2. Smaller amounts of N_2F_4 and N_2F_2 are also present in the products. The absolute conversion of nitrogen into fluorides is relatively low in these systems, only about 1%. Yields of nitrogen fluorides increase at higher discharge powers and larger F:N ratios. Pyrolysis of CHF_3 in the N_2 plasma jet leads to formation of such fluorocarbons as CF_4 (57.7%), C_2F_6 (18.2%), C_2F_4 (12.9%), and C_3F_8 (3.4%). The major nitrogen-containing organic fluorine compound in this case is CF_3CN – 7.8%. Thermodynamic quasi-equilibrium in the mixture C:N:F = 1:4:4 at atmospheric pressure and temperature of 4000 K gives major fluorocarbons: CF_2 radicals and C_2F_2; the major nitrogen-containing compounds at 2500–4200 K are FCN and CN radicals. The maximum concentration of FCN is 1 vol %, and it is reached at 3500 K. The maximum concentration of CN radicals is 5 vol %, and it is reached at temperature 4500 K. Data related to plasma pyrolysis of fluorocarbons in a nitrogen plasma jet and by thermal reactions with NO and HCN (Bronfine, 1970) are summarized in Table 9–7.

9.4.5. Thermal Plasma Pyrolysis of Chlorofluorocarbons

Chlorofluorocarbons (CFCs) are associated with high ozone-destroying and global warming potential, which determines interest in their destruction by plasmas (see Chapter 11). Most CFC decomposition processes are performed in thermal plasmas with addition of hydrogen and oxygen into the plasma gas (see, for example, Bjornson, 1966a,b; Begin, 1970; Badie, Flamant, & Granier, 1997, Brozek, Hrabovsky, & Kopecky, 1997; Coulibaly et al., 1997; Park & Cha, 1997; Rutberg et al., 1997; Murphy et al., 2002; Ponelis & van der Walt, 2003). When the process is carried out in Ar plasma jets at powers of 2–10 kW without adding O_2 and N_2, the main products of pyrolysis of chlorofluoromethane are chlorofluoroethane compounds, benzene derivatives, and so on. Examples of plasma pyrolysis of CFCs without addition of H_2 and O_2 are presented in Table 9–7. Thermal plasma pyrolysis involving bromine-containing organic compounds is also presented in the table in a couple of examples. It is interesting to discuss the effect of hydrogen and oxygen on the plasma-chemical process for CFC pyrolysis. Introducing hydrogen improves thermal conductivity of the plasma, which increases the efficiency of the process (Coulibaly et al., 1997). The most important function of hydrogen is its involvement in irreversible reactions with fluorine and chlorine-containing radicals and formation of HF and HCl, which are products of the pyrolysis. Introducing oxygen helps avoid soot formation during pyrolysis. Relatively small amounts

Table 9–7. Thermal Plasma Pyrolysis of Organic Fluorine Compounds

Reagent	Pressure Torr	Power, kW	Reagent flow, mol/s; composition	Products	Plasma-chemical reactor type
CF_4	35–760	0.1–30	0.02–3.7	C_2F_4, C_2F_6, C_3F_6	High-intensity arc, cold wall quenching
C_3F_8	32	0.4	1.0	C_2F_4, CF_4, C_2F_4, C_2F_6, C_3F_6	Carbon electrode arc with quenching
C_3F_6	40	0.4	1.4	C_2F_4, CF_4, C_2F_4, C_2F_6, C_3F_8	Carbon electrode arc with quenching
C_8F_{18}	10	0.4	0.01	C_2F_4, CF_4	Carbon electrode arc with quenching
F_2 + (C, vap.)	2	0.07	Const. vol.	CF_4, C_2F_4, C_3F_6	Carbon electrode arc with quenching
$CF_4 + Cl_2$	55	0.5	Const. vol.	CF_4, CF_2Cl_2, CF_3Cl, C_2F_6, $(CF_2Cl)_2$	Carbon electrode arc with Cl_2 injection
$(CF_2Cl)_2$	45	0.5	0.1	C_2F_4, CF_4, CF_3Cl, CF_2Cl_2	Carbon electrode arc with quenching
C_2F_5Cl	45	0.5	0.1	C_2F_4, CF_4, CF_3Cl, C_2F_6	Carbon electrode arc with quenching
$C_2F_2Cl_2$	45	0.5	0.07	C_2F_4, CF_4, CF_3Cl, C_2F_6, $(CF_2Cl)_2$	Carbon electrode arc with quenching
C_2F_3Cl	45	0.5	0.08	C_2F_4, CF_4, CF_2Cl_2	Carbon electrode arc with quenching
$CF_4 + CCl_4$	760	3.0	–	CCl_2F_2	RF-ICP discharge in argon
$CF_4 + C_6H_6$	20	1.3	9 : 1	C_2F_4, CF_4, C_2F_6, HF	Carbon electrode arc with quenching
$CF_4 + CH_4$	20	1.3	3 : 1	C_2F_4, CF_4, C_2F_6, HF	Carbon electrode arc with quenching
$CF_4 + C_2H_2$	20	1.3	9 : 1	C_2F_4, CF_4, C_2F_6, HF	Carbon electrode arc with quenching
$CF_4 + NO$	20–65	0.7–2.0	0.04, 0.08	C_2F_4, CF_4, C_2F_6, N_2, CO, COF_2, NF_3, NO, N_2O	Carbon electrode arc with quenching
$CF_4 + O_2$	20	0.75	0.06, 0.06	C_2F_4, CF_4, C_2F_6, CO	Carbon electrode arc with quenching
CF_3COOH	8	0.42	0.007	C_2F_4, CF_4, C_2F_6, C_3F_8, HF, CO	Carbon electrode arc with quenching
$CF_4 + Ar$	250	0.8	0.02, 0.2	C_2F_4, C_2F_6, C_3F_6, higher fluorocarbons, tar	Microwave discharge with graphite outside
$CF_4 + N_2$	760	20	35, 18	NF_3, CF_3NF_2, N_2F_4, C_2F_6, N_2F_2, C_2F_4	N_2 plasma jet with CF_4 injection
$COF_2 + NO$	28	2.3	0.17, 0.10	NF_3, N_2F_2, CF_4, CO, N_2O, NO	Carbon electrode arc discharge with cold wall quenching
$CF_4 + C_2N_2$	20	1.4	2 : 1	C_2F_4, CF_4, FCN, CF_3CN, C_2N_2, N_2	Carbon electrode arc discharge with cold wall quenching
$CF_4 + HCN$	20	1.4	2 : 1	HF, N2, C_2F_4, CF_4	Carbon electrode arc discharge with cold wall quenching
N_2 + cyclo- $(-N{=}CF-)_3$	60	1.7	0.19, 0.19	FCN, C_2F_4, CF_4, C_2F_6, C_2N_2, $N{=}C{-}C{=}C{-}C{=}N$	Carbon electrode arc discharge with cold wall quenching
$C_2F_4 + Br_2$	100–500	0.4	–	Fluoro-α, ω- dibromo alkanes, fluoroalkanes, 1-bromo-fluoroalkanes	Carbon electrode arc discharge with cold wall quenching
$C_2F_4 + C_2F_4Br_2$	100–500	0.4	–	1-bromo-fluoroalkanes	Carbon electrode arc discharge with cold wall quenching
$C_2F_4Br_2$	100–500	0.4	–	$CFBr_3$, C_2F_3Br, $Br_2(CF_2)_4$, mixture of olephines	Carbon electrode arc discharge with cold wall quenching

of oxygen lead mostly to the formation of CO, and larger oxygen admixtures result in the formation of CO_2. When oxygen admixtures are relatively high, the concentration of CO in products can be increased with respect to CO_2 by increasing the flow rate of hydrogen (Park & Cha, 1997). Generally, it can be concluded that the degree of perfluoro compound (PFC) destruction and product composition are essentially controlled by the amounts of H_2 and O_2 added to the plasma gas. Hydrogen and oxygen can be introduced into the thermal plasma in the form of steam (Watanabe, Taira, & Takeuchi, 2005). In this case CFCs are injected in a direct-current (DC) steam plasma torch operating at a power of about 1.5 kW. The system provides the complete destruction of CFC compounds with a high degree of fluorine recovery – about 95%. The destruction degree of the PFCs in thermal plasma can be very high (99–99.99%). However, by-products of the pyrolysis should be carefully analyzed. Although CFCs are usually non-toxic; products of their pyrolysis, especially in the presence of oxygen (such as fluorophosgene [COF_2] phosgene [$COCl_2$], and COClF); and are very toxic. Their elimination from the products of plasma pyrolysis requires special measures – for example, plasma organization in a spouted bed (Coulibaly et al., 1997).

9.4.6. Non-Thermal Plasma Conversion of CFCs and Other Plasma Processes with Halogen-Containing Organic Compounds

While most of the CFC conversion processes are done in thermal discharges, non-equilibrium plasmas are also applied for this purpose (see, for example, Suhr, 1974; Genet et al., 1997; Barkhudarov et al., 1997; Foglein et al., 2003; Opalinska & Opalska, 2003; Oda, Han, & Ono, 2005; Pushkarev et al., 2005). An advantage of non-thermal plasma approaches is their significantly higher energy efficiency in destruction and conversion of CFC compounds diluted in other gases, primarily in air. Electrons and active species of non-equilibrium plasma systems are able to destroy the highly diluted CFC compounds without heating the whole gas, which makes the energy cost of the process significantly lower. For example, application of a DBD to dry atmospheric air with 100–300 ppm of trichloroethylene (C_2HCl_3) requires an energy input of only about 60 J/L for a destruction efficiency of C_2HCl_3 up to 80%. Different types of non-thermal discharges have been applied for the destruction of CFC compounds, specifically low-pressure RF capacitively coupled plasma (CCP) and glow discharges, moderate-pressure (about 100 Torr) microwave and atmospheric-pressure sliding surface discharges, as well as pulsed electron beams. An important problem of all non-equilibrium systems is the wide variety of by-products. While non-equilibrium plasma stimulates some elementary conversion processes, many intermediate products are not completely converted into desirable final products because of low gas temperature in the system. For example, decomposition of CF_4 in the presence of oxygen leads not only to the formation of CO and CO_2 but also to the significant production of highly toxic phosgenes, COF_2 and COF. Phosgene yields can be reduced, in particular, by applying calcium oxide particles that trap fluorine in the form of CaF_2 (Ayoubi, Gonnord, & Amouroux, 1995). Application of pulsed electron beams to the conversion of CF_4 results in the formation of nanoscale carbon particles (Pushkarev et al., 2005), which affect the phosgene yield. Among other non-thermal plasma processes with halogen-containing organic compounds (Rusanov & Fridman, 1984), there are also dehydration reactions of organic acids,

$$Cl_2 + HCOOH \rightarrow 2HCl + CO_2, \qquad (9\text{--}39)$$

chlorine–bromine exchange reactions in conversion of organic bromine and chlorine compounds,

$$Cl_2 + CCl_3Br \rightarrow BrCl + CCl_4, \qquad (9\text{--}40)$$

and halogen attachment to hydrocarbons, specifically the formation of dibromoethylene from C_2H_2:

$$C_2H_2 + Br_2 \rightarrow C_2H_2Br_2 \tag{9-41}$$

Numerous synthetic reactions of this kind are reviewed by Baddur and Timmins (1970), McTaggart (1967), and Andreev (1953).

9.5. PLASMA SYNTHESIS OF ALDEHYDES, ALCOHOLS, ORGANIC ACIDS, AND OTHER OXYGEN-CONTAINING ORGANIC COMPOUNDS

9.5.1. Non-Thermal Plasma Direct Synthesis of Methanol from Methane and Carbon Dioxide

The reaction of methane and carbon dioxide in thermal plasma conditions leads mostly to the production of syngas. While non-thermal plasma can also be applied for effective conversion of the CH_4–CO_2 mixture into syngas (see Chapter 10), non-equilibrium plasma conditions can also lead to the direct formation of methanol in the following process (Rusanov & Fridman, 1984):

$$CH_4 + CO_2 \rightarrow CH_3OH + CO, \quad \Delta H = 1.6\,\text{eV/mol.} \tag{9-42}$$

This plasma-chemical process is promoted in strongly non-equilibrium conditions, when the vibrational temperature of CO_2 molecules significantly exceeds their translational gas temperature ($T_v \gg T_0$). Vibrational excitation of CO_2 molecules stimulates the chain reaction of methanol production:

$$CH_3 + CO_2^*(v) \rightarrow CO + CH_3O, \quad \Delta H \approx E_a = 1.5\,\text{eV/mol,} \tag{9-43}$$

$$CH_3O + CH_4 \rightarrow CH_3OH + CH_3, \quad \Delta H = 0.1\,\text{eV/mol,} \quad E_a \approx 0.3\,\text{eV/mol.} \tag{9-44}$$

Reactions between CH_3O radicals also contribute to synthesis in the CH_4–CO_2 mixture:

$$CH_3O + CH_3O \rightarrow CH_3OH + CH_2O. \tag{9-45}$$

Although the fast elementary reaction produces valuable products – methanol and formaldehyde – it decreases the yield of methanol because of chain termination (9–43) and (9–44). Direct production of methanol and formaldehyde in the CH_4–CO_2 mixture (9–43)–(9–45) is possible only at very low gas temperatures. A translational temperature increase leads to CH_3OH and CH_2O decomposition in the system and to formation of the syngas (CO, H_2) as major reaction products. The kinetics of process (9–43)–(9–45) in the CH_4–CO_2 mixture in strongly non-equilibrium plasma is analyzed by Peters and Pranschke (1930); Gerasimov, Gorogankin, and Gusev (1979); and Lukianov, Simonov, and Mogaisky (1973). The conversion degree of the direct production of methanol and formaldehyde in the CH_4–CO_2 mixture is limited.

9.5.2. Non-Thermal Plasma Direct Synthesis of Methanol from Methane and Water Vapor

Similarly to the aforementioned reactions in the CH_4–CO_2 mixture, the reaction of methane and water vapor in thermal plasma leads to the production of syngas (H_2, CO). Non-thermal plasma is also usually applied to produce the H_2–CO mixture (see Chapter 10). As was demonstrated by Okazaki and Kishida (1999), application of non-thermal plasma permits the direct production of methanol from methane:

$$CH_4 + H_2O \rightarrow CH_3OH + H_2. \tag{9-46}$$

These experiments were carried out in an atmospheric-pressure DBD supported by short pulses (duration 400 ns, repetition frequency 500 Hz). Typical plasma-chemical parameters of the system were peak voltage, 3 kV; residence time, 50 s; gas temperature, 393 K; and specific energy input, 11 J/cm^3. The methanol yield was about 1% at a water vapor concentration of about 50%. The energy cost of methanol production from methane was relatively high in this case – about 200 eV/mol CH_3OH. In contrast to the process with CO_2 molecules (9–42), reactions (9–46) with H_2O molecules cannot be effectively stimulated by vibrational excitation, which results in a lower energy efficiency for methanol production (9–46). According to Okazaki and Kishida (1999), reaction (9–46) proceeds via dissociation of water and methane molecules with formation of CH_3 and OH radicals, which leads then to methanol production. The effectiveness of the process can be improved by choosing plasma conditions that stimulate the dissociation of water and methane molecules with the formation of CH_3 and OH radicals. The dissociation processes proceeding through electronic excitation of the molecules can be stimulated by increasing the voltage and, therefore, the electric field and electron temperature in the discharge. The dissociation of CH_4 and H_2O can also be stimulated by diluting the CH_4–H_2O mixture with noble gases. The dilution effectively increases the electron temperature at a fixed peak voltage. Also the dilution with noble gases results in their intensive electronic excitation with further transfer of the electronic excitation energy to dissociation of CH_4–H_2O molecules and production of methanol. Experiments of Wang and Xu (2005) with non-thermal DBD plasma demonstrate these effects. The percentage of methanol in products can be increased to 50% and the yield of methanol can reach 10% when the ratio of dilution of the CH_4–H_2O mixture with argon is 1:4 and peak voltage is 45 kV. At lower voltages, a dilution ratio of 1:4 increases the yield of methanol to about 6%. An additional increase of the methanol yield from the CH_4–H_2O mixture can be achieved by combination of DBD with an Ni catalyst (Nozaki, 2002; Kado et al., 2003).

9.5.3. Production of Formaldehyde (CH_2O) by CH_4 Oxidation in Thermal and Non-Thermal Plasmas

Although partial oxidation of methane in thermal plasma leads mostly to the generation of syngas (see Chapter 10), special quenching can also result in a significant production of formaldehyde (CH_2O):

$$CH_4 + O_2 \rightarrow CH_2O + H_2O. \tag{9–47}$$

Experiments with formaldehyde production by methane oxidation in an argon thermal plasma jet were carried out by Ovsiannikov, Polak, and Yudin (1968) and Ovsiannikov and Polak (1965). The major products of the plasma-chemical process were H_2, CO, CH_2O, CO_2, and H_2O. The highest yield of CH_2O was 3.5% with respect to the initial methane concentration. This yield has been achieved at reaction temperatures of 1600–1700 K. The plasma-chemical process (9–47) can be stimulated in non-equilibrium discharges by vibrational excitation of oxygen molecules (Rusanov & Fridman, 1984). Formaldehyde formation proceeds via the following chain mechanism:

$$CH_3 + O_2^*(v) \rightarrow CH_2O + OH, \quad E_a = 0.8 \, eV/mol, \tag{9–48}$$

$$OH + CH_4 \rightarrow CH_3 + H_2O, \quad E_a = 0.2 \, eV/mol. \tag{9–49}$$

Selective completion of the chain reaction (9–48)–(9–49) for formaldehyde production is possible only at low translational temperatures and vibrational temperatures of oxygen kept in some specific range ($T_v > T_0$). When the O_2 vibrational temperature is not high, reaction (9–48) is suppressed by the reaction

$$CH_3 + O_2 \rightarrow HCO + H_2O, \tag{9–50}$$

which terminates the chain mechanism for CH_2O production, (9–48)–(9–49). On the other hand, when the vibrational temperature of oxygen is too high, reaction (9–48) of CH_2O production is suppressed by

$$CH_3 + O_2^*(v) \rightarrow CH_3O + O. \tag{9–51}$$

This reaction together with (9–49) and the following elementary processes leads to branching of the chain mechanism:

$$O + CH_4 \rightarrow CH_3 + OH, \tag{9–52}$$

$$CH_3O + CH_3O \rightarrow CH_3OH + CH_2O. \tag{9–53}$$

The chain mechanism (9–49) and (9–51)–(9–53) leads to co-production of formaldehyde and methanol and is characterized in general by low selectivity. The selectivity of the plasma production of formaldehyde even at low translational temperature is limited by the fast reaction for CH_2O destruction:

$$CH_2O + OH \rightarrow HCO + H_2O. \tag{9–54}$$

The energy efficiency of formaldehyde production (9–47) in non-equilibrium plasma is limited by fast non-adiabatic VT relaxation of molecular oxygen (see Section 2.6.3).

9.5.4. Non-Thermal Plasma Oxidation of Methane and Other Hydrocarbons with Production of Methanol and Other Organic Compounds

Somewhat similarly to the aforementioned synthesis of formaldehyde, the oxidation of methane in non-thermal plasma conditions can be optimized for the production of methanol:

$$CH_4 + 1/2O_2 \rightarrow CH_3OH. \tag{9–55}$$

This process has been demonstrated in atmospheric-pressure pulsed DBD by Okazaki et al. (1995). The main characteristics of the discharge are peak voltage, 15 kV; frequency of pulses, 250 Hz; pulsed current and pulsed voltage duration, about 100 ns, average DBD power, 0.5 W; initial gas composition, $CH_4:O_2 = 92:8$; and residence time, 9 s. The yield of methanol reached 2.4%; the selectivity of CH_3OH production from CH_4 (percentage of methanol in converted products) was as high as 33%. Similar methanol yield (2.5%) in the process (9–55) (2.5%) has been achieved using atmospheric-pressure pulsed corona (Gong et al., 1997). Higher methanol yields can be achieved by combining non-thermal plasma with catalysts and recirculation of products (Okazaki et al., 1995). Among other hydrocarbon oxidation processes, there is also the production of oxides of C_2–C_{10} olephines by oxidation of the relevant olephines with CO_2 (Burleson & Jates, 1967a,b), H_2O (Ruppel et al., 1970), and O_2–N_2 and N_2O (Weisbeek & Hullstrung, 1970). The processes have been performed in DBD plasma at pressures up to 3.5 atm, peak electric fields of 30–120 kV/cm, wall temperatures of 15–100°C, residence time of 30–100 s, and frequencies of 30–60 Hz. The yield of the olephine oxides in the systems is 20–30%, and the energy cost of the products is about 10 kWh/kg. A similar plasma-chemical oxidation of paraphenes with CO_2, analyzed by Burleson and Jates (1967a,b), leads to the formation of alcohols with a corresponding number of carbon atoms.

9.5.5. Non-Thermal Plasma Synthesis of Aldehydes, Alcohols, and Organic Acids in Mixtures of Carbon Oxides with Hydrogen: Organic Synthesis in CO_2–H_2O Mixture

A hydrogen mixture with carbon monoxide is widely used for conventional catalytic synthesis; the mixture is called syngas. Non-thermal plasma processing of the CO–H_2 mixture leads to direct synthesis of formaldehyde:

$$H_2 + CO \rightarrow CH_2O, \quad \Delta H \approx 1\,kcal/mol. \tag{9-56}$$

This process has been carried out in different non-thermal discharges (Losanitch & Jovitschitsch, 1897; Lob, 1906; Sahasrabudhey & Kolyanasundaram, 1948; Sahasrabudhey & Deshpande, 1950, 1951). Although reaction (9–57) is thermoneutral, the energy cost of CH_2O production in the experiments is quite high – about 70 eV/mol. The maximum yield of formaldehyde is 7%, and it is reached in the initially hydrogen-rich mixtures (up to 75% of H_2). Lower energy costs of the process are possible, when synthesis of formaldehyde is stimulated by vibrational excitation of CO and H_2 molecules (Givotov et al., 1984b). When the syngas CO–H_2 mixture is treated in non-equilibrium plasma with effective vibrational excitation of both molecules (CO, H_2), a sufficient increase in the specific energy input leads to partial conversion of formaldehyde into methanol (Givotov et al., 1984b):

$$H_2 + CH_2O \rightarrow CH_3OH, \quad \Delta H = -1\,eV/mol. \tag{9-57}$$

Starting not from the syngas CO–H_2 but from higher initially oxidized mixtures, CO–H_2O or CO_2–H_2, plasma-chemical synthesis leads to the formation of formic acid:

$$CO + H_2O \rightarrow HCOOH, \tag{9-58}$$

$$CO_2 + H_2 \rightarrow HCOOH. \tag{9-59}$$

These processes were demonstrated in different non-equilibrium plasma-chemical systems (Hemptinne, 1897; Losanitch & Jovitschitsch, 1897; Lob, 1906; Losanitch, 1911; Lind, 1923; Briner & Hoefer, 1940). It is interesting to compare hydrogenization of carbon dioxide (9–59) with that of carbon disulfide (CS_2), which has been done in non-thermal plasma conditions and leads to the production of acetylene and sulfur:

$$CS_2 + 1/2H_2 \rightarrow 1/2C_2H_2 + S_2. \tag{9-60}$$

An interesting non-equilibrium process for plasma-chemical synthesis of formic acid from a mixture of CO_2 and H_2O was first demonstrated more than 100 years ago by Losanitch and Jovitschitsch (1897):

$$CO_2 + H_2O \rightarrow HCOOH + 1/2O_2. \tag{9-61}$$

This process can be considered the first plasma-chemical stage of a two-step process for hydrogen production from water. The second stage in this case is formic acid decomposition with formation of hydrogen and carbon dioxide:

$$HCOOH \rightarrow H_2 + CO_2. \tag{9-62}$$

The carbon dioxide produced from formic acid is then recycled back to the plasma-chemical synthesis (9–61). The cycle (9–61) and (9–62) of hydrogen production from water is especially interesting because the plasma-chemical stage (9–61) does not require thermodynamically significant energy input. Most of the energy required for hydrogen production from water should be spent in this cycle during decomposition of formic acid in the second stage (9–62), which proceeds at a relatively low temperature. Therefore, the cycle (9–61) and (9–62) of hydrogen production from water can be performed using mostly low-temperature heat, if energy efficiency of the plasma process (9–61) is sufficiently high (Novikov, Bochin, & Romanovsky, 1978; Rusanov & Fridman, 1984). The cycle (9–61) and (9–62) of hydrogen production from water is somewhat similar to the natural process of photosynthesis. CO_2 and H_2O molecules are first converted into a relatively large organic molecule, HCOOH, with the production of oxygen without consuming significant energy but using selective and catalytic properties of non-thermal plasma. Then the relatively large molecule dissociates, consuming significant energy but at low temperatures.

9.5.6. Non-Thermal Plasma Production of Methane and Acetylene from Syngas ($CO–H_2$)

Methane and acetylene can be effectively produced from syngas ($CO–H_2$) in non-thermal plasma conditions. Syngas-based production of methane is an exothermic process:

$$CO + 3H_2 \rightarrow CH_4 + H_2O, \quad \Delta H = -2\,eV/mol. \tag{9–63}$$

The total plasma-chemical process of methane synthesis (9–63) can be considered a continuation of the carbon monoxide hydrogenization sequence, which starts with the production of formaldehyde from syngas (9–56), then proceeds to production of methanol from formaldehyde (9–57), and finally leads to the production of methane (9–62) in the following reaction:

$$CH_3OH + H_2 \rightarrow CH_4 + H_2O, \quad \Delta H = -1\,eV/mol. \tag{9–64}$$

The entire carbon monoxide hydrogenization sequence from CO to methane can be effectively stimulated by vibrational excitation of hydrogen molecules in non-equilibrium plasma conditions (Rusanov & Fridman, 1984). Methane formed in process (9–63) can be partially converted into acetylene, which determines CH_4 and C_2H_2 as the major products of the non-thermal plasma treatment of syngas. The synthetic process (9–63) in non-equilibrium microwave discharge at moderate pressures (10–50 Torr) has been described by McTaggart (1967). To reach a higher conversion degree, water produced in the reaction has been continuously frozen out. As a result, 80% of carbon monoxide has been converted in the non-thermal plasma experiments into methane with some additional small amount of acetylene. Effective production of methane and acetylene from syngas has also been demonstrated in glow discharges (Fisher & Peters, 1931), DBDs (Lunt, 1925), and spark discharges (Lefebvre & Overbeke, 1934).

The endothermic reverse process of (9–63), referred to as steam reforming of methane, can be effectively stimulated in plasma. The steam reforming of methane can be accomplished using any other energy source, particularly a high-temperature nuclear reactor. For this reason, the cycle including direct and reverse chemical reactions (9–63),

$$CH_4 + H_2O \Leftrightarrow CO + 3H_2, \tag{9–65}$$

plays an important role in long-distance heat transportation systems from high-temperature gaseous nuclear reactors (Ponomarev-Stepnoy, Protsenko, & Stoliarevsky, 1979). In these systems, heat generated by the nuclear reactor provides conversion of methane into syngas, which is then transported to the customer. At the destination point, the syngas is converted back to methane, releasing heat. The role of plasma in the system is to stimulate exothermic process (9–63) without using catalysts, which are not effective at relatively high temperatures.

9.6. PLASMA-CHEMICAL POLYMERIZATION OF HYDROCARBONS: FORMATION OF THIN POLYMER FILMS

9.6.1. General Features of Plasma Polymerization

The formation of high-molecular products (polymers) from initial low-molecular substances (monomers) in non-thermal, mostly low-pressure (0.01–10 Torr), discharges is usually referred to as plasma polymerization. Formation of high-molecular-weight products (polymerization) proceeds either on solid surfaces contacting with plasma, which results in growth of polymer films, or in the plasma volume, which results in production of polymer powders or some forms of polymer macroparticles. In contrast to conventional polymerization which requires the use of specific monomers, the application of plasma permits the polymerization to start from practically any organic compound. Plasma-stimulated deposition of polymer films is of significant interest due to the applications of these films to surface processing and formation of thin dielectric and protective films, which is especially important in microelectronics as well as in biology and medicine. Numerous research efforts have been focused on the investigation of plasma polymerization. Almost a century-old publication of Linder and Davis (1931) describes the plasma synthesis of 57 different polymers. Reviews on plasma polymerization can be found in Hollahan and Bell (1974), Havens (1976), Bradley (1970), Vinogradov and Ivanov (1977), Tkachuk and Kolotyrkin (1977), Yasuda (1985), Vinogradov (1986), Biederman and Osada (1992), Inagaki (1996), D'Agostino et al. (2003), Oehr (2005), and Sardella et al. (2005).

Characteristics of plasma polymerization processes essentially depend on the specific power of non-equilibrium discharges, gas pressure, and dilution degree of the initial hydrocarbons in noble gases. At low values of the specific power ($\ll 0.1$ W/cm^3), which are typical for discharges in mixtures highly diluted with noble gases, the translational gas temperature is usually relatively low ($T_0 < 450$ K). In this case, the initial volume dissociation of hydrocarbons is mainly due to non-equilibrium processes stimulated by direct electron impact. At higher levels of specific power (about 0.1 W/cm^3), which is typical for discharges in non-diluted hydrocarbons, translational gas temperatures can exceed 500 K. In this case, thermal decomposition of hydrocarbons and their reactions with atomic hydrogen make larger contribution to polymerization kinetics. Also the higher temperatures lead to higher volume concentration of heavier gas-phase hydrocarbons and to the acceleration of plasma polymerization. An increase of pressure above 3–10 Torr in the discharge slows down diffusion of the hydrocarbon radicals to the reactor walls and stimulates their volume reactions and recombination. Nevertheless, interesting experiments with plasma polymerization in non-thermal atmospheric-pressure discharges have been carried out recently.

9.6.2. General Aspects of Mechanisms and Kinetics of Plasma Polymerization

Numerous experiments have focused on the kinetics of plasma polymerization in low-pressure non-equilibrium plasma. For example, experiments with plasma polymerization of benzene, toluene, ethyl benzene, and styrene in glow discharges (Gilman, Kolotyrkin, & Tunitsky, 1970) demonstrate an exponential decrease of the polymer deposition rate with temperature, which indicates the contribution of absorption kinetics to the polymerization process (see kinetic details following). Similar experiments of Tusov et al. (1975) show the significant contribution of charged particles in the deposition rate, while contribution of ultraviolet (UV) radiation, as well as excited and chemically active neutrals, is also essential. Which plasma components – ions or radicals – mostly dominate polymerization is actually still an open question (Vinogradov, Ivanov, & Polak, 1978, 1979a,b). Consider as an example the glow discharge plasma polymerization of cyclohexane, analyzed by Slovetsky (1981). In this system, the film deposition rate linearly grows with cyclohexane concentration in the initial mixture with noble gases, whereas all plasma parameters including the charged particles flux to the surface remain the same. It shows that gradual cyclohexane ion attachment does not contribute to the polymer film deposition. Kinetic analysis results in the following polymerization mechanism in this case. First, cyclohexane dissociates in the plasma volume to form a cyclohexane radical and atomic hydrogen. The radicals diffuse to the walls and recombine there with open bonds formed on the polymer film surface by plasma particle bombardment. Further bombardment leads to more bond breaking and reorganization of C–C and C–H bonds on the polymer surface. Hydrogen is then desorbed from the surface, while free surface bonds recombine between one another and with gas-phase radicals, creating a highly cross-linked polymer structure. A similar mechanism describes the plasma polymerization of some other hydrocarbons diluted in noble gases (Alekperov, 1979).

Development of the general plasma polymerization model is restricted by an essentially non-linear contribution of different plasma components in the process. Plasma particles are able to simultaneously build and destroy thin polymer films. Nevertheless, some models are able to describe a wide variety of experiments. The plasma polymerization model developed by Osipov and Folmanis (1973) is one of those. According to this model, charged particles activate adsorbed molecules, allowing then to cross-link with the polymer film. The rate of polymer film deposition from a non-equilibrium low-pressure discharge (number of deposited molecules per unit time) can be calculated in this case as

$$w_d = \frac{\Phi}{1 + \frac{1}{\sigma j}\left(\frac{1}{\tau} + a\Phi\right)}, \tag{9–66}$$

where Φ is the flux of the polymer creating molecules to the surface, j is the flux of charged particles to the surface, σ is the cross-section for activation of the adsorbed molecules by charged particles, τ is the effective lifetime of the adsorbed molecule, and a is the surface area occupied by an absorbed molecule. When fluxes to the surface of neutral and charged particles are not very high,

$$\sigma j \tau \ll 1, \quad a\Phi\tau \ll 1, \tag{9–67}$$

and the rate of polymer film deposition (9–66) can be rewritten, taking into account the adsorption time dependence on surface temperature as (Roginsky, 1948)

$$w_d \propto \Phi j \cdot \sigma \exp\left(\frac{E_a^a + \Delta H_a}{T_s}\right). \tag{9–68}$$

Here E_a^a and ΔH_a are activation energy and enthalpy of the adsorption relevant to plasma polymerization and T_s is the surface temperature. Formula (9–68) interprets the plasma polymerization kinetics, when the film deposition rate is proportional to fluxes of both neutral and charged particles. It explains the exponential acceleration of the polymer film deposition rate in plasma with reduction of surface temperature (Gilman et al., 1970; Tusov et al., 1975). This kinetic effect is due to surface stabilization of intermediate products at lower temperatures, which accelerate the polymerization rate. Polymer film growth rates in non-thermal plasma vary over a large range, between 1 nm/s and 1 μ/s. These values are time-averaged; instantaneous film growth rates can be higher.

9.6.3. Initiation of Polymerization by Dissociation of Hydrocarbons in Plasma Volume

Consider the kinetics of the plasma process starting with decomposition of hydrocarbons in low-pressure glow discharges, which initiates polymerization (Yensen, Bell, & Soong, 1983; Ivanov & Epstein, 1984; Ivanov, 1989; Ivanov, Rytova, & Soldatova, 1990a; Ivanov et al., 1990b). If the initial concentration of hydrocarbons mixed with an inert gas exceeds 10–20 vol %, dissociation of hydrocarbons is mostly due to electronic excitation by direct

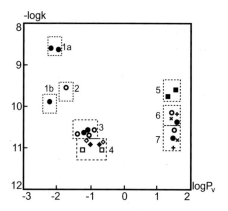

Figure 9–17. Rate coefficients of dissociation of hydrocarbons by direct electron impact in glow discharges: effective experimental values (k_{diss}) and simulated partial values (k_{ed}). Pressures: 13 Pa ($\bullet$), 29 Pa ($\square$), 59 Pa ($\blacklozenge$), 133 Pa ($\blacksquare$), 266 Pa ($\circ$), 400 Pa (X), 530 Pa ($\blacksquare$), 660 Pa ($+$), 2660 Pa ($\diamond$). Effective coefficients of dissociation (k_{diss}): (1a) cis-C_6H_{12} (cis-C_5H_{10} and C_6H_6 are 30%–40% lower, cis-$C_6H_{11}CH_3$ and n-C_6H_{14} are twice lower), DC glow discharge in Ar + 1% hydrocarbons; (3) C_2H_6, RF glow discharge in ethane; (4) CH_4, RF glow discharge in methane; (5) CH_4 (C_2H_4 and C_2H_6 are 10% higher, n-C_5H_{12} are more than 10% lower), RF glow discharge in hydrocarbons; (6) C_6H_6, microwave discharge in C_6H_6 + H_2. Simulated partial values of the dissociation rate coefficient (k_{ed}): (1b) cis-C_6H_{12}, DC glow discharge in Ar + 1% hydrocarbons; (2) CH_4, DC glow discharge in Ar + 1% hydrocarbons. Rate coefficients are given in cm^3/s, specific power P_v is presented in W/cm^3.

electron impact (see Sections 2.5.5 and 3.6.3). When the concentration of hydrocarbons is less that 3 vol %, electronically excited atoms of inert gases and hydrogen atoms start making an essential contribution to the dissociation of hydrocarbons. Ion–molecular processes also can make a contribution to the dissociation at very low hydrocarbon concentrations. If gas temperature in the non-thermal discharges exceeds 500 K, additional decomposition of hydrocarbons is due to their reactions with atomic hydrogen, which can double the plasma dissociation rate of hydrocarbons.

Vibrational excitation of hydrocarbons stimulates dissociation if the ionization degree is sufficiently high; see formula (3.201). For low-pressure glow discharges in hydrocarbons, the criterion of significant contribution of vibrational excitation in dissociation can be presented as (Ivanov, 2000)

$$A \cdot P_v / p^2 \gg 1, \tag{9–69}$$

where P_v is the specific power in the discharge, in W/cm^3; p is the gas pressure, in Pa; and A is a numeric factor, which equals 10^4 for discharges in CH_4 and decreases for discharges in higher hydrocarbons. When the pressure is about 1000 Pa, a significant contribution of vibrational excitation requires very high values of specific power exceeding 100 W/cm^3. At lower pressures, power requirements are not as strong, but essential losses of vibrational excitation can be related to heterogeneous relaxation (see Section 2.6.3). Effective rate coefficients for dissociation of different hydrocarbons in different low-pressure non-thermal discharges (glow, RF, and microwave discharges) are summarized in Fig. 9–17 as a function of specific discharge power P_v (Ivanov, 2000). This dissociation process results in the formation of active hydrocarbon radicals, which are responsible for growth of the thin polymer film. An increase of the specific discharge power between 0.1 and 10 W/cm^3 leads to an increase of the effective dissociation coefficient due to the increase of gas temperature and reactions of atomic hydrogen with hydrocarbons. The figure also illustrates that dilution (99%) with argon leads to higher values of the effective dissociation coefficient (10^{-9} cm^3/s), which is due to additional dissociations in collisions with electronically excited atoms of the inert gas.

9.6.4. Heterogeneous Mechanisms of Plasma-Chemical Polymerization of C_1/C_2 Hydrocarbons

Plasma polymerization, in contrast to conventional polymerization, permits a polymer to be built up starting from any organic compound. Building up a polymer in plasma does not require any specific chemical behavior of monomers, like opening of double bonds, but is based on the attachment of chemically active species (radicals, atoms, non-saturated molecules, etc.) generated in plasma beforehand. Such a mechanism of plasma polymerization based on preliminary "plasma activation of monomers" is sometimes referred to as "stepwise." The heterogeneous mechanism of polymerization is quite complicated in this case. Next we discuss two examples, for which a homogenous mechanism is better investigated: plasma polymerization of C_1/C_2 hydrocarbons and plasma-initiated chain polymerization. In the specific case of plasma polymerization of C_1/C_2 hydrocarbons, mechanisms of heterogeneous stages have been well investigated (Ivanov, 2000). The key heterogeneous process initializing the plasma polymerization is the formation of free bonds near the polymer film surface (depth up to several monolayers), which are called **centers of polymer growth**. Center formation is mostly due to recombination of electrons with ions of hydrocarbons, which proceeds by electron tunneling through the potential barrier of the dielectric polymer film. The center of polymer growth is usually related to a broken C–H bond; the probability of breaking C–C and C=C bonds is much lower. Major growth of the polymer films is due to the attachment of radicals and non-saturated hydrocarbons produced in the plasma volume to the centers of polymer growth.

The main mechanism of polymer film growth is accompanied by a wide variety of other heterogeneous processes that also contribute to the plasma polymerization of C_1/C_2 hydrocarbons, specifically the formation of the centers of polymer growth as a result of recombination of electrons with ions of inert gases resulting in energy transfer to the hydrocarbons and formation of a free bond. On the other hand, not only radicals and non-saturated molecular hydrocarbons but also hydrocarbon ions and ion radicals produced in plasma can be attached to the centers of polymer growth (and even to the polymer macromolecules). UV radiation from non-thermal plasma also leads to breaking bonds in the macromolecules of hydrocarbons. It is absorbed, however, not in the near-surface region (several monolayers thin), where polymer film growth actually takes place; the UV radiation generated in plasma is absorbed in a much thicker layer of up to several microns, where it leads not to polymer film growth but to cross-linking of the macromolecules. Detailed kinetics of the plasma-chemical polymerization of C_1/C_2 hydrocarbons in low-pressure non-equilibrium discharges is described by Ivanov (2000). Analysis of the heterogeneous mechanisms also indicates that plasma polymerization includes two competing processes: plasma-stimulated polymer film growth and plasma-stimulated etching of the polymer film (see Section 8.2). Generally, a "more active" plasma interaction with the surface leads to domination by etching. Sometimes it is possible to have domination of a polymer film growth in one part of a reactor, while etching dominates in another part of the same reactor. The effective growth of polymer films requires choosing plasma-chemical conditions that suppress the etching. The contribution of etching becomes especially important when the macromolecules include an oxygen group –O– in the major polymer chain, and when the macromolecules include any oxygen-containing side groups.

9.6.5. Plasma-Initiated Chain Polymerization: Mechanisms of Plasma Polymerization of Methyl Methacrylate

Non-thermal plasma can be applied not only for the stepwise polymerization discussed earlier, but also for effective stimulation of more conventional chain polymerization processes. Plasma-initiated polymerization of methyl methacrylate (MMA) with production of practically important polymer polymethyl methacrylate (PMMA) is a good example of such processes (Ponomarev, 1996, 2000). MMA is a quite large organic molecule: $CH_2=C(CH_3)-C(=O)-O-CH_3$, creating polymer (PMMA) by opening of the C=C double bond. Conventional chain propagation reactions of MMA polymerization can be represented by the elementary MMA attachment to a radical $R(\bullet)$:

$$R(\bullet) + CH_2 = C(CH_3)-C(=O)-O-CH_3$$
$$\rightarrow (RH_2)C-C(\bullet, CH_3)-C(=O)-O-CH_3. \qquad (9\text{–}70)$$

Plasma initiation of the chain polymerization is due to formation of a primary free radical $R(\bullet)$, starting the traditional scheme (9–70), and by formation of positive or negative ion radicals, which are also capable of initiating the MMA polymerization. The primary free radical $R(\bullet)$ as well as the charged centers of polymer growth are formed from the absorbed monomers by electron/ion bombardment and UV radiation from plasma. Formation of a positive ion radical from an adsorbed MMA molecule on the surface under electron/ion bombardment and UV radiation can be schematically shown as the ionization process

$$CH_2 = C(R_1)-R_2 \rightarrow CH_2(\bullet)-C^+(R_1)-R_2 + e. \qquad (9\text{–}71)$$

The positive ion radical then initiates a sequence of attachments of further and further MMA molecules in the ion–molecular chain propagation reactions:

$$CH_2(\bullet)-C^+(R_1)-R_2 + CH_2 = C(R_1)-R_2$$
$$\rightarrow CH_2(\bullet) - (R_1)C(R_2) - CH_2 - C^+(R_1)-R_2. \qquad (9\text{–}72)$$

Formation of a negative ion radical (which is also a center of polymer growth) from an adsorbed MMA molecule on the surface is due to direct electron attachment:

$$CH_2 = C(R_1)-R_2 + e \rightarrow CH_2(\bullet)-C^-(R_1)-R_2. \qquad (9\text{–}73)$$

Similarly to process (9–72), the negative ion radical operating as a center of polymer growth also initiates a sequence of attachment processes of further MMA molecules in the ion–molecular chain-propagation reactions involving the negative ions:

$$CH_2(\bullet)-C^-(R_1)-R_2 + CH_2 = C(R_1)-R_2$$
$$\rightarrow CH_2(\bullet) - (R_1)C(R_2) - CH_2-C^-(R_1)-R_2. \qquad (9\text{–}74)$$

To deposit the PMMA thin film on a substrate of interest, the partial pressure of MMA should be high enough to provide sufficient MMA concentration in the absorbed surface layer. Such a requirement is common for all plasma-initiated chain polymerization processes. Discharge power should not be high in the plasma-initiated chain polymerization to minimize conversion of MMA (or other monomers) in the plasma volume. The role of plasma should be limited in this case to the generation on the surface of relatively low concentrations of active centers initiating chain polymerization in the layer of absorbed monomers.

9.6.6. Plasma-Initiated Graft Polymerization

Polymer film deposition on polymer substrates or other surfaces can be accomplished by graft polymerization. In this case, chain polymerization of a monomer proceeds on a polymer (or other material) substrate preliminarily treated in non-thermal plasma. A common approach to such graft polymerization is based on the preliminary polymer substrate treatment in O_2-containing plasma, which forms organic radicals $R(\bullet)$ on the surface and converts them into organic peroxides. Formation of the organic peroxide compounds is the chain process, which starts with the direct attachment of molecular oxygen and formation of an organic peroxide radical:

$$R(\bullet) + O_2 \rightarrow R-O-O(\bullet). \qquad (9\text{–}75)$$

Further propagation of the plasma-initiated chain leads to production of the organic peroxides and restoration of organic radicals:

$$R - O–O(\bullet) + RH \rightarrow ROOH + R(\bullet), \qquad (9–76)$$

$$R - O–O(\bullet) + RH \rightarrow ROOR_1 + R_2(\bullet). \qquad (9–77)$$

The polymer substrate activated this way in plasma (by attachment of the surface peroxide groups) is able to initiate the graft polymerization of gas-phase or liquid-phase monomers. To initiate the graft polymerization process, the substrate should first be heated up to dissociate the organic peroxide on the surface (9.76) and (9.77) and to form active organic radicals:

$$ROOH \rightarrow RO(\bullet) + (\bullet)OH, \quad ROOR_1 \rightarrow RO(\bullet) + R_1O(\bullet). \qquad (9–78)$$

The RO radicals then function as the centers of polymer growth. They initiate the graft polymerization, which in the specific case of MMA polymerization into PMMA is the sequence of attachment reactions of the monomer to the center of polymer growth:

$$RO(\bullet) + CH_2 = C(CH_3)–C(=O)–O–CH_3$$
$$\rightarrow RO–CH_2–C(\bullet, CH_3)–C(=O) - O–CH_3 \qquad (9–79)$$

Grafting of monomers with special functional groups significantly changes the surface properties of initial polymer substrates and allows the creation of new special compounds. Such an approach has been effectively applied to create new types of immobilized catalysts on a polymer base. In particular, polyethylene powder with specific surface area $2 \, m^2/g$ has been activated in oxygen-containing plasma for further graft polymerization of acrylic and methacrylic acids from the gas phase. Then vanadium (V), titanium (Ti), and cobalt were chemically deposited on the powder to produce highly effective immobilized metal catalysts. Such plasma-activated powders were successfully used as catalysts for ethylene polymerization into polyethylene (PE), as bifunctional catalysts of C_2H_4 dimerization and C_2H_4–C_4H_8 co-polymerization, and for other catalytic applications, particularly the non-thermal plasma-initiated graft polymerization of vinyl monomers on the surface of polytetrafluoroethylene (PTFE), which significantly changes the PTFE surface properties. The PTFE film is first activated in non-thermal plasma of a low-pressure discharge and then treated in a special liquid monomer that significantly enhances its adhesion to different materials and especially to steel (Ponomarev, 2000).

9.6.7. Formation of Polymer Macroparticles in Volume of Non-Thermal Plasma in Hydrocarbons

Sometimes, when the residence time of hydrocarbons in the discharge volume is relatively long, the polymer film growth on substrates, reactor walls, and the electrode is accompanied by the formation of polymer powder in the plasma volume. Polymer macroparticles of the powder are deposited on different surfaces but also can be incorporated into the growing polymer film. The typical size of spheroidal macroparticles is $0.1 \, \mu m$ and later. Although mechanisms of growth of the polymer powder in the plasma volume are similar to those of polymer films on surfaces, some specific features are related in this case to the formation of precursors of the macroparticles. The precursors of macroparticles are usually sufficiently large hydrocarbon molecules formed in plasma. The minimal size of a macromolecule to be considered a macroparticle is determined by the ability of the particle to be effectively negatively charged in plasma (see Section 8.8). In the case of large hydrocarbon molecules, the ability to be charged usually requires the macroparticle sizes to exceed 5 nm. While the growth rate of the plasma polymer film on substrate surfaces usually varies between 1 nm/s

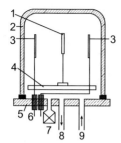

Figure 9–18. General schematic of the DC plasma polymerization system with internal electrodes: (1) substrate; (2) reactor walls; (3) electrodes; (4) screen; (5) baseplate; (6) high-voltage wiring; (7) manometer; (8) pumping; (9) inlet for a polymer-forming gas.

and 1 μ/s, the growth rate of the polymer macroparticles can be significantly faster (10^{-8}–10^{-4}cm/s). Therefore, formation of a macroparticle larger than 1 μm typically requires more than 1 s of residence time. The polymer macroparticles leave the discharge zone through different channels. Relatively large macroparticles (5–10 μm) simply fall down to the bottom of the plasma-chemical reactor due to gravity. While gas flow drags the polymer powder, stagnation zones lead to longer residence time and to an increase of size and concentration of the macroparticles. The polymer macroparticles are usually charged in plasma, which essentially influences their evolution and their dynamics (see Section 8.8). Factors that prevent significant formation of the polymer macroparticles include the high flow rate of hydrocarbons through the discharge active zone, the absence of stagnation zones there, and the limitation of the film growth rate.

9.6.8. Plasma-Chemical Reactors for Deposition of Thin Polymer Films

Configurations of reactors applied for plasma polymerization can be quite different. They can be based on low-pressure DC glow discharge, and discharges over a wide variety of frequencies from industrial (50–60 Hz) to RF and microwave frequencies. The DC and industrial-frequency alternating-current (AC) discharges use special electrodes located inside the plasma-chemical reactor, whereas RF and microwave discharges are usually electrodeless. The most conventional configurations of the plasma polymerization reactors are illustrated as follows (Ponomarev, 2000). A schematic of a conventional reactor for plasma deposition of the polymer films in a low-pressure DC glow discharge is presented in Fig. 9–18. The typical voltage is hundreds of volts and the typical distance between parallel planar electrodes is 3–5 cm. The substrate is located between the electrodes. Typical pressure in the discharge varies in the range 30–300 Pa. A typical plasma polymerization reactor based on a low-pressure RF-ICP discharge is shown in Fig. 9–19. Parameters of the electrodeless discharge are somewhat similar to those of the aforementioned DC glow discharge. Polymer film deposition can be done in this case either in the active non-thermal plasma zone located inside of the inductive coil or on the substrate outside of the RF-ICP discharge in the special chamber for film deposition. A schematic of a plasma-chemical reactor for plasma-initiated chain polymerization (Section 9.6.5) is presented in Fig. 9–20. In this system, the glow discharge zone is located sufficiently far from the substrate to avoid intensive interaction of the active plasma component with the growing polymer film. As was explained in Section 9.6.5, such a regime is optimal for plasma-initiated chain polymerization. The system has been applied, in particular, to polymerization of MMA and deposition of thin PMMA film on the surface of semiconducting materials. The PMMA films are characterized by sufficient electric resistivity and high lithographic resolution (about 0.10–0.15 μm).

9.6.9. Some Specific Properties of Plasma-Polymerized Films

The most specific property of plasma-polymerized films is a high concentration of free radicals in the films and a large number of cross-links between macromolecules. The

9.6. Plasma-Chemical Polymerization of Hydrocarbons

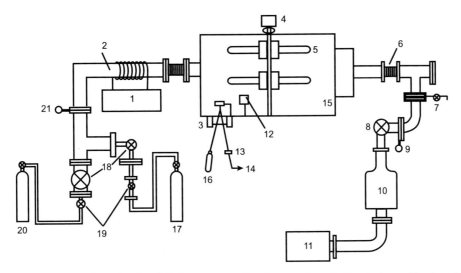

Figure 9–19. Schematic of an RF plasma system for plasma polymerization (polymer film deposition): (1) RF generator; (2) plasma zone; (3) heated window; (4) engine; (5) rotating substrates; (6) flexible connection; (7) air valve; (8) vacuum valve; (9) manometer; (10) cooled trap; (11) vacuum pump; (12) quartz gauge for measurements of the film thickness; (13) photodiode; (14) registration system; (15) polymer deposition chamber; (16) laser; (17) argon tank; (18) valve; (19) dose valve; (20) tank with a manometer; (21) manometer.

concentration of free radicals can be actually very high, up to 10^{19}–10^{20} spin/g. The cross-linkage immobilizes the free radicals, significantly slowing down their recombination and chemical reactions. The slowly developing chemical processes with the free radicals result in "slow but sure" changing of gas permeability, electric characteristics, and other physical and chemical properties of the plasma-polymerized films, which is usually referred to as the **aging effect**. Plasma-polymerized films are often characterized by high internal stresses, up to 5–$7 \cdot 10^7$ N/m^2. In contrast to conventional polymerization, internal stresses in the plasma-polymerized films are related to extension of the material. The effect is due to intensive insertion of free-radical fragments between already deposited macromolecules with simultaneous cross-linking during the polymer film growth. The solubility of plasma-polymerized films in water and organic solvents is usually very low because of very strong cross-linkage between macromolecules. It should be mentioned that if the film growth rate is very high and the formed macromolecules are relatively short, the film can be dissolved in organic solvents and even in water. Generally, however, when polymer film growth rates are 0.1 nm/s and less, the plasma-polymerized films are highly cross-linked and not soluble in either water or organic solvents. The majority of plasma-polymerized films are characterized

Figure 9–20. General schematic of a plasma-chemical reactor for polymer film formation due to plasma-initiated chain polymerization: (1) glass reactor; (2) cooled substrate; (3) cooling system; (4) electrodes; (5) discharge zone; (6) inlet system for polymer-forming gases (for example, MMA and Ar); (6) pumping system.

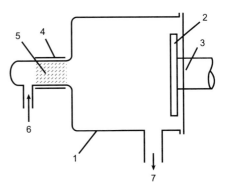

Table 9–8. Dielectric Properties of Plasma-Polymerized Films in Comparison with Conventional Polymers at Temperature 20°C and Frequency 1 kHz

Polymer	Dielectric permittivity ε, conventional polymer	Dielectric permittivity ε, plasma polymer	Dielectric loss tangent $\tan \delta$, conventional polymer	Dielectric loss tangent $\tan \delta$, plasma polymer
Polyethylene	2.3	3.57	$1–2 \cdot 10^{-4}$	$5–30 \cdot 10^{-4}$
Polystyrene	2.55	2.67–3.3	$2 \cdot 10^{-4}$	$1–3 \cdot 10^{-3}$
Polyisobutylene	2.2–2.3	3.0	$2 \cdot 10^{-4}$	$9 \cdot 10^{-3}$
Polytetrafluoroethylene	2.0	–	$1–2 \cdot 10^{-4}$	$25 \cdot 10^{-4}$
Hexamethyldisiloxane (silicon rubber)	3.2	2.5–5.23	$4 \cdot 10^{-3}$	$1–10 \cdot 10^{-3}$

by high thermal stability. As an example, a thin polymer film deposited from a non-thermal low-pressure discharge in methane remains stable and does not lose weight after being treated in argon at temperatures up to 1100 K and in air at temperatures up to 800 K.

The wettability of plasma-polymerized films is related to their surface energy and depends on the type of plasma gas used for polymerization. For example, polymer films formed from fluorocarbon and organic silicon compounds are characterized by low surface energy and low wettability. Plasma-polymerized hydrocarbon films usually have high wettability. It is important that their wettability is higher than that of their counterparts formed without plasma. The effect is due to oxygen-containing groups usually incorporated into the plasma-polymerized films. These groups usually make a significant contribution to increased surface energy and wettability. Thin plasma-polymerized films are also specific in selective permeability for different gases. A large number of cross-links makes this film somewhat like a molecular sieve, creating interest for applications of such films as membranes for gas separation. Ultrathin plasma-polymerized films with thickness of 0.1 μm and smaller are usually characterized by very strong adhesion to a substrate; the adhesion of relatively thick films is significantly lower. Also the highest adhesion is usually achieved at lower rates of polymer film deposition. As an example, ultrathin plasma-polymerized films with thickness below 0.2 μm have an adhesion to aluminum substrate exceeding 1500 N/cm^2 (Ponomarev, 2000).

9.6.10. Electric Properties of Plasma-Polymerized Films

The electric properties of plasma-polymerized films are especially important in connection with their applications as dielectrics in microelectronics. Major dielectric properties of plasma-polymerized films (dielectric permittivity ε and dielectric loss tangent $\tan \delta$) are shown in Table 9–8 in comparison with characteristics of polymer films produced in a conventional manner from the same monomers. It can be seen from the table that the dielectric permittivity ε is slightly higher and the dielectric loss tangent, $\tan \delta$, is significantly higher in the case of plasma-polymerized films. This effect occurs because the plasma-polymerized films always include some polar groups. The dielectric properties of plasma-polymerized films are affected by an aging effect and are also sensitive to humidity. The aging effect means in this case the dielectric properties change with time, which is due to a high concentration of free radicals in the films, and slow reactions of the radicals with oxygen from air lead to the formation of new oxygen-containing polar groups. The dielectric properties can be stabilized by special treatment of the freshly deposited films, partially decreasing the concentration of free radicals. It can be achieved by heating the films in vacuum or by their additional treatment in hydrogen plasma.

Table 9–9. Electric Conductivity of Thin Plasma-Polymerized Films and Its Temperature Dependence

Initial substance used for plasma polymerization	Electric conductivity at temperature 150°C, σ, ohm^{-1} cm^{-1}	Electric conductivity at temperature 250°C, σ, ohm^{-1} cm^{-1}	Activation energy, E_a, eV
Naphthalene	$9 \cdot 10^{-16}$	$2.7 \cdot 10^{-13}$	1.1
Styrene	$6 \cdot 10^{-16}$	$9 \cdot 10^{-14}$	1.2
n-Xylol	$5 \cdot 10^{-17}$	$1.5 \cdot 10^{-13}$	1.8
Cyclopentadiene	$1.0 \cdot 10^{-16}$	$1.2 \cdot 10^{-13}$	1.5
Hexamethylbenzene	$7 \cdot 10^{-17}$	$7 \cdot 10^{-14}$	1.5
Ethylene oxide	$4 \cdot 10^{-16}$	$1.6 \cdot 10^{-13}$	1.1
Methoxinaphthalene	$1.1 \cdot 10^{-16}$	$7 \cdot 10^{-14}$	1.5
Thiocarbamide	$3.3 \cdot 10^{-16}$	$4 \cdot 10^{-13}$	1.7
Chlorobenzene	$8 \cdot 10^{-17}$	$1.9 \cdot 10^{-14}$	1.4
Picoline	$2.2 \cdot 10^{-14}$	$6 \cdot 10^{-12}$	1.1
N-nitrosodiphenylamine	$8 \cdot 10^{-15}$	$3 \cdot 10^{-12}$	1.2
n-Toluidine	$7 \cdot 10^{-16}$	$2.3 \cdot 10^{-12}$	1.5
Aniline	$2.8 \cdot 10^{-16}$	$1.4 \cdot 10^{-12}$	1.8
n-Nitrotoluene	$5 \cdot 10^{-16}$	$2.5 \cdot 10^{-13}$	1.2
Diphenylselenide	$3 \cdot 10^{-18}$	$8 \cdot 10^{-13}$	1.5
Diphenylmercury	$2.8 \cdot 10^{-15}$	$2.7 \cdot 10^{-13}$	0.85
Ferrocene	$2.7 \cdot 10^{-13}$	$4.5 \cdot 10^{-12}$	0.55
Benzene-selenol	$2.5 \cdot 10^{-14}$	$7 \cdot 10^{-12}$	1.1
Hexa-n-butyl-tin	$1.5 \cdot 10^{-15}$	$7 \cdot 10^{-13}$	1.1
Tetracyanoethylene	$1.8 \cdot 10^{-13}$	$5 \cdot 10^{-12}$	0.6
Malononitrile	$3 \cdot 10^{-14}$	$1.8 \cdot 10^{-12}$	0.75
Thianthrene	$1.5 \cdot 10^{-14}$	$1.6 \cdot 10^{-12}$	0.85
Thiophene	$6 \cdot 10^{-14}$	$3 \cdot 10^{-12}$	0.75
Thioacetamide	$8 \cdot 10^{-14}$	$9 \cdot 10^{-12}$	0.85

Resistance of plasma-polymerized films with respect to electric breakdown can be quite high. This resistance increases at relatively low specific powers of the non-thermal discharges applied for polymer film deposition. As an example, plasma-polymerized fluorocyclobutane film is characterized by breakdown electric fields: $5 \cdot 10^6$ V/cm for a film thickness of 150 nm, $6 \cdot 10^6$ V/cm for a film thickness of 100 nm, and $7–8 \cdot 10^6$ V/cm for a film thickness of 75 nm. To compare, breakdown of a conventional PTFE film with thickness 0.1–0.2 mm requires only $0.4–0.8 \cdot 10^6$ V/cm. Electric conductivity of the plasma-polymerized films is low and strongly depends on temperature with typical activation energy of about 1 eV. Table 9–9 presents the electric conductivities of some specific plasma-polymerized films measured at temperatures of 150 and 250°C as well as the corresponding activation energies. The conductivity of the thin films generally does not follow Ohm's law; therefore, data in the table are summarized for a specific film thickness around 1 μm and voltage 1.4 V (Ponomarev, 2000). Plasma-polymerized films usually have high photo-electric conductivity, which is due to effective photo-generation of electric charge carriers in the material of the polymer films. Oxygen-containing groups and free radicals in the material of the films are able to accept free electrons and therefore act as centers for photo-generation of electric charges and photo-electric conductivity.

9.6.11. Some Specific Applications of Plasma-Polymerized Film Deposition

Applications of thin plasma-polymerized films in microelectronics, gas separation, and catalysis, such as new types of immobilized polymer-based catalysts, have been already mentioned. Applications of the films as insulators and protective layers in thin-film capacitors in different areas of electronics are of interest. Multi-layer capacitors produced by this plasma-chemical method are characterized by high values of specific capacitance – more than $0.1\ \mu F/cm^2$, high resistance to electric breakdown, and high stability of parameters in a wide range of frequencies and temperatures. Another important application of plasma-polymerized films is deposition of thin (<100 nm) corrosion-protective coatings of metal surfaces. A single plasma-chemical reactor is used in this case to clean the surface, to deposit the protective polymer film, and, if necessary, to additionally plasma-modify the film, which is especially interesting for treatment of large-area steel surfaces. The film is formed in this process by plasma polymerization of hexamethyldisiloxane (HMDSO). The HMDSO plasma-polymerized 16–25 nm coating effectively protects steel surfaces against corrosion during several days at a temperature of 65°C and relative humidity of 85%. Similar plasma coatings are also effective in protecting magnetic recording systems. For effective plasma coating of steel surfaces, the surfaces should be preliminarily treated in oxygen plasma and then argon–hydrogen plasma. The mechanism providing corrosion protection of metals by the plasma-polymerized films is mostly related not to the barrier effect of the film itself but to formation on the metal surface of stable Fe–O–Si groups preventing corrosion. It explains why nanoscale plasma coatings (about 20 nm thick) are so effective in suppressing corrosion. Deposition of special plasma-polymerized films makes materials that are difficult to glue together using conventional approaches much more adhesive to the glue. This method has been successfully applied for gluing of PTFE and PE with aluminum, stainless steel, PTFE, and PE.

Plasma-polymerized films are used for changing the surface friction coefficient and other tribological properties of different materials, especially rubber. The elastomer-steel friction coefficient can be significantly decreased to 0.3–0.4 by deposition on the elastomer surface of plasma-polymerized fluorocarbons. Typical glow-discharge plasma reactors used for this application have a volume of about 0.7 m^3 and AC power of about 1 kW. An interesting application of plasma-polymerized HMDS thin films is related to surface modification by active carbon, which is further used for absorption of hazardous vapors and gases. Active carbon absorbs water vapor during its storage period, which reduces its ability to later absorb hazardous vapors and gases. Plasma deposition of HMDS-based thin films decreases water vapor absorption in this case more than two-fold. Absorption of CCl$_4$ decreases after the plasma treatment by less than 10%. Other interesting applications of plasma-polymerized thin films are considered by Ponomarev (1996, 2000).

9.7. INTERACTION OF NON-THERMAL PLASMA WITH POLYMER SURFACES: FUNDAMENTALS OF PLASMA MODIFICATION OF POLYMERS

9.7.1. Plasma Treatment of Polymer Surfaces

Non-thermal plasma treatment of polymers leads to significant changes of their surface properties, in particular surface energy, wettability, adhesion, surface electric resistance, dielectric loss tangent, dielectric permittivity, catalytic activity, tribological parameters, gas absorption, and permeability characteristics. Plasma treatment and modification of polymer surfaces are widely used today in numerous applications, from painting of textiles and printing on synthetic wrapping materials to treatment of photographic materials and in microelectronic fabrication. Some of these applications will be discussed in the following sections after preliminary consideration of the fundamentals of non-thermal plasma interaction with polymer surfaces. Plasma treatment of polymers can be performed in non-thermal plasma at both low and high pressures. Low-pressure non-thermal discharges are usually applied for polymer surface chemical functionalization and for "specific and accurate" chemical modification, as in the application for functionalization of photographic materials and in microelectronics. At the same time, high-pressure non-thermal discharges are usually sufficient for less specific treatment of polymers directed to surface cleaning, changing wettability, and so on. More and more applications that traditionally required low-pressure discharges now use non-thermal atmospheric-pressure discharges (Roth, 2001). Taking into account the significant practical interest, numerous research efforts have been focused recently on investigation of plasma treatment of polymer surfaces, similar to the case of plasma polymerization. Reviews and interesting materials on the subject can be found in particular in Boenig (1988), D'Agostino (1990), Yasuda (1985), Kramer, Yeh, & Yasuda (1989), Biederman & Osada (1992), Liston (1993), Lipin et al. (1994), Gilman & Potapov (1995), Garbassi, Morra, & Occhiello (1994), Ebdon (1995), Inagaki (1996), Ratner (1995), Chan (1996), Maximov, Gorberg, & Titov (1997), Arefi-Khonsari et al. (2001), Hocker (2002), D'Agostino et al. (2003), Wertheimer, Dennler, & Guimond (2003), Tatoulian et al. (2004), Oehr (2005), and Sardella et al. (2005).

9.7.2. Major Initial Chemical Products Created on Polymer Surfaces During Their Interaction with Non-Thermal Plasma

Chemical processes in thin polymer surface layers are stimulated by all major plasma components, especially by electrons, ions, excited particles, atoms, radicals, and UV radiation. The major primary products of plasma polymer treatment are free radicals, non-saturated organic compounds, cross-links between polymer macromolecules, products of destruction of the polymer chains, and gas-phase products (mostly molecular hydrogen). Processes for the formation of radicals on the polymer surface under plasma treatment, which are due to electron impact and UV radiation, are related to breaking of R–H and C–C bonds in polymer macromolecules:

$$RH \xrightarrow{e,\hbar w} R(\bullet) + H, \quad RH \xrightarrow{e,\hbar w} R_1(\bullet) + R_2(\bullet). \tag{9–80}$$

Direct formation of non-saturated organic compounds with the double bonds on the surface of a polymer, which are treated by non-thermal plasma, can be illustrated in similar simple cases as

$$RH \xrightarrow{e,\hbar w} R_1-CH{=}CH-R_2. \tag{9–81}$$

Secondary reactions of atomic hydrogen usually lead to the formation of molecular hydrogen through different mechanisms, including recombination and hydrogen transfer with the polymer macromolecule:

$$H + H \to H_2, \quad H + RH \to R(\bullet) + H_2. \tag{9–82}$$

The secondary reaction of atomic hydrogen with organic radical R can result not only in recombination but also in simultaneous formation of molecular hydrogen and a double bond in the organic macromolecule, which can be illustrated as

$$H + R(\bullet) \to R_1-CH{=}CH-R_2. \tag{9–83}$$

When the plasma gas contains oxygen, the free organic radical R generated by non-thermal plasma treatment on polymer surfaces (9–80) very effectively attaches molecular oxygen from the gas phase, forming active organic peroxide radicals:

$$R(\bullet) + O_2 \to R-O-O-. \tag{9–84}$$

The RO_2 peroxide radicals formed on the polymer surface by treatment in non-thermal plasma systems are able to initiate different important chemical surface processes. The simplest processes started by the RO_2 radicals are related to the formation of hydro-organic peroxide and other peroxide compounds on the surface of polymers:

$$R–O–O– + RH \rightarrow R–O–O–H + R(\bullet), \qquad (9–85)$$

$$R–O–O– + RH \rightarrow R–O–O–R_1 + R_2(\bullet). \qquad (9–86)$$

Reactions (9–85) and (9–86) together with attachment process (9–84) create the chain reaction for formation of hydro-organic peroxide and other peroxide compounds on the surface of polymers under treatment by oxygen-containing plasma. The formation of chemical products takes place in the relatively thin surface layers of the polymer because energies of plasma electrons and ions are quite limited and coefficients of extinction of UV radiation in polymers are high. Electrons make a significant contribution to the modification of polymer materials; this has been demonstrated in particular in the specific cases of plasma treatment of polystyrene films (Avasov, Bagirov, & Malin, 1969) and biological objects (Bochin, Rusanov, & Lukianchuk, 1976a,b; Rusanov & Fridman, 1978). In the specific conditions of the aforementioned experiments, non-thermal plasma electrons penetrate into the polymers to a depth of about $x_0 \approx 2\,\mu m$. The characteristic depth of modification of polymer material and formation of cross-links between macromolecules can be estimated according to Rusanov and Fridman (1978a,b) and Bamford and Ward (1961) as

$$x^2 \approx x_0^2(\ln \sigma_{i0} N - \ln \ln K), \qquad (9–87)$$

where σ_{i0} is the effective cross-section of modification of the macromolecules (in the case of bio-macromolecules it can be estimated as $\sigma_{i0} \approx 10^{-9}\,cm^2$; Dubinin, 1969); $N(cm^{-2})$ is the total time-integrated flux of electrons to the polymer surface; and $K \gg 1$ is the degree of polymer modification. Assuming, for example, $N = 10^{13}\,cm^{-2}$ and $K = e$, the characteristic depth for modification of a polymer material according to (9–87) is about 6 μm.

9.7.3. Kinetics of Formation of Main Chemical Products in Process of Polyethylene Treatment in Pulsed RF Discharges

Systematic kinetic investigations of the chemical modification of polymer surfaces in non-thermal plasma were carried out by Ponomarev (1982), Vasilets, Tikhomirov, and Ponomarev (1979), Ponomarev and Vasilets (2000), and Vasilets (2005), using as an example the treatment of PE in low-pressure strongly non-equilibrium RF discharges. The pulsed RF discharge in these experiments is characterized by voltage amplitude 20–22 kV, pulse duration 5 μs, RF in the pulse 0.5 MHz, repetition frequency of the pulses 50 Hz, and average specific discharge power 0.007 W/cm^3. High-density PE was used in powder form with specific surface area 2 m^2/g or in the form of a polymer film. Air, helium, and hydrogen were used as plasma gases. The gas pressure was changed over a wide range, between 10^{-2} and 50 Torr. Kinetic curves showing the formation of free radicals in PE at different plasma pressures are presented in Fig. 9–21 (Vasilets, 2005). All the curves demonstrate linear growth at first and then saturation. The saturation level of free radical concentration significantly decreases at higher pressures, when the reduced electric field E/p is relatively low and the effective depth of plasma treatment on the polymer is less than the size of the polymer particles. At very low pressures (10–100 mTorr), the depth of the plasma polymer treatment is $d \approx 20$–$30\,\mu m$. At relatively high pressures (≥ 5 Torr), the treatment depth is much shorter, $d < 2\,\mu m$.

The kinetics of free radical formation in PE is determined by the relative contribution of plasma electrons and UV radiation from plasma. The saturation level of the free radical concentration in PE films treated by only UV radiation ($\lambda > 160$ nm) from the plasma doesn't depend on discharge pressure in the range 1–20 Torr and doesn't depend on thickness of the polymer film in the range 2–110 μm. The saturation value of the free radical concentration

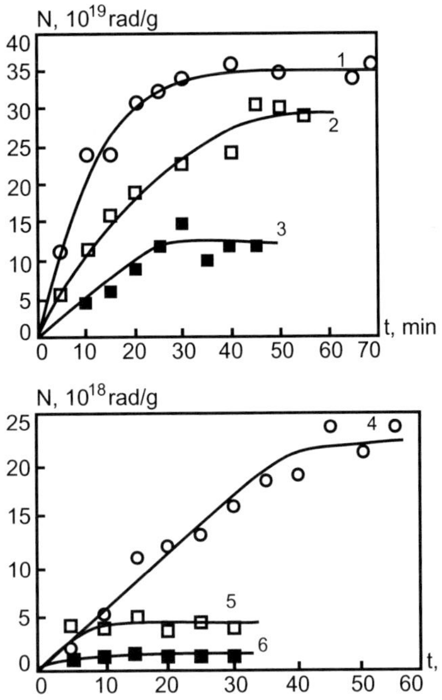

Figure 9–21. Accumulation of free radicals in polyethylene powder (specific surface 2 m^2/g) in the pulsed RF discharge with pressure: (1) 0.01–0.1 Torr, (2) 1 Torr, (3) 5 Torr, (4) 10 Torr, (5) 20 Torr, (6) 50 Torr.

is about $2.2 \cdot 10^{17}$ radicals/g, which is much below that corresponding to total plasma treatment at very low pressures (10–100 mTorr). These results lead to the conclusion that free radical formation in PE under pulsed RF discharge treatment is dominated by the contribution of UV radiation at higher pressures, whereas at very low pressures (10–100 mTorr) it is mostly due to the contribution of plasma electrons. Figure 9–22 illustrates the formation of molecular hydrogen, *trans*-vinylene bonds, and free radicals in PE powder under treatment in pulsed RF discharge in air at very low pressure (10 mTorr) and temperature (77 K; Vasilets, 2005). Initial formation rates of intermolecular cross-links and double bonds are close and exceed those of free radicals three-fold, meaning that formation of cross-links and double bonds in this case does not proceed through free radicals. The depth of the layer, where the intermolecular cross-links and double bonds are formed at very low pressures (10–100 mTorr), is about (2–4 μm), which is much lower than the depth of formation of free radicals under these conditions.

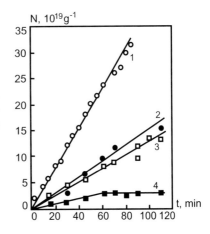

Figure 9–22. Kinetic curves of formation of (1) molecular hydrogen H$_2$, (2) intermolecular cross-links; (3) *trans*-vinylene bondings; (4) free radicals during treatment of polyethylene powder in pulsed discharge in air, pressure 0.01 Torr, temperature 77 K.

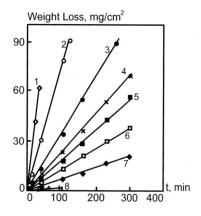

Figure 9–23. Time evolution of mass losses for different polymers treated by oxygen plasma: (1) polysulphone, (2) polypropylene, (3) low-density polyethylene, (4) polyethylene-glycol-terephthalate, (5) polystyrene, (6) polytetra fluorethylene (PTFE), (7) resin based on natural caoutchouc.

9.7.4. Kinetics of Polyethylene Treatment in Continuous RF Discharge

Treatment of high-density PE film and powder with specific surface area 2 m^2/g has also been performed using low-pressure continuous RF discharge (Vasilets et al., 1979; Ponomarev, 1982; Ponomarev & Vasilets, 2000; Vasilets, 2005). The voltage amplitude in this case was 0.5–1.2 kV, frequency was 1.2 MHz, and specific discharge power was 0.04–0.35 W/cm^3. Similarly to the case of pulsed discharges, air, He, and H$_2$ were used as plasma gases in the continuous RF discharge at pressures between 10^{-2} and 50 Torr. Continuous RF discharge in comparison with the pulsed discharge at the same power and pressure is characterized by lower electric field. The lower electric fields in the continuous discharge result in a lower fraction of energetic electrons, a higher density of charged particles, and a higher flux of positive ions to the polymer surface. The higher flux of positive ions leads to significant etching of the polymer surfaces in this case. The kinetic curves of free radical formation in PE in continuous RF discharge are similar to those in pulsed RF discharge (Fig. 9–21), but treatment depths are different. Lower electron energies, less intensive UV radiation, and significant ion etching of the polymer surfaces in the continuous RF discharge result in smaller treatment depths of polymer surfaces. The depth of formation of free radicals is 0.27 µm at a specific power of 0.35 W/cm^3 and decreases to 0.15 and 0.07 µm at lower specific powers of 0.14 and 0.04 W/cm^3. For the same reason mentioned earlier, PE treatment in continuous RF discharge is characterized by slower formation of *trans*-vinylene bonds and thinner surface layers where they are formed.

9.7.5. Non-Thermal Plasma Etching of Polymer Materials

Etching of polymer materials, like etching in general (Section 8.2), is provided in plasma by two mechanisms: physical sputtering and chemical etching. Physical sputtering is due to simple ion bombardment, while chemical etching is due to surface reactions and gasification of polymers, particularly provided by atomic oxygen, fluorine, ozone, and electronically excited oxygen molecules O$_2$($^1\Delta_g$). Typical examples of plasma gases applied for polymer etching are CF$_4$ and CF$_4$–O$_2$ mixture, which are widely used in microelectronics for cleaning substrates of organic deposits and for etching of photoresist layers. Treatment of polymers in oxygen-containing plasma initially leads to the formation on the surface of the following specific oxygen-containing chemical groups: –C=O, –C(R)–O–H, –C(R)–O–O–H, –C(R$_1$)–O–O–C(R$_2$)–. Further interaction of the polymer with non-thermal oxygen-containing plasma can result in further oxidation, formation of CO$_2$ and H$_2$O, and their transition to the gas phase. Such a process can be interpreted as chemical etching of polymer oxygen-containing plasma. Etching rates of different polymers are very different. Fluorocarbon polymers, for example PTFE, are characterized by the slowest etching rates in plasma. Etching of hydrocarbon polymers is much faster. Heteroatoms

Table 9–10. Etching Rates of Different Polymers in Non-Thermal Low-Pressure Plasma of Helium, Nitrogen, and Oxygen

Polymer	He plasma, etch rate $g/(cm^2 \cdot h)$	N_2 plasma, etch rate $g/(cm^2 \cdot h)$	O_2 plasma, etch rate $g/(cm^2 \cdot h)$
Polytetrafluoroethylene (PTFE)	0.2	0.73	5.6
Poly-(tetrafluoroethylene– hexafluoropropylene) (PTFE/HFP)	0.17	0.45	3.4
Polyvinyl difluoride (PVDF)	0.83	0.91	10.6
Polyethylene (PE)	0.70	0.90	4.2

and especially oxygen-containing groups are the least resistive to plasma etching. Rates of polymer mass losses during etching in low-pressure oxygen plasma of an RF discharge are illustrated in Fig. 9–23 (Ponomarev & Vasilets, 2000; Vasilets, 2005). Branching of a polymer chain usually leads to higher etching rates, whereas cross-links between macromolecules usually decrease the etching rate. Etching rates are different for amorphous and crystal regions of a polymer, which results in changing the polymer surface morphology and can lead to the formation of porous surface structure after long plasma treatment. Etching of the amorphous regions is usually faster, which means the crystal phase can dominate after plasma treatment. This etching effect is applied, in particular, to increase the durability of polymer fibers. Characteristic plasma etching rates of some specific polymers in helium, nitrogen, and oxygen are shown in Table 9–10 (Gilman, 2000).

9.7.6. Contribution of Electrons and Ultraviolet Radiation in the Chemical Effect of Plasma Treatment of Polymer Materials

The formation of chemical products of plasma polymer treatment – radicals, double bonds, cross-links, hydrogen, and so forth – or, in other words, the chemical effect of plasma treatment is due to several plasma components. Usually the primary plasma components active in high-depth polymer treatment are divided into two groups, electrons and UV radiation, which can penetrate relatively far into the surface of polymer material. The contribution of these two deep-penetrating plasma components to polymer treatment is discussed below (Ponomarev, Vasilets, & Talrose, 2002; Vasilets, 2005). The phrase "contribution of electrons" sometimes also assumes the contribution of secondary factors related to the primary plasma electrons such as gas-phase radicals, excited atoms and molecules, and so on. Interaction of the chemically active heavy particles with polymers is usually limited, however, to very thin surface layers (except the effect of ozone and other relatively stable neutrals) and is discussed in particular in the next section. "UV radiation" includes the contribution of vacuum-UV (VUV) radiation as well as softer UV radiation. VUV radiation is characterized by short wavelengths (110–180 nm) and very high photon energies sufficient for electronic excitation and direct dissociation of chemical bonds. The high chemical activity of VUV radiation leads, on the other hand, to very short absorption lengths in this case. Effective transfer of this type of radiation across the relatively long distance is possible only in low-pressure gases, which explains the name "vacuum ultraviolet." Softer UV radiation is characterized by longer wavelengths, exceeding 200 nm, and therefore lower photon energies. Softer UV radiation is much less chemically active but is able to penetrate much deeper into the polymer film, which is important for some applications. As an example, treatment of PE by VUV radiation of a low-pressure non-thermal RF discharge results in the significant formation of intermolecular cross-links. VUV radiation is separated from other plasma components able to modify the polymer surface by using a special LiF window.

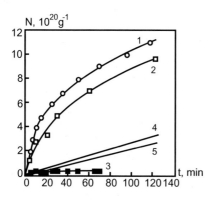

Figure 9–24. Accumulation of H₂ (1, 4), *trans*-vinylene double bonds (2, 5), and free radicals (3) in polyethylene powder (specific surface 2 m²/g) due to direct treatment in the pulsed RF discharge in air at pressure 0.01 Torr and due to only vacuum ultraviolet (VUV) radiation of the discharge (4, 5).

Softer UV radiation from a mercury lamp (typical wavelength 253.7 nm) does not produce the intermolecular cross-links in PE.

Kinetic curves for the formation of molecular hydrogen (H₂), *trans*-vinylene bonds, and free radicals in PE powder (with specific surface 2 m²/g) by its treatment in non-thermal pulsed RF discharge in an inert gas (Section 9.7.3) are presented in Fig. 9–24 (Vasilets, 2005). The figure compares the total chemical effect of the discharge with partial contribution of VUV radiation. Initially ($t < 30$ min), VUV contribution toward the formation of molecular hydrogen and *trans*-vinylene bonds is negligible, but it becomes significant for longer treatment times ($t < 60$ min). The effect of heavy particles is negligible in this case. Analysis of the PE treatment by continuous non-thermal RF discharge (Section 9.7.4) shows that the formation of H₂ and *trans*-vinylene bonds at the depth 1–2 μm after more than 60 min of the plasma–polymer interaction is dominated by VUV in the range 140–160 nm. Plasma electrons at the same time make a significant contribution toward the formation of free radicals in the PE powder (Ponomarev & Vasilets, 2000). It should be mentioned that the effect of plasma-generated active heavy particles can be neglected when considering depths of 1–2 μm. The non-thermal low-pressure plasma treatment of PTFE leads to the formation in PTFE of two types of radicals: those formed by detachment of an F atom, $-CF_2-CF(\bullet)-CF_2-$, and those formed by breaking a macromolecule, $-CF_2-CF_2(\bullet)$. During the initial period of the treatment ($t <$ min), formation of radicals is mostly due to the contribution of plasma electrons; later the partial contribution of VUV (140–155 nm) becomes significant (Ponomarev & Vasilets, 2000; Ponomarev et at., 2002). Quantum yields of the free radical formation in this case are $3 \cdot 10^{-3}$ for VUV with wavelength 147 nm, and $5 \cdot 10^{-3}$ for VUV with wavelength 123 nm. The depth of the effective formation of free radicals in PTFE is about 0.6 μm.

9.7.7. Interaction of Atoms, Molecules, and Other Chemically Active Heavy Particles Generated in Non-Thermal Plasma with Polymer Materials: Plasma-Chemical Oxidation of Polymer Surfaces

The interaction of plasma-activated heavy particles with polymers plays an important role in plasma–polymer treatment but is usually localized on the surface in only few molecular monolayers. Probably only ozone is an exception because of its ability to effectively diffuse and penetrate deeper into the polymer. Molecular and atomic particles provide etching of polymer surface layers as well as form new functional groups, significantly changing the surface characteristics of the polymer. The volume properties of the polymer materials remain the same. The interaction of heavy particles with polymer surfaces is very special in different plasma gases; therefore, examples of the polymer interaction with some specific gases are to be discussed next. We start with plasma-chemical oxidation of polymer surfaces, which is widely used today in different industrial areas. Interaction with oxygen-containing plasma results in the formation of polar groups on the polymer surfaces, which leads to growth of the polymer surface energy and significant increase of the polymers' wettability and adhesion to metals and different organic compounds. Photo-electronic spectra of plasma-treated PE show that, first of all, plasma-chemical oxidation leads to formation on the polymer surfaces of $-C(-)-O-$ bonds. These bonds correspond to such specific groups as peroxides, alcohols, ethers, and epoxies. Second in the row are $-C(-)=O$ bonds, which are typical for aldehydes and ketones. The least probable are $-C(-O-)=O$ bonds, corresponding to the carboxyl-acidic groups. Plasma-chemical oxidation always includes the simultaneous formation of oxygen-containing surface groups and surface etching. As a result, the polymer surface oxidation degree essentially depends on the polymer composition and structure and can be significantly varied by discharge power, plasma parameters, and treatment time. It is possible to avoid the etching of oxygen-containing groups by preliminary treatment of the polymer in plasma of inert gases and following contact of the activated surface with oxygen-containing gases outside of the discharge.

9.7.8. Plasma-Chemical Nitrogenation of Polymer Surfaces

The interaction of nitrogen and nitrogen compounds (especially NH_3) with non-thermal plasma results, in particular, in the formation of amine groups ($R-NH_2$), amide groups (R_1-NH-R_2), and imine groups ($(R_1, R_2)C=N-H$) on the polymer surfaces. These groups promote surface metallization or adhesion of different materials to the polymers (Liston, Martinu, & Wertheimer, 1994; Grace et al., 1996; Gerenser et al., 2000). The non-thermal plasma nitrogenation approach has been used, for example, for promoting silver adhesion to polyethylene terephthalate (PET) (Spahn & Gerenser, 1994). Another practically important example is plasma nitrogenation of polyester webs to promote adhesion of gelatin-containing layers, which are used in the production of photographic film (Grace, 1995). Non-thermal nitrogen plasma is similarly effective in promoting adhesion on the surface of polyethylene-2,6-naphthalate (Gerenser et al., 2000). In the preceding examples, effective nitrogenation has been achieved in low-frequency RF-CCP discharge in N_2 (Conti et al., 2001), and nitrogen has been incorporated into the polymers in the form of amine and amide groups.

The types of nitrogen-containing groups incorporated into the polymer surface essentially depend on the plasma gas. For example, NH_3 plasma treatment of polystyrene leads mostly to the formation of amine groups (NH_2) on the polymer surface, whereas N_2 plasma in similar conditions does not produce amine groups. Treatment of polymers in N_2 plasma results in a more significant formation of imine groups ($C=N-H$) on the surface. Experiments with low-pressure non-equilibrium microwave discharges in molecular nitrogen (N_2) indicate that the total surface concentration of nitrogen-containing groups can be very high, reaching up to 40 atomic %. Ammonia (NH_3) plasma in similar conditions results in a lower surface concentration of the nitrogen-containing groups (Ponomarev & Vasilets, 2000). Fluorine-containing polymers are the most resistive to nitrogenation and oxidation in non-thermal plasma conditions. For example, treatment of PTFE in N_2 or NH_3 plasma usually leads to the formation of not more than 6 atomic % of the nitrogen-containing groups in a thin 10 nm surface layer. Plasma nitrogenation of polymer surfaces strongly promotes further oxidation of the surfaces in atmospheric air. For example, nitrogenation of PE in N_2 plasma and following contact of the surface with atmospheric air results in an oxygen concentration in PE of about 8 atomic % in the 10 nm layer. A similar procedure in NH_3 plasma leads to an oxygen concentration in PE of about 4–6 atomic % in the 10 nm surface layer. Surface modification of polymers in nitrogen-containing plasmas is widely used to improve the biocompatibility of polymer materials, which will be discussed later with regard to plasma medicine. For example, amine groups formed in plasma on polymer surfaces provide effective immobilization of heparin and albumin on the surfaces.

Biocompatibility and adhesion of different cells to the polystyrene surface is significantly enhanced by the formation of amine groups on the surface during treatment in NH_3 plasma. Similarly, the amine groups created in NH_3 plasma on the surface of PTFE provide adhesion of collagen.

9.7.9. Plasma-Chemical Fluorination of Polymer Surfaces

The interaction of fluorine-containing gases with non-thermal plasma leads to a decrease of surface energy for hydrocarbon-based polymers and makes these polymer surfaces hydrophobic, which is widely used for practical applications. The interaction of hydrocarbon polymer materials with fluorine-containing plasmas results in the formation of different surface groups, especially C–F, CF_2, CF_3, and C–CF. The interaction of CF_4 plasma with hydrocarbon-based polymer materials leads mostly to the formation of C–F and CF_2 groups on the polymer surface, while interaction with CF_3H plasma mostly leads to the formation on the surface of C–CF and CF_3 groups. Treatment of polymer surfaces with fluorine-containing plasmas stimulates three groups of processes simultaneously: formation of the fluorine-containing groups (C–F, CF_2, CF_3, and C–CF), polymer etching, and plasma polymerization. The relative contribution of these three processes strongly depends on the relative concentration in plasma of CF and CF_2 radicals on the one hand, and F atoms on the other hand. The CF and CF_2 radicals are building blocks for plasma polymerization, whereas atomic fluorine is responsible for etching and formation of the fluorine-containing surface groups. The relative concentration in plasma of CF and CF_2 radicals and F atoms depends on the type of applied plasma gas. Typical fluorine-containing gases applied for polymer treatment are CF_4, C_2F_6, C_2F_4, C_3F_8, and CF_3Cl. Generally, higher C:F ratios in the initial plasma gases lead to an increase of relative concentration of CF and CF_2 radicals in volume with respect to F atoms. Therefore, higher C:F ratios are favorable for plasma polymerization, whereas lower C:F ratios are favorable for polymer etching and formation of fluorine-containing surface groups. An example of the application of plasma fluorination of polymers is the plasma treatment of PMMA-based contact lenses to minimize their interaction with the eye's tissues. Plasma fluorination of polymers is also used to increase their durability and to decrease the friction coefficient.

9.7.10. Synergetic Effect of Plasma-Generated Active Atomic/Molecular Particles and UV Radiation During Plasma Interaction with Polymers

Plasma interaction with polymers, and especially biopolymers, is a multi-stage and multi-channel process, which is often characterized by strongly non-linear kinetics. The contribution of different plasma components such as atoms, radicals, active and excited molecules, charged particles, and UV radiation into the polymer treatment process is not simply cumulative but is essentially synergetic. Therefore, the answer to the simple question of which plasma component dominates the plasma–polymer treatment process often cannot be so simple and unambiguous. Two or more plasma components can make a synergistic contribution into the total process. An example is the non-thermal plasma etching of PE and PVC in low-pressure RF discharges in oxygen (Vasilets, 2005). Generally, etching of polymers in low-pressure oxygen plasma is mostly due to atomic oxygen, electronically excited molecular oxygen, and UV radiation. Experiments with these plasma components individually give total etching rates after summation about three times lower than the combined contribution of the components applied together. Ponomarev and Vasilets (2000) claim that PE and PVC etching is a synergetic effect of atomic particles and UV radiation. A similar situation probably takes place in the plasma treatment of biopolymers and specifically in plasma sterilization processes, where UV radiation, radicals, and chemically active and electronically excited molecules make essentially non-linear synergistic contributions.

9.7.11. Aging Effect in Plasma-Treated Polymers

The composition and space distribution of products of plasma treatment of polymer materials can keep changing long after the plasma treatment process is finished. The phenomenon is referred to as the aging effect in plasma-treated polymers. Four major mechanisms of the aging effect can be pointed out:

- Re-orientation and shift of the polar groups formed on the polymer surface inside of the polymer material due to thermodynamic relaxation;
- Diffusion of the low-molecular-mass admixtures and oligomers from the volume of the polymer material to the polymer surface;
- Diffusion to the polymer surface of the low-molecular-mass products formed during the plasma treatment in the relatively thick surface layer; and
- Post-plasma treatment reactions of free radicals and other plasma-generated active species and groups between themselves and with the environment.

Aging of hydrocarbon-based polymer materials treated in oxygen plasma is mostly due to the reorientation and shift of the polar peroxide groups formed on the polymer surface inside of the polymer, which is related to thermodynamic relaxation. The wettability contact angle in H_2O decreases several times immediately after treatment because of formation of the polar peroxide groups. Then the wettability contact angle starts increasing and because of the aging effect can almost return to the initial value after several days of storage in atmospheric air. Aging of hydrocarbon-based polymers treated in nitrogen is mostly due to the post-processing reactions of nitrogen-containing surface groups with the environment. As an example, the major effect of plasma nitrogenation of PE in N_2 discharges is related to the formation of imine groups ($(R_1, R_2)C=N-H$) on the polymer surface. Storage of the plasma-treated polymer in atmospheric air results first of all in the hydrolysis of the imine groups:

$$(R_1, R_2)C=N-H + H_2O \rightarrow (R_1, R_2)C=O + NH_3. \qquad (9-88)$$

Longer storage in atmospheric air results in additional reactions of nitrogen incorporated into polymer with atmospheric water:

$$R_1-CH=N-R_2 + H_2O \rightarrow R_1-CH=O + H_2N-R_1. \qquad (9-89)$$

Polypropylene is characterized by the strongest aging effect. PET is affected by aging a little less. The most durable with respect to aging are PE and polyimide. Generally, a higher level of crystallinity of polymers leads to their stronger durability with respect to aging after plasma treatment.

9.8. APPLICATIONS OF PLASMA MODIFICATION OF POLYMER SURFACES

9.8.1. Plasma Modification of Wettability of Polymer Surfaces

The most important result of the plasma treatment of polymers, which are produced on an industrial scale, is the change in their wettability and adhesion characteristics. As was discussed earlier, plasma treatment can make polymers more hydrophilic as well as more hydrophobic. Both effects are widely used for practical applications. The change of wettability is usually characterized experimentally by the **contact angle** θ, which is formed on the solid surface along the linear solid–liquid borderline of air (see Fig. 9–25). An increase of

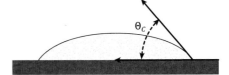

Figure 9–25. Contact angle θ_c of a liquid sample.

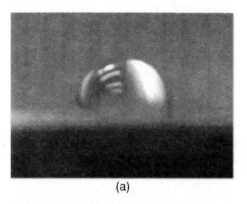 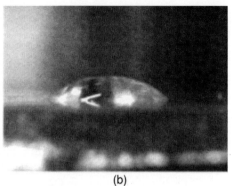

(a)   (b)

Figure 9–26. Increase of wettability and relevant change of the contact angle as a result of surface treatment by non-thermal atmospheric plasma.

wettability or making a polymer more hydrophilic leads to a decrease of the contact angle. An example illustrating improvement of wettability and decrease of the contact angle as a result of plasma treatment of a polymer is shown in Fig. 9–26. Change of the wettability contact angles with respect to water and glycerin for different polymer materials after treatment in a low-frequency (50 Hz) low-pressure AC discharge in air is presented in Table 9–11 (Gilman, 2000). The treatment of polymers in oxygen-containing plasma leads to a decrease of contact angle and therefore to an increase of wettability. The wettability increase effect is related to plasma-stimulated formation of polar peroxide groups on the polymer surface as discussed in Section 9.7.7.

Changes in contact angles for plasma-modified polymer surfaces depend on the applied plasma gas and conditions of the plasma treatment. The application of discharges in air, oxygen, nitrogen, and ammonia transforms initially hydrophobic surfaces into hydrophilic surfaces. Application of discharges in fluorine-containing gases such as tetrafluoroethylene, perfluoropropane, and octafluorocyclobutane provides significant enhancement of hydrophobic properties of polymer surfaces. An increase of gas pressure, discharge current, and plasma treatment time lead to reduction of the contact angle θ and to an enhancement of wettability. Major enhancements of hydrophilic properties usually occur during the initial treatment time of 30–120 s (see Fig. 9–27). The plasma-enhanced hydrophilic properties become less strong with time (aging effect). Maintaining the enhanced hydrophilic properties over a long time period is important in practical applications. Plasma treatment of polypropylene permits high wettability maintenance (contact angle $\theta < 60°$) for 30 days, and plasma treatment of polyimide permits high wettability maintenance (contact angle

Table 9–11. Change of Contact Angles with Respect to Water and Glycerin due to Treatment of Different Polymers in Low-Pressure Low-Frequency AC Plasma in Air

Polymer	Contact angle θ (degr.), water: before treatment	Contact angle θ (degr.), water: after treatment	Contact angle θ (degr.), glycerin: before treatment	Contact angle θ (degr.), glycerin: after treatment
Polypropylene (PP)	92	46	78	39
Polyimide (PI)	76	13	58	6
Poly-(tetrafluoroethylene–hexafluoropropylene) (PTFE/HFP)	111	85	90	66

9.8. Applications of Plasma Modification of Polymer Surfaces

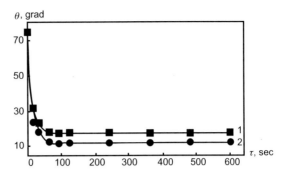

Figure 9–27. Typical dependence of water/PI contact angle (θ) on plasma treatment time. Air plasma, 50 Hz, current (1) 50 mA, (2) 100 mA.

$\theta < 50°$) for 12 months. The aging effect is illustrated in Fig. 9–28 for the case of plasma-treated polyimide fluoroplastic film. As one can see from the figure, restoration of the wettability contact angle θ takes place mostly during the first 10 days, and then the aging process slows down.

9.8.2. Plasma Enhancement of Adhesion of Polymer Surfaces: Metallization of Polymer Surfaces

Plasma treatment cleans the polymer surface and makes it highly hydrophilic, which results in an enhancement of adhesion properties of the polymer materials and, as a result, in a wide variety of different relevant applications. The process is widely used in practice for gluing polymers in different combinations and fabrication of composite materials with special mechanical and chemical properties. In particular, plasma treatment-enhanced adhesion of fibers to binding materials significantly improves the characteristics of composites. Plasma technology can be applied for fabrication of high-temperature-resistant electric insulation. Contact properties of polymer surfaces are often improved in practical applications by plasma deposition of special polymer films characterized by strong contact properties with respect to the adhesive material as well as to the major polymer (see Section 9.6). Plasma enhancement of polymer surface adhesion can be considered a generalization of the plasma stimulation of wettability. Plasma enhancement of adhesion is widely used in various large-scale industrial applications. Increased adhesion energy of different polymers with respect to water and glycerin in a low-frequency (50 Hz) low-pressure AC discharge in air is illustrated in Table 9–12 (Gilman, 2000). As one can see from the table, plasma can significantly stimulate the adhesion process. An increase of the adhesion energy of polymer materials by their treatment in oxygen-containing plasma is related to an increase of surface energy for the plasma-treated polymers. Change of **surface energy** of different polymers after treatment in the same plasma of low-frequency (50 Hz) low-pressure AC discharge in air is presented in Table 9–13 (Gilman, 2000). Data in the table illustrate the growth of the total surface energy of the polymers as a result of treatment in oxygen-containing non-thermal

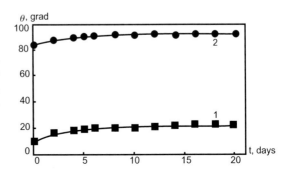

Figure 9–28. Typical dependence of water contact angle (θ) on storage time in normal air conditions. Plasma-modified film: (1) laminated polyimide fluoroplast film treated in air plasma 50 Hz; (2) TFE fluoroplast.

Table 9–12. Change of Adhesion Energy with Respect to Water and Glycerin due to Treatment of Different Polymers in Low-Pressure Low-Frequency AC Plasma in Air

Polymer	Adhesion energy to water, mJ/m^2, before treatment	Adhesion energy to water, mJ/m^2, after treatment	Adhesion energy to glycerin, mJ/m^2, before treatment	Adhesion energy to glycerin, mJ/m^2, after treatment
Polypropylene (PP)	70.3	123.4	76.6	112.7
Polyimide (PI)	91.6	144.0	100.6	126.4
Poly-(tetrafluoroethylene–hexafluoropropylene) (PTFE/HFP)	46.7	79.14	63.4	83.0

plasma. The table also shows how plasma modification changes two components of the surface energy: polar and dispersion.

The **polar component of the surface energy** characterizes polar interactions between the surface of the polymer material and the working fluid or surface film. This component is determined by the presence of polar groups, electric charges, and free radicals on the polymer surface. In contrast, the **dispersion component of the surface energy** characterizes the dispersion interaction between the surface and working liquid or surface film. This component is determined by the roughness, unevenness, and branching level of the polymer surface. The polar component of surface energy grows during treatment in oxygen-containing plasma. This effect is due to the formation of polar groups, especially peroxide groups, on the polymer surface, as discussed in Section 9.7.7. In contrast, the dispersion component of the surface energy, which is relatively large before plasma treatment, can be decreased after plasma modification. The plasma-enhanced adhesion of polymer surfaces permits effective metallization of surfaces using such conventional methods as vacuum thermal and magnetron spraying, deposition by decomposition of organic metal compounds, and so on. A comparison of adhesion of vacuum thermally sprayed thin aluminum (Al) films on the surface of different polymer materials treated or not treated in plasma is presented in Table 9–14 together with relevant data on the wettability contact angles with respect to water (Gilman, 2000). The table shows the results of polymer treatment in non-thermal discharge plasmas generated in different gases: helium (He), oxygen (O$_2$), and O$_2$–CF$_4$

Table 9–13. Change of Total Surface Energy and Its Components (Polar One and Dispersion One) due to Treatment of Different Polymers in Low-Pressure Low-Frequency AC Plasma in Air

Polymer	Surface energy, total, mJ/m^2	Surface energy, dispersion component, mJ/m^2	Surface energy, polar component, mJ/m^2
Polypropylene (PP), before treatment	29.3	26.1	3.2
Polypropylene (PP), after treatment	52.5	19.2	33.3
Polyimide (PI), before treatment	45.73	40.7	5.03
Polyimide (PI), after treatment	71.55	17.5	54.05
Poly-(tetrafluoroethylene–hexafluoropropylene) (PTFE/HFP), before treatment	30.19	30.09	0.10
Poly-(tetrafluoroethylene–hexafluoropropylene) (PTFE/HFP), after treatment	37.83	35.04	2.79

Table 9–14. Adhesion of Plasma-Treated Industrial Polymer Films with Respect to Sprayed Aluminum Layers; Wettability Contact Angles of the Plasma-Treated Industrial Polymer Films

Polymer	Contact angle θ(degrees)/ adhesion, before treatment	Contact angle θ(degrees)/ adhesion, O_2 plasma	Contact angle θ(degrees)/ adhesion, CF_4–O_2 plasma	Contact angle θ(degrees)/ adhesion, He plasma
Polycarbonate (PC)	72/B	39/A	<15/A	37/A
Polyether sulphone (PSU)	70/A	25/A	<15/A	26/A
Polyethylene fluoride (PEF)	66/C	29/A	30/A	29/A
Polyvinyl difluoride (PVDF)	71/C	40/A	70/A	57/A
Polypropylene (PP)	98/C	40/A	72/A	53/A
Polyethylene (PE)	90/C	–	20/A	50/A
Polypropylene (PP)	83/B	15/A	<15/A	26/A

Note: (Adhesion is characterized as: A – very strong, B – strong, C – medium strong).

mixture. Plasma treatment of all considered industrial polymer materials leads to a significant improvement of adhesion to the vacuum thermally sprayed thin aluminum (Al) films and to a decrease of the wettability contact angles.

9.8.3. Plasma Modification of Polymer Fibers and Polymer Membranes

Polymers are used in composite materials as a dispersed phase (fibers or powders) as well as a matrix phase. In both cases, adhesion between the phases can be significantly improved by treatment in non-thermal plasma. Some examples of non-thermal plasma modification of polymer fibers (Gilman, 2000), which significantly change their properties for specific technological applications, are summarized in Table 9–15. Plasma modification of porous and non-porous polymer substrates allows the production of composite membranes for gas separation, pervaporation (separation of liquids by evaporation through the membrane), and water cleaning by reverse osmosis. Plasma modification approaches differ for porous and non-porous substrates. Treatment of porous substrates is focused on the porous size reduction due to cross-linkage of the polymer surface in air, O_2, or inert gas discharges; activation of the substrate surface followed by grafting; and deposition of thin polymer film (<1 μm) on the porous substrate surface or on the special adhesive sublayer. Some examples of plasma modification of porous membranes (Gilman, 2000) for different applications are summarized in Table 9–16. Plasma treatment of non-porous membranes can be focused on functionalization, hydrophilization, and cross-linking of the polymer surfaces in plasma of air, O_2, N_2, NH_3, and so on, on plasma deposition of thin polymer films with preliminary surface activation; and on grafting on a preliminarily plasma-activated membrane surface. Some examples of plasma modification of non-porous membranes (Gilman, 2000) for different applications are summarized in Table 9–17. Generally, the plasma modification of polymer membranes permits one to achieve excellent results in separation of gases, water purification from salts, as well as in concentrating different organic and inorganic compounds. As a practical example it can be mentioned that a plasma-modified polyacrylonitrile membrane with 98% salt separation efficiency is used, in particular, in a device to clean 2 m^3 of water (initial pH = 1–10) per day. Characteristics of plasma-modified membrane systems will be considered in a special section.

9.8.4. Plasma Treatment of Textile Fibers: Treatment of Wool

A non-thermal plasma approach plays an important and multi-functional role in the treatment of natural as well as man-made textile materials (Maximov, 2000; Hocker, 2002). The contribution of plasma technology is not limited to the well-known and widely used positive

Table 9–15. Modification of Properties of Polymer Fibers after Plasma Treatment

Polymer fiber	Plasma gas	Plasma treatment conditions	Modification of fiber properties
Polyethylene terephthalate (PET)	Acrylic acid (AA), hexamethyl-disilazane, air	RF discharge (13.6 MHz), 13.3 Pa, 20–40 W	Reduction of combustibility, hydrophilization: initial contact angle $\theta = 57°$, final $\theta = 40°$
Polyethylene terephthalate (PET)	Tetrachloroethylene (TECE), trichloroethylene (TCE)	RF discharge (13.6 MHz), 13.3 Pa, 20–40 W	Reduction of combustibility, hydrophilization: initial contact angle $\theta = 57°$, final $\theta = 46°$
Polyethylene (PE), Kevlar	Oxygen (O_2)	RF discharge (13.6 MHz), 133 Pa, 20–40 W, 5–60 min	Adhesion enhancement to epoxy matrix: PE-fiber 118% increase, Kevlar-fiber 45% increase
Polyethylene (PE)	Acrylic acid (AA)	RF discharge (13.6 MHz), 6.6 Pa, 30–70 W, 5–40 min	Adhesion enhancement to epoxy matrix. Hydrophilization: initial contact angle $\theta = 65°$, final contact angle $\theta = 8°$
Kevlar	Ar, O_2, NH_3	RF discharge (13.6 MHz), 7–13.3 Pa, 25 W, 2–8 s	20% Adhesion enhancement to polycarbonate (PC) matrix
Cellulose acetate (CA)	Air	RF discharge (13.6 MHz), 13.3 Pa, 100 W	Adhesion enhancement to matrix from polypropylene (PP), polystyrene (PS), chlorinated polyethylene (PE)
Polyethylene	Oxygen (O_2)	RF discharge (13.6 MHz), 17.3 Pa, 67 W, 5 min	Adhesion enhancement to epoxy matrix

plasma effect on dyeing and printing of textiles. Plasma is also effectively used for more specific treatment of natural fibers, including enhancement of shrink-resistance of wool and selective oxidation of lignin in cellulose, which transforms the lignin in a water-soluble form for further extraction. We start this and following chapters with a consideration of some specific representative examples of non-thermal plasma treatment of wool, cotton,

Table 9–16. Plasma Modification of Porous Polymer Membranes

Membrane	Plasma gas	Treatment conditions	Properties of treated membranes
Polysulfone (PSU)	Ar	RF discharge (13.6 MHz), 6.5 Pa, 30 s, sublayer of silicon polymer Sylgard 184	Increase of gas-separation coefficient for O_2–N_2 mixture from 1.2 to 4.6
Polystyrene (PS)	Butyronitrile	RF discharge (13.6 MHz), 4 Pa, 25–60 W, sublayer	Increase of gas-separation coefficient for H_2–CH_4 mixture from 1 to 15
Polyacrylonitrile (PAN)	He, Ar, O_2	AC discharge (60 Hz), 13.3 Pa, 350 W, 120 s	Water purification from salts, effectiveness 98%
Polypropylene (PP), Celgard 2400	air	RF discharge (13.6 MHz), 2.7 Pa, 10 W, 60 s, MAC-grafting in water solution (5%, 70°C, 2 h)	Pervaporation of ethanol–water mixture, selectivity with respect to water 10%
Polysulfone (PSU)	C_2F_4(TFE), C_3F_8, C_3F_6	RF discharge (13.6 MHz), 30–100 W, 230 min	Pervaporation of ethanol–water mixture, separation factor 10, productivity 1 kg/(m^{2}h)

9.8. Applications of Plasma Modification of Polymer Surfaces

Table 9–17. Plasma Modification of Non-Porous Polymer Membranes

Membrane	Plasma gas	Treatment conditions	Properties of treated membranes
Polydimethylsiloxane (PDMS)	Ar	RF discharge (13.6 MHz), 35 W, 6 min	Increase of gas-separation coefficient for O_2–N_2 mixture from 2 to 5.4, for CO_2–CH_4 mixture from 2 to 20
SC (silicon-based membrane)	Methacryl-, benzo-, crotono-nitrile	RF discharge (13.6 MHz), 133 Pa, 100 W, 60 s	Increase of gas-separation coefficient for H_2–CH_4 mixture from 0.8 to 40
Polyphenylene oxide (PPO)	Cyanogen Bromide	RF discharge (13.6 MHz), 67 Pa, 100 W, 30 s	Increase of gas-separation coefficient for H_2–CH_4 mixture from 23 to 297
Nylon 4	Ar, N_2, O_2	RF discharge (13.6 MHz), 13.3 Pa, 50 W, AC, VP-grafting in solution	Water purification from salts, effectiveness 94%
Polyvinyl-trimethylsilane (PVTMS)	Air	AC discharge (50 Hz), 40 W, 60 s	Pervaporation of water solution of HNO_3, selectivity coefficient is 0.99 for 3N-solution HNO_3, productivity 2 kg/h

and synthetic textile fibers. To discuss plasma treatment of wool it should be mentioned that the cuticle cells in wool overlap each other to create a directional frictional coefficient, and the very surface is highly hydrophobic. The hydrophobic behavior leads in aqueous medium to aggregation of the fibers, which move to their root, resulting in felting and shrinkage of the wool. A significant effect of wool treatment in oxygen-containing (in particular, air) plasma is due to the oxidation and partial removal of the hydrophobic lipid layer on the very surface of wool. This applies both to the adhering external lipids and to the covalently bound 18-methyl-eicosanoic acid. Thus, the plasma-induced decrease of the hydrophobic behavior results in shrink resistance. The effect can be achieved in both low-pressure and atmospheric-pressure non-thermal plasma discharges.

Another non-thermal plasma effect, which can also be achieved in both low-pressure and atmospheric-pressure non-thermal discharges, is related to a significant reduction of the cross-link density of the exocuticle layer in wool. The exocuticle is the layer located below the fatty acid layer of the very surface, which is called the epicuticle. The exocuticle layer is highly cross-linked via disulfide bridges. Treatment of the wool in oxygen-containing plasma leads to oxidation and breaking of the disulfide bonds, which results in a significant reduction of the cross-link density and improvement of the wool properties (Hocker, 1995). Protein loss after even intensive plasma treatment and extraction is very low (about 0.05%) because of the surface-oriented nature of the plasma treatment. The specific surface area of the wool is significantly increased as a result of plasma treatment from about 0.1 to 0.35 m^2/g. Due to the surface-directed nature of the plasma treatment, the tenacity of the fibers is only slightly influenced. Thus, plasma modification of the wool surface leads to a decrease in the shrinkage behavior of the wool top. The felting density of the wool top before spinning decreases from more than 0.2 g/cm^3 to less than 0.1 g/cm^3. Especially strong shrinkage resistance can be achieved by additional resin coverage of the plasma-treated fiber surface. This combined plasma–resin procedure leads to the formation of a smooth surface with reduced scale height and shrinkage of about 1% after 50 simulated washing cycles. To

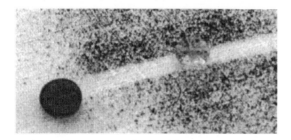

Figure 9–29. Water drop–induced dye removal from an HMDSO-plasma-treated cotton fabric.

compare, the area felting shrinkage of untreated wool is 69%, and of plasma-treated wool without resin is 21%. As the wool is oxidized during treatment in oxygen-containing non-thermal plasma, the hydrophobic behavior of its surface is changed to become increasingly hydrophilic. It results in additional advantages of plasma treatment related to improved dyeing kinetics, enhanced depth of shade, and improved bath exhaustion (Thomas et al., 1992; Hocher et al., 1994; Kusters et al., 1995; Thomas & Hocker, 1995).

9.8.5. Plasma Treatment of Textile Fibers: Treatment of Cotton and Synthetic Textiles and the Lotus Effect

Treatment of cotton and synthetic textile fibers can be performed in non-thermal plasma under atmospheric-pressure as well as under reduced pressure depending on the specific modification needs of the materials (Maximov, 2000; Hocker, 2002). First consider some physical and chemical features of plasma treatment of cotton fibers. Similar to wool treatment, the non-thermal plasma treatment of cotton in oxygen-containing gases leads to a significant increase of specific surface area of cotton. On the other hand, treatment of cotton using HMDSO plasma results in a strong hydrophobization effect. It leads to formation of smooth surfaces with increased contact angle, about 130° with respect to water. The strong hydrophobization effect of the cotton fiber can be similarly achieved using non-thermal hexafluoroethane plasma, whereby fluorine is effectively incorporated into the fiber (Hocker, 2002). Plasma-induced hydrophobization of cotton fabric in conjunction with increased specific surface area leads to an interesting and practically important effect. Water droplets are able to effectively remove dirt particles from the surface of the cotton fabric. This phenomenon is illustrated in Fig. 9–29 for the case of HMDSO-plasma-treated cotton fabric (Hocker, 2002) and is usually referred to as the **Lotus effect**. Thus, the highly hydrophobic plasma-treated surface of cotton with specific plasma-modified surface topography is extremely dust- and dirt-repellant in contact with water. As an important consequence, the plasma-treated surface also becomes repellant to bacteria and fungi. The effect is relevant not only to cotton fiber but to some other materials as well.

Effective treatment of synthetic fibers can be achieved by using atmospheric-pressure plasma of DBD or corona. As an example, treatment of polypropylene (PP) in these discharges significantly increases the hydrophilicity of the surface. In particular, the contact angle with respect to water can be decreased from 90° to 55°; even after 2 weeks, the contact angle remains about 60°. The hydrophilicity increase is due to a plasma-induced high oxygen-to-carbon ratio, which is significant even at the tenth surface layer of the polymer. The non-equilibrium atmospheric-pressure plasma intensification of the PP surface hydrophilicity can be additionally enhanced by using maleic acid anhydride as an assisting reagent. Incorporation of oxygen into the polymer fiber surface is permanent in this case, and a contact angle with respect to water can be decreased to 42°. PET fibers are used for enforcing the PE matrix in the process of fabrication of polymer composite materials. In this case, non-equilibrium ethylene plasma treatment of the PET fibers significantly increases their adhesion strength to the PE matrix. Another example of effective plasma treatment

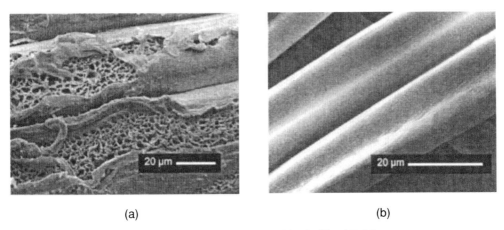

(a)   (b)

Figure 9–30. SEM pictures of Nomex fibers after exposition in diluted H_2SO_4.

of synthetic textiles is treatment of polyaramid textile fibers, such as Nomex, which are considered to be high-performance fibers but are prone to hydrolysis. Treatment of the fiber in hexafluoroethane–hydrogen plasma creates a diffusion barrier layer on the surface that is resistant to hydrolysis. The fiber becomes resistant even to 85% H_2SO_4. Contact of the treated fibers with sulfuric acid for 20 h at room temperature leaves the fibers completely intact whereas conventional fluorocarbon finishing under the given conditions causes significant damage (see Fig. 9–30). The damage includes significant shrinkage and loss of properties of the Nomex polyaramid textile fibers (Hocker, 2002).

9.8.6. Specific Conditions and Results of Non-Thermal Plasma Treatment of Textiles

The effect of applying different plasma gases for the plasma treatment of textiles (Maximov, 2000) is summarized in Table 9–18. The same textile treatment can be achieved using different plasma gases. This effect can be explained by taking into account two factors. First, most of the plasma treatment processes are not chemically specific and are related to an integral change of surface energy. Chemical peculiarities of the plasma gas are not so important in this case. Second, plasma treatment of polymers usually results in the production of gases with a flow rate close to that of the initial plasma gas. Thus, the polymer is treated in plasma including not only the initial plasma gas but also the gas generated during the treatment process. Different specific plasma-chemical processes for treatment of textiles are organized in non-equilibrium discharges under different discharge conditions, including pressures, discharge currents and powers, flow rates of applied plasma gases, and treatment times. Although similar results can be achieved generally in different discharges at different conditions, the most typical ones applied to achieve specific properties of textiles and related materials (Maximov, 2000) are summarized in Table 9–19.

9.8.7. Plasma-Chemical Processes for Final Fabric Treatment

Industrial plasma-chemical reactors used in final fabric treatment process a large amount of wet materials and sustain stable operation of uniform plasma in large volumes. The reactors are able to treat fabric widths of about 2 m on a roll of diameter 1–1.4 m. Discharge powers are 100–200 kW, treatment rates are 10–100 m/min, current densities are 1–3 mA/cm^2, and pressures are 50–150 Pa. The plasma-chemical reactors are used for different applications related to final fabric treatment. Specifically we can mention treatment of wool fabric before printing. The fabric is treated in this technological process in air plasma after smoothing and drying. Subsequent printing has better quality than in the traditional wet chemical approach

Table 9–18. Application of Different Plasma Gases for Plasma Treatment of Textiles

Plasma gas	Results of plasma treatment	Treated polymer materials
Inert gases: He, Ne, Ar	1. Enhancement of hydrophilicity	Wool, cotton, polyamide (PA) (nylon), polyethylene (PE), polyethylene terephthalate (PET)
	2. Decrease of gas permeability of polymer films	Polyethylene (PE), polypropylene (PP), polystyrene (PS)
	3. Suppressing of losses of plastificator	Polyvinylchloride (PVC)
	4. Enhancement of polymer film tearing strength	Polyethylene (PE), polypropylene (PP), polystyrene (PS)
	5. Adhesion properties enhancement	Polyether
	6. Suppressing of surface gloss	Polyvinylchloride (PVC)
	7. Stimulation of antistatic properties	Polyethylene-terephthalate (PET)
Air, oxygen	1. Enhancement of wettability and adhesion properties for gluing fabrics	Cotton, flax, wool, lavsan
	2. Suppressing of shrinkage and felting, enhancement of stability with respect to fabric wear-out	Wool
	3. Enhancement of wettability and intensification of soaking	Rough cotton fabric
	4. Enhancement of adhesion for vacuum metallization and gluing polymers with metals	Polyethylene, polypropylene, poly-oxymethylene (acetal), polyethylene terephthalate (PET), polytetrafluoroethylene (PTFE), polyimide (PI), polycarbonate (PC)
	5. Decrease of gas permeability of polymer films	Polyethylene (PE), polypropylene (PP), polystyrene (PS)
	6. Enhancement of polymer film tearing strength	Polyethylene (PE), polypropylene (PP), polystyrene (PS)
	7. Suppressing of surface gloss	Polyvinylchloride (PVC)
	8. Stimulation of antistatic properties	Lavsan fabric
Silanes (silicon hydrides)	1. Decrease of gas permeability of polymer films	Polyethylene (PE), polypropylene (PP), polystyrene (PS)
	2. Enhancement of polymer film tearing strength	Polyethylene (PE), polypropylene (PP), polystyrene (PS)
Hydrocarbons: methane, ethane, butane, etc.	1. Decrease of gas permeability of polymer films	Polyethylene (PE), polypropylene (PP), polystyrene (PS)
	2. Enhancement of polymer film tearing strength	Polyethylene (PE), polypropylene (PP), polystyrene (PS)
	3. Suppressing of losses of plastificator	Polyvinylchloride (PVC)
$CO_2 + Ar$	1. Modification of surface morphology	Carbon fabrics
CO_2	1. Suppressing of surface gloss	Polyvinylchloride (PVC)
	2. Wettability enhancement	Cotton, wool, lavsan fabric
NO, NO_2, CO	1. Suppressing of surface gloss	Polyvinylchloride (PVC)
CO, H_2O	1. Enhancement of adhesion	Lavsan films
N_2	1. Enhancement of adhesion	Polyether
	2. Suppressing of losses of plastificator	Polyvinylchloride (PVC)
	3. Decrease of gas permeability of polymer films	Polyethylene (PE), polypropylene (PP), polystyrene (PS)
	4. Enhancement of polymer film tearing strength	Polyethylene (PE), polypropylene (PP), polystyrene (PS)
	5. Wettability enhancement	Cotton, wool, lavsan fabric
	6. Suppressing of surface gloss	Polyvinylchloride (PVC)

Table 9–18 (*cont.*)

Plasma gas	Results of plasma treatment	Treated polymer materials
Amines	1. Enhancement of axisymmetric fiber strength	Aromatic polyamides
	2. Fiber wettability enhancement	Aromatic polyamides
	3. Enhancement of fiber adhesion	Aromatic polyamides
	4. Stimulation of fiber's antistatic properties	Aromatic polyamides
Alkylamines	1. Surface activation of fibers, films, and fabrics for further fabrication of epoxy- and other composites	Aromatic polyamides, polytetrafluoroethylene (PTFE)
NH_3 (ammonia)	1. Wettability enhancement	Aromatic polyamides, wool, polyether, rough cotton fabric
	2. Suppressing of losses of plastificator	Polyvinylchloride (PVC)
	3. Enhancement of adhesion	Polyoxymethylene (acetal), polytetrafluoroethylene (PTFE), polycarbonate (PC), polyacrylate, polyphenyleneoxide
Hydrogen	1. Wettability enhancement	Cotton, wool, lavsan fabric
	2. Suppressing of surface gloss	Polyvinylchloride (PVC)
	3. Surface flattening	Polyvinylchloride (PVC)
$H_2 + Ar$	1. Enhancement of adhesion	Carbon fibers
Chlorine	1. Suppressing of surface gloss	Polyvinylchloride (PVC)
CF_4, $CF_4 + O_2$	1. Enhancement of wettability of plastic surfaces	Polyethylene (PE), polyoxymethylene (acetal), polystyrene (PS), polypropylene (PP), polytetrafluoroethylene (PTFE), polymethylmethacrylate (PMMA), polyimide (PI)
NF_3, BF_3, SiF_4	1. Surface activation for further liquid treatment	Silk, wool, cotton
CF_4, C_2F_4, $CHClF_2$, CF_3Br, CF_3Cl, CHF_3	1. Surface activation with changing of wettability	Polyethylene (PE), polystyrene (PS), polypropylene (PP)
CF_4, CHF_3	1. Improvement of dyeing properties	Polytetrafluoroethylene (PTFE), polyethyleneterephthalate (PET)
	2. Suppressing of losses of plastificator	Wool, polyvinylchloride (PVC)
Fluorocarbons/ H_2 mixture, CF_4, C_2F_6, SF_6	1. Decrease of surface friction	Polyethers
	2. Stimulation of antistatic properties	Polyethers
	3. Enhancement of hydrophobicity	Polyethers
	4. Enhancement of weather resistance	Polyethers

and does not require chlorination, treatment in tin salt solution, and washing. Another application is related to pre-treatment of rough cotton and mixed fabric (cotton/polyether, cotton/nylon) before dyeing. The water absorption ability of the plasma-treated fabric is sufficiently high to exclude the otherwise required process of alkaline boiling of the fabric. This feature is especially important for cotton/polyether and cotton/nylon mixed fabrics. Plasma treatment permits the production of fabrics that cannot be manufactured using other technologies. An example is the production of cotton/polyether lace fabric; the semi-transparency of the lace fabric is achieved by special plasma modification of the cotton component of the mixed fabric with subsequent partial elimination of the material during alkaline boiling and bleaching.

Table 9–19. Non-Equilibrium Discharge Conditions Typical for Achieving Specific Results of Plasma-Chemical Treatment of Textiles and Related Materials

Specific textile treatment	Discharge	Power, current	Plasma gas	Pressure	Flow rate	Other
Surface activation of fabrics and polymer films	RF, 1 kHz–40 MHz	0.1–1 W/cm^2	Air	1.33–266 Pa		
Smoothing of mixed fabric (cotton, polyether)	RF, 13.56 MHz	200 W	Air, O_2	133 Pa	50 cm^3/s	Treatment time 29 min
Enhancement of adhesion properties of polyethylene films	Glow	50–200 W	O_2, $O_2 + CF_4$	100 Pa		Treatment time 3–10 min
Enhancement of adsorption properties and durability of polyether and polyamide fibers and knitted wear	Glow	50–100 mA	Air	0.133–1.33 Pa		
Enhancement of adhesion properties of polypropylene surface	RF	70–1000 W	O_2	66–133 Pa	10-400 cm^3/min	Treatment time 3–60 s
Suppressing of losses of plasticator from PVC-films, enhancement of hydrophobicity, oil and weather resistance	Glow		H_2/C_nF_{2n+2} n = 1, 2, 3; admixture of CO, Ar, He, CO_2, N_2	6.65–1330 Pa		
Enhancement of adhesion properties of polyether	Glow	>100 W	Ar, N_2, He and their mixtures	0.13–133 Pa		
Stimulation of antistatic properties of polymers and decrease of friction by deposition (and etching) of polymers	RF, 13.56 MHz	500 W	vapors of saturated hydro-carbons	0.001–5 Pa		Treatment time 10 min
Stimulation of antistatic properties of polyether textiles	RF, 100 KHz–13.56 MHz	Ar, He: 0.1—30 W/cm^2; CO_2: 0.1–5 W/cm^2	Ar, He, CO_2	Ar, He: 0.1–133 Pa; CO_2: 0.66–66 Pa		Surface activation for further grafting antistatic materials
Enhancement of adhesion and printing properties of polypropylene surface	Glow		O_2	10–1330 Pa		Treatment time 5 s–15 min
Enhancement of wettability of polymethylmethacrylate (PMMA) surface	RF, 13.56 MHz	1 W/cm^2	NH_3	13.3 Pa		Treatment time 200 s
Acceleration of drying and wettability enhancement of silk fabric	RF, 18 MHz	100 W	N_2			Treatment time < 90 min
Enhancement of non-crumple and water-repulsion properties of cellulose; reduction of cellulose solubility in basic solutions	RF, 1–30 MHz	20–80 W	NH_3	133–665 Pa	1–5 cm^3/min	Treatment time 2 h
Enhancement of wettability to dyeing solution and acceleration of drying of cotton fabrics	RF, 13.56 MHz	40 W	Ar	13.3 Pa	6.5 cm^3/min	Treatment time < 90 min
Enhancement or wettability of polyether fabrics	Glow, 1–10 kHz	1–1.5 mA/cm^2	N_2, CO_2, NH_3, air	66 Pa		

Application	Discharge, frequency	Power	Gas	Pressure	Parameters
Improvement of water-absorption and dirt-resistive properties of polyether and mixed cotton–polyether fabrics	RF, 1 and 13.56 MHz	0.41–1.52 mA/cm²	N_2, O_2, CO_2, air	133 Pa	
Enhancement of adhesion properties of polyamide fibers	Microwave, 2.45 GGz	100–700 W	Ar, N_2, NH_3	0.7–133 Pa	Activation for allyl-amine grafting; Treatment time 5–60 s
Reduction of wool felting	RF, 13.56 MHz	30–100 W	Air, O_2, N_2, H_2, CO_2, He, NH_3	266–532 Pa	Treatment time 0.7–1.5 s
Wettability enhancement and smoothing of flax fabric	Glow	250 mA	Air	130 Pa	
Lavsan metallization in production of wrapping materials and information carriers	Corona		Air, N_2, Ar, N_2+NH_3		
Metallization of polypropylene	Glow, 70 kHz	100 mA	Air	133–800 Pa	Voltage, 1 kV
Metallization of fluoroplast	Glow	0.06–0.17 mA/cm²	Air	53–80 Pa	Voltage, 700–900 V
Fluoroplast gluing with epoxy	RF, 13.56 MHz	4 W/cm²	Ar	0.67 Pa	Plasma activation; treatment time 30 s
Enhancement of adhesion of polyethylene film for gluing sliding ski surface	Glow, 50 Hz	500 W	Air		Voltage 1.5 kV; film width 6 cm, film velocity 2 m/min
Polymer film preparation to thermal pressing	Corona	0.005 W/cm²	Air	1 atm	
Improvement of dyeing properties of lavsan	RF, 13.56 MHz	300 W	O_2	27–200 Pa	Treatment time 10 min
Preparation of automobile's bumper to painting	RF	500 W	O_2	133 Pa	Treatment time 30 s
Intensification of dyeing of nap fabrics produced from polyether fibers	RF, 13.56 MHz	700 W	Air	13 Pa	Treatment time 5 min
Enhancement of elasticity and dyeing properties of synthetic fibers	Glow	2–5 kW		13–133 Pa	Treatment time 5–10 min
Reduction of shrinkage, felting, and wearing of wool, cotton, and mohair	Corona, 2 kHz		Air	1 atm	Voltage 17.5 kV
Enhancement of water-repulsive properties of fabrics	RF, 13.56 MHz	0.025 W/cm²	CF_4	13–200 Pa	
Strengthening of gluing connection of siliconized cloak fabric with lining materials	DBD		air	1 atm	
Enhancement of plastic adhesion at metallization	RF	25–150 W	O_2	8–67 Pa	
Enhancement of adhesion of arylox-polymer at metallization	RF, 27.12 MHz		N_2	267 Pa	Treatment time 5–20 min

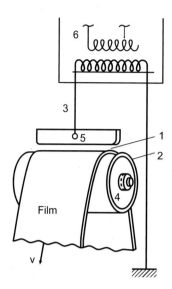

Figure 9–31. General schematic of a system for surface modification of polymer films based on a surface barrier discharge: (1) air gap; (2) metallic roller covered with a dielectric; (3) high-voltage wiring; (4) ground system; (5) electrode; (6) high-voltage power supply.

9.8.8. Plasma-Chemical Treatment of Plastics, Rubber Materials, and Special Polymer Films

Industrial equipment has been developed for the plasma-chemical surface modification of rolled materials, different films, and special products from plastics, rubber, paper, metal, textiles, different fabrics, and so forth for protection of the materials and preparation for dyeing, varnishing, gluing, antifriction treatment, and so on (Maximov, 2000). Plasma-chemical equipment can be operated in this case either continuously or periodically and is used either for plasma activation and coating or just for plasma activation of surfaces as a pretreatment phase of the whole technology. Non-thermal discharges applied in the equipment are DBDs and corona discharges operating at atmospheric-pressure, or glow discharges operating at low pressures. A DBD system for surface modification of polymer films shown in Fig. 9–31 can be considered an example of the atmospheric-pressure plasma systems. The system is able to activate about 10 m/min of polymer band with width up to 1.6 m. It is applied in particular for treatment of PE, PP, and PET. A glow discharge system for surface modification of polymer films shown in Fig. 9–32 can be considered an example of low-pressure plasma systems. The discharge power in the system is 50 kW, voltage is 1.4–1.6 kV, current is 0.3–0.35 A, and gas pressure is 30 Pa. The system is able to activate about 3 m/min of polymer band with width up to 60 mm. It has been specifically designed for activation of high-density PE applied as a gliding ski surface. A similar glow discharge system has been applied for 7 m/min pretreatment of fluorine-based film before gluing; the width of the glow discharge–treated films can be 0.6 m or more. The discharge system has also been used for activation of fluoroplastic insulation of 1–4 mm wires with treatment rate 5 m/min. Plasma-chemical surface modification of rolled materials and different films from plastics, rubber, paper, metal, textiles, and different fabrics has been also performed using industrial-scale low-pressure RF and microwave discharges. The power of the plasma

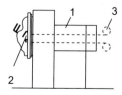

Figure 9–32. General schematic of a reactor for activation of polymer films by an industrial frequency AC glow discharge: (1) gas control system; (2) treatment zone; (3) film-moving system.

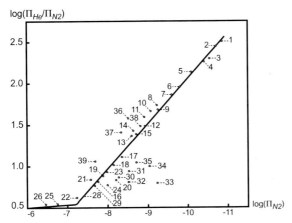

Figure 9–33. Selectivity of different polymer membranes to He–N$_2$ separation as a function of nitrogen permeability (Π, in cm^3/(cm^2 × atm × s)): (1) polyvinylidene chloride; (2, 4) polyethylene terephthalate; (3) polyvinylfluoride; (5) polyvinylchloride; (6) polyamide; (7) plastified polyvinylidene chloride; (8) cellulose nitrate; (9) polypropylene; (10) fluoroplast (26); (11) co-polymer of isoprene (74%) and acryl-nitryl (26%); (12, 18, 20) different co-polymers of butadiene and acryl-nitryl; (13) polyacrylate; (14) polycarbonate; (15) polyisobutylene; (16) butyl latex; (17) co-polymer of vinyl chloride and vinyl acetate; (19, 37) butyl acetate of cellulose; (21) polyethylene vinyl acetate; (22) polybutadiene; (23) special polymer SKI-3; (24) natural latex; (25) nitryl silicon latex; (26) dimethyl silicon latex; (27) special polymer SKS-30; (28) special polymer SKMS-50; (29) special polymer SKMS-30; (30, 34, 35) high-density, medium-density, and low-density polyethylene; (31) polyethylene with 5% soot; (32) co-polymer of ethylene (90%) and propylene (10%); (33) co-polymer of ethylene (96.5%) and vinyl acetate (3.5%); (36) triacetate of cellulose; (38) acetate cellulose; (39) polystyrene.

systems is as high as 300 kW, treatment rate is 60 m/min, and the width of the treated band is as much as 1.55 m. Such systems have been also applied for the treatment of different-shaped products and powders. More detailed information on the subject can be found in Lipin et al. (1994), Dai, Kviz, & Denning (2005), Maximov, Gorberg, &Titov (1997), and Pichal et al. (2004, 2005).

9.9. PLASMA MODIFICATION OF GAS-SEPARATING POLYMER MEMBRANES

9.9.1. Application of Polymer Membranes for Gas Separation: Enhancement of Polymer Membrane Selectivity by Plasma Polymerization and by Plasma Modification of Polymer Surfaces

Polymer membranes and specifically those treated in plasma are widely used in the separation of gases (Matsuura, 1993; Paul & Yampolskii, 1993; Biederman, 2004). Figure 9–33 presents the selectivity of different polymer materials $S(\Pi_{He}/\Pi_{N_2})$ as a function of their permeability with respect to separation of helium and nitrogen, which is a practically important process in He extraction technologies. The dimensionless selectivity S is determined here as a ratio of the membrane permeabilities Π, which are coefficients of proportionality between gas flow rate across a unit area and pressure gradient (cm^3 · cm)/(cm^2 · s · atm). The $S(\Pi_{He}/\Pi_{N_2})$ dependence in Fig. 9–33 demonstrates that an increase in selectivity of a polymer material is always related to a decrease in its permeability (Reitlinger, 1974; Krykin & Timashev, 1988). The creation of polymer membranes with high permeability and simultaneously high selectivity for gas separation becomes possible via fabrication of asymmetric multi-layer polymer membranes. Such asymmetric membranes consist of a high-permeability and low-selectivity mechanically strong substrate layer, which is covered

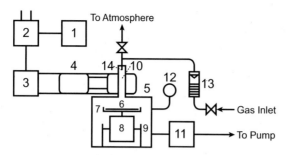

Figure 9–34. General schematic of the microwave installation for plasma treatment of gas-separating polymer membranes: (1) modulation system; (2) power supply; (3) microwave magnetron; (4) plasmatron; (5) chamber for polymer membrane treatment; (6) polymer membrane; (7) substrate; (8) electric engine; (9) system of vertical displacement of the membrane; (10) quartz discharge tube; (11) liquid nitrogen gas-trapping system; (12) manometer; (13) gas-flow meter.

by thin high-selectivity and low-permeability gas separating layer. Taking into account that the mechanical strength of the asymmetric membrane is provided by the substrate layer, the gas separating layer can be made very thin to maintain high multi-layer membrane permeability (Hwang, Kammermeyer, & Weissberger, 1984). Plasma polymerization is used in this case for deposition of the thin and highly selective polymer layer of the asymmetric membranes. The application of low-pressure RF discharges allows the deposition of layers to be uniform with thicknesses from 30 to 300 nm. The plasma-polymerized films have sufficient adhesion with the high-permeability substrate layer and permit the fabrication of gas-separation asymmetric membranes with sufficiently good characteristics.

A more effective approach to fabrication of asymmetric polymer membranes for gas separation is related not to plasma polymerization but to plasma surface modification of the polymer membranes. A high-permeability and low-selectivity polymer membrane is used in this case as a substrate. Plasma modification of a surface of the membrane leads to high selectivity of its thin surface layer, creating an asymmetric high-permeability membrane effective for gas separation. Among methods of surface modification of polymer membranes, treatment by non-thermal plasma and by UV radiation permit modification of super-thin surface layers, which is especially important in the case of gas-separating polymer membranes (Leonenkova et al., 1987). Treatment by strongly non-equilibrium plasma as well as by VUV radiation changes the permeability of the very thin surface layers, about 10–30 nm (Dorofeev & Skurat, 1978; Yasuda, 1985; Dorofeev, Skurat, & Talrose, 1987). Plasma modification of polymer surfaces is able to provide such treatment uniformly for large areas exceeding 30 m^2. The permeability change during plasma modification of polymer membrane surfaces is different for different gases, which leads to selectivity of the polymer membranes and to effective separation of gases. Plasma modification of polymer surfaces permits the fabrication of asymmetric polymer membranes with high permeability and simultaneous high selectivity for gas separation (Arbatsky et al., 1988). Values of polymer membrane selectivity achieved by plasma treatment exceed those of any other polymer materials and approach the values typical for inorganic materials.

9.9.2. Microwave Plasma System for Surface Modification of Gas-Separating Polymer Membranes

Microwave plasma modification of the surfaces of siloxane membranes (particularly lestosil and polycarbosil membranes) and acetate cellulose membranes is an example of plasma fabrication of asymmetric highly selective gas-separating polymer membranes (Arbatsky et al., 1988, 1990). A schematic of the microwave plasma installation is shown in Fig. 9–34.

Table 9–20. Permeability of Polymer Membranes with Respect to Different Individual Gases

Membrane parameters	Lestosil	Acetate-cellulose	Polycarbosil
Permeability of He, 10^{-4} cm^3 cm/cm^2 s atm	50–70	200–400	1.0–1.2
Permeability of CH_4, 10^{-4} cm^3 cm/cm^2 s atm	110–150	3–6	2.3–2.8
Permeability of N_2, 10^{-4} cm^3 cm/cm^2 s atm	35–50	2.5–5	0.6
Permeability of O_2, 10^{-4} cm^3 cm/cm^2 s atm	90–130	12–25	1.2
Permeability of H_2, 10^{-4} cm^3 cm/cm^2 s atm	90–120	180–360	1.8
Permeability of CO_2, 10^{-4} cm^3 cm/cm^2 s atm	300–400	120–240	6–7
Thickness of separating layer, μm	2–3	0.1	50

A magnetron with frequency 2.45 GHz and power 2.5 kW is applied as a microwave source. In the pulse-periodic regime, pulsed microwave power is 1–3.5 kW. The average microwave power is controlled by the pulse repetition frequency. The minimal average power to sustain the discharge is about 10 W. The microwave discharge is sustained in a quartz tube of a plasmatron with symmetric microwave irradiation (Fig. 9–34). Air as well as different atomic and molecular plasma gases have been used with a flow rate of about 40 cm^3/s, and the typical gas pressure was above 0.5 Torr. The distance between the microwave discharge and polymer film was 4–12 cm in these experiments; the typical treated polymer surface area was 40 cm^2. Polymer membrane permeability was measured using the Barrer method (Barrer, 1939; Buckner & Schrickerb, 1960; Becker, 1964; Linowitzki, 1971). The non-thermal plasma treatment was applied in the experiments to polymer films from polycarbosil and to asymmetric membranes from lestosil and acetate cellulose. Permeabilities of the membranes with respect to individual gases are presented in Table 9–20 in the form of permeability coefficients, which (as mentioned earlier) are coefficients of proportionality between gas flow rate across a unit area and pressure gradient $(cm^3 \cdot cm)/(cm^2 \cdot s \cdot atm)$. The initial selectivity of the membranes for separation of different pairs of gases is shown in Table 9–21. The tables demonstrate that all initial membranes chosen in the experiments have high permeability coefficients but low selectivity for separation of practically important gases. It should be mentioned that siloxane polymer materials are generally characterized by the highest permeability coefficients between polymer materials applied for gas separation (see Fig. 9–33).

9.9.3. Influence of Non-Thermal Discharge Treatment Parameters on Permeability of Plasma-Modified Gas-Separating Polymer Membranes

Among all plasma gases, oxygen and oxygen compounds are the most effective in changing the permeability of polymer membranes. In the microwave plasma system described earlier (Arbatsky et al., 1988, 1990), application of inert gases (He, Ar, Kr, Xe) and molecular nitrogen did not result in any significant changes in membrane permeability with respect to any penetrating gases for all considered discharge power levels and treatment times. At the same time, the application of oxygen, oxygen mixtures with all of the aforementioned gases,

Table 9–21. Selectivity of Polymer Membranes with Respect to Separation of Different Pairs of Gases

Pairs of gases	Lestosil	Acetate-cellulose	Polycarbosil
Selectivity for He–CH_4 separation	0.40–0.50	40–70	0.42–0.45
Selectivity for N_2–O_2 separation	0.40–0.50	3–5	0.5
Selectivity for H_2–CO_2 separation	0.23–0.29	1.5	0.27

Π, cm^3/(cm^2 · atm · sec)

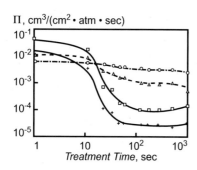

Figure 9–35. Permeability of the lestosil polymer membrane with respect to He, H$_2$, CO$_2$, and CH$_4$ as function of microwave plasma treatment time: ○, Π_{He}; △, Π_{H_2}; □, Π_{CO_2}; +, Π_{CH_4}; microwave pulse power 2 kW; pulsing period 2 ms; pulse duration 100 μs; gas pressure in the microwave discharge chamber 2 Torr; nitrogen/oxygen ratio in the plasma gas N$_2$:O$_2$ = 4:1; flow rate of the plasma gas 40 cm^3/s.

or triatomic oxygen-containing gases (H$_2$O, CO$_2$) leads to significant permeability changes, which indicates the key role of plasma-activated oxygen on permeability characteristics of polymer membranes. Figures 9–35–9–38 present changes in permeabilities of different polymer membranes as functions of microwave plasma treatment time at specific discharge conditions. Figures 9–35 and 9–36 show permeabilities of plasma-treated lestosil with respect to different penetrating gases. Similar characteristics of microwave plasma treatment of acetate cellulose are presented in Fig. 9–37; the treatment of polycarbosil is characterized in Fig. 9–38.

Treatment in non-thermal plasma can lead to either a decrease or an increase of permeability of the gas-separating polymer membranes. For example, plasma treatment of asymmetric membranes from acetate cellulose leads to the destruction of a thin dense surface layer, which results in a significant permeability increase (see Fig. 9–37). Siloxane polymers are not etched during microwave plasma treatment; therefore, surface modification results in this case mostly in the densification of a surface layer due to and other effects. This effect leads to a decrease of permeability of the siloxane membranes after plasma treatment (see Figs. 9–35, 9–36, and 9–38). Longer treatment times of siloxane membranes saturate the decrease of their permeability, which is due to the presence of pores in the membranes. While plasma treatment reduces the permeability of the dense siloxane, it cannot affect gas penetration through the pores in the material. As a result, longer treatment does not decrease permeability below some level, which can be seen in Figs. 9–35, 9–36, and 9–38. In good agreement with common sense, the equality of fluxes through the pores and plasma-treated dense polymer layers occurs faster (after shorter plasma treatment time) for larger penetrating molecules (CO$_2$, CH$_4$, N$_2$, O$_2$) than for smaller ones (He, H$_2$). Although the absolute value of gas permeability (total gas flows) through a dense polymer membrane is proportional to the pressure change across the membrane ($\Pi \propto \Delta p$), the laminar flow through the pores results in parabolic dependence: $\Pi \propto (\Delta p)^2$ (Osipov, 1982). This difference permits one to separate the contribution of porous penetration from

Π, cm^3/(cm^2 · atm · sec)

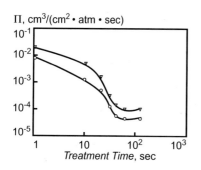

Figure 9–36. Permeability of the lestosil polymer membrane with respect to O$_2$, N$_2$, and CH$_4$ as function of microwave plasma treatment time: ▽, Π_{O_2}; ○, Π_{N_2}: microwave pulse power 2 kW; pulsing period 2 ms; pulse duration 100 μs; gas pressure in the microwave discharge chamber 2 Torr; nitrogen/oxygen ratio in the plasma gas N$_2$:O$_2$ = 4:1; flow rate of the plasma gas 40 cm^3/s.

Figure 9–37. Permeability of the polymer membrane from acetate cellulose with respect to He, H_2, CO_2, and CH_4 as function of microwave plasma treatment time: o, Π_{He}; □, Π_{H_2}; Δ, Π_{CO_2}; +, Π_{CH_4}; microwave pulse power 2 kW; pulsing period 2 ms; pulse duration 100 μs; gas pressure in the microwave discharge chamber 2 Torr; nitrogen/oxygen ratio in the plasma gas N_2:O_2 = 4:1; flow rate of the plasma gas 40 cm^3/s.

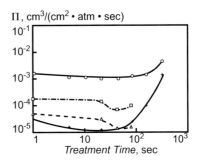

Π, cm^3/(cm^2 · atm · sec)

Treatment Time, sec

the total permeability of polymer membranes. Flow rates through two different lestosil membranes are shown in Fig. 9–39 before and after plasma treatment. The parabolic dependence $\Pi \propto (\Delta p)^2$ is easier to observe for plasma-treated membranes where the linear permeability component is suppressed.

9.9.4. Plasma Enhancement of Selectivity of Gas-Separating Polymer Membranes

The selectivity of gas-separating polymer membranes can be significantly increased by treatment in a microwave plasma (Section 9.9.2), which is illustrated in Figs. 9–40, and 9–41. Figure 9–40 shows the dependence of selectivity of a lestosil membrane with respect to different pairs of gases as a function of plasma treatment time. Similar dependences for acetate cellulose and polycarbosil are presented in Fig. 9–41. It is important that whereas the selectivity increases for separation of He–CH_4 and H_2–CO_2 mixtures, permeabilities of the plasma-treated membranes with respect to helium and hydrogen remain almost as high as before the plasma treatment. Because of the presence of micropores, the maximum selectivity of membranes is determined not only by the plasma treatment process but also by the quality of the initial polymer material, specifically by the concentration and sizes of the pores. A decrease of selectivity for higher treatment times (Figs. 9–40, and 9–41) is due to the fact that permeability for smaller molecules (like hydrogen) keeps reducing, while permeability for bigger molecules (like methane) is already determined by the gas penetration through pores and remains almost constant. To determine the maximum selectivity of plasma-treated membranes, experiments (Arbatsky et al., 1988) have been carried out with special 20 μm poreless dimethylsiloxane membranes with an initial He–CH_4 selectivity of 0.5, close to that of lestosil (see Table 9–21). Microwave plasma treatment results in this case in a selectivity increase of up to $5 \cdot 10^4$, whereas reduction of He permeability is only 1.5- to 2-fold. A similar selectivity increase (up to 10^5-fold) has also been achieved by plasma treatment of relatively thick polydimethylsiloxane (PDMS) membranes.

Figure 9–38. Permeability of the polycarbosil polymer membrane with respect to He and CH_4 as function of microwave plasma treatment time: o, Π_{He}; +, Π_{CH_4}; microwave pulse power 2 kW; pulsing period 0.6 ms; pulse duration 70 μs; gas pressure in the microwave discharge chamber 1 Torr; nitrogen/oxygen ratio in the plasma gas N_2:O_2 = 4:1; flow rate of the plasma gas 40 cm^3/s.

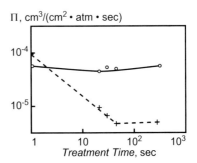

Π, cm^3/(cm^2 · atm · sec)

Treatment Time, sec

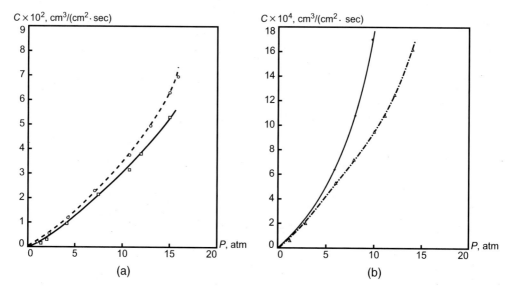

Figure 9–39. Nitrogen flow through the lestosil polymer membrane as function of pressure drop across the membrane at different pore concentration: (a) initial non-treated specimen; (b) the same specimen after 120 s of treatment in microwave plasma. Microwave pulse power 2 kW; pulsing period 2 ms; pulse duration $100 \, \mu$s; gas pressure in the microwave discharge chamber 2 Torr; nitrogen/oxygen ratio in the plasma gas $N_2:O_2 = 4:1$; flow rate of the plasma gas $40 \, cm^3$/s.

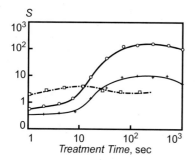

Figure 9–40. Selectivity of the lestosil polymer membrane with respect to separation of gases as a function of microwave plasma treatment time: ○, $S(\Pi_{He}/\Pi_{CO_2})$; +, $S(\Pi_{H_2}/\Pi_{CO_2})$; □, $S(\Pi_{O_2}/\Pi_{N_2})$; microwave pulse power 2 kW; pulsing period 2 ms; pulse duration $100 \, \mu$s; gas pressure in the microwave discharge chamber 2 Torr; nitrogen/oxygen ratio in the plasma gas $N_2:O_2 = 4:1$; flow rate of the plasma gas $40 \, cm^3$/s.

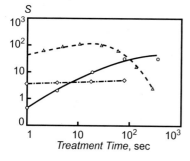

Figure 9–41. Gas-separation selectivity of the acetate cellulose polymer membrane ($\triangle$, $S(\Pi_{He}/\Pi_{CH_4})$; $\Diamond$, $S(\Pi_{H_2}/\Pi_{CO_2})$) and polycarbosil polymer membrane (○, $S(\Pi_{He}/\Pi_{CH_4})$) as function of microwave plasma treatment time. In the case of acetate cellulose treatment: microwave pulse power 2 kW; pulsing period 2 ms; pulse duration $100 \, \mu$s; gas pressure in the microwave discharge chamber 2 Torr; nitrogen/oxygen ratio in the plasma gas $N_2:O_2 = 4:1$; flow rate of the plasma gas $40 \, cm^3$/s. In the case of polycarbosil treatment: microwave pulse power 2 kW; pulsing period 0.6 ms; pulse duration $70 \, \mu$s; gas pressure in the microwave discharge chamber 1 Torr; nitrogen/oxygen ratio in the plasma gas $N_2:O_2 = 4:1$; flow rate of the plasma gas $40 \, cm^3$/s.

9.9. Plasma Modification of Gas-Separating Polymer Membranes

Figure 9–42. Gas-separation selectivity of the acetate cellulose polymer membrane $S(\Pi_{He}/\Pi_{CH_4})$ as function of the microwave discharge energy (dose) deposited during total duration of treatment: microwave pulse power 2 kW; pulse duration 100 μs; gas pressure in the microwave discharge chamber 2 Torr; nitrogen/oxygen ratio in the plasma gas $N_2:O_2 = 4:1$; flow rate of the plasma gas 40 cm^3/s. The same dose can be achieved at different values of average microwave power: +, 10 W (pulsing period 20 ms); $\circ$, 20 W (pulsing period 10 ms); $\square$, 100 W (pulsing period 2 ms).

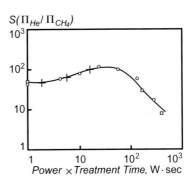

Although plasma treatment of polymer membranes is generally affected by the aging effect (Section 9.7.11), the plasma enhancement of gas-separating membrane selectivity remains stable for a long period of time, exceeding at least 100 days of storage in air. The stability to aging is due to the chemical and structural stability of compounds formed on the polymer surface during plasma treatment which enhances the gas-separating properties of the membranes. The permeability of the plasma-modified membranes and, hence, their selectivity depends not on the discharge power and treatment time separately but on their product, which is the discharge energy input. This so-called **dose effect** is illustrated in Fig. 9–42, where He–CH$_4$ selectivity of the acetate cellulose membrane is shown as a function of dose at different values of average power and treatment time. The dose effect is due to the fact that the permeability of the plasma-modified membranes is determined by surface reactions and the density of active oxygen species, which is proportional to the specific energy input in the microwave discharge. The rate and effectiveness of plasma treatment of the polymer membranes essentially depend on the concentration of oxygen in the plasma gas. The rate of permeability change of a gas-separating membrane is usually proportional to the molar fraction of oxygen in the plasma gas (Arbatsky et al., 1988). As an example, the selectivity of an acetate cellulose membrane with respect to He–CH$_4$ separation is shown in Fig. 9–43 as a function of relative concentration of oxygen (%) in N_2–O_2 mixture used as plasma gas and the treatment time in the microwave discharge.

9.9.5. Chemical and Structural Modification of Surface Layers of Gas-Separating Polymer Membranes by Microwave Plasma Treatment

The microwave plasma treatment of gas-separating polymer membranes, described in Section 9.9.2, leads to chemical and structural changes resulting in significant improvement in the selectivity of the membranes (Section 9.9.4) without any significant decrease of their permeability with respect to one gas component. These chemical and structural polymer surface changes have been analyzed using X-ray photo-electron spectroscopy (XPES) in

Figure 9–43. Gas-separation selectivity of the acetate cellulose polymer membrane $S(\Pi_{He}/\Pi_{CH_4})$ as function of oxygen density in the plasma gas [O_2] and microwave plasma treatment time: $\lozenge$, 100% O_2; +, 20% O_2 and 80% N_2; $\circ$, (0.1% O_2 and 99.9% N_2). The membrane treatment regime: microwave pulse power 2 kW; pulsing period 2 ms; pulse duration 100 μs; gas pressure in the microwave discharge chamber 2 Torr; nitrogen/oxygen ratio in the plasma gas $N_2:O_2 = 4:1$; flow rate of the plasma gas 40 cm^3/s.

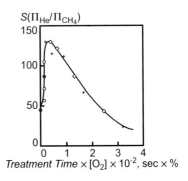

Table 9–22. Composition of Surface Layer Including Concentration of Cross-Links for Initial and Plasma-Treated Lestosil Membrane

Treatment time, s	0	5	20	80	320	1290
C, Atomic concentration, %	52	35	36	30	30	22
O, Atomic concentration, %	23	38	39	42	42	47
Si, Atomic concentration, %	25	26	27	28	29	31
N, Atomic concentration, %	0.0	0.1	0.3	0.4	0.3	0.4
C:Si Atomic concentration ratio	2.08	1.36	1.33	1.07	1.03	0.70
O:Si Atomic concentration ratio	0.92	1.48	1.45	1.50	1.43	1.58
Cross-link density, cm^{-3}	0	$1.2 \cdot 10^{21}$	$1.3 \cdot 10^{21}$	$1.8 \cdot 10^{21}$	$2.1 \cdot 10^{21}$	$2.6 \cdot 10^{21}$

the specific case of plasma treatment of lestosil, which is a block-copolymer of PDMS and polymethylsilsesquioxane (Clark & Peeling, 1976; Gorelova et al., 1980; Arbatsky et al., 1988). The concentration of cross-links and the chemical composition of the lestosil surface layer treated in plasma are presented in Table 9–22 as a function of treatment time. The data presented in the table correspond to the following parameters of the plasma-chemical system: pulsed power of magnetron microwave source, 2 kW; pulse duration, 100 μs, interval between the pulses, 2 ms, gas pressure, 2 Torr; plasma-gas composition, $O_2:N_2 = 1:4$; and gas flow rate, 40 cm^3/s. Longer plasma treatment leads to a decrease of the surface concentration of carbon atoms, a significant increase of oxygen concentration on the polymer surface, and some increase of silicon concentration. The concentration of nitrogen on the surface is very low, which indicates that chemical reactions of plasma-activated nitrogen with the lestosil surface are negligible. Analysis of XPES spectra indicates that lestosil treatment in oxygen-containing plasma leads to cross-linking of the linear siloxane. The structure of the treated surface includes such fragments as $(CH_3)_2–Si(–O–)_2$, $(CH_3)–Si(–O–)_3$, and $Si(–O–)_4$. Cross-linking occurs mostly due to extraction of methyl groups with the formation of interchain siloxane bonds. Initial plasma treatment leads to replacement of methyl groups of PDMS by oxygen atoms. Then reactions of the plasma-activated oxygen result mostly in extraction of methyl groups and formation of siloxane bonds. The surface concentration of oxygen atoms with respect to silicon atoms remains constant in this case.

9.9.6. Theoretical Model of Modification of Polymer Membrane Surfaces in After-Glow of Oxygen-Containing Plasma of Non-Polymerizing Gases: Lame Equation

As was found from XPES spectral analysis (Section 9.9.5), the plasma-induced selectivity enhancement of polymer membranes is mostly due to cross-linking of linear macromolecular chains into the polymer network. The cross-linking, in turn, is mostly due to oxygen-excited molecules and atoms penetrating into the membrane. The high density of the cross-links leads to a decrease in membrane permeability, especially with respect to larger penetrating molecules, which results in selectivity enhancement. A theoretical model describing the selectivity enhancement of polymer membranes due to the polymer cross-linking induced by plasma treatment has been developed by Arbatsky et al. (1988, 1990). The polymer membrane in this model is considered a potential field of mechanical forces with respect to the penetrating molecules. The model is based on the following three assumptions:

1. Change in selectivity is due to the formation of cross-links or other local quasi-rigid areas in the polymer matrix. The surface density of the cross-links or other local quasi-rigid areas is proportional to the energy input in the discharge during the entire period of plasma treatment.

2. The polymer membrane material is considered to be a uniform elastic medium with Young modulus E. The cross-links and other local quasi-rigid areas are presented in the model as immobile solid globules dispersed in the elastic medium.
3. There are through-pores in the membrane, but most of the gas penetration across the non-treated membrane is due to diffusion through the polymer material. As permeability is reduced by plasma treatment and formation of cross-links, the gas flow through the pores becomes a major fraction of the total permeability.

The elastic medium representing the polymer material in the absence of shearing can be described by the **Lame equation** (Landau & Lifshitz, 1986):

$$\lambda \Delta \vec{u} + \vec{f} = 0 \tag{9-90}$$

where λ is the Lame coefficient of elasticity, $\vec{u}$ is the displacement of medium from its equilibrium position, and $\vec{f}$ is an external force acting on the unit volume. The Lame vector equation (9–90) resembles the scalar Poisson equation in electrostatics (Lifshitz, Landau, & Pitaevskii, 1984), which allows us to use the analogy with electrostatics to describe the behavior of a perturbed elastic medium. This elasticity/electrostatics similarity approach (Feynman, Leighton, & Sands, 2005) will be applied to describe the influence of plasma treatment and cross-links on the permeability and selectivity of gas-separating polymer membranes.

9.9.7. Elasticity/Electrostatics Similarity Approach to Permeability of Plasma-Treated Polymer Membranes

To describe the influence of non-equilibrium plasma-treatment and cross-links on the permeability and selectivity of the gas-separating polymer membranes, let us first calculate the energy required for formation in an unbounded elastic medium of a hollow sphere with radius R corresponding to the radius of a penetrating molecule. Solving the Lame equation (9–90) in the absence of external forces ($\vec{f} = 0$) and for radial displacement $u_r = R$, the compression energy of the elastic medium can be found in this case as (Landau & Lifshitz, 1986)

$$W_0 = \frac{E}{2} \int_R^\infty \left(\frac{\partial u_r}{\partial r}\right)^2 4\pi r^2 dr = 2\pi E R^3. \tag{9-91}$$

This relation corresponds in electrostatics to the electric field energy of a charged conductive sphere. The displacement of elastic medium in this case is an analog of electrostatic potential, while electric field and electric charge correspond, respectively, to $\sqrt{E}(\partial u_r/\partial r)$ and $4\pi\sqrt{E}R^2$ (Feynman et al., 2005; here the Gaussian system of units is chosen). Developing the elasticity/electrostatics similarity (Section 9.9.6), the interaction energy of repulsion between two molecules with radii R_1 and R_2 penetrating in the elastic medium can be expressed as

$$W_1 = c_1 \frac{E R_1^2 R_2^2}{r}, \tag{9-92}$$

where r is the distance between the two penetrating molecules and c_1 is a numerical constant of the order of unity. Similarly, the interaction energy of repulsion inside the polymer matrix between a penetrating molecule of radius R and a rigid globule of radius R_c, representing a cross-link or another local quasi-rigid area, can be calculated as

$$W_2 = c_2 \frac{E R^4 R_c}{r^2 (1 - R_c^2/r^2)}. \tag{9-93}$$

663

In this relation, C_2 is another numerical constant of the order of unity and r is the distance between the penetrating molecule and the center of the sphere. In this case, the elasticity/electrostatics similarity leads to an electrostatic problem for determination of the interaction energy between a point charge q (penetrating molecule) and a grounded spherical conductor (cross-link; Feynman et al., 2005). To make zero potential on the sphere of radius R_c, it can be equivalently replaced by a point charge $q_1 = q \cdot (R_c/r)$ shifted on the distance $x = R_c^2/r$ from center of the sphere (Lifshitz et al., 1984). The presence of a cross-link (effective size R_c) significantly affects the motion of a penetrating molecule and, therefore, the permeability of a polymer membrane when the penetrating molecule moves sufficiently close to the cross-link and the energy of their repulsion exceeds temperature T. The critical distance r_D between the penetrating gas molecule and the cross-link (when the energy of their repulsion exceeds temperature T) is similar to the Debye radius (Section 3.7) in the framework of the elasticity/electrostatics analogy. This critical distance can be calculated by assuming $R \approx R_c \ll r$ as follows:

$$r_D = \sqrt{\frac{c_2 E R^5}{T}}. \tag{9–94}$$

The relation between the density of the cross-links, n, and average distance L between them can be estimated as $L = 1/n^{1/3}$. If the quasi-Debye radius is short with respect to distance between the cross-links ($L \gg r_D$), the motion of the penetrating molecules is only slightly perturbed by the cross-links. The density of the cross-links should be relatively low in this case:

$$n \ll \left(\frac{T}{c_2 E R^5}\right)^{\frac{3}{2}}. \tag{9–95}$$

The permeability P of the membrane is decreased only slightly in this case ($r_D \ll L$) by plasma treatment and formation of the cross-links, which can be expressed as:

$$P = P_0\left[1 - \left(\frac{r_D}{L}\right)^2\right], \tag{9–96}$$

where P_0 is the permeability of the initial membrane before plasma treatment and without any cross-links ($n = 0$). At a higher density n of the cross-links, $r_D > L$ (inequality opposite to (9–95)), and the energy of interaction between penetrating molecules and the cross-links exceeds temperature everywhere in the volume of the polymer membrane. The permeability P of the polymer membrane is significantly decreased in this case ($r_D > L$) by plasma treatment and formation of the cross-links and can be expressed by the following exponential relation:

$$P = P_0 \cdot \exp\left(-\frac{c_2 E R^4 R_c n^{2/3}}{T}\right). \tag{9–97}$$

Thus, relations (9–96) and (9–97) describe the influence of plasma treatment and formation of the cross-links on the permeability of polymer membranes. The size of the penetrating molecules R very strongly affects the permeability rate (9–97), which explains the strong effect of plasma treatment on the selectivity of the gas-separating membranes.

9.9.8. Effect of Cross-Link's Mobility and Clusterization on Permeability of Plasma-Treated Polymer Membranes

In deriving relations (9–96) and (9–97) for permeability of the plasma-treated polymer membranes, it was assumed that the rigid globules R_c immersed in the polymer and representing the cross-links are immobile. A more accurate consideration requires taking into account that as a result of repulsion between a penetrating molecule and the quasi-rigid globule, the latter can be slightly shifted, mitigating the repulsion (Arbatsky et al., 1988, 1990). A relevant derivation in the framework of the elasticity/electrostatics analogy requires consideration of the charge–dipole interaction. In the case of non-interconnected quasi-rigid areas, such consideration leads to the following expression for the energy of repulsion between penetrating molecules of radius R and cross-links (Arbatsky et al., 1988, 1990):

$$\Delta W = W_0 \cdot \gamma \cdot \frac{R}{L}, \tag{9--98}$$

where γ is the volume fraction of the elastic medium occupied by the quasi-rigid globules, representing the cross-links; W_0 is a normalizing energy factor, which characterizes the initial energy of the elastic medium; and L is a distance between the quasi-rigid globules. If the concentration of the quasi-rigid globules representing the cross-links is relatively high, they are interconnected into clusters of cross-links with a larger size, R_{cl}. An expression for the energy of repulsion between penetrating molecules of radius R and the clusters of cross-links can be presented in this case as (Arbatsky et al., 1988, 1990)

$$\Delta W = W_0 \cdot \gamma^{4/3} \cdot \frac{R}{R_{cl}}. \tag{9--99}$$

Thus, the interaction energy between penetrating molecules and cross-links increases with the growth of density of cross-links, which is characterized by the factor γ. Energy ΔW should be added simultaneously to the chemical potential of the molecules penetrating through the polymer, and to the activation energy of diffusion. An expression for the permeability of polymers with clusters of cross-links immersed inside can be presented according to Arbatsky et al. (1988, 1990) as

$$P = P_0 \exp\left(-\frac{2\pi E R^3}{T}\gamma^{4/3}\right), \tag{9--100}$$

where P_0 is the permeability of a non-treated membrane, R is the size of a penetrating molecule, and E is the Young modulus. Relation (9–100) is valid when the factor γ is small and the quasi-rigid globules are isolated. When the volume fraction of the elastic medium occupied by the quasi-rigid globules γ increases, effective clusters become bigger and bigger. As soon as the factor γ reaches its critical value, γ_{cr} (percolation limit, for spherical clusters $\gamma_{cr} = 0.16$), the "infinite cluster" is formed (Kirpatrick, 1977). In this case, although the clusters occupied only 16% of the polymer, cross-links are actually interconnected, forming one large quasi-rigid matrix. Similar to (9–98) and (9–99), the energy ΔW of interaction of a penetrating molecule (size R) with the matrix can be expressed as

$$\Delta W = c \cdot W_0 \cdot \frac{R}{l}, \tag{9--101}$$

where c is a constant of the order of unity and l is a correlation radius of the quasi-rigid matrix. The correlation radius l of the percolating macrocluster (the quasi-rigid matrix) can be expressed according to the Shklovsky-de Jean model (Shklovsky & Efros, 1979) near the percolation threshold as

$$l = \frac{R_c}{|\gamma - \gamma_{cr}|^\nu}, \tag{9--102}$$

where R_c is a characteristic size of a cross-link; $\nu = 1.33$ in the two-dimensional case and $\nu = 0.9$ in the three-dimensional case. The fraction of the globules $A(\gamma)$ belonging to the quasi-rigid matrix can be expressed near the percolation limit as (Shklovsky & Efros, 1979)

$$A(\gamma) = A_0 \cdot (\gamma - \gamma_{cr})^\beta, \tag{9--103}$$

where A_0 is a constant of the order of unity; $\beta = 0.14$ in the two-dimentional case and $\beta = 0.4$ in the three-dimentional case. The expression for gas permeability of the plasma-treated gas-separating polymer membranes near the percolation threshold ($\gamma \geq \gamma_{cr}$) can be presented as (Arbatsky et al., 1988, 1990)

$$
P = P_0 \cdot A_0 (\gamma - \gamma_{cr})^\beta \exp \left[- \frac{c \cdot 2\pi E R^4 (\gamma - \gamma_{cr})^\nu}{R_c T} \right]
$$

$$
+ P_0 \cdot [1 - A_0 \cdot (\gamma - \gamma_{cr})^\beta] \cdot \exp \left(- \frac{2\pi E R^3}{T} \cdot \gamma^{4/3} \right). \qquad (9\text{--}104)
$$

The first term describes the contribution of quasi-rigid globules comprising the infinite cluster of cross-links; the second term corresponds to the contribution of all other cross-links forming smaller clusters.

9.9.9. Modeling of Selectivity of Plasma-Treated Gas-Separating Polymer Membranes

Relation (9–104) demonstrates a strong exponential dependence of permeability of the plasma-treated membranes on the size of the penetrating molecules. Penetration of larger molecules is usually strongly suppressed by non-thermal plasma treatment of the membrane surface, which leads to significant plasma enhancement of the membrane selectivity, determined as the ratio of permeabilities with respect to two gases characterized by molecular sizes R_1 and R_2. The effect of plasma treatment is determined by specific energy input, or dose (Section 9.9.4), which is proportional to the factor γ – the volume fraction of the elastic medium occupied by the quasi-rigid globules, representing the cross-links. Based on the formulas from Section 9.9.8, the plasma enhancement of selectivity can be expressed by the following two relations. The first one describes the case of relatively low doses and low values of factor γ ($\gamma < \gamma_{cr} = 0.16$), when clusters of cross-links are not interconnected:

$$
S = S_0 \cdot \exp \left(\frac{2\pi E \left(R_2^3 - R_1^3 \right) \gamma^{4/3}}{T} \right). \qquad (9\text{--}105)
$$

The second formula describes the case of relatively high specific energy inputs and doses, which results in high values of the factor γ ($\gamma > \gamma_{cr} = 0.16$) when clusters of cross-links are interconnected in a quasi-rigid matrix:

$$
\frac{S}{S_0} = \frac{A_0 (\gamma - \gamma_{cr})^\beta \exp \left[- \frac{2\pi E R_1^4 (\gamma - \gamma_{cr})^\nu}{R_c T} \right] + [1 - A_0 (\gamma - \gamma_{cr})^\beta] \exp \left(- \frac{2\pi E R_1^3 \gamma^{4/3}}{T} \right)}{A_0 (\gamma - \gamma_{cr})^\beta \exp \left[- \frac{2\pi E R_2^4 (\gamma - \gamma_{cr})^\nu}{R_c T} \right] + [1 - A_0 (\gamma - \gamma_{cr})^\beta] \exp \left(- \frac{2\pi E R_2^3 \gamma^{4/3}}{T} \right)},
$$
$$
(9\text{--}106)
$$

where S_0 is the selectivity of the polymer membrane before plasma treatment. Relation (9–105) shows that the selectivity starts increasing exponentially when the specific energy input and factor γ exceed the threshold value:

$$
\gamma_{th} = \left[\frac{T}{2\pi E \left(R_2^3 - R_1^3 \right)} \right]^{\frac{3}{4}}. \qquad (9\text{--}107)
$$

Thus, plasma treatment makes the thin surface layer of the polymer membrane more stiff, which significantly decreases the permeability with respect to larger molecules and results in an enhancement of selectivity in gas separation. The increase in stiffness of the

9.9. Plasma Modification of Gas-Separating Polymer Membranes

Figure 9–44. Gas-separation selectivity of the lestosil polymer membrane $S(\Pi_{He}/\Pi_{CH_4})$ as function of microwave plasma treatment duration: o, experimental results; solid line, theory. The membrane treatment regime: microwave pulse power 2 kW; pulsing period 2 ms; pulse duration 100 μs; gas pressure in the microwave discharge chamber 2 Torr; nitrogen/oxygen ratio in the plasma gas $N_2:O_2 = 4:1$; flow rate of the plasma gas 40 cm³/s.

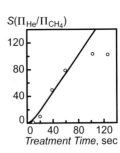

polymer surface after plasma treatment can be characterized by an effective increase of the Young modulus:

$$E_{\text{eff}} = E \cdot \left(1 + \gamma^{4/3}\frac{R}{R_{cl}}\right). \tag{9–108}$$

Relations (9–105) and (9–106) are in good agreement with experimental data on microwave plasma treatment of lestosil membranes (Sections 9.9.2–9.9.4), which is demonstrated in Fig. 9–44 for the case of He–CH₄ gas separation. The polymer membrane parameters used for the modeling are $E = 5 \cdot 10^6$ N/m², $T = 300$ K, $R(CH_4) = 0.41$ nm, $R(He) = 0.215$ nm, $R_{cl} = 1$ nm. Based on experimental measurements (Section 9.9.5), the time evolution of the cross-link density can be characterized as $\gamma \propto t^k (k \approx 0.6)$.

9.9.10. Effect of Initial Membrane Porosity on Selectivity of Plasma-Treated Gas-Separating Polymer Membranes

As seen from Fig. 9–44, the maximum lestosil selectivity for He–CH₄ separation achieved in the microwave plasma (Sections 9.9.2–9.9.4) is 120. Generally, the presence of a maximum in the dependence of selectivity on plasma treatment time is due to the porosity of the initial membranes (Section 9.9.3). Although plasma treatment reduces the permeability of the dense polymer, it cannot affect gas penetration through the pores in the material. As a result, longer plasma treatment does not decrease permeability below some level (see also Figs. 9–35, Fig. 9–36, and 9–38), which actually explains the presence of the maximum in the dependence of selectivity on plasma treatment time. The selectivity of the plasma-treated polymer membranes with through-pores, in which gas flows in a Knudsen or Poiseuille fashion depending on the pore diameter, can be calculated at low values of the factor γ ($\gamma < \gamma_{cr} = 0.16$) as (Arbatsky et al., 1988, 1990)

$$S = \frac{Q_{01}\exp\left(-\beta_1\gamma^{4/3}\right)\cdot(n_{1b} - n_{1a}) + K_1(n_{1b} - n_{1a}) + \Pi_1\left(n_{1b}^2 - n_{1a}^2\right)}{Q_{02}\exp\left(-\beta_2\gamma^{4/3}\right)\cdot(n_{2b} - n_{2a}) + K_2(n_{2b} - n_{2a}) + \Pi_2\left(n_{2b}^2 - n_{2a}^2\right)}, \tag{9–109}$$

where the subscripts 1 and 2 ($i = 1, 2$) refer to different penetrating gas components; $\beta_i = 2\pi E R_i^3/T$; n_{ib} and n_{ia} are partial gas densities from opposite sides of the membrane; $Q_{0i}(n_{ib} - n_{ia})$ is the flux of the penetrating molecules through a dense polymer membrane material before treatment ($\gamma = 0$); $K_i(n_{ib} - n_{ia})$ is Knudsen flux through pores; $\Pi_i(n_{ib}^2 - n_{ia}^2)$ is the Poiseuille flux through pores; $K_1/K_2 = \sqrt{M_2/M_1}$; $\Pi_1/\Pi_2 = \mu_2/\mu_1$; M_i are molecular masses; and μ_i are coefficients of viscosity. If the concentration of the through-pores is low and the permeability of a membrane is due to a dense polymer layer, then an increase in the cross-link density (assuming $R_1 < R_2$, $\beta_1 < \beta_2$) leads to a rapid decrease of the polymer membrane permeability with respect to larger molecules (R_2). The final value of the permeability is determined by the flux of the molecules (R_2) through the pores and does not depend any more on plasma treatment. Maximum selectivity of the membrane is achieved in this case at the value of the specific energy input corresponding to the factor γ:

$$\gamma_m^{4/3} = \frac{1}{\beta_2} \ln\left(\frac{\beta_2}{\beta_1} \cdot \frac{1}{b}\right), \tag{9-110}$$

where $b = S_0[K_2 + \Pi_2(n_{2b} - n_{2a})]/Q_{01}$. The maximum value of the selectivity of the plasma-treated polymer membrane corresponding to $\gamma = \gamma_{max}$ is

$$S_{max}(\gamma_m) = \frac{Q_{01}}{K_2 + \Pi_2(n_{2b} - n_{2a})}. \tag{9-111}$$

Thus, the maximum value of the selectivity is determined by the ratio of the flow of smaller molecules through dense polymer material and through the pores to the flow of larger molecules to the pores only. A further increase of the specific energy input of the plasma treatment ($\gamma > \gamma_m$) leads to a slow decrease of the membrane selectivity due to a slow decrease of polymer material permeability with respect to smaller molecules. The initial porosity of the polymer membranes essentially limits the maximum value of their selectivity to gas separation. When the lestosil membrane is especially prepared without pores, its selectivity with respect to He–CH_4 separation exceeds 1000 after microwave plasma treatment (Sections 9.9.2–9.9.5; Arbatsky et al., 1988, 1990).

9.10. PLASMA-CHEMICAL SYNTHESIS OF DIAMOND FILMS

9.10.1. General Features of Diamond-Film Production and Deposition in Plasma

Non-equilibrium plasma is effectively used in the production of synthetic diamonds, specifically diamond films, which are widely used in modern technology. Diamond is a metastable form of carbon at normal conditions; therefore, it can be synthesized in thermodynamic quasi-equilibrium conditions only at very high pressures (150 kbar) and temperatures (about 3000°C). The application of special catalysts decreases diamond production pressures to less than 70 kbar and temperatures to less than 1500°C, which are still quite extreme conditions. Plasma provides non-equilibrium conditions that permit effective diamond synthesis at atmospheric and lower pressures from different carbon compounds including CH_4, C_2H_2, CO, CO_2, C_3H_8, CH_3OH, C_2H_5OH, CH_3Cl, CF_4, CCl_4, and others (Deriagin & Fedoseev, 1977; Prelas, Popovici, & Bigelow, 1997; Asmussen & Reinhard, 2002). High electron temperature and relatively low gas temperature in non-equilibrium plasma provide intensive dissociation of the carbon compounds and carbon-atom deposition on special substrates. The non-equilibrium plasma systems also provide a high concentration of hydrogen atoms, usually by dissociation of specially added molecular hydrogen. The hydrogen atoms mostly etch graphite from the deposited film, leaving mostly diamond. To provide better selective deposition of diamond films, the substrates (made, for example, from silicon, metals, or ceramics) are treated to create diamond nucleation centers. Typical substrate temperatures for effective deposition of diamond films are 1000–1300 K.

The influence of the initial C/H/O composition of the carbon compounds used for non-equilibrium diamond synthesis on the composition of the deposited films is usually illustrated by the so-called **Bachmann triangle**, shown in Fig. 9–45. The Bachmann triangle summarizes experimental data and indicates the C/H/O compositions corresponding to deposition of non-diamond films, diamond films, or no deposition at all (which corresponds to domination of etching over deposition). Deposition of diamond films from the gas phase is typically performed at pressures of 10–100 Torr in a mixture of hydrogen with about 1–5% methane. The plasma-chemical conversion of carbon compounds can lead not only to the formation of diamond films but also to so-called diamond-like carbon (DLC). DLC films are

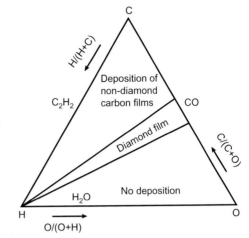

Figure 9–45. Bachmann triangle.

hard amorphous films with a significant fraction of sp^3-hybridized carbon atoms and which can contain a significant amount of hydrogen. Depending on deposition conditions, DLC films can be fully amorphous or contain diamond crystallites. Plasma-chemical production of the DLC films as well as their characteristics is described by Robertson (2002). DLC films are not called "diamond" unless a full three-dimentional crystalline lattice of diamond is proven. The physical and chemical properties of DLC films are often not as attractive as those of diamond films. On the other hand, the production of DLC films is generally easier and requires lower substrate temperature, gas pressure, hydrogen fraction in the initial mixture, and a lower level of gas-phase activation, which attracts interest to their large-scale industrial application as protective, anticorrosive, and other special coatings.

9.10.2. Different Discharge Systems Applied for Synthesis of Diamond Films

The most conventional plasma-chemical systems for synthesizing diamond films from gas-phase carbon compounds are based on the application of **microwave discharges**. The important advantage of these systems is the purity of the produced diamond films, which is due to an absence of electrodes and sustaining plasma only near the substrate, quite far from the reactor walls. Microwave discharge reactors are built either in microwave resonators or in microwave waveguides. A schematic of the microwave plasma reactor for diamond film synthesis is shown in Fig. 9–46 (Mankelevich & Suetin, 2000). Magnetron generators with frequency 2.45 GHz and power in the range between 100 W and 10 kW are usually used as relevant microwave sources. Waveguides with internal cross section 90 × 45 mm or 72 × 34 mm are usually applied for microwave power transfer from the magnetron generator to the discharge chamber. The microwave discharge chamber is usually made from stainless steel with a special quartz window transparent to microwave radiation and separating the plasma reactor from ambient atmosphere. Deposition of diamond films from the gas phase is performed at pressures of 10–100 Torr in a mixture of hydrogen with about 1–5% methane. The growth rate of the diamond films is about 10 μm/h on the substrate with 5–10 cm diameter. The price of diamond films produced in microwave discharge plasma is about $160 per carat (Mankelevich & Suetin, 2000).

Diamond films can also be effectively synthesized in **low-pressure DC glow discharges**, where the film growth rate can be relatively high, exceeding 30 μm/h. The substrate in this case is located on the grounded anode and heated up to the required temperature by the discharge itself. Hydrogen with 1–10% methane is injected into the low-pressure plasma reactor through the cathode. Although spectral measurements show that the gas temperature

Organic and Polymer Plasma Chemistry

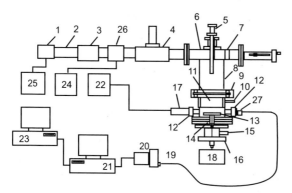

Figure 9–46. General schematic of an experimental microwave discharge plasma system for plasma-initiated deposition of diamond films: (1) magnetron; (2) waveguide system; (3) ferrite-circulation system; (4) attenuator; (5) tuning system; (6) microwave coupling system; (7) piston; (8) microwave mode converter; (9) separating glass; (10) gas inlet system; (11) vacuum chamber volume; (12) windows; (13) substrate support; (14, 16) sample displacement system; (15) pumping port; (17) optical pyrometer; (18) external displacement system; (19) monochromator; (20) video camera; (21) multi-channel spectra analyzer; (22) temperature block; (23) computer; (24) measurements of incident and reflected microwave power; (25) magnetron power supply; (26) electric measurement bridge; light-transport water/PI contact angle (θ) sport system.

in the system is about 2000 K, the plasma-chemical process of diamond film synthesis in low-pressure DC glow discharges is strongly non-equilibrium and similar to that in microwave discharges. The most popular laboratory deposition method, especially for the case of DLC, is **RF discharge** plasma-enhanced chemical vapor deposition (Koidl et al., 1990; Catherine, 1991; Wild & Koidl, 1991; Martinu et al., 1993; Weiler et al., 1996; Zarrabian, Fourches-Coulon, & Turban, 1997; He et al., 1999; Weiler, Lang, & Robertson, 1998; Conway et al., 2000). The reactor usually consists of two electrodes of a different area; the RF power is capacitively coupled to the smaller electrode, on which the substrate is mounted. The other electrode, typically including the reactor walls, is grounded. For DLC deposition, the plasma should be operated at the lowest pressures in order to maximize the ion-to-radical ratio. A review of the discharges used for DLC film deposition can be found in Robertson (2002).

The highest deposition rates of the diamond films (1 mm/h) are achieved in plasma-chemical reactors based on **thermal arc discharges.** In contrast to the previous systems, the discharge is clearly thermal and close to quasi-equilibrium conditions; gas temperature is about 3000–5000 K. The diameter of the deposited films is usually not large (about 3–5 cm), and the deposition uniformity is not as good as in the previously discussed discharge systems. To increase the diameter of the deposited film to 10 cm or more, and to improve the deposition uniformity, several arc discharge plasmatrons are applied simultaneously. Typical gas pressures in the arc discharges applied for diamond film synthesis are in the range 250–760 Torr, while pressure in the reactor is about ten times lower (20–50 Torr). A plasma jet with velocities of 100–300 m/s formed in a narrow nozzle is directed from the arc discharge zone into the reactor, where the substrate is located. The reacting H_2–CH_4 mixture is usually diluted in the discharge zone with 90% argon or other inert gas. The carbon-containing gas component (for example, methane) is sometimes not injected to the arc but admixed to the Ar–H_2 plasma jet in the reactor zone. Effective deposition of diamond films requires fast gas cooling and quenching in the substrate area. Taking into account that the thickness of the boundary layer near the substrate is inversely proportional to the square root of the jet velocity, the transsonic jet velocities are usually applied to provide effective deposition of diamond films in thermal plasma systems.

9.10.3. Non-Equilibrium Discharge Conditions and Gas-Phase Plasma-Chemical Processes in the Systems Applied for Synthesis of Diamond Films

Non-equilibrium microwave and glow discharges applied for the production of diamond films have similar plasma parameters: the gas temperature is about 2000–3000 K and equal to the ion temperature, the average electron energy is 1–2 eV with a strongly non-Maxwellian electron energy distribution function, the electron density is about 10^{11}–10^{12} cm^{-3}, the gas pressure is moderately low (50–150 Torr), and the reduced electric field is $E/n_0 = 35$–50 Td $= 3.5$–$5 \cdot 10^{-16}$ V $\cdot$ cm^2. DC glow discharges are usually stratified in this case. The plasma gas applied in the diamond film deposition systems is a mixture of molecular hydrogen with a small admixture of hydrocarbons. Non-equilibrium gas-phase processes lead to the formation of acetylene, ethylene, and such atoms and radicals as H, CH$_3$, CH$_2$, CH, C, and so forth. The active species formed in the gas phase are deposited on the surface, where they participate in a sophisticated sequence of such processes as adsorption/desorption, recombination, surface diffusion, and surface chemical reactions, which result in the formation of the diamond film. Atomic hydrogen predominantly etches non-diamond carbon structures, promoting selective production of the diamond films. The discharge is usually first initiated in pure hydrogen to clean the substrate. To provide effective breakdown, the initial pressure is 10–30 Torr. Then the pressure is increased to about 50–150 Torr, which leads to the growth of gas temperature up to the required 2000–3000 K. While plasma is sustained at $E/n_0 = 40$ Td in pure hydrogen, most of the discharge energy (63%) goes to vibrational excitation, 17% goes to rotational excitation, 11% goes to dissociation of molecular hydrogen and formation of H atoms, 7.5% goes to elastic collision and direct heating, and 1.5% goes to ionization and electronic excitation not related to dissociation. At $T_0 = 2000$–3000 K, vibrational relaxation is relatively fast and most of the vibrational and translation energy of molecular hydrogen are converted into translational, that is, to heating of the reacting gas mixture. Atomic hydrogen, which plays an important role in the formation of diamond films, is generated as a result of thermal dissociation

$$H_2 + H_2 \rightarrow H + H + H_2, \tag{9–112}$$

as well as dissociation of hydrogen by electron impact through excitation of electronically excited states:

$$e + H_2 \rightarrow H_2^* + e \rightarrow H + H + e. \tag{9–113}$$

A major contribution is provided in this case by electronically excited states of hydrogen molecules with relatively low excitation thresholds: 8.9, 11.75, and 11.8 eV. Thermal dissociation (9–112) dominates the formation of atomic hydrogen only at very high translational gas temperatures exceeding 2750–2850 K. Atomic hydrogen formed in the thermal (9–112) and non-thermal (9–113) processes initiates conversion of hydrocarbons and the formation of radicals (CH$_x$, $x < 4$), which are the gas-phase precursors of the diamond film. A typical gas-phase composition created in non-equilibrium discharge is presented in Table 9–23 (Mankelevich & Suetin, 2000) for the plasma system reduced electric field $E/n_0 = 37$ Td, electron density $2.65 \cdot 10^{11}$ cm^{-3}, gas temperature 2370 K, and density of the H$_2$–CH$_4$ gas mixture, $5.4 \cdot 10^{17}$ cm^{-3}. The major active gas-phase species created in the non-equilibrium plasma that provide the diamond film growth are atomic hydrogen (H), methyl radical (CH$_3$), and acetylene (C$_2$H$_2$). Oxygen is sometimes added to the initial gas mixture applied for deposition of diamond films. It leads, in particular, to the formation of O atoms and OH radicals in reactions involving electron impact and in reactions of atomic hydrogen (which is present in plasma with relatively high concentration; see Table 9–23) with molecular oxygen:

$$H + O_2 \rightarrow OH + O. \tag{9–114}$$

Table 9–23. Typical Gas-Phase Composition in Non-Equilibrium Discharge System for Diamond Film Production in H_2–CH_4 Mixture

Component	Molar fraction, %	Component	Molar fraction, %
H	2.44	C_2H	0.00413
CH_3	0.48	C_2H_6	0.000173
C_2H_2	2.93	C_2H_4	0.0146
CH_2	0.0135	C_2H_5	0.000516
CH_2 (singlet)	0.00133	C_2H_3	0.0023
CH	0.00105	CH_4	1.01
C	0.00108	H_2	93.1

The addition of oxygen also results in the formation of CO in the gas phase. Further redistribution of radicals in the system is influenced in this case by the following reactions: $O + H_2 \Leftrightarrow H + OH$, $OH + H_2 \Leftrightarrow H + H_2O$, and $H_2 + CO \rightarrow H + HCO$. Deposition of diamond films can be also performed using CO as a major gas-phase source of carbon atoms (see, for example, Aithal & Subramaniam, 2002). The formation of carbon atoms in this case is due to different non-equilibrium mechanisms of CO disproportioning (see Section 5.7), as well as by direct dissociation of carbon monoxide by electron impact. The thermal mechanism is not effective in non-equilibrium systems, because it requires temperatures exceeding 3000 K which are not present in the non-thermal plasma discharges under consideration.

9.10.4. Surface Chemical Processes of Diamond-Film Growth in Plasma

The most conventional non-equilibrium plasma-chemical systems that produce diamond films use H_2–CH_4 mixture as a feed gas. Plasma activation of this mixture leads to the gas-phase formation of hydrogen atoms, methyl radicals (CH_3), and acetylene (C_2H_2), which play a major role in further film growth. Transport of the gas-phase active species to the substrate is mostly provided by diffusion. The substrate is usually made from metal, silicon, or ceramics and is specially treated to create diamond nucleation centers. The temperature of the substrate is sustained at the level of 1000–1300 K to provide effective diamond synthesis. The synthesis of diamond films is provided by numerous elementary surface reactions. Four chemical reactions in particular describe the most general kinetic features of the process. First of all, surface recombination of atomic hydrogen from the gas phase into molecular hydrogen returns back to the gas phase:

$$C\text{–}H + H \rightarrow (C\text{–}) + H_2, \quad k = 1.1 \cdot 10^{-10} \exp\left(-\frac{3430\,\text{K}}{T}\right), \text{cm}^3/\text{s}, \quad (9\text{–}115)$$

where C–H is a surface group representing a hydrogen atom chemisorbed on the carbon structure, (C–) is a surface carbon atom with a free bond, and T is the substrate temperature. The surface reaction rate coefficients are given in cm^3/s, taking into account that volume densities of the gas-phase reacting components are measured in cm^{-3}, whereas surface densities of the deposited groups and surface reaction rates are considered with respect to unit area. Reaction (9–115) can be partially balanced by reverse reaction (9–115) of surface dissociation of molecular hydrogen, which covers the free carbon bond and create an H atom. While elementary recombination reaction (9–115) eliminates hydrogen from the surface, the following process of adsorption of atomic hydrogen covers the open carbon bond without growth of the diamond film:

$$(C\text{–}) + H \rightarrow C\text{–}H, \quad k = 3.3 \cdot 10^{-11} \text{cm}^{-3}. \quad (9\text{–}116)$$

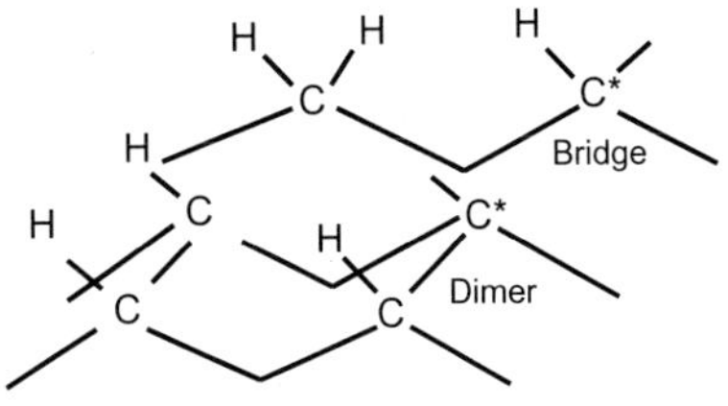

Figure 9–47. Different surface centers on (100) diamond plane.

This fast barrierless process is also partially balanced by the reverse reaction of desorption of an H atom from the surface group C–H. A major contribution to the actual deposition of the diamond film can be attributed to the initial chemisorption of methyl radicals (CH_3) and acetylene (C_2H_2) to open carbon bonds on the surface of the film:

$$(C-) + CH_3 \rightarrow C\text{-}CH_3, \quad k = 8.3 \cdot 10^{-12}\,cm^{-3}, \tag{9–117}$$

$$(C-) + C_2H_2 \rightarrow C\text{-}C_2H_2, \quad k = 4.0 \cdot 10^{-13}\,cm^{-3}. \tag{9–118}$$

After the primary chemisorption, (9–117) and (9–118), of the carbon components, a sequence of surface reactions incorporates the carbon atoms into the growing layer of the diamond crystal structure, which is discussed in the next section (see also Farouk, Nagayama, & Lee, 1995, 1996, 1997; Farouk et al., 1998; Farouk & Bera, 1999; Robertson, 2002).

9.10.5. Kinetics of Diamond-Film Growth

Kinetics of the surface reactions leading to crystal growth in the diamond film can be considered by analysis of growth of the (100) diamond plane from the plasma-generated CH_x ($x = 0$–3) components, which is illustrated in Fig. 9–47. First, it should be mentioned that a monohydrated diamond surface is thermodynamically more stable than a dihydrated surface. On such surfaces, at temperatures conventional for diamond film growth, the fraction of the surface centers (C–) with a hydrogen free bond is about 10–20%, while the fraction of the hydrogen-covered (C–H) centers is about 80–90%. Losses of hydrogen atoms due to their recombination through surface reactions (9–115) and (9–116) at conventional substrate temperatures of about 1200 K are characterized by a relatively high probability, in the range 10–20% (Mankelevich & Suetin, 2000). The major active carbon compound, generated in non-equilibrium H_2–CH_4 plasma, is the methyl radical. Its concentration far exceeds that of other CH_x compounds and that of acetylene (C_2H_2). When the CH_3 radical from the gas phase reacts with the aforementioned surface, it attaches to the surface in the form of CH_2, releasing a hydrogen atom. Attachment of the CH_2 group to the diamond surface occurs at the (C–) center of the dimer carbon pair, which is illustrated in Fig. 9–47. The attachment process proceeds through breaking of the dimer bond and incorporation of the CH_2 group in the form of a "bridge" between the two carbon atoms of the dimer pair. Formation of the bridges results in the creation of a new layer of the diamond lattice. Formation of the bridges above the dimer carbon pair proceeds very effectively and is characterized by a low-energy barrier. However, when vacancies above the dimer carbon pairs are occupied, growth of the new diamond layer slows down due to the fact that further incorporation of the CH_x groups between the dimers in the through site centers and between the bridge sites is kinetically limited by high-energy barriers. A significant contribution to the completion of the building of the new diamond layer is made by surface migration of the CH_2 groups, which leads to effective filling of the intermediate empty spots between the bridges. The attachment of acetylene (C_2H_2) to the diamond film surface (9–118) also makes an essential contribution to the growth of the (100) diamond plane. The process is dominated by surface migration of the $C=CH_2$ groups (Mankelevich & Suetin, 2000). Although the rate coefficient for C_2H_2

673

attachment to the diamond surface about 20 times lower than that of the methyl radicals, the total contribution of acetylene in the diamond film growth can be significant because its density in plasma volume can exceed that of CH_3 by about 10–100-fold. Limitation of the substrate temperature to the optimal range of 1000–1300 K is due to the following kinetically competing effects. A temperature increase above 1300 K is limited by intensive desorption of CH_3 radicals from the surface, which is a reverse process with respect to attachment (9–117). A temperature decrease below 1000 K leads to essential slowing down of various surface reactions required for the film growth. Also the lower temperature values lead to significant reduction of the CH_3 radical density in the gas phase over the substrate due to a faster three-body recombination of the methyl radicals with hydrogen atoms.

The key property of the diamond (and DLC) films is sp^3 bonding. The plasma deposition process which promotes the sp^3 bonding and, therefore, diamond synthesis is the ion bombardment (Fallon et al., 1993; Robertson, 1993a,b, 1994a,b; Lifshitz, 1996, 1999; McKenzie, 1996). The highest sp^3 fractions are formed by energetic C^+ ions with energies of around 100 eV. As noted by McKenzie, Muller, and Pailthorpe (1991) and McKenzie (1993), sp^3-bonded graphite occupies 50% more volume than sp^3-bonded diamond. Therefore, energetic ions, penetrating into the film, create compressive stress and stabilize the diamond phase (Davis, 1993). The process is essentially a subsurface one, and usually is referred to as low-energy subsurface implantation or simply **subplantation**. It has been shown that subplantation creates a metastable increase in density of the film, which tends to cause the bonding to change to sp^3 and promote synthesis of diamond crystal structure (Robertson, 1993a,b, 1994a,b, 1997; Silva, 1998).

PROBLEMS AND CONCEPT QUESTIONS

9–1. Pyrolysis of Hydrocarbons in Thermal Plasma. Using data from Tables 9–1–9–3 regarding the discharge power and energy cost of $C_2H_2 + C_2H_4$ synthesis, find the thermal plasma system where the absolute co-production of acetylene and ethylene is the largest. Which system represents the most energy-effective approach to co-production of acetylene and ethylene in thermal plasma?

9–2. Methane Conversion into Acetylene in Non-Equilibrium Plasma Conditions. Compare relations (9–14) and (9–15) to estimate the relative accuracy of logarithmic approximation (9–15) for calculations of the CH_4 dissociation rate coefficient at $T_v > T_0$. Why can't the logarithmic approximation be applied to describe methane conversion into acetylene in strongly non-equilibrium plasma-chemical system, when the vibrational temperature significantly exceeds the translational one?

9–3. Non-Thermal Plasma Synthesis of Formic Acid in CO_2–H_2O Mixture. Determine the minimum energy efficiency of the plasma-chemical HCOOH synthesis in the CO_2–H_2O mixture (9–60) required for effective hydrogen production in the double-step cycle (9–60) and (9–61). Assume that thermodynamically about 70% of the total energy required for hydrogen production from water should be consumed in this case for decomposition of formic acid (9–61) to form hydrogen and to recycle carbon dioxide back to the plasma process.

9–4. Mechanism and Kinetics of Plasma Polymerization. Using kinetic relation (9–66), find the plasma polymerization regimes (derive criteria for the regimes) when the polymer film deposition rate depends on and is proportional to (1) only the flux of charged particles, (2) only the flux of relevant neutral species, and (3) fluxes of both active neutral and charged particles.

9–5. Temperature Dependence of Electric Conductivity of Plasma-Polymerized Films. The electric conductivity of plasma-polymerized films is low and strongly depends on temperature following the Arrhenius formula $\sigma = \sigma_0 \exp(-E_a/T_0)$. Comparing values of electric conductivity presented in Table 9–9 for two temperatures, calculate activation energies E_a and pre-exponential factors σ_0 for plasma-polymerized naphthalene, styrene, ethylene oxide, and chlorobenzene. Taking into account the high concentration of radicals in the films, give your interpretation of why such different polymers have relatively close values of activation energy (around 1 eV).

9–6. Depth of Plasma Modification of Polymer Surfaces. Analyze the double logarithmic dependence of the modification depth x on the polymer modification degree (9–87). Based on the analysis, explain the sharp front of plasma modification of polymer materials. Compare for illustration the depth of polymer treatment corresponding to its slight modification ($K = e$) and very significant modification ($K = 1000$). Assume a total flux of electrons $N = 10^{13}\ \mathrm{cm}^{-2}$ and effective cross-section of modification of the macromolecules, $\sigma_{i0} \approx 10^{-9}\ \mathrm{cm}^2$.

9–7. Permeability of Plasma-Treated Gas-Separating Polymer Membranes. Analyzing relation (9–97), compare the influence of sizes of penetrating molecules and cross-links on the permeability of plasma-treated gas-separating polymer membranes. Why do these two sizes appear in relation (9–97) for gas permeability through the membrane in a non-symmetric way with significantly different powers?

9–8. Threshold Effect of Plasma Treatment on Selectivity of Gas-Separating Polymer Membranes. Using relation (9–107), calculate the threshold of the γ-factor (which characterizes the volume fraction of the elastic medium occupied by the quasi-rigid globules, representing the cross-links) corresponding to the beginning of exponential growth of selectivity of a gas-separating membrane. Take numerical values of parameters specifically for lestosil, which can be found in the end of Section 9.9.9.

9–9. Effect of Through Pores on Selectivity of Plasma-Treated Polymer Membranes Applied for Gas Separation. Using relation (9–109), find out the maximum value of selectivity of the plasma-treated polymer membranes considered as a function of the factor γ, which characterizes the specific energy input and density of cross-links.

9–10. Contribution of Methyl Radicals (CH_3) and Acetylene (C_2H_2) in Non-Equilibrium Plasma-Chemical Deposition of Diamond Films. Taking into account the reaction rate coefficients of attachment of methyl radicals (9–117) and acetylene (9–118) to the diamond film surface as well as typical molar fractions of CH_3 radicals and C_2H_2 (see Table 9–23), calculate the relative contribution of these major active carbon-containing chemical species to the deposition of diamond films from a H_2–CH_4 mixture in non-equilibrium plasma conditions.

10

Plasma-Chemical Fuel Conversion and Hydrogen Production

10.1. PLASMA-CHEMICAL CONVERSION OF METHANE, ETHANE, PROPANE, AND NATURAL GAS INTO SYNGAS (CO–H$_2$) AND OTHER HYDROGEN-RICH MIXTURES

10.1.1. General Features of Plasma-Assisted Production of Hydrogen from Hydrocarbons: Plasma Catalysis

Industrial demand for H$_2$ and H$_2$-rich gases is growing, resulting in the development of the hydrogen energy concept in energy production and transportation (Hoffmann & Harkin, 2002; Ewing, 2004). Hydrogen production plays a key role in the development of fuel cell technology and should be especially noted (Busby, 2005). Also the conversion of natural gas into syngas (CO–H$_2$ mixture) plays an important role in natural gas liquefaction and numerous processes of organic synthesis (Smith, 2001). Conventional thermo-catalytic technology for H$_2$ production is limited by relatively low specific productivity, high metal capacity, and the large equipment size, which becomes especially important in the case of small- and medium-scale H$_2$-generation systems. The application of plasma in the production of hydrogen and syngas creates an attractive alternative to conventional thermo-catalytic technologies (Deminsky et al., 2002). Plasma stimulation of hydrogen-rich gas production results in a significant increase of specific productivity and reduces the capital and operational costs of the process. The use of traditional catalysis is often limited by a time delay based on heating the catalyst to the required high temperature, which is especially critical in transportation applications such as hydrogen production from hydrocarbons on board a vehicle. The plasma method of hydrogen production has almost no inertia, which makes the plasma approach very attractive to the automotive industry. The application of non-thermal plasma for fuel conversion and hydrogen production is especially effective because plasma is used not as a source of energy but as a non-equilibrium generator of radicals and charged and excited particles. These cold plasma-generated active species can lead to long-chain reactions of fuel conversion and hydrogen production. The energy required for fuel conversion and hydrogen production can be provided mostly by chemical energy of reagents and low-temperature heat in non-thermal plasma. The plasma-generated active species just stimulate this process and contribute only a very small fraction (on the level of only a couple percent) of the total process energy. This effect is usually referred to as **plasma catalysis**. Different kinetic mechanisms can be responsible for plasma catalysis, including the generation of atoms, radicals, and excited and charged species. Charged and excited particles are able to stimulate chain processes, which cannot be accomplished in conventional chemistry. Plasma charged particles are especially important when taking into account exothermic ion–molecular reactions that proceed without activation barriers. This mechanism of plasma catalysis has been discussed in Section 2.2.6. Generally speaking,

Table 10–1. Energy Characteristics of Some Specific Plasma-Chemical Processes of Fuel Conversion and Hydrogen Production

Fuel conversion/hydrogen production process	Enthalpy, ΔH, eV/mol	Process temperature, T, K	Energy for heating, ΔQ, eV/mol	Combustion enthalpy of initial fuel, eV/mol
Pyrolysis	All endothermic			
$CH_4 \rightarrow C(s) + 2H_2$	0.9	1000–1100	0.39	−8
$C_3H_8 \rightarrow 3C(s) + 4H_2$	1.3	1000–1100	0.9	−20.4
$C_8H_{16} \rightarrow 8C(s) + 8H_2$	12.5	1000–1100	3.9	−51
Steam reforming	All endothermic			
$C(s) + H_2O \rightarrow CO + H_2$	1.3	900–1000	0.43	−3.9
$CH_4 + H_2O \rightarrow CO + 3H_2$	2.2	1000–1100	0.56	−8
$C_3H_8 + 3H_2O \rightarrow 3CO + 7H_2$	5.4	1000–1100	0.79	−20.4
$C_8H_{16} + 8H_2O \rightarrow 8CO + 16H_2$	12.5	1000–1150	3.9	−51
$CH_3OH + H_2O \rightarrow CO_2 + 3H_2$	0.7	500–600	0.33	−6.7
Partial oxidation	All exothermic			
$C(s) + 1/2\,O_2 \rightarrow CO$	−1.1		0.16	−3.9
$CH_4 + 1/2\,O_2 \rightarrow CO + 2H_2$	−0.2	1100–1500	0.34	−8
$C_3H_8 + 3/2\,O_2 \rightarrow 3CO + 4H_2$	−2.1	1100–1600	0.45	−20.4
$C_8H_{16} + 4\,O_2 \rightarrow 4CO + 8H_2$	−9.9	1100–1900	3.9	−51

the effect of plasma catalysis should not be confused with the plasma–catalytic approach to fuel conversion and hydrogen production, when plasma and catalytic technologies are combined in series to increase an effective process. This latter approach can also be useful in fuel conversion technologies and will be discussed later in this chapter.

Production of H_2-rich gases using plasma can be divided into three different approaches: pyrolysis (direct decomposition) processes with the production of carbon and hydrogen, steam (or CO_2) reforming of hydrocarbons with the production of syngas, and partial oxidation of hydrocarbons usually with the production of syngas. The energy characteristics of the processes are summarized in Table 10–1 (Deminsky et al., 2002), which indicates the enthalpy of the processes, the combustion enthalpy of initial fuel, the temperature required for thermal organization of the process, and the heat energy required to reach the temperature of the process. Fuel conversion and hydrogen production processes were examined both in thermal and non-thermal plasmas. Numerous experiments have been focused on the thermal plasma approach to the technology (see, for example, Alexeev, Souris, & Shorin, 1971, 1972; Bromberg et al., 2000; Brown, 2001). The degree of conversion for hydrocarbons into syngas is fairly high in these systems; the energy cost is about 13 MJ/kg H_2 (0.27 eV/mol) for catalytic air–steam oxidation, and 100–130 MJ/kg H_2 (2.1–2.7 eV/mol) for non-catalytic steam oxidation (Bromberg et al., 1999a,b; Deminsky et al., 2002).

For the case of partial oxidation of methane in quasi-thermal discharges, the energy cost of hydrogen production can be decreased to 7 MJ/kg H_2 (0.15 eV/mol; Bromberg et al., 2000). These energy costs are still higher than that of conventional thermo-catalytic processes, and to become competitive with conventional systems would require heat recuperation. The aforementioned energy costs are given with respect to a molecule of syngas (CO–H_2). It is assumed that CO in the plasma-produced syngas is converted into hydrogen by the conventional water–gas shift reaction:

$$CO + H_2O \rightarrow CO_2 + H_2, \quad \Delta H = -0.4 \text{ eV/mol.} \tag{10–1}$$

Table 10–2. Plasma-Catalytic Effect of Decomposition and Conversion of Different Fuels in Non-Equilibrium Pulsed Microwave Discharge

Plasma process	Plasma fraction in total energy	Conversion degree increase by plasma	Plasma energy cost of H_2 production, eV/mol
Methane Decomposition	5%	7% → 18%	0.1
Ethane Decomposition	5%	19% → 26%	0.25
Ethanol Decomposition	5%	23% → 62%	0.1
Methane Steam Conversion	5%	10% → 16%	0.35
Ethanol Steam Conversion	1%	41% → 58%	0.1

The shift reaction usually requires temperatures of about 400°C and the application of chromium-promoted iron oxide catalysts. A significant decrease in CH_4 partial oxidation energy cost becomes possible with the application of non-thermal plasma catalysis, which can be illustrated using experiments with non-thermal pulsed microwave discharges as an example. The hydrocarbons have been preheated in these experiments to relatively low temperatures of about 700–1000 K and then treated with pulsed non-equilibrium microwave discharge. Even a very small addition of plasma energy (5% or less from the total energy) leads to a significant increase in the degree of conversion and a sharp decrease of the overall hydrogen energy cost (Deminsky et al., 2002). This specific plasma-catalytic effect of decomposition and conversion of different fuels is illustrated in Table 10–2.

10.1.2. Syngas Production by Partial Oxidation of Methane in Different Non-Equilibrium Plasma Discharges, Application of Gliding Arc Stabilized in Reverse Vortex (Tornado) Flow

Partial oxidation of methane is a slightly exothermic process:

$$CH_4 + 1/2O_2 \rightarrow CO + 2H_2, \quad \Delta H = -0.2 \text{ eV/mol}. \quad (10–2)$$

Air is often used in plasma-chemical systems as a source of oxygen; the oxygen-to-carbon ratio in the stoichiometric mixture is $[O]/[C] = 1$. Taking into account the exothermicity of the partial oxidation (10–2), the process can be realized using plasma catalysis, that is, with very low energy input and through stimulation of the low-temperature long-chain reactions by selective generation of radicals and charged and excited particles. Atmospheric- and near-atmospheric-pressure non-equilibrium discharges can be effective for this purpose, including pulsed corona discharges (Mutaf-Yardimci et al., 1998; Okumoto et al., 1999, 2001; Mutaf-Yardimci, 2001), microwave discharges (Deminsky et al., 2002; Tsai et al., 2005), and gliding arc discharges (Czernichowski, 2001; Iskenderova, et al., 2001; Czernichowski & Wesolovska, 2003; Kalra, 2004; Kalra, Gutsol, & Fridman, 2005); the process in lower-pressure non-equilibrium radiofrequency (RF) and pulsed discharges has been demonstrated by Tsai and Hsieh (2004), Da Silva et al. (2006), and Yao, Nakayama, and Suzuki (2001a,b). While high-energy-efficiency plasma catalysis of the partial oxidation process (10–2) is mostly relevant to the non-thermal discharges, industrial applications of the technology usually require relatively high productivity of syngas and, therefore, a relatively high power of the applied discharges. The gliding arc discharges are well suited for this due to their ability to remain strongly non-equilibrium and mostly non-thermal even at relatively high power levels (see Chapter 4 for more details; Fridman & Kennedy, 2004). The most practically attractive yield and energy efficiency characteristics of the CH_4 partial oxidation process (10–2) and syngas production have been achieved by using the non-equilibrium gliding arc discharge stabilized at atmospheric-pressure in the reverse vortex (tornado) flow (Kalra, 2004; Kalra et al., 2005; also see Section 4.3.11). Before

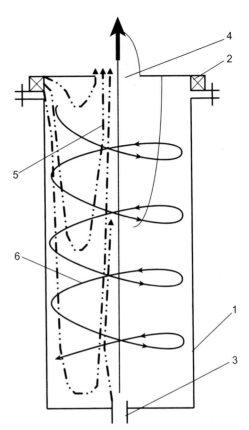

Figure 10–1. Schematic of a reverse vortex flow system: (1) cylindrical chamber walls; (2) major tangential gas flow inlet; (3) axial gas flow inlet; (4) axial flow exit (from the same side as major tangential inlet); (5) dashed lines representing the flow streamlines on the axial plane; (6) solid arrows representing three-dimentional sketch of rotating flow near the walls.

analyzing the energy efficiency and yield characteristics of the process, let us first discuss the organization of the plasma-chemical reactor for this case.

A schematic of the reverse vortex (tornado) flow is shown in Fig. 10–1. To obtain the reverse vortex flow, pressurized gas enters the cylindrical volume tangentially through the inlet (2) and exits (4) from the top of the volume on the same side that the tangential gas entry is placed. The diameter of the exit is considerably smaller than that of the cylindrical vessel. Arrows in Fig. 9.48 show the flow direction: the solid arrows (6) represent a three-dimensional sketch of rotating flow in the region near the wall, and the dotted arrows (5) show the streamlines on the axial plane. The "tornado" flow provides high gas velocities necessary for motion of the non-equilibrium gliding arc and effective heat and mass exchange at the central zone of the plasma caused by the fast radial migration of the turbulent microvolumes decelerated near the tube walls. The design of the reactor stabilizing the non-equilibrium gliding arc inside the reverse vortex flow is presented in Fig. 10–2 (Kalra, 2004; Kalra et al., 2005). The reactor is made of a quartz tube with internal diameter 40 mm and length 150 mm (1), which encloses a cylindrical volume (2). At one end of the tube, a swirl generator with tangential inlet holes injects air or reacting gases. The gas outlet (5) is on the same side as the swirl generator, setting up the reverse vortex flow. An additional axial inlet (4) is located at the bottom end of the cylindrical reactor. Typical swirl velocities vary from 10 to 60 m/s; axial velocities are from 2 to 10 m/s. A gliding arc requires electrodes to be in the "plane" of the gas flow. In the case of reverse vortex, this "plane" is cylindrical and parallel to the walls. Thus, the spiral cathode (6, in Fig. 10–2b) is placed inside of the cylindrical volume coaxially with the discharge tube. The helix angle of the spiral electrode is made identical to the flow pattern to avoid its disturbance. The spiral and ring electrodes (6 and 8, in Fig. 10–2) are connected to the high-voltage power supply, while the flat circular disk electrode (7) is

679

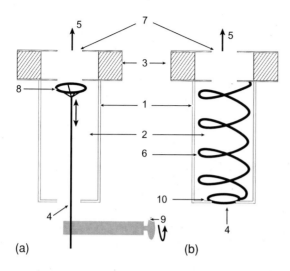

Figure 10–2. Reverse vortex flow system for gliding arc discharge stabilization: (a) configuration with a movable ring electrode; (b) configuration with a spiral electrode. In the figures, (1) quartz tube; (2) cylindrical reactor volume; (3) swirl generator with tangential inlet holes; (4) additional axial gas flow inlet; (5) gas flow exit; (6) spiral electrode; (7) flat circular disk electrode; (8 and 10) ring electrodes; (9) the ring electrode position control.

used as the ground. In the spiral electrode configuration, initial breakdown starts between the top circular disk electrode and the tip of the spiral electrode. This gliding arc configuration can be interpreted as diverging electrodes in circular geometry. At the opposite end of the spiral, a ring electrode (10) is placed. This electrode is smaller in diameter than the spiral one; therefore, when the arc is fully elongated it can be stabilized when attached to the ring. A similar gliding arc column can be created by using a simple ring cathode, which can be moved away from the anode (Fig. 10–2a). Initially in this configuration the cathode ring is about 3 mm away from the anode disk (7) to ignite the arc. Due to the reverse vortex flow, the arc rotates on the ring electrode, remaining more or less constant in length. When the cathode ring is moved away from the anode, the gliding arc is elongated to reach the non-equilibrium plasma regime optimal for partial oxidation of natural gas or other hydrocarbons (see Fig. 10–3). Experimental data on partial oxidation of CH_4 by atmospheric air in a reverse-vortex (tornado) non-equilibrium gliding arc discharge is to be discussed later.

Figure 10–3. Photo of the gliding arc "tornado" discharge in configuration with the movable ring electrode (high power, but you can touch it because of effective heat insulation of the walls by the reverse vortex flow).

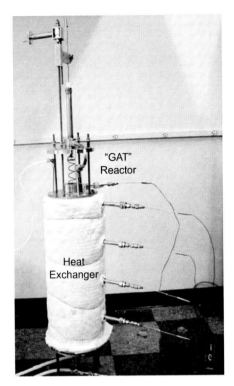

Figure 10–4. Schematic of the gliding arc tornado (GAT) reactor with a heat exchanger for effective fuel conversion into syngas.

10.1.3. Plasma Catalysis for Syngas Production by Partial Oxidation of Methane in Non-Equilibrium Gliding Arc Stabilized in Reverse Vortex (Tornado) Flow

The methane partial-oxidation process (10–2) for a methane–air mixture is illustrated in Fig. 10–4, which consists of the gliding-arc-tornado (GAT) reactor, heat exchanger for internal heat recuperation, power supply (10 kV, 1 A), and instrumentation and data acquisition elements (Kalra, 2004; Kalra et al., 2005). The GAT reactor is made from quartz, includes a movable circular electrode, and has volume 0.2 L. Methane and air are injected into the GAT reactor through the inlet tubes connected to a well-insulated counterflow heat exchanger. The non-equilibrium plasma reactor operates in fuel-rich conditions ([O]:[C] ≈ 1; see chemical formula (10–2)). Gases enter the system at room temperature. After preheating in the heat exchanger, gases flow into the plasma reactor, typically at 700 K, and leave it at 1050 K; the final outlet temperature from the heat exchanger is 450 K. The methane-to-syngas conversion as a function of the [O]:[C] ratio in the inflow is presented in Fig. 10–5. The conversion rises at first with the [O]:[C] ratio due to an increase of the process temperature, but at [O]:[C] > 1.35 the conversion starts decreasing because of CO_2 and water formation. Maximum methane-to-syngas conversion in the non-equilibrium plasma system is quite high at 80–85% compared to 60%, which is the maximum for the thermal processes without application of non-thermal plasma (see Fig. 10–5). The absolute production rate of the syngas in the system is about 1 standard L/s with less than 50 ms residence time inside the reactor: Most syngas is produced directly in the GAT reactor; no more than 5% of the total conversion takes place in the heat exchanger. It should be noted that long and continuous operation of the methane partial-oxidation process in the non-thermal plasma of a gliding arc resulted in no erosion or corrosion of the electrode surface. Electric power input in the non-equilibrium GAT plasma is relatively low (about 200 W), indicating low electric energy cost of the partial-oxidation process and its plasma-catalytic nature.

681

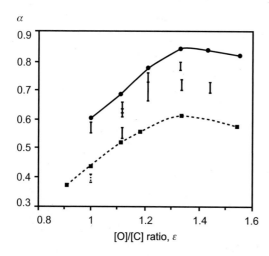

Figure 10–5. Methane conversion to syngas as a function of [O]:[C] ratio. Solid lines with circles represent the kinetic simulation results; solid bars represent the experimental results; dotted line and bar represent kinetic simulation and experimental results without plasma.

The electric energy cost of syngas production as function of the [O]:[C] ratio is shown in Fig. 10–6. Similar to the methane-to-syngas conversion dependence (Fig. 10–5), electric energy efficiency is also optimal at [O]:[C] = 1.3–1.4. The minimum electric energy cost is quite low, only about 0.09 kWh/m^3 of syngas. Taking into account that the total energy of 1 m^3 of syngas is about 3 kWh, only 3% of the energy is consumed in the form of electricity; everything else is chemical or thermal energy of the initial products. Non-equilibrium GAT plasma mainly stimulates the process kinetically by operating as a catalyst and does not contribute significant energy to partial oxidation. Therefore, the process can be qualified as plasma catalysis. The typical composition of the products is shown in Fig. 10–7 for different [O]:[C] ratios and different temperatures at the GAT reactor inlet. Partial oxidation of methane in the non-equilibrium gliding arc completely avoids soot formation in the relatively rich mixtures considered earlier and presented in Figs. 10–5–10–7. Kinetic analysis shows that syngas production in the system can be divided into two stages: first, the short combustion stage, and second, the longer reforming stage. During the combustion stage, densities of atomic oxygen and OH radicals are relatively large, which results in the kinetics resembling that of combustion. This phase is highly exothermic and, therefore, gas temperature grows. The first stage ends when no more molecular oxygen is available. The second stage proceeds in the absence of oxygen, where the rest of the methane should be oxidized and converted into syngas (H$_2$–CO mixture) by water vapor and CO$_2$ formed during the first combustion phase. The reforming reactions are endothermic, causing an observable temperature decrease and slowing down the partial-oxidation process. Detailed

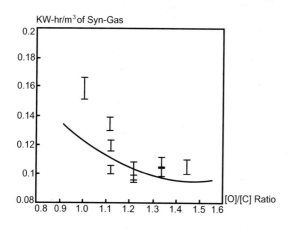

Figure 10–6. Electric energy cost of syngas production as function of [O]:[C] ratio.

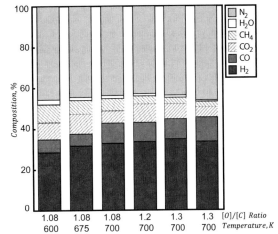

Figure 10–7. Outflow composition at different [O]:[C] ratios and inlet temperatures.

kinetic modeling of the non-equilibrium plasma-chemical process can be found in Kalra (2004). The methane partial-oxidation process (10–2) produces syngas with H_2:CO ratio 2:1. If for a specific practical application (i.e., fuel cells) it is preferable to produce mostly hydrogen, water vapor can be added to stimulate the shift reaction (10–1), converting CO into hydrogen (Ouni et al., 2005a,b). The more desirable outcome is if the H_2O reacts not only with CO, shifting products to hydrogen, but also with methane to increase syngas yields and specifically hydrogen yields. This is feasible given the exothermic nature of the partial-oxidation process (10–2). Generally, the application of non-equilibrium gliding arc plasma stabilized in reverse-vortex (tornado) flow can achieve very attractive parameters for methane conversion into syngas (H_2–CO) by partial oxidation. The plasma-assisted process allows for almost complete conversion without the use of any conventional catalysts. The system operates at relatively low temperature levels (about 300–400°C) with plasma providing only 2–3% of the total process energy.

10.1.4. Non-Equilibrium Plasma-Catalytic Syngas Production from Mixtures of Methane with Water Vapor

Syngas can be produced in the endothermic process of methane oxidation by water vapor, often referred to as the **steam reforming of methane**:

$$CH_4 + H_2O \rightarrow CO + 3H_2, \quad \Delta H = 2.2 \, \text{eV/mol}. \quad (10-3)$$

Steam reforming can be effectively carried out in plasma and especially in non-equilibrium discharges, including microwave (Givotov et al., 2005) and pulsed corona discharges (Mutaf-Yardimci et al., 1998, 1999; Mutaf-Yardimci, 2001). In contrast to methane partial oxidation (10–2), steam reforming is endothermic and, therefore, requires significant energy consumption. To illustrate, the energy cost of the steam reforming (10–3) in atmospheric-pressure thermal plasma is presented in Fig. 10–8 as a function of specific energy input in different quenching modes (Nester et al., 1988). Minimum values of syngas energy cost – 0.59 eV/mol H_2–CO (ideal quenching) and 0.69 eV/mol H_2–CO (absolute quenching) – are relatively high when the energy should be provided in the form of electric energy. Again, the situation can be improved by using non-thermal discharges in the plasma catalysis regime as discussed in Section 10.1.1. The plasma-catalytic approach for the case of endothermic reactions means most of the energy required for the process is provided in the form of reasonably low-temperature heat and not as electric energy going through the electric discharge. Plasma catalysis of the endothermic process (10–3) and

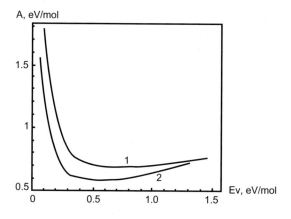

Figure 10–8. Energy cost of production of a syngas molecule (H_2–CO mixture) in the thermal plasma process of steam reforming of methane, $CH_4 + H_2O \rightarrow 3H_2 + CO$: (1) absolute quenching; (2) ideal quenching.

other hydrocarbon conversion reactions supported by a combination of low-temperature heat and plasma energy has been demonstrated in particular by using pulsed microwave discharge (Givotov et al., 2005). The initial products in these experiments with flow rates of 50–200 cm^3/s at atmospheric-pressure are preheated to 360–570°C and then treated in the pulsed microwave discharge (frequency of 9 GHz, power in the pulses up to 100 kW, pulse duration 0.3–1μs, pulse repetition frequency of 1 kHz, average power up to 30–100 W). In the experiments with steam reforming of methane (10–3), the preheating temperature is 500–570°C, the ratio of the microwave power and heating power is $P_p/P_T = 5$–10% , and the ratio of flow rates of CH_4:H_2O varies from 1:1 to 1:2.

Gas-phase products are syngas (H_2, CO) and some carbon dioxide (CO_2). The molar fraction of molecular hydrogen produced is presented in Fig. 10–9 as a function of specific energy input to the microwave discharge. For comparison purposes, the thermal production of hydrogen due to only the thermal preheating effect is presented in the same figure. The added microwave discharge energy, which is only a small fraction of the total energy ($P_p/P_T = 5$–10%), decreases the energy cost of the process from 30 to 60%, which proves the plasma-catalytic nature. The plasma-chemical syngas production process (10–3) is accompanied in the experiments with some soot production. The fraction of solid carbon (soot) in the products decreases from 8 to 1% by increasing the preheating temperature from 500 to 570°C. Similar non-equilibrium plasma-chemical approaches to steam reforming of methane have been developed using non-equilibrium pulsed corona, sparks, and gliding arc discharges (Fridman et al., 1999; Czernichowski, 1994; Mutaf-Yardimci, 2001; Sekine et al., 2004). In some of the experiments, process (10–3) did not require preheating. Methane reformation can be quite intensive in this case; however, the plasma-catalytic effect is absent, which results in significant consumption of electric energy. Effective reduction in

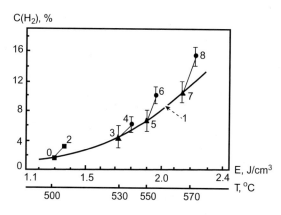

Figure 10–9. Concentration of H_2 for the process of methane conversion in mixture with water vapor as function of specific energy input. Curve 1 represents experiments with purely thermal energy input; points 2, 4, 6, and 8 represent experiments with a discharge.

Table 10–3. Conversion and Energy-Efficiency Characteristics of Methane Reforming into Syngas in Gliding Arc Discharge in CO_2–CH_4 Mixture

CO_2:CH_4 initial molar ratio	Specific energy input, $\frac{MJ}{m^3}(\frac{KWh}{m^3})$	Energy efficiency, %	Moles of CO_2 produced from 100 moles of inlet gas	Moles of H_2O produced from 100 moles of inlet gas	Moles of CO produced from 100 moles of inlet gas
0.55	3.36 (0.93)	23	28	1	13
1.13	3.40 (0.94)	33	42	3	19
2.17	2.79 (0.78)	42	55	5	22

the methane-reforming energy cost can be achieved by adding air (or oxygen) to the initial mixture. Production of syngas in this case would be due to the combination of endothermic steam reformation (10–3) with exothermic partial oxidation (10–2) in non-equilibrium plasma.

10.1.5. Non-Equilibrium Plasma-Chemical Syngas Production from Mixtures of Methane with Carbon Dioxide

In the non-equilibrium pulsed corona and gliding arc discharges, methane reformation was also analyzed in mixture with carbon dioxide (Czernichowski, 1994; Mutaf-Yardimci et al., 1998, 1999; Fridman et al., 1999; Mutaf-Yardimci, 2001):

$$CH_4 + CO_2 \rightarrow 2CO + 2H_2, \quad \Delta H = 2.6 \text{ eV/mol.} \tag{10-4}$$

The process is similar to steam reforming of methane (10–3), although it is more endothermic. In contrast to the steam reforming, the CO_2 reforming can be more effectively stimulated by vibrational excitation of the reagents (Rusanov & Fridman, 1984), taking into account the completely gas-phase nature of the reaction and effectiveness at sustaining vibrational–translational non-equilibrium in carbon dioxide (see Sections 5.2 and 5.3). The plasma-catalytic effect in non-equilibrium methane CO_2 reformation (10–4) by sustaining the process with a combination of plasma energy and low-temperature preheating was demonstrated by Mutaf-Yardimci (2001) using an atmospheric-pressure pulsed corona discharge.

Plasma processing of CO_2–CH_4 mixtures at atmospheric-pressure using the non-equilibrium gliding arc has been studied by Lesueur, Czernichowski, and Chapelle (1994). Gliding arc power in these experiments was up to 3 kW, CO_2–CH_4 molar ratios were from 0.55 to 2.17, and flow rates of the premixed CO_2–CH_4 mixture varied from 1.8 to 4.7 standard m^3/h. The major conversion and energy characteristics of the process are summarized in Table 10–3. The maximum energy efficiency of syngas production (the ratio of the reaction enthalpy and actual energy expenses) reached as high as 42% in these experiments. The table also shows the number of moles of CO_2, H_2O, and CO produced from 100 moles of inlet gas; the corresponding amount of hydrogen can be estimated from there. Plasma-chemical methane reformation by CO_2 (10–4) has also been investigated in the three-phase atmospheric-pressure gliding arc (Czernichowski, 1994). Typical results are presented in Table 10–4 at different values of process temperature in comparison with the relevant results of theoretical modeling of the process. The results correspond to a 2 kW gliding discharge power, and CO_2–CH_4 molar ratios varied in these systems from 0.5 to 2. Similar to the previous table, concentrations of major products of the reforming process are presented in the form of number of moles produced per 100 moles of the inlet gas. Generally, the application of non-equilibrium plasma allows for methane reformation with CO_2 at a relatively

Table 10–4. Experimentally Measured and Theoretically Calculated Concentrations of Main Products of Methane Reforming into Syngas in Gliding Arc Discharge in CO_2–CH_4 Mixture

CO_2:CH_4 initial molar ratio	CH_4 (exp, th) concentration in products (mol)	CO_2 (exp, th) concentration in products (mol)	CO (exp, th) concentration in products (mol)	H_2 (exp, th) concentration in products (mol)	H_2O (exp, th) concentration in products (mol)	Process temperature, K
0.5, experimental	44	19	9	12	0.2	810
0.5, theoretical	47	17	12	14	0.7	810
1.0, experimental	40	38	9	11	1.0	750
1.0, theoretical	41	38	12	11	2.9	750
1.5, experimental	33	46	9	9	2.8	750
1.5, theoretical	32	46	12	6	3.2	750
2.0, experimental	26	52	10	7	3.6	750
2.0, theoretical	26	52	13	6	3.5	750

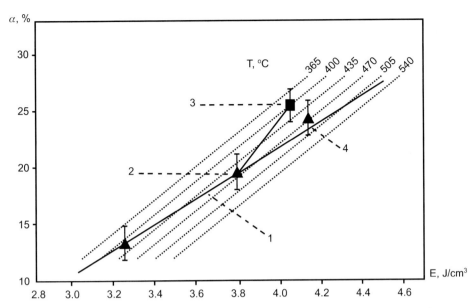

Figure 10–10. Conversion degree of ethane into hydrogen as a function of total specific energy input (preheating and energy input in the discharge): α, conversion degree of ethane into hydrogen; T, temperature.

low temperature (about 800 K) without using any conventional catalysts. Although the energy efficiency of the reformation exceeds 40%, the total consumption of electric energy can be relatively high in the process. Plasma catalysis allows for a significant decrease in the consumption of electric energy. The plasma-chemical process (10–4) consumes CO_2 during the fuel conversion, which can be attractive from the perspectives of environmental control. Produced syngas is not supposed to be burned in this case but rather used for further chemical synthesis.

10.1.6. Plasma-Catalytic Direct Decomposition (Pyrolysis) of Ethane in Atmospheric-Pressure Microwave Discharges

Direct decomposition (pyrolysis) of hydrocarbons with production of soot (or carbon) and hydrogen can also be qualified as fuel conversion processes. Whereas hydrocarbon combustion results in the production and exhaust of CO_2, pyrolysis produces hydrogen leading to CO_2-free clean combustion. The carbon produced in the process should be not burned, but rather used for chemical purposes. Hydrogen production by direct decomposition of hydrocarbons has been discussed in Section 9.1.7. Now we consider the effect of plasma catalysis on these processes. The experiments were performed in atmospheric-pressure non-equilibrium pulsed microwave discharge with controlled gas preheating as described in Section 10.1.4 (Givotov et al., 2005). Typically, microwave discharge power was on the level of 5% with respect to the total power introduced in the system. Figure 10–10 illustrates the plasma catalysis of hydrogen production in the direct decomposition (pyrolysis) of ethane:

$$C_2H_6 \rightarrow 2C(s) + 3H_2, \quad \Delta H = 0.85 \text{ eV/mol}. \tag{10–5}$$

The chart shows the dependence of ethane conversion into hydrogen, α (relative decrease of the C_2H_6 amount), on the total specific energy input, ε (including preheating and energy input in the discharge). Isothermal curves (from $T = 365$ to $T = 540°C$) represent functions $\alpha(\varepsilon)$ at fixed process temperature. Line segment 2–4 in the figure show the experimental dependence of $\alpha(\varepsilon)$ when energy is introduced in the system as purely thermal energy, which is spent on both temperature increase and conversion increase (curve

1 is theoretical dependence $\alpha(\varepsilon)$ for thermal processes). The introduction of thermal energy (by either additional external heating or thermal plasma heating) leads to temperature increases from $T = 470°C$ at point 2 to about $T = 500°C$ at point 4; conversion grows from $\alpha = 19\%$ to $\alpha = 22–23\%$, respectively. The experimental line segment 2–3 represents the plasma-catalytic effect of non-equilibrium discharge on the gas initially preheated at point 2 to 470°C ($\varepsilon = 3.75$ J/cm^3). A specific energy input in the discharge of $\Delta\varepsilon \approx 0.2$ eV/mol (which is about only 5% of the total energy in the system) results in the conversion growing from $\alpha = 19\%$ at point 2 to $\alpha = 25–26\%$ at point 3. Gas temperature decreases from 470°C at point 2 to about 400°C at point 3. Thus, intensive plasma production of the active species results in such a significant production of hydrogen that not only the plasma energy but also the thermal energy of the preheated gas is used for this purpose. This is a clear demonstration of plasma catalysis in the ethane decomposition process. Details on the chemical composition of products of the non-equilibrium plasma-chemical process can be found in Rusanov et al. (1997) and Babaritsky et al. (1999a,b).

The energy cost of the additional hydrogen produced in process 2–3 (Fig. 10–10) by the microwave discharge specific energy input $\Delta\varepsilon \approx 0.2$ eV/mol is only 0.26 eV/mol H$_2$. Taking into account the specific composition of gas-phase products in these experimental conditions (47% H$_2$, 20% C$_2$H$_4$, 3% CH$_4$, 30% non-reacted C$_2$H$_6$), the enthalpy of the process is 0.65 eV/mol H$_2$ and it can be concluded that the energy efficiency is 250% (much exceeding 100%). This plasma-catalytic effect doesn't interdict the First or Second Laws of thermodynamics. The rest of the energy required for hydrogen production is taken from the thermal reservoir (preheated gas), somewhat similar to heat pumps. The total energy cost of hydrogen production, including both preheating thermal energy and discharge energy, also decreases in the case of plasma catalysis. Point 2 in Fig. 10–10 shows that thermal decomposition of ethane without plasma catalysis results in a hydrogen energy cost of 1.5 eV/mol H$_2$. Point 3, representing a combination of plasma with preheating, corresponds to a hydrogen energy cost of 1.2 eV/mol H$_2$. Thus, non-equilibrium plasma (with power only 5% of total thermal power of the system) results in a 30% reduction of the total energy cost of hydrogen produced in the atmospheric-pressure pulsed microwave discharge. Carbon has been experimentally produced in the process partially in the form of soot C(s) and partially in the form of gas-phase compounds (C(g): CH$_4$ and C$_2$H$_4$). The atomic ratio C(s):C(g), with and without discharge, was about the same (Givotov et al., 2005): 44% at point 3 and 37% at point 2 (Fig. 10–10).

10.1.7. Plasma Catalysis in the Process of Hydrogen Production by Direct Decomposition (Pyrolysis) of Methane

Similar to the pyrolysis of ethane (10–5), effective plasma catalysis takes place in the direct plasma decomposition of methane with the production of soot and hydrogen:

$$CH_4 \rightarrow C(s) + 2H_2, \quad \Delta H = 0.7 \text{ eV/mol.} \tag{10–6}$$

The plasma-catalytic effect has been observed in an atmospheric-pressure non-equilibrium pulsed microwave discharge with preheating (see Section 10.1.4; Givotov et al., 2005). Methane flow rates in the system were 30–250 cm^3/s and preheating temperatures of methane were 400–600°C. Experimental results of pyrolysis are presented in Fig. 10–11 (similar to Fig. 10–10) as a chart showing the dependence of the methane conversion, α, on the total specific energy input, ε, and including preheating energy and energy input in the discharge.

Line 1 in Fig. 10–11 represents the thermodynamic equilibrium conversion of methane as a function of specific energy input (enthalpy). Line 2 in the figure represents experimental results achieved by thermal preheating without discharge, which leads to growth of methane conversion and increase of temperature at the same time. Isothermal lines 4, 5, and others

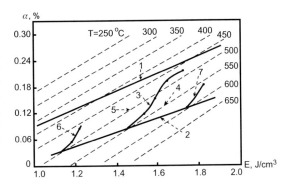

Figure 10–11. Plasma conversion of methane: results of experiments and simulations presented as the methane conversion dependence on specific energy input: (1) thermodynamic calculations of the quasi-equilibrium states; (2) experiments with purely thermal energy input; (3, 6, 7) experiments with a discharge; (4, 5) theoretical lines representing isotherms (fixed temperature T).

represent the $\alpha(\varepsilon)$ dependence at fixed temperatures in the range between 250 and 650°C. Curves 3, 6, and 7 represent preheating to temperatures 550, 500, and 600°C, respectively, followed by the application of the microwave discharge at different powers. Similar to ethane pyrolysis, curves 3, 6, and 7 show that the application of non-equilibrium plasma results in a temperature decrease, meaning that a small amount of plasma energy ($< 20\%$) initiates intensive contribution of CH_4 thermal energy into the endothermic pyrolysis (10–6). Non-equilibrium plasma shifts the conversion curves closer to thermodynamic equilibrium (line 1), demonstrating the effect of plasma catalysis. The energy cost of hydrogen production (10–6) is shown in Fig. 10–12 as a function of specific energy input in plasma in the regime corresponding to point 3 of Fig. 10–11. As one can see from the upper curve of the figure, adding only 15% plasma energy (with respect to thermal energy) allows for a two-fold decrease of the energy cost of hydrogen production. The lower curve on Fig. 10–12 represents the electric (plasma) energy cost of hydrogen production, which is lower than the enthalpy of the process (10–6). Specifically when the fraction of plasma energy is 10%, the temperature during the process decreases from 550 to 480°C, and the electric (plasma) energy cost of hydrogen production is only 0.2 eV/mol H_2 (see (10–6) to compare). This proves that most of the required energy is provided by the thermal reservoir, while the effect of plasma is mostly catalytic.

10.1.8. Mechanism of Plasma Catalysis of Direct CH_4 Decomposition in Non-Equilibrium Discharges

Figure 10–11 shows that the application of non-equilibrium plasma effectively increases methane conversion (from curve 2 to curve 1), which approaches the thermodynamic limit. Thus, non-equilibrium plasma operates in this case as a catalyst, which explains why the

Figure 10–12. Energy cost of the plasma-chemical pyrolysis of methane as a function of the specific plasma energy input: upper curve, total energy cost of the process with respect to the total produced hydrogen; lower curve, electrical energy cost with respect to additional hydrogen produced by the discharge.

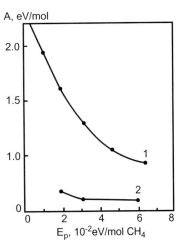

process is called "plasma catalytic". The temperature of the preheated methane decreases during the plasma-catalytic process, because a fraction of the thermal reservoir energy is spent to increase methane conversion to hydrogen in the endothermic reaction (10–6). Different mechanisms of the non-equilibrium plasma-catalytic effect in direct methane decomposition (10–6) have been analyzed by Givotov, Potapkin, and Rusanov (2005), including process stimulation by radicals and excited atoms and molecules, non-uniform temperature distribution in the discharge, stimulation of CH_4 decomposition on the surface of carbon clusters, and finally process stimulation by the ion–molecular Winchester mechanism (see Section 2.2.6). Comparison with experimental data shows that the ion–molecular Winchester mechanism makes the biggest contribution to plasma catalysis of methane decomposition with hydrogen production (10–6). Although the role of the ion–molecular reactions of negative ions has been discussed, it is concluded that the CH_3^+ ion radical stimulates the most effective chain of methane conversion into carbon clusters and hydrogen. The initial generation of CH_3^+ ion radicals is mostly due to dissociative ionization by electron impact:

$$e + CH_4 \rightarrow CH_3^+ + H + e + e. \tag{10–7}$$

Hydrogen production and carbon cluster formation then proceeds by the [CH_3^+ ion radical reaction with methane, leading to incorporation of a CH_2 group into the growing ionic complex and releasing an H_2 molecule:

$$CH_3^+ + CH_4 \rightarrow C_2H_5^+ + H_2, \quad k = 7.24 \cdot 10^{-14} cm^3/s. \tag{10–8}$$

Further growth of the ion clusters (to sizes of about C_{20}) are related to the sequence of the following attachments of CH_2 groups accompanied by hydrogen production:

$$C_nH_{2n+1}^+ + CH_4 \rightarrow C_{n+1}H_{2n+3}^+ + H_2(n > 1), \quad k = 6 \cdot 10^{-13} \exp\left(-\frac{600\,K}{T}\right) cm^3/s. \tag{10–9}$$

Methane molecules can be directly attached to the C_nH_{2n+1} ion clusters:

$$C_nH_{2n+1}^+ + CH_4 \rightarrow [C_{n+1}H_{2n+5}^+]^*, \quad k \approx 6 \cdot 10^{-9} cm^3/s. \tag{10–10}$$

The excited complexes $C_{n+1}H_{2n+5}^+$ are unstable with respect to decomposition, rearrangements, and hydrogen desorption (10–9). Termination of the sequence of reactions is due to recombination of charged particles:

$$e + C_{n+1}H_{2n+3}^+ \rightarrow C_{n+1}H_{2n+2} + H,$$
$$e + CH_3^+ \rightarrow CH_2 + H, \quad k = 2 \cdot 10^{-7}\left(\frac{300\,K}{T}\right)^{1.5} cm^3/s. \tag{10–11}$$

The Winchester mechanism (10–7)–(10–11) of methane conversion into hydrogen with growth of the carbon cluster (Section 2.2.6) can be interpreted as a chain reaction with multiple uses of a positive ion. Decrease of the ionization potentials in the sequence $C_nH_{2n+1}^+$ makes chain propagation (10–9) exothermic (Winchester mechanism) and fast because the ion–molecular reactions have no significant activation barriers. Additional information regarding the kinetics of ion–molecular reactions in methane and its derivatives can be found, in particular, in Hamlet and Moss (1969); Davis and Libby (1966); Sheridan, Greer, and Libby (1972); Olah and Schlosberg (1968); Adams and Babcock (2001); and Harioka and Kebarle (1976, 1977).

10.1.9. Plasma-Chemical Conversion of Propane, Propane–Butane Mixtures, and Other Gaseous Hydrocarbons to Syngas and Other Hydrogen-Rich Mixtures

Plasma-chemical conversion of gaseous hydrocarbons into syngas and hydrogen-rich mixtures is not limited to methane, natural gas, or ethane. Effective conversion and

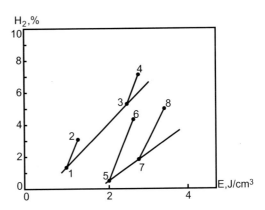

Figure 10–13. Concentration of hydrogen (H_2) during plasma-chemical conversion of the propane–butane mixture as function of specific energy input. Curves 1–3 and 5–7 correspond to experiments with only thermal energy input; points 2, 4, 6, and 8 correspond to experiments with the discharge.

plasma-catalytic effects have been observed in the conversion of other gaseous and relatively light hydrocarbons.

Partial oxidation of propane has been investigated in particular in atmospheric-pressure gliding arc discharges (Czernichowski, Czernichowski, & Wesolowska, 2003a,b) and in transitional arcs (Bromberg, Hadidi, & Cohn, 2005). The experiments with a 0.1 kW gliding arc were performed with a propane–air mixture. Based on the gas input/output analysis and mass balance, the process can be presented as follows:

$$C_3H_8 + 1.85\,O_2 \rightarrow 2.7\,CO + 3.6\,H_2 + 0.3\,CO_2 + 0.4\,H_2O. \qquad (10\text{–}12)$$

The O/C ratio in the process is 1.23, which is close to the ideal partial oxidation reaction. The process produces nitrogen-diluted syngas containing up to 45% of $H_2 + CO$. The output syngas flow rate is 2.7 standard m^3/h, which is equivalent to 8.6 kW of output power. In terms of plasma catalysis, discharge power in this case (0.1 kW) is only 1.2% of the total power of the produced syngas. Reaction (10–12) proceeds with the total absence of soot.

The gliding arc discharge has also been applied for **partial oxidation of cyclohexane (C_6H_{12}), heptane (C_7H_{17}), and toluene (C_7H_8)** (Czernichowski, Czernichowski, & Czernichowski, 2002). The O/C ratio in these experiments was 0.9–1.2, preheating temperature was 295–345°C, and gliding arc power was about 0.3 kW. Flow rates of the produced syngas in these processes are up to 1.6 standard m^3/h, with an output syngas power of 5.3 kW. In terms of plasma catalysis, the discharge power (0.3 kW) contributes 5.7% of the total process power in the produced syngas. The processes had no soot formation even in the case of the syngas production from such cyclic organic compounds as toluene, which is highly likely to be converted to soot.

Partial oxidation of natural gas in plasma is not much different from that of methane. However, this process has some specific features related in particular to the presence of sulfur compounds (i.e., H_2S). The subject has been investigated and discussed by Czernichowski, Czernichowski, and Wesolowska (2004). It was shown that partial oxidation of hydrocarbons with production of syngas is high-sulfur tolerant. Plasma-catalytic **pyrolysis of a propane–butane mixture** was investigated in a non-equilibrium atmospheric-pressure pulsed microwave discharge (see Section 10.1.4; Givotov et al., 2005). Experimental results on hydrogen production (H_2 concentration in products) for the plasma treatment of the C_3H_8:$C_4H_{10} = 1$:1 mixture preheated to 360–420°C are presented in Fig. 10–13. Curves 1–3 and 5–7 on the figure represent experiments with pure thermal heating of the mixture (at two different conditions). In this case, the plasma-catalytic effect results in the faster-growing dependence of hydrogen concentration on specific energy input when the energy is introduced by plasma (paths 1–2, 3–4, 5–6, and 7–8). Plasma catalysis also leads to an increase of the total conversion degree of the propane–butane mixture, which increases

691

three-fold, while plasma adds only 12% to the total power. The major products of pyrolysis in this regime are hydrogen, ethylene, propylene, methane, and propane. Soot formation is not significant: only 7% of carbon atoms in products are in the form of soot; the rest is in the gas phase.

10.2. PLASMA-CHEMICAL REFORMING OF LIQUID FUELS INTO SYNGAS (CO–H$_2$): ON-BOARD GENERATION OF HYDROGEN-RICH GASES FOR INTERNAL COMBUSTION ENGINE VEHICLES

10.2.1. Specific Applications of Plasma-Chemical Reforming of Liquid Automotive Fuels: On-Board Generation of Hydrogen-Rich Gases

Liquid fuels, such as gasoline, diesel, biodiesel, diesel oil, kerosene, ethanol, and so forth can be effectively converted into syngas (CO–H$_2$) by plasma-assisted partial oxidation or steam–air conversion, similar to the gas conversion processes considered in Section 10.1. Such technologies of hydrogen-rich gas production can be applied in different industrial schemes, but they are especially interesting for transportation systems. The applications for transportation systems focus on either development of hydrogen filling stations or development of devices for on-board generation of hydrogen-rich gases. The systems for on-board generation of hydrogen-rich gases should meet two crucial requirements: the first is compactness and low weight of the equipment; the second is no inertia of the process or, in other words, the possibility of fast start. Both requirements can be met by using plasma stimulation of the fuel conversion processes. Plasma systems are characterized by specific productivity exceeding that of conventional catalytic systems by up to 1000-fold (Rusanov & Fridman, 1984) and have no technological inertia related to preheating. The plasma-chemical fuel-conversion processes are sulfur resistant (in contrast to catalysis), which is important when taking into account the composition of the automotive fuels. All of these facts explain why plasma-chemical systems attract significant attention today for application in on-board generation of hydrogen-rich gases from liquid automotive fuels.

Plasma-stimulated on-board generation of hydrogen-rich gases can be used in a number of applications to improve environmental quality and reduce total fuel, especially petroleum fuel consumption in internal combustion engine vehicles. Four major applications can be specifically mentioned in this regard. The first is the production of relevant hydrogen-rich fuel for fuel cells providing on-board electricity for automotive vehicles (Czernichowski, Czernichowski, & Wesolowska, 2004). Second, the generated syngas can be directly used in internal combustion engines as an admixture to the conventional fuel–air mixture to reduce toxic exhausts and improve the major engine characteristics (Bromberg et al., 1999a,b). Third, the on-board syngas plasma generated from liquid automotive fuel can be effectively applied for aftertreatment of particulates and NO$_x$ emissions. This is important because the engine in-cylinder techniques alone are currently not sufficient to meet stringent emission regulations especially for diesel vehicles (Bromberg et al., 2001, 2003, 2004). Finally, the syngas produced on board the diesel vehicles can be effectively used for regeneration of the diesel particulate filter (DPF; Bromberg et al., 2004). The effective conversion of liquid fuels (gasoline, diesel, biodiesel, diesel oil, kerosene, ethanol, etc.) into syngas (CO–H$_2$) by plasma-assisted partial oxidation or steam–air conversion becomes possible because of the plasma catalysis and because it is similar to the gas-conversion processes (Section 10.1). Consumption of the electric (plasma) energy in plasma-catalytic systems is only a small fraction of the total energy of the produced syngas. As a reminder, plasma catalysis is understood here as plasma stimulation of the process (see Section 10.1). It should be mentioned that liquid fuels can also be converted into syngas by the simple combination of plasma technology with conventional catalysis, which is to be discussed later in this section.

10.2. Plasma-Chemical Reforming of Liquid Fuels into Syngas

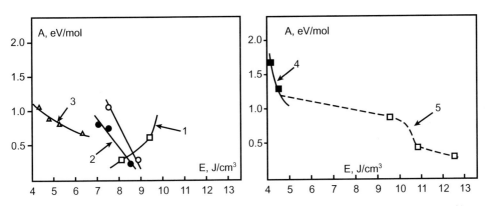

Figure 10–14. Plasma (electric) energy cost of syngas production during plasma conversion of kerosene as function of specific energy input in the system: (1) partial oxidation; (2) steam-oxygen conversion with preheating of air; (3) steam-oxygen conversion with excess water; (4) steam-conversion in low-temperature regime; (5) steam conversion in high temperature regime.

10.2.2. Plasma-Catalytic Steam Conversion and Partial Oxidation of Kerosene for Syngas Production

Effective plasma-catalytic conversion of such liquid fuel as kerosene (total composition $C_{10.78}H_{22.33}$, for simplicity $C_{11}H_{22}$; with major components undecane ($C_{11}H_{24}$), 21%; decane ($C_{10}H_{22}$), 14%; and dodecane ($C_{12}H_{26}$), 14%) into syngas has been achieved, in particular, in the non-equilibrium atmospheric-pressure pulsed microwave discharge, described in Section 10.1.4 (Givotov et al., 2005). The process has been investigated in regimes of partial oxidation and steam reformation with different fractions of fuel, air, and water vapor. The endothermic process of kerosene steam reformation can be represented by the chemical formula

$$C_{11}H_{22} + 11H_2O \rightarrow 11CO + 22H_2. \tag{10–13}$$

The exothermic plasma-stimulated process of partial oxidation of kerosene can be represented as

$$C_{11}H_{22} + 5.5\,O_2(+N_2) \rightarrow 11CO + 11H_2(+N_2). \tag{10–14}$$

The actual kerosene conversion is a combination of reactions (10–13) and (10–14); that is, as a steam–air conversion process. The more oxygen introduced in the reaction, the higher the process exothermicity and the higher the process temperature. Temperature control is especially important because of possible soot formation when the temperature of the fuel-rich mixture in thermodynamic quasi equilibrium is below some critical level. This minimum quasi-equilibrium temperature for steam–air reforming of kerosene is about 1170 K. The kerosene steam–air reforming stoichiometry corresponding to the minimum temperature providing syngas production without soot formation is

$$28.05\,C_{11}H_{22} + 112.4\,O_2 + 414\,N_2 + 83.78\,H_2O$$
$$\rightarrow 308.57\,CO + 392.35\,H_2 + 414\,N_2. \tag{10–15}$$

In practice, the critical steam/air ratio for soot formation can be lower because of possible heat losses. Essentially, a non-equilibrium process and significant over-equilibrium density of active species can help to avoid soot formation at lower temperatures and higher steam/air ratios. Preheating and evaporation of the liquid fuel can be effectively incorporated into either the plasma system or in a special recuperative heat exchange system. The electric (plasma) energy cost of syngas production by conversion of kerosene is presented in Fig. 10–14 as a function of the total specific energy input, including electric and thermal

693

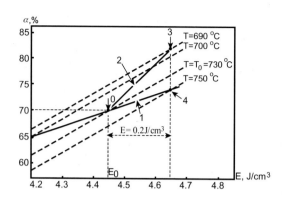

Figure 10–15. Illustration of the plasma catalysis in the steam conversion of kerosene. Additional plasma energy input $\Delta E = 0.2\,\text{J/cm}^3$ (pathway 0–3) leads not to an increase, but to a decrease of temperature in the system. Additional thermal energy input of the equivalent value (pathway 1–4) results in lower conversion degree and increase of temperature in the system. Curve 2 represents experiments with the plasma-catalytic microwave discharge.

energy consumption, particularly in the process involving preheating and evaporation of initial liquid fuel and other reagents (Givotov et al., 2005). Two regimes result in the relatively low energy cost of the process (about 0.2 eV/mol CO–H_2; see Fig. 10–14). Those regimes are (1) high-temperature steam conversion of kerosene in plasma with thermal energy input of about 10 J/cm^3 provided by preheating of initial reagents, and (2) plasma-stimulated kerosene partial oxidation, where the required energy is provided by the exothermicity of the reaction. The effect of plasma catalysis in the steam conversion process of kerosene is illustrated in Fig. 10–15, similar to Figs. 10.10 and 10.11. The effect reveals itself in a decrease of gas temperature, when a small amount of plasma energy is introduced in the system and the conversion degree grows significantly.

A prototype of the on-board, plasma-catalytic, partial oxidation converter of kerosene into syngas based on the application of non-equilibrium microwave discharge was developed by Babaritsky et al. (2003). A schematic of the converter is illustrated in Fig. 10–16. The microwave plasma-igniter is combined in this case with the main conversion reactor (Babaritsky et al., 2003; Givotov et al., 2005). The major parameters of the on-board prototype are 20 m^3/h syngas production, 10 L/h kerosene consumption, 10 L/s atmospheric-pressure air flow rate, and 3 kW microwave discharge power. The degree of the kerosene gasification was 100%; the exit gas composition was as follows: 19.85% hydrogen, 54.25% nitrogen, 20.16% carbon monoxide (CO), 2.48% carbon dioxide (CO_2), 2.42% methane (CH_4), 0.1% ethylene (C_2H_4), and 0.74% acetylene (C_2H_2). A smaller on-board partial oxidation plasma converter of kerosene into syngas (CO–H_2) can be created using a magnetron from a microwave oven. Such a magnetron is small, relatively cheap (about \$25), and quite efficient (about 70% energy efficiency). The parameters for such a small-size converter are 3–7 standard m^3/h syngas production; total mass less than 10 kg for the converter including the reactor, microwave generator, waveguide system, and internal heat exchanger; below 1 kW microwave generator power; and the following product composition: syngas > 40.0%, nitrogen < 60.0%, carbon dioxide (CO_2), < 1.5%, and $CH_4 + C_2H_4 + C_2H_2 < 0.5\%$.

10.2.3. Plasma-Catalytic Conversion of Ethanol with Production of Syngas

The effect of plasma catalysis has been demonstrated for the conversion of ethanol (C_2H_5OH) into syngas using non-equilibrium atmospheric-pressure pulsed microwave discharge (described in Section 10.1.4; Givotov et al., 2005) and atmospheric-pressure transitional arc discharge (Bromberg et al., 2004). The process was investigated in different regimes including direct conversion, steam reforming, and partial oxidation. Plasma catalysis of the ethanol conversion to syngas is illustrated in Fig. 10–17 by the dependence of the conversion degree on specific energy input in the pulsed microwave discharge (Givotov et al., 2005). Although electric (plasma) energy in this case (see Fig. 10–17) is about 5% or less, the plasma-catalytic effect is fairly strong. The conversion degree of ethanol increases from

10.2. Plasma-Chemical Reforming of Liquid Fuels into Syngas

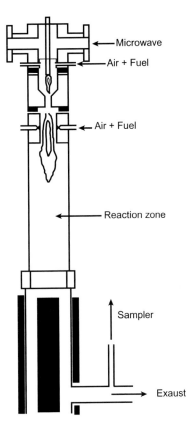

Figure 10–16. Schematic of the plasma-catalytic converter of liquid fuels into syngas, including a microwave discharge and a mixer-reactor.

23 to 62%, while the thermal reservoir temperature decreases from 410 to 370°C. Curve 1 in the figure shows (for comparison) the thermal effect of preheating without application of non-equilibrium plasma, which clearly demonstrates the effect of plasma catalysis.

Direct plasma-catalytic decomposition of ethanol with formation of syngas can be achieved in non-equilibrium conditions at low levels of power and gas temperature as a mostly gas-phase process:

$$C_2H_5OH \rightarrow CO + H_2 + CH_4. \qquad (10\text{–}16)$$

Higher intensity of the process leads to the formation of soot and hydrogen:

$$C_2H_5OH \rightarrow CO + 3H_2 + C_{(s)}. \qquad (10\text{–}17)$$

Figure 10–17. Plasma-catalytic conversion degree of ethanol as a function of specific energy input. Curve 1 corresponds to an experiment with only thermal energy input; point 2 corresponds to an experiment with the microwave discharge; dashed lines correspond to the isothermal conditions.

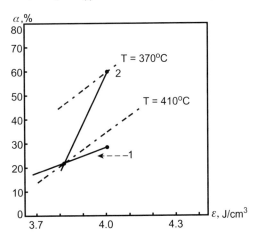

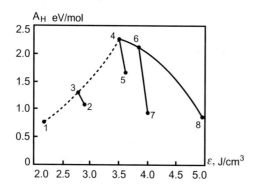

Figure 10–18. Energy cost of the plasma conversion of ethanol vapor into hydrogen: (4, 6, 8) only preliminary heating of the ethanol; (3, 1) addition of oxygen (4% and 7.2% respectively) with simultaneous reduction of preheating; (2, 5, 7) additional energy input provided by the plasma discharge.

The actual process is a combination of the gas-phase and heterogeneous reactions mentioned earlier, with the additional formation of water. The typical product composition observed in the experiments with pulsed microwave discharge in methanol is (molar fractions in gas phase) 55% hydrogen, 12% carbon monoxide (CO), 10% methane, 20% water, 1% carbon dioxide, 1% ethane, and 1% ethylene. The solid-to-gas ratio (C(s):C(g)) is close to 1 in this case, in good agreement with formula (10–17).

The plasma-stimulated steam conversion of ethanol with the production of syngas is less endothermic in the production of one syngas (CO or H_2) molecule (Sekine et al., 2004):

$$C_2H_5OH + H_2O \rightarrow CO + 4H_2, \quad \Delta H = 2.6 \text{ eV/mol}. \tag{10–18}$$

In experiments with steam conversion of ethanol in the aforementioned non-equilibrium pulsed microwave discharge considered (Givotov et al., 2005), the solid-to-gas ratio (C(s):C(g)) is about 0.5 (compare with formula (10–17)), which is half of that for the case without water vapor at similar conditions. The increase of oxidizer concentration leads to the avoidance of soot formation. When the steam flow rate is 2.5 times that of ethanol, the typical composition of the gas-phase products is 45% hydrogen, 17% carbon monoxide (CO), 3% carbon dioxide (CO_2), 1% ethane (C_2H_6), 3% ethylene (C_2H_4), 18% methane (CH_4), and 18% additionally produced water. In the microwave experiments, the preheating temperature is 410°C and electric (plasma) power is only 1% with respect to the total thermal power. The plasma-catalytic effect is quite strong: the conversion degree of ethanol increases in the presence of plasma from 41 to 58%, while the temperature decreases from 410 to 400°C. Introducing equivalent additional power in the form of heat leads to growth of the conversion degree to only 47% and a temperature increase to 415°C. A small addition of air to the ethanol allows for a significant decrease of the required preheating in the partial oxidation regime for the microwave experiments. For example, adding 4% of oxygen into a 97 cm³/s flow of ethanol results in a temperature increase from 390 to 415°C and a C_2H_5OH conversion degree increase from 26 to 48%. Preheating power is decreased in this case from 340 to 285 W. Application of the 10 W non-equilibrium discharge leads to a further increase of the C_2H_5OH conversion degree to 52% and a temperature decrease to 405°C.

The plasma-catalytic reduction of the total energy cost (preheating plus discharge) of ethanol conversion and syngas production is illustrated in Fig. 10–18. The decrease of energy cost at higher thermal energy inputs (points 4, 6, and 8) is due to growth of the ethanol conversion degree at higher temperatures. Reduction of the energy cost at lower thermal energy inputs (points 1 and 3) is related to the addition of oxygen. Points 2, 5, and 7 demonstrate the effect of plasma catalysis on the energy cost: the application of plasma with only 5% of the total power decreases the energy cost of ethanol conversion into syngas more than twice. It should be mentioned that the electric (plasma) energy cost of one additionally produced syngas (CO or H_2) molecule in this process is below 0.1 eV. The high-efficiency

10.2. Plasma-Chemical Reforming of Liquid Fuels into Syngas

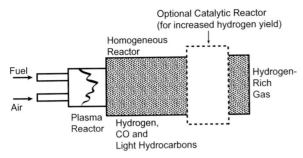

Figure 10–19. General schematic of the plasma fuel reformer.

conversion of ethanol for on-board production of syngas has also been demonstrated using this atmospheric-pressure transitional arc discharge (Bromberg et al., 2004). In addition to homogeneous plasma reformation, both combined plasma and conventional catalytic reforming have been utilized in the system, which results in an additional increase of the hydrogen yield. The application of a catalyst also accelerates the plasma reformation of ethanol in the transitional arc (Bromberg et al., 2004).

10.2.4. Plasma-Stimulated Reforming of Diesel Fuel and Diesel Oils into Syngas

Effective plasma-stimulated partial oxidation of diesel fuel has been demonstrated in atmospheric-pressure transitional arcs (Bromberg et al., 2001, 2004) and atmospheric-pressure gliding arcs (Czernichowski et al., 2004). A schematic of the plasmatron fuel reformer using transitional arcs (Bromberg et al., 2001, 2004) is shown in Fig. 10–19. The fuel is injected from the top of the plasmatron through a nozzle. Directly downstream from the nozzle there is an orifice to provide air. The high air velocity further atomizes the diesel fuel into a fine mist. To prevent the diesel fuel from striking the solid surfaces inside the plasmatron and producing soot by pyrolysis, wall air is introduced through an annular gap between the electrode and the atomization air plug. This air creates a sheath along the inner walls of the plasmatron and prevents the diesel fuel from hitting the wall. A third air stream, the plasma air, is introduced with a large tangential component in the gap between the two cylindrical electrodes where the electric discharge takes place. The tangential motion of the air prevents the plasma from attaching to one spot on the electrodes by moving the discharge and its roots somewhat similar to the case of gliding arc discharges. The air also pushes the discharge into the main volume of the plasmatron. The plasma-chemical reactor can also be supplied with the conventional catalytic element for further conversion of plasma-generated light hydrocarbons into hydrogen. The described diesel fuel reformer has been used on a bus to produce H_2 for NO_x exhaust aftertreatment (Bromberg et al., 2001, 2004). The diesel fuel flow rate in the plasma converter is up to 2 g/s (corresponding fuel power 90 kW), which results in a production of up to 1.5 standard L/s of H_2 without use of catalysts. In a typical reforming regime, the diesel fuel flow rate is 0.8 g/s (corresponding fuel power 36 kW), and the production of hydrogen is 20 standard L/min without use of catalysts. The parameters of the diesel fuel reformer in this case are 200 W electric power; O:C ratio of 1.2; and 65% energy efficiency of the diesel fuel conversion into hydrogen, carbon monoxide (CO), and light hydrocarbons. It is important that absolutely no soot is generated in the process. The composition (in vol %) of the gas-phase products is 7.6% H_2, 1.3% O_2, 64% N_2, 2.4% CH_4, 13.0% CO, 4.4% CO_2, 2.2% C_2H_4, 0.0% C_2H_2, and 7.1% H_2O. For on-board NO_x catalyst regeneration with the generated hydrogen, the plasmatron was turned on only for a few seconds at half-minute intervals. For diesel particulate filter application, the plasmatron should be operated for a few minutes every few hours. Tests of on-board applications of the plasma-stimulated diesel fuel reformation into syngas performed by ArvinMeritor are described by Khadiya (2004).

Plasma-stimulated partial oxidation of dirty diesel oils was studied using the gliding arc discharge by Czernichowski, Czernichowski, and Czernichowski (2002) and Czernichowski, Czernichowski, and Wesolowska (2004). A short description of the fuel conversion reactor can be found in the next section. A 280 W gliding arc was applied to treat commercial Canadian road diesel oil characterized by a density of 846 kg/m^3, 20 cm^3/min flow rate, and 300 ppm sulfur content. The feed product was mixed with air at atmospheric pressure and a 91 standard L/min flow rate, providing an O:C atomic ratio of 1.4. No soot is produced in this regime. The composition of products after the complete plasma conversion of the diesel oil is (in dry vol %) 19% CO, 14% H_2, 62% N_2, 1.5% CH_4, 0.04% C_2H_4, and 3.3% CO_2. The total gas flow output in the system is 6.9 standard m^3/h, and the output of syngas ($CO-H_2$) is 2.3 standard m^3/h. The electric energy cost for production of syngas ($CO-H_2$) from the dirty diesel fuel is 0.12 kW/m^3. The chemical power of the plasma-produced syngas is 7.4 kW, which means that the electric (plasma) energy spent in the process is less than 4% with respect to total energy. Thus, the reforming process, like almost all of those considered earlier, can be qualified as plasma catalytic. Sulfur contained in the diesel fuel is converted in this process into H_2S and can be removed by conventional methods – for example, by a ZnO bed.

10.2.5. Plasma-Stimulated Reforming of Gasoline into Syngas

Effective conversion of gasoline 95 into syngas by partial oxidation has been demonstrated in atmospheric-pressure non-equilibrium gliding arc discharges (Czernichowski et al., 2002, 2004). The process was explored in relatively small 0.6–2 L tubular reactors containing two, three, or six knife-shaped electrodes. Gasoline as a liquid fuel (can be slightly preheated) is mixed in a simple T-connector with air (also either cold or preheated), and then blown into the interelectrode space and treated in non-equilibrium gliding arc plasma (see Section 4.3.8). Then the plasma-treated mixture enters a post-plasma zone filled with metal rods (Ni or other common metal), where total gasoline conversion is achieved. The gliding arc fuel reforming system has been used not only for partial oxidation of gasoline, but for the conversion of many different liquid fuels and other hydrocarbons into syngas (Czernichowski et al., 2002, 2004). In particular, the 440 W gliding arc was applied to gasoline conversion into syngas. Commercial unleaded French gasoline 95 under treatment contained 27% aromatics, 15% olefins, and 58% saturated hydrocarbons; the density of the treated gasoline was 728 kg/m^3 and the gasoline flow rate was 22 cm^3/min. The feed product was mixed with atmospheric-spressure air at a flow rate of 77 standard L/min, providing an O:C atomic ratio of 1.2. No soot was produced in this regime. The composition of the products after complete plasma conversion of gasoline 95 is (in dry vol %) 21% CO, 19% H_2, 55% N_2, 2.4% CH_4, and 3% CO_2. The total gas flow output in the gasoline conversion system is 6.6 standard m^3/h, and the output of syngas ($CO-H_2$) is 2.6 standard m^3/h. The electric energy cost of the production of syngas from gasoline 95 is 0.17 kW/m^3. The chemical power of the produced syngas is 8.4 kW, which means that electric energy spent in the process is about 5% with respect to total energy. Thus, gasoline 95 reformation can be qualified as plasma catalytic.

10.2.6. Plasma-Stimulated Reforming of Aviation Fuels into Syngas

A plasma-chemical reactor, based on the atmospheric-pressure gliding arc (see Section 10.2.5), has been used for syngas production by partial oxidation of liquid aviation fuels (Czernichowski et al., 2004). Specifically, the analyzed aviation fuels were **JP-8** and **heavy naphtha** with composition close to jet fuel. The initial temperature in the system varied from room temperature (no preheating) to a maximum preheating at $516°C$. Refractory granules of 5–10 mm size dump into the post-plasma zone of the JP-8 aviation fuel treatment reactor to reach better conversion efficiency. Characteristics of the treated JP-8 aviation fuel are density of 796 kg/m^3, 13.9 w% hydrogen content, 15.3 vol % total aromatics, 433 weight ppm total sulfur, an initial boiling point of $150°C$, a final boiling point of $252°C$, and a net heat of combustion of 43.2 MJ/kg. The explored range of the JP-8 aviation fuel flow is 0.25–0.48 kg/h and air flow rate is 23–41 standard L/min. The typical preheated air temperature is up to $516°C$, and typical preheating of the JP-8 aviation fuel temperature is up to $235°C$. The gliding arc discharge power is in the 45–90 W range. The described plasma treatment of aviation fuel results in the following major output data: bottom reactor temperature is in the $389–723°C$ range; syngas exit temperature is $253–488°C$; the dry content of the gas-phase products (in vol %) is 10.1–15.6% hydrogen (H_2), 14.8–19.4% carbon monoxide (CO), 25.5–35.0% total syngas (CO–H_2), 58.5–66.6% nitrogen (N_2), 0.70–0.80% argon (Ar), 2.8–4.8% carbon dioxide (CO_2), 0.7–2.8% methane (CH_4), 0.06–2.8% ethylene (C_2H_4), and 0.00–0.19% ethane (C_2H_6). No oxygen, acetylene (C_2H_2), C_{3+} hydrocarbons, soot, or tar were found in the products of the plasma-stimulated conversion. The flow rate of the output syngas is in the 9.6–18.1 standard L/min range. The output syngas power is 1.8–3.6 kW, and total output gas fuel power is 2.2–4.4 kW. Thus, the consumed electric power of the gliding arc discharge is only 2% from the total chemical power of the produced fuel, and 2.5% from the power of the produced syngas (17–31% of the total energy is still converted into heat). The plasma-stimulated JP-8 aviation fuel conversion has been effectively integrated with solid-oxide fuel cells (SOFCs; Czernichowski et al., 2004). In the integrated system, the JP-8 aviation fuel has been converted into an H_2–CO–CH_4 mixture with a quality sufficient for effective operation of the SOFCs. One kilogram of the initial fuel leads to production of 8.2–9.4 kWh of electric energy.

10.2.7. Plasma-Stimulated Partial Oxidation Reforming of Renewable Biomass: Biodiesel

Plasma-catalytic reforming of renewable biomass is important, not only for the generation of syngas for further production of effective liquid fuels and for electricity production in fuel cells but also for the on-board production of hydrogen-rich gases for reducing petroleum consumption in automotive transportation (Bromberg et al., 2004). The most abundant renewable fuels are biodiesel and ethanol (Shapouri, Duffield, & Wang, 2002; Piementel & Patzek, 2005). Ethanol can be obtained from cellulosic materials, corn, or sugar; Biodiesel can be obtained from a variety of seeds. The renewable fuels can be effectively converted into syngas, particularly in the production of hydrogen-rich gases to feed fuel cells. Plasma-stimulated conversion of ethanol has been already discussed in Section 10.2.3. Plasma-catalytic reforming of biodiesel fuel into syngas has been achieved in atmospheric-pressure transitional arc discharge (Bromberg, Hadidi, & Cohn, 2005a,b; see Section 10.2.4). In experiments with atmospheric-pressure transitional arc discharge, the flow rate of the biodiesel is about 0.5 g/s. Variation in the air flow allows for a range in the O:C ratio from 1.05 to 1.8. The hydrogen concentration in the products is relatively low at lower values of O:C. The H_2 concentration doubles (from 5 to 10% on a molar basis) when the O:C ratio increases from 1.3 to 1.7. CO concentration in products follows the same tendency, growing from 8 to 16% on a molar basis when the O:C ratio increases from 1.3 to 1.7. The energy efficiency of the plasma-chemical process is conventionally determined as the ratio of the heating value of the reforming products to the heating value of the initial biodiesel fuel. The energy efficiency is poor (about 20%) at low O:C ratios. At higher O:C ratios, the energy efficiency grows, reaching 60% when O:C = 1.5 and 66% when O:C = 1.7. Soot is not produced in the plasma-stimulated biodiesel fuel conversion process.

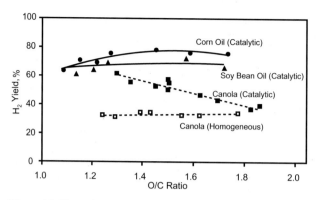

Figure 10–20. Hydrogen yield for catalytic and homogeneous (non-catalytic) reformation of corn, canola, and soybean oil.

10.2.8. Plasma-Stimulated Partial Oxidation Reforming of Bio-Oils and Other Renewable Biomass into Syngas

An atmospheric-pressure transitional arc plasma system, similar to the one considered earlier and described in Section 10.2.4, has been used for conversion of different bio-oils, including corn oil and such vegetable oils as soy and canola, into syngas (Bromberg et al., 2004). Hydrogen yield from partial oxidation oil reformation is presented in Fig. 10–20 as a function of the O:C ratio. The hydrogen yield is defined in this case as the ratio of the mass flow rates of hydrogen in the reforming gases (products) to that of hydrogen bound in the initial bio-oils. The experimental results are presented for plasma-chemical conversion of the bio-oils without any conventional catalysts as well as for the plasma-chemical conversion process combined with conventional catalysis to increase the hydrogen yield (see Section 10.2.4). The initial biofuel flow rate in the experiments is about 0.4 g/s, which corresponds to about 17 kW of chemical power. The transitional arc discharge power is about 200 W. As seen in Fig. 10–20, the hydrogen yields for reformation of corn and soybean oils are similar in the presence of a catalyst. The hydrogen yield for canola oil in the presence of a catalyst is lower than that for corn and soybean oils, especially at higher O:C ratios. Volume fractions of hydrogen produced by plasma-stimulated reformation of corn, soybean, and canola oils in the presence of the catalyst are shown in Fig. 10–21 as a function of the O:C ratio. The effect of combining plasma stimulation with a conventional catalyst is illustrated in Table 10–5 for the partial oxidation of unrefined soybean oil (Bromberg et al., 2004). In these experiments, the power of the atmospheric-pressure transitional arc discharge is about 200 W, flow rate of the bio-oil is 0.37 g/s (corresponding chemical power 16 kW) and the O:C ratio is 1.08. The catalyst in this process doubles the production of hydrogen and significantly increases the production of CO; soot generation is minimal.

Similar experiments with partial oxidation of renewable biomass to produce syngas have been performed using atmospheric-pressure gliding arc discharge without the application

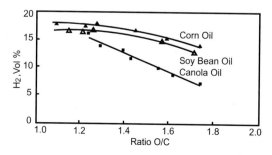

Figure 10–21. Hydrogen concentration (vol %) for catalytic reformation of corn, soybean, and canola oil.

Table 10–5. Composition of Products of Plasma: Reforming of Unrefined Soybean Oil with and without Additional Catalytic Stage (O:C ratio 1.08)

Product of reforming	Plasma w/o catalyst, vol %	Plasma with catalyst, vol %
Hydrogen, H_2	8.1%	15.5%
Carbon monoxide, CO	11.1%	16.6%
Carbon dioxide, CO_2	6.7%	5.0%
Nitrogen, N_2	57%	53%
Methane, CH_4	1.2%	1.0%
Ethylene, C_2H_4	5.1%	2.8%
Acetylene, C_2H_2	0.1%	0.0%
Oxygen, O_2	0.6%	0.6%
Water, H_2O	9.8%	5.3%

of additional catalysts (Czernichowski et al., 2002, 2004; see Section 10.2.5). Effective syngas production without any soot formation has been demonstrated in particular for ethyl alcohol and saturated sugar solution, which opened opportunities for conversion of such farm products as molasses, residual and waste oils, and glycerin. Renewable biomass conversion into syngas using gliding arc discharge has been demonstrated for reforming French commercial rapeseed oil (Czernichowski et al., 2004). The gliding arc with power 200–300 W and without preheating or conventional catalysts led to complete conversion of the rapeseed oil with no production of any soot or coke. The flow rate of the produced syngas is 3.5 standard m^3/h, which corresponds to 11 kW of chemical power. The air flow rate in the experiments was in the range 35–102 standard L/min, while the rapeseed oil input flow rate was between 11 and 30 mL/min. The concentration of nitrogen in the products was about 55–60 vol % (dry basis) depending on the O:C ratio. The typical mass balance of the plasma-chemical process can be presented in the form

$$C_{18.1}H_{34.1}O_2 + 10.53\,O_2 \rightarrow 15.09\,CO + 12.54\,H_2$$
$$+ 2.31\,CO_2 + 0.47\,CH_4 + 0.11\,C_2H_4 + 3.35\,H_2O. \quad (10\text{–}19)$$

The corresponding O:C ratio is 1.16, taking into account that the only oxygen introduced in the reaction is in the form of air. The energy efficiency of the process, defined as the ratio of produced syngas power to the chemical power of the initial biomass and electric power consumption, is about 70%. The product temperature is about 850°C and can be recuperated.

10.3. COMBINED PLASMA–CATALYTIC PRODUCTION OF HYDROGEN BY PARTIAL OXIDATION OF HYDROCARBON FUELS

10.3.1. Combined Plasma–Catalytic Approach Versus Plasma Catalysis in Processes of Hydrogen Production by Partial Oxidation of Hydrocarbons

The plasma-chemical processes of fuel conversion and hydrogen production considered in the Sections 10.1 and 10.2 are often referred to as plasma catalysis. The concept of plasma catalysis in general has been discussed in Section 10.1.1. Plasma catalysis means the application of plasma *instead of* conventional catalysts to stimulate the process without providing major energy input. Although they sound similar, the combined plasma–catalytic approach is quite different from plasma catalysis. The combined plasma–catalytic approach implies application of both plasma (possibly plasma-catalytic) and conventional catalytic systems to convert hydrocarbons to hydrogen-rich gases. The combined plasma–catalytic system can be arranged in different configurations, which are always intended to achieve

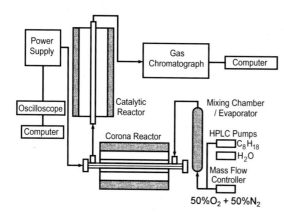

Figure 10–22. General schematic of the combined plasma–catalytic system.

synergy of the plasma and catalysis. One configuration, which can be referred to as **plasma preprocessing**, means that incoming gas is first processed by the plasma reactor before treatment in the catalytic reactor. Plasma preprocessing aims to enrich the gas with reactive species and then process it by the catalyst, thus enhancing the kinetics of hydrogen production. Another configuration, which can be referred to as **plasma postprocessing**, means that plasma is used for treating the exhaust gases from a catalytic unit. The postprocessing operation is intended to complete the reformation process and to destroy unwanted by-products generated during traditional catalytic reactions. Some examples of plasma–catalysis combinations in the production of hydrogen have been already discussed earlier (Bromberg et al., 2004; for example see 10–5). Consistent consideration has been accomplished by Sobacchi et al. (2002) by investigation of partial oxidation of isooctane (trimethylpentane, C_8H_{18}):

$$C_8H_{18} + 3.7(O_2 + 3.76N_2) \rightarrow \text{ products } (H_2, CO, \text{others}). \quad (10\text{--}20)$$

Different separate (stand-alone) and combined methods were applied for reformation: (1) a stand-alone catalytic reactor, (2) a stand-alone non-equilibrium atmospheric-pressure plasma reactor, (3) a combined plasma–catalytic reactor in the plasma preprocessing configuration, and (4) a combined plasma–catalytic reactor in the plasma postprocessing configuration. Water was added for the shift reaction ($CO + H_2O \rightarrow H_2 + CO_2$) to replace CO with hydrogen. The isooctane (trimethylpentane, C_8H_{18}), while being a simple chemical substance, is a liquid fuel which is often used to represent gasoline.

10.3.2. Pulsed-Corona-Based Combined Plasma–Catalytic System for Reforming of Hydrocarbon Fuel and Production of Hydrogen-Rich Gases

A schematic of the combined plasma–catalytic reactor designed and used to investigate the production of hydrogen-rich gases by partial oxidation of isooctane (10–20) is presented in Fig. 10–22 (Sobacchi et al., 2002). The plasma section of the combined system includes a pulsed corona discharge, generated in a wire-into-cylinder coaxial electrode configuration. The inner high-voltage electrode is a 0.5 mm diameter Inconel wire. The outer electrode of the corona is a 1.2-m-long cylindrical stainless steel tube with 22.2 mm internal diameter. The central 0.9-m-long section of the reactor is placed inside a high-temperature furnace, which can be stabilized at temperatures as high as 1200°C. For the experiments, the pulsed corona was kept at a constant temperature of 400°C with accuracy better than 20°C. Power is supplied by a thyrotron, providing pulses of 100 ns duration with rise time 10 ns and voltage up to 20 kV. The power is varied from 1 to 20 W by controlling the pulse amplitude and frequency (from 0.2 to 2 kHz). The catalytic section of the combined system is a 50-cm-long, 1.3-cm-diameter stainless steel reactor placed inside a tubular furnace that provides the required heat. The furnace has been stabilized at temperatures from 600 to 800°C with accuracy better than 20°C; 3 cm³ of a special catalyst developed at the Argonne National

10.3. Combined Plasma–Catalytic Production of Hydrogen

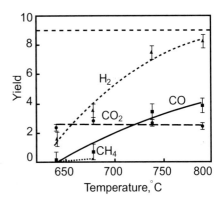

Figure 10–23. Gaseous product yields as function of temperature for the stand-alone catalytic reformation. Maximum possible hydrogen yield from isooctane is 9, as shown by the straight dashed line.

Laboratory (Ahmed, Kumar, & Krumpelt, 1997; Ahmed, 1998) is used in the catalytic reactor. Inlet gas is a 1:1 atmospheric-pressure O_2–N_2 mixture (O_2 and N_2 flow rates are 22 standard cm^3/ min). Two liquids are supplied to the system: isooctane fuel with flow rate 0.04 mL/min and water with flow rate 0.04 mL/min. To evaporate the liquid components and to properly mix the different chemicals, a stainless steel mixing chamber, heated up to $190°C$ and filled with glass beads, is placed before the reactor. Separate experiments were run to characterize the following: the stand-alone catalytic reactor, the stand-alone non-equilibrium atmospheric-pressure pulsed corona plasma reactor, the combined plasma–catalytic reactor in the plasma preprocessing configuration, and finally the combined plasma–catalytic reactor in the plasma postprocessing configuration. Product yields are defined as the ratio of the number of moles of a given compound X to the number of moles of isooctane:

$$\eta_x = [X]/[C_8H_{18}]. \qquad (10\text{–}21)$$

Thus, the maximum possible hydrogen yield from isooctane is $\eta_{H_2} = 9$, and the maximum possible carbon monoxide (CO) yield from isooctane is $\eta_{CO} = 8$.

10.3.3. Catalytic Partial Oxidation Reforming of Isooctane
The first set of the preceding experiments with isooctane reformation (Section 10.3.2) was performed using the catalytic reactor by itself (Sobacchi et al., 2002) to provide a baseline for comparison with the results for combined plasma–catalytic processing. The catalytic reactor temperature varied from 630 to $800°C$. The results are presented in Fig. 10–23 (Ahmed, 1998; Sobacchi et al., 2002) in the form of the process yield (for different reaction products, see (10–21)) dependence on temperature. The catalytic process results in a high level of conversion of hydrocarbons into hydrogen only at the upper level of investigated temperatures. The hydrogen yield of the isooctane reformation process drops sharply with the temperature decrease. At low temperatures, a significant amount of isooctane remains unprocessed, and also some heavy organic acids are formed. The optimal temperatures are relatively high, around $800°C$. Therefore, one of the major goals of combination of the conventional catalytic system with the non-equilibrium plasma reactor is reduction of the required temperature level.

10.3.4. Partial Oxidation Reforming of Isooctane Stimulated by Non-Equilibrium Atmospheric-Pressure Pulsed Corona Discharge
The second set of the preceding experiments with isooctane (C_8H_{18}) partial oxidation reformation (Section 10.3.2) was performed using atmospheric-pressure pulsed corona discharge by itself (Sobacchi et al., 2002). These "pure" plasma experiments provide another baseline to compare the results for combined plasma–catalytic processing. The results are presented in Fig. 10–24 in the form of process yield dependence on the corona discharge power. Gas

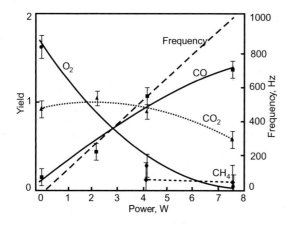

Figure 10–24. Yields of gaseous products and pulse frequency as functions of discharge power for the stand-alone corona reforming.

temperature in the plasma reactor was sustained at $400°C$. At this temperature, an initial degree of isooctane conversion is observed in the corona reactor even with zero discharge power (see Fig. 10–24). As a result of the conversion process, the oxygen content is reduced and carbon dioxide is formed. The effect of the corona on initial preheating-related C_8H_{18} conversion is to initiate further oxidation of the fuel into CO and CO_2. The consumption of O_2 due to oxidation reactions is completed when the corona pulse frequency reaches 1000 Hz, which corresponds to an average discharge power of 8 W. Reduction of CO_2 and increase of CO concentrations in the reformation products accompany the pulsed corona–stimulated oxidation of isooctane (see Fig. 10–24), which indicates a significant change in the main oxidation pathways in the presence of plasma. No H_2 is produced, but some methane is detected for high corona powers. Intermediate partial oxidation products with high molecular weights, such as heavy organic acids, have also been generated at low discharge power levels. The liquid solution collected after pulsed corona treatment shows an acidic pH.

10.3.5. Reforming of Isooctane and Hydrogen Production in Pulsed-Corona-Based Combined Plasma–Catalytic System

In the next two sections, the reforming process is to be considered using a combination of the catalytic and plasma approaches in the two possible system configurations: plasma preprocessing and plasma postprocessing (see Section 10.3.1). Similar to the experiments considered in Sections 10.3.3 and 10.3.4, the temperature of the catalytic reactor in the combined system varied from 630 to $800°C$. The pulse frequency of the corona discharge was set to 256 Hz. Results of Sobacchi et al. (2002) are summarized in Figs. 10–25–10–28. Product yield as function of catalytic reactor temperature is shown in Fig. 10–25 for the plasma postprocessing configuration and in Fig. 10–26 for the plasma preprocessing configuration. Figure 10–27 presents a comparison among the hydrogen yields achieved in the individual catalytic treatment process, the plasma preprocessing case, and in the plasma postprocessing configuration. Finally, power inputs for the two different configurations of the combined plasma–catalytic system are shown in Fig. 10–28. An electric power input of 1 W in the processed stream is equivalent to a temperature increase of $350°C$. At the same time, the thermal effect of the discharge on the system temperature and as a result on the hydrogen generation is negligible because of effective heat exchange and rigorous temperature control. Hydrogen production from isooctane is augmented by the plasma treatment in both postprocessing and preprocessing configurations. However, the two configurations differ in terms of hydrogen yield and power consumption. The enhancement of the hydrogen yield is particularly significant at relatively low temperatures. For example, at a temperature of $630°C$, the

10.3. Combined Plasma–Catalytic Production of Hydrogen

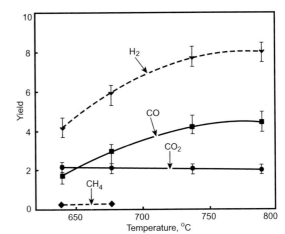

Figure 10–25. Gaseous product yields as a function of temperature for plasma postprocessing.

conversion to hydrogen is increased 2.5-fold due to the effect of the discharge. The effect of plasma on the conversion to hydrogen decreases at higher temperatures and becomes almost negligible at temperatures of about 800°C. An increase of CO concentration accompanies a higher degree of conversion to hydrogen. The increase of CO production is especially significant at low temperatures when the contribution of plasma-chemical processes makes a difference (in comparison to the strongly temperature-sensitive reactions of thermal catalysis). The output carbon monoxide concentration for all analyzed conditions was higher than 3%.

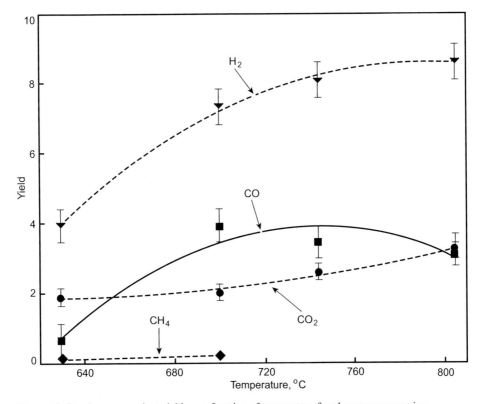

Figure 10–26. Gaseous product yields as a function of temperature for plasma preprocessing.

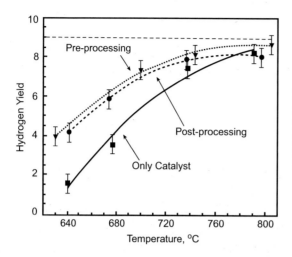

Figure 10–27. Hydrogen yield as function of temperature for various system configurations: plasma preprocessing, plasma postprocessing, pure catalysis. Maximum possible hydrogen yield from isooctane equals 9 and is represented by a straight dashed line.

10.3.6. Comparison of Isooctane Reforming in Plasma Preprocessing and Plasma Postprocessing Configurations of the Combined Plasma–Catalytic System

Experiments with the combined plasma–catalytic system allow for a comparison of the effectiveness of hydrogen production by partial oxidation reformation of isooctane using the two major configurations of the system: plasma preprocessing and plasma postprocessing (see Figs. 10–25 to 10–28; Sobacchi et al., 2002). Plasma preprocessing seems to be the optimal configuration of the combined plasma–catalytic system. For all tested temperatures, this configuration resulted in larger hydrogen production than plasma postprocessing or the catalytic treatment alone. Moreover, at higher temperatures, when the production of hydrogen is highest, plasma preprocessing reduces the concentration of CO, which is oxidized to CO_2. The final concentration of CO at these high temperatures is lower than that for the other configurations. The better conversion of isooctane into hydrogen in the preprocessing configuration with respect to the stand-alone operation of the catalytic unit suggests that the catalyst acts more effectively on the intermediate oxidized compounds produced by the pulsed corona discharge than on the initial gas mixture. Plasma stimulates the kinetically suppressed and slowest initial stages of oxidation of isooctane, whereas the catalyst supports the kinetically "easier" reactions of further oxidation of the intermediate oxidized compounds to syngas. This effect is most evident and seen from the figures for the catalytic reactor temperatures below 750°C. The same pulse frequency results in higher

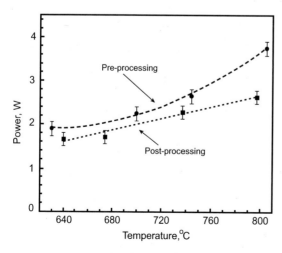

Figure 10–28. Power consumption as a function of temperature for the plasma postprocessing and plasma preprocessing configurations. Pulse frequency 256 Hz.

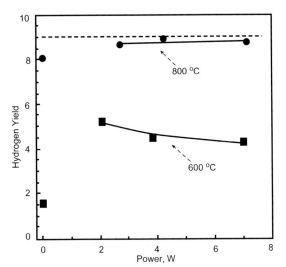

Figure 10–29. Variation of hydrogen yield with corona power input in the plasma preprocessing configuration. Catalytic reactor temperatures are 600 and 800°C.

power loads in the pulsed corona discharge operating in the preprocessing configuration of the combined system. Moreover, the power input in the preprocessing operation has a higher sensitivity to the catalytic reactor temperature. These effects are mostly due to a difference in the reactant compositions and pressure variations inside the plasma-chemical partial oxidation reactor.

For the postprocessing configuration of the combined system, plasma is formed in the gases produced by the catalytic conversion process, when the oxidation processes are mostly completed. On the other hand, in the case of the preprocessing configuration, the gas composition in the non-equilibrium plasma reactor is closer to the initial composition, with only some preliminary level of fuel oxidation into CO and CO_2. The oxidation itself depends on the power input in the discharge. The pressure inside the pulsed corona is different for plasma preprocessing and plasma postprocessing (higher in the preprocessing configuration). This pressure increase is mostly due to the pressure drop in the catalytic reactor. Higher catalytic reactor temperatures produce higher pressure drops, which changes the reduced electric field, E/N, in the corona discharge. In the postprocessing configuration, this pressure rise does not influence corona operation. Overall, it has been found that the variation of chemical composition leads to a stronger effect on the energy-per-pulse value than the recorded variations of the reactor pressure, which is about 0.2 atm. Hydrogen yield in the case of the plasma preprocessing configuration is shown in Fig. 10–29 as a function of power of the pulsed corona discharge. The catalytic reactor temperature was fixed in these experiments at 600 or 800°C. For high catalytic reactor temperatures, the power variation of the pulsed corona does not influence the level of hydrogen production. For lower catalytic reactor temperatures, the hydrogen yield decreases even with the discharge power. For both temperatures, a combination of the catalytic treatment with relatively low frequency and, hence, lower power pulsed corona discharge seems to be optimal in terms of H_2 yield and power consumption of the combined plasma–catalytic reformation.

10.4. PLASMA-CHEMICAL CONVERSION OF COAL: MECHANISMS, KINETICS, AND THERMODYNAMICS

10.4.1. Coal and Its Composition, Structure, and Conversion to Other Fuels

Coal is the fuel most able to cover world deficiencies in oil and natural gas. Wide application of coal is, however, somewhat limited by its transportation and ecological "non-purity" when used directly as a fuel, which is related to its significant formation of ashes, sulfur compounds, nitrogen oxides, and so forth. This motivates the development of new and more effective technologies for coal conversion into other fuels. Such technologies are focused on coal gasification with production of syngas or gaseous hydrocarbon fuels, as well as on direct coal liquefaction with production of liquid fuels (Williams, 2000). Both thermal and non-thermal plasma discharges can be used to convert coal into other fuels. The benefits of plasma application in these technologies is based on the high selectivity of the plasma-chemical processes, the high efficiency of conversion of different types of coal including those of low quality, relative simplicity of the process control, and significant reduction in the production of ashes, sulfur, and nitrogen oxides (see, for example, Xie et al., 2002; He et al., 2004; Messerle, 2004).

Before considering the coal conversion processes, it is helpful to overview the typical chemical composition and structure of coal. The major chemical elements that form coal are carbon, hydrogen, oxygen, sulfur, and nitrogen. Coal also contains moisture and mineral components. The age of coal determines its degree of **metamorphism**: the oldest coals have the highest degree of metamorphism and usually have the highest quality. Low-quality (low-metamorphism-degree) coals have about 65% carbon (% total organic mass), whereas the high-quality (high-metamorphism-degree) coals can have up to 91% carbon. The mass fraction of hydrogen in the same sequence of growing degree of metamorphism decreases from 8 to 4%. The mass fraction of oxygen decreases in the metamorphism sequence from 30 to 2%. All types of coal contain 0.5–2% nitrogen and 0.5–3% sulfur. The concentration of mineral components varies in a wide range, from 1–3% to 10–30%. Most organic mass of coal is a rigid irregular three-dimensional polymer. The remaining organic mass is made of mobile monomolecular or slightly polymerized substances which are immobilized in pores of the rigid skeleton (Lipovich, 1988). Oxygen contained in coal is mostly bound in hydroxyl, carbonyl, and carboxyl groups. The concentration of phenol-hydroxyl groups decreases with higher concentrations of carbon in coal and almost disappears when the carbon mass fraction reaches 89%. Sulfur is contained in coal in both organic (30–50%) and inorganic (50–70%) compounds in the form of disulfides and sulfates. Disulfides are mostly represented in coal by pyrite (FeS_2); sulfates are mostly represented by gypsum ($CaSO_4$) and ferrous sulfate ($FeSO_4$). The total amount of sulfates in coal is low and usually does not exceed 0.1 mass %. Nitrogen is incorporated only in the organic fraction of coal, mostly in heterocyclic compounds or as a bridge between two carbocyclic compounds. Coal is characterized by high porosity. The pores can be divided into three classes (Gavalas & Willks, 1980): nanopores with sizes below 30 nm, intermediate pores with size 30–300 nm, and macropores with size 0.3–3 μm. The nanopores have a very large surface area. In these pores most of the coal destruction products are generated during thermal treatment. Diffusion of chemicals during adsorption and reactions also takes place through nanopores. The intermediate pores provide major channels for the transport of reagents and products in and out of the coal particles. The fraction of macropores in the coal is relatively low.

10.4.2. Thermal Conversion of Coal

Most kinetic data regarding the thermal conversion of coal are known for relatively low temperatures (≤ 1200 K) and heating rates ($\leq 10^4$ K/s), which will be discussed in this section. Kinetic peculiarities related to higher temperatures and higher coal heating rates are less available and will be considered in Section 10.4.7. Heating of coal in an inert atmosphere to temperatures of 620–670 K leads to its decomposition with the formation of primary volatile destruction products and coke. Formation and release of the volatile

Table 10–6. Arrhenius Parameters of Kinetics of Releasing of Different Functional Groups and Their Decomposition Products from Coal During Its Pyrolysis

Functional groups, decomposition products	Strength of bond	Pre-exponential factor, A, s^{-1}	Activation energy, E_a, kcal/mol
CO_2	Weak	$1.0 \cdot 10^{21}$	76.7 ± 4.0
CO_2	Strong	$2.4 \cdot 10^{11}$	57.6 ± 2.0
H_2O	–	$1.7 \cdot 10^{14}$	60.0 ± 3.0
CO	Weak	$1.7 \cdot 10^{11}$	50.0 ± 5.0
CO	Strong	$2.5 \cdot 10^{7}$	46.0 ± 4.6
HCN	Weak	5400	17.7
HCN	Strong	$7.0 \cdot 10^{7}$	64.0
NH_3	–	$1.2 \cdot 10^{12}$	54.6 ± 6.0
CH_x (from aliphatic groups)	–	$1.7 \cdot 10^{14}$	60.0 ± 3.0
CH_4	–	$1.2 \cdot 10^{12}$	54.6 ± 6.0
H (from aromatic group)	–	$1.6 \cdot 10^{7}$	46.0 ± 4.6

products continues until the temperature is about 1220 K. The volatile compounds include water vapor, hydrogen, methane, carbon oxides, hydrocarbon gases, and heavy organic products (tars). The fraction of the volatile compounds varies in different types of coal and generally is about 15–40%. The amount of the volatile compounds is highest in the younger coals with the lowest degree of metamorphism. Anthracite, the highest-quality hard coal with the highest degree of metamorphism, releases the minimum amount of volatile compounds. The fraction of heavy organic products (tars) in volatile compounds is high (40–80%) and increases with the rate of coal heating.

Coal heating in an inert atmosphere also significantly changes the residual solid fraction. First, its plasticity grows and the microparticles agglomerate. Then the solid microparticles become rigid again, forming the porous structure of the coke. It should be mentioned that the volatile compounds are not contained in the initial coal. Formation of the volatile compounds, as well as formation of coke, is the result of multiple chemical reactions taking place during the coal heating. For this reason the terms "formation of volatile compounds," "cracking of coal," and "pyrolysis of coal," which describe the thermal conversion of coal, are actually synonymous.

Coal structure can be considered a combination of different functional chemical groups. Some of the groups during pyrolysis are detached and decomposed with the formation of relatively small volatile products. Other functional groups are released without decomposition, which results in tar generation. Coals with different degrees of metamorphism are characterized by different structures and concentrations of the functional groups. Nevertheless, the kinetics of release of functional groups and products of their decomposition from different coals can be described more or less by universal Arrhenius parameters summarized in Table 10–6 (Solomon et al., 1982). Note that some products can be formed by breaking either the weak or strong chemical bond with the skeleton or the rest of the functional group, which results in two different pre-exponential factors and activation energies for those components (see Table 10–6). Release of the functional groups and products of their decomposition can be described by zero kinetic order. This is applicable when the characteristic size of the coal particles exceeds some critical value that depends on the heating rate. The critical diameter of the coal particles changes from 2 to 0.2 mm, when heating rate grows from 10^2 to 10^4 K/s. When the coal particles are large enough (greater than the critical value), the relevant chemical reactions take place mostly on the surface and are controlled by heat transfer, which results in the zero kinetic order of the processes presented in Table 10–6 (Koch, Juntgen, & Peters, 1969).

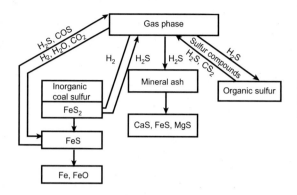

Figure 10–30. Schematic of conversion of sulfur compounds of coal in inert and reduction medium.

10.4.3. Transformations of Sulfur-Containing Compounds During Thermal Conversion of Coal

Sulfur is mostly contained in coal as pyrite (FeS_2) and organic compounds (30–50%). Thermal treatment of coal leads to the decomposition of pyrite and organic sulfur compounds and the formation of mostly H_2S, which then in turn reacts with the organic mass of coal and ash components. A general schematic of the transformations of sulfur-containing compounds during pyrolysis in an inert or reducing atmosphere is illustrated in Fig. 10–30. Decomposition of pyrite (FeS_2) starts at temperatures exceeding 820 K and leads to the formation of elemental sulfur and iron sulfide (FeS). In the presence of hydrogen (H_2) formed during coal pyrolysis, the pyrite is more intensively converted into iron sulfide and hydrogen sulfide at temperatures of about 770 K:

$$FeS_2 + H_2 \rightarrow FeS + H_2S. \tag{10–22}$$

Pyrite and iron FeS also react with carbon oxides, CO and CO_2 (formed by the pyrolysis), at temperatures around 1100 K, which results in the formation of COS:

$$FeS_2 + CO \rightarrow FeS + COS, \tag{10–23}$$

$$FeS + CO_2 \rightarrow FeO + COS. \tag{10–24}$$

The reactions of pyrite and iron sulfide (10–22)–(10–24) are limited by diffusion of the gas-phase reagents across the formed surface layer of iron. Therefore, relevant conversion degrees usually remain below 30% (Attar, 1978). At relatively high temperatures (exceeding 1270 K), carbon reduces pyrite and iron sulfide, forming metallic iron and CS_2:

$$2FeS + C \rightarrow 2Fe + CS_2, \tag{10–25}$$

$$FeS_2 + C \rightarrow Fe + CS_2. \tag{10–26}$$

The kinetic parameters of some reactions of pyrite (FeS_2) and iron sulfide (FeS) contained in coal with gas-phase reagents are summarized in Table 10–7 (Schwab & Phillinis, 1974; Bjerle et al., 1982). The Arrhenius parameters of the two-phase processes are given, assuming zero kinetic order of the reactions. Organic sulfur in coal is contained mostly in thiophenes (40–70%) and sulfides (60–30%). Thiophenes are thermally stable and their decomposition requires temperatures at about 1100 K. For this reason, thiophenes are the major sulfur-containing products of coal pyrolysis. Sulfides are presented in coal as aryl sulfides (about 50%), cyclic sulfides (about 30%), and aliphatic sulfides (about 20%). Their thermal decomposition leads mostly to gas-phase production of H_2S and CS_2, which then reacts with the organic and inorganic fraction of coal. Hydrogen sulfide reacts with the mineral compounds of coal:

$$MO + H_2S \rightarrow MS + H_2O, \tag{10–27}$$

$$MCO_3 + H_2S \rightarrow MS + H_2O + CO_2. \tag{10–28}$$

Table 10–7. Kinetic Parameters of Pyrite and Iron Sulfide Reactions in Coal with Gas-Phase Reagents

Reactions	Pre-exponential factor, A, s^{-1}	Activation energy, E_a, kcal/mol
$FeS_2 + H_2 \rightarrow FeS + H_2S$	$5 \cdot 10^4$	28.8–33.1
$FeS + CO_2 \rightarrow FeO + COS$	31	31.0
$FeS + H_2 \rightarrow Fe + H_2S$	$9 \cdot 10^2$	17.2
$FeS_2 + O_2 \rightarrow$ products	$5 \cdot 10^{-5}$	0
$FeS + H_2O \rightarrow FeO + H_2S$	10^{-4}	16.8
$FeS + O_2 \rightarrow$ products	$3 \cdot 10^{-5}$	0

Due to these reactions, a significant amount of sulfur remains in coke after the thermal treatment.

10.4.4. Transformations of Nitrogen-Containing Compounds During Thermal Conversion of Coal

Nitrogen is contained almost completely in organic compounds of coal, mostly in aromatic rings. Release of nitrogen in the gas phase during pyrolysis does not start immediately; it begins after the preliminary formation of about 10% of volatile compounds not containing nitrogen. The volatile nitrogen-containing products of pyrolysis are NH_3, HCN, N_2, and such tar components as pyridine, pyrol, nitryl, carbosol, chinoline, and indole. The Arrhenius kinetic (zero-order) parameters of formation of some volatile nitrogen compounds during coal pyrolysis were shown in Table 10–6. Generally, formation rates of HCN and tar are close at 800 K, whereas ammonia production in this case is about 10-fold slower. At higher temperatures (about 1000 K), the rate of tar formation is 10 times faster than that of HCN and ammonia. Some fraction of HCN is strongly bound in the coke skeleton and remains there even at higher temperatures (see Table 10–6). The amount of nitrogen released from coal in the form of volatile compounds depends on the pyrolysis temperature, heating rate, and type of the coal. A temperature increase from 770 to 1170 K leads to growth of the nitrogen fraction released into the gas phase from 20 to 80% (Fine et al., 1974). The rest of the nitrogen is fixed in the rigid coal skeleton by very strong chemical bonds (see Table 10–6). It remains in the coke even during pyrolysis at very high temperatures. Conversion of the nitrogen, strongly bound in the coke skeleton, into the gas phase becomes possible only in the processes of complete gasification of coke using, for example, steam and air.

10.4.5. Thermodynamic Analysis of Coal Conversion in Thermal Plasma

Thermal plasma coal conversion takes place in conditions close to quasi equilibrium; therefore, thermodynamic analysis can be effectively used to describe the process (Venugopalan et al., 1980; Messerle et al., 1985; Tselishchev & Abaev, 1987). Messerle et al. (1985) described thermal plasma gasification of Turgai coal by water vapor. The thermodynamic analysis was performed in this case using 1 atm pressure and a C:O mass ratio of 0.75, which corresponds to a mixture of 1000 kg of dry Turgai coal with 360 kg of water vapor. The elemental composition (mass fraction) of the steam–coal mixture under the thermal plasma treatment is presented in Table 10–8. Results of the thermodynamic analysis of the coal gasification in thermal plasma by steam are presented in Figs. 10–31–10–33. The thermodynamic analysis shows that with gasification of the organic coal fraction at temperatures below 2000 K, the gas-phase products mostly (93%) consist of syngas (CO–H_2). Other products, including acetylene, nitrogen, and cyan- and sulfur-containing compounds, comprise altogether less than 1% at $T = 1800$ K. The concentration of CO_2 and H_2O is less than 0.01%. Sulfur is represented in gas phase by three major compounds: H_2S at temperatures

Table 10–8. Elemental Composition of Typical Steam–Coal (Turgai Coal) Mixture under Thermal Plasma Treatment

Element present in coal	Mass fraction, %	Element present in coal	Mass fraction, %
C, carbon	35.43	Al, aluminum	2.90
O, oxygen	47.24	Fe, iron	1.06
H, hydrogen	5.69	Ca, calcium	0.44
N, nitrogen	0.53	Mg, magnesium	0.32
S, sulfur	0.76	Ti, titanium	0.30
Si, silicon	5.22		

of 1000–1875 K, CS_2 at temperatures of 1875–3950 K, and S at temperatures exceeding 3950 K. The fraction of condensed substances in the mixture remains almost fixed at temperatures below 1400 K. At higher temperatures of about 3100 K, the major components of the mineral fraction of coal are gasified. The mineral fraction components are reduced in these conditions to elements: silicon (Si), aluminum (Al), iron (Fe), calcium (Ca), and titanium (Ti). The energy cost of Turgai coal gasification is shown in Fig. 10–33 and grows gradually with temperature. The coal gasification degree, shown in the figure, jumps up sharply at temperatures close to 1800 K. This typical effect for the high-mineral-content coals is due to the decomposition of SiO and the formation of carbon monoxide.

10.4.6. Kinetic Phases of Coal Conversion in Thermal Plasma

Two-phase chemical processes involving plasma gas and solid fuel particles are complicated and include destruction of the solid fuel with release of volatile products, gas-phase

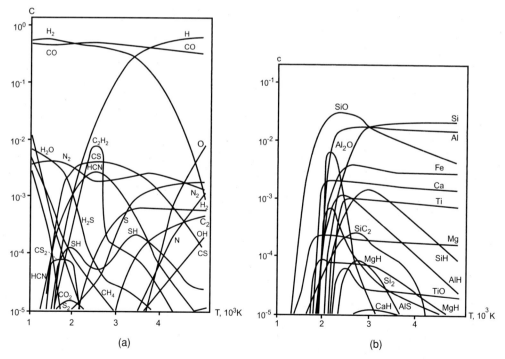

(a)   (b)

Figure 10–31. Quasi-equilibrium gas-phase composition at the gasification of low-quality coal as a function of temperature: (a) organic fraction; (b) mineral fraction. In the figure, c volume fraction of a component, T, average temperature.

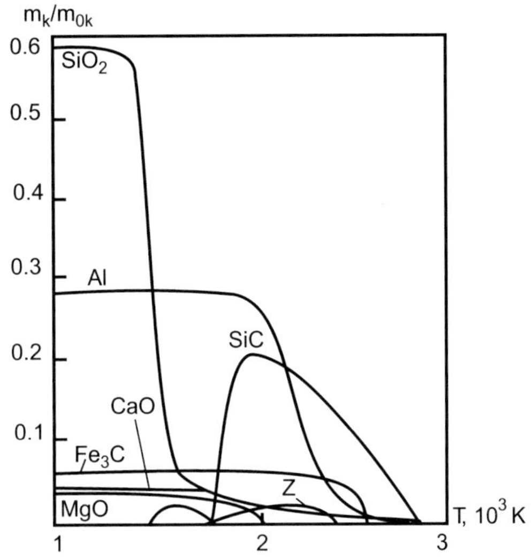

10.4. Plasma-Chemical Conversion of Coal

Figure 10–32. Mass fraction of specific condensed components of coal gasification as a function of temperature: $m_{ok} = 281$ kg, initial amount of the condensed components of coal; m_k, mass of the components at gasification temperature T.

reactions of the volatile compounds, and reactions between the gas-phase and solid-phase components. Relevant plasma-chemical reactors are characterized by very high temperature and velocity gradients, reaching in the radial direction 10^4 K/cm and $500 \frac{cm/s}{cm}$. Obviously, heat and mass transfer make a significant contribution into the kinetics in such systems. To clarify the plasma-chemical kinetics, the overall coal conversion is conventionally divided into five major phases:

1. **Heating the coal powder particles** to temperatures required to start the chemical processes of coal decomposition;
2. **Destruction of the coal powder particles** with formation of the volatile compounds and the solid-state residual product, coke;
3. **Chemical transformations of the volatile products** from the initial coal destruction in the high-temperature conditions of thermal plasma;
4. **Heating the coke particles** to temperatures required for reactions of coke with gas-phase products of pyrolysis; and
5. **Gas–solid interphase chemical reactions** of the coke with volatile compounds and the chemically active fraction of the plasma gas.

The second, third, and fifth phases are chemical and have their special localization in the reactor. The destruction of coal (phase 2) takes place in the initial reactor zone; whereas the reactions of coal with gas-phase products (phase 5) take place in the end of the reactor, where temperatures are sufficiently high. The second and third phases begin simultaneously. Their reaction rates are determined, however, by different temperatures. Destruction of the coal powder (phase 2) is controlled by the temperature of the solid particles, whereas the

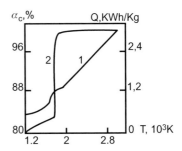

Figure 10–33. (1) Specific energy cost and (2) degree of the coal gasification as function of temperature.

transformation of the volatile products (phase 3) is controlled by the much higher temperature of the plasma gas. For this reason, the kinetic processes of the second and third phases of coal conversion are practically independent. The second phase (initial destruction of coal powder) is similar for different types of plasma-chemical coal conversion technologies, because the coal destruction is not sensitive to the type of plasma gas. In contrast, the kinetics of the third and fifth phases depends on the nature of the plasma gas and, therefore, varies with different specific processes of plasma-chemical coal conversion.

10.4.7. Kinetic Analysis of Thermal Plasma Conversion of Coal: Kinetic Features of the Major Phases of Coal Conversion in Plasma

The kinetics of the thermal conversion of coal at relatively low temperatures (≤ 1200 K) and relatively low rates of heating ($\leq 10^4$ K/s) have been discussed in Sections 10.2–10.4. There is less kinetic information related to higher temperatures and higher rates of coal heating. Some of the results obtained at higher temperatures are considered next. It was shown by Beiers et al. (1985) that the rate of plasma pyrolysis of coal (40 mass % of volatiles in dry organic fraction) does not depend on the plasma gas temperature in the temperature range from 1560 to 2180 K. The effect is explained by taking into account that the formation of volatile compounds starts for this case at a temperature of about 1220 K, which remains fixed during release of the volatiles. The degree of formation of the volatile compounds does depend on temperature. Less than 20% of carbon is converted into gas-phase products at 1690 K, while at 2220 K the degree of formation of volatiles grows to 40%. Most volatiles released at lower temperatures are determined to be tar, which is decomposed to lighter gas-phase compounds at higher temperatures. Generally, the kinetic analysis shows that coal pyrolysis at any temperature results in the formation of the same primary products, which then react in the gas phase. In other words, the formation mechanism of the volatile compounds during coal pyrolysis remains mainly the same at lower temperatures and at higher temperatures of thermal plasma (Beiers et al., 1985). As has been shown by Lesinski et al. (1985), the primary destruction of coal (phase 2 of the thermal plasma process) does not depend on the nature of the plasma gas. The formation of volatile compounds is so intensive, and their flux from the coal particles is so high, that it prevents effective interaction of the particles with the plasma gas. For the third phase, the volatile compounds react with very-high-temperature plasma gases. The thermal plasma of the inert gases and hydrogen stimulates decomposition of the volatile hydrocarbons from their initially relatively large structure to simple ones, mostly through the formation of acetylene. The application of nitrogen plasma leads to the formation of hydrogen cyanide (HCN) and cyanogen ($(CN)_2$). During this stage, application of the water vapor plasma results in the formation of hydrogen, carbon monoxide (CO), and carbon dioxide (CO_2). The next chemical stage is the interaction of coke with high-temperature gas components (phase 5). This stage is kinetically dominated by the following complex (non-elementary) reactions of solid carbon with different gases:

$$C_{(s)} + CO_2 \rightarrow CO + CO, \tag{10–29}$$

$$C_{(s)} + H_2O \rightarrow CO + H_2, \tag{10–30}$$

$$C_{(s)} + 2H_2 \rightarrow CH_4, \tag{10–31}$$

$$C_{(s)} + O_2 \rightarrow CO_2, \tag{10–32}$$

$$2C_{(s)} + O_2 \rightarrow 2CO. \tag{10–33}$$

According to Laurendeau (1978), rate coefficients of the first four reactions at 800 K and 0.1 atm are related as $1:3:(3 \cdot 10^{-3}):(10^5)$. Thus, the coke reaction with oxygen is the fastest (if O_2 is present in the gas phase), whereas the coke reaction with hydrogen producing methane is the slowest. The activation energy of the coke oxidation by CO_2 (10–29) is in the 13–62 kcal/mol range. The activation energy of the coke oxidation by H_2O (10–30) is higher: 32.5–84 kcal/mol. The ratio of the rate coefficients of reactions (10–33) and (10–32) leading to production of CO and CO_2 can be expressed by the Arrhenius relation,

$$k_{(10.33)}/k_{(10.32)} = A \exp(-E/RT), \tag{10–34}$$

with the Arrhenius parameters dependent on pressure. At lower pressures, $A = 10^{2.5}$ and $E = 6$–9 kcal/mol, while at higher pressures, $A = 10^{3.5}$ and $E = 12$–19 kcal/mol (Laurendeau, 1978). According to this relation, production of CO by means of coke oxidation is about 10 times faster than production of CO_2 at temperatures greater than 1200 K. At temperatures below 700 K, the production of CO by means of coke oxidation is slower than that of CO_2. The last group of reactions taking place during high-temperature coal conversion in thermal plasma is the carbon interaction with mineral components of ash. These reactions result in the formation of silicon, silicon carbide, aluminum, iron-carbide, iron-silicon compounds, and so on. At temperatures exceeding 1300 K, silicon dioxide is converted to the gas–phase compound SiO. A maximum yield of condensed-phase SiC is achieved at 2000 K; then it is converted into the gas phase in the form of SiC_2 (Messerle et al., 1985). At temperatures exceeding 3100 K, all oxides of the mineral fraction of coke get reduced to their corresponding elements. Detailed modeling of the multi-stage process of coal conversion in thermal plasma is described by Zhukov et al. (1990).

10.4.8. Coal Conversion in Non-Thermal Plasma

Coal conversion (especially partial oxidation with the production of syngas) using non-thermal plasma has been demonstrated in different strongly non-equilibrium discharges including microwave, glow, corona, DBD, and gliding arc discharges. Some of the processes are described in Section 10.5 in relation to specific discharge systems applied to coal conversion in plasma (see also Zhukov et al., 1990; Boulanov et al., 2003). An important advantage of coal gasification at relatively low temperatures is related to the simplification of accompanying technologies. This is especially true for the separation of fly ash, which is much easier to perform at lower gas product temperatures. Coal gasification in non-thermal plasma has all the typical features of plasma catalysis (see Section 10.1.1). Non-thermal plasma is not a major source of energy, but it is an accelerator or a catalyst of the process. Chemically active neutral and charged particles effectively generated in strongly non-equilibrium plasma conditions stimulate most of the gas-phase and gas-surface reactions comprising coal gasification. Coal gasification in non-thermal plasma, although very practically attractive, is much more kinetically complex than that of thermal plasma and, therefore, much less developed (Zhukov et al., 1990).

In analyzing the major stages of coal conversion (see Section 10.4.6), it is clear that some require heating of the solid particles without any non-thermal alternatives. The chemically active neutral and charged particles, generated in non-thermal plasma, are able to stimulate gas-surface reactions, but not all of those are required for gasification. As a consequence, non-equilibrium discharges, which are able to supply not only active plasma species but also some controlled level of translational temperature, have certain advantages for this case. As an application example of this type of transitional "warm" discharges, non-equilibrium gliding arc discharge can be effective for the coal conversion processes. It should be pointed out (see Section 7.1.6) that non-thermal plasma can provide additional intensive mechanisms of surface heating required for the stimulation of surface reactions leading to coal gasification. These non-equilibrium mechanisms are mostly due to surface recombination and surface relaxation processes, which are often more intensive on the surfaces than in the plasma volume. Some specific approaches to coal conversion in non-thermal plasma conditions are to be discussed in the next section.

Table 10–9. Conversion of Coal in Thermal Plasma of Argon, Hydrogen, and Their Mixtures

Discharge power, kW	Plasma gas	Coal feeding rate, g/min	Carbon conversion from coal to C_2H_2	Energy cost, kWh/kg C_2H_2	Process and flow organization
7	Ar	0.05	37	2500	Axial
6	Ar	1.0	35	–	Radial
4–8	Ar	6.0	10–20	100	Radial
1.7–4.9	Ar	8.3–22.4	2.6–15.4	50–190	Radial
4.2	Ar–H_2	0.05	37	–	Radial
13.6	Ar–H_2	0.05	74	>6000	Axial
4.3	H_2	0.08	18	–	Radial
30	H_2	80–275	7–18	15.5	Ring System
250–350	H_2	1000–3300	20	11.0	–

10.5. THERMAL AND NON-THERMAL PLASMA-CHEMICAL SYSTEMS FOR COAL CONVERSION

10.5.1. General Characteristics of Coal Conversion in Thermal Plasma Jets

In coal conversion in a thermal plasma jet, the coal particles are not introduced directly in the thermal plasma or, for example, an electric arc. Here the coal particles are introduced in the high-temperature jet produced in the thermal discharge. Coal conversion in high-temperature plasma jets is characterized by high selectivity. Products contain only substances stable at high-temperatures, such as H_2, C_2H_2, HCN, CO, and CO_2. Higher hydrocarbons and tar are practically absent in the products. The specific product composition is determined by the type of coal and by the plasma gas. The temperature of the gas and coal particles has much less influence on the product composition during coal conversion in thermal plasma jets. Product yield is determined in this case by the type of coal and the degree of heating of the coal particles by the high-temperature plasma gas. The degree of heating of the particles depends on their size, the type of the plasma-chemical reactor, and the degree of mixing of the solid particles with the hot plasma gas. The specific processes of coal conversion in thermal plasma jets are to be classified according to the type of plasma gas, starting with pyrolysis in plasma-heated argon, hydrogen, and their mixtures. This type of plasma-chemical coal conversion has been widely investigated using arc discharges with power mostly up to 100 kW, and with either radial (Bond et al., 1963; Graves, Kawa, & Hiteshue, 1966; Bond, Ladner, & McConnel, 1966a,b; Chakravartty, Dutta, & Lahiri, 1976) or axial (Nicholson & Littlewood, 1972) supply of the plasma gases.

10.5.2. Thermal Plasma Jet Pyrolysis of Coal in Argon, Hydrogen, and Their Mixtures: Plasma Jet Production of Acetylene from Coal

The Ar–H_2 plasma jet pyrolysis of coals with a high degree of metamorphism leads to the formation of a single major gas-phase product: acetylene (C_2H_2). The addition of hydrogen does not change many characteristics of the pyrolysis; it only increases the yield of C_2H_2. Some major characteristics of Ar–H_2 plasma pyrolysis of coals are summarized in Table 10–9 (Zhukov et al., 1990). For coal pyrolysis in an argon plasma jet, an increase of the coal feeding rate leads to a decrease of the conversion degree of carbon from coal to acetylene and to a decrease of the energy cost for the formation of acetylene (see the table). Acetylene yield is proportional to the mass fraction of the volatile compounds in coal. This effect is illustrated in Fig. 10–34. When coal particles are supplied along the radius, the maximum yield of acetylene from coals containing 50% volatiles is 20%

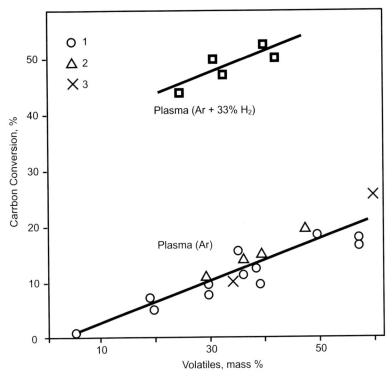

Figure 10–34. Conversion of coal carbon into acetylene at different values of mass fraction of volatile coal components (points 1, 2, and 3 correspond to different experimental systems, but are in good correlation among themselves).

(Bond et al., 1966a,b; Chakravartty et al., 1976). Axial injection of coal particles through a hollow cathode leads to a maximum yield of C_2H_2 that is close to the total amount of volatile compounds in the coal (Nicholson & Littlewood, 1972). In contrast to the conversion of coals with high degrees of metamorphism, thermal plasma treatment of lower-quality coals results in a decrease of acetylene production and an increase of CO and H_2 generation with growth of the total amount of volatiles in the coals. Generally, plasma conversion of lower-quality coals with lower degrees of metamorphism is more effective, not in the production of acetylene but for syngas, the CO–H_2 mixture. Addition of 10% hydrogen into the Ar plasma results in roughly a doubling of the acetylene yield in the products of pyrolysis (Bond et al., 1966a,b; Nicholson & Littlewood, 1972). The plasma pyrolysis of coal in pure H_2 is characterized by growth of the acetylene yield with an increase in the amount volatiles in the coals up to 34%, which is shown in Fig. 10–35 (Bittner, Baumann, & Peucker, 1981). Acetylene yield is close to the total amount of volatiles, and its concentration in the gas is about 5–9%. Note that the acetylene yield in plasma pyrolysis of coal is practically independent of the amount of mineral components in the coal (Chakravartty et al., 1976).

The release of volatile compounds starts in the H_2 plasma when the temperature of the coal particles reaches about 1220 K (Beiers et al., 1985). This critical temperature of volatiles formation is almost independent of the plasma-gas temperature. The temperature of the coal particles remains fixed during the intensive generation of volatile compounds, which prevents effective heat exchange between the particles and the plasma gas. Coal-to-carbon conversion, therefore, also remains almost fixed at different plasma-gas temperatures. While C_2H_2 is a major product of plasma coal pyrolysis in Ar and H_2, conventional by-products are

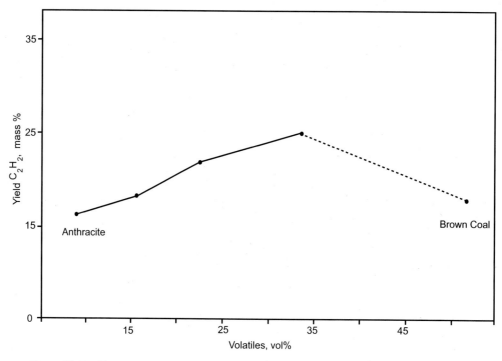

Figure 10–35. Plasma conversion of coal into acetylene in hydrogen plasma. Yield of acetylene as a function of mass fraction of volatile coal components.

C_1–C_4 hydrocarbons, CO, CO_2, sulfur and nitrogen compounds, soot, and coke. The yield of the by-products depends on the type of coal and conditions of the pyrolysis. The by-product composition and some typical parameters of coal pyrolysis in a hydrogen plasma with a vortex and magnetic arc stabilization are summarized in Table 10–10 (Peuckert & Muller, 1985; Muller, Peuckert, & Krebs, 1987). Mechanisms of C_2H_2 formation from the light and heavy primary products of coal destruction in thermal plasma jets have been analyzed by

Table 10–10. Plasma Pyrolysis of Coal in Hydrogen Plasma Using Electric Arcs with Vortex and Magnetic Stabilization

Process parameter	Vortex stabilization	Magnetic stabilization
Discharge power, kW	250–350	400
Hydrogen flow rate, standard m^3/h	60–200	200
Pressure, atm	0.2–10	0.2–10
Yield C_2H_2, mass %	30–40	20
Energy cost, kWh/kg C_2H_2	11.0	–
Yield of C_4H_2 (in kg) per 1 ton of C_2H_2	30–40	34
Yield of C_2H_4 (in kg) per 1 ton of C_2H_2	40–80	15
Yield of CH_4 (in kg) per 1 ton of C_2H_2	50–110	36
Yield of C_6H_6 (in kg) per 1 ton of C_2H_2	20	–
Yield of H_2 (in kg) per 1 ton of C_2H_2	20–100	65
Yield of CO (in kg) per 1 ton of C_2H_2	400–700	550
Yield of CO_2 (in kg) per 1 ton of C_2H_2	20–30	20
Yield of N_2 (in kg) per 1 ton of C_2H_2	–	34
Yield of H_2S, COS, CS_2 (in kg) per 1 ton of C_2H_2	30–40	–
Yield of HCN (in kg) per 1 ton of C_2H_2	30–40	–

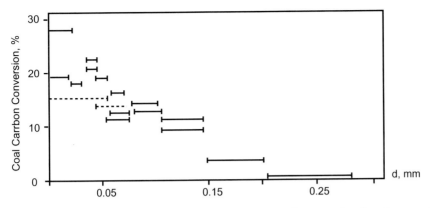

Figure 10–36. Coal carbon conversion into acetylene in argon plasma as a function of average diameter of coal particles.

Beiers et al. (1985). In these systems, most of the acetylene was not produced from primary methane released from coal, but came from hydroaromatic compounds as well as aromatic hydrocarbons including benzene, toluene, ethyl benzene, and anthracite oil.

10.5.3. Heating of Coal Particles and Acetylene Quenching During Pyrolysis of Coal in Argon and Hydrogen Plasma Jets

Acetylene is formed during plasma jet pyrolysis from volatile compounds released from solid coal particles. Therefore, the acetylene yield is determined by the completeness of volatile release, which is related to the rate and depth of heating the particles, effectiveness of mixing the particles with the plasma gas, size distribution of particles, temperature, and time of the reaction. Heating the coal particles is a major factor determining the completeness of releasing volatile compounds from coal. The release of volatile compounds in plasma jets is often incomplete, although the plasma-gas temperatures significantly exceed the required values of about 1000–1200 K. For example, pyrolysis of coal in a hydrogen plasma jet at a temperature of 1560–2180 K leaves 6.6–7.7% of volatiles in coke (Beiers et al., 1985). According to Graves et al. (1966), pyrolysis of 44 μm coal particles, initially containing 37.2% volatiles, in an argon plasma jet with temperatures of 4000–8000 K leaves 12.3–17.4% of volatile compounds in the coke. Not completing the volatile formation can be caused by some particles missing the hot zone or that their residence time is insufficient. Coal particles smaller than 60 μm can be heated to 1200 K with ideal mixing during a time interval quite sufficient for maximum production of acetylene (Bittner et al., 1981). For this reason, missing the hot zone is probably the main cause for the incomplete release of volatile compounds from the coal particles. The second factor important for heating and pyrolysis of the coal particles is their size, which is illustrated in Fig. 10–36. Particles with diameters exceeding 0.2 mm are difficult to heat and pyrolize (Bond et al., 1966a,b; Chakravartty et al., 1976). The best results are achieved with coal particle sizes in the range 10–50 μm. Particles smaller than 30 μm have a tendency to agglomerate; they are also pushed out from the plasma jet by thermophoretic forces (Sheer, Korman, & Dougherty, 1979).

Coal pyrolysis often leads to intensive formation of small and mostly spherical particles with sizes of 60–90 nm (Kozlova & Kukhto, 1976; Shimansky, 1979). They originate from fast plasma heating of the coal particles and intensive formation of volatile compounds inside of them. This leads to an explosive decay of the coal particles into very small sizes. Another source of the small particles is the formation of soot in the plasma system from the preliminarily formed acetylene (Razina et al., 1973). Other factors affecting heating and pyrolysis of coal particles are the plasma-gas temperature and residence time of the

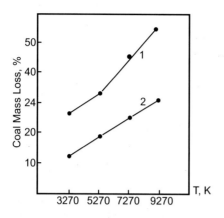

Figure 10–37. Mass losses by coal powder ($d \approx 0.1$ mm) as a function of temperature during pyrolysis in argon plasma; residence time: (1) 46.6 ms; (2) 23.5 m.

particles in the reactor. As was already mentioned, the plasma-gas temperature does not much affect the pyrolysis, which is illustrated in Fig. 10–37. The influence of residence time at a fixed temperature is somewhat stronger. Doubling the residence time results in halving the amount of volatiles in the plasma-treated coal particles. To summarize, the major factors determining the completeness of heating, release of volatiles, and formation of acetylene in the process of coal pyrolysis in argon–hydrogen plasma jets are the size of the coal particles and the effectiveness of getting them into the hot zone of the plasma jet.

Plasma pyrolysis of coal in Ar–H_2 jets with C_2H_2 generation requires fast product quenching. The highest yields of acetylene can be achieved when the quenching is in the initial zone of the reactor (see Table 10–11; Chakravartty et al., 1976). As an example, the maximum yield of acetylene in the pyrolysis process of 60 μm coal particles in a hydrogen plasma jet is achieved at a residence time of $4 \cdot 10^{-4}$ s with a 20 cm reactor length (Bittner et al., 1981). The maximum yield of acetylene requires quenching with a product cooling rate of about 10^7 K/s. Mixing of the products with cold hydrogen is effectively used as such a quenching process. The two-stage quenching process, which includes quenching of the products by gaseous hydrocarbons in the first stage and water quenching in the second stage, is also effective in this case (Zhukov et al., 1990). To avoid problems with the quenching of acetylene, coal can be pyrolyzed in hydrogen plasma in the presence of CaO particles (Kim et al., 1979). In this case, the process results in the formation of calcium carbide and CO:

$$CaO + 3C \rightarrow CaC_2 + CO(g). \tag{10–35}$$

The pyrolysis products are quite stable in this case. This plasma process also has an advantage compared to the conventional process described by the same chemical formula (10–35). The conventional process requires the application of coke, whereas the plasma-chemical one starts directly from coal. In the experimental system of plasma production of calcium carbide (CaC_2), the size of the coal particles is 8–44 μm and the molar ratio between carbon (of coal) and CaO is in the range 1–3. The experimentally achieved conversion degree in the optimal regime of the process is 90% for carbon and 70% for CaO. Achieving higher

Table 10–11. Acetylene Yield During Coal Pyrolysis in Argon Plasma Jet Depending on Location of Quenching Zone in the Reactor

Mass % of volatile compounds in initial coal	Mass % of C_2H_2 in products; quenching in initial (upper) zone	Mass % of C_2H_2 in products; quenching in middle zone	Mass % of C_2H_2 in products; quenching in final (lower) zone
37.1	13.5	6.5	5.0
44.6	20.0	8.0	7.5

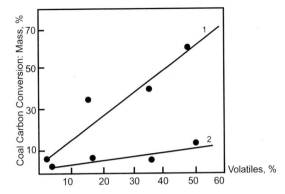

Figure 10–38. Yields of (1) HCN and (2) acetylene during pyrolysis of coal in nitrogen plasma as a function of percentage of volatiles in the coal.

conversion degrees for CaO is limited by the formation of a calcium carbide film on the surface of CaO particles, which is not permeable to carbon fragments.

10.5.4. Pyrolysis of Coal in Thermal Nitrogen Plasma Jet with Co-Production of Acetylene and Hydrogen Cyanide

Pyrolysis of coal effectively occurs in a plasma jet of nitrogen or Ar–N_2 mixture (Ar + 10% N_2). The major products of pyrolysis are acetylene, hydrogen cyanide (HCN), and CO. When the coal pyrolysis occurs in a plasma jet of pure nitrogen, the major product of pyrolysis is HCN, whereas acetylene dominates if the amount of nitrogen in the Ar–N_2 mixture is relatively small. The yield of HCN in the pyrolysis in the pure nitrogen plasma jet is close to the amount of volatile compounds in the coal. Acetylene yield in this case is about 3–5%. Generally, yields of acetylene and HCN in coal pyrolysis in a nitrogen plasma jet are proportional to the amount of volatile compounds in the treated coal, which is illustrated in Fig. 10–38 (Bond et al., 1966a,b; Zhukov et al., 1990).

10.5.5. Coal Gasification in a Thermal Plasma Jet of Water Vapor

Generally the gasification of any carbon-based solid fuel can be described by the formula

$$C(s) + H_2O \rightarrow CO + H_2. \tag{10–36}$$

This process requires high-temperatures for reasonable reaction rates, and it is very endothermic. To provide necessary heat into the system, a portion of the carbon should be burnt. Heat in this case is released by an exothermic combustion process:

$$C(s) + O_2 \rightarrow CO_2, \tag{10–37}$$

$$2 \cdot C(s) + O_2 \rightarrow 2 \cdot CO. \tag{10–38}$$

To provide the required energy balance for reactions (10–36)–(10.38), the amount of CO_2 in the products of conventional gasification is relatively high. The conventional industrial coal gasification processes of Coppers-Totzec, Winkler, and Lurgi produce 12.6, 15.7, and 30.6%, respectively, of CO_2 (Zhukov et al., 1990). The application of plasma jets provides an independent source of energy, thereby opening wide possibilities for process control and significantly diminishing the production of CO_2. Coal conversion in a plasma jet of water vapor results in the production in the gas phase of mostly (about 95%) syngas, the H_2–CO mixture. Usually, syngas contains more hydrogen than carbon monoxide (H_2:CO > 1). The conversion products also contain CO_2, usually about 3%, and some traces of methane and hydrogen sulfide. The yield of CO_2 is much smaller in plasma than in conventional technologies. Also the plasma conversion of coal does not produce tars, phenols, or polycyclic hydrocarbons, which decreases the costs of syngas purification and makes plasma technology more preferential from an ecology standpoint. Coal gasification with water vapor

Table 10–12. Typical Material Balance of Plasma-Chemical Coal Gasification Process in DC Plasma Jet by Water Vapor

Moisture contained in coal	8.2 mass %
Inorganic compounds contained in coal	32.2 mass %
Volatile compounds with respect to organic mass of coal	58.9 mass %
Carbon (C) contained in coal before gasification	57.7 mass %
Hydrogen (H) contained in coal before gasification	6.2 mass %
Sulfur (S) contained in coal before gasification	4.2 mass %
Nitrogen (N) contained in coal before gasification	1.1 mass %
Oxygen (O) contained in coal before gasification	31.0 mass %
Electric discharge power	22–25 kW
Water vapor flow rate	0.4–0.8 g/s
Coal feeding rate	0.3–0.6 g/s
Argon flow rate	0.3–0.6 g/s
Amount of dry gas-phase products generated from 1.05 kg of coal and 1.45 kg of water vapor	2.45 standard m_3
Amount of syngas (H_2–CO) generated from 1.05 kg of coal and 1.45 kg of water vapor	1.90 standard m^3

in atmospheric-pressure thermal plasma jets is usually based on one of the following two approaches. In the first approach, the cold coal powder is injected into the atmospheric-pressure plasma jet of water vapor (Dobal, 1978; Sheer et al., 1979; Krukovsky & Kolobova, 1981, 1982; Kolobova, 1983). In the second approach, both the coal particles and the water vapor are heated in plasma jets of some other gases (Herlitz & Santen, 1984; Camacho, 1985).

10.5.6. Coal Gasification by H_2O and Syngas Production in Thermal Plasma Jets: Application of Steam Plasma Jets and Plasma Jets of Other Gases

The injection of cold coal powder in an atmospheric-pressure plasma jet of water vapor was investigated by Dobal (1978), Kolobova (1983), Krukovsky and Kolobova (1981, 1982), and Sheer, Korman, and Dougherty (1979). For example, in the plasma system of Sheer, Korman, and Dougherty (1979), coal particles were injected in the plasma through a channel in the cathode. The system has a discharge power of 120 kW, consumption of electric energy of 3.3 kWh per kg of coal, coal particle size of 50–80 μm, and coal-particle residence time in the water vapor plasma jet of about 1 s. In this plasma system, 40–50% of the carbon contained in the coal is gasified; the rest of the carbon is transformed into very small particles with sizes of about 0.1 μm. In the experiments of Sheer, Korman, and Dougherty (1979), the product composition in a wide range of flow rates is 62 mass % H_2, 26 mass % CO, and 12 mass % CO_2. The gasification degree of coal, α_C, increases with growth of the amount of volatile compounds in the coal (Krukovsky & Kolobova, 1982). At the same time, the coal gasification degree exceeds the amount of volatiles in the coal. This means that the plasma treatment of coal provides not only gasification of the volatile compounds but also gasification of coke (Krukovsky et al., 1987). A typical material balance of plasma-chemical coal gasification is shown in Table 10–12. Simple calculations based on the data in the table prove that plasma treatment gasifies not only volatile compounds but also coke. The degree of utilization of water vapor, α_{H_2O}, as well as the gasification degree of carbon in the coal, α_C, depends on the relation between H_2O and C in the initial reagents of the gasification process. At higher values of the H_2O:C ratio, the gasification degree of carbon in the coal, α_C, grows, as does the amount of CO_2 in the products. At lower H_2O:C ratios, the gasification degree of carbon in the coal, α_C, is smaller but the utilization of water vapor, α_{H_2O}, becomes higher.

Table 10–13. Energy Balance of Plasma-Chemical Coal Gasification by Water Vapor with Heating the Reagents in Thermal Plasma Jets of Other Gases

Energy capacity of the coal introduced for gasification	6.815 kcal/kg
Energy of water vapor introduced in the system	0.910 kcal/kg
Electric energy introduced by plasma jet	1.700 kcal/kg
Introduced electric energy in coal equivalent (taking into account efficiency of production of electric energy)	4.987 kcal/kg
Total energy introduced into the system	9.425 kcal/kg
Effective total energy introduced into the system (taking into account efficiency of production of electric energy)	12.712 kcal/kg
Energy capacity of the produced gas	6.985 kcal/kg
Recuperated heat transferred to the gas	1.065 kcal/kg
Heat of gas from economizer	0.220 kcal/kg
Heat of the produced inorganic solid exhausts	0.100 kcal/kg
Energy of non-converted coal and other non-recuperated heat	1.055 kcal/kg
Total energy output balanced with introduced energy	9.425 kcal/kg
Energy efficiency of plasma gasification of coal (without taking into account efficiency of production of electric energy)	74%
Energy efficiency of plasma gasification of coal (taking into account efficiency of production of electric energy)	35%

Note: All energies are calculated with respect to 1 kg of coal introduced into the system.

With a deficiency of water vapor, some carbon from the coal remains in the products in the form of the technical carbon with relatively high specific surface area of 350–400 m^2/g. Such by-products are of interest for the production of different rubber-based materials (Kolobova, 1983).

The first approach considered earlier for plasma gasification of coal is based on coal particle injection into plasma jets of water vapor. The second approach is based on co-injection of coal and water into plasma jets of other gases. The plasma jets heat coal and water to 1570–1680 K. These temperatures are sufficient for the melting of ash, which is extracted from the system as a liquid. If, for example, the applied coal consists of 75 mass % of C, 4 mass % of H, 9 mass % of O, 1 mass % of S, 1 mass % of N, and 10 mass % of ashes, then the produced gas would consist of 54.6% H$_2$, 42.8% CO, 2% CO$_2$, 0.2% H$_2$S, and 0.2% N$_2$. The energy capacity of the produced gas is 12.4 MJ/m^3 (Zhukov et al., 1990). Such a gasification scheme consumes 65% of the total energy in the form of coal and 35% of the energy in the form of electricity. Coal gasification technology with such a high electric energy consumption becomes practical only in the presence of a cheap source of electric energy. In the system developed by Camacho (1985), coke and volatile compounds of coal produced and heated in nitrogen plasma are gasified by water vapor. The temperature inside of the reactor is 1270–1370 K, the wall temperature is 450 K, and the temperature of the exit gases is 1070 K. Conversion in the plasma system of coal with an elemental composition of 65 mass % C, 4.56 mass % H, 11.24 mass % N, 1.24 mass % S, and generally containing 7.92% moisture, 30.82% volatile compounds, and 19.99% ashes results in the production of gas with composition 53 vol % H$_2$, 25 vol % CO, 12 vol % N$_2$, 9 vol % CO$_2$, and 1–2 vol % CH$_4$. The pyrite sulfur leaves the system with the solid inorganic exhaust, while organic sulfur leads to production of some H$_2$S in the gas phase. Consumption of electric energy in the system is relatively high – about 25% of the energy capacity of gasified coal. The energy balance of such a coal gasification system based on heating the coal and water in thermal plasma jets of other gases is illustrated in Table 10–13 (Zhukov et al., 1990; Camacho, 1985).

10.5.7. Coal Gasification in Steam–Oxygen and Air Plasma Jets

Energy consumption during coal gasification in atmospheric-pressure plasma jets is significantly decreased by adding oxygen to the water vapor (Khudiakov et al., 1982; Herlitz & Santen, 1984; Kolev & Georgiev, 1987). In the system developed by Kolev and Georgiev (1987), the average temperature in the plasma jet is 3420 K and decreases to 2840 K after the injection of coal. The degree of conversion of the carbon from the coal is 96% in this steam–oxygen gasification system. The composition of the produced gas is 39 vol % H_2, 54 vol % CO, and 7 vol % CO_2. Another process of steam–oxygen coal gasification in plasma jets was developed by Herlitz and Santen (1984). In the first stage, coal powder with particles smaller than 0.1 mm is gasified in a plasma jet of water vapor and oxygen at a temperature of about 1670 K. It results in the production of syngas and 10–15% of CO_2 and H_2O. The gas produced in the first stage then reacts in the second stage with preheated coke to generate more syngas. In this process 1 ton of coal, 317 kg of water vapor, and 500 m^3 of oxygen lead to the production of gas containing 61.8 vol % CO, 34.9 vol % H_2, 2.6 vol % of CO_2, 0.3 vol % H_2S, and 0.4 vol % N_2. The consumption of electric energy in the process is relatively small – 7% from the total consumed energy (compare with 10–13). Gasification of 1000 tons of coal in such a plasma-chemical process requires only 35 MW of electric energy. Thus, coal gasification in the steam–oxygen and steam–air plasma jets is more energy effective and more similar to the plasma-catalytic systems (see Section 10.1.1).

10.5.8. Conversion of Coal Directly in Electric Arcs

In contrast to the processes where coal is injected in thermal plasma jets, coal also can be effectively converted directly in atmospheric-pressure thermal plasma of electric arcs. Coal is introduced directly in the thermal arc discharge with the plasma gas. Therefore, the coal is introduced in the very-high-temperature zone of the thermal plasma discharge in this approach, and the heating rate of the coal particles is also very high. Such thermal plasma reactors are usually not used for the conversion of gaseous and liquid hydrocarbons into ethylene, acetylene, or other valuable products. High temperatures and high concentrations of charged, excited, and chemically active species inside of the arc result in too much cracking of the gaseous and liquid hydrocarbons and the production of some undesirable products. In contrast to the treatment of gaseous and liquid hydrocarbons, the direct treatment of coal in electric arcs allows for better results regarding product yield and their energy costs. Better efficiency of the direct thermal plasma treatment of coal has been demonstrated, both in the case of coal pyrolysis to produce acetylene and in the case of coal gasification by water vapor with the production of syngas (Zhukov et al., 1990). As an example, industrial pyrolysis of coal in a hydrogen plasma with the production of acetylene is done by AVCO using the direct approach. The advantage of the direct use of an electric arc for coal conversion can be explained by taking into account that in the high-temperature arc zone the coal is only heated to temperatures required for releasing the volatiles. The release of the volatiles itself and their further cracking occurs outside of the highest temperature zone. The direct approach maximally uses the thermal arc to heat up the coal particles, while the produced gas-phase products are protected from the undesirable influence of very high-temperatures. Specific applications of the direct plasma approach both for coal pyrolysis and gasification are considered later.

10.5.9. Direct Pyrolysis of Coal with Production of Acetylene (C_2H_2) in Arc Plasma of Argon and Hydrogen

In early investigations of direct coal pyrolysis, coal powder was used as a consumable anode (Amman et al., 1964; Krukonis, Gannon, & Modell, 1974). The use of hydrogen as a quenching gas in these experiments increases the yield of acetylene two- to three-fold compared to quenching with argon. Krukonis, Gannon, and Modell (1974) proved that the use of

Table 10–14. Direct Coal Pyrolysis with Production of Acetylene in Different Thermal Plasma Systems Based on Arc Discharges in Hydrogen

Parameters of plasma-chemical process	Arc discharge with consumable anode	Arc with magnetic stabilization	Pilot arc discharge with coal injection
H$_2$ flow rate, standard m^3/hour	–	41	25.5–60
Coal feeding rate, g/min	200	650	225–900
Conversion degree of carbon from coal, %	5–12	35	35
Acetylene C$_2$H$_2$, vol %	3–4	10	15
Energy cost of acetylene production, kWh/kg C$_2$H$_2$	37–75	10	10
Discharge power, kW	–	80–100	50–180

hydrogen as a quenching gas in the system did not increase the C$_2$H$_2$ yield due to direct reactions of coal and hydrogen:

$$2 \cdot C(\text{coal}) + H_2 \rightarrow C_2H_2. \tag{10–1}$$

Due to intensive decomposition of the initially generated hydrocarbons, the yield of the products and their energy cost were not good in these early direct plasma-chemical experiments. Relevant results obtained in electric arcs with a consumable anode are summarized in Table 10–14 (Zhukov et al., 1990). Further developments of the atmoelectric arc systems for direct coal pyrolysis include injection of coal into the arc discharge zone with the hydrogen flow, as well as magnetic stabilization of the arc applied to direct pyrolysis. As seen from Table 10–14, such modifications significantly improve the acetylene yield and energy cost of the process. Axial injection of coal with an argon flow through a hole in the cathode results in an acetylene yield of 23 vol % at a temperature of 3900 K, whereas radial injection into the arc at a temperature of 5900 K leads to a lower acetylene yield of about 10% (Razina et al., 1976). Kinetic analysis of direct coal pyrolysis shows that improvement of the product yield and energy cost with the application of hydrogen as a plasma gas cannot be explained only by changes in the physical characteristics of plasma gases; it is mostly a chemical effect. C$_2$H$_2$, as a major product of the pyrolysis, is many times formed, decomposed, and formed again in the systems. The addition of hydrogen suppresses the decomposition of acetylene, which results in improvements of the product yield and energy cost.

10.5.10. Direct Gasification of Coal with Production of Syngas (H$_2$–CO) in Electric Arc Plasma of Water Vapor

Direct coal gasification with water vapor in an atmospheric-pressure electric arc discharge reaches higher coal conversions than in the case of the gasification process in plasma jets (see Sections 10.5.5 and 10.5.6). This effect has been demonstrated in the electric arc reactor with 75 kW power and magnetic stabilization of the arc (Sakipov, Messerle, & Ibraev, 1986; Vdovenko, Ibraev, & Messerle, 1987). The following parameters of the plasma-chemical process were achieved in experiments with average temperatures of 2500–3400 K, coal gasification degree $\alpha_C = 68.9–93.5\%$, degree of conversion of sulfur into gas phase of $\alpha_S = 69.0–98.8\%$, and syngas yield (H$_2$–CO) of 68.9–95.6%. It should be mentioned that values of the conversion degrees, α_C and α_S, grow for smaller sizes of coal particles. Effective direct plasma gasification in electric arcs can also be achieved by treatment of water–coal suspensions (Ryabinin, Orumbaeva, & Toboyakov, 1987). Reagents are inherently well mixed in such an approach, which helps to achieve high process efficiency. This

Table 10–15. Typical Composition of Coals Treated in Microwave Discharge System

Compounds of coal	Anthracite coal	Bituminous coal A	Bituminous coal B	Lignite coal A	Lignite coal B
Carbon (C)	91.05 mass %	89.55 mass %	81.75 mass %	66.45 mass %	67.0 mass %
Hydrogen (H)	2.5 mass %	4.65 mass %	5.55 mass %	5.4 mass %	4.1 mass %
Nitrogen (N)	0.95 mass %	1.25 mass %	0.7 mass %	0.3 mass %	1.2 mass %
Sulfur (S)	0.83 mass %	0.81 mass %	0.97 mass %	1.4 mass %	0.5 mass %
Oxygen (O)	3.0 mass %	2.3 mass %	7.9 mass %	22.8 mass %	17.5 mass %
Ash	1.8 mass %	1.5 mass %	2.05 mass %	3.6 mass %	9.7 mass %
Volatiles	6.1 mass %	20.2 mass %	39.2 mass %	44.0 mass %	–

process significantly reduces the energy consumption related to the preparation of coal for gasification, allows for a process pressure exceeding atmospheric conditions, and avoids the compression stage of syngas generated in the process.

10.5.11. Coal Conversion in Non-Equilibrium Plasma of Microwave Discharges

In contrast to the aforementioned thermal plasma discharges generally close to thermodynamic quasi equilibrium, the microwave discharges to be considered in the following two sections are essentially far from thermodynamic equilibrium. The electron temperature in these discharges is about 3 eV and greatly exceeds the translational temperature ($T_e \gg T_0$). This leads to an extremely high super-equilibrium density of chemically active species, which significantly accelerates most of the coal conversion processes. Although the translational gas temperature is much lower than that of the electrons, it can still be high in these systems (up to 2000–3000 K and higher). This is why microwave discharges used for coal conversion are not always "completely non-thermal" and are sometimes classified as "warm" discharges. The conversion of different coals in different gases was investigated in detail using microwave discharges with a frequency of 2.45 GHz and voltages of 35–50 V. The experiments were done in static conditions (Fu & Blaustein, 1968, 1969; Fu, Blaustein, & Wender, 1971; Fu, Blaustein, & Sharkey, 1972) as well as in a gas flow (Fu, 1971; Fu et al., 1971; Nishida & Berkowitz, 1972a,b). In contrast to the thermal plasma systems discussed in the preceding sections, the experiments with non-equilibrium microwave discharges were preferred at relatively low pressures. The pressure of gases added to the system for coal conversion is in the range 600–6000 Pa. Obviously, treatment of the coal leads to release of the volatile compounds and to gasification of the solid particles. In the flow systems, gases and gas mixtures (Ar, Ar + H$_2$, Ar + CO$_2$, Ar + H$_2$O, N$_2$, N$_2$ + H$_2$) were introduced with a flow rate of 1–10 cm^3/min. Coal in the systems was located in a vertical reactor tube on a grid. To account for the conductivity of the coal particles, they were placed either in the microwave zone itself or immediately below the zone along the gas flow. The residence time

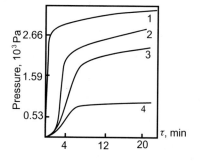

Figure 10–39. Release of volatile compounds from (1) lignite, from bituminous coal with (2) high and (3) low amount of volatiles, and from (4) anthracite.

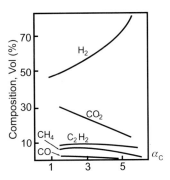

Figure 10–40. Composition of treatment products of bituminous coal with high concentration of volatiles in a microwave discharge at different levels of gasification.

of the coal in the microwave system was varied from minutes to hours. Typical compositions of coal treated in the microwave discharge system are shown in Table 10–15. The release of volatile compounds is mostly characterized using some induction period (Fu & Blaustein, 1969), which can be seen in Fig. 10–39. Generation of the volatile compounds in static conditions leads to sharp pressure increases followed by saturation. Even with an increase of pressure, the non-equilibrium microwave system stays in the low-pressure regime (see Chapter 4 and Fridman & Kennedy, 2004).

Coal conversion in non-equilibrium low-pressure plasma can be divided into three stages. The first stage is related to the induction period, when tar is formed on the reactor walls. Formation of volatiles starts in the second stage as a result of partial decomposition of the tar. The second stage starts when the pressure in the system reaches 70–130 Pa. Concentrations of electrons, ions, atoms, radicals, and other chemically active species become relatively high and this results in the acceleration of coal decomposition. The formation rate of the volatile compounds increases to the amount of volatiles in the treated coal. The quantity of generated gases is comparable to that of thermal coal decomposition at 1200–1300 K. The composition of the gas-phase products formed in non-equilibrium plasma (see Figs. 10–40 and 10–41 and Table 10–16) is quite different from that of the thermal systems (Fu & Blaustein, 1969). Specifically, products of coal conversion in non-equilibrium microwave discharge plasma contain much more H_2, CO, and C_2H_2, but less CH_4 and CO_2 with respect to the case of thermal treatment of coal. The third stage of the process is gasification of the residual coke, which proceeds very slowly in the microwave discharge. It has been demonstrated that the amount of carbon in hydrocarbon gas products can essentially exceed (by more than 60%) the amount of carbon in the volatile compounds of coal. This provides for the effective gasification of the residual coke during non-equilibrium

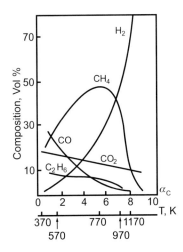

Figure 10–41. Composition of products of conversion of coal at different temperatures and degrees of gasification.

Table 10–16. Yield of Products for Coal Conversion in Microwave Discharge, Static Regime (in 10^{-4} mol/g of coal)

Products of coal conversion	Bituminous coal, treatment 30 min, without Ar	Bituminous coal, treatment 30 min, with Ar (1300 Pa)	Lignite coal, treatment 10 min, without Ar	Lignite coal, treatment 20 min, with Ar (1300 Pa)
H_2O	30	14	75	43
CO	3.6	6.7	0.6	9.8
CO_2	0.7	1.3	11.9	14.2
H_2	3.2	17.6	0.9	12.9
CH_4	0.5	0.4	0.3	0.2
C_2H_2	48.4	85.2	50.2	38.2
C_2H_4	3.6	6.1	1.0	0.8
C_2H_6	9.4	5.4	3.2	1.3
C_3H_6	3.8	0.9	0.6	0.3
C_3H_8	2.7	0.7	1.0	0.3
C (from coal) converted into gas hydrocarbons	21.7 mass %	30.4 mass %	22.8 mass %	19.3 mass %

microwave plasma pyrolysis. Generally, the microwave plasma pyrolysis of coal results in a lower conversion degree into acetylene than in the case of the application of a thermal arc discharge using the same plasma gases of Ar or Ar–H_2 (Zhukov et al., 1990). Analysis of coal pyrolysis in a continuous-flow microwave discharge plasma (Fu et al., 1971) shows that an increase of the H:C ratio in the treated coal leads to an increase of the gasification degree of the coal. For example, the continuous-flow microwave discharge plasma treatment of bituminous coals with low H:C ratio and a high fraction of volatile compounds results in a gasification degree of 36.2%. In similar discharge conditions, the continuous-flow, non-equilibrium microwave plasma treatment of kerogen with a higher H:C ratio results in a much higher gasification degree of 69.1%. The maximum conversion degree in the system is usually achieved after 2–4 min of plasma treatment and does not increase after that time. The composition of products and the conversion degrees achieved in the coal pyrolysis in continuous-flow microwave discharge in the different gases Ar, Ar–H_2, and Ar–CO_2 (Fu et al., 1971) are illustrated in Table 10–17. The table shows that application of the Ar–H_2 mixture results in higher yields of hydrocarbons, whereas application of the Ar–CO_2 mixture leads to higher yields of CO and higher degrees of coal gasification.

10.5.12. Coal Conversion in Non-Equilibrium Microwave Discharges Containing Water Vapor or Nitrogen

Gasification of coal in microwave discharges in Ar–H_2O mixtures has been studied at relatively low pressures (600–6000 Pa), similar to those described in the previous section. Coal gasification in this case results in a higher degree of coal conversion as well as in higher yields of hydrocarbons and CO with respect to a similar discharge in pure argon (Fu & Blaustein, 1969; see Table 10–18). This can be explained through the fact that the atoms O and OH radicals from water better "open" the coal structure, while hydrogen from the water promotes additional gasification. The major gas-phase product in these conditions is acetylene and the yield dependence on the amount of volatiles is similar to that for the Ar–H_2 mixture. Gasification of coal in a microwave discharge in molecular nitrogen was also studied in the relatively low-pressure range of 600–6000 Pa under conditions similar to those described earlier. The gasification process in nitrogen leads to significant formation

Table 10–17. Yield of Products (10^{-4} mol/g of coal) for Lignin Coal Conversion in Microwave Discharge in Different Gases, Continuous Flow Regime (treatment time 30 min)

Products of coal conversion	Plasma gas: Ar, 5 Torr	Plasma gas: Ar:H_2 = 1:4, 25 Torr	Plasma gas: Ar:CO_2 = 1:4, 25 Torr
C_2H_2	7.5	9.9	2.4
C_2H_4	4.1	2.1	0.3
C_2H_6	0.7	0.6	0.2
C_3H_8	0.1	0.1	<0.1
C_4H_2	0.6	0.5	0.1
C_6H_6	0.2	0.2	<0.1
CO_2	17.1	14.6	31.1
HCN	1.3	1.3	0.6
H_2O	3.9	4.8	21.4
Mass losses of coal, mass %	42.5	43.7	51.0

of C_2N_2, CH_3CN, and C_2H_5CN (Fu et al., 1972; Nishida & Berkowitz, 1972). Formation of the compounds is attributed to super-equilibrium concentrations of atomic nitrogen, vibrationally excited nitrogen molecules, and electronically excited metastable nitrogen $N_2(A^3\Sigma_u^+)$. Ignition of the discharge in static conditions (without gas flow) leads to rapid pressure growth in the reactor with saturation reached after 40–60 s. Pressure growth in the reactor does not depend on the initial nitrogen pressure (in the range 600–6000 Pa). In these conditions, the microwave plasma conversion of bituminous coal typically leads to the formation of $1.8 \cdot 10^{-2}$ mol of gas-phase products per gram of coal. Yields of gasification products and the degree of coal conversion in the microwave discharge in nitrogen under the static regime with continuous frozen products (Fu et al., 1972; Zhukov et al., 1990) is shown in Table 10–19.

Gasification of coal in a continuous-flow microwave discharge in nitrogen was investigated by Nishida and Berkowitz (1972). The conditions for this non-equilibrium plasma system are 300 Pa nitrogen pressure and 5 cm^3/min N_2 flow rate; the major gasification products are HCN and C_2N_2. The rate of C_2N_2 formation is 0.1–0.2 with respect to that of HCN. An increase in the degree of metamorphism of the coal (and amount of carbon in the coal) does not affect the yield of C_2N_2, whereas the yield of HCN significantly decreases

Table 10–18. Yield of Products (10^{-4} mol/g of coal) for Microwave Plasma Conversion of Coal in Ar–H_2O Mixtures (static regime, treatment time 1 min)

Products of coal conversion	H_2O pressure 7.3 Pa, Ar pressure 5.0 Pa	H_2O pressure 10.0 Pa, Ar pressure 7.7 Pa	H_2O pressure 40.5 Pa, Ar pressure 8.2 Pa	H_2O pressure 11.0 Pa, Ar pressure 8.8 Pa
H_2	45.4	20.9	56.4	89.0
CH_4	0.6	1.1	2.0	1.5
C_2H_2	10.4	13.6	24.4	10.0
CO	42.4	25.8	44.8	87.2
CO_2	0.8	0.5	2.0	5.0
C (coal) conversion to all gas products, mass %	8.7	7.6	14.8	20.9
C (coal) conversion to hydrocarbon gas products, mass %	3.0	4.0	7.9	4.1

Table 10–19. Yield of Products (10^{-4} mol/g of coal) for Microwave Plasma Conversion of Coal in Molecular Nitrogen (static regime, treatment time 26 min, initial pressure 4000 Pa)

Products of coal conversion	Yield of products, 10^{-4} mol/g, C (coal) conversion degree, mass %
HCN	232
C_2H_2	28
C_2H_6	<0.1
C_3H_6	0.1
C_3H_8	<0.1
C_4H_2	<0.1
CO_2	1.6
H_2O	5.2
C (coal) conversion to all gas products, mass %	42.5

from 30 to 1–2 cm^3/g. The application of nitrogen and hydrogen together as a plasma gas leads to a 10–50-fold increase of the HCN production rate to 4–5 cm^3/min per gram of coal. The N_2:H_2 ratio does not signicantly affect the rate and selectivity of product formation. Kinetic analysis of the coal gasification in non-equilibrium N_2 plasma shows that HCN is formed mostly from the non-aromatic coal fractions and C_2N_2 from the aromatic fractions. Rates of formation of HCN and C_2N_2 are proportional to the coal surface area, which proves the surface nature of the reactions. The diffusion of nitrogen inside the coal particles does not make a significant contribution to the formation of HCN and C_2N_2.

10.5.13. Coal Conversion in Low-Pressure Glow and Other Strongly Non-Equilibrium Non-Thermal Discharges

In the aforementioned non-equilibrium microwave discharges, the electron temperature (about 3 eV) is much higher than the gas temperature, but the gas temperature is still elevated (up to 2000–3000 K). Therefore, the microwave discharges cannot generally be qualified as non-thermal. In contrast, the low-pressure glow and other strongly non-equilibrium discharges, to be discussed next, are characterized by room or slightly elevated temperatures and can be qualified as non-thermal. Sanada and Berkowitz (1969) considered coal gasification in a non-equilibrium, high-voltage, 120 Pa discharge in hydrogen. The major product of the cold plasma process is methane. The CH_4 formation rate at a 30 mA discharge current increases four-fold (from 0.1 to 0.4 cm^3/min per gram of coal) when the voltage is increased from 5 to 6 kV. A pressure increase from 10 to 130 Pa leads to an increase of the CH_4 generation rate with saturation. A typical composition of the gasification products can be characterized as CH_4:C_2H_2:C_2H_6:C_3H_8 = 150:4:1:1. Hydrocarbons C_2–C_4 are mostly generated from non-aromatic fraction of coal.

Kinetic analysis shows that the CH_4 formation process in the preceding system takes place on the coal surface. Hydrogen atoms, generated in plasma volume, react with CH_3- and CH_2-groups on the surface and form methane. Formation of higher hydrocarbons takes place with aliphatic and cyclic hydrocarbons released from the coal reacting with atomic hydrogen created in the plasma (Sanada & Berkowitz, 1969; Kobayashi & Berkowitz, 1971). Scott and Venugopalan (1976) discuss the conversion of bituminous coal in electrodeless discharge with 26 kV voltage, 2.6 mA current, and 60 Hz frequency. The process takes place in the system at room temperature. Major products of the cold plasma-chemical coal treatment are hydrogen and carbon monoxide, with a small amount of methane. The addition of hydrogen in the discharge zone stimulates the generation of methane. EPR analysis shows that plasma treatment increases the spin concentration in coal by 2.5-fold, which indicates a radical-based mechanism for coal gasification (Scott, Chu, & Venugopalan, 1976).

Rusakova and Eremin (1975a,b) analyzed non-thermal plasma gasification of anthracite by carbon dioxide in a low-pressure glow discharge with a 100 mA direct current. The stationary concentration of carbon monoxide is achieved in these experiments after about an hour of coal treatment. The maximum CO concentration is 93.2% of the initial CO_2, the pressure is 5340 Pa, and the coal location is on the cathode. In similar conditions, if the coal sample is located on the anode of the glow discharge, the maximum CO concentration is a little lower at 87%. Formation of CO in the glow discharge system is due in particular to the interaction of CO_2^+ ions with coal located on the cathode. This leads to secondary electron emission and CO_2 surface dissociation with the formation of CO gas and chemisorbed atomic oxygen. The oxygen atom chemisorbed on the coal surface potentially leads to the formation of an additional CO molecule. Generally the low temperature of coal surfaces in the considered systems prevents high rates of coal gasification. Optimization of such systems require the application of non-equilibrium surface heating (see Section 7.1.6), or a combination of cold non-equilibrium discharges with external control of the coal temperature.

10.5.14. Plasma-Chemical Coal Conversion in Corona and Dielectric Barrier Discharges

Gasification and direct liquefaction of coal in strongly non-equilibrium, non-thermal, atmospheric-pressure discharges is interesting due to its simplicity and potential for practical application. DBDs, atmospheric-pressure glow discharges, as well as pulsed corona discharges can all be used for coal gasification and direct liquefaction. Effective large-scale operation of the relatively cold atmospheric-pressure plasma processes is still challenging and requires a better understanding of the surface reaction kinetics and multi-phase non-thermal plasma fluid dynamics. Coal gasification and liquefaction in a 20 kV corona discharge was analyzed in particular by Dibelius et al. (1964) and Zhukov et al. (1990). The coal powder in these experiments was located between the electrodes and hydrogen was used as a plasma gas. Hydrogen atoms formed in the discharge attach to non-saturated aromatic bonds, which leads to hydrogenization of coal and its conversion to liquid-phase and gas-phase hydrocarbons. Yields of liquid fuels in the system are similar to those in thermal plasma, but without the formation of tar. Electric conductivity of the initial coal particles in experiments with atmospheric-pressure corona discharge was relatively low. However, the coal conductivity grew during the plasma treatment, which resulted at some point in short-circuiting and extinguishing the discharge. The problem can be solved by locating the coal particles outside the discharge, which was considered by Boulanov et al. (2003).

Boulanov et al. (2003) proposed experiments with low-temperature plasma of DBDs and gliding arcs to convert coal into gas and liquid fuel with relatively low process outlet temperature (below 450°C). Two working approaches were considered. In the first one, plasma is used as a volumetric catalytic agent producing radicals and ions inside the reactor. Produced active particles then promote chemical reactions of coal gasification at a temperature of 450°C or less. These temperatures are sufficiently low to simplify cleaning the produced gases. In the second approach, a low-temperature gliding arc plasma is used to produce hydrogen atoms from natural gas. The hydrogen atoms increase the H:C ratio in coal by reacting with char. An increase of the H:C ratio to 2:1 from the typical 0.8 stimulates production of liquid fuel from coal.

10.6. ENERGY AND HYDROGEN PRODUCTION FROM HYDROCARBONS WITH CARBON BONDING IN SOLID SUBOXIDES AND WITHOUT CO_2 EMISSION

10.6.1. Highly Ecological Hydrogen Production by Partial Oxidation of Hydrocarbons without CO_2 Emission: Plasma Generation of Carbon Suboxides

One of the key questions of the modern energy crisis is whether it is possible to extract energy from hydrocarbon feedstock without CO_2 emission into the atmosphere. Considerable efforts are focused on answering this question by CO_2 sequestration, dissolving CO_2 into ocean water, or pumping CO_2 into the salt caverns or old gas and oil fields. The application of non-equilibrium plasma gives a completely different option for energy generation, using coal for example, without CO_2 production (Fridman et al., 2006), which is to be discussed next. As a background, there exist carbon oxides other than CO and CO_2. These are carbon suboxides, such as C_3O_2, that can be polymerized and then as a polymer form chemically and thermodynamically stable substances (Greenwood & Earnshaw, 1984). Vast amounts of carbon on Earth exist as a similar polymer substance – the humic acids, which are a major organic component of soils. Conversion of hydrocarbons into humic acids with an effective energy release process may appear to be an advantageous approach for the modern energy industry. The conversion of solid hydrocarbons like coal and biomass into carbon suboxides would solve many important problems simultaneously. In addition to energy generation without CO_2 emission, this process would eliminate problems related to sulfur and heavy metal component emission, ash utilization, and the addition of inorganic components into organic fertilizers.

Oxidation of hydrocarbons in quasi-equilibrium conditions usually leads to the formation of CO and CO_2. Non-equilibrium processes and specifically strongly non-equilibrium plasma can be effective in the production of carbon suboxides (Fridman et al., 2006), opening the possibility for highly ecological hydrogen production technology from different hydrocarbons such as natural gas, biomass, and coal without CO_2 emissions. This approach, using non-equilibrium plasma conditions, raises key questions to be adressed: (1) How much energy may be obtained in such a process, and what will be the quality of this energy? The quality of the extracted energy is of importance since it should then be effectively converted into other forms (i.e., mechanical, electrical, etc.). (2) What would the energy-generating technological process look like if the polymer carbon suboxides are produced simultaneously? This question is relevant first with respect to coal, biomass, and natural gas as the most abundant hydrocarbon feedstocks.

10.6.2. Thermodynamics of the Conversion of Hydrocarbons into Hydrogen with Production of Carbon Suboxides and without CO_2 Emission

Carbon suboxide (C_3O_2) is a foul-smelling lachrymatory gas. Its molecules are linear and symmetric with structure that can be represented as $O=C=C=C=O$. The suboxide is stable, but at 25°C it polymerizes to form a highly colored solid material with a polycyclic six-membered lactone structure (Balauff et al., 2004), shown in Fig. 10–42. The carbon suboxide is the acid anhydride of malonic acid, and it slowly reacts with water to produce that acid. It can be stored at a pressure of a few Torr, but under standard conditions C_3O_2 forms a yellow, red, or brown polymer $(C_3O_2)_n$ (ruby-red above 100°C, violet at 400°C, and it decomposes into carbon at 500°C). The formation energy for the gas-phase carbon suboxide C_3O_2 is -97.6 kJ/mol (Chase, 1998), or -95.4 kJ/mol (Balauff et al., 2004). The formation energy for liquid polymerized carbon suboxide $(C_3O_2)_n$ is -58.8 kJ per one mole of carbon, and the formation energy for the solid polymerized carbon suboxide $(C_3O_2)_n$ is about -112 kJ per one mole of carbon (Kybett et al., 1965; McDougall & Kilpatrick, 1965; Kondratiev, 1974).

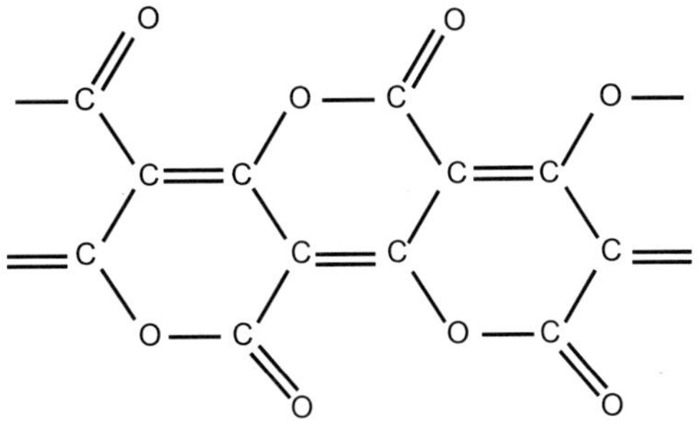

10.6. Energy and Hydrogen Production from Hydrocarbons

Figure 10–42. Basic structure of $(C_3O_2)_n$.

This and similar polymer substances are major organic components of soils called humic acids, which are black solids with approximately 55–60% carbon, 2–3% hydrogen, and the remainder predominantly oxygen. The chemical composition of humic substances can vary but its basic component is $(C_3O_2)_n$. Because of composition variations in natural humic substances (coal and biomass), the data for pure substances are used here for estimates: graphite instead of coal, methane instead of natural gas, polymerized suboxide $(C_3O_2)_n$ instead of humic substances, and cellulose, $(C_6H_{10}O_5)_n$, with a formation energy of -962 kJ per mole of monomer instead of biomass (Perks & Liebman, 2000). The heat release that accompanies burning coal obviously depends on the coal composition. However, to give specific numeric values one can illustrate the process by oxidation of carbon in the form of graphite:

$$C(s) + O_2 \rightarrow CO_2(g), \quad \Delta H = -393.5 \text{ kJ/mol} \approx -3.9 \text{ eV/mol}. \quad (10\text{–}40)$$

Under the natural conditions of weathering, slow oxidation of coal occurs with the formation of carbon suboxide and no, or very little, carbon dioxide is generated (Davidson, 1990). The heat release accompanying coal oxidation to the polymerized substance can be described as

$$C(s) + (1/3n)O_2 \rightarrow (C_3O_2)_n(s), \quad \Delta H = -112 \text{ kJ/mol} \approx -1.1 \text{ eV/mol}. \quad (10\text{–}41)$$

Carbon oxidation to a suboxide phase still provides considerable heat release but, in contrast to burning, it occurs without any CO_2 emission. Taking into account the essential hydrogen content in coal, water will also be produced during the process, and the heat release of the oxidation can, therefore, be higher than the enthalpy just shown. It should also be emphasized that suboxides can be further utilized as fertilizers and/or in biochemical technologies that do not generate large-scale CO_2 emission. The oxidation of coal to suboxide (10–41) is a very slow process at low temperatures (Davidson, 1990). It takes many hours or days, which prevents effective use of the process in energy generation. Acceleration of the process by means of quasi-equilibrium temperature increase is impossible, because a temperature increase above 300°C results in the formation of gas-phase carbon oxides, CO and CO_2 (Davidson, 1990). The process becomes possible in non-equilibrium conditions, specifically by applying non-thermal plasma systems. The energy of carbon oxidation to suboxide can also be extracted in the chemical form. For example, water vapor can be added to the oxygen, or air, to make the oxidation process (10–41) thermoneutral:

$$2.16\, C(s) + 0.22\, O_2 + H_2O \rightarrow 0.72/n(C_3O_2)_n(s) + H_2, \quad \Delta H \approx 0. \quad (10\text{–}42)$$

Coal contains hydrogen (typically 0.5–1 hydrogen atoms per carbon atom), so the process (10–42) with coal would result in a higher hydrogen yield. Additional high-quality energy can be obtained if a part of the oxygen in process (10–42) is substituted with water vapor and

733

a low-temperature waste heat, which is discussed in the next section. Changing from coal to natural gas as a feedstock, complete CH_4 oxidation is seen to be strongly exothermic:

$$CH_4 + 2O_2 \rightarrow CO_2 + 2H_2O, \quad \Delta H = -802.5\,\text{kJ/mol} \approx -8.0\,\text{eV/mol}. \quad (10\text{--}43)$$

Replacing the complete oxidation process by partial oxidation with the formation of polymerized solid suboxides and without CO_2 emission leads to the reaction

$$CH_4 + 1/3\,O_2 \rightarrow (1/3n)(C_3O_2)_n(s) + 2H_2, \quad \Delta H = -37.4\,\text{kJ/mol}. \quad (10\text{--}44)$$

The addition of a small amount of water leads to the thermoneutral process

$$1.3\,CH_4 + 1/3\,O_2 + 0.2\,H_2O \rightarrow (1.3/3n)(C_3O_2)_n(s) + 2.8\,H_2, \quad \Delta H \approx 0. \quad (10\text{--}45)$$

The LCV of the produced hydrogen is about 520.8 kJ per mole of CH_4. Using the LCV of methane, 802.5 kJ/mol, which can be released in combustion, the conversion of CH_4 into hydrogen without CO_2 emission (10–44) has a 65% efficiency. This efficiency is higher than that of the aforementioned coal conversion and is due to the high H_2 content of methane. The rest of the energy (in this case about 35%) stays in the form of suboxides, which can then be used as organic fertilizer. Additional high-quality energy in the form of hydrogen can be obtained if part of the oxygen in reaction (10–45) is substituted by a low-temperature waste heat. It is interesting to compare CH_4 conversion into polymerized carbon suboxides (10–44) with the direct CH_4 conversion into carbon and H_2, which was performed in thermal plasma by Steinberg (2000):

$$CH_4 \rightarrow C(s) + 2H_2, \quad \Delta H = 74.6\,\text{kJ/mol}. \quad (10\text{--}46)$$

Though this plasma process leads to CH_4 conversion to H_2 with no CO_2 emission, the thermodynamic efficiency here is 51% – lower than in the production of suboxides. The endothermic Steinberg process requires significant electrical or concentrated solar energy (Hirsch & Steinfeld, 2004).

10.6.3. Plasma-Chemical Conversion of Methane and Coal into Carbon Suboxide

The polymerized carbon suboxide was produced from methane by plasma-assisted oxidation in a "super-rich" mixture with equivalence ratio exceeding 4 (Kalra, Gutsol, & Fridman, 2005). Pictures of the plasma reactor quartz tubes with a deposited film of polymerized carbon suboxide are shown in Fig. 10–43. A similar product was formed in a non-equilibrium microwave discharge in CO (see Sections 5.7.2 and 5.7.3). The formation of polymerized C_3O_2 in non-thermal plasma was also reported by D'Amico and Smith (1977). Experiments show that the formation of polymerized carbon suboxides is possible even at relatively high-temperatures (about 500°C) with an oxygen deficit. Non-equilibrium plasma effectively stimulates the process. Separation of the polymerized carbon suboxides from gaseous reagents and products is a separate technological task, which can be accomplished by $(C_3O_2)_n$ condensation on solid-phase material passing through the reactor (e.g., on the surface of sand particles) or even on coal particle residues. In the latter case it is possible to have thermoneutral hydrogen production (Fridman et al., 2006):

$$4.9\,CH_4 + 3.3\,C(s) + 6.3/3,\,O_2 + 2.6/3\,H_2O$$
$$\rightarrow (8.2/3n)(C_3O_2)_n(s) + 33.2/3\,H_2, \quad \Delta H \approx 0. \quad (10\text{--}47)$$

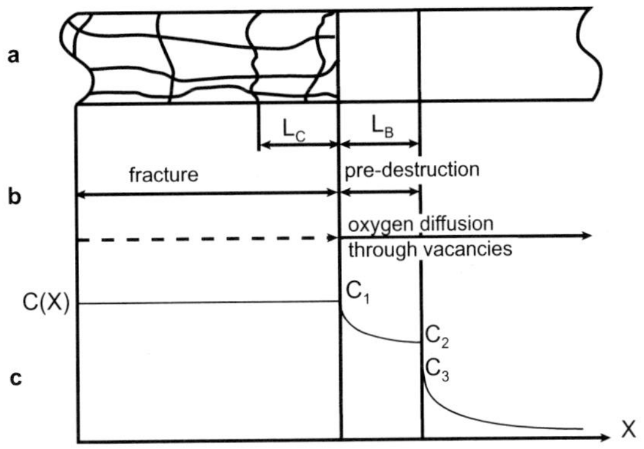

Figure 10–43. Deposited film of the polymerized carbon suboxide on the inner surface of the plasma reactor quartz tube.

10.6.4. Mechanochemical Mechanism of Partial Oxidation of Coal with Formation of Suboxides

During conventional high-temperature coal combustion, new portions of solid fuel start reacting (10–40) after removal of the gaseous products (CO_2 and CO) from the surface of the reagent. This is not the case during coal conversion into polymerized suboxides because both the feedstock and the product are solid. The kinetic mechanism of the process is complicated and includes a sequence of mechanochemical reactions to be considered after some remarks regarding coal oxidation mechanisms in general. Oxidation of coal at low temperatures starts with the absorption of oxygen at the surface with further propagation into the volume by means of diffusion. At temperatures above 70°C, this leads to the formation of alkali-soluble products, or humic acids. Under natural conditions the oxidation process lasts from 80 to 100 h and leads to degradation of the coal substance. At temperatures in the 150–250°C range the acidic products are stable, but at higher temperatures the reaction quickens and becomes strongly exothermic – the coal reaches the ignition point and begins to burn. With an oxygen deficiency and moderate temperatures, slow oxidation occurs with a condensed-phase product. Acceleration of the slow oxidation process by a few orders of magnitude can be achieved in non-equilibrium plasma. During low-temperature coal oxidation, the temperature slowly rises. If it reaches a threshold of 80°–120°C, a steady reaction resulting in the production of gaseous products such as carbon dioxide ensues. In this temperature range, the oxidation rate increases and becomes dependent on the structural properties of coal, such as porosity and presence of microdefects, microcracks, and so forth. At this point the combination of mechanical and chemical properties of coal determines the oxidation kinetics. The absorption of oxygen gives further rise to microdefects, which then interact to create pores, microcracks, and so on. This enhances oxygen diffusion and creates a positive feedback loop for propagation of the coal degradation zone. The mechanochemical oxidation mechanism is important for coal conversion into carbon suboxides because the product material is solid and blocks the transport of oxygen. A similar mechanism plays a key role in the catastrophic breakaway steam–zirconium oxidation that occurs in nuclear reactors during severe accidents (Baranov et al., 1991). The breakaway phenomenon is characterized by a transition from parabolic to linear time dependence of the oxidation-front propagation inside the solid material. Oxygen diffusion deep in the coal triggers dissolution of the oxide film and propagation of the chemical oxidation reaction front. This chemical reaction, in turn, causes oxygen to transition to a bound state, which results in a local increase of molar volume, nonconformity of the reacting components and product structures, and mechanical stresses. Growth of the elastic stresses is accompanied by phase transformations in the oxide layer that finally result in the formation of a scale defective structure, which is a system of interconnected cracks and pores. The scale degradation ensures fast migration

735

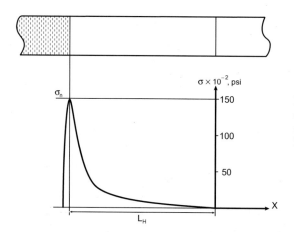

Figure 10–44. Breakaway oxidation wave front structure: morphological changes in the oxide, non-homogeneous stress distribution in oxide layer.

of the gas through macroscopic defects and enhanced oxygen diffusion through the oxide. The formation of a permeable system of cracks and/or pores makes the oxidation process autocatalytic. The key feature of this autocatalysis is that it occurs due to morphological and structural changes in the oxidized coal, and these changes are caused by the elastic stresses affected by the topochemical reaction (Baranov et al., 1991).

10.6.5. Kinetics of Mechanochemical Partial Oxidation of Coal to Carbon Suboxides

The conceptual basis of breakaway oxidation, viewed as a degradation of the oxide film resulting from a combined action of diffusion, chemical reaction, structural rearrangement, and its consequent destruction, allows for the construction of the kinetic model for coal conversion under plasma treatment into carbon suboxide (Fridman et al., 2006). Diffusion-driven propagation of the oxidation reaction front is analyzed by using the self-similar approximation for the diffusion equation solution for the concentration of oxygen in coal. After that, by invoking fracture mechanics concepts, the dependence of the breakaway oxidation rate on the thermodynamic parameters (e.g., temperature) and elastic and morphological oxide characteristics are explicitly calculated. The structure of the oxidation front and the oxygen concentration distribution is illustrated in Fig. 10–44 for the relevant temperatures of $150°–250°C$ for the natural slow oxidation of coal. For analysis of the oxidation front, only diffusion and chemical processes should be considered. Surfaces where the destruction begins separate the compact oxide from the scale permeable to gaseous oxygen. Microscopic pores and cracks nucleate and grow in the predestruction zone (Fig. 10–44). In the destruction zone, macroscopic interaction and the growth of pores and cracks takes place. Macroscopic consequences of the oxide degradation (i.e., enhanced oxygen penetration through the permeable oxide scale) can be phenomenologically delineated by the equation

$$\frac{\partial c}{\partial t} = D\frac{\partial^2 c}{\partial t^2} + S(c, x, \xi_r - \xi_f), \tag{10–48}$$

where $c(x)$ is the oxygen concentration in the film, $D = D_0 \exp\{-\frac{E_a}{RT}\}$ is the oxygen diffusion coefficient, $S(c, x, \xi_r - \xi_f)$ is the oxygen source term, and ξ_r and ξ_f are the coordinates of the reaction and degradation fronts. The equation can be rewritten in terms of the self-similar variable $\xi = x + Vt$:

$$V\frac{\partial c}{\partial \xi} = D\frac{\partial^2 c}{\partial \xi^2} + S(c, \xi). \tag{10–49}$$

The oxygen source term for equation (10–49) can be introduced as (Baranov et al., 1991)

$$S(c, \xi) = \frac{c_0 D \psi}{L_{\mathrm{B}}^2} \exp \left[-\frac{c_2}{c(\xi - L_{\mathrm{B}})} - \frac{c(\xi)}{c_1} \right], \qquad (10\text{–}50)$$

where $J_0 = c_0 D \psi / L^2{}_{\mathrm{B}}$ is the diffusion flux through the oxide barrier of width L_{B}, and ψ is the volume fraction of the interconnected permeable cracks or pores. The exponential form of the factor provides a rapid fall-off of the source power outside the compact region of the deteriorated material. This property of nonlinear source uses asymptotic methods of analysis. The solution of the homogeneous equation with the source absent ($S = 0$) and the parameter V characterizing the front propagation velocity takes the form

$$c(\xi) = c_1 \exp \left(\xi \frac{V}{D} \right). \qquad (10\text{–}51)$$

The non-homogeneous equation (10–49) with the source taken to the first approximation is

$$\frac{\partial}{\partial \xi} \left[V c(\xi) - D \frac{\partial c}{\partial \xi} \right] = J_0 \exp \left\{ -\frac{c_2}{c_1} \exp \left[\frac{(L_{\mathrm{B}} - \xi)}{D} V \right] \right\}. \qquad (10\text{–}52)$$

The front velocity can then be found by matching the solutions for the regions $\xi < 0$ and $\xi > 0$:

$$V = \frac{D}{L_{\mathrm{B}}} \ln \left[\frac{c_1}{c_2} \ln \left(\frac{c_0 D^2 \psi}{L_{\mathrm{B}}^2 V^2 c_2} \right) \right]. \qquad (10\text{–}53)$$

A weak dependence of the volume fraction ψ of the interconnected cracks and pores allows the approximation

$$V \approx \frac{D}{L_{\mathrm{B}}} \ln \frac{c_1}{c_2}, \qquad (10\text{–}54)$$

which represents the velocity V of the steady-state, self-similar, diffusion-driven front of the oxidation reaction, supported by the degradation front staying a distance L_{B} behind the oxidation reaction front. The kinetic details regarding propagation of the breakaway oxidation front and conversion front of coal into carbon suboxide in plasma conditions can be found in Baranov et al. (1991) and Fridman et al. (2006).

10.6.6. Biomass Conversion into Hydrogen with the Production of Carbon Suboxides and Without CO_2 Emission

Cellulose $(C_6H_{10}O_5)_n$ was used to model biomass conversion into hydrogen and carbon suboxide without CO_2 emissions (Fridman et al., 2006). Non-equilibrium plasma can be applied to stimulate the process. Under natural anaerobic conditions, the conversion of biomass into an effective energy carrier proceeds over millennia. In order to use biomass as an effective renewable source of energy, the conversion process should occur at a much faster pace. The conventional combustion of dry biomass can be presented as

$$(1/n)(C_6H_{10}O_5)_n + 6O_2 \rightarrow 6CO_2 + 5H_2O, \quad \Delta H = -2608 \text{ kJ/mol.} \qquad (10\text{–}55)$$

Regarding prevention of the greenhouse effect, biomass releases more energy than coal but less than natural gas: the production of energy per CO_2 mole is 434.7 kJ for cellulose, 393.5 kJ for carbon, and 802.5 kJ for methane. Similar to coal, the use of biomass causes many technical problems because of residual ash formation. Therefore, most research efforts concerning biomass use are directed toward the production of high-quality biofuels (i.e., methanol, biodiesel, etc.). Nevertheless, in all methods of processing for use as an energy source, the final products of biomass with complete conversion are water, CO_2, and some residual ash. This raises a general question: What is the advantage of using biomass compared to methane (which is a biomass conversion by-product) from the standpoint of preventing the greenhouse effect? This is especially important since we produce more CO_2 per unit energy from biomass because the energy quality of biomass is lower than the energy quality of methane. A positive result can be achieved if during the biomass conversion process part of the carbon produced is turned into solid-phase products. In this case, the conversion of biomass into carbon suboxides for organic fertilizer is a very attractive choice. This process is easier from an engineering point of view than the conversion of coal, because dry biomass has both porous structure and a significant oxygen content. Therefore, no limitation on the diffusion process can be expected. Moreover, cellulose structure is quite similar to the structure of $(C_3O_2)_n$, which simplifies the process kinetics. The excess of oxygen in this case converts part of the hydrogen into water:

$$(1/n)(C_6H_{10}O_5)_n \rightarrow 2/n(C_3O_2)_n(s) + 4H_2 + H_2O, \quad \Delta H = 496.2 \text{ kJ/mol}. \quad (10\text{–}56)$$

To compensate for the energy demand of the endothermic reaction, about half of the hydrogen produced by this process is oxidized:

$$(1/n)(C_6H_{10}O_5)_n + O_2 \rightarrow 2/n(C_3O_2)_n(s) + 2H_2 + 3H_2O, \quad \Delta H \approx 0, \quad (10\text{–}57)$$

or the use of additional low-temperature heat is necessary. The effectiveness of biomass conversion into hydrogen without CO_2 emissions still requires additional research efforts (Fridman et al., 2006). The presented approach suggests a significant shift of the paradigm of hydrocarbon feedstock use as an energy source. It allows a substantial portion of the hydrocarbon chemical energy to be released in the form of hydrogen or electricity, together with simultaneous bonding of carbon into polymerized carbon suboxides that can be used as organic fertilizers. In addition to energy generation without CO_2 emission, the benefits of the approach include elimination of ash accumulation, heavy metal, and sulfur component emission; and addition of microelements (e.g., N, S, P, K, Na, etc.) into organic fertilizers.

10.7. HYDROGEN SULFIDE DECOMPOSITION IN PLASMA WITH PRODUCTION OF HYDROGEN AND SULFUR: TECHNOLOGICAL ASPECTS OF PLASMA-CHEMICAL HYDROGEN PRODUCTION

10.7.1. H_2S Dissociation in Plasma with Production of Hydrogen and Elemental Sulfur and Its Industrial Applications

Hydrogen sulfide dissociation for the production of hydrogen and elemental sulfur is a slightly endothermic, almost thermoneutral process:

$$H_2S \rightarrow H_2 + S_{solid}, \quad \Delta H = 0.2 \text{ eV/mol}. \quad (10\text{–}58)$$

Although the enthalpy required for H_2S decomposition (10–58) is theoretically very low, a practical version of the process with low energy cost in quasi-equilibrium thermal conditions is not possible. When the temperature is increased to stimulate the dissociation, sulfur is produced not in the condensed phase, but as a gas (S_8), which requires more energy:

$$H_2S \rightarrow H_2 + 1/8S_8, \quad \Delta H = 0.3 \text{ eV/mol}. \quad (10\text{–}59)$$

Because the enthalpy of the gas-phase sulfur process (10–59) exceeds that of the solid-phase sulfur process (10–58), the required temperature to stimulate dissociation of hydrogen sulfide is higher. Continuing along these lines, the higher temperature required for the decomposition results in the production of sulfur in the form of the dimer S_2, and not in the form S_8. As a result, quasi-equilibrium thermal H_2S decomposition with the production of

10.7. Hydrogen Sulfide Decomposition in Plasma

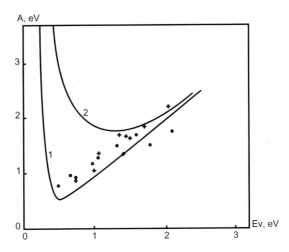

Figure 10–45. Energy cost of H_2S dissociation as a function of specific energy input: (1) non-equilibrium process, modeling; (2) minimal energy cost in quasi-equilibrium process, modeling. Experiments: •, microwave discharge, 50 kW; +, RF discharge, 4 kW.

hydrogen and elemental sulfur actually requires much more energy than can be expected from the chemical formula (10–58) and is actually

$$H_2S \rightarrow H_2 + 1/2\,S_2, \quad \Delta H = 0.9\,\text{eV/mol}. \tag{10–60}$$

This decomposition takes place in the high-temperature thermal plasma zone, which is followed by clusterization and condensation of sulfur in the lower-temperature zones usually on the discharge periphery:

$$1/2\,S_2 \rightarrow S_6, S_8 \rightarrow S_{\text{solid}}, \quad \Delta H = -0.7\,\text{eV/mol}. \tag{10–61}$$

Because the energy released during recombination (10–61) cannot easily be returned to the high-temperature zone of quasi-equilibrium thermal plasma systems, the energy cost of hydrogen production in such systems remains high. The minimal energy cost of the process (10–58) in quasi-equilibrium thermal systems with ideal quenching (see Section 3.6.6) is 1.8. eV/mol, which corresponds to a maximum energy efficiency of only 22%. The minimal energy cost can be significantly decreased and energy efficiency significantly increased in a non-equilibrium plasma, where the essential deviation from equilibrium is due to a forced selectivity of the transfer processes. Such plasma systems with fast gas rotation and strong centrifugal separation of the produced plasma clusters are discussed next. The selective transfer of the sulfur clusters to the discharge periphery produces hydrogen and sulfur with a minimal energy cost of 0.5 eV/mol. The energy efficiency reaches 40% in such plasma-chemical systems. The H_2S conversion into hydrogen and sulfur can also have a relatively high level of conversion in the non-equilibrium processes. Special experiments in the relevant non-equilibrium microwave and RF discharges with strong centrifugal effect at moderate and atmospheric-pressures give the minimal energy cost value of about 0.7–0.8 eV/mol (Rusanov & Fridman, 1984; Balebanov et al., 1985; Nester, Rusanov, & Fridman, 1985; Krasheninnikov et al., 1986; Balebanov et al., 1989; Harkness & Doctor, 1993). The optimal reaction temperature in these non-equilibrium systems is about 1150 K, and the effective clusterization temperature is 850 K. Experimental and modeling results illustrating the plasma-chemical process of hydrogen sulfide decomposition with production of hydrogen and elemental sulfur, as well as the centrifugal effect on its efficiency, are summarized in Fig. 10–45. The low energy cost and high-conversion characteristics of the H_2S decomposition process (10–58) achieved in the non-equilibrium, high-pressure plasma attracts significant industrial interest, especially in natural gas processing and in oil refining. Some natural gas from major sources contains 25% or more of H_2S, which is usually

739

separated from the natural gas and converted into sulfur in the Claus process (Speight & Ozum, 2001). The traditional Claus process is exothermic, converting H_2S into sulfur and water,

$$H_2S + 1/2\,O_2 \rightarrow S(s) + H_2O, \tag{10-62}$$

and actually loses H_2 production potential. The industrial application of plasma technology (10–58) in natural gas processing produces sulfur and saves the hydrogen with minimal energy expenditures. The application of plasma technology (10–58) in oil refineries relates to hydrogen recycling in the hydrogen desulfurization process of crude oil. Hydrogen used for oil desulfurization is conventionally produced in refineries by partial oxidation of the oil. The hydrogen desulfurization of oil leads to H_2 conversion into H_2S, which is then conventionally transferred into sulfur in the Claus process (10–62). New hydrogen is continuously produced in conventional refineries to purify the crude oil. Plasma technology (10–58) allows for recycling the hydrogen, making the hydrogen desulfurization of crude oil more effective and ecological (Harkness & Doctor, 1993). More details on the industrial applications of non-equilibrium plasma technology in natural gas treatment and oil refining are discussed in the next section.

10.7.2. Application of Microwave, Radiofrequency, and Arc Discharges for H_2S Dissociation with Production of Hydrogen and Elemental Sulfur

Non-equilibrium microwave and RF capacitively coupled plasma (CCP) discharges operating at moderate and atmospheric-pressures with a strong centrifugal effect have been applied to the process of H_2S dissociation with the production of hydrogen and elemental sulfur (Rusanov & Fridman, 1984; Balebanov et al., 1985; Nester, Rusanov, & Fridman, 1985; Krasheninnikov et al., 1986; Balebanov et al., 1989; Harkness & Doctor, 1993). Laboratory experiments were carried out at pressures of 60–200 Torr, powers of 1–2 kW and higher, and H_2S flow rates of 150–400 standard cm^3/s and higher. Larger-scale experiments were accomplished at pressures up to atmospheric, at a power level up to 1 MW, and flow rates up to several thousand standard m^3 per hour. The key feature of the reactors is the injection of the feed gas into the plasma zone tangentially at very high velocities. This provides a strong centrifugal effect, separating sulfur clusters from other products. The centrifugal effect is responsible for a significant decrease of the process energy cost, which is discussed below. The optimal reaction temperature in the non-equilibrium microwave and RF-CCP systems is near 1150 K, and the effective non-equilibrium sulfur clusterization temperature is about 850 K.

The microwave and RF-CCP discharge systems applied to H_2S dissociation are somewhat similar to those applied to CO_2 dissociation and described in Sections 5.5.1 and 5.5.6. In these experimental systems, strong centrifugal effects providing effective product separation result in conversion rates close to 100% and record low values of energy cost (about 0.7–0.8 eV/mol) for production of hydrogen and elemental sulfur (10–58). The energy efficiency characteristics of the process (10–58) in the microwave and RF-CCP discharge systems are shown in Fig. 10–45 as a function of specific energy input in the discharges. Similar experiments focused on H_2S dissociation with the production of hydrogen, and elemental sulfur (10–58) was explored in thermal arc discharges (Grachev et al., 1990; Khrikulov et al., 1992). The arc discharge power in this atmospheric-pressure plasma-chemical system is 10 kW. The hydrogen sulfide gas flow rate is above 10 standard m^3/h. Significant improvement of the process energy efficiency due to fast gas rotation and centrifugal separation of products was also demonstrated in these experiments. The maximum degree of H_2S conversion (10–58) in these experiments is close to 100%, while the minimum energy cost (1–1.2 eV/mol) is slightly higher than that achieved in the microwave and RF-CCP discharges. Experiments with non-equilibrium, atmospheric-pressure, gliding arc

10.7. Hydrogen Sulfide Decomposition in Plasma

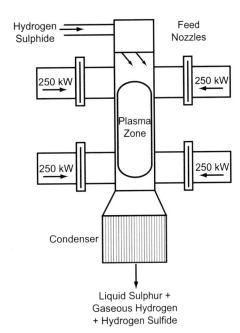

Figure 10–46. Schematic of the 1 MW plasma-chemical reactor for H_2S treatment in Orenburg.

discharges also result in an energy efficiency higher than that in the microwave and RF-CCP discharges (Czernichowski, 1994; Fridman et al., 1999). It is expected that stabilizing the gliding arc discharges in the reverse vortex (tornado) flow should decrease the energy cost of H_2S dissociation and hydrogen production. Although the energy efficiency of the process is slightly worse in the arc discharges, their practical application can be justified by taking into account the possibility of using much higher levels of power and flow rate of H_2S in the process. Dissociation of H_2S with production of H_2 and sulfur in the gliding arc tornado has been demonstrated in Drexel Plasma Institute.

10.7.3. Technological Aspects of Plasma-Chemical Dissociation of Hydrogen Sulfide with Production of Hydrogen and Elemental Sulfur

An industrial-level plasma-chemical facility was built in the Orenburg Gas Plant (Orenburg, Russia) for treatment of H_2S from natural gas (Balebanov et al., 1989; Harkness & Doctor, 1993). The four-port microwave discharge configuration used in the facility is illustrated in Fig. 10–46. The plasma-chemical reactor is a short section of circular waveguide with four ports for injecting the microwave energy. The microwave power is produced by four separate 250 kW magnetron generators and combined in the plasma-chemical reactor to yield up to a total of 1 MW microwave power. Orenburg natural gas contains a large fraction of sour or acid gas (H_2S–CO_2 mixture). The acid-gas mixture is extracted from crude natural gas and then treated in the plasma-chemical reactor. The gas flow through the plasma reactor reaches several thousands of standard m^3/h, and the pressure is as high as 1.3 atm. Orenburg acid gas, produced at the Orenburg natural gas purification plant, typically contains 40% CO_2. Energy efficiency and H_2S conversion characteristics of the process were discussed in the preceding section and the energy cost of the process is summarized in Fig. 10–45. The end of the reactor volume is defined by the sulfur condenser (Fig. 10–46), which brings the temperature of the reactor effluent to between 130 and 135°C.

The plasma-chemical reactor is a major component, but not the sole part, of the complicated technological scheme required to produce pure hydrogen and sulfur from hydrogen sulfide. A simple flow process diagram for the Orenburg experimental facility, including sulfur recovery and polymer membrane gas separation, is shown in Fig. 10–47. Polymer membrane technologies were effectively applied in this case for separation of the H_2S-containing

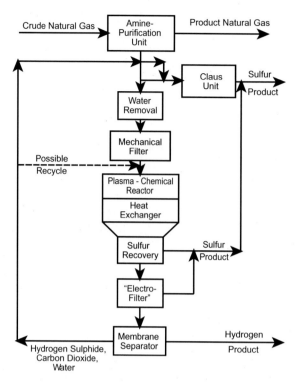

Figure 10–47. Flow sheet for the H_2S treatment Orenburg experimental plasma facility.

gases (Amirkhanov et al., 1998, 2001). An important advantage of polymer membrane technologies is their compatibility with the high specific productivity of the plasma-chemical reactor. Other gas separation devices are usually much larger than the plasma devices and have the same productivity, which diminishes the advantages of the highly productive and very compact plasma-chemical technology as a whole (see Section 10.7.8). An industrial version of the Orenburg plasma-chemical system is presented in Fig. 10–48 (Harkness & Doctor, 1993). One process line shown in Fig. 10–48 (marked I) starts with the acid-gas stream from the amine-purification unit, which is used to separate the acid gas from crude natural gas. The acid-gas stream contains a relatively wet H_2S–CO_2 mixture with particulates. The first step is the removal of water and particulates from the mixture. The second step is the separation of dry sour gas into H_2S and CO_2 using the special polymer membrane technology (membrane unit 1 in the figure; see Amirkhanov et al., 1998, 2001). Then H_2S is dissociated in the plasma-chemical reactor to produce hydrogen and sulfur. An alternative process line shown in Fig. 10–48 (marked II) starts with crude natural gas without the preliminary amine purification used in the previous approach. The special polymer membrane technology is used in this case to extract the acid-gas (H_2S–CO_2) mixture from crude natural gas (membrane unit 3; Amirkhanov et al., 1998, 2001). The acid-gas (H_2S–CO_2) mixture is then directed to membrane unit 1 to separate H_2S for further plasma-chemical decomposition into hydrogen and sulfur similar to the previous process line. Commercial quantities of helium could also be recovered in this case from methane by using the subsequent polymer membrane gas-separation unit 4 (see Fig. 10–48). Traces of H_2S in the CO_2 product gas from polymer membrane gas-separation unit 1 are removed by an ultraviolet (UV) treatment/filtration step, where H_2S is dissociated and the solid sulfur particles are removed mechanically. The H_2S stream passes through the plasma reactor and dissociates as discussed in the previous section. Then the bulk of the sulfur is condensed and

10.7. Hydrogen Sulfide Decomposition in Plasma

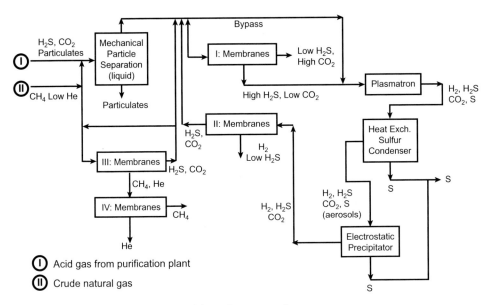

Figure 10–48. Industrial version of the H_2S treatment plant.

separated, and the residual sulfur aerosols are removed by an electrostatic precipitator. The gaseous stream H_2–H_2S–CO_2 goes to polymer membrane gas-separation unit 2 to extract the hydrogen product (see Fig. 10–48). The stream, rich in hydrogen sulfide, is recycled to membrane unit 1 and then to the plasma-chemical reactor for dissociation. Traces of H_2S in the hydrogen product are removed with the aforementioned UV treatment/filtration technology.

Similar plasma treatment of H_2S waste from crude natural gas was proposed at Argonne National Laboratory and is illustrated in Fig. 10–49 (Harkness & Doctor, 1993). Acid gas from an amine-purification system containing H_2S, CO_2, and H_2O is combined with recycled H_2S and then fed into the plasma-chemical reactor, which is connected to a microwave generator. The process stream leaving the reactor contains mostly H_2, sulfur vapor and liquid, unconverted H_2S and CO_2, water, and some impurities such as COS and CS_2. The sulfur product is cooled and collected in a scrubber. Available process energy is recoverable at this point. The scrubbing step removes the bulk of the sulfur and some of the sulfur-containing

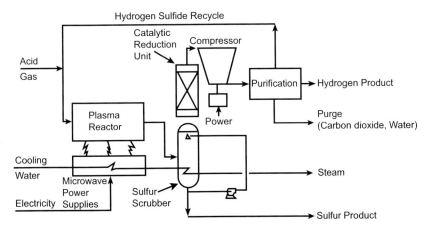

Figure 10–49. Argonne plasma process for the treatment of hydrogen sulfide waste.

impurities. Remaining traces of the undesirable compounds can be removed by an internal catalytic reducing unit or TGCU. Sulfur compounds are converted into H_2S by using a small amount of hydrogen and then recycled back to the plasma-chemical reactor. The purified gas-phase products are compressed and sent through a separation step to recover the hydrogen product, recycle the unconverted H_2S, and purge water and CO_2. Economic analysis of the technology shows its advantages with respect to the Claus Sulfur Recovery Unit (CSRU) and TGCU (Daniels, Harkness, & Doctor, 1992). Effective plasma dissociation in oil refining of H_2S into H_2 and sulfur was demonstrated using the microwave plasma system at a power level of up to 50 kW with productivity of 50 standard m^3 of H_2 per hour at a refinery near L'vov in the Ukraine (Harkness & Doctor, 1993). The plasma unit was applied to hydrogen recycling in the hydrogen desulfurization of crude oil (see Section 10.7.1).

10.7.4. Kinetics of H_2S Decomposition in Plasma

As discussed earlier, the most effective plasma-chemical decomposition of hydrogen sulfide (10–58) occurs when the process starts with thermal decomposition (10–60) and is followed by sulfur clusterization (10–61) and product separation in a centrifugal field. Consider the kinetics of the process beginning with thermal decomposition (Potapkin et al., 1988). The elementary endothermic reaction initiating the total process (10–58) is the thermal dissociation of hydrogen sulfide:

$$H_2S + M \rightarrow H + HS + M. \tag{10–63}$$

Atomic hydrogen formed in the dissociation process then reacts with and decomposes another H_2S molecule and forms molecular hydrogen in the fast exothermic process

$$H + H_2S \rightarrow H_2 + HS, \quad k = 1.3 \cdot 10^{-11}\ cm^3/s \cdot \exp(-860\ K/T). \tag{10–64}$$

Active radical HS generated in reactions (10–63) and (10–64) react with each other and form molecular hydrogen and sulfur in the following fast exothermic reaction:

$$HS + HS \rightarrow H_2 + S_2, \quad k = 2.13 \cdot 10^{-10}\ cm^3/s. \tag{10–65}$$

The HS radicals can also be converted into the products H_2 and S_2 via the following sequence of two fast exothermic elementary reactions:

$$HS + HS \rightarrow H_2S + S, \quad k = 1.2 \cdot 10^{-11}\ cm^3/s, \tag{10–66}$$

$$H_2S + S \rightarrow H_2 + S_2, \quad k = 10^{-11},\ cm^3/s \cdot \exp(-2500\ K/T). \tag{10–67}$$

Kinetic curves illustrating H_2S decomposition (10–58) in a plasma-chemical reactor at a temperature of 1700 K and pressure of 20 kPa are shown in Fig. 10–50. The curves were calculated by taking into account the detailed kinetic mechanism of the thermal process (Potapkin et al., 1988). The decomposition time in the process is about 5 ms. The final composition of the products in Fig. 10–50 corresponds to the thermodynamic equilibrium of the mixture. Kinetic curves characterizing quenching of the decomposition products at different cooling rates (in the range 10^1–10^7 K/s) are presented in Fig. 10–51 (Potapkin et al., 1988). It is assumed in the detailed kinetic modeling that the initial temperature in the thermal plasma system is 1500 K. The molar fractions of the decomposition products are shown in Fig. 10–51 as functions of the final temperature. Complete quenching of H_2S dissociation products is achieved at relatively low cooling rates exceeding 10^4 K/s. The requirements for quenching rates in thermal plasma decomposition of H_2S are a couple of orders of magnitude lower than those of other plasma decomposition processes (see Chapters 5 and 7). This positive quenching effect essentially facilitates plasma-chemical decomposition of H_2S with the production of hydrogen and elemental sulfur and has a simple explanation. In the process mechanism (10–63)–(10–67), only the first reaction (10–63) of the primary

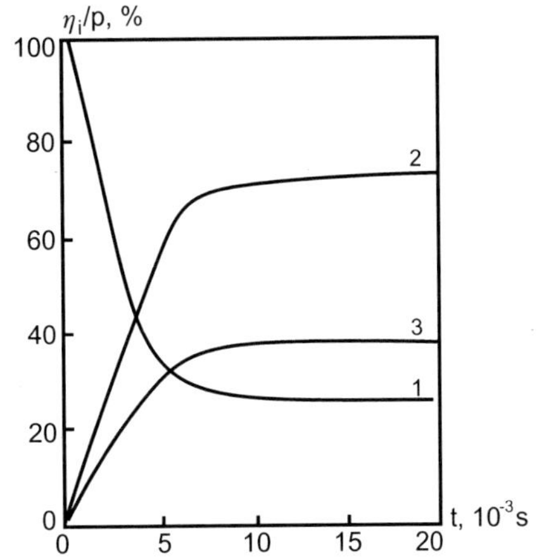

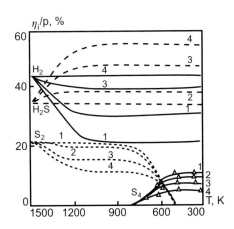

Figure 10–50. Kinetic curves (molar fraction vs. time) for dissociation of H_2S at temperature 1700 K: (1) H_2S, (2) H_2, (3) S_2.

H_2S dissociation is endothermic. The secondary reactions (10–64)–(10–67), converting H and HS into molecular hydrogen and sulfur, are strongly exothermic. As a result, the reverse reactions to secondary processes (10–64)–(10–67) are essentially endothermic and therefore relatively slow at the quenching temperatures. The relative slow kinetics of the reverse reactions explains the low requirements for the quenching rates ($\geq 10^4$ K/s) in plasma-chemical H_2S decomposition. The energy cost of hydrogen production in the thermal decomposition of H_2S, found from the kinetic analysis (Potapkin et al., 1988), is presented in Fig. 10–52 as a function of the process temperature (or the specific energy input) at different quenching rates and in the case of ideal quenching (Potapkin et al., 1984). It is seen from the figure that relatively low quenching rates ($\geq 10^4$ K/s) are sufficient to reach the minimum value of the energy cost for the thermal plasma conditions (about 1.8 eV/mol). Quenching rates of 10^7 K/s and earlier lead to the absolute quenching mode (the effective conversion of intermediate radicals into H_2) in the entire range of the process temperatures and specific energy inputs.

10.7.5. Non-Equilibrium Clusterization in a Centrifugal Field and Its Effect on H_2S Decomposition in Plasma with Production of Hydrogen and Condensed-Phase Elemental Sulfur

The clusterization process (10–61), accompanying the H_2S dissociation, can be essentially affected by fast plasma rotation (in the RF and microwave discharges, the tangential velocities v_φ can be close to the speed of sound). Centrifugal forces proportional to the cluster mass

Figure 10–51. Molar fraction of the H_2S dissociation products (H_2, S_2, S_4, H_2S) after quenching from initial temperature 1500 K at different quenching rates: (1) 10^4 K/s and above; (2) 10^3 K/s; (3) 10^2 K/s; (4) 10 K/s.

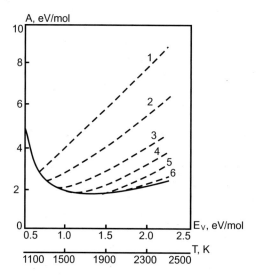

Figure 10–52. Specific energy cost of production of hydrogen (eV/mol) by thermal plasma dissociation of H_2S as a function of specific energy input (and heating temperature) at different quenching rates: (1) 10 K/s; (2) 100 K/s; (3) 10^3 K/s; (4) 10^4 K/s; (5) 10^5 K/s; (6) 10^6 K/s; (7) 10^7 K/s. Solid line represents the case of ideal quenching.

push the particles to the discharge periphery faster than the heat transfer occurs. This can result in a significant shift of chemical equilibrium toward product formation. This effect provides product separation in the processes and decreases its energy cost significantly below the minimum of the quasi-equilibrium thermal process (compare Figs. 10–45 and 10–52). In contrast to conventional condensation (Becker & Doring, 1935; Zeldovich, 1942; Frenkel, 1945), cluster formation in the fast-rotating plasma can occur at temperatures much exceeding the condensation point, which will be considered in this section. Clusters A_n, consisting of n molecules, can be generally formed or destroyed in the following processes:

$$A_{n1} + A_{n-n1} \Leftrightarrow A_n, \quad A_{n-k} + A_{m+k} \Leftrightarrow A_n + A_m. \tag{10–68}$$

In H_2S decomposition, A_n represents the sulfur clusters, which diffuse and drift in the centrifugal field. Consider the cluster size n as a continuous coordinate (varying from 1 to ∞). Another coordinate is the radial one, x, varying from 0 to the radius of the discharge tube, R. Clusterization in the centrifugal field can be considered an evolution of the distribution function $f(n, x, t)$ in the space (n, x) of cluster sizes. Dissociation of H_2S takes into account a source A (sulfur atoms) that is located at the point $x = 0$. Thus, evolution of the cluster distribution function can be described by the continuity equation in the space (n, x) of cluster sizes (Macheret, Rusanov, & Fridman, 1985):

$$\frac{\partial f(n, x, t)}{\partial t} + \frac{\partial j_n}{\partial n} + \frac{\partial j_x}{\partial x} = 0, \tag{10–69}$$

where j_n is the flux of clusters along the axis of their sizes n, and j_x is the radial flux of the particles. Including non-isothermal diffusion and centrifugal drift, the radial flux, j_x, can be presented as

$$j_x = -D_x \left(\frac{\partial f}{\partial x} + \frac{f}{T} \frac{\partial T}{\partial x} - \frac{nmv_\varphi^2}{xT} f \right), \tag{10–70}$$

where D_x is the diffusion coefficient and m is the mass of A. At temperatures above the condensation temperature, diffusion in the space of cluster sizes is mostly due to atom attachment and detachment:

$$A_{n-1} + A \Leftrightarrow A_n, \tag{10–71}$$

which leads to the flux of clusters along the axis of their sizes n given in the following linear Fokker-Plank form (Rusanov, Fridman, & Macheret, 1985):

$$j_n = -D_n \left(\frac{\partial f}{\partial n} - \frac{\partial \ln f_0}{\partial n} f \right), \tag{10–72}$$

where D_n is the diffusion coefficient along n, and $f_0(n)$ is the quasi-equilibrium distribution function. A simple solution of equation (10–69) can be obtained assuming fluxes j_n and j_x are equal to zero:

$$f(n, x) \propto \frac{T(x = 0)}{T(x)} f_0(x) \exp \left[n \int \frac{m v_\varphi^2 dx}{x T(x)} \right]. \tag{10–73}$$

This distribution does not satisfy the typical boundary conditions in the space of cluster sizes. Also the characteristic establishment time of the distribution $f(n, x)$ along n (about n^2/D_n) is much shorter for smaller clusters ($n < 10^2$–10^3) than the characteristic time along $x (R^2/D_x)$. For this reason, it should be assumed in (10–69) that the distribution $f(n, x)$ is determined by fast diffusion along the n axis, and by the centrifugal drift along the radius x. In the equations that follow, R is the discharge radius, D_x is the diffusion coefficient, and D_n is the diffusion coefficient in the space of cluster sizes:

$$D_n = k_0[A] \exp \left(-\frac{E_a}{T} \right); \tag{10–74}$$

k_0 is the rate coefficient of gas-kinetic collisions, $[A]$ is the concentration of molecules A, and E_a is the activation energy of the direct process (10–71). Assuming that E_a does not depend on n, and that the diffusion coefficient D_x is independent of n and x, the derivative of the centrifugal drift is

$$\frac{\partial}{\partial x} \left[D_x \frac{m v_\varphi^2}{x T} n f \right] \approx \frac{D_x}{R^2} \frac{m v_\varphi^2}{T} n f. \tag{10–75}$$

In this case, the kinetic equation for the non-equilibrium cluster distribution $f(n, x)$ can be derived by taking into account diffusion along the n axis and centrifugal drift along the radius x as

$$\frac{\partial^2 f}{\partial n^2} - \frac{\partial \ln f_0}{\partial n} \frac{\partial f}{\partial n} - \left(\frac{\partial^2 \ln f_0}{\partial n^2} + \frac{D_x}{D_n R^2} \frac{m v_\varphi^2}{T} n \right) f = 0. \tag{10–76}$$

The solution of this kinetic equation for $f(n)$ can be found as (Rusanov, Fridman, & Macheret, 1985)

$$f(n) = \frac{\text{const}}{\sqrt[4]{1 + 4bn_0^2 n}} \exp \left[-\frac{n}{2n_0} - \frac{(1 + 4bn_0^2 n)^{3/2}}{12bn_0^3} \right], \tag{10–77}$$

where n_0 is the exponential parameter of the cluster-size distribution non-disturbed by centrifugal effect $f_0(n) \propto \exp(-n/n_0)$; and the centrifugal parameter b is determined as

$$b = \frac{D_x}{D_n R^2} \frac{m v_\varphi^2}{T}. \tag{10–78}$$

The distribution $f(n)$ (10–77) is close to the quasi-equilibrium distribution, $f_0(n) \propto \exp(-n/n_0)$, at relatively small cluster sizes of $n < 1/4bn_0^2$. At relatively large cluster sizes of $n > 1/4bn_0^2$, the distribution function falls because of intensive centrifugal losses:

$$f(n) \propto \exp \left(-\frac{2\sqrt{b}}{3} n^{3/2} \right). \tag{10–79}$$

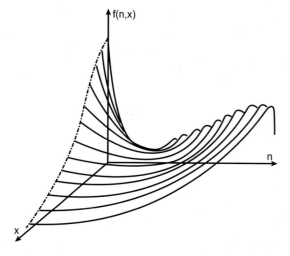

Figure 10–53. Typical cluster distribution in the (n, x) cluster size space.

The cluster size can then be found as

$$\langle n \rangle = n_0, \quad \text{if} \quad n_0 > 0, \tag{10–80}$$

$$\langle n \rangle = 1/4bn_0^2, \quad \text{if} \quad n_0 < 0. \tag{10–81}$$

An example of the non-equilibrium cluster size distribution $f(n, x)$ in the centrifugal field is shown in Fig. 10–53. To further analyze the influence of centrifugal effect on the energy efficiency of H_2S decomposition in plasma, now consider the effect of the centrifugal forces on the average cluster sizes.

10.7.6. Influence of the Centrifugal Field on Average Cluster Sizes: Centrifugal Effect Criterion for Energy Efficiency of H_2S Decomposition in Plasma

The average size of the sulfur clusters produced by the plasma-chemical dissociation of H_2S and moving to the discharge periphery from a radius x can be determined (Rusanov, Fridman, & Macheret, 1985) based on the flux $j(x, n)$ in the combined radius–cluster size space (see the previous section) as

$$\langle n \rangle_f = \int_0^\infty n \cdot j(x, n)\, dn \Big/ \int_0^\infty j(x, n)\, dn. \tag{10–82}$$

Using the relations from Section 10.7.5 for any cluster sizes, including small ones ($n_0 \ll 1$), requires a replacement of the equilibrium average size parameter n_0 by the effective one, $\tilde{n}_0$:

$$\frac{1}{\tilde{n}_0} = 1 - \exp\left(\frac{\partial \ln f_0}{\partial n}\right), \quad \frac{1}{n_0} = -\frac{\partial \ln f_0}{\partial n}. \tag{10–83}$$

Integration (10–82) with the flux (10–70) determines the cluster size in the centrifugal field as

$$\langle n \rangle_f = \frac{\tilde{n}_0 + \alpha n_m + \dfrac{m v_\varphi^2}{T}(2\tilde{n}_0^2 + \alpha n_m^2)}{1 + \dfrac{m v_\varphi^2}{T}(1 + \alpha n_m)}. \tag{10–84}$$

In this relation, $\alpha = [A_{n_m}]/[A]$ is the ratio of concentration of "magic" clusters ($n = n_m$, for sulfur it can be taken as $n_m = 8$) to the concentration of molecules A. Average sizes of clusters (10–84) moving from the radius x with temperature $T(x)$ in the absence of centrifugal forces ($\frac{mv_\varphi^2}{T} \to 0$) are given by

$$\langle n \rangle_f = \tilde{n}_0(T) + \alpha(T)n_m. \tag{10–85}$$

At relatively high temperatures, $\alpha(T) \ll n_m^{-1}$ and, therefore, $\langle n \rangle_f \approx \tilde{n}_0 \approx 1$, which means that primarily individual molecules move to the discharge periphery. The fraction of the magic clusters ($n_m = 8$) increases at lower temperatures. Large clusters with sizes $n \geq n_m$ begin dominating in the flux to the discharge periphery only when temperature exceeds the condensation point. Centrifugal forces stimulate the domination of large clusters in the flux of particles to the discharge periphery, even at temperatures essentially exceeding the condensation point. If the centrifugal effect is strong enough,

$$\frac{mv_\varphi^2}{T} \cdot \alpha(T) \cdot n_m^2 > 1, \tag{10–86}$$

then according to equation (10–84), the average cluster sizes become relatively large:

$$\langle n \rangle_f \approx \frac{mv_\varphi^2}{T} \cdot \alpha(T) \cdot n_m^2 \gg 1, \quad \text{if} \quad \frac{mv_\varphi^2}{T} \cdot \alpha(T) \cdot n_m < 1, \tag{10–87}$$

$$\langle n \rangle_f \approx n_m, \quad \text{if} \quad \frac{mv_\varphi^2}{T} \cdot \alpha(T) \cdot n_m > 1. \tag{10–88}$$

Criterion (10–86) determines the temperature when there are more large clusters than molecules moving to the discharge periphery. The higher the value of the centrifugal factor $\frac{mv_\varphi^2}{T}$, the more this critical temperature exceeds the condensation point. For example, the condensation temperature of sulfur is $T_c \approx 550\,\text{K}$ at 0.1 atm. At high values of the centrifugal factor, $\frac{mv_\varphi^2}{T}$, when tangential velocity is close to the speed of sound, the effective clusterization temperature reaches the high value of 850 K. The magic clusters for this case are sulfur compounds S_6 and S_8. Transfer phenomena do not affect the maximum energy efficiency of plasma processes when there are no external forces and the Lewis number is close to unity (Fridman & Kennedy, 2004). A strong increase of energy efficiency is achieved in the centrifugal field if the molecular mass of the products essentially exceeds that of other components. Then the fraction of products moving from the discharge can exceed the relevant fraction of heat, which results in a decrease of product energy cost with respect to the minimal value for quasi-equilibrium (1.8 eV/mol; see Fig. 10–45). If plasma rotation is sufficiently fast and criterion (10–86) is satisfied, selective transfer of the sulfur clusters to the discharge periphery produces hydrogen and sulfur with a minimal energy cost of 0.5 eV/mol (see Sections 10.7.1 and 10.7.2). Note that the lowest energy cost (0.5 eV/mol) can be achieved only if the strong centrifugal effect (10–86) occurs in the discharge zone, where the dissociation process takes place. If the required high rotation velocities are only in the lower-temperature clusterization zones, the minimal energy cost is higher at 1.15 eV/mol.

10.7.7. Effect of Additives (CO_2, O_2, and Hydrocarbons) on Plasma-Chemical Decomposition of H_2S

It was mentioned earlier (Section 10.7.1) that hydrogen sulfide is present in natural gas as sour or acid-gas mixture (H_2S–CO_2). The previously considered technologies (Section 10.7.3) mostly assumed preliminary separation of the acid gas before plasma-chemical dissociation of H_2S (10–58) and the production of hydrogen and sulfur. In some cases, however, it can be preferable to treat acid gas in plasma directly without the preliminary

A, eV/mol

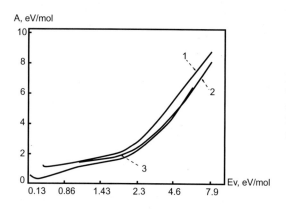

Figure 10–54. Energy cost of hydrogen formation in the plasma-chemical process of H_2S dissociation in the mixture $H_2S + 1/32\,O_2$: (1) absolute quenching mode; (2) ideal quenching mode; (3) super-ideal quenching mode.

separation step (Asisov et al., 1985; Harkness & Doctor, 1993). This endothermic plasma process can be presented as

$$H_2S + 2CO_2 \rightarrow H_2 + 2CO + SO_2, \quad \Delta H = 3.2\,\mathrm{eV/mol}\,H_2S. \qquad (10\text{–}89)$$

The enthalpy of production of a syngas molecule in this case is about 1 eV/mol, which is higher than in process (10–58). In the experiments of Asisov et al. (1985) using non-equilibrium moderate-pressure microwave discharge, it is important that syngas and SO_2 are the only products (no water, sulfur, or sulfur-carbons are produced). Although the energy cost of hydrogen production in plasma-chemical H_2S decomposition (10–58) is relatively low, it can be further decreased by a small addition of oxygen. As an example, consider the plasma-chemical process (Nester et al., 1988)

$$H_2S + (1/32)O_2 \rightarrow (15/16)H_2 + S(s) + (1/16)H_2O, \quad \Delta H = 0.05\,\mathrm{eV/mol}. \qquad (10\text{–}90)$$

Comparison of reactions (10–58) and (10–90) shows that the addition of only 3% of oxygen (and therefore burning 6% of the produced hydrogen) reduces the required process enthalpy four-fold. The energy cost of hydrogen production in atmospheric-pressure plasma is presented in Fig. 10–54 as a function of specific energy input for different quenching modes. In the fast-rotating plasma (Section 10.7.6), the process leads to the super-ideal quenching mode and a decrease of the minimal energy cost to 0.33 eV/mol (compared to 0.5 eV/mol in the case of decomposition of pure H_2S; see Sections 10.7.1, 10.7.2, and 10.7.6). The hydrogen sulfide extracted from natural gas and from crude oil desulfurization gases sometimes contains up to 10% of hydrocarbons in industrial conditions. The hydrocarbons in this case react with the sulfur produced in process (10–58), forming CS_2. In other words, the presence of hydrocarbons replaces sulfur with CS_2 in the products of H_2S dissociation. This plasma-chemical process has been analyzed by Balebanov et al. (1989) in microwave discharge in mixtures of H_2S with 0–20% propane (C_3H_8). The experiments were carried out at microwave powers up to 3 kW, 0.1–1 atm pressures, and gas flow rates up to 3 standard m^3/h. The major results of the microwave experiments in the H_2S–C_3H_8 mixtures, where sulfur is replaced by sulfur–carbon compounds, are summarized in Fig. 10–55. Fig.10–55 shows that a relatively small amount of propane (C_3H_8, 10 vol %) results in the complete conversion of sulfur into CS_2 and other sulfur–carbon compounds. However, according to a simple mass balance, the complete replacement of sulfur in the H_2S dissociation products by CS_2 requires significantly more than 10% propane:

$$H_2S + (1/6)C_3H_8 \rightarrow (7/3)H_2 + (1/2)CS_2. \qquad (10\text{–}91)$$

Balebanov et al. (1988) explain this effect by using special non-equilibrium co-clusterization of carbon and sulfur in the system. The replacement of sulfur by sulfur–carbon

10.7. Hydrogen Sulfide Decomposition in Plasma

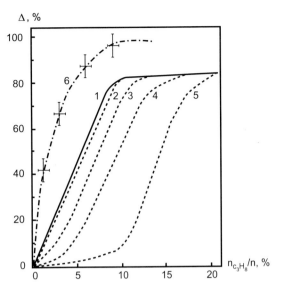

Figure 10–55. Amount of sulfur bound in the compounds with carbon as a function of molar fraction of propane in the mixture $H_2S + C_3H_8$: (1) without addition of oxygen; (2) 1% of oxygen; (3) 3% of oxygen; (4) 7% of oxygen; (5) 15% of oxygen; (1)–(5) kinetic modeling results; (6) experiments in the mixture $H_2S + C_3H_8$.

compounds can be suppressed by the addition of oxygen in the mixture (see Fig. 10–55). When the molar fraction of C_3H_8 is 10%, sulfur can be kept as a product by adding 15% oxygen. This addition of 15% oxygen also leads to a reduction of energy cost of hydrogen production by a factor of 3 (Balebanov et al., 1988).

10.7.8. Technological Aspects of H_2 Production from Water in Double-Step and Multi-Step Plasma-Chemical Cycles

Plasma technologies for hydrogen production, discussed in Section 10.7.3, are not limited to the plasma itself. Products should be separated, additionally purified, and so forth to integrate plasma-chemical processes into the more general technologies. Taking into account the high specific productivities of plasma devices (productivity per unit volume of equipment), system integration can be especially challenging. It can be challenging to find conventional chemical engineering equipment (product separators, catalytic reactors, heat exchangers, etc.) consistent with the process intensity and specific productivity of plasma reactors. If system integration and technological compatibility are not solved, plasma-chemical technology for hydrogen production could be symbolized by a dinosaur with a small head representing the compact and highly productive plasma unit itself, and a massive tail representing the auxiliary (tail) slow chemical processes. The problem was mentioned earlier regarding product separation in hydrogen production by means of H_2S dissociation (Section 10.7.3). In this section, the important technological problem will be addressed shortly in connection with plasma-chemical cycles of hydrogen production from water. A highly energy-effective, two-step, technological cycle of H_2 production from water using a nuclear power plant as the electric energy source was mentioned in Section 5.1.1. A simplified technological scheme of the plasma-chemical cycle is presented in Fig. 10–56 (Rusanov & Fridman, 1984). Electricity produced during the nighttime hours at a nuclear power plant (off-peak electricity) is the source of electric energy for the plasma-chemical reactor, where CO_2 is dissociated in strongly non-equilibrium conditions with the production of oxygen and carbon monoxide:

$$CO_2 \rightarrow CO + 1/2O_2. \qquad (10\text{–}92)$$

This plasma-chemical process in non-equilibrium plasma conditions was considered in detail in Chapter 5. The energy efficiency of the plasma-chemical process reaches 90%;

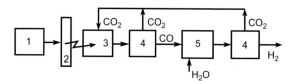

Figure 10–56. Technological scheme of the plasma-chemical double-stage process of hydrogen production from water: (1) nuclear energy power source; (2) electric power supply system; (3) plasma-chemical reactor; (4) gas separation system; (5) steam conversion shift reactor.

the energy cost of CO production is about 3.2 kWh/m^3. Specific productivity (productivity per unit volume of the reactor) in this case is very high at 0.03–0.1 standard L/(s ·cm^3) (Givotov et al., 1981). The produced CO–O$_2$–CO$_2$ mixture is then separated, with the CO$_2$ recirculating back to the plasma-chemical reactor. Oxygen is extracted at this stage from the mixture as a product. Special gas-separating polymer membranes are used to achieve specific productivity comparable with that of a plasma reactor. In the second chemical stage, the produced carbon monoxide is converted into hydrogen in the exothermic shift reaction

$$CO + H_2O \rightarrow CO_2 + H_2. \tag{10–93}$$

Conventional catalysis using the shift reaction is characterized by a relatively low specific productivity. Better productivity was achieved by an electrochemical method applied to the shift reaction. The final stage of the technology is based again on using polymer membranes for separation of gas products and recirculation of CO$_2$. A detailed description of the two-step plasma-chemical technology of hydrogen from water can be found in Belousov, Legasov, and Rusanov (1977) and Legasov et al. (1988). Of the other plasma-chemical cycles of hydrogen production from water, we should note those based on plasma-chemical decomposition of HBr with the production of hydrogen and bromine:

$$HBr \rightarrow \frac{1}{2}H_2 + \frac{1}{2}Br_2. \tag{10–94}$$

This endothermic plasma-chemical process was considered in Section 5.11.1. The process requires super-ideal quenching and separation of the products by fast plasma rotation similar to the case of plasma-chemical H$_2$S dissociation (see Sections 10.7.5 and 10.7.6, and Fig. 5–76). The plasma-chemical stage (10–94), using off-peak electricity, can be a key step in the following HBr cycles of hydrogen production from water. The first cycle, proposed by Parker and Clapper (2001), applies the bromine produced in plasma (10–94) for reaction with water:

$$Br_2 + H_2O \rightarrow 2HBr + O_2. \tag{10–95}$$

Hydrogen bromide produced in the reaction goes to the plasma reactor (10–94) to close the cycle of hydrogen and oxygen production of water. Reaction (10–95) is endothermic; therefore, Parker and Clapper (2001) propose to use solar energy to complete the energy balance. As an exothermic alternative for Br$_2$ conversion into HBr, Parker and Clapper (2001) propose a slightly exothermic process using CH$_4$ and converting it to carbon (without CO$_2$ emission):

$$2Br_2 + CH_4 \rightarrow 4HBr + C(s). \tag{10–96}$$

Reaction (10–95) can be an exothermic one through the addition of methane to the bromine–water mixture: $4Br_2 + 2H_2O + CH_4 \rightarrow 8HBr + CO_2$. However, this hydrogen production technology does lead to significant CO$_2$ emission. An interesting HBr cycle for H$_2$ production from water was proposed by Doctor (2000) and Doctor, Matonis, and

Wade (2003, 2004). In this approach, bromine generated in the plasma by decomposition of hydrogen bromide (10–94) reacts with calcium oxide (CaO) to produce calcium bromide and oxygen:

$$Br_2 + CaO \rightarrow CaBr_2 + 1/2\, O_2. \tag{10–97}$$

This second step of the H_2 production cycle proceeds best at a temperature of 823 K. The third step is conversion of calcium bromide ($CaBr_2$) from the second stage (10–97) into calcium oxide (CaO):

$$CaBr_2 + H_2O \rightarrow CaO + 2HBr. \tag{10–98}$$

This reaction recovers CaO to be used in the second stage of the cycle (10–97), and recovers HBr to be used in the first plasma-chemical stage of the cycle (10–94). Thus, the three stages (10–94), (10–97), and (10–98) form a three-step technological cycle of hydrogen production from water built on the plasma-chemical process (10–94). Note that the third stage, (10–98), is essentially endothermic with $\Delta H = 1.9$ eV/mol and activation energy exceeding 2 eV/mol. Thermal stimulation of the process requires relatively high-temperatures of about 1000 K. The process was investigated at Argonne National Laboratory by Doctor (2000) and Doctor, Matonis, Wade (2003, 2004). Thus, effective use of plasma-chemical technologies for hydrogen production generally requires significant efforts in building a complete technology with all the individual components consistent with high-performance characteristics of a plasma reactor.

PROBLEMS AND CONCEPT QUESTIONS

10–1. Methane Reforming into Syngas in Non-Equilibrium Discharges in CO_2–CH_4 Mixtures. Table 10–3 presents the number of moles of CO_2, H_2O, and CO produced from 100 moles of CO_2–CH_4 mixtures during methane reformation into syngas (CO–H_2) in a non-equilibrium gliding arc discharge. Assuming reformation (10–4), estimate the corresponding amount of hydrogen produced in the plasma-chemical process from 100 moles of the inlet gas at three different initial mixture compositions.

10–2. Plasma Catalysis of Hydrogen Production by Direct Decomposition (Pyrolysis) of Ethane. Interpreting the plasma-catalytic effect of ethane decomposition and hydrogen production illustrated in Fig. 10–10, explain why the application of thermal plasma results in an increase of gas temperature, while application of non-equilibrium plasma results in gas cooling and additional hydrogen production. Compare the thermodynamics of these systems with that of refrigerators and heat pumps.

10–3. Plasma Catalysis of Direct Methane Decomposition in Non-Equilibrium Discharges. Assuming the mechanism (10–7)–(10–11) for plasma catalysis of methane decomposition, estimate the chain length of the ion–molecular Winchester process. In other words, estimate the number of CH_4 molecules that can be converted into molecular hydrogen and growing clusters in this chain reaction using one positive ion before its recombination. Assume in your calculations that translational gas temperature in the non-equilibrium discharge is 550°C.

10–4. Plasma-Stimulated Partial Oxidation of Liquid Fuel into Syngas (CO–H_2). Explain why the total energy efficiency of syngas production by partial oxidation of liquid fuels is always less than 100%, even in the ideal plasma process (for example, the efficiency is 65% for diesel fuel conversion; see Section 10.2.4). A fraction of energy is always converted into heat in the processes. Calculate the maximum energy efficiency for conversion of kerosene into syngas, assuming that the plasma-stimulated process follows the simplified chemical formula (10–14).

10–5. Thermal Plasma Conversion of Coal. Compare production of CO and CO_2 by oxidation of coke in reactions (10–32) and (10–33) in thermal plasma conditions. Taking into account relation (10–34) between the coke oxidation rate coefficients, calculate temperatures at which CO and CO_2 production becomes equal at relatively low and relatively high pressures.

10–6. Gasification of Coal by Water Vapor in Thermal Plasma Jet. Prove that thermal plasma treatment of coal with water vapor provides not only gasification of the volatile compounds of the coal but also gasification of coke. Use the material balance data shown in Table 10–12.

10–7. Coal Conversion into Hydrogen and Carbon Suboxides Without CO_2 Emission. Use relation (10–54) to estimate the front velocity of solid-state (coal) oxidation at low temperatures. How does the oxidation front velocity depend on temperature? Which effects determine the temperature limit for acceleration of the oxidation front? In which way does non-equilibrium plasma stimulate the process?

10–8. Effective Clusterization Temperature as Function of Centrifugal Factor: mv_φ^2/T. At a 0.1 atm pressure and high centrifugal factor of $mv_\varphi^2/T \approx 1$, the effective clusterization temperature in sulfur (formation of S_6 and S_8) is about 850 K. Using the condensation temperature of sulfur, $T_c \approx 550$ K, at 0.1 atm, estimate the effective clusterization temperature of sulfur as a function of the centrifugal factor.

Plasma Chemistry in Energy Systems and Environmental Control

11.1. PLASMA IGNITION AND STABILIZATION OF FLAMES

11.1.1. General Features of Plasma-Assisted Ignition and Combustion

Spark ignition is one of the oldest applications of plasma, known and successfully applied for thousands of years. Even in the automotive industry, spark ignition has been applied for more than a hundred years. Nevertheless, other plasma discharges, especially, non-thermal discharges, have been attracting more and more attention for use in ignition and stabilization of flames. An example, in this regard, is the non-thermal plasma ignition of fuel–air mixtures at moderate pressures and high velocities, including ignition in supersonic flows, plasma enhancement of combustion at atmospheric pressure, and stimulation of combustion of lean mixtures (Anikin et al., 2005; Starikovskaia, 2006). Numerous investigations have been focused on plasma ignition and stabilization of flames. The effectiveness of spark ignition relies on the essential non-uniformity of the thermal plasma of spark discharges and, therefore, restrictions of the system geometry (see, for example, Thiele, Warnatz, & Maas, 2000). Relevant application of thermal arc discharges, related in particular to hypersonic flows, has been analyzed, for example, by Takita (2002) and Matveev et al. (2005). Initiation of flame by a short-pulse thermal discharge and a conventional arc has been investigated in CH_4–air mixtures using the time-resolved interferometry by Maly and Vogel (1979). The ignition effect of gliding arc discharges, which generates non-thermal plasma but also can result in some controlled heating, has been analyzed by Ombrello et al. (2006a,b). The use of such strongly non-equilibrium discharges as pulsed corona for the ignition of fuel–oxidizer mixtures has been described in particular by Wang et al. (2005), of a microwave discharge by Brovkin and Kolesnichenko (1995), of pulsed nanosecond sparks by Pancheshnyi et al. (2005), of pulsed nanosecond dielectric barrier discharges (DBDs) by Anikin et al. (2003), and of volume nanosecond discharges by Bozhenkov, Starikovskaia, and Starikovskii (2003) and Lou et al. (2006). Some of these discharges are to be considered in detail in the following sections in connection with their efficiency in the plasma ignition and stabilization of combustion. The effect of a spark with different currents in a primary electric circuit on the explosion limits of an H_2–O_2 mixture was demonstrated long ago (Gorchakov & Lavrov, 1934; Nalbanjan, 1934; Dubovitsky, 1935). The plasma ignition effect observed in these classical experiments is summarized in Fig. 11–1. The curve marked "0 A" (no current) in the figure corresponds to the conventional H_2–O_2 explosion limit (so-called Z-curve). Relatively high current discharges significantly shift the Z-curve to lower temperatures and pressures. The hydrodynamics of spark ignition development (including nanosecond sparks) has been investigated by Chomiak (1979), Haley and Smy (1989), and Reinmann and Akram (1997). Spark ignition starts with rapid formation of an ignition

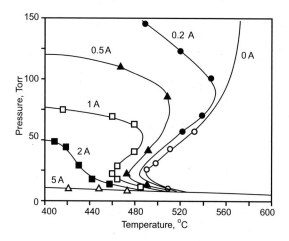

Figure 11–1. Illustration of extension of explosion limits of the H_2–O_2 mixture due to the plasma stimulation of ignition. Different curves correspond to different currents in a primary electric circuit.

kernel, and generation of a uniform, small-scale turbulence flow field near the ignition point is able to improve the repeatability of the plasma ignition process.

Two qualitatively different mechanisms, thermal and non-thermal, make contributions to plasma-assisted ignition and combustion. The thermal mechanism implies that plasma as an energy source provides a temperature increase (usually local), which results in exponential acceleration of the elementary reactions of dissociation, chain propagation of fuel oxidation, and, finally, ignition. The locally ignited mixture then propagates, establishing a stable combustion process. The thermal ignition mechanism does not require the special contribution of specific active species produced in plasma but is based mostly on the plasma-related temperature increase. Application of thermal plasma discharges like coronas and arcs leads usually to thermal ignition. Thermal plasma ignition is the key mechanism of spark ignition in internal combustion engines. Strongly non-equilibrium non-thermal discharges (like DBD, corona, gliding arc, etc.) also are able to increase gas temperature and stimulate the thermal ignition mechanism. Non-equilibrium discharges, however, can also trigger a much more effective non-thermal ignition mechanism. The non-thermal ignition mechanism implies that not temperature but specific plasma-generated active species – radicals, excited atoms and molecules, charged particles, and so on – stimulate the chain reactions of fuel oxidation leading to the ignition of the fuel–oxidizer mixture. The role of atoms and radicals in non-thermal plasma ignition has been analyzed in particular by Anikin et al. (2003); the contribution of ion chemistry has been considered in particular by Williams et al. (2004), and the influence of excited atoms and molecules has been discussed in particular by Starik and Titova (2005). Even relatively small concentrations of atoms, radicals, and other active species (molar fraction 10^{-5}–10^{-3}) are sufficient to initiate combustion (Starikovskaia, 2006). The temperature increase in the non-thermal discharges corresponding to this amount of active species is usually not more than 300 K. The non-thermal plasma ignition mechanism has significant advantages over the thermal mechanism, because it requires much less energy and can provide more uniform and well-controlled ignition and stabilization of flames. Simply, the non-thermal mechanism is able to use the plasma energy selectively, only in the active plasma species relevant to the flame ignition and stabilization, whereas the thermal ignition mechanism distributes plasma energy over all degrees of freedom, and most of those are irrelevant to flame ignition and stabilization. The relative uniformity of non-thermal plasma ignition keeps it far from the detonation condition, which is important for many applications (Starikovskaia, 2006). Truly uniform non-thermal plasma ignition and flame stabilization, however, usually require very high power input in the gas flow; thus, a reasonable compromise (see the following sections) between energy cost and uniformity of the plasma ignition process should always be chosen.

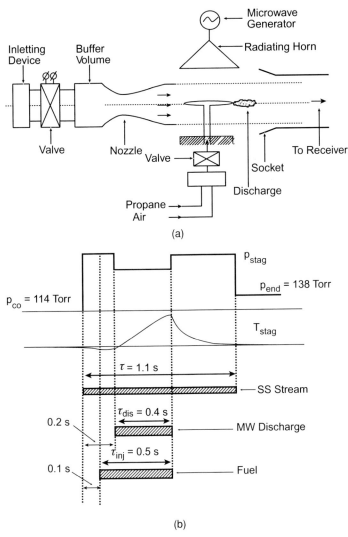

Figure 11–2. General schematic of the experimental setup for (a) the hypersonic microwave ignition; and (b) the time evolution of stagnation pressure and temperature in the system.

11.1.2. Experiments with Plasma Ignition of Supersonic Flows

Plasma is effective in ignition and combustion control in fast and specifically supersonic flows. It is used in important applications for aircraft engines, where problems of reduction in ignition delay time, high-altitude (low-pressure) flameholding, and extension of the flame blow-off limits are of crucial importance. It is especially important for use in supersonic combustion ram (scram) engines. The plasma ignition of supersonic flows is especially challenging because normal flame propagation velocity is always less than the speed of sound. Leonov et al. (2006) analyzed the applicability for ignition, and advantages and disadvantages of different discharges including arcs, microwave discharges, atmospheric-pressure glow (APG) discharges, DBDs, nanosecond pulsed discharges, and laser breakdowns. Most of the experiments with plasma ignition of fast flows, however, have been carried out in microwave and radiofrequency (RF) discharges, which are considered later. A schematic and major characteristics of a setup for microwave plasma ignition of supersonic and subsonic flows of air–propane mixtures (Esakov et al., 2006) are shown in Fig. 11–2. Air flows into the chamber through a supersonic nozzle and fuel or fuel–air mixture is injected (injection time τ_{ing}) through internal tubes in the pylon and in the microwave vibrator; the stagnation pressure in the supersonic system is about hundreds of Torr. Sub-critical microwave discharge (attached to the surface, see Fig. 11–2) with power up to 200 W is used for ignition of the mixture. Ignition and combustion of the propane–air mixtures is observed in the system at gas velocities up to 500 m/s, which corresponds to Mach number $M = 2$.

Plasma ignition of the supersonic propane–air mixture using RF and pulsed discharges has been demonstrated by Klimov et al. (2006). The RF discharge has frequencies of 0.5–1.2 MHz, voltage amplitude up to 60 kV, and mean electric power 0.1–10 kW. The pulsed discharge in the experiments is combined with a direct current (DC) discharge: DC current is 1 A, pulse duration is 1–10 μs, pulsed voltage amplitude is up to 60 kV, repetition frequency is 10 kHz, and mean electric power is about 1 kW. Ignition of the supersonic propane–air mixture with stagnation pressure of about 1 atm is observed in the system at Mach numbers $M < 2$. The RF discharge plasma providing the effective supersonic flow ignition in these experiments is close to thermodynamic equilibrium with all temperatures on the level of 4000–5000 K. The filamentary mode of the RF discharge is unable to provide effective ignition of the supersonic flow of the propane–air mixture. It is important to point out that the combustion process in the supersonic system is almost complete. The effective plasma ignition of supersonic flow ($M = 2$) of the propane–air mixture using a microwave discharge has also been demonstrated by Shibkov et al. (2006). They additionally compared plasma ignition provided by non-equilibrium discharges characterized by different values of reduced electric field E/n_0. They analyzed a DC discharge ($E/n_0 \approx 10$–30 Td), a pulse-periodic transversal electrode discharge ($E/n_0 \approx 30$–70 Td), a free-localized microwave discharge ($E/n_0 \approx 70$–120 Td), and a surface microwave discharge ($E/n_0 \approx 100$–200 Td). It has been shown that discharges with higher reduced electric fields (providing higher densities of chemically active species) give shorter induction times for combustion, which indicates the non-thermal nature of plasma ignition in the systems.

11.1.3. Non-Equilibrium Plasma Ignition of Fast and Transonic Flows: Low-Temperature Fuel Oxidation Versus Ignition

The high efficiency of plasma-assisted ignition of premixed fuel–air mixtures at regimes close to transonic flow (Mach numbers $M \approx 0.8$–0.9) has been demonstrated in the experiments of Lou et al. (2006) and Chintala et al. (2005, 2006). The uniform diffuse RF capacitively coupled plasma (CCP) discharge at pressures of 60–130 Torr has been applied by Chintala et al. (2005) to ignite large-volume transonic flows of air premixed with methane, ethylene, or CO. Flame stabilization by the non-equilibrium RF discharge is provided at significantly higher flow velocities and lower pressures than in the case of the application of thermal discharges: arcs and sparks. A schematic of the RF-CCP discharge plasma ignition system operating at frequency 13.6 MHz and power 200–500 W (Chintala et al., 2005, 2006) is shown in Fig. 11–3. The air–fuel mixture enters the rectangular combustion chamber through a flow straightener and passes between two electrode blocks. It should be mentioned that the electrodes have been rounded at the edges to avoid high local electric fields and possible "hot spots" influencing ignition. The electrodes were separated from the gas flow by dielectric barriers to avoid secondary emission from the electrode surfaces. The gas temperature has been characterized using fourier transform infrared spectroscopy from the rotationally resolved R-branch of the CO fundamental band $1 \rightarrow 0$ (CO was especially added to the flow). The flow temperatures are summarized in Fig. 11–4. The shaded areas in the figure show the range of pressures where the flow ignition has been by adding a stoichiometric amount of methane or ethylene. It is important that the flow temperatures before ignition in the system are much lower than the auto-ignition temperature at the given pressure. For example, the auto-ignition temperature for ethylene at these conditions is

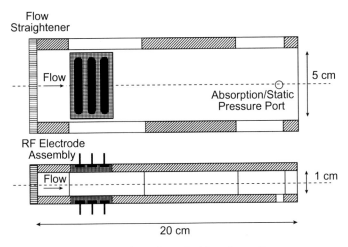

Figure 11–3. Schematic of RF plasma ignition of a gas flow.

600–700°C (compare to Fig. 11–4; see also the section below on the theoretical analysis of the plasma ignition). Application of the RF plasma results in a significant decrease in ignition delay time from several seconds to milliseconds.

Non-equilibrium plasma is able to stimulate low-temperature oxidation of the fuel without ignition of the combustible mixture. Such a regime has been observed, for example, in experiments with non-equilibrium RF-CCP discharge at low equivalent ratios ($\phi < 0.75$–0.8, lean mixtures). A similar effect of plasma-stimulated fuel oxidation without ignition has also been observed in experiments with strongly non-equilibrium nanosecond discharges (Anikin, Starikovskaia, & Starikovskii, 2004; Lou et al., 2006). The effect of low-temperature oxidation of fuel without ignition can be explained by the generation of O, H, and OH radicals in the non-equilibrium RF discharge. The active atoms and radicals generated in the discharge partially oxidize the fuel, which leads to a certain increase of temperature. If the temperature increase is not sufficient, low-temperature oxidation takes place

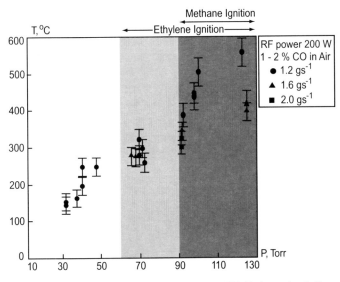

Figure 11–4. Flow temperatures in a transverse RF discharge in air flow at different pressures and mass flow rates (RF power is 200 W).

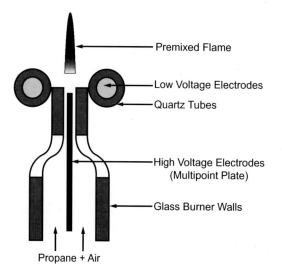

Figure 11–5. General schematic of the propane–air burner sustained by a nanosecond pulsed DBD.

without ignition. If the temperature increase is significant, it can stimulate thermal ignition. The kinetic comparison of such an ignition mechanism with "completely" non-thermal ignition is discussed in particular regarding the theoretical analysis of plasma ignition below the auto-ignition limit.

11.1.4. Plasma Sustaining of Combustion in Low-Speed Gas Flows

Non-equilibrium plasma has proven to be effective in sustaining combustion in low-speed (3–100 m/s) turbulent or laminar atmospheric-pressure gas flows, which are employed in industrial burners. For example, the application of pulsed nanosecond DBD allows doubling of the blow-off velocity of the premixed propane–air flame (Anikin et al., 2003). The required discharge power in this case is less than 1% with respect to the burner power. The application of pulsed nanosecond DBD also allows for burning of lean propane–air mixtures without flame deceleration. A schematic of the nanosecond DBD system for sustaining a flame is shown in Fig. 11–5. Voltage pulses (12.5 kV in the cable, 25 kV on the electrode) are provided in the system with repetition frequency 1.2 kHz and duration 75 ns. Flow velocity in the burner is a few meters per second. The blow-off velocities of the premixed propane–air flame are summarized in Fig. 11–6 as a function of the equivalence ratio of the mixture with and without application of plasma at different polarities and pulsation regimes (Anikin et al.,

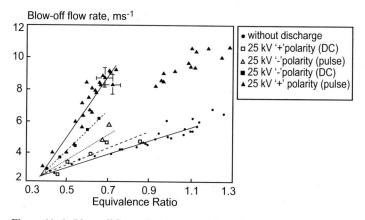

Figure 11–6. Blow-off flow velocities as function of equivalence ratio at different discharge parameters.

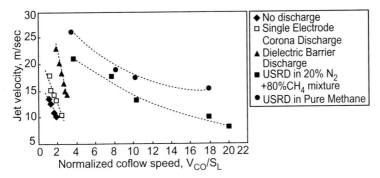

Figure 11–7. Enhancement of the lift-off jet velocity as a function of normalized co-flow speed: S_L is the laminar flame speed; SECD, single-electrode corona discharge; USRD, ultrashort repetitively pulsed discharge.

2003). The figure demonstrates the effect of doubling the blow-off velocity by application of the strongly non-equilibrium nanosecond DBD. Similar results on the plasma stabilization of the premixed lean propane–air flames have been obtained by Galley et al. (2005) also using a nanosecond pulsed discharge but without a dielectric barrier.

Plasma stabilization of a lifted jet diffusion (non-premixed) flame in co-flow configuration at atmospheric pressure has been demonstrated by Leonov et al. (2006). Three different types of discharges have been used for stabilization of the non-premixed lifted flame, namely, a single-electrode corona discharge, a DBD, and a nanosecond pulsed discharge between two electrodes. The DBD and corona are powered by an AC power supply with an open circuit peak voltage of 20 kV and frequency range 25–35 kHz. The nanosecond discharge is generated by pulses of 10 ns duration, peak voltage up to 10 kV, and repetition frequency up to 100 kHz. The efficiency of the three non-thermal discharges for flame stabilization is compared in Fig. 11–7 by plotting a liftoff jet velocity on a normalized co-flow speed. Presented liftoff jet velocities correspond to the input peak-to-peak voltage 9.35 kV and frequency 30 kHz in the case of the corona discharge and DBD, and 6 kV and 15 kHz, respectively, in the case of the nanosecond pulsed discharge. It is seen from Fig. 11–7 that the nanosecond pulsed discharge can increase the flame stability limit 10-fold, whereas the effect of the DBD discharge is less pronounced. The power consumption of all three discharges is less than or about 0.1% of the chemical power. Taking into account that flame itself is plasma, the flame can be controlled by sub-breakdown electric fields. Control of the hydrocarbon flame by sub-breakdown microwave radiation with frequency 2.45 MHz has been demonstrated by Zaidi et al. (2006). Even relatively low absorption of microwave radiation in a laminar flame leads to a significant effect in the flame speed increase without changing the flow velocity structure and without causing electric breakdown. The stretched flame speed increases from 33.7 to 45.3 cm/s as long as the incident microwave power is below 1.2 kW. At the 1.2 kW incident microwave power level, there is no electric breakdown of the gas and only a few watts of electromagnetic radiation are absorbed by the flame. In these conditions, however, the flame speed increases up to 34.5%.

11.1.5. Kinetic Features of Plasma-Assisted Ignition and Combustion

Interesting kinetic conclusions can be derived from a comparison of quasi-thermal spark ignition and ignition by strongly non-equilibrium nanosecond pulse corona discharge performed by Liu, Ronney, and Gundersen (2003) and Liu et al. (2004). The pulsed corona discharge has been generated in these experiments in the wire-cylinder configuration using nanosecond pulses with voltage 30–55 kV, duration about 150 ns, and maximum pulse energy 1.8 J. A comparison of a pulsed corona and a common spark plug ignition has been done by measuring the time delay for ignition of a C_8H_{18}–air mixture. It has been shown that, at the same equivalence ratio values, the nanosecond discharge gives at least three to five times shorter ignition delay. Also, the nanosecond pulsed corona permits ignition at equivalence ratio 0.8, which is impossible for a spark discharge at the same conditions. A kinetic comparison of the spark ignition and ignition by the nanosecond pulse corona indicates significant advantages of the non-thermal atmospheric-pressure ignition mechanism (see Section 11.1.1). The kinetic peculiarities of separate fuel activation by non-equilibrium plasma were analyzed by Rosocha et al. (2004). DBD has been used in these experiments to activate propane before its mixing with air and stabilizing the combustion process. The time delay between the plasma activation of propane and the fuel–air mixing is sufficient for recombination of most radicals in the atmospheric-pressure system. Thus, it can be concluded that the plasma-stabilization effect should be attributed in this case to the long-lived excited and chemically active species, as well as to a possible temperature effect in the discharge. Interesting kinetic information has been obtained by investigating the ignition of propane–air mixtures at elevated pressures using a high-voltage repetitively pulsed nanosecond discharge (pulse voltage about 10 kV, duration 10 ns, frequency 30 kHz) in a point-to-plate 1.5 mm gap (Pancheshnyi et al., 2005). The energy input in the system depends on the type of discharge in the system: corona occurs at lower currents, providing an energy input not exceeding 0.2 mJ/pulse; nanosecond spark occurs at higher currents, leading to energy inputs greater than 1 mJ/pulse. It has been demonstrated that energy in the streamer phase is not sufficient to ignite the mixture, whereas energies deposited during the phase of developed breakdown are sufficient for ignition of the mixture. Increased pressure leads to a decrease of the minimum energy required for ignition but makes the delay time longer. If the energy of a single pulse is not sufficient for ignition, a sequence of pulses has a cumulative effect on the ignition. Charged species recombine faster than the recombination and relaxation of active neutrals. Therefore, each following breakdown is only slightly affected by the previous discharge, while chemically active neutrals are accumulated in the discharge gap from pulse to pulse, leading to significant decrease in the ignition time.

Kinetic studies of auto-ignition are commonly carried out using the shock-wave technique. Using this technique for kinetic studies of strongly non-equilibrium plasma ignition with a spatially uniform fast ionization wave has been proposed by Kof and Starikovskii (1996). The nanosecond fast-ionization-wave discharge is spatially uniform in large volumes at voltage amplitudes as high as tens or hundreds of kilovolts. Due to the large difference between the time for nanosecond discharge development (< 50 ns) and typical ignition time (0.01–1 ms), these processes are separated in the system, which simplifies kinetic analysis. Detailed experiments with plasma-assisted ignition of H_2–O_2 and CH_4–O_2 mixtures using a fast ionization wave have been carried out by Bozhenkov, Starikovskaia, and Starikovskii (2003) and Starikovskaia et al. (2004). Some typical results of the experiments are summarized in Figs. 11–8 and 11–9. The experiments clearly demonstrated a pronounced (up to 600 K) temperature shift in the ignition threshold, and the ignition delay time has been decreased by an order of magnitude and more. Understanding the kinetic mechanisms of plasma-assisted ignition and combustion requires a detailed experimental characterization of the composition of chemically active plasma, and detailed kinetic modeling of the system. Different spectroscopic characterization experiments have been carried out by Ganguly and Parish (2004), Marcum, Parish and Ganguly (2004), Brown, Forlines, and Ganguly (2006), and Starikovskaia (2006). The diagnostic analysis permits the evaluation of densities of such radicals as CH, C_2, and H, as well as rotational and vibrational temperatures of nitrogen in the plasma ignition system.

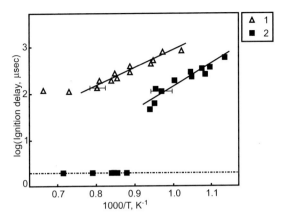

Figure 11–8. Ignition delay time in the mixture $H_2:O_2:N_2:Ar = 6:3:11:80$ presented as a function of gas temperature: (1) auto-ignition; (2) ignition by a nanosecond 160 kV discharge.

11.1.6. Combined Non-Thermal/Quasi-Thermal Mechanism of Flame Ignition and Stabilization: "Zebra" Ignition and Application of Non-Equilibrium Magnetic Gliding Arc Discharges

Non-thermal mechanisms can be more effective in ignition and stabilization of flames. First of all, it is because non-thermal discharges are able to provide energy selectively only into the channels relevant to stimulation of combustion. It should be mentioned, however, that a certain level of translational temperature (or in other words, direct heating) can also be helpful in stimulating combustion. Anyway some increase of translational temperature is inherent to non-equilibrium discharges and can be provided extremely quickly (see Sections 3.4–3.6). It attracts interest to the application of combined (hybrid) non-thermal/quasi-thermal discharges with significantly super-equilibrium generation of active plasma species and at the same time with a controlled level of translational temperature (Fridman et al., 2007a,b). The non-equilibrium gliding arc (see Sections 4.3.7–4.3.11, 10.1.2, and 10.1.3) is an example of such discharges, which are especially promising for ignition and stabilization of flames. The non-equilibrium gliding arc discharge is also interesting for flame ignition and stabilization because of its periodic structure in space and time (see Sections 4.3.7–4.3.11). Homogeneous non-equilibrium plasma activation of flames leads to power requirements that can become extremely high, sometimes in excess of several MJ/m³ (Fridman et al., 2007a,b; see also Section 11.2). The application of non-equilibrium gliding arcs helps provide excitation of not all the combustible gas flow but to ignite the gas only in some well-organized time-space structured array, which decreases energy consumption but still avoids unstable detonation-related effects on the fast gas flow during combustion (Fridman et al., 2007a,b). The time-space structured array of plasma-assisted ignition, sometimes

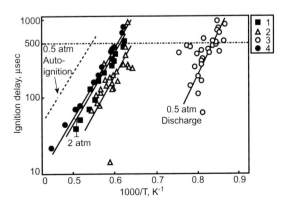

Figure 11–9. Ignition delay time in the mixture $CH_4:O_2:N_2:Ar = 1:4:15:80$ presented as a function of gas temperature: (1) auto-ignition at 2 atm; (2) ignition with the discharge at 2 atm; (3) pressure 0.5 atm, ignition with the discharge; dashed line, auto-ignition at 0.5 atm (calculated); (4) auto-ignition at 2 atm (calculated).

Figure 11–10. Photographs of the magnetic gliding arc device: front, top, and isometric views.

referred to as **"zebra" ignition** (Fridman et al., 2007; see also Section 11.2), permits a significant decrease of the non-equilibrium plasma power required for ignition of large-volume flows, particularly in the case of scramjet engines.

The magnetically stabilized non-equilibrium gliding arc (MGA) driven by Lorentz force and fast rotating in a magnetic field is convenient for effective flame ignition and stabilization, which has been demonstrated by special experiments with a counterflow burner (Ombrello et al., 2006a,b; Fridman et al., 2007a,b). The MGA apparatus, as seen in Fig. 11–10, is composed of stainless steel inner and outer electrodes, which serve as the cathode (high voltage) and anode (grounded), respectively. A wire is fixed to the cathode that is kept separated from the outer anode by approximately 2 mm at the smallest gap (Fig. 11–11). The spiral wire gets progressively closer to the cathode (inner electrode) where it is attached, which is at the largest gap between the two electrodes. In addition, there is a magnetic field in the discharge region produced by an external donut-shaped permanent ceramic magnet as seen in Fig. 11–10. The direction of the magnetic field determines the direction of the rotation of the arc. The magnetic field strength (about 0.1 T) as well as the plasma current determines the rotation frequency of the gliding arc. When a high enough voltage is applied, there is an initial breakdown of the gas at the shortest gap between electrodes and quasi-thermal arc plasma is established. The arc then rotates in the magnetic field and elongates as the distance between the spiraled wire and the outer electrode increases (as seen in Fig. 11–11). The increased length of the arc results in transition to a non-equilibrium plasma leading to rapid cooling to temperatures < 2000 K, as well as to an increase in electric field and average electron temperatures (> 1 eV). Once the arc reaches the cylindrical inner electrode (Fig. 11–11), it is stabilized as a non-equilibrium transitional discharge. A top view of this plasma disk can be seen in Fig. 11–12. The plasma arc rotation frequency ranges from approximately 20 to 50 Hz (for a current range of 30 to 90 mA) and only decreases by a few percent when the flow rate is increased (Fig. 11–13). The increased rotation frequency comes from the higher current input (and, hence, higher power addition), forcing the arc to rotate faster in the fixed magnetic field. When the arc reaches the largest gap, it remains at a

(a)　　　　　(b)

Figure 11–11. Gliding arc plasma system: (a) top view of the system, (b) side view of the central electrode.

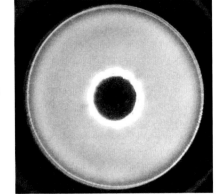

Figure 11–12. Top view of the magnetic gliding arc plasma disk.

fixed axial position with a constant length as it rotates. The gliding arc sustains itself during this period of continuous rotation between electrodes. This is because it moves repeatedly through a medium that was previously ionized. It therefore requires less power input to remain at that position (stabilize) as opposed to quenching and reinitiating at the smallest electrode gap, as it does in the flat geometric configuration. Since the rotation frequency of the arc is fast when compared to the gas velocity, there is quasi-uniform activation of the flow. Energy is deposited in these conditions not to all gas flow, which corresponds to the aforementioned energy-efficient "zebra" ignition regime. Typical gliding arc current in the ignition experiments is 30–40 mA, power is 30–40 W, gas temperature in the arc is about 1600–2000 K (from N_2 spectrum), vibrational temperature of N_2 in the non-equilibrium discharge is about 4000 K. Reduced electric field in the positive column of the MGA in the experiments is about 20–30 Td, electron temperature is about 1 eV, and plasma density is $1–2 \cdot 10^{12}$ cm^{-3}.

11.1.7. Magnetic Gliding Arc Discharge Ignition of Counterflow Flame

The effectiveness of non-equilibrium MGA has been demonstrated in particular by ignition of the counterflow flame (Ombrello et al., 2006a,b; Fridman et al., 2007a,b). A schematic of the system is shown in Fig. 11–14. It consisted of two converging nozzles of 15 mm in diameter, separated by 13 mm. The upper nozzle is water cooled. The gas treated by the MGA is air and it is exited from the lower nozzle, while nitrogen-diluted methane flows through the upper nozzle. To isolate the flame from the ambient air and disturbances, a nitrogen co-flow is used, emanating from a 0.75 mm annular slit around the circumference of each nozzle exit. The velocity of the curtain is maintained at or below the exit speed of the nozzle to minimize diffusion into the stream. The flame is established on the upstream air side of the stagnation plane. The use of this system allowed for the examination of

Figure 11–13. Frequency of the arc rotation for two air flow rates.

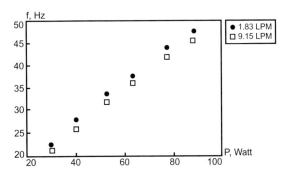

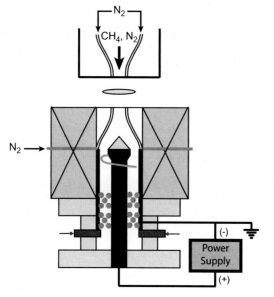

Figure 11–14. Counterflow burner with integrated plasma system: Magnetic gliding arc discharge stimulation of counterflow flame.

strain rates at extinction. To find the extinction limits, the flame is first established with the bypass fully open (the lowest strained flame). Then each bypass is slowly closed, while maintaining the stagnation plane at a fixed position. During this process the flame moves closer to the stagnation plane between the two nozzles, therefore decreasing the residence mixing time as well as increasing the strain rate until flame extinction. Pictures with and without plasma power addition of the nitrogen-diluted methane–air counterflow flames are shown in Fig. 11–15 (Ombrello et al., 2006a,b), where the one-dimentional structure of the flame is seen. The flame with plasma activation of the air has a larger luminous zone with a distinct white and orange coloring as compared to the typical bright blue of the non-activated flame. To quantify the combustion characteristics, Ombrello et al. (2006a,b) measured the extinction limits, planar Rayleigh scattering for temperature profiles, OH planar laser-induced fluorescence (PLIF) for absolute OH number density, and numerical computations with radical addition. The extinction limits for the counterflow flames were observed with and without plasma power addition. The strain rates for both situations were increased, causing the flame to move toward the stagnation plane, continuously increasing the heat loss and decreasing the residence time for reaction completion, eventually leading to extinction. The measured values for the strain rates at extinction can be seen in Fig. 11–15 for three different combinations of nitrogen-diluted methane–air counterflow diffusion

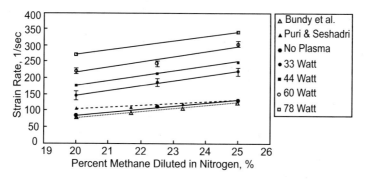

Figure 11–15. Effects of plasma power addition on the strain rates at extinction for different levels of nitrogen dilution.

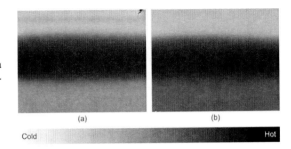

Figure 11–16. Rayleigh scattering images at a strain rate of 98.6 s⁻¹: (a) no plasma power addition; (b) 33 W plasma power addition.

flames for various levels of plasma power. The extinction limits with no plasma power addition agree well with values reported by Puri and Seshadri (1986) and Bundy, Hammins, and Lee (2003). It was observed that using only 78 W of plasma power input gave ~220% increase in the extinction strain rate. The MGA power input is less than 6% of the flame power, but the plasma activation of air leads to a significant extinction limit enhancement.

Planar Rayleigh scattering thermometry has been performed on the counterflow flame for various strain rates. The laser sheet was passed through the diameter of the disk-shaped counterflow flame between the nozzles of the burner. Two sample images can be seen in Fig. 11–16: (a) with no plasma and (b) with 33 W of plasma power addition. The background has been subtracted and the air reference divided into each image. The observed intensity is a function of scattering cross sections and number density distributions between the nozzles. The flame is located in the darkest region of these images (transition from dark to light depicts descending values of temperature and increased scattering). The nitrogen-diluted methane mixture is entering the field of view from the top of the images and the air (MGA side) from the bottom of the images. The one-dimentional flame structure along the flow direction is observed even with the use of the MGA. This result greatly simplified the coupling between the flow field and the plasma–flame interaction. By comparison of the two images it is possible to differentiate them based on the extents of darkness, indicative of less scattering and higher temperatures. The region below the flame in panel b (plasma on) is darker than that in panel a (plasma off) below the flame. In each case, an averaged temperature from a 10-pixel-wide "stripe" is taken from each image. The intensity observed by the ICCD camera is a function of the scattering cross section and number density change between the two nozzles. By referencing those intensities to that of air at room temperature, the ratios have been found (Fig. 11–17) for the plasma power off and on. Using refractory data (Gardiner, Hidaka, & Tanzawa, 1981), the variation in scattering cross sections gives temperature profiles in the burner (Fig. 11–18). The density of OH radicals has been measured in the counterflow burner for different plasma powers and strain rates using PLIF (Fig. 11–19). The ignition experiments have also been performed by Ombrello et al. (2006a,b) using a

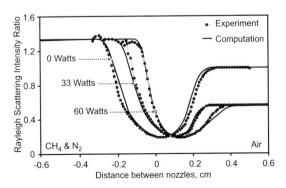

Figure 11–17. Comparison of the Rayleigh scattering intensity ratio profiles corresponding to plasma power: 0 W, 33 W, 60 W. Strain rates are 87.1 s⁻¹, 98.6 s⁻¹, and 298.5 s⁻¹, respectively.

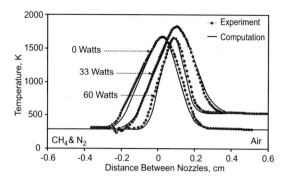

Figure 11–18. Comparison of the temperature profiles corresponding to plasma power: 0 W, 33 W, 60 W. Strain rates are 87.1 sec^{-1}, 98.6 sec^{-1}, and 298.5 s^{-1}, respectively.

preheater together with the MGA on the upstream air side of the counterflow burner. A silicon carbide heater placed upstream of the MGA effectively raised the air temperature to allow for hydrogen ignition. The MGA was turned on and the heater power (and, hence, the air temperature) raised until the hydrogen diffusion flame ignited. The ignition temperatures observed using the preheater and MGA are presented in Fig. 11–20 in comparison with the corresponding ignition temperatures observed by using only a preheater (Fotache et al., 1995). The ignition temperatures obtained with plasma assistance (at low energy addition) are significantly lower than that of just heated air, demonstrating the contribution of the non-thermal ignition mechanism in the combined non-thermal/quasi-thermal discharge (see Section 11.1.6).

11.1.8. Plasma Ignition and Stabilization of Combustion of Pulverized Coal: Application for Boiler Furnaces

Plasma is effectively applied for ignition and stabilization of combustion of not only gaseous or evaporated fuels but also solid fuels. This application is especially important for the ignition of low-quality coals in power plant boilers. The conventional combustion of low-quality coals requires the addition of up to 21–25% natural gas and heavy oils, which is effective neither economically nor technologically (Ibragimov et al., 1983). Although intensive spark discharges can be used for the ignition of the coal powder (Gubin & Dik, 1986), thermal arcs prove to be very effective (Blackburn, 1980; Verbovetsky & Kotler, 1984). Air heated in a thermal arc discharge has a very high temperature and active species density, which are sufficient to provide ignition of coal powder more effectively than the conventionally used burning of natural gas and heavy oil. This process was first demonstrated on an industrial scale in a power plant boiler in Marietta, Ohio, involving the burning of bituminous coal and producing a total of 200 MW of electric power. The energy efficiency of the thermal arc igniter varied in this case in the range 45–68%. Stable ignition requires

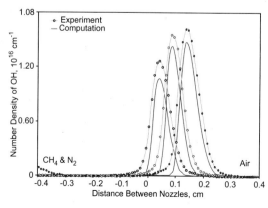

Figure 11–19. Comparison of the OH number density distributions corresponding to plasma power: 0 W, 48 W, 78 W. Strain rates are 83.3 s^{-1}, 127.7 s^{-1}, and 183.0 s^{-1}, respectively.

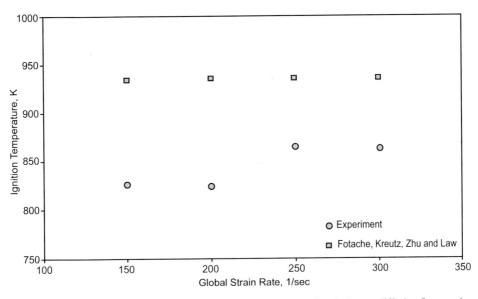

Figure 11–20. Comparison of preheating temperatures of the counterflow hydrogen diffusion flames using preheater versus the preheater and gliding arc activation of the air stream.

arc power of 109 kW and exit plasma jet temperature of 5700 K. A single arc discharge provides stable ignition of 1.5 tons of pulverized coal per hour under these conditions. The fraction of plasma power applied for ignition is only 1% of the combustion power. To compare, the application of heavy oil in the same system required about 6% of the total combustion power.

Effective industrial-scale plasma ignition of coals with a high fraction of volatiles has been achieved by Ibragimov, Marchenko, and Tuvalbaev (1987); Messerle and Sakipov (1988), and Sakipov et al. (1987b). A thermal arc applied for ignition has power 16.8–22.1 kW and is able to increase air temperature to 500 K before it mixes with pulverized coal. The plasma is effective in ignition of 300–600 kg of coal per hour. The effectiveness of the igniter can be characterized by the ratio of the arc power to the power of the coal burner, which is 0.75–1%. Better efficiency can be achieved when the arc is organized directly in the air–coal mixture (Sakipov et al., 1987b). The ratio of the arc power to the power of the coal burner can be decreased in this case to 0.32%. More challenging is the plasma ignition of coals containing a low amount of volatiles, especially anthracites with a percentage of volatiles below 5%. Effective ignition can be achieved in this case by special thermo-chemical coal preparation before combustion (Solonenko, 1986, 1996; Messerle & Sakipov, 1988). This approach assumes thermal plasma heating of the pulverized coal premixed with some fraction of air to temperatures sufficient for release of volatiles and partial gasification of the coke. It results in a total generation of combustible gases on the level of 30%, which actually corresponds to the characteristics of highly reactive coals. This approach permits the conversion of low-reactive coals into a double-component coke-combustible gas mixture, which can then be conventionally ignited without any additional chemical admixtures. The precombustion plasma treatment of coal requires about 0.6 kWh of electric energy spent in the thermal arc plasma per 1 kg of coal, heating of the air–coal mixture to about 930 K, and air consumption on the level of 670 kg per 1000 kg of coal (Sakipov, Imankulov, & Messerle, 1987b; Messerle & Sakipov, 1988). It is important that the precombustion plasma treatment of coal permits a decrease in the effective combustion temperature and, therefore, reduces the amount of nitrogen oxides in the combustion exhaust gases by about 20%. This effect can be explained because the precombustion plasma treatment of coal makes the coal combustion a double-stage process (air is added in two steps), which always leads to lower temperatures and less NO_x generation (Law, 2006). Additional information on the subject can be found in the review of Karpenko and Messerle (2000).

11.2. MECHANISMS AND KINETICS OF PLASMA-STIMULATED COMBUSTION

11.2.1. Contribution of Different Plasma-Generated Chemically Active Species in Non-Equilibrium Plasma Ignition and Stabilization of Flames

Various particles generated in non-equilibrium plasma make contributions to the stimulation of combustion. In non-equilibrium discharges with relatively low reduced electric fields E/n_0, low-energy excitation of molecules such as vibrational excitation and excitation of lower electronic states can be essential in the stimulation of combustion. Lukhovitskii, Starik, and Titova (2005) investigated the reaction kinetics of a hydrogen–air mixture with a small ($< 1\%$) addition of ozone molecules (O_3), vibrationally excited by CO_2 laser radiation at 9.7 µm. The vibrational excitation of ozone results in 10–1000-fold increase of ignition efficiency in comparison to equivalent local gas heating by IR radiation. The same group proposed a kinetic mechanism whereby electronically excited oxygen molecules, $O_2(a^1\Delta_g)$ and $O_2(b^1\Sigma_g^+)$, make a significant contribution to stimulation of CH_4–O_2 combustion (Starik & Titova, 2005). The excitation of $O_2(a^1\Delta_g)$ and $O_2(b^1\Sigma_g^+)$ by laser radiation at wavelengths of 1268 and 762 nm leads to ignition of the methane–air mixture at low temperatures. For example, laser excitation of $O_2(b^1\Sigma_g^+)$ with energy input $10\ \text{J/cm}^2$ permits ignition of the methane–air mixture at room temperature.

Experiments with non-equilibrium plasmas characterized by high reduced electric fields indicate a significant role of the electronic excitation in high energy levels (Kossyi et al., 1992), which results in dissociation either through direct electron impact or by the following collisions:

$$N_2^* + O_2 \rightarrow N_2 + O + O. \tag{11-1}$$

The kinetics of plasma ignition of H_2–O_2–Ar mixtures has been investigated by Bozhenkov, Starikovskaia, and Starikovskii (2003) by assuming a key role for H and O atoms. A similar analysis for methane–air mixtures, additionally taking into account electronically excited species, at different initial pressures and temperatures has been performed by Starikovskii (2003). In many cases, but not always (see the following sections), the role of a non-equilibrium discharge in the ignition of hydrocarbon–air and CO–air mixtures can be explained by a branching mechanism that multiplies the primary radicals generated in the non-equilibrium plasma (Chintala et al., 2006). The role of the non-equilibrium discharge in low-temperature oxidation of different alkanes, starting from CH_4 to C_6H_{14}, can be reduced to that of the supplier of the active radicals (Anikin et al., 2004), while further oxidation takes place via well-known low-temperature oxidation mechanisms (see, for example, Von Elbe & Lewis, 1942). It is also known that plasma-generated ions can play an important role in ignition kinetics (see Williams et al., 2004). For example, addition of 10^{-4} molar fraction of positive ions (NO^+) to the atmospheric-pressure mixture i-C_8H_{18}:O_2:NO:Ar $=$ 1.4:17.6:1:80 significantly shifts the ignition delay time.

11.2.2. Numerical Analysis of Contribution of Plasma-Generated Radicals to Stimulate Ignition

Combustion and ignition kinetics for hydrogen and different hydrocarbons is well known and facilitates the numerical analysis of plasma stimulation, assuminig that the major contribution of plasma is limited to the generation of radicals. Such an efficiency analysis of the plasma-generated O atoms in ignition of H_2–O_2 mixture at atmospheric pressure and different initial temperatures has been carried out, in particular, by Starikovskaia et al. (2006). The minimal density of O atoms leading to ignition has been determined in the analysis. At lower temperatures, according to the modeling, the main effect leading to ignition acceleration is the fast gas heating due to the recombination of radicals. The injection of O atoms becomes more and more effective starting from several hundreds of kelvins, when the atoms not only lead to heating but also initiate chemical chains. The analysis also indicates a possible contribution of chain oxidation mechanisms involving ions as well as electronically and vibrationally excited species.

Campbell and Egolfopoulos (2005) investigated the kinetic paths to radical induced ignition of methane–air mixtures for atmospheric pressure and temperatures from 300 K. They analyzed the kinetic evolution to the ignition by injecting CH_3, CH_2, CH, and C (up to a few percent). CH radicals oxidized to HCO are the most effective in stimulating ignition through a chain mechanism. The CH_2 radicals react with H producing CH and, thus, stimulating the ignition. The contribution of carbon atoms (C) is limited to the gas heating due to their recombination with O_2 leading to production of CO and CH_3. The impact of addition of radicals (O, H, N, NO) on the extinction limit of H_2 and CH_4 flames in a perfectly stirred reactor has been analyzed by Takita and Ju (2006). The contribution of radicals at the extinction limit does not differ much. The effect of radicals on the extinction limit is not as strong as that on the ignition. At low inlet temperatures, the radicals initially added to the mixture are quickly quenched; therefore, the combustion enhancement is mostly due to the thermal effect provided by a recombination of the radicals. The effect of radical addition on the extinction limit becomes much stronger at higher inlet temperatures.

11.2.3. Possibility of Plasma-Stimulated Ignition Below the Auto-Ignition Limit: Conventional Kinetic Mechanisms of Explosion of Hydrogen and Hydrocarbons

Recent experiments, reviewed in particular by Haas (2006), have shown non-thermal plasma-stimulated ignition in H_2, CH_4, and C_2H_4 at temperatures apparently below their respective auto-ignition thresholds. This means that non-thermal plasma is able to accelerate the ignition not only where it is possible according to traditional chemical kinetics, but also below the auto-ignition limit where production of new radicals is suppressed by their recombination and ignition is considered impossible in the framework of conventional combustion kinetics (Williams, 1985). To analyze this important plasma effect, let us first discuss explosion limits for fuel–oxidizer gas mixtures without plasma. A large number of fuel–oxidizer mixtures share qualitatively similar ignition characteristics with respect to branched-chain ignition. A characteristic S-shaped curve – the auto-ignition or explosion limit – divides the plane described by pressure versus temperature into a region of explosion and no explosion (see Fig. 11–21). Most hydrocarbons exhibit more complicated behavior (e.g., lobes of multi-stage ignition, zones for single and multiple cool flames not leading to explosion) as a consequence of their particular oxidation kinetics (see, for instance, Barat, Cullis, & Pollard, 1972; Vanpee, 1993); however, the general S shape remains.

The explosion limit demarcates the locus of initial temperature and pressure conditions where termination and branching reaction rates are balanced, leading to a non-explosive reaction. Since the rate of termination of radical chains exceeds that of chain branching below and/or to the left of the limit, it is generally held in combustion science that the mixture cannot ignite at these conditions without thermal stimulus to raise at least local temperatures to the explosion limit. To non-thermally ignite a mixture at a temperature below the threshold, it follows then that either the termination rate must be reduced or the branching rate must be accelerated to extend the region of auto-ignition in the temperature–pressure plane to lower temperatures.

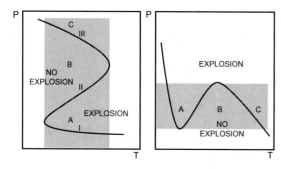

Figure 11–21. Explosion limit diagram characteristic of (a) hydrogen and (b) hydrocarbons.

Auto-ignition of H_2–O_2 is governed by three explosion limits (I–III in Fig. 11–21a). Below the first explosion limit (I), chain-carrying radicals and heat diffuse to vessel walls where their chains terminate faster than branching occurs, inhibiting the overall reaction rate explosion. Above I, in region A, radical chains branch explosively for initial pressures up to the second limit, II, at which point effective termination reaction rates again exceed branching rates. In this case, the rate of the pressure-dependent three-body reaction

$$H + O_2 + M \rightarrow HO_2 + M, \tag{11–2}$$

eclipses the rate of the branching reactions:

$$H + O_2 \rightarrow OH + O, \tag{11–3}$$

$$O + H_2 \rightarrow H + OH. \tag{11–4}$$

The hydroperoxy radical (HO_2) terminates the atomic H branching chain since it is a relatively unreactive radical at the temperatures corresponding to limit II. The plasma extension of the auto-ignition limits in the vicinity of limit II appears to depend on either recovering active radicals (H, O, or OH) from HO_2 or preventing HO_2 formation (11–2). At higher pressures corresponding to limit III, this is the case; the reaction

$$HO_2 + H_2 \rightarrow H_2O_2 + H, \tag{11–5}$$

recovers chain carriers from HO_2, once again enabling explosion.

Hydrocarbon auto-ignition differs from hydrogen auto-ignition in that reactions competing with chain branching are a function of temperature rather than pressure. Glassman (1996) attributes region B of Fig. 11–21b to competition between reactions:

$$R + O_2 \rightarrow \text{olefin} + HO_2, \tag{11–6}$$

$$R + O_2 \rightarrow RO_2, \tag{11–7}$$

where R is any alkyl radical (except methyl). The former reaction only serves to propagate the chain, whereas the latter contributes to branching after subsequent steps:

$$RO_2 + R^1H \rightarrow ROOH + R^1, \quad ROOH \rightarrow RO + OH, \tag{11–8}$$

or

$$RO_2 \rightarrow R^1CHO + R^2O, \quad R^1CHO + O_2 \rightarrow R^1CO + HO_2. \tag{11–9}$$

At the lowest temperatures, chain branching pathway (11–7) predominates, but termination rates exceed branching rates, at least up to region A. At temperatures corresponding to region B, the non-branching step (11–6) predominates, preventing explosion; however, reaction self-heating causes explosive runaway to occur at temperatures corresponding to that in region C. The methyl radical primarily reacts with molecular oxygen at low and

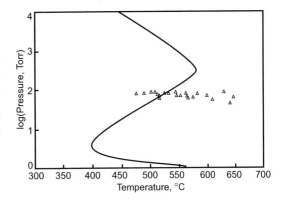

Figure 11–22. Ignition in H_2–O_2 mixture stimulated by the fast ionization wave. Solid curve is for H_2–O_2 without diluents; recorded ignition pressures have been adjusted by removing partial pressure of He. Temperatures and pressures are those after passage of the reflected shock, but before the fast-ionization-wave energy deposition.

intermediate temperatures to form CH_3OO or formaldehyde (CH_2O) and OH (Zhu, Hsu, & Lin, 2001; Herbon et al., 2005), with both pathways potentially leading to branching. This may be why methane does not exhibit a region like B in its explosion curve. Thus, according to traditional combustion kinetics, auto-ignition is a critical (threshold-type) phenomenon. It becomes possible only at such pressures and temperatures when multiplication of radicals overcomes recombination and quenching, leading to explosive growth of their density. From this point of view, the non-thermal plasma generation of radicals without influencing temperature should not result in ignition, because the radicals would be quenched anyway. Experiments discussed in the next two sections demonstrate the possibility of non-thermal plasma ignition below the auto-ignition temperature threshold. This effect, which has significant potential for applications, can be obviously explained by the contribution of specific non-thermal plasma species including charged and excited atoms and molecules, which is also to be discussed.

11.2.4. Plasma Ignition in H_2–O_2–He Mixtures

The subthreshold auto-ignition of a stoichiometric hydrogen–oxygen mixture ($12H_2$: $6O_2$:$82He$) has been observed by Bozhenkov, Starikovskaia, and Starikovskii (2003), using shockwaves to bring the mixture to near-threshold temperature and the fast ionization wave (FIW) as single-pulse plasma stimulation for the gas mixture. Ignition delay times were measured as the time between initiation of the $\sim$40 ns duration FIW pulse and the emission of the $A^2\Sigma \rightarrow X^2\Pi$ transition band for electronically excited OH radical (306 nm). None of the measured ignition delay times were faster than 10^{-4} s, and many ignition results suggested ignition below the explosion limit for stoichiometric H_2–O_2 mixtures (Fig. 11–22). The theoretical explosion limit corresponds to an infinite ignition delay time, although practically it is determined as corresponding to ignition delays of many seconds or minutes. This makes the subthreshold ignition delay times on the order of 10^{-4} s all the more remarkable because isochrones for finite ignition delays lie at higher temperatures (i.e., to the right) of the conventional explosion limit in Fig. 11–22. Each order-of-magnitude change in ignition delay time corresponds to an increase of many tens of Kelvins. Experimental results of Bozhenkov, Starikovskaia, and Starikovskii (2003) are very tractable for modeling as a consequence of the time evolution of each experiment. At the observation point, shockwave-related events occur up to triggering the FIW pulse. This pulse concludes on the order of $4 \cdot 10^{-8}$ s later, while plasma and ignition chemistry leading to OH* emission occurs on the ignition delay timescale of $\sim$$10^{-4}$ s. It follows that the original gas composition at the temperature and pressure behind the reflected shock can be considered initial FIW conditions, and the resulting effects of the FIW discharge can be considered initial conditions for plasma-stimulated ignition chemistry. This essentially divorces the FIW discharge and ignition processes from each other.

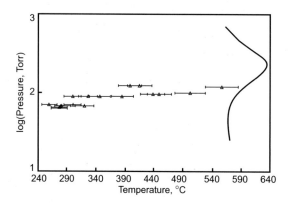

Figure 11–23. RF plasma-stimulated ignition in stoichiometric C_2H_4–air mixture at test section flow rates from 1.2 to 2.0 g/s. Solid curve is the conventional explosion limit for 5% C_2H_4–air mixture ($\varphi = 0.75$). Ignition temperatures are inferred from plasma heating of a CO marker–air flow at different flow rates and pressures.

11.2.5. Plasma Ignition in Hydrocarbon–Air Mixtures

RF plasma-stimulated subthreshold volumetric ignition of premixed methane–air and ethylene–air flows in a supersonic combustion test section has been demonstrated by Chintala et al. (2005, 2006). The nanosecond pulsed plasma-stimulated subthreshold ignition of ethylene–air flows and substantial degree of flameless oxidation of methane–air flows have been achieved in the same apparatus (Lou et al., 2006, 2007). The test section used in the experiments is a 20-cm-long rectangular channel with dielectric-covered electrodes as the walls of the first several centimeters of the apparatus, giving an effective plasma volume of ~15–20 cm³. The RF experiments used a 13.56 MHz CCP to excite flows of room air doped with 1–2% CO, which acted as an IR-active thermometry species. For a given flow rate of CO marker–air mixture subjected to 200 W RF plasma stimulation, flow heating appears to increase linearly with pressure over the 30–125 Torr range. Subthreshold temperature ignition has been reported in stoichiometric ethylene–air (Fig. 11–23) and methane–air flows (Fig. 11–24). Ignition is not obtained in tests below 60 Torr, which corresponds to post-heating temperatures lower than 523 K. Experiments that vary equivalence ratio while holding power input, pressure, and mass flow rate constant show nearly 100% flameless oxidation of C_2H_4 and 70% flameless oxidation of CH_4 due to the plasma stimulation of lean flows. O and H atoms and OH radicals stimulated by plasma contribute to non-thermal chain branching and ignition of lean flows, while near-stoichiometric mixtures attain enough temperature to achieve thermal runaway and attain thermal ignition.

Similar experiments have been conducted in the same test section by Lou et al. (2006, 2007) using a pulsed DBD plasma. The system has been driven by a pulsed power supply at 30–50 kHz and 9 kV voltage peaks giving 6 mJ per 50 ns pulse. The results of this work parallel results of the RF plasma-stimulated ignition in CH_4 and C_2H_4. Ignition of CH_4 is not observed under any conditions tested, but up to 60% of it has been destroyed in flameless

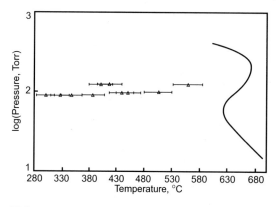

Figure 11–24. RF plasma-stimulated ignition in stoichiometric CH_4–air mixture at test section flow rates from 1.2 to 2.0 g/s. Solid curve is the conventional explosion limit for stoichiometric CH_4–O_2 mixture, which differs only slightly to that of CH_4–air relative to the reported temperature depressions. Ignition temperatures are inferred from plasma heating of a CO marker–air flow at different flow rates and pressures.

oxidation at lean equivalence. Flameless oxidation at lean equivalence ratio φ and ignition at near-stoichiometric equivalence ratio φ has been observed in C_2H_4–air mixtures, further supporting the claims that combustion radicals like H, OH, and O are produced by plasma, with ignition occurring only when heat generated by the exothermic oxidation reactions leads to thermal ignition. At power 240 W for $\varphi = 1.2$ of the C_2H_4–air mixture, ignition temperatures as low as $140°C$ have been recorded, which is many hundreds of kelvins below the conventional ignition threshold. To summarize, experimental results presented in the preceding two sections demonstrate ignition from tens to hundreds of degrees below the auto-ignition temperature threshold, as well as a substantial degree of flameless oxidation for some conditions. Chintala et al. (2005, 2006) and Lou et al. (2006) inferred plasma heating of their gas flows using a surrogate air–CO mixture, which demonstrates that non-thermal chain branching (explosion) occurs under plasma stimulation in the temperature–pressure (T–P) range where conventional combustion chemistry, dealing with temperature and radicals, predicts only slow reaction (i.e., branching radical termination rate exceeds branching rate). The experiments provide data permitting the analysis of different mechanisms of the subthreshold plasma ignition.

11.2.6. Analysis of Subthreshold Plasma Ignition Initiated Thermally: The "Bootstrap" Effect

Conventional kinetic schemes for auto-ignition and combustion (see Section 11.2.3) in general explain the phenomena almost exclusively through the interactions of radicals and temperature. For a potentially explosive mixture at a point in T–P space initially below the explosion limit, an increase in temperature up to or beyond the limit will result in a rate of branching exceeding that of termination and an ensuing explosion. It follows that the subthreshold ignition data presented in Sections 11.2.4 and 11.2.5 could be explained by Joule heating or fast relaxation of excited plasma species. Consider, as an example, an isothermal well-mixed isobaric system undergoing a simplified net exothermic illustrative reaction:

$$F + R \rightarrow R + R + \Delta Q, \quad k_{rxn} = A\exp(-E_a/T), \quad (11\text{--}10)$$

where the component F represents fuel species and R corresponds to oxidizer radicals; ΔQ is the heat liberated from the reaction; k_{rxn} is the reaction rate coefficient in the Arrhenius form. For a reaction mixture with an invariant initial concentration of fuel $[F]_0$ and even one F radical, it follows that, if R does not recombine on the walls or otherwise terminates, the reaction will proceed with the increasing rate

$$\frac{d[R]}{dt} = 2\omega_{rxn}(T), \quad \omega_{rxn} = k_{rxn}(T)[F]_0 \exp(2k_{rxn}[F]_0 t). \quad (11\text{--}11)$$

Initial conditions for the general branching explosion reaction are

$$\frac{d[R]}{dt} = 2\omega_{rxn}(T), \text{ with } [R](t = 0) = [R]_0, \quad (11\text{--}12)$$

which in a non-isothermal case can lead to subthreshold explosion through thermal chain branching if $[R]_0$ is above the critical threshold, though more complicated radical injection forcing functions leading to explosion are also conceivable. At subthreshold temperatures, $\omega_{rxn}(T)$ is negative, which is consistent with radical termination rate exceeding branching rate. However, exothermic recombination of the initial radical pool $[R]_0$, less any heat losses, can heat the mixture to temperatures where $\omega_{rxn}(T)$ is positive and effects an explosion, which is illustrated in Fig. 11–25. Since non-thermal plasmas generate significant amounts of radicals, this radical-thermal "bootstrap" to ignition may explain the experimental subthreshold ignition results.

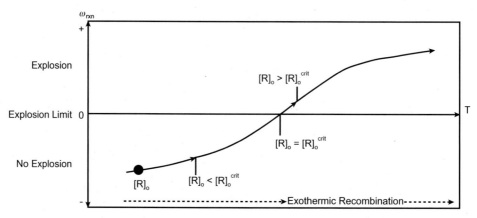

Figure 11–25. Radical-thermal bootstrap can achieve ignition if exothermic recombination of the initial concentration of radicals can lift the temperature to the explosion limit.

Kinetic modeling with impulsive thermal and/or radical injection analyzes the possibility of explanation of the subthreshold experimental results by effect of temperature, radicals, or a combination of both (see, for example, Haas, 2007). Bozhenkov, Starikovskaia, and Starikovskii (2003) give the energy per FIW pulse ΔQ in the range 1–50 mJ/cm^3. A conservative assumption of $\Delta Q = 50$ mJ/cm^3 gives temperature increases due to instantaneous plasma energy thermalization of 340–485 K, placing all of the apparent subthreshold ignition data well into the explosion region (see Fig. 11–26). Thus, some subthreshold plasma ignition results can be explained by thermal effects. Plasma heating through multiple channels, including relaxation of excited species and recombination of radicals, is able sometimes to boost the combustion mixture from subthreshold to above-threshold conditions. Short-length oxidation chains (below the auto-ignition limit) leading to additional heating can accelerate the bootstrap effect of thermal ignition stimulated in non-thermal plasmas.

11.2.7. Subthreshold Ignition Initiated by Plasma-Generated Radicals
Although subthreshold plasma ignition can be explained sometimes thermally (see previous section), several experiments cannot be reduced to only a temperature effect. For example, in the subthreshold plasma ignition experiments of Chintala et al. (2005, 2006), actual thermalization due to plasma has been especially accounted for. Thus, it is interesting to analyze the contribution of plasma-generated radicals to the phenomenon, which can be done in particular based on the modeling mentioned in Section 11.2.6 (Haas, 2006). In the model, the initial concentration of radicals $[R]_0$, which could explain auto-ignition

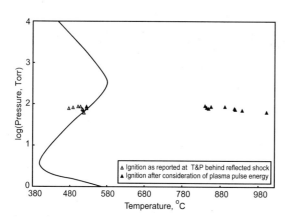

Figure 11–26. Interpretation of the subthreshold ignition data of Bozhenkov, Starikovskaia, and Starikovskii (2003) by fast thermalization of plasma energy input.

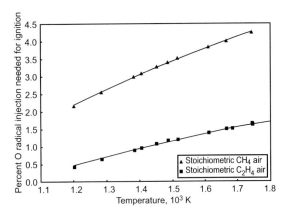

Figure 11–27. Temperature dependence of O atoms injection into $\varphi = 1$ CH$_4$–air and C$_2$H$_4$–air mixtures necessary for 5 ms ignition delay.

results in ethylene and methane in Figs. 11–23 and 11–24, were calculated to the nearest 10 ppm. Impulsive injection was considered, whereas plasma energy in the experiments was deposited over the flow's residence time in the plasma volume. Residence time in the test section was a round 3–5 ms, so modeling in this study aimed to find the [R]$_0$ which corresponded to a 5 ms ignition delay time. According to Lou et al. (2007), the oxygen-containing radicals O and OH generated by the plasma are responsible for ethylene destruction. Screening of both OH and H radicals in the stoichiometric C$_2$H$_4$–air system showed that, at the temperatures and pressures tested, neither achieved the 5 ms ignition delay target at lower concentrations than O. That is, atomic O is a more effective radical in achieving ignition and can be selected as the radical for radical injection modeling.

The kinetic analysis of ignition data of Chintala et al. (2005, 2006) for O atoms injected into methane and ethylene show that initial concentration [O]$_0$ between 2.1 and 4.3% for methane and 0.4 and 1.8% for ethylene is necessary to achieve the 5 ms ignition delay, depending primarily on temperature (see Fig. 11–27). This is the same general level of dissociation reported by Lavid et al. (1994) for subthreshold, excimer laser–stimulated photochemical ignition of H$_2$–O$_2$. The radical-thermal bootstrap effect on temperature and [O] for achieving 5 ms ignition by radical injection of 1.1% atomic oxygen (O) into $\varphi = 1$ C$_2$H$_4$–air mixture is demonstrated in Fig. 11–28 for the initial temperature and pressure of 416°C and 125 Torr, respectively. Much of the O initially injected into the mixture exothermically recombines in the first 100 μs, contributing to a temperature increase of 266 K to 682°C (955 K). For most of the rest of the 5 ms ignition delay period, the temperature increases substantially more slowly as the radical pool necessary for ignition becomes established. An important feature of the radical injection data is whether or not the experimental specific energy input required to generate [O]$_0$ greatly exceeds the model prediction; if this is the case, the subthreshold ignition data may be explicable by radical injection. Chintala

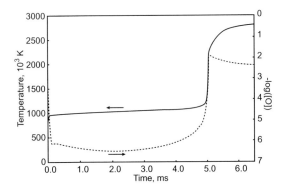

Figure 11–28. Time evolution of temperature and density of atomic oxygen [O] demonstrating the radical-thermal bootstrap effect.

et al. (2005, 2006) estimated plasma energy deposition of 0.04–0.05 eV/mol of mixed gas. The dissociation energy for O_2 is 2.56 eV per O atom generated, so as an example, for a 1% O injection requirement, about 0.026 eV/mol must be ideally deposited into O_2 dissociation. Modeling in methane–air mixture requires 2.1–4.2% O injection for ignition, which under the assumption of ideality corresponds to energy depositions of 0.055–0.11 eV/mol to produce the necessary $[O]_0$, not including energy lost to heating the gas. Hence, the results of Chintala et al. (2005, 2006) in methane cannot be explained by either thermal or radical injection, even when allowing for the shortcomings of modeling assumptions. For ethylene–air mixtures, the required ideal energy deposition range is 0.011–0.047 eV/mol, which can possibly be explained by the experimental energy deposition. In actuality, much of the energy deposited by plasma will not ideally dissociate O_2 into atomic oxygen, but rather will be captured in other degrees of freedom (e.g., ionization, vibration, side reactions, fast relaxation, etc.). The actual energy cost to generate O radicals can be described by a G-factor – the number of particles produced per 100 eV energy input. The disposition of deposited energy depends on the reduced electric field and gas composition, so while O_2 is very efficient at dissociating in non-thermal plasmas at 3 or 4 eV mean electron energy, still more energy is deposited into N_2 vibration, up to $\sim$5 eV (Penetrante, 1997). In the case of the ethylene data of Chintala et al. (2005, 2006), the G-factor thermalization can be accounted for in the flow temperature measurements. Thus, for the experiments with plasma ignition of CH_4–air mixture, the concentrations of O atoms necessary to meet the 5 ms ignition delay target correspond to energy greater than that deposited by the plasma, even under the assumption of 100% efficient energy deposition into O_2 dissociation. For experiments in C_2H_4, calculations show that energy for ideal O_2 dissociation is available from the plasma; however, upon considering non-ideality, it appears the plasma energy may not even account for flow heating. For the experimental data considered, the subthreshold ignition appears to be unexplained by the radical-thermal mechanism in the case of CH_4, whereas evidence for the phenomenon remains tenuous in the case of C_2H_4 and H_2. Conservative assumptions of this analysis show that ignition may be explained by the temperature or radicals in these cases, but experimental uncertainties render this conclusion uncertain.

11.2.8. Subthreshold Ignition Initiated by Plasma-Generated Excited Species

Plasma-stimulated ignition experiments (Sections 11.2.4 and 11.2.5) may not be able to be explained by purely radical-thermal means, so the participation of other plasma species remains tenable. Such plasma species (for example, charged and excited particles) influence the radical-thermal pathway leading to explosion, either by singly or simultaneously increasing the number of initial active centers ($[R]_0$), reducing the rate of chain termination, or increasing the rate of chain branching. Of these, the latter mechanisms of sustaining or multiplying chains are attractive as they may lead to positive values for $\omega_{rxn}(T)$ at lower temperatures (Haas, 2006). Important reactions between H atoms with oxygen include (see Section 11.2.3)

$$H + O_2 + M \rightarrow HO_2 + M, \qquad (11\text{--}13)$$

which establishes the second explosion limit of hydrogen, and the branching reaction

$$H + O_2 \rightarrow OH + O. \qquad (11\text{--}14)$$

Transition state theory suggests that reactions (11–13) and (11–14) stem from the same process, illustrated in Fig. 11–29 (Westmoreland et al., 1986). Reactants associate to a metastable state, HO_2^*, and temporarily store 205 kJ/mol of energy of the H–O_2 bond in mostly vibrational modes. In hydroperoxy radical formation, the third body M collisionally stabilizes the HO_2^* complex into the stable HO_2 potential energy well. Failing stabilization, HO_2^* may decompose back to H and O_2, while at higher temperatures, the energy barrier

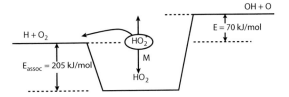

Figure 11–29. Potential energy diagram for the reaction path profile $H + O_2 \rightarrow$ products.

about 70 kJ/mol for the channel (11–14) is increasingly likely to be overcome, leading to chain branching. In order to overcome the second explosion limit, at least a couple of reaction schemes that recover active radicals OH and O (which further react to regenerate H) from HO_2 are imaginable. In the first scheme, an active plasma particle, whether vibrationally or electronically excited, collides with stable HO_2, imparting the roughly 2.87 eV per HO_2 to lift the system out of the potential energy well to products OH and O. Note that HO_2 destruction back to reactants H and O_2 requires less energy but does not lead directly to branching. An alternative, energetically less costly scheme would be

$$H + O_2 \rightarrow HO_2^*, \tag{11–15}$$

$$HO_2^* + P^* \rightarrow OH + O + P. \tag{11–16}$$

where P^* represents an active plasma particle. In this case, the HO_2^* activated complex is collisionally destabilized to the extent of the 0.73 eV needed to lift its energy to form the branching products. Possible contributions of the two schemes of the subthreshold plasma ignition, including destruction of HO_2 by the plasma-excited species, are to be discussed in the next section starting from the (11–15)–(11–16) mechanism.

11.2.9. Contribution of Plasma-Excited Molecules into Suppressing HO_2 Formation During Subthreshold Plasma Ignition of Hydrogen

Non-equilibrium plasma can stimulate the ignition of hydrogen through multiple channels of electronic and vibrational excitation. To demonstrate the effect, kinetic modeling (Haas, 2006) can combine them into the effective vibrational excitation of N_2 molecules admixed to the stoichiometric H_2–O_2 mixture with composition similar to that of (Bozhenkov, Starikovskaia, and Starikovskii, 2003) $12H_2$:$6O_2$:$82N_2$. The second limit of the auto-ignition (Turns, 2000; see also Section 11.2.3) can be compared in this case to the second limits in otherwise congruent systems where the vibrational temperature of N_2 is far from equilibrium with the rest of the mixture $T_v > T_0$ (see Fig. 11–30). In the kinetic modeling, an additional reaction, involving a vibrationally excited N_2 molecule, is added to the conventional mechanism:

$$H + O_2 + N_2^*(vib) \rightarrow OH + O + N_2. \tag{11–17}$$

The reaction rate coefficient for the plasma-stimulated three-body reaction can be expressed as

$$k_{rxn}(T_v) = Z \cdot \exp(-\Delta E / T_v), \tag{11–18}$$

where the pre-exponential factor is calculated from the three-body collision model (Benson, 1960) as $Z = 10^{15}$ cm^6/mol^2 s; ΔE is the necessary energy to elevate HO_2^* to the OH + O level (see Fig. 11–29). Fig. 11–30 shows the ignition threshold depression due to the effect of vibrational excitation in plasma on the order of 50–100 K relative to the equilibrium curve depending on pressure and the non-equilibrium vibrational temperature T_{vib}. The vibrationally excited N_2^* molecules have been considered as an example of the plasma-ignition stimulants since they are least prone to relaxation among common gases, obviating the assumption of long-lived N_2^* for 15 isochrones under the assumption of a single plasma

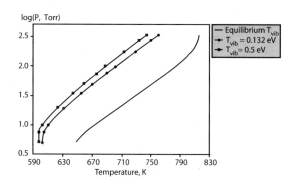

Figure 11–30. Explosion limits (15 s isochrone) in the mixture $12H_2:6O_2:82N_2$ as a function of vibrational temperature T_{vib} for N_2.

pulse at zero time. Results of modeling for 1 s isochrones are presented in Fig. 11–31. The 1 s isochrones display the same left-shifting saturation to a limit curve as seen with the 15 s isochrones, with maximum ignition temperature depression of just over 100 K. Whereas the 15 s isochrones appear relatively evenly spaced, the 1 s isochrones have taken on characteristics of both the first and second limits. Second-limit portions of these curves retain the even spacing noted in the 15 sec isochrones, ranging from $\sim$80 to 108 K temperature depression in the case of N_2^* T_v of 0.5 eV. At pressures lower than 30 Torr, the rates of three-body collisions forming HO_2 and involving N_2^* are sufficiently slow that reasonably regular branching through H, O, and OH dominate, establishing a first limit (albeit excluding wall effects). The relatively minor contribution of the destruction of HO_2^* by N_2^* due to low three-body collision rates explains why the first limit curves for $T_v = 0.132$ eV and $T_v = 0.5$ eV nearly overlap at low pressures, as well as why both non-equilibrium curves lie closer to the equilibrium one as pressure decreases.

Kinetic curves for the H and HO_2 radicals, with and without non-equilibrium vibrational temperatures, were at 698 K and 100 Torr (in the second limit region) and are illustrated in Fig. 11–32 to show relative rates of branching through H and reaction suppression through HO_2. For vibrational–translation (VT) equilibrium, [H] remains nearly flat, while [HO_2] remains 2.5 orders higher. This is to be expected since the point (698 K, 100 Torr) lies well to the left of even the 15 s ignition isochrone (Fig. 11–32) in the steady reaction regime. The generation of H atoms is due to the reactions

$$O + H_2 \rightarrow H + OH, \quad OH + H_2 \rightarrow H + H_2O, \tag{11–19}$$

so the flat [H] profile suggests that the rate of H scavenged to HO_2 outcompetes branching. The kinetic curves for $T_v = 0.5$ eV are markedly different. The sharp peak in [H] for the case of non-equilibrium vibrational temperature at $\sim$0.4 s corresponds to ignition, which

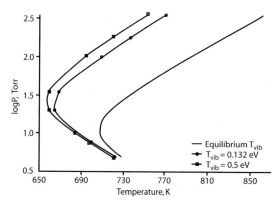

Figure 11–31. Explosion limits (1 s isochrone) in the mixture $12H_2:6O_2:82N_2$ as a function of vibrational temperature T_{vib} for N_2.

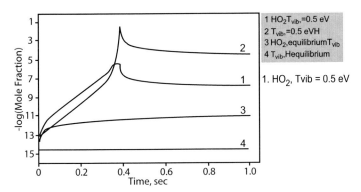

Figure 11–32. Radical concentration histories for H and HO_2 at 698 K and 100 torr, with and without non-equilibrium vibrational excitation.

clearly illustrates the effect of the added reaction involving N_2^*. Since reaction (11–17) effectively skips over formation of HO_2 to recover O and OH (and subsequently H), the fact that $[HO_2]$ exceeds $[H]$ by over an order of magnitude for much of the ignition delay may seem counterintuitive. However, upon careful analysis, this should be expected since, under the steady-state approximation, it is clear that the rate of formation of HO_2 through conventional equilibrium three-body collisions is essentially dependent on $[H]$. The greater concentrations of $[H]$ ultimately liberated through (11–17) feed back to higher concentrations of HO_2. Adding reaction (11–17) into the H–O kinetic mechanism apparently decreases ignition temperatures to around 100 K below the ignition threshold for the gas composition considered that is, a large excess of vibrationally excited N_2. These results should phenomenologically apply for H_2–O_2 mixtures with different equivalence ratios and concentrations of both gases. A similar effect also takes place when not vibrationally but rather electronically excited species suppress formation of HO_2 and therefore stimulate ignition. To apply the preceding kinetic data to the more general plasma excitation of atoms and molecules, vibrational temperature T_v should be replaced by a general excitation temperature.

11.2.10. Subthreshold Plasma Ignition of Hydrogen Stimulated by Excited Molecules Through Dissociation of HO_2

Excited molecules can stimulate the ignition of hydrogen more effectively when they not only suppress formation of HO_2, as discussed earlier (11–17), but also collisionally dissociate the HO_2 already formed. For the gas composition $12H_2$:$6O_2$:$82N_2$, reaction (11–17) acting alone results in a second-limit ignition temperature depression of $\sim$100 K (see Figs. 11–30 and 11–31). Moreover, HO_2 concentrations exceed H concentrations for much of the ignition period (Fig. 11–32). Dissociation of HO_2 by collision with excited particles (for example, vibrationally excited N_2 molecules) permits ignition temperature depressions to be achieved on the order of several hundred degrees, corresponding to experimental data presented in Figs. 11–30 and 11–24. The process can be modeled by introducing an additional reaction (Haas, 2006):

$$HO_2 + N_2^* \rightarrow OH + O + N_2. \qquad (11\text{–}20)$$

Similarly to the case of non-equilibrium reaction (11–17) assuming vibrational excitation of the nitrogen molecules, the reaction rate of dissociation process (11–20) can be expressed as

$$k_{rxn}(T_v) = Z \cdot \exp(-\Delta E / T_v), \quad Z = 10^{14} \text{ cm}^3/\text{mol} \cdot \text{s}. \qquad (11\text{–}21)$$

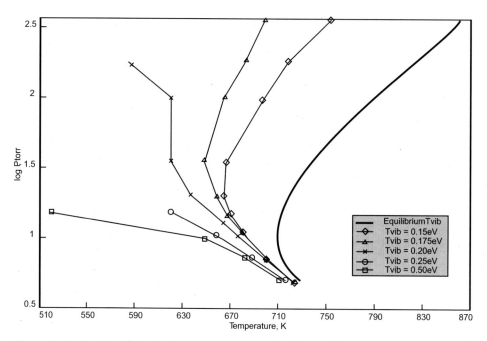

Figure 11–33. Explosion limits (1 s isochrone) in the mixture $12H_2:6O_2:82N_2$ as a function of vibrational temperature of nitrogen N_2^* T_{vib}, taking into account decomposition of HO_2 in collisions with the excited molecules.

The energy barrier in this process is higher: $\Delta E \approx 2.87\,eV/mol$ (see Fig. 11–29). The results of kinetic modeling (Haas, 2006) of ignition of the stoichiometric mixture $12H_2:6O_2:82N_2$ taking into account the non-equilibrium HO_2 dissociation (11–20) are shown in Fig. 11–33. The ignition threshold depression behavior is similar to that shown in Fig. 11–29 up to T_v of about 0.2 eV, when the dissociation (11–20) is still relatively slow. The effect of reaction (11–20) in this regime is a 100 K leftward shift in the threshold at vibrational temperatures lower than 0.5 eV (like in the case of reaction (11–17) acting alone). At $T_v \approx 0.2\,eV$, the second explosion limit seems to disappear and is replaced by an inflection, and, at higher vibrational temperatures, ignition temperatures plummet several hundred kelvins. The 1 s ignition isochrone at higher vibrational temperature $T_v = 0.5\,eV$ and lower gas temperatures (down to 450 K) is shown in Fig. 11–34. It should be mentioned that only minor effects of HO_2 destruction on the ignition curve are seen at lower pressures, where HO_2 formation is slow. At pressures of about 15 Torr, which corresponds to the toe of the HO_2-defined ignition peninsula at equilibrium conditions, the effects of reaction (11–20) clearly cause a significant ignition threshold temperature depression; temperature reductions increase to nearly 320 K (at a minimum) at 100 Torr and roughly 387 K at 350 Torr. Kinetic curves for H and HO_2 radicals for ignition stimulated by reactions (11–17) and (11–20) at various degrees of vibrational temperature non-equilibrium are presented in Fig. 11–35. Gas temperature and pressure are fixed at 698 K and 100 Torr to remain consistent with the results presented in Fig. 11–32. The case of slow reaction at equilibrium vibrational temperature has been discussed previously, while the HO_2 and H histories at $T_v = 0.15\,eV$, where [HO_2] substantially exceeds [H], is reminiscent of the $T_v = 0.50\,eV$ result presented in Fig. 11–32 and is due to the dominance of reaction (11–17). At $T_v = 0.20\,eV$, [H] and [HO_2] are virtually the same over the ignition time, and at $T_v = 0.50\,eV$, [H] exceeds that of [HO_2] by two to three orders of magnitude until ignition. These results further support the conclusion that at low vibrational temperatures reaction (11–17) dominates, whereas

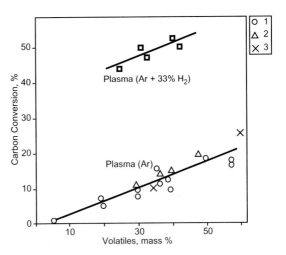

Figure 11–34. Explosion limits (1 s isochrone) in the mixture $12H_2:6O_2:82N_2$ at lower gas temperatures as a function of vibrational temperature of nitrogen N_2^* T_{vib}, taking into account decomposition of HO_2 in collisions with the excited molecules.

at higher vibrational temperatures the role of reaction (11–20) becomes important and can shift the ignition threshold beyond the limit curve defined by (11–17) acting alone. In short, at high enough vibrational temperatures, the combined action of non-equilibrium reactions (11–17) and (11–20) appears to effectively remove the second explosion limit for the H_2–O_2 mixtures. The presented kinetic data can be applied to plasma stimulation of ignition not only due to vibrational excitation but due to more general plasma excitation of atoms and molecules. To describe the effect in a more general way, vibrational temperature T_v should be replaced by excitation temperature.

11.2.11. Subthreshold Plasma Ignition of Ethylene Stimulated by Excited Molecules Effect of NO

Low-temperature ignition in hydrocarbons proceeds through pathways less dependent on free H radicals. Nevertheless, releasing active radicals by dissociation of the passive radical HO_2 can also significantly shift ignition curves similarly to the case of explosion of H_2–O_2 mixtures discussed in the preceding sections. The passive radical HO_2 is produced, in particular, in processes such as the initiation reaction in ethylene ignition,

$$C_2H_4 + O_2 \rightarrow C_2H_3 + HO_2, \tag{11–22}$$

and in the initiation and chain-branching steps in methane ignition,

$$CH_4 + O_2 \rightarrow CH_3 + HO_2, \tag{11–23}$$

$$CH_2O + O_2 \rightarrow HCO + HO_2. \tag{11–24}$$

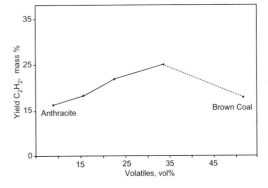

Figure 11–35. Kinetic curves for concentrations of atomic hydrogen H and HO_2 radicals at temperature 698 K and pressure 100 Torr, with and without non-equilibrium vibrational excitation.

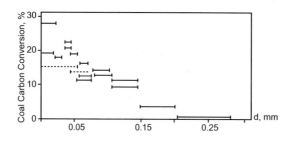

Figure 11–36. Explosion limits (1 s isochrone) in ethylene–air mixture as a function of vibrational temperature of nitrogen molecules.

Thus, the non-equilibrium reaction (11–20) for HO_2 destruction as well as (11–17) to reduce reversion of H back to HO_2 are equally as applicable to hydrocarbon ignition as to hydrogen ignition, which has been considered earlier. Modeling involving both reactions of non-equilibrium vibrationally excited nitrogen molecules has been conducted by Haas (2006) for ethylene ignition in air. Ignition temperature depression behavior as a function of vibrational temperature of nitrogen parallel to the hydrogen system (at similar conditions) has been observed (see Fig. 11–36). At vibrational temperatures about $T_v \approx 0.2\,eV$, just as in the case of H_2 ignition, the curve ceases to reverse curvature and the ignition temperature limit appears to monotonically decrease with increasing pressure.

Not only excited molecules but also some chemically active radicals and molecules produced with relatively high concentrations in plasma conditions can convert passive HO_2 into active radicals stimulating ignition and combustion processes. This can be illustrated by plasma production of NO, which is able to suppress the chain-termination processes by the reaction

$$NO + HO_2 \rightarrow NO_2 + OH.$$
$$H + NO_2 \rightarrow NO + OH$$

This plasma stimulated chain reaction converting HO_2 into active radicals H and OH can significantly decrease the ignition limit. It takes place not only in the case of ethylene but in the ignition of H_2 and different hydrocarbons.

11.2.12. Contribution of Ions in the Subthreshold Plasma Ignition

Ions can play a very productive role in plasma ignition through the so-called processes of ion catalysis or plasma catalysis, which have been discussed earlier regarding plasma conversion of fuels (see Section 10.1.1). Among the different mechanisms of ion catalysis in plasma ignition, the one providing low-temperature subthreshold oxidation of hydrogen is of particular interest. The chain termination in the hydrogen–oxygen mixture is recombination (see Section 11.2.3):

$$H + O_2 + M \rightarrow HO_2 + M, \tag{11–25}$$

which converts an active H atom into a passive HO_2 radical. Conventional ignition is impossible when reaction (11–25) suppresses the chain-branching reaction $H + O_2 \rightarrow OH + O$. The passive radical HO_2 is traditionally considered a dead end of chain propagation. It should be mentioned that the reaction

$$HO_2 + H_2 \rightarrow H_2O + OH \tag{11–26}$$

is exothermic ($\Delta H = -192.7$ kJ/mol) and releases a very active OH radical, which could serve to propagate the hydrogen oxidation chain reaction. Exothermic reaction (11–26) does not however, make any significant contribution into the hydrogen oxidation kinetics because of its high activation energy $E_a \sim 100$ kJ/mol. This activation energy can be effectively eliminated if process (11–26) is organized in the framework of ion catalysis as an ion–molecular reaction. The exothermic chemical reactions with ions do not have any activation energy; see Section 2.2.6. If hydrogen in reaction (11–26) is bonded into an ion cluster such as MH_2^+, the hydrogen oxidation chain can propagate, while the electrical charges survive and are able to participate in the following new reactions:

$$M^+ + H_2 \rightarrow MH_2^+, \qquad (11\text{–}27)$$

$$HO_2 + MH_2^+ \rightarrow H_2O + OH + M^+. \qquad (11\text{–}28)$$

Data on the ion–molecular reactions in the H_2–O_2 mixture related to its ignition can be found in Shibkov and Konstantinovskij (2005). The ion–molecular sequence (11–27) and (11–28) is equivalent to reaction (11–26) converting the passive HO_2 radical into the active OH, which further reacts with hydrogen:

$$OH + H_2 \rightarrow H_2O + H, \qquad (11\text{–}29)$$

forming the chain reaction (11–25), ((11–27), (11–28), (11–29). The chain mechanism (11–25), (11–27) (11–28) (11–29) does not lead to branching. The presence of ions makes it possible to oxidize hydrogen into water through the chain reaction but without branching and, therefore, without any specific threshold. Thus, non-equilibrium plasma operates in this case exactly like a conventional low-temperature oxidation catalyst.

Another ionic mechanism of low-temperature hydrogen oxidation in non-equilibrium plasma conditions has been proposed by Starikovskii (2000). The chain mechanism in this case contains more reactions, including formation and destruction of negative ions:

$$e + O_2 + M \rightarrow O_2^- + M, \qquad (11\text{–}30)$$

$$O_2^- + H_2 \rightarrow OH^- + OH, \qquad (11\text{–}31)$$

$$OH + H_2 \rightarrow H_2O + H, \qquad (11\text{–}32)$$

$$H + O_2 + M \rightarrow HO_2 + M, \qquad (11\text{–}33)$$

$$OH^- + H \rightarrow H_2O + e, \qquad (11\text{–}34)$$

$$OH^- + HO_2 \rightarrow H_2O + O_2 + e. \qquad (11\text{–}35)$$

While the first four reactions of the mechanism can proceed relatively fast, the last two reactions, which are responsible for the electron detachment, require collision of two partners having low concentration and, therefore, proceed relatively slowly. Kinetic competitition of the two last reactions of electron detachment with fast recombination processes determines the effectiveness of the mechanism. Similar ionic mechanisms can stimulate ignition and combustion of hydrocarbons. Generally, although the contribution of ions into plasma ignition and especially the subthreshold ignition can be quite significant, understanding the ionic mechanisms in the plasma-assisted combustion processes still requires additional kinetic investigations.

11.2.13. Energy Efficiency of Plasma-Assisted Combustion in Ram/Scramjet Engines

Plasma stimulation is very promising for application in the ram- and scramjet engines (see, for example, Macheret, Shneider, & Miles, 2005). The development of ram/scramjet engines for hypersonic airbreathing vehicles is associated with a number of challenges, including ignition, flameholding, and spreading (Jacobsen et al., 2003). In Mach 4–8 flights with dynamic pressures of 0.5–1 atm, the combustor inlet conditions are static pressures 0.5–2 atm, static temperatures 400–1000 K, and flow velocity 1200–2400 m/s. Hydrocarbon fuel is injected at Mach numbers around 4 as a liquid, and at higher Mach numbers as a gas. Mixing in the scramjets is helped somewhat by the flow turbulence. The auto-ignition delay exceeds 10 ms for hydrocarbon fuels at pressure of about 1 atm and temperature 600–1200 K, which results in an unacceptably long ignition region at the scramjet flow velocities. Thus, forced ignition is required, which can be provided by plasmas. Different approaches to ignition require different power consumption, often unacceptably high, which will be analyzed next. Significant efforts have focused on volumetric (space-uniform) plasma ignition to remove the flame spread problem. Gas is preheated in the ram/scramjet combustors due to its compression, which facilitates plasma ignition. Ignition at temperatures of 1000–1400 K depends on the plasma generation of the pool of active radicals. The molar fraction of plasma-generated active radicals ($\alpha = 1 - 3 \cdot 10^{-3}$) permits the ignition delay time to be shortened to about $10\,\mu s$ and thus the ignition length to about a few centimeters (Williams et al., 2001). The specific energy input in the discharge per molecule required to achieve the initial fraction α of the active radical can be calculated as (Macheret, Shneider, and Miles, 2005)

$$E_v = \alpha \cdot W = \frac{\alpha e^2}{m k_{en} k_{rad}} (E/n_0)^2, \qquad (11\text{--}36)$$

where W is the energy cost of production of one radical; E/n_0 is the reduced electric field; k_{en} is the gas-kinetic rate coefficient for the electron–neutral collisions; k_{rad} is the effective rate coefficient for production of radicals (for example, through dissociation) in the electron–neutral collisions; and e and m are an electron charge and mass. In the conventional low-electric-field non-equilibrium discharges in molecular gases under consideration, electron temperature is $T_e \approx 1\,eV$ and reduced electric field is $E/n_0 \approx 30\,Td$. Most of the discharge energy in such conditions goes into the vibrational excitation of molecules (see Section 2.5.6), and the energy cost of a radical is relatively high, at least $W \approx 100\,eV$. Then, according to (11–36), the specific energy input required to achieve the molar fraction of plasma-generated active radicals ($\alpha = 1$–$3 \cdot 10^{-3}$) and to shorten the ignition delay time to about $10\,\mu s$ is rather high and equals 0.1–0.3 eV/mol. This specific energy input per molecule coresponds to a very high macroscopic energy input of 0.3–1 MJ/kg, which is comparable with the total flow enthalpy of 1.5–2.5 MJ/kg at Mach numbers of 5–7. In power requirements, this translates to more than 100 MW per square meter of a combustor cross section, which is clearly unacceptable (Macheret, Shneider, & Miles, 2005).

The energy requirements for ignition in the ram- and scramjet engines can be reduced by application of plasma systems with higher electron energies, including electron beams and electric discharges such as high-voltage pulse discharges with high values of reduced electric field, E/n_0. Using electron beams or the high-voltage nanosecond pulse discharges with megahertz repetition rate operating at $E/n_0 \approx 300$–$1000\,Td$ permits the effective energy of a radical production to be reduced to about $W \approx 15\,eV$. Then, according to (11–36), the specific energy input required to achieve the molar fraction of plasma-generated active radicals ($\alpha = 1$–$3 \cdot 10^{-3}$) and to shorten the ignition delay time to about $10\,\mu s$ decreases to values a bit lower: 0.015–0.05 eV/mol. It corresponds to the macroscopic energy input 0.05–0.15 MJ/kg. This energy is still very high and is equivalent to at least several percent of the total flow enthalpy. Assuming that 50–100 μs ignition delay time (0.1–0.2 m ignition length) is still acceptable for the ram- and scramjet engines, and the static temperature is 1000–1400 K, then a mole fraction $\alpha \approx 10^{-4}$ of the plasma-generated radical can be sufficient (Williams et al., 2001). Then the minimum required value of the specific energy input during application of the high-electron-energy plasma system can be reduced to about 10^{-3} eV/mol, which corresponds to the power budget of several megawatts per square meter. This power budget is still quite high because the power needs to be generated on board and delivered to the flow. The problems discussed earlier related to volumetric quasi-uniform plasma ignition attract more attention to space- and time-periodic plasma ignition systems.

Examples of such space- and time-periodic plasma ignition systems include the zebra igniters considered in Section 11.1.6, and the multi-spot ignition system proposed by Macheret, Shneider, and Miles (2005). These volumetric but not completely homogeneous plasma systems require much less energy and power budget because they do not cover all the flow uniformly, but their volumetric nature still helps avoid significant flow perturbations typical for local igniters.

11.3. ION AND PLASMA THRUSTERS

11.3.1. General Features of Electric Propulsion: Ion and Plasma Thrusters

Launching of a spacecraft to an orbit requires powerful high-thrust rocket engines based on the use of liquid or solid fuel. Attitude control, precision spacecraft control, and low-thrust maneuvers can be effectively provided by ion and plasma thrusters, which are high-specific-impulse, low-power electric propulsion systems (Grishin, Leskov, & Kozlov, 1975; Morozov, 1978; Tajmar, 2004; Jahn, 2006). The useful load of a spacecraft is very expensive ($10,000–20,000 per kg) and should be spent carefully. The electric propulsion systems are able to decrease the required mass of the spacecraft, because they are characterized by very high **specific impulse**, I_{sp}, which is the impulse (change of momentum) per unit mass of propellant. The specific impulse is determined by exit velocity of the propellant and measured in meters per second. The I_{sp} is often presented not per unit mass but per unit weight of propellant; then it is $g = 9.8 \, \text{m/s}^2$ times less, and measured in seconds. The specific impulse I_{sp} is used as a measure of the efficiency of the propulsion system. In practice, the specific impulses of real engines, vary somewhat with both altitude and thrust; nevertheless, I_{sp} is a useful value to compare engines, much like "miles per gallon" is used for cars. The specific impulse of rocket engines based on liquid and solid fuels does not exceed 4500–5000 m/s (450–500 s, if presented per unit weight), which is limited by neutral gas velocities at high temperatures. Much higher exit velocities can be achieved by forced electromagnetic acceleration of the propellant, using accelerators of charged particles or plasma. Such electric propulsion principles lead to the creation of ion and plasma thrusters with very high values of the specific impulse. The high values of the specific impulse of plasma and ion thrusters permit increased spacecraft loads (Jahn, 2006). A schematic of the electric rocket engine is presented in Fig. 11–37. A distinctive feature of the electric rocket engine is the necessity of a special on-board electric energy generator, which actually determines the limit for the total thrust of the engines. Thrusts exceeding 50–100 N already require an unacceptable mass of on-board electric energy generators. As an example, thrust of 100 N and specific impulse $I_{sp} = 40 \, \text{km/s}$ correspond to jet power of an electric rocket engine of 2 MW. The typical required power of the on-board electric energy generator in this case is 2.8 MW, and the mass of the generator is 86 tons (Grishin, 2000). The electric rocket engines cannot be applied for launching spacecraft and can be used only in zero-gravity conditions. The acceleration provided by electric rocket engines is usually in the

Figure 11–37. General components of an electric rocket engine.

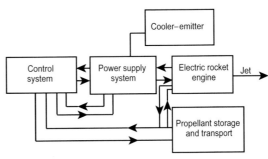

range 10^{-4}–10^{-2} m/s^2. To reach velocities of tens of kilometers per second, conventional for spacecrafts, the electric rocket engines would require several thousand hours.

11.3.2. Optimal Specific Impulse of an Electric Rocket Engine

Each flight of a spacecraft with an electric rocket engine (ion or plasma thruster) has an optimal specific impulse, which minimizes the initial mass of the spacecraft and cost of its launching to an orbit. It distinguishes the electric propulsion based on the ion or plasma thrusters from the rocket engines using liquid or solid fuel, for which efficiency always increases with the specific impulse. To find the optimal specific impulse for an electric rocket engine, consider a mass model of a spacecraft with an electric rocket engine. Initial mass M_0 of such a spacecraft at low (near-Earth) orbit includes the useful load M_{use}, mass of the on-board electric generator and the electric rocket engine $M_{electric}$, mass of the propellant (working fluid) M_{wf}, mass of the fuel reservoirs M_{fr}, and mass of the construction elements M_{const}:

$$M_0 = M_{use} + M_{electric} + M_{wf} + M_{fr} + M_{const}. \tag{11–37}$$

The mass of an electric rocket engine, increases linearly with the total consumed power of the engine, $P_{electric}$, where the coefficient of proportionality $\gamma_{electric}$ can be estimated today as 0.03 kg/W:

$$M_{electric} = \gamma_{electric} \cdot P_{electric}. \tag{11–38}$$

The mass of the propellant (working fluid) can be calculated based on the thrust F, operation time of the engine, τ, and the specific impulse I_{sp}:

$$M_{wf} = \frac{F \cdot \tau}{I_{sp}}. \tag{11–39}$$

The jet power of the propellant, P_j, leaving the engine is determined by thrust and specific impulse:

$$P_j = F I_{sp}/2. \tag{11–40}$$

Also, the jet power of the propellant, P_j, is related to the total consumed power of the engine, $P_{electric}$, through the thrust efficiency η_{thrust}, which can be numerically estimated as $\eta_{thrust} = 0.7$:

$$P_{elecrtic} - P_j = \eta_{thrust} \cdot P_{electric}. \tag{11–41}$$

The relation between mass of the fuel reservoir and mass of the propellant (working fluid) is conventionally described using the coefficient γ_{fr} characterizing relative mass of the fuel reservoir:

$$M_{wf} - M_{fr} = \gamma_{fr} \cdot M_{wf}. \tag{11–42}$$

A typical numerical value of the relative fuel reservoir mass is $\gamma_{fr} = 0.2$. Combination of equations (11–37)–(11–42) results in the following relation between the initial mass M_0 of a spacecraft at low (near-Earth) orbit and its specific impulse I_{sp}:

$$M_0 = M_{use} + \frac{\gamma_{electric} F I_{sp}}{2\eta_{thrust}} + \frac{F\tau}{I_{sp}}(1 + \gamma_{fr}) + M_{const}. \tag{11–43}$$

Relation (11–43) shows that high specific impulses permit, on the one hand, decreasing initial spacecraft mass through reduction of the mass of propellant (working fluid). On the other hand, however, the higher specific impulses lead to an increase of the initial spacecraft mass because of the required larger mass of the on-board electric generator and the electric

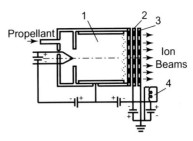

Figure 11–38. General schematic of an ion thruster: (1) source of ions; (2) accelerating electrode; (3) external electrode; (4) neutralizer or cathode compensator.

rocket engine. Relation (11–43) permits one to find the optimal value of the specific impulse I_{sp}^{opt} minimizing $dM_0/dI_{sp} = 0$, the initial mass of a spacecraft:

$$I_{sp}^{opt} = \sqrt{2\eta_{thrust}\tau(1 + \gamma_{fr})/\gamma_{electric}}. \tag{11–44}$$

Thus, optimal values of the specific impulse do not depend on thrust and are mostly determined by operation time of the electric rocket engine, or simply on the flight time of the spacecraft. For example, the flight to Mars (when $\tau \approx 1$ year) requires in optimum $I_{sp} \approx 40$ km/s. The parameters of electric rocket engines are expected to be improved in the near future to the values $\gamma_{electric} = 0.02$ kg/W, $\gamma_{fr} = 0.1$, and $\eta_{thrust} = 0.8$. For the improved parameters, the same flight to Mars requires an optimum of $I_{sp} \approx 50$ km/s. Flights to Jupiter and Saturn ($\tau \approx 3$–4 years) require an optimum of $I_{sp} \approx 90$–100 km/s. Optimal values of the specific impulse of ion or plasma thrusters for short near-Earth flights and for corrections of orbits can be estimated as $I_{sp}^{opt} = 15$–30 km/s. The propellant (working fluid) velocities in electric propulsion systems (ion and plasma thrusters) typically expected today are in the range 15–100 km/s (Grishin, 2000).

11.3.3. Electric Rocket Engines Based on Ion Thrusters

An ion thruster is an electric rocket engine where the propellant (working fluid) is represented by positive ions accelerated in a constant electric field. Ion thrusters were developed historically before plasma thrusters mostly because of the relative simplicity of generation and acceleration of ions in these engines. A schematic of an ion thruster is shown in Fig. 11–38. An external electrode and a neutralizer (cathode compensator) are grounded on the spacecraft body. An ion source is sustained under positive potential; the accelerating electrode is sustained under negative potential, and the external electrode has a zero potential (grounded). Such distribution of electric potential prevents electrons from the neutralizer from getting to the accelerating electrode and the ion source. The accelerated ions leave the thruster and go out into the environment, entraining electrons from the neutralizer along with them. Therefore, although the accelerated ions leave the thruster, the electric potential of the spacecraft does not grow, and the space charge of the ion beams is neutralized. The ion thrusters can be based on different methods of ion generation. Most modern ion thrusters use gas-discharge ion sources, where ionization occurs due to electron impact. A schematic of the gas-discharge ion sources is shown in Fig. 11–39. Another approach to ion generation is based on surface ionization; a schematic of such ion sources is presented

Figure 11–39. General schematic of a gas-discharge ion source: (1) cathode; (2) anode; (3) discharge chamber; (4) electromagnetic coil; (5) forming electrode; (6) propellant inlet.

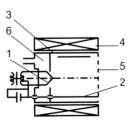

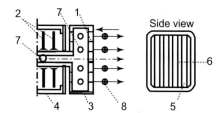

Figure 11–40. Schematic of an ion source with surface ionization: (1) porous ionizer; (2) heater; (3) chamber of ionizer; (4) thermal screen; (5) porous surface; (6) non-porous surface; (7) atom; (8) ion.

in Fig. 11–40. In the surface ionization ion sources, vapors of cesium characterized by low ionization potential are ionized by penetration through a porous tungsten membrane characterized by high work function. It should be mentioned that electric rocket engines similar to the ion thrusters can be organized by accelerating not ions but nanometer-size charged liquid droplets. Such electric engines are referred to as **colloidal thrusters**. The acceleration zone of ion thrusters lies between the ion source and the external electrodes and is filled with ions of the same polarity, which creates a significant space charge limiting the operational characteristics of the electric rocket engines. Better characteristics can be achieved in plasma thrusters using quasi-neutral plasma as a propellant.

11.3.4. Classification of Plasma Thrusters: Electrothermal Plasma Thrusters

Quasi-neutral plasma with a high ionization degree (usually 90–95%) is a propellant (working fluid) in plasma thrusters. Electrons neutralize the positive space charge of ions in the accelerating channel and thus improve the characteristics of the thrusters. Both electrons and ions are accelerated in plasma thrusters to velocities of around 100 km/s. Most of the energy in this case is accumulated obviously in ions (energy of an electron accelerated to such velocity is only 0.03 eV); thus, the major task of the plasma thrusters, similar to that of ion thrusters, is ion acceleration. The motion of an ion in the completely ionized rarefied plasma can be described (see Fridman & Kennedy, 2004) as

$$M\frac{d\vec{v}_i}{dt} = e\vec{E} - \frac{\nabla p_i}{n_{e,i}} - \frac{1}{\sigma}\vec{j} + e[\vec{v}_i \times \vec{B}], \qquad (11\text{–}45)$$

where, $\vec{v}_i$, e, and M are the velocity, charge, and mass of an ion; $n_{e,i}$ and σ are the number density and conductivity of plasma; $\vec{E}$ and $\vec{B}$ are the electric and magnetic fields; ∇p_i is the gradient of the ionic pressure; and $\vec{j}$ is the current density. Only the first three forces on the right-hand side of the equation are able to increase the ion energy and in these terms to accelerate the ion. Acceleration of the ions by the force $e\vec{E}$ is called "electrostatic"; by the force $\nabla p_i/n_{e,i}$ is called "gasdynamic" or "thermal"; and by the force $\vec{j}/\sigma$ is called "ohmic" or acceleration by the "electronic wind." Taking into account different mechanisms of ion acceleration, plasma thrusters are usually subdivided into three categories: electrothermal plasma thrusters, electrostatic plasma thrusters, and magneto-plasma-dynamic plasma thrusters, which are discussed separately later. In **electrothermal plasma thrusters**, plasma as a propellant is generated in a thermal gas discharge (usually either thermal arc discharge or thermal RF inductively coupled plasma discharge) and then accelerated in a de Laval nozzle similarly to the case of conventional rocket engines using liquid fuel. An advantage of electrothermal plasma thrusters with respect to conventional liquid fuel propulsion systems is due to the possibility of using the light gas, hydrogen, as a propellant, and also due to the possibility of heating the gas in a discharge to extremely high temperatures, which leads to high exit velocities of the propellant and therefore high values of the specific impulse.

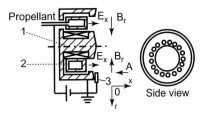

Figure 11–41. General schematic of an electrostatic thruster with closed drift of electrons: (1) magnet; (2) anode/vapor-distributor; (3) cathode-compensator; (B), magnetic field; E, electric field.

11.3.5. Electrostatic Plasma Thrusters

Ion acceleration in electrostatic plasma thrusters is provided by an electric field generated in the plasma in a circular gap between poles of a magnet due to the sharp decrease of the transverse electron mobility in the magnetic field (see Fig. 11–41). The magnetic field in the thrusters is chosen to magnetize plasma electrons but leave the ions non-magnetized (see Kennedy & Fridman, 2004). When the anode has a typical potential (0.3–3 kV), and the magnet and external cathode compensator are sustained under zero potential, a symmetrical discharge occurs in the circular gap with strong axial electric field. Drift of the magnetized electrons is in the tangential directions, while ions are not constrained by the magnetic field and are accelerated in the axial directions. Because electrons are trapped in the layer of ion acceleration and they are not able to exit the propulsion system together with ions, a cathode compensator should be placed at the exit from the ion-acceleration channel. In the cathode compensator, some fraction of electrons compensates the ion beam similarly to the case of ion thrusters. Another fraction of electrons goes to the ion-acceleration channel and sustains the electric discharge. Neutral atoms of the propellant are injected into the accelerating channel and are ionized there by the rotating magnetized electrons. Collisions of electrons with heavy particles break their tangential drift and result in a shift of electrons in the direction toward the anode providing the axial electron current to the anode. Ions formed by ionization of the propellant (working fluid) are accelerated in the electric field and exit the propulsion system. Thus, electrostatic plasma thrusters have a combined layer for both ionization and acceleration of the propellant. In contrast to the ion thrusters, ion acceleration in the electrostatic plasma thrusters takes place inside the cloud of electrons. It compensates space charge of the ions and improves the characteristics of the engine. Electrostatic plasma thrusters can be constructionally organized in two different ways. The so-called **stationary thruster** has a long ionization and acceleration layer. Another configuration, the so-called **thruster with an anode layer**, has a narrow ionization and acceleration layer adjacent to the anode.

11.3.6. Magneto-Plasma-Dynamic Thrusters

In magneto-plasma-dynamic thrusters, ions are accelerated by electrons through electron–ion friction. The magnetic field in such a thruster can be induced either by its own discharge current or by external magnetic systems. The effectiveness of ion acceleration in the thruster with a magnetic field induced by its own discharge current is higher at higher levels of electron current, which is sustained in this case usually in the range 1–10 kA. Such thrusters, for this reason, are usually referred to as **high-current thrusters**. Electrons in the thrusters leave the system together with ions; therefore, a special cathode neutralizer is not required in these systems. A schematic of a high-current thruster is shown in Fig. 11–42. The thruster consists of a thermionic cathode, through which vapor propellant comes to the engine partially ionized, an external ring anode with the shape of the de Laval nozzle, and an insulator. A cathode in the high-current thrusters is shorter than an anode. Therefore, plasma confinement to the axis of a thruster can be effectively achieved in these kinds of electric engines.

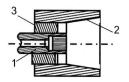

Figure 11–42. General schematic of a high-current magneto-plasma-dynamic thruster: (1) thermionic cathode; (2) ring-anode; (3) insulator.

11.3.7. Pulsed Plasma Thrusters

Propellant and electric energy are introduced in the pulsed plasma thrusters, not continuously but by separate pulses with a duration in the millisecond range. The pulsed thrusters include capacitors to collect sufficient energy and a system for periodic ignition of the discharge. The most effective of the pulsed thrusters are so-called **erosion thrusters**, which are illustrated in Fig. 11–43. Propellant is located in the thruster between the anode and the cathode in the form of solid or paste-like dielectric and works also as an insulator. Operation of the pulsed plasma thruster is initiated by an additional capacitor (see Fig. 11–43), generating initial plasma going into the accelerating space between the major cathode and anode, where the voltage and energy of the major bank of capacitors is ready to be discharged. Plasma generation in the vicinity of the dielectric initiates surface discharge and evaporation of the dielectric. Interaction of the discharge current with its own tangential magnetic field leads to plasma acceleration in the coaxial system. When the major bank of capacitors is discharged, formation of the vaporized propellant stops automatically, which terminates a pulse. Plasma thrusters have been only briefly discussed in this section. Much more detailed information can be found, for example, in Bouchoule et al. (2001), Jahn (2006), Tajmar (2004), Morozov (1978), Grishin, Leskov, and Kozlov (1975), and Grishin (2000).

11.4. PLASMA APPLICATIONS IN HIGH-SPEED AERODYNAMICS

11.4.1. Plasma Interaction with High-Speed Flows and Shocks

It has been known for several decades that weak ionization of gases results in large changes in the standoff distance ahead of a blunt body in ballistic tunnels, in reduced drag, and in modifications of traveling shocks. These interesting plasma effects are of great practical importance for high-speed aerodynamics and flight control, and are briefly discussed. It is clear that energy addition to the flow results in an increase in the local sound speed that leads to expected modifications of the flow and changes to the pressure distribution around a vehicle due to the decrease in local Mach number. Intensive research, however, has recently been focused on determining specific strong plasma effects influencing high-speed flows and shocks, which are the most attractive for applications. It has been demonstrated that, although the heating in many cases is global, experiments with positive columns, DBDs, and focused microwave plasmas can produce localized special energy deposition effects more attractive for energy efficiency in flow control (Bletzinger et al., 2005). Numerous schemes have been proposed recently for modifying and controlling the flow around a hypersonic vehicle. These schemes include novel approaches for plasma generation, magneto-hydrodynamic

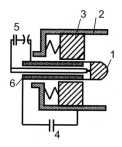

Figure 11–43. General schematic of an erosion pulsed plasma thruster: (1) cathode; (2) anode; (3) propellant; (4) main battery of capacitors; (5) battery of capacitors of the ignition system; (6) igniting needle.

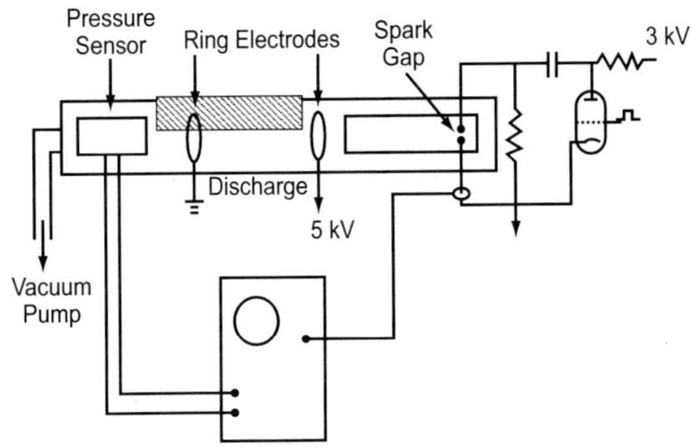

Figure 11–44. General schematic of the plasma-aerodynamic experimental setup.

(MHD) flow control and power generation, and other purely thermal approaches. Two major physical effects of heat addition to the flow field lead to drag reduction at supersonic speeds: (1) reduction of density in front of the body due to the temperature increase (that assumes either constant heating or a pulsed heating source having time to equilibrate in pressure before impinging on the surface) and (2) coupling of the low-density wake from the heated zone with the flow field around the body, which can lead to a dramatically different flow field (the effect is very large for blunt bodies, changing their flow field into something more akin to the conical flow). Efficiency of the effect grows significantly with Mach number. Let us begin to consider plasma effects in high-speed aerodynamics with early impressive experiments regarding specific plasma influence on shockwave structure and velocity.

11.4.2. Plasma Effects on Shockwave Structure and Velocity

Klimov et al. (1982) were first to investigate the propagation of shockwaves in low-pressure plasmas, and they found that the observations could not completely be explained via plasma-induced thermal gradients. A schematic of the experiment is illustrated in Fig. 11–44. Ring electrodes were placed in a 1-m-long, 10×10 cm^2 cross-sectional channel, and the length of the plasma column was varied from 30 to 160 mm. A pulsed electric discharge was used to launch the shockwave at one end; a fast-response-piezoelectric transducer for measuring the shock pressure was placed at the other end. Shock velocity in the system with a neutral gas was 740 m/s, and reached 1200–1300 m/s velocity 1–3 μs after entering the plasma. The shock front broadened from a temporal width in the neutral gas of 2 μs to a width of 8–10 μs in the plasma. The calculated increased velocity due to the temperature gradient at the plasma boundary was only 900 m/s versus the measured 1200–1300 m/s. The effect can be explained by the release of stored vibrational energy. With a similar experimental system, but using the afterglow of the discharge, Klimov et al. (1982) measured the shock velocity and the shock amplitude as a function of the time after the discharge was switched off (see Fig. 11–45). Further experiments of Klimov and Mishin (1990), Gorshkov et al. (1987), and Klimov et al. (1989) provide additional evidence that the plasma influence on the shockwave structure and velocity is essentially stronger than expected, assuming that plasma affects only thermal gradients.

11.4.3. Plasma Aerodynamic Effects in Ballistic Range Tests

The flight of simple-shaped gunlaunch projectiles (spheres, cones, and stepped cylinders) through steady DC discharges was analyzed by Mishin, Serov, and Yavor (1991), and Mishin (1997). The shock standoff distance significantly increases when the sphere is flown through a steady DC discharge region. Experiments were conducted at gas pressures of 40–50 Torr with discharge current densities of 20–50 mA/cm^2, degrees of ionization 10^{-5}–10^{-6},

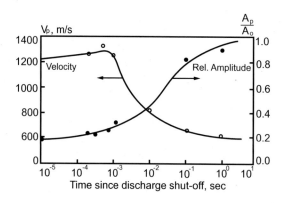

Figure 11–45. Velocity and relative amplitude of the shockwave in the decaying plasma as functions of delay time since switch-off of the discharge. A_p and A_0 are the amplitudes of the shockwaves in air and in the plasma, respectively.

electron densities 10^{11}–10^{12} cm^{-3}, and electron temperatures 1–4 eV. For projectile velocities in the range of 1200–2400 m/s, the shock standoff distance was observed to increase by a factor of 1.6–2.2 compared with air at the measured gas temperature. Ballistic range tests of Bedin and Mishin (1995) permitted measurement of the drag of a sphere flying through gas discharge plasmas. The experiments were conducted using 15 mm polyethylene spheres fired at speeds of 200–1350 m/s in air at 15 Torr. An RF plasma generator operating at 440 kHz was used to produce plasma with an ionization degree 10^{-6} over a length of 3 m. Gas temperature in the discharge was 1140 K, and electron temperature was 2–4 eV. Anomalous shock standoff distances were observed for spheres within the discharge zone at distances far from the thermal non-uniformities near the entrance to the plasma. Time-of-flight measurements have been made to determine the drag coefficient of the sphere (see Fig. 11–46). The measurements indicate a substantial decrease in the drag coefficient at subsonic speeds, with a small increase at supersonic speeds. Numerous further experimental and theoretical investigations were focused on understanding the plasma aerodynamic effects, observed in ballistic range tests, and especially on mechanisms leading to the anomalous shock standoff distances (see, for example, Serov & Yavor, 1995; Lowry et al., 1999a,b; Candler et al., 2002). The gas discharges used in ballistic ranges result in significant energy tied up in the vibrational modes; therefore, the flow field around the sphere can be altered significantly if the energy relaxes back into translational and rotational modes within the shock layer. The role of vibrational energy relaxation in the shock layer has been analyzed, and it has been shown that its contribution to the anomalous shock standoff distances is usually less significant than that of thermal non-uniformities.

Shockwave propagation in plasmas leads to polarization of the plasmas and specific electric effects (see, for example, Lieberman & Velikovich, 1985; Kolesnikov, 2001; Ishiguro, Kamimura, & Sato, 1995; Hershkowitz, 1985; Raadu, 1989; Maciel & Allen, 1989;

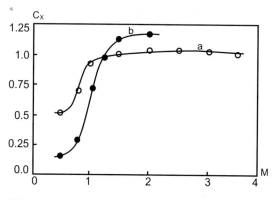

Figure 11–46. Drag coefficient (measured values) as functions of Mach number for: (a) flight through air without a discharge and (b) flight through the weakly ionized gas.

Williamson & Ganguly, 2001; Bletzinger & Ganguly, 1999; Ganguly, Bletzinger, & Garscadden, 1997). Some flow kinetic energy can be converted into electrical energy through the charge separation. A traveling strong double layer can be formed near the shock front in the case of shock propagations through a non-equilibrium ($T_e \gg T_0$) glow discharge plasma. The existence of a strong double layer with a significant potential jump leads to local electron heating, excitation, ionization, and local gas heating. The last effect causes shock front broadening and velocity changes in addition to the effect of overall discharge heating. Optical, microwave, and electrical measurements give a picture of the plasma–shockwave interactions and evidence for the presence of strong excitation (and ionization) enhancement in a narrow region propagating with the shockwave for initially weakly ionized plasmas, evidence of the formation of strong double layers in collisional shockwave propagation in non-equilibrium plasmas. The formed double layer is a transient moving layer able to support a strong space-charge layer with a potential jump exceeding 10 kTe.

11.4.4. Global Thermal Effects: Diffuse Discharges
The energy of electric discharges through different channels is finally transferred to the temperature of the neutral gas components and results in spatial temperature gradients, which modifies shockwaves. In a cylindrical discharge tube this process results in a radial temperature distribution with the maximum and, in turn, a neutral density minimum at the tube axis. Similarly there is a temperature and density gradient at the axial discharge. At some plasma conditions, the global thermal effect can dominate over any local gas heating caused by the shock-induced polarization of the plasma. Good agreement between two-dimensional theory and measurements on a cylindrical shock tube with a discharge section has been achieved by Voinovich et al. (1991) for both the shock velocity and the shock profile. The shock profile, after the initial change from planar to curved shock, doesn't vary further when traversing a length of 13 tube radii. Macheret et al. (2001a), using their carefully measured temperature profiles for a discharge in argon with 1% N_2 at 50 Torr pressure and a 10–40 mA discharge current, calculated shock profiles which increased in curvature along the path of propagation. Experimental results of Macheret et al. (2001a) and White et al. (2002) show that, for plasmas with low mean electron energy and Debye length smaller than the shock thickness, the global thermal effects dominate over other local energy-dissipation processes.

11.4.5. High-Speed Aerodynamic Effects of Filamentary Discharges
Filamentary discharges, in particular those created using high-frequency discharges, microwave discharges, or injected small aperture e-beams are applied for supersonic and hypersonic drag reduction. High-frequency pulsed discharges have been investigated for the application by Bityurin et al. (1999), Klimov et al. (1999), and Leonov, Bityurin, and Kolesnichenko (2001a). Models with both well-shaped aerodynamic forebodies and a $30°$ cone cylinder have been tested at Mach 2. Models with diameters between 40 and 80 mm had a sharp electrode at the tip to concentrate the electric field. A 5 kW power supply operating with a carrier frequency of 13–16 MHz and 100 Hz amplitude modulation has been used. Both diffuse and thin streamer-like discharges have been observed. With a diffuse discharge, little change in the shock position was observed, but the drag decreased by approximately 6% with an energy efficiency near 100%. With the filamentary discharges, Schlieren photographs have been obtained, showing the interaction of the plasma filaments with the bowshockwave, but negligible change in drag has been observed. Single filaments have been found to be created with a frequency between 1 and 10 kHz, grew with an average speed of 5–10 km/s, and had a lifetime of 10–100 μs. The total length of the filaments produced was approximately 5 cm. Leonov et al. (1999) investigated high-frequency discharges with multiple filaments produced simultaneously, which leads to effective drag reduction

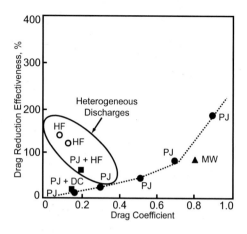

Figure 11–47. Drag reduction effectiveness for heterogeneous discharges (PJ, plasma jet; HF, high-frequency discharge; MW, microwave).

(see Fig. 11–47). The effectiveness of drag reduction is shown in the figure as a function of the baseline model drag coefficient. The effectiveness is defined as the drag power savings (drag decrease multiplied by freestream velocity) divided by the electrical input power. The energy efficiency of a heterogeneous discharge can be high even for aerodynamic shapes.

Kolesnichenko et al. (2001, 2003) have investigated the aerodynamic effects of a microwave discharge upstream of blunt cylinders, spheres, and cones in supersonic flow. Measurements of the microwave filamentary discharge show the existence of a central high-temperature filament surrounded by a halo of ionized gas. The filament, which is formed in 1–2 ns and has a radial dimension of less than 3×10^{-3} cm, has a specific energy of approximately 7 eV per particle with a peak electron density $n_e \approx 5 \times 10^{16}$ cm^{-3}. The temperature rise in the filament is 2800 K with a heating rate of 2000–3000 K/μs. The surrounding halo is formed at a slower rate with a temperature rise of 100 K/μs and formation time of 1–2 μs and possesses a radial dimension of approximately 3 mm. In the halo region, the specific energy is approximately 2 J/cm^3 atm with a peak electron density $n_e = 6 \times 10^{13}$ cm^{-3}. Time-resolved measurements show that the discharge significantly alters the bowshock structure by coupling the thermal gradients induced by the discharge with the bowshockwave to produce vorticity. The bowshock propagates forward into the zone heated by the discharge, leading to vortices produced at the interface between the hot and cold zones. A single hot filament on the centerline is most efficient for reducing aerodynamic drag. Heterogeneous discharges upstream of a two-dimensional 5° wedge have been described numerically by Wilkerson, Van Wie, and Cybyk (2003). The calculations were conducted to investigate the effects of modest energy deposition levels on the drag reduction of a 5° wedge at Mach numbers between 2 and 10 using both constant-pressure and constant-density energy deposition models (see Fig. 11–48). The constant-pressure model shows increasing drag reduction energy efficiency η_D with increasing Mach number, which

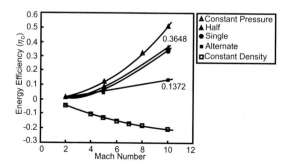

Figure 11–48. Drag reduction energy efficiency for small energy additions.

is partially due to the mass removal from the freestream. The constant-density model results in negative η_D values (i.e., drag increases), which become more negative as the Mach number increases. The effects of reduced energy deposition have been evaluated by investigating three discharge configurations: (1) removal of every other filament, (2) calculation of a single filament on the centerline, and (3) calculation of the effect of a half-thickness filament on the centerline. The drag reduction effectiveness is shown as a function of freestream Mach number for the cases with reduced energy deposition and compared with the baseline case with constant-pressure and constant-density energy deposition models. The energy deposition is reduced and concentrated near the centerline of the wedge; the energy efficiency approaches the trend produced with the constant-pressure deposition model. The energy should be concentrated on the centerline and located far enough upstream for the pressure equilibration that occurs after energy deposition to be completed prior to heated air impacting the body.

11.4.6. Aerodynamic Effects of Surface and Dielectric Barrier Discharges: Aerodynamic Plasma Actuators

Surface discharges permit rapid and selective heating of a boundary layer at or slightly off the surface, as well as creating plasma zones for effective MHD interactions (see, for example, Shibkov et al., 2000, 2001; Leonov et al., 2001b; Grossman, Cybyk, & Van Wie, 2003). The surface discharges for supersonic and hypersonic aerodynamic applications have been effectively organized using surface microwave discharges, DC and pulsed discharges between surface-mounted electrodes and internal volumes, and sliding discharges. Leonov et al. (2001b) investigated the use of a surface discharge between the surface-mounted electrodes to reduce the friction drag force of a flat plate. At transonic flow speeds with a local static pressure of 500 Torr, application of the 15 kW discharge results in gas heating sufficient to produce a temperature of about 1500 K. A portion of the flat plate downstream of the discharge was isolated and mounted on a force balance. The drag measurements provided evidence that the friction drag downstream of the discharge is significantly reduced (20–30%). Surface discharges are effective for adding thermal energy to a boundary layer. For speeds below approximately Mach 6, the creation of a hot bounder layer stabilizes the boundary layer so that transition can be delayed. Above Mach 6, alternative transition modes become dominant. Levin and Larin (2003) have calculated the reduction in surface skin friction coefficient on a flat plate at Mach 3 due to a surface discharge modeled as an energy deposition in a small volume above the surface. The discharges have been shown to impact the skin friction well downstream of the discharge region, and energy efficiencies approaching 50% have been calculated for Mach 3.

Asymmetric DBDs, with one electrode located inside (beneath) the barrier and another mounted on top of the dielectric barrier and shifted aside, stimulate intensive "ion wind" along the barrier surface, influence boundary layers, and can be used as aerodynamic actuators (see, for example, Boeuf et al., 2007; Likhanskii et al., 2007). Pioneering experiments in this directons were carried out by R. Roth and his group. Aerodynamic plasma actuators have lately attracted significant interest in numerous applications for flow control above different surfaces (aircraft wings, turbine blades, etc.) The ion wind pushes neutral gas with not very high velocities (typically 5–10 m/s at atmospheric pressure). However, even those values of the DBD-stimulated gas velocities near the surface are able to make significant changes in the flow near the surface, especially when the flow speed is not fast. The major contribution to the ion-wind drag in DBD plasma actuators is mostly due not to the DBD streamers themselves but to the ion motion between the streamers when ions move along the surface from one electrode to another, similarly to the case of corona discharges (see Sections 4.5.1–4.5.4). During the phase when the electrode mounted on the barrier is positive, the positive ions create the wind and drag the flow. It is interesting that, during the following phase when the electrode mounted on the barrier is negative, the negative ions formed by electron attachment to oxygen (in air) make a significant contribution to total ion wind and drag the flow in the same direction (Likhanskii et al., 2007). Application of nano-second pulse DBD permitted A. Starikousky and his group to achieve strong effect at very high speeds (M = 0.7). The effect is due in this case to ion wind but to fast heating of the boundary layer and fosmation of a shock wave.

11.4.7. Plasma Application for Inlet Shock Control: Magneto-Hydrodynamics in Flow Control and Power Extraction

Plasma aerodynamic flow-control can be applied for controlling shock positions within an inlet so as to reduce or eliminate the complexity of a variable geometry inlet. The plasma flow-control techniques are usually based on the use of Lorentz force while minimizing the role of heat addition. MHD flow control has been suggested, in particular, for use in control of the captured flow rate, shock positioning, compression ratio, and local shockwave/boundary-layer interactions for scramjet inlets (Kopchenov, Vatazhin, & Gouskov, 1999; Shang, Gaitonde, & Canupp, 1999; Vatazhin, Kopchenov, & Gouskov, 1999; Bobashev & Golovachov, 2000; Hoffman, Damevin & Dietikey, 2000; Vatazhin & Kopchenov, 2000; Munipalli & Shankar, 2001; Poggie & Gaitonde, 2001). Generation of adequate electric conductivity is a requirement for the development of an MHD inlet flow control. For blunt vehicles, significant electron number densities can occur naturally in the viscous shock layer, but the scramjet inlets are long and slender with small flow deflection angles; therefore, the electric conductivity for MHD interaction requires an additional source of ionization. The most energy-efficient technique for this purpose is the use of an electron beam. The electron beam can be operated at moderate energy fluxes with injection through foil windows or at high energy fluxes with injection through aerodynamic windows. The effect of inlet flow control on the overall performance of a scramjet engine has been analyzed by Brichkin, Kuranov, and Sheikin (1999). It was shown that at Mach 6 the use of an MHD flow-control system can increase the captured flow rate between 25 and 37%, which leads to an increase in the specific impulse of 10–12% and to a thrust increase between 40 and 50%. Analysis of the inlet performance shows that MHD flow control can be used to modify the shock structure in an inlet, although operation in the flow expansion mode is not efficient. Macheret, Shneider, and Miles (2000) have examined the feasibility of using an external MHD system with electron-beam ionization for control of a scramjet inlet. They show that the power required for ionization is less than the power generated within the MHD system for inlet operation at Mach 8 and 30 km altitude operating conditions.

Plasma-initiated MHD effects can also be used to reduce heat transfer. Analysis of the hypersonic flow over a hemisphere shows that the stagnation-point heat transfer could be reduced by approximately 25% at Mach 5 with a magnetic interaction parameter of 6. Reduction in heat transfer at a blunt leading edge can be achieved using current flows along the leading edge to produce a circumferential magnetic field. The resulting $j \times B$ force leads to a deceleration of the flow approaching the leading edge. Another concept is on-board power extraction from a scramjet flow, which has been considered both as energy bypass and as auxiliary power generation (Fraishtadt, Kuranov, & Sheikin, 1998; Burakhanov et al., 2000; Macheret, Schneider, & Miles, 2000, 2001a,b; Brichkin, Kuranov, & Sheikin, 2001; Park, Bogdanoff, & Mehta, 2001; Slavin et al., 2001). The energy bypass involves extraction of flow energy upstream of the combustion using an MHD power extraction system with re-introduction of the power into the propulsive stream downstream of the combustor using an MHD flow acceleration system. Energy bypass results in the possibility of organizing the combustion, process at a lower effective velocity. Auxiliary MHD power generation can be used for plasma aerodynamic applications, plasma-assisted combustion, or

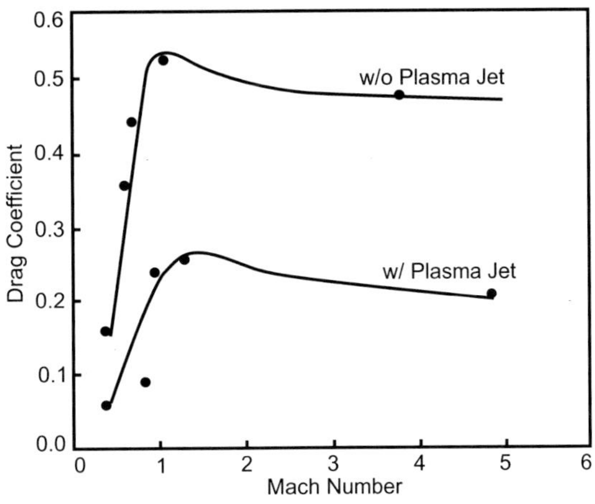

Figure 11–49. Drag coefficient (measured values) for a cone-cylinder model with and without plasma jet injection.

non-equilibrium ionization systems. Macheret, Schneider, and Miles (2000) have analyzed an MHD power extraction system located in the forward portion of a vehicle on flow turning surfaces at low angles, with an emphasis on extraction of the maximum possible energy for bypass applications. Electron beams with kiloelectron volt-range energies have been used for generating adequate conductivity. Calculations have been conducted for an MHD channel with an in-flow cross section of 25×25 cm^2, length 3 m, and magnetic field 7 T. In this system, 8–12% of the flow kinetic energy is converted to electric power at low altitudes at Mach numbers between 4 and 10, whereas up to 33% can be converted at high altitudes.

11.4.8. Plasma Jet Injection in High-Speed Aerodynamics
The reduction of drag on a body at supersonic speeds can also be achieved by injection of plasma from an on-board generator (see, for example, Ganiev et al., 1998, 2000). Plasma in this case can be generated by a thermal arc discharge providing plasma heating to about 6000 K prior to its injection through a Laval nozzle at the leading edge of a 30° cone-cylinder model. Decreases in the drag on the model between two- and four-fold have been achieved at speeds between Mach 0.4 and 5.0 as illustrated in Fig. 11–49. The reduction of drag coefficient is due to a modification of the flowfield over the cone with the plasma jet injection resulting in an effectively lower-angle cone. The effect has been analyzed by injection of hot solid-fuel combustion products and hot helium. The effect has also been demonstrated by Shang (2002) and Fomin et al. (2002). More detail can be found in a review by Bletzinger et al. (2005), and in a special issue of the *Journal of Propulsion and Power* (2007, vol. 22), edited by S. Macheret.

11.5. MAGNETO-HYDRODYNAMIC GENERATORS AND OTHER PLASMA SYSTEMS OF POWER ELECTRONICS

11.5.1. Plasma Power Electronics
Low-temperature plasma can be effectively applied to produce electricity from high-temperature heat energy "directly," without any special mechanical devices, and without any rotating or linearly moving mechanical parts such as rotors, pistons, and so forth. The major plasma power electronic systems of the direct production of electricity are MHD generators and thermionic converters. MHD generators use fossil fuels as their primary energy source and create thermal plasma through combustion. In the future, nuclear energy is to be used to

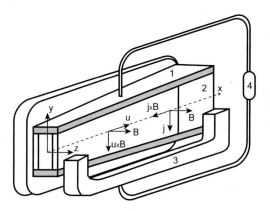

Figure 11–50. Schematic of an MHD generator channel: (1) electrode wall; (2) insulator wall; (3) magnetic coil; (4) load.

produce plasma in MHD generators. Plasma flow moving in an MHD channel works against the magnetic field generating electricity. The MHD approach can be performed not only in stationary systems but also in pulse generators of electricity. MHD pulse generators use explosives for short-duration thermal plasma generation; the pulse duration varies between $100\,\mu s$ and $100\,s$. Thermionic converters (or thermo-electronic converters) are based on electric current generation by means of the thermionic emission from a hot cathode (called an emitter) through a vacuum gap to a colder electrode (called a collector). An interest in thermionic converters is their application in nuclear power systems for autonomous electricity production without any mechanical devices and moving parts. Other systems of plasma electronics are plasma commutation devices; see Section 11.5.6.

11.5.2. Plasma MHD Generators in Power Electronics: Different Types of MHD Generators

Application of plasma as a working fluid in MHD generators permits conversion of both thermal and kinetic energy of the working fluid into electric energy. Plasma motion with velocity $\vec{u}$ in magnetic field $\vec{B}$ results in the formation of electromotive force $\vec{u} \times \vec{B}$. When electrodes connected to an external electric circuit are located in the plasma flow, the electromotive force $\vec{u} \times \vec{B}$ leads to electric current with density $\vec{j}$. A schematic of a channel of an MHD generator is shown in Fig. 11–50. In contrast to a conventional turbine-based electric generator, the MHD generator has no moving parts, which permits a significant increase of the working fluid temperature. Depending on the type of working fluid, MHD generators can be divided into two classes: open cycle and closed cycle. Open-cycle MHD generators use high-temperature combustion products as the working fluid. Closed-cycle MHD generators operate with inert gases (for example, in the case of nuclear energy sources).

With respect to the method of induction of electric field, MHD generators can be divided into two other classes: conductive and inductive MHD generators. **Conductive** MHD generators are characterized by potential electric field $\vec{E} = -\nabla\varphi$; and electric current generated in a working fluid goes through an external load (see Fig. 11–50). Although conductive MHD generators can operate with AC and DC, they always include electrodes. The electric field in **inductive** MHD generators is induced by alternating magnetic field ($\vec{E} = -\frac{\partial \vec{A}}{\partial t}$, $curl\,\vec{A} = \vec{B}$). The electric current generated in the system can be closed inside of the working fluid. The magnetic field in the inductive generators can be used in different ways, separating the systems with the running magnetic field and the transformer-based systems. Further consideration will be mostly focused on the wider applied conductive type of MHD generators.

11.5.3. Major Electric and Thermodynamic Characteristics of MHD Generators

Major electric characteristics of MHD generators are power generated in the volume of working fluid, $P = \int \vec{j} \vec{E} dV$, and local energy efficiency, determined as the ratio of the power density $\vec{j} \vec{E}$ to the work of electromagnetic force per unit time, $(\vec{j} \times \vec{B}) \cdot \vec{u}$:

$$\eta = \frac{\vec{j} \cdot \vec{E}}{(\vec{j} \times \vec{B}) \cdot \vec{u}}. \tag{11–46}$$

The power density $\vec{j} \vec{E}$ and the work of electromagnetic force per unit time are related by the equation of energy conservation:

$$\vec{j} \cdot \vec{E} = (\vec{j} \times \vec{B}) \cdot \vec{u} + j^2/\sigma. \tag{11–47}$$

The energy efficiency is maximized by reduction of the Joule heating j^2/σ, where σ is the plasma conductivity. The thermodynamic characteristics of MHD generators are coefficient of enthalpy conversion, η_P; and internal energy efficiency, η_{0i}. η_P is determined as the ratio of the power ($P = \int \vec{j} \vec{E} dV$) to the flux of stagnation enthalpy of the working fluid at the inlet of an MHD generator:

$$\eta_P = \frac{P}{\dot{m} \cdot h_{01}}, \tag{11–48}$$

where $\dot{m}$ is the mass flow rate of the working fluid, and h_{01} is the stagnation enthalpy of unit mass of the working fluid at the inlet of an MHD generator. The internal energy efficiency η_{0i} is determined as the ratio of the power ($P = \int \vec{j} \vec{E} dV$) to the change of the stagnation enthalpy fluxes between inlet and exit of an MHD generator at the fixed drop in the stagnation pressure:

$$\eta_{0i} = \frac{P}{\dot{m}(h_{01} - h_{02})}. \tag{11–49}$$

The local internal energy efficiency can be determined at any specific point as the local derivative:

$$\eta_0 = \frac{1}{\dot{m}} \cdot \frac{dP}{dh_0}. \tag{11–50}$$

The relation between the local internal energy efficiency η_0 (11–50) and the local energy efficiency η (11–46) can be derived from the equations of mass, momentum, and energy conservation:

$$\eta_0 = \eta \cdot \frac{1}{1 + \dfrac{\gamma - 1}{2} M^2(1 - \eta)}, \tag{11–51}$$

where M is the Mach number of the working fluid. Equation (11–51) reflects the influence of the Joule dissipation incoming at higher Mach numbers on the local internal energy efficiency η_0. The dependence between the local electric characteristics and flow characteristics in MHD generators is controlled by Ohm's law in the presence of a magnetic field. In the case of a linear conductive MHD generator with rectangular channel, flow velocity in the x-direction, magnetic field in the z-direction, and uniform flow $j_z = 0$ (see Fig. 11–50), the relation between the local electric and flow characteristics is (Medin, 2000)

$$\vec{j}\vec{E} = -\frac{\sigma u^2 B^2 \eta(1 - \eta)}{1 + (j_x/j_y)^2}. \tag{11–52}$$

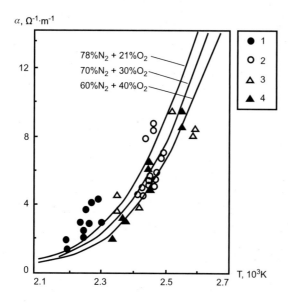

Figure 11–51. Electric conductivity of plasma created by products of combustion of natural gas in air enriched by oxygen at pressure 0.8 atm. Curves represents different simulations. Experiments: (1) 78% N_2, 21% O_2, 1% additives; (2) 70% N_2, 30 % O_2; (3) 60% N_2, 40 % O_2; (4) 50% N_2, 50 % O_2.

The maximum value of the power density can be achieved at $\eta = 0.5$, $j_x/j_y = 0$, maximum value of the magnetic field, and an optimal value of the Mach number. The optimal value of Mach number, M_{opt}, is due to non-monotonic σu^2 dependence on flow velocity u at fixed stagnation flow parameters. In monoatomic gases usually $M_{opt} < 1$, whereas in molecular gases $M_{opt} < 1$.

11.5.4. Electric Conductivity of Working Fluid in Plasma MHD Generators

The electric conductivity of plasma as the working fluid is one of the critical parameters of MHD generators. To increase the electric conductivity, small amounts of alkaline metals (K, Cs, etc.), characterized by low ionization potentials, are usually added to the working fluid. Figure 11–51 illustrates the electric conductivity of the products of combustion of natural gas with sodium addition (1%), which are used as the working fluid for an MHD generator (Medin, 2000). Electric conductivity of thermal plasma strongly exponentially depends on temperature (see Section 3.1.1). In the case of molecular gases, plasma in MHD generator is not far from the thermodynamic quasi equilibrium, which imposes strong limitations on operational temperature in the MHD generators to reach a necessary level of electric conductivity. For example, MHD generators, using the products of combustion of natural gas with alkaline-metal additives as working fluid, operate with exit temperatures always exceeding 2000 K. The operating temperature of MHD generators can be significantly decreased by using plasma of inert gases as a working fluid. Such a situation takes place in closed-cycle MHD generators (when heating of the working fluid is provided, for example, from a nuclear power source). A large contribution to the ionization of inert gases in this case is made by non-equilibrium ionization by electric fields present in the system. The non-equilibrium ionization of inert gases requires lower values of electric fields (see

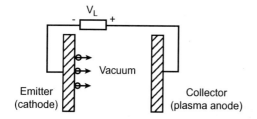

Figure 11–52. General schematic of a thermionic converter.

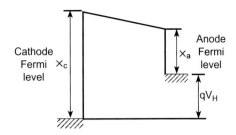

Figure 11–53. Electron energy diagram for a thermionic converter.

Chapters 2 and 3), which makes the effect of operating temperature reduction relevant mostly for closed-cycle MHD generators. MHD generators of electricity have been only briefly discussed here. Much more information regarding plasma physics, chemistry, and engineering of MHD generators can be found in special reviews and books on the subject (Vatazhin, Lubimov, & Regirer, 1970; Yantovsky & Tolmach, 1972; Medin, 2000; Decher, 2006; Sutton & Sherman, 2006).

11.5.5. Plasma Thermionic Converters of Thermal Energy into Electricity: Plasma Chemistry of Cesium

Thermionic converters are simple systems for direct conversion of heat energy to electricity. The thermionic converters can be easily combined with nuclear power sources and with solar power sources, which attracts particular interest to their application in spacecraft. A schematic of a thermionic converter is shown in Fig. 11–52. A thermionic converter is a flat or cylindrical diode consisting of a hot cathode to emit electrons and a cold anode to collect electrons. If the work function of a cathode, U_C, exceeds that of an anode, U_A, then connecting the electrodes (and equalizing the Fermi levels of cathode and anode) leads to the formation of an electric field in the interelectrode gap, which pushes electrons from cathode to anode (see Fig. 11–53). In a vacuum thermionic converter, the electron current remains constant until the load voltage V_L does not exceed the contact difference of potentials: $V_L \leq U_C - U_r$. At larger load voltages, the external circuit current decreases, because a fraction of the emitted electrons returns via the electric field back to the cathode. Ideal current-voltage characteristics of a vacuum thermionic converter are shown in Fig. 11–54. Generated power is larger at the highest values of cathode work function and the lowest values of anode work function. To provide high thermionic current from a cathode with high work function, the cathode temperature should be sufficiently high. Usually such special metals as W, Mo, Re, Nb, and their alloys, characterized by high melting temperatures, are used for manufacturing the high-temperature cathodes. Work functions of the cathode are usually in the range 4.4–5.1 eV. They operate at temperatures exceeding 2800 K to provide a thermionic current density on the level of 10 A/cm². Anodes are usually coated by special thin films that reduce their work function. An important role in the thermionic converters is played by the addition of cesium, which is characterized by very low work function (1.8 eV) and

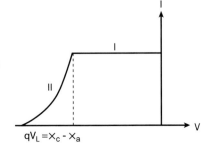

Figure 11–54. Ideal current-voltage characteristic of a thermionic converter.

ionization potential (3.89 eV); it makes a thermionic converter work as a plasma-chemical reactor. Cesium plays a multi-functional role in thermionic converters. First, adsorption of cesium on the high-melting-temperature cathode surface changes operation of the cathode in the so-called film-cathode regime. In this regime, the work function of the cathode can be controlled by flux of the Cs atoms to the surface and the surface temperature. The cathode work function in the film-cathode regime can be varied from the highest value, corresponding to the work function of the initial cathode material, to the lowest value, corresponding to a the work function of cesium. A second effect of Cs addition is related to a decrease of anode work function to the low values required for high effectiveness of thermionic converters. The reduction of the anode work function occurs due to formation of a Cs film on the surface of a cold anode. Finally, the third positive effect of cesium is related to the generation of a significant concentration of Cs^+ ions in the gap, which results in neutralization of the electron space charge there and growth of the converter effectiveness. Cesium atoms have very low ionization potential (3.89 eV) and are easily ionized in the gap by different ionization mechanisms. More information can be found in Dugev (2000), Stakhanov (1968), Baksht, Dugev, and Martsinovsky (1973), and Hatsopoulos and Gyftopoulos (1973, 1979).

11.5.6. Gas-Discharge Commutation Devices

The application of electric discharges as switches and communication devices has already been briefly discussed in Section 4.3.11 (Fig. 4–58) in relation to gliding-arc discharges. Gas discharges are also widely applied as pulsed commutation devices, specifically for the generation of pulses with current amplitude exceeding 100 A and pulse voltages from several to hundreds of kilovolts (see, for example, Bochkov & Korolev, 2000). Gas-discharge commutation devices can be classified into groups, either according to gas pressure or to their position on the Paschen curve (see Section 4.1.1). **Thyratrones, tasitrons** with hot or cold cathodes, and **pseudo-spark discharges** operate at pressures of 1–100 Pa, which corresponds to the left branch of the Paschen curve. Vacuum commutating discharges with plasma generation due to erosion of electrodes and discharges with liquid (usually Hg) cathode can also be classified in the lower-pressure group. In contrast to that, commutating spark discharges can be classified as high-pressure gas-discharge commutation devices. These commutation devices operate at pressures from 100 Pa to several atmospheres in conditions corresponding to a minimum of the Paschen curve or its right-hand branch. Some characteristic parameters (maximum achieved values) of the gas-discharge commutating devices are summarized in Table 11–1 (Bochkov & Korolev, 2000). More information on the subject can be found in Schaefer, Kristiansen, and Guenther (1990), Guenther, Kristiansen, and Martin (1987), and Bochkov & Korolev (2000).

11.6. PLASMA CHEMISTRY IN LASERS AND LIGHT SOURCES

11.6.1. Classification of Lasers: Inversion Mechanisms in Gas and Plasma Lasers and Lasers on Self-Limited Transitions

Strongly non-equilibrium gas-discharge plasmas ($T_e \gg T_0$) significantly overpopulate highly excited states of atoms, molecules, and ions and, therefore, can be effectively used as an active medium for lasers. Physics, engineering, and applications of gas-discharge lasers are described in numerous books and reviews (Svelto, 2004; Silfvast, 2004; Gudzenko & Yakovlenko, 1978; McDaniel & Nighan, 1982; Yakovlenko, 2000). Strongly non-equilibrium plasma systems with electron temperature significantly exceeding that of neutrals permit intensive excitation of an upper working level, which leads to lasing. The lower working level, however, is usually also excited in the plasma system; therefore, lasing requires special plasma-kinetic mechanisms to provide effective inversion in the population of upper and lower working levels. Thus, effective lasing requires special kinetic

Table 11–1. Maximum Parameters of the Gas-Discharge Commutating Devices

Maximum values of parameters	Spark discharge commutating systems	Vacuum discharge commutating systems	Thyratrons, tasitrons with hot/cold cathodes
Operating voltage, kV	10,000	70	200
Commutated current, kA	1000	500	100
Commutation rate, A/s	$> 10^{12}$	10^{10}	$5 \cdot 10^{11}$
Commutated charge in a pulse, C	500	100	0.5
Pulse frequency, kHz	0.5	1	300
Deviation of starting moments with respect to initiating pulse, ns	10	100	0.1
Lifetime as total commutated charge, C	10^4–10^5	10^6	10^5–10^8

organization of the laser as a plasma-chemical system. Lasers in gas discharges can be classified in different ways. For example, lasers can be classified according to the type of working transition into those on transitions between electronically excited states (to be discussed in the beginning of Section 11.5), and those on transitions between vibrationally excited states (to be discussed later in Section 11.5). When laser generation takes place during ionization (in an active discharge zone), they are usually called gas lasers. If generation occurs in the recombining plasma (that is not in an active discharge phase), the lasers are usually referred to as plasma lasers. Let us begin by considering lasers in strongly non-equilibrium gas discharges on the transitions between electronically excited states of atoms, molecules, and ions. The most successful organization of the inversion in lasers on electronic transitions has been achieved in the so-called plasma-gas lasers on self-limited transitions. The self-limited transitions are those to a metastable state "a," which is operating as a lower working level of the laser system. The upper working level, "b," is usually a resonant one that can be easily excited in a strongly non-equilibrium plasma ($T_e \gg T_0$). The lasers on self-limited transition efficiently operate on the front of current pulses, in pulse-periodic regimes. The optimal time of the pulse front should be shorter than the inversion time, that is, less than 10^{-8}–10^{-6} s. The pulsed self-limited lasers only provide intensive excitation of upper levels, whereas deactivation of the lower level is not as important because of the pulsed-periodic operation. The advantages of self-limited lasers are mostly related to their simplicity in lasing of atomic and ionic spectral lines in the visible optical range. The average power of the lasers is limited by the pulse-periodic mode of their operation.

11.6.2. Pulse-Periodic Self-Limited Lasers on Metal Vapors and on Molecular Transitions

Gas-discharge lasers using metal vapors can be considered an example of lasers on self-limited transitions (between different electronically excited states of atoms). The highest average and pulse power have been achieved for the case of copper vapors (lasing wavelengths $\lambda = 510.5$ nm and $\lambda = 578.2$ nm). These lasers are able to provide an average radiation power exceeding 100 W from one active element. Between other gas-discharge lasers on metal vapors, we can point out those in lead vapor ($\lambda = 722.9$ nm), barium vapor ($\lambda = 1499.9$ nm), europium vapor ($\lambda = 1759.6$ nm), and gold vapor ($\lambda = 627.8$ nm and $\lambda = 312.3$ nm).

A gas-discharge laser operating on the vapor of a single metal is able to generate radiation at two or more wavelengths. If two metals are loaded in the same discharge tube, it is possible to have simultaneous generation of transitions from the different metals. The operating temperatures of metal vapor lasers are high. The lasers using copper vapors require a temperature around 1500°C. The operation temperature can be decreased by almost half by making the lasers in metal halides. The presence of halogens in the laser systems also increases the time of useful excitation. Lasers based on self-limited transitions of vapors of copper, lead, and gold are widely applied because of their high average power and stable operation in the pulse-periodic regime with pulse frequencies exceeding 10 kHz. The highest pulse power and the shortest wavelength generation of the lasers are achieved using the electronic transitions of diatomic molecules. The lowest generation wavelength in such systems, 116 nm, has been achieved in lasers on H_2 molecules. The highest pulsed power (24 MW) at pulse duration 10 ns has been achieved on self-limited transitions of the second positive system of nitrogen molecules. The repetition frequency in the lasers is typically about 100 Hz; energy efficiency is about 0.1%.

11.6.3. Quasi-Stationary Inversion in Collisional Gas-Discharge Lasers, Excitation by Long-Lifetime Particles and Radiative Deactivation

Quasi-stationary inversion mechanisms permit laser generation to be longer than the lifetime of the working levels. It requires not only population of the upper lasing level but also deactivation of the lower working level. To decrease the excitation of the lower working level, the upper level can be excited by resonant energy-exchange collisions with particles of a buffer gas, which are plasma-excited to long-lifetime states. The molar fraction of the buffer gas usually exceeds that of the working gas in order to initially transfer most of the energy of plasma electrons to the buffer. Excitation transfer from the buffer to the working gas can be organized either through energy-transfer relaxation collision of neutral particles or through charge exchange processes. For example, the He–Ne laser is based on excitation of upper levels of Ne in energy-exchange relaxation collisions with metastable electronically excited levels of helium (buffer). As other examples, lasing from the ionic transitions of cadmium and zinc can be effectively sustained in mixture with a helium buffer. Excitation energy transfer occurs in this case through the charge transfer processes, which is usually organized in the recombination regime. Inversion and lasing can also be achieved by the so-called radiative deactivation, if radiative decay time of a lower working level is essentially shorter than that of an upper working level. Such a type of inversion is not common for most atoms and ions with simple configurations, because the rapidly decaying lower working levels are usually also very effective for excitation. The inversion mechanism based on the radiative deactivation can be achieved, however, in the case of ions with complex configuration.

11.6.4. Ionic Gas-Discharge Lasers of Low Pressure: Argon and He–Ne Lasers

The radiative mechanism of deactivation of the lower level determines the inversion and operation of low-pressure ionic gas-discharge lasers. Effective lasing has been achieved in numerous ionic transitions of such atoms as Ne, Ar, Kr, Xe, Cl, I, O, N, Hg, C, Si, S, Cd, Zn, and P. The most powerful low-pressure laser today is the argon laser, operating on the radiative mechanism of deactivation of the lower working level. The lifetime of the upper working levels in the Ar laser is about 10^{-8} s; the lifetime of the lower levels having radiative transition to the ground state is about 10^{-9} s, which actually leads to population inversion and lasing. The Ar lasers usually are organized in low-pressure discharge tubes with a diameter of 1–5 mm, current density of 1–$2 \cdot kA/cm^2$, and energy efficiency of about 0.1%. The highest power of the argon laser in the blue-green range of wavelengths is 0.5 kW. The widely used He–Ne laser operates based on the population inversion sustained by excitation of the upper working levels of neon atoms Ne(4s, 5s) in collisions with helium atoms (buffer) preliminarily excited in plasma to the metastable states $He(2^3S)$ and $He(2^1S)$ by plasma electrons. Inversion in He–Ne lasers is also essentially supported by radiative deactivation of the lower working levels. Lasing is achieved in the system on more than 200 spectral lines; the most used from those are $\lambda_1 = 3.39\,\mu m$, $\lambda_2 = 0.633\,\mu m$, and $\lambda_3 = 1.15\,\mu m$. The excitation of the laser is usually organized in the low-pressure positive column of a glow discharge with typical current densities 100–200 mA/cm^2. Optimal electron density in the gas-discharge lasers is about 10^{11} cm^{-3}; the typical density of neon is $2 \cdot 10^{15}$ cm^{-3} and the typical density of helium is $2 \cdot 10^{16}$ cm^{-3}.

11.6.5. Inversion Mechanisms in Plasma Recombination Regime: Plasma Lasers

In most of the gas lasers using transition between electronically excited states considered earlier, lasing takes place during ionization in the active discharge zone. In contrast to that, generation in the so-called plasma lasers takes place during plasma recombination. The inversion can be sustained as a quasi-stationary state in the overcooled recombining plasma, when a significant fraction of electrons appear on the upper working level after recombination and sequence of gradual relaxation processes. Effective lasing in this case also requires deactivation of the lower working levels, which can be achieved by means of radiative deactivation, or deactivation in collisions with electrons or heavy particles. Radiative deactivation of the lower working levels takes place in many metal vapor and inert gas lasers. This mechanism is important in sustaining inversion in short-wavelength lasers using transitions of multi-charged ions. It should be mentioned that the energy characteristics of lasers based on the radiative deactivation of the lower working levels are limited in both gas lasers (ionization regime) and plasma lasers (recombination regime). When electron density is relatively high ($n_e > 10^{12}$–10^{13} cm^{-3}), effective deactivation of the lower working levels can be provided via electron impact. This deactivation mechanism is typical for conditions when plasma recombination processes dominates over ionization. The deactivation of the lower working levels in collisions with electrons is most effective for the levels located close to the ground state. Deactivation of lower working levels and inversion population in dense gases can be provided by collisions with heavy neutral particles. A significant role can be played in such laser systems by Penning deactivation. Electronically excited particles in these systems, corresponding to the lower working level, lose their activation in ionizing collisions with particles of a specially added gas characterized by low ionization potential (see Section 2.1.11). The Penning deactivation effect is limited in the active phases of conventional gas discharges because a significant presence of an additive with low ionization potential leads to its intensive ionization and restriction of ionization of the major gas. Therefore, some special ionization approaches, such as an external ionizer (for example, electron beams), need to be applied in this case. Effective Penning deactivation is organized, in particular, in lasers using transitions of neon and helium. As an example of non-Penning mechanisms of deactivation of the lower working levels, we can point out fast chemical reactions (especially the harpoon reactions) from these levels.

11.6.6. Plasma Lasers Using Electronic Transitions: He–Cd, He–Zn, He–Sr, and Penning Lasers

Plasma He–Cd lasers operate at high pressures with inversion based on radiative deactivation of the lower working level. When excitation of plasma laser is provided by electron beams or by charged products of nuclear reactions, effective lasing has been achieved on the following electronic transitions of the singly charged cadmium ions: $4d^9 5s^2\,{}^2D_{5/2} \rightarrow 4d^{10} 5p^2 P_{3/2}(\lambda = 441.6\,nm)$, $4d^{10} 4f^2 F_{5/2,7/2} \rightarrow 4d^{10} d^2 D_{3/2,5/2}(\lambda = 533.7\,nm$ and 537.8 nm). The recombination regime lasing (radiative deactivation of the lower working level) on the cadmium ion transition $4d^9 5s^2\,^2D_{3/2} \rightarrow 4d^{10} 5p\,^2 P_{1/2}(\lambda = 325.0\,\text{nm})$ in the He–Cd mixture of high pressure has been achieved only by nanosecond electron-beam excitation and on the front of the microsecond electron-beam excitation. Quasi-stationary generation is achieved in the system only by adding to the mixture a small amount of electronegative CCl$_4$ gas.

Another recombination-regime lasing (radiative deactivation of lower working level) on the cadmium ion transitions $6p^2 P_{1/2,3/2} \rightarrow 6s^2 S_{1/2}(\lambda = 325.0\,\text{nm}, \lambda = 853.1\,\text{nm})$ in He–Cd mixture at high pressure has been achieved by nuclear excitation with fragments from nuclear fission of ^{235}U. The generation power is not high in the system (36 W), and energy efficiency is also low (0.02%). The highest generation power (1 kW) and energy efficiency (0.4–0.7%) have been achieved by nuclear excitation in the quasi-stationary regime at wavelength $\lambda = 441.6\,\text{nm}$. Generation in an He–Zn laser has been achieved on the following electronic transitions of a Zn atom: $3d^9 4s^2\,^2D_{5/2} \rightarrow 4p^2\,P_{3/2}(\lambda = 747.9\,\text{nm})$, $5d^2 D_{5/2} \rightarrow 5p^2 P_{3/2}(\lambda = 610.25\,\text{nm})$. Lasing has also been observed in an He–Zn mixture at wavelengths 491.2, 492.4, 589.4, and 758.8 nm. Laser generation at wavelength 747.9 nm with power 62 W has been organized by quasi-stationary nuclear excitation; lasing at wavelength 610.25 nm with power 50 W has been organized by nonstationary nanosecond excitation using an electron beam. Population inversion based on the deactivation of the lower working level in collision with electrons takes place in an He–Sr plasma laser. The laser operates in the recombination regime on electronic transitions of strontium ions with wavelengths of 430.5 and 416.2 nm. Excitation of the He–Sr mixture can be effectively organized in this case in the afterglow of a pulse discharge. The average power of the laser is 4 W and energy efficiency is 0.12%.

Penning plasma lasers, operating at high pressures and based on the Penning deactivation of lower working levels, have been organized in mixtures of inert gases. The lasers provide high power generation in the visible range with energy efficiency below 1%. The Ne Penning laser is generated in He–Ne–Ar, He–Ne–Ar–H$_2$, Ne–Ar, He–Ne–Kr, He–Ne–Xe, Ne–Kr, He–Ne–H$_2$, and Ne–H$_2$ mixtures using electronic transition with wavelengths 585.3, 626.7, 633.4, 703.2, 724.5, and 743.9 nm. Excitation of the mixtures in the Ne Penning laser can be organized for use in electron beams, special electric discharges, and also nuclear excitation. A penning plasma laser using electronic transition of He ($\lambda = 706.5\,\text{nm}$) has been organized, in particular, in the afterglow of high-voltage (50–170 kV) pulsed (10–20 ns) discharge in an He–H$_2$ mixture with pressure 15 Torr. The lasing duration in the system is around 1 μs. Using helium and neon excited in a discharge with admixtures of NF$_3$ results in laser generation of an electronic transition in the helium ($\lambda = 706.5\,\text{nm}$) and neon ($\lambda = 585.3\,\text{nm}$). The addition of electronegative NF$_3$ gas to the plasma laser system provides effective deactivation of the lower working level and also increases the effectiveness of the excitation.

11.6.7. Plasma Lasers Based on Atomic Transitions of Xe and on Transitions of Multi-Charged Ions

Among the dense gas plasma lasers, lasers using electronic transitions of xenon atoms in Ar–Xe, He–Ar–Xe, and He–Xe mixtures have the lowest excitation thresholds and highest energy efficiency. The energy of a radiation pulse in this case reaches several hundreds of Joules, when the lasing is initiated by an electron beam. The high-pressure Xe plasma laser intensively radiates on wavelengths 1.73, 2.03, 2.63, 2.65, 3.37, and 3.51 μm. The most powerful and most effective lasing is achieved in the Ar–Xe mixture; the generation corresponds to electronic transition in Xe with $\lambda = 1.73\,\mu$m. Excitation of the upper working levels of Xe occurs as a result of four major processes: (1) dissociative recombination of molecular ions ArXe$^+$ and Xe$_2{}^+$, (2) EE relaxation energy transfer from Ar* to Xe,

(3) three-body recombination of Xe^+ ions, and (4) direct electron impact excitation from the 6s states. The deactivation of lower working levels of Xe is mostly due to collisions with atoms of He, Ar, and Xe together with the radiative decay. Maximum energy efficiency of the Xe laser is achieved at atmospheric pressure by electron-beam excitation in the mixture; Ar:Xe $= 200:1$ is 4.5%. The current density in this system is 0.7 A/cm^2, and the duration of the excitation pulse is 4.5 μs. The interest in plasma lasers using electronic transitions of multi-charged ions is due to the high energy of the transitions and possibilities of achieving effective lasing at very low wavelengths. Effective lasing in the short-wavelength range is very challenging because spontaneous radiation from the upper working levels significantly grows with frequency ($\propto \omega^2$), whereas cross sections of the induced photo-transitions decrease with frequency (see Fridman & Kennedy, 2004). Examining the lasing based on the transitions of multi-charged ions, we can point out the powerful amplification of the Ne-like ion SeXXV (wavelengths 20.63 and 20.93 nm). The radiation energy in the pulse was about 3 mJ, and radiation power was several megawatts. The lasing has been achieved in the decaying plasma created by a powerful laser pulse. Radiation of this kind has also been observed from the transition of ions with charge 30 by excitation with radiation of a nuclear explosion. The contracting plasma of a capillary discharge has also been used for the excitation of lasers using transitions of multi-charged ions. Lasing has been observed in a relatively narrow range of plasma parameters: plasma density about 10^{19} cm^{-3} and electron temperature 70–90 eV. The duration of amplification in the system was about 1 ns, and the amplification coefficient was 0.6 cm^{-1}. The generation observed in the capillary discharge was on the transition of an Ne-like ion of argon ArIX ($\lambda = 46.9$ nm), and on the transition of another Ne-like ion of sulfur, SVII ($\lambda = 60.8$ nm). The maximum radiation energy in the system was 0.1 mJ at a power level of 80 kW. Energy accumulated in a capacitor was about 100 J, and the energy efficiency of the laser was relatively low ($\approx 10^{-4}$%).

11.6.8. Excimer Lasers

Chemical compounds stable in an electronically excited state and easily dissociating in a ground state are called "exciplex" (short for "excited complex"). An exciplex built from identical atoms or fragments is usually referred to as an excimer, which stands for "excited dimer." Sometimes the term eximer is generalized to describe any kind of dimers stable only in electronically excited states. The exciplex molecules used in lasers usually contain atoms of inert gases, which don't form stable compounds in a ground state (with rare exceptions like XeF_2). The excimer lasers radiate using photo-dissociative electronic transitions from a stable electronically excited state to a dissociating ground state. The spectral width of photo-dissociative transitions is large, which causes major difficulties with lasing in the systems. Electron-beam excitation of liquid xenon results in the amplification of a photo-dissociative transition, which demonstrates the excimer lasing effect. Electron beams also have been used for laser excitation in compressed inert gases. Effective lasing has been achieved this way for such dimers as $Xe_2^*(\lambda = 172$ nm$)$ $Kr_2^*(\lambda = 146$ nm$)$, and $Ar_2^*(\lambda = 126$ nm$)$. The most effective are eximer lasers on halides of inert gases: $KrF^*(\lambda = 250$ nm$)$, $XeF^*(\lambda = 350$ nm$)$, $XeCl^*(\lambda = 308$ nm$)$, and $ArF^*(\lambda = 193$ nm$)$.

Special interest in excimer lasers is due to the high energy density of their radiation (up to 40 kJ/m^3) at high lasing energy efficiency (up to 9–15%). Electron-beam excitation of the eximer lasers permits the operation of systems with large volumes of active medium (about 1 m^3). Excimer lasers are also especially interesting because of their ability to generate radiation over a wide range of wavelengths, from the visible to vacuum ultraviolet (VUV). Using electron-beam excitation, the highest energy efficiency of lasing has been achieved in KrF systems at high power density (2–7 MW/cm^3) of excitation of the high pressure (≈ 3 atm) gas mixture. From a liter of active medium, 40–50 J of laser radiation has been generated in the systems; the total energy reached 6 kJ. Excitation of the excimer Xe–Cl and Xe–F lasers has also been achieved using high-energy radiation from nuclear explosions. The excimer lasers using halides of inert gases are today the most powerful UV lasers. Much more information on the subject can be found in the book by Basting and Marowsky (2005).

11.6.9. Gas-Discharge Lasers Using Vibrational–Rotational Transitions: CO_2 Lasers

Lasing based on the vibrational–rotational transitions in the IR and submillimeter range has been achieved using gas-discharge excitation of different molecular gases, including CO_2, CO, N_2O, SO_2, CS_2, COS, NH_3, HCN, ICN, H_2O, and C_2H_2. An important advantage of such lasers is due to the fact that for many molecules the majority of the discharge energy can be transferred to molecular vibration (see Section 3.6.2), which leads to their high energy efficiency and high lasing power. The gas-discharge CO_2 laser is particularly advantageous from this point of view; its efficiency can exceed 10%, and its power in a continuous regime is the record high and can exceed 100 kW (McDaniel & Nighan, 1982). CO_2 lasers, both in continuous and pulsed regimes, are usually operated in a CO_2–N_2–He mixture (typical composition 1:1:8). Lasing occurs on the transition $00^01 \rightarrow 10^00$ from more energetic asymmetric vibrations of CO_2 to a lower-lying energy level corresponding to the deformational oscillations of CO_2 (Section 5.2.1). Nitrogen molecules (as well as CO_2 molecules) are very effectively vibrationally excited by direct electron impact in strongly non-equilibrium discharges. The vibration of N_2 molecules is close to resonance with the asymmetric oscillation mode of CO_2 and effectively transfers energy there through vibrational–vibrational (VV) exchange. The N_2 molecules actually accumulate energy and then transfer it to the lasing asymmetric oscillations of CO_2. At optimal plasma, 80% of the total discharge energy can be transferred specifically to vibration of N_2 and the asymmetric mode of CO_2 responsible for lasing, which results in the high energy efficiency and power of the CO_2 lasers. Atomic helium increases heat conductivity, decreases gas temperature, and deactivates the deformational CO_2 oscillations, which increases the lasing efficiency.

Conventional gas-discharge CO_2 lasers are performed in low-pressure (3–10 Torr) DC discharges (current density $j = 10$–$50 \, \text{mA/cm}^2$, tube diameter 1–4 cm) in CO_2:N_2:He = 1:1:8 mixtures. Typical values of the reduced electric field in the discharges are $E/n_0 = 20$–$80 \, \text{Td} \, (2$–$8 \cdot 10^{-16} \, \text{V} \cdot \text{cm}^2)$, which corresponds to electron temperatures of 1.8–2.2 eV. The gas temperature in optimal lasing conditions should not exceed about 600 K. Electron densities in CO_2 gas-discharge laser systems can be calculated using the semi-empirical formula proposed by Novgorodov, Sviridov, and Sobolev (1971), and Judd (1974):

$$n_e(\text{cm}^3) \approx 0.5\text{–}1 \cdot 10^9 \, j(\text{mA/cm}^2). \qquad (11\text{–}53)$$

Effective lasing in the CO_2–N_2–He mixture has also been achieved in strongly non-equilibrium pulsed discharges at high pressures, up to 100 atm (Basov, Belenov, & Danilychev, 1972). The energy efficiency of the lasers is very high – up to 30%. The specific energies of CO_2 laser radiation in pulses can be very high, reaching about 0.1 J/cm³. Taking into account the high level of vibrational excitation in the CO_2 laser mixtures, it is clear that plasma-chemical reactions (particularly CO_2 dissociation with formation of CO and O_2) take place in the gas discharges in parallel with the lasing. These processes have been discussed in detail in Chapter 5. The CO_2 dissociation degree in laser systems can reach 20–50%, strongly changing the characteristics and effectiveness of the lasers (Gasilevich, Ivanov, & Lotkova, 1969; Wiegang, Fowler, & Benda, 1970; Wiegand & Nighan, 1973). To avoid the intensive CO_2 dissociation through vibrational excitation of the molecule, in accordance with criteria discussed in Section 5.3.5, specific energy input E_v in the discharge system should be lower than the critical value of about 1 J per standard cm^3. It should be mentioned that, even below the E_v limit, CO_2 dissociation occurs by direct electron impact through electronic excitation and dissociative attachment (typical rate coefficient 0.6–$2.4 \cdot 10^{-10}$ cm^3/s), which restricts CO_2 residence time in the active discharge zone.

11.6.10. Gas-Discharge Lasers Using Vibrational–Rotational Transitions: CO Lasers

CO lasers operate in CO–He or CO–N_2–He mixtures using the vibrational–rotational transitions of the P-branch without absolute population inversion of vibrational states. Non-equilibrium vibrational distribution functions (Treanor distributions) with overpopulation of highly excited vibrational states are effectively organized in strongly non-equilibrium gas discharges ($T_e \gg T_v \gg T_0$) and have been discussed in Sections 3.1.8, 3.4.3, and 3.4.4. Actually the Treanor effect in CO is the major kinetic mechanism of vibrational population inversion and lasing in the CO–He and CO–N_2–He mixtures. Generation occurs in the lasers usually from several vibrational transitions, $v \rightarrow v - 1 (v = 6$–$15)$, with wavelength in the range $\lambda = 5$–$5.8 \,\mu m$. Continuous CO lasers usually work in low-pressure strongly non-equilibrium discharges in mixtures of CO:He = 1:10/1:30 with parameters close to those of CO_2 lasers. Typical gas discharge parameters are pressure, 3–20 Torr; current density, $(2$–$10) \cdot 10^9$ cm^{-3}; internal diameter of a long discharge tube, 1–4 cm; and discharge power, up to 500 W/m. The powers of continuous CO lasers are very high and only slightly below those of the CO_2 lasers. The energy efficiencies of CO lasers can be even higher than those of the CO_2 lasers.

Similar to the case of CO_2 lasers, high levels of vibrational excitation in CO laser mixtures can lead to plasma-chemical processes with formation of O, CN, CO_2, C_3O_2, and so forth occurring in parallel with lasing and strongly changing the characteristics and effectiveness of the lasers (Taieb & Legay, 1970; Mikaberidze, Ochkin, & Sobolev, 1972). To avoid the intensive CO conversion stimulated by vibrational excitation of the molecules (see Section 5.7.2), the specific energy input E_v in the discharge should be limited by the critical value of about 1 J per standard cm^3. Below the E_v limit, plasma-chemical process in the CO laser mixture are mostly due to direct electron impact processes and can be partially suppressed by restriction of CO residence time in the active discharge zone as well as by addition to the laser mixture of about 7% molecular oxygen. The major factor that limits characteristics of the gas-discharge CO lasers is gas heating, which decreases the level of overpopulation of highly excited vibrational states (Treanor effect; see Sections 3.1.8, 3.4.3, and 3.4.4). High-pressure pulsed discharges in a CO laser mixture, limiting values of the specific energy input, have some advantages from this point of view (Lacina & Mann, 1972; Mann, Rice, & Eguchi, 1974). The analysis of such system shows possibilities of reaching lasing pulses with specific energies of about 1 J/cm^3atm and energy efficiencies of about 60% (Basov et al., 1977a).

11.6.11. Plasma Stimulation of Chemical Lasers

Population inversion in chemical lasers is due to fast exothermic chemical processes releasing energy into internal degrees of freedom, usually in molecular vibration. For effective lasing, the rate of the chemical reactions should exceed that of the corresponding relaxation processes. The most relevant reactions for this type of lasers are those leading to the formation of hydrogen halides, because they release 30–70% of the energy into vibrational energy of the products and create absolute inversion in the population of several vibrational levels. The first chemical laser was organized in an H_2–Cl_2 mixture (Kasper & Pimentel, 1965), where the chain reaction generating vibrationally excited HCl* molecules can be initiated using pulsed discharges or pulsed photolysis in the following sequence:

$$Cl_2 + e, \hbar\omega \rightarrow Cl + Cl, \tag{11-54}$$

$$Cl + H_2 \rightarrow HCl + H, \tag{11-55}$$

$$H + Cl_2 \rightarrow HCl^* + Cl. \tag{11-56}$$

Most chemical lasers are organized in pulse regimes, although some of those are effective in continuous operation. Some major initial mixtures and exothermic chemical reactions that provide effective lasing are summarized in Table 11–2 (Oraevsky, 1974; Gordiets, Osipov, & Shelepin, 1980). The most known of the chemical lasers is the HF laser. The lasers presented in Table 11–2 have many common features. They are usually organized in a preliminary prepared mixture with total pressure in the range 1–100 Torr (in some cases at much higher pressures). External energy sources, such as pulsed discharges, electron beams, or energetic photon sources, create active chemical species, which are able then to react with a laser-mixture component, producing a vibrationally excited molecular product. Inversion in the vibrational population of the molecules results in lasing in the chemical laser systems. To provide not pulsed (which is common) but continuous generation in chemical lasers, fast mixing of the chemical reagents and fast exchange of products are required. The continuous regimes of chemical lasers, therefore, require fast flow velocities (10^4–10^5 cm/s) across the generation zone, gas discharge, electron beam, and so on (Gordiets, Osipov, & Shelepin, 1980).

11.6.12. Energy Efficiency of Chemical Lasers: Chemical Lasers with Excitation Transfer

Effective lasing in chemical lasers require sustaining the population inversion of vibrational levels of products provided by selective energy release in the exothermic reactions. Unfortunately, relatively fast vibrational relaxation of the molecules/products of the exothermic reactions (through VT- and VV-relaxation processes; see Section 2.6) results in the suppression of inversion during the initial stages of the exothermic reaction when only a small fraction of the reagents is consumed. The effect is especially severe in the case of lasing of the diatomic molecule/products of the exothermic reactions. The suppressing of inversion terminates the lasing, leading to low utilization efficiency of energy accumulated in chemical reagents, and finally to low energy efficiency of the chemical lasers. As an example, we can mention the characteristics of a pulsed chemical laser in mixture $H_2:F_2:MoF_6:He = 1:1:0.25:40$, operating at a total pressure of 600 Torr. The laser has quite high exit parameters: a pulse power of 12 kW and pulse energy of 0.07 J. Energy efficiency of the laser, however, is very low – only 0.2% – due to the aforementioned relaxation effect (Hess, 1972). This negative effect can be avoided if generation is achieved on vibrational transitions of polyatomic molecule/products, or if vibrational energy of the diatomic molecule/products is transferred to the especially admixed polyatomic molecules, which are then lasing on transitions between different vibrational modes (Dzidzoev, Platonenko, & Khokhlov, 1970; Gordiets, Osipov, & Shelepin, 1980). Such lasers are sometimes referred to as chemical lasers with excitation transfer. Usually, CO_2 molecules are used as the accumulators of vibrational energy in chemical lasers with excitation transfer. Vibrational energy is transferred in the chemical lasers with excitation transfer from vibrationally excited diatomic products to the CO_2 molecules, which then generate on the vibrational intermode transition $00^01 \rightarrow 10^00(\lambda = 10.6\,\mu m)$. Chemical reagents and the processes of chemical excitation in this type of chemical laser are shown in Table 11–3 (Oraevsky, 1974; Gordiets, Osipov, & Shelepin, 1980). Between the chemical lasers with excitation transfer, the O_2^*–I laser should

Table 11–2. Initial Mixtures and Exothermic Chemical Reactions Providing Effective Generation in Chemical Lasers

Initial mixture	Chemical reaction of the chemical laser excitation	Lasing wavelength, μm
$H_2 + Cl_2$	$H + Cl_2 \rightarrow HCl^* + Cl$	About 3.7
$D_2 + Cl_2$	$D + Cl_2 \rightarrow DCl^* + Cl$	About 5.1
$HD + Cl_2$	$H + Cl_2 \rightarrow HCl^* + Cl$	About 3.7
	$D + Cl_2 \rightarrow DCl^* + Cl$	About 5.1
$H_2 + Br_2$	$H + Br_2 \rightarrow HBr^* + Br$	Rotational transitions
$D_2 + Br_2$	$D + Br_2 \rightarrow DBr^* + Br$	Rotational transitions
$CF_4 + H_2$	$F + H_2 \rightarrow HF^* + H$	About 2.7
$CF_4 + D_2$	$F + D_2 \rightarrow DF^* + D$	About 4.2
$CF_4 + CH_3Cl$	$F + CH_3Cl \rightarrow HF^* + CH_2Cl$	About 2.7
$NOCl + H_2$	$H + NOCl \rightarrow HCl^* + NO$	About 3.7
$CS_2 + O_2$	$CS + O \rightarrow CO* + S$	5.1–5.7
$CS_2 + SO_2$	$CS + O \rightarrow CO* + S$	5.1–5.7
$NF_2H + H_2$	$F + RH \rightarrow HF^* + R$	About 2.7
$NF_2H + RH$	$F + RH \rightarrow HF^* + R$	About 2.7
$D_2 + F_2$	Chain reaction:	About 4.2
	$F + D_2 \rightarrow DF^* + D$	About 4.2
	$D + F_2 \rightarrow DF^* + F$	
$D_2 + NF_3$	$F + D_2 \rightarrow DF^* + D$	About 4.2
$D_2 + N_2F_4$	$F + D_2 \rightarrow DF^* + D$	About 4.2
$D_2 + FNO_2$	$F + D_2 \rightarrow DF^* + D$	About 4.2
$MoF_6 + H_2$	$F + H_2 \rightarrow HF^* + H$	About 2.7
$OF_2 + H_2$	$F + H_2 \rightarrow HF^* + H$	About 2.7
$Cl_2 + HI$	$H + Cl_2 \rightarrow HCl^* + Cl$	About 3.7
$Cl_2 + HBr$	$Cl + HBr \rightarrow HCl^* + Br$	About 3.7
$H_2 + SbF_6$	$F + H_2 \rightarrow HF^* + H$	About 2.7
$H_2 + XeF_4$	$F + H_2 \rightarrow HF^* + H$	About 2.7
$CH_4 + XeF_4$	$F + H_2 \rightarrow HF^* + H$	About 2.7
$CH_4 + XeF_6$	$F + H_2 \rightarrow HF^* + H$	About 2.7
$H_2 + UF_6$	$F + H_2 \rightarrow HF^* + H$	About 2.7
$CH_4 + UF_6$	$F + CH_4 \rightarrow HF^* + CH_3$	About 2.7
$C_2H_6 + UF_6$	$F + C_2H_6 \rightarrow HF^* + C_2H_5$	About 2.7
$C_3H_8 + UF_6$	$F + C_3H_8 \rightarrow HF^* + C_3H_7$	About 2.7
$CCl_3H + UF_6$	$F + CCl_3H \rightarrow HF^* + CCl_3$	About 2.7
$CF_3H + UF_6$	$F + CF_3H \rightarrow HF^* + CF_3$	About 2.7
$H_2 + SF_6$	$F + H_2 \rightarrow HF^* + H$	About 2.7
$HBr + SF_6$	$F + HBr \rightarrow HF^* + Br$	About 2.7
$H_2 + F_2$	Chain reaction:	2.7–3.2
	$F + H_2 \rightarrow HF^* + H$	
	$H + F_2 \rightarrow HF^* + F$	

be especially pointed out because it is the first chemical laser generating on the electronic transitions. The laser is based on the excitation of a metastable electronically excited state of oxygen, $O_2(^1\Delta_g)$, which can be generated in particular in the exothermic chemical reaction between Cl_2 and H_2O_2 in alkaline solution. The concentration of singlet oxygen $O_2(^1\Delta_g)$ in O_2 discharges can exceed 5–10%. When the ratio $[O_2(^1\Delta_g)]/[O_2(^3\Sigma_g^-)]$ is large enough, the singlet oxygen molecules $O_2(^1\Delta_g)$ can effectively transfer excitation to iodine atoms. The excitation transfer leads to absolute inversion on the electronic transition in iodine $I(^2P_{1/2} \rightarrow {}^2P_{2/3})$ and lasing with wavelength $\lambda = 1.315\ \mu$m (McDermott et al., 1978).

Table 11–3. Reagents and Major Reactions in the Chemical Lasers with Excitation Transition

Initial reagents	Excitation transfer	Laser excitation processes
$HN_3 + CO_2 + He$	$N_2^* \rightarrow CO_2$	$NH_3 \rightarrow N_2^* + H_2$
		$N_2^* + CO_2 \rightarrow N_2 + CO_2^*$
$HI + Cl_2 + CO_2$	$HCl^* \rightarrow CO_2$	$Cl + HI \rightarrow HCl^* + I$
		$HCl^* + CO_2 \rightarrow HCl + CO_2^*$
$D_2 + F_2 + CO_2$	$DF^* \rightarrow CO_2$	$F + D_2 \rightarrow DF^* + D$
		$DF^* + CO_2 \rightarrow DF + CO_2^*$
$D_2 + F_2 + CO_2 + He + MoF_6$	$DF^* \rightarrow CO_2$	$F + D_2 \rightarrow DF^* + D$
		$D + F_2 \rightarrow DF^* + F$
		$DF^* + CO_2 \rightarrow DF + CO_2^*$
$D_2 + O_3 + CO_2$	$OD^* \rightarrow CO_2$	$O(^1D) + D_2 \rightarrow OD^* + D$
		$D + O_3 \rightarrow OD^* + O_2$
		$OD^* + CO_2 \rightarrow OD + CO_2^*$

11.6.13. Plasma Sources of Radiation with High Spectral Brightness

High spectral brightness radiation is usually defined as radiation characterized by brightness exceeding that of an absolute blackbody with temperature 20,000 K. An absolute blackbody with temperature 20,000 K generates about 85% radiation in the UV and VUV ranges. Thus, radiation sources of high spectral brightness are supposed to radiate in the UV and VUV ranges, usually with high power ($\geq 1 \, MW/cm^2$). The radiation sources of high spectral brightness are usually divided into the following four groups.

1. **Laser sources** are based on the stimulated radiation mechanism (Fridman & Kennedy, 2004) and generate monochromatic radiation with low angular divergence, which results in extremely high brightness corresponding to the absolute blackbody temperatures exceeding 10^6 K. The laser light sources were discussed earlier in Section 11.6.
2. **Luminescent sources** are based on the non-equilibrium spontaneous radiation of atoms, ions, and molecules (Fridman & Kennedy, 2004), which are excited by strongly non-equilibrium discharges, electron beams, and soforth (as an example, we can point out excimer VUV lamps on the dimers of inert gases excited by relativistic electron beams).
3. **Synchrotron (ondulator) sources** are based on the bremsstrahlung radiation of the relativistic electrons in continuous magnetic fields (synchrotron sources) and space-time-modulated magnetic fields (ondulator sources).
4. **Thermal plasma sources** are based on the radiation of quasi-equilibrium plasma (see Fridman & Kennedy, 2004) heated to very high temperatures. Taking into account their wide area of application, let's consider these high-spectral-brightness radiation sources in more detail.

Thermal plasma light sources are effective radiation devices in generating short-wavelength (UV and VUV) radiation (Protasov, 2000). The emission spectrum of the absolute blackbody with temperature T has a radiation maximum at the frequency $\hbar\omega_{max} \approx 2.8 \cdot T$. Thermal plasma radiation cannot exceed that of the absolute blackbody at the plasma temperature. Therefore, fixing the required spectral emission interval determines the required values of the thermal plasma temperature. Specifically, for effective emission in UV and VUV spectral ranges, the thermal plasma temperature should be in the interval from 2 to 15 eV. Plasma in high-brightness light sources is usually characterized by volume from tens to hundreds of cubic centimeters, and by high electron density (10^{18}–10^{20} cm^{-3}). The following three methods are applied for generation and heating of thermal plasma for high-brightness radiation sources:

1. **Joule (ohmic) heating** is provided by electric current of high amplitude (10^4–$10^5 A$) and high density (10^4–10^6 A/cm^2). The method of plasma heating is the most widely used in high-brightness radiation sources. To reach the brightness temperature in the range 20,000–40,000 K, the power of thermal discharges with Joule heating is very high (3–45 MW/cm or 4–60 MW/cm^3).

2. **Plasma-dynamic heating** is based on the thermalization of kinetic energy of a fast-moving gas or plasma flow. A major advantage of this method is energy transfer from heavy plasma particles to heavy plasma particles, which is more effective at high temperatures than electron–ion energy transfer in the case of the Joule heating.

3. **Radiative heating** is provided by the interaction of highly concentrated radiation fluxes with materials. Plasma, generated as a result of such interaction, can be in some conditions a powerful source of thermal radiation in the range of VUV and X-rays. The method is very effective, in particular, when power laser radiation is focused on a target. The conversion of powerful coherent laser radiation into high-temperature thermal radiation can reach 50–70% in such systems. More details on the subject can be found in books and reviews by Protasov (2000), Alexandrov and Rukhadze (1976), and Rokhlin (1991).

11.6.14. Mercury-Containing and Mercury-Free Plasma Lamps

Plasma is widely used today in both mercury-containing and mercury-free electric lamps. The search for efficient and environmentally friendly substitutes for mercury today is a key issue in many plasma lighting technologies such as automotive lamps and fluorescent lighting systems. As an example, the use of Ne–Xe-based low-pressure discharges is an interesting solution for the excitation of phosphors (Capdeville et al., 1999; Sarroukh et al., 2001). The development of a pulse power supply, generating high-repetition current pulses in the microsecond time range, allows a three- to four-fold enhancement of both the luminous flux and the efficiency of mercury-free plasma lamps to be achieved (Robert et al., 2005). Thus, the development of mercury-free plasma lamps is quite successful. Overwhelming environmental effects from the use of electric lighting are from the power generation over the life of the lamps. These effects depend on the power source and include atmospheric emissions of mercury, carbon dioxide, nitrous oxide and sulfur dioxide, noise, and radioactive waste. Efficient lighting systems that use mercury-containing lamps will typically reduce these effects by at least 75% because of their lower power requirements. Power generation effects dwarf any environmental emissions which might occur during the manufacture or proper disposal of mercury-containing lamps. Generally, mercury-free electric lamps cannot be substituted for mercury-containing lamps because of incompatibilities of light output, shape, color, life, electrical characteristics, and excessive heat, or because their increased energy consumption may violate energy codes and overload electrical circuits. Let us briefly discuss mercury-containing and mercury-free plasma lighting in relation to specific types of these widely used lamps.

1. Fluorescent lamps. No viable replacement has yet been discovered for mercury in general-purpose fluorescent lamps. Mercury-free xenon-based fluorescent discharges are available in a flat-panel format, suitable for back lighting of liquid crystal displays, but the efficiency is approximately 30% of a normal mercury-based fluorescent lamp; therefore, this technology is environmentally counterproductive for general lighting applications.

2. High-intensity discharge (HID) lamps. HID lamps are used in street lighting, flood-lighting, and industrial and some commercial applications. There are better prospects for mercury-free HID lamps. Mercury-free high-pressure sodium lamps are available up to powers of 150 watts, with some higher wattages under development. This has been achieved by reengineering the arc tube geometry and fill pressure. The lamps will retrofit into existing sockets and operate on existing ballasts. Metal halide lamps without mercury present a greater challenge. These lamps may not be a "screw-in replacement" for existing types. Maintaining a stable discharge is challenging and the availability of mercury-free metal halide products is still several years away. The high-pressure sulfur lamp is fundamentally mercury-free but is unstable and requires forced cooling. Lamps which have been marketed so far are high-wattage (at least 1 kW), and they require coupling to a lighting distribution system such as a light pipe. The overall system efficiency is lower than an equivalent fluorescent or HID system, especially if the greenish color is corrected by means of a filter.

3. Low-pressure sodium lamps. These types of lamps are mercury-free plasma light sources, characterized by orange appearance and very poor color. Although very efficient in photometric terms, their visual efficiency in typical outdoor (street lighting) applications is less than that of other lamps. All colors are rendered in shades of brown or gray, making recognition of people and vehicles very difficult. The lamp contains sodium in sufficient quantities to fail tests for reactivity and ignitability.

4. Light-emitting diodes (LEDs). LEDs are mercury-free plasma light sources that have long lifetimes. They are very bright and lend themselves to use in signs, traffic signals, and some accent and display lighting. At present, their overall light output is much less than that of fluorescent or HID lamps, despite their brightness. They are not currently suitable for most general illumination purposes, but this is a new technology that is evolving very rapidly. More details on the subject can be found in Manheimer, Sugiyama, and Stix (1996), Robert et al. (2005a,b), and Protasov (2000).

11.6.15. Plasma Display Panels and Plasma TV

Plasma TV is not the largest plasma application today, but surely the most publicly known. The flat plasma display is a major competitor among several flat-panel display technologies, competing for the potentially enormous high-definition television market. The strongly non-equilibrium plasma in a display panel is much like the plasma in fluorescent lamps (see previous section). To be more exact, this plasma is generated by tiny surface DBDs, (see Section 4.5.3). A plasma display panel (PDP) is essentially a collection of numerous very small DBD lamps, each a few tenths of a millimeter in size. If we look closely, it is easy to distinguish the individual PDP cells – the tiny color elements of red, green, and blue light that together form what is called a pixel. The DBD arranged on the dielectric surface in a PDP cell is illustrated in Fig. 11–55. Similarly to fluorescent lamps, the light we see from the PDP does not come directly from the plasma, but rather from the phosphor coatings on the inside walls of the cells when they are exposed to UV radiation emitted by the plasma. Because each cell emits its own light, a PDP is called an "emissive display." This contrasts with the familiar liquid crystal display (LCD), a type of flat display in which the light comes from a plasma lamp behind the liquid crystal, which has arrays of small switches controlling where light is allowed to pass through. It is interesting that LCD TV can also be called plasma TV according to the type of light source used in that system. Thus, all modern TVs are "plasma" based. The DBD surface discharge, illustrated in Fig. 11–55, operates in xenon mixed with a buffer gas, which consists usually of neon or helium to optimize UV emission. The operating conditions of the display (gas composition, pressure, voltage,

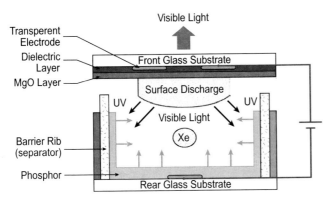

Figure 11–55. Plasma cell cross section of a typical plasma display panel.

geometry, etc.) represent a compromise, taking into account performance requirements such as low-voltage operation, long life, high brightness, and high contrast.

The plasma display consists of two parallel glass plates separated by a precise spacing of some tenths of a millimeter and sealed around the edges. The space between the plates is filled with a mixture of rare gases at a pressure somewhat less than one atmosphere. Parallel stripes of transparent conducting material (electrodes) with a width of about a tenth of a millimeter are deposited on each plate, with the stripes on one plate perpendicular to those on the other. Thus, the stripes operate as "electrodes" to which voltage is applied. The intersections of the rows of electrodes on one side and the columns of electrodes on the opposite glass plate define the individual color elements – or cells – of a PDP. For high-quality color images it is important to keep the UV radiation from passing between cells. To isolate the individual cells, barriers are created on the inside surface of one of the plates before sealing. Honeycomb-like structures and other shapes have been used. The red, green, and blue phosphors are deposited inside these structures. The plasma in each individual cell can be turned on and off rapidly enough to produce a high-quality moving picture. To help turn the individual cells on and off, there are two electrodes on one side and a third electrode on the opposite side of each cell. PDPs in modern plasma TVs consists of several million DBD cells which are switched at a rate that creates 60 TV picture frames per second. PDPs are manufactured with diagonals as large as 80 inches and with a thickness of 3–4 inches. More information can be found in Meunier, Belenguer, and Boeuf (1995), Boeuf (2003), and Boeuf et al. (1997).

11.7. NON-THERMAL PLASMA IN ENVIRONMENTAL CONTROL: CLEANING EXHAUST GAS OF SO₂ AND NOₓ

11.7.1. Industrial SO₂ Emissions and Plasma Effectiveness of Cleaning Them

Industrial SO_2 emissions cause acid rain and result in serious environmental disasters. The world's most severe acid rain regions today are Northern America, Europe, and China. The major sources of SO_2 emissions are coal-burning power plants (for example, in China, they produce about 26 million tons of the hazardous SO_2, which is almost 40% of the total SO_2 emissions), steelworks, non-ferrous metallurgical plants, as well as oil refineries and natural gas purification plants. The SO_2 emissions in air are usually high-volume and low-concentration: the SO_2 fraction is usually on the level of hundreds of ppm, while total flow of polluted air in one system can reach a million cubic meters per hour (Pu & Woskov, 1996). Oxidation of SO_2 in air to SO_3 results in the rapid formation sulfuric acid (H_2SO_4). The kinetics of this process is limited by very low rates of natural oxidation of SO_2 in low-temperature air conditions. Therefore, the formation of SO_3 and sulfuric acid from the exhausted SO_2 occurs not immediately in the stacks of the industrial or power plants (burning sulfur-containing fuel) but later on in the clouds, which ultimately results in acid rain. Non-thermal plasma can be used to stimulate oxidation of SO_2 into SO_3 inside the stack. This permits collection of the sulfur oxides in the form of sulfates, for example, in the form of a fertilizer, $(NH_4)_2SO_4$, if ammonia (NH_3) is admixed to the plasma-assisted oxidation products (Pu & Woskov, 1996). An important advantage of the non-thermal plasma method is the possibility of simultaneous exhaust gas cleaning from SO_2 and NO_x. Taking into account the very large exhaust gas volumes to be cleaned from SO_2, the non-thermal atmospheric-pressure plasma systems mostly applied for this purpose are based on electron beams and pulsed corona discharges (Tokunaga & Suzuki, 1981; Leonhardt, 1984; Svedchikov & Belousova, 1988; Baranchikov et al., 1990a,b, 1992; Chae, Desiaterik, & Amirov, 1996; Jiandong et al., 1996; Mattachini, Sani, & Trebbi, 1996). The crucial question to be addressed is the energy cost of the plasma cleaning process, which strongly depends on the treatment dose rate (current density in the case of electron beams), air humidity, temperature, and other factors. As mentioned, non-thermal plasma just stimulates exothermic oxidation of SO_2 to SO_3, which then is removed chemically. Nevertheless, the process can be quite energy expensive. At low current densities ($j < 10^{-5}$ A/cm^2) conventional for many experimental systems, the energy cost of the desulfurization process is usually high, on the level of 10 eV per SO_2 molecule. In this case, cleaning exhaust of a 300 MW power plant containing 0.1% (1000 ppm) of SO_2 requires about 12 MW of plasma power. Even when such an energy cost of plasma exhaust cleaning is acceptable, the use of such powerful electron beams or pulsed coronas is questionable. Decreasing the energy cost requires the use of very specific plasma parameters, particularly dose rates and current densities in the case of electron beams (Baranchikov et al., 1990a,b, 1992), which will be discussed next.

11.7.2. Plasma-Chemical SO_2 Oxidation to SO_3 in Air and Exhaust Gas Cleaning Using Relativistic Electron Beams

The application of relativistic electron beams (electron energies $\geq$ 300–500 keV) is attractive for exhaust gas cleaning in large ducts, because the propagation length of high-energy electrons in atmospheric air can be estimated as 1 m per 1 MeV (Fridman & Kennedy, 2004). Also the high-energy electrons are characterized by the lowest energy cost of formation of charged particles (about 30 eV per an electron–ion pair), which permits a decreased energy cost of SO_2 oxidation in air stimulated by the chain ion–molecular reactions (see Sections 11.7.4 and 11.7.5). Electron-beam-generated plasmas provide effective SO_2 oxidation to SO_3, which is removed afterward using conventional chemical methods (in particular, admixture of NH_3 and filtration of $(NH_4)_2SO_4$). As an example, electron-beam experiments of Baranchikov et al. (1990a,b, 1992) were carried out in an SO_2–air mixture with molar composition: SO_2 1%, H_2O 3.4%, O_2 19.6%, and N_2 76% at a total pressure of 740 Torr and temperature 298 K. A cylindrical reactor in the experiments had volume 800 mL and diameter 80 mm. Experiments were carried out with pulse-periodic beams and single-pulse beams. In the case of pulse-periodic beams (pulse duration 100 ns, repetition frequency 0.1 kHz) with electron energy 300 keV, current densities varied from 1 to 10 A/cm^2. In the case of single-pulse beams, the current density varied from 10 mA/cm^2 to 10 A/cm^2, corresponding to the specific energy input in gas in the range 10^{-4}–$3 \cdot 10^{-2}$ J/cm^3. The energy cost of SO_2 oxidation into SO_3 in the electron-beam experiments of Baranchikov et al. (1990a,b, 1992) is presented in Fig. 11–56 as a function of the electron-beam current density. In the high-current density range ($j_{eb} \geq 0.1$ A/cm^2), the energy cost of SO_2 oxidation is proportional to the square root of the relativistic electron-beam current density:

$$A_{SO_2} \propto \sqrt{j_{eb}}. \tag{11–57}$$

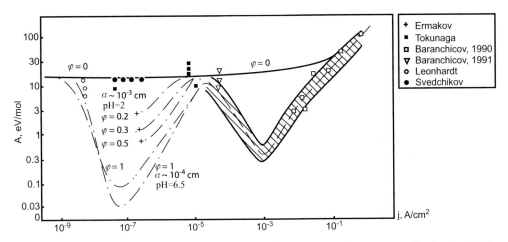

Figure 11–56. Energy cost of SO_2 oxidation as a function of electron beam current density. Dashed line shows energy cost decrease by the cluster effect; dotted line shows energy cost decrease by the effect of droplets.

The minimum achieved value of the energy cost, A_{SO_2}, without taking into account oxidation in droplets is 2–3 eV/mol SO_2. Including oxidation in the droplets the minimum achieved oxidation energy cost is 0.8 eV/mol SO_2. The low oxidation energy costs are due to plasma stimulation of chain oxidation mechanisms at relatively high values of electron-beam current density. Electron beams with lower current densities ($\leq 10^{-4}$ A/cm^2) are unable to initiate chain oxidation, which results in an oxidation energy cost of about 10 eV/mol; see Fig. 11–56 (Tokunaga & Suzuki, 1981; Washino, Tokunaga, Hashimoto, 1984; Leonhardt, 1984; Svedchikov & Belousova, 1988).

11.7.3. SO_2 Oxidation in Air to SO_3 Using Continuous and Pulsed Corona Discharges

Corona discharges, and especially pulsed corona discharges (see Sections 4.5.1, and 4.5.2), are also effective in SO_2 oxidation to SO_3 and therefore in the exhaust gas cleaning of sulfur oxides (Baranchikov et al., 1992; Chae, Desiaterik, & Amirov, 1996; Jiandong et al., 1996; Mattachini, Sani, & Trebbi, 1996). Baranchikov et al. (1992) investigated SO_2 oxidation in air using both continuous and pulsed corona discharges in a system similar to the one described in the previous section. The continuous corona has been generated by applying a constant 20 kV voltage to the central hexahedron electrode. The pulsed corona has been sustained by negative or positive voltage pulses (30–70 kV) with duration 50–200 ns and repetition frequency 50 Hz. The energy cost of the SO_2 oxidation process in continuous and pulsed corona discharges (Baranchikov et al., 1992) is also presented in Fig. 11–56 as a function of current density (dose rate). The minimum energy cost for the pulsed coronas is 3–5 eV/mol SO_2, and 10 eV/mol SO_2 for the continuous coronas. Thus, the minimum SO_2 oxidation energy costs are close for the pulsed corona discharges and electron beams. The SO_2 oxidation energy cost dependence on the current density (dose rate) has similar behavior for electron beams and continuous and pulsed corona discharges. At low current densities ($\leq 10^{-4}$ A/cm^2), the energy cost of SO_2 oxidation to SO_3 remains almost constant at the level of 10 eV/mol SO_2, whereas at high current densities (0.1 A/cm^2 and higher), the energy cost follows equation (11–57) with an expected minimum at 10–100 mA/cm^2. The low values of SO_2 oxidation energy cost at current densities in the range 10–100 mA/cm^2 are explained by plasma stimulation of the chain mechanism of SO_2 oxidation, which is to be discussed next. There is no chain mechanism for SO_2 oxidation to SO_3 in air in conventional gas-phase chemistry, which results in the aforementioned relatively high energy cost of the oxidation. Even ozone is not effective in gas-phase SO_2 oxidation. The chain oxidation of SO_2 is possible, however, in liquid phase, and even in small clusters, which can be stimulated in non-thermal plasma.

11.7.4. Plasma-Stimulated Liquid-Phase Chain Oxidation of SO_2 in Droplets

Atomic oxygen and other reactive oxygen species generated in non-thermal air plasma with an energy cost of about 10 eV per particle are able to oxidize SO_2 to SO_3 with the same energy cost, which probably explains experiments with low current densities (see Fig. 11–56). The SO_2 oxidation energy costs of 1–3 eV/mol, observed in plasma at higher current densities and discussed earlier, can be explained only by the plasma stimulation of chain oxidation processes. There is no chain mechanism of SO_2 oxidation to SO_3 in conventional gas-phase chemistry (without a catalyst). Plasma is able to stimulate chain oxidation of SO_2 either in liquid droplets or in ionic clusters generated in non-thermal discharges in humid air, which is to be discussed here and in Section 11.7.5. Plasma-stimulated liquid-phase chain oxidation, of SO_2 into SO_3 occurs in droplets formed in non-thermal discharges by water condensation around H_2SO_4, ions, and other active species. When the cold plasma is generated in atmospheric air containing sulfur compounds, the active species and especially sulfuric acid generated in the system immediately lead to water condensation and formation of mist. The presence of water droplets permits organization of the further chain process of SO_2 oxidation to SO_3 and sulfuric acid in liquid phase. Liquid-phase chain oxidation of SO_2 into SO_3 and sulfuric acid has been investigated by several groups including Calvet (1985), Penkett and Jones (1979), Daniel and Jacob (1986), and Deminsky et al. (1990). The oxidation mechanism in droplets depends on their acidity, which can be presented through the following kinetic schemes:

1. **Acidic Droplets ($pH < 6.5$).** The major ion formed in this case in water solution by gaseous SO_2 is the HSO_3^- ion:

$$SO_2(g) + H_2O \rightarrow HSO_3^- + H^+. \tag{11–58}$$

The product of oxidation in this case is the sulfuric acid ion, HSO_4^-. HSO_3^- oxidation in the acidic solution into HSO_4^- is a chain process starting from the chain initiation through the HSO_3^- decomposition in collision with an active particle M (in particular an OH radical):

$$M + HSO_3^- \rightarrow MH + SO_3^-. \tag{11–59}$$

Chain propagation in the acidic solution starts with attachment of oxygen and proceeds through active sulfur ion radicals SO_3^-, SO_4^-, and SO_5^-:

$$SO_3^- + O_2 \rightarrow SO_5^-, \quad k = 2.5 \cdot 10^{-12}\, cm^3/s, \tag{11–60}$$

$$SO_5^- + HSO_3^- \rightarrow SO_4^- + HSO_4^-, \quad k = 1.7 \cdot 10^{-16}\, cm^3/s, \tag{11–61}$$

$$SO_5^- + SO_5^- \rightarrow 2SO_4^- + O_2, \quad k = 10^{-12}\, cm^3/s, \tag{11–62}$$

$$SO_4^- + HSO_3^- \rightarrow HSO_4^- + SO_3^-, \quad k = 3.3 \cdot 10^{-12}\, cm^3/s, \tag{11–63}$$

$$SO_5^- + HSO_3^- \rightarrow SO_3^- + HSO_5^-, \quad k = 4.2 \cdot 10^{-17}\, cm^3/s, \tag{11–64}$$

$$HSO_5^- + HSO_3^- \rightarrow 2SO_4^{2-} + 2H^+, \quad k = 2.0 \cdot 10^{-14}\, cm^3/s. \tag{11–65}$$

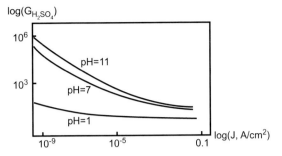

Figure 11–57. Dependence of the radiation yield of SO$_2$ liquid-phase oxidation on the current density and acidity.

Termination of chain SO$_2$ oxidation is due to recombination and destruction of the active sulfur ion radicals SO$_3^-$, SO$_4^-$, and SO$_5^-$:

$$SO_3^- + SO_3^- \rightarrow S_2O_6^{2-}, \quad k = 1.7 \cdot 10^{-12}\, cm^3/s, \tag{11–66}$$

$$SO_5^- + SO_5^- \rightarrow S_2O_8^{2-} + O_2, \quad k = 3.3 \cdot 10^{-13}\, cm^3/s, \tag{11–67}$$

$$SO_3^-, SO_4^-, SO_5^- + admixture \rightarrow destruction. \tag{11–68}$$

2. Neutral and Basic Droplets (pH > 6.5). The major ion formed in this case in water solution by gaseous SO$_2$ is the SO$_3^{2-}$ ion:

$$SO_2(g) + H_2O \rightarrow SO_3^{2-} + 2H^+. \tag{11–69}$$

The product of SO$_2$ oxidation in this case is the sulfuric acid ion, SO$_4^{2-}$. SO$_3^{2-}$ oxidation in neutral and basic solutions into the sulfuric acid ion (SO$_4^{2-}$) is also a chain process. It starts from the chain initiation (through the SO$_3^{2-}$ decomposition in collision with an active particle M, in particular an OH radical) and formation of an active SO$_3^-$ radical similarly to (11–59):

$$M + SO_3^{2-} \rightarrow MH^- + SO_3^-. \tag{11–70}$$

Chain propagation in the neutral and basic solutions also starts with attachment of oxygen (11–60), and after that (similarly to acidic solutions) also proceeds through the active sulfur ion radicals SO$_3^-$, SO$_4^-$, and SO$_5^-$:

$$SO_3^- + O_2 \rightarrow SO_5^-, \quad k = 2.5 \cdot 10^{-12}\, cm^3/s, \tag{11–71}$$

$$SO_5^- + SO_3^{2-} \rightarrow SO_4^{2-} + SO_4^-, \quad k = 5 \cdot 10^{-14}\, cm^3/s, \tag{11–72}$$

$$SO_4^- + SO_3^{2-} \rightarrow SO_4^{2-} + SO_3^-, \quad k = 3.3 \cdot 10^{-12}\, cm^3/s, \tag{11–73}$$

$$SO_5^- + SO_3^{2-} \rightarrow SO_5^{2-} + SO_3^-, \quad k = 1.7 \cdot 10^{-14}\, cm^3/s, \tag{11–74}$$

$$SO_5^{2-} + SO_3^{2-} \rightarrow 2SO_4^{2-}, \quad k = 2 \cdot 10^{-14}\, cm^3/s. \tag{11–75}$$

The chain termination reactions in the case of neutral and basic droplets are mostly the same as in the case of acidic droplets (11–66)–(11–68). The SO$_3^-$ ion radical dominates the chain propagation in cases of both low and high acidity. The SO$_3^-$ ion radicals can be significantly replaced by SO$_5^-$ ion radicals in propagation of the oxidation chain reaction, when oxygen (O$_2$) is in excess in the liquid droplets. The radiation yield of the SO$_2$ liquid-phase oxidation (number of oxidized SO$_2$ molecules per 100 eV of plasma energy) is presented in Fig. 11–57 as a function of the current density and acidity (Potapkin et al., 1993, 1995). The radiation yield is very high (up to 10^6, which corresponds to the chain length 10^5) for the case of basic droplets, and relatively low (about 100, which corresponds to the chain length 10) for the case of acidic droplets. It can be explained taking into account the much faster kinetics of HSO_3^- chain oxidation with respect to that of SO_3^{2-}. Fig. 11–57 also illustrates a decrease of the radiation yield of SO_2 liquid-phase oxidation with current density (dose rate of the plasma treatment), which is due to acceleration of the chain-termination rate as the square of density of charged particles (11–66), and (11–67). The SO_2 oxidation chain length can be very high at low dose rates (up to 10^5), which was first observed by Backstrom (1934). The energy-efficient experiments with SO_2 oxidation by relatively high-current electron beams and pulsed corona discharges (Baranchikov et al., 1990a,b, 1992), described in Sections 11.7.2 and 11.7.3, cannot be explained by the liquid-phase chain mechanism (compare Figs. 11–56 and 11–57). The high current and high dose rate plasma-chemical systems can provide the SO_2 chain oxidation in clusters, which is to be discussed next.

11.7.5. Plasma-Catalytic Chain Oxidation of SO_2 in Clusters

The plasma-catalytic chain oxidation of SO_2 in clusters is not limited by dose rate or acidity, which therefore explains the high energy efficiency of the process in high-current electron-beam systems and pulsed corona discharges (Baranchikov, 1990a,b, 1992; Sections 11.7.2, and 11.7.3). The long-chain oxidation process is provided in this case by the complex ions (clusters), which catalyze SO_2 oxidation similar to the general plasma-catalytic mechanism described in Section 10.1.1. The plasma-stimulated cluster mechanism of chain SO_2 oxidation in air (Fehsenfeld, 1974; Virin et al., 1978; Laudavala & Moroussi, 1985; Potapkin et al., 1993, 1995) starts with electron-beam or pulsed corona generation of negative oxygen ions by the electron attachment to molecular oxygen (which is usually faster in the system than attachment to SO_2 and water clusters):

$$e + O_2 + M \rightarrow O_2^- + M, \quad k = 3 \cdot 10^{-31} \text{ cm}^6/\text{s}. \tag{11–76}$$

The simple oxygen ions then attach to water molecules, creating the negative complex ions (ion clusters) in the stepwise clusterization:

$$O_2^-(H_2O)_n + H_2O + M \rightarrow O_2^-(H_2O)_{n+1} + M, \quad k = 3–6 \cdot 10^{-28} \text{ cm}^6/\text{s}. \tag{11–77}$$

Attachment of an SO_2 molecule leads to reconstruction of the ion cluster and formation of the most thermodynamically stable cluster core (Fehsenfeld, 1974):

$$O_2^-(H_2O)_n + SO_2 \rightarrow O_2^-(H_2O)_n SO_2 \rightarrow O_2^- SO_2(H_2O)_n \rightarrow SO_2^- O_2(H_2O)_n. \tag{11–78}$$

The core ion of the cluster (SO_4^-) is characterized by the peroxide structure and is a strong oxidant (Murtabekov, 1980; Lovejoy et al., 1987). The attachment of a second SO_2 molecule to the cluster can lead to strongly exothermic oxidation with formation of sulfuric acid ($\Delta H \approx -2\text{eV}$) and subsequent detachment of an electron from the acidic cluster:

$$SO_2^- O_2(H_2O)_n + SO_2 \rightarrow SO_2^- \cdot O_2 \cdot SO_2(H_2O)_n \rightarrow H_2SO_4 \cdot H_2SO_4 \cdot (H_2O)_n + e. \tag{11–79}$$

Release of the electron during the formation of sulfuric acid results in chain propagation of the SO_2 oxidation chain reaction mechanism (11–76)–(11–79). We should note that the SO_4^- ion, playing the key role in the chain propagation, exists in two configurations: a high-energy active peroxide isomer and a low-energy non-active cyclic isomer (Fehsenfeld, 1974; Lovejoy et al., 1987). Plasma-excited species in air promote O_2 penetration into the core of the cluster and stimulate formation of the chemically active isomer of the SO_4^- ion. The contribution of excited species to the formation of the chemically active isomer of the SO_4^- ion explains the suppression of the chain oxidation mechanism at low current densities and low dose rates (see Fig. 11–56), when concentration of the excited species is low. An increase of the chain oxidation energy cost at higher currents (Fig. 11–56) is due, as was mentioned, to domination in this case of recombination and chain-termination processes (proportional to the square of the concentration of active charged particles) over the chain-propagation reactions (linear with respect to the concentration of active charged particles). The two effects explain the optimal value of the current density (or dose rate) on the level of about 10^{-3} A/cm^2, resulting in an energy cost of SO$_2$ oxidation close to 1 eV/mol SO$_2$.

11.7.6. Simplified Mechanism and Energy Balance of the Plasma-Catalytic Chain Oxidation of SO$_2$ in Clusters

To analyze the energy balance and efficiency of plasma cleaning of SO$_2$ in exhaust gases, it is convenient to simplify the chain oxidation mechanism considered earlier to the kinetics of simple ions, neutrals, and excited molecules (Potapkin, Rusanov, & Fridman, 1989; Baranchikov et al., 1992). The plasma-assisted SO$_2$ chain oxidation mechanism starts with formation of negative ions (O$_2^-$) through the three-body electron attachment process (11–76). Then the SO$_2^-$ ions are formed due to a fast exothermic charge exchange process:

$$O_2^- + SO_2 \rightarrow O_2 + SO_2^-. \tag{11–80}$$

The SO$_2^-$ ions are then converted into the chemically active peroxide configuration of SO$_4^-$ ions in collisions with excited oxygen molecules:

$$SO_2^- + O_2^* \rightarrow SO_4^-. \tag{11–81}$$

The formation of oxidation products (SO$_3$) takes place in the reactions of the active SO$_4^-$ ions with the excited oxygen molecules:

$$SO_4^- + O_2^* \rightarrow SO_3 + O_3^-. \tag{11–82}$$

Restoration of electrons and chain propagation is provided by ion–molecular reactions and associative detachment, producing additional sulfuric acid:

$$O_3^- + SO_2 \rightarrow SO_3^- + O_2, \tag{11–83}$$

$$SO_3^- + H_2O \rightarrow H_2SO_4 + e. \tag{11–84}$$

With an excess of water vapor, ion–molecular processes (11–82) and (11–84) are limiting reactions of the chain oxidation mechanism, which permits us to express the energy cost of the SO$_2$ oxidation in steady-state conditions as (Baranchikov et al., 1992)

$$A = \frac{W_i}{\frac{4G_v}{G_i \beta} + \delta}, \tag{11–85}$$

$$\beta = \left(1 + \frac{n_1}{n_i} + \frac{n_i}{n_2}\right) \cdot \left\{1 + \frac{1}{1 - \frac{4n_i/n_2}{\left(1 + \frac{n_1}{n_i} + \frac{n_i}{n_2}\right)^2}}\right\}. \tag{11–86}$$

In these relations, n_i is the steady-state ion density generated in atmospheric air (density n_0) by, for example, a relativistic electron beam (current density j, electron velocity c, electron charge e), which can be calculated as

$$n_i = \sqrt{\frac{k_{ib} j n_0 G_i}{k_r^{ii} e c}}; \tag{11–87}$$

characteristic ion concentrations, n_1, and n_2, are determined by the expressions

$$n_1 = k_{\text{VT}} n_0 / k_{(11.82)},\qquad (11\text{--}88)$$

$$n_2 = k_{(11.84)} n_{\text{SO}_2} \frac{k_{\text{r}}^{\text{ii}} G_{\text{v}}}{G_{\text{i}}};\qquad (11\text{--}89)$$

G_{i} and G_{v} are the radiative yields of ions and vibrationally excited molecules; k_{ib}, k_{r}^{ii}, and k_{VT} are rate coefficients of relativistic electron-beam ionization, ion–ion recombination, and VT relaxation; W_{i} is energy cost of generation of an ion. The factor δ characterizes non-chain mechanisms of the plasma-stimulated SO_2 oxidation in air, including

$$SO_2 + O \rightarrow SO_3, \quad SO_3 + H_2O \rightarrow H_2SO_4,\qquad (11\text{--}90)$$

$$SO_2 + OH \rightarrow HSO_3, \quad HSO_3 + OH \rightarrow H_2SO_4.\qquad (11\text{--}91)$$

The factor δ can be calculated using the degradation cascade method (Fridman & Kennedy, 2004) and doesn't make a significant contribution to the total oxidation energy cost (11–85). As seen from (11–85) when $n_1 < n_2$ (which is relevant to the experimental conditions corresponding to SO_2 oxidation stimulated by the relativistic electron beams and pulsed coronas; Baranchikov et al., 1990a,b, 1992), the SO_2 oxidation energy cost can be significantly lower than the energy cost of an ion (W_{i}) because of the long oxidation chain length. The minimum value of SO_2 oxidation energy cost can be calculated in this case as

$$A^{\text{min}} = \frac{G_{\text{i}} W_{\text{i}}}{2 G_{\text{v}}} = \frac{100\,\text{eV}}{G_{\text{v}}},\qquad (11\text{--}92)$$

which is achieved at the optimal electron-beam current densities (or the corresponding dose rates) resulting in an optimal value of ion density:

$$n_{\text{i}} = \sqrt{n_1 n_2}.\qquad (11\text{--}93)$$

Taking into account that the G_{v} factor can be in the range 100–300, the simplified kinetic model considered quite satisfactory describes the experimental dependence $A(j)$ shown in Fig. 11–56. Specifically, the model explains minimal values of the energy cost of the plasma-stimulated SO_2 oxidation to SO_3 in air, 0.3–1 eV/mol SO_2 (Baranchikov et al., 1992).

11.7.7. Plasma-Stimulated Combined Oxidation of NO_x and SO_2 in Air: Simultaneous Industrial Exhaust Gas Cleaning of Nitrogen and Sulfur Oxides

An important advantage of plasma-assisted exhaust gas cleaning, especially in the case of power plant exhaust, is due to the possibility of simultaneous oxidation of SO_2 and NO_x to sulfuric and nitric acids. The products can then be collected in the form of non-soluble sulfates and nitrates, for example, in the form of the fertilizers $(NH_4)_2SO_4$ and $(NH_4)NO_3$ if ammonia (NH_3) is admixed to the plasma-assisted oxidation products (Pu & Woskov, 1996). The mechanisms of simultaneous NO_x and SO_2 oxidation stimulated by electron beams have been investigated in numerous experiments and simulations (Matzing, 1989a,b; Namba, 1990; Paur, 1991; Potapkin et al., 1995). It is interesting that simultaneous NO and SO_2 oxidation appears to be more effective than their plasma-assisted individual oxidation in air. In the presence of SO_2, the radiative yield of NO oxidation to NO_2 and nitric acid can exceed 30 (energy cost below 3 eV/mol NO). The low-energy-cost effect of simultaneous NO and SO_2 oxidation can be explained, in particular, by the chain co-oxidation mechanism, propagating through the radicals OH and HO_2:

$$HO_2 + NO \rightarrow OH + NO_2,\qquad (11\text{--}94)$$

$$OH + SO_2 \rightarrow SO_3 + H, \quad H + O_2 + M \rightarrow HO_2 + M.\qquad (11\text{--}95)$$

It should be mentioned that the gas-phase chain process, (11–94) and (11–95), is limited by the formation of a certain concentration of NO_2, when the plasma-generated atomic nitrogen causes significant destruction of the product in the following reverse reactions (Matzing, 1989a,b)

$$N + NO_2 \rightarrow 2NO, \quad N + NO_2 \rightarrow N_2O + O. \quad (11\text{–}96)$$

To suppress the reverse reactions, Matzing (1989a,b) proposed multi-stage plasma treatment with intermediate removal of the oxidation products. More interesting quenching effects occur in the plasma-generated droplets (see Section 11.7.4). In such droplets at certain pH conditions, it is possible not only to stabilize NO_2 by formation of nitric acid in the solution but also to reduce NO_2 to molecular nitrogen (N_2) with simultaneous oxidation of sulfur from S(IV) to S(VI)

$$NO_2 + 2HSO_3^- \rightarrow 1/2\,N_2 + 2HSO_4^-. \quad (11\text{–}97)$$

More kinetic details regarding the simultaneous plasma-chemical cleaning of industrial exhaust gases of sulfur and nitrogen oxides using relativistic electron beams can be found, in particular, in Potapkin et al. (1993, 1995). Discussing the non-thermal plasma treatment of large-volume exhausts, we should point out the interesting concept of the "**radical shower,**" applied in particular to NO_x-removal flow processing (Yan et al., 2001; Chang, 2003; Wu et al., 2005). The radical shower approach implies that plasma treats directly only a portion of the total flow (or even separate gas) producing active species (in particular, radicals), which then treat the total gas flow as a "shower." Although the approach is strongly limited by mixing efficiency of the plasma-treated and non-plasma treated gases, radical shower gas cleaning can be very energy effective especially in the case of very high flow rates of the exhaust gas streams. Because the radical shower approach is often organized using corona discharges, it is sometimes referred to as the corona radical shower.

11.7.8. Plasma-Assisted After Treatment of Automotive Exhaust: Kinetic Mechanism of Double-Stage Plasma-Catalytic NO_x and Hydrocarbon Remediation

Lean-burn car engines are in wide use because of their higher efficiency (more miles per gallon) and lower relative CO_2 emission. Combustion of the lean mixtures results, on the other hand, in larger nitrogen oxide emissions, which then require their effective remediation. Such after-treatment of automotive exhaust can be effectively organized using the strongly non-equilibrium plasma of, in particular, pulsed corona discharges (Pu & Woskov, 1996; Hammer & Broer, 1998; Hoard & Servati, 1998; Penetrante et al., 1998; Puchkarev, Roth, & Gunderson, 1998; Slone et al., 1998). Lean-burn engines operate under oxidizing conditions, when conventional three-way catalysts are ineffective in controlling emissions of NO_x through stimulating its reduction by hydrocarbons. Concentrations of hydrocarbons are low in the oxidizing conditions of the lean-burn engines, which is good for engine efficiency but makes the three-way catalyst ineffective in controlling NO_x emission. Application of plasma provides effective NO_x-removal aftertreatment even under oxidizing conditions, simultaneously with burning out the residual hydrocarbons (Hoard & Servati, 1998). Although plasma-generated atomic nitrogen is able to reduce NO to molecular nitrogen in the barrierless fast elementary reaction (see Section 6.3.6)

$$N + NO \rightarrow N_2 + O, \quad (11\text{–}98)$$

plasma in lean-combustion exhaust (containing a significant amount of oxygen) mostly stimulates not reduction but oxidation of NO to NO_2 (Penetrante et al., 1998). The plasma-stimulated NO-to-NO_2 conversion becomes especially effective in the presence of hydrocarbons (Penetrante et al., 1998). Complete NO-to-NO_2 conversion permits effective NO_2 reduction to molecular nitrogen (N_2) subsequently by using special catalysts. Thus, plasma-catalytic NO-reduction aftertreatment means in this case the actual combination of plasma and catalytic technologies to provide double-stage automotive exhaust cleaning (Hoard & Servati, 1998). The plasma-chemical phase of the double-stage exhaust treatment process is organized in the pulsed corona or similar strongly non-equilibrium atmospheric-pressure discharge (see Section 4.5) in an automotive exhaust mixture containing conventional combustion products as well as NO, hydrocarbons, and residual oxygen (O_2). The major purpose of the plasma-chemical phase is the complete conversion of NO to NO_2 with simultaneous partial destruction of hydrocarbons. The oxidation process starts in the plasma with dissociation of molecular oxygen (O_2) and generation of atomic oxygen through electronic excitation and dissociative attachment:

$$e + O_2 \rightarrow O + O + e, \quad e + O_2 \rightarrow O + O^-. \tag{11-99}$$

Atomic oxygen then quickly reacts with the hydrocarbons (which can be represented for simplicity as propylene, C_3H_6), stimulating initial destruction of the hydrocarbons with the production of hydrogen atoms, methyl radicals, and HO_2 radicals:

$$O + C_3H_6 \rightarrow CH_2CO + CH_3 + H, \tag{11-100}$$

$$O + C_3H_6 \rightarrow CH_3CHCO + H + H, \tag{11-101}$$

$$H + O_2 + M \rightarrow HO_2 + M. \tag{11-102}$$

Intermediate hydrocarbon produced by the propylene destruction further react with molecular oxygen, producing additional HO_2 radicals:

$$CH_2OH + O_2 \rightarrow CH_2O + HO_2, \tag{11-103}$$

$$CH_3O + O_2 \rightarrow CH_2O + HO_2, \tag{11-104}$$

$$HCO + O_2 \rightarrow CO + HO_2. \tag{11-105}$$

Plasma-generated HO_2 radicals are the major particles responsible for effective plasma-chemical conversion of NO into NO_2 in the systems under consideration:

$$NO + HO_2 \rightarrow NO_2 + OH. \tag{11-106}$$

The OH radicals produced in the system are effective in both destruction of hydrocarbons,

$$OH + C_3H_6 \rightarrow C_3H_6OH, \tag{11-107}$$

$$OH + C_3H_6 \rightarrow C_3H_5 + H_2O, \tag{11-108}$$

and in NO_x conversion into acids relevant for further catalytic reduction in the second (catalytic) phase of the double-stage automotive exhaust aftertreatment:

$$NO + OH + M \rightarrow HNO_2 + M, \tag{11-109}$$

$$NO_2 + OH \rightarrow HNO_3 + M. \tag{11-110}$$

After plasma conversion of NO into NO_2 and the acids, the double-stage aftertreatment is completed by NO_2 reduction to nitrogen and burning out the residual hydrocarbons in the catalytic stage of the process:

$$NO_2 + \text{Hydrocarbons} + \text{Catalyst} \rightarrow N_2 + CO_2 + H_2O + \text{Catalyst}. \tag{11-111}$$

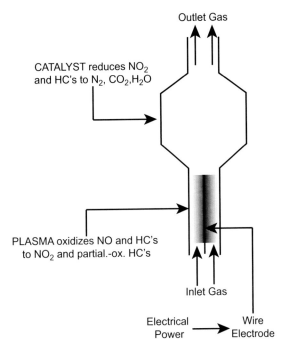

Figure 11–58. An embodiment of the plasma-assisted catalytic reduction process (the same result has been achieved in experiments when the catalyst has been placed inside the plasma reactor).

Metal oxides or metal-covered zeolites (such as Na-ZSM-5, Cu-ZSM-5, Ce-ZSM-5, H-ZSM-5, Al_2O_3, ZrO_2, and Ga_2O_3) can be applied as catalysts for the process. Some additional information regarding the catalysts will be given in the next section. Generally, there are three key features in the plasma-catalytic reduction of NO_x:

1. The plasma oxidation process is only partially complete in this approach: NO is mostly oxidized to NO_2, not to nitric acid. The plasma also produces partially oxygenated hydrocarbons but does not oxidize them to CO_2 and H_2O. For some catalysts, the partially oxygenated hydrocarbons are much more effective in comparison with the original hydrocarbons in reducing NO_x to N_2.

2. The plasma oxidation process is selective. In the specially chosen discharge regimes (compare with Section 11.7.7), NO is oxidized to NO_2, while SO_2 that is present can be kept from oxidizing to SO_3. It makes the plasma-catalytic approach more tolerant to the sulfur content compared to conventional lean-NO_x technologies.

3. Using plasma to change the composition of NO_x from NO to NO_2 permits a new class of catalysts to be used that are potentially more durable and more active than conventional lean-NO_x catalysts. More details can be found in the literature (Hoard & Servati, 1998; Penetrante et al., 1998).

11.7.9. Plasma-Assisted Catalytic Reduction of NO_x in Automotive Exhaust Using Pulsed Corona Discharge: Cleaning of Diesel Engine Exhaust

The double-stage system for plasma-catalytic NO_x reduction in automotive exhaust, built using a pulsed corona discharge (Penetrante et al., 1998), is illustrated in Fig. 11–58. The typical parameters of pulsed corona discharges used for the application are maximum pulse voltage, 20 kV; pulse duration, 100 ns; pulse repetition frequency, 10 kHz; voltage rise rate, 1–2 kV/ns; and average power of a pulsed power supply, about 1 kW. The non-equilibrium plasma experiments show that the effectiveness of NO oxidation to NO_2 at relevant temperatures (about 300°C) significantly increases in the presence of hydrocarbons

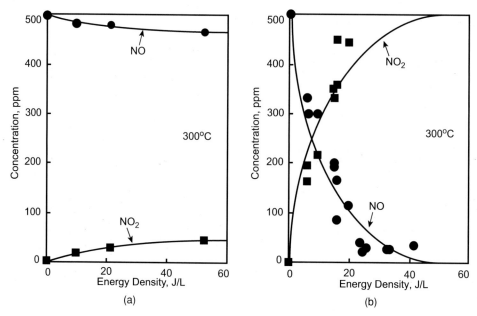

Figure 11–59. Effect of propene on the plasma oxidation of NO at 300°C. Plasma processing of 500 ppm NO in 10% O_2, balance N_2: (a) without propene; (b) with 1000 ppm propene.

(see Fig. 11–59a and b). Without hydrocarbons, plasma-generated atomic oxygen (11–99) intensively reacts with NO_2 in the reverse process:

$$O + NO_2 \rightarrow NO + O_2. \qquad (11\text{–}112)$$

The presence of hydrocarbons makes the atomic oxygen mostly reacting with them in the elementary oxidation processes (11.100 and 11.101) preventing the destruction of the product (NO_2) in a wide temperature range (see Fig. 11–59). The effectiveness of the combined plasma-catalytic process is obviously very sensitive to the choice of catalyst. γ–Al_2O_3 is one of the good choices, because this selective catalytic reduction catalyst is much more active in NO_2 reduction than NO. Some other good choices have been mentioned in the previous section. Combining pulsed corona plasma with the γ–Al_2O_3 catalyst can provide NO_x reduction efficiencies much higher than those achieved by the conventional approach of loading a metal on γ–Al_2O_3. Fig. 11–60a shows the NO reduction to N_2 using the γ–Al_2O_3 catalyst. The temperature operating window occurs at high temperature and is narrow. The addition of 2 wt % of Ag to γ–Al_2O_3 stimulates NO reduction at lower temperatures (Fig. 11–60b). When the input gas contains NO_2 instead of NO, the NO_x reduction activity over γ–Al_2O_3 increases dramatically in the wide range of temperatures (see Fig. 11–60c). Plasma-assisted catalytic reduction of NO_x in the exhaust from a Cummins B5.9 diesel engine is illustrated in Fig. 11–61 (Penetrante et al., 1998). The specific system of 0.5 L volume consists of a pulsed corona reactor packed with γ–Al_2O_3 pellets. A Cummins B5.9 diesel engine running with a 95 kW load exhausts about 600 ppm of NO_x at temperatures of 350–400°C. Propylene was used in these experiments as the hydrocarbon reductant with a Cl/NO_x ratio of 5. Fig. 11–59 shows the amount of NO_x reduction at space velocities of 12,000–18,000 L/h. The NO_x reduction strongly increases at higher specific energy inputs (J/L). Complete plasma-catalytic cleaning of the automotive exhausts containing 200–300 ppm of NO_x typically requires about 30 J/L of pulsed corona plasma energy input, which corresponds to an energy cost of about 30 eV/mol NO (compare with

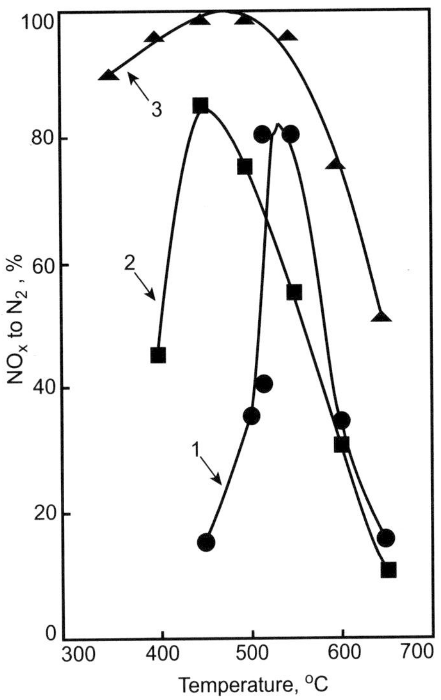

Figure 11–60. NO$_x$ reduction to N$_2$ as function of temperature. (a) NO over γ–Al$_2$O$_3$; NO or NO$_2$ over 2 wt% Ag/Al$_2$O$_3$; (c) NO$_2$ over γ–Al$_2$O$_3$. Catalyst weight 0.25 g. Dry gas feed: 1000 ppm NO or NO$_2$; 1000 ppm C$_3$H$_6$, 6% O$_2$; balance helium at 100 mL/min.

Fig. 11–59). The typical exhaust gas flow rate per unit engine power can be estimated as 10 L/min per 1 kW of engine power. A car with a power of about 200 hp, or about 150 kW, exhausts gas with a maximum flow rate of about 1,500 L/min. It corresponds to minimum power of the pulsed corona discharge on the level of 750 W, required for the effective plasma-catalytic exhaust treatment from nitrogen oxides.

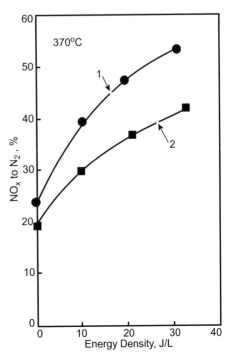

Figure 11–61. Plasma-assisted catalytic reduction of NO$_x$ at 370°C in a pulsed corona plasma reactor packed with γ–Al$_2$O$_3$ pellets. Total plasma and catalyst reactor volume is 0.5 L. The NO$_x$ reduction is shown as a function of the energy density input to the plasma. A Commins B5.9 diesel engine running with a 95 kW load was used as the source of NO$_x$. Propene reductant ratio C1/NO$_x$ = 5. Space velocity in the system: (a) 12,000 L/h, (b) 18,000 L/h.

11.8. NON-THERMAL PLASMA TREATMENT OF VOLATILE ORGANIC COMPOUND EMISSIONS, AND SOME OTHER PLASMA-ECOLOGICAL TECHNOLOGIES

11.8.1. General Features of the Non-Thermal Plasma Treatment of Volatile Organic Compound Emissions

Volatile organic compounds (VOCs) form a class of air pollutants that has been addressed by environmental regulations in the past few decades due to the pollutants' toxicity and contribution to global warming mechanisms (Hewitt, 1999; Hunter & Oyama, 2000; Lippmann, 2000). The control of air pollution from dilute large-volume sources, such as paint spray booths, paper mills, pharmaceutical, and food and wood processing plants, is a challenging problem. Conventional technologies, such as carbon adsorption/solvent recovery or catalytic/thermal oxidation – widely used in industry – and regenerative thermal incineration and especially regenerative thermal oxidation (RTO) systems, have too high an annual cost and are not economical for large gas flow rates (50,000–250,000 scfm) and low VOC concentrations (<100 ppm; see Paur, 1999; Hunter & Oyama, 2000). Among the emerging low-temperature VOC treatment technologies are low-temperature catalysis, biofiltration, and non-thermal plasma, which is considered in the next section. Catalysts easily suffer from plugging, fouling, or poisoning by particulates and non-VOC materials in the exhaust stream, which results in a high maintenance cost (Bertelsen, 1992). The major disadvantage of biofilters is their large specific footprint, typically 100–400 square feet per 1,000 scfm of treated gas (Leson & Dharmavaram, 1995). Biofilter systems and filter materials may also require costly maintenance and replacement. The application of non-thermal plasma methods (particularly pulsed coronas, DBDs, and electron beams) solves most of the problems typical for alternative VOC treatment methods. The application of non-thermal plasma requires, however, decreased energy cost similar to the case of plasma desulfurization and NO_x reduction processes (Section 11.7). Plasma approaches usually become more energy efficient and competitive with other cleaning methods when stream volumes are large and concentration of pollutants is small. Non-thermal atmospheric-pressure plasma systems that have been investigated for VOC abatement include, in particular, pulsed corona (Yamamoto et al., 1992, 1993; Hsiao et al., 1995; Sobacchi et al., 2003), DBD (Evans et al., 1993; Rosocha et al., 1993; Neely et al., 1993; Clothiaux, Koropchak & Moore, 1984; Fraser, Fee & Sheinson, 1985; Fraser & Sheinson, 1986), surface discharge (Oda et al., 1991; Masuda, 1993), gliding arc discharge (Mutaf-Yardimci et al., 1998; Czernichowski, 1994; Fridman et al., 1999), microwave discharge (Bailin et al., 1975), and electron-beam plasma (Penetrante et al., 1996a,b). The highest energy efficiency of the cleaning of large-volume air exhaust from low-concentration VOCs has been achieved in plasma systems with the highest electron energies (such as electron beams and pulsed coronas), where the energy cost of generation of oxidation active species (such as OH) is lowest. Non-thermal plasma systems have been successfully applied for cleaning of high-volume low-concentration (HVLC) VOC exhausts, particularly from volatile hydrocarbons (acetone, methanol, pinene, etc.), sulfur-containing compounds (dimethyl sulfide, H_2S, etc.), and chlorine-containing compounds (vinyl chloride, trichloroethylene, trichloroethane, carbon tetrachloride, etc.). Some of the processes are specifically discussed in the next section.

11.8.2. Mechanisms and Energy Balance of the Non-Thermal Plasma Treatment of VOC Emissions: Treatment of Exhaust Gases from Paper Mills and Wood Processing Plants

Electric energy cost is one of the crucial factors characterizing the plasma treatment HVLC VOC exhausts. To analyze the energy cost of VOC treatment in non-thermal plasma, we should point out that the mechanism of plasma oxidation of hydrocarbons can be generally interpreted as the low-temperature burning out of the VOCs in air to preferentially CO_2 and H_2O using different plasma-generated active oxidizers (OH, atomic oxygen, electronically excited oxygen, ozone, etc.), but especially OH radicals. The non-thermal plasma oxidation (burning out) of a small admixture of volatile hydrocarbons in the air stream can be generally outlined by the following illustrative kinetic mechanism.

Between different plasma-generated active oxidizers (OH, atomic oxygen, electronically excited oxygen, ozone, etc.) responsible for the cold burning out of VOC, OH radicals play a major role. The formation of OH radicals in air plasma is due to air humidity and is provided by numerous channels, from which it is interesting to point out a selective one starting with the charge exchange from any positive air ions M^+ to water ions, H_2O^+:

$$M^+ + H_2O \rightarrow M + H_2O^+. \tag{11–113}$$

The H_2O^+ ions, formed through the charge exchange, then react with water molecules to produce active OH radicals in the fast ion–molecular reaction

$$H_2O^+ + H_2O \rightarrow H_3O^+ + OH. \tag{11–114}$$

Taking into account that the ionization potential of the water molecules is relatively low, most of the positive ions initially formed in air have a tendency for charge exchange (11–113). Therefore, the discharge energy initially distributed over different air components is somewhat selectively localized on the ionization of water and selective production of OH radicals. It explains the relatively low energy price of the production of OH radicals in air plasma: 10–30 eV/radical, although the fraction of water molecules in air is not large (Gutsol, Tak, & Fridman, 2005). Representing a hydrocarbon VOC molecule as RH, where R is a relevant organic group, the oxidation of the molecule starts with the elementary reaction of its dehydrogenization:

$$OH + RH \rightarrow R + H_2O. \tag{11–115}$$

Almost immediate attachment of molecular oxygen to the active organic radical R results in the formation of an organic peroxide radical:

$$R + O_2 \rightarrow RO_2. \tag{11–116}$$

The peroxide radical RO_2 is able then to react with another saturated hydrocarbon VOC molecule RH, forming saturated organic peroxide (RO_2H) and propagating the chain mechanism (11–116) and (11–117) of RH oxidation in air:

$$RO_2 + RH \rightarrow RO_2H + R. \tag{11–117}$$

Depending on the temperature and chemical composition, peroxides RO_2 and RO_2H are further oxidized up to CO_2 and H_2O with or without additional consumption of OH and other plasma-generated oxidizers. Summarizing the mechanism, the energy cost corresponding to the VOC treatment process is on the level of 10–30 eV per molecule of the pollutant RH (Gutsol, Tak, & Fridman, 2005). This energy (10–30 eV/mol) is relatively high, but the total energy consumption can be small because of the very low concentration of pollutants. A major conventional approach to VOC control, RTO, consumes about 0.1 eV/mol, but this energy is calculated per molecule of air. Therefore, energy consumption in plasma becomes lower than that of conventional RTO, when the VOC concentration in air is below 0.3–1% (3,000–10,000 ppm). In the case of plasma treatment of large-volume exhausts, the "radical shower" approach can be effective (see Section 11.7.7). The radical shower approach implies that plasma treats only a portion of the total flow generating active species, which then clean the whole gas. While it has been discussed earlier regarding NO_x reduction and desulfurization processes (Yan et al., 2001; Chang, 2003; Wu et al., 2005), it can be similarly useful for VOC removal. Exhaust gases of paper mills (especially the brownstock

Table 11–4. Characteristics of the Major HVLC VOC Exhaust Streams of Paper Mills and Wood Processing Plants, Range of the Parameters Tested for Plasma Treatment Using Pulse Corona Discharges

Stream composition, humidity, and temperature	Paper mills, brownstock washer ventilation stream	Wood processing plants, strandboard press ventilation stream	Wood processing plants, strandboard dryer ventilation stream	Range of parameters, tested in pulsed corona plasma treatment
Methanol, CH_3OH, ppm	83	25	8	5–1000
Acetone, CH_3COCH_3, ppm	3	1	28	5–1000
α-pinene, ppm	209	4	16	150–800
Dimethyl Sulfide, $S(CH_3)_2$, ppm	2	–	–	5–1000
Relative Humidity, %	100	70	90	0–100
Temperature, °C	43	37	93	25–220

washer ventilation gases) and wood processing plants (especially strandboard press and dryer ventilation gases) are good examples of HVLC VOC streams, that can be effectively cleaned by non-thermal plasma (Harkness & Fridman, 1999). Compositions, temperature, and humidity of the streams are presented in Table 11–4. The same table shows a range of stream parameters tested in the experiments with VOC treatment in pulsed corona discharges (Sobacchi et al., 2003; Gutsol et al., 2005).

11.8.3. Removal of Acetone and Methanol from Air Using Pulsed Corona Discharge

Before discussing plasma cleaning of large-volume VOC exhaust streams of paper mills and wood processing plants using a powerful pulsed corona system (Gutsol et al., 2005), consider laboratory-scale experiments using pulsed corona (wire-in-cylinder coaxial electrode configuration) with power 1–20 W, voltage pulse amplitude 9–12 kV, pulse repetition rate 0.2–2 kHz, pulse duration 100 ns, and rise time 10 ns (Sobacchi et al., 2003). Typical atmospheric-pressure gas flow through the plasma system was 2 slm; the corresponding residence time was about 13 s. The pulsed corona reactor was placed inside a tubular furnace

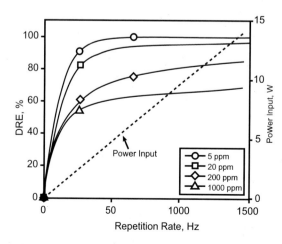

Figure 11–62. Destruction-removal efficiency (DRE) and electric power input as a function of the pulse repetition rate for treatment of acetone (diluted in air) in pulsed corona discharge. Initial acetone concentrations are 5, 20, 200, and 1000 ppm.

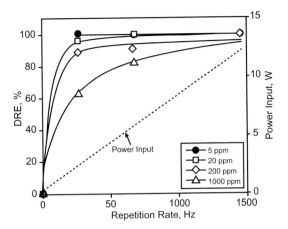

Figure 11–63. Destruction-removal efficiency (DRE) and electric power input as a function of the pulse repetition rate for treatment of methanol (diluted in air) in pulsed corona discharge. Initial methanol concentrations are 5, 20, 200, and 1000 ppm.

to sustain fixed gas temperature between 25 and 220°C. In the experiments with acetone and methanol removal, the initial concentration of both CH_3COCH_3 and CH_3OH varied between 5 and 1000 ppm (see Table 11–4), temperature was maintained constant at 200°C, and the air flow humidity corresponded to saturation at about 55°C. Destruction and removal efficiency (DRE – the mole percentage of VOC removed with respect to initial amount) of acetone and discharge power are shown in Fig. 11–62 as a function of the pulse repetition rate. As shown, the discharge power is proportional to the pulse repetition rate. The DRE of acetone depends strongly on the initial composition and on the corona power. Higher power levels result in higher DRE; increase of the initial acetone concentration requires higher power to reach the same level of DRE. No organic by-products were detected at initial acetone concentrations of 5 and 20 ppm. Treatment of the air stream with 200 and 1000 ppm of acetone leads to the formation of very small amounts of methanol as a by-product ($\ll 10$ ppm). A similar dependence of the DRE of methanol on the discharge power in the pulsed corona system at 200°C is presented in Fig. 11–63. No organic by-products or CO have been observed during the CH_3OH treatment; the methanol has been entirely converted into CO_2. The methanol DRE dependence on specific energy input (SEI) in the discharge is shown in Fig. 11–64. Summarizing the results, the relatively low energy cost of treatment (about 20 Wh per m^3 of air) can be sufficient for 98% of DRE, which is well within acceptable limits for the industrial applications.

11.8.4. Removal of Dimethyl Sulfide from Air Using Pulsed Corona Discharge

Dimethyl sulfide (DMS), $S(CH_3)_2$, represents sulfur-containing VOCs, which are air pollutants of major concern in different industrial exhausts, particularly including the brownstock

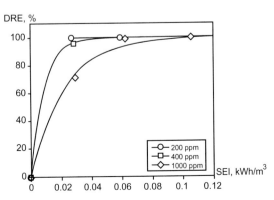

Figure 11–64. DRE of methanol as a function of specific energy input (SEI) in pulsed corona discharge. Initial methanol concentrations are 200, 400, and 1000 ppm.

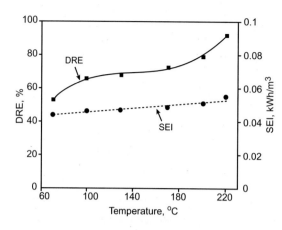

Figure 11–65. DRE of dimethyl sulfide (DMS) as a function of SEI in pulsed corona discharge (SEI values are related to temperature).

ventilation gases of paper mills (see Table 11–4). DMS is highly toxic, flammable, and emits a foul odor; the DMS odor threshold is on the level of 1 ppb. While the typical DMS concentration in brownstock ventilation gases is 2 ppm (11–4), it reaches 1700 ppm (0.17 vol %) in the "worst-case-scenario" exhausts. Effective plasma removal of DMS has been demonstrated in pulsed corona discharges described in the previous section (Sobacchi et al., 2003; Gutsol et al., 2005). The DRE of DMS is presented in Fig. 11–65 as function of the SEI and temperature. The products of non-thermal plasma DMS removal are CO_2, H_2O, and SO_2. At the lower initial DMS concentrations, no DMS has been detected in aftertreatment gases. For higher initial concentrations, DRE exceeding 98% can be achieved at fairly low levels of SEI; an SEI of about 50 Wh/m^3 is sufficient for processing the highest DMS concentrations (Sobacchi et al., 2003). The only by-product of the DMS treatment is methanol, its concentration in the aftertreatment air is shown in Table 11–5 depending on initial DMS concentration and the SEI. The mechanism of plasma removal of DMS from air starts with oxidation provided by plasma-generated oxidizers and especially OH (Turnipseed, Barone, & Ravishenkara, 1996a,b; Sun, Sato, & Clements, 1997; see also Section 11.8.2):

$$OH + CH_3SCH_3 \rightarrow CH_3S + CH_3OH, \tag{11–118}$$

$$OH + CH_3SCH_3 \rightarrow H_2O + CH_3SCH_2. \tag{11–119}$$

The CH_3S radical is an intermediate particle in the oxidation of DMS. The CH_3S radical is formed either directly in elementary reaction (11–118) or through the transformation of CH_3SCH_2 radicals produced in (11–119) into corresponding peroxide radicals by means of attachment of molecular oxygen:

$$CH_3SCH_2 + O_2 + M \rightarrow CH_3SCH_2OO + M. \tag{11–120}$$

Table 11–5. Concentration of Methanol as a By-Product of the Plasma Destruction of Dimethyl Sulfide

Initial DMS concentration, ppm	SEI, 27 Wh/m^3	SEI, 60 Wh/m^3	SEI, 100 Wh/m^3
5	0	0	0
20	0	0	0
200	5	4	2
400	14	12	8
1000	21	28	37

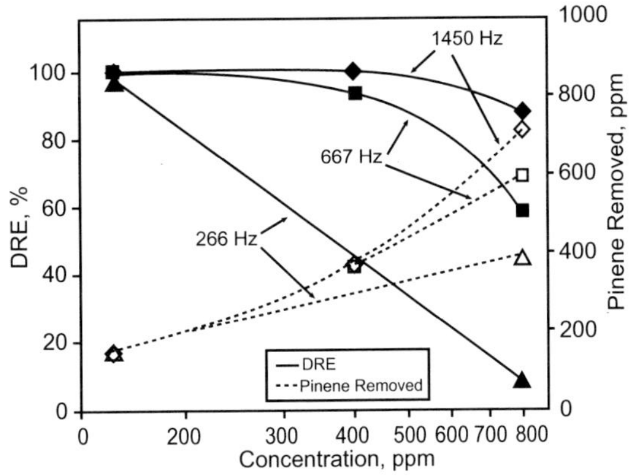

Figure 11–66. DRE and actual amount of α-pinene removed in pulsed corona discharge as a function of initial concentration of α-pinene.

The peroxide radical CH_3SCH_2O is then transformed into the radical CH_3SCH_2O by reactions with different active species but especially by reactions with NO (even when the concentration of NO is relatively very low):

$$CH_3SCH_2OO + NO \rightarrow CH_3SCH_2O + NO_2. \qquad (11\text{–}121)$$

The major intermediate particles of DMS oxidation – the CH_3S radicals – are then generated by decomposition of the CH_3SCH_2O radicals with the additional formation of formaldehyde molecules as by-products:

$$CH_3SCH_2O + M \rightarrow CH_3S + CH_2O. \qquad (11\text{–}122)$$

Further oxidation of the major intermediate active particles (CH_3S) leads to formation of CO_2, H_2O, and SO_2 as final products. It should be mentioned that another primary product of the DMS reaction with plasma-generated OH is the DMS $\cdot$ OH adduct:

$$OH + CH_3SCH_3 (+M) \rightarrow OH \cdot CH_3SCH_3(+M). \qquad (11\text{–}123)$$

The DMS $\cdot$ OH adduct then effectively reacts with molecular oxygen to generate dimethyl sulfoxide ($CH_3S(O)CH_3$) and dimethyl sulfone ($CH_3S(O_2)CH_3$):

$$OH \cdot CH_3SCH_3 + O_2 \rightarrow HO_2 + CH_3S(O)CH_3, \qquad (11\text{–}124)$$

$$OH \cdot CH_3SCH_3 + O_2 \rightarrow OH + CH_3S(O_2)CH_3. \qquad (11\text{–}125)$$

Both dimethyl sulfoxide ($CH_3S(O)CH_3$) and dimethyl sulfone ($CH_3S(O_2)CH_3$) are known to rapidly react with OH (which leads to effective further oxidation), and in the presence of liquid water to be very likely removed by absorption in the liquid phase. The effective absorption of dimethyl sulfoxide and dimethyl sulfone in water is due to their high solubility; the Henri constant is $K_H > 5 \cdot 10^4$ for both organic sulfoxides. Therefore, effective DMS removal from air can be provided not only by its complete oxidation to CO_2, H_2O, and SO_2, but also by means of formation of highly soluble compounds following scrubbing. Such a no-by-product approach to VOC removal in the wet pulse corona will be considered later in Section 11.8.

11.8.5. Removal of α-Pinene from Air Using Pulsed Corona Discharge; Plasma Treatment of Exhaust Gas Mixtures

VOC emissions of paper mills and wood processing plants usually contain a significant amount of terpenes, especially α-pinene ($C_{10}H_{16}$; see Table 11–4). Effective plasma removal of α-pinene ($C_{10}H_{16}$) has been demonstrated in pulsed corona discharges described in Section 11.8.3 at temperatures of 70–200°C (Sobacchi et al., 2003), which is illustrated in Fig. 11–66 for different pulse frequencies corresponding to different levels of discharge power (the SEI is 20, 50, and 100 Wh/m^3 for the frequencies 266, 667, and 1450 Hz).

Table 11–6. Characteristics of Plasma Removal of 400 ppm of α-Pinene from Air Stream Using Pulsed Corona with Pulse Repetition Rate 667 Hz

Plasma VOC treatment, process characteristics	Gas temperature, 70°C	Gas temperature, 100°C	Gas temperature, 130°C	Gas temperature, 170°C	Gas temperature, 200°C
Specific energy input, (SEI), Wh/m^3	47	48	48.5	49	50
Destruction and removal efficiency (DRE) %	76.6	81.5	84	89.7	96.4
Treatment energy cost per 1 kg of α-pinene, kWh/kg	25.5	24.2	23.7	22.5	21.4
Initial by-product: acetone, ppm	44	55	59	88	189

A major initial by-product of α-pinene plasma removal is acetone, which is shown in Table 11–6 together with the DRE dependence on SEI and gas temperature. Overall the α-pinene plasma removal tests demonstrate that 98% DRE for the typical initial VOC concentration of 200 ppm can be reached at quite low SEI values into the air at 25 Wh/m^3. Treatment efficiency and especially energy cost of the exhaust treatment can be enhanced at higher humidity (Sobacchi et al., 2003), which is discussed next.

Decomposition of VOCs in some specific exhaust mixtures related to paper mills and wood processing plants is illustrated in Table 11–7 at two values of the specific energy input: 90 and 190 Wh/m^3 (Sobacchi et al., 2003). As seen from the table, the presence in the exhaust gases of large molecules of α-pinene decreases the DREs of other components, which is due to the formation of methanol and acetone as treatment by-products.

11.8.6. Treatment of Paper Mill Exhaust Gases Using Wet Pulsed Corona Discharge

The significant reduction of VOC-removal energy cost as well as the complete elimination of plasma-treatment by-products can be achieved by application of the so-called wet or spray pulsed corona discharges, which is a combination of pulsed corona discharge with a scrubber (Gutsol et al., 2005). General schematics of the wet and spray configurations of the pulsed corona discharges (which are both often referred to the wet pulsed corona) are

Table 11–7. Decomposition of VOC Exhaust Mixtures Related to Paper Mills and Wood Processing Plants at Different Values of the Pulsed Corona Power Input

Input and DRE	Methanol	Acetone	Dimethyl sulfide (DMS)	α-Pinene
Initial fraction, ppm (SEI = 90 Wh/m^3)	100	20	20	0
DRE, % (SEI = 90 Wh/m^3)	100	91	100	–
Initial fraction, ppm (SEI = 190 Wh/m^3)	100	20	20	0
DRE, % (SEI = 190 Wh/m^3)	100	99.3	100	–
Initial fraction, ppm (SEI = 190 Wh/m^3)	100	20	20	200
DRE, % (SEI = 190 Wh/m^3)	86.5	14.5	83.8	96.5

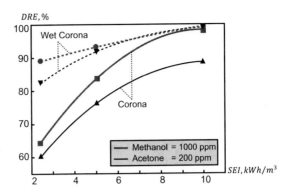

Figure 11–67. Comparison of the plasma-stimulated VOC removal efficiency using wet pulsed corona and regular (dry) pulsed corona. DRE of methanol (initial concentration 1000 ppm) and acetone (initial concentration 200 ppm) are presented in both types of corona as function of SEI. Water flow rates in the spray of the wet pulsed corona are: 0.4 mL/min for methanol removal, and 1 mL/min for acetone removal.

shown in Fig. 4–95 (see Section 4.5.2). First of all, water droplets (or film) effectively absorb soluble VOCs (such as methanol) present in the exhaust gases, which obviously simplifies the task for plasma treatment. Non-soluble organic compounds (in particular, non-polar hydrocarbons) cannot be directly removed by water scrubbing. Their plasma treatment (see mechanism (11–115)–(11–117) and discussion in Section 11.8.2), however, immediately leads to the formation of soluble peroxides (ROOH) and peroxide radicals (RO_2) which can be effectively removed by water scrubbing. Thus, plasma cleaning of VOC emissions is provided in this plasma approach not by complete oxidation of the organic compounds to CO_2 and H_2O, but only by plasma-induced conversion of the non-polar non-soluble compounds into soluble ones, which are then removed by water scrubbing.

Complete VOC oxidation to CO_2 and H_2O requires much more plasma-generated oxidizers (such as OH) than conversion of the non-polar non-soluble compounds (RH) into soluble ones (ROOH or ROO), where a single OH radical can be sufficient (see mechanism (11–115)–(11–117). It results in much lower energy requirements for VOC treatment in a pulsed corona discharge, which is illustrated, for example, in Fig. 11–67. The by-products of plasma VOC treatment are usually soluble; therefore, the application of wet pulsed corona permits removal of all of them from the gas flow. In the particular case of plasma cleaning of sulfur-containing exhausts (such as DMS; see Section 11.8.4), sulfur oxides and polar by-products of oxidation are also effectively removed from the air flow in the form of relevant acidic solutions. Plasma-stimulated oxidation of organic compounds continues even when they are absorbed in water droplets, which increases the absorbing capacity of the water droplets with respect to the same initial VOC. Therefore, water consumption for the VOC exhaust cleaning approach is very low. The major requirements and characteristics of the wet pulsed corona treatment of brownstock washer ventilation gases from paper mills (see Table 11–4) are summarized in Table 11–8 (Gutsol et al., 2005). Although the VOC removal parameters presented in the table are quite impressive, the organic compounds in

Table 11–8. Major Requirements and Characteristics of the Wet Pulsed Corona Treatment of the Brownstock Washer Ventilation Gases from Paper Mills

Total Destruction-Removal Efficiency of the VOC Mixture	98–100%
Plasma Treatment By-Products in Gas-Phase	0% (no by-products)
Energy Cost (Regular Conditions)	10 Wh/m³
Energy Cost ("Worst-Case Scenario": additional 1700 ppm of DMS)	20 Wh/m³
Power Requirements for VOC removal from 25,000 scfm Brownstock Washer Ventilation Stream	500 kW
Water Requirements for VOC removal from 25,000 scfm Brownstock Washer Ventilation Stream	25 gal/min

Figure 11–68. Mobile Environmental Laboratory built in Drexel University carrying precise exhaust gas characterization facility as well as 10 kW wet (spray) pulsed corona discharge for treatment of large-volume low-concentration exhaust gases containing VOCs.

this approach are not completely removed but just converted from the gas to the liquid phase. Such a solution of VOC treatment is, however, very much acceptable for the paper industry, where the amount of polluted water from other sources is large, and effective water cleaning should be organized anyway.

A mobile plasma laboratory for VOC removal from the exhaust streams typical for paper mills and wood processing plants (see Table 11–4) has been built at Drexel University (Fig. 11–68, see Gutsol et al., 2005) based on 10 kW wet (spray) pulsed corona discharge. A photograph of the discharge from one of the six windows is presented in Fig. 11–69; all

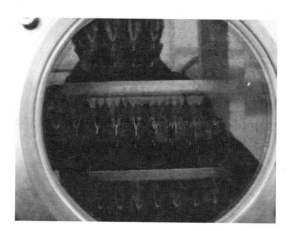

Figure 11–69. Photo of the 10-kW pulsed corona discharge (organized in the water-spray configuration) through a window of the Mobile Environmental Laboratory.

Figure 11–70. Photo of the Mobile Environmental Laboratory demonstrating all six access windows to the 10-kW pulsed corona discharge system applied for treatment of large-volume low-concentration exhaust gases containing VOCs.

six windows showing the discharge can be seen in Fig. 11–70. The pneumatic and hydraulic scheme of the system is shown in Fig. 11–71, and organization of the pilot plant is illustrated in Fig. 11–72. Exhaust gas in the pilot plant first goes to a scrubber where it is mostly washed out from soluble VOCs. After that, the exhaust gas is directed to the wet pulsed corona discharge, where non-soluble VOCs are converted into soluble compounds to be scrubbed out by water spray in the chamber. Final air refining is provided by the mist separator, finalizing the exhaust cleaning process. The pilot plant is mounted on a trailer (see Figs. 11–68 and 11–70) providing opportunity for industrial field experiments. The mobile plasma laboratory for VOC removal includes not only the gas cleaning system (Figs. 11–71 and 11–72) but also special exhaust gas characterization capability for gas diagnostics before and after plasma treatment. It should be especially mentioned that the mobile plasma pilot plant for VOC removal has been effectively applied not only in combination with the scrubber, but also for "dry" cleaning (without water shower) of industrial exhaust gases.

11.8.7. Non-Thermal Plasma Control of Diluted Large-Volume Emissions of Chlorine-Containing VOCs

Non-thermal atmospheric-pressure plasma systems, including those based on electron beams, are effectively applied in the destruction of different chlorine-containing VOCs including vinyl chloride (Slater & Douglas-Hamilton, 1981), trichloroethylene (Matthews et al., 1993; Scheytt et al., 1993; Vitale et al., 1995), and trichloroethane (Vitale et al., 1995), and carbon tetrachloride (Bromberg et al., 1993; Koch et al., 1993; Penetrante et al., 1995, 1996a,b). The major specific plasma systems used for the application are electron beams, pulsed corona, and DBD (see, for example, Penetrante & Schultheis, 1993; Hadidi et al., 1996; Penetrante et al., 1996a,b). Considering as an example the destruction of carbon tetrachloride (CCl_4) diluted in air, a major contribution to the process kinetics according to Penetrante et al. (1996a,b) is provided by plasma-generated OH radicals, O and N atoms, as

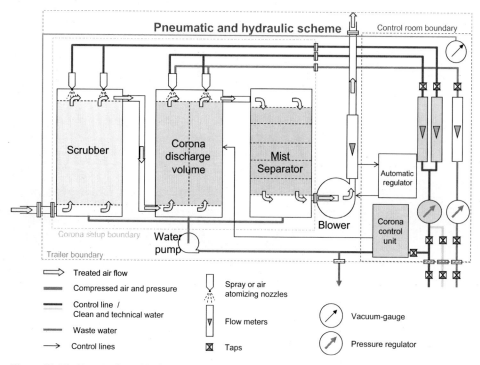

Figure 11–71. Pneumatic and hydraulic scheme of the Mobile Environmental Laboratory showing initial scrubbing chamber, wet pulsed corona chamber, final mist separation unit, as well as air and water flow channels.

well as by direct electron-impact destruction through dissociative attachment. The radical-induced decomposition starts with the formation of OH radicals and O and N atoms:

$$e + O_2 \rightarrow O(^3P) + O(^1D) + e, \quad O(^1D) + H_2O \rightarrow OH + OH, \quad (11\text{–}126)$$

$$e + N_2 \rightarrow N(^4S) + N(^2D) + e. \quad (11\text{–}127)$$

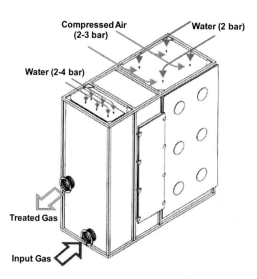

Figure 11–72. Constructional organization of the exhaust gas treatment unit of the Mobile Environmental Laboratory.

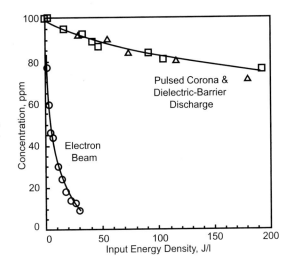

Figure 11–73. Electron beam, pulsed corona, and DBD processing of 100 ppm carbon tetrachloride in dry air at 25°C.

Then oxidation of CCl_4 is initiated by its reactions with O and OH:

$$OH + CCl_4 \rightarrow HOCl + CCl_3, \qquad (11-128)$$

$$O(^3P) + CCl_4 \rightarrow ClO + CCl_3. \qquad (11-129)$$

N atoms generated in reactions (11–127) promote, in particular, reduction of dichloroethylene:

$$N(^4S) + CH_2Cl_2 \rightarrow NH + CHCl_2. \qquad (11-130)$$

The major initial reaction of CCl_4 decomposition through dissociative attachment starts with the following elementary process (Penetrante et al., 1996a,b):

$$e + CCl_4 \rightarrow CCl_3 + Cl^-. \qquad (11-131)$$

Fig. 11–73 shows a comparison between electron-beam, pulsed corona, and DBD processing of 100 ppm CCl_4 in dry air at 25°C (Penetrante et al., 1996a,b). The major products of plasma processing of CCl_4 in air are Cl_2, $COCl_2$, and HCl. These products can be easily removed from the gas stream; they dissolve and/or dissociate in aqueous solutions and combine with $NaHCO_3$ in a scrubber solution to form NaCl. Similar approaches can be applied as performed for product removal from air streams during the plasma cleaning of other diluted chlorine-containing VOC exhausts.

The plasma-stimulated decomposition of CCl_4 is strongly dependent on the dissociative attachment (11–131a,b), consuming an electron per each decomposition (Bromberg et al., 1993; Koch et al., 1993; Penetrante et al., 1995, 1996a,b). Therefore, the energy cost of the CCl_4 decomposition is determined by the energy cost of ionization. The ionization energy cost is lower at higher electron energies in a plasma system. The lowest ionization energy cost can be achieved in plasma generated by electron beams (about 30 eV per anion–electron pair). To compare, the ionization energy cost in pulsed corona and DBD is usually on the level of hundreds of electron volts, which explains the higher energy efficiency of electron beams in the destruction of carbon tetrachloride (see Fig. 11–73).

The advantages of electron beams versus pulsed corona discharges have also been demonstrated (Penetrante et al., 1996a,b) in the plasma treatment of 100 ppm of methylene chloride ($C_2H_2Cl_2$) in dry air at 25°C (see Fig. 11–74), and in the plasma treatment of 100 ppm of TCE (trichloroethylene, C_2HCl_3) in dry air at 25°C (see Fig. 11–75). Reactions

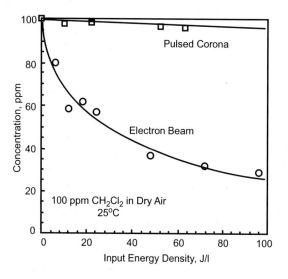

Figure 11–74. Electron beam and pulsed corona processing of 100 ppm CH_2Cl_2 in dry air at 25°C.

with N and O atoms (see previous reactions) make significant contributions in the destruction of these chlorine-containing VOCs. However, higher energies of plasma electrons typical for electron beams permit a decrease in the energy cost of not only ionization but also the generation of atoms, which explains the advantages of electron beams in the destruction of methylene chloride and TCE (Figs. 11–74 and 11–75). We should mention that major by-products of the plasma destruction of TCE diluted in air are dichloroacetyl chloride, phosgene, hydrochloric acid, as well as some amount of CO and CO_2. Plasma destruction of TCE in air requires significantly less energy than that of methylene chloride ($C_2H_2Cl_2$) and CCl_4 (compare Figs. 11–73–11–75). This can be explained by the plasma-stimulated chain mechanism of TCE destruction, with the chain propagation provided by Cl atoms. Among other plasma-treated diluted streams with Cl-containing VOCs, we should point out trichloroethane (Penetrante et al., 1996a,b). Non-thermal plasma destruction of different diluted streams with chlorine-containing VOCs has been investigated by several other research groups, including Oda, Han, and Ono (2005) and Futamura and Yamamoto (1996).

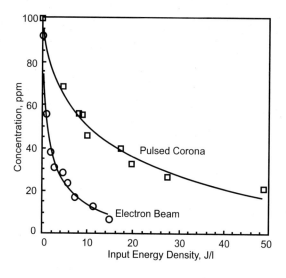

Figure 11–75. Electron beam and pulsed corona processing of 100 ppm TCE (trichloroethylene) in dry air at 25°C.

11.8.8. Non-Thermal Plasma Removal of Elemental Mercury from Coal-Fired Power Plants and Other Industrial Offgases

Elemental mercury and mercury compounds are extremely toxic, volatile, and difficult to remove from offgases. Hg removal can be effectively organized, however, using non-thermal plasma. Major sources of mercury are combustion facilities (coal-fired power plants), municipal solid waste incinerators, hazardous waste incinerators, and medical waste incinerators, which alltogether account for 87% of total Hg emissions. The U.S. Enviornmental Protection Agency established a standard regulating Hg concentration to 40 $\mu g/m^3$ (about 4.5 ppb). Typical offgas from incineration or a coal-fired boiler contains from 100 to 1,000 $\mu g/m^3$ (0.1–1 ppm) of mercury in both elemental and oxidized form. While oxidized forms of mercury (Hg(II), primarily $HgCl_2$ and HgO) can be removed by wet scrubbers, elemental mercury is chemically inert, insoluble in water solutions, and therefore difficult to remove. A significant portion (10–30%) of the total mercury load in coal-fired power plant offgases stays in the elemental form (Lee et al., 2004).

Typically, a combination of different removal techniques is required today to remove oxidized and elemental mercury. Oxidized mercury is typically removed by wet scrubbing; the basic pH normally used in the scrubbers (to increase efficiency of removal of acidic gases such as SO_2 and NO_2) promotes reduction of mercury into elemental form followed by its reentering the offgas. Conventional Hg removal technologies based on adsorption, scrubbing, and electrostatic precipitation (ESP) are not effective and also generate secondary wastes. Mercury capture in conventional control devices is limited to 27% in the cold-side ESP, 4% in the hot-side ESP, and 58% in the fabric filters (Pavlish et al., 2003). The application of non-thermal plasma permits significant improvement in the effectiveness of mercury capture, especially in the case of elemental mercury. Initial investigations of Hg removal from flue gas using a pulsed-corona enhanced dry ESP were performed by Masuda et al. (1987). The pulsed corona reactor of cylindrical coaxial geometry was powered by a 55 kV peak voltage with 50–250 Hz frequency. The initial Hg concentration in air or simulated flue gas was 500 $\mu g/cm^3$. The experiments showed a high (over 90%) removal efficiency of mercury in pulsed and DC coronas of both polarities. The removal efficiency increased with the decrease of gas temperature (from about 350°C to 100°C) and it was higher in simulated flue gas than in air. Negative polarity discharges performed better. A removal efficiency close to 99% was achieved in the negative pulsed corona regime at an energy input level of an ordinary ESP (about 0.3 kJ/m^3).

Experiments on mercury removal by pulsed corona were also performed by Helfritch, Harmon, and Feldman (1996). Elemental mercury vapor was introduced into the simulated flue gas at a concentration of 30 $\mu g/cm^3$. Pulsed corona was initiated with 27 kV pulses at 400 Hz frequency. Hg removal efficiency of 100 was achieved at an energy efficiency of 10 w/scfm (5.9 $kWh/1,000 m^3$, or 21.2 kJ/m^3). It should be mentioned for comparison that 500 W/scfm (1,060 kJ/m^3) is a typical value for energy production by coal-fired electric power plants. Powerspan Corp. developed the process named Electro-Catalytic Oxidation (ECO; Alix, Neister, & McLarnon, 1997), which combines DBD with WESP (wet ESP). The process is capable of oxidizing and removing 80–90% of mercury at initial concentrations in the offgas of ~26 $\mu g/cm^3$ (out of which ~6 $\mu g/cm^3$ is in elemental form). The DBD reactor is operated in this case at a power level of 18 W/scfm (~37 kJ/m^3) at a flue gas flow rate of 1,500 scfm. MSE Technology Applications, Inc. performed tests on removal of mercury from air and simulated flue gases using a capillary corona reactor (Babko-Malyi et al., 2000). In this reactor, oxygen and water vapor have been injected through an electrode into the offgas (at 0.1% flow rate of the main gas). At initial concentrations of elemental mercury (30–500 $\mu g/cm^3$), removal efficiencies of up to 97% have been achieved at an energy input of 6 kJ/m^3.

11.8.9. Mechanism of Non-Thermal Plasma Removal of Elemental Mercury from Exhaust Gases

Generally, Hg removal from offgases is quite challenging, mostly because mercury is present in offgases in both elemental (Hg^0) and oxidized ($Hg(II)$) forms. These two forms of mercury atoms present in air require different capture approaches. Oxidized mercury (HgO, built from $Hg(II)$) can be transferred into water solution and separated from the offgas by using conventional scrubber or WESP technologies. Elemental mercury, on the other hand, is very difficult to remove because it is not soluble in water and therefore easily escapes scrubber systems. As an example, wet and dry flue gas desulfurization systems can remove 80–90% of oxidized mercury (II) but cannot remove elemental mercury (Hg^0) at all. Plasma stimulation of mercury removal is mostly based on plasma's ability to oxidize elemental mercury (Hg^0) into Hg (II), which then can be effectively removed using the aforementioned conventional technologies. Oxidation of elemental mercury (Hg^0) by molecular oxygen has a very high activation barrier (above 3 eV/mol) and, therefore, can be completely neglected not only at room temperature but even at elevated temperatures (Hall, Schager, & Weesmaa, 1995). Thus, oxidation of mercury requires the participation of stronger oxidizers. The mechanism of plasma-stimulated mercury oxidation starts with the generation in air of such strong oxidizers as atomic oxygen and electronically excited molecular oxygen, ozone, and OH radicals. The mechanism and kinetics of generation of O atoms, electronically excited oxygen, and ozone (O_3) have been discussed in Section 6.5; the formation of OH radicals can be due to the interaction of water with atomic oxygen and ozone:

$$O + H_2O \rightarrow OH + OH, \quad O_3 + H_2O \rightarrow 2OH + O_2, \quad (11\text{–}132)$$

as well as the intensive and highly selective sequence of elementary reactions (11–113, and 11–114) based on charge transfer to water molecules. Elemental mercury (Hg^0) can be oxidized in plasma into Hg(II) in homogeneous elementary reactions with electronically excited metastable oxygen molecules:

$$Hg^0 + O_2(A^3\Sigma_u^+) \rightarrow HgO + O. \quad (11\text{–}133)$$

Another reaction path is oxidation of Hg^0 by ozone. The reaction has been experimentally studied at low reagent concentration (30 ppbV of ozone) and characterized by relatively slow kinetics (Tokos et al., 1998):

$$Hg^0 + O_3(A^3\Sigma_u^+) \rightarrow HgO + O_2, \quad k = 3 \cdot 10^{-20}\,\text{cm}^3/\text{s}. \quad (11\text{–}134)$$

Oxidation of Hg^0 by peroxide (H_2O_2) also takes place in plasma but is characterized by relatively slow kinetics similar to (11–134). The presence of water clusters, droplets, and microsurfaces (such as fly ash) can significantly accelerate the oxidation of Hg^0, in particular due to the following surface reaction (Babko-Malyi et al., 2000):

$$Hg^0 + 2OH \rightarrow Hg^{2+} + 2OH^-. \quad (11\text{–}135)$$

In liquid or cluster phases, Hg^{2+} is stable (in particular at low pH) but also may be transferred into other ionic forms of Hg(II) (Hg_2^{2+}, $HgCl^+$, $HgCl_4^{2-}$, etc.; Babko-Malyi et al., 2000). Hg(II) dissolved or captured in the clusters can then be removed from the system using conventional technologies.

The energy cost of mercury oxidation and removal through the aforementioned non-thermal plasma mechanisms is actually determined by the energy cost of plasma generation of strong oxidizers (electronically excited oxygen, OH radicals, etc.). Strongly non-equilibrium DBDs – used in the Powerspan (2004) ECO process, for example – requires 10 Wh/m^3 of combustion offgases. The application of non-thermal discharges with special additive injection is able to provide higher selectivity toward the removal of mercury and decreases the energy cost of the process (Babko-Malyi et al., 2000).

11.8.10. Plasma Decomposition of Freons (Chlorofluorocarbons) and Other Waste Treatment Processes Organized in Thermal and Transitional Discharges

Thermal and transitional (quasi-thermal) plasma systems can be effectively applied for decomposition, incineration, and reforming of different concentrated industrial and municipal wastes. The decomposition of freons (chlorofluorocarbons) [CFCs], can be considered an example of such waste treatment processes. Freons have been widely used in cooling systems, in the production of insulation foams and aerosols, and as solvents, especially in the electronic industry. The total amount of freons industrially produced and applied is very large. Exhaust from freons can penetrate to the upper atmosphere, destroying the ozone layer. Solar UV radiation dissociates CFC-producing Cl atoms, which stimulates a chain reaction of ozone destruction (a single chlorine atom is able to destroy up to 10^6 ozone molecules). For these reasons, CFC production has been prohibited by the Montreal Protocol since 1987. Any remaining stocks of freons by now have been destroyed. Decomposition of the freons in thermal plasma jets (Murphy, 1997; Murphy et al., 2001) and under direct action of the electric discharges (Kohchi, Adachi, & Nakagawa, 1996; Nakagawa, Adachi & Kohchi, 1996; Wang et al., 1999; Jasinski et al., 2002a,b) is an effective practical approach to CFC destruction.

Most of plasma CFC destruction processes are carried out in the presence of oxygen (Wang et al., 1999), air and/or water vapor (Kohchi et al., 1996; Jasinski et al., 2000), hydrogen and oxygen (Nakagawa et al., 1996), hydrogen (Wang et al., 1999), and nitrogen (Jasinski et al., 2002a,b). An Australian company, SRL Plasma Ltd., decomposes halocarbon compounds with a destruction degree of 99.99% in argon plasma jets in the presence of oxygen and water vapor, or water vapor alone (Murphy, 1997; Murphy et al., 2001). The major decomposition products under such conditions are CO_2, HCl, and HF. The presence of water vapor significantly reduces the production of other "unwanted" products such as other freons, dioxins, and furans. Although CFC destruction effectiveness in thermal plasma jets is excellent, this technology is characterized by significant heat losses and relatively high electric energy cost. The electric energy cost of plasma CFC destruction can be reduced by performing the process not in a thermal plasma jet but directly in the active discharge zone (Kohchi et al., 1996; Nakagawa et al., 1996; Wang et al., 1999; Jasinski et al., 2002a,b). Direct decomposition of freons in plasma has been organized at atmospheric pressure (Jasinski et al., 2000, 2002a,b) as well as at reduced pressures (Kohchi et al., 1996; Nakagawa et al., 1996; Wang et al., 1999); the concentration of freons varied from very low (several dozen ppm) in diluted mixtures to very high (several dozen percent) in non-diluted exhausts.

Effective decomposition of $CFCl_3$, CF_2Cl_2, and CHF_2Cl was achieved in the transitional regimes of gliding arc discharges in the presence of water vapor (Czernichowski et al., 1994) and oxygen (Opalska, Opalinska, & Ochman, 2002a; Opalska et al., 2002b). An admixture of H_2 to freons during their treatment in gliding arc discharges limits products to simple aliphatic hydrocarbons, completely avoiding the formation of chlorinated dioxins and furans or phosgene, which can be generated in oxidative or redox media (Czernichowski et al., 1994). Effective CFC destruction by steam plasma generated in an atmospheric-pressure DC discharge has also been demonstrated by Watanabe, Taira, and Takeuchi (2005). Similar to freons, plasmas can be effectively applied for reducing perfluorinated compound (PFC; for example, CF_4, SF_6) emissions from chamber cleaning and dielectric etch tools in electronics (Kabouzi et al., 2005). Although the plasma PFC abatement is mostly organized today at relatively low pressures, more and more attention is being given to the use of atmospheric-pressure discharges for this application.

Thermal plasmas can be effective in compaction and destruction of liquid and solid hazardous wastes (Watanabe, 2003), including treatment of radioactive wastes by plasma incineration and vitrification for final disposal (Tzeng et al., 1998), plasma compaction of radioactive waste (in particular, ion exchange resin; Nezu, Morishima, & Watanabe, 2003), and complete incineration of medical and municipal wastes (Dmitriev et al., 2003). Thermal plasmas are effective in fly ash treatment and detoxification (Sakano, Tanaka, & Watanabe, 2001). In this case, useful metals and other materials can be recovered from the fly ash separately due to the difference in condensation temperatures. A similar approach has been successfully applied for treatment of Li ion battery waste using transferred arc plasma (Silla & Munz, 2005). From products of the plasma battery waste treatment, the metal phase (Containing 84% Co, 10% Cu, 4% Fe, and 2% Ni) can be recovered separately due to the mentioned condensation temperature effect. Thermal plasma treatment of such hydrocarbon-based waste as wood, car tires, some municipal wastes, and others can be carried out not only as incineration but also as reformation leading to syngas production, which was especially discussed in Chapter 10 (see also Rutberg et al., 2005). The large-scale environmental application of plasma is related to water purification and cleaning of exhaust water, which is considered in particular in the next chapter in relation to biomedical aspects of water treatment.

PROBLEMS AND CONCEPT QUESTIONS

11–1. Radical-Thermal "Bootstrap" Effect in the Subthreshold Plasma-Ignition of Hydrogen. Estimate how long a non-branching chain of radical-stimulated oxidation in stoichiometric H_2–O_2 mixture should be to provide the radical-thermal "bootstrap" ignition 200 K below the temperature threshold of auto-ignition. Assume that the initial concentration of the plasma-generated radicals is low, molar fraction is 10^{-4}, and neglect direct plasma heating during the non-equilibrium plasma ignition.

11–2. Contribution of Vibrationally and Electronically Excited Molecules into Plasma-Stimulated Ignition of H_2–Air Mixtures. Estimate the relative contribution of nitrogen molecules, vibrationally and electronically excited in non-equilibrium plasma, in the stimulation of ignition of H_2–air mixtures through reactions (11–17) and (11–20). Compare the possible contribution to the ignition of electronically and vibrationally excited nitrogen molecules with the contribution of electronically and vibrationally excited oxygen molecules.

11–3. Contribution of Ions in the Subthreshold Plasma Ignition of Hydrogen. Analyze the non-branching chain mechanism (11–25), (11–27), (11–28), (11–29) of the plasma-stimulated quasi-catalytic low-temperature oxidation of H_2. Why does the conventional gas-phase oxidation of hydrogen have a threshold nature (explosion/no explosion; see Section 11.2.3), whereas the quasi-catalytic chain mechanism (11–25), (11–27), (11–28), (11–29) is able to occur in a wide range of pressures and temperatures without any specific thresholds?

11–4. Energy Requirements for Plasma Ignition in Ram/Scramjet Engines. Analyze equation (11–36) to explain why the specific energy input E_v required for a non-equilibrium plasma ignition in ram- and scramjet engines decreases when using discharge systems operating at higher values of the reduced electric field E/n_0.

11–5. Optimal Specific Impulse of Electric Propulsion Systems (Ion and Plasma Thrusters). Based on relation (11–43) between the initial mass M_0 of a spacecraft at low (near-Earth) orbit and its specific impulse I_{sp}, derive formula (11–44) for the optimal value of the specific impulse, I_{sp}^{opt}. Explain why the optimal value of the specific impulse of the ion and plasma thrusters, I_{sp}^{opt}, does not depend on the value of their thrust and is mostly determined by the duration of space flight.

11–6. Electric and Thermodynamic Characteristics of Magneto-Hydrodynamic Generators. Using equations of mass, momentum, and energy conservation and neglecting heat and friction losses, derive relation (11–51) between the local internal energy efficiency η_0 (11–50) and the local energy efficiency η(11–46) of an MHD generator.

11–7. Discharge Parameters and Plasma-Chemical Processes in CO$_2$ Lasers. Using equation (11–53), determine the typical values of electron density in conventional stationary low-pressure discharge CO$_2$ laser systems. Based on the values of electron density and the relevant reaction rate coefficient, estimate the time of CO$_2$ dissociation in the system through dissociative attachment and electronic excitation by direct electron impact.

11–8. Plasma Cleaning of SO$_2$-Containing Exhaust Gases; Plasma-Stimulated Liquid-Phase Chain Oxidation of SO$_2$ in Droplets. Analyzing kinetic mechanisms (11–58)–(11–68) and (11–69)–(11–75), compare the effectiveness of liquid-phase SO$_2$ oxidation to sulfuric acid (H$_2$SO$_4$) in acidic (pH $<$ 6.5) and neutral/basic (pH $>$ 6.5) droplets. Check your predictions with the data on radiation yield of the SO$_2$ oxidation presented in Fig. 11–57.

11–9. Plasma Cleaning of SO$_2$-Containing Exhaust Gases; Plasma-Stimulated Chain Oxidation of SO$_2$ in Ionic Clusters. Analyzing the SO$_2$ oxidation energy cost dependence on the ion density (11–85) and (11–86), find the minimum value of the energy cost, A^{min} ($n_1 < n_2$), and the optimal value of the ion density (and related electron-beam current density) when the minimum energy cost A^{min} can be achieved. Compare your results with equations (11–92), and (11–93), and numerically with experimental data presented in Fig. 11–56.

11–10. VOC Removal from Exhaust Gases Using Wet Pulsed Corona Discharge. Explain why the combination of non-thermal plasma treatment with water scrubbing permits a significant decrease of energy cost of VOC removal from polluted gas streams? Why is it that a single OH radical can be sufficient to convert a non-soluble organic molecule into a soluble one?

12

Plasma Biology and Plasma Medicine

Plasma biology and medicine are rapidly growing new areas of non-thermal plasma science and engineering. Not so long ago, biomedical applications of non-thermal plasma were mostly focused on surface sterilization as well as treatment of different surfaces to control their compatibility with biomaterials. Now, plasma is also applied in solving novel sophisticated problems of tissue engineering, sterilization of reusable heat-sensitive medical instruments, and of large-volume air and water streams. Recently, non-thermal plasmas have also been directly applied in medicine, in treatment of living tissues, including sterilization and healing of wounds, blood coagulation, and treatment of skin deceases. Such applications are to be discussed, starting with the biological and finishing with the medical.

12.1. NON-THERMAL PLASMA STERILIZATION OF DIFFERENT SURFACES: MECHANISMS OF PLASMA STERILIZATION

Non-thermal plasma is an effective source of active species and factors, such as radicals, ions, excited atoms and molecules, UV radiation and so on, which are able to deactivate, kill, or even completely disintegrate bacteria, viruses, and other micro-organisms without any significant temperature effects. It attracts interest to plasma applications in sterilization and disinfection. Disinfection usually implies a couple-order-of-magnitude reduction of populations of micro-organisms, while sterilization usually requires at least a 10^4–10^5-fold reduction in the number of micro-organisms. This section focuses on plasma sterilization and disinfection of different non-living surfaces (such as, for example, medical instruments and equipment). The following sections focus on sterilization of air and water streams, as well as on sterilization of living tissues (such as, for example, human skin and organs in vivo). This section will start with separate discussions regarding plasma sterilization using low-pressure and atmospheric-pressure discharges, which will be followed by an analysis of mechanisms and kinetics of plasma sterilization.

12.1.1. Application of Low-Pressure Plasma for Biological Sterilization

Earlier non-thermal plasma sterilization experiments were carried out mostly at low gas pressures. It should also be mentioned that initial low-pressure plasma sterilization methods implied the application of gas mixtures containing components with germicidal properties such as H_2O_2 and aldehydes (Boucher, 1980; Jacobs & Lin, 1987). Advantages of plasma sterilization are revealed when it uses gases (such as air, He–air, He–O_2, N_2–O_2) without germicidal properties of their own, which become biocidal only when the plasma is ignited. Such studies, motivated by the decontamination of interplanetary space probes and sterilization of medical tools, have been performed using radiofrequency (RF) and microwave low-pressure discharges in oxygen and O_2–N_2 mixtures (Moreau et al., 2000; Moisan et al.,

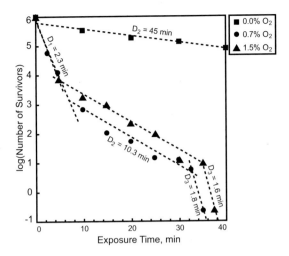

Figure 12–1. Survival curves for *B. subtilis* spores exposed to a N_2/O_2 discharge afterglow (field frequency is 2450 MHz). Gas pressure in the sterilization chamber is 5 Torr. Gas flow is one standard L/min with some percentage of oxygen added.

2001; Bol'shakov et al., 2004). The effect of low-pressure RF oxygen plasma on bacteria has been investigated by Bol'shakov et al. (2004) in both inductively coupled plasma (ICP) and capacitively coupled plasma (CCP) discharges. The ICP provided better efficiency in destroying biological matter due to higher electron and ion densities in this mode. High densities of atomic and electronically excited oxygen in synergy with ultraviolet (UV) photons induced chemical degradation of the biological materials followed by volatilization of the decomposition products (CO_2, CO, etc.). DNA degradation was evaluated for both ICP and CCP modes of low-pressure non-thermal RF plasma. It was found that at the same power the ICP discharge destroyed over 70% of supercoiled DNA in 5s, whereas only 50% of DNA was destroyed in the same conditions in the CCP discharge.

The effect on bacteria of low-pressure N_2–O_2 afterglow plasma generated using a surfatron source driven by microwave power with frequencies 0.915 and 2.450 GHz (Moisan & Zakrzewski, 1991) has been investigated by Moreau et al. (2000) and Moisan et al. (2001). The **survival curves** (**colony forming units**, [cfu] versus treatment time) in the experiments with *Bacillus subtilis* spores exhibited three inactivation phases, which can be seen in Fig. 12–1 (Philip et al., 2002). The first phase, which exhibited the shortest **D-value** (decimal value, or time required to reduce an original concentration of micro-organisms by 90% – one log_{10} reduction), corresponds to the action of UV radiation on isolated spores or on the first layer of stacked spores. The second phase, which is characterized by the slowest kinetics, can be attributed to a slow erosion process by active species (such as atomic oxygen, O). The third phase starts after the outer spore debris are cleared during phase 2, hence allowing UV to hit the genetic material of the still-living spores. The D-value of this phase is close to that of the first phase. The survival curves of the low-pressure plasma inactivation of *B. subtilis* spores have only two phases (Moisan et al., 2001). The radical debris removal stage is not required, and UV radiation can dominate both phases. The first phase is similar to that described in the previous paragraph (see Fig. 12–1). The second phase represents in this case spores that are shielded by others and require longer irradiation time to accumulate a lethal UV dose. The key contribution of UV radiation in the experiments is supported by the fact that at low UV intensity there is a lag time before inactivation. A minimum UV dose should be achieved before irreversible damage to the DNA strands occurs (Moisan et al., 2001). Practical applications of low-pressure non-thermal plasmas for surface sterilization can be related to treatment of medical instruments or equipment without the use of high-temperature and aggressive chemicals, as well as disinfection in low-pressure extraterrestrial conditions, for example on Mars or other planets.

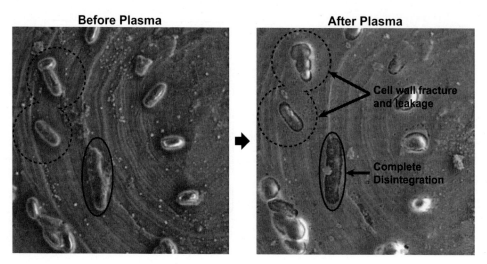

Figure 12–2. *B. subtilis* (left) before and (right) after treatment with DBD plasma (120 s, 0.8 W/cm^2).

12.1.2. Inactivation of Micro-Organisms by Non-Equilibrium High-Pressure Plasma

Non-thermal atmospheric-pressure discharges, particularly in air, are very effective and convenient in deactivation of micro-organisms. Atmospheric-pressure non-thermal discharges in air are characterized by a high density of strongly reactive oxidizing species and are able not only to kill micro-organisms without any notable heating of the substrate, but also to destroy and decompose the micro-organism. This ability of complete but cold destruction of micro-organisms in atmospheric-pressure plasmas is attractive for complete sterilization of spacecrafts in the framework of the planetary protection program. atmospheric-pressure cold dielectric barrier discharge (DBD) plasma has been demonstrated as an effective tool, particularly, in the destruction of such resistive micro-organisms as *Bacillus subtilis* (spores), *Bacillus anthracis* (anthrax spores), and *Deinococcus radiodurans* (micro-organisms surviving strong radiation of nuclear materials). The effective sterilization (5-log reduction) and significant destruction (cell wall fracture and leakage as well as complete disintegration) of *Bacillus subtilis* spores in DBD air plasma (120 s treatment, D-value 24 s) is illustrated in Fig. 12–2. The range of applications of plasma sterilization at atmospheric pressure is wide, from medical instruments and spacecraft to different food products, which has been reviewed, in particular, by Fridman (2003).

The mechanisms and kinetics of sterilization processes are more sophisticated in the high-pressure strongly non-equilibrium plasmas, because of the significant contribution of collisional gas-phase processes at higher pressures and therefore a wider variety of active species involved in the sterilization kinetics. In particular, the germicidal effect of non-thermal atmospheric-pressure plasma is characterized by different shapes of the survival curves (see previous section) depending on the type of micro-organism, the type of the medium supporting the micro-organisms, and the method of exposure (Laroussi, 2005). The "single slope" survival curves (one-line curves) for inactivation of the bacterial strains have been observed in atmospheric-pressure plasma sterilization by Herrmann et al. (1999), Laroussi et al. (1999) and Yamamoto, Nishioka, and Sadakata (2001), which is illustrated in Fig. 12–3. The D-values range in this case from 4 s to 5 min. Two-slope survival curves (two consecutive lines with different slopes) have been observed in atmospheric-pressure plasma by Kelly-Wintenberg et al. (1998), Laroussi, Alexeff, and Kang (2000), and Montie, Kelly-Wintenberg, and Roth (2000). The D-value of the second line (D$_2$) is smaller (shorter

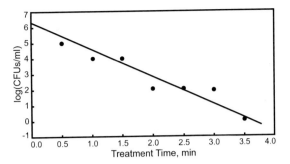

Figure 12–3. Single-phase surviving curve: *E. coli* on Luria-Bertani broth exposed to a He–air plasma generated by DBD at atmospheric pressure.

time) in these systems than the D-value of the first line (D_1). The D_1-value is dependent on the species being treated, and the D_2-value is dependent on the type of surface supporting the micro-organisms. The two-slope survival curve can be explained by taking into account that the active plasma species during the first phase react with the outer membrane of the cells, inducing damaging alterations. After this process has sufficiently advanced, the reactive species can quickly cause cell death, resulting in a rapid second phase. Multi-slope survivor curves (three kinetic phases or more) were also observed in sterilization by non-thermal atmospheric-pressure plasmas. This type of kinetics can be illustrated by deactivation of *Pseudomonas aeuroginosa* on a filter exposed to He–air DBD plasma (see Fig. 12–4; Laroussi, Alexeff, & Kang, 2000). Interpretation of the multi-slope kinetic effect can be similar in this case to that described in the previous section regarding sterilization by low-pressure plasma (see Fig. 12–3). Thus, the bacterial inactivation kinetics reveals the complexity of mechanisms of sterilization by non-equilibrium high-pressure plasmas. Several factors can impact the killing process: type of bacteria, type of medium in which the cells are seeded, number of cell layers, type of exposure, contribution of UV, operating gas mixture, and so on. Generally, if UV plays an important or dominant role, the survival curves tend to exhibit a first rapid phase (small D-value) followed by a second slower phase (outer layer of bacteria is deactivated by UV during the fast first phase; then slower erosion of the bacteria proceeds during the longer second phase). When UV does not play a significant role, such as in the case of air plasma (where the contribution of active oxidizing species is essential), the single-phase survival curves are mainly observed.

12.1.3. Plasma Species and Factors Active for Sterilization: Direct Effect of Charged Particles

Plasma sterilization is a quite complicated process determined by multiple plasma species and factors, including charged and excited species, reactive neutrals, and UV radiation. The contribution of these factors differs in different types and regimes of non-equilibrium plasma discharges; also it differs from the point of view of induced biological pathways. The synergistic nature of the interaction between the plasma factors is important in sterilization similarly to the plasma treatment of polymers (see Section 9.7). First consider the

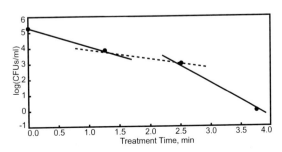

Figure 12–4. Multi-phase surviving curve: *Pseudomonas aeuroginosa* on filter exposed to He–air plasma generated by a DBD at atmospheric pressure.

contribution of charged particles (electrons and ions) in plasma sterilization. The charged particles play a major role in sterilization, especially in the case of so-called direct plasma treatment, when plasma is in direct contact with a micro-organism and, furthermore, electric current can cross the sterilization zone. Advantages and higher effectiveness of the direct (vs. indirect) plasma sterilization and key contribution of charged particles in this case will be discussed in Section 12.2. The contribution of charged particles into sterilization can be divided into three effects: direct chemical effect of electrons and ions, direct effect of ion bombardment, and effect of electric field induced by the charged particles. The first two (direct) sterilization are to be discussed in this section, and the third will be considered in Section 12.1.4 in relation to general effects of electric fields.

1a. Chemical effect of electrons. Electrons make a significant contribution to modification of polymers and especially biopolymers (see Section 9.7.2). The effectiveness of electrons in sterilization is related in particular to a quite significant depth of their penetration into the polymers, $x_0 \approx 2 \mu m$ (Avasov, Bagirov, & Malin, 1969; Bochin, Rusanov, & Lukianchuk, 1976; Rusanov & Fridman, 1978a,b). Characteristic depth of inactivation of bioorganisms (sterilization) provided by plasma electrons has been estimated by Rusanov and Fridman (1978) and Bamford and Ward (1961) as

$$x^2 \approx x_0^2 (\ln \sigma_{io} N - \ln \ln K), \qquad (12\text{--}1)$$

where $\sigma_{io} \approx 10^{-9} cm^2$ is the electron-impact biodeactivation cross section (Dubinin, 1969); $N(cm^{-2})$ is the total time-integrated flux of electrons to the surface of a bioorganism; and $K \gg 1$ is the destruction degree. Assuming, for example, $N = 10^{13} cm^{-2}$ and $K = e$, the characteristic biodeactivation depth provided by the plasma electrons is about $6 \mu m$. The significance of electrons with respect to other charged particles (ions) in sterilization is also due to the possibility that they provide relatively high electron flux to the treated surface, especially in the case of direct plasma treatment (see Section 12.2). Charged particles compete in sterilization effectiveness with reactive neutral species (OH, ozone, etc.) characterized by about 10^5-fold higher number density in non-thermal plasma. Electrons with very high drift velocities (10^7 cm/s and higher) are able nevertheless to provide fluxes higher than that of the reactive neutral species. In electronegative gases (in particular, air), it requires the application of discharges (for example, DBD) operating in the "breakdown regimes," when ionization rates exceed those of electron attachment, preventing significant formation of negative ions O_2^- (see Section 4.1).

Plasma sterilization is characterized by quite significant depth of penetration, which means that deactivation of bioorganisms occurs under a layer of water, biosolution, and so on. From this point of view (and also remembering that the micro-organisms themselves contain a lot of water), it is interesting to analyze the behavior of plasma electrons when they reach and immerse in the water surface, becoming hydrated (aqueous) electrons e_{aq}. The **hydrated electrons** are surrounded by polar water molecules and remain quite stable, providing the deep penetration of the electrons in water. They react with different admixtures in water, in particular with biomolecules and with peroxide (converting it into OH and OH$^-$, which is the Fenton reaction). Between the most important processes, a hydrated electron e_{aq} is able to convert oxygen dissolved in water into **superoxide**, the highly reactive O_2^- ion radical:

$$e_{aq} + O_2(H_2O) \rightarrow O_2^-(H_2O). \qquad (12\text{--}2)$$

The superoxide is a precursor to other strong oxidizing agents, including singlet oxygen and peroxynitrite. It is interesting to mention that certain cells in the human body produce superoxide as an antibiotic "weapon" to kill invading micro-organisms. The superoxide also acts as a signaling molecule needed to regulate cellular processes. In biological systems with a sufficient level of acidity, the superoxide is converted into hydrogen peroxide and oxygen in a reaction called the **superoxide dismutation**:

$$2O_2^- + 2H^+ \rightarrow H_2O_2 + O_2. \tag{12-3}$$

The superoxide dismutation can be spontaneous or can be catalyzed and therefore significantly accelerated by the enzyme superoxide dismutase (SOD). The superoxide not only generates hydrogen peroxide (H_2O_2) but also stimulates its conversion into OH radicals, which are actually extremely strong oxidizers very effective in sterilization through a chain oxidation mechanism (to be especially discussed in Section 12.1.5). The H_2O_2 conversion into OH, known as the **Fenton reaction**, proceeds as a redox process provided by oxidation of metal ions (for example, Fe^{2+}):

$$H_2O_2 + Fe^{2+} \rightarrow OH + OH^- + Fe^{3+}. \tag{12-4}$$

Restoration of the Fe^{2+} ions takes place in the Fe^{3+} reduction process provided by the superoxide ion radicals, O_2^-:

$$Fe^{3+} + O_2^- \rightarrow Fe^{2+} + O_2. \tag{12-5}$$

Superoxide can also react in water with nitric oxide (NO) producing peroxynitrate ($OONO^-$), which is another highly reactive oxidizing ion radical:

$$O_2^- + NO \rightarrow OONO^-. \tag{12-6}$$

To summarize, electrons are able to provide significant sterilization effects even deep under the surface of a liquid. In water, plasma electrons become the hydrated electrons e_{aq} and the superoxide O_2^- ion radicals (12–2), which are then converted into H_2O_2 (by dismutation (12–3)) and such extremely strong oxidants and as hydroxyl (OH) and peroxynitrate ($OONO^-$). These oxidants are very effective in deactivation of micro-organisms.

1b. Chemical Effect of Plasma Ions. Taking into account the balance of charges on the treated surfaces, the effect of ambipolar diffusion and acceleration of ions in the sheaths and fluxes of positive ions to the treated surface can be significant (although usually lower than fluxes of reactive neutral species, and drift-related pulsed fluxes of electrons in the "breakdown" regimes, see earlier discussion). Fluxes of negative ions are not so high with respect to those of positive ions typically because of negative charge of the surfaces and not as strong fields in the anode sheaths. The major negative ion in non-thermal air plasma is O_2^-. When these ions are immersed in water, they become superoxides (O_2^-) and further form H_2O_2 and OH according to mechanism (12–3 to 12–5), similarly to the case of sterilization provided by the chemical effect of plasma electrons (see previous discussion). As mentioned, the fluxes of negative ions from plasma are typically relatively low, which somewhat decreases their contribution to sterilization. The chemical effect of positive plasma ions M^+ (for example, N_2^+) on micro-organisms can be illustrated by the effect of positive plasma ions M^+ on the water surface (which represents either the biological solution where the micro-organisms are located, or the micro-organisms themselves). Interaction of the plasma ions M^+ with water molecules starts with fast charge transfer processes to the water molecules, characterized by relatively low ionization potential:

$$M^+ + H_2O \rightarrow M + H_2O^+. \tag{12-7}$$

Water ions then rapidly react with water molecules, producing H_3O^+ (which makes the water acidic) and OH radicals, playing an important role in sterilization:

$$H_2O^+ + H_2O \rightarrow H_3O^+ + OH. \tag{12-8}$$

The positive plasma ions make water acidic. Hydrated electrons can stabilize the plasma-stimulated decrease of pH (see reactions (12.2 to 12.5)), which requires, however, special conditions. Thus, non-thermal plasma interaction with water conventionally decreases pH, making the water slightly acidic. The contribution of both electrons and ions (positive and negative) in one way or another leads to the effective generation of OH and other highly reactive oxidants resulting in intensive oxidation of biomaterials and a strong sterilization effect. It should be mentioned that ions also provide significant catalytic effect and affect multiple biochemical channels.

2. Effect of Ion Bombardment. The deactivation of micro-organisms can be due to the destruction of the lipid layer of their membranes or other membrane damage by ion bombardment from plasma. This effect is especially important in the case of direct plasma treatment of micro-organisms. Although the ion bombardment makes a bigger contribution in the low-pressure plasmas when the ion energies are quite high (see Section 8.2), the effect of ion bombardment in atmospheric-pressure discharges with relatively low ion energies can also be significant. The average energy of ions $\langle \varepsilon_i \rangle$ in the ion bombardment process in high-pressure non-equilibrium ($T_e \gg T_0$) plasmas can be estimated as energy received by the ions in electric field E during their displacement on the mean free path $eE\lambda$ (Fridman & Kennedy, 2004):

$$\langle \varepsilon_i \rangle \approx T_0 + T_e \sqrt{\delta}, \tag{12–9}$$

where T_e, T_0 are electron and gas temperatures in plasma; δ is the average fraction of energy lost by electrons in the electron–neutral collisions. Taking into account that $\delta \approx 0.1$ in molecular gases, average energies of the ion bombardment even in high-pressure plasma systems can reach 0.3–0.7 eV. Although these energies are not sufficient to break strong chemical bonds, they are high enough to break the hydrogen bonds responsible for the integrity of lipid layers forming cellular membranes. Thus, in addition to chemical effects, ions are able to provide deactivation of micro-organisms through mechanical effects of bombardment (and possibly related osmotic-pressure effects).

12.1.4. Plasma Species and Factors Active for Sterilization: Effects of Electric Fields, Particularly Related to Charged Plasma Particles

The direct effect of externally applied electric fields on sterilization and deactivation of micro-organisms (electroporation) in plasma can usually be neglected (Neumann, Sowers, & Jordan, 2001). At the same time, the effects of electric fields related to collective motion and deposition of charged particles can be significant. According to Mendis, Rosenberg, and Azam (2000) and Laroussi, Mendis, and Rosenberg (2003), deposited charged particles can play a significant role in the rupture of the outer membrane of bacterial cells. It was shown that electrostatic forces caused by charge accumulation on the outer surface of the cell membrane could overcome the tensile strength of the membrane and cause its rupture. Charged bacterial cells with micrometer sizes experience large electrostatic repulsive forces proportional to the square of the charging potential Φ and inversely proportional to the square of the radius of the cells curvature, r. The charging (floating) potential Φ is negative and depends on the ratio of the ion mass to the electron mass. The condition for disruption of the bacterial cell membrane can be expressed in this case as (Mendis, Rosenberg, & Azam, 2000)

$$|\Phi| > 0.2 \cdot \sqrt{r \Delta F_t}, \tag{12–10}$$

where Δ is the thickness of the membrane, and F_t is its tensile strength. This mechanism can be relevant for gram-negative bacteria with irregular surface membranes. These irregularities with small radii of curvatures can result in localized high electrostatic forces (Laroussi, Mendis, & Rosenberg, 2003). High strengths of electric fields and electroporation

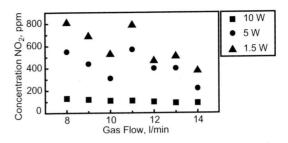

Figure 12–5. NO$_2$ concentration generated in air DBD as a function of flow rate at three power levels: 1.5, 5, and 10 W.

effects can be achieved not only because of the charge deposition on irregular surfaces of gram-negative bacteria but also due to the collective and strongly localized motion of the charged particles, especially in streamers of DBDs, (see Sections 4.1.2, and 4.1.3). Strengths of electric fields in the DBD streamers significantly exceed those of the externally applied electric fields (see equation (4.12)). Therefore, while the externally applied electric fields in the atmospheric-pressure discharges are not sufficient for electroporation and sterilization, electric fields in the streamers and electric fields of ions are able, in principle, to make a contribution in plasma deactivation of micro-organisms.

12.1.5. Plasma Species and Factors Active for Sterilization: Effect of Reactive Neutral Species

Reactive neutral species generated in non-thermal plasmas make a significant contribution into sterilization and more generally into the treatment of biomaterials, especially at high pressures. Strongly non-equilibrium air plasmas, for example, are excellent sources of reactive oxygen-based and nitrogen-based species, such as O, O$_2$ ($^1\Delta_g$), O$_3$, OH, NO, NO$_2$, and so on. Some of the neutrals (such as active atoms, radicals O and OH, as well as excited species O$_2$ ($^1\Delta_g$) are extremely reactive, but have very short lifetimes and are effective for sterilization only inside or in close vicinity of plasma. Other neutrals (such as saturated molecules (O$_3$) and passive radicals (NO, NO$_2$)) are not so extremely reactive, but have a relatively long lifetime and can be effective for treatment of biomaterials far from the plasma zone where they have been generated. The kinetics of air-plasma generation of such major reactive neutrals as O$_3$, O, OH, NO, and NO$_2$ has been analyzed in Sections 6.1 and 6.5 regarding the plasma synthesis of ozone and nitrogen oxides. Concentrations of O$_3$ and NO$_2$ in DBD air plasma applied for sterilization have been measured by Laroussi and Leipold (2004) and Minayeva and Laroussi (2004). The concentrations of O$_3$ and NO$_2$ are presented in Figs. 12–5 and 12–6 as functions of the discharge power and air flow rate.

The germicidal effects of ozone are caused, in particular, by its interference with the cellular respiration system. Generally, the oxygen-based and nitrogen-based plasma-generated reactive neutral species have strong oxidative effects on the outer structures of cells. Cell membranes are made of lipid bilayers, an important component of which is unsaturated fatty acids. The unsaturated fatty acids give the membrane a gel-like nature, which allows transport of the biochemical agents across the membrane. Since unsaturated fatty acids are susceptible to OH attacks (Laroussi & Leipold, 2004), the presence of hydroxyl radicals can compromise the function of the membrane lipids whose role is to act as a barrier against the transport of ions and polar compounds in and out of the cells (Bettleheiem & March, 1995). Protein molecules, which are basically linear chains of amino acids, are susceptible to oxidation by atomic oxygen, OH, or metastable electronically excited oxygen molecules. Proteins play the role of gateways that control the passage of various macromolecules in and out of cells. In the case when bacteria are of the gram-positive type, they are able to form spores, which are highly resistive states of cells. Spores are made of several coats surrounding a genetic core. These coats are also made of proteins susceptible to chemical

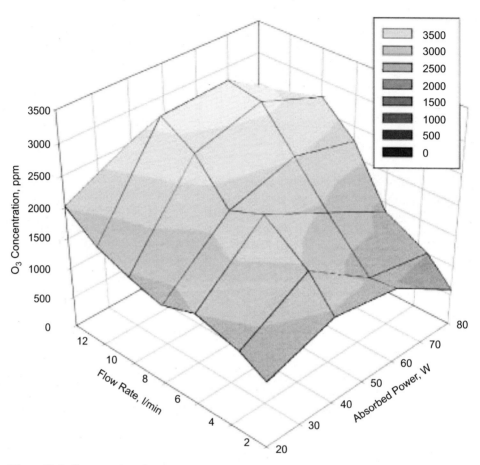

Figure 12–6. Concentration of ozone generated in a DBD in air as function of air flow rate and dissipated discharge power.

attack of reactive neutrals. Therefore, the reactive neutral species generated by air plasmas are expected to greatly compromise the integrity of the walls, coats, and membranes of the cells of micro-organisms.

Extremely high reactivity in biochemical reactions is provided by hydroxyl radicals (OH) generated in plasma. The OH radicals easily pass through membranes and cannot be kept out of cells. Hydroxyl radical damage is so fast that it is only "diffusion rate-limited." Hydroxyl attaches to an organic substrate R (which could be, for example, a fatty acid), forming a hydroxylated adduct-radical ROH:

$$OH + R \rightarrow ROH \, (radical). \tag{12–11}$$

The adduct-radical ROH can be further oxidized by metal ions or oxygen, resulting in an oxidized but stable product:

$$ROH \, (radical) + Fe^{3+} \rightarrow HOR + Fe^{2+} + H^+, \tag{12–12}$$

$$ROH \, (radical) + O_2 \rightarrow HOR + O_2^- + H^+. \tag{12–13}$$

Two adduct-radicals ROH can also react with each other, forming a stable, cross-linked-but-oxidized product:

$$ROH \, (radical) \, H + ROH \, (radical) \, H \rightarrow R - R + 2H_2O. \tag{12–14}$$

A significant part of OH oxidation of bioorganic materials is due to the initial dehydrogenization of the substrate organic molecules RH:

$$OH + RH \rightarrow R + H_2O. \tag{12–15}$$

The resulting oxidized substrate bioorganic compound R is a radical and reacts with molecules in chain processes. It could react with oxygen to produce a peroxyl radical, R-OO-:

$$R + O_2 \rightarrow R - OO - . \tag{12–16}$$

The peroxyl radical is highly reactive and can react with another organic substrate, RH, propagating the chain oxidation mechanism:

$$(R - OO-) + RH \rightarrow ROOH + R. \tag{12–17}$$

This type of biochemical chain reaction of OH is common in the oxidative damage of fatty acids and other lipids and demonstrates why radicals such as the hydroxyl radical can cause so much more damage than one might have expected. Plasma-generated ozone is at least partially dissociated when immersed into water, liquid-based, or liquid-like biomaterial:

$$(O_3)_{aq} \rightarrow (O_2)_{aq} + O_{aq}. \tag{12–18}$$

Atomic oxygen created in the process intensively reacts with the bioorganic molecules RH, producing R radicals and OH radicals:

$$O + RH \rightarrow R + OH, \tag{12–19}$$

which then stimulate the same biochemical chain oxidation mechanism (12–15)–(12–17). Thus, ozone (O_3) can actually be added to the intensive OH-based biochemical oxidation.

Similar damage caused by hydroxyl radicals and other reactive oxygen species can occur in proteins and with nucleic acids (mainly DNA). Proteins are highly susceptible to oxidative damage, particularly at sites where sulfur-containing amino acids are found. DNA can be oxidatively damaged at both the nucleic bases (the individual molecules that make up the genetic code) and at the sugars that link the bases. Oxidative damage of DNA results in degradation of the bases, breaking of the DNA strands by oxidation of the sugar linkages, or cross-linking of DNA to protein (a form of damage particularly difficult for the cell to repair). Although all cells have some capability of repairing oxidative damage to proteins and DNA, excess damage can cause mutations or cell death (Christophersen et al., 1991). Thus, non-thermal atmospheric plasma-generated reactive neutral species, particularly OH and O_3, are very effective in deactivation of micro-organisms, and generally in the treatment of biomaterials. The reactive neutrals, especially OH, can be not only generated in plasma and transported to the surface of biomaterials (as assumed in this section) but also produced in the biomaterials from plasma-generated charged particles (see Section 12.1.4). In the case of direct plasma treatment with sufficiently high electric fields (see the next section), the flux of charged particles can dominate in the production of reactive neutrals in the biomaterial. Otherwise, the reactive neutrals are mostly generated in plasma and then transported to the surface of the biomaterials. Numerical kinetic data on the effect of O_3 and OH on deactivation of different micro-organisms are to be summarized later in connection with plasma sterilization of air streams (Section 12.3).

Significant sterilization effect can be provided by H_2O_2 plasma generated in water. Conventional DBD produces in water more than 10–100 μM of H_2O_2, which are able to penetrate into cells and stimulate double-strand breaks of DNA after seconds of treatment. Direct treatment of water by air-plasma can lead to higher H_2O_2 concentrations.

12.1.6. Plasma Species and Factors Active for Sterilization: Effects of Heat

The first factor that usually comes to mind when discussing sterilization is the thermal effect. Many conventional sterilization methods are based on the use of either moist heat or dry heat with typical treatment time of about 1 h (Block, 1992). Moist heat sterilization is usually organized in autoclaves at a typical temperature of $121°C$ and atmospheric-pressure. Dry heat sterilization requires temperatures close to $170°C$. Most non-thermal plasma discharges operate at lower temperatures (from room temperature to about $70°C$), which does not provide thermal sterilization (Laroussi & Leipold, 2004). Neglecting the effect of heat in non-thermal plasma sterilization, we should keep in mind that even strongly non-equilibrium discharges can be characterized by elevated temperatures in some localized intervals in space or in time. For example, non-equilibrium gliding arc discharges (see Sections 4.3.8–4.3.11) are cold on average, but they can provide significant short-term local increase of gas temperature exceeding $1000°C$. Such short-time local but intensive heating makes a significant contribution to sterilization, providing in particular not only the killing of micro-organisms also but their complete disintegration. Even analyzing sterilization in DBD, traditionally considered non-thermal, the thermal effect should be sometimes also taken into account. It can be related to the DBD microdischarge channels (see Section 4.5.3), where temperature can locally reach several hundreds of degrees. Overheating of the microdischarge channels occurs if they are relatively thick due to non-uniformity of electrodes or special conditions of the electrodes (wetness, dirt, surface conductivity, etc.). These problems are especially important when living tissue is used as one of DBD electrodes, which will be discussed later regarding plasma medicine. Although the thermal sterilization effect of non-thermal plasma discharges is not significant by itself, the strong contribution of, for example, radicals or charged species discussed earlier can be additionally enhanced by even a small temperature increase. Generally, plasma sterilization is a complicated process where even a minor factor can make an important synergetic contribution in combination with the major plasma sterilization factors. Local overheating of plasma discharges can lead to the formation of shockwaves and to the **shockwave sterilization effect**. This effect can be significant for discharges in water and related water sterilization (see Section 12.4). To summarize experimental results, the heat effect is not a major contributor to the sterilization effect in non-thermal and strongly non-equilibrium plasmas.

12.1.7. Plasma Species and Factors Active for Sterilization: Effect of Ultraviolet Radiation

Plasma is a source of UV with different wavelengths, which can be effective in sterilization. Four relevant ranges of UV wavelengths specifically can be indicated: vacuum ultraviolet (VUV) radiation between 10 and 100 nm, UV-C radiation between 100 and 280 nm, UV-B radiation between 280 and 315 nm, and UV-A radiation between 315 and 400 nm. VUV photons have high energy sufficient for breaking chemical bonds. Their efficiency in sterilization is limited, however, by very short penetration depth. The efficiency of UV-A/B photons in sterilization is limited by low energy of the photons. UV-C photons have energy sufficient for significant reconstruction of organic molecules and are characterized by sufficiently large penetration depth at the same time, which makes the UV-C photons most effective in direct sterilization processes. UV radiation in the 200–300 nm wavelength range with doses of several mJ/cm^2 causes lethal damage to cells. There are several biological mechanisms of UV deactivation of bioorganisms. Among specific UV effects on bacterial cells there is dimerization of thymine bases in bacterial DNA strands. The dimerization of thymine bases inhibits the ability of bacteria to replicate properly (Norman, 1954). The contribution of UV radiation into the total plasma sterilization effect essentially depends on the effectiveness of the specific plasma to radiate in the UV-C range.

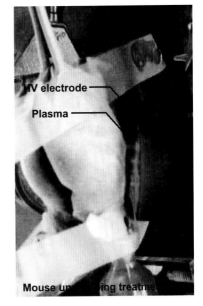

Figure 12–7. FE-DBD direct treatment of living tissue: animal directly treated in plasma for up to 10 min remains healthy and no tissue damage is observed visually or microscopically immediately after and up to 2 weeks following the treatment.

Low-pressure plasma discharges are able to provide significant UV radiation in the range of wavelengths effective in sterilization. It explains the important contribution of UV radiation in plasma sterilization at low pressures, which has been discussed in Section 12.1.1. Non-thermal atmospheric-pressure plasmas are typically not very effective sources of high-energy UV radiation. According to Laroussi and Leipold (2004), no significant UV emission occurs in non-thermal atmospheric-pressure plasmas below 285 nm, which reduces the UV contribution to sterilization in such plasmas. Even when direct UV contribution to sterilization is negligible, it can be important in synergy with other factors like radicals and charged particles. Such an effect has been discussed earlier in relation to non-thermal plasma treatment of polymers (see Section 9.7). Kinetic data on the effect of UV on deactivation of different micro-organisms are to be summarized later in connection with plasma sterilization of air streams (see Section 12.3).

12.2. EFFECTS OF ATMOSPHERIC-PRESSURE AIR PLASMA ON BACTERIA AND CELLS: DIRECT VERSUS INDIRECT TREATMENT, SURFACE VERSUS IN-DEPTH TREATMENT, AND APOPTOSIS VERSUS NECROSIS

12.2.1. Direct and Indirect Effects of Non-Thermal Plasma on Bacteria

Non-thermal plasma is effective in killing parasites, bacteria, fungi, and viruses on living tissue (Fridman et al., 2006, 2007; Lu & Laroussi, 2006; Raiser & Zenker, 2006; Sakiyama & Graves, 2006; Sladek, Baede, & Stoffels, 2006; Stoffels, 2006). The sterilization of living tissue can be subdivided into two approaches, classified as indirect and direct. The *indirect treatment* uses a jet of products (plasma afterglow) generated in a remotely located plasma discharge. The *direct treatment*, by contrast, uses the tissue itself as an electrode that participates in creating the plasma discharge, as illustrated in Fig. 12–7. Plasma in this case is contained between the quartz surface of the powered high-voltage electrode and the surface of bacteria or tissue being treated. The distinction between direct and indirect treatment is not only related to the proximity of the plasma and the tissue; the direct contact with plasma brings charged energetic particles to the plasma–tissue interface. By contrast, no charged particles are usually taken out of the plasma region by a jet even if the plasma region is located only a fraction of a millimeter away. Fridman et al. (2007) have demonstrated that the

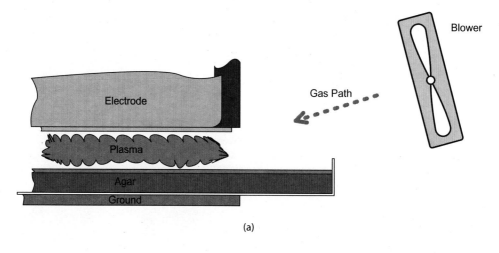

(a)

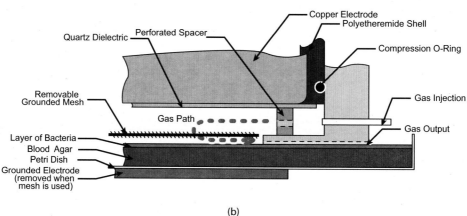

(b)

Figure 12–8. Direct and indirect plasma experimental setups: grounded mesh is removed for direct plasma treatment and placed back for indirect treatment (left) or air is blown through plasma to carry out long-lived species (right).

direct treatment can achieve sterilization much faster without any thermal effects, proving that the effect of charged particles in sterilization (Section 12.1.3) can be significantly stronger than that of the long-living active neutrals (Section 12.1.5) and longer-wavelength UV radiation (VUV is absorbed by air at atmospheric-pressure within microns).

The significant advantages of direct plasma treatment (Fridman et al., 2007) have been demonstrated by comparison of two types of experiments with DBD discharges. In one set of experiments, a surface covered by bacteria was employed as one DBD electrode; a second smaller-area DBD electrode was placed over it. Areas of sterilization with and without air flow parallel to the surface have been compared. In another set of experiments, a surface covered by bacteria was employed as one DBD electrode; the rate of sterilization in the system was compared to the case when the surface was separated from the DBD by a grounded mesh electrode (see Fig. 12–8). Bacterial samples for the experiments were collected from de-identified skin samples from human cadavers. Bacteria from the skin samples consisted of staphylococci, streptococci, and *Candida* species of yeast. They were transferred onto the blood agar plate, cultured, and diluted in 10X PBS to approximately 10^9 cfu/mL. To quantify sterilization efficiency, 20 μL drops of 10^9 cfu/mL bacteria were placed on the agar surface, left to dry for 5 min, and finally treated by plasma. This volume

Table 12–1. Bacteria Sterilization Results (in cfu/mL) in Experiments Comparing Direct and Indirect Plasma Treatment Using Dielectric Barrier Discharge

Original bacteria concentration	Bacteria concentration after 5 s of DBD treatment	Bacteria concentration after 10 s of DBD treatment	Bacteria concentration after 15 s of DBD treatment
10^9	850 ± 183	9 ± 3	4 ± 4
10^8	22 ± 5	5 ± 5	0 ± 0
10^7	6 ± 6	0 ± 0	0 ± 0

was selected as it spread to ~ 1 cm^2 over the agar surface; thus, the area covered by the bacterial sample drop was entirely within the area covered by the insulated plasma electrode. Following treatment the drop was spread over the entire agar surface and incubated in air at 37°C for 24 h. Bacterial colonies (CFUs) were then counted and the results are presented in Table 12–1. There is a clearly visible difference in agar appearance between untreated, partially sterilized, and completely sterilized agar. Based on this difference, it is reasonable to classify the appearance of the bacterial surface into five categories: (1) untreated (10^9 cfu/mL), (2) partially disinfected (10^9 to 10^7 cfu/mL), (3) disinfected (10^7 to 10^4 cfu/mL), (4) partially sterilized (1,000 to 10 cfu/mL), and (5) completely sterilized (zero cfu/mL). To quantify the extent of sterilization, 1 mL of 10^9 cfu/mL was poured over the entire agar surface. These samples were left to dry for 3 h and then treated by plasma. Here plasma only covered a portion of the petri dish, while bacteria covered the entire dish. Following the 24 h incubation period, the extent of sterilization was clearly visible – areas where bacteria were killed looked like uncontaminated agar while areas that received no treatment changed color and appearance significantly as bacteria grew there. The complete sterilization area is easiest to identify – it corresponds to the agar area that is completely clear from bacteria, which can be seen in Fig. 12–9. Partial sterilization is also relatively easy to define since the number of CFUs is relatively small and can be counted. Disinfection is more difficult to assess because of the difficulty of counting the large number of CFUs. The complete sterilization gradually fades into untreated areas, forming a "gray-scale" fade that gradually increases from 0 cfu/mL in the sterile zone to 10^9 cfu/mL in the untreated zone.

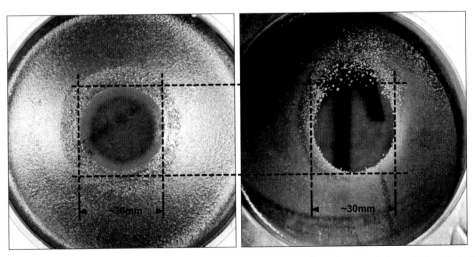

Figure 12–9. Using a blower to shift sterilization region does not affect plasma and shows little effect of the afterglow: 15 s of treatment with (a) blower off and (b) blower on (air flows up).

12.2.2. Two Experiments Proving Higher Effectiveness of Direct Plasma Treatment of Bacteria

The experimental system developed to compare the effectiveness of direct and indirect plasma treatment of micro-organisms (Fridman et al., 2007) is illustrated in Fig. 12–8. An agar-covered petri dish with bacteria spread on its surface from the earlier prepared samples of 10^9 cfu/mL concentration was used to test the sterilization. Specifically, 1 mL of such bacterial sample was transferred onto the blood agar, spread over the entire plate, and dried for 3 h prior to the plasma treatment. After the plasma treatment, the samples were incubated in air at $37°C$ for 24 h to assess the ability of the DBD plasma to sterilize. Control experiments were performed with the gas flow only. Bacteria were completely unaffected by flowing room air in up to 30 min. Placed on a metal substrate, the agar dish with bacteria on its surface acts as one of the DBD electrodes – bacteria are at the plasma–agar interface. The other smaller-area DBD electrode is made of a 2.5-cm-diameter solid copper disc separated from the plasma by a 1-mm-thick quartz dielectric. The plasma gap, or the distance between quartz–plasma and agar–plasma interfaces, was set to 1.5 mm. Alternating polarity sinusoidal voltage of 35 kV magnitude (peak to peak) and 12 kHz frequency was applied between the copper electrode and the aluminum substrate under the agar petri dish. The flow speed for this experiment was measured between the electrode and the substrate, directly outside of the electrode, to be 0.8 ± 0.3 m/s. Electrical and calorimetric measurements indicated a DBD power of 0.8 ± 0.2 W/cm^2. When there is no air flow parallel to the surface, the area directly under the second smaller electrode is exposed simultaneously to charged particles, UV, and the plasma afterglow containing active neutrals such as ozone (O_3), nitric oxide (NO), hydroxyl radicals (OH), and other excited molecular and atomic species. In the presence of the air flow, the area under the second smaller electrode experiences practically the same conditions as without the flow. However, in the presence of flow the plasma afterglow will also impinge on the areas of the substrate that are not directly under the second smaller electrode. Results comparing the sterilization with and without flow are illustrated in Fig. 12–9. Complete sterilization by direct plasma treatment under the insulated electrode is achieved in 15 s. If air is passed parallel to the substrate through the plasma region, the disinfection area shifts by several millimeters, but a "tail" of disinfection appears. Even though complete sterilization is not observed in this tail, different degrees of disinfection are evident. Locations closer to the plasma region have greater degree of disinfection. This suggests that plasma "afterglow," or the gas carried through and outside of plasma, does contain active species and radicals that have bactericidal effects, although direct plasma is substantially more potent as a sterilizing agent. Complete sterilization, however, did not extend to the region reachable by the plasma afterglow in as much as 5 min of treatment. This clearly demonstrates that plasma-generated active neutrals are much less effective than direct plasma where charged particles participate in the process.

First experiments (described earlier) separated the effects of plasma afterglow from those of charged particles and UV. In the second type of experiment (described next), the plasma afterglow together with UV is compared against direct plasma, also involving charged particles. The direct treatment is obtained similarly to the first type of experiments. However, instead of creating airflow parallel to the substrate, provisions were made to allow introduction of air into the discharge gap (Fig. 12–8). To obtain indirect treatment including UV effects, the agar surface covered by bacteria was again employed as one DBD electrode; however, this surface was separated from the DBD by a mesh electrode, as illustrated on the right in Fig. 12–8. With a grounded mesh in place, bacteria are no longer at the plasma interface, as plasma is bounded by the grounded mesh. The mesh is placed 1 mm above the surface of the agar, which permits only the passage of UV, radicals, and other longer-living particles created in the plasma gap between the mesh and the quartz-insulated electrode above it. The second electrode used with the mesh was the same electrode that was employed

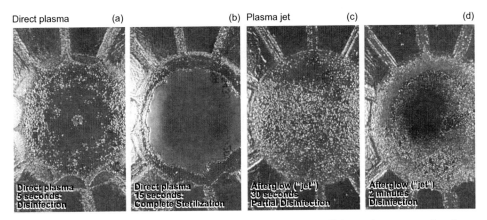

Figure 12–10. Direct application of plasma yields better sterilization efficiency than treatment by plasma afterglow: (a) 5 s and (b) 15 s of direct plasma compared with (c) 30 s and (d) 2 min of plasma jet.

in the direct treatment setup illustrated on the left in Fig. 12–8. The setup used to introduce air into the discharge gap is illustrated on the right in Fig. 12–8. The air was introduced into the discharge gap to enhance extraction of the plasma afterglow through the mesh.

A comparison of direct (without the mesh) and indirect (with the mesh) plasma sterilization is shown in Fig. 12–10. It is clearly visible that plasma that comes in direct contact with bacteria is able to sterilize significantly faster than an afterglow or jet. Only 5 s of direct plasma treatment results in the appearance of a sterilization region (spot near the center). Complete sterilization occurs within 15 s when direct treatment is employed and only partial disinfection can be achieved within the same time frame with indirect treatment. Over 5 min of indirect treatment is required to achieve sterilization results similar to direct treatment obtained within 15 s. Therefore, even with substantial UV radiation, indirect plasma treatment is substantially weaker than the direct treatment provided by charged particles (see Section 12.1.3).

Comparing direct DBD plasma effect with radiation biology, we should mention that they are quite similar (including possible double-strand breaks in DNA). Typical DBD plasma doses after seconds of treatment are about 10^6 Gy (J/kg). Similar treatment effect in the case of γ-radiolysis requires $0.5 \div 1$ Gy. Thus, plasma is $\sim 10^6$ (million) times less destructive with respect to non-repairable DNA damages.

12.2.3. Surface Versus In-Depth Plasma Sterilization: Penetration of DBD Treatment into Fluid for Biomedical Applications

Effective plasma treatment of bacteria, cells, and other biomaterials is possible not only if they are in contact with plasma but even when they are separated from plasma and protected by a layer of intermediate substance, in particular submerged into a culture medium (Kieft et al., 2004, 2005; Fridman et al., 2007). Detachment of mammalian vascular cells was observed after indirect treatment by a non-thermal atmospheric-pressure "plasma needle" when the thickness of the liquid layer covering the cells was about 0.1 mm (Kieft et al., 2005). Effective direct treatment of cells that are covered by up to a half a millimeter of cell culture fluid has been demonstrated by Fridman et al. (2007), which is to be discussed next. Experiments were carried out using DBD with submerged melanoma cell culture as illustrated in Fig. 12–11. The discharge gap between the bottom of the 1-mm-thick quartz plate covering the copper electrode and the top surface of the fluid was set to 1.5 mm. AC voltage of 35 kV and 12 kHz was applied to the copper electrode, while the aluminum petri dish remained at a floating potential to mimic the conditions that exist when this discharge is used to treat living tissue. The surface power density of the DBD was

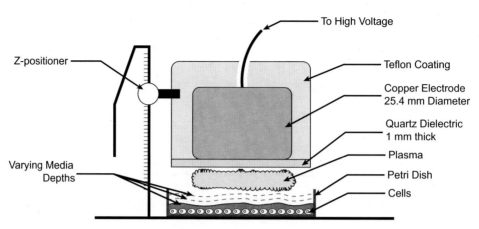

Figure 12–11. Schematic of FE-DBD treatment of melanoma cells.

0.8 ± 0.2 W/cm^2. Melanoma cells were cultured on the surface of an aluminum dish. Cells were allowed to grow while the dish was kept within an incubator. The original growth medium was then removed and a measured volume of fresh culture medium was added while the cells remained attached to the bottom surface. The volume of the culture medium was measured to obtain the desired depth, which is varied in these experiments from 0 to 0.33 mm (Fig. 12–11). The submerged cells were placed into the DBD treatment setup and treated by plasma for periods of time ranging from 0 to 30 s. Immediately after the plasma treatment the culture medium used during the treatment was replaced with a fresh medium. The cells were then analyzed with a Trypan Blue exclusion test to determine their viability through the cellular membrane integrity. Trypan Blue only enters cells through permeabilized cytoplasmic membranes. Fig. 12–12 shows percentages of affected cells and treatment times. No effect on the cells is observed in control experiments where plasma treatment does not occur. Without the medium protecting the cells, cell wall fracture is achieved within 10 s of treatment and all cells take up the ink (Trypan Blue), indicating that the cell wall has been compromised and the cell is dead. Even though the DBD plasma is not

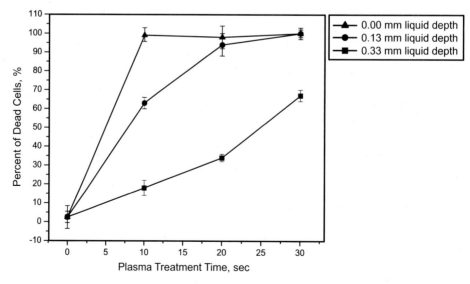

Figure 12–12. Percent of dead cells (cells that take up Trypan blue) after treatment at different liquid depths. The thickness of the cell layer at the bottom of the dish is about 30 μm.

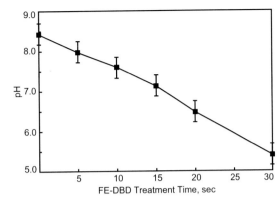

Figure 12–13. pH change of the buffer medium under FE-DBD treatment.

uniform, all the cells are affected by the treatment in as little as 10 s. When a layer of media is introduced which covers these cells, the number of compromised cells decreases but it is clear that the plasma treatment is permeating these cells through a layer of liquid cell growth medium.

The DBD plasma treatment leads to notable changes in the pH of the culture medium (see Fig. 12–13). It can be suspected as a cause of deep penetration of the plasma sterilization effect, which is, however, not the case. Special control experiments were carried out to understand the role of the changing chemistry of the culture medium on the melanoma cells. In these experiments, the grown cell culture was transferred into the culture medium pretreated by the DBD in a manner identical to the one used to treat submerged cell culture. Changes in the pH of the culture medium were identical to the ones found in experiments with the submerged cells. As before, the cells were placed into the fresh culture medium after remaining in the pretreated culture medium for the amount of time equal to the plasma pretreatment time. Similarly to the control (untreated) cells, cells transferred into the culture medium pretreated by the plasma in this fashion and then placed into the fresh medium did not take up Trypan Blue. The results of placing cells in acidified media temporarily indicate that the DBD plasma does not affect the cells simply through changes in the chemical composition of the culture medium.

12.2.4. Apoptosis Versus Necrosis in Plasma Treatment of Cells: Sublethal Plasma Treatment Effects

Non-thermal plasma can stimulate both apoptosis and necrosis in treated cells. Apoptosis, or programmed cell death, is a complex biochemical process of controlled self-destruction of a cell in a multicellular organism (Johnstone, Ruefli, & Lowe, 2002; Silverthorn et al., 2004). This process plays an important role in maintaining tissue homeostasis, fetal development, immune cell "education," development, and aging. Examples of apoptosis that occur during normal body processes include the formation of the outer layer of skin, the inner mucosal lining of the intestine, and the endometrial lining of the uterus, which is sloughed off during menstruation. During apoptosis, cellular macromolecules are digested into smaller fragments in a controlled fashion, and ultimately the cell collapses into smaller intact fragments that can be removed by phagocytosis without damaging the surrounding cells or causing inflammation. In contrast, during necrosis, also termed "accidental cell death," the cell bursts and the cellular contents spill out into the extracellular space, which can cause inflammation. Necrosis is induced by cellular injury, for example, extreme changes in osmotic pressure or heat, that lead to adenosine tri-phosphate (ATP) depletion of the cell. Apoptosis and necrosis effects of plasma treatment on mammalian cells have been studied, in particular, by Stoffels (2003). Chinese hamster ovarian cells were used as a model. The cells were exposed to an RF small-volume cold plasma generated around the tip of a needle-shaped electrode (**plasma needle**). Necrosis (cell death due to catastrophic injury) occurs in the plasma needle treatment for powers greater than 0.2 W and exposure times longer than 10 s. The cell membranes are damaged and the cytoplasm is released. Lower doses of exposure lead to apoptosis. If the power level and exposure time were reduced significantly to about 50 mW and 1 s, the Chinese hamster ovarian cells partly detached from the sample, took a more rounded shape, and did not undergo apoptosis (Stoffels, 2003).

The sublethal plasma effect on bacterial cells has been demonstrated by Laroussi, Richardson, and Dobbs (2002). When a sublethal plasma exposure is administered to bacterial cells, a change in their metabolic behavior could occur. To demonstrate these effects, Laroussi, Richardson, and Dobbs (2002) addressed the biochemical impacts of plasma on *Escherichia coli* with sole carbon substrate utilization (SCSU) experiments, using Biolog (Hayward, CA) GN2TM 96-well microtiter plates. The purpose of the SCSU experiments was to determine if exposure to plasma altered the heterotrophic pathways of the bacteria. It was presumed that any changes in metabolism would be indicative of plasma-induced changes in cell function. The Biolog GN2 plate was comprised of a control well and 95 other wells, each containing a different carbon substrate. Color development of a redox dye present in each well indicated utilization of that particular substrate by the inoculated bacteria. The 95 substrates were dominated by amino acids, carbohydrates, and carboxylic acids. The plasma exposure caused an increase of utilization of some substrates and a decrease of utilization in others, indicating noticeable changes in the corresponding enzyme activities without causing any lethal impact on the cells. Plasma-stimulated apoptotic behavior in melanoma skin cancer cell lines has been demonstrated by Fridman et al. (2007). This effect has important potential in cancer treatment because the cancer cells frequently acquire the ability to block apoptosis and thus are more resistant to chemotherapeutic drugs. Therefore, the effect is to be discussed later in this chapter regarding plasma medical treatment of melanoma skin cancer.

12.3. NON-THERMAL PLASMA STERILIZATION OF AIR STREAMS: KINETICS OF PLASMA INACTIVATION OF BIOLOGICAL MICRO-ORGANISMS

12.3.1. General Features of Plasma Inactivation of Airborne Bacteria

In comparison with plasma surface sterilization (see Sections 12.1 and 12.2) and water sterilization (to be considered in Section 12.4), only a few plasma researchers have been focused on air decontamination. Most of them have been successful only when coupling plasma technology with high-efficiency particulate air (HEPA) filters to both trap and kill micro-organisms (Jaisinghani, 1999; Gadri, 2000; Kelly-Wintenberg, 2000). Usually such coupling of filtration and plasma systems means trapping the micro-organisms with the filter followed by sterilization of the filter using non-thermal plasma discharges. The downside of relying on HEPA filters is that they have limited efficiency at trapping submicron-sized airborne micro-organisms (Harstad, 1969) and they also cause significant pressure losses in heating, ventilation, and air conditioning (HVAC) systems, giving rise to higher energy and maintenance costs. An experimental facility, the Pathogen Detection and Remediation Facility (PDRF), was designed by Gallagher et al. (2005) to perform air decontamination experiments using a dielectric barrier grating discharge (DBGD). The PDRF combines a plasma device with a laboratory-scale ventilation system with bio-aerosol sampling capabilities. Experiments with PDRF show that direct contact of the bio-aerosol and the DBGD plasma with very short duration can cause an approximate 2-log reduction (97%) in culturable *E. coli* with a ∼5-log reduction (99.999%) measured in the 2 min following exposure. Fast treatment times within plasma are due to a high air flow rate and high velocity of the bio-aerosol particles in flight and a small discharge length of the DBGD device, which results in a residence treatment time of approximately 1 ms. Thus, direct plasma treatment

12.3. Non-Thermal Plasma Sterilization of Air Streams

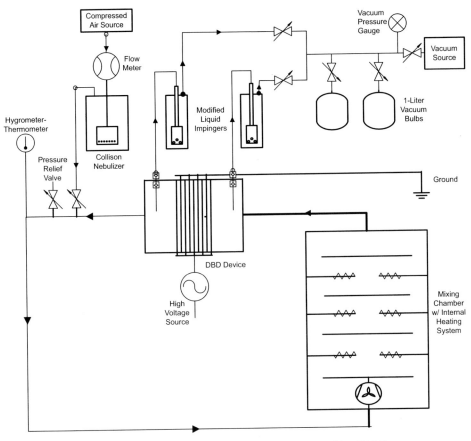

Figure 12–14. Schematic of the Pathogen Detection and Remediation Facility (PDRF).

leads to fast and effective deactivation of airborne bacteria similarly to the aforementioned plasma surface sterilization.

12.3.2. Pathogen Detection and Remediation Facility for Plasma Sterilization of Air Streams

The PDRF system, as mentioned, is a bio-aerosol treatment facility that provides a recirculating air flow environment for DBD air decontamination experiments (Gallagher et al., 2005). The bio-aerosol can be treated with repeated passes through the same plasma discharge. The sealed, recirculating system allows for complete control over relative humidity (RH), which is important because even small fluctuations in RH have been shown to significantly decrease the survivability of airborne bacteria (Hatch & Dimmick, 1966). The PDRF is a plug flow reactor, where air flow is turbulent so that radial variation of the bacterial concentration is minimized. Figure 12–14 shows a general schematic of the PDRF. The PDRF system has a quite large total volume of 250 L and operates at high air flow rates (~25 L/s or greater), which are typical for indoor ventilation systems. The system has an inlet with attached collison nebulizer for bio-aerosol generation and two air-sampling ports connected to a vacuum air-sampling system. The system also has a large-volume barrel that contains a series of aluminum baffle plates and a variable-speed centrifugal blower motor that drives the air through the DBGD treatment chamber. The PDRF system residence time – that is, the time for one bio-aerosol particle to make one complete revolution through the whole system – is approximately 10 s.

Figure 12–15. Liquid impinger for bioaerosol sampling.

An important element of the PDRF is the **air-sampling system**. To separate the decon-tamination effect of direct exposure to DBD plasma from the remote exposure of ozone and other long-lived chemical species that can interact with bio-aerosols downstream of the discharge, a sampling method was devised by Gallagher et al. (2005) so that air samples are taken just before and after bio-aerosol passes through the discharge area. As there are only two sample ports, each set of two air samples measures the change in viability of bacteria on a "per-pass" basis through the discharge. For each of the subsequent sets of air samples, the sample taken "before plasma" can give a measurement of the change in viability due to the effect of ozone from the previous "after plasma" sample. Liquid impingement was the chosen air-sampling method because it minimizes desiccation stress on the bacteria by directly depositing them into a buffered saline solution. Liquid impingers operate by drawing a sample of air through an inlet tube submerged in a solution, thereby causing the air stream to strike the liquid bed, trapping aerosols in the solution through forces of inertia. The AGI-30 is the most commonly used liquid impinger, which contains a critical orifice that limits the maximum air-sampling rate to 12.5 L/min. The vacuum air sampling system was designed in PDRF to take as large a volume of air sample as possible (∼1 L) in the shortest period of time (∼1 s) so as not to significantly disturb the flow inside the system. To accommodate this high air-sampling rate, the AGI-30 impinger was modified by replacing the standard critical orifice with a hollow tip with several jet ports. Figure 12–15 shows an image of the modified air samplers. The overall efficiency of this type of sampler was in

Figure 12–16. Dielectric barrier grating discharge (DBGD) with air sterilization chamber.

the range $6 \pm 3\%$. To be injected into the PDRF, the bacterial culture was placed into a BGI 24-jet collison nebulizer operating at 40 psi for a period of 45 s (nebulizing rate: 1.1 mL/min). The DBGD device was then switched on for a period of 10 s and during this time the first two air samples were taken in sequential order: before and after passing through plasma. The discharge time of 10 s is used because this time is just enough for taking two samples and is equal to the time of the bio-aerosol to make one complete revolution in the PDRF system. Therefore, measurement of the decontamination effectiveness of the DBGD is made on a per-pass basis and this ensures that each bio-aerosol particle has been treated once by the discharge. The next set of air samples was taken approximately 2 min later because this amount of time is required to remove and replace the air samplers with the next set of presterilized samplers. This process was repeated again until the typical number of air samples, usually six, was achieved. Each of the presterilized air samplers was initially filled with 30 ml of sterile PBS solution and after sampling each solution was transferred to a sterile 50 mL centrifuge tube for assaying. The concentration of *E. coli* in PBS solution used for nebulization was on average 10^8 cfu/mL.

Detailed balance of the plasma-treated airborne bacteria was done in the experiments of Galagher et al. (2005) using flow cytometry. The flow-cytometric measurements were made using a flow cytometer with 488 nm excitation from an argon-ion laser at 15 mW. Fluorochromes with a high affinity for nucleic acid SYBR Green I and propidium iodide (PI; molecular probes) were used for flow cytometry. The SYBR Green I, a green fluorescent nucleic acid stain, has been shown to stain living and dead gram-positive and gram-negative bacteria (Porter et al., 1993). PI is a red fluorescent dye that intercalates with dsDNA and only enters permeabilized cytoplasmic membranes (Gregory et al., 2001).

12.3.3. Special DBD Configuration – the Dielectric Barrier Grating Discharge – Applied in PDRF for Plasma Sterilization of Air Streams

A special DBD configuration, called the DBGD, has been applied in the PDRF to provide effective direct sterilization of air flow. The DBGD consists of a thin plane of wires with equally spaced air gaps of 1.5 mm. The high-voltage electrodes are 1-mm-diameter copper wire shielded with a quartz capillary dielectric that has an approximate wall thickness of 0.5 mm. The total area of the DBGD including electrodes is 214.5 cm^2 and without electrodes is 91.5 cm^2. Figure 12–16 shows an image of the DBGD device, which has two air sample ports located at a distance of 10 cm from each side of the discharge so that bio-aerosol can be sampled right before and after it enters the discharge. When the PDRF system is operated at a flow rate of 25 L/s, the air velocity inside the DBGD chamber is 2.74 m/s and the residence time of treatment, that is, the duration of one bio-aerosol particle (containing one *E. coli*

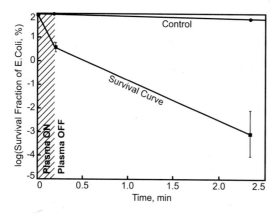

Figure 12–17. Culture test results of DBD-treated *E. coli* in PDRF system in four replicate trials. The gray shaded areas indicate the time when the DBD was ignited. Control experiments are also shown.

bacterium) passing through the DBGD device is approximately 0.73 ms. The DBGD device is operated using a power supply that delivers a distinct sinusoidal current-voltage waveform with a very fast rise time that nearly simulates a true square wave pulse. The period between pulses is approximately 600 μs, peak-to-peak voltage is 28 kV, and pulsed current is nearly 50 A (peak-to-peak value). The average power of the discharge is approximately 330 W and, considering the discharge area of 91 cm^2, the power density is 3.6 W/cm^2. The majority of power is discharged in the very short duration of the pulse itself, which has a period of 77 μs and average pulse power of 2600 W. Since the residence time of a bio-aerosol particle passing through the discharge area is 0.73 ms and the period between pulses is 0.6 ms, each bio-aerosol particle that passes through the DBGD device experiences about 1 pulse of DBD power. The typical concentration of bio-aerosol is approximately 5.10^5 bacteria per liter of air, which translates to approximately 9.10^3 bacteria within the cross section of the discharge area at any given time (in each 2-mm-wide cross section of flow passing through the DBGD).

12.3.4. Rapid and Direct Plasma Deactivation of Airborne Bacteria in the PDRF

The culture test results from four replicate trials of DBGD-treated *E. coli* bio-aerosol in the PDRF system are shown in Fig. 12–17 (Gallagher et al., 2005). Control experiments show a negligible change in the surviving fraction of *E. coli* over the total experimental period of 6 min. In the DBGD-treated trials, an approximate 1.5-log reduction (97%) in the surviving fraction of *E. coli* was measured between samples 1 and 2 (before and after plasma treatment). It is important to note that the plasma was only ignited for a short period during each set of samples, which is noted by the gray shaded areas in Fig. 12–17. A second decrease in the surviving fraction of *E. coli* is shown between samples 2 and 3, which occurred in the time between plasma treatments. In this second decline, the remaining culturable bacteria (3% from the initial number) decreased by an additional 99.95% (3.5 logs) when the plasma discharge was switched off. Additionally samples 4, 5, and 6 taken after the second (sample 4) and before and after the third "pass" (samples 5 and 6) through plasma showed no culturable *E. coli* in the probes (results not shown), which means that after the second and third passes through plasma the number of culturable *E. coli* was less than the detection limit (10^{-4}%).

Flow cytometry has been employed to detect the presence *E. coli* in each of six air samples taken during experiments. Where colony-counting techniques are limited to detecting only culturable (i.e., visibly growing) bacteria, flow cytometry is capable of detecting the physical presence of bacteria in a sample regardless of culturability. Flow cytometry analysis of air samples taken during the trials indicates that the total number of bacteria (both active and inactive) remains almost constant; therefore, the DBGD device is not acting as an electrostatic precipitator and the concentration of bio-aerosol particles remains undisturbed for the duration of each experiment. Flow cytometry results also showed that bacterial outer membranes of *E. coli* were not disintegrated with up to three passes of direct exposure in the DBGD plasma device. However, culture tests demonstrated a 97% reduction in culturable *E. coli* with a millisecond exposure time in DBGD (one pass through discharge) and a subsequent 3.5-log reduction in the 2 min following treatment. The direct plasma exposure time of 0.73 ms (per pass) allows enough time for bacteria to be attacked by all active plasma factors: charged particles, UV radiation, OH radicals, atomic oxygen, and ozone, which is one explanation for the initial 97% reduction in culturability. Subsequent remote exposure to ozone in the 2 min following direct plasma treatment, as well as the high sensitivity of airborne bacteria to slight atmospheric changes, may account for the remaining 3.5-log reduction. Thus, direct plasma treatment of air flow permits its rapid sterilization (much faster than corresponding indirect plasma treatment), similarly to the surface sterilization (see Sections 12.2.1 and 12.2.2). This conclusion has been supported by experiments with the PDRF where some small number of the DBGD wires has been eliminated (Gallagher et al., 2005). Some relatively small fraction of the air stream was not been treated directly in plasma in this case (but only indirectly), which resulted in a much higher number of bacteria surviving the treatment or, in other words, in significant decrease of the sterilization efficiency.

12.3.5. Phenomenological Kinetic Model of Non-Thermal Plasma Sterilization of Air Streams

Sterilization of airborne micro-organisms by plasma-generated UV and active chemical species (atoms, radicals, excited and charged particles, as well as active stable molecules like O_3, etc.) can be described in terms of conventional gas-phase chemical kinetics. The sterilization process is considered in this case as an elementary reaction, which occurs with some probability during a "collision" of the micro-organisms and relevant plasma-generated active species. The rate of the "elementary sterilization reaction" can be described in such a chemical kinetic approach in terms of conventional reaction rate coefficients. Such an approach to plasma sterilization kinetics, proposed by Gangoli et al. (2005), is a phenomenological one, because it cannot describe detailed biochemical pathways of the very complicated process of deactivation of micro-organisms. However, the phenomenological plasma sterilization kinetics permits generalization of the experimental kinetic data obtained in different discharge systems (with different concentrations of different relevant active plasma components) applied to deactivation of different types of micro-organisms. Rate coefficients for plasma sterilization (initially focused only on ozone, OH radicals, and UV) have been calculated by Gangoli et al. (2005) through analysis of empirical data (available in the food and water decontamination literature) that resulted from many experiments in which ozone and UV radiation were used to destroy a variety of bacteria, viruses, and spores in various liquids and surfaces Broadwater, 1973; Burelson, 1975; Khadre, Yousef, & Kim, 2001. The rate coefficients k have been calculated assuming second kinetic order of the micro-organism deactivation:

$$\frac{d[M]}{dt} = -k \cdot [M] \cdot [A],$$
(12–20)

where d[*M*]/dt is the change of concentration of viable micro-organisms with time, [*M*] is the concentration of micro-organisms, and [*A*] is the concentration of active chemical species

Table 12–2. Phenomenological Reaction Rate Coefficients for Plasma Sterilization of Bacteria (*E. coli*) using Ozone O_3, Hydroxyl Radicals (OH), and UV Radiation

Plasma-generated active species and radiation	Phenomenological reaction rate coefficients
Ozone (O_3)	$1.5 \cdot 10^{-16}$ (cm³/s)
Hydroxyl radical (OH)	$3.6 \cdot 10^{-13}$ (cm³/s)
UV radiation	$3.8 \cdot 10^{-3}$ (cm²/μJ)

A. The reaction rate coefficient for interaction of *A* with the micro-organism, k_A, is defined in this case as

$$k_A = \frac{\ln\left(\frac{1}{S}\right)}{[A] \cdot t}, \qquad (12\text{–}21)$$

where S is the surviving fraction of micro-organism M, and t is treatment time. Using the same approach, the model accounts for the destructive effects of plasma-generated UV on each individual micro-organism by assigning the reaction rate constants k_{UV} for each micro-organism. Taking into account that the density of active species [*A*] in equations (12–20) and (12–21) are replaced in the case of UV radiation with energy flux [*J*] usually measured in μJ/(s · cm²), the sterilization reaction rate coefficient is measured in the case of UV radiation not in cm³/s (which is conventional for the second-order reactions) but in cm²/μJ (see Table 12–2). Regarding hydroxyl radicals, Von Gunten (2003) showed the inactivation rate of pathogens by OH radicals in their system to effectively be about 10^6–10^9 times faster than that of O_3. Recently the interaction between OH radicals and *E. coli* were carefully studied by Cho et al. (2004). It was found that the *CT* value (product of characteristic concentration *C* and time of exposure, *T*) corresponding to 2 log reduction of *E. coli* inactivation by OH is approximately $0.8 \cdot 10^{-5}$ mg · min/L. In comparison, the $O_3 - E.~coli$ interaction *CT* value ($4 \cdot 10^{-2}$ mg · min/L) is higher. Summarizing experimental data, it can be taken for the kinetic analysis that the inactivation ability of OH exceeds that of O_3 by 10^3–10^4-fold (Gangoli et al., 2005). Table 12–2 summarizes the phenomenological reaction rate coefficients for the inactivation of *E. coli* by O_3, OH, and UV radiation.

12.3.6. Kinetics and Mechanisms of Rapid Plasma Deactivation of Airborne Bacteria in the PDRF

In the framework of the aforementioned kinetic model of plasma sterilization, the following formula describes the rate of inactivation of airborne micro-organisms inside a DBD from the combination of effects of ozone, hydroxyl, and UV radiation:

$$\frac{d[M]}{dt} = -k_{O_3} \cdot [M] \cdot [O_3] - k_{UV} \cdot [M] \cdot [I] - k_{OH} \cdot [M] \cdot [OH], \qquad (12\text{–}22)$$

where [*M*] is the concentration of micro-organisms; [O_3] and [OH] are the concentrations of the active chemical species; [I] is the energy flux of UV radiation; k_{O3}, k_{OH}, and k_{UV} are the corresponding reaction rate coefficients (see Table 12–2). Since the DBD is operated in air, typical concentrations of ozone range from 100 ppm ($3 \cdot 10^{15}$ cm⁻³, see Sections 6.5, 6.6). The G-factor for OH formation (number of OH radicals produced per 100 eV energy of the discharge) is about 0.5 (Penetrante et al., 1997), which corresponds in the PDRF conditions (DBD power 330 W, 25 L/s flow rate) to an OH concentration about 15 ppm ($4 \cdot 10^{14}$ cm⁻³). The UV radiation intensity in the wavelength range 200–300 nm is below 50 μW/cm².

12.3. Non-Thermal Plasma Sterilization of Air Streams

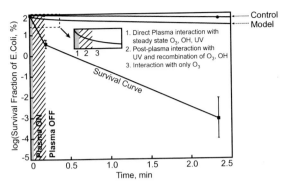

Figure 12–18. Comparison of modeling results with experimental and control results. In the figure: discrepancy between the modeling and experimental results due to consideration of only a few active agents from plasma. Inset: a detailed explanation of various sections in the graph. As shown in the inset figure, modeling regions are divided into: (1) in plasma, where *E. coli* encounters steady-state O_3, OH, and UV concentrations and intensities; (2) post-plasma interaction, where *E. coli* interacts with the above three species in the downstream flow as they recombine (O_3 and OH) or go through extinction (UV); (3) *E. coli* interacts only with long living O_3.

To be consistent with the transient nature of the PDRF experiments, the changes in **CFU** numbers of *E. coli* for a given volume (cm^3) are tracked as polluted air travels through the experimental system. Three zones are considered based on differential interaction modes of *E. coli* (shown in Fig. 12–18): (1) direct plasma interaction (DPI), (2) post-plasma interaction (PPI), and (3) interaction with only O_3 (IO). During DPI, the concentrations of O_3, OH, and the UV intensity are assumed at a constant steady-state value. DPI only lasts for about 1 ms due to the HVAC conditions of operation (25 L/s). The *E. coli* then goes through PPI, where it interacts with recombining O_3 and OH species and decreasing UV intensity (absorption of UV radiation by ozone, cross section -10^{-17} cm^2; Fix et al., 2002). Ozone and OH radical deactivation reactions considered in the PPI simulation are limited to

$$OH + O_3 \rightarrow O_2 + HO_2, \quad OH + OH + M \rightarrow H_2O_2 + M, \quad O_3 + M \rightarrow O_2 + O + M.$$
$$(12\text{–}23)$$

According to the phenomenological kinetic modeling, OH radical interaction time with pathogens is about 600 μs outside plasma, which is reasonable when compared to experiments of Ono and Oda, (2001). The IO part of the model is used to describe the situation when plasma is turned "off" in experiments. During IO, since there is no production of new OH or UV, the *E. coli* interacts with only remnant ozone for the duration of 2 min (the time between sets of air samples taken during experiments). Keeping in mind the aforementioned considerations, the predicted variation of concentration of active species outside plasma with length from the discharge zone is shown in Fig. 12–18. The simulation results depicted in Fig. 12–18 predict a 14% decrease in viable *E. coli* during one pass of direct contact time with DBGD plasma. However, during further interaction with ozone (outside DBGD plasma zone) through the recirculating air flow system, a total 36% inactivation is predicted for treatment time of about 10 s (DPI, 1 ms; PPI, 10 s; shaded area in Fig. 12–18). This result, obviously, is not in agreement with the PDRF experiments, where a 97% decrease has been achieved during the same times and modes of treatment. The disagreement is surely due to the fact that the model only considers the effect of OH, UV radiation, and O_3 as the primary active components of plasma, when it is known that there are additional active species present (especially charged particles, making significant contribution into direct plasma treatment of micro-organisms; see Sections 12.1.3, 12.2.1, and 12.2.2). Considering only UV radiation and the most aggressive plasma neutral species

OH and O_3 cannot explain the initial 1.5 log reduction during 1 ms treatment in DBGD plasma and subsequent 4 log reduction (after 2 min treatment by remnant species), as was observed experimentally (Ganguli et al., 2005). Active species, such as, first of all, charged species (ions and electrons) and other radicals (atomic oxygen and nitric oxides), contribute to the inactivation process and may account for the remainder of the observed bactericidal activity of DBGD plasma.

12.4. PLASMA CLEANING AND STERILIZATION OF WATER: SPECIAL DISCHARGES IN LIQUID WATER APPLIED FOR ITS CLEANING AND STERILIZATION

12.4.1. Needs and General Features of Plasma Water Treatment: Water Disinfection Using UV Radiation, Ozone, or Pulsed Electric Fields

Currently, an estimated 1.1 billion people are unable to acquire safe drinking water, which determines the need for improved methods of water treatment. Contaminated water can be attributed to a number of factors, including chemical fouling, inadequate treatment, and deficient or failing water treatment and distribution systems. An additional important cause of contamination is the presence of untreated bacteria and viruses within the water. As estimated by the U.S. Environmental Protection Agency (EPA), nearly 35% of all deaths in developing countries are related directly to contaminated water. The increased presence of *Escherichia coli*, along with various other bacteria within some areas of the United States, has been a cause for national concern. In an effort to inactivate these bacteria, successful experiments and commercial applications of chemical treatments, UV radiation, and ozone injection units have been developed and implemented for potable water delivery systems. The experimental success and commercialization of these water treatment methods are not, however, without deficiencies. With regard to human consumption, chemical treatments such as chlorination can render potable water toxic. UV radiation and ozone injection have also been proven to be two practical methods of bacterial inactivation in water, but the effectiveness of such methods largely depends upon adherence to regimented maintenance schedules. Plasma methods effectively combining the contribution of UV radiation, active chemicals, and high electric fields (see Section 12.1) are considered, therefore, to be a very effective approach to water treatment (Locke et al., 2006; Fridman, Gutsol, & Cho, 2007). Before considering direct application of plasma to water treatment (which is a major goal of this section), we will discuss briefly the independent application of UV radiation, active chemicals, and high electric fields for deactivation of micro-organisms in water.

Pulsed electric fields can be mentioned as a novel approach to water treatment. The electric field associated with this technology is not strong enough (membrane potential of more than 1 V can kill a bacterium) to initiate electrical breakdown in water. The deactivation of micro-organisms is due to **electroporation**, which is the creation of holes in cell membranes. It means that plasma-originated electric fields (for example, those in DBD streamers) can be sufficient for electroporation (see Section 12.1.4). The energy for the 2-log reduction of number of viable cells is high, about 30,000 J/L (Gahl & Markl, 1996; Katsuki et al., 2002).

UV radiation (plasma-generated) has proven to be effective in decontamination processes and is gaining popularity, particularly in European countries, because chlorination leaves undesirable by-products in water. Most bacteria and viruses require relatively low UV dosages for inactivation, which is usually in a range of $2000-6000 \,\mu W \cdot s/cm^2$ for a 90% kill rate. For example, *E. coli* requires a dosage of $3000 \,\mu W \cdot s/cm^2$ for a 90% reduction (Wolfe, 1990). Cryptosporidium, which shows an extreme resistance to chlorine, requires a UV dosage greater than $82,000 \,\mu W \cdot s/cm^2$. The criteria for the acceptability of UV disinfecting units include a minimum dosage of $16,000 \,\mu W \cdot s/cm^2$ and a maximum water penetration depth of approximately 7.5 cm. UV radiation in the wavelength range from 240 to 280 nm causes irreparable damage to the nucleic acid of micro-organisms. The most potent wavelength of UV radiation for DNA damage is approximately 260 nm. The total energy cost of the UV water treatment is also quite large, similar to the case of pulsed electric fields.

Ozonation (O_3 is also plasma-generated; see Sections 6.5 and 6.6) is one of the powerful methods of water treatment. Ozone gas, generated in plasma, is bubbled into a contaminated solution and dissolves in it. The ozone is chemically active and is capable of efficiently inactivating micro-organisms at a level comparable to chlorine. Achieving a 4-log reduction at 20°C with an ozone concentration of 0.16 mg/L requires an exposure time of 0.1 min (Anpilov et al., 2001). At higher temperatures and pH levels, ozone tends to rapidly decay and requires more exposure time. Plasma discharges, especially DBD (see Section 6.5), have been used for the production of ozone in the past several decades to kill micro-organisms in water. Ozone has a lifetime of approximately 10–60 min, which varies depending on pressure, temperature, and humidity of surrounding conditions. Because of the relatively long O_3 lifetime, ozone gas can be produced in air or oxygen plasma and then stored in a tank and finally injected into water. The bactericidal effect of O_3 in water is usually characterized by the *Ct*-factor, defined as the product of ozone concentration C [mg/L] and the required time t [min] to disinfect a micro-organism in water. For example, for *Ditylum brightwelli*, an important ballast water species, the *Ct* value was 50 mg-min/L. In other words, if the ozone concentration is 2 mg/L, it takes 25 min of contact time to disinfect this organism in ballast water (Dragsund, Andersen, & Johannessen, 2001). The energy efficiency of ozonation is additionally limited by O_3 losses during storage and transportation.

Thus, ozone and UV radiation generated in remote plasma sources are effective means of water cleaning and sterilization. If plasma is organized not remotely but directly in water, the effectiveness of the treatment due to plasma-generated UV radiation and active chemicals can be much higher. The organization of plasma inside of water also leads to an additional significant contribution of short-living active species (electronically excited molecules, active radicals like OH, O, etc.), charged particles, and plasma-related strong electric fields into cleaning and sterilization (Sun, Sato, & Clements, 1997; Locke et al., 2006; Fridman, Gutsol, & Cho, 2007). While direct water treatment by plasma generated in the water can be very effective (see following discussion), initiation and sustaining of plasma in water (where the mean free path of electrons is very short) is more complicated than in the gas phase, which is to be discussed in the following section.

12.4.2. Electrical Discharges in Water

Electric discharges are conventionally organized in gases (see Chapter 4). The electric breakdown of liquids is limited by their high density and short mean free path of electrons, and, therefore, requires very high electric fields E/n_0 (see Paschen curves, Section 4.1.1). Nevertheless, the breakdown of liquids can be performed not at the extremely high electric fields required by Paschen curves but at those only slightly exceeding breakdown fields in atmospheric-pressure molecular gases. The effect can be explained by different electrically induced mechanisms of formation of microvoids and quasi "cracks" in the liquids. Some of the mechanisms have been discussed in Section 8.7 regarding micro-arc oxidation, (see also Section 12.4.3). The discharges in liquids have been reviewed by Fridman, Gutsol, and Cho (2007); and Locke et al. (2006). They can be sustained in water by pulsed high-voltage power supplies. The discharges in water usually start from sharp electrodes. If the discharge does not reach the second electrode it can be interpreted as pulsed corona; branches of such a discharge are referred to as streamers. If a streamer reaches the opposite electrode a spark is formed. If the current through the spark is high (above 1 kA), this spark is called a pulsed arc. Various electrode geometries have been used for plasma generation in water for the purpose of water treatment. Two simple geometries are shown in Fig. 12–19, which

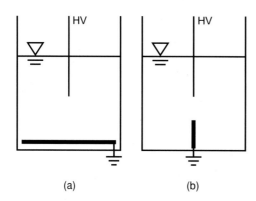

Figure 12–19. Two configurations for the plasma discharges in water.

(a) (b)

are a point-to-plane geometry and a point-to-point geometry. The former is often used for pulsed corona discharges, whereas the latter is often used for pulsed arc systems. One of the concerns in the use of the point-to-plane geometry is the adverse effect of the tip erosion. Sunka et al. (1999) pointed out that the very sharp tip anode would be quickly eroded by the discharge and one had to find some compromise between the optimum sharp anode construction and its lifetime for extended operations. Sunka et al. (1999) proposed a coaxial reactor, which consisted of a 6-mm-diameter stainless steel wire anode and a long stainless steel tubule cathode of 30 mm inside diameter. In particular, the anode wire was spray-coated by a thin (0.2–0.3 mm) layer of porous ceramics whose electric conductivity was about 1–3 μS/cm. The ceramic layer increased the electric field strength on the anode wire surface due to the redistribution of the field inside the interelectrode space during the prebreakdown stage, thus generating a large number of discharge channels (i.e., hundreds of streamers) during each voltage pulse. Another concern in the use of pulsed discharges is the limitation posed by the electrical conductivity of water on the production of such discharges (Sunka et al., 1999). In the case of a low electric conductivity of water (<10 μS/cm), the range of the applied voltage that can produce a corona discharge without sparking is very narrow. In the case of the high electric conductivity of water (>400 μS/cm), streamers become short and the efficiency of radical production decreases, and a denser and cooler plasma is generated. In general, the production of OH radicals and O atoms is more efficient at water conductivity less than 100 μS/cm. Al-Arainy, Jayaram, and Cross (1996) pointed out that, for the case of tap water, bulk heating was one of the problems in the use of corona discharges. They reported that, at a frequency of 213 Hz (i.e., a relatively high frequency), the temperature of the treated water rose from 20 to 55°C in 20 min, indicating a significant power loss and extra loading to the pulse generator.

12.4.3. Mechanisms and Characteristics of Plasma Discharges in Water

Mechanisms of plasma discharges and breakdowns in liquids (specifically in water) can be classified into two groups: the first presents the breakdown in water as a sequence of a bubble process and an electronic process, and the second divides the process into a partial discharge and a fully developed discharge such as arc or spark (Akiyama, 2000). In the first approach, the bubble process starts from a microbubble which is formed by the vaporization of liquid from local heating in the strong electric field region at the tips of electrodes. The bubble grows, and an electrical breakdown takes place within the bubble. In this case, the cavitation mechanism is suggested to explain the slow bush-like streamers (Beroural, 1993; Beroural et al. 1998). The appearance of bright spots is delayed from the onset of the voltage, and the delay time tends to be greater for smaller voltages. The time lag to water breakdown increases with increasing pressure, supporting the bubble mechanism in a submicrosecond discharge formation in water (Jones & Kunhardt, 1994, 1995). The

Table 12–3. Major Characteristics of Pulsed Corona, Pulsed Arc, and Pulsed Spark Discharges in Liquid Phase

Pulsed corona in water	Pulsed arc in water	Pulsed spark in water
Streamer-like channels.	Current is transferred by electrons.	Similar to pulsed arc, except for short pulse durations and low er temperature.
Streamer-channels do not propagate across the entire electrode gap, i.e., partial electrical discharge.	Quasi-thermal plasma.	
	Arc generates strong shock waves within cavitation zone.	
Streamer length in order of cm, channel width ~10–20 μm.	High-current filamentous channel bridges the electrode gap.	Pulsed spark is faster than pulsed arc, i.e., strong shock waves are produced.
Electric current is transferred mostly by ions.	Gas in channel (bubble) is ionized.	
Non-thermal plasma.	High UV emission and radical density.	Plasma temperatures in spark is around a few thousand kelvin.
Weak to moderate UV generation.		
Relatively weak shock waves.	A smaller gap between two electrodes of ~5 mm is needed than that in pulsed corona.	
Water treatment area is limited at a narrow region near the corona.		
Pulse energy: a few joules per pulse, often less than 1 J per pulse.	Large discharges pulse energy (greater than 1 kJ per pulse).	
Frequency is about 100–1000 Hz.	Large current (about 100 A), peak current greater than 1000 A.	
Relatively low current, i.e., peak current is less than 100 A.	Electric field intensity at the tip of electrode is 0.1–10 kV/cm.	
Electric field intensity at the tip of electrode of 100–10,000 kV/cm.	Voltage rise time is 1–10 μs.	
A fast-rising voltage on the order of 1 ns, but less than 100 ns.	Pulse duration ~20 ms. Temperature exceeds 10,000 K.	

time to form the bubbles is about 3–15 ns, depending on the electric field and pressure (Akiyama, 2000). The influence of the water electrical conductivity on this regime of the discharges is small (Akiyama, 2000). Bulk heating via ionic current does not contribute to the initiation of the breakdown. The power necessary to evaporate the water during streamer propagation can be estimated using the streamer velocity, the size of the streamer, and the heat of vaporization (Lisitsyn et al., 1999). Using a streamer radius of 31.6 μm, a power of 2.17 kW was estimated to be released into a single streamer to ensure its propagation in the form of vapor channels. In the case of multiple streamers, the required power can be estimated by multiplying the number of visible streamers to the power calculated for a single streamer. In the electronic process, electron injection and drift in liquid take place at the cathode, while hole injection through a resonance tunneling mechanism occurs at the anode (Katsuki et al., 2002). In the electronic process, breakdown occurs when an electron makes a suitable number of ionizing collisions in its transit across the breakdown gap.

According to the second approach to the mechanisms of electrical discharges in water, they are divided into partial electrical discharges and arc and spark discharge, which is illustrated in Table 12–3 (Locke et al., 2006; Sato, Ohgiyama, & Clements, 1996; Sugiarto, Ohshima, & Sato, 2002; Sun, Sato, Clements, 1999; Sugiarto et al., 2003; Manolache, Shamamian, & Denes, 2004; Ching et al., 2001; Ching, Colussi, & Hoffmann, 2003; Robinson, Ham, & Balaster, 1973; Robinson, 1973). In the partial discharges, the current is mostly transferred by ions. For the case of high-electrical-conductivity water, a large discharge current flows, resulting in a shortening of the streamer length due to the faster compensation of the space charge electric fields on the head of the streamer. Subsequently, a higher power density in the channel is obtained, resulting in a higher plasma temperature, a higher UV radiation, and generation of acoustic waves. In the arc or spark discharges, the current is transferred by electrons. The high current heats a small volume of plasma in the gap between the two electrodes, generating quasi-thermal plasma. When a high-voltage, high-current discharge takes place between two submerged electrodes, a large part of the energy is consumed in the formation of a thermal plasma channel. This channel emits UV radiation and its expansion against the surrounding water generates intense shockwaves (Sunka et al., 1999). The shockwave can directly interact with the micro-organism, causing it to explode. Alternatively, the pressure waves can dissociate micro-organism colonies within the liquid, thus increasing their exposure to inactivation factors. For the corona discharge in water, the shockwaves are weak or moderate, whereas for the pulsed arc or spark the shockwaves are strong.

12.4.4. Physical Kinetics of Water Breakdown

The critical breakdown condition for a gas is described by the Paschen curve (Section 4.1.1), from which one can calculate the breakdown voltage for air, for example. A value of 30 kV/cm is a well-accepted breakdown voltage of air at 1 atm. When one attemps to produce direct plasma discharges in water, it could be expected that a much greater breakdown voltage on the order of 30,000 kV/cm might be needed due to the density difference between air and water. A large body of experimental data on the breakdown voltage in water shows, however, that without special precautions this voltage is of the same magnitude as for gases. This interesting and practically important effect can be explained by taking into account the fast formation of gas channels in the body of water under the influence of the applied high voltage (see previous section). When formed, the gas channels give the space for the gas breakdown inside of the body of water, which explains why the voltage required for water breakdown is of the same magnitude as for gases. The gas channels can be formed by development and electric expansion of gas bubbles already existing in water as well as by additional formation of the vapor channel through fast local heating and evaporation. We are going to focus on the second mechanism, which is usually referred to as the **thermal breakdown** (Fridman, Gutsol, and Cho, 2007). When a voltage pulse is applied to water, it induces a current and redistribution of the electric field. Due to the dielectric nature of water, an electric double layer is formed near the electrode, which results in the localization of the applied electric field. This electric field can become high enough for the formation of a narrow conductive channel, which is heated up by electric current to temperatures of about 10,000 K. Thermal plasma generated in the channel is rapidly expanded and ejected from the narrow channel into water, forming a plasma bubble. The energy required to form and sustain the plasma bubble is provided by Joule heating in the narrow conductive channel in water. The physical nature of thermal breakdown can be related to thermal instability of local leakage currents through water with respect to the Joule overheating. If the leakage current is slightly higher at one point, the Joule heating and, hence, temperature also grow there. The temperature increase results in a significant growth of local conductivity and leakage current. Exponential temperature growth to several thousand degrees at a local point leads to the formation of the narrow plasma channel in water, which determines the thermal breakdown. The thermal breakdown is a critical thermo-electric phenomenon taking place at the applied voltages exceeding a certain threshold value, when heat release in the conductive channel cannot be compensated by heat transfer losses to the surroundings.

The thermal condition of water is constant during the breakdown; water stays liquid away from the discharge with the thermal conductivity about 0.68 W/mK. When Joule heating between the two electrodes is larger than a threshold value, an instability can occur, resulting in instant evaporation and a subsequent thermal breakdown. When Joule heating is smaller than a threshold value, nothing happens but electrolysis and the breakdown never takes place. The thermal breakdown instability is characterized by the instability increment showing frequency of its development (Fridman, Gutsol, & Cho, 2007):

$$\Omega = \left[\frac{\sigma_o E^2}{\rho C_p T_0} \right] \frac{E_a}{T_0} - D \frac{1}{R_o^2}, \qquad (12\text{--}24)$$

where σ_0 is the water conductivity, E_a is the Arrhenius activation energy for the water conductivity, E is the electric field, ρC_p is the specific heat per unit volume, T_0 is the temperature, R_0 is the radius of the breakdown channel, and $D \approx 1.5 \cdot 10^{-7}\,\text{m}^2/\text{s}$ is the thermal diffusivity of water. When the increment Ω is greater than zero, the perturbed temperature exponentially increases with time, resulting in thermal explosion; when Ω is less than zero, the perturbed temperature exponentially decreases with time, resulting in the steady-state condition. For the plasma discharge in water, the breakdown voltage in the channel with length L can be estimated based on equation (11.24) as (Fridman, Gutsol, & Cho, 2007)

$$V \geq \sqrt{\frac{DC_p \rho T_o^2}{\sigma_o E_a}} \cdot \frac{L}{R_0}. \qquad (12\text{--}25)$$

The breakdown voltage increases with L/R_o. Assuming $L/R_0 = 1000$, it is about 30 kV. It should be mentioned that recent investigations indicate the possibility of "cold" water breakdown in nanosecond pulsed discharges. The effect can be provided in particular by electrostatic water "cracking."

12.4.5. Experimental Applications of Pulsed Plasma Discharges for Water Treatment

The application of electrical pulses in the microsecond range to biological cells has been investigated in detail by Schoenbach et al. (1997), Joshi, Qian, & Schoenbach (2002), Katsuki et al. (2002), Abou-Ghazala et al. (2002), and Joshi et al. (2002). They used a point-to-plane geometry to generate pulsed-corona discharges for bacterial (*E. coli* or *Bacillus subtilis*) decontamination of water with 600 ns, 120 kV square wave pulses. The wire electrode was made of a tungsten wire with $75\,\mu\text{m}$ diameter, 2 cm apart from a plane electrode. The concentration of *E. coli* could be reduced by three orders of magnitude after applying eight corona pulses to the contaminated water with corresponding energy expenditure of 10 J/cm^3 (10 kJ/L). Plasma pulses cause the accumulation of electrical charges at the cell membrane, shielding the interior of the cell from the external electrical fields. Since typical charging times for the mammalian cell membrane are on the order of 1 μs, these microsecond pulses do not penetrate into cells. Hence, shorter pulses in the nanosecond range can penetrate the entire cell, nucleus, and organelles, and affect cell functions, thus disinfecting them. High-voltage pulse generators were used to apply nanosecond pulses as high as 40 kV to small test chambers. Biological cells held in liquid suspension in the cuvettes were placed between two electrodes for pulsing. The power density was up to 10^9 W/cm^3, but the energy density was rather low (less than 10 J/cm^3, a value that could slightly increase the temperature of the suspension by approximately 2°C).

Heesch et al. (2000) also applied pulsed electric fields and pulsed corona discharges to inactivate micro-organisms in water. They used four different types of plasma treatment configurations: a perpendicular water flow over two wire electrodes, a parallel water flow along two electrodes, air bubbling through a hollow needle electrode toward a ring electrode, and a wire cylinder. They used 100 kV pulses (producing a maximum of 70 kV/cm electric field) with a 10 ns rising time with 150 ns pulse duration at a maximum rate of 1000 pulses per second. The pulse energy varied between 0.5 and 3 J/pulse and an average pulse power

Figure 12–20. Point-to-plane plasma discharge system for pulsed corona discharge and spark in water.

was 1.5 kW with 80% efficiency. Micro-organism inactivation was found to be 85 kJ/L per one-log reduction for *Pseudomonas flurescens* and 500 kJ/L per one-log reduction for spores of *Bacillis sereus*. It was demonstrated that corona directly applied to water was more efficient than pulsed electric fields. With direct corona, Heesch et al. (2000) achieved 25 kJ/L per one-log reduction for both gram-positive and gram-negative bacteria. Pulsed plasma discharges in water have been applied for the sterilization and removal of organic compounds such as dyes by Sato, Ohgiyama, and Clements (1996), Sugiarto, Ohshima, and Sato (2002), Sun, Sato, Clements (1999), and Sugiarto et al. (2003). The streamer discharge was produced from a point-to-plane electrode, where a platinum wire in a range of 0.2–1 mm in diameter was used for the point electrode, which was positioned 1–5 cm from the ground plane electrode.

Lisitsyn et al. (1999) used a dense medium plasma reactor for the disinfection of various waters. The plasma reactor consisted of an upper electrode rotating at a range of 500–5000 rpm and a hollow conical cross-sectional end piece. An advantage of the rotating electrode was that the rotating action spatially homogenized the multiple microarcs, activating a larger effective volume of water. In addition, spinning the upper electrode also simultaneously pumped fresh water and vapors into the discharge zone. Water sterilization in the electrohydraulic (arc) discharge reactor was investigated by Ching et al. (2001) and Ching, Colussi, and Hoffmann (2003). A typical operational condition included a discharge of 135 mF capacitor bank stored energy at 5–10 kV through a 4 mm electrode gap within 40 μs with a peak current of 90 kA. They studied the survival of *E. coli* in aqueous media exposed to the aforementioned electrohydraulic discharges. They reported the disinfection of 3 L of a $4 \cdot 10^7$ cfu/mL *E. coli* suspension in 0.01 M PBS at pH 7.4 by 50 consecutive electrohydraulic discharges. It was demonstrated that UV radiation emitted from the electrohydraulic discharge was the lethal agent that inactivated *E. coli* colonies rather than the thermal/pressure shocks or the active chemical species. More data regarding efficiency of thermal and non-thermal discharges for water cleaning and sterilization can be found in Fridman, Gutsol, and Cho (2007) and Locke et al. (2006).

12.4.6. Energy-Effective Water Treatment Using Pulsed Spark Discharges

The energy cost of water treatment is one of the most critical parameters of the plasma process. The highest energy efficiency of water sterilization has been achieved using pulsed spark discharge (Campbell et al., 2006; Fridman, Gutsol, & Cho, 2007), organized in the point-to-plane electrode configuration (see Fig. 12–20). Variance in the plasma from corona to spark discharge was observed in these experiments to be dependent on the gap distance

12.4. Plasma Cleaning and Sterilization of Water

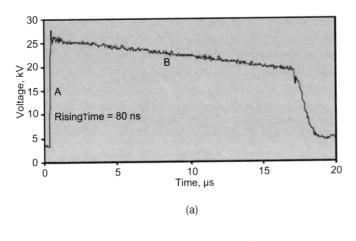

(a)

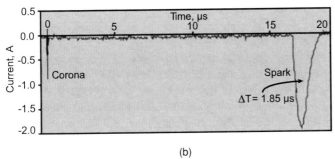

(b)

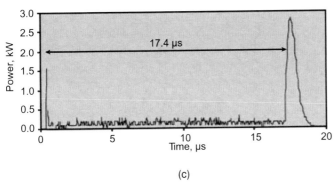

(c)

Figure 12–21. Voltage, current, and power profiles measured using an oscilloscope during a typical pulsed spark test.

measured from the anode to the grounded cathode. A stainless steel electrode (0.18 mm) was encased in silicon residing in a hollow Teflon tube, providing the necessary insulation for the electrodes. The electrode extended approximately 1.6 mm from the bottom of the glass tube, providing a region for spark discharge initiation. The critical distance between electrodes for the corona-to-spark transition was about 50 mm. Interelectrode distance greater than 50 mm resulted in corona discharge, whereas a distance less than 50 mm resulted in spark discharge. Figure 12–21 presents voltage, current, and power profiles during a pulsed spark test. The initial steep rise in the voltage profile indicates the time moment of breakdown in the spark gap, after which the voltage linearly decreased with time over the next 17 μs due to a long delay while the corona was formed and transferred to a spark. The rate of

Table 12–4. *E. coli* Bacterial Concentrations in Water
Following the Pulsed Spark Discharge Treatment of Water

Experiment 1 Number of pulses	Cells/mL Run A	Cells/mL Run B	Cells/mL Run C
0	$1.84 \cdot 10^8$	$1.76 \cdot 10^8$	$1.60 \cdot 10^8$
20	$1.01 \cdot 10^8$	$7.40 \cdot 10^7$	–
40	$1.54 \cdot 10^7$	$1.01 \cdot 10^7$	$1.85 \cdot 10^7$
60	$1.67 \cdot 10^6$	$4.20 \cdot 10^6$	$3.20 \cdot 10^6$
80	$2.10 \cdot 10^5$	$4.80 \cdot 10^5$	$3.30 \cdot 10^5$
100	$9.30 \cdot 10^4$	$2.43 \cdot 10^4$	$3.05 \cdot 10^4$
Experiment 2 Number of pulses	Cells/mL Run A	Cells/mL Run B	Cells/mL Run C
0	$2.05 \cdot 10^6$	$2.04 \cdot 10^6$	$1.96 \cdot 10^6$
50	$8.70 \cdot 10^3$	$1.52 \cdot 10^4$	$1.58 \cdot 10^4$

the voltage drop over time depends on the capacitance. The current and power profiles show the corresponding histories, which show initially sharp peaks and then very gradual changes over the next 17 μs. The duration of the initial peak was about 70 ns. At $t \cong 17\ \mu$s, there was a sudden drop in the voltage, indicating the onset of a spark or the moment of channel appearance, which was accompanied by sharp changes in both the current and power profiles. The duration of the spark was approximately 2 μs, which was much longer than the duration of the corona.

Plasma sterilization data collected by treatment of *E. coli* in the described pulsed spark discharge (Campbell et al., 2006) are presented in Table 12–4 for two different initial conditions. When the initial cell count was high ($1.8 \cdot 10^8$ cfu/mL), the spark discharge could produce a 4-log reduction at 100 pulses and 2-log reduction at about 65–70 pulses. When the initial cell count was at an intermediate level ($2 \cdot 10^6$ cfu/mL), the spark discharge produced a 2-log reduction at 50 pulses. Taking into account the energy of one pulse and volume of water, the minimum energy per 1 L of water for 1-log reduction in *E. coli* concentration is as low as 77 J/L. It should be mentioned that similar but much more detailed experiments with spark discharge treatment of *E. coli* in water at lower bacterial concentrations (Arjunan et al., 2007) results in similar very low values of the plasma sterilization energy cost (about 100 J/L). Table 12–5 compares major plasmas applied for water sterilization. The properties of spark discharges are beneficial for water treatment. The spark discharges in water investigated by Campbell et al. (2006) and Arjunan et al. (2007) require very low power in comparison to other systems: arcs, which are too hot and localized; and coronas, which are too weak to generate UV effective for water sterilization. To achieve a 4-log reduction for a typical household flow rate of 6 gpm (22.7 L/min), the electrical power requirement is only 120 W. The pulsed spark discharges indicate a potential to accommodate a 1000 gpm (3786 L/min) water flow rate, while retaining the capability of achieving a 4-log reduction in biological contaminant at a power of only 20 kW.

12.5. PLASMA-ASSISTED TISSUE ENGINEERING

12.5.1. Plasma-Assisted Regulation of Biological Properties of Medical Polymer Materials

Plasma methods are widely applied today in different aspects of tissue engineering. First consider plasma control of the biological properties of medical polymers and **biocompatibility**.

Table 12–5. Comparison of Major Discharges Used in Plasma-Based Water Treatment

Water treatment parameters	Pulsed arc discharges	Spark discharges	Pulsed corona discharges
Energy per liter for 1-log reduction in *E. coli*, J/L	860	77	30,000–150,000
Power requirement for water consumption at 6 gpm, kW	0.326	0.029	11.4–56.8
Power requirement for water consumption at 1,000 gpm, kW	54.3	4.9	1892.7–9463.5
Power available in small power system (10 · 10 · 10 cm), kW	30	10	0.3
Maximum water throughput based on the above power, gpm	553	2058	0.03–0.16

The main requirement imposed on all polymer biomaterials applied in medicine is a combination of their desired physicochemical and physicomechanical characteristics with biocompatibility. Depending on particular applications, the biocompatibility of polymers can include various requirements, which can sometimes be contradictory to each other. Thus, in the case of artificial vessels, drainages, intraocular lenses, biosensors, or catheters, the interaction of the polymer with a biological medium should be minimized for the reliable operation of the corresponding device after implantation. In contrast, in the majority of orthopedic applications, the active interaction and fusion of an implant with a tissue is required. General requirements imposed on all medical polymers consist in non-toxicity and stability.

The body response to a polymer implant mainly depends on its surface properties: chemical composition, structure, and morphology. Physical techniques are required for regulating the biological properties of polymer materials. These techniques should vary the physicochemical, structural, and functional properties of surfaces over a wide range without affecting bulk characteristics such as strength, elasticity, transmission factor, refractive index, and electrophysical parameters. Such a multi-purpose and multi-functional technique is treatment in a low-temperature gas-discharge plasma, which is widely used for modifying the surface characteristics of polymers (Chu et al., 2002; Poncin-Epaillard & Legeay, 2003; Detomazo et al., 2005; Lopez et al., 2005; see also Sections 9.7 and 9.8).

The cross-linking effect under the action of inert gas plasmas can be used for the immobilization of biocompatible low-molecular-weight compounds and various functional groups on polymer surfaces. For example, treatment in argon plasma was used for grafting the Pluronic^{TM} 120 copolymer, which is highly biocompatible, onto the surface of PTFE (Vasilets et al., 2002). The Pluronic ^{TM} 120 triblock copolymer (PEO/PP/PEO) was initially supported onto the polymer surface by physisorption from solution, and then grafted by treatment in low-pressure argon plasma. A similar procedure was used for the covalent immobilization of sulfo groups on the surface of PE (Terlinger, Feijen, & Hoffman, 1993). Plasma etching changes the polymer surface morphology, which plays an important role in biocompatibility. For example, surface smoothing due to plasma etching upon treatment in oxygen- and fluorine-containing plasmas positively affects hemocompatibility because it decreases the probability of thrombosis at surface irregularities in a bloodstream. The surface smoothing of a PMMA lens without changes in its bulk optical characteristics is a positive consequence of plasma treatment in a CF_4 discharge (Eloy et al., 1993).

Since most biological fluids are aqueous solutions or contain water, the wettability or hydrophilicity of polymer surfaces is of paramount importance for biocompatibility. According to the current **hypothesis of complementarity** (Sevastianov, 1991), in order to improve hemocompatibility, it is necessary to minimize not only the average interfacial surface energy at the material–blood interface but also at every point on the surface; that is, the character of free energy distribution at the interface of the biomaterial with an adsorbed protein layer should be taken into consideration. In other words, the hydrophilic and hydrophobic regions of a surface that contacts with blood and analogous regions of an adsorbed protein molecule should be complementary in order for the product to be highly hemocompatible. The hydrophilicity and surface energy of biomaterials can be varied over wide ranges by generating various surface polar groups with the use of treatment of medical polymers in oxygen plasma (Sevastianov, 1991), oxygen-containing CO_2 (Baidarovtsev, Vasilets, & Ponomarev, 1985), H_2O (Simon et al., 1996), or SO_2 (Klee & Hocker, 1999) plasmas, or nitrogen-containing (N_2, NH_3) plasmas (Ramires et al., 2000). Thereby, the processes of protein adsorption and cell adhesion can be regulated. Thus, plasma treatment is effective in regulating adhesion of medical polymers (making their surfaces adhesive or repulsive to certain biological agents), as well as controlling their biocompatibility. More details can be found in Vasilets, Kuznetsov, and Sevastianov (2006) and Poncin-Epaillard and Legeay (2003).

12.5.2. Plasma-Assisted Attachment and Proliferation of Bone Cells on Polymer Scaffolds

The significant effect of non-thermal plasma treatment of polymer scaffolds on not only the attachment of bone cells but also on their further biological activity, specifically on their proliferation, has been demonstrated by Yildirim et al. (2007). The experiment in the atmospheric-pressure DBD plasma addressed attachment and proliferation of osteoblasts cultured over poly-ε-caprolactone (PCL) scaffolds. Traditional bone-grafting procedures due to pathological conditions, trauma, or congenital deformities have disadvantages like graft rejection, donor site morbidity, and disease transmission. Tissue engineering of bone is increasingly becoming the treatment of choice among surgeons to alleviate these problems. Favorable cell-substrate interaction during the early stage of cell seeding is one of the most desirable features of tissue engineering. The ability of bone cells to produce an osteoid matrix on the scaffold can be affected by the quality of the cell–scaffold interaction. Plasma treatment of the scaffolds significantly improves this interaction. PCL is an aliphatic polyester with a promising potential in tissue engineering because of its biocompatibility and mechanical properties. However, the surface hydrophobicity works against PCL when it comes to cell attachment. Plasma makes the PCL surface hydrophilic, which leads to successful formation of tissue constructs and cell proliferation, differentiation, and new tissue ingrowth. The attachment phase of the anchor-dependent cells starts with formation of a cell adhesive protein layer from serum containing media on the substrate at the cell–material interface. The cells get attached to these absorbed proteins via cell-surface adhesion receptors like integrins and extracellular matrix (ECM) proteins including fibronectin, vitronectin, fibrinogen, and collagen that have a cell-binding domain containing the arginine-glycine-aspartate (RGD) sequence. Interactions between the proteins and the surface determine the residence time of the initial attachment, thereby influencing the cell proliferation and differentiation capacity on contact biomaterial. Attachment and proliferation of osteoblasts, which are anchor-dependent cells, can be effectively controlled by oxygen plasma treatment which increases the cell affinity of the material by binding the cell adhesion proteins (Anselme, 2000; Marxer, 2002; Xu, 2007). Experiments of Yildirim et al. (2007) investigated the influence of physicochemical properties of atmospheric-pressure oxygen pulsed DBD plasma-treated PCL on early time points of osteoblastic cell adhesion and proliferation. For this purpose, PCL scaffolds were treated with oxygen-based microsecond pulsed DBD plasma with different exposure time; then mouse osteoblast cells (7F2) were cultured on treated PCL scaffolds for 15 h. The plasma-treated PCL surfaces were

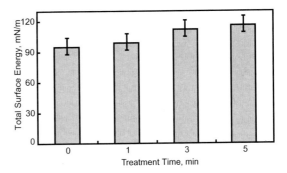

Figure 12–22. Variation in total solid surface energy of PCL with oxygen plasma treatment time.

characterized in terms of the surface energy, and topography, via the Owens-Went method, and atomic force microscopy (AFM). The mouse osteoblast cell adhesion and proliferation were characterized by live/dead cell viability assay and the AlamarBlue[TM] assay. Results of the experiments are to be discussed in the next section.

12.5.3. DBD Plasma Effect on Attachment and Proliferation of Osteoblasts Cultured over Poly-ε-Caprolactone Scaffolds

The experiments of Yildirim et al. (2007) first consider effect of atmospheric-pressure oxygen DBD plasma on surface hydrophilicity and surface energy of the PCL scaffold by contact angle measurements (see Section 9.8.1). The contact angles on untreated and treated PCL scaffolds were measured by sessile drop technique using ultrapure water, diiodomethane, and glycerol as probe liquids. This permits the variation with DBD treatment time (at power about 0.1 W/cm^2) of total solid surface energy of PCL to be calculated (see Fig. 12–22) as well as dispersive and polar components of the solid surface energy (see Fig. 12–23). The total surface energy of PCL increased with prolonged oxygen plasma treatment time from 95 mN/m for untreated to 115 mN/m for a 5-min oxygen plasma-treated scaffold. While the dispersive component was not changed much, the polar component increased significantly, which demonstrates that the major contribution to the increment of total solid surface energy is due to the formation of polar (mostly peroxide) groups on the polymer surface during its treatment in atmospheric-pressure non-thermal oxygen plasma (see Section 9.8.1). The effect of oxygen plasma treatment on PCL microstructure was characterized in the Yildirim et al. (2007) experiments using AFM. The average roughness of untreated film was 13 nm and the surface pattern was relatively smooth. As a result of 1-min and 3-min DBD plasma treatment, the roughness was increased to 20 and 26 nm, respectively. For the 5-min treatment

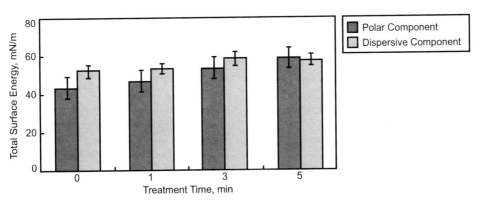

Figure 12–23. Variation in dispersive and polar components of the total solid surface energy of PCL with oxygen plasma treatment time.

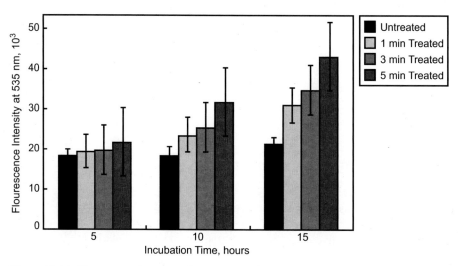

Figure 12–24. Fluorescence intensity of mouse osteoblast cells cultured on untreated and treated PCL scaffolds up to 15 h.

time, the change in surface morphology became obvious by large peaks in nanoscale with the roughness increase to 39 nm. Thus results show that with prolonged treatment time the mean surface roughness is increased almost three-fold. It is believed that the change in surface morphology is the result of the ion bombardment and selective destruction of the polymer surface layer (Chan, Ko, & Hiraoka, 1996).

The metabolic activity and the morphology of mouse osteoblast cells on plasma treated and untreated PCL scaffold were examined in the Yildirim et al. (2007) experiments by the live/dead cell vitality assay and the AlamarBlue[TM] assay. On the untreated PCL scaffolds, cell proliferation, cell attachment, and cell spreading with culturing time were significantly lower as compared to oxygen plasma-treated PCL scaffolds. The proliferation of mouse osteoblast cells for a 15-hour culture period on treated and untreated (control) PCL scaffolds is compared in Figs. 12–24 and 12–25. The mouse osteoblasts on untreated PCL scaffolds did not show significant metabolic activity during the incubation time. In contrast, a higher

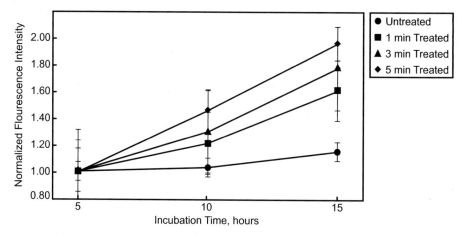

Figure 12–25. Normalized fluorescence intensity of mouse osteoblast cells according to 5-h values for different plasma treatment durations.

degree of cell attachment and proliferation can be seen on the plasma-treated surface. The viable cell count was also increased with prolonged oxygen-plasma treatment. The highest improvement for proliferation rate was observed for the 5-min treated sample. Normalized fluorescence intensity data (Fig. 12–25) show that, after a 15-h incubation period, the cell proliferation rate on the 5-min treated samples increased 90% from the beginning of the experiment. From the fluorescence images, it was observed that the mouse osteoblast cells are hardly attached on untreated scaffolds. After a 15-h incubation period, cells were hardly spread out on the 1-min treated scaffold. With the increased plasma treatment time from 1 to 5 min, osteoblastic adhesions improved. On the 5-min oxygen plasma-treated samples a confluent mouse osteoblast cell layer was observed. A similar trend was found for cell proliferation results given in Fig. 12–25. Thus, oxygen-plasma-treated PCL scaffolds are much more favorable than untreated ones for cell attachment due to the higher hydrophilicity and increased roughness.

12.5.4. Controlling of Stem Cell Behavior on Non-Thermal Plasma Modified Polymer Surfaces

Plasma treatment of polymers is effective not only in regulating attachment and proliferation of cells as discussed earlier, but also in controlling the much more sophisticated behavior of biological systems on treated surfaces. An example is the selective inhibition of type X collagen expression in human mesenchymal stem cell (MSC) differentiation on polymer substrates surface modified by low-pressure RF-CCP discharge plasma (see Section 4.4.5). Such plasma-assisted tissue engineering experiments were carried out by Nelea et al. (2005). Recent evidence indicates that a major drawback of current cartilage- and disc-tissue engineering is that human MSCs rapidly express type X collagen, which is a marker of chondrocyte hypertrophy associated with endochondral ossification. Some studies have attempted to use growth factors to inhibit type X collagen expression, but none to date have addressed the possible effect of the substratum on chondrocyte hypertrophy. Nelea et al. (2005) examined the growth and differentiation potential of human MSCs cultured on two polymer types, polypropylene and nylon-6, both of which have been surface-modified by low-pressure RF-CCP discharge plasma treatment in ammonia gas. Plasma treatments were performed in a cylindrical aluminum/steel chamber (Bullett et al., 2004). High-purity ammonia gas was admitted to the chamber at an operating pressure of 300 mTorr (40 Pa). RF-CCP (13.56 MHz) was generated at the 10-cm-diameter powered electrode in the center of the chamber with the walls acting as the grounded electrode. The samples were placed on the powered electrode, after which 30-s plasma treatments were performed at 20 W of RF power. The negative DC bias (250 V) developed on the powered electrode led to ion bombardment of the polymer surface during exposure to the plasma. Cultures in the Nelea et al. (2005) experiments were performed for up to 14 days in Dulbecco's modified Eagle medium plus 10% fetal bovine serum. Commercial polystyrene culture dishes were used as the control. Reverse transcriptase polymerase chain reaction was used to assess the expression of types I, II, and X collagens and aggrecan using gene-specific primers. Glyceraldehyde-3-phosphate dehydrogenase was used as a housekeeping gene. Types I and X collagens, as well as aggrecan, were found to be constitutively expressed by human MSCs on polystyrene culture dishes. Whereas both untreated and treated nylon-6 partially inhibited type X collagen expression, treated polypropylene almost completely inhibited its expression. These results indicate that plasma-treated polypropylene or nylon-6 may be a suitable surface for inducing MSCs to a disc-like phenotype for tissue engineering of intervertebral discs in which hypertrophy is suppressed. The Nelea et al. (2005) experiments indicate the effectiveness of plasma polymer treatment in controlling not only simple surface properties but also the very specific biological behavior of cells on treated surfaces.

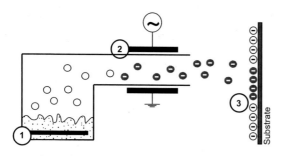

Figure 12–26. Plasma bioprinter: (1) droplet atomizer; (2) DBD plasma reactor; (3) charge droplet deposition substrate.

12.5.5. Plasma-Assisted Bio-Active Liquid Microxerography, Plasma Bioprinter

Biochemical patterning (bioprinting), which allows for microscale resolution on non-planar substrates, has been demonstrated by Fridman et al. (2003, 2005). Utilizing this method, biomolecules (including DNA, proteins, and enzymes) can be delivered to charged locations on surfaces by charged water buffer droplets. Charging of water droplets has been accomplished by using atmospheric DBD stabilized in the presence of a high concentration of micron-size water droplets. The patterning of biomolecules on surfaces has many applications ranging from biosensors, used in genetic discovery and monitoring of dangerous toxins, to tissue engineering constructs, where surfaces control tissue assembly or adhesion of cells. Most methods of biochemical patterning are suitable only for planar surfaces. In addition, micro- and nanoscale patterning often rely on complex sequences of lithography-based process steps. Plasma provides the biochemical substance patterning, or printing, allowing micro- and even nanoscale resolution on planar and non-planar substrates. It implies printing droplets of buffer containing biomolecules (including DNA), peptides, and cells. The method includes the creation of droplets with relevant molecules or cells in their respective buffer solutions, then DBD charging of the droplets, and finally delivering them onto the substrate with charge prewritten onto the substrate via conventional xerography (micron resolution) or via charge stamping (nanometer resolution); see Fig. 12–26. The key issue of plasma-assisted bioprinting is the protection of the biological agents contained in the buffer droplets from their plasma-based deactivation (see Section 12.2.3). Survival of the biological agents in plasma can be achieved by choosing relatively low specific DBD power and special composition of the protective buffer solutions.

12.6. ANIMAL AND HUMAN LIVING TISSUE STERILIZATION

12.6.1. Direct Plasma Medicine: Floating-Electrode Dielectric Barrier Discharge

Direct plasma treatment can be very effective in sterilization and other aspects of tissue treatment. The direct plasma treatment implies that living tissue itself is used as one of the electrodes and directly participates in active plasma discharge processes. Figure 12–7 illustrates direct plasma treatment (sterilization) of skin of a live mouse. DBD plasma is generated in this case between the quartz-covered high-voltage electrode and the mouse as a second electrode. Direct application of the high-voltage (10–40 kV) non-thermal plasma discharges in atmospheric air to treat live animals and people requires high level of safety precautions. Non-damaging treatment are the crucial issues of plasma medicine. The discharge current should be limited below the values permitted for treatment of living tissue. Moreover, discharge itself should be homogeneous enough to avoid local damage and discomfort. The creation of special atmospheric discharges effectively solving these problems is an important challenge for plasma medicine.

Figure 12–27. Floating-electrode dielectric barrier discharge: one electrode is protected by a dielectric barrier, another electrode is living tissue to be treated.

Fridman et al. (2006) especially developed for this purpose the floating-electrode DBD (FE-DBD), which operates under conditions where one of the electrodes is a dielectric-protected powered electrode and the second active electrode is a human or animal skin or organ – without human or animal skin or a tissue surface present the discharge does not ignite. In the FE-DBD setup, the second electrode (a human, for example) is not grounded and remains at a floating potential. The discharge ignites when the powered electrode approaches the surface to be treated at a distance (discharge gap) of less than about 3 mm, depending on the form, duration, and polarity of the driving voltage. A simple schematic of the FE-DBD (Fridman et al., 2006) was illustrated in Fig. 12–11. A typical value of plasma power in initial experiments was kept at about 3–5 W and power density was 0.5–1 W/cm^2. Further development of the FE-DBD is related to shape optimization of the applied voltage to minimize DBD non-uniformities, and the related possible damaging effects. The best results have been achieved by organization of the FE-DBD in the pulsed mode with pulse duration below 30–100 ns (Ayan et al., 2007), which results in the no-streamer discharge regime, sufficient uniformity, and the possibility of the non-damaging direct plasma treatment even when the second electrode is living tissue and therefore wet and essentially non-uniform. As soon as the atmospheric discharge is safe, it can be applied directly to the human body, as illustrated in Fig. 12–27. Thus, highly intensive non-thermal plasma devices can be directly applied to living animal or human tissue for different medical and cosmetic treatments.

12.6.2. Direct Plasma-Medical Sterilization of Living Tissue Using FE-DBD Plasma

Sterilization of living animal or human tissue with minimal or no damage to this tissue is of importance in a hospital setting. Chemical sterilization does not always offer a solution. For example, transporting chemicals for sterilization becomes a major logistics problem in a military setting, while use of chemicals for sterilization of open wounds, ulcers, or burns is not possible due to the extent of damage they cause to punctured tissues and organs. Non-thermal atmospheric-pressure plasma is non-damaging to the animal and human skin but quite a potent disinfecting and sterilizing agent (Fridman et al., 2006), which will be discussed later. Human tissue sterilization has been investigated by Fridman et al. (2006). Bacteria in this case were a mix of "skin flora" – bacteria collected from cadaver skin containing *Staphylococcus*, *Streptococcus*, and yeast. Direct FE-DBD plasma sterilization leads roughly to a 6-log reduction in bacterial load in 5 s of treatment (Table 12–6). A similar level of skin flora sterilization using the indirect DBD approach (see Section 12.2.2) requires 120 s and longer of plasma treatment at the same level of discharge power. Sterilization of the mix of bacteria collected from cadaver skin containing *Staphylococcus*, *Streptococcus*, and yeast occurred in the experiments generally after 4 s of treatment in most cases and 6 s in a few cases. Thus, non-thermal atmospheric plasma, especially when it is applied

Table 12–6. Human Skin Flora Sterilization by FE-DBD Plasma

Original concentration	5 s of FE-DBD	10 s of FE-DBD	15 s of FE-DBD
10^9	850 ± 183	9 ± 3	4 ± 4
10^8	22 ± 5	5 ± 5	0 ± 0
10^7	6 ± 6	0 ± 0	0 ± 0

Note: Concentrations of bacteria are given in cfu/mL

directly, is an effective tool for sterilization of living tissue. It opens interesting possibilities for non-thermal plasma applications in medicine, including presurgical patient treatment, sterilization of catheters (at points of contact with human body), sterilization of wounds and burns, as well as treatment of internal organs in gastroenterology.

12.6.3. Non-Damage (Toxicity) Analysis of Direct Plasma Treatment of Living Tissue

Plasma is an excellent sterilization tool for different surfaces. The key question for direct plasma skin sterilization is if the skin remains intact after the treatment. Moreover, the problem of non-damage is a key issue in all plasma medicine. A topical treatment which damages the tissue surface would not be acceptable to the medical community; therefore, cadaver tissue was first tested and then escalating skin toxicity trials were carried out on SKH1 hairless mice in the FE-DBD experiments of Fridman et al. (2006). Cadaver tissue in these experiments was treated by FE-DBD plasma for up to 5 min without any visible or microscopic change in the tissue, as was verified with tissue sectioning and staining via the Hematoxylin & Eosin (H&E) procedure, which is illustrated in Fig. 12–28. Based on the knowledge that FE-DBD plasma has non-damaging regimes, an animal model was tested by Fridman et al. (2007). In an SKH1 mouse model, the skin treatment was carried out at varying doses to locate damaging power/time (dose) combination, and skin damage was

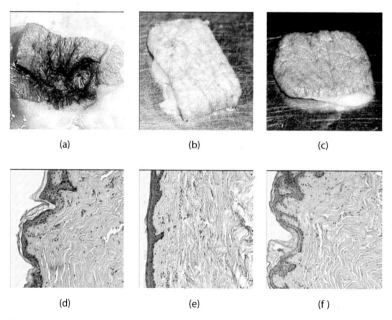

(a) (b) (c)

(d) (e) (f)

Figure 12–28. Photos (top) and tissue histology (bottom) of cadaver skin samples after FE-DBD treatment: control (left), 15 s of treatment (center), and 5 min of treatment (right) – no visible damage is detected.

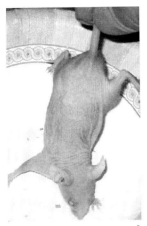
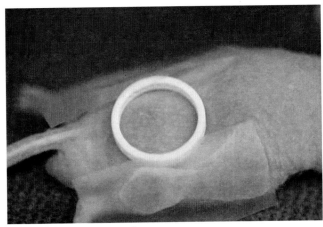

40 seconds at 1.4 Watt/cm^2 40 seconds at 1.4 Watt/cm^2
2 weeks post-treatment Immediately after treatment

(a) (b)

Figure 12–29. Animal remains fine after a reasonably high plasma dose (more than 10 times higher than needed for skin sterilization). Dark circle on the left animal's back shows the area within which the animal was treated by FE-DBD plasma.

analyzed in two stages. First, the animal was treated at what was deemed to be a toxic (damaging) dose based on trials with cadaver skin tissue. Once the dose where the damage was visible was located, a new animal was treated at a lower dose. If no damage was observed at that dose, two more animals were treated and if no damage was observed in all three the dose was deemed "maximum acceptable dose." Once the maximum dose was located, three animals were treated at that dose and left alive under close observation for 2 weeks. Based on the experimental matrix, a dose of 10 min at 0.6 W/cm^2 was deemed the maximum acceptable prolonged treatment and a dose of 40 s at 2.3 W/cm^2 was deemed the maximum acceptable high-power treatment. Histological (microscopic) comparison of control SKH1 skin samples with toxic and non-toxic plasma doses shows regions where plasma dose is fairly high while the animal remains unaffected (Fig. 12–29, animal after the treatment; Fig. 12–30, histological samples). Of note is that sterilization was achieved at 2–4 s at high-power treatment of 0.8 ± 0.2 W/cm^2 and at 10 ± 4 s at half that power. The variation in time necessary for sterilization is attributed to the initial contamination level of the animal (as for cadaver tissue); in other words, some skin samples are simply cleaner than others.

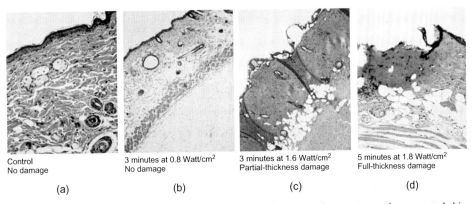

Control 3 minutes at 0.8 Watt/cm^2 3 minutes at 1.6 Watt/cm^2 5 minutes at 1.8 Watt/cm^2
No damage No damage Partial-thickness damage Full-thickness damage

(a) (b) (c) (d)

Figure 12–30. Histology of toxic and non-toxic to SKH1 skin plasma doses, compared to untreated skin.

891

(a)   (b)

Figure 12–31. Blood drop treated by FE-DBD: 15 s of FE-DBD (left) and control (right); photo was taken 1 min after the drops were placed on brushed stainless steel substrate.

Following the investigation in mice, an investigation was carried out in pigs, achieving the same results. The toxicity due to FE-DBD treatment of living tissue depends not only on the treatment dose (discharge power and treatment duration), but also strongly on the shape of the voltage applied to the discharge. Pulsing of the DBDs can significantly decrease their damaging ability. An application of nanosecond pulses completely prevents the formation of streamers and therefore DBD microdischarges, which helps decrease toxicity of the direct plasma-medical treatment of living tissue (Ayan et al., 2007).

12.7. NON-THERMAL PLASMA-ASSISTED BLOOD COAGULATION

12.7.1. General Features of Plasma-Assisted Blood Coagulation

Blood coagulation is an important issue in medicine, in particular regarding wound treatment. Thermal plasma has been traditionally used for this application in the form of the so-called cauterization devices: argon plasma coagulators (APCs), argon beam coagulators, and so forth. In these devices, widely used in particular in surgery, plasma is just a source of local high-temperature heating, which cauterizes (actually cooks) the blood. Recent developments of the effective non-thermal plasma-medical systems permit effective blood coagulation to be achieved without any thermal effects. In such systems, which will be discussed later, coagulation is achieved through non-thermal plasma stimulation of specific natural mechanisms in blood without any "cooking" and damaging of surrounding tissue (Fridman et al., 2006). Both coagulating the blood and preventing the coagulation could be needed, depending on the specific application. For example, in wound treatment one would want to close the wound and sterilize the surrounding surface. Flowing blood, in that case, would prevent wound closure and create the possibility of re-introduction of bacteria into the wound. Where blood coagulation would be detrimental is, for example, in sterilization of stored blood in blood banks. There, a potential exists for blood to contain or to have somehow acquired bacterial, fungal, or viral infection which needs to be removed for this blood to be usable. Here, of course, the treatment cannot coagulate the blood. Thus, clearly, an understanding of the mechanisms of blood coagulation by non-thermal plasma is needed (Fridman et al., 2006, 2007).

12.7.2. Experiments with Non-Thermal Atmospheric-Pressure Plasma-Assisted In Vitro Blood Coagulation

FE-DBD plasma was experimentally confirmed to significantly hasten blood coagulation in vitro (Fridman et al., 2006, 2007). Visually, a drop of blood drawn from a healthy donor and left on a stainless steel surface coagulates on its own in about 15 min, while a similar drop treated for 15 s by FE-DBD plasma coagulates in under 1 min (Fig. 12–31). FE-DBD treatment of cuts on organs leads to similar results where blood is coagulated without any visible or microscopic tissue damage. Figure 12–32 shows a human spleen treated by FE-DBD for

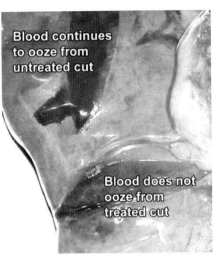

Figure 12–32. Thirty seconds of FE-DBD treatment of human spleen: blood coagulates without tissue damage. Top cut: blood continues to ooze from an untreated area; bottom cut: blood coagulates while the wound remains wet.

30 s – blood is coagulated and the tissue surrounding the treatment area looks "cooked"; however, the cut remains at room temperature (even after 5 min of FE-DBD treatment) and the wound remains wet, which could potentially decrease healing time. Additionally, a significant change in blood plasma protein concentrations is observed after treatment by plasma of blood plasma samples from healthy patients, patients with hemophilia, and blood samples with various anticoagulants. Anticoagulants, like sodium heparin or sodium citrate, are designed to bind various ions or molecules in the coagulation cascade, thus controlling coagulation rate or preventing it altogether. Analysis of changes in concentration of blood proteins and clotting factors indicates that FE-DBD aids in promoting the advancement of blood coagulation or, in other words, plasma is able to catalyze the biochemical processes taking place during blood coagulation (Fridman et al., 2005, 2006, 2007).

12.7.3. In Vivo Blood Coagulation Using FE-DBD Plasma

Plasma stimulation of in vivo blood coagulation has been demonstrated by Fridman et al. (2007) in experiments with live hairless SKH1 mice. FE-DBD plasma treatment for 15 s can coagulate blood at the surface of a cut saphenous vein (Fig. 12–33) and vein of a mouse. In these experiments only the ability of direct non-thermal plasma treatment to coagulate blood was tested and the animal was not left alive to test improvement in healing times. A

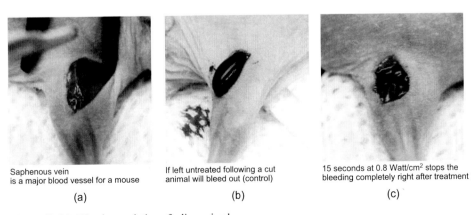

Saphenous vein is a major blood vessel for a mouse

(a)

If left untreated following a cut animal will bleed out (control)

(b)

15 seconds at 0.8 Watt/cm^2 stops the bleeding completely right after treatment

(c)

Figure 12–33. Blood coagulation of a live animal.

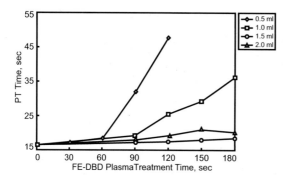

Figure 12–34. Prothrombin (PT) time for blood samples of different volumes with the same surface area of FE-DBD treatment.

full in vivo investigation of ability of plasma to hasten wound healing through sterilization and blood coagulation is discussed in Fridman et al. (2007).

12.7.4. Mechanisms of Non-Thermal Plasma-Assisted Blood Coagulation

Detailed biochemical pathways of non-thermal plasma-stimulated blood coagulation remain largely unclear. Several possible mechanisms, however, were investigated (Fridman et al., 2006; Kalghatgi et al., 2007). First and most important, it was demonstrated that direct non-thermal plasma can trigger natural, rather than thermally induced, coagulation processes. Second, it was observed that the release of calcium ions and change of blood pH level, which could be responsible for coagulation, is insignificant. Instead, the evidence points to the selective action of direct non-thermal plasma on blood proteins involved in natural coagulation processes. Mechanisms of plasma interaction with blood can be deduced from the facts observed in the experiments with FE-DBD plasma: (1) plasma can coagulate both normal and anticoagulated blood, but the rate of coagulation depends on the anticoagulant used; (2) plasma is able to alter the ionic strength of the solution and change its pH, but normal and anticoagulated blood buffers these changes even after long treatment time; (3) plasma changes the natural concentration of clotting factors significantly, thus promoting coagulation; (4) plasma effects are non-thermal and are not related to gas temperature or the temperature at the surface of blood; (5) plasma is able to promote platelet activation and formation of fibrin filaments, even in anticoagulated blood. Some more details follow.

1. Anticoagulants like sodium heparin bind thrombin in the coagulation cascade, thus slowing coagulation; while sodium citrate or ethylene diamine tetraacetic acid (EDTA) are designed to bind calcium, an important factor in the cascade, thereby preventing coagulation altogether. Plasma treatment promotes visible coagulation in blood with all of the these anticoagulants.

2. The initial plasma coagulation hypothesis was focused on increasing the concentration of Ca^{2+}, which is an important factor in the coagulation cascade. It was suggested that plasma stimulates generation of Ca^{2+} through the redox mechanism $[Ca^{2+}R^{2-}]+$ $H^+_{(H_2O)} \xrightleftharpoons[k_{-Ca}]{k_{Cal}} [H^+R^{2-}]_{(H_2O)} + Ca^{2+}_{(H_2O)}$ provided by hydrogen ions produced in blood in a sequence of ion–molecular processes induced by plasma ions. The validity of this hypothesis was tested experimentally by measuring Ca^{2+} concentrations in plasma-treated anticoagulated whole blood using a calcium-selective microelectrode. The calcium concentration was measured immediately after plasma treatment and remained almost constant for up to 30 s of treatment and then increased slightly for prolonged treatment times of 60 and 120 s. Although plasma is capable of coagulating anticoagulated blood within 15 s, no significant change occurs in calcium-ion concentration during the typical time of blood coagulation in discharge-treated blood. In vivo, the pH of blood is maintained in a very narrow range of 7.35–7.45 by various physiological processes. The change in pH by plasma treatment

12.7. Non-Thermal Plasma-Assisted Blood Coagulation

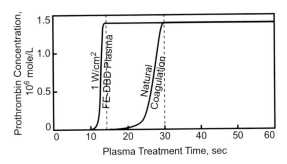

Figure 12–35. Prothrombin kinetics: two-fold decrease in clot formation time with plasma treatment.

(about 0.1 after 30 s) is less than the natural variation of pH, which indicates that the coagulation is not due to a pH change in blood.

3. FE-DBD treatment of whole blood samples changes concentrations of proteins participating in the coagulation cascade. Plasma treatment is shown to "consume" coagulation factors (proteins and enzymes) and a visible film is formed on the surface of the treated samples. An increase in sample volume and keeping the surface area fixed decrease the effect, indicating that plasma treatment initiates clot formation at the surface, not in the volume (Fig. 12–34). A corresponding kinetic model of plasma-assisted blood coagulation indicates a two-fold decrease in clot formation time with plasma treatment (Fig. 12–35).

4. When the surface of blood is protected by small thin aluminum foil, which prevents contact between blood and FE-DBD plasma but transfers all the heat generated by the plasma, no influence on blood is observed, which proves the non-thermal mechanism of plasma-stimulated blood coagulation.

5. The final step in the natural biological process of blood coagulation is the production of thrombin, which converts fibrinogen into fibrin monomers that polymerize to form fibrin microfilaments. FE-DBD plasma treatment of fibrinogen solution in physiological medium coagulates it, which is confirmed visually through a change in the color of the solution (from clear to milky-white) and through dynamic light scattering. Of note is that plasma does not influence fibrinogen through a pH or temperature change. FE-DBD treatment, however, is unable to polymerize albumin (directly not participating in the coagulation cascade); no change in its behavior is observed either visually or through dynamic light scattering (DLS). Thus, non-thermal plasma selectively affects proteins (specifically, fibrinogen) participating in the natural coagulation mechanism.

To assess the plasma influence on protein activity, compared with the plasma influence on the protein itself, trypsin (pretreated with L-1-tosylamido-2-phenylethyl chloromethyl ketone [TPCK] to inhibit contaminating chymotrypsin activity without affecting trypsin activity) was treated by plasma for up to 2 min and its total protein weight and protein activity analyzed via fluorescence spectroscopy. The total protein weight, or the amount of protein in the treated solution, remains practically intact after up to 90 s of treatment (Fig. 12–36),

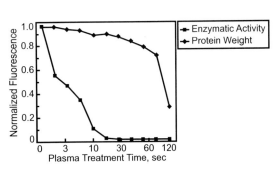

Figure 12–36. Total protein weight compared to enzymatic activity of Trypsin following plasma treatment.

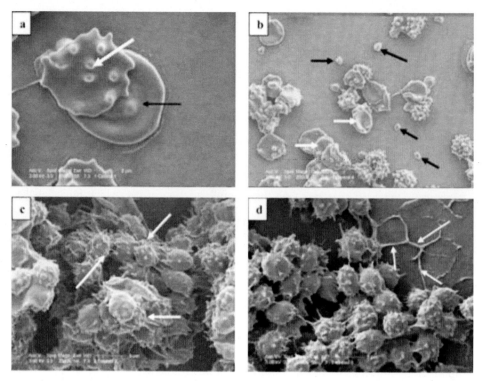

Figure 12–37. SEM images of untreated (a,b) and treated (c,d) anti-coagulated whole blood: (a) whole blood (control) showing single activated platelet (white arrow) on a red blood cell (black arrow); (b) whole blood (control) showing many non-activated platelets (black arrows) and intact red blood cells (white arrows); (c) whole blood (treated) showing extensive platelet activation (pseudopodia formation) and platelet aggregation (white arrows); and (d) whole blood (treated) showing platelet aggregation and fibrin filament formation (white arrows).

whereas the enzymatic (catalytic) activity of this protein drops to nearly zero after 10–15 s of treatment. Similar behavior is also observed for albumin. This effect also proves that the plasma effect on proteins is not just destructive but quite selective and "natural." Morphological examination of the clot layer by scanning electron microscopy (SEM) further proves that plasma does not "cook" blood but initiates and enhances "natural" sequences of blood coagulation processes. Activation followed by aggregation, of platelets is the initial step in the coagulation cascade and conversion of fibrinogen into fibrin is the final step in the coagulation cascade. Figure 12–37 shows extensive platelet activation, platelet aggregation, and fibrin formation following FE-DBD plasma treatment.

12.8. PLASMA-ASSISTED WOUND HEALING AND TISSUE REGENERATION

12.8.1. Discharge Systems for Air-Plasma Surgery and Nitrogen Oxide (NO) Therapy

The effective use of plasma in surgery was first demonstrated in the 1960s: the plasma afterglow jet of an inert gas was applied for tissue sectioning with instant blood coagulation. Because of that, plasma-surgical devices acquired the long-standing name of "plasma scalpels" in hospitals (see, for example, Glover et al., 1978). The significant advancement in plasma surgery of wound healing and tissue regeneration is due to the development of the "plazon" system based on the jet of hot air plasma that is rapidly quenched and provides

12.8. Plasma-Assisted Wound Healing and Tissue Regeneration

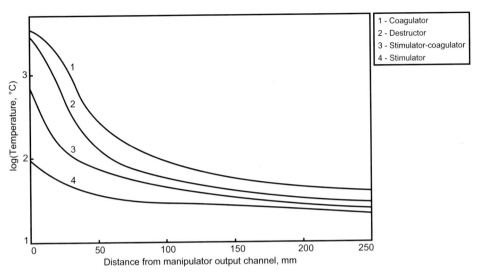

Figure 12–38. Temperature in the center of the gas flow jet for different manipulators.

relatively high NO concentration with significant therapeutic effect (Shekhter et al., 1998; Pekshev, 2001). This plasma device is used for rapid coagulation and sterilization of wound surfaces, destruction and desiccation of dead tissue and pathologic growths, dissection of biological tissues with the plasma jet, and also stimulation of regenerative processes and wound healing by NO-containing gas flow with temperatures of 20–40°C. The plazon generators (Shekhter et al., 1998; Pekshev, 2001) are DC arcs with different configurations of exit channels corresponding to the different applications (blood coagulation, tissue destruction, therapeutic manipulation/stimulation). The main and common elements of the system construction are the liquid-cooled cathode, intra-electrode insert, and anode. Atmospheric air enters the manipulator through the built-in microcompressor, passes through the plasma arc, heats up and thus accelerates, and exits through the hole in the anode of the plasma-generating module. The plasma temperature at the anode exit differs in different configurations of the device, corresponding to different medical applications (see Fig. 12–38). Farther away from the anode, the temperature drops rapidly, and at 30–50 mm from the anode the flow is composed simply of the warm gas and plasma-generated NO. The nitrogen oxide content in the gas flow is mainly determined by the quenching rate (see Section 6.1). The necessary quenching rate for effective operation of the medical device is about 10^7–10^8 K/s. Commonly, the cooling rate of plasma jets is on the order of $\sim 10^6$ K/s. Thus, to achieve the cooling rate of $\sim 10^7$–10^8 K/s, it is necessary to utilize additional cooling of the plasma jet, which has been achieved by special construction of the plasma nozzles. The therapeutic manipulator/stimulator configuration of the plazon discharge system is used solely for therapeutic treatment by exogenic NO. The principle difference of this manipulator is that the air-plasma jet does not freely exit into the atmosphere, but rather it exits the anode into the two-step cooling system, the gas channels of which are created in a maze scheme to force-cool the jet by the liquid circulating from the cooling system. This construction allows for obtaining NO-containing gas flow (NO-CGF) with sufficiently low temperature and optimal concentration of NO molecules, which makes it possible to apply this manipulator for treatment of external body surfaces by using the 150-mm-length cooling hose (temperature of NO-CGF at the exit ~ 36°C). Of course, NO content in the gas flow depends on the distance from the exit channel (Fig. 12–39). Additionally, for laparoscopic operation, a special manipulator of 350 mm length and 10 mm diameter is utilized. The possible operating

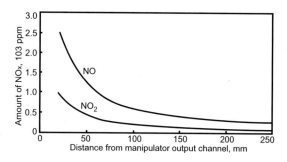

Figure 12–39. NO and NO$_2$ content at the center of the gas jet.

regimes of the apparatus are defined by the characteristics of the gas flow exiting from the manipulator, the main parameters of which are its temperature and the NO content. The first group of regimes, free-flowing plasma offgas exiting the manipulator, and the second group of regimes, treatment of biotissues by completely cooled (20°C) NO-CFG – to obtain which manipulator is connected to the internal gas cooler – achieve delivery of NO-CGF to biotissues through a silicone tube with an attached tip of 130 or 390 mm length and exit channel diameter of 0.7 mm. This allows not only direct treatment of the bio tissues by NO, but also to deliver it to a pathologic center through drainage tubes, puncture needles, or any endoscopic devices (gastroscope, broncoscope, cystoscope, rectascope, etc).

12.8.2. Medical Use of Plasma-Generated Exogenic NO

The Nobel prize in medicine and biology was awarded in 1998 to R.F. Furchgott, L.J. Ignarro, and F. Murad for investigation of the function of NO as a signal molecule. In humans, NO serves a multitude of essential biological functions – it regulates blood vessel tone (via relaxation of flat epithelial cells) and blood coagulation, immune system and early apoptosis, neural communication and memory, relaxation of flat bronchial and gastrointestinal muscles, and hormonal and sex functions; NO offers also antimicrobial and antitumor defense. In pathology, NO plays a major role in adaptation, stress, tumor growth, immunodeficiency, cardiovascular, liver, and gastrointestinal tract diseases, which explains the wide possibilities of plasma-generated exogenic NO in medical applications. The importance of exogenic NO in infection and inflammation processes is also well studied and is linked with antimicrobial effects; stimulation of macrophages; induction of cytokines, T-lymphocytes, and many immunoglobulins; interaction with oxygen radicals; and influence on microcirculation, cytotoxic, and cytoprotective in different conditions. During inflammation macrophages and some other cells (i.e., aibroblasts, epithelial cells, etc.) produce NO via inducible NO synthase (iNOS) in quantities significantly (two orders of magnitude) greater than normal when NO is formed via constructional NOS: endothelial (eNOS) and neuronal (nNOS). Exogenic NO is also crucial in trauma wound processes. The activity of iNOS grows substantially in trauma wounds, burn wound tissues, bone fracture site tissues, and others in the inflammatory and proliferation phases of the healing process. Activation of iNOS was also discovered in the cultivation of wound fibroblasts. Macrophage activation in a wound, cytokine synthesis and proliferation of fibroblasts, epithelization, and wound healing processes are all linked with the activity levels of iNOS. In animal models, the injection of iNOS inhibitors disrupts all of these processes and especially the synthesis of collagen, whereas NO synthesis promoters accelerate these processes. Animals with iNOS deficiency demonstrate a significant decrease in wound healing rate; however, this can be reversed by injection of the iNOS gene. In complicated wound models, for example in experimentally induced diabetes, protein deficiency, injection of corticosteroids or immunosuppressants, and also in patients with tropic ulcers, lowered activity of iNOS is usually discovered, which correlates to slowed healing processes. The exogenic delivery of

NO donors (nitrogen-containing compounds) to the wound promotes healing processes in animals with complicated wounds and in animals with inhibited iNOS. This knowledge, coupled with theoretical and experimental data on NO generation in air plasmas, served as a basis for a series of biomedical experiments focused on the use of plasma-generated exogenic NO delivered directly to the pathologic site for control of inflammatory processes and increase in the rate of wound healing.

12.8.3. Experimental Investigations of NO Effect on Wound Healing and Inflammatory Processes

EPR spectroscopy was utilized to investigate the dynamics of the level of endogenic and exogenic NO in wound tissues and in organs in an animal model (70 rats; Shekhter et al., 2005). An NO "trap," diethyl thiocarbamate (DETC), was injected into rats with full thickness flat wound of 300 mm^2 area 5 days prior to EPR analysis. Following euthanasia, samples were collected from the animals: blood, and granular tissue from the bottom of the wound and from internal organs (heart, liver, kidney, and the small intestine). For a portion of the animals, on day 5 following the initial wound introduction, the wound surface was treated by NO-CGF (500 ppm). Without the NO treatment, the results indicate a high content of endogenic NO in wound tissues (10.3 ± 2.3 μM). The liver of the animals with the wound contained 2.3 ± 1.4 μM of DETC-iron-mononitrosyl complex (IMNC), while the control group (without the wound) contained only 0.06 ± 0.002 μM. Animals without the wound were used for investigation of penetration capability of gaseous exogenic NO through undamaged tissues of the abdominal wall. Treatment by NO-CGF was performed for 60 and 180 s. A linear dependence of the amount of DETC-IMNC produced in the liver and blood of the animal on the NO-containing gas treatment time was observed. A maximum signal was registered in the bowels of the animal 2 min after the 180 s treatment, 2.6 times higher than in the control group. In the heart, liver, and kidney the difference was 1.7-fold. These results are indicative of the ability of the exogenic NO molecules to penetrate the undamaged tissues.

A more complex relationship was observed in treatment by exogenic NO of the wound tissues. If the animal was euthanized 30–40 min following the treatment, then NO content in wound tissue and blood was observed to increase 9–11-fold more than in the case of the 2 min interval. This is probably due to the formation of peroxinitrite, which can be formed through NO reacting with superoxide anions (O_2^-), as it is known that superoxide levels are increased in the organism during inflammatory processes. In response to the oxidative stress, the organism mobilizes the antioxidant defense mechanisms first via the increase in the levels of reducing agents (thioles, ascorbate, etc.) and then via activation of synthesis of antioxidant enzymes. Following wound treatment by exogenic NO (30–40 min later), activation of the first cascade of antioxidant defense allowed for a significant decrease in the level of superoxide anions. This considerably decreases the destructive influence on DETC-IMNC and the nitrosyl complexes of the hemoproteins, which leads to an increase in their concentration as is detected by the EPR method. Additionally, activation of NOS by the increase in endogenic NO cannot be neglected. It partially explains the discovered phenomena of stimulation of wound development processes via the influence of exogenic NO, when there is a deficiency of endogenic NO or excess of free radicals, including superoxide.

In experiments on the cornea of rabbits, the mucous membrane of the cavity of the mouth of hamsters, and on the meninx membrane of rats, via lifetime biomicroscopy it was found that the effect of the expansion of the opening of the microvessels under the influence of exogenic NO (500 ppm) lasts with varying intensity up to 10–12 h, whereas the lifetime of NO molecules is no more than 10–15 s (Shekhter et al., 1998, 2005; Vanin, 1998). The experiments serve as additional evidence that a single application of exogenic

NO initiates a cycle of cascade reactions, including biosysnthesis endogenic NO, which leads to a long-lasting effect and explains the successes of NO therapy.

The action of exogenic NO on the cellular cultures of human fibroblasts and rat nervous cells was studied by Shekhter et al. (1998, 2005), and Ghaffari et al. (2005). Single treatment by plasma-generated NO of the cell cultures significantly increases (2.5-fold) the cell proliferation rate via the increase of DNA synthesis (tested by inclusion of C^{14} thymidine) and to a lesser extent (1.5-fold) an increase of protein synthesis by the cells (tested by inclusion of C^{14} aminoacids). As expected, the stimulating effect is dose-dependent. The action of exogenic NO on the phagocytic activity of the cultured wound macrophages from the washings of trophic human ulcers, studied via photochemiluminescence (Krotovskii et al., 2002), revealed that a maximum increase in the luminous intensity (1.95-fold in comparison with control) testifies to the activation of the proteolytic enzymes of macrophages under the effect of NO-CGF. A statistically significant increase in fluorescence of macrophages was observed in less than 24 h following a 30-s treatment.

In vitro investigation of the influence of NO-CGF on *Escherichia coli*, *Staphylococcus aureus*, *Pseudomonas aeruginosa*, *Proteus vulgaris*, and *Candida albicans*, which are typically associated with many hospital infections, showed that 75 s of treatment by NO-CGF significantly decreases viable CFUs, 80 s practically removes them all, and no growth is detected at all following 90 s of treatment (Shulutko, Antropova, & Kryuger, 2004). The major mechanisms of NO influence on various pathologic processes can be summarized as follows (Shekhter et al., 1998, 2005):

1. Direct bactericidal effect (through formation of peroxynitrite in the reaction: NO + O_2^- → ONOO$^-$).
2. Induction of the phagocytosis of bacteria and necrotic detrite by neutrophils and by macrophages.
3. Inhibition of free oxygen radicals, which exert pathogenic influence, and also possible activation of the antioxidant protection.
4. Normalization of microcirculation due to vasodilatation, anti-aggregation and anticoagulant properties of NO, that improves vascular trophicity and nutrient exchange.
5. Improvement of nerve conductance.
6. Regulation of immune deficiencies, which are common in wound pathology.
7. Secretion of cytokines by the activated macrophages, which increase fibroblast proliferation, angiogenesis factors, chemokines, in particular, monocyte chemoattractant protein (MCP-1), G-protein, nuclear factor κB (NFkB), and other biologically active factors which regulate wound healing and inflammatory processes.
8. Direct induction of proliferation of fibroblasts and synthesis by them of proteins.
9. Increase in or regulation of collagen synthesis.
10. Regulation of apoptosis in remodeling granular and fibrous tissues.
11. Influence on the proliferation of keratinocytes and thus on the epithelization of the wound.

12.8.4. Clinical Aspects of Use of Air Plasma and Exogenic NO in Treatment of Wound Pathologies

The application of air plasma and exogenic NO in the treatment of trophic ulcers of vascular etiology in 318 patients showed high efficiency of NO-therapy in the treatment of venous and arterial trophic ulcers of lower extremities with an area from 6 to 200 cm^2 (Shekhter et al., 1998, 2005). For assessment of the effectiveness of plasma NO therapy, clinical and planimetric indices were analyzed in the course of the process of sanitation and epithelization of ulcers, a bacteriological study of discharge from the ulcer, cytological study of exudate, a histopathological study of biopsies from the boundary of a trophic ulcer, the indices of

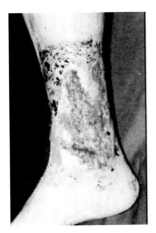
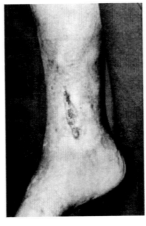
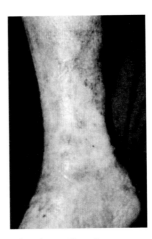

Before treatment   21st day of NO-therapy   After 2 months of
                          (10 seances)   NO-therapy

Figure 12–40. Dynamics of the healing of venous trophic ulcer during the plasma NO therapy.

microcirculation (according to the data obtained by laser Doppler flowmetry [LDF]), and transcutaneous partial pressure of oxygen (pO_2). In the main groups of observations trophic ulcers were processed in the regime of NO therapy (500 and 300 ppm), or prior to beginning the therapy the ulcer surface was treated in the regime of coagulation until the evaporation of necrotic debris. Following initial treatment, the wounds were treated for 10–30 days in the NO-therapy regime. In the control group, proteolytic and antimicrobial drugs were used in the phase of exudation and necrosis, and wound coatings in the phase of tissue regeneration and epithelization.

Planimetric observation of the dynamics of decrease in trophic ulcer area showed that, on average, traditional treatment methods applied in the control group lead to 0.7% per day decrease, while those in the experimental group lead to 1.7% per day. Cleansing of ulcers from necrosis and exudate, and the appearance of granulation and boundary epithelization, were accelerated with NO therapy an average of 2.5-fold. The time to final healing was reduced 2.5- to 4-fold depending on the initial ulcer size (Fig. 12–40). Larger ulcers tended to close faster than smaller ones. LDF investigation of microcirculation in the tissues of trophic ulcers showed that, following NO therapy, there was normalization of pathologic changes in the amplitude–frequency characteristics of the microvasculature and activation of regulatory mechanisms. By 14–18 days the average index of microcirculation, value of root-mean-square deviation, coefficient of variation, and index of fluctuation of microcirculation approached those of the symmetrical sections of healthy skin. In the control group the disturbances of microcirculation remained. Against the background of treatment, normalization of the level of transcutaneous partial pressure of oxygen (TpO_2) happened at a higher rate in the experimental group than in the control group, especially at the NO concentration of 500 ppm (Fig. 12–41). A bacteriological study of wound discharge from trophic ulcers showed that in the experimental group, against the background, NO therapy (especially in combination with the preliminary coagulation of ulcerous surface) reduced the degree of bacterial seeding (microbial associations) and already by days 7–14 went below the critical level necessary for maintaining the infectious process in the wound (Fig. 12–42).

Using plasma-generated NO for local treatment of ulcerous and necrotic tissues in patients with diabetes (diabetic foot ulcer) has been demonstrated by Shulutko, Antropova, and Kryuger (2004). Patients were selected for this study following 2 months of unsuccessful

901

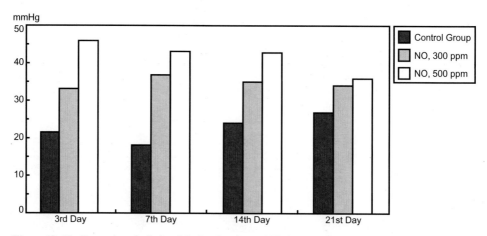

Figure 12–41. Dynamics of pO_2 level during the plasma NO therapy of venous trophic ulcers.

treatments by the state-of-the-art techniques. Already from the first few sessions the difference was evident: the inflammatory reaction was clearly reduced, patients reported a decrease in pain, and cleansing of the ulcer surface was clearly visible. Following 10 sessions, most patients expressed positive healing dynamics: ulcer size decreased to one-third to one-fourth of the original size. LDF markers, pO_2, and bacteriological investigation all showed a positive dynamic. In patients with relatively small-sized ulcers (initial diameter less than 1 cm), full epithelization occurred after six to eight NO treatment sessions. The period of stationary treatment and full clinical recovery of patients was noticeably shortened (on average by 2.3-fold). In the cases of large ulcerating wounds, the necessity for amputation decreased 1.9-fold (Fig. 12–43). The effectiveness of exogenic NO and air plasma in healing pyro-inflammatory diseases of soft tissues has been demonstrated by studying 520 patients with purulent wounds of different etiology and 104 patients with the phlegmonous-necrotic form of erysipelatous inflammation (Lipatov et al., 2002). By day 5 of therapy, wounds on most of the patients in the experimental group (90%), contrary to the control group, were clear of necrotic tissue, and the wounds began to be covered by bright spots of granular tissue. Microbial infestation of the wound tissue had lowered from 10^{6-8} cfu/g of tissue to 10^{1-2}. Data from complex analysis of microcirculation (LDF, pO_2) showed significant repair in the microvasculature and blood flow in the wound tissues in most of the patients in the experimental group. The predominant types of cytograms were regenerative and regenerative-inflammatory with a notable increase in fibroblast proliferation – on average $18.5 \pm 3.1\%$. Large suppurated wounds, for example suppurated burn wounds (Fig. 12–44), by day 7–10 of treatment were clear of the pyonecrotic exudate and were beginning to be covered by granular tissue; in other words, these wounds were ready for dermautoplasty.

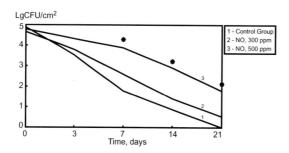

Figure 12–42. Dynamics of bacterial contamination of trophic ulcers during the plasma NO therapy (* statistical significance, $p < 0.01$).

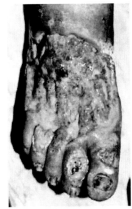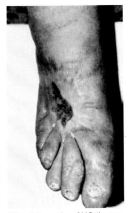

Figure 12–43. Plasma-based treatment of the extensive pyonecrotic ulcer of the foot (neuroischemic form of the syndrome of diabetic foot).

Before treatment   After 4.5 months of NO-therapy
                          (3 courses; 12 seances per course)

The effectiveness of plasma NO therapy is most apparent with the treatment of the pyonecrotic form of erysipelatous inflammation – patients who are considered the most severe cases of purulent surgery departments (Lipatov et al., 2002). The combination of surgical preparation of extensive pyonecrotic centers and local NO therapy allowed the majority of the patients with phlegmonous-necrotic erysipelas during 12–14 days of treatment to liquidate heavy pyonecrotic process and to create conditions for completion of reparative procedures.

The plasma NO treatment has been also successfully applied in surgical oncology (Kabisov et al., 2000; Reshetov et al., 2000). Interoperative treatment in the coagulation regime of the plazon system ensures ablation, and considerably decreases blood plasma and whole blood losses from extensive wound surfaces as a result of thin-film formation over the wound surface, consisting of coagulative necrotic tissue. As a result of plasma NO therapy of the postoperative wounds, a significant decrease in inflammation is observed along with stimulated proliferation of granular tissue and epithelization. This effect is observed independently of the location of the wound on the body and also of the plastic material used. An additional positive benefit of this treatment is the prophylactic treatment of the local relapses of the tumor, which allows for a wide application of this method in oncoplastic surgeries. The effectiveness of NO therapy in treatment of early and late radiation reactions allows for the surgeon to carry out a full course in radiation therapy in 88% of the patients. Treatment of radiation tissue fibrosis also yields a statistically significant improvement, confirmed in

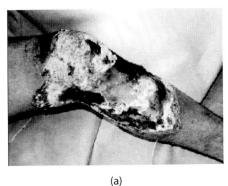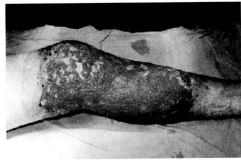

(a)   (b)

Figure 12–44. Healing dynamics of the festered burn wound in process of the plasma NO therapy: (a) prior to the beginning of treatment; (b) after five sessions of the therapy.

morphological investigation of these tissues. Plasma NO therapy is successfully used both for the preventive maintenance of the formation of postoperative hypertrophic and keloid scars, and for the treatment of already formed scars, including softening of the scar tissue, decrease of fibrosis, and preventive maintenance of their relapse with surgical removal.

12.8.5. Air Plasma and Exogenic NO in Treatment of Inflammatory and Destructive Illnesses

Possibility of directing of the plasma-generated NO-containing gas flows through puncture needles, vent lines, and endoscopic instruments, and also the inhalation method of action considerably enlarges prospects for the plasma NO-therapy in treatment of the ulcero-necrotic, erosive and inflammatory processes in the pleural and abdominal cavities, lungs, stomach and bowels, ENT organs (purulent sinusitis, purulent otitis media, paratonsillar abscesses), etc. Effectiveness of the plasma NO-therapy is already at present shown with a number of diseases in gynecology, traumatology, stomatology, ophthalomology, otorhino-laryngology, dermatology, gastroenterology, etc. Some specific relevant medical applications of the plasma system are summarized below (Shekhter et al., 1998, 2005).

Pulmonology. Strong effect is demonstrated in the treatment of plural empyema via insufflation of the NO-CGF into the cavity of the pleura through the vent lines (Shulutko, Antropova, & Kryuger, 2004). Therapy in treatment of 60 patients with pleural empyema showed stimulative and regulative influence on the development of the wound tissues. Acceleration of the purification of pleural cavity from the micro-organisms and the debris, stimulation of phagocytosis, and normalization of microcirculation accelerate the passage of the phase of inflammation during wound regeneration, which leads to a significant decrease in the drainage time for all patient categories in the experimental group, as compared to control, and to the reduction in the hospitalization time. The inhalation application in treatment of patients with the complex chronic unspecific inflammatory lung diseases led to the clearly expressed positive dynamics of the endoscopic picture of the tracheobronchial tree: decrease of the degree of inflammatory changes in the mucous membrane of bronchi, reduction in the quantity and the normalization of the nature of contents of the respiratory tract. During the study of biopsies of mucosa of bronchi it was verified for all cases that the liquidation or the considerable decrease of inflammatory changes occurred, in addition to a complete or partial restoration of the morphological structure of the bronchi.

Phthisiology. Plasma NO-therapy together with the specific treatment was used in patients with infiltrative and fibrous-cavernous pulmonary tuberculosis via NO insuffla-tion through the bronchoscope or cavernostomy for cavernous tuberculosis, through the vent line with tubercular pleurisy or empyema. Through 8–10 therapy sessions, a signifi-cant acceleration of healing of the cavities, tubercular bronchitis, and pleurisy is achieved (Seeger, 2005).

Traumatology and Orthopedics. Positive effect has been demonstrated for treat-ment of patients with the infected, prolongedly not healing wounds after sequestrectomy, osteomyelitic blowholes, etc. (Shekhter et al., 2005; Efimenko et al., 2005). It was shown that following the NO-therapy the bacterial load of wounds and blowholes was significantly reduced, inflammatory manifestations were reduced in the wound and the surrounding tis-sues, cleansing from the necrotic mass advanced, and active granulations appeared. All the participants of this study were previously treated by state-of-the-art methods for a long time without apparent success. Morphological investigations in these clinical observations showed that already by 3–4 plasma NO-therapy sessions a significant reduction in the infec-tion was observed, weakening microcirculatory disorders and signs of inflammation were reduced, proliferation and differentiation of fibroblasts and angiogenesis were evident, and an increase in the granular tissue and the cicatrization of the wound were apparent. This therapy is also being used for treatment of open fractures.

Gynecology. Effectiveness of the plasma NO-therapy in combination with the coagulation by air plasma is shown with treatment of patients with purulent inflammation of appendages of the womb (Davydov et al., 2002, 2004; Kuchukhidze et al., 2004). In surgery, where the abdominal cavity was opened, the purulent wound was treated by air plasma in the coagulation regime, and then at a later time, the plasma NO-therapy was achieved by remote action through the front abdominal wall and vagina. With operational laparoscopy after dissection and sanitation of purulent center the region of surgical incision and the organs of small basin were treated by NO-CGF, which was delivered locally through the aspiration tube. The plasma NO-therapy was also continued in the post-operation period. The use of NO-CGF in surgical and therapeutic regimes aided the rapid decrease in the microbial load, decrease in swelling, lowered risk of post-operative bleeding, rapid development of reparative processes, and overall time the patients remained in the hospital was decreased by 6–8 days on average. The NO-CGF was also used in organ-saving surgical operations on the womb, the uterine pipes, and the ovary.

Dentistry. Effectiveness of the plasma NO-therapy has been demonstrated on the chronic gingivitis. After the first session of the therapy, gum bleeding ceases, after 1–2 weeks normalization of tissue and regional blood flow in the tissues of periodontium (Grigorian et al., 2001). Normalization of cytological signs was observed in 2–3 months, however in the control group normalization was not observed at all. Utilization of NO-CGF in surgical intervention of generalized paradontitis (both in intra-operative and post-operative NO-therapy) showed that normalization of clinical and cytological signs occurs by the 7th day in the experimental group, while only on the 14th day in the control group. Complications were not observed in the experimental group, unlike in the control group where they did occur.

Maxillofacial Surgery. The plasma NO-therapy is used to accelerate the healing of postoperative wounds and preventive maintenance of the formation of hypertrophic and keloid scars, treatment of the formed scars, treatment of pyonecrotic processes (abscesses, phlegmon, etc.). With the latter, preliminary coagulation of purulent centers is sometimes utilized (Shekhter et al., 2005).

Ophthalmology. Treatment by NO-CGF (300 ppm) does not result in altering and/or toxic reaction, and does not cause changes in the intraocular pressure and morphological changes in the tissues of the eye, but considerably accelerates the healing of wounds and burns of the cornea. The therapy was then used in the clinic for the effective treatment of burns, erosions and injuries of cornea, and burn ischemia of conjunctiva (Chesnokova et al., 2003).

Otorhinolaryngology. Effectiveness is demonstrated on treatment of scar stenoses of larynx and trachea, chronic tonsillitis, relapsing nose hemorrhages, chronic and sharp rhinitis, pharyngitis, maxillary sinusitis, otitis, polypous purulent etmoidita, and other ENT pathologies (Golubovskii et al., 2004).

Dermatology. NO-therapy is effectively used with the treatment of psoriasis, eczemas, dermatitis, ulcerous injuries with local and systemic angiitises, scleroderma, red flat lishchaya, and a number of other skin illnesses (Zaitsev, 2003).

Gastroenterology. For treatment of chronic ulcers and erosions of stomach and duodenum, and blowholes of small intestine, the therapy is delivered through endoscopic instruments (Chernechovskaia et al., 2004). Stomach ulcers heal twice as fast as in the control group. The proliferating activity of the epithelium according to the data of the immunomorphology of biopsies is strengthened 7.8 times.

Purulent Peritonitis. In the case of the purulent peritonitis caused by the diseases of the organs of the abdominal cavity, the effect is achieved by the direct treatment of peritoneum, and in the postoperative period, cooled NO-CGF is delivered through the vent lines (Efimenko et al., 2005). NO carries bactericidal action, stimulates microcirculation and lymphodrainage, normalizes the indices of cellular and humoral immunity, dilutes the inflammatory process, and serves as a factor of the preventive maintenance of sealings in the abdominal cavity.

12.9. NON-THERMAL PLASMA TREATMENT OF SKIN DISEASES

12.9.1. Non-Thermal Plasma Treatment of Melanoma Skin Cancer

The FE-DBD plasma treatment (see section 12.6.1) is shown to initiate apoptosis in Melanoma cancer cell lines, that is a threshold at which plasma treatment does not cause immediate necrosis but initiates a cascade of biochemical processes leading to cell death many hours after the treatment (Fridman et al., 2007). Melanoma cells, treated by plasma at doses below those required for cell destruction, survive the plasma treatment but develop apoptosis many hours post treatment and die (disintegrate) by themselves gracefully. This could potentially be an intriguing approach for cancer treatment, especially if by manipulation of plasma parameters the treatment could be made selective to cancerous cells over healthy cells, as was demonstrated before for bacteria vs. healthy cells (Fridman et al., 2006).

The concept of apoptosis is discussed in section 12.2.4. Cellular macromolecules during apoptosis are digested into smaller fragments in a controlled fashion, and ultimately the cell collapses without damaging the surrounding cells or causing inflammation. With cancer cells, however, a problem arises with apoptosis as the tumor cells frequently "learn" how to turn off apoptosis as one of the processes they employ in evading the immune system and surviving under unfavorable conditions. A way to target apoptosis development only in specific areas of the body is needed, and can be achieved by the non-thermal plasma treatment. Melanoma cancer cell line (ATCC A2058) was prepared in the Fridman et al. (2007) experiments to the total concentration of $\sim 1.5 \cdot 10^6$ per dish. On days 3–5 of cell development, the cells were treated with the FE-DBD plasma for 5, 10, 15, 20, or 30 s. The distance from the electrode surface to the fluid surface was 3 ± 0.5 mm. After the treatment media was removed from the dishes, the culture was allowed to propagate further by adding 2 mL of the fresh media or harvested by trypsinization for further testing. Trypan Blue exclusion test was performed at different time periods after treatment: immediately following treatment, 1, 3, 24, 48, and 72 h following treatment.

Another group of experiments was performed, testing cells for the onset of apoptosis. For this set of experiments, cells were treated by plasma for 5 and 15 s. Following treatment, cells were harvested at 3, 24, 48 and 72 h after treatment. Two apoptosis assays were then performed: TUNEL assay which detects DNA breaks indicative of a late onset of apoptosis and cell's final preparations to disintegrate; and Anexin staining which detects "early" and "late" stages of apoptosis development where the structure of cell's membrane begins to change. These two biochemical fluorescence-based staining techniques, coupled with the careful analysis of the cell lifecycle are together indicative of FE-DBD plasma's ability to initiate apoptosis development in these cells. Melanoma cell growth patterns were noted to assess "background" cell death through lack of nutrition, cell age, or the influence of aluminum substrate on the cell's life cycle. Figure 12–45 demonstrates cell survival numbers after 5, 10, 20, and 30 s of treatment compared to control analyzed by Trypan blue exclusion test. Total cell numbers are normalized to 1 (100%) to account for cell growth between the counting sessions: controls are set to 100% and cell viability is expressed as percent to control to allow for comparison between experiments. It is of no great surprise that FE-DBD plasma is able to kill cells; what is unusual is that 24 hours following treatment the total number of cells continues to decrease significantly (see Fig. 12–45).

It is important to distinguish between cell death by "poisoning" of the growth media the cells are in and the effect of direct plasma on these cells. To assess the difference, growth media was treated for up to 120 s by plasma separately and then the cells were placed in this acidified media. Cells did not appear to react negatively to the acidified media. Additionally,

12.9. Non-Thermal Plasma Treatment of Skin Diseases

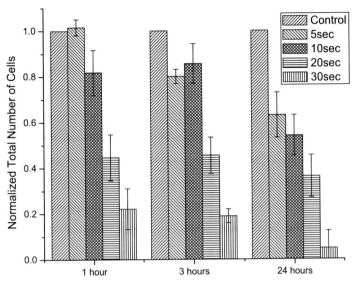

Figure 12–45. Results of FE-DBD treatment of melanoma cancer cells: control, 5, 10, 20, and 30 s, counted 1, 3, and 24 h post-treatment.

cell inactivation under varying depths of growth media was investigated: FE-DBD is able to inactivate cells under as much as 1 mm of cell growth media, though the time to achieve same inactivation as without media increases (Fridman et al., 2007).

The general trend observed in treated cells is that they continue to die for days after the treatment. Figure 12–46 presents an observation of groups of cells treated by plasma for 5 s and observed for a 3-day period following treatment. An emergent pattern appears where growth rate of treated cells is impaired as well as the number of inactivated cells grows substantially. Figure 12–46 shows the percentage of inactivated (dead) cells among treated and untreated populations. It was observed that 5 s of plasma treatment does not inactivate cells immediately; however, cell growth slows down significantly, and the number of dead cells increases 24 h after treatment, which is indicative of cell death occurring long after the treatment. To analyze whether those plasma-treated cells that survive the initial insult die through an apoptosis-like process, TUNEL assays were performed. Cells treated for 5 s were then incubated and stained for DNA fracture 24 h later. Following the TUNEL assay procedure, it was observed that a significant percentage of these cells exhibit apoptotic behavior as is evident from Fig. 12–47. Apoptosis develops 24 h following treatment, where 25.5% of cells are present in the treated group, compared with 2.2% in

Figure 12–46. Results of observation of plasma treated and untreated cells for a 3-day period: total number of cells before and after FE-DBD treatment. Treatment time: 5 s.

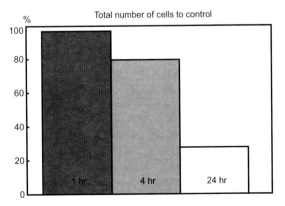

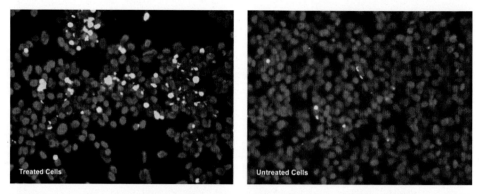

Figure 12–47. Images of plasma treated (left) and untreated (right) melanoma cancer cells stained following TUNEL assay protocol. All cells are stained blue (darker circles) and apoptotic cells are also stained green (bright spots).

the control group. As time progresses, even more cells undergo apoptosis, further reaching 72.8% of apoptotic cells in the treatment group vs. 3.2% in the control group 72 h following treatment.

Thus FE-DBD can kill Melanoma skin cancer cells through necrosis at higher treatment doses (15 s and over at 1.4 W/cm^2) which are still below the threshold of damaging healthy tissue (Fridman et al., 2006). Very low doses of FE-DBD (5 s at 0.8 W/cm^2 of plasma treatment), where no cell necrosis was observed, were shown to initiate apoptotic behavior in Melanoma cancer cells. Apoptotic behavior was deduced from the fact that treated cells do not initially die but stop growth and die 12 to 24 h following treatment, while untreated cells continue to grow and proliferate. Apoptotic behavior was confirmed through DeadEnd^{TM} Fluorometric TUNEL System apoptosis staining with subsequent flow-cytometry. It was shown that the plasma treatment initiates this behavior in cells not through poisoning of the growth media in which the cells reside or through interaction with the aluminum dishes the cells reside in, but through direct interaction with the cells (Fridman et al., 2007).

12.9.2. Non-Thermal Plasma Treatment of Cutaneous Leishmaniasis

Direct non-thermal FE-DBD plasma treatment (see section 12.6.1) is inherently a surface phenomenon and can be effectively applied to topical wounds and diseases. Cutaneous Leishmaniasis (C.L.) is a good example of such plasma treatment, specifically in the case of the Post-Kala-azar Dermal Leishmaniasis (PKDL). PKDL is a topical disease, and is a growing concern with 200 million people at risk and 500,000 cases of Leishmania per year. Few options are available today to treat the Cutaneous Leishmaniasis cases. Aside from surgical removal of the infected and surrounding tissue, there are two investigational drugs administered by the Center for Disease Control and Prevention (CDC): Sodium antimony gluconate (Pentostam) and Amphotericin B (AmBisome). Both are very expensive, requiring frequent (in some cases daily for 2 months) visits to a trained physician with intravenous injections of the drug, and both are associated with reports of adverse side effects. Secondary infection of ulcers with skin flora is also common and must be treated. If prolonged, the Cutaneous Leishmaniasis can transform to visceral Leishmaniasis – a systemic disease where parasites enter the blood stream and settle in vital organs. Ultra Violet (UV) radiation treatments of C.L. have been applied but the reported results indicate slow inactivation rates and high dose requirements which in turn may cause tissue damage. Non-thermal plasma is considered as an important possible solution to the medical problem (Fridman, 2007).

Leishmania promastigotes (parasites) have been treated at various FE-DBD plasma doses (see section 12.6.1) and separately human Macrophage cultures have been treated to

12.9. Non-Thermal Plasma Treatment of Skin Diseases

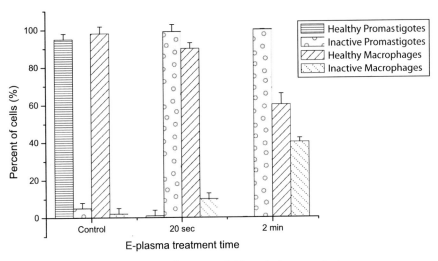

Figure 12–48. Plasma inactivation of cutaneous leishmaniasis promastigotes.

assess both the difference in inactivation rates between two different cell lines and the dose required to inactivate the parasite (Fridman et al., 2007). About 20–30% of macrophages are inactivated in 2 minutes of plasma treatment, while 100% of promastigotes are inactivated in 20 s (Fig. 12–48). Even though apoptosis was not tested in the promastigotes, they did exhibit behavior indicative of it. For this study, promastigotes of Leishmania Major were used. Interestingly, C.L. promastigotes exhibit apoptosis-like behavior in a similar way that cancer cells do, though apoptosis is not possible in this type of cells. The promastigotes take longer than 24 h to grow and duplicate, so it is no surprise that number of alive parasites is not increasing 24 h post treatment as can be seen from Fig. 12–49 (Fridman et al., 2007). What is more interesting is that the number of metabolically inactive ("dead," "non-viable," or simply "not moving") parasites decreases 24 h post-treatment. This is indicative of the

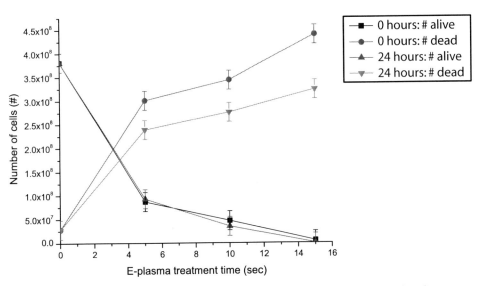

Figure 12–49. Apoptosis-like behavior of cutaneous leishmaniasis promastigotes following plasma treatment.

Table 12–7. Dependence of Current Density and Ion Concentration of the Xenon Plasma Flow in the Medical Microplasma System on the Distance from the Nozzle

Distance, mm	1	1.3	2	3
Current density, mA/cm^2	2040	2000	240	60
Velocity, cm/s	$2 \cdot 10^4$	$1.8 \cdot 10^4$	$1.2 \cdot 10^4$	$7 \cdot 10^3$
Ion concentration, cm^{-3}	$6.4 \cdot 10^{14}$	$3.7 \cdot 10^{14}$	$1.5 \cdot 10^{14}$	$5.3 \cdot 10^{13}$

fact that the parasites that were inactivated continued to "kill" themselves and disintegrated in the 24 h period following the treatment.

12.9.3. Non-Equilibrium Plasma Treatment of Corneal Infections

A special micro-plasma system has been developed for local medical treatment of skin diseases, and especially for treatment of corneal infections (Gostev & Dobrynin, 2006). The device allows generation of plasma flows with average gas temperature not exceeding 30–40°C. It consists of a coaxial cathode and needle-like anode, which is fixed in metal capillary. The gas is fed through the capillary to the discharge gap. The anode is connected with the positive lead of the power source and the cathode is grounded. The discharge appears on the nozzle output if the pressure on the nozzle input is higher than the atmospheric-pressure. The discharge is a specific plasma sphere with the diameter ~4 mm, atmospheric air or xenon are fed through the capillary at 0.2–0.5 atm, the discharge voltage is 1–3 kV, the pulse duration is about 50 μs, the total power was kept on the order of 1–2 W.

The medical micro-plasma operated in Xe radiates intensively in the UV range, and operated in air it generates excited oxygen species, ozone, oxides of nitrogen, and OH radicals (Gostev & Dobrynin, 2006). Both regimes have bactericidal effects and air plasma is able also to aid in tissue regeneration via the NO-therapy mechanisms discussed above in sections 12.7 and 12.8. UV-radiation of Xe plasma in this case is: UVA (315–400 nm) 180 μW/cm^2, UVB (280–315 nm) 180 μW/cm^2, and UVC (200–280 nm) 330 μW/cm^2. UV-radiation of air plasma is: UVA (315–400 nm) 53 μW/cm^2, UVB (280–315 nm) 25 μW/cm^2, and UVC (200–280 nm) 90 μW/cm^2. Results of probe measurements of current density, velocity, and ion concentration at different distances from the exit nozzle are shown in Table 12–7.

The ability of this medical micro-plasma system to sterilize surface has been demonstrated by Misyn et al. (2000). Staphylococcus culture in liquid media (~$2 \cdot 10^6$ cfu/ml) have been treated by the air plasma plume of 3 mm diameter, incubated for 24 h, and counted (see Table 12–8). A 6-log reduction in viable bacteria is achieved in 25 s of treatment; however the sterilization efficiency drops off with increasing volume of liquid which inhibits UV penetration and diffusion of active species generated in plasma. Nevertheless, the micro-plasma system should be a good solution for treatment of living human and animal skin as the bacteria are normally at much lower concentrations on skin ($<10^5$ cfu/cm^2).

Table 12–8. Staphylococcus Inactivation by Air Microplasma System

Culture volume, mL	Plasma exposure time, s			
	0 (control)	25	50	100
1	$2 \cdot 10^6$ cfu	0 cfu	0 cfu	0 cfu
2	$4 \cdot 10^6$ cfu	25 cfu	0 cfu	0 cfu
3	$6 \cdot 10^6$ cfu	$1 \cdot 10^6$ cfu	680 cfu	460 cfu

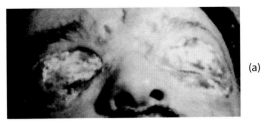

(a)

Figure 12–50. Result of six sessions of plasma treatment of the complicated ulcerous eyelid wound.

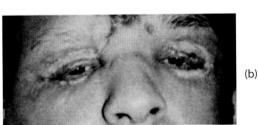

(b)

A series of in vitro experiments on bacterial cultures and in vivo experiments on rabbit eyes (Misyn et al., 2000) affirm the strong bactericidal effect of the micro-discharge with minimal and reversible changes, if any, in biological tissues, even in such delicate tissues as the cornea. During the investigation of plasma treatment of ulcerous dermatitis of rabbit cornea two important observations were made: 1) plasma treatment has a pronounced and immediate bactericidal effect, and 2) the treatment has an effect on wound pathology and the rate of tissue regeneration and the wound healing process.

These results offered a strong ground for application of the medical micro-plasma system for treatment of human patients with complicated ulcerous eyelid wounds, which is shown in Fig. 12–50 (Misyn & Gostev, 2000). Necrotic phlegm on the surface of the upper eyelid was treated by air plasma plume of 3 mm diameter for 5 s once every few days. By the 5th day of treatment (two 5-second plasma treatment sessions) the eyelid edema and inflammation were reduced; and by the 6th day (third session) the treated area was free of edema and inflammation and a rose granular tissue appeared. Three more plasma treatments were administered (six total), and the patient was discharged from the hospital six days following the last treatment (Fig. 12–50). The micro-plasma treatment is being further developed for stimulation of reparative processes in various topical wounds, tropic ulcers, chronic inflammatory complications, and other diseases of soft tissues and mucous membrane (Misyn & Gostev, 2000).

12.9.4. Remarks on the Non-Thermal Plasma-Medical Treatment of Skin

Thus non-thermal plasma was shown to be able to initiate, promote, control, and catalyze various complex behaviors and responses in biological organisms. More importantly, it was shown that this plasma can be tuned to achieve the desired medical effect, especially in medical sterilization and treatment of different kinds of skin diseases. Wound healing and tissue regeneration was achieved following various types of plasma treatment of a multitude of wound pathologies. The non-thermal plasma was shown to be non-destructive to tissue, safe, and effective in inactivation of various parasites and foreign organisms.

Treatment of various skin diseases by thermal plasmas and related afterglow plasma systems have been employed in a hospital setting for a long time; which, however, can be quite destructive as is evident from Table 12–9. Depending on thermal capacity of the plasma flow there can be pyrolysis, coagulation, and destruction of biological tissue during this contact. In treatment of especially sensitive areas, for example, the cornea, temperature should not exceed 43°C. Non-thermal plasma treatment opens in this perspective a whole

Table 12–9. Effect of Temperature on Tissue

Effect on tissue	Temperature, °C
Absence of irreversible modifications	37–43
Tissue division (hypostasis)	45–48
Tissue welding, protein denaturation	45–60
Tissue coagulation, necrosis, dehydration	60–100
Tissue water evaporation	100
Tissue pyrolysis, burning off	100–300
Carbonization of the tissue solid components	>200
Tissue solid components evaporation	>300

new area of opportunities of plasma treatment without any damages and undesirable medical consequences. It is generally clear that Plasma Medicine is an emerging field with a great potential. Obviously, numerous open questions remain to be answered. Though plasma is shown to be selective and specific in medical treatment, mechanisms of this selectivity remain largely unknown. Mutagenic effects of repeated treatment also need to be analyzed in detail. In dermatological applications, for example, a patient can be subjected to a series of repeat treatments potentially spanning months before the desired effect is achieved (i.e., tattoo removal or acne treatment).

PROBLEMS AND CONCEPT QUESTIONS

12–1. Contribution of the Electron Flux from Non-Thermal Plasma into Deactivation of Biological Micro-Organisms. Based on the equation (12–1) describing contribution of the non-thermal plasma electrons into sterilization, derive corresponding formula for the survival curve. Compare the curve related to contribution of plasma electrons with experimental survival curves for atmospheric-pressure discharges presented in Section 12.2.

12–2. Effect of Non-Thermal Atmospheric Pressure Air Plasma Contacting Water Surface on the Water Acidity. Analyzing reactions (12–2–8) describing interaction of electrons and ions of air plasma with water, explain why the non-thermal plasma treatment of water usually decreases pH and makes the water slightly acidic. Find out conditions, when plasma treatment does not make the water acidic or at least stabilizes the plasma-stimulated decrease of pH.

12–3. Contribution of Ion Bombardment to Sterilization Effect of Non-Thermal Plasmas. Compare average energies of ion bombardment and related effect of deactivation of micro-organisms in high pressure non-thermal plasmas organized in molecular and monatomic gases. Take into account that from one hand δ-factors in molecular gases are up to 1000 times greater than in monatomic ones; while electron temperatures at the same strengths of electric fields are essentially higher in monatomic gases.

12–4. Contribution of Reactive Neutral Species to Sterilization Effect of Non-Thermal Plasma. Compare contribution to sterilization of OH-radicals generated in plasma and then transported to the surface of biomaterials (see section 12.5) with that of OH-radicals produced inside of the biomaterials from plasma-generated charged particles (see section 12.1.4). Analyze rates of OH-production by surface bombardment with electrons and with positive and negative ions. What is the effect of electric field in the non-thermal atmospheric-pressure air discharges on the balance between the channels of OH production?

12–5. Kinetic Model of Non-Thermal Plasma Sterilization of Air Flow. Based on kinetic data presented in Table 12–1 and on comparison of experimental and simulation data on direct plasma deactivation of *E. coli* bacteria in air stream using a DBD discharge (Fig. 12–18), estimate reaction rate coefficient of plasma sterilization attributed to effect of charged particles. Assume that plasma density in the discharge is $n_e = 10^{12}$ cm^{-3}, and that the charged particles are responsible for the discrepancy between experimental and simulation results presented in the Fig. 12–18.

12–6. Discharge in Water Applied for the Direct Water Treatment. Analyzing equation (12–24) for the increment of thermal breakdown instability in water and typical parameters of the breakdown channel given in the section 12.4.4, estimate characteristic time required for development of the pulsed plasma discharge used for direct water treatment.

References

Ablekov, V.K., Denisov, Yu.N., Lubchenko, F.N. (1982), Gas-Dynamic Laser Handbook, Mashinostroenie (Machinery), Moscow.

Abolentsev, V.A., Korobtsev, S.V., Medvedev, D.D., Shiryaevsky, V.L. (1992), Euro Physics Conference, ESCAMPIG-92, vol. 16F, p. 396.

Abou-Ghazala, A., Katsuki, S., Schoenbach, K.H., Moreira, K.R. (2002), IEEE Trans Plasma Sci., vol. 30, p. 1449.

Achasov, O.V., Ragosin, D.S. (1986), Rate Coefficients of Vibrational Energy Exchange in Laser Active Medium of CO_2 Gas-Dynamic Laser with Additions of O_2, H_2, H_2O and CO, Preprint no. 16, A.V. Lykov Institute of Heat and Mass Transfer, Minsk.

Adamovich, I., Saupe, S., Grassi, M.J., Schulz, O., Macheret, S., Rich, J.W. (1993), Chem. Phys., vol. 174, p. 219.

Adams N.G., Babcock, L.M., eds. (2001), Advances in Gas Phase Ion Chemistry, vol. 4, Elsevier Science.

Affinito, J., Parson, R.R. (1984), J. Vac. Sci. Technol., vol. A2, p. 1275.

Ageeva, N.P., Novikov, G.I., Raddatis, V.K., Ruanov, V.D., Fridman, A. (1986), Sov. Phys., High Energy Chem. (Khimia Vysokikh Energij), vol. 20, p. 284.

Ahmed, I., Bergman, T.L. (1999), Therm. Spray Technol., vol. 9, p. 215.

Ahmed, S. (1998), "Catalytic Partial Oxidation Reforming of Hydrocarbon Fuels," 1998 Fuel Cell Seminar, California, p. 242, Elsevier, N.Y.

Ahmed, S., Kumar, R., Krumpelt, M. (1997), "Gasoline to Hydrogen – a New Route for Fuel Cells," Electric & Hybrid Vehicle Technology 97, p. 77, UK and Int. Press, UKIP, L.

Ahmedganov, R.A., Bykov, Yu.V., Kim, A.V., Fridman, A. (1986), J. Inst. Appl. Phys., vol. 147, p. 20.

Aithal, S.M., Subramaniam, V. (2002), "Laser-Assisted and Optical Pumping Techniques for Diamond Synthesis," in Diamond Films Handbook, Asmussen, J., and Reinhard, D. (eds.), Marcel Dekker, New York.

Akashi, K., Ishizuka, R., Egami, I. (1972), J. Mining Metall. Inst. Jpn, vol. 88, no. 1081, p. 885.

Akishev, Y., Grushin, M., Kochetov, I., Karal'nik, V., Napartovich, A., Trushkin, N. (2005), Plasma Source Sci. Tech., vol. 14, p. 18.

Akiyama, H. (2000), IEEE Trans. Dielectr. Elec. Insul., vol. 7, p. 646.

Akulintsev, V.M., Gorshunov, V.M., Neschimenko, Y.P. (1977), Sov. Phys., J. Appl. Mech. Techn. Phys., vol. 18, p. 593.

Akulintsev, V.M., Gorshunov, V.M., Neschimenko, Y.P. (1983), Sov. Phys., J. Appl. Mech. Techn. Phys., vol. 24, p. 1.

Al-Arainy, A.A., Jayaram, S., Cross, J.D. (1996), 12th Int. Conf. on Conduction and Breakdown in Dielectric Liquids, Rome, Italy.

Aldea, E., Peeters, P., de Vries, H., Van De Sanden, M.C.M. (2005), Surf. Coat. Technol., vol. 200, p. 46.

Aldea E., de Vries, H., Van De Sanden, M.C.M. (2006), 3rd Int. Workshop on Microplasmas (IWM), Greifswald, Germany, p. 96.

Alekperov, G.A. (1979), "Plasma-Chemical Polymerization Processes," Ph.D. Dissertation, Institute of Petrochemical Synthesis, USSR Academy of Sciences, Moscow.

Alexandrov, A.F., Rukhadze, A.A. (1976), Physics of High-Current Gas-Discharge Light Sources, AtomIzdat, Moscow.

Alexandrov, A.P. (1978), in Nuclear – Hydrogen Energy and Technology, Legasov, V.A. (ed.), vol. 1, p. 5, Atom-Izdat, Moscow.

Alexandrov, N.L. (1981), in Plasma Chemistry, Smirnov, B.M. (ed.), vol. 8, Energoizdat, Moscow.

Alexandrov, N.L., Konchakov, A.M., Son, E.E. (1978a), Sov. Phys., Plasma Phys. (Phisica Plasmy), vol. 4, p. 169.

Alexandrov, N.L., Konchakov, A.M., Son, E.E. (1978b), Sov. Phys., Plasma Phys. (Phisica Plasmy), vol. 4, p. 1182.

References

Alexeev, A.M., Ivanov, A.A., Starykh (1979), 4th International Symposium on Plasma Chemistry, ISPC-4, vol. 2, p. 427, Zurich, Switzerland.

Alexeev, N.V., Souris, A.L., Shorin, S.N. (1971), in Theory and Practice of Gas Combustion, Nedra (ed.), vol. 5, Leningrad, Russia.

Alexeev, N.V., Souris, A.L., Shorin, S.N. (1972), in Low-Temperature Plasma in Technology of Inorganic Compounds, Polak, L.S. (ed.), Nauka (Science), Novosibirsk, Russia.

Alexeeva, L.P., Bobrov, A.A., Lysov, G.V. (1977), 7th USSR Symposium on Low Temperature Plasma Generators, p. 152, Alma-Ata.

Alexeeva, L.P., Bobrov, A.A., Lysov, G.V. (1979), 3rd USSR Symposium on Plasma Chemistry, vol. 2, p. 157, Nauka (Science), Moscow.

Alexeff, I. (2001), "Direct Current Energy Discharge System," U.S. Patent no. 6,232,732, May 15.

Alexeff, I., Laroussi, M., Kang, W., Alikafesh, A. (1999), IEEE Int. Conf. on Plasma Science, Monterey, CA, 4P21, p. 208.

Aliev, Z.G., Emelianov, Yu.M., Babayan, V.G. (1971), in Chemistry and Physics of Low-Temperature Plasma, p. 189, Moscow State University, Moscow.

Alix, F.R., Neister, S.E., McLarnon, C.R. (1997), U.S. Patent no. 5,871,703.

Alkhasov, G.D. (1971), Sov. Phys., J. Techn. Phys., vol. 41, p. 2513.

Allis, W.P. (1960), Nuclear Fusion, D. Van Nostrand, Princeton, NJ.

Amberger, E., Druminsky, M., Dietze, W. (1968), J. Less-Common Metals, vol. 14, no. 4, p. 466.

Amberger, E., Druminsky, M., Ploog, K. (1970), Electr. Technol., vol. 1, p. 133.

Ambrazevicius, A. (1983), Heat Transfer During Quenching of Gases, Mokslas Publ., Vilnius, Lithuania.

Amirkhanov, D.M., Kotenko, A.A., Rusanov, V.D., Tul'skii, M.N. (1998), Polym. Sci. A, vol. 40, p. 206.

Amirkhanov, D.M., Kotenko, A.A., Tul'skii, M.N., Chelyak, M.M. (2001), Fibre Chem., vol. 33, p. 67.

Amman, P.R., Baddour, R.F., Johnston, M.M., Timmins, R.S., Mix, J.W. (1964), Chem. Eng. Progr., vol. 60, p. 52.

Anders, A., Anders, S., Brown, I.G., Dickinson, M.R., MacGill, R.A. (1994), J. Vac. Sci. Technol. B., vol. 12, p. 815.

Andreev, D.N. (1953), Organic Synthesis in Electric Discharges, Publishing House of the USSR Academy of Sciences, Moscow.

Andreev, E.A., Nikitin, E.E. (1976), in Plasma Chemistry, vol. 3, Smirnov, B.M., (ed.), Atom-Izdat, Moscow.

Andreev, Yu.P. (1964), Sov. Phys., J. Phys. Chem., vol. 38, p. 794.

Andrews, T., Tait, P.G. (1860), Phil. Trans. Roy. Soc., London, vol. 150, p. 113.

Anikin, N.B., Mintoussov, E.I., Pancheshnyi, S.V., Roupassov, D.V., Sych, V.E., Starikovskii, A.Yu. (2003), "Nonequilibrium Plasma and Its Application for Combustion and Hypersonic Flow Conrol," 41st AIAA Aerospace Sciences Meeting and Exhibit, January 9–12, Reno, NV, AIAA Paper 2003–1053.

Anikin, N.B., Starikovskaia, S.M., Starikovsky, A.Yu. (2004), Plasma Phys. Rep., vol. 30, p. 1028.

Anikin, N.B., Bozhenkov, S.A., Zatsepin, D.B., Mintoussov, E.I., Pancheshnyi, S.B., Starikovskaia, S.M., Starikovsky, A.Yu. (2005), in Encyclopedia of Low-Temperature Plasma, vol. VIII-I, Chemistry of Low-Temperature Plasma, Lebedev, Yu.A., Plate, N.A., Fortov, V.E. (eds.), YANUS-K, Moscow.

Anpilov, A.M., Barkhudarov, E.M, Bark, Y.B., Zadiraka, Y.V., Christofi, N., Kozlov, Y.N., Kossyi, I.A., Kop'ev, V.A., Silakov, V.P., Taktakishvili, M.I., Temchin, S.M. (2001), J. Phys., D: Appl. Phys., vol. 34, p. 993.

Anselme, K. (2000), Biomaterials, vol. 21, p. 667.

Arbatsky, A.E., Vakar, A.K., Vorobiev, A.B., Golubev, A.V., Gorelova, M.M., Krasheninnikov, E.G., Levin, V.Yu., Liventsov, V.V., Fridman, A., Rusanov, V.D. (1988), Modification of Polymer Films Using Microwave Plasma After-Glow, Kurchatov Institute of Atomic Energy, CNII-Atom-Inorm, vol. 4722/7, Moscow.

Arbatsky, A.E., Vakar, A.K., Golubev, A.V., Krasheninnikov, Liventsov, V.V., Macheret, S.O., Rusanov, V.D., Fridman, A. (1990), Sov. Phys., High Energy Chemistry (Khimia Vysokikh Energij), vol. 24, p. 256.

Arefi-Khonsari, F., Kurdi, J., Tatoulian, M., Amouroux, J. (2001), Surface & Coatings Technology, vol. 142–444, p. 437.

Arjunan, K.P., Gutsol, A., Vasilets, V., Fridman, A. (2007), 18th Int. Symposium on Plasma Chemistry (ISPC-18), Kyoto, Japan.

Arkhipenko, V.I., Zgirovskii, S.M., Kirillov, A.A., Simonchik, L.V. (2002), Plasma Physics Reports, vol. 28, no. 10, p. 858.

Arkhipenko, V.I., Kirillov, A.A., Simonchik, L.V., Zgirouski, S.M. (2005), Plasma Sources Sci. Technol., vol. 14, p. 757.

Arshinov, A.A., Musin, A.K. (1958a), Sov. Phys., Doklady, vol. 118, p. 461.

Arshinov, A.A., Musin, A.K. (1958b), Sov. Phys., Doklady, vol. 120, p. 747.

Arshinov, A.A., Musin, A.K. (1959), Sov. Phys., J. Phys. Chem., vol. 33, p. 2241.

Arshinov, A.A., Musin, A.K. (1962), Sov. Phys., Radio Engineering and Electronics, vol. 7, p. 890.

Artukhov, A.A. (1976), Sov. Phys., High Energy Chemistry (Khimia Vysokikh Energij), vol. 10, p. 513.

Artukhov, A.A., Legasov, V.A., Makeev, G.N. (1977), Sov. Phys., High Energy Chemistry (Khimia Vysokikh Energij), vol. 11, p. 88.

References

Artukhov, I.M., Bezmozgin, E.S., Vinokurov, H.L., Klimenko, V.L., Prigogin, N.S. (1969), in Thermocatalytic Methods of Conversion of Hydrocarbons, Chimia (Chemistry), Leningrad, Russia.

Ash, R.L., Dowler, W.L., Varsi, G. (1978), Acta Astronautica, vol. 5, p. 705.

Ashurly, Z.I., Babayan, V.G., Emelianov, Yu.M., Morozova, N.P., Fedotova, T.A. (1971), in Chemistry and Physics of Low-Temperature Plasma, Moscow State University, Moscow, p. 121.

Asinovsky, E.I., Vasilyak, L.M. (2001), Encyclopedia of Low Temperature Plasma, Fortov, V.E. (ed.), vol. 2, p. 234, Nauka (Science), Moscow.

Asisov, R.I. (1980), "Plasma-Chemical Experiments with Non-Equilibrium Microwave Discharges of Moderate Pressure with Magnetic Field," Ph.D. Dissertation, Kurchatov Institute of Atomic Energy, Moscow.

Asisov, R.I., Givotov, V.K., Rusanov, V.D. (1977), 3rd International Symposium on Plasma Chemistry, Limoge, vol. 1, p. 51.

Asisov, R.I., Givotov, V.K., Rusanov, V.D., Fridman, A. (1980), Sov. Phys., High Energy Chemistry (Khimia Vysokikh Energij), vol. 14, p. 366.

Asisov, R.I., Givotov., V.K., Krasheninnikov, E.G., Krotov, M.F., Potapkin, B.V., Rusanov, V.D., Fridman, A. (1981a), 5th International Symposium on Plasma Chemistry, Edinburgh, vol. 1, p. 52.

Asisov, R.I., Givotov., V.K., Krasheninnikov, E.G., Potapkin, B.V., Rusanov, V.D., Fridman, A. (1981b), 5th International Symposium on Plasma Chemistry, Edinburgh, vol. 2, p. 774.

Asisov, R.I., Givotov., V.K., Krasheninnikov, E.G., Potapkin, B.V., Rusanov, V.D., Fridman, A. (1983), Sov. Phys., Doklady, vol. 271, p. 94.

Asisov, R.I., Vakar, A.K., Givotov, V.K., Krasheninnikov, E.G., Rusanov, V.D., Fridman, A. (1985a), International Journal of Hydrogen Energy, vol. 10, p. 475.

Asisov, R.I., Vakar, A.K., Givotov, V.K., Krasheninnikov, E.G., Rusanov, V.D., Fridman, A. (1985b), Sov. Phys., Journal of Technical Physics, vol. 55, p. 79.

Asisov, R.I., Potapkin, B.V., Rusanov, V.D., Fridman, A. (1986), Sov. Phys., Doklady, vol. 286, p. 1143.

Asmussen, J., Reinhard, D.K. (2002), Diamond Film Handbook, CRC Press, Boca Raton, FL.

Atamanov, V.M., Ivanov, A.A., Nikiforov, V.A. (1979a), Sov. Phys., J. Tech. Phys., vol. 49, p. 2311.

Atamanov, V.M., Ivanov, A.A., Nikiforov, V.A. (1979b), Sov. Phys., Plasma Physics (Fizika Plazmy), vol. 5, p. 663.

Atkins, P. (1998), Physical Chemistry, Freeman, New York.

Attar, A. (1978), Fuel, vol. 57, p. 201.

Aulsloos, P. (1968), Fundamental Processes in Radiation Chemistry, Wiley Interscience Publ., N.Y.

Avasov, S.A., Bagirov, M.A., Malin, V.P. (1969), Kinetics and Mechanisms of Polymer Reactions, Hungarian Academy of Sciences Budapest, Hungary.

Avramenko, L.I., Kolesnikov, R.V. (1967), Sov. Phys., News of Academy of Sciences (Izvestia), Chemistry, vol. 8, p. 996.

Ayan, H., Fridman, G., Friedman, G., Gutsol, A., Fridman, A. (2007), 18th Int. Symposium on Plasma Chemistry (ISPC-18), Kyoto, Japan.

Ayoubi, S.A., Gonnord, M.F., Amouroux, J. (1995), Int. Symposium on Environmental Technologies, Atlanta, GA.

Babaeba, N., Naidis, G. (1996), J. Phys. D: Applied Physics, vol. 29, p. 2423.

Babaritsky, A.I., Ivanov, A.A., Shapkin, V.V. (1977), in Plasma Chemistry, vol. 4, Smirnov, B.M. (ed.), Atom-Izdat, Moscow.

Babaritsky, A.I., Diomkin, S.A., Givotov, V.K., Makarenko, V.G., Nester, S.A., Rusanov, V.D., Fridman, A. (1991), Non-Equilibrium Approach to Methane Conversion into Acetylene in Microwave Discharge, Kurchatov Institute of Atomic Energy, vol. 5350/12, Moscow.

Babaritsky, A.I., Baranov, E.I., Deminsky, M.A., Demkin, S.A., Givotov, V.K., Potapkin, B.V., Potekhin, S.I., Rusanov, V.D., Ryazantsev, E.I., Etivan, C. (1999a), Sov. Phys., High Energy Chemistry (Khimia Vysokikh Energij), vol. 33, p. 59.

Babaritsky, A.I., Baranov, E.I., Demkin, S.A., Givotov, V.K., Potapkin, B.V., Potekhin, S.I., Rusanov, V.D., Ryazantsev, E.I., Etivan, C. (1999b), Sov. Phys., High Energy Chemistry (Khimia Vysokikh Energij), vol. 33, p. 458.

Babaritsky, A.I., Baranov, E.I., Bibikov, M.B., Demkin, S.A., Deminsky, M.A., Givotov, V.K., Lysov, G.V., Konovalov, G.M., Moskovsky, A.S., Potapkin, B.V., Rusanov, V.D., Smirnov, R.V., Strelkova, M.I., Fedotov, N.G., Chebankov, F.N. (2003), Kurchtov Institute of Atomic Energy, RSC-KI, vol. 63/0213, Moscow.

Babayan, S.E. (2000), "Deposition of Materials with an Atmospheric Pressure Plasma," Ph.D. Dissertation, University of California at Los Angeles, CA.

Baber, S.C., Dean, A.M. (1974), J. Chem. Phys., vol. 60, p. 307.

Babko-Malyi, S., Battleson, D., Ray, I., Buckley, W.P., Reynolds, J. (2000), "Mercury Removal from Combustion Flue Gases by the Plasma-Enhanced Electrostatic Precipitator," Air Quality II: Mercury Trace Elements, and Particular Matter Conference, Proceedings, McLean, VA.

Backstrom, H. (1934), Z. Phys. Chem., vol. B25, p. 99.

Baddur, R. (1967), U.S. Patent no. 3,333,927.

Baddur, R., Dandus, P. (1970), in Application of Plasma in Chemical Processes, Polak, L.S. (ed.), Mir, Moscow.

References

Baddur, R., Timmins, R. (1970), in Application of Plasma in Chemical Processes, Polak, L.S. (ed.), p. 255, Mir, Moscow.

Badie, J.M., Flamant, J., Granier, B. (1997), 13th Int. Symposium on Plasma Chemistry (ISPC-13), Beijing, vol. 4, p. 1748.

Bai, M., Zhang, Z., Gao, H., Bai, X. (2005), 17th International Symposium on Plasma Chemistry (ISPC-17), p. 895, Toronto, Canada.

Baidarovtsev, Yu.P., Vasilets, V.N., Ponomarev, A.N. (1985), Sov. Phys., Chemical Physics, vol. 4, p. 89.

Bailin, L.J., Sibert, M.E., Jonas, L.A., Bell, A.T. (1975), Environ. Sci. Tech., vol. 9, p. 254.

Baksht, F.G., Dugev, G.A., Martsinovsky, A.M. (1973), Thermionic Converters and Low-Temperature Plasma, Nauka (Science), Moscow.

Balebanov, A.V., Butylin, B.A., Givitov, V.K., Matolich, R.M., Macheret, S.O., Novikov, G.I., Potapkin, B.V., Rusanov, V.D., Fridman, A. (1985a), Sov. Phys., Doklady, vol. 283, p. 657.

Balebanov, A.V., Butylin, B.A., Givitov, V.K., Matolich, R.M., Macheret, S.O., Novikov, G.I., Potapkin, B.V., Rusanov, V.D., Fridman, A. (1985b), Journal of Nuclear Science and Technology, Nuclear-Hydrogen Energy, vol. 3, p. 46.

Balebanov, A.V., Givitov, V.K., Krasheninnikov, E.G., Nester, S.A., Potapkin, B.V., Rusanov, V.D., Samarin, A.E., Fridman, A., Shulakova, E.V. (1989), Sov. Phys., High Energy Chemistry (Khimia Vysokikh Energij), vol. 5, p. 440.

Ballauff, M., Li, L., Rosenfeldt, S., Dingenouts, N., Beck, J., Krieger-Beck, P. (2004), "Analysis of Polycarbon Suboxide by Small-Angle X-ray Scattering," Ed. Angew. Chem. Int., p. 43.

Baltin, L.M., Batenin, V.M. (1970), Plasmatron Applications in Spectroscopy, Ilim, Frunze (Bishkek), Kyrgyzstan.

Bamford, C.H., Ward, J.C. (1961), Polymer, vol. 2, p. 277.

Baranchicov, E.I., Belenky, G.S., Deminsky, M.A., Denisenko, V.P., Potapkin, B.V., Rusanov, V.D., Fridman, A. (1990a), "Plasma-Catalytic SO$_2$ Oxidation in Air," Kurchatov Institute of Atomic Energy, vol. IAE-5256/12, Moscow.

Baranchicov, E.I., Denisenko, V.P., Potapkin, B.V., Rusanov, V.D., Fridman, A. (1990b), Sov. Phys., Doklady, vol. 339, p. 1081.

Baranchicov, E.I., Belenky, G.S., Deminsky, M.A., Dorovsky, E.M., Erastov, E.M., Kochetov, V.A., Maslenikov, D.D., Potapkin, B.V., Rusanov, V.D., Severny, V.V., Fridman, A. (1992), Radiat. Phys. Chem., vol. 40, p. 287.

Baránková, H., Bárdos, L. (2002), Catalysis Today, vol. 72, p. 237.

Baranov, I.E., Fridman, A., Kirillov, I.A., Rusanov, V.D. (1991), Breakaway Oxidation Effect and Its Influence on Severe Nuclear Accidents, Kurchatov Institute of Atomic Energy, Moscow.

Barat, P., Cullis, C.F., Pollard, R.T. (1972), Proceedings of the Royal Society of London, A, vol. 329, p. 433.

Barcicki, J., Myrdzik, J. (1974), Roshniki Chemii, Annales Sco. Chem. Polonium, vol. 48, no. 3, p. 539.

Barkhudarov, E.M., Gritsinin, S.I., Korchagina, E.G., Kossyi, I.A., Misakyan, M.A., Silakov, V.P., Tarasova, N.M., Temchin, S.M. (1997), 14th Int. Symposium on Plasma Chemistry (ISPC-14), Prague, Czech Repulic, vol. 5, p. 2563.

Barnes, M.S., Keller, J.H., Forster, J.C., O'Neil, J.A., Coutlas, D.K. (1992), Phys. Rev. Lett., vol. 68, p. 313.

Barnett, C.F. (1989), in A Physicist's Desk Reference, Anderson, H.L. (ed.), American Institute of Physics, New York.

Barone, M.E., Graves, D.B. (1995), J. Appl. Phys., vol. 77, p. 1263.

Baronov, G.S., Bronnikov, D.K., Fridman, A., Potapkin, B.V., Rusanov, V.D., Varfolomeev, A.A., Zasavitsky, A.A. (1989), J. Phys. B., vol. 22, p. 2903.

Barrer, R.M. (1939), Trans. Faraday Soc., vol. 35, p. 628.

Barry, T.J., Bayliss, R.K., Lay, L.A. (1968), Journal of Material Science, vol. 3, p. 229.

Bartlett, A.H., Castro, R.G., Butt, D.P., Kung, H., Petrovic, J.J., Zurecki, Z. (1996), Industrial Heating, vol. LXIII, no. 1, p. 33.

Baselyan, E.M., Raizer, Yu.P. (1997), Spark Discharge, Moscow Institute of Physics and Technology, Moscow.

Bashkirov, Y.A., Medvedev, S.A. (1968), "Application of a Radio-Frequency Electrodeless Discharge for Synthesis of Niobium-Vanadium Intermetallic Compounds," in Low Temperature Plasma Generators, p. 501, Energia (Energy), Moscow.

Basov, N.G., Belenov, E.M., Danilychev, V.A. (1972), Publication no. 56, Physical Institute of the USSR Academy of Sciences, Moscow.

Basov, N.G., Belenov, E.M., Danilychev, V.A, Panteleev, V.I. (1973), Sov. Phys., Journal of Experimental and Theoretical Physics, vol. 64, p. 108.

Basov, N.G., Dolinina, V.I., Suchkov, A.F., Urin, B.M. (1977a), Sov. Phys., Quantum Electronics, vol. 4, p. 776.

Basov, N.G., Danilychev, V.A., Panteleev, V.I. (1977b), Sov. Phys., Doklady, vol. 233, p. 6.

Basov, N.G., Danilychev, V.A, Lobanov, A.N. (1978a), Theoretical and Experimental Investigations of Nitrogen Oxide Synthesis in Non-Thermal Electron-Ionization Plasma, Physical Institute of Academy of Sciences, Moscow.

References

Basov, N.G., Danilychev, V.A., Panteleev, V.I. (1978b), Sov. Phys., High Energy Chemistry (Khimia Vysokikh Energij), vol. 3, p. 266.

Basting, B., Marowsky, G., ed. (2005), Excimer Laser Technology, Springer, Berlin.

Batenin, V., Klimovsky, I.I., Lysov, G.V., Troizky, V.N. (1988), Microwave Plasma Generators: Physics, Engineering, Applications, EnergoAtomIzdat, Moscow.

Baulch, D.L., Drysdele, D.D., Horne, D.J., Lloyd A.C. (1976), Evaluated Kinetic Data for High Temperature Reactions, vol. 3, Butterworth, London.

Becker, K. (1964), Kunststoffe, vol. 54, p. 155.

Becker, K.H., Kogelschatz, U., Schonbach, K.H., Barker, R.J. (2005), Non-Equilibrium Air Plasmas at Atmospheric Pressure, Series in Plasma Physics, IOP Publ., Bristol.

Becker, R., Doring, W. (1935), "Condensation Theory," Annnalen der Physik, vol. 24, p. 719.

Bedin, A.P., Mishin, G.I. (1995), Sov. Phys., Tech. Phys. Letters (Pis'ma v JTF), vol. 21, p. 5.

Begin, K. (1970), in Application of Plasma in Chemical Processes, p. 49, Mir (World), Moscow.

Beiers, H.-J., Baumann, H., Bittner, D., Klein, I. (1985), 7th Int. Symposium on Plasma Chemistry (ISPC-7), Eindhoven, no. B-2-2, p. 232.

Bekefi, G., ed. (1976), Principles of Laser Plasma, John Wiley & Sons, New York.

Belenguer, Ph., Blondeau, J.P., Boufendi, L., Toogood, L.M., Plain, A., Laure, C., Bouchoule, A., Boeuf, J.P. (1992), Phys. Rev. A, vol. 46, p. 7923.

Belenov, E.M., Markin, E.P., Oraevsky, A.N., Romanenko, V.I. (1973), Sov. Phys., Journal of Experimental and Theoretical Physics, Letters (Pis'ma JETF) vol. 18, p. 116.

Beliaev, S.T., Budker, G.I. (1958), Plasma Physics and Problems of the Controlled Thermo-Nuclear Reactions, Academy of Sciences of the USSR, vol. 3.

Belousov, I.G., Legasov, V.A., Rusanov, V.D. (1977), Journal of Nuclear Science and Technology, Nuclear-Hydrogen Energy, vol. 2, no. 3, p. 158.

Belousov, I.G., Krasnoshtanov, V.F., Rusanov, V.D., Fridman, A. (1979), Journal of Nuclear Science and Technology, Nuclear-Hydrogen Energy, vol. 1, no. 5, p. 43.

Ben Gardi, R., Roth, J.R., Montie, T.C., Kelly-Wintenberg, K., Tsai, P.P.-Y., Helfritch, D.J., Feldman, P., Sherman, D.M., Karakaya, F., Chen, Z. (2000), Surface Coatings Technology, vol. 131, p. 528.

Bennett, W.H. (1934), "Magnetically self-focusing streams," Phys. Rev., vol. 45, p. 890.

Ben-Shaul, A., Levine, R.D., Bernstein, R.B. (1974), J. Chem. Phys., vol. 61, p. 4937.

Benson, S.W. (1960), The Foundations of Chemical Kinetics, McGraw-Hill, New York.

Benson, S.W., Axworthy, A.E. (1957), J. Chem. Phys., vol. 26, p. 1718.

Berg, S., Blom, H.O., Moradi, M., Nender, C., Larsson, T. (1989), Journal of Vacuum Science and Technology, vol. A7, p. 1225.

Bergman, R.C., Homicz, G.F., Rich, J.W., Wolk, G.L. (1983), J. Chem. Phys., vol. 78, p. 1281.

Bernius, M.T., Chutjian, A. (1989), Rev. Sci. Imstrum., vol. 60, p. 779.

Beroural, A. (1993), J. Appl. Phys., vol. 73, p. 4528.

Beroural, A., Zahn, M., Badent, A., Kist, K., Schwabe, A.J., Yamashita, H., Yamazawa, K., Danikas, M., Chadband, W.G., Torshin, Y. (1998), IEEE Electrical Magazine, vol. 14, p. 6.

Bertelsen, B. (1992), Catalytic Control of VOC Emissions, Manufactures of Emission Controls Association, Washington, DC.

Bettleheiem, F.A., March, J. (1995), Introduction to General, Organic and Biochemistry, 4th edition, Saunders College Publ., Orlando, FL.

Bezmelnitsyn, V.N., Legasov, V.A., Chaivanov, B.B. (1977), Sov. Phys., Doklady (Reports of USSR Academy of Sciences), vol. 235, p. 96.

Bezmelnitsyn, V.N., Siniansky, V.F., Chaivanov, B.B. (1979), Plasma Chemistry, vol. 6, Smirnov, B.M. (ed.), Atom-Izdat, Moscow.

Biberman, L.M., Vorobiev, V.S., Yakubov, I.T. (1979), Adv. in Phys. Sci., (Uspehi Phys. Nauk), vol. 128, p. 233.

Biederman, H. (2004), Plasma Polymer Films, Imperial College Press, London.

Biederman, H., Osada, Y. (1992), Plasma Polymerization Process, Elsevier, Amsterdam.

Bilgic, A.M. (2000a), J. Anal. At. Spectrom., vol. 15, p. 579.

Bilgic, A.M. (2000b), Plasma Sources Science and Technology, vol. 9, p. 1.

Billing, G.D. (1986), Non-Equilibrium Vibrational Kinetics, Topics of Current Physics 39, Capitelli, M. (ed.), Springer-Verlag, Berlin.

Biondi, M.A. (1976), Principles of Laser Plasma, Bekefi, G. (ed.), John Wiley & Sons, New York.

Birmingham, J.G., Moore, R.R. (1990), "Reactive Bed Plasma Air Purification," U.S. Patent no. 4,954,320.

Birukov, A.S., Gordiets, B.F., Shelepin, L.A. (1968), Sov. Phys., Journal of Experimental and Theoretical Physics, vol. 55, p. 1456.

Birukov, A.S., Kulagin, Yu.A., Shelepin, L.A. (1977), Sov. Phys., Journal of Technical Physics, vol. 47, p. 332.

Bittencourt, J.A. (2004), Fundamentals of Plasma Physics, Springer, New York, Heidelberg.

Bittner, D., Baumann, H., Peucker, C. (1981), Erdol & Kohle Erdgas Petrochemie, vol. 34, p. 237.

References

Bityurin, V., Klimov, V., Leonov, S., Pashina, A., Skvortsov, V., Cain, T. (1999), AIAA paper 99-4856.

Bjerle, J., Eklund, H., Linne, M., Svensson, O. (1982), Ind. and Eng. Chemical Process Des. and Development, vol. 21, p. 141.

Bjerre, A., Nikitin, E.E. (1967), Chem. Phys. Lett., vol. 1, p. 179.

Bjornson, D. (1966a), U.S. Patent no. 3,471,546.

Bjornson, D. (1966b), U.S. Patent no. 3,494,974.

Bjornson, D. (1968), U.S. Patent no. 3,409,403.

Blackburn, P.R. (1980), "Ignition of Pulverized Coal with Arc-Heated Air," Energy, vol. 4, p. 98.

Blauer, J.A., Nickerson, G.R. (1974), "A Survey of Vibrational Relaxation Rate Data for Processes Important to CO_2-N_2-H_2O Infrared Plume Radiation," American Institute of Aeronautics and Astronautics, New York, p. 74.

Bletzinger, P., Ganguly, B.N. (1999), Phys. Rev. Letters, vol. A258, p. 342.

Bletzinger, P., Ganguly, B.N., Van Wie, D., Garscadden, A. (2005), J. Phys. D: Applied Physics, vol. 38, p. 33.

Blinov, L.M., Volod'ko, V.V., Gontarev, G.G., Lysov, G.V., Polak. L.S. (1969), Low-Temperature Plasma Generators, Polak, L.S. (ed.), Energia (Energy), Moscow.

Bloch, F., Bradbury, N. (1935), Phys. Rev., vol. 48, p. 689.

Block, S.S. (1992), "Sterilization," in Encyclopedia of Microbiology, vol. 4, Lederberg, J. (ed.), Academic Press, San Diego, p. 92.

Bobashev, S.V., Golovachov, Yu.P. (2000), Technical Report 807577, JHU/APL.

Bobrov, A.A., Kudrevatova, O.V., Lysov, G.V. (1979a), Sov. Phys., Electronic Engineering (Electronnaja Technica), Microwave Electronics, no. 5, p. 45.

Bobrov, A.A., Kudrevatova, O.V., Lysov, G.V. (1979b), 3rd USSR Symposium on Plasma Chemistry, vol. 2, p. 172., Nauka (Science), Moscow.

Bobylev, V.M., Ivashov, A.A., Krutiansky, M.M. (1976), "Operation Conditions of Electric Arc Plasma Installation for Reduction of Cobalt Oxide," in Plasma Processes in Metallurgy and Technology of Inorganic Materials, A.A. Baikov Institute of Metallurgy of USSR Academy of Sciences, Nauka (Science), Moscow.

Bochin, V.P., Rusanov, V.D., Lukianchuk, A.N. (1976a), Biophysics of Microbes and Bioengineering, Leningrad, Russia, p. 93.

Bochin, V.P., Rusanov, V.D., Lukianchuk, A.N. (1976b), 2nd Int. School on Physics and Engineering of Low-Temperature Plasma, vol. 1, Minsk, Belarus, p. 20.

Bochin, V.P., Legasov, V.A., Rusanov, V.D., Fridman, A., Sholin, G.V. (1977a), Journal of Nuclear Science and Technology, Nuclear-Hydrogen Energy, vol. 1, no. 2, p. 55.

Bochin, V.P., Legasov, V.A., Rusanov, V.D., Fridman, A., Sholin, G.V. (1977b), Journal of Nuclear Science and Technology, Nuclear-Hydrogen Energy, vol. 2, no. 3, p. 93.

Bochin, V.P., Legasov, V.A., Rusanov, V.D., Fridman, A., Sholin, G.V. (1978a), Nuclear–Hydrogen Energy and Technology, Legasov, V.A. (ed.), vol. 1, Atom-Izdat, Moscow.

Bochin, V.P., Legasov, V.A., Rusanov, V.D., Fridman, A., Sholin, G.V. (1978b), 2nd World Conference on Hydrogen Energy, Zurich, vol. 3, p. 1183.

Bochin, V.P., Legasov, V.A., Rusanov, V.D., Fridman, A., Sholin, G.V. (1979), Nuclear–Hydrogen Energy and Technology, Legasov, V.A. (ed.), vol. 2, Atom-Izdat, Moscow.

Bochin, V.P., Krasnoshtanov, V.F., Rusanov, V.D., Fridman, A. (1983), Journal of Nuclear Science and Technology, Nuclear-Hydrogen Energy, vol. 2, no. 15, p.12.

Bochkov, V.D., Korolev, Yu.D. (2000), "Pulsed Gas-Discharge Commutation Devices," in Encyclopedia of Low-Temperature Plasma, Fortov, V.E. (ed.), vol. 4, Nauka (Science), Moscow, p. 203.

Boenig, H.V. (1988), Fundamentals of Plasma Chemistry and Technology, Technomic Publ., Lancaster, PA.

Boeuf, J.P. (1992), Phys. Rev. A, vol. 46, p. 7910.

Boeuf, J.P. (2003), J. Phys. D: Appl. Phys., vol. 36, p. 53.

Boeuf, J.P., Marode, E. (1982), J. Phys., D, Appl. Phys., vol. 15, p. 2169.

Boeuf, J.P., Punset, C., Hirech, A., Doyeux, H. (1997), J. Physique IV, 7, p. C4-3.

Boeuf, J.P., Lagmich, Y., Callegari, T., Pitchford, L. (2007), "EHD Force in Dielectric Barrier Discharge: Parametric Study an Influence of Negative Ions," 45th AIAA Aerospace Science Meeting and Exhibit, Reno, NV, AIAA paper 2007-0183.

Bohm, C., Perrin, J. (1991), J. Phys. D: Appl. Phys., vol. 24, p. 865.

Bolotov, A.V., Glikman, M.A., Isikov, V.S., Kopylov, Yu.M. (1976a), "Reduction of Metal Oxides Using Electric Arc Heating of Nitrogen-Hydrogen Mixture," in Plasma Processes in Metallurgy and Technology of Inorganic Materials, A.A. Baikov Institute of Metallurgy of USSR Academy of Sciences, Nauka (Science), Moscow.

Bolotov, A.V., Isikov, V.S., Filkov, M.N. (1976b), "Reduction of Molybdenum Tri-Oxide in Fluidized Bed Using Electric-Arc Hydrogen Heater," in Plasma Processes in Metallurgy and Technology of Inorganic Materials, A.A. Baikov Institute of Metallurgy of USSR Academy of Sciences, Nauka (Science), Moscow.

Bol'shakov, A.A., Cruden, B.A., Mogul, R., Rao, M.V.V.S., Sharma, A.P., Khare, B.N., Meyyappan, M. (2004), AIAA Journal, vol. 42, p. 823.

References

Bond, R.L., Galbraight, I.F., Ladner, W.R., McConnel, G.I.T. (1963), "Production of Acetylene from Coal Using a Plasma Jet," Nature, vol. 200, p. 1313.

Bond, R.L., Ladner, W.R., McConnel, G.I.T. (1966a), "Reaction of Coals under Conditions of High Energy Input and High Temperature," Adv. Chem. Ser., vol. 55, p. 650.

Bond, R.L., Ladner, W.R., McConnel, G.I.T. (1966b), "Reaction of Coals in Plasma Jet," Fuel, vol. 45, p. 381.

Bonhoeffer, K. (1924), Ztschr. Phys. Chem., vol. 113, p. 199.

Borisov, A.A., De Luca, L., Merzhanov, A. (2002), "Self-Propagating High-Temperature Synthesis of Materials" Combustion Science and Technology, Taylor & Francis, New York.

Borodulia, V.A. (1973), Electrothermal Fluidized Bed, Nauka I Technika (Science and Technology), Minsk, Belarus.

Boswell, R.W. (1970), Plasma Phys. and Controlled Fusion, vol. 26, p. 1147.

Boucher, R.M. (1980), U.S. Patent no. 4,207,286.

Bouchoule, A. (1993), Physics World, vol. 47, August.

Bouchoule, A., ed. (1999), Dusty Plasmas: Physics, Chemistry and Technological Impacts in Plasma Processing, John Wiley & Sons, New York.

Bouchoule, A., Boufendi, L. (1993), Plasma Sources Sci. Technol., vol. 2, p. 204.

Bouchoule, A., Boufendi, L., Blondeau, J.Ph., Plain, A., Laure, C. (1991), J. Appl. Phys., vol. 70, p. 1991.

Bouchoule, A., Boufendi, L., Blondeau, J.Ph., Plain, A., Laure, C. (1992), Appl. Phys. Lett., vol. 60, p. 169.

Bouchoule, A., Boufendi, L., Blondeau, J.Ph., Plain, A., Laure, C. (1993), J. Appl. Phys., vol. 73, p. 2160.

Bouchoule, A., Cadiou, A., Heron, A., Dudeck, M., Lyszyk, M. (2001), "An Overview of the French Research Program on Plasma Thrusters for Space Applications," Contributions to Plasma Physics, vol. 41, no. 6, p. 573.

Boufendi, L. (1994), Ph.D. Thesis, Faculte de Science, GREMI, University of Orleans, France.

Boufendi, L., Bouchoule, A. (1994), Plasma Sources Sci. Technol., vol. 3, p. 70.

Boufendi, L., Hermann, J., Stoffels, E., Stoffels, W., De Giorgi, M.L., Bouchoule, A. (1994), J. Appl. Phys., vol. 76, p. 148.

Bougaenko, L.T., Kuzmin, M.G., Polak, L.S. (1993), High Energy Chemistry, Ellis Horwood Series in Physical Chemistry, Horwood Publ., UK.

Boulanov, D., Romanets, A., Saveliev, A.V., Kennedy, L.A. (2003), Plasma Catalytic Coal Conversion, University of Illinois at Chicago, COE, Chicago.

Boulos, M.I. (1985), Pure and Applied Chemistry, vol. 57, p. 1321.

Boulos, M. I. (1992), Journal of Thermal Spray Technology, vol. 1, no. 1, p. 33.

Boulos, M. I. (1999), 2nd United Thermal Spray Conference, Düsseldorf, Germany.

Boulos, M.I., Fauchais, P., Vardelle, A., Pfender, E. (1993), Plasma Spraying: Theory and Applications, Suryanarayanan (ed.), World Scientific, Singapore.

Boulos, M.I., Fauchais, P., Pfender, E. (1994), Thermal Plasmas. Fundamentals and Applications, Plenum Press, New York, London.

Boyd, T.J.M., Sanderson J.J. (2003), Physics of Plasmas, Cambridge University Press.

Bozhenkov, S.A., Starikovskaia, S.M., Starikovskii, A.Yu. (2003), Combustion and Flames, vol. 133, p. 133.

Bradley, A. (1970), Ind. and Eng. Chem. Prod. Research and Development, vol. 9, p. 101.

Bragin, V.E., Matukhin, V.D. (1979), 3rd USSR Symposium on Plasma Chemistry, Nauka (Science), Moscow, vol. 2, p. 184.

Braginsky, S.I. (1958), Sov. Phys., Journal of Theoretical and Experimental Physics, vol. 34, p. 1548.

Braginsky, S.I. (1963), Problems of Plasma Theory, Leontovich, M.A. (ed.), Atomizdat, Moscow, vol. 1, p. 183.

Braun, D., Pietsch, G. (1991), 10th International Symposium on Plasma Chemistry, (ISPC-10), vol. 1, p. 6, Bochum, Germany.

Braun, D., Kuchler, U., Pietsch, G. (1988), Pure and Appl. Chem., vol. 60, p. 741.

Braun, D., Kuchler, U., Pietsch, G. (1991), J. Phys. D: Appl. Phys., vol. 20, p. 564.

Braun, D., Gibalov, V., Pietsch, G. (1992), Plasma Sources Sci. Technol., vol. 1, p. 166.

Brau, C.A. (1972), Physica, vol. 58, p. 533.

Bray, C.A., Jonkman, R.M. (1970), J. Chem. Phys., vol. 52, p. 477.

Bray, K.N.C. (1968), J. Phys. B: Atomic and Molecular Physics, vol. 1, p. 705.

Brewer, A.K., Westhaver, J.W. (1930), J. Phys. Chem., vol. 34, p. 1280.

Brichkin, D.I., Kuranov, A.L., Sheikin, E.G. (1999), AIAA paper 1999-4969.

Brichkin, D.I., Kuranov, A.L., Sheikin, E.G. (2001), AIAA paper 2001-0381.

Briner, E., Hoefer, H. (1940), Helv. Chim. Acta, vol. 23, p. 800.

Briner, E., Ricca, M. (1955a), Compt. Rendu, vol. 240, p. 2470.

Briner, E., Ricca, M. (1955b), Helv. Chim. Acta, vol. 38, p. 329.

Broadwater, W.T. (1973), Applied Microbiology, vol. 26, p. 391.

Bromberg, L., Cohn, D.R., Koch, M., Patrick, R.M. (1993), Phys. Lett., A, vol. 173, p. 293.

Bromberg, L., Cohn, D.R., Rabinovich, A., Alexeev, N. (1999a), International Journal of Hydrogen Energy, vol. 24, p. 1131.

921

References

Bromberg, L., Cohn, D.R., Rabinovich, A., Surma, J.E., Virben, J. (1999b), International Journal of Hydrogen Energy, vol. 24, p. 341.

Bromberg, L., Cohn, D.R., Rabinovich, A., Alexeev, N., Samochin, A., Ramprasad, R., Tamhankar, S. (2000), International Journal of Hydrogen Energy, vol. 25, p. 1157.

Bromberg, L., Cohn, D.R., Rabinovich, A., Heywood, J. (2001), International Journal of Hydrogen Energy, vol. 26, p. 1115.

Bromberg, L., Crane, S., Rabinovich, A., Kong, Y., Cohn, D.R., Heywood, J., Alexeev, N., Samochin, A. (2003), Diesel Engine Emissions Reduction (DEER) Workshop, Newport, RI.

Bromberg, L., Cohn, D.R., Hadidi, K., Heywood, J., Rabinovich, A. (2004), Diesel Engine Emission Reduction (DEER) Workshop, Coronado, CA.

Bromberg, L., Hadidi, K., Cohn, D.R. (2005a), Experimental Investigation of Plasma Assisted Reforming of Propane, Massachusetts Institute of Technology, Plasma Science and Fusion Center, Publ. PSFC/JA-05-15, Cambridge, MA.

Bromberg, L., Hadidi, K., Cohn, D.R. (2005b), Plasmatron Reformation of Renewable Fuels, Massachusetts Institute of Technology, Plasma Science and Fusion Center, Publ. PSFC/JA-05-3, Cambridge, MA.

Bronfine, B. (1970), Application of Plasma in Chemical Processes, Mir (World), Moscow, p. 182.

Bronin, S.Ya., Kolobov, V.M. (1983), Sov. Phys., Plasma Physics (Fizika Plazmy), vol. 9, p. 1088.

Brovkin, V.G., Kolesnichenko, Yu.F. (1995), J. Moscow Phys. Soc., vol. 5, p. 23.

Brown, I.G. (1989), The Physics and Technology of Ion Sources, Wiley, New York.

Brown, L.F. (2001), International Journal of Hydrogen Energy, vol. 26, p. 381.

Brown, M.S., Forlines, R.A., Ganguly, B.N. (2006), "Dynamics of Hydrocarbon-Based Pulsed DC Discharge for Ignition Application," 44th AIAA Aerospace Science Meeting and Exhibit, Reno, NV, AIAA paper 2006-611.

Brozek, V., Hrabovsky, M., Kopecky, V. (1997), 13th Int. Symposium on Plasma Chemistry (ISPC-13), Beijing, vol. 4, p. 1735.

Buckner, N., Schrickerb, B. (1960), Kunststoffe, vol. 50, p. 156.

Budko, D.Ya., Emetz, Yu.P., Repa, I.I. (1981), Sov. Phys., Thermal Physics of High Temperatures (Teplophysica Vysokikh Temperatur), vol. 19, p. 697.

Bugaev, V.A., Kukhta, A.V. (1979), 5th USSR Conference on Physics and Technology of Low Temperature Plasma, Kiev, vol. 1, p. 114.

Bullett, N.A., Bullett, D.P., Truica-Marasecu, F., Lerouge, S., Mwale, F., Wertheimer, M.R. (2004), Appl. Surf. Science, vol. 235, p. 395.

Bundy, M., Hammins, A., Lee, K.Y. (2003), Combustion and Flame, vol. 133, p. 299.

Burakhanov, B., Likhachev, A., Medin, S., Novikov, V., Okunev, V., Rickman, V. (2000), AIAA paper 2000-0614.

Burelson, G.R. (1975), Applied Microbiology, vol. 29, p. 340.

Burleson, J.C., Jates, W.F. (1967a), U.S. Patent no. 3,475,308.

Burleson, J.C., Jates, W.F. (1967b), U.S. Patent no. 3,497,436.

Burov, I.S., Mosse, A.L., Pechkovsky, V.V., Chelnokov, A.A. (1970), "Plasma-Chemical Process of P_2O_5 Production from Natural Phosphates," in Physics, Engineering and Applications of Low-Temperature Plasma, Nauka (Science), Alma-Ata, Kazakhstan.

Burov, I.S., Vilk, Yu.N., Ershov, V.A., Mosse, A.L., Pechkovsky, V.V., Teterevkov, A.I. (1973), Sov. Phys., High Energy Chemistry (Khimia Vysokikh Energij), vol. 6, p. 1.

Busby, R.L. (2005), Hydrogen and Fuel Cells: A Comprehensive Guide, Pennwell Books.

Bussmann, M., Mostaghimi, J., Chandra, S. (1999), Phys. Fluids, vol. 11, p. 1406.

Butterbaugh, J.W., Baston, L.D., Sawin, H.H. (1990), Journal of Vacuum Science and Technology, vol. A8, no. 2, p. 916.

Butterbaugh, J.W., Gray, D.C., Sawin, H.H. (1991), Journal of Vacuum Science and Technology, vol. B9, p. 3.

Butylkin, Yu.P., Polak, L.S., Slovetsky (1978), Sov. Phys., High Energy Chemistry (Khimia Vysokikh Energij), vol. 12, p. 526.

Butylkin, Yu.P., Grinenko, A.A., Ermin, Yu.S. (1979), Sov. Phys., High Energy Chemistry (Khimia Vysokikh Energij), vol. 13, p. 545.

Butylkin, Yu.P., Givotov, V.K., Krasheninnikov, E.G., Rusanov, V.D., Fridman, A. (1981), Sov. Phys., Journal of Technical Physics, vol. 51, p. 925.

Cabannes, F., Chapelle, J. (1971), Reactions under Plasma Conditions, vol. 1, Wiley Interscience, New York.

Cabot Corp. (1967), U.S. Patent no. 1,202,587.

Cacciatore, M., Capitelli, M. (1981), Chem. Phys., vol. 55, p. 67.

Cacciatore, M., Capitelli, M., De Benedictis, S., Dilonardo, M., Gorse, C. (1986), Non-Equilibrium Vibrational Kinetics, Topics in Current Physics, vol. 39, Capitelli, M., (ed.), Springer-Verlag, Berlin.

Cale, T.S., Raupp, G.B., Gandy, T.H. (1992), Journal of Vacuum Science and Technology, vol. A10, p. 1128.

Calderazzo, F. (1970), Organic Synthesis via Metal Carbonyls, Mir, Moscow.

Calvet, J.G. (1985), SO_2, NO and NO_2 Oxidation Mechanism: Atmospheric Consideration, vol. 3, Butterworths.

References

Camacho, S.L. (1985), 7th Int. Symposium on Plasma Chemistry (ISPC-7), Einhoven, Netherlands, no. B-1-2, p. 188.

Campbell, C.A., Snyder, F., Szarko, V., Yelk, E., Zanolini, J., Cho, Y., Gutsol, A., Fridman, A. (2006), "Water Treatment Using Plasma Technology," SD Report, Drexel University, Philadelphia, PA.

Campbell, C.S., Egolfopoulos, F.N. (2005), Combustion Science and Technology, vol. 177, p. 2275.

Candler, G.V., Kelley, J.D., Macheret, S.O., Shneider M.N., Adamovich, I. (2002), AIAA Journal, vol. 40, p. 1803.

Capdeville, H., Guillot, Ph., Pitchford, L.C., Galy, J., Brunet, H. (1999), XXIVth Int. Conf. on Phenomena in Ionized Gases (ICPIG-24), vol. 3, p. 5.

Capezzutto, P., Cramarossa, F., D'Agostino, R. (1975), J. Chm. Phys., vol. 79, p. 1487.

Capezzutto, P., Cramarossa, F., D'Agostino, R. (1976), J. Chem. Phys., vol. 80, p. 882.

Capitelli, M. (1986), Non-Equilibrium Vibrational Kinetics, Topics in Current Physics 39, Springer-Verlag.

Capitelli, M. (2000), Plasma Kinetics in Atmospheric Gases, Springer Series on Atomic, Optical and Plasma Physics, vol. 31, Springer-Verlag.

Capitelli, M. (2001), 25th International Conference on Phenomena in Ionized Gases (ICPIG-25), Nagoya, Japan, vol. 4, p. 1.

Capitelli, M., Dilonardo, M. (1977), Chem. Phys., vol. 20, p. 417.

Carl, D.A., Hess, D.W., Lieberman, M.A., Nguen, T.D., Gronsky, R. (1991), J. Appl. Phys., vol. 70, p. 3301.

Carleton, H.E., Oxley, J.H., Hall, E.E., Blochler, J.M. (1970), 2nd Int. Conf. on Vapor Deposition, Blochler, J.M., and Withers, I.C. (eds.), Electrochemical Society, New York.

Carr, A.K. (1997), "Increase of the Surface Energy of Metal and Polymeric Surfaces Using the One Atmosphere Uniform Flow Discharge Plasma," MS Thesis, University of Tennessee.

Case, L.E. (1967), "Arc Torch Reduction of Metal Halides," U.S. Patent no. 3,320,145.

Castro, R.G. (1992), Materials Science and Engineering, vol. A155, p. 101.

Castro, R.G., Kung, H., Stanek, P.W. (1994), Materials Science and Engineering, vol. A185, p. 65.

Catherine, Y. (1991), "Diamond and Diamond-like Carbon Thin Films," NATO ASI, vol. 266, p. 193.

Chae, J.O., Desiaterik, Yu.N., Amirov, R.H. (1996), Int. Workshop on Plasma Technologies for Pollution Control and Waste Treatment, MIT, Cambridge, MA.

Chakravartty, S.H., Dutta, D., Lahiri, A. (1976), "Reaction of Coals under Plasma Conditions: Direct Production of Acetylene from Coal," Energy, vol. 56, p. 43.

Chan, C.M., Ko, T.M., Hiraoka, H. (1996), Surf. Sci. Rep., vol. 24, p. 3.

Chan, J.K. (1996), Surf. Sci. Rep., vol. 24, no. 1/2, p. 1.

Chang, B.S. (2003), J. Electrostat., vol. 57, p. 313.

Chase, J.D. (1974), "Method of Making Sub-Micron Particles of Metal Carbides of Controlled Size," American Cyanamid Corp., U.S. Patent no. 3,839,542.

Chase M.W. (1998), NIST-JANAF Themochemical Tables, 4th Edition, J. Phys. Chem. Ref. Data, Monograph 9.

Chen, F.F. (1991), Plasma Phys. and Controlled Fusion, vol. 33, p. 339.

Chen, F.F. (2006), Introduction to Plasma Physics and Controlled Fusion: Plasma Physics, 2nd Edition, Plenum Press, New York.

Chen F.F., Chang, J.P. (2003), Lecture Note on Principles of Plasma Processing, Springer-Verlag, Berlin, New York.

Chen, H. C., Duan, Z., Heberlein J. V. R., Pfender, E. (1996), 9th National Thermal Spray Conf., ASM Int'l, Cincinnati, OH, p. 553.

Chernekhovskaia, N.E., Shishlo, V.K., Svistunov, B.D., Svistunova, A.S. (2004), IVth Army Medical Conference on Intensive Therapy and Profilactic Treatments of Surgical Infections, Moscow, Russia.

Chernyi, G.G., Losev, S.A., Macheret, S.O., Potapkin, B.V. (2002), "Physical and Chemical Processes and Gas Dynamics: Cross- Sections and Rate Constants," Progress in Astronautics and Aeronautics, vol. 196, American Institute of Aeronautics & Astronautics, Reston, VA.

Chesnokov, E.N., Panfilov, V.N. (1981), J. Theoret. Exper. Chem., vol. 17, p. 699.

Chesnokova, N.V., Gundorova, R.A., Krvasha, O.I., Bakov, V.P., Davydova, N.G. (2003), News of Russian Academy of Medical Sciences, vol. 5, p. 40.

Cheung, N.W. (1991), Nuclear Instrum. Methods, vol. 55, p. 811.

Ching, W. K., Colussi, A.J., Sun, H.J., Nealsonm, K.H., Hoffmann, M.R. (2001), Environ. Sci. Technol., vol. 35, p. 4139.

Ching, W.K., Colussi, A.J., Hoffmann, M.R. (2003), Environmental Science & Technology, vol. 37, p. 4901.

Chintala, N., Meyer, R., Hicks, A., Bao, A., Rich, J.W., Lempert, W.R., Adamovich, I.V. (2005), J. Propul. Power, vol. 21, p. 583.

Chintala, N., Bao, A., Lou, G., Adamovich, I.V. (2006), Combustion and Flame, vol. 144, p. 744.

Chirokov, A. (2005), "Stability of Atmospheric Pressure Glow Discharges," Ph.D. Dissertation, Drexel University, Philadelphia, PA.

Chirokov, A., Gutsol, A., Fridman, A., Sieber, K.D., Grace, J.M., Robinson, K.S. (2004), Plasma Sources Sci. Technol., vol. 13, p. 623.

References

Chirokov, A., Gutsol, A., Fridman, A., Sieber, K.D., Grace, J.M., Robinson, K.S. (2006), Plasma Chemistry and Plasma Processing, vol. 26, p. 127.

Chizhikov, D.M., Tsvetkov, Yu.V., Tagirov, I.K. (1970), "Special Features of Reduction Processes in Thermal Plasma Conditions," in Mechanism and Kinetics of Reduction of Metals, Samarin, A.M. (ed.), Nauka (Science), Moscow.

Chizhikov, D.M., Deyneka, S.S., Tribunsky, L.M., Gorunkov, V.V. (1976), "Application of Plasma Processes in Technology of Production of Powder Materials," in Non-Ferrous Metallurgy, Nauka (Science), Moscow, p. 126.

Cho, M., Chung, H., Choi, W., Yoon, J. (2004), Water Research, vol. 38, p. 1069.

Choi, S., Kushner, M. (1993), J. Appl. Phys., vol. 74, p. 853.

Chomiak, J. (1979), 17th Symposium on Combustion, Pittsburgh, PA, p. 255.

Chraska, P., Hrabovsky, M. (1992), 13th Int. Thermal Spray Conf., ASM Int., Orlando, FL, p. 81.

Chraska, P., Dubsky, J., Kolman, B., Ilavsky, J., Forman, J. (1992), Journal of Thermal Spray Technology, vol. 1, no. 4, p. 301.

Christman, K., Schaber, O., Ertl, G., Neuman, M. (1974), J. Chem. Phys., vol. 60, p. 4528.

Christophersen, A.G., Jun, H., Jørgensen, K., Skibsted, L.H. (1991), Z. Lebensm. Unters. Forsch., vol. 192, p. 433.

Chu, P.K., Chen, J.Y., Wang, L.P., Huang, N. (2002), Mater. Sci. Eng., vol. R36, p. 143.

Chyin, V.I. (2002), Condensed Matter Physics, vol. 5, p. 429.

Clark, D.T., Peeling, J.J. (1976), Journal Pol. Sci.–Pol. Chem., vol. 14, p. 543.

Clark, J. (1970), Application of Plasma in Chemical Processes, Polak L.S. (ed.), Mir (World), Moscow, p. 152.

Clothiaux, E.J., Koropchak, J.A., Moore, R.R. (1984), Plasma Chemistry and Plasma Processing, vol. 4, p. 15.

Cobine, J.D. (1958), Gaseous Conductors, Dover, New York.

Coburn, J.W., Winters, H.F. (1979), J. Vac. Sci. Technol., vol. 16, p. 391.

Conrad, J.R. (1987), J. Appl. Phys., vol. 62, p. 777.

Conrad, J.R. (1988), "Method and Apparatus for Plasma Source Ion Implantation," U.S. Patent no. 4,764,394.

Conrad, J.R., Dodd, R.A., Han, S., Madapura, M., Scheuer, J., Sridharam, K., Worzala, F.J. (1990), Journal of Vacuum Science and Technology, vol. A8, p. 3146.

Conti, S., Fridman, A., Raoux, S. (1999), 14th International Symposium on Plasma Chemistry (ISPC-14), vol. 5, p. 2581, Prague, Czech Republic.

Conti, S., Porshnev, P.I., Fridman, A., Kennedy, L.A., Grace, J.M., Sieber, K.D., Freeman, D.R., Robinson, K.S. (2001), Experimental Thermal and Fluid Science, vol. 24, p. 79.

Conway, N.M.J., Ferrari, A.C., Flewitt, A.J., Robertson, J., Milne, W.I., Tagliaferro, A., Beyer, W. (2000), Diamond Related Materials, vol. 9, p. 765.

Cook, J.M., Ibbotson, D.E., Foo, P.D., Flamm, D.L. (1990), Journal of Vacuum Science and Technology, vol. A8, p. 1820.

Coppinger, E.A., Johnson, B.M. (1969), "Production of Uranium and Plutonium Carbides and Nitrides," U.S. Patent no. 3,697,436.

Cormier, J.-M., Richard, F., Chapelle, J., Dudemaine, M. (1993), Proceedings of the 2nd International Conference on Elect. Contacts, Arcs, Apparatus and Applications, p. 40.

Corvin, K.K., Corrigan, S.J. (1969), J. Chem. Phys., vol. 50, p. 2570.

Coulibaly, K., Genet, F., Renou-Gonnord, M.F., Amouroux, J. (1997), 13th Int. Symposium on Plasma Chemistry (ISPC-13), Beijing, vol. 4, p. 1721.

Cromwell, W.E., Manley, T.C. (1959), Adv. Chem. Ser., vol. 21, p. 304.

Croquesel, E., Gherardi, N., Martin, S., Massines, F. (2000), HAKONE-VII, 7th Int. Symp. on High Pressure Low Temperature Plasma Chem., vol. 1, p. 88.

Crowford, O.H. (1967), Mol. Phys., vol. 13, p. 181.

Crutzen, P.J. (1970), Quart. J. R. Met. Soc., vol. 96, p. 320.

Cuelleron, G., Crusiat, B. (1974), "High Purity Boron Production by Hydrogen Reduction of Boron in Inductive Plasma Discharge," in Boron: Production, Structure and Properties, 4th Int. Boron Symposium, Nauka (Science), Moscow.

Czernichowski, A. (1994), Pure and Applied Chemistry, vol. 66, no. 6, p. 1301.

Czernichowski, A. (2001), Oil & Gas Science and Technology, Revue de l'IFP, vol. 56, no. 2, p. 181.

Czernichowski, A., Wesolovska, K. (2003), First Int. Conf. on Fuel Cell Science, Engineering and Technology, Rochester, New York, p. 181.

Czernichowski, A., Opalinska, T., Czernichowski, P., Lesueur, H. (1994), French Patent no. 94.10738.

Czernichowski, A., Czernichowski, M., Czernichowski, P. (2002), France-Deutschland Fuel Cell Conference: Materials, Engineering, Systems, Applications, Forbach, France, p. 322.

Czernichowski, A., Czernichowski, M., Wesolowska, K. (2003a), "Glidarc-Assisted Production of Synthesis Gas from Rapeseed Oil," 1st European Hydrogen Energy Conference, Grenoble, France, Publ. CP1, p. 67.

Czernichowski, A., Czernichowski, M., Wesolowska, K. (2003b), 1st Int. Conference on Fuel Cell Science, Engineering and Technology, Rochester, New York, p. 175.

References

Czernichowski, A., Czernichowski, M., Wesolowska, K. (2004), 2nd Int. Conference on Fuel Cell Science, Engineering and Technology, Rochester, New York, p. 75.

D'Agostino, R. (1990), Plasma Deposition and Etching of Polymers, Academic Press, Boston.

D'Agostino, R., Favia, P., Oehr, C., Wertheimer, M.R. (2003), Plasma Processes and Polymers, Wiley-VCH, Berlin.

Dai, X.J., Kviz, L., Denning, R.J. (2005), 17th International Symposium on Plasma Chemistry (ISPC-17), Toronto, Canada, p. 644.

Daith, F.S., Marcio, J.A. (1967), "Method of Contacting Slag with a Reducing Arc Atmosphere to Reduce the Metal Oxides Contained Therein," U.S. Patent no. 3,347,766.

Dalaine, V., Cormier, J.-M., Pellerin, S., Lefaucheux, P. (1998), J. Appl. Phys., vol. 84, p. 1215.

Dale, W. (1974), U.S. Patent no. 3,848,068.

D'Amico, K.M., Smith, A.L.C. (1977), J. Phys. D: Appl. Phys., vol. 10, p. 261.

Daniel, J., Jacob, J. (1986), J. Geophys. Res., vol. 91, p. 9807.

Darken, L.S., Gurry, R.V. (1960), Physical Chemistry of Metals, Metallurgy, Moscow.

Da Silva, C.F., Ishikawa, T., Santos, S., Alves, C., Martinelli, A.E. (2006), International Journal of Hydrogen Energy, vol. 31, p. 49.

Daugherty, J.E., Porteuos, R.K., Kilgore, M.D., Graves, D.B. (1992), J. Appl. Phys., vol. 72, p. 3934.

Daugherty, J.E., Porteuos, R.K., Kilgore, M.D., Graves, D.B. (1993a), J. Appl. Phys., vol. 73, p. 1619.

Daugherty, J.E., R.K.Porteuos, Kilgore, M.D., Graves, D.B. (1993b), J. Appl. Phys., vol. 73, p. 7195.

Davidson, R.M. (1990), Natural Oxidation of Coal, IEACR/29, International Energy Agency, Coal Research Center, University of Kentucky.

Davis, C.A. (1993), Thin Solid Films, vol. 226, p. 30.

Davis, D., Libby, F.J. (1966), J. Amer. Chem. Soc., vol. 45, p. 4481.

Davydov, A.I., Strijakov, A.N., Pekshev, A.V., Kuchukhidze, S.T., Klihdukhov, I.A. (2002), Problems of Obstetrics and Gynaecology (OB/GYN), vol. 1, no. 2, p. 57.

Davydov, A.I., Kuchukhidze, S.T., Shekhter, A.B., Khanin, A.G., Pekshev, A.V., Pankratov, A.V. (2004), Problems of Gynecology, Obstetrics and Perinatology, vol. 3, no. 4, p. 12.

Dawson, G.A., Winn, W.P. (1965), Zs. fur. Phys., vol. 183, p. 159.

Decher, R. (2006), Direct Energy Conversion: Fundamentals of Electric Power Production, Oxford University Press.

Delcea, L.B., Herrington, D. S. (1995), Why Axial Feed and Why AxiJet, Metcon Thermal Spray, Abbotsford, BC, Canada.

Delluc, G., Mariaux, G., Vardelle, A., Fauchais, P., Pateyron, B. (2003), 16th Int. Symposium on Plasma Chemistry (ISPC-16), D'Agostino, R. (ed.), University of Bari, Italy.

Dembovsky, V. (1985), Plasma Metallurgy: Principles, Elsevier Science.

Deminsky, M.A., Potapkin, B.V., Rusanov, V.D., Fridman, A. (1990), Possibility of SO_2 Chain Oxidation in Heterogeneous Air Stream by Relativistic Electron Beams, Kurchatov Institute of Atomic Energy, vol. IAE-5260/12, Moscow.

Deminsky, M.A., Potapkin, B.V., Cormier, J.-M., Richard, F., Bouchoule, A., Rusanov, V.D. (1997), Sov. Phys., Doklady, vol. 42, p. 337.

Deminsky, M., Givotov, V., Potapkin, B., Rusanov, V. (2002), Pure and Applied Chemistry, vol. 74, p. 413.

Demura, A.V., Macheret, S.O., Fridman, A. (1981), 5th Int. Symposium on Plasma Chemistry (ISPC-5), Edinburgh, vol. 1, p. 52.

Demura, A., Macheret, S.O., Fridman, A. (1984), Sov. Phys. Doklady, vol. 275, p. 603.

Denaro, A.R., Owens, P.A., Crawshaw, A. (1968), Eur. Polymer J., vol. 4, p. 93.

Deriagin, B.V., Fedoseev, D.B. (1977), Growth of Diamond and Graphite from Gas Phase, Nauka (Science), Moscow.

Detomazo, L., Gristina, R., Favia, P., d'Agostino, R. (2005), 17th International Symposium on Plasma Chemistry (ISPC-17), Toronto, Canada, p. 1103.

DeVink, I.A. (1970), Silicates Ind., vol. 35, no. 12, p. 308.

DeVink, I.A. (1971a), Silicates Ind., vol. 36, no. 1, p. 17.

DeVink, I.A. (1971b), Silicates Ind., vol. 36, no. 2, p. 49.

Diana, M., Russo, G., de Marino, S. (1974), "Boron Trichloride Reduction to Boron in H_2 Plasma Reactor," Chimica Industria (Chemical Industry), Italy, vol. 56, no. 8, p. 544.

Dibelius, N.R., Fraser, J.C., Kawahafa, M., Dyle, C.D. (1964), Chem. Eng. and Process., vol. 60, p. 41.

Diskina, P.M., Koval, V.V., Petrishchev, V.A. (1979), 3rd USSR Symposium on Plasma Chemistry, vol. 1., Nauka (Science), Moscow, p. 44.

Djermoune, D., Samson, S., Marode, E., Segur, P. (1995), 2nd Int. Conference on Gas Discharges and Their Applications (GD-95) II-484, vol. 2, Tokyo, p. 484.

Dmitriev, S., Gnedenko, V., Pegaz, D., Suris, A. (2003), 16th Int. Symposium on Plasma Chemistry (ISPC-16), Taormina, Italy, p. 763.

References

Dmitrieva, I.K., Zenevich, V.A. (1983), "Calculation of Rate Coefficients of Atomic Oxygen Oxidation of Nitrogen Molecules at Different Levels of Vibrational Excitationo. Theoretical–Informational Approximation," A.V. Lykov Institute of Heat and Mass Exchange, vol. 10, Minsk, Belarus.

Dmitrieva, I.K., Zenevich, V.A. (1984), Sov. Phys., Chemical Physics (Khimichekaja Fizika), vol. 3, p. 1075.

Dobal, V. (1978), Chemical Technology of Fuels (KhTT), vol. 3, p. 97.

Dobkin, S.V., Son, E.E. (1982), Sov. Phys., Thermal Physics of High Temperatures, vol. 20, p. 1081.

Dobretsov, L.N., Gomounova, M.V. (1966), Emission Electronics, Nauka (Science), Moscow.

Doctor, R.D. (2000), "Hydrogen Production in Thermochemical Calcium-Bromine Water Splitting Cycle," Argonne National Laboratory, Argonne, IL.

Doctor, R.D., Matonis, D.T., Wade, D.C. (2003), "Hydrogen Generation Using a Calcium Bromine Thermochemical Water Splitting Cycle," Symposium on Nuclear Energy and the Hydrogen Economy, MIT, Cambridge, MA.

Doctor, R.D., Matonis, D.T., Wade, D.C. (2004), Nuclear Production of Hydrogen, NEA, no. 5308.

Donelly, I.J., Watterson, P.A. (1989), J. Phys. D., vol. 22, p. 90.

Dong-Yan, W., Xin-Can, W., Xi, C. (2002), Surf. Coat. Technol., vol. 171, p. 149.

Dorofeev, Yu.I., Skurat, V.E. (1978), Sov. Phys., Doklady, Reports of USSR Academy of Sciences, vol. 243, p. 1479.

Dorofeev, Yu.I., Skurat, V.E., Talrose, V.L. (1987), Sov. Phys., Doklady, Reports of USSR Academy of Sciences, vol. 292, p. 654.

Drabkina, S.I. (1951), Sov. Phys., Journal of Theoretical and Experimental Physics, vol. 21, p. 473.

Dragsund, E., Andersen, A.B., Johannessen B.O. (2001), "Ballast Water Treatment by Ozonation," 1st Int. Ballast Water Treatment, R & D Symposium, Global Monograph Series no. 5, IMO, London, p. 21.

Drawin, H.W. (1968), Ztschr., Bd. 211, S. 404.

Drawin, H.W. (1969), Ztschr., Bd. 255, S. 470.

Drawin, H.W. (1972), "The Thermodynamic Properties of the Equilibrium and Non-Equilibrium States of Plasma," in Reactions under Plasma Conditions, Venugopalan, M. (ed.), Wiley, New York.

Dreizin, E.L. (2004), Second International Workshop on Microplasmas, Proceedings, p. 23.

Dresselhaus, M.S., Dresselhaus, G., Eklund, P.C. (1996), Science of Fullerenes and Carbon Nanotubes: Their Properties and Applications, Academic Press.

Dresvin, S.V. (1977), Physics and Technology of Low Temperature Plasmas, Iowa State University Press, Ames, IA.

Dresvin, S.V. (1991), Theory and Simulations of RF Discharges, Energo-Atom-Izdat, Leningrad, Russia.

Dresvin, S.V., Ivanov, D.V. (2004), Mathematical Modeling of RF-Discharges, St. Petersburg State Polytechnical University, St. Petersburg.

Dresvin, S.V., Ivanov, D.V. (2006), Mathematical Modeling of RF-Discharges. Electromagnetic Problems, St. Petersburg State Polytechnical University, St. Petersburg.

Dresvin, S.V., Shi, N.K., Ivanov, D.V. (2006), Mathematical Modeling of RF-Discharges. Plasma Dynamics, St. Petersburg State Polytechnical University, St. Petersburg.

Dubinin, N.P. (1969), Modern Problems of Radiation Genetics, Atom-Izdat, Moscow.

Dubovitsky, F. (1935), Acta Physicochimica, URSS, vol. 2, p. 361.

Dubrovin, V.Yu. (1983), "Ionization Processes and Dissociation of Water Molecule in Low Pressure Plasma," Ph.D. Dissertation, Kurchatov Institute of Atomic Energy, Moscow.

Dubrovin, V.Yu., Maximov, A.I. (1980), Sov. Phys., High Energy Chemistry (Khimia Vysokikh Energij), vol. 14, p. 54.

Dubrovin, V.Yu., Maximov, A.I. (1981), Sov. Phys., News of Higher Education (Izvestia Vuzov), Chemistry and Chemical Technology, vol. 14, p. 252.

Dubrovin, V.Yu., Maximov, A.I. (1982), Sov. Phys., High Energy Chemistry (Khimia Vysokikh Energij), vol. 16, p. 92.

Dubrovin, V.Yu., Maximov, A.I., Rybkin, V.V. (1979), Sov. Phys., News of Higher Education (Izvestia Vuzov), Chemistry and Chemical Technology, vol. 22, p. 1469.

Dubrovin, V.Yu., Maximov, A.I., Menagarishvili, V.M. (1980), Sov. Phys., Plasma Physics (Fizika Plazmy), vol. 6, p. 888.

Dugev, G.A. (2000), "Plasma Thermionic Converters of Thermal Energy into Electricity," in Encyclopedia of Low-Temperature Plasma, Fortov, V.E. (ed.), vol. 4, Nauka (Science), Moscow, p. 203.

Dundas, P.H., Thorpe, M.L. (1974), Chemical Engineering Progr., vol. 66, no. 10, p. 66.

Dussoubs, B. (1998), "3D Modeling of the Plasma Spray Process," Ph.D. Thesis, University of Limoge, France.

Dzantiev, B.G., Ermakov, A.N., Popov, V.N. (1979), Journal of Nuclear Science and Technology, Nuclear-Hydrogen Energy, vol. 1, no. 5, p. 89.

Dzidzoev, M.S., Platonenko, V.T., Khokhlov, R.V. (1970), Sov. Phys., Advances in Physical Sciences (Uspekhi), vol. 100, p. 641.

Ebdon, J.R. (1995), New Methods of Polymer Synthesis, Blackie, Glasgow, UK.

Edelson, D. (1953), Industrial and Chemical Engineering, vol. 45, p. 2094.

References

Efimenko, N.A., Hrupkin, V.I., Marahonich, L.A., Iashenko, V.I., Moskalenko, V.I., Lykianenko, E.V. (2005), Journal of Military Medicine (Voenno-Meditsinskii Jurnal), vol. 5, p. 51.

Einbinder, H. (1957), J. Chem. Phys., vol. 26, p. 948.

Eletsky, A.V. (1997), Adv. in Phys. Sci. (Uspekhi Physicheskikh Nauk), vol. 167, p. 945.

Eletsky, A.V. (2000a), Adv. in Phys. Sci. (Physics-Uspekhi), vol. 43, no. 2, p. 111.

Eletsky, A.V. (2000b), "Production of Fullerenes and Nano-Tubes in Low-Temperature Plasma," in Encyclopedia of Low-Temperature Plasma, Fortov, V.E. (ed.), vol. 4, Nauka (Science), Moscow.

Eletsky, A.V., Smirnov, B.M. (1995), Adv. in Phys. Sci. (Uspekhi Physicheskikh Nauk), vol. 165, p. 977.

Eletsky, A.V., Zaretsky, N.P. (1981), Sov. Phys., Doklady, vol. 260, p. 591.

Eletsky, A.V., Palkina, L.A., Smirnov, B.M. (1975), Transport Phenomena in Weakly Ionized Plasma, Atom-Izdat, Moscow.

Eliasson, B., Kogelschatz, U. (1986), J. Chem. Phys., vol. 83, p. 279.

Eliasson, B., Kogelschatz, U. (1991), IEEE Transactions on Plasma Science, vol. 19, p. 309.

Eliasson, B., Kogelschatz, U., Baessler, P. (1984), J. Phys. B: At. Mol. Opt. Phys., vol. 17, p. 797.

Eliasson, B., Hirth, M., Kogelschatz, U. (1987), J. Phys. D: Applied Physics, vol. 20, p. 1421.

Eloy, R., Parrat, D., Duc, T.M., Legeay, G., Bechetoille, A. (1993), J. Cataract and Refract. Surg., vol. 19, p. 364.

Emelianov, Yu.M. (1971), Chemistry and Physics of Low-Temperature Plasma, Moscow State University, Moscow, p. 255.

Emelianov, Yu.M., Filippov, Yu.V. (1960), Sov. Phys., Journal of Physical Chemistry, vol. 34, p. 2841.

Enduskin, P.N., Polak, L.S., Volodin N.L., Uglev, V.N. (1969), Sov. Phys., High Energy Chemistry (Khimia Vysokikh Energij), vol. 3, p. 202.

Engelgardt, A.G., Felps, A.V., Risk, G.G. (1964), Phys. Rev., vol. 135, p. 1566.

Eremin, E.N., Maltsev, A.N. (1956), Sov. Phys., Journal of Physical Chemistry, vol. 30, p. 1615.

Eremin, E.N., Vasiliev, S.S., Kobosev, N.I. (1936), Sov. Phys., Journal of Physical Chemistry, vol. 18, p. 814.

Esakov, I.I., Grachev, L.P., Khodataev, K.V., Vinogradov, V.A., Van Wie, D.M. (2006), "Efficiency of Propane-Air Mixture Combustion Assisted by Deeply Undercritical MW Discharge in Cold High-Speed Airflow," 44th AIAA Aerospace Science Meeting and Exhibit, Reno, Nevada, AIAA paper 2006-1212.

Essoltani, A., Proulx, P., Boulos, M.I., Gleizes, A. (1991), Thermal Plasma Processing, ASME, vol. 161, p. 145.

Este, G., Westwood, W.D. (1984), J. Vac. Sci. Technol., vol. A6, no. 3, p. 1845.

Este, G., Westwood, W.D. (1988), J. Vac. Sci. Technol., vol. A2, p. 1238.

Evans, A.W., Wynne, W.G., Marinowsky, C.W. (1968), Verfahren zur Herstellung von Feinteiligen Siliciumkarbid, U.K. Patent no. 1283813.

Evans, D., Rosocha, L.A., Anderson, G.K., Coogan, J.J., Kushner, M.J. (1993), J. Appl. Phys., vol. 74, p. 5378.

Evseev, A.V., Eletsky, A.V., Palkina, L.A. (1979), Journal of Nuclear Science and Technology, Nuclear- Hydrogen Energy, vol. 1, no. 5, p. 30.

Ewing, R.A. (2004), Hydrogen: Hot Stuff Cool Science, Pixyjack Press.

Eyring, H., Lin, S.H., Lin, S.M. (1980), Basic Chemical Kinetics, John Wiley & Sons, New York.

Fallon, P.J., Veerasamy, V.S., Davis, C.A., Robertson, J., Amaratunga, G.A.J., Milne, W.I., Koskinen, J. (1993), Phys. Rev. B, vol. 48, p. 4777.

Farhat, S., Hinkov, I., Scott, C.D. (2002), Journal of Nano-Science and Nano-Technology, vol. 13, p. 377.

Farmer, A., Bogdan, J.R. (1969), "Method of Production of Titanium Dioxide," U.S. Patent no. 3,674,383.

Farouk, B., Bera, K. (1999), IEEE International Conference on Plasma Sciences, Monterey, CA, p. 147.

Farouk, B., Nagayama, K., Lee, Y.H. (1995), 12th Int. Symposium on Plasma Chemistry (ISPC-12), vol. 4, p. 2247, Minneapolis, MN.

Farouk, B., Nagayama, K., Lee, Y.H. (1996), IEEE International Conference on Plasma Sciences, p. 183, Boston.

Farouk, B., Nagayama, K., Lee, Y.H. (1997), Int. Conference on Fluid Engineering, Vol. II, JSME Centennial Grand Congress, Tokyo, p. 977.

Farouk, B., Bera, K., Yi, W.J., Lee, Y.H. (1998), IEEE International Conference on Plasma Sciences, Raleigh, NC, p. 268.

Farouk, T., Farouk, B., Staack, D., Gutsol, A., Fridman, A. (2006), Plasma Sources Science and Technology, vol. 15, p. 676.

Fauchais, P., (ed.) (1997), Progress in Plasma Processing of Materials, Athens, Begell House Publishers.

Fauchais, P. (2004), J. Phys. D: Appl. Phys., vol. 37, p. 86.

Faushais, P., Vardelle, A., Dusoubs, B. (2001), J. Thermal Spray Technol., vol. 10, p. 44.

Fehsenfeld, F.C. (1974), J. Chem. Phys., vol. 61, p. 3181.

Fehsenfeld, F.C., Evenson, K.M., Broida, H.C. (1965), Rev. Scientific Instrum., vol. 36, p. 294.

Feldman L.C., Mayer, J.W. (1986), Fundamentals of Surface and Thin Film Analysis, North-Holland, New York.

Felten, A., Bittencourt, C., Azioune, A., Pireau, J.J. (2005), 17th Int. Symposium on Plasma Chemistry (ISPC-17), p. 780, Toronto.

Fendler, E., Henne, R., Lang, M. (1995), 8th National Thermal Spray Conf., Houston, TX, ASM Int'l, p. 533.

References

Fey, M.G., Kemeny, G.A. (1973), Method of Direct Ore Reduction Using a Short Gap Arc Heater," U.S. Patent no. 3,765,870.

Feynman, R.P., Leighton, R.B., Sands, M. (2005), Feynman Lectures on Physics, 2nd Edition, Addison Wesley Publishing.

Filippov, Yu.V., Popovich, M.P. (1971), Chemistry and Physics of Low-Temperature Plasma, p. 271, Moscow State University, Moscow.

Filippov, Yu.V., Vendillo, V.P. (1961), Sov. Phys., Journal of Physical Chemistry, vol. 35, p. 624.

Filippov, Yu.V., Emelianov, Yu.M., Semiokhin, I.A. (1968), Modern Problems of Physical Chemistry, vol. 2, p. 76, Moscow State University, Moscow.

Fine, D.H., Slater, S.M., Sarofim, A.F., Williams, G.C. (1974), Fuel, vol. 53, p. 120.

Finkelburg, W., Maecker, H. (1956), Elektrische Bogen und Thermisches Plasma, Handbuch der Physik, vol. 22, S. 254.

Firestone, R.F. (1957), J. Am. Chem. Soc., vol. 79, p. 5593.

Fisher, F., Peters, K. (1931), Brennst.-Chem., vol. 12, p. 268.

Fix, A., Wirth, M., Meister, A., Efret, G., Pesch, M., Weidauer, D. (2002), Appl. Phys. B, vol. 75, p. 153.

Flamm, D.L. (1989), Plasma Etching: An Introduction, Manos, D.M., and Flamm, D.L., (eds.), Academic, New York.

Flamm, D.L. (1990), "Mechanisms of Silicon Etching in Fluorine- and Chlorine Containing Plasmas," Report UCB/ERL M90/41, College of Engineering, University of California, Berkley.

Flamm, D.L., Herb, G.K. (1989), Plasma Etching: An Introduction, Manos, D.M., and Flamm, D.L. (eds.), Academic, New York.

Fleischer, K., Wielage, B., Müller, M., Henne, R., Borck, V. (1999), 2nd United Thermal Spray Conference, Düsseldorf, Germany, p. 608.

Foest, R., Schmidt, M., Becker, K. (2006), Int. J. Mass Spectrom., vol. 248, p. 87.

Foglein, K.A., Szabo, P.T., Dombi, A., Szepvolgyi, J. (2003), Plasma Chemistry and Plasma Processing, vol. 23, p. 4.

Fomin, V.M., Cherepanov, A.N. (1995), Plasma-Chemical Synthesis of Ultrafine Powders and Their Application for Modification of Metals and Alloys, Low-Temperature Plasma Series, Gukov, M.F. (ed.), vol. 12, Nauka (Science), Novosibirsk, Russia.

Fomin, V.M., Maslov, A.A., Malmuth, N.D., Fomichev, V.P., Shashkin, A.P., Korotaeva, T.A., Shipyiuk, A.M., Pozdnyakov, G.A. (2002), AIAA J., vol. 40, p. 1170.

Fominsky, L.P. (1978), Sov. Phys., Journal of Technical Physics, vol. 48, p. 436.

Forrester, A.T. (1988), Large Ion Beams: Fundamentals of Generation and Propagation, John Wiley, New York.

Fortov, V.E., Yakubov, I.T. (1994), Non-Ideal Plasma, Energo-Atom-Izdat, Moscow.

Fotache, C.G., Kreutz, T.G., Zhu, D.L., Law, C.K. (1995), Combust. Sci. Technol., vol. 109, p. 373.

Fraishtadt, V.L., Kuranov, A.L., Sheikin, E.G. (1998), J. Techn. Phys., vol. 43, p. 1309.

Frank-Kamenetsky, D.A. (1971), Diffusion and Heat Transfer in Chemical Kinetics, Nauka (Science), Moscow.

Frank-Kamenetsky, M.D., Lukashin, A.V. (1975), Adv. Phys. Sci. (Uspekhi Physicheskikh Nauk), vol. 116, p. 193.

Fraser, M.E., Sheinson, R.S. (1985), Plasma Chem. Plasma Process., vol. 6, p. 27.

Fraser, M.E., Fee, D.A., Sheinson, R.S. (1985), Plasma Chem. Plasma Process., vol. 5, p. 163.

Frenkel, Ya.I. (1945), Kinetic Theory of Liquids, Academy of Sciences of USSR, Moscow.

Frenkel, Ya.I. (1975), Kinetic Theory of Gases and Liquids, Nauka (Science), Leningrad.

Freton, P., Gonzalez, J.J., Gleizes, A. (2000), J. Phys. D: Appl. Phys., vol. 33, p. 2442.

Fridman, A. (1976a), Ionization Processes in Heterogeneous Media, Moscow Institute of Physics and Technology, Moscow.

Fridman, A. (1976b), "Water Vapor Decomposition in Non-Self-Sustained Atmospheric Pressure Glow Discharge Supported by High-Current Relativistic Electron Beam," M.S. Dissertation, Moscow Institute of Physics and Technology, Moscow.

Fridman, A. (1986), "Physical Kinetics of Chemically Active Plasma," Dr. Sci. Dissertation, Kurchatov Institute of Atomic Energy, Moscow.

Fridman, A. (2003), 16th Int. Symposium on Plasma Chemistry (ISPC-16), Taormina, Italy, p. 8.

Fridman, A. (2007), 1st Int. Conf. on Plasma Medicine (ICPM-1), Corpus Christi, TX.

Fridman, A., Kennedy, L.A. (2004), Plasma Physics and Engineering, Taylor & Francis, New York, London.

Fridman, A., Rusanov, V.D. (1994), Pure Appl. Chem., vol. 66, no. 6, p. 1267.

Fridman, A., Czernichowski, A., Chapelle, J., Cormier, J.-M., Lesueur, H., Stevefelt, J. (1993), 11th Int. Symposium on Plasma Chemistry (ISPC-11), Loughborough, UK.

Fridman, A., Czernichowski, A., Chapelle, J., Cormier, J.-M., Lesuer, H., Stevefelt, J. (1994), J. Physique, p. 1449.

Fridman, A., Potapkin, B.V., Strelkova, M.I., Czernichowski, A., Charamel, A., Gorius, A. (1995), 12th Int. Symposium on Plasma Chemistry (ISPC-12), Minneapolis, MN, vol. 2, p. 683.

References

Fridman, A., Boufendi, L., Bouchoule, A., Hbid, T., Potapkin, B. (1996), J. Appl. Phys., vol. 79, p. 1303.

Fridman, A., Nester, S., Yardimci, O., Saveliev, A., Kennedy, L. (1997), 13th Int. Symposium on Plasma Chemistry (ISPC-13), Beijing.

Fridman, A., Nester, S., Kennedy, L., Saveliev, A., Mutaf-Yardimci, O. (1999), Progress in Energy and Combustion Science, vol. 25, p. 211.

Fridman, A., Chirokov, A., Gutsol, A. (2005), J. Physics D: Appl. Phys., vol. 38, pp. 1–24.

Fridman, A., Gutsol, A., Dolgopolsky, A., Stessel, E. (2006), Energy Fuels, vol. 20, p. 1242.

Fridman, A., Gutsol, A., Cho, Y. (2007a), Advance in Heat Transfer, Fridman, A., and Cho, Y. (eds.), vol. 40.

Fridman, A., Gutsol, A., Gangoli, S., Ju, Y., Ombrello, T. (2007b), J. Propul. Power, Macheret, S. (ed.), vol. 22.

Fridman, G., Friedman, G., Gutsol, A., Fridman, A. (2003), 16th Int. Symposium on Plasma Chemistry (ISPC-16), p. 703, Taormina, Italy.

Fridman, G., Li, M., Lelkes, P.I., Friedman, G., Fridman, A., Gutsol, A. (2005a), IEEE Trans. Plasma Sci., vol. 33, p. 1061.

Fridman, G., Peddinghaus, L., Fridman, A., Balasubramanian, M., Gutsol, A., Friedman, G. (2005b), 17th Int. Symposium on Plasma Chemistry (ISPC-17), p. 1066, Toronto, Canada.

Fridman, G., Peddinghaus, L., Ayan, H., Fridman, A., Balasubramanian, M., Gutsol, A., Brooks, A., Friedman, G. (2006), Plasma Chemistry and Plasma Processing, vol. 26, p. 425.

Fridman, G., Brooks, A., Balasubramanian, M., Fridman, A., Gutsol, A., Vasilets, V., Halim Ayan, Friedman, G. (2007a), Plasma Processing and Polymers, vol. 4.

Fridman, G., Friedman, G., Gutsol, A., Fridman, A. (2007b), 18th Int. Symposium on Plasma Chemistry (ISPC-18), Kyoto, Japan.

Fridman, G., Shereshevsky, A., Jost, M.M., Brooks, A., Fridman, A., Gutsol, A., Vasilets, V., Friedman, G. (2007c), Plasma Chemistry and Plasma Processing, vol. 27, p. 163.

Friedlander, S.K. (2000), "Smoke, Dust and Haze: Fundamentals of Aerosol Dynamics," Topics in Chemical Engineering, Oxford University Press.

Frolova, I.B., Minkin, V.M., Frolov, V.A. (1976), "Treatment of Dispersed Oxides of Iron, Chromium and Nickel by Reducing Gases Heated Up in Electric Arc Discharge," in Plasma Processes in Metallurgy and Technology of Inorganic Materials, A.A. Baikov Institute of Metallurgy of USSR Academy of Sciences, Nauka (Science), Moscow.

Fu, Y.C. (1971), "Gasification of Fossil Fuels in Microwave Discharge in Argon," Chemistry and Industry, no. 31, p. 896.

Fu, Y.C., Blaustein, B.D. (1968), Fuel, vol. 47, p. 463.

Fu, Y.C., Blaustein, B.D. (1969), Ind. Eng. Chem. Process Des. Dev., vol. 8, p. 257.

Fu, Y.C., Blaustein, B.D., Wender, I. (1971), Chem. Eng. Progr. Symposium Ser., vol. 67, no. 111, p. 47.

Fu, Y.C., Blaustein, B.D., Sharkey, A.G. (1972), Fuel, vol. 51, p. 308.

Funahashi, H. (1953), Journal of Society of Organic Synthetic Chemistry of Japan (Yuki Gosei Kagaku), vol. 11, p. 262.

Funke, V.F., Klementiev, A.A., Kosukhin, V.V. (1969), Sov. Phys., Powder Metallurgy (Poroshkovaja Metallurgija), vol. 12, no. 33, p. 39.

Futamura, S., Yamamoto, T. (1996), IEEE Trans. Ind. Appl., vol. 32, p. 1969.

Gadri, R.B. (2000), Surf. Coat. Technol., vol. 131, p. 528.

Gahl T., Markl, H. (1996), Appl. Microbiol. Biotechnol., vol. 45, p. 148.

Galkin, N.P. (1961), Chemistry and Technology of Fluorine-Compounds of Uranium, Gos-Atom-Izdat, Moscow.

Gallagher, M.J., Friedman, G., Dolgopolsky, A., Gutsol, A., Fridman, A. (2005), 17th Int. Symposium on Plasma Chemistry (ISPC-17), p. 1056, Toronto, Canada.

Galley, D., Pilla, G., Lacoste, D., Ducruix, S., Lacas, F., Veynante, D., Laux, C.O. (2005), "Plasma-Enhanced Combustion of a Lean Premixed Air-Propane Turbulent Flame Using a Nanosecond Repetitively Pulsed Plasma," 43d AIAA Aerospace Sciences Meeting and Exhibit, 10–13 January 2003, Reno, NV, AIAA Paper 2005-1193.

Gallimberti, I. (1972), J. Phys. D: Applied Phys., vol. 5, p. 2179.

Gangoli, S., Gallagher, M., Dolgopolsky, A., Gutsol, A., Fridman, A. (2005), 17th Int. Symposium on Plasma Chemistry (ISPC-17), p. 1111, Toronto, Canada.

Ganguly, B.N., Parish, J.W. (2004), Appl. Phys. Lett., vol. 84, p. 4953.

Ganguly, B.N., Bletzinger, P., Garscadden, A. (1997), Phys. Lett. A, vol. 230, p. 218.

Ganiev, Yu.Ch., Gordeev, V.P., Krasilnikov, A.V., Lagutin, V.I., Otmennikov, V.N., Panasenko, A.V. (1998), 2nd Weekly Ionized Gas Workshop, Proceedings, Norfolk, MD.

Ganiev, Yu.Ch., Gordeev, V.P., Krasilnikov, A.V., Lagutin, V.I., Otmennikov, V.N., Panasenko, A.V. (2000), J. Thermophys. Heat Transfer, vol. 14, p. 10.

Ganz, S.N., Krasnokutsky, Yu.I. (1971), Chemical Technology, Kharkov State University, vol. 19, p. 18.

Ganz, S.N., Parkhomenko, V.D. (1976), Plasma Production of Nitrogen Compounds, Vysshaja Shkola (Higher Education), Kiev, Ukraine.

References

Ganz, S.N., Parkhomenko, V.D., Krasnokutsky, Yu.I. (1969a), Generators of Low-Temperature Plasma, Energia (Energy), Moscow.

Ganz, S.N., Parkhomenko, V.D., Krasnokutsky, Yu.I. (1969b), Chemical Technology, Kharkov State University, vol. 5, pp. 356, 518.

Garbassi, F., Morra, M., Occhiello, E. (1994), Polymer Surfaces, From Physics to Technology, Wiley & Sons, New York.

Gardi, B.R. (1999), IEEE Trans. Plasma Sci., vol. 27, p. 36.

Gardiner, W.C., Jr., Hidaka, Y., Tanzawa, T. (1981), Combust. Flame, vol. 40, p. 213.

Garscadden, A. (1994), Pure Appl. Chem., vol. 66, p. 1319.

Garsky, I.M., Givotov, V.K., Ivanov, V.A. (1981), J. Nucl. Sci. Technol., Nucl. Hydrog. Ener., vol. 1, no. 9, p. 44.

Gasilevich, E.S., Ivanov, V.A., Lotkova, E.N. (1969), J. Tech. Phys., vol. 39, p. 126.

Gauyacq, J.P. (1985), J. Phys., B, vol. 18, p. 1859.

Gavalas, G.R., Willks, K. (1980), AIChE J., vol. 26, p. 201.

Gelb, A., Kim S.K. (1971), J. Chem. Phys., vol. 55, p. 10.

Generalov, N.A., Zimakov, V.P., Kozlov, G.I., Masyukov, V.A., Raizer, Yu.P. (1970), Sov. Phys., Lett. J. Exp. Theoret. Phys. (Pis'ma v JETPh), vol. 11, p. 447.

Generalov, N.A., Zimakov, V.P., Kozlov, G.I., Masyukov, V.A., Raizer, Yu.P. (1971), Sov. Phys., J. Exp. Theoret. Phys., vol. 61, p. 1444.

Genet, F., Coulibaly, K., Cavvadias, S., Gonnord, M.F., Amouroux, J. (1997), 13th Int. Symposium on Plasma Chemistry (ISPC-13), Beijing, vol. 4, p. 1740.

Gerasimov, G.A., Gorogankin, E.V., Gusev, E.V. (1979), 3rd USSR Symposium on Plasma Chemistry, vol. 1, p. 179, Nauka (Science), Moscow.

Gerenser, L.J., Grace, J.M., Apai, G., Thompon, P.M. (2000), Surf. Interf. Anal., vol. 29, p. 12.

Gerjoy, E., Stein, S. (1955), Phys. Rev., vol. 95, p. 1971.

Gershenson, Yu., Rosenstein, V., Umansky, S. (1977), Plasma Chemistry-4, Smirnov, B.M. (ed.), Atomizdat, Moscow.

Gershuni, S.Sh., Souris, A.L., Shorin, S.N., Alekseev, N.V. (1975), Sov. Phys., High Energy Chem. (Khimia Vysokikh Energij), vol. 1, p. 1(1).

Ghaffari, A., Neil, D.H., Ardakani, A., Road, J., Ghahary, A., Miller, C.C. (2005), Nitric Oxide-Biology and Chemistry, vol. 12, p. 129.

Gianchandani, Y.B., Wilson, C.G. (2004), U.S. Patent no. 6,827,870 B1.

Gibalov, V.I., Samoilovich, V.G., Filippov, Yu.V. (1981), Sov. Phys., J. Phys. Chem. (JFCh), vol. 55, p. 471.

Gibbs, W.E., McLeary, R. (1971), Phys. Lett. A, vol. 37, p. 229.

Gibson, J.O., Weidman, R. (1963), Chem. Eng. Progr., vol. 59, no. 9, p. 53.

Gill, P., Webb, C.E. (1977), J. Phys. D, Appl. Phys., vol. 10, p. 229.

Gilles, H.L., Clump, C.W. (1969), Ind. Eng. Chem. Process Des. Dev., vol. 9, p. 191.

Gilles, H.L., Clump, C.W. (1970), Ind. Eng. Chem. Process Des. Dev., vol. 9, no. 2, p. 194.

Gilman, A.B. (2000), "Interaction of Chemically Active Plasma with Surfaces of Synthetic Materials," in Encyclopedia of Low-Temperature Plasma, Fortov, V.E. (ed.), vol. 4, p. 393, Nauka (Science), Moscow.

Gilman, A.B., Potapov, V.K. (1995), Appl. Phys., vols. 3–4, p. 14.

Gilman, A.B., Kolotyrkin, Ya.M., Tunitsky, N.N. (1970), Kinet. Catal. (Kinetika i Kataliz), vol. 11, p. 1267.

Ginzburg, V.L. (1960), Propagation of Electromagnetic Waves in Plasma, FizMat Giz, Moscow.

Girshick, S. L. (1993), Diamond Related Mat., vol. 2, p. 1090.

Girshick, S.L. (2005), 17th Int. Symposium on Plasma Chemistry (ISPC-17), p. 7, Toronto, Canada.

Girshick, S.L., Kortshagen, U.R., Bhandarkar, U.V., Swihart, M.T. (1999), Pure Appl. Chem., vol. 71, p. 1871.

Girshick, S.L., Bhandarkar, U.V., Swihart, M.T., Kortshagen, U.R. (2000), J. Phys., D: Appl. Phys., vol. 33, p. 2731.

Givotov, V.K., Krasheninnikov, E.G., Krotov, M.F., Rusanov, V.D., Fridman, A. (1980a), 3rd World Hydrogen Energy Conference, Tokyo, p. 92.

Givotov, V.K., Krasheninnikov, E.G., Rusanov, V.D., Fridman, A. (1980b), 3rd World Hydrogen Energy Conference, Tokyo, p. 84.

Givotov, V.K., Fridman, A., Krotov, M.F., Krasheninnikov, E.G., Patrushev, B.I., Rusanov, V.D., Sholin, G.V. (1981a), Int. J. Hydrogen Energy, vol. 6, p. 441.

Givotov, V.K., Krotov, M.F., Krasheninnikov, E.G., Rusanov, V.D., Fridman, A. (1981b), Int. J. Hydrogen Energy, vol. 6, p. 441.

Givotov, V.K., Malkov, S.Yu., Krasheninnikov, E.G., Krotov, M.F., Rusanov, V.D., Fridman, A. (1981c), J. Nucl. Sci. Technol., Nucl.-Hydrogen Energy, vol. 2, no. 9, p. 36.

Givotov, V.K., Vakar, A.K., Krasheninnikov, E.G., Rusanov, V.D., Fridman, A. (1981d), 5th Int. Symposium on Plasma Chemistry, Edinburgh, vol. 1, p. 52.

Givotov, V.K., Vakar, A.K., Krasheninnikov, E.G., Rusanov, V.D., Fridman, A. (1981e), 15th Int. Conference on Phenomena in Ionized Gases (ICPIG-15), Minsk, vol. 1, p. 99.

References

Givotov, V.K., Pakhomov, V.P., Raddatis, V.K. (1982a), 2nd USSR Conference on New Perspectives in Sulfur Production, vol. 1, p. 111, NIITechChem, Cherkassy, Ukraine.

Givotov, V.K., Rusanov, V.D., Fridman, A. (1982b), Plasma Chemistry, vol. 9, Smirnov, B.M. (ed.), Atom-Izdat, Moscow.

Givotov, V.K., Asisov, R.I., Bezmenov, F.V., Butylkin, Yu.P., Rusanov, V.D., Fridman, A. (1983a), J. Nucl. Sci. Technol., Nucl.-Hydrogen Energy, vol. 1, no. 14, p. 46.

Givotov, V.K., Kalachev, I.A., Krasheninnikov, E.G. (1983b), Spectral Methods of Diagnostics of Plasma-Chemical Non-Uniform Microwave Discharge in CO_2, Kurchatov Institute of Atomic Energy, Moscow, vol. 3704/7.

Givotov, V.K., Malkov, S.Yu., Rusanov, V.D., Fridman, A. (1983c), J. Nucl. Sci. Technol., Nucl.-Hydrogen Energy, vol. 1, no. 14, p. 52.

Givotov, V.K., Malkov, S.Yu., Potapkin, B.V., Rusanov, V.D., Samarin, A.E., Fridman, A. (1984a), Sov. Phys., High Energy Chem. (Khimia Vysokikh Energij), vol. 18, p. 252.

Givotov, V.K., Rusanov, V.D., Fridman, A. (1984b), Plasma Chemistry, vol. 11, Smirnov, B.M. (ed.), Atom-Izdat, Moscow.

Givotov, V.K., Levitsky, A.A., Macheret, S.O., Fridman, A. (1985a), Sov. Phys., High Energy Chem. (Khimia Vysokikh Energij), vol. 19, p. 530.

Givotov, V.K., Rusanov, V.D., Fridman, A. (1985b), Diagnostics of Non-Equilibrium Chemically Active Plasma, Energo-Atom-Izdat, Moscow.

Givotov, V.K., Potapkin, B.V., Rusanov, V.D. (2005), "Low-Temperature Plasma Chemistry," in Encyclopedia of Low-Temperature Plasma, vol. 8.1, Lebedev, Yu.A., Plate, N.A., Fortov, V.E. (eds.), p. 4, Nauka (Science), Moscow.

Gladisch, H. (1962), Chim. Ind., vol. 88, p. 471.

Gladisch, H. (1969), Chem. Ingr. Techn., vol. 4, p. 204.

Glassman, I. (1996), Combustion, 3rd Edition, Academic Press, San Diego, CA.

Glasstone, S., Edlung, M.C. (1952), Nuclear Reactor Theory, D. van Nostrand, Princeton, NJ.

Glover, J., Bendick, P., Link, W. (1978), Curr. Probl. Surg., vol. 15, no. 1, p. 1.

Godyak, V.A. (1971), Sov. Phys., J. Tech. Phys. (JTPh), vol. 41, p. 1361.

Godyak, V.A. (1975), Sov. Phys., Plasma Phys. (Phizika Plasmy), vol. 2, p. 141.

Godyak, V.A. (1986), Soviet Radio Frequency Discharge Research, Delphic Associates, Falls Church, VA.

Godyak, V.A., Piejak, R.B. (1990a), Phys. Rev. Lett., vol. 65, p. 996.

Godyak, V.A., Piejak, R.B. (1990b), J. Vac. Sci. Technol., vol. A8, p. 3833.

Godyak, V.A., Piejak, R.B., Alexandrovich, B.M. (1991), IEEE Trans. Plasma Sci., vol. 19, p. 660.

Goetschel, C.T. (1969), J. Am. Chem. Soc., vol. 91, p. 4702.

Gold, R.G., Sandall, W.R., Cheplick, P.G., Mac-Rae, D.R. (1975), Int. Round Table on Study and Applications of Transport Phenomena in Thermal Plasmas, pp. 1–8, Odeillo, France.

Goldberger, W.M. (1966a), Chem. Eng., vol. 73, no. 6, p. 173.

Goldberger, W.M. (1966b), Chem. Eng., vol. 73, no. 7, p. 125.

Goldman, M., Goldman, N. (1978), Gaseous Electronics, Hirsh, M.N., Oscam H.J. (eds.), Academic Press, New York.

Golubev, A.V., Krasheninnikov, E.G., Tishchenko, E.A. (1981), 15th Int. Conference on Phenomena in Ionized Gases (ICPIG-15), Minsk, vol. 1, p. 620.

Golubovskii, Yu.B., Maiorov, V.A., Behnke, J., Behnke J.F. (2002a), J. Phys. D: Appl. Phys., vol. 35, p. 751.

Golubovskii, Yu.B., Maiorov, V.A., Behnke, J., Behnke J.F. (2002b), J. Phys. D: Appl. Phys., vol. 36, p. 39.

Golubovskii, G.A., Zenger, V.G., Nasedkin, A.N., Prokofieva, E.I., Borovtsov, A.A. (2004), IVth Army Medical Conference on Intensive Therapy and Profilactic Treatments of Surgical Infections, Moscow, Russia.

Gomi, Sh. (1964), Petrol Refiner, vol. 43, no. 11, p. 165.

Gong, W., Zhu, A., Zhou, J., Shi, H., Ruan, G., Zhang, B. (1997), 13th Int. Symposium on Plasma Chemistry (ISPC-13), Beijing, vol. 4, p. 1578.

Gorchakov, G., Lavrov, F. (1934), Acta Phisicochim., URSS, vol. 1, p. 139.

Gordeyenya, E.A., Matveyev, A.A. (1994), Plasma Sources Sci. Technol., vol. 3, p. 575.

Gordiets, B., Zhdanok, S. (1986), Non-Equilibrium Vibrational Kinetics, Capitelli, M. (ed.), Topics in Current Physics 39, Springer-Verlag.

Gordiets, B.F., Osipov, A.I., Stupochenko, E.B., Shelepin, L.A. (1972), Adv. Phys. Sci., Uspehi Phys. Nauk, vol. 108, p. 655.

Gordiets, B.F., Osipov, A.I., Shelepin, L.A. (1980), Kinetic Processes in Gases and Molecular Lasers, Nauka (Science), Moscow.

Gorelova, M.M., Levin, Yu.V., Dubchak, I.L., Zhdanov, A.A., Makarova, L.I., Storozhuk, I.P. (1980), High-Molecular Compounds, vol. 14, p. 543.

Gorov, Yu.M. (1989), Kinetics of Reactions in Industrial Plasma Chemistry, p. 82, Chimia (Chemistry), Moscow.

Gorshkov, V.A., Klimov, A.I., Mishin, G.I., Fedotov, A.B., Yavor, I.P. (1987), Sov. Phys., J. Techn. Phys. (JTF), vol. 32, p. 1138.

References

Gostev, V., Dobrynino. D. (2006), Medical Microplasmatron, 3rd Int. Workshop on Microplasmas (IWM-2006), Greifswald, Germany.

Grabis, Ya.G., Miller, T.N., Heidemane, G.M., Raitsis, M. (1971), Plasma-Chemistry-71, p. 142, Polak, L.S. (ed.), Institute of Petrochemical Synthesis of USSR Academy of Sciences, Moscow.

Grabis, Ya.G., Miller, T.N., Heidemane, G.M., Tenas, A. (1975), 2nd USSR Symposium on Plasma Chemistry, vol. 1, p. 164., Riga, Latvia.

Grace, J.M. (1995), U.S. Patent no. 5,425,980.

Grace, J.M., Freeman, D.R., Corts, R., Kozel, W. (1996), J. Vac. Sci. Technol., vol. A14, no. 3, p. 727.

Grachev, V.G., Krotov, M.F., Potapkin, B.V., Rusanov, V.D., Fridman, A., Khrikulov, V.V. (1992), Hydrogen Sulfide Dissociation in the Arc Discharge, Kurchatov Institute of Atomic Energy, vol. IAE-5149/6, Moscow.

Granovsky, V.L. (1971), Electric Current in Gas, Steady Current, Nauka (Science), Moscow.

Graves, R.D., Kawa, W., Hiteshue, R.W. (1966), Ind. Eng. Chem. Process. Des. Dev., vol. 5, p. 59.

Greenwood, N.N., Earnshaw, A. (1984), Chemistry of the Elements, Pergamon.

Gregory, G., Citterio, S., Labra, M., Sgorbati, S., Brown, S., Denis, M. (2001), Appl. Environ. Microbiol., vol. 67, p. 4462.

Griem, H.R. (1974), Spectral Broadening by Plasma, Academic Press, New York, London.

Grigorieva, T.A., Levitsky, A.A., Macheret, S.O., Polak, L.S., Fridman, A. (1984), Sov. Phys., High Energy Chem. (Khimia Vysokikh Energij), vol. 18, p. 336.

Grishin, S.D. (2000), "Ion and Plasma Rocket Engines," in Encyclopedia of Low-Temperature Plasma, Fortov, V.E. (ed.), vol. 4, p. 291, Nauka (Science), Moscow.

Grishin, S.D., Leskov, L.V., Kozlov, N.P. (1975), Electric-Propulsion Rocket Engines, Machinery (Mashinostroenie), Moscow.

Gross, B., Grycz, B., Miklossy, K. (1969), Plasma Technology, Elsevier, New York.

Gross, R.W.F., Wesner, F. (1973), Appl. Phys. Lett., vol. 23, p. 559.

Grossman, K.R., Cybyk, B., Van Wie, D. (2003), AIAA paper 2003-0057.

Gromov, B.V., Sudarikov, B.N., Savelieva, V.I., Rakov, E.G., Zaitsev, V.A. (1971), Chemical Engineering of Radiated Nuclear Fuel, Shevchenko, V.B. (ed.), Atom-Izdat, Moscow.

Gubin, E.I., Dik, I.G. (1986), Sov. Phys., Phys. Combust. Expl. (FGV), vol. 22, p. 10.

Gudzenko, L.I., Yakovlenko, S.I. (1978), Plasma Lasers, Atomizdat, Moscow.

Guenther, A., Kristiansen, M., Martin, T. (1987), Opening Switches, Plenum, New York.

Guertin, J.P., Christe, K.O., Pavlath, A.E. (1966), Inorg. Chem., vol. 5, p. 1921.

Guliaev, G.V., Polak, L.S. (1965), Kinetics and Thermodynamics of Chemical Reactions in Low-Temperature Plasma, Polak, L.S. (ed.), p. 72, Nauka (Science), Moscow.

Guliaev, G.V., Kozlov, G.I., Polak, L.S. (1965), Kinetics and Thermodynamics of Chemical Reactions in Low Temperature Plasma, Polak, L.S. (ed.), p. 132–150, Nauka (Science), Moscow.

Gurin, A.A., Chernova, N.I. (1985), Sov. Phys., Plasma Phys. (Physika Plasmy), vol. 11, p. 244.

Gutmann, V. (1955), Monatsch. Chem., Bd. 86, S. 765.

Gutsol, A. (1995), Sov. Phys., High Energy Chem. (Chimia Vysokikh Energii), vol. 29, p. 373.

Gutsol, A.F. (1997) Successes Phys. Sci., vol. 40, no. 6, p. 639.

Gutsol, A., Fridman, A. (2001), XXVth Int. Conf. on Phenomena in Ionized Gases (ICPIG-25), Nagoya, Japan, vol. 1, p. 55.

Gutsol, A.F., Potapkin, B.V., Rusanov, V.D, Fridman, A. (1984), 4th USSR Symposium on Plasma Chemistry, vol. 1, p. 105, Dnepropetrovsk.

Gutsol, A.F., Givotov, V.K., Korobtsev, S.V., Rusanov, V.D., Fridman, A. (1985a), J. Nucl. Sci. Technol., Nucl.-Hydrogen Energy, vol. 1, no. 9, p. 39.

Gutsol, A.F., Givotov, V.K., Malkov, S.Yu., Rusanov, V.D., Fridman, A. (1985b), Sov. Phys., High Energy Chem., (Khimia Vysokikh Energij), vol. 19, p. 69.

Gutsol, A.F., Givotov, V.K., Potapkin, B.V., Rusanov, V.D., Fridman, A. (1990), Sov. Phys., J. Techn. Phys. (JTF), vol. 60, p. 62.

Gutsol, A.F., Givotov, V.K., Potapkin, B.V., Rusanov, V.D., Fridman, A. (1992), Sov. Phys., High Energy Chem. (Khimia Vysokikh Energij), vol. 26, p. 361.

Gutsol, A., Larjo, J., Hernberg, R. (1999), XIVth Int. Symposium on Plasma Chemistry (ISPC-14), Prague, vol. 1, p. 227.

Gutsol, A., Tak, G., Fridman, A. (2005), 17th Int. Symposium on Plasma Chemistry (ISPC-17), Toronto, Canada, p. 1128.

Haas, F.M. (2006), "Plausibility Analysis of Branching Radical Reclamation from HO_2 in Plasma Stimulated Ignition below the Autoignition Temperature Threshold," Thesis, Drexel University.

Hadidi, K., Cohn, D.R., Vitale, S., Bromberg, L. (1996), Proceedings of the International Workshop on Plasma Technologies for Pollution Control and Waste Treatment, Pu, Y.K., Woskov, P.P. (eds.), p. 166, Beijing.

Haley, R.F., Smy, P.R. (1989), J. Phys., D: Appl. Phys., vol. 22, p. 258.

References

Hall, B., Schager, P., Weesmaa, J. (1995), Chemosphere, vol. 30, p. 611.

Hamblym, S.M.L., Reuben, B.G., Thompson, S. (1972), Spec. Ceram., vol. 5, p. 147.

Hamlet, P., Moss, J. (1969), J. Am. Chem. Soc., vol. 91, p. 258.

Hammer, T., Broer, S. (1998), Plasma Exhaust Aftertreatment, Society of Automotive Engineers, Warrendale, PA.

Hancock, R. (1973), "Production of Titanium Carbide," U.K. Patent no. 1309660.

Harbec, D., Meunier, J.-L., Guo, L., Gauvin, R. (2005a), 17th Int. Symposium on Plasma Chemistry (ISPC-17), p. 774, Toronto, Canada.

Harbec, D., Meunier, J.-L., Guo, L., Gauvin, R., El Mallah, N. (2005b), J. Phys. D: Appl. Phys., vol. 37, p. 2121.

Harioka, K, Kebarle P. (1976), J. Am. Chem. Soc., vol. 98, p. 6119.

Harioka, K, Kebarle P. (1977), J. Am. Chem. Soc., vol. 99, p. 366.

Harkness, J.B.L., Doctor, R.D. (1993), "Plasma-Chemical Treatment of Hydrogen Sulfide in Natural Gas Processing," Argonne National Laboratory, Gas Research Institute, GRI-93/0118, Chicago, IL.

Harkness, J.B.L., Fridman, A. (1999), The Technical and Economic Feasibility of Using Low-Temperature Plasmas to Treat Gaseous Emissions from Pulp Mills and Wood Product Plants, National Council of the Paper Industry for Air and Stream Improvement, NCASI, Research Triangle Park, NC.

Harris, P.J.F. (2002), Carbon Nanotubes and Related Structures, Cambridge University Press.

Harris, V., Holmgren, J.D., Korman, S. (1959), J. Electrochem. Soc., vol. 106, no. 10, p. 874.

Harstad, J.B. (1969), Am. Ind. Hyg. Assoc. J., vol. 106, p. 874.

Hartech, P., S. Doodes (1957), J. Chem. Phys., vol. 26, p. 1727.

Hassan, H.A., Deese, J.E. (1976), Phys. Fluids, vol. 19, p. 2005.

Hassouni, K., Gicquel, A., Capitelli, M. (1998), Chem. Phys. Lett., vol. 290, p. 502.

Hassouni, K., Gicquel, A., Capitelli, M. (1999a), Phys. Rev., vol. 59, p. 3741.

Hassouni, K., Gicquel, A., Capitelli, M., Loureiro (1999b), Plasma Sources Sci. Technol., vol. 8, p. 494.

Hatch, M.T., Dimmick, R.L. (1966), Bacteriol. Rev., vol. 30, p. 3.

Hatsopoulos, G.N., Gyftopoulos, E.P. (1973), Thermionic Energy Conversion, Vol. 1: Processes and Devices, MIT Press, Cambridge, MA.

Hatsopoulos, G.N., Gyftopoulos, E.P. (1979), Thermionic Energy Conversion, Vol. 2: Theory, Technology and Applications, MIT Press, Cambridge, MA.

Havens, M.R. (1976), J. Vac. Sci. Technol., vol. 13, p. 575.

Hayashi, M. (1987), Swarm Studies and Inelastic Electron-Molecular Collisions, Pitchford, L.C., McKoy, B.V., Chutjan, A., Trajmar, S. (eds.), Springer, New York.

He, X., Ma, T., Qiu, J., Sun, T., Zhao, Z., Zhou, Y., Zhang, J. (2004), Plasma Sources Sci. Technol., vol. 13, p. 446.

He, X.M., Lee, S.T., Bello, L., Cheung, A., Lee, C.S. (1999), J. Mat. Res., vol. 14, p. 1055.

Heath, W.O., Birmingham, J.G. (1995), PNL-SA-25844, Pacific Northwest Laboratory, Annual Meeting of the American Nuclear Society, Philadelphia, PA.

Heberlein, J. (2002), Pure Appl. Chem., vol. 74, p. 327.

Heesch, E.J.M., Pemen, A.J.M., Huijbrechts, A.H.J., van der Laan, P.C.T., Pissinski, K.J., Zanstra, G.J., de Jong, P. (2000), IEEE Trans. Plasma Sci., vol. 28, p. 137.

Hegeler, F., Akyiama, H. (1997), IEEE Trans. Plasma Sci., vol. 25, p. 1158.

Heidt, L.J., Forbes, G.S. (1934), J. Am. Chem. Soc., vol. 56, p. 1671.

Heilmann, R. (1963), Rev. Chem. Ind., vol. 533, p. 134.

Helfritch, D.J., Harmon, G., Feldman, P. (1996), Emerging Solutions to VOC and Air Toxics Conference, A&WMA, Proceedings, p. 277.

Hemptinne, A. (1897), Acad. Roy. Med. Belg., vol. 34, p. 269.

Henne, R., Kayser, A., Borck, V., Schiller, G. (1992), 13th Int. Thermal Spray Conf., Orlando, FL, ASM Int., p. 817.

Henne, R., Borck, V., Müller, M., Ruckdäschel, R., Schiller, G. (1999), 2nd United Thermal Spray Conference, Dusseldorf, Germany, p. 598.

Herbon, J.T., Hanson, R.K., Bowman, C.T., Golden, D.M. (2005), Proc. Combust. Inst., vol. 30, p. 955.

Herlitz, H., Santen, S. (1984), Chem. Stosow., vol. 28, p. 49.

Herman, K., Schmidt, H. (1971), Production of Olephines and Acetylene in Chemical Industry, 8-World Petrolium Congress, p. 60, Moscow.

Herrmann, H.W., Henins, I., Park, J., Selwyn, G.S. (1999), Phys. Plasma, vol. 6, p. 2284.

Herron, J.T. (1999), J. Phys. Chem. Ref. Data, vol. 28, p. 1453.

Hershkowitz, N. (1985), Space Sci. Rev., vol. 41, p. 351.

Herzenberg, A. (1968), J. Phys. B: Atom. Molec. Phys., vol. 1, pp. 548–553.

Hess, L.D. (1972), J. Appl. Phys., vol. 43, p. 1173.

Hewitt, C.N. (1999), Reactive Hydrocarbons in the Atmosphere, Academic Press, San Diego, CA.

Hill, A.E. (1971), Appl. Phys. Lett., vol. 18, p. 194.

Hinazumi, H., Hosoya, M., Mitsui, T. (1973), J. Phys. D: Appl. Phys., vol. 21, p. 1973.

References

Hirayama, C., Maniero, D.A. (1967), U.S. Patent no. 3,460,902.

Hirsch, D., Steinfeld, A. (2004), Int. J. Hydrogen Energy, vol. 29, p. 47.

Hirsh, M.N., Oscam, H.J., eds. (1978), Gaseous Electronic, Vol. 1, Electrical Discharges, Academic Press, New York.

Hoard, J., Servati, H., eds. (1998), Plasma Exhaust Aftertreatment, Society of Automotive Engineers (SAE), Warrendale, PA.

Hocher, H., Thomas, H., Kusters, A., Herrling, J. (1994), Melliand Textilber., vol. 75, pp. 506, 510, 512.

Hocker, H. (1995), Int. Text. Bull. Veredlung, vol. 41, p. 18.

Hocker, H. (2002), Pure Appl. Chem., vol. 74, p. 423.

Hoffman, K.A., Damevin, H.M., Dietikey, J.F. (2000), AIAA paper 2000-2259.

Hoffmann, P., Harkin, T. (2002), Tomorrow's Energy: Hydrogen, Fuel Cells, and the Prospects for a Cleaner Planet, MIT Press.

Hollahan, J.R., Bell, A.T. (1974), Techniques and Applications of Plasma Chemistry, Wiley, New York.

Hollenstein, Ch., Howling, A.A., Dorier, J.L., Dutta, J., Sansonnens, L. (1993), NATO Advanced Research Workshop, Formation, Transport and Consequences of Particles in Plasma, France.

Honda, Y., Tochikubo, F., Watanabe, T. (2001), 25th Int. Conf. on Phenomena in Ionized Gases (ICPIG-25), vol. 4, p. 37, Nagoya, Japan.

Hood, J.L. (1980), Int. Conf. on Gas Discharges and their Applications, Edinburgh, p. 86.

Hopwood, J., Guarnieri, C.R., Whitehair, S.J., Cuomo, J.J. (1993a), J. Vac. Sci. Technol., vol. A11, p. 147.

Hopwood, J., Guarnieri, C.R., Whitehair, S.J., Cuomo, J.J. (1993b), J. Vac. Sci. Technol., vol. A11, p. 152.

Hopwood, J.A., Iza, F. (2004a), U.S. Patent no. 0,164,682 A1.

Hopwood, J.A., Iza, F. (2004b), U.S. Patent no. 5,942,855.

Horvath, M., Bilitzky, L., Hattner, J. (1985), Ozone, Science, Budapest.

Howling, A.A., Hollenstein, Ch., Paris, P.J. (1991), Appl. Phys. Lett., vol. 59, p. 1409.

Howling, A.A., Dorier, J.L., Hollenstein, Ch. (1993a), Appl. Phys. Lett., vol. 62, p. 1341.

Howling, A.A., Sansonnens, L., Dorrier, J.L., Hollenstein, Ch. (1993b), J. Phys. D: Appl. Phys., vol. 26, p. 1003.

Hseuh, H.C., Chou, T.S., Christianson, C.A. (1985), J. Vac. Sci. Technol., vol. A3, p. 518.

Hsiao, M.C., Meritt, B.T., Penetrante, B.M., Vogtlin, G.E., Wallman, P.H. (1995), J. Appl. Phys., vol. 78, p. 3451.

Huang, P.C., Heberlein, J., Pfender, E. (1995), Plasma Chem. Plasma Process., vol. 15, p. 27.

Hunter, P., Oyama, S.T. (2000), Control of Volatile Organic Emissions: Conventional and Emerging Technologies, John Wiley & Sons, New York.

Huxley, L.G.H., Crompton, R.W. (1974), The Diffusion and Drift of Electron in Gases, Wiley, New York.

Hwang, S.T., Kammermeyer, K., Weissberger, A. (1984), Membranes in Separation, Techniques of Chemistry Series, Krieger.

Ibragimov, M.Kh., Bashilov, V.A., Marchenko, E.M., Tuvalbaev, V.G. (1983), Energy and Technology of Fuel Usage, VZPI, Moscow.

Ibragimov, M.Kh., Marchenko, E.M., Tuvalbaev, V.G. (1987), Energetics Electricity, no. 1, p. 1.

Ichiki, T., Koidesawa, T., Horiike, Y. (2003), Plasma Sources Sci. Technol., vol. 12, p. 16.

Ichiki, T., Taura, R., Horiike, Y. (2004), J. Appl. Phys., vol. 95, p. 35.

Il'in, D.T., Eremin, E.N. (1962), Sov. Phys., J. Phys. Chem., vol. 36, p. 1560; vol. 36, p. 2222.

Il'in, D.T., Eremin, E.N. (1965), Sov. Phys., J. Appl. Chem., vol. 38, p. 12.

Il'in, D.T., Eremin, E.N. (1967), Low-Temperature Plasma, Polak, L.S. (ed.), p. 576, Mir (World), Moscow.

Il'in, D.T., Eremin, E.N. (1968), Modern Problems of Physical Chemistry, Moscow State University, vol. 2, p. 61, Moscow.

Inagaki, N. (1996), Plasma Surface Modification and Plasma Polymerization, CRC Press, Boca Raton, FL.

Inocuti, M., Kim, Y., Platzman, R.L. (1967), Phys. Rev., vol. 164, p. 55.

Inoue, E., Sugino, K. (1959), Adv. Chem. Ser, vol. 21, p. 313.

Intesarova, E.I., Kondratiev, V.N. (1967), Sov. Phys., News of USSR Academy of Sciences (Izvestia), Chemistry, p. 2440.

Isaev. I.F., Legasov, V.A., Rakhimbabaev, Ya.R. (1975), 2nd USSR Symposium en plasma chemistry, vol. 1, p. 184.

Isbarn, G., Ederer, H.J., Ebert, K.H. (1981), "The Thermal Decomposition of n-Hexane," in Mechanism and Simulation in Modeling of Chemical Reaction Systems, Ebert, K.H. (ed.), p. 235, Springer-Verlag, Berlin, Heidelberg, New York.

Ishiguro, S., Kamimura, T., Sato, T. (1995), Phys. Fluids, vol. 28, p. 2100.

Ishizuka, K. (1958), "Titanium Reduction in Glow Discharge," U.S. Patent no. 2,860,094.

Iskenderova, K. (2003), "Cleaning Process in High-Density Plasma Chemical Vapor Deposition Reactor," Ph.D. Dissertation, Drexel University, Philadelphia, PA.

Iskenderova, K., Porshnev, P., Gutsol, A., Saveliev, A., Fridman, A., Kennedy, L., Rufael, T. (2001), 15th International Symposium on Plasma Chemistry (ISPC-15), Orleans, France, p. 2849.

Ivanov, A.A. (1975), Sov., Phys., Plasma Phys. (Fizika Plasmy), vol. 1, p. 147.

References

Ivanov, A.A., Nikiforov, V.A. (1978), Plasma Chemistry, vol. 5, Smirnov, B.M. (ed.), Atom-Izdat, Moscow.

Ivanov, A.A., Soboleva, T.K. (1978), Non-Equilibrium Plasma Chemistry, Atom-Izdat, Moscow.

Ivanov, A.A., Starykh, V.V., Filiushkin, V.V. (1979), 3rd USSR Symposium on Plasma Chemistry, vol. 1, p. 70, Nauka (Science), Moscow.

Ivanov, A.V. (2000), "Plasma Production of Oxide Materials," in Encyclopedia of Low-Temperature Plasma, Fortov, V.E. (ed.), vol. 4, p. 354, Nauka (Science), Moscow.

Ivanov, O.R., Matytsin, A.P. (1973), Sov. Phys., High Energy Chem. (Khimia Vysokikh Energij), vol. 23, p. 81.

Ivanov, Yu.A. (1989), Sov. Phys., High Energy Chemistry (Khimia Vysokikh Energij), vol. 5, p. 382.

Ivanov, Yu.A. (2000), "Mechanisms of Plasma-Chemical Polymerization of Hydrocarbons," in Encyclopedia of Low-Temperature Plasma, Fortov, V.E. (ed.), vol. 3, p. 330, Nauka (Science), Moscow.

Ivanov, Yu.A., Epstein, I.L. (1984), Sov. Phys., High Energy Chem. (Khimia Vysokikh Energij), vol. 18, p. 461.

Ivanov, Yu.A., Polak, L.S., Slovetsky, D.I. (1971), Sov. Phys., High Energy Chem. (Khimia Vysokikh Energij), vol. 5, p. 382.

Ivanov, Yu.A., Rytova, N.M., Soldatova, I.V. (1990a), Plasma Chemistry–90, Polak, L.S. (ed.), Institute of Petrochemical Synthesis, Nauka (Science), Moscow.

Ivanov, Yu.A., Rytova, N.M., Timakin, V.N., Epstein, I.L. (1990b), Sov. Phys., High Energy Chem. (Khimia Vysokikh Energij), vol. 24, p. 460.

Ivanov, Yu.A., Rytova, N.M., Timakin, V.N., Epstein, I.L. (1990c), Sov. Phys., High Energy Chem. (Khimia Vysokikh Energij), vol. 24, p. 541.

Iza, F., Hopwood, J. (2003), IEEE Trans. Plasma Sci., vol. 31 p. 782.

Iza, F., Hopwood, J. (2004), IEEE Trans. Plasma Sci., vol. 32, p. 2.

Jackson, M., Rairden, J., Smith, J., Smith, R.W. (1981), J. Metals, vol. 1, p. 146.

Jacobs, P.T., Lin, S.M. (1987), U.S. Patent no. 4,643,876.

Jacobsen, L., Carter, C., Baurle, R., Jackson, T. (2003), 41st AIAA Aerospace Sciences Meeting and Exhibit, 9–12 January 2003, Reno, NV, AIAA paper 2003-871.

Jahn, R.G. (2006), Physics of Electric Propulsion, Dover Publications.

Jaisinghani, R. (1999), Bactericidal Properties of Electrically Enhanced HEPA Filtration and a Bioburden Case Study, InterPhex Conference, New York.

Jasinski, M., Szczucki, P., Dors, M., Mizeraczyk, J., Lubanski, M., Zakrzewski, Z. (2000), Int. Symp. High Pressure Low Temperature Plasma Chemistry, HACONE VII, Griefswald, Germany, p. 496.

Jasinski, M., Mizeraczyk, J., Zakrzewski, Z., Chang, J.S. (2002a), Czech. J. Phys., vol. 52, D, p. 743.

Jasinski, M., Mizeraczyk, J., Zakrzewski, Z., Ohkubo, T., Chang, J.S. (2002b), J. Phys. D: Appl. Phys., vol. 35, p. 1.

Javan, A., Bennett, W.R., Herriot, D.R. (1961), Phys. Rev. Lett., vol. 6, p. 106.

Jellum, G., Graves, D., Daugherty, J.E. (1991), J. Appl. Phys., vol. 69, p. 6923.

Jiandong, L., Xianfu, X., Yao, S., Tianen, T. (1996), Int. Workshop on Plasma Technologies for Pollution Control and Waste Treatment, MIT, Cambridge, MA.

Jiang, X.L., Tiwari, R., Gitzhofer, F., Boulos, M.I. (1993), J. Therm. Spray Technol., vol. 2, no. 3, p. 265.

Johnson, T.P., Dowson, A.L., Ward, R.M. (1998), Scripta Materialia, vol. 39, no. 6, p. 783.

Johnston, H.S. (1992), Anno. Rev. Phys. Chem., vol. 43, p. 1.

Johnstone, R.W., Ruefli, A.A., Lowe, S.W. (2002), Cell, vol. 108, p. 153.

Jones, H.M., Kunhardt, E.E. (1994), IEEE Trans. Dielectric Insul., vol. 1, p. 1016.

Jones, H.M., Kunhardt, E.E. (1995), J. Appl. Phys, vol. 77, p. 795.

Joshi, R.P., Hu, Q., Schoenbach, K.H., Beebe, S.J. (2002a), IEEE Trans. Plasma Sci., vol. 30, p. 1536.

Joshi, R.P., Qian, J., Schoenbach, K.H. (2002b), J. Appl. Phys., vol. 92, p. 6245.

Jozef, R., Sherman, D.M. (2005), J. Phys. D: Appl. Phys., vol. 38, p. 547.

Judd, O.P. (1974), J. Appl. Phys., vol. 45, p. 4572.

Kabisov, R.K., Sokolov, V.V., Shekhter, A.B., Pekshev, A.V., Maneilova, M.V. (2000), Russ. J. Oncol., vol. 1, p. 24.

Kabouzi, Y., Nantel-Valiquette, M., Moisan, M., Rostaing, J.C. (2005), 17th Int. Symposium on Plasma Chemistry (ISPC-17), Toronto, Canada, p. 1121.

Kado, S., Sekine, Y., Muto, N., Nozaki, T., Okazaki, K. (2003), 16th Int. Symposium on Plasma Chemistry (ISPC-16), Orleans, France, p. 569.

Kadomtsev, B.B. (1976), The Collective Phenomena in Plasma, Nauka (Science), Moscow.

Kalghatgi, S., Cooper, M., Fridman, G., Peddinghaus, L., Nagaraj, G., Balasubramanian, M., Vasilets, V.N., Gutsol, A., Fridman, A., Friedman, G. (2007), Mechanism of Blood Coagulation by Nonthermal Atmospheric Pressure Dielectric Barrier Discharge Plasma, 9th Annual RISC, Drexel University, Philadelphia, PA.

Kalinnikov, V.T., Gutsol, A. (1997), Sov. Phys., Doklady, vol. 353(42), p. 469(179).

Kalinnikov, V.T., Gutsol, A. (1999), Sov. Phys., Therm. Phys. High Temp., vol. 37, p. 194(172).

Kalra, C.S. (2004), "Gliding Arc in Tornado and its Application for Plasma-Catalytic Partial Oxidation of Methane," Thesis, Drexel University, Philadelphia, PA.

References

Kalra., C.S., Gutsol, A.F., Fridman, A. (2005), IEEE Trans. Plasma Sci., vol. 33, p. 32.

Kanazawa, S., Kogoma, M., Moriwaki, T., Okazaki, S. (1987), 8th Int. Symposium on Plasma Chemistry (ISPC-8), Tokyo, Japan, vol. 3, p. 1839.

Kanazawa, S., Kogoma, M., Moriwaki, T., Okazaki, S. (1988), J. Phys. D, vol. 21, p. 838.

Kapitsa, P.L. (1969), Sov. Phys., J. Exp. Theoret. Phys., vol. 57, p. 1801.

Karachevtsev, G.V., Fridman, A. (1974), Sov. Phys., J. Techn. Phys., vol. 44, p. 2388.

Karachevtsev, G.V., Fridman, A. (1975), 4th USSR Conference on Physics of Low Temperature Plasma, vol. 2, p. 26, Kiev, Ukraine.

Karachevtsev, G.V., Fridman, A. (1975), Sov. Phys., Therm. Phys. High Temp., vol. 15, p. 922.

Karanassios, V. (2005), U.S. Patent no. 0,195,393 A1.

Karpenko, E.I., Messerle, V.E. (2000), "Plasma – Energetic Technologies of Utililation of Fuels," in Encyclopedia of Low-Temperature Plasma, Fortov, V.E. (ed.), vol. 4, p. 359, Nauka (Science), Moscow.

Kasper, J.V.V., Pimentel, G.C. (1965), Phys. Rev. Lett., vol. 14, p. 352.

Kassel, L.S. (1932), J. Am. Chem. Soc., vol. 54, p. 3949.

Kastenmeier, B.E., Matsuo, P.J., Oehrein, G.S., Langan, J.G. (1998), J. Vac. Sci. Technol., vol. A, no. 16, p. 2047.

Kastenmeier, B.E., P.J., Oehrein, G.S., Langan, J.G., Entley, W.R. (2000), J. Vac. Sci. Technol., vol. A, no. 18, p. 2102.

Katsuki, S., Akiyama, H., Abou-Ghazala, A., Schoenbach, K.H. (2002a), IEEE Trans. Dielectrics Electrical Insul., vol. 9, p. 498.

Katsuki S., Moreira, K., Dobbs, F., Joshi, Schoenbach, K.H. (2002b), IEEE J., no. 8, p. 648.

Kaufman, F., (1964) Anno. Geophys., vol. 20, p. 106.

Keebler, P.F. (1990), "A Large Volume Microwave Plasma Facility for Plasma Ion Implantation Studies," M.S. Thesis, University of Tennessee, Knoxville, TN.

Keebler, P.F., Roth, J.R., Buchanan R.A., Lee, I.-S. (1989), APS Bull., vol. 34, p. 2020.

Kelly-Wintenberg, K. (2000), IEEE Trans. Plasma Sci., vol. 28.

Kelly-Wintenberg, K., Montie, T.C., Brickman, C., Roth, J.R., Carr, A.K., Sorge, K., Wadworth, L.C., Tsai, P.P.Y. (1998), J. Ind. Microbiol. Biotechnol., vol. 20, p. 69.

Kennedy, L., Fridman, A., Saveliev, A., Nester, S. (1997), APS Bull., vol. 41, no. 9, p. 1828.

Khadiya, N. (2004), "A Fast Start-up On-Board Diesel Fuel Reformer for NO_x Trap Regeneration and Desulfation," Desel Engine Exhaust Reduction Workshop, San Diego, CA.

Khadre, M.A., Yousef, A.E., Kim, J.G. (2001), J. Food Sci., vol. 66, p. 9.

Khare, S.P. (1966), Phys. Rev., vol. 149, p. 33.

Khare, S.P. (1967), Phys. Rev., vol. 157, p. 107.

Kharitonov, D.Yu., Gogish-Klushin, S.Yu., Novikov, G.I. (1987), News Belarus Acad. Sci., Chem., vol. 6, p. 105.

Kharitonov, D.Yu., Gutsevich, E.I., Novikov, G.I., Fridman, A. (1988), Mechanism of Pulsed Electrolytic-Spark Oxidation of Aluminum in Concentrated Sulfuric Acid, Kurchatov Institute of Atomic Energy, vol. 4705/13, Atom-Inform, Moscow.

Khrikulov, V.V., Grachev, V.G., Krotov, M.F., Potapkin, B.V., Rusanov, V.D., Fridman, A. (1992), High Energy Chem. (Khimia Vysokikh Energij), vol. 26, p. 371.

Khudiakov, G.N., Tselishchev, P.A., Bogacheva, T.M., Abayev, G.G. (1982), "Application of Plasma Gasification of Coals in Energetics," in Economy and Enhancement of Energy Efficiency of Usage of Fuel Resources, p. 84, Energy Institute, ENIN, Moscow.

Kieft, I.E., Broers, J.L.V., Caubet-Hilloutou, V., Slaaf, D.W., Ramaekers, F.C.S., Stoffels, E. (2004a), Bioelectro-magnetics, vol. 25, p. 362.

Kieft, I.E., v.d. Laan, E.P., Stoffels, E. (2004b), New J. Phys., vol. 6, p. 149.

Kieft, I.E., Darios, D., Roks, A.J.M., Stoffels, E. (2005), IEEE Trans. Plasma Sci., vol. 33, p. 771.

Kim, C.S., Baddour, R.F., Howard, J.B., Meissner, H.P. (1979), Ind. Eng. Chem. Process Des. Dev., vol. 18, p. 323.

Kim, G.J., Iza, F., Lee, J.K. (2006), J. Phys. D: Appl. Phys., vol. 39, no. 10.

Kim, T., Suh, S.-M., Girshick, S.L., Zachariah, M.R., McMurry, P.H., Rassel, R.M., Shen, Z., Campbell, S.A. (2002), J. Vac. Sci. Technol., vol. A no. 20, p. 413.

Kirillin, V.A, Sheindlin, A.E. (1971), MHD – Generators as a Method of Electric Energy Production, Energia (Energy), Moscow.

Kirk, R.W. (1974), Techniques and Applications of Plasma Chemistry, Bell, A.T., and Hollohan, J.F. (eds.), p. 347, Wiley Inter-Science, New York, London.

Kirpatrick, S. (1977), Theory and Properties of Disordered Materials, Bonch-Bruevich, V.L. (ed.), Mir (World), Moscow.

Kitayama, J., Kuzumoto, M. (1999), J. Phys. D: Appl. Phys., vol. 32, p. 3032.

Kiyshy, U. (1967), U.S. Patent no. 3,513,014.

Klee, D., Hocker, H. (1999), Adv. Polym. Sci., vol. 49, p. 1.

Klemenc, A., Hinterberger, H., Hofer, H. (1937), Z. Elektrochem., vol. 43, p. 261.

References

Klimov, A., Bitiurin, V., Moralev, I., Tolkunov, B., Nikitin, A., Velichko, A., Bilera, I. (2006), "Non-Premixed Plasma-Assisted Combustion of Hydrocarbon Fuel in High Speed Airflow," 44th AIAA Aerospace Science Meeting and Exhibit, Reno, NV, AIAA paper 2006-617.

Klimov, A.I., Mishin, G.I. (1990), Sov. Phys., Tech. Phys. Lett. (Pis'ma v JTF), vol. 16, p. 960.

Klimov, A.I., Koblov, A.N., Mishin, G.I., Serov, Yu.L., Yavor, I.P. (1982), Sov. Phys., Tech. Phys. Lett. (Pis'ma v JTF), vol. 8, p. 192.

Klimov, A.I., Mishin, G.I., Fedotov, A.B., Shakhovatov, V.A. (1989), Sov. Phys., Tech. Phys. Lett. (Pis'ma v JTF), vol. 15, p. 800.

Klimov, A.I., Leonov, S., Pashina, A., Skvortsov, V., Cain, T., Tmofeev, B. (1999), AIAA paper 99-4857.

Klimov, V.D., Legasov, V.A., Chaivanov, B.B. (1968a), Sov. Phys., Atom. Energy (Atomnaja Energia), vol. 25, p. 244.

Klimov, V.D., Legasov, V.A., Chaivanov, B.B. (1968b), Sov. Phys., News USSR Acad. Sci. (Vestnik), vol. 7, p. 104.

Klingbeil, R.D., Tidman, A., Fernsler, R.F. (1972), Phys. Fluids, vol. 15, p. 1969.

Knight, R., Smith, R., Apelian, D. (1991), Int. Mat. Rev., vol. 36, no. 6, p. 221.

Knight, R., Smith, R. (1998), "Thermal Spray Forming of Materials," ASM Int., vol. 7, p. 408.

Knight, R. (2005), Thermal Spray: Past, Present and Future, 17th International Symposium on Plasma Chemistry (ISPC-17), p. 913, Toronto, Canada.

Knizhnik, A.A., Potapkin, B.V., Medvedev, D.D., Korobtsev, S.V., Shiyaevskii, V.L., Rusanov, V.D. (1999a), Techn. Phys., vol. 365, p. 336.

Knizhnik, A.A., Potapkin, B.V., Medvedev, D.D., Korobtsev, S.V., Shiyaevskii, V.L., Rusanov, V.D. (1999b), 14th International Symposium on Plasma Chemistry (ISPC-14), vol. 5, p. 2319, Prague, Czech Republic.

Knypers, A.D. Hopman, H.J. (1988), J. Appl. Phys., vol. 63, p. 1899.

Knypers, A.D., Hopman, H.J. (1990), J. Appl. Phys., vol. 67, p. 1229.

Kobayashi, K., Berkowitz, N. (1971), Fuel, vol. 50, p. 254.

Kobsev, Yu.N., Khudiakov, G.N. (1971), Plasma Chemistry - 71, Polak, L.S. (ed.), p. 155, Institute of Petrochemical Synthesis, USSR Academy of Sciences, Moscow.

Koch, M., Cohn, D.R., Patrick, R.M., Schuetze, M.P., Bromberg, L., Reily, D., Thomas, P. (1993), Phys. Lett. A, vol. 184, p. 109.

Koch, V., Juntgen, H., Peters, W., (1969), Brennst. Chem, vol. 50, p. 369.

Kochetov, I.V., Pevgov, V.G., Polak, L.S., Slovetsky, D.I. (1979), Plasma-Chemical Processes, Polak, L.S. (ed.), Nauka (Science), Moscow.

Kof, L.M., Starikovskii, A.Yu. (1996), Int. Symposium on Combustion, Naples, Italy, Abstracts of WIP papers, p. 406.

Koga, K., Matsuoka, Y., Tanaka, K., Shiratani, M., Watanabe, Y. (2000), Appl. Phys. Lett., vol. 77, p. 196.

Kogelschatz, U. (1988), "Advanced Ozone Generation," in Process Technologies for Water Treatment, Stucki, S. (ed.), p. 87, Plenum Press, New York.

Kogelschatz, U. (1992), 10th Int. Conf. on Gas Discharges and Their Applications (GD-92), vol. 2, p. 972, Swansea, UK.

Kogelschatz, U. (1999), International Ozone Symposium, Basle, p. 253.

Kogelschatz, U. (2003), Plasma Chem. Plasma Process., vol. 23, no. 1, p. 1.

Kogelschatz, U., Eliasson, B. (1995), "Ozone Generation and Applications," in Handbook of Electrostatic Processes, Chang, J.S., Kelly, A.J., and Crowley, J.M. (eds.), Chap. 26, p. 581, Marcel Dekker, New York.

Kogelschatz, U., Eliasson, B., Hirth, M. (1987), J. Phys. D: Appl. Phys., vol. 20, p. 1421.

Kogelschatz, U., Eliasson, B., Egli, W. (1997), J. Physique IV, vol. 7, Colloque C4, p. 4.

Kohchi, A., Adachi, S., Nakagawa, Y. (1996), Jpn. J. Appl. Phys., vol. 35, p. 2326.

Koidl, P., Wagner, C., Dischler, B., Wagner, J., Ramsteiner, M. (1990), Mat. Sci. Forum, vol. 52, p. 41.

Kolesnichenko, Yu.F., Matukhin, V.D., Muraviov, V.F., Smaznov, S.I. (1979), Sov. Phys., Doklady, (Reports of USSR Academy of Sciences), vol. 246, p. 1091.

Kolesnichenko, Yu.F., Brovkin, V.G., Leonov, S.B., Krylov, A.A., Lashkov, V.A., Mashek, I.Ch., Gorynya, A.A., Ryvkin, M.I. (2001), AIAA paper 2001-0345.

Kolesnichenko, Yu.F., Brovkin, V.G., Azarova, O.A., Gruditsky, V.G., Lashkov, V.A., Mashek, I.Ch. (2003), AIAA paper 2003-0361.

Kolesnikov, A.F. (2001), 32nd Plasma Dynamics and Laser Conference, Anaheim, CA, AAIA paper 2001-2871.

Kolev, K., Georgiev, I. (1987), 8th Int. Symposium on Plasma Chemistry (ISPC-8), Tokyo, Japan, no. P055, p. 678.

Kolman, D., Volenik, K. (2002), Plasma Chem. Plasma Process., vol. 22, p. 437.

Kolman, D., Heberlein, J., Pfender, E., Young, R. (1996), Plasma Chem. Plasma Process., vol. 16., p. 57.

Kolobova, E.A. (1983), Chem. Technol. Fuels (KhTT), vol. 2, p. 91.

Komachi, K. (1992), J. Vac. Sci. Technol., vol. A11, p. 164.

References

Komkov, I.D., Kruchinin, A.M., Nosikov, A.V., Sevostianov, A.N. (1976), " Plasma Furnace for Carbothermic Reduction of Rare and Refractory Metals," in Plasma Processes in Metallurgy and Technology of Inorganic Materials, A.A. Baikov Institute of Metallurgy of USSR Academy of Sciences, Nauka (Science), Moscow.

Komolov, S.A. (1992), Total Current Spectroscopy of Surfaces, Gordon & Breach.

Kondratiev, V.N. (1971), Rate Constants of Gas-Phase Reactions, Nauka (Science), Moscow.

Kondratiev, V.N. (1974), Chemical Bonding Energies, Enthalpy of Chemical Processes, Ionization Potentials and Electron Affinities, Nauka (Science), Moscow.

Kondratiev, V.N., Nikitin, E.E. (1981), Chemical Processes in Gases, John Wiley & Sons, New York.

Konuma, M. (1992), Film Deposition by Plasma Techniques, Springer, New York.

Kopchenov, V., Vatazhin, A., Gouskov, O. (1999), AIAA paper 99-4971.

Koriagin, V.S., Suris, A.L., Troitsky, V.N., Shorin, S.N. (1973), Sov. Phys., High Energy Chem., (Khimia Vysokikh Energij), vol. 7, p. 215.

Kornilov, A.A., Lamikhov, L.K., Barashkina, S.A. (1976a), "Reduction of Oxides of Tungsten and Molybdenum in Nitrogen Plasma," in Plasma Processes in Metallurgy and Technology of Inorganic Materials, A.A. Baikov Institute of Metallurgy of USSR Academy of Sciences, Nauka (Science), Moscow.

Kornilov, A.A., Lamikhov, L.K., Glukhov, V.P. (1976b), "Silicon Oxides Reduction by Propane-Butane in Nitrogen Plasma," in Plasma Processes in Metallurgy and Technology of Inorganic Materials, A.A. Baikov Institute of Metallurgy of USSR Academy of Sciences, Nauka (Science), Moscow.

Korobtsev, S. (1982), "Experimental Investigations of Plasma-Chemical Processes in a Reactor with Fast Rotating Plasma," Ph.D. Dissertation, Kurchatov Institute of Atomic Energy, Moscow.

Korobtsev, S., Kosinova, T.A., Rakhimbabaev, Ya.R., Rusanov, V.D. (1982), Sov. Phys., Lett. J. Techn. Phys. (Pis'ma v JTF), vol. 8, p. 397.

Korobtsev, S., Medvedev, D., Rusanov, V.D., Shiryaevski (1997), 13th Int. Symposium on Plasma Chemistry (ISPC-13), vol. 2, p. 755, Peking University Press, Beijing.

Korolev, Yu.D., Mesiatz, G.A. (1982), Field Emission and Explosive Processes in Gas Discharge, Nauka (Science), Novosibirsk.

Korzec, D., Rapp, J., Theirich, D., Engemann, J. (1994), J. Vac. Sci. Technol., vol. A, no. 12, p. 369.

Kossyi, I.A., Kostinsky, A.Yu., Matveyev, A.A., Silakov, V.P. (1992), Plasma Sources Sci. Technol., vol. 1, p. 207.

Kouprine, A., Gitzhofer, F., Boulos, M., Fridman, A. (2003), Plasma Chem. Plasma Process., vol. 24, p. 189.

Kouprine, A., Gitzhofer, F., Boulos, M., Veres, T. (2005), 17th Int. Symposium on Plasma Chemistry (ISPC-17), p. 786, Toronto, Canada.

Koval, V.V., Petrishchev, V.A., Khmelevoy, P.S. (1979), 3rd USSR Symposium on Plasma Chemistry, vol. 2., p. 192, Nauka (Science), Moscow.

Kozlov, G.I., Khudiakov, G.N., Kobsev, Yu.N. (1967), Petrochemistry, vol. 7, p. 224.

Kozlov, G.I., Khudiakov, G.N., Kobsev, Yu.N. (1974), Plasma-Chemical Process of Natural Gas Conversion into Acetylene, Hydrogen and Soot, Institute of Problem of Mechanics of USSR Academy of Sciences, ENIN, vol. 48, Moscow.

Kozlov, S.V., Kosinova, T.A., Rakhimbabaev, Ya.R., Rusanov, V.D. (1979), 3rd USSR Symposium on Plasma Chemistry, vol. 1, p. 195, Nauka (Science), Moscow.

Kozlova, S.G., Kukhto, V.A. (1976), Chem. Technol. Fuel (KhTT), vol. 2, p. 82.

Kramer, P.W., Yeh Y.-S., Yasuda H. (1989), J. Membrane Sci., p. 461.

Krapukhin, V.V., Korolev, E.A. (1968), Sov. Phys., Higher Educ. News (Izvestia VUZov), Non-Ferrous Metall., no. 6, p. 268.

Krasheninnikov, E.G. (1981), "Experimental Investigations of Non-Equilibrium Plasma-Chemical Processes in Elevated Pressure Microwave Discharge," Ph.D. Dissertation, Kurchatov Institute of Atomic Energy, Moscow.

Krasheninnikov, E.G., Rusanov, V.D., Saniuk, S.V., Fridman, A. (1986), Sov. Phys., J. Techn. Phys., vol. 56, p. 1104.

Krasheninnikov, S.I. (1980), "Electron Beam Interaction with Chemically Active Plasma," Ph.D. Dissertation, Kurchatov Institute of Atomic Energy, Moscow.

Krasnov, A.I., Zilberberg, V.G., Sharivker, S.Yu. (1970), Low-Temperature Plasma in Metallurgy, Metallurgia (Metallurgy), Moscow.

Krieger, R.L., Gatti, R., Schumacher, H.J. (1966), Phys. Chem., Neue Folge, vol. 51, p. 240.

Krotovskii, G.S., Pekshev, A.V., Zudin, A.M., Uchkin, I.G., Atkova, E.O., Mosesov, A.G. (2002), Cardiovasc. Surg. (Grudnaia i serdechno-sosudistaia hirurgia), vol. 1, p. 37.

Kruchinin, A.M., Orlov, G.I., Komkov, I.D. (1976), "Vacuum High Current Discharge with Hot Plasma Hollow Cathode as a Source of Electron-Plasma Heating in Melting Vacuum Electric Furnaces," in Plasma Processes in Metallurgy and Technology of Inorganic Materials, A.A. Baikov Institute of Metallurgy of USSR Academy of Sciences, Nauka (Science), Moscow.

Krukonis, V.J., Gannon, R.E., Modell, M. (1974), Adv. Chem., vol. 131, p. 29.

Krukovsky, V.K., Kolobova, E.A. (1981), "Gasification of Coal in Water Vapor Plasma," in Theory and Technology of Liquid, Gaseous and Synthetic Fuels, p. 71, IGI, Moscow.

References

Krukovsky, V.K., Kolobova, E.A. (1982), "Influence of Yield of Volatiles on Interaction of the Fuel Carbon with Water Vapor," in Conversion of Coal into Liquid and Gaseous Fuels, p. 83, IGI, Moscow.

Krukovsky, V.K., Kolobova, E.A., Lubchanska, L.I., Nikishkov, B.V. (1987), "Complex Plasma-Chemical Conversion of Solid Hydrocarbon Fuel in Water Vapor," in Plasma Gasification and Pyrolysis of Low-Quality Coals, p. 81, Energy Insitute, ENIN, Moscow.

Krykhtina, L.M., Maltsev, A.N., Eremin, E.N. (1966), Sov. Phys., J. Phys. Chem. (JFCh), vol. 40, p. 2784.

Krykin, M.A., Timashev, S.F. (1988), J. High-Mol. Compounds, no. 1, p. 21.

Kubanek, G.R., Chevalier, P., Cauvin, W.H. (1968), Can. J. Chem. Eng., vol. 46, p. 101.

Kuchukhidze, S.T., Klihdukhov, I.A., Bakhtiarov, K.R., Pankratov, V.V. (2004), Probl. Obstet. Gynaecol. (OB/GYN), vol. 3, no. 2, p. 76-82.

Kudinov, V.V., Pekshev, P.Yu., Belashchenko, V.I., Solonenko, O.P., Safiulin, V.A. (1990), Coating Spraying by Plasma, Nauka (Science), Moscow.

Kulagin, I.D., Lubimov, V.K., Marin, K.G. (1967), Sov. Phys., Phys. Chem. Mat. Treatment, FChOM, no. 2, p. 38.

Kulikovsky, A.A. (1994), J. Phys. D: Appl. Phys., vol. 27, p. 2556.

Kulikovsky, A.A. (1997a), IEEE Trans. Plasma Sci., vol. 25, p. 439.

Kulikovsky, A.A. (1997b), 23rd International Symposium on Phenomena in Ionized Gases (ICPIG-23), vol. 1, p. 258, Toulouse, France.

Kulikovsky, A.A. (2001), IEEE Trans. Plasma Sci., vol. 29, p. 313.

Kummler, R (1977), J. Phys. Chem., vol. 81, p. 2451.

Kunchardt, E. E., Tzeng, Y. (1988), Phys. Rev., vol. A38, p. 1410.

Kureha Corporation (1958), U.K. Patent no. 1068793.

Kurenkov, V.V., Gorogankin, E.V., Klementieva, L.L., Panteleev, V.I. (1979), 3rd USSR Symposium on Plasma Chemistry, vol. 1, p. 95, Nauka (Science), Moscow.

Kurochkin, Yu.V., Polak, L.S., Pustogarov A.V. (1978), Sov. Phys., Therm. Phys. High Temp., TVT, vol. 16, p. 1167.

Kushner, M.J. (1988), J. Appl. Phys., vol. 63, p. 2532.

Kushner, M.J. (2004), J. Appl. Phys., vol. 95, no. 3, p. 846.

Kushner M.J. (2005), J. Phys. D: Appl. Phys., vol. 38, p. 1633.

Kuster, H. (1931), Brennst. Chem., vol. 12, p. 329.

Kusters, A., Herrling, J., Thomas, H., Hocker, H. (1995), 9th Int. Wool Text. Res. Conf., Biella, vol. 2, p. 403.

Kusz, J. (1978), Plasma Generation on Ferroelectric Surface, PWN, Warsaw-Wroclaw, Poland.

Kutlin, A.P., Larionov, V.V., Yarosh, A.M. (1980), Sov. Phys., J. Phys. Chem., JFCh, vol. 54, p. 3086.

Kuznetzov, N.M. (1971), J. Techn. Exp. Chem. (Tekh. Eksp. Khim., TEKh), vol. 7, p. 22.

Kuznetsov, N.M., Raizer, Yu. P. (1965), Sov. Phys., J. Appl. Mech. Techn. Phys., PMTF, vol. 4, p. 10.

Kuzumoto, M., Tabata, Y., Yagi, S. (1996), Trans. IEE Jpn, vol. 116A, p. 121.

Kybett, B.D., Johnson, G.K., Barker, C.K., Margrave, J.L. (1965), J. Phys. Chem., vol. 69, p. 3603.

Lacina, W.B., Mann, M.M. (1972), Appl. Phys. Lett., vol. 21, p. 224.

Lacour, B., Vannier, C. (1987), J. Appl. Phys., vol. 38, p. 5244.

Lafferty, J.M. (ed.), (1980), Vacuum Arcs, Theory and Applications, Wiley, New York.

Laimer, J., Haslinger, S., Meissl, W., Hell, J., Störi, H. (2005a), Vacuum, vol. 79, p. 209.

Laimer, J., Haslinger, S., Meissl, W., Hell, J., Störi, H. (2005b), 17th Int. Symposium on Plasma Chemistry (ISPC-17), p. 330, Toronto, Canada.

Landau, L.D. (1997), Quantum Mechanics, Butterworth-Heinemann, 3rd edition.

Landau, L.D., Lifshitz, E.M. (1980), Statistical Physics, Part 1, Butterworth–Heinemann, Oxford.

Landau, L.D., Lifshitz, E.M. (1981a), Quantum Mechanics: Non-Relativistic Theory, Butterworth–Heinemann, Oxford.

Landau, L.D., Lifshitz, E.M. (1981b), Physical Kinetics, Course of Theoretical Physics, Butterworth–Heinemann, Oxford.

Landau, L.D., Lifshitz, E.M. (1986), Theory of Elasticity. Theoretical Physics, vol. 6, 3rd Edition (co-authored with Kosevich, A.M., and Pitaevskii, L.P.), Butterworth–Heinemann, Oxford.

Landau, L.D., Teller, E. (1936), Phys. Ztschr. Sow., Bd.10, p. 34.

Lange, H., Huczko, A., Sioda, M., Pacheco, M., Razafinimanana, M., Gleizes, A. (2002), Plasma Chem. Plasma Process., vol. 22, p. 523.

Langmuir, I. (1928), Proc. Nat. Acad. Sci. USA, vol. 14, p. 628.

Langmuir, I. (1961), The Collected Works, vol. 3, Pergamon, p. 107.

Laroussi, M. (2005), Plasma Process. Polym., vol. 2, p. 391.

Laroussi, M., Leipold, F. (2004), Int. J. Mass Spectrom., vol. 233, p. 81.

Laroussi, M., Sayler, G., Galscock, B., McCurdy, B., Pearce, M., Bright, N., Malott, C. (1999), IEEE Trans. Plasma Sci., vol. 27, p. 34.

Laroussi, M., Alexeff, I., Kang, W. (2000), IEEE Trans. Plasma Science, vol. 28, p. 184.

Laroussi, M., Alexeff, I., Richardson, P., Dyer, F.F. (2002a), IEEE Trans. Plasma Sci., vol. 30, no.1, p. 159.

References

Laroussi, M., Richardson, J.P., Dobbs, F.C. (2002b), Appl. Phys. Lett., vol. 81, p. 772.

Laroussi, M., Mendis, D.A., Rosenberg, M. (2003), New J. Phys., vol. 5, p. 41.

Lathman, B. (1967), U.S. Patent, no. 3,344,051.

Laudavala, W.K., Moroussi, J.W. (1985), J. Phys. D: Appl. Phys., vol. 14, p. 2015.

Laurendeau, N.M. (1978), Prog. Energy Combust. Sci., vol. 4, p. 221.

Lavid, M., Nachshon, Y., Gulati, S. K., Stevens, J.G. (1994), Combust. Sci. Technol., vol. 96, p. 231.

Law, C.K. (2006), Combustion Physics, Cambridge University Press, New York, London.

Lebedev, O.G., Murashov, M.V., Prusakov, V.N. (1972), Sov. Phys., High Energy Chem. (Khimia Vysokikh Energij), vol. 6, p. 533.

Lee, S.J., Seo, Y.C., Jurng, J., Hong, J.H., Park, J.W., Hyun, J.E., Lee, T.G. (2004), Sci. Total Environ., vol. 325, p. 155.

Lefebvre, H., Overbeke, M. (1934), C.R. Acad. Sci., vol. 198, p. 172.

Legasov, V.A., Ed. (1978), Nuclear – Hydrogen Energy and Technology, vol. 1, Atom-Izdat, Moscow.

Legasov, V.A., Ed. (1988), Nuclear – Hydrogen Energy and Technology, vol. 8, Energo-Atom-Izdat, Moscow.

Legasov, V.A., Belousov, I.G., Rusanov, V.D., Fridman, A. (1977a), J. Nucl. Sci. Technol., Nuclear-Hydrogen Energy, vol. 2, no. 3, p. 158.

Legasov, V.A., Rusanov, V.D., Fridman, A., Sholin, G.V. (1977b), 3rd Int. Symposium on Plasma Chemistry (ISPC-3), vol. 5, p. 18, Limoge, France.

Legasov, V.A., Bochin, V.P., Rusanov, V.D., Sholin, G.V., Fridman, A. (1978a), 2nd World Hydrogen Energy Conference, Zurich, vol. 3, p. 1183.

Legasov, V.A., Givotov, V.K., Krasheninnikov, E.G., Rusanov, V.D., Fridman, A. (1978b), Sov. Phys., Doklady (Reports of USSR Academy of Sciences), vol. 238, p. 66.

Legasov, V.A., Rusanov, V.D., Fridman, A. (1978c), Plasma Chemistry, vol. 5, Smirnov, B.M. (ed.), Atom-Izdat, Moscow.

Legasov, V.A., Vakar, A.K., Denisenko, V.P., Maximov, G.P., Rusanov, V.D., Fridman, A., Sholin, G.V. (1978d), Sov. Phys., Doklady (Reports of USSR Academy of Sciences), vol. 243, p. 323.

Legasov, V.A., Ezhov, V.K., Potapkin, B.V., Rusanov, V.D., Fridman, A., Sholin, G.V., Bochin, V.P. (1980), Sov. Phys., Doklady (Reports of USSR Academy of Sciences), vol. 251, p. 845.

Legasov, V.A., Asisov, R.I., Butylkin, Yu.P., Rusanov, V.D., Fridman, A. (1981a), J. Nucl. Sci. Technol., Nuclear-Hydrogen Energy, vol. 1, no. 8, p. 13.

Legasov, V.A., Belousov, I.G., Givotov, V.K., Rusanov, V.D., Fridman, A. (1981b), J. Nucl. Sci. Technol., Nuclear-Hydrogen Energy, vol. 1, no. 8, p. 3.

Legasov, V.A., Smirnov, B.M., Chaivanov, B.B. (1982), Plasma Chemistry, vol. 9, Smirnov, B.M. (ed.), Atom-Izdat, Moscow.

Legasov, V.A., Asisov, R.I., Butylkin, Yu.P., Potapkin, B.V., Rusanov, V.D., Fridman, A. (1983a), Nuclear-Hydrogen Energy and Technology - 5, Legasov, V.A. (ed.), p. 71, Energo-Atom-Izdat, Moscow.

Legasov, V.A., Belousov, I.G., Givotov, V.K., Rusanov, V.D., Fridman, A. (1983b), Nuclear-Hydrogen Energy and Technology, vol. 5, Legasov, V.A. (ed.), Energo-Atom-Izdat, Moscow, p. 71.

Legasov, V.A., Belousov, I.G., Givotov, V.K., Krasheninnikov, E.G., Krotov, M.F., Patrushev, B.I., Rusanov, V.D., Fridman, A., Sholin, G.V. (1988), Nuclear – Hydrogen Energy and Technology, Legasov, V.A. (ed.), vol. 8, Energo-Atom-Izdat, Moscow.

Leonenkova, E.G., Smirnov, S.E., Shishkova, I.I., Tarasov, A.V. (1987), 11th USSR Conference on Membrane Methode of Separation of Mixtures, D.I. Mendeleev Moscow Institute of Chemical Technology, vol. 2, p. 88.

Leonhardt, J.W. (1984), Radiat. Phys. Chem., vol. 24, p. 167.

Leonov, A.V., Yarantsev, D.A., Napartovich, A.P., Kochetov, I.V. (2006), "Plasma-Assisted Ignition and Flame-holding in High-Speed Flows," 44th AIAA Aerospace Science Meeting and Exhibit, Reno, NV, AIAA paper 2006-563.

Leonov, S., Nebolsin, V., Shilov, V., Timofeev, B., Kozlov, A. (1999), "Perspectives of MHD and Plasma Technologies in Aerospace Applications," AIAA paper, March.

Leonov, S., Bityurin, V., Kolesnichenko, Yu. (2001a), AIAA paper 2001-0493.

Leonov, S., Bityurin, V., Savishenko, N., Yuriev, A., Gromov, V. (2001b), AIAA paper 2001-0640.

Le Roy, R.L. (1969), J. Phys. Chem., vol. 73, p. 4338.

Lesinski, J., Baronnet, J.M., Meillot, E., Debbagh-Nour, G. (1985), 7th Int. Symposium on Plasma Chemistry (ISPC-7), no. P-3-3, p. 261.

Leson, G., Dharmavaram, S. (1995), "A Status Overview of Biological Air Pollution Control," 88th Annual Meeting & Exhibition of the Air and Waste Management Association, San Antonio, TX, paper no. 95-MP 9A.01.

Lesueur, H., Czernichowski, A., Chapelle, J. (1990), J. de Physique, Colloque C5, p. 51.

Leung, K.N., Taylor, G.R., Barrick, J.M., Paul, S.L., Kribel. R.E. (1976), Phys. Lett., vol. 57A, p. 145.

Leutner, H.W. (1962), Eng. Chem. Process Des. Dev., vol. 1, p. 166.

Leutner, H.W. (1963), Eng. Chem. Process Des. Dev., vol. 2, p. 315.

Levin, A., Larin, O.B. (2003), AIAA paper 2003-0036.

References

Levine, R.D., Bernstein R.B. (1975), "Thermodynamic Approach to Collision Processes," in Modern Theoretical Chemistry, vol. 3, Dynamics of Molecular Collisions, Miller, W.H. (ed.), Plenum, New York.

Levitsky, A.A. (1978a), "Kinetics and Mechanisms of Chemical Reactions, Investigated by Means of Mathematical Modeling," Ph.D. Dissertation, Institute of Petrochemical Synthesis, USSR Academy of Sciences, Moscow.

Levitsky, A.A. (1978b), "Investigation of Elementary Reaction $O + N_2$ Using Method of Classical Trajectories," in New Aspects of Petrochemical Synthesis, Institute of Petrochemical Synthesis of USSR Academy of Sciences, Moscow.

Levitsky, A.A., Polak L.S. (1980), Sov. Phys., High Energy Chem., (Khimia Vysokikh Energij), vol. 14, p. 3.

Levitsky, A.A., Macheret, S.O., Fridman, A., (1983a), High Energy Chem. (Khimia Vysokikh Energij), vol. 17, p. 625.

Levitsky, A.A., Macheret, S.O., Fridman, A. (1983b), "Kinetic Modeling of Plasma-Chemical Processes Stimulated by Vibrational Excitation," in Chemical Reactions in Non-Equilibrium Plasma, Polak, L.S. (ed.), p. 3, Nauka (Science), Moscow.

Levitsky, A.A., Macheret, S.O., Polak, L.S., Fridman, A. (1983c), Sov. Phys., High Energy Chem. (Khimia Vysokikh Energij), vol. 17, p. 625.

Levitsky, S.M. (1957), Sov. Phys., J. Techn. Phys., JTPh, vol. 27, p. 970, 1001.

Lewis, B., von Elbe, G. (1987), Combustion, Flames and Explosion of Gases, 3rd edition, Academic Press, Orlando, FL, London.

Li, H.P., Chen, X. (2001), Thin Solid Films, vol. 380, p. 175.

Liakhovich, A.P. (1974), Elec. Ind. Electrotherm. Process., vol. 8, no. 144, p. 21.

Lieberman, M.A. (1989a), IEEE Trans. Plasma Sci., vol. 17, p. 338.

Lieberman, M.A. (1989b), J. Appl. Phys., vol. 65, p. 4168.

Lieberman, M.A. (1989c), J. Appl. Phys., vol. 66, p. 2926.

Lieberman, M.A., Gottscho, R.A. (1994), Physics of Thin Films, vol. 18, Francombe, M.H., and Vossen, J.L. (eds.), Academic Press, New York.

Lieberman, M.A., Lichtenberg, A.J. (1994), Principles of Plasma Discharges and Material Processing, John Wiley & Sons, New York.

Lieberman, M.A., Velikovich, A.L. (1985), Physics of Shock Waves in Gases and Plasmas, Springer, B., New York.

Lieberman, M.A., Lichtenberg, A.J., Savas, S.E. (1991), IEEE Trans. Plasma Sci., vol. 19, p. 189.

Lifshitz, A. (1974), J. Chem. Phys., vol. 61, p. 2478.

Lifshitz, E.M., Landau, L.D., Pitaevskii, L.P. (1984), Electrodynamics of Continuous Media. Course of Theoretical Physics, vol. 8, 2nd Edition, Butterworth-Heinemann, Oxford.

Lifshitz, Y. (1996), Diam. Relat. Mater., vol. 5, p. 388.

Lifshitz, Y. (1999), Diam. Relat. Mater., vol. 8, p. 1659.

Likalter, A.A. (1975a), Sov. Phys., Quant. Electr. (Kvantovaja Electronika), vol. 2, p. 2399.

Likalter, A.A. (1975b), Sov. Phys., J. Appl. Mech. Techn. Phys., JPMTF, vol. 3, p. 8.

Likalter, A.A. (1976), Sov. Phys., J. Appl. Mech. Techn. Phys., JPMTF, vol. 4, p. 3.

Likalter, A.A., Naidis, G.V. (1981), Plasma Chemistry - 8, Smirnov, B.M. (ed.), Energoizdat, Moscow.

Likhanskii, A., Shneider, M., Macheret, S., Miles, R. (2007), "Optimization of Dielectric Barrier Discharge Plasma Actuators Driven by Repetitive Nanosecond Pulses," 45th AIAA Aerospace Science Meeting and Exhibit, Reno, NV, AIAA paper 2007-0183.

Lin, I. (1985), J. Appl. Phys., vol. 58, p. 2981.

Lind, S.C. (1923), Trans. Am. Electrochem. Soc., vol. 44, p. 68.

Linder, W., Davis, S. (1931), J. Phys. Chem, vol. 35, p. 3649.

Linowitzki, V. (1971), Kunststoffe, vol. 61, p. 188.

Lipatov, K.V., Kanorskii, I.D., Shekhter, A.B., Emelianov, A.Y. (2002a), Ann. Surg., vol. 1, p. 58.

Lipatov, K.V., Sopromadze, M.A., Shekhter, A.B., Emelyanov, A.Y., Grachev, S.V. (2002b), Surgery, vol. 2, p. 188.

Lipin, Yu.V., Rogachev, A.V., Sydorsky, S.S., Kharitonov, V.V. (1994), Technology of Vacuum Metallization of Polymer Materials, Belorussian Academy of Engineering and Technology, Gomel, Belarus.

Lipovich, V.L., Ed. (1988), Chemistry and Conversion of Coal, Khimia (Chemistry), Moscow, Russia.

Lippmann, M., (2000), Environmental Toxicants: Human Exposures and Their Health Effects, John Wiley & Sons, New York.

Lisitsyn, I.V., Nomiyama, H., Katsuki, S., Akiyama, H. (1999), Rev. Sci. Instr., vol. 70, p. 3457.

Liston E.M. (1993), J. Adhes. Sci. Technol., vol. 7, p. 1091.

Liston, E.M., Martinu, L., Wertheimer, M.R. (1994), Plasma Surface Modification of Polymers: Relevance to Adhesion, Strobel, M., Lyons, C., and Mittal, K.L. (eds.), VSP, Netherlands.

Liu, J., Ronney, P.D, Gundersen, M.A. (2003), 3rd Joint Meeting of the US Sections of the Combustion Institute, Chicago, IL, paper B-25.

References

Liu, J., Wang, F., Lee, L., Ronney, P.D, Gundersen, M.A. (2004), "Effect of Fuel Type on Flame Ignition by Transient Plasma Discharges," 42nd AIAA Aerospace Sciences Meeting and Exhibit, Reno, NV, AIAA paper 2004.

Liventsov, V.V., Rusanov, V.D., Fridman, A. (1983), Sov. Phys., J. Techn. Phys. Lett., Pis'ma v JTF, vol. 7, p. 163.

Liventsov, V.V., Rusanov, V.D., Fridman, A. (1984a), Sov. Phys., Doklady (Reports of USSR Academy of Sciences), vol. 275, p. 1392.

Liventsov, V.V., Rusanov, V.D., Fridman, A. (1984b), Energy Efficiency of CO_2 – Dissociation in Nonthermal Plasma in Approximation of Quasi-Equilibrium of Vibrational Modes, Kurchatov Institute of Atomic Energy, Moscow, vol. 4201/6.

Lob, W. (1906), Z. Electrochem., vol. 12, p. 282.

Locke B.R., Sato, M., Sunka, P., Hoffmann, M.R., Chang, J.S. (2006), Ind. Eng. Chem. Res., vol. 45, p. 882.

Loeb, L.B. (1960), Basic Processes of Gaseous Electronics, University of California Press, Berkeley.

Loeb, L.B. (1965), Electrical Coronas – Their Basic Physical Mechanisms, University of California Press, Berkeley.

Long, B.H., Wu, H.H., Long, B.Y., Wang, J.B., Wang, N.D., Lu, X.Y., Jin, Z.S., Bai, Y.Z. (2005), J. Phys. D: Appl. Phys., vol. 38, p. 3491.

Longfield, J.E. (1969), BRD Patent no. 2053206, Germany.

Lopez, L.C., Gristina, R., Favia, P., d'Agostino, R. (2005), 17th Int. Symposium on Plasma Chemistry (ISPC-17), p. 1078, Toronto, Canada.

Losanitch, S.M. (1911), Ber. (Beratung), vol. 44, p. 312.

Losanitch, S.M., Jovitschitsch, M.Z. (1897), Ber. (Beratung), vol. 30, p. 135.

Losev, S.A. (1977), Gas-Dynamic Lasers, Nauka (Science), Moscow.

Losev, S.A., Generalov, N.A. (1961), Sov. Phys., Doklady (Reports of USSR Academy of Sciences), vol. 141, p. 1072.

Losev, S.A., Shatalov, O.P., Yalovik, M.S. (1970), Sov. Phys., Doklady, vol. 195, p. 585.

Losev, S.A., Sergievska, A.L., Rusanov, V.D., Fridman, A., Macheret, S.O (1996), Sov. Phys. Doklady (Reports of USSR Academy of Sciences), vol. 346, p. 192.

Lou, G., Bao, A., Nishihara, M., Keshav, S., Utkin, Y.G., Adamovich, I.V. (2006), "Ignition of Premixed Hydrocarbon-Air Flows by Repetitively Pulsed, Nano-Second Pulsed Duration Introduction Plasma," 44th AIAA Aerospace Sciences Meeting and Exhibit, January 2006, Reno, NV, AIAA paper 2006-1215.

Lou, G., Bao, A., Nishihara, M., Keshav, S., Utkin, Y.G., Adamovich, I.V. (2007), Proc. Combust. Inst., vol. 31.

Lovejoy, E.R., Wang, N.S., Howard, C.J. (1987), J. Phys. Chem., vol. 91, p. 5749.

Lowke, J.J., D'Alessandro, F. (2003), J. Phys. D: Appl. Phys., vol. 36, p. 2673.

Lowry, H., Smith, M., Sherrouse, P., Felderman, J., Drake, J., Bauer, M., Pruitt, D., Keefer, D. (1999a), 3rd Weakly Ionized Gas Workshop (Norfolk, November 1999), AIAA paper 99-4822.

Lowry, H., Stepanek, C., Crosswy, L., Sherrouse, P., Smith, M., Price, L., Ruyten, W., Felderman, J. (1999b), AIAA paper 90-0600.

Lozansky, E.D., Firsov, O.B. (1975), Theory of Sparks, Atomizdat, Moscow.

Lu, X.P., Laroussi, M. (2006), J. Appl. Phys., vol. 100, p. 302.

Lugscheider, E., Rass, I. (1992), J. Therm. Spray Technol., vol. 1, no. 1, p. 49.

Lugscheider, E., Born, K., Wagner, N. (1995), 2nd Plasma-Technik Symp., Lucerne, Switzerland, vol. 1, p. 221.

Lugscheider, E., Jungklaus, H., Schwier, G., Mathesius, H., Heinrich, P. (1995), 8th National Thermal Spray Conf., ASM Int'l, Houston, TX, p. 333.

Lukhovitskii, B.I., Starik, A.M., Titova, N.S. (2005), Combust. Explo. Shock Waves, vol. 41, p. 386.

Lukianov, V.B., Simonov, E.F., Mogaisky, A.M. (1973), Sov. Phys., J. Phys. Chem., JFCh, vol. 47, p. 2838.

Lukiyanchuk, I.V., Rudnev, V.S., Andenko, N.A., Kaidalova, T.A., Panin, E.S., Gordienko, P.S. (2002), Russ. J. Appl. Chem., vol. 75, p. 573.

Lukyanova, A.B., Rahimov, A.T., Suetin, N.B. (1990), Sov. Phys., Plasma Physics (Physica Plasmy), vol. 16, p. 1367.

Lukyanova, A.B., Rahimov, A.T., Suetin, N.B. (1991), Sov. Phys., Plasma Physics (Physica Plasmy), vol. 17, p. 1012.

Lunt, R.W. (1925), Proc. Roy. Soc. London A, vol. 108, p. 172.

Lytkin, A.Ya., Pendrakovsky, V.T., Varapaeva V.G. (1969), Sov. Phys., Nitrog. Ind., vol. 7, p. 569.

Lytkin, A.Ya., Zyrichev, N.A., Pendrakovsky, V.T. (1970), Physics, Technology and Applications of Low Temperature Plasma, Science (Nauka), Alma-Ata, Kyrgyzstan.

Lyubimov, G.A., Rachovsky, V.I. (1978), Adv. Phys. Sci., Uspekhi Phys. Nauk, vol. 125, p. 665.

MacDonald, A.D., Tetenbaum, S.J. (1978), Gaseous Electronics, Hirsh, M.N. Oskam, H.J. (eds.), vol. 1, Electrical Discharges, Academic Press, New York.

Macheret, S.O., Rusanov, V.D., Fridman, A., Sholin, G.V. (1978), Sov. Phys., Lett. J. Techn. Phys., Pis'ma v JTF, vol. 4, p. 28.

Macheret, S.O., Rusanov, V.D., Fridman, A., Sholin, G.V. (1980a), Sov. Phys., Doklady, vol. 255, p. 98.

Macheret, S.O., Rusanov, V.D., Fridman, A., Sholin, G.V. (1980b), Sov. Phys., J. Techn. Phys., vol. 50, p. 705.

References

Macheret, S.O., Rusanov, V.D., Fridman, A. (1984), Sov. Phys., Doklady, vol. 276, p. 1420.

Macheret, S.O., Fridman, A., Adamovich, I.V., Rich, J.W., Treanor, C.E. (1994), Mechanism of Non-Equilibrium Dissociation of Diatomic Molecules, AIAA paper 94-1984.

Macheret, S.O., Shneider, M.N., Miles, R.B. (2000), 2nd Workshop on Magneto-Plasma-Aerodynamics in Aerospace Applications, Moscow.

Macheret, S.O., Ionikh, Y.Z., Chemysheva, N.V., Yalin, A.P., Martinelli, L., Miles, R. (2001a), Phys. Fluids, vol. 13, p. 2693.

Macheret, S.O., Shneider, M.N., Miles, R.B. (2001b), AIAA paper 2001-0492.

Macheret, S.O., Shneider, M.N., Miles, R.B. (2001c), AIAA paper 2001-0795.

Macheret, S.O., Shneider, M.N., Miles, R.B. (2005), 36th AIAA Plasmadynamics and Lasers Conference, Toronto, Canada, AIAA-2005-5371.

Maciel, H.S., Allen, J.E. (1989), J. Plasma Phys., vol. 42, p. 321.

MacKinnon, J.M., Reuben, B.G. (1975), J. Electrochem. Soc., vol. 122, no. 6, p. 806.

MacKinnon, J.M., Wickens, A.J. (1973), Chem. Ind., vol. 16, p. 800.

Madou, M.J. (2002), Fundamentals of Microfabrication, CRC Press, Boca Raton, FL.

Maeg, S., Geshi, O. (1971), Elec. Eng. Jpn, vol. 91, p. 116.

Makeev, G.N., Sokolov, V.B., Chaivnov, B.B. (1975), Plasma Chemistry, vol. 2, Smirnov, B.M. (ed.), Atom-Izdat, Moscow.

Maker, P.D., Terhune, R.W., Savage, C.M. (1964), Quantum Electronics III, Grivet, P., Bloembergen, N. (eds.), Columbia University Press, New York.

Maksimenko, A.P., Tverdokhlebov, V.P. (1964), Sov. Phys., News Higher Educ. (Izvestia VUZov), Phys., vol. 1, p. 84.

Malkov, S.Yu. (1981), "Dissociation of Water Vapor in Non-Equilibrium Low-Temperature Plasma of Microwave Discharges," Ph.D. Dissertation, Moscow Institute of Physics and Technology, Moscow.

Maltsev, A.I., Eremin, E.N., Ivanter, V.L. (1967), Sov. Phys., J. Phys. Chem., vol. 41, p. 1190.

Maly, R., Vogel, M. (1979), 17th Symposium on Combustion, Pittsburgh, PA, p. 821.

Mamedov, Sh.S. (1979), Trudy FIAN, Physical Institute of Academy of Sciences, vol. 107, p. 3.

Mandich, M.L., Reents, W.D., Jr. (1992), J. Chem. Phys., vol. 96, p. 4233.

Mangano, J.A., Jacob, J.H. (1975), Appl. Phys. Lett., vol. 27, p. 495.

Mangolini, L., Thimsen, E., Kortshagen, U. (2005), 17th Int. Symposium on Plasma Chemistry (ISPC-17), p. 770, Toronto, Canada.

Manheimer, W., Sugiyama, L.E., Stix, T.H., eds. (1996), Plasma Science and the Environment, Springer, Berlin, Heidelberg, New York.

Maniero, D.A., Kienast, P.F., Hirayama, C. (1966), Westinghouse Eng., vol. 26, p. 66.

Mankelevich, Yu.A., Suetin, N.V. (2000), "Physical and Chemical Processes in Plasma Reactors for Deposition of Diamond Films," in Encyclopedia of Low-Temperature Plasma, Fortov, V.E. (ed.), vol. 4, p. 404, Nauka (Science), Moscow.

Mann, M.M., Rice, D.K., Eguchi, R.G. (1974), IEEE J. Quant. Electr., vol. QE-10, p. 682.

Manning, D. (1971), Kinetics of Atomic Diffusion in Crystals, Mir, Moscow.

Manolache, S., Shamamian, V., Denes, F. (2004), J. Environo. Eng., vol. 130, p. 17.

Manos, D.M., Flamm, D.L. (1989), Plasma Etching, Academic New York.

Marcum, S.D., Parish, J.W., Ganguly, B.N. (2004), J. Propul. Power, vol. 20, p. 360.

Margeloff, J.B. (1966), U.S. Patent no. 3,254,958 (6/7/1966).

Margolin, A.D., Mishchenko, A.V., Shmelev, V.M. (1980), Sov. Phys., High Energy Chem. (Khimia Vysokikh Energij), vol. 14, p. 162.

Markovsky, L.Ya, Lvova, V.I., Kondrashev, Ya.D. (1958), Chemistry of Boron and Its Compounds, p. 36, Gos-Chim-Izdat, Moscow.

Marrone, P.V., Treanor, C.E. (1963), Phys. Fluids, vol. 6, p. 1215.

Martinu, L., Raveh, A., Boutard, D., Houle, S., Wertheimer, M. (1993), Diam. Rel. Mater., vol. 2, p. 673.

Marxer, C.G. (2002), "Protein and Cell Absorption: Topographical Dependency and Adlayer Viscoelastic Properties Determined with Oscillation Amplitude of Quartz Resonator," Department of Physics, University of Fribourg, Fribourg, Switzerland.

Massey, H.S.W. (1969), Electronic and Ionic Phenomena, vol. 2, Oxford University Press, UK.

Massey, H.S.W. (1976), Negative Ions, Cambridge University Press, Cambridge.

Massey, H.S.W., Burhop, E., Gilbody, H.B. (1974), Electron and Ion Impact Phenomena, Clarendon, Oxford, UK.

Massines, F., Gouda, G. (1998), J. Phys. D: Appl. Phys., vol. 31, no. 24, p. 3411.

Masuda, S. (1993), Nonthermal Plasma Techniques for Pollution Control: Part B – Electron Beam and Electrical Discharge Processing, Penetrante, B.M., Schultheis, S.E. (eds.), Springer-Verlag, Berlin.

Masuda, S., Wu, Y., Urabe, T., Ono, Y. (1987), 3rd Int. Conf. on Electrostatic Precipitation, Abano-Padova, Italy.

Masuda, S., Akutsu, K., Kuroda, M., Awatsu, Y., Shibuya, Y. (1988), IEEE Trans. Ind. Appl., vol. 24, p. 223.

Matossian, J.N. (1994), J. Vac. Sci. Technol., vol. 12, p. 850.

References

Matossian, J.N., Goebel, D.M. (1992), 34th Annual APS Division of Plasma Physics, Seattle, WA.

Matsumoto, O. (1973), Ist International Symposium on Plasma Chemistry (ISPC-I), paper 5-5, Kiel, Germany.

Matsumoto, O., Abe, K. (1975), Denki Kagaku, vol. 43, no. 2, p. 75.

Matsumoto, O., Hayakawa, Y. (1966), J. Electrochem. Soc. Jpn, vol. 34, no. 4, p. 216.

Matsuoka, M., Ono, K. (1988), J. Vac. Sci. Technol., vol. A6, p. 25.

Matsuura, T. (1993), Synthetic Membranes and Membrane Separation Processes, CRC Press, Boca Raton, FL.

Mattachini, F., Sani, E., Trebbi, G. (1996), Int. Workshop on Plasma Technologies for Pollution Control and Waste Treatment, MIT, Cambridge, MA.

Matthews, S.M., Boegel, A.J., Loftis, J.A., Caufield, R.A., Mincher, B.J., Meikrantz, D.H. (1993), Radiat. Phys. Chem., vol. 42, p. 689.

Matveev, I., Matveeva, S., Gutsol, A., Fridman, A. (2005), 43d AIAA Aerospace Sciences Meeting and Exhibit, 10–13 January 2005, Reno, NV, AIAA paper 2005-1191.

Matzing, H. (1989a), Chemical Kinetics of Flue Gas Cleaning by Electron Beam, Keruforschunfzektrum, KFK, 4494.

Matzing, H. (1989b), Radiat. Phys. Chem., vol. 33, p. 81.

Maximov, A.I. (2000), Encyclopedia of Low-Temperature Plasma, Fortov, V.E. (ed.), vol. 4, p. 399, Nauka (Science), Moscow.

Maximov, A.I., Sizov, V.D., Chebotarenko, O.B. (1970), Sov. Phys., Electron. (Electronika), vol. 2, no. 18, p. 92.

Maximov, A.I., Polak, L.S., Sergienko, A.F., Slovetsky, D.I. (1979), Sov. Phys., High Energy Chem. (Khimia Vysokikh Energij), vol. 13, p. 165.

Maximov, A.I., Gorberg, B.L., Titov, V.A. (1997), "Possibilities and Problems of Plasma Treatment of Fabrics and Polymer Materials," in Textile Chemistry – Theory, Technology, Equipment, Moryganov, A.P. (ed.), NOVA Science Publishers, New York.

Mayer, H. (1962), Ber. Dtsch. Keram. Ges., vol. 39., no. 2, p. 115.

McCarthy, R.L. (1954), J. Chem. Phys., vol. 22, p. 1360.

McCaughey, M.J., Kushner, M.J. (1989), J. Appl. Phys., vol. 65, p. 186.

McDaniel, E.W. (1964), Collision Phenomena in Ionized Gases, Wiley, New York.

McDaniel, E.W. (1989), Atomic Collisions: Electron and Photon Projectiles, Wiley, New York.

McDaniel, E.W., Mason, E.A. (1973), The Mobility and Diffusion of Ions in Gases, Wiley, New York.

McDaniel, E.W., Nighan, W.L. (1982), "Gas Lasers," Applied Atomic Collision Physics, vol. 3, Academic Press, New York, London.

McDermott, W.E., Pchelkin, N.R., Benard, D.J., Bousek, R.R. (1978), Appl. Phys. Lett. vol. 32, p. 469.

McDougall, L.A., Kilpatrick, J.E. (1965), J. Chem. Phys., vol. 42, p. 2311.

McEwan, M.J., Phillips, L.F. (1975), Chemistry of the Atmosphere, Wiley, New York.

McHenry, M.E., Majetich, S.A., Artman, J.O., DeGraef, M., Staley, S.W. (1994), Phys. Rev., B, vol. 49, p. 11358.

McKenzie, D.R. (1993), J. Vac. Sci. Technol., vol. B 11, p. 1928.

McKenzie, D.R. (1996), Rep. Prog. Phys., vol. 59, p. 1611.

McKenzie, D.R., Muller, D., Pailthorpe, B.A. (1991), Phys. Rev. Lett., vol. 67, p. 773.

McTaggart, F. (1963), Nature, vol. 199, p. 399.

McTaggart, F. (1964), Nature, vol. 201, p. 1320.

McTaggart, F. (1965a), Austral. J. Chem., vol. 18, p. 937.

McTaggart, F. (1965b), "Plasma Production of Pure Metals in Microwave Discharges," Australia Patent no. 66043/65.

McTaggart, F. (1967), Plasma Chemistry in Electric Discharges, Topics in Inorganic and General Chemistry, Elsevier.

Medin, S.A. (2000), "Plasma MHD-Denerators in Power Electronics," in Encyclopedia of Low Temperature Plasma, Fortov, V.E. (ed.), vol. 4, p. 154, Nauka (Science), Moscow.

Meek, J.M., Craggs, J.D. (1978), Electrical Breakdown of Gases, Wiley, New York.

Meeks, E., Larson, R.S., Vosen, S.R., Shon, J.W. (1997), J. Electrochem. Soc., vol. 144, p. 1.

Melik-Aslanova, T.A., Rusanov, V.D., Shilnikov, V.I., Potapkin, B.V., Fridman, A., Givotov, V.K., Gutsol, A.F. (1985), Sov. Phys., Doklady (Reports of Azerbaijan Academy of Sciences), vol. 10, p. 44.

Mendis, D.A., Rosenberg, M., Azam, F. (2000), IEEE Trans. Plasma Sci., vol. 28, p. 1304.

Meschi, D.S., Meyers, R.J. (1956), J. Am. Chem. Soc., vol. 78, p. 6220.

Messerle, V.E. (2004), High Energy Chem. (Khimia Vysokikh Energij), vol. 38, p. 35.

Messerle, V.E., Sakipov, Z.B. (1988), Chem. Solid Fuels (Khimia Tverdykh Topliv), vol. 4, p. 123.

Messerle, V.E., Sakipov, Z.B., Siniarev, G.B., Trusov, B.G. (1985), Sov. Phys., High Energy Chem. (Khimia Vysokikh Energij), vol. 19, p. 160.

Metel, A.S., Nastukha, A.I. (1977), Sov. Phys., High Energy Chem. (Khimia Vysokikh Energij), vol. 11, p. 366.

Meubus, P. (1973), Can. J. Chem. Eng., vol. 51, no. 4, p. 440.

Meunier, J., Belenguer, Ph., Boeuf, P. (1995), J. Appl. Phys., vol. 78, p. 731.

Meyer, S., Gorges, R., Kreisel, G. (2004a), Thin Solid Films, vol. 450, p. 276.

References

Meyer, S., Gorges, R., Kreisel, G. (2004b), Electrochim. Acta, vol. 49, p. 3319.

Mezdrikov, O.A. (1968), Electrical Methods of Volumetric Granulometry, Energia (Energy), Leningrad.

Migdal, A.B. (1981), Qualitative Methods in Quantum Mechanics, Nauka (Science), Moscow.

Mikaberidze, A.A., Ochkin, V.N., Sobolev, N.N. (1972), J. Techn. Phys. (JTF), vol. 42, p. 1464.

Milko, V.I., Gukov, M.F., Isay, P.P. (1973), "Method of Production of Titanium Dioxide," Patent USSR (avtorskoe svid.), no. 322960, Discoveries and Inventions, vol. 46, p. 84.

Miller, T.N. (1971a), Low-Temperature Plasma in Technology of Inorganic Substances, Nauka (Science), Novosibirsk, Russia.

Miller, T.N. (1971b), Plasma-Chemistry-71, p. 140, Polak, L.S. (ed.), Institute of Petrochemical Synthesis of USSR Academy of Sciences, Moscow.

Miller, T.N., Polak, L.S., Raitsis, L.S (1980), Sov. Phys., High Energy Chem. (Khimia Vysokikh Energij), vol. 14, p. 168.

Minayeva, O., Hopwood, J. (2003), J. Appl. Phys., vol. 94, p. 2821.

Minayeva, O., Laroussi, M. (2004), Proc. IEEE Int. Conf. on Plasma Science (ICOPS), IEEE Press, Baltimore, MD, p. 122.

Mishin, G.I. (1997), AIAA Aerospace Sciences Meeting, AIAA paper 1997-2298.

Mishin, G.I., Serov, Yu.L., Yavor, I.P. (1991), Sov. Phys., J. Techn. Phys. (JTF), vol. 17, p. 413.

Misyn, F.A., Gostev, V.A. (2000), "Cold Plasma Application in Eyelid Phlegmon Curing," in Diagnostics and Treatment of Infectious Diseases, Petrozavodsk University, Petrozavodsk, Russia.

Misyn, F., Besedin, E., Gostev, V., and Komkova, O. (2000a), "Experimental Studying of Bactericidal Action of Cold Plasma," in Diagnostics and Treatment of Infectious Diseases, Petrozavodsk University, Petrozavodsk, Russia.

Misyn, F.A., Besedin, E.V., Komkova, O.P., Gostev, V.A. (2000b), "Experimental Investigation of Bactericidal Influence of Cold Plasma and Its Interaction with Cornea," in Diagnostics and Treatment of Infectious Diseases, Petrozavodsk University, Petrozavodsk, Russia.

Misyn, F.A., Besedin, E.V., Obraztsova, A.M., Gostev, V.A. (2000c), "Experimental Curing of Bacterial Ulcerous Keratitis with Cold Plasma," in Diagnostics and Treatment of Infectious Diseases, Petrozavodsk University, Petrozavodsk, Russia.

Mitscuio, M., Yasio, K. (1968), J. Chem. Soc. Jpn, Ind. Chem., vol. 71, p. 492.

Mitscuio, M., Yasio, K. (1970), J. Chem. Soc. Jpn, Ind. Chem., vol. 73, p. 306.

Mitsuda, Y., Yoshida, T., Akashi, K. (1989), Rev. Sci. Instrum., vol. 60, p. 249.

Mnatsakanyan, A.Kh., Naidis, G.V. (1992), Institute of High Temperatures, Russian Academy of Sciences, vol. I-334.

Modest M.F. (1993), Radiative Heat Transfer, McGraw-Hill, New York.

Modinos, A. (1984), Field, Thermionic and Secondary Electron Emission Spectroscopy, Plenum Press, New York.

Mogab, C.J., Adams, A.C., Flamm, D.L. (1979), J. Appl. Phys., vol. 49, p. 3796.

Mohanty, M., Smith, R.W., Knight, R., Chen, W.L.T., Heberlein, J.V.R. (1996), 9th National Thermal Spray Conf., ASM Int'l, Cincinnati, OH, p. 967.

Moisan, M., Zakrzewski, Z. (1991), J. Phys. D: Appl. Phys., vol. 24, p. 1025.

Moisan, M., Barbeau, Moreau, S., J., Pelletier, Tabrizian, M., Yahia, L.H. (2001), Int. J. Pharm., vol. 226, p. 1.

Molchanov Yu.S., Starik, A.N. (1984), "Kinetics of Vibrational Energy Exchange in Hydrocarbon Combustion Products during Gas Expansion in Supersonic Nozzles," Central Institute of Aviation Motors, Preprint # 10160, Moscow.

Molinary, E. (1974), Pure Appl. Chem., vol. 39, p. 343.

Montie, T.C., Kelly-Wintenberg, K., Roth, J.R. (2000), IEEE Trans. Plasma Sci., vol. 28, p. 41.

Moon, S.Y., Choe, W., Kang, B.K. (2004), Appl. Phys. Lett., vol. 84, p. 188.

Moreau, C., Gougeon, P., Burgess, A., Ross, D. (1995), 8th National Thermal Spray Conf., ASM Int'l., Houston, TX, p. 141.

Moreau, S., Moisan, M., Barbeau, J., Pelletier, J., Ricard, A. (2000), J. Appl. Phys., vol. 88, p. 1166.

Morgan, R.A. (1985), Plasma Etching in Semiconductor Fabrication, Elsevier, New York.

Morozov, A.I. (1978), Physical Basis of the Space Electric-Jet Engines, AtomIzdat, Moscow.

Mort, J., Jansen, F. (1986), Plasma Deposited Thin Films, Pub. Chemical Rubber Company, Boca Raton, FL.

Moruzzi, J.L., Kiermasz, A., Eccleston, W. (1982), Plasma Phys., vol. 24, p. 605.

Moskalev, B.I. (1969), The Hollow Cathode Discharge, Energy, Moscow.

Mosse, A.L., Burov, I.S. (1980), Treatment of Dispersed Materials in Plasma Reactors, Nauka I Technica (Science and Engineering), Minsk, Belarus.

Mosse, A.L., Pechkovsky, V.V. (1973), Application of Low-Temperature Plasma in Technology of Inorganic Materials, Nauka I Technica (Science and Engineering), Minsk, Belarus.

Mosse, A.L., Zabrodin, V.K., Krylova, I.A. (1971), Sov. Phys., Eng.-Phys. J. (IFJ), vol. 20, p. 3.

Mosse, A.L., Zabrodin, V.K., Krylova, I.A. (1973), Sov. Phys., High Energy Chem. (Khimia Vysokikh Energij), vol. 7, p. 491.

References

Moukhametshina, Z.B., Chekmarev, A.M., Givotov, V.K., Kalachev, I.A., Fridman, A. (1986a), Sov. Phys., J. Techn. Phys. (JTF), vol. 56, p. 757.

Moukhametshina, Z.B., Chekmarev, A.M., Givotov, V.K., Kalachev, I.A., Fridman, A. (1986b), Sov. Phys., High Energy Chem. (Khimia Vysokikh Energij), vol. 20, p. 354.

Muller, M., Heimann, R.B., Gitzhofer, F, Boulos, M.I. (1995), 8th National Thermal Spray Conf., ASM Int'l, Houston, TX, p. 567.

Muller, R., Peuckert, G., Krebs, W. (1987), 8th Int. Symposium on Plasma Chemistry (ISPC-8), Tokyo, Japan, no. P060, p. 707.

Müller, U., Lugscheider, E. (1992), 3rd Int. Conf. on New Diamond Science and Technology (Diamond '92), Heidelberg, FRG, COMST.

Munipalli, R., Shankar, V. (2001), AIAA paper 2001-0198.

Munir, Z. A., Anselmi-Tamburini, V. (1989), Mater. Sci. Rep., vol. 3, no. 7, p. 277.

Murdoch, H.D., Hamblyn, S.M.L. (1967), French Patent, no. 1,582,154.

Murdoch, H.D., Hamblyn, S.M.L. (1972), "Chemical Process and Apparatus Utilizing Plasma," U.S. Patent no. 3,625,846.

Murphy, A.B. (1997), J. Proc. Roy. Soc. NSW, vol. 130, p. 93.

Murphy, A.B., McAlister, T., Farmer, A.J.D., Horrigan E.C. (2001), 15th Int. Symposium on Plasma Chemistry (ISPC-16), Orleans, France, p. 619.

Murphy, A.B., Farmer, A.J.D., Horrigan, E.C., McAllister, T. (2002), Plasma Chem. Plasma Proces., vol. 22, p. 371.

Murtabekov, M.B. (1980), Investigations of Multi-Component Systems, Nauka (Science), Moscow.

Musin, A.K. (1974), Ionization Processes in Heterogeneous Gas Plasma, Moscow State University, Moscow.

Mutaf-Yardimci, O. (2001), "Plasma-Catalysis in Hydrocarbon Processing by Using Non-Equilibrium Plasma Discharges," Ph.D. Dissertation, University of Illinois at Chicago, Chicago, IL.

Mutaf-Yardimci, O., Kennedy, L., Saveliev, A., Fridman, A. (1998), "Plasma Exhaust Aftertreatment," SAE, SP-1395, p. 1.

Mutaf-Yardimci, O., Saveliev, A.V., Fridman, A., Kennedy, L.A. (1999), Int. J. Hydrog. Energy, vol. 23, p. 1109.

Naidis, G.V. (1997), J. Phys. D: Appl. Phys., vol. 30, p. 1214.

Nakagawa, Y., Adachi, S., Kohchi, A. (1996), Jpno. J. Appl. Phys., vol. 35, p. 2808.

Nalbandjan, A. (1934), Acta Physicochem, URSS, vol. 1, p. 305.

Namba, H. (1990), Appl. Radiat. Isotope, vol. 41, p. 569.

Natanson, G. (1959), Sov. Phys., J. Techn. Phys., vol. 29, p. 1373.

Naville, A.A., Guye, C.E. (1904), French Patent No. 350120.

Neely, W.C., Newhouse, E.I., Clothiaux, E.J., Gross, C.A. (1993), Non-Thermal Plasma Techniques for Pollution Control: Part B – Electron Beam and Electrical Discharge Processing, Penetrante, B.M., Schultheis, S.E. (eds.), Springer-Verlag, Berlin.

Nelea, V., Luo, L., Demers, C.N., Antoniou, J., Petit, A., Lerouge, S., Wertheimer, M.R., Mwale, F. (2005), J. Biomed. Res., vol. 75A, p. 216.

Nelson, L.Y., Saunder, A.W., Harwey, A.B. (1971), J. Chem. Phys., vol. 55, p. 5127.

Nemchinsky, V.A. (1996), Wel. J., vol. 75, p. 388.

Nemchinsky, V.A. (1997), J. Phys. D: Appl. Phys., vol. 30, p. 2566.

Nemchinsky, V.A. (1998), J. Phys. D: Appl. Phys., vol. 31, p. 3102.

Nemchinsky, V.A. (2002), IEEE Trans. Plasma Sci., vol. 30, no. 6, p. 2113.

Nemchinsky, V.A. (2003), J. Phys. D: Appl. Phys., vol. 36, p. 1573.

Nemchinsky, V.A., Showalter, M.S. (2003), J. Phys. D: Appl. Phys., vol. 36, p. 704.

Nemchinsky, V.A., Severance, W.S., Showalter, M.S. (1999), J. Phys. D: Appl. Phys., vol. 32, p. 1364.

Nester, S., Demura, A.V., Fridman, A. (1983), Disproportioning of the Vibrationally Excited CO-Molecules, Kurchatov Institute of Atomic Energy, vol. 3518/6, Moscow.

Nester, S., Rusanov, V.D., Fridman, A. (1985), Dissociation of Hydrogen Sulfide in Plasma with Additives, Kurchatov Institute of Atomic Energy, vol. 4223/6, Moscow.

Nester, S., Potapkin, B.V., Levitsky, A.A., Rusanov, V.D., Trusov, B.G., Fridman, A. (1988), Kinetic and Statistical Modeling of Chemical Reactions in Gas Discharges, Central Research Institute of Informatics in Nuclear Science and Industry (CNII Atom Inform), Moscow.

Neuenschwander, E., Schnet, K., Scheller, W. (1965), U.S. Patent nos. 3,429,661 and 3,545,922.

Neumann, E., Sowers, A.E., Jordan, C.A., eds., (2001), Electroporation and Electrofusion in Cell Biology, Springer-Verlag, Berlin, Heidelberg.

Neumann, K.K. (1964), Erdol Kohle, vol. 17, no. 9, p. 708.

Neuschutz, D. (2000), High Temp. Mater. Process., vol. 4, p. 127.

Newman, I.E., Watts, J.A. (1960), J. Am. Chem. Soc., vol. 82, p. 2113.

Nezu, A., Morishima, T., Watanabe, T. (2003), Thin Solid Films, vol. 435, p. 335.

Niazi, K., Lichtenberg, A.J., Lieberman, M.A., Flamm, D.L. (2004), Plasma Source Sci. Technol., vol. 3, p. 482.

Nicholson, R., Littlewood, K. (1972), Nature, vol. 236, p. 397.

References

Nikerov, V.A., Sholin, G.V. (1978), Sov. Phys., Plasma Phys., vol. 4, p. 1256.

Nikerov, V.A., Sholin, G.V. (1985), Kinetics of Degradation Processes, Energo-Atom-Izdat, Moscow.

Nikiforov, V.A. (1979), "Experiments with Plasma-Beam Discharge; Chemically Active Plasma," Ph.D. Dissertation, Kurchatov Institute of Atomic Energy, Moscow.

Nikitin, E.E. (1970), Theory of Elementary Atom-Molecular Processes in Gases, Khimia (Chemistry), Moscow.

Nikitin, E.E. (1984), Theory of Slow Atomic Collisions, Springer Series in Chemical Physics, Springer-Verlag, Berlin, Heidelberg.

Nikitin, E.E., Osipov, A.I. (1977), Vibrational Relaxation in Gases, VINITI, Moscow.

Nikitin, E.E., Umansky, S.Ya. (1971), Sov. Phys., Doklady (Reports of USSR Academy of Sciences), vol. 196, p. 145.

Nikitin, I.V., Rosolovsky, V.Ya. (1970), Adv. Chem. (Uspekhi Khimii), vol. 39, p. 1161, Nauka (Science), Moscow.

Nikitin, I.V., Rosolovsky, V.Ya. (1971), Adv. Chem. (Uspekhi Khimii), vol. 40, p. 1913, Nauka (Science), Moscow.

Nishida, S., Berkowitz, N. (1972a), Nature, vol. 236, p. 267.

Nishida, S., Berkowitz, N. (1972b), Nature, vol. 236, p. 262.

Norman, A. (1954), J. Cell. Comp. Physiol., vol. 44, p. 1.

Novgorodov, M.Z., Sviridov, A.G., Sobolev, N.N. (1971), Sov. Phys., J. Techn. Phys., vol. 41, p. 752.

Novikov, G.I., Bochin, V.P., Romanovsky, M.K. (1978), Nuclear – Hydrogen Energy and Technology, Legasov, V.A. (ed.), vol. 1, Atom-Izdat, Moscow, p. 231.

Novobrantsev, I.V., Starostin, A.N. (1974), Sov. Phys., J. Appl. Mech. Techn. Phys., no. 2, p. 164.

Nozaki, T. (2002), 55th Gaseous Electronics Conference (GEC-55), Minneapolis, MN.

Oda, T., Takahashi, T., Nakano, H., Masuda, S. (1991), 1991 IEEE Industrial Application Society Meeting, p. 734, Dearborn, MI.

Oda, T., Han, S.B., Ono, R. (2005), 17th Int. Symposium on Plasma Chemistry (ISPC-17), Toronto, Canada, p. 1180.

Oehr, C. (2005), 17th Int. Symposium on Plasma Chemistry (ISPC-17), Toronto, Canada, p. 8.

Oehrlein, G.S., Hua X., Stolz, C. (2005), U.S. Patent no. 0051517 A1.

Ogryzlo, E.A., Ibbotson, D.E., Flamm, D.L., Mucha, J.A. (1990), J. Appl. Phys., vol. 67, p. 3115.

Okano, H., Yamazaki, T., Horiike, Y. (1982), Solid State Technol., vol. 25, p. 166.

Okazaki, K., Kishida, T. (1999), 14th Int. Symposium on Plasma Chemistry (ISPC-14), Prague, Czech Republic, vol. 5, p. 2283.

Okazaki, K., Nozaki, T., Uemitsu, Y., Hijikata, K. (1995), 12th Int. Symposium on Plasma Chemistry (ISPC-12), Minneapolis, MN, vol. 2, p. 581.

Okazaki, S., Kogoma, M. (1993), J. Photopolym. Sci. Technol., vol. 6, p. 339.

Okumoto, M., Su, Z.Z., Katsura, S., Mizuno, A. (1999), IEEE Trans. Ind. Appl., vol. 35, p. 1205.

Okumoto, M., Kim, H.H., Takashima, K., Katsura, S., Mizuno, A. (2001), IEEE Trans. Ind. Appl., vol. 37, p. 1618.

Okumura, T., Saitoh, M. (2004), U.S. Patent no. 0157447 A1.

Okumura, T., Kimura, T., Yashiro, Y., Sato, K., Saitoh, M. (2004), U.S. Patent no. 0075396 A1.

Olah, A.G., Schlosberg, R.H. (1968), J. Am. Chem. Soc., vol. 90, p. 2726.

Ombrello, T., Qin, X., Ju, Y., Gangoli, S., Gutsol, A., Fridman, A. (2006a), 44th AIAA Aerospace Sciences Meeting and Exhibit, 9–12 January 2006, Reno, NV, AIAA paper 2006-1214.

Ombrello, T., Qin, X., Ju, Y., Gutsol, A., Fridman, A., Carter, C. (2006b), AIAA J., vol. 44, no. 1, p. 142.

Ono, R., Oda, T. (2001), IEEE Trans. Ind. Appl., vol. 37, p. 3.

Opalinska, T., Opalska, A. (2003), 15th Int. Symposium on Plasma Chemistry (ISPC-16), Orleans, France, p. 594.

Opalinska, T., Szymanski, A. (1996), Contrib. Plasma Phys., vol. 36, p. 63.

Opalska, A., Opalinska, T., Ochman, P. (2002a), Acta Agrophys., vol. 80, p. 367.

Opalska, A., Opalinska, T., Polaczek, J., Ochman, P. (2002b), Int. Symp. on High Pressure Low Temperature Plasma Chemistry, HAKONE VIII, Puhajarve, Estonia, p. 191.

Oraevsky, A.N. (1974), Sov. Phys., High Energy Chem. (Khimia Vysokikh Energij), vol. 8, p. 3.

Orbach, H. (1966), U.S. Patent, no. 3,288,696.

Orlikovsky, A.A. (2000), "Plasma in Sub-Micron Electronic Technology," in Encyclopedia of Low Temperature Plasma, Fortov, V.E. (ed.), vol. 4, p. 370, Nauka (Science), Moscow.

Ormonde S. (1975), Rev. Modern Phys., vol. 47, p. 193.

Osherov, V.I. (1965), Sov. Phys., J. Exper. Theoret. Phys., vol. 49, p. 1156.

Osipov, K.A., Folmanis, G.E. (1973), Film Deposition Using Low-Temperature Plasma and Ion Beams, Nauka (Science), Moscow.

Osipov, O.A. (1982), High-Mol. Compounds, vol. 24(A), p. 1322.

Ouni, F., El Ahmar, E., Aubry, O., Met, C., Khacef, A., Cormier, J.-M. (2005a), 17th Int. Symposium on Plasma Chemistry (ISPC-17), Toronto, Canada, p. 307.

Ouni, F., Khacef, A., Aubry, O., Met, C., Cormier, J.-M. (2005b), 17th Int. Symposium on Plasma Chemistry (ISPC-17), Toronto, Canada, p. 884.

References

Ovsiannikov, A.A., Polak, L.S. (1965), Kinetics and Thermodynamics of Chemical Reactions in Low-Temperature Plasma, Polak, L.S. (ed.), Nauka (Science), Moscow, p. 118.

Ovsiannikov, A.A., Polak, L.S., Yudin, V.I. (1968), Sov. Phys., High Energy Chem. (Khimia Vysokikh Energij), vol. 2, p. 5.

Pancheshnyi, S., Lacoste, D.A., Bourdon, A., Laux, C.O. (2005), 17th Int. Symposium on Plasma Chemistry (ISPC-17), Toronto, Canada, p. 1025.

Panteleev, V.I. (1978), "Electro-Ionization Synthesis of Chemical Compounds," Ph.D. Dissertation, Lebedev Physical Institute of USSR Academy of Sciences, Moscow.

Park, C. (1987), "Assessment of Two-Temperature Kinetic Model for Ionizing Air," AIAA paper 87-1574.

Park, C., Bogdanov, D.W., Mehta, U.B. (2001), AIAA paper 2001-0972.

Park, D.-W., Cha, W.-B. (1997), 13th Int. Symposium on Plasma Chemistry (ISPC-13), Beijing, vol. 4, p. 1764.

Park, J., Henins, I., Herrmann, H.W., Selwyn, G.S., Hicks, R.F. (2001), J. Appl. Phys., vol. 89, no. 1.

Park, M., Chang, D., Woo, M., Nam, G., Lee, S. (1998), Plasma Exhaust Aftertreatment, Society of Automotive Engineers, Warrendale, PA.

Park S.-J., Chen, J., Liu, C., Eden, J.G. (2001), Appl. Phys. Lett., vol. 78, p. 419.

Parker, J. (1959), Phys. Fluids, vol. 2, p. 449.

Parker, J., Stephens, R.R. (1973), J. Quant. Electron., vol. 9, p. 643.

Parker, R., Clapper, W. (2001), Proceedings of the 2001 DOE Hydrogen Program Review, NREL/CP-570-30535.

Parkhomenko, V.D., Rudenko, A.I., Ganz, S.N. (1969), Sov. Phys., High Energy Chem. (Khimia Vysokikh Energij), vol. 3, p. 274.

Pasandideh-Fard, M., Bhola, R., Chandra, S., Mostaghimi, J. (1998), Int. J. Heat Mass Transf., vol. 41, p. 2229.

Pasandideh-Fard, M., Pershin, V., Chandra, S., Mostaghimi, J. (2002), J. Therm. Spray Technol., vol. 11, p. 206.

Patel, C.K.N. (1964), Phys. Rev. Lett., vol. 13, p. 617.

Pathria, R.K. (1996), Statistical Mechanics, Butterworth-Heinemann, 2nd edition.

Patrushev, B.I., Rykunov, G.V., Spector, A.M. (1977), 3rd Int. Symposium on Plasma Chemistry (ISPC-3), vol. 2, p. 18, Limoge, France.

Paul, D.R., Yampolskii, Yu.P. (1993), Polymeric Gas Separation Membranes, CRC Press, FL.

Paur, H.-R. (1991), European Aerosol Conference, Proceedings, KFK, Karlsruhe, Germany.

Paur, H.-R. (1999), Nonthermal Plasma Techniques for Pollution Control, Penetrante, B.M., Schultheis, S.E. (eds.), Springer-Verlag, Berlin, p. 77.

Pavlish, J.H., Sondreal, E.A., Mann, M.D., Olson, E.S., Galbreath, K.C., Laudal, D.L., Benson, S.A. (2003), Fuel Process. Technol., vol. 82, p. 2003.

Pawlowski, L. (1995), The Science and Engineering of Thermal Spray Coatings, Wiley, New York.

Pearson, T.G., Robinson, P.L., Stoddart, E.M. (1933), Proc. R. Soc., A Lond., vol. 142, p. 275.

Pekshev, A.V. (2001), Stomatol., vol. 1, p. 80.

Pellerin, S., Cormier, J.-M., Richard, F., Musiol, K., Chapelle, J. (1996), J. Phys. D: Appl. Phys., vol. 29, p. 726.

Penetrante, B.M. (1997), Plasma Source Sci. Technol., vol. 6, p. 251.

Penetrante, B.M., Schultheis, S.E., eds. (1993), Non-Thermal Plasma Techniques for Pollution Control: Part B – Electron Beam and Electrical Discharge Processing, Springer-Verlag, Berlin.

Penetrante, B.M., Hsiao, M.C., Meritt, B.T., Vogtlin, G.E., Wallman, P.H., Kuthi, A., Butkhart, C.P., Bayless, J.R. (1995), Phys. Lett. A, vol. 209, p. 69.

Penetrante, B.M., Hsiao, M.C., Bardsley, J.N., Meritt, B.T., Vogtlin, G.E., Wallman, P.H., Kuthi, A., Butkhart, C.P., Bayless, J.R. (1996a), Pure Appl. Chem., vol. 68, p. 1868.

Penetrante, B.M., Hsiao, M.C., Bardsley, J.N., Meritt, B.T., Vogtlin, G.E., Wallman, P.H., Kuthi, A., Butkhart, C.P., Bayless, J.R. (1996b), Proceedings of the International Workshop on Plasma Technologies for Pollution Control and Waste Treatment, Pu, Y.K., Woskov, P.P. (eds.), p. 99, Beijing.

Penetrante, B.M., Hsiao, M.C., Bardsley, J.N., Meritt, B.T., Vogtlin, G.E., Kuthi, A., Butkhart, C.P., Bayless, J.R. (1997), Plasma Source Sci. Technol., vol. 6, p. 251.

Penetrante, B.M., Brusasco R.M., Meritt, B.T., Pitz, W.J., Vogtlin, G.E., Kung, M.C., Kung, H.H., Wan, C.Z., Voss K.E. (1998), Plasma Exhaust Aftertreatment, Society of Automotive Engineers, Warrendale, PA.

Penetrante, B.M., Brusasco, R.M., Merritt, B.T., Vogtlin, G.E. (1999), Pure Appl. Chemi., vol. 71, p. 1829.

Penkett, S.A., Jones, R. (1979), Atmos. Environ., vol. 13, p. 123.

Penning, F.M. (1936), Physica, vol. 3, p. 873.

Penning, F.M. (1937), Physica, vol. 4, p. 71.

Penning, F.M., Moubis, J.H.A. (1937), Physica, vol. 4, p. 1190.

Perks, H.M., Liebman, J.F. (2000), J. Struct. Chem., vol. 11, p. 325.

Perrin, J. (1991), J. Non-Cryst. Solids, vol. 137, p. 639.

Perrin, J. (1993), J. Phys. D: Appl. Phys., vol. 26, p. 1662.

Perrin, J., Bohm, C., Etemadi, R., Lloret, A. (1993), NATO Advanced Research Workshop "Formation, Transport and Consequences of Particles in Plasma," France.

References

Peters, K., Kuster H. (1931), Brennstoff-Chemie, vol. 12, p. 122.

Peters, K., Pranschke (1930), Brennstoff-Chemie, vol. 11, pp. 239, 473.

Peuckert, C., Muller, R. (1985), 7th Int. Symposium on Plasma Chemistry (ISPC-7), Eindhoven, no. P-3-5, p. 274.

Peyrous, R. (1990), Ozone Sci. Eng., vol. 12, p. 19.

Peyrous, R., Pignolet, P., Held, B. (1989), J. Phys. D: Appl. Phys., vol. 22, p. 1658.

Pfender, E., Lee, Y.C. (1985), Plasma Chem. Plasma Process., vol. 5, p. 391.

Philip, N., Saoudi, B., Crevier, M.C., Moisan, M., Barbeau, J., Pelletier, J. (2002), IEEE Trans. Plasma Sci., vol. 30, p. 1429.

Pichal, J., Koller, J., Aubrecht, L., Vatuna, T., Spatenka, P., Wiener, J. (2004), Czechoslovak J. Phys., vol. 54, p. 54.

Pichal, J., Koller, J., Vatuna, T., Aubrecht, L., Spatenka, P. (2005), 17th Int. Symposium on Plasma Chemistry (ISPC-17), Toronto, Canada, p. 712.

Pico, C.A., Lieberman, M.A., Cheung, N.W. (1992), J. Electrono. Mater., vol. 21, p. 75.

Piementel, D., Patzek, T.W. (2005), Nat. Resources Res., vol. 14, no. 1.

Pietsh, E.Z. (1933), Electrochem., vol. 39, p. 577.

Pikaev, A.K. (1965), Pulse Radiolysis of Water and Water Solutions, Science (Nauka), Moscow.

Piley, M.E., Matzen, M.K. (1975), J. Chem. Phys., vol. 63, p. 4787.

Pitaevsky, L.I. (1962), Sov. Phys., J. Exper. Theoret. Phys., JETF, vol. 42, p. 1326.

Plonjes, E., Palm, P., Rich, W.J., Adamovich, J. (2001), 32nd AIAA Plasmadynamics and Lasers Conference, AIAA paper 2001-3008, Anaheim, CA.

Plumb, I.C., Ryan, K.R. (1986), Plasma Chem. Plasma Process., vol. 6, p. 205.

Poeschel, R.L., Beattle, J.R., Robinson, P.A., Ward, J.W. (1979), 14th Int. Electric Propulsion Conf., Princeton, NJ, paper 79-2052.

Poggie, J., Gaitonde, D.V. (2001), AIAA paper 2001-0196.

Polak, L.S. (1965), Kinetics and Thermodynamics of Chemical in Low-Temperature Plasma, Nauka (Science), Moscow.

Polak, L.S. (1971), "Production of Olephines and Acetylene in Chemical Industry," 8-World Petrolium Congress, p. 40, Moscow.

Polak, L.S., Shchipachev, V.S. (1965), Kinetics and Thermodynamics of Chemical Reactions in Low Temperature Plasma, Polak, L.S. (ed.), p. 151–166, Nauka (Science), Moscow.

Polak, L.S., Valibekov, Yu.V., Guliaev, G.V., Bolotov, G.M. (1969), News Tadjik Acad. Sci., vol. 1, no. 31, p. 43.

Polak, L.S., Parkhomenko, V.D., Melnikov, B.I., Ganz, S.N. (1973), Sov. Phys., High Energy Chem. (Khimia Vysokikh Energij), vol. 7, pp. 303, 408.

Polak, L.S., Ovsiannikov, A.A., Slovetsky, D.I., Vurzel, F.B. (1975a), Theoretical and Applied Plasma Chemistry, Nauka (Science), Moscow.

Polak, L.S., Slovetsky, D.I., Todesaite, R.D. (1975b), Sov. Phys., High Energy Chem. (Khimia Vysokikh Energij), vol. 9, p. 142.

Polak, L.S., Slovetsky, D.I., Butylkin, Yu.P. (1977), Carbon Dioxide Dissociation in Electric Discharges: Arc Discharge, Institute of Petrochemical Synthesis, USSR Academy of Sciences, Moscow.

Polishchuk, A., Rusanov, V.D., Fridman, A. (1980), Theoret. Exp. Chem., vol. 16, p. 232.

Poluektov, N.P., Efremov, N.P. (1998), J. Phys. D., vol. 31, p. 988.

Poncin-Epaillard, F., Legeay, G., J. (2003), Biomat. Sci. Polym. Ed., vol. 14, p. 1005.

Ponelis, A.A., van der Walt, I.J. (2003), 16th Int. Symposium on Plasma Chemistry (ISPC-16), Taormina, Italy, p. 733.

Ponomarev, A.N. (1982), Heat and Mass Transfer in Plasma-Chemical Processes, Nauka (Science), Minsk, Belarus, vol. 1, p. 137.

Ponomarev, A.N. (1996), News Russ. Acad. Sci., vol. 6, p. 78.

Ponomarev, A.N. (2000), Encyclopedia of Low-Temperature Plasma, Fortov, V.E. (ed.), vol. 4, p. 386, Nauka (Science), Moscow.

Ponomarev, A.N., Vasilets, V.N. (2000), Encyclopedia of Low-Temperature Plasma, Fortov, V.E. (ed.), vol. 3, p. 374, Nauka (Science), Moscow.

Ponomarev, A.N., Vasilets, V.N., Talrose, R.V. (2002), Chem. Phys. (Russ. J.), vol. 21, no. 4, p. 96.

Ponomarev-Stepnoy, N.N., Protsenko, A.N., Stoliarevsky, A.Ya. (1979), Nuclear-Hydrogen Energy and Technology, Legasov, V.A. (ed.), vol. 2, p. 184, Atom-Izdat, Moscow.

Poplavko, Yu.M. (1980), Physics of Dielectrics, Vysshaya Shkola (Higher Education), Kiev.

Porteous, B.K., Graves, D.B. (1991), IEEE Trans. Plasma Sci., vol. 19, p. 204.

Porteous, B.K., Hbid, T., Boufendi, L., Fridman, A., Potapkin, B.V., Bouchoule, A. (1994), ESCAMPIG-12, Euro-Physics Conference Abs., Noordwijkerhout, Netherlands, 18E, p. 83.

Porter, J., Edwards, C., Morgan, J.A.W., Pickup, R.W. (1993), Appl. Environ. Microbiol., vol. 59, p. 3327.

Potapkin, B.V. (1986), "Kinetics and Fluid Mechanics of Non-Equilibrium Plasma-Chemical Processes in Fast Flows," Ph.D. Dissertation, Kurchatov Institute of Atomic Energy, Moscow.

References

Potapkin, B.V., Rusanov, V.D., Samarin, A.E., Fridman, A. (1980), Sov. Phys., High Energy Chem. (Khimia Vysokikh Energij), vol. 14, p. 547.

Potapkin, B.V., Rusanov, V.D., Fridman, A. (1983a), Sov. Phys., High Energy Chem. (Khimia Vysokikh Energij), vol. 17, p. 528.

Potapkin, B.V., Rusanov, V.D., Samarin, A.E., Fridman, A. (1983b), Sov. Phys., High Energy Chem. (Khimia Vysokikh Energij), vol. 17, p. 1251.

Potapkin, B.V., Rusanov, V.D., Fridman A. (1984a), Sov. Phys., High Energy Chem. (Khimia Vysokikh Energij), vol. 18, p. 252.

Potapkin, B.V., Rusanov, V.D., Fridman A. (1984b), 4th USSR Symposium on Plasma Chemistry, Dnepropetrovsk, Ukraine, vol. 1, p. 214.

Potapkin, B.V., Rusanov, V.D., Strelkova, M.I., Fridman A. (1984c), 4th USSR Symposium on Plasma Chemistry, Dnepropetrovsk, vol. 1, p. 214.

Potapkin, B.V., Rusanov, V.D., Fridman A. (1985), Non-equilibrium Effects in Plasma, Provided by Selectivity of Transfer Processes, Kurchatov Institute of Atomic Energy, vol. 4219/6, Moscow.

Potapkin, B.V., Rusanov, V.D., Fridman A. (1987), Sov. Phys., High Energy Chem. (Khimia Vysokikh Energij), vol. 21, p. 452.

Potapkin, B.V., Rusanov, V.D., Strelkova, M.I., Fridman A. (1988), Sov. Phys., High Energy Chem. (Khimia Vysokikh Energij), vol. 22, p. 537.

Potapkin, B.V., Rusanov, V.D., Fridman, A. (1989), Sov. Phys., Doklady (Reports of USSR Academy of Sciences), vol. 308, p. 897.

Potapkin, B.V., Deminsky, M., Fridman, A., Rusanov, V.D. (1993), Non-Thermal Plasma Techniques for Pollution Control, NATO ASI Series, vol. G 34, Part A, Penetrante, B.M., Schultheis, S.E. (eds.), Springer-Verlag, Berlin.

Potapkin, B.V., Deminsky, M., Fridman, A., Rusanov, V.D. (1995), Radiat. Phys. Chem., vol. 45, p. 1081.

Powel, R.A. (1984), Dry Etching for Microelectronics, Elsevier, New York.

Powerspan (2004), Mercury Removal in Nonthermal Plasma Based Multi-Pollutant Control Technology for Utility Boilers, DE-FC26-01NT41182.

Prelas, M.A., Popovici, G., Bigelow, L.K. (1997), Handbook on Industrial Diamonds and Diamond Films, CRC Press, FL.

Protasov, S.S. (2000), "Low-Temperature Plasma in Photon Energy Sources," in Encyclopedia of Low-Temperature Plasma, Fortov, V.E. (ed.), vol. 4, p. 231, Nauka (Science), Moscow.

Protasov, Yu.S. (2000), "Plasma Sources of Radiation with High Spectral Brightness," in Encyclopedia of Low-Temperature Plasma, Fortov, V.E. (ed.), vol. 4, p. 232, Nauka (Science), Moscow.

Provorov, A.S., Chebotaev, V.P. (1977), Gas Lasers, Soloukhin, R.I. (ed.), p. 174, Nauka (Science), Novosibirsk.

Prusakov, V.N., Sokolov, V.B. (1971), Sov. Phys., Atomic Energy (Atomnaja Energia), vol. 31, p. 259.

Pu, Y.K., Woskov, P.P. (1996), Int. Workshop on Plasma Technologies for Pollution Control and Waste Treatment, Ed., MIT, Cambridge, MA.

Puchkarev, V., Roth, G., Gunderson, M. (1998), Plasma Exhaust Aftertreatment, Society of Automotive Engineers, Warrendale, PA.

Puri, I.K., Seshadri, K. (1986), Combust. Flame, vol. 65, p. 137.

Pushkarev, A., Remnev, G., Vlasov, V., Sosnovsky, S., Ponomarev, D. (2005), 17th Int. Symposium on Plasma Chemistry (ISPC-17), Toronto, Canada, p. 909.

Qu, X. (2007), Biomaterials, vol. 28, p. 9.

Raadu, M.A. (1989), Phys. Rep., vol. 178, p. 25.

Rabkina, A.L., Borisov, P.A., Ponomareva, V.T. (1971a), Chem. Ind., no. 1, p. 35.

Rabkina, A.L., Borisov, P.A., Ponomareva, V.T. (1971b), J. Econ., Org. and Contr. Petrochem. Ind., no. 2, p. 1.

Rabkina, A.L., Borisov, P.A., Ponomareva, V.T. (1971c), Plasma Cemistry – 71, Polak, L.S. (ed.), p. 44, Institute of Petrochemical Synthesis of USSR Academy of Sciences, Moscow.

Radford, H. (1964), J. Chem. Phys., vol. 40, p. 2732.

Raether, H. (1964), Electron Avalanches and Breakdown in Gases, Butterworth, London.

Raghavachari, K. (1992), J. Chem. Phys., vol. 96, p. 4440.

Rains, R.K., Kadlek, R.H. (1970), Metallurg. Trans., vol. 1, no. 6, p. 1501.

Raiser, J., Zenker, M. (2006), J. Phys. D: Appl. Phys., vol. 39, p. 3520.

Raitsis, M.B., Raitsis, L.S., Miller, T.N. (1976), News Latvian Acad. Sci., Chem., p. 261.

Raizer, Yu.P. (1970), Sov. Phys., Lett. J. Exper. Theoret. Phys., vol. 11, p. 195.

Raizer, Yu.P. (1972a), Sov. Phys., Therm. Phys. High Temp., vol. 10, p. 1152.

Raizer, Yu.P. (1972b), Sov. Phys., Adv. Phys. Sci., Uspehi Phys. Nauk, vol. 108, p. 429.

Raizer, Yu.P. (1974), Laser Spark and Discharge Propagation, Nauka (Science), Moscow.

Raizer, Yu.P. (1977), Laser-Induced Discharge Phenomena, Consultants Bureau, New York.

Raizer, Yu.P. (1991), Gas Discharge Physics, Springer, Berlin, Heidelberg, New York.

Raizer, Yu.P., Shneider, M., Yatsenko, N.A. (1995), Radio-Frequency Capacitive Discharges, Nauka (Science), Moscow, and CRC Press, New York.

References

Ramires, P.A., Mirenghi, L., Romano, A.R., Palumbo, F., Nicolardy, G.M. (2000), J. Biomed. Mater. Res., vol. 51, p. 535.

Raoux, S., Cheung, D., Fodor, M., Taylor, N., Fairbairn, K. (1997), Plasma Source Sci. Technol., vol. 6, p. 405.

Raoux, S., Tanaka, T., Bhan, M., Ponnekanti, H., Seamons, M., Deacon, T., Xia, L.Q., Pham, F., Silvetti, D., Cheung, D., Fairbairn, K., Johnson, A., Pierce, R., Langan, J. (1999), J. Vac. Sci. Technol., vol. B17, p. 477.

Rapp, D. (1965), J. Chem. Phys., vol. 43, p. 316.

Rapp, D., Francis, W.E. (1962), J. Chem. Phys., vol. 37, p. 2631.

Ratner, B. (1995), Biosensors Bioelectr., vol. 10, p. 797.

Razina, G.N., Kaftanov, S.V., Fedoseev, S.D., Kleimenov, V.V. (1973), Investigation of High Temperature Pyrolysis of Low-Quality Coals, Adv. of Moscow Institute of Chemical Engineering (Mendeleev University), vol. 74, p. 69, Moscow.

Razina, G.N., Fedoseev, S.D., Gvozdarev, V., Staroverov, V.A. (1976), "Influence of Gas-Dynamic Factors on the Gas-Phase Products of Coal Pyrolysis in Conditions of Plasma Temperatures," in Modern Processes of Conversion of Solid Fuels, Adv. of Moscow Institute of Chemical Engineering (Mendeleev University), vol. 91, p. 94, Moscow.

Razumovsky, S.D., Zaikov, G.E. (1974), Ozone and its Reactions with Organic Compounds, Nauka (Science), Moscow.

Redsten, A.M., Sridharan, K., Worzala, F.J., Conrad, J.R. (1992), J. Mater. Process. Technol., vol. 30, p. 253.

Reents, W.D. Jr., Mandich, M.L. (1992), J. Chem. Phys., vol. 96, p. 4449.

Reich, S., Thomsen, C., Maultzsch, J. (2004), Carbon Nanotubes: Basic Concepts and Physical Properties, Wiley-VCH.

Reinmann, R., Akram, M. (1997), J. Phys. D: Appl. Phys., vol. 30, p. 1125.

Reitlinger, S.A. (1974), Permeability of Polymer Materials, Khimia (Chemistry), Moscow.

Reshetov, I.V., Kabisov, R.K., Shekhter, A.B., Pekshev, A.V., Maneilova, M.V. (2000), Ann. Plastic, Reconstr. Aesth. Surg., vol. 4, p. 24.

Rich, J.W., Bergman, R.C. (1986), Isotope Separation by Vibration–Vibration Pumping, Topics of Current Physics-39, Capitelli, M. (ed.), Springer-Verlag, Berlin.

Richard, F., Cormier, J.-M., Pellerin, S., Chapelle, J. (1996), J. Appl. Phys., vol. 79, no. 5, p. 2245.

Ridge, M.I., Howson, R.P. (1982), Thin Solid Films, vol. 96, p. 113.

Robert, E., Cachoncinlle, C., Viladrosa, R., Dozias, S., Point, S., Pouvesle, J.M. (2005a), 17th Int. Symp. on Plasma Chemistry (ISPC-17), Toronto, Canada, p. 1166.

Robert, E., Saroukh, H., Cachoncinlle, C., Viladrosa, R., Hochet, V., Eddanoui, S., Pouvesle, J.M. (2005b), Pure Appl. Chem., vol. 77, p. 1789.

Robertson, G.D., Baily, N.A. (1965), Bull. Am. Phys. Soc., vol. 2, p. 709.

Robertson, J. (1993a), Diam. Rel. Mater., vol. 2, p. 984.

Robertson, J. (1993b), Philos. Trans. R. Soc., vol. A342, p. 277.

Robertson, J. (1994a), Pure Appl. Chem., vol. 66, p. 1789.

Robertson, J. (1994b), Diam. Rel. Mater., vol. 3, p. 361.

Robertson, J. (1997), Radiat. Effect, vol. 142, p. 63.

Robertson, J. (2002), Mater. Sci. Eng., vol. R37, p. 129.

Robertson, J. (2004), Mater. Today, October.

Robinson, J.W. (1973), J. Appl. Phys., vol. 44, p. 76.

Robinson, J.W., Ham, M., Balaster, A.N. (1973), J. Appl. Phys., vol. 44, p. 72.

Roginsky, S.Z. (1948), Adsorption and Catalysis on Non-Uniform Surfaces, Publ. by Academy of Sciences of USSR, Moscow.

Rokhlin, G.N. (1991), Gas-Discharge Light Sources, EnergoAtomIzdat, Moscow.

Rosocha, L.A., Anderson, G.K., Bechtold, L.A., Coogan, J.J., Heck H.G., Kang, M., McCulla, W.H., Tennant, R.A., Wantuck, P.J. (1993), Non-Thermal Plasma Techniques for Pollution Control: Part B – Electron Beam and Electrical Discharge Processing, Penetrante, B.M., Schultheis, S.E. (eds.), Springer-Verlag, Berlin.

Rosocha, L.A., Platts, D., Coates, D.M., Stange, S. (2004), Proceedings of the First International Workshop on Plasma-Assisted Combustion (PAC-1), Washington.

Roth, J.R. (1966), Rev. Sci. Instrum., vol. 37, p. 1100.

Roth, J.R. (1973a), IEEE Trans. Plasma Sci., vol. 1, p. 34.

Roth, J.R. (1973b), Plasma Phys., vol. 15, p. 995.

Roth, J.R. (1995), Industrial Plasma Engineering, vol. 1, Principles, Institute of Physics Publishing, Bristol, Philadelphia.

Roth, J.R. (2001), Industrial Plasma Engineering, vol. 2, Applications for Nonthermal Plasma Processing, Institute of Physics Publishing, Bristol, Philadelphia.

Roth, J.R. (2006), The One Atmosphere Uniform Glow Discharge Plasma (OAUGDP) – a Platform Technology for the 21st Century, Plenary talk at the 33rd IEEE International Conference on Plasma Science (ICOPS-2006), Traverse City, MI.

References

Rozovsky, M.O. (1972), Sov. Phys., J. Appl. Mech. Techn. Phys., PMTPh, no. 6, p. 176.

Ruehrwein, R. (1960), J. Chem. Phys., vol. 64, p. 1317.

Ruppel, Th. C., Mossbauer, P.F., Ferrer, M.F., Beinstock, D. (1970), Ind. Eng. Chem. Product. Res. Dev., vol. 9, no. 3, p. 369.

Rusakova, L.A., Eremin, E.N. (1975a), Sov. Phys., J. Phys. Chem. (Jornal Physicheskoj Chimii), vol. 49, p. 752.

Rusakova, L.A., Eremin, E.N. (1975b), Sov. Phys., J. Phys. Chem. (Jornal Physicheskoj Chimii), vol. 49, p. 1031.

Rusanov, V.D., Fridman, A., (1976), Sov. Phys., Doklady (Reports of USSR Academy of Sciences), vol. 231, p. 1109.

Rusanov, V.D., Fridman, A. (1978a), Sov. Phys., J. Phys. Chem. (JFCh), vol. 52, p. 92.

Rusanov, V.D., Fridman, A. (1978b), Sov. Phys., Lett. J. Techn. Phys. (Pis'ma v JTF), vol. 4, p. 28.

Rusanov, V.D., Fridman, A. (1984), Physics of Chemically Active Plasma, Nauka (Science), Moscow.

Rusanov, V.D., Fridman, A., Sholin, G.V. (1977), Sov. Phys., Doklady (Reports of USSR Academy of Sciences), vol. 237, p. 1338.

Rusanov, V.D., Fridman, A., Sholin, G.V. (1978), Plasma Chemistry, vol. 5, Smirnov, B.M. (ed.), Atom-Izdat, Moscow.

Rusanov, V.D., Fridman, A., Sholin, G.V. (1979a), Sov. Phys., J. Techn. Phys., vol. 49, p. 554.

Rusanov, V.D., Fridman, A., Sholin, G.V. (1979b), Sov. Phys., J. Techn. Phys., vol. 49, p. 2169.

Rusanov, V.D., Potapkin, B.V., Fridman, A., Sholin, G.V. (1979c), 3rd USSR – Symposium on Plasma Chemistry, Nauka (Science), Moscow, vol. 1, p. 121.

Rusanov, V.D., Fridman, A., Sholin, G.V. (1981), Adv. Phys. Sci. (Uspekhi Physicheskikh Nauk), vol. 134, p. 185.

Rusanov, V.D., Fridman, A., Sholin, G.V. (1982), Heat and Mass Transfer in Plasma-Chemical Processes, vol. 1, p. 137, Nauka (Science), Minsk.

Rusanov, V.D., Fridman, A., Macheret, S.O. (1985a), Sov. Phys., Doklady (Reports of USSR Academy of Sciences), vol. 283, p. 590.

Rusanov, V.D., Fridman, A., Sholin, G.V., Potapkin, B.V. (1985b), Vibrational Kinetics and Reactions of Polyatomic Molecules in Non-Equilibrium Systems, Kurchatov Institute of Atomic Energy, Moscow, vol. 4201/6.

Rusanov, V.D., Fridman, A., Sholin, G.V. (1986), Non-Equilibrium Vibrational Kinetics, Topics of Current Physics, vol. 39, Capitelli, M. (ed.), Springer-Verlag, Berlin.

Rusanov, V.D., Petrusev, A.S., Potapkin, B.V., Fridman, A., Czernichowski, A., Chapelle, J. (1993), Sov. Phys., Doklady, vol. 332, no. 6, p. 306.

Rusanov, V.D., Babaritsky, A.I., Baranov, E.I., Demkin, S.A., Givotov, V.K., Potapkin, B.V., Ryazantsev, E.I., Etivan, C. (1997), Sov. Phys., Doklady (Reports of Russian Academy of Sciences), vol. 354, p. 1.

Rutberg, F.G., Safronov, A.A., Goryachev, V.L., Ufimtsev, A.A., Polovtsev, I. (1997), 13th Int. Symposium on Plasma Chemistry (ISPC-13), Beijing, China, vol. 4, p. 1727.

Rutberg, F.G., Bratsev, A.N., Popov, V.E., Ufimtsev, A.A. (2005), 17th Int. Symposium on Plasma Chemistry (ISPC-13), Toronto, Canada, p. 1162.

Rutherford, P.H., Goldston, R.J. (1995), Introduction to Plasma Physics, Inst. of Phys. Pub.

Ryabinin, V.P., Orumbaeva, B.K., Toboyakov, B.O. (1987), "Technological Scheme of Direct Plasma Gasification of Water-Coal Suspensions," in Plasma Gasification and Pyrolysis of Low-Quality Coals, p. 101, Energy Institute, ENIN, Moscow.

Ryan, N.W. (1966), U.S. Patent no. 3,420,632.

Rykalin, N.N., Sorokin, L.M. (1987), Metallurgical RF-Plasmatron: Electro- and Gas-Dynamics, Nauka (Science), Moscow.

Rykalin, N.N., Tsvetkov Yu.V., Petrunichev, V.A., Glushko, I.K. (1970), Physics, Engineering and Application of Low-Temperature Plasma, p. 602, Nauka (Science), Alma-Ata, Kazakhstan.

Rykalin, N.N., Petrunichev, V.A., Kulagin, I.D., Sorokin, L.M., Koroleva, E.B., Gugniak, A.B. (1973), Plasma Processes in Metallurgy and Technology of Inorganic Materials, p. 220, Nauka (Science), Moscow.

Rykalin, N.N., Ogurtsov, S.V., Kulagin, I.D. (1975), Sov. Phys., Phys. Eng. Mater. Treatment, vol. 1, p. 154.

Sahasrabudhey, R.S., Deshpande, S.M. (1950), Proc. Indian Acad. Sci., vol. 31, p. 317.

Sahasrabudhey, R.S., Deshpande, S.M. (1951), J. Indian Chem. Soc., vol. 28, p. 377.

Sahasrabudhey, R.S., Kolyanasundaram, A. (1948), Proc. Indian Acad. Sci., A, vol. 27, p. 366.

Sahoo, P., Lewis, T. F. (1992), 13th Int. Thermal Spray Conf., ASM Int'l., Orlando, FL, p. 75.

Saidane, K., Razafinimanana, M., Lange, H, Huczko, A., Baltas, M., Gleizes, A., Meunier, J.-L. (2004), J. Phys. D: Appl. Phys., vol. 39, p. 232.

Sakai, O., Kishimoto, Y., Tachibana, K. (2005), J. Phys. D: Appl. Phys., vol. 38, p. 431.

Sakano, M., Tanaka, M., Watanabe, T. (2001), Thin Solid Films, vol. 386, no. 2, p. 189.

Sakharov, B.S., Lukianychev, Yu.L. (1974), Sov. Phys., Non-Ferr. Metals (Tsvetnye Metally), vol. 4, p. 43.

Sakharov, B.A., Marin, K.G., Lubimov, V.K. (1968), Sov. Phys., News of USSR Academy of Sciences (Izvestia), Inorg. Mater., vol. 4, p. 2035.

Sakipov, Z.B., Messerle, V.E., Ibraev, Sh. (1986), Sov. Phys., High Energy Chem. (Khimia Vysokikh Energij), vol. 20, p. 61.

References

Sakipov, Z.B., Imankulov, E.R., Messerle, V.E. (1987a), Effectiveness of Burning Low-Quality Coals in Boilers, YVTI, Gorlovka, Ukraine.

Sakipov, Z.B., Riabinin, V.P., Seytimov, T.I., Imankulov, E.R. (1987b), Plasma Gasification and Pyrolysis of Low-Quality Coals, p. 90, Energy Institute (ENIN), Moscow.

Sakiyama, Y., Graves, D.B. (2006), J. Phys. D: Appl. Phys., vol. 39, p. 3451.

Samaranayake, W.J.M., Miyahara, Y., Namihira, T., Katsuki, S., Sakugava, T., Hackam, T., Akyiama, H. (2000), IEEE Trans. Dielect. Electrical Insul., vol. 7, p. 254.

Samoilovich, V.G., Filippov, Yu.V. (1964), Sov. Phys., J. Phys. Chem., vol. 38, p. 2712.

Samoilovich, V.G., Gibalov, V.I., Kozlov, K.V. (1989), Physical Chemistry of Barrier Discharge, Nauka (Science), Moscow.

Sampath, S. (1993), 1993 Powder Metallurgy World Congress, Kyoto, Japan, Japan Soc. Powder and Powder Metallurgy, p. 401.

Sampath, S. Herman, H., Rangaswamy, S. (1987), 1st National Thermal Spraying Conference, ASM Int'l., Orlando, FL, p. 47.

Samuilov, E.V. (1966a), Sov. Phys., Doklady, vol. 166, p. 1397.

Samuilov, E.V. (1966b), Sov. Phys., Therm. Phys. High Temp., vol. 4, p. 143.

Samuilov, E.V. (1966c), Sov. Phys., Therm. Phys. High Temp., vol. 4, p. 753.

Samuilov. E.V. (1967), Properties of Gases at High Temperatures, Nauka (Science), Moscow, p. 3.

Samuilov, E.V. (1968), Physical Gas Dynamics of Ionized and Chemically Reacting Gases, Nauka (Science), Moscow, p. 3.

Samuilov, E.V. (1973), Thermo-Physical Properties of Gases, Nauka (Science), Moscow, p. 153.

Sanada, Y., Berkowitz, N. (1969), Fuel, vol. 48, p. 375.

Sardella, E., Gristina, R., Gilliland, D., Ceccone, G., Senesi, G.S., Rossi, F., d'Agostino, R., Favio, P. (2005), 17th Int. Symposium on Plasma Chemistry (ISPC-17), Toronto, Canada, p. 608.

Sarroukh, H., Robert, E., Cachoncinlle, C., Viladrosa, R., Pouvesle, J.M. (2001), 9th Int. Symp. on the Science and Technology of Light Sources, p. 353.

Sato, M., Ohgiyama, T., Clements, J.S. (1996), IEEE Trans. Ind. Appl., vol. 32, p. 106.

Sayasov, Yu.S. (1958), Sov. Phys., Doklady, vol. 122, p. 848.

Schaefer, G., Kristiansen, M., Guenther, A. (1991), "Gas Discharge Closing Switches," Springer, Berlin.

Schallus, E., Gotz, A. (1962), U.S. Patent no. 2,616,534, 25218.

Scheuer, J.T., Shamim, M., Conrad, J.R (1990), J. Appl. Phys., vol. 67, p. 1241.

Scheytt, H., Esrom, H., Prager, L., Mehnert, R., von Sontag, C. (1993), Non-Thermal Plasma Techniques for Pollution Control: Part B – Electron Beam and Electrical Discharge Processing, Penetrante, B.M., Schultheis, S.E. (eds.), Springer-Verlag, Berlin.

Schiller, S., Beister, G., Buedke, E., Becker, H.J., Schmidt, H. (1982), Thin Solid Films, vol. 96, p. 113.

Schoenbach, K.H., Verhappen, R., Tessnow, T., Peterkin, F.E., Byszewski, W.W. (1996), Appl. Phys. Lett., vol. 68, p. 13.

Schoenbach, K.H., Peterkin, F.E., Alden, R.W., Beebe, S.J. (1997a), IEEE Trans. Plasma Sci., vol. 25, p. 284.

Schoenbach, K.H., El-Habachi, A., Shi, W., Ciocca, M. (1997b), Plasma Sources Sci. Technol., vol. 6, p. 468.

Schoenbach, K.H., Moselhy, M., Shi, W. (2004), Plasma Sources Sci. Technol., vol. 13, p. 177.

Schonbein, C.F. (1840), Poggendorff's Anno. Phys. Chem., vol. 50, p. 616.

Schultz, G.J. (1976), "Vibrational Excitation of Molecules by Electron Impact at Low Energies," in Principles of Laser Plasma, Bekefi, G. (ed.), John Wiley & Sons, New York.

Schultz, J.F., Wulf, O.R. (1940), J. Am. Chem. Soc., vol. 62, p. 2980.

Schumacher, H.J. (1957), Z. Phys. Chem., vol. 13, p. 353.

Schütze, A., Jeong, J.Y., Babayan, S.E., Park, J., Selwyn, G.S., Hicks, R.F. (1998), IEEE Trans. Plasma Sci., vol. 26, p. 1685.

Schwab, G.-M., Phillinis, J. (1974), J. Am. Chem. Soc., vol. 69, p. 2588.

Schwartz, R.N., Slawsky, Z.I., Herzfeld, K.F. (1952), J. Chem. Phys., vol. 20, p. 1591.

Scott, T.W., Venugopalan, M. (1976), Nature, vol. 262, p. 48.

Scott, T.W., Chu, K.-C., Venugopalan, M. (1976), EPR Studies of Plasma Treated Coal, Chemical Industry, London, p. 739.

Secrest, D. (1973), Annu. Rev. Phys. Chem., vol. 24, p. 379.

Seeger, W. (2005), Deutsche Medizinische Wochenschrift, vol. 130, nos. 25–26, p. 1543.

Sekine, Y., Urasaki, K., Kado, S., Matsukata, M., Kikuchi, E. (2004), Energy Fuels, vol. 18, p. 455.

Selwyn, G.S., Heidenreich, J.E., Haller, K.L. (1990), Appl. Phys. Lett., vol. 57, p. 1876.

Selwyn, G.S., Park, J., Herrmann, H., Snyder, H., Henins, I., Hicks, R.F. Babayan, S.E., Jeong, J.Y., Tu, V.J. (1999), Adv. Appl. Plasma Sci., vol. 2, p. 111.

Sennewald, K., Schallus, E., Pohl, F. (1963), Chem.-Ingr.-Techn., vol. 35, p. 1.

Sergeev, P.A., Slovetsky, D.I. (1979), Plasma Chemistry-1979, III-Symposium on Plasma Chemistry, p. 132, Nauka (Science), Moscow.

References

Sergievska, A.L., Kovach, E.A., Losen, S.A. (1995), "Mathematical Modeling in Physical and Chemical Kinetics," Ed. Moscow State University, Institute of Mechanics.

Serov, Yu.L., Yavor, I.P. (1995), Sov. Phys., J. Techn. Phys., vol. 40, p. 248.

Sevastianov, V.I. (1991), High-Performance Biomaterials, Technomic, PA.

Shahin, M.M. (1971), Reactions under Plasma Conditions, Venugopalan, M. (ed.), vol. 2, p. 237, John Wiley & Sons, New York.

Sham, W., Spenceley, R., Teetzel, F. (1975), "Production of Uranium Tetrafluoride," U.S. Patent no. 2,898,187.

Shang, J.S. (2002), AIAA J., vol. 40, p. 1178.

Shang, J.S., Gaitonde, D.V., Canupp, P.W. (1999), AIAA paper 99-4903.

Shapouri, H., Duffield, J.A., Wang, M. (2002), The Energy Balance of Corn Ethanol: An Update, USDA, Office of Energy Policy and New Uses, Agricultural Economics, Report no. 813.

Shaub, W.M., Nibler, J.W., Harwey, A.B. (1977), J. Chem. Phys., vol. 67, p. 1883.

Shchuryak, E.V. (1976), Sov. Phys., J. Exper. Theoret. Phys., vol. 71, p. 2039.

Sheer, C. (1961), U.S. Patent no. 3,009,783.

Sheer, C., Korman, S., Dougherty, T.J. (1979), 4th Int. Conference on Plasma Chemistry (ISPC-4), Zurich, p. 277.

Shekhter, A.B. (1935), Chemical Reaction in Electric Discharge, ONTI, Moscow, Leningrad.

Shekhter, A.B., Kabisov, R.K., Pekshev, A.V., Kozlov, N.P., Perov, Y.L. (1998), Bull. Exper. Biol. Med., vol. 126, no. 8, p. 829.

Shekhter, A.B., Serezhenkov, V.A., Rudenko, T.G., Pekshev, A.V., Vanin, A.F. (2005), Nitric Oxide Biol. Chem., vol. 12, p. 210.

Sheppard, R.S., Wilson, W.L. (1972), "Preparation of Titanium Carbide," U.S. Patent no. 3,661,523.

Sheridan, M., Greer, E., Libby, W.F. (1972), J. Am. Chem. Soc., vol. 94, p. 2614.

Shevchenko, S.A., Zverev, A.G., Pavlov, S.A. (1974), "Mechanism of Formation of Metastable Crystal Structures of Fine Powders in Low-Temperature Plasma Conditions," in Problems of Chemistry and Chemical Engineering, vol. 35, Higher Education (Vyscha Schola), Charkov, Ukraine.

Shi, D., Wang, S.X., van Ooij, W.J., Want, L.M. (2001), Appl. Phys. Lett., vol. 78, p. 1243.

Shi, J. J., Kong, M.G. (2005), IEEE Trans. Plasma Sci., vol. 33, no. 2, p. 624.

Shibkov, V.M., Konstantinovskij, R.S. (2005), Kinetical Model of Ignition of Hydrogen-Oxygen Mixture under Conditions of Non-equilibrium Plasma of the Gas Discharge, AIAA paper 2005-987.

Shibkov, V.M., Chernikov, A.V., Chernikov, V.A., Ershov, A.P., Shibkova, L.V., Timofeev, I.B., Vinogradov, D.A., Voskanyan, A.V. (2000), 2nd Workshop on Magneto-Plasma-Aerodynamics in Aerospace Applications, Moscow, Russia.

Shibkov, V.M., Chernikov, V.A., Ershov, Dvinin, S.A., Raffoul, Ch.N., A.P., Shibkova, L.V., Timofeev, I.B., Van Wie, D.M., Vinogradov, D.A., Voskanyan, A.V. (2001), AIAA paper 2001-3087.

Shibkov, V.M., Aleksandrov, A.F., Chernikov, V.A., Ershov, A.P., Konstantinovskij, R.S., Zlobin, V.V. (2006), 44th AIAA Aerospace Science Meeting and Exhibit, Reno, NV, AIAA paper 2006-1216.

Shimansky, A. (1979), Chem. Technol. Fuels (KhTT), vol. 1, p. 84.

Shklovsky, B.I., Efros, A.L. (1979), Electronic Properties of Doped Semiconductors, Nauka (Science), Moscow.

Shmakin, Yu.A., Marusin, V.V. (1970), Application of Low-Temperature Plasma in Technology of Inorganic Materials and Powder Metallurgy, Nauka (Science), Novosibirsk, Russia.

Sholin, G.V., Nikerov, V.A., Rusanov, V.D. (1980), J. Phys., vol. 41, C9, p. 305.

Shuler, K. Weber, J. (1954), J. Chem. Phys., vol. 22, p. 3.

Shulutko, A.M., Antropova, N.V., Kryuger, Y.A. (2004), Surgery, vol. 12, p. 43.

Siemens, W. (1857), Poggendorff's Anno. Phys. Chem., vol. 102, p. 66.

Silbiger, J., Schnell, C. (1968), Swiss Patent no. 472337.

Silfvast, W.T. (2004), Laser Fundamentals, Cambridge University Press, New York, London.

Silla, V., Munz, R. (2005), 17th Int. Symposium on Plasma Chemistry (ISPC-17), Toronto, Canada, Abstracts and Full-Paper CD, p. 1210.

Silva, S.R.P., Ed. (1998), Amorphous Carbon: State of the Art, World Scientific, Singapore.

Silverthorn, D.U., Garrison, C.W., Silverthorn, A.C., Johnson, B.R. (2004), Human Physiology, an Integrated Approach, 3rd ed., Benjamin-Cummings Publishing Company.

Simek, M., Clupek, M. (2002), J. Phys. D: Appl. Phys., vol. 35, p. 1171.

Simon, F., Hermel, G., Lunkwitz, D., Werner, C., Eichhorn, K., Jacobasch, H.J. (1996), Macromol. Symp., vol. 103, p. 243.

Skanavi, G. (1958), Physics of Dielectrics at High Electric Fields, Fizmatgiz, Moscow.

Sladek, R.E.J., Baede, T.A., Stoffels, E. (2006), IEEE Trans. Plasma Sci., vol. 34, p. 1325.

Slater, R.C., Douglas-Hamilton, D.H. (1981), J. Appl. Phys., vol. 52, p. 5820.

Slavin, V.S., Gavrilov, V.M., Zelinsky, N.I., Bozhkov, A.R. (2001), AIAA J. Propul. Power, vol. 17, January/February.

Slone, R., Ramavajjala, M., Palekar, V., Pushkarev, V. (1998), Plasma Exhaust Aftertreatment, Society of Automotive Engineers, Warrendale, PA.

References

Slovetsky, D.I. (1977), "Kinetics and Mechanisms of Physical and Chemical Processes in Non-Equilibrium Plasma-Chemical Systems," Dr. Sci. Dissertation, Institute of Petro-Chemical Synthesis, USSR Academy of Sciences.

Slovetsky, D.I. (1980), Mechanisms of Chemical Reactions in Non-Equilibrium Plasma, Nauka (Science), Moscow.

Slovetsky, D.I. (1981), Plasma Chemistry, vol. 8, Smirnov, B.M. (ed.), Energo-Izdat, Moscow, p. 181.

Slovetsky, D.I., Todesaite, R.D. (1973a), Sov. Phys., High Energy Chem. (Khimia Vysokikh Energij), vol. 7, p. 291.

Slovetsky, D.I., Todesaite, R.D. (1973b), Sov. Phys., High Energy Chem. (Khimia Vysokikh Energij), vol. 7, p. 297.

Smirnov, A.S. (2000), Encyclopedia of Low Temperature Plasma, Fortov, V.E. (ed.), vol. 2, p. 67, Nauka (Science), Moscow.

Smirnov, B.M. (1968), Atomic Collisions and Elementary Processes in Plasma, Atomizdat, Moscow.

Smirnov, B.M. (1974), Ions and Excited Atoms in Plasma, Atomizdat, Moscow.

Smirnov, B.M. (1982a), Introduction to Plasma Physics, Nauka (Science), Moscow.

Smirnov, B.M. (1982b), Negative Ions, McGraw-Hill, New York.

Smirnov, B.M. (2001), Physics of Ionized Gases, John Wiley & Sons, New York.

Smirnov, B.M., Chibisov, M.I. (1965), Sov. Phys., J. Exper. Theoret. Phys., vol. 49, p. 841.

Smith, D.L. (1995), Thin-Film Deposition: Principles and Practice, McGraw-Hill, New York.

Smith, M.B. (2001), Organic Synthesis, McGraw-Hill, New York.

Smith, R.W. (1993), Powder Metall. Int. (PMI), Springer-Verlag, vol. 25, no. 1, p. 9.

Smith, R.W., Kangutkar, K., Drossman, R., Krepski, R. (1992), 13th Int. Thermal Spray Conference, ASM Int'l., Orlando, FL, p. 653.

Smullin, L.D., Chorney, P. (1958), Proc. IRE, vol. 46, p. 360.

Sobacchi, M.G., Saveliev, A.V., Fridman, A., Kennedy, L.A., Ahmed, S., Krause, T. (2002), Int. J. Hydrog. Energy, vol. 27, p. 635.

Sobacchi, M.G., Saveliev, A.V., Fridman, A., Gutsol, A., Kennedy, L.A. (2003), Plasma Chem. Plasma Process., vol. 23, p. 347.

Sobolev, V.V., Guilmany, J.P., Mart, A.J. (1999), J. Mater. Process. Technol., vol. 87, p. 37.

Sodha, M.S. (1963), Brit. J. Appl. Phys., vol. 14, p. 172.

Solomon, P.R., Hamblen, D.J., Carangelo, R.M., Krause, J.L. (1982), 19th Int. Combustion Symposium, p. 1139.

Solonenko, O.P. (1986), News of Siberian Branch of the USSR Academy of Sciences, Eng. Sci., vol. 4, no. 1, p. 136.

Solonenko, O.P. (1996), Thermal Plasma and New Materials Technology, Cambridge Int. Science Publishing, UK.

Sommerer, T.J., Barnes, M.S., Keller, J.H., McCaughey, M.J., Kushner, M. (1991), Appl. Phys. Lett., vol. 59, p. 638.

Souris, A.L., Shorin, S.N. (1969), Sov. Phys., High Energy Chem. (Khimia Vysokikh Energij), vol. 3, p. 2.

Spahn, R.G., Gerenser, L.J. (1994), U.S. Patent no. 5,324,414.

Spears, K., Kampf, R., Robinson, T. (1988), J. Phys. Chem., vol. 92, p. 5297.

Speight, J.G., Ozum, B. (2001), Petrolium Refining Processes, CRC Press, Boca Raton, FL.

Spence, P.D., Roth, J.R. (1994), 21st IEEE Int. Conf. on Plasma Science, Santa Fe, NM, p. 97.

Spence, P.D., Keebler, P.F., Freeland, M.S., Roth, J.R. (1991), 10th Int. Symposium on Plasma Chemistry (ISPC-10), Proc. Workshop on Industrial Plasma Appl. and Eng., Bochum, Germany.

Spitzer, L. (1941), Astrophys. J., vol. 93, p. 369.

Spitzer, L. (1944), Astrophys. J., vol. 107, p. 6.

Staack, D., Farouk, B., Gutsol, A., Fridman, A. (2005a), Plasma Source Sci. Technol., vol. 14, p. 700.

Staack, D., Farouk, B., Gutsol, A., Fridman A. (2005b), 17th Int. Symposium on Plasma Chemistry (ISPC-17), Toronto, Canada, Abstracts and Full-Paper CD, p. 281.

Staack, D., Farouk, B., Gutsol, A., Fridman A. (2006), Plasma Sources Sci. Techn., vol. 15, p. 818.

Staack, D. (2008), "Atmospheric Pressure Microplasmas and Their Application to Thin Film Deposition," PhD. Dissertation, Drexel University, Philadelphia, PA.

Stachorska, D. (1965), J. Chem. Phys., vol. 42, p. 1887.

Stakhanov, I.P. (1968), Plasma Thermionic Converters, AtomIzdat, Moscow.

Starik, A.M., Titova, N.S. (2005), Combust., Explos. Shock Waves, vol. 40, p. 499.

Starikovskaia, S.M. (2006), J. Phys. D: Appl. Phys., vol. 39, p. 265.

Starikovskaia, S.M., Kukaev, E.N., Kuksin, A.Yu., Nudnova, M.N., Starikovskii, A.Yu. (2004), Combust. Flame, vol. 139, p. 177.

Starikovskaia, S.M., Anikin, N.V., Kosarev, I.N., Popov, N.A., Starikovskii, A.Yu. (2006), 44th AIAA Aerospace Science Meeting and Exhibit, Reno, NV, AIAA, paper 2006-616.

Starikovskii, A.Yu. (2000), Plasma Phys. Rep., vol. 26, p. 701.

Starikovskii, A.Yu. (2003), Combust. Explo. Shock Waves, vol. 39, p. 619.

Starke, A. (1923), Z. Elektrochem., vol. 29, p. 358.

Steffens, H. D., Hohle, M.-H. (1980), 9th Int. Thermal Spraying Conf., Nederlands Institut voor Lastechniek, The Hague, p. 420.

References

Steiger, R.A., Wilson, W.L. (1974), "Preparation of Monotungsten Carbide," U.S. Patent no. 3,848,062.

Steinberg, M. (2000), "Decarbonization and Sequestration for Mitigating Global Warming," Int. Symp. "Deep Sea & CO, 2000," SRI, Mitaka, Tokyo, pp. 4-2-1, 4-2-6.

Stevens, J.E., Huang, Y.C., Larecki, R.L., Cecci, J.L. (1992), J.Vac. Sci. Technol., vol. A10, p. 1270.

Stewart, R.A., Lieberman, M.A. (1991), J. Appl. Phys., vol. 70, p. 3481.

Stewart, R.L. (1967), U.S. Patent nos. 3,586,613, 3/31/1967.

Stoffels, E. (2003), Proceedings Gaseous Electrono. Conference, AIP, San Francisco, CA, p. 16.

Stoffels, E. (2006), J. Phys. D: Appl. Phys., vol. 39, p. 16.

Stokes, C.S. (1971), Reactions under Plasma Conditions, Venugopalan, M. (ed.), vol. 2, p. 259, John Wiley & Sons, New York.

Storer, J., Haynes, J.H. (1994), Welding Manual: Basics of Gas, ARC, MIG, TIG and Plasma Welding and Cutting, MBI Publishing Company.

Stout, P.J., Kushner, M.J. (1993), J. Vac. Sci. Technol., vol. A11, p. 2562.

Struckman, C.K., Kusse, B.R. (1993), J. Appl. Phys., vol. 74, p. 3658.

Stueber, G.J., Clarke, S.A., Bernstein, E.R., Raoux, S., Tanaka, T., Porshnev, P. (2003), J. Phys. Chem., vol. A107, p. 7775.

Su, T., Bowers, M.T. (1973), Int. J. Mass Spectrom. Ion Phys., vol. 12, p. 347.

Su, T., Bowers, M.T. (1975), Int. J. Mass Spectrom. Ion Phys., vol. 17, p. 221.

Sugano, T. (1985), Applications of Plasma Processes to VLSI Technology, Wiley, New York.

Sugiarto, A.T., Ito, S., Ohshima, T., Sato, M., Skalny, J.D. (2003), J. Electrostatics, vol. 58, p. 135.

Sugiarto, A.T., Ohshima, T., Sato, M. (2002), Thin Solid Films, vol. 407, p. 174.

Suh, S.-M., Girshick, S.L., Kortshagen, U.R., Zachariah, M.R. (2003), J. Vac. Sci. Technol., vol. A21, p. 251.

Suhr, H. (1973), Fortschr. Chem. Forsch., vol. 36, p. 39.

Suhr, H. (1974), Techniques and Applications of Plasma Chemistry, Hollahan, J.R., Bell, A.T. (eds.), Wiley-Interscience, p. 57.

Sulzmann, K.J.P., Myer, B.F., Bartle, E.R. (1965), J. Chem. Phys., vol. 42, p. 3969.

Sun, B., Sato, M., Clements, J.S. (1997), J. Electrostatics, vol. 39, p. 189.

Sun, B., Sato, M., Clements, J.S. (1999), J. Phys. D: Appl. Phys., vol. 32, p. 1908.

Sunka, P., Babicky, V., Clupek, M., Lukes, P., Simek, M., Schmidt J., Cernak M. (1999), Plasma Source Sci. Technol, vol. 8, p. 258.

Sutton, G.W., Sherman, A. (2006), Engineering Magnetohydrodynamics, Dover.

Svedchikov, A.P., Belousova, E.V. (1988), Radiat. Phys. Chem., vol. 31, p. 15.

Svelto, O. (2004), Principles of Lasers, Springer, 4th ed.

Swaney, L.R. (1968a), France Patent no. 3,485,858.

Swaney, L.R. (1968b), U.S. Patent no. 3,812,239.

Swaney, L.R. (1974), "Preparation of Submicron Titanium Carbide," U.S. Patent no. 3,812,239.

Taieb, G., Legay, F. (1970), Can. J. Phys., vol. 48, p. 1956.

Tajmar, M. (2004), Advanced Space Propulsion Systems, Springer, Berlin.

Takita, K. (2002), Combust. Flame, vol. 128, p. 301.

Takita, K., Ju, Y. (2006), 44th AIAA Aerospace Science Meeting and Exhibit, Reno, NV, AIAA paper 2006-1209.

Talrose, V.L. (1952a), Ph.D. Dissertation, N.N. Semenov Institute of Chemical Physics, Moscow.

Talrose, V.L. (1952b), Sov. Phys., Doklady (Reports of USSR Academy of Sciences), vol. 86, p. 909.

Talrose, V.L., Vinogradov, P.S., Larino. I.K. (1979), Gas Phase Ion Chemistry, Bowers, M.(ed.), Academic Press, vol. 1, p. 305.

Tareev, B.M. (1978), Physics of Dielectric Materials, Mir (World), Moscow.

Tarras, P., Musil, J., Bardos, L., Fric, Z. (1981), 15th Int. Conference on Phenomena in Ionized Gases (ICPIG-15), vol. 1, p. 381, Minsk.

Tatoulian, M., Bouloussa, O., Moriere, F., Arefi-Khonsari, F., Amouroux, J., Rondelez, F. (2004), Langmuir, vol. 20, p. 10481.

Taylor, R.L., Bitterman, S. (1969), Rev. Mod. Phys., vol. 41, p. 26.

Terlinger, J.G.A., Feijen, J., Hoffman, A.S. (1993), J. Colloid Interf. Sci., vol. 155, p. 55.

Tesner, P.A. (1972), Formation of Carbon from Gas-Phase Hydrocarbons, Chimia (Chemistry), Moscow.

Tesner, P.A., Altshuller, B.N. (1969a), Gas Ind., vol. 6, p. 4.

Tesner, P.A., Altshuller, B.N. (1969b), Sov. Phys., Doklady (Reports of USSR Academy of Sciences), vol. 187, p. 1100.

Thiele, M., Warnatz, J., Maas, U. (2000), Combust. Theor. Model., vol. 4, p. 413.

Thiyagarajan, M., Alexeff, I., Parameswaran, S., Beebe, S. (2005), IEEE Trans. Plasma Sci., vol. 33, no. 2, p. 322.

Thomas, H., Hocker, H. (1995), 9th Int. Wool Text. Res. Conf., Biella, vol. 4, p. 351.

Thomas, H., Herrling, J., Rakowski, W., Kaufmann, R., Hocker, H. (1992), DWI Reports, vol. 111, p. 315.

Thomson, J.J. (1912) Philos. Mag., vol. 23, p. 449.

Thomson, J.J. (1924) Philos. Mag., vol. 47, p. 337.

References

Thornton, J.A. (1986), J. Vac. Sci. Technol., vol. A4, p. 3059.

Thorpe, M.L. (1971), Chem. Eng. News, vol. 49, no. 35, p. 20.

Thorpe, M.L., Wilks, P.H. (1971), Chem. Eng., p. 117.

Thursfield, G., Davies, G.J. (1974), Trans. Inst. Chem. Eng., vol. 52, p. 237.

Timmins, R., Amman, P. (1970), Plasma Applications in Chemical Processes, Polak, L.S. (ed.), Mir (World), Moscow.

Tira, J.S. (1987), SAMPE J., vol. 18, p. 18.

Tisone, G.C., Branscome, L.M. (1968), Phys. Rev., vol. 170, p. 169.

Tkachuk, B.V., Kolotyrkin, V.M. (1977), Production of Thin Polymer Films from Gas Phase, Khimia (Chemistry), Moscow.

Tokos, J., Hall, B., Calhoun, J., Prestbo, E. (1998), Atmos. Environ., vol. 32, p. 823.

Tokunaga, O., Suzuki, N. (1981), Radiat. Phys. Chem., vol. 24, p. 145.

Tomita, S., Hikita, M. (2000), Chem. Phys. Lett., vol. 316, p. 361.

Treanor, C.E., Rich, I.W., Rehm, R.G. (1968), J. Chem. Phys., vol. 48, p. 1798.

Trivelpiece, A.W., Gould, R.W. (1959), J. Appl. Phys., vol. 30, p. 1784.

Troitsky, V.N., Aivasov, M.I., Kuznetsov, V.M., Koriagin, V.S. (1971a), Low-Temperature Plasma in Technology of Inorganic Substances, p. 19, Nauka (Science), Novosibirsk, Russia.

Troitsky, V.N., Grebtsov, B.M., Aivasov, M.I., Koriagin, V.S. (1971b), Plasma Chemistry 71, p. 145, Institute of Petrochemical Synthesis of USSR Academy of Sciences, Moscow.

Troitsky, V.N., Grebtsov, B.M., Aivasov, M.I. (1973), Powder Metall., vol. 11, p. 6.

Tsai, C., Nelson, J., Gerberich, W., Heberlein, J. V. R., Pfender, E. (1992), 3rd Int. Conf. on New Diamond Science and Technology (Diamond '92), Heidelberg, FRG, COMST.

Tsai, C.-H., Hsieh, T.-H. (2004), Ind. Eng. Chem. Res., vol. 43, p. 4043.

Tsai, C.-H., Hsieh, T.-H., Shih, M., Huang, Y.J., Wei, T.C. (2005), Environ. Energy Eng., AIChE J., vol. 51, p. 2853.

Tselishchev, P.A., Abaev, G.G. (1987), Plasma Gasification and Pyrolysis of Low-Quality Coals, Energy Institute, ENIN, Moscow.

Tsvetkov, Yu.V., Panfilov, S.A. (1980), Low-Temperature Plasma in Reduction Processes, Nauka (Science), Moscow.

Tsvetkov, Yu.V., Panfilov, S.A., Kliachko, L.I. (1976), "Production of Tungsten Powder in Hydrogen Plasma Jet, and Investigations of Properties of the Powder," in Plasma Processes in Metallurgy and Technology of Inorganic Materials, A.A. Baikov Institute of Metallurgy of USSR Academy of Sciences, Nauka (Science), Moscow.

Tumanov, Yu.N. (1975), 2nd USSR Symposium on Plasma Chemistry, vol. 1, p. 238, Riga, Latvia.

Tumanov, Yu.N. (1976), Sov. Phys., Nucl. Energ. (Atomnaya Energia), vol. 39, no. 6, p. 424.

Tumanov, Yu.N. (1981), Electro-Thermal Reactions in Modern Chemical Technology and Metallurgy, Energo-Atom-Izdat, Moscow.

Tumanov, Yu.N. (1989), Low-Temperature Plasma and High-Frequency Electromagnetic Fields in Production of Nuclear Energy Related Materials, Energo-Atom-Izdat, Moscow.

Tumanov, Yu.N., Galkin, N.P. (1972), Sov. Phys., Nucl. Energ. (Atomnaya Energia), vol. 32, no.1, p. 21.

Tumanov, Yu.N., Galkin, N.P. (1974), Sov. Phys., Nuclear Energy (Atomnaya Energia), vol. 34, no. 4, p. 340.

Turnipseed, A.A., Barone, S.B., Ravishenkara, A.R. (1996a), J. Phys. Chem., vol. 100, p. 14,694.

Turnipseed, A.A., Barone, S.B., Ravishenkara, A.R. (1996b), J. Phys. Chem., vol. 100, p. 14,703.

Turns, S.R. (2000), Introduction to Combustion: Concepts and Applications, 2nd ed., McGraw-Hill, Boston.

Tusov, L.S., Kulikov, V.V., Kolotyrkin, Ya.M., Tunitsky, N.N. (1975), 2nd USSR Symposium on Plasma Chemistry, vol. 1, p. 234, Riga, Latvia.

Tylko, J.K. (1973), Ist Int. Symposium on Plasma Chemistry, pp. 2–7, Kiel, Germany.

Tylko, J.K. (1974), "High Temperature Treatment of Material," U.S. Patent no. 8,783,167.

Tzeng, C.-C, Kuo, Y.-Y., Huang, T.-F., Lin, D.-L., Yu, Y.-J. (1998), J. Haz. Mater., vol. 58, p. 207.

Uehara, T. (1999), Adhesion Promotion Techniques, Marcel Dekker, New York.

Ultee, C.J. (1977), Chem. Phys. Lett., vol. 46, p. 366.

Uman, M.A. (1964), Introduction to Plasma Physics, McGraw-Hill, New York.

Urbas, A.D. (1978a), "Kinetics and Mechanisms of Chemical Processes in Glow Discharges in Nitrogen, Hydrogen, Ammonia and Their Mixtures," Ph. D. Dissertation, Institute of Petrochemical Synthesis of USSR Academy of Sciences, Moscow.

Urbas, A.D. (1978b), Modern Problems of Physical Chemistry, Karpov Institute of Physical Chemistry, Moscow.

Usov, L.N, Borisenko, A.I. (1965), Application of Plasma for Production of High-Temperature Coatings, Nauka (Science), Moscow.

Vahedi, V., Lieberman, M.A., Alves, M.V., Verboncoeur, J.P., Birdsall, C.K. (1991), J. Appl. Phys., vol. 69, p. 2008.

Vahedi, V., Stewart, R.A., Lieberman, M.A. (1993), J. Vac. Sci. Technol., vol. A11, p. 1275.

Vakar, A.K., Denisenko, V.P., Rusanov, V.D. (1977), 3rd Int. Symposium on Plasma Chemistry (ISPC-3), vol. 1, p. 10, Limoge, France.

Vakar, A.K., Denisenko, V.P., Maksimov, G.P., Rusanov, V.D., Fridman, A. (1978a), 3rd USSR Symposium on High Current Electronics, Tomsk, Russia, p. 31.

Vakar, A.K., Maximov, G.P., Denisenko, V.P., Rusanov, V.D., Fridman, A., Sholin, G.V. (1978b), 3rd USSR Symposium on High Current Pulse Electronics, Academy of Sciences of the USSR, Tomsk, Russia, p. 31.

Vakar, A.K., Givotov, V.K., Krasheninnikov, E.G., Fridman, A. (1981a), Sov. Phys., J. Techn. Phys. Lett., vol. 7, p. 996.

Vakar, A.K., Givotov, V.K., Krasheninnikov, E.G., Rusanov, V.D., Fridman, A. (1981b), 15th Int. Symposium on Phenomena in Ionized Gases (ICPIG-15), VINITI, vol. 1, p. 323, Minsk.

Vakar, A.K., Krasheninnikov, E.G., Rusanov, V.D. (1981c), Plasma-Chemical Process of CO_2 – Dissociation in Non-Equilibrium Microwave Discharge, Kurchatov Institute of Atomic Energy, Moscow, vol. 3466/7.

Vakar, A.K., Krasheninnikov, E.G., Tischenko, E.A. (1984), IVth Symposium on Plasma Chemistry, Dnepropetrovsk, Ukraine, p. 35.

Valibekov, Yu.V., Volodin, N.L., Vurzel, F.B., Gutman, B.E., Polak, L.S., Enduskin, P.N., Epstein, I.L. (1975), Sov. Phys., High Energy Chem. (Khimia Vysokikh Energij), vol. 9, p. 1(1).

VanDrumpt, J.D. (1972), Ind. Eng. Chem. Fund., vol. 11, no. 4, p. 594.

Vanin, A.F. (1998), Biochemistry, vol. 63, p. 867.

Vanpee, M. (1993), Combust. Sci. Technol., vol. 93, p. 363.

Vardelle, A. (2005), IEEE International Conference on Plasma Science (ICOPS-2005), Monterey, CA, p. 86.

Vardelle, M., Vardelle, A., Li, K.-I., Fauchais, P., Themelis, N.J. (1996), Pure Appl. Chem., vol. 68, p. 1893.

Vasilets, V.N. (2005), "Modification of Physical Chemical and Biological Properties of Polymer Materials Using Gas-Discharge Plasma and Vacuum Ultraviolet Radiation," Dr. Sci. Dissertation, Institute of Energy Problems of Chemical Physics, Russian Academy of Sciences, Moscow.

Vasilets, V.N., Tikhomirov, L.A., Ponomarev, A.N. (1979), Sov. Phys., High Energy Chem. (Khimia Vysokikh Energiy), vol. 13, p. 475.

Vasilets, V.N., Werner, C., Hermel, G., Pleul, D., Nitschke, M., Menning, A., Janke, A., Simon, F. (2002), J. Adhes. Sci. Technol., vol. 16, p. 1855.

Vasilets, V.N., Kuznetsov, A.V., Sevastianov, V.I. (2006), High Energy Chem. (Khimia Vysokikh Energiy), vol. 40, p. 105.

Vasilyak, L.M., Kostuchenko, S.V., Kurdyavtsev, N.N., Filugin, I.V. (1994), Adv. Phys. Sci. (Uspehi Phys. Nauk), vol. 164, p. 263.

Vatazhin, A.B., Kopchenov, V.I. (2000), Scramjet Propulsion, Chapter 14, Curran, E.T., Murphy, S.N.B. (eds.), AIAA Progress in Aeronautics and Astronautics Series, vol. 189.

Vatazhin, A.B., Lubimov, G.A, Regirer, S.A. (1970), Magneto-Hydrodynamic Flows in Channels, Nauka (Science), Moscow.

Vatazhin, A.B., Kopchenov, V.I., Gouskov, O.V. (1999), AIAA paper 99-4972.

Vdovenko, M.I., Ibraev, Sh.Sh., Messerle, V.E. (1987), "Plasma Gasification of Coals with Utilization of Mineral Fraction," in Plasma Gasification and Pyrolysis of Low Quality Coals, p. 59, Energy Institute, ENIN, Moscow.

Venugopalan, M., Roychowdhury, U.K., Chan, K., Pool, M. (1980), "Plasma Chemistry of Fossil Fuels," in Topics in Current Chemistry, Springer-Verlag, Heidelberg, New York.

Veprek, S. (1972a), J. Chem. Phys., vol. 57, p. 952.

Veprek, S. (1972b), J. Crystal Growth, vol. 17, p. 101.

Veprek, S., Venugopalan, M. (1980), Plasma Chemistry, vol. III, Springer, New York.

Veprek, S., Veprek- Heijman, M.G. (1990), Appl. Phys. Lett., vol. 56, p. 1766.

Veprek, S., Veprek- Heijman, M.G. (1991), Plasma Chem. Plasma Process., vol. 11, p. 323.

Verbovetsky, E.Kh., Kotler, V.R. (1984), Thermal Power Plants, Energo-Atom-Izdat, Moscow.

Vinogradov, G.K. (1986), Sov. Phys., High Energy Chem. (Khimia Vysokikh Energiy), vol. 20, no. 3, p. 195.

Vinogradov, G.K., Ivanov, Yu.A. (1977), in Chemical Reactions in Low-Temperature Plasma, Polak, L.S. (ed.), p. 27, Institute of Petrochemical Synthesis of USSR Academy of Sciences, Moscow.

Vinogradov, G.K., Ivanov, Yu.A., Polak, L.S. (1978), Sov. Phys., High Energy Chem. (Khimia Vysokikh Energiy), vol. 12, p. 542.

Vinogradov, G.K., Ivanov, Yu.A., Polak, L.S. (1979a), Sov. Phys., High Energy Chem. (Khimia Vysokikh Energiy), vol. 13, p. 84.

Vinogradov, G.K., Ivanov, Yu.A., Polak, L.S. (1979b), Plasma-Chemical Processes, Polak, L.S. (ed.), Nauka (Science), Moscow, p. 64.

Virin, L., Dgagaspanian, R., Karachevtsev, G., Potapov, V., Talrose V. (1978), Ion-Molecular Reactions in Gases, Nauka, Moscow.

Vishnevska, V.A., Konstant, Z.A., Miller, T.N., Vaivad, A.Ya. (1969), News Acad. Sci. Latvia, Chem., vol. 1, p. 69.

Vitale, S.A., Hadidi, K., Cohn, D.R., Falkos, P., Bromberg, L. (1995), I&EC Special Symposium of the American Chemical Society, Atlanta, GA.

Vitello, P.A., Penetrante B.M., Bardsley J. N. (1994), Phys. Rev., vol. E49, p. 5574.

Vitruk, P.P., Baker, H.J., Hall, D.R. (1992), J. Phys. D: Appl. Phys., vol. 25, p. 1767.

References

Voinovich, P.A., Ershov, A.P., Ponomareva, S.E., Shibkov, V.M. (1991), Phys. High Temp., vol. 29, p. 468.

Volchenok, V.I., Egorov, I.P. (1976), Sov. Phys., J. Techn. Phys., vol. 46, p. 2541.

Volchenok, V.I., Komarov, V.N., Kuprianov, S.E., Stukanog, V.I. (1980), Plasma Chemistry, vol. 7, Smirnov, B.M. (ed.), AtomIzdat, Moscow.

Volenik, K., Hanonsek, F., Chraska, P., Ilavsky, J., Neufuss, K. (1999), Mater. Sci. Eng., vol. A272, p. 199.

Volodin, N.L., Polak, L.S., Enduskin, P.N., Levenson, R.I., Krugly, S.M., Diatlov, V.T. (1969), U.K Patent no. 1316668.

Volodin, N.L., Vurzel, F.B., Polak, L.S., Enduskin, P.N. (1970a), Physics, Technology and Applications of Low-Temperature Plasma, Polak, L.S. (ed.), pp. 576, 583, Alma-Ata, Kazakhstan.

Volodin, N.L., Vurzel, F.B., Polak, L.S., Enduskin, P.N. (1970b), Problems of Petrochemical Synthesis, Polak, L.S. (ed.), pp. 11, 24, Ufa, Russia.

Volodin, N.L., Vurzel, F.B., Diatlov, V.T., Polak, L.S., Shmykov, Yu.I., Enduskin, P.N. (1971a), Sov. Phys., High Energy Chem. (Khimia Vysokikh Energij), vol. 5, p. 3.

Volodin, N.L., Vurzel, F.B., Polak, L.S., Enduskin, P.N. (1971b), Plasma Chemistry-71, Polak, L.S. (ed.), p. 157, Institute of Petrochemical Synthesis of USSR Academy of Sciences, Moscow.

Von Elbe, G., Lewis, B. (1942), J. Chem. Phys., vol. 10, p. 366.

von Engel, A. (1994), American Vacuum Society Classics, Springer-Verlag, Berlin.

von Engel, A., Steenbeck, M. (1934), Elektrische Gasentladungeno. Ihre Physik und Technik, vol. II, Springer, Berlin.

Von Gunten, U. (2003), Water Res., vol. 37, p. 1469.

Vossen, J.L., Kern (1978), Thin Films Processes, Academic, New York.

Vossen, J.L., Kern (1991), Thin Films Processes II, Academic, New York.

Vurzel, F.B. (1970), Application of Plasma in Chemical Processes, Polak, L.S. (ed.), Chapter 10, Mir (World), Moscow.

Vurzel, F.B., Nazarov, N.F. (2000), "Plasma-Chemical Treatment of Powder Materials and Coating Formation," in Encyclopedia of Low-Temperature Plasma, Fortov, V.E. (ed.), vol. 4, p. 349.

Vurzel, F.B., Polak, L.S. (1965), Kinetics and Thermodynamics of Chemical Reactions in Low Temperature Plasma, Polak, L.S. (ed.), p. 101, Nauka (Science), Moscow.

Vurzel, F.B., Dolgopolov, N.N., Maksimov, A.I., Polak, L.S., Fridman, V.I. (1965), Kinetics and Thermodynamics of Chemical Reactions in Low Temperature Plasma, Polak, L.S. (ed.), p. 223, Nauka (Science), Moscow.

Vurzel, F.B., Polak, L.S., Shchipachev, V.S. (1967), Sov. Phys., High Energy Chem. (Khimia Vysokikh Energij), vol. 1, p. 268.

Vurzel, F.B., Valibekov, Yu.V., Gutman, B.E., Polak, L.S. (1973), Sov. Phys., High Energy Chem. (Khimia Vysokikh Energij), vol. 7, p. 3.

Vurzel, F.B., Polak, L.S., Epstein, I.L. (1975), Sov. Phys., High Energy Chem. (Khimia Vysokikh Energij), vol. 9, p. 2(2).

Vuskovic, L., Ash, R.L., Shi, Z., Popovic, S., Dinh, T. (1997), J. Aerospace, SAE Trans., Section 1, vol. 106, p. 1041.

Wadehra, J.M. (1986), Non-Equilibrium Vibrational Kinetics, Topics in Current Physics, vol. 39, Capitelli, M. (ed.), Springer-Verlag, Berlin.

Walder, B., Hancock, R. (1973), Ist Int. Symposium on Plasma Chemistry (ISPC-1), paper 5-6, Kiel, Germany.

Waldie, B. (1973), Trans. Inst. Chem. Eng., vol. 48, no. 3, p. 90.

Walker, P.A., Rusinko, F., Austin, L.G. (1963), Carbon Reactions with Gases, Izdatelstvo Inostrannoj Literatury, Moscow.

Wan, Y.P., Fincke, J.R., Sampath, S., Prasad, V., Herman, H. (2002), Int. J. Heat Mass Transf., vol. 45, p. 1007.

Wang, B.W., Xu, G.H. (2005), 17th Int. Symposium on Plasma Chemistry (ISPC-17), Toronto, Canada, p. 851.

Wang, F., Liu, J.B., Sinibaldi, J., Brophy, C., Kuthi, A., Jiang, C., Ronney, P., Gundersen, M.A. (2005), IEEE Trans. Plasma Sci., vol. 33, p. 326.

Wang, X., Li, C., Lu, M., Pu, Y. (2003), Plasma Sources Sci. Technol., vol. 12, p. 358.

Wang, Y.F., Lee, W.J., Chen, C.Y., Hsieh, L.T. (1999), Ind. Eng. Chem. Res., vol. 38, p. 3199.

Washino, M., Tokunaga, O., Hashimoto, S. (1984), IEEE-SM-194/706.

Watanabe, T. (2003), 16th Int. Symposium on Plasma Chemistry (ISPC-16), p. 734, Taormina, Italy.

Watanabe, T., Taira, T., Takeuchi, A. (2005), 17th Int. Symposium on Plasma Chemistry (ISPC-17), Toronto, Canada, p. 1153.

Watanabe, Y., Shiratani, M., Kubo, Y., Ogava, I., Ogi, S. (1988), Appl. Phys. Lett., vol. 53 , p. 1263.

Watanabe, Y., Shiratani, M., Fukuzava, T., Koga, K. (2000), J. Tech. Phys., vol. 41, p. 505.

Watanabe, Y., Shiratani, M., Koga, K. (2001), XXV Int. Symposium on Phenomena in Ionized Gases (ICPIG-25), vol. 2, p. 15.

Weiler, M., Sattel, S., Giessen, T., Jung, K., Ehrhardt, H., Veerasamy, V.S., Robertson, J. (1996), Phys. Rev., B., vol. 53, p. 1594.

Weiler, M., Lang, K., Robertson, J. (1998), Appl. Phys. Lett., vol. 72, p. 1314.

References

Weisbeek, R., Hullstrung, D. (1970), Chem. Ingr. Techn., vol. 42, no. 21, p. 1302.

Wendt, A.E., Lieberman, M.A. (1993), 2nd Workshop on High Density Plasmas and Applications, AVS Topical Conference, San Francisco, CA.

Wertheimer, M.R., Bailon, J.P. (1977), 3rd Int. Symposium on Plasma Chemistry (ISPC-3), vol. 5, p. 23, Limoge, France.

Wertheimer, M.R., Dennler, G., Guimond, S. (2003), 16th Int. Symposium on Plasma Chemistry (ISPC-16), p. 11, Taormina, Italy.

West, A.R. (1999), Basic Solid State Chemistry, John Wiley & Sons, New York.

Westfall, L. J. (1987), 1st National Thermal Spray Conference, ASM Int'l., Orlando, FL, p. 417.

Westmoreland, P.R., Howard, J.B., Longwell, J.P., Dean, A.M. (1986), AIChE J., vol. 32, p. 1971.

Wiegand, W.J., Nighan, W.L. (1973), Phys. Lett., vol. 22, p. 583.

Wiegand, W.J., Fowler, M.C., Benda, J.A. (1970), Appl. Phys. Lett., vol. 16, p. 237.

White, A.D. (1959), J. Appl. Phys., vol. 30, p. 711.

White, A.R., Palm, P., Plonjes, E., Subramaniam, V., Adamovich, I.V. (2002), J. Appl. Phys., vol. 91, p. 2604.

Whittier, J.S. (1976), J. Appl. Phys., vol. 47, p. 3542.

Wild, C., Koidl, P. (1991), J. Appl. Phys., vol. 69, p. 2909.

Wilhelm, H.A., MacClusky, J.K. (1969), J. Metals, vol. 21, no. 12, p. 51.

Wilkerson, J.T., Van Wie, D.M., Cybyk, B.Z. (2003), AIAA paper 2003-0526.

Wilks, P.H. (1973a), Chem. Ind., no. 18, p. 891.

Wilks, P.H. (1973b), 1st Int. Symposium on Plasma Chemistry (ISPC-1), Kiel, Germany (BRD), paper 4-1.

Wilks, P.H. (1975), Chem. Eng., vol. 82, no. 25, p. 56.

Wilks, P.H., Ravinder, P., Grant, C.L. (1972), Chem. Eng. (Progr.), vol. 68, no. 4, p. 82.

Williams, A. (2000), Combustion and Gasification of Coal, Taylor & Francis, New York, London.

Williams, F.A. (1985), Combustion Theory, 2nd edition, Benjamin/Cummings Publishing, Menlo Park, CA.

Williams, S., Midey, A.J., Arnold, S.T., Miller, T., Bench, P.M., Dressler, R.A., Chiu, Y.H., Levandier, D.J. (2001), 32nd AIAA Plasmadynamics and Lasers Conference, Anaheim, CA, AIAA paper 2001-2873.

Williams, S., Popovic, S., Vuskovich, L., Carter, C., Jacobson, L., Kuo, S., Bivolaru, D., Corera, S., Kahandawala, M., Sidhu, S. (2004), 42nd AIAA Aerospace Sciences Meeting and Exhibit, 5–8 January 2004, Reno, NV, AIAA paper 2004-1012.

Williamson, J.M., Ganguly, B.N. (2001), Phys. Rev., vol. E64, p. 403.

Williamson, R.L., Fincke, J.R., Chang, C.H. (2000), Plasma Chem. Plasma Process., vol. 20, p. 299.

Willis, G., Saryend, W.J., Marlow, D.M. (1969), J. Appl. Phys., vol. 50, p. 68.

Wilson, J., Hao-Line, C., Fyfe, W. (1973), J. Appl. Phys., vol. 44, p. 5447.

Wilson, R.G., Brewer, G.R. (1973), Ion Beams, with Applications to Ion Implantation, Wiley, New York.

Wilson, W.L., Strain, F., Hockie, H.H. (1965), U.S. Patent no. 3,443,897.

Winters, H.F., Coburn, J.W. (1992), Surface Science Rep., vol. 14, p. 161.

Winters, H.F., Haarer, D. (1987), Phys. Rev. B, vol. 36, p. 6613.

Winters, H.F., Haarer, D. (1988), Phys. Rev. B, vol. 37, p. 10379.

Wolfe, R.L. (1990), Environo. Sci. Tech., vol. 24, p. 768.

Wolfkovich, S.I., Asiev, R.G. (1965), Sov. Phys., Doklady (Reports of USSR Academy of Sciences), vol. 162, p. 6.

Wu, D., Outlaw, R.A., Ash, R.L. (1993), J. Appl. Phys., vol. 74, p. 4990.

Wu, Z.L., Gao, X., Luo, Z.Y., Wei, E.Z., Zhang, Y.S., Zhang, J.Z., Ni, M.J., Cen, K.F. (2005), Energy Fuels, vol. 19, p. 2279.

Wutzke, S. A., Pfender, E., Eckert, E. R. G. (1984), AIAA J., vol. 6, no. 8, p. 1474.

Xiaohui, Y., Raja, L.L. (2003), IEEE Trans. Plasma Sci., vol. 31, p. 495.

Xie, K.-C., Lu Y.-K., Tian, Y.-J., Wang D.-Z. (2002), Energy Sources, vol. 24, p. 1093.

Xu, X. P., Kushner, M. J. (1998), J. Appl. Phys., vol. 84, p. 4153.

Xue, S., Proulx, P., Boulos, M.I. (2001), J. Phys. D: Appl. Phys., vol. 34, p. 1897.

Xue, S., Proulx, P., Boulos, M.I. (2002a), J. Phys. D: Appl. Phys., vol. 35, p. 1123.

Xue, S., Proulx, P., Boulos, M.I. (2002b), J. Phys. D: Appl. Phys., vol. 35, p. 1131.

Xue, S., Proulx, P., Boulos, M.I. (2003a), Plasma Chem. Plasma Process., vol. 23, p. 245.

Xue, S., Proulx, P., Boulos, M.I. (2003b), Thermal Spray 2003: Advancing the Science and Applying the Technology, vol. 1, Moreau, C., Marple, B. (eds.), ASM Int. Materials Park, OH, p. 993.

Yagi, S., Tanaka, M. (1979), J. Phys. D: Appl. Phys., vol. 12, p. 1509.

Yakovlenko, S.I. (2000), "Gas and Plasma Lasers," in Encyclopedia of Low-Temperature Plasma, Fortov, V.E. (ed.), vol. 4, p. 262, Nauka (Science), Moscow.

Yamamoto, T., Ramanathan, K., Lawless, P.A., Ensor, D.S., Newsome, J.R., Plaks, N., Ramsey, G.H., Vogel, C.A., Hamel, L. (1992), IEEE Trans. Ind. Appl., vol. 28, p. 528.

Yamamoto, T., Lawless, P.A., Owen, M.K., Ensor, D.S., Boss, C. (1993), Non-Thermal Plasma Techniques for Pollution Control: Part B – Electron Beam and Electrical Discharge Processing, Penetrante, B.M., Schultheis, S.E. (eds), Springer-Verlag, Berlin.

References

Yamamoto, M., Nishioka, M., Sadakata, M. (2001), 15th Int. Symposium on Plasma Chemistry (ISPC-15), Orleans, France, vol. 2, p. 743.

Yamanouchi, T., Horie, H. (1952), J. Phys. Soc. Jpn, vol. 7, p. 52.

Yan, K., Yamamoto, T., Kanazawa, S., Ohkubo, T., Nomoto, Y., Chang, J.S. (2001), IEEE Trans. Ind. Appl., vol. 37, p. 1499.

Yang, X., Moravej, M., Nowling, G.R., Babayan, S.E., Panelon, J.,Chang, J.P., Hicks, R.F. (2005), Plasma Sources Sci. Technol., vol. 14, p. 314.

Yantovsky, E.I., Tolmach, I.M. (1972), Magnetohydrodynamic Generators, Nauka (Science), Moscow.

Yao, S.L., Nakayama, A., Suzuki, E. (2001a), AIChE J., vol. 47, p. 413.

Yao, S.L., Nakayama, A., Suzuki, E. (2001b), AIChE J., vol. 47, p. 419.

Yasuda, H. (1978), J. Poly. Sci., vol. 16, p. 199.

Yasuda, H. (1985), Plasma Polymerization, Academic Press, Orlando, FL.

Yasui, T., Akitani, T. (1965), Japnese Patent no. 02215/69.

Yatsenko, N.A. (1981), Sov. Phys., J. Techn. Phys., vol. 51, p. 1195.

Yensen, R.J., Bell, A.T., Soong, D.S. (1983), Plasma Chem. Plasma Process., vol. 3, no. 2, pp. 139, 163.

Yeom, G.Y., Thornton, J.A., Kushner, M.J. (1989), J. Appl. Phys., vol. 65, p. 3816.

Yerokhin, A.L., Snizhko, L.O., Gurevina, N.L., Leyland, A., Pilkington, A., Matthews, A. (2003), J. Phys. D: Appl. Phys., vol. 36, p. 2110.

Yildirim, E.D., Ayan, H., Vasilets, V.N., Fridman, A., Guceri, S., Sun, W. (2007), Plasma Process. Polym., vol. 4.

Yoshida, K., Tagashira, H. (1987), Symp. on High-Pressure Low-Temperature Plasma-Chemistry, HAKONE, Japan, p. 29.

Yoshida, T. (2005), 17th International Symposium on Plasma Chemistry (ISPC-17), p. 9, Toronto, Canada.

Young, R.A., Black, G., Slanger, T.G. (1968), J. Chem. Phys., vol. 49, p. 4758.

Yoyce, R.J., Sterling, H.F., Alexander, J.H. (1968), Solid Films, vol. 1, no. 6, p. 481.

Yukhimchuk, S.A., Shinka, V.P. (1976), "Plasma Reduction of Silicon Oxides by Hydrogen and Natural Gas," in Plasma Processes in Metallurgy and Technology of Inorganic Materials, A.A. Baikov Institute of Metallurgy of USSR Academy of Sciences, Nauka (Science), Moscow.

Zaidi, S., Stockman, E., Qin, X., Zhao, Z., Macheret, S., Ju, Y., Miles, R.B., Sullivan, D.J., Kline, J.F. (2006), 44th AIAA Aerospace Science Meeting and Exhibit, Reno, NV, AIAA paper 2006-1217.

Zaitsev, V.M. (2003), Russ. J. Otorinolaringol., vol. 1, p. 58.

Zaitsev, V.V. (1978), Sov. Phys., High Energy Chem. (Khimia Vysokikh Energij), vol. 12, p. 516.

Zaitsev, V.V. (1983), Physics and Engineering of Non-Equilibrium Processes, Ivanovo State University, Ivanovo, Russia.

Zarrabian, M., Fourches-Coulon, N., Turban, G. (1997), Appl. Phys. Lett., vol. 70, p. 2535.

Zaslavskii, G.M., Chirikov, B.V. (1971), Adv. Phys. Sci. (Uspekhi Physicheskikh Nauk), vol. 105, p. 3.

Zeldovich, Ya.B. (1942), Sov. Phys., J. Exper. Theoret. Phys., vol. 12, p. 525.

Zeldovich, Ya.B. (1947), Acta Physicochim. USSR, no. 4, p. 577.

Zeldovich, Ya.B., Raizer Yu.P. (1966), Physics of Shock Waves and High Temperature Hydrodynamic Phenomena, Academic Press, New York.

Zeldovich, Ya.B., Sadovnikov, P.Ya., Frank-Kamenetsky, D.A. (1947), Nitrogen Oxidation during Combustion, Ed. by USSR Academy of Sciences, Moscow.

Zhang, B., Liu, B., Renault, T., Girshick, S.L., Zachariach, M.R. (2005), 17th Int. Symposium on Plasma Chemistry, (ISPC-17), p. 782, Toronto, Canada.

Zhang, T., Gawne, D.T., Lin, B. (2000), Surf. Coat. Technol., vol. 132, p. 233.

Zhdanok, S.A., Soloukhin, R.I. (1982), Sov. Phys., Lett. J. Techn. Phys., vol. 8, p. 295.

Zhu, H., Lau, Y. C., Pfender, E. (1991), Mater. Sci. Eng., vol. A139, p. 352.

Zhu, R., Hsu, C.-C., Lin, M.C. (2001), J. Chem. Phys., vol. 115, p. 195.

Zhukov, M.F., Solonenko, O.P. (1990), High Temperature Dusted Jets in the Powder Material Processing, Institute of Thermophysics, Siberian Branch of Russian Academy of Sciences, Novosibirsk.

Zhukov, M.F., Kalinenko, R.A., Levitsky, A.A., Polak, L.S. (1990), Plasma-Chemical Conversion of Coal, Nauka (Science), Moscow.

Zyrichev, N.A. (1968), Technological Gas Production (Fixation of Atmospheric Nitrogen), State Institute of Nitrogen Industry (GIAP), Moscow.

Zyrichev, N.A., Kushnir, V.S., Poliakov, O.V., Cherkasky, V.S. (1979), 3rd USSR Symposium on Plasma Chemistry, vol. 1, p. 187, Nauka (Science), Moscow.

Zyrichev, N.A., Kulish, S.M., Rusanov, V.D. (1984), CO_2 – Dissociation in Supersonic Plasma-Chemical Reactor, Kurchatov Institute of Atomic Energy, vol. 4045/6, Moscow.

Index

absorption of radiation in the continuum, 113
accommodation coefficient, 72
acidic behavior of plasma, 30
activation energy of ion-molecular reactions, 31
actuators, 797
adiabatic principle, 21
aerodynamic plasma effect
  actuators, 797
  asymmetric DBD, 797
  ballistic range tests, 793
  dielectric barrier discharge, DBD, 797
  drag reduction, 796
  filamentary discharges, 795
  high-speed flows, 792
  influencing boundary layers, 797
  inlet shock control, 798
  jet injection, 799
  shocks, 792
  shock wave structure and velocity, 793
  surface discharge, 797
aerodynamics, 792
aerosols
  electric conductivity of thermally ionized, 53
  photo-ionization, 46
  photo-ionization by continuous-spectrum radiation, 49
  photo-ionization by monochromatic radiation, 46
  thermal ionization, 51
Ag-hydride direct synthesis, 489, 490
aging effect, 629, 641
Al nanopowder production, 579
Al-nitride synthesis from chloride, 473
$AlCl_3$
  direct decomposition, 457
  hydrogen reduction, 448
  oxidation, 486
AlF_3 hydrogen reduction, 448
AlI_3 hydrogen reduction, 448
Al_2O_3
  direct thermal decomposition, 434
  reduction, 423

Alfven velocity
  magneto-hydrodynamic, 150
  plasma centrifuges, 185
anode
  dark space, 176
  glow, 176
  layer of arc discharges, 189
anodization, 545
apoptosis, 865, 908, 909
arc discharge
  anode layer, 189
  cathode layer, 189
  classification, 187
  coaxial flow-stabilized, 199
  current-voltage characteristics, 187
  flow-stabilized linear, 199
  free-burning linear, 197
  gliding, 200
  high-pressure, 189
  hot cathode spot, 188, 191
  hot thermionic cathode, 188
  jet, 200
  low-pressure, 189
  magnetically stabilized, 200
  non-transferred, 200
  obstructed, 199
  plasma torch, 200
  rotating, 200
  segmented wall-stabilized, 199
  transferred, 199
  transpiration-stabilized, 199
  vacuum, 188
  voltaic, 187
  vortex-stabilized, 200
  wall-stabilized linear, 199
  wall-stabilized with unitary anode, 199
$AsCl_3$
  direct decomposition, 461
  hydrogen reduction, 448
As-hydride direct synthesis, 489
Aston dark space, 176
atmospheric pressure glow, APG
  atmospheric pressure plasma jet mode, 245
  DBD mode, 242

atmospheric pressure glow, APG (*cont.*)
  electronically stabilized, 244
  one-atmosphere uniform glow discharge plasma, 243
  resistive barrier discharge mode, 242
atmospheric pressure plasma jet, APPJ
  discharge, 245
  stability, 247
aurora borealis, 2
auto-displacement voltage, 222
automotive exhaust cleaning
  catalytic reduction of NO_x, 827
  diesel engine, 827
  principles, 825
  pulsed corona discharge, 827
avalanche, 160
aviation fuel reforming into syngas, 698

Bachmann triangle, 668
back ionization wave, 241
$BaCl_2$ direct decomposition, 455
B_4C synthesis from BCl_3, 476
BCl_3
  hydrogen reduction in non-thermal plasma, 448
  hydrogen reduction in thermal plasma, 441
B-hydrides
  direct synthesis, 489
  synthesis, 492
B-nitride synthesis from chloride, 473
Be-hydride direct synthesis, 489
$BeCl_2$ direct decomposition, 452
Bennett relation, 148
Bethe-Bloch formula, 20
BF_3 direct decomposition, 457
Bi-carbide solid-phase synthesis, 475
biocompatibility, 882
biodiesel reforming into syngas, 699
biomass conversion into H_2
  production of C_3O_2, 737
  without CO_2 emission, 737
bio-oil reforming into syngas, 700
bioprinter, 888
Bloch-Bradbury mechanism, 34
blood coagulation
  general features, 892
  in vitro, 892
  in vivo, 893
  mechanisms, 894
  prothrombin kinetics, 895
  trypsin effect, 895
Boltzmann distribution, 19, 92
bone cells
  attachment, 884
  proliferation, 884
Bouguer law, 154
Br_2 dissociation, 347
breakaway oxidation wave, 736
breakdown
  gas, 157
  thermal, 878
  water, 878

Breit-Wigner formula, 54
butyl chloride pyrolysis, 612

Ca-carbide synthesis, 477
$CaBr_2$ direct decomposition, 453
$Ca_5F(PO_4)_3$ (fluoroapatite) treatment, 487
canola oil reforming into syngas, 700
$Ca_3(PO_4)_2$ (tricalcium phosphate) treatment, 487
carbon nanotubes synthesis, 581
carbon suboxide (C_3O_2), 732
$CaSO_4$ non-equilibrium decomposition, 427
capacitively coupled plasma, CCP, 210
capillary plasma electrode, 255
carbothermic reduction, 429
cathode
  dark space, 176
  glow, 176
  layer of arc discharges, 189
  spot, 181
CCl_4 destruction, 841
centrifuges, 108
CF_4
  conversion to C_2F_4, 613
  pyrolysis, 615
  remote plasma sources, 533
C_2F_6 remote plasma sources, 533
C_3F_6 pyrolysis, 615
C_3F_8 pyrolysis, 615
C_8F_{18} pyrolysis, 615
$C_2F_2Br_4$ pyrolysis, 615
$CFCl_3$ decomposition, 845
CF_2Cl_2 decomposition, 845
CF_2Cl_2 pyrolysis, 615
$C_2F_2Cl_2$ pyrolysis, 615
C_2F_3Cl pyrolysis, 615
C_2F_5Cl pyrolysis, 615
CF_3CN synthesis, 614
CF_3COOH pyrolysis, 615
CF_3NH_2 synthesis, 614
CH_4 conversion with CO_2, 685
CH_4 partial oxidation
  gliding arc tornado, 678
  reverse vortex flow, 679
  syngas production, 678
CH_4 pyrolysis
  plasma catalysis, 688
  plasma catalytic mechanism, 689
  process, 589
CH_4 steam reforming, 683
C_2H_2 production from coal
  particle size effect, 719
  pyrolysis in H_2, Ar jets, 716
  volatile components effect, 717
  acetylene quenching, 719
C_2H_2 production from hydrocarbons
  microwave discharge, 598
  process 589, 598
  stimulated by vibrational excitation, 602
C_2H_4 production, 589
C_2H_6 pyrolysis, 687

Index

C_3H_8
  partial oxidation, 691
  pyrolysis, 593
C_6H_6 pyrolysis, 593
C_6H_{12} (cyclohexane) partial oxidation, 691
C_7H_8 (toluene) partial oxidation, 691
C_7H_{17} (heptane) partial oxidation, 691
C_8H_{18} (isooctane) partial oxidation, 703
channel model of an arc, 197
charge transfer
  electron tunneling effect, 28
  ion-atomic, 28
  non-resonant, 28
  resonant, 28
$C_2H_2Br_2$ synthesis, 617
C_2HCl_3 destruction, 841
$C_2H_2Cl_2$ destruction, 841
C_2H_5CN (ethyl cyanide) synthesis, 611
CHF_2Cl decomposition, 845
chlorofluorocarbons (CFC) decomposition, 845
chemical lasers, 811
Child law of space-charge-limited current, 145
chirality, 582
$C_6H_5NH_2$ synthesis, 611
CH_2O (formaldehyde) synthesis, 617
CH_3OH (methanol) synthesis, 617
C_2H_5OH (ethanol) conversion to syngas, 693
$(CH_3)_2S$ (dimethyl sulfide) oxidation, 833
Cl_2 dissociation, 347
cleaning
  active, 540
  CVD reactors, 531
  etching reactors, 531
  passive, 539
ClF_5 synthesis, 406
clusterization in centrifugal field
  average cluster size, 748
  cluster size distribution, 746
  effect on energy efficiency, 749
  non-equilibrium effect, 745
CN synthesis, 412
C_2N_2 (dicyanogen) synthesis, 604
CO disproportioning, 314
CO laser, 811
CO_2 dissociation
  applied aspects, 259
  capacitively coupled RF discharge, 300
  complete with production of C and O_2, 314
  corona discharge, 304
  characteristic time scales, 282
  chemical mechanism, 278
  discharge sustained by electron beam, 302
  discharge sustained by UV radiation, 302
  energy efficiency, 259
  energy efficiency in non-equilibrium plasma, 288
  energy losses to product excitation, 286
  experimental results, 260
  gasdynamic stimulation, 309
  glow discharges, 302
  hollow cathode discharge, 303
  hyperbolic decrease of energy efficiency, 268

  inductively coupled RF discharge, 299
  ionization degree regimes, 285
  microwave discharge of moderate pressure, 293
  microwave discharge with electron cyclotron
    resonance, 316
  non-self-sustained discharges, 302
  non-thermal mechanisms, 259
  physical kinetics, 268
  plasma-beam discharge, 317
  products explosion, 292
  products stability, 293
  pulsed glow discharge, 303
  radiofrequency discharges, 299
  radiolysis by relativistic electron beams, 310
  stimulated by dissociative attachment of electrons,
    267
  stimulated by electronic excitation, 265
  stimulated by vibrational excitation, 263
  supersonic discharges, 304
  supersonic microwave discharge, 308
  thermal mechanisms, 259, 262
  energy balance, 276
CO_2
  emission, 732
  laser, 810
$CO-N_2$ mixture chemistry, 609
C_3O_2 production
  from coal, 734
  from methane, 734
  mechanochemical mechanism, 735
  mechanochemical oxidation kinetics, 736
coagulation of nanoparticles, 575
coal
  composition and structure, 707
  direct gasification into C_2H_2, 725
  direct gasification into syngas, 725
  gasification by H_2O, 721
  gasification in steam-air jets, 724
  gasification in steam-oxygen jets, 724
  heating, 713
  inorganic fraction, 708
  metamorphism, 708
  organic fraction, 708
  particle destruction, 713
  pyrolysis in Ar and H_2, 716
  pyrolysis in N_2, 721
  porosity, 708
  steam reforming, 721
  Turgai, 712
coal conversion
  coke kinetics, 713
  corona discharge, 731
  dielectric barrier discharge, DBD, 731
  directly in arc, 724
  glow discharge, 730
  kinetics, 707, 712
  mechanisms, 707
  microwave discharge, 726, 728
  non-thermal plasma, 715, 726
  phases, 713
  pyrolysis, 708

coal conversion (*cont.*)
  thermal, 708
  thermal plasma jets, 716
  thermodynamics, 707, 711
  transformation of nitrogen compounds, 711
  transformation of sulfur compounds, 710
  volatiles, 713
coating
  Al in sulfuric acid, 559
  electrolytic spark oxidation, 557
  microarc oxidation, 557
coaxial-hollow micro-DBD, 254
$CoCl_2$ direct decomposition, 465
collisions
  Coulomb, 15
  elastic, 13
  inelastic, 13
  ion-molecular polarization, 26
  superelastic, 13
colony forming units, cfu, 849
combustion
  blow-off velocities, 760
  kinetics, 761
  low-speed gas flows, 760
  plasma-assisted, 755
  pulverized coal, 768
  ram-jet engine, 785
  scram-jet engine, 785
  stabilization by magnetic gliding arc, 765
  stabilization by plasma, 755
combined plasma-catalytic partial oxidation of hydrocarbons
  plasma postprocessing, 702
  plasma preprocessing, 702
complementarity, 883
conductivity in crossed electric and magnetic fields, 107
conductivity of plasma, 104, 106
contact angle, 641
continuity equation, 109
corn reforming into syngas, 700
corona
  igniting electric field, 233
  negative, 233
  positive, 233
cotton treatment, 648
Coulomb coupling parameter, 141
Cr-oxide reduction, 424
CrB_2 synthesis, 492
$CrCl_3$ direct decomposition, 466
$Cr(CO)_6$ synthesis, 491
CrO_2Cl_2 oxidation, 486
cross section of elementary process, 13
crude oil pyrolysis, 595
Cs plasma chemistry, 803
current-voltage characteristic
  cathode layer, 179
  CCP-discharge, moderate pressure, 219
  DC discharges, 177
  load line, 177

cutting, 493
cylindrical coil RF discharge, 224

dark-to-glow transition, 179
dark Townsend discharge, 177
Debye
  radius, 110, 141
  shielding, 141
decomposition processes, 259
densification of powders, 495
deposition
  amorphous Si, 542
  non-conformal, 545
  plasma-enhanced chemical vapor, 541
  reactive sputter, 549
  silane, 543
  Si_3N_4, 547
  SiO_2, 545
  sputter, 548
  thin films, 542
diamond film synthesis, 668
dichloroethane pyrolysis, 612
dielectric barrier discharge (DBD)
  atmospheric pressure glow mode, 241
  coaxial-hollow micro-, 254
  configurations, 238
  cylindrical, 239
  memory effect, 165, 241
  packed-bed, 239
  planar, 239
  self-organization, 170
  structure, 166
  surface, 239
diesel
  fuel reforming into syngas, 697
  oil reforming into syngas, 697
diffusion
  ambipolar, 110
  electrons, 109
  ions, 109
discharge
  arc, 7
  atmospheric pressure, 233
  atmospheric pressure glow, 241, 242
  controlled by diffusion of charges, 172
  controlled by electron attachment, 172
  controlled by electron-ion recombination, 171
  controlled by surface recombination, 170
  controlled by volume recombination, 170
  corona, 6, 233
  dark, 177
  dielectric barrier, 7, 237
  electric, 157
  electromagnetic surface wave, 230
  electron cyclotron resonance, ECR, 229
  ferroelectric, 240
  floating-electrode, 7
  floating-electrode dielectric-barrier discharge, FE-DBD, 888
  gliding arc, 7
  glow, 6

Index

helicon, 229
micro-arc, 253
micro-glow discharge, 249
micro-hollow-cathode discharge, 251
microwave, 209, 231
modes α and γ, 218
non-self-sustained, 302
plasma-beam, 317
pseudo-spark, 804
pulsed corona, 234
radiofrequency, RF, 209
spark, 240
steady-state regimes, 170
tasitron, 804
thyratrone, 804
"tornado" (reverse vortex), 7
water, 875
wave-heated low-pressure, 229
dissociation factor Z, non-equilibrium, 86
dissociation of molecules
diatomic, 341
direct electron impact, 61
equilibrium, 94
hydrogen halides, 341
in plasma, 54
mechanisms, 61
stimulated by translational energy, 86
dissociative reduction, 420
distribution of electron energy in non-thermal
discharges, 63
dissociative attachment to water molecules, 323
DNA damage
double-strand breaks, 863
non-repairable, 863
drift in crossed electric and magnetic fields,
107
Druyvesteyn distribution function, 101
dusty plasma
kinetics, 566
silane discharge, 566

EEDF controlled by vibrational excitation, 103
efficiency α of excitation energy in overcoming the
activation barrier, 79, 81
Einbinder formula, 51
Einstein relation, 109
elastic scattering, 15
electric
propulsion, 787
rocket engine, 787
rocket engine, optimal specific impulse, 788
electromagnetic surface wave discharge, 230
electromagnetic wave
plasma absorption and reflection, 154
propagation in plasma, 153
electron affinity, 37
electron attachment
direct, 35
dissociative, 31
radiative, 35
three-body, 33

electron beam
non-relativistic, 20
relativistic, 20
electron-cyclotron resonance, ECR, 229
electron detachment
associative, 35
electron impact, 35
collisions with excited particles, 35
electron energy distribution function (EEDF), 12, 100
electron-plasma furnace, 431
electronegative gas, 31
electronic excitation energy
relaxation, 76
transfer, 78
electronic excitation in plasma
electron impact, 59
rate coefficient, 59
electrons, 12
electroporation, 874
electrostatic plasma oscillations, 145
Elenbaas-Heller equation, 193
emission
explosive electron, 192
field, 43, 187
of electrons, 15, 42
photo-electron, 46
secondary electron, 45
secondary electron-electron, 46
secondary ion-electron, 45
thermionic, 42, 187
thermionic field, 43
emission and absorption of radiation, 112
endohedrals, 583
energy balance of plasma-chemical processes, 132
energy transfer equations, 137
energy balance of CO_2 dissociation
critical specific energy input, 281
critical vibrational temperature, 281
equations, 276
one-temperature approach, 280
two-temperature approach, 277
energy efficiency
chemical factor, 137
excitation factor, 136
influence of transfer phenomena, 139
non-equilibrium processes, 132
processes stimulated by electronic excitation, 134
processes stimulated by dissociative attachment,
134
processes stimulated by vibrational excitation, 133
relaxation factor, 136
quasi-equilibrium processes, 132, 137
Engel-Steenbeck relation, 172, 173
environmental control, 817
etching
Al, 529
anisotropic, 511, 513
discharge kinetics, 520
discharges, 519
doping effect, 524
F/C ratio, 527

etching (*cont.*)
  gas composition, 523
  isotropic, 511
  ion energy-driven, 516
  ion-enhanced inhibitor, 516
  kinetics, 517
  over-, 513
  photoresist, 531
  polymers, 636
  principles, 510
  pure chemical, 515
  rate, 511
  refractory metals, 530
  selectivity, 511, 513
  Si in CF_4, 525
  Si with Cl-atoms, 528
  Si with F-atoms, 523
  Si_3N_4, 529
  SiO_2 with CF_x radicals, 529
  SiO_2 with F-atoms, 529
  sputtering, 514
excitation of neutrals, 54
exhaust gas cleaning
  α- pinene, 835
  acetone, 832
  automotive exhaust, 825
  chlorine-containing VOC, 839
  continuous corona, 819
  dimethyl sulfide, $(CH_3)_2S$, 833
  Hg, 843
  methanol, 832
  mobile laboratory, 838
  NO_x, 817, 824
  paper mills, 830
  pulsed corona, 819, 832
  relativistic electron beams, 818
  SO_2, 817, 824
  volatile organic compounds, VOC, 830
  wood processing plants, 830
explosion limit
  H_2-O_2 mixture, 756
  plasma control, 756

F-atoms recombination
  surface, 535
  volume, 536
F-atoms transportation from remote plasma source, 538
F_2 dissociation, 344
Faraday dark space, 176
FCN synthesis, 614
$FeCl_3$ direct decomposition, 465
Fenton reaction, 853
Fe_2O_3 reduction, 417
$FeO \cdot TiO_2$ carbothermic reduction, 431
ferroelectric DBD discharge, 240
FeS_2 (pyrite) and FeS conversion, 710
Fick's law, 109
Flamm formulas, 523
floating potential, 143

fluxes (VV, VT) of excited molecules in energy space, 115
Fokker-Planck kinetic equation, 100, 114
fourth state of matter, 1
Fowler-Nordheim formula, 43
Frank-Condon principle, 17
Franklin's experiments, 2
freons decomposition, 845
Fridman approximation for vibrational excitation, 56
Fridman-Macheret α-model, 79, 81
fullerenes synthesis, 581, 583

$GaCl_3$ direct decomposition, 457
Ga-hydride direct synthesis, 489
gas condensate pyrolysis, 594
gas discharge, 5
gas-phase decomposition in multi-phase technologies, 336
gas-separating polymer membranes
  dose effect, 661
  effect of cross-link's clusterization, 664
  effect of cross-link's mobility, 664
  effect of membrane porosity, 667
  H_2-CO_2, 657
  He-CH_4, 657
  modeling, 662
  modification of surface layers, 661
  N_2-O_2, 657
  permeability control, 657
  plasma modification, 655
  selectivity enhancement, 659
gasoil pyrolysis, 594
gasoline
  pyrolysis, 594
  reforming into syngas, 698
$GeCl_4$ hydrogen reduction, 445
GeF_4 direct decomposition, 459
Ge-hydride direct synthesis, 489
GeO_2 direct plasma decomposition, 439
G-factor
  concept, 311
  nitrogen, 386
  oxygen, 387
gliding arc
  breakdown, 203
  equilibrium stage, 203
  equilibrium-to-non-equilibrium transition, 204
  fluidized bed, 207
  non-equilibrium stage, 203
  rotating in magnetic field, 207
  stability, 205
  stabilization in reverse-vortex (tornado) flow, 207
  switchgears, 207
  tornado, 207
glow discharge
  abnormal, 181
  anode layer, 175
  cathode layer, 175
  configurations, 175
  energy efficiency, 186

Index

hollow-cathode, 175, 183
longitudinal organization, 175
magnetron, 174
normal, 179
obstructed, 181
Penning, 184
positive column, 175
structure, 174
subnormal, 181
transverse organization, 175

H_2 dissociation, 345
Hall effect, 149
HBr dissociation, 341
HCl
dissociation, 343
laser, 811
oxidation, 413
HCN production
from coal, 721
process, 605
HCOOH (formic acid) synthesis, 620
heat flux potential, 194
heavy naphtha reforming into syngas, 698
helical resonator discharge, 228
helicon discharge, 229
hemocompatibility, 884
heterogeneous stabilization of products, 317
hexachlorane pyrolysis, 612
HF dissociation, 343
Hf-nitride synthesis from chloride, 473
Hf reduction from chloride, 443
$HfCl_4$
direct decomposition, 462
hydrogen reduction, 446
HfO_2 non-equilibrium reduction, 426
Hg removal from offgases, 843
HI dissociation, 343
high-density plasma, HDP, 224, 520
high frequency
dielectric permittivity of plasma, 151
plasma conductivity, 151
H_2O dissociation
applied aspects, 318
attachment/detachment chain reaction, 324
cooling rate effect, 326
energy efficiency, 320
glow discharge, 322
kinetics, 319
microwave discharge, 331
plasma radiolysis, 335
product explosion limit, 321
product stability, 322
production of H_2O_2, 334
quenching modes, 325
stimulated by dissociative attachment, 322
stimulated by vibrational excitation, 319
super-ideal quenching, 326
supersonic discharge, 334
thermal plasma, 325
H_2O_2 chemistry, 857

hollow cathode discharge
fused, 253
general features, 183
Lidsky capillary, 184
"hot" atoms
chemical reactions, 123
VT relaxation, 122
H_2S direct synthesis, 488
H_2S dissociation
arc discharge, 740
centrifugal effect on energy cost, 748
effect of CO_2, 749
effect of hydrocarbons, 749
effect of O_2, 749
energy cost, 739
gliding arc tornado discharge, 741
industrial applications, 738
kinetics, 744
microwave discharge, 740
production of H_2 and S, 738
radiofrequency discharge, 740
S clusterization in centrifugal field, 745
technological schemes, 741
hydrated electrons, 852
hydrocarbons
elimination, 603
oxidation without CO_2 emission, 732
hydrogen production
combined plasma-catalytic, 701
double-step CO_2 cycle, 751
cycles, 751
from CO_2-H_2O mixture, 328
from H_2S, 738
from hydrocarbons, 597, 676
from water, 318
HBr cycle, 752

ICP discharge
helical resonator, 228
low-pressure regime, 226
moderate-pressure regime, 225
parallel plate reactor, 227
planar coil, 226
ignition
below autoignition limit, 771
bootstrap effect, 775
comparison with fuel oxidation, 758
contribution of different species, 770
contribution of excited species, 778
contribution of ions, 784
contribution of NO, 783
contribution of radicals, 770, 776
counterflow flame, 765
delay, 763
H_2-O_2, 771
H_2-O_2-He mixtures, 773
HO_2 formation suppressing, 779, 781
hydrocarbons, 771
kinetics, 761
magnetic gliding arc, 763
microwave, 757

ignition (*cont.*)
  nanosecond pulsed DBD, 760
  non-thermal mechanism, 763
  pulverized coal, 768
  radiofrequency plasma, 759
  subthreshold plasma, 775
  supersonic flows, 757
  thermal mechanism, 763
  transonic flows, 758
  zebra, 763
implantation
  ion, 551
  ion-beam, 551
  plasma-immersion ion, 551, 553
In-hydride direct synthesis, 489
inactivation of microorganisms, 850
inflammatory and destructive illnesses
  dentistry, 905
  dermatology, 905
  gastroenterology, 905
  gynecology, 905
  maxillofacial surgery, 905
  ophthalmology, 905
  orthopedics, 904
  otorhinolaryngology, 905
  phthisiology, 904
  pulmonology, 904
  purulent peritonitis, 905
  traumatology, 904
In_2O_3 direct plasma decomposition, 439
inductively coupled plasma, ICP
  high-density, 224
  non-thermal, 224
  thermal, 209,
inorganic gas-phase decomposition, 259
integrated circuit fabrication, 510
intermetallic compound synthesis, 493
intermolecular VV relaxation, 74
ion
  conversion, 23
  drift, 109
  energy, 109
  negative, 13
  positive, 12
  thruster, 787
ion-cluster
  growth, 31
  negative, 31
  positive, 31
ion-ion recombination
  binary collisions, 38
  three-body, 39
ion-molecular reactions, 26, 29
ionization
  associative, 21
  collision of heavy particles, 14, 21
  collision of vibrationally excited molecules, 22
  dissociative, 17
  direct, by electron impact, 14, 16
  electron beams, 20
  heterogeneous, 42

high-energy electrons, 20
  Penning, 21
  photo-, 15, 20
  potential, 16
  processes, 12
  stepwise, by electron impact, 14, 18
  surface, 15
isomerization of hydrocarbons, 603
isotopic mixture
  coefficient of selectivity, 125
  reverse kinetic effect, 125
  Treanor formula, 125

Jacob's ladder, 201
jet engine stimulation
  ram-jets, 785
  scram jets, 785
jet fuel reforming into syngas, 698
jet lift-off velocity, 761
Joule heating, 106
JP-8 fuel reforming into syngas, 698

Kassel mechanism, 589, 600
kerosene conversion to syngas, 693
KI direct decomposition, 452
kinetic equation
  Fokker-Planck for EEDF, 100
  Fokker-Plank for vibrational distribution functions,
    114
  gas mixtures, 124
kinetics
  CO disproportioning, 314
  gas mixtures, 124
  macroscopic, 14
  microscopic, 14
  NO synthesis, 367
  plasma-chemical, 92
  population of electronically excited states, 120
  reactions of vibrationally excited molecules, 124
  relaxation of vibrationally excited molecules, 124
kinetics of dissociation of vibrationally excited CO_2
  asymmetric and symmetric modes, 268
  distribution function, 270
  elementary reaction rates, 275
  one-temperature approximation, 272
  super-sonic flow, 306
  two-temperature approximation, 272, 274
  vibrational quasi continuum, 271
KrF_2 sysnthesis
  barrier discharge, 401
  glow discharge, 401
  kinetics in Kr-matrix, 400
  microwave discharge, 402
  photo-chemical, 401
  volume mechanism, 399
  surface stabilization mechanism, 400

Lame equation, 662
lamps
  fluorescent, 815
  high-intensity discharge (HID), 816

Index

light-emitting diode (LED), 816
mercury-containing, 815
mercury-free, 815
Landau-Teller formula, 69, 131
Landau-Zener formula, 364
Langevin
cross-section, 27
model, 39
polarization capture, 26
relation, 52
rate coefficient, 26
scattering, 27
Langmuir frequency, 145
Langmuir-Rideal mechanism, 535
Larmor radius, 150
lasers
Ar, 806
classification, 804
C_2H_2, 810
chemical, 811
CO, 810, 813
CO_2, 810
COS, 810
CS_2, 810
DCl, 813
DF, 813
electronic transitions, 807
excimer, 809
gas-discharge, 810
HBr, 813
HCl, 812
HCN, 810
He-Cd, 807
He-Ne, 806
He-Sr, 806
He-Zn, 806
HF, 813
H_2O, 810
inversion in plasma recombination, 807
inversion mechanisms, 804
ICN, 810
ionic gas-discharge, 806
molecular transitions, 805
multi-charged ions, 806
NH_3, 810
N_2O, 810
Penning, 807
plasma, 804, 807
pulse-periodic self-limited, 805
radiative deactivation, 806
self-limited transitions, 804
SO_2, 810
quasi-stationary inversion, 806
UV, 808
vibrational-rotational transitions, 810
Xe transition, 808
laser spark, 241
lavsan fabric treatment, 650
Le Roy formula, 81
Lichtenberg figures, 164, 240
light-emitting diode (LED), 816

light sources, 804
LiH_2 synthesis, 83
LiI
direct decomposition in thermal plasma, 451
direct decomposition in non-thermal plasma, 469
living tissue sterilization
animal tests, 891
direct plasma medicine, 888
floating-electrode dielectric-barrier discharge, FE-DBD, 888
histology, 890
human skin flora, 890
non-damage analysis, 890
toxicity analysis, 890
loading effect, 522
local thermodynamic equilibrium (LTE), 4
Losev model, 88
Losev formula, 131
Lotus effect, 648

Macheret-Fridman model, 86
magnetic
field frozen in plasma, 146
lines tension, 148
mirror effect, 186
pressure, 147
Reynolds number, 150
viscosity, 147
magnetically enhanced reactive ion etcher, MERIE, 222, 519
magneto-hydrodynamic equations, 147
magneto-hydrodynamic
flow control, 798
generators, MHD, 799
power electronics, 799
power extraction, 798
principles, 146
magnetron discharge
coplanar configuration, 186
parallel plate electrodes, 185
radiofrequency, 222
sputtering configuration, 185
Margenau distribution function, 101
Marrone-Treanor model, 88
mass transfer equations, 137
Massey parameter, 21, 68
Maxwell-Boltzmann distribution function, 13
Maxwellian distribution for plasma electrons, 102
mean electron energy, 13
mean free path, 14
medical polymer materials, 882
Meek
breakdown condition, 163
criterion, 163
melting, 495
memory effect, 165, 241
metal halides dissociation kinetics, 470
metallic cylinder model, 209
methyl methacrylate (MMA) plasma polymerization, 625
MgI_2 direct decomposition, 453

$Mg(NO_3)_2$ aqueous solution and melt decomposition, 483
MgO production, 483
MHD generators
  conductive, 800
  electric characteristics, 801
  inductive, 800
  thermodynamic characteristics, 801
micro-arc discharge, 253
microdeposition in atmospheric pressure
    microdischarges, 564, 565
microdischarge
  arrays, 252
  DBD, 164
  general features, 247
  glow, 248
  ICP jet, 256
  interaction, 166
  kilohertz-frequency range, 254
  memory effect, 165
  microwave, 257
  Monte Carlo modeling, 167
  quasi-repulsion, 167
  remnant, 165
  RF-, 255
  self-organized pattern, 168, 252
  streamer, 165
  structure, 252
microetching in atmospheric pressure
    microdischarges, 564
microfabrication in atmospheric pressure
    microdischarges, 563
micro-glow discharge
  current-voltage characteristics, 249
  images, 249
micro-hollow-cathode discharge, 251
micropatterning in atmospheric pressure
    microdischarges, 563
microwave
  CO_2 discharge, 296
  discharge, 211
  discharge in air, 213, 374
  discharge in water vapor, 331
  discharge pressure regimes, 232
  ECR discharge, vibrational temperature, 376
  electron cyclotron resonance (ECR), 374
  F_2 discharge, 402
  low-pressure discharge, 229
  microdischarge, 257
  moderate-pressure discharge, 231
  modes, 212
  plasma torch, 213
  radial plasmatron, 214
  rectangular-to-circular mode converter, 214
  resonator, 212
  thermal plasma generation, 211
microxerography, 888
minimum power principle, 181, 195
miniaturized atmospheric-pressure ICP jet, 256
mixtures of gases, vibrational kinetics, 124

$Mn_2(CO)_{10}$ synthesis, 491
MnO production, 478
$MnSiO_3$ (rhodonite) decomposition, 478
Mo reduction from chloride, 443
$MoCl_5$ hydrogen reduction, 446
$Mo(CO)_6$ synthesis, 491
modification of polymers, 632
MoF_6 direct decomposition, 467
MoO_3
  non-equilibrium reduction, 426
  thermal plasma reduction, 422
mobility of electrons, 106

N_2 dissociation, 347
NaCl
  direct decomposition in non-thermal plasma, 469
  direct decomposition in thermal plasma, 451
nanocomposites, 580
nanoparticles, 566
nanoparticles coating, 581
nanotube
  armchair, 582
  chirality, 582
  functionalization, 586
  multi-wall, 582
  Russian doll, 582
  scroll, 582
  single-wall, 582
  synthesis, 584
  zigzag, 582
nanotubes synthesis, 581
natural gas partial oxidation, 691
Nb-nitride synthesis from chloride, 473
$NbCl_5$
  direct decomposition, 463
  hydrogen reduction, 442
Nb_2O_5 carbothermic reduction, 430
necrosis, 865
negative ion
  destruction, 35
  thermal destruction, 37
NF_3
  dissociation, 534
  plasma chemistry, 404
  remote plasma sources, 533
  synthesis, 404, 614
NF_4AsF_6 synthesis, 405
NH_3 (ammonia) dissociation, 337
NH_3 (ammonia) synthesis, 408
N_2H_4 (hydrazine) synthesis, 406
Ni-extraction from serpentine minerals, 482
Ni-oxide reduction, 424
$NiCl_2$ direct decomposition, 466
nitrogen oxides synthesis, 355
NO synthesis
  adiabatic channel, 361
  air and N_2-O_2 mixtures, 355
  applied aspects, 355
  arc discharge, 380
  discharge sustained by electron beam, 378

Index

ECR microwave discharge, 377
Electron-beam discharge, 379
effect of hot H-atoms, 372
elementary reaction, 361
energy balance, 367
energy efficiency, 356, 359, 370
formation of N_2O complex, 363
kinetics, 367
microwave discharge, 374
non-adiabatic channel, 363
non-self-sustained discharge, 378
non-thermal mechanisms, 356
product stability, 371, 373
reaction path profile, 361
stimulated by electronic excitation, 357
stimulated by ion-molecular reactions, 358
stimulated by vibrational excitation, 356
technology, 381
thermal mechanisms, 358
Zeldovich mechanism, 356
N_2O destruction and conversion, 340
NO_x oxidation
  air plasma, 824
  simultaneous with SO_2, 824
non-equilibrium
  evaporation, 426
  surface heating, 425
non-ideality parameter, 141
non-self-sustained discharges
  sustained by electron beam, 302, 378
  sustained by UV radiation, 302
normal
  cathode potential drop, 182
  current density, 182
  thickness of cathode layer, 182

O_2 dissociation, 347
O_3 (ozone) biochemistry, 857
O_3 (ozone) synthesis
  dielectric barrier discharge, DBD, 392
  discharge poisoning effect, 387
  effect of H_2 and hydrocarbons, 390
  effect of halogens, 391
  energy cost, 383
  energy efficiency, 383
  from air, 386
  from oxygen, 383
  industrial application, 382
  kinetic curves, 384, 387
  large production installations, 392
  optimum DBD strength, 385
  planar DBD, 394
  pulsed corona discharge, 395
  surface DBD, 394
  temperature effect, 388
  water vapor effect, 389
O_2F_2 synthesis, 403
O_4F_2 synthesis, 404
Ohm's law, 106
  generalized, 149

on-board fuel reforming, 692
one-atmosphere uniform glow discharge plasma,
  OAUGDP, 243
optical
  breakdown, 214
  discharge, 211, 214
  thermal plasma generation, 211
optically thick plasma, 113
optically thin plasma, 113
organic
  chlorine compounds pyrolysis, 612
  fluorine compounds pyrolysis, 613
osteoblasts
  attachment, 885
  proliferation, 885
ozonation, 875

P-hydride direct synthesis, 489
P-oxidation, 488
P_6N_6 synthesis, 409
partition functions, 96
Park model, 87
Parker formula, 76
partial oxidation of hydrocarbons, 677
Paschen curve, 158
pathogen detection and remediation facility, PDRF,
  867
Pb-carbide solid-phase synthesis, 475
PbH_4 direct synthesis, 488
Peek formula, 233
Perfluoro compounds, PFC
pinch
  effect, 147
  Z-, 148
plasma
  assisted chemonuclear reactor, 313
  biology, 848
  centrifuges, 108, 185
  capacitively coupled, 210
  catalysis, 31, 603, 676
  chemistry without electricity, 309
  cleaning, 539
  communication devices, 804
  completely ionized, 3
  components, 8
  cutting, 493
  densification, 495
  density, 3
  diffusion across magnetic field, 149
  display panels, PDP, 816
  electrodynamics, 140
  etching, 510
  fluidized bed, 431
  frequency, 145
  ideal, 140
  ignition and stabilization of flames, 755
  inductively coupled, 209
  lasers, 804
  medicine, 848
  melting, 495

Index

plasma (*cont.*)
  metallurgy, 417
  nano-technology, 499, 566
  needle, 255, 865
  non-equilibrium, 4
  non-equilibrium statistics, 97
  non-ideal, 141
  non-thermal, 4
  polarization, 141
  polyethylene treatment kinetics, 634
  polymerization, 622
  pyrolysis, 591
  radiolysis, 310
  reduction of metal oxides, 417
  solar, 4
  sources, 5
  spheroidization, 495
  spraying, 499
  statistics, 92
  sterilization, 848
  surface processing, 499
  surgery, 896
  temperature, 3
  thermal, 4
  thermodynamics, 92
  thruster, 787
  torch, 200
  TV, 816
  two-temperature statistics and thermodynamics, 97
  weakly ionized, 3
  welding, 494
plasma-chemical reaction
  elementary, 12
  frequency, 14
  mechanism, 12
  macro-kinetics, 14
  micro-kinetics, 14
  positive ions, 22
  rate, 14
  rate coefficient, 14
  two excited molecules, 315
plastics treatment, 654
Pointing vector, 154
polar dissociation, 34
polarizability of atoms and molecules, 26
polymer
  adhesion enhancement, 643
  fiber modification, 645
  film deposition, 628
  growth centers, 625
  membrane modification, 645
  surface energy, 643
  surface fluorination, 640
  surface metallization, 643
  surface nitrogenization, 639
  surface oxidation, 638
polymerization
  graft, 626
  heterogeneous, 625
  initiation, 623

kinetics, 622
methyl methacrylate, MMA, 625
population of electronically excited states, 120
positive column, 182
  high pressure arcs, 193
  current-voltage characteristic, 193
predissociation, 63
presheath, 143
probability of elementary process, 14
propane-butane mixture partial oxidation, 691
pulsed corona
  discharge, 234
  pin-to-plate configuration, 236
  spray, 237
  wet, 237
  wire-cylinder configuration, 236

quantum yield, 46
quasi-equilibrium of vibrational modes, 273
quasi-equilibrium plasma energy efficiency, 288
quasi-equilibrium thermal plasma mechanisms, 262
quenching of acetylene, 719
quenching of CO_2 dissociation products
  energy efficiency, 289
  kinetics, 288
  non-equilibrium effects, 290
  super-ideal mode, 290
quenching of products
  absolute, 137
  energy efficiency, 288
  ideal, 137, 288
  super-ideal, 137, 288

radiation sources of high spectral brightness
  laser, 814
  luminescent, 814
  synchrotron (ondulator), 814
  thermal plasma, 814
radiation
  biology, 863
  dose, 863
  transfer in plasma, 113
radical shower, 825
radiofrequency discharges
  capacitive coupling, 210, 215
  coupling circuits, 216
  high-density, 224
  inductive coupling, 209, 215, 224
  low pressure, 219
  magnetron, 222
  moderate pressure, 216
  modes α and γ, 218
  non-thermal, 215
  thermal, 209
radiofrequency (RF)
  bias, 224
  diode, 519
  triode, 519

radiolysis in tracks of nuclear fission fragments, 311, 335
Ramsauer effect, 103
rare metals carbothermic reduction, 430
reactions of excited molecules
  α-model, 79
  macrokinetics, 129
  rate coefficient, 79
  theoretical-informational approach, 79
  through intermediate complexes, 83
reactive ion etcher, RIE, 519
recombination
  dissociative, 22
  electron-ion, 22
  ion-ion, 37
  radiative electron-ion, 23, 25
  three-body electron-ion, 23, 25
reduced electric field, 18, 104
refractory metals, 420, 430
remelting, 495
remote plasma sources, 531
renewable biomass reforming into syngas, 699
resistive barrier discharge, 242
reverse vortex flow, 679
rotational excitation of molecules, 58
rotational relaxation, 76
rubber materials treatment, 654
Rutherford formula, 16

Saha equation, 25, 92, 94
$SbCl_5$
  direct decomposition, 461
  hydrogen reduction, 448
$ScCl_3$ hydrogen reduction, 448
ScF_3 hydrogen reduction, 448
Schottky effect, 42
secondary ion-electron emission
  induced by metastable atoms, 46
  Penning mechanism, 46
  potential mechanism, 46
SF_6 direct decomposition, 470
sheath, 4, 142
  Bohm criterion, 143
  Child law, 145
  DC, 142
  high voltage, 141
  matrix, 144
Si cluster growth, 568
Si-hydrides direct synthesis, 489
Si-nitride synthesis from chloride, 473
Si-oxide reduction, 424
Si-oxynitride synthesis from SiH_4, 474
Si quantum dots synthesis, 579
$SiCl_4$
  direct decomposition, 459
  hydrogen reduction, 445
SiC synthesis from SiO_2, 477
SiF_4 steam conversion, 484
SiH_4 oxidation to SiO_2, 475
silane plasma, 567, 572

similarity principle of concentrations and temperature fields, 140
Si_3N_4 synthesis from SiH_4, 474
single excited state approach, 98
SiO production, 432
SiO_2
  decomposition, 432
  film growth, 547
  non-equilibrium reduction, 426
  production from SiF_4, 484
skin
  effect, 146
  layer, 146
skin diseases
  corneal infections, 910
  Cuteneous Leishmaniasis, 908
  melanoma skin cancer, 906
  ulcerous eyelid wound, 911
sliding surface
  corona, 239
  spark, 239
$SnCl_4$ hydrogen reduction, 448
SnH_4 direct synthesis, 488
SO_2
  dissociation, 338
  oxidation, 414, 818
  oxidation in clusters, 822
  oxidation in droplets, 820
Sommerfeld formula, 42
soot production, 597
soybean oil reforming into syngas, 700
spark breakdown mechanism, 159
spark discharges, 240
specific energy input, 135
spectral density of plasma emission, 113
spheroidization of powders, 495
spray
  air plasma, 500
  ceramics, 504
  composites, 503, 505
  controlled atmosphere plasma, 500
  DC-arc plasma, 500
  functionally gradient materials, 503, 508
  intermetallics, 504
  layered materials, 503
  low-pressure plasma, 500
  metals, 503
  modeling, 510
  monolithic materials, 503
  radiofrequency plasma, 502
  reactive plasma, 507
  shrouded plasma, 500
  underwater plasma, 500
  vacuum plasma, 500
sputtering, 514
stabilization of plasma
  forward-vortex, 207
  reverse-vortex (tornado), 207
statistical distribution over vibrational-rotational states, 93
statistical theory of chemical processes, 83

statistics of vibrationally excited molecules, 99
steam reforming of hydrocarbons, 677
Steenbeck minimum power principle, 181, 195
Steenbeck-Raizer arc channel model, 196
stem cells, 887
sterilization
 air streams, 866
 airborne bacteria, 866
 atmospheric pressure plasma, 859
 B. subtilis spores, 849
 chemical effect of electrons, 852
 chemical effect of ions, 853
 colony forming units, cfu, 849
 D-value, 849
 dielectric-barrier discharge, 860
 dielectric barrier grating discharge, 869
 direct plasma effect, 851
 E. coli, 866, 872
 effect of electric fields, 854
 effect of heat, 858
 effect of reactive neutral species, 855
 effect of UV radiation, 858, 874
 H_2O_2 effect, 857
 human skin flora, 890
 in-depth, 863
 indirect plasma effect, 859
 kinetic model, 871
 living tissue, 889
 low-pressure plasma, 848
 melanoma cells, 864
 ozone effect, 857
 pH effect, 865
 pulsed electric fields, 874
 staphylococci, 860
 streptococci, 860
 surfaces, 848
 survival curve, 849
 water, 874
 yeast, 860
stochastic heating effect, 221
streamer, 157, 160
 anode-directed, 161
 cathode-directed, 161
 filament, 165
 ideally conducting channel, 164
 interaction, 166
 Monte Carlo modeling, 167
 negative, 161
 positive, 161
 propagation models, 163
 quasi–self-sustained, 163
sublethal plasma effect, 865
subplantation, 674
sulfur gasification by CO_2, 409
superconducting composite production, 483
superoxide, O_2^-, 852
superoxide dismutation, SOD, 853
supersonic plasma flow
 choking, 308
 critical heat release, 308

surface discharge, 239
surface energy
 dispersion component, 644
 polar component, 644
synthesis
 $CO-N_2$ mixture, 412
 F-oxidizers, 402
 inorganic carbides, 472
 inorganic nitrides, 472
 self-propagating high-temperature, 506

Ta-carbide synthesis, 477
Ta-hydride direct synthesis, 489
Ta-nitride synthesis from chloride, 473
Ta reduction from chloride, 443
$TaCl_5$ hydrogen reduction, 446
temperature of electrons, 3, 104
textile fiber treatment, 645, 648
thermal conductivity
 in plasma, 111
 non-equilibrium, 112
 related to dissociation of molecules, 112
 Treanor effect in vibrational energy transfer, 112
thermal
 breakdown of dielectrics, 561
 spray technologies, 499
thermionic converter, 803
thermodynamic equilibrium
 complete, CTE, 95
 local, LTE, 95
thermodynamic functions, 95
thermodynamic probability, 92
Thomson
 formula, 16
 theory, 39
thrusters
 colloidal, 790
 electronic wind, 790
 electrostatic plasma, 791
 electrothermal plasma, 790
 erosion, 792
 high-current, 791
 ion, 787, 789
 magneto-plasma-dynamic, 791
 plasma, 787
 pulsed plasma, 792
 specific impulse, 787
 stationary, 791
 with an ion layer, 791
Ti-carbide solid-phase synthesis, 475
TiB_2 synthesis, 492
TiC
 synthesis from $TiCl_4$, 477
 synthesis from TiO_2 and CH_4, 475
$TiCl_4$
 direct decomposition in non-thermal plasma, 469
 direct decomposition in thermal plasma, 461
 hydrogen reduction in non-thermal plasma, 448
 hydrogen reduction in thermal plasma, 445
 oxidation, 485, 486
 oxidation with CO_2, 486

Index

TiN synthesis from TiCl$_4$
TiO$_2$
  non-equilibrium reduction, 426
  production, 485
tissue engineering, 882
tornado discharges, 207
Townsend
  breakdown mechanism, 158
  coefficient β, 159
  formula, 157
  ionization coefficient α, 157
tracks of nuclear fission fragments
  ionization degree, 313
  plasma features, 311
transition $\alpha-\gamma$ in dusty plasma, 569, 577
translational energy distribution function of neutrals, 122
trapping effect, 568, 573
Treanor distribution function, 99
two-fluid magneto-hydrodynamics, 149

U-carbide synthesis, 477
U$_3$O$_8$ production, 482, 484
U production
  direct decomposition of hexafluoride, 449
  hydrogen reduction of hexafluoride, 442
  carbothermic reduction of oxides, 429
UF$_4$ calcium reduction, 443
UF$_6$
  direct decomposition, 449
  hydrogen reduction, 442
  steam conversion into U$_3$O$_8$, 484
Unsold-Kramers formula, 112
UO$_2$(NO$_3$)$_2$ (uranyl nitrate) aqueous solution decomposition, 482

VB$_2$ synthesis, 492
VCl$_4$ direct decomposition, 463
vibrational distribution
  Brau, 120
  Bray, 120
  excitation regimes, 117
  functions, 99, 114
  Gordiets, 119
  hyperbolic plateau, 118
  inverse population, 119
  Treanor, 119
vibrational excitation of molecules
  boomerang resonance, 55
  electron impact, 54
  long-lifetime resonances, 56
  rate coefficients, 56
  semi-empirical Fridman approximation, 56
  short-lifetime resonances, 55
vibrational energy
  losses due to non-resonant VV exchange, 132
  losses due to VT relaxation, 131
  transfer, 72

vibrational-translational (VT) relaxation
  anharmonic oscillators, 70
  collisions with atoms and radicals, 71
  fast non-adiabatic, 71
  formation of long-life complexes, 71
  harmonic oscillators, 68
  heterogeneous, 72
  Landau-Teller rate coefficient, 69
  slow adiabatic mechanism, 67
  symmetrical exchange reactions, 71
vibrational-vibrational (VV) relaxation
  resonant, 72
  non-resonant, 74
  intermolecular, 74
vibronic terms, 81, 363
vinyl chloride production, 596, 611
V$_2$O$_3$ direct plasma decomposition, 437
V$_2$O$_5$ direct plasma decomposition, 436
Voronoi polyhedra, 169

W-carbide synthesis, 477
W reduction from chloride, 443
water
  dissociation in plasma, 318
  treatment, 879
welding
  keyhole mode, 495
  medium current mode, 495
  microplasma mode, 495
  technology, 494
wettability
  improvement, 541, 641
  patterning, 566
Winchester mechanism, 31, 570
WCl$_6$ hydrogen reduction, 446
W(CO)$_6$ synthesis, 491
WF$_6$ direct decomposition, 467
WO$_3$
  non-equilibrium reduction, 426
  thermal plasma reduction, 421
wool treatment, 645, 650
work function, 42
wound treatment
  burns, 903
  diabetic ulcer, 903
  exogenic NO, 898
  fundamentals, 896
  inflammatory processes, 899
  NO therapy, 896
  ulcerous eyelid wound, 911
  venous trophic ulcer, 901

Xe-F, Xe-Cl lasers, 810
XeF$_2$, XeF$_4$, XeF$_6$ synthesis, 406
Xe-transition plasma laser, 808
Xenon fluoride synthesis, 405

yield of CO$_2$ dissociation in tracks of nuclear fission fragments, 312
yield of hydrogen production in tracks of nuclear fission fragments, 336

Index

yield of plasma radiolysis, 312
yield of water dissociation in tracks of nuclear fission
    fragments, 335

Z-factor of non-equilibrium dissociation, 86
zebra plasma ignition, 763
Zeldovich/Frank-Kamenetsky rule, 5
Zeldovich mechanism, 356
Zr-carbide solid-phase synthesis, 475
Zr-nitride synthesis from chloride, 473
Zr reduction from chloride, 443
ZrB_2 synthesis, 492
$ZrBr_4$ hydrogen reduction, 448

$ZrCl_4$
    direct decomposition in thermal plasma, 462
    hydrogen reduction in non-thermal plasma,
        448
    hydrogen reduction in thermal plasma, 446
ZrF_4 hydrogen reduction, 448
ZrI_4
    hydrogen reduction, 448
    direct decomposition, 469
ZrO_2
    non-equilibrium reduction, 426
    production, 477
$ZrSiO_4$ (zircon sand) decomposition, 477